Quick Guide to Inference

	THINK			**SHOW**			**TELL?**
Inference about?	One group or two?	Procedure	Model	Parameter	Estimate	SE	Chapter
Proportions	One sample	1-Proportion z-Interval	z	p	$\hat{p}$	$\sqrt{\dfrac{\hat{p}\hat{q}}{n}}$	18
		1-Proportion z-Test				$\sqrt{\dfrac{p_0 q_0}{n}}$	19
	Two independent groups	2-Proportion z-Interval	z	$p_1 - p_2$	$\hat{p}_1 - \hat{p}_2$	$\sqrt{\dfrac{\hat{p}_1 \hat{q}_1}{n_1} + \dfrac{\hat{p}_2 \hat{q}_2}{n_2}}$	22
		2-Proportion z-Test				$\sqrt{\dfrac{\hat{p}\hat{q}}{n_1} + \dfrac{\hat{p}\hat{q}}{n_2}},\ \hat{p} = \dfrac{y_1 + y_2}{n_1 + n_2}$	22
Means	One sample	t-Interval t-Test	t df $= n-1$	μ	$\bar{y}$	$\dfrac{s}{\sqrt{n}}$	20
	Two independent groups	2-Sample t-Test 2-Sample t-Interval	t df from technology	$\mu_1 - \mu_2$	$\bar{y}_1 - \bar{y}_2$	$\sqrt{\dfrac{s_1^2}{n_1} + \dfrac{s_2^2}{n_2}}$	22
	n Matched pairs	Paired t-Test Paired t-Interval	t df $= n-1$	μ_d	$\bar{d}$	$\dfrac{s_d}{\sqrt{n}}$	23
Distributions (one categorical variable)	One sample	Goodness of fit	χ^2 df $=$ cells $- 1$			$\sum \dfrac{(\text{Obs} - \text{Exp})^2}{\text{Exp}}$	24
	Many independent groups	Homogeneity χ^2 Test	χ^2 df $= (r-1)(c-1)$				
Independence (two categorical variables)	One sample	Independence χ^2 Test					
Association (two quantitative variables)	One sample	Linear Regression t-Test or Confidence Interval for β	t df $= n-2$	β_1	b_1	$\dfrac{s_e}{s_x \sqrt{n-1}}$ (compute with technology)	25
		Confidence Interval for μ_ν		μ_ν	$\hat{y}_\nu$	$\sqrt{SE^2(b_1) \cdot (x_\nu - \bar{x})^2 + \dfrac{s_e^2}{n}}$	
		Prediction Interval for y_ν		y_ν	$\hat{y}_\nu$	$\sqrt{SE^2(b_1) \cdot (x_\nu - \bar{x})^2 + \dfrac{s_e^2}{n} + s_e^2}$	
Inference about?	One group or two?	Procedure	Model	Parameter	Estimate	SE	Chapter

Stats: Data and Models

edition 4

Richard D. De Veaux
Williams College

Paul F. Velleman
Cornell University

David E. Bock
Cornell University

PEARSON

Boston Columbus Indianapolis New York San Francisco Amsterdam Cape Town
Dubai London Madrid Milan Munich Paris Montréal Toronto Delhi
Mexico City São Paulo Sydney Hong Kong Seoul Singapore Taipei Tokyo

Editor-in-Chief: *Deirdre Lynch*
Acquisitions Editor: *Patrick Barbera*
Editorial Assistants: *Salena Casha and Justin Billing*
Program Manager: *Chere Bemelmans*
Project Manager: *Shannon Steed*
Program Management Team Lead: *Marianne Stepanian*
Project Management Team Lead: *Christina Lepre*
Media Producer: *Stephanie Green*
TestGen Content Manager: *John Flanagan*
MathXL Content Developer: *Bob Carroll*
Senior Marketing Manager: *Erin Kelly*
Senior Author Support/Technology Specialist: *Joe Vetere*
Rights and Permissions Project Manager: *Diahanne Lucas*
Procurement Specialist: *Carol Melville*
Associate Director of Design: *Andrea Nix*
Senior Designer: *Barbara Atkinson*
Text Design: *Studio Montage*
Production Management, Composition, and Illustrations: *Lumina Datamatics, Inc.*
Cover Design: *Studio Montage/Barbara Atkinson*
Cover Image: *liravega/Shutterstock*

Copyright © 2016, 2012, 2008 by Pearson Education, Inc. All Rights Reserved. Printed in the United States of America. This publication is protected by copyright, and permission should be obtained from the publisher prior to any prohibited reproduction, storage in a retrieval system, or transmission in any form or by any means, electronic, mechanical, photocopying, recording, or otherwise. For information regarding permissions, request forms and the appropriate contacts within the Pearson Education Global Rights & Permissions department, please visit www.pearsoned.com/permissions/.

Acknowledgements of third party content appear on pages A-51–A-52, which constitutes an extension of this copyright page.

PEARSON, ALWAYS LEARNING, MyStatLab, MathXL are exclusive trademarks in the U.S. and/or other countries owned by Pearson Education, Inc. or its affiliates.

Unless otherwise indicated herein, any third-party trademarks that may appear in this work are the property of their respective owners and any references to third-party trademarks, logos or other trade dress are for demonstrative or descriptive purposes only. Such references are not intended to imply any sponsorship, endorsement, authorization, or promotion of Pearson's products by the owners of such marks, or any relationship between the owner and Pearson Education, Inc. or its affiliates, authors, licensees or distributors.

MICROSOFT AND/OR ITS RESPECTIVE SUPPLIERS MAKE NO REPRESENTATIONS ABOUT THE SUITABILITY OF THE INFORMATION CONTAINED IN THE DOCUMENTS AND RELATED GRAPHICS PUBLISHED AS PART OF THE SERVICES FOR ANY PURPOSE. ALL SUCH DOCUMENTS AND RELATED GRAPHICS ARE PROVIDED "AS IS" WITHOUT WARRANTY OF ANY KIND. MICROSOFT AND/OR ITS RESPECTIVE SUPPLIERS HEREBY DISCLAIM ALL WARRANTIES AND CONDITIONS WITH REGARD TO THIS INFORMATION, INCLUDING ALL WARRANTIES AND CONDITIONS OF MERCHANTABILITY, WHETHER EXPRESS, IMPLIED OR STATUTORY, FITNESS FOR A PARTICULAR PURPOSE, TITLE AND NON-INFRINGEMENT. IN NO EVENT SHALL MICROSOFT AND/OR ITS RESPECTIVE SUPPLIERS BE LIABLE FOR ANY SPECIAL, INDIRECT OR CONSEQUENTIAL DAMAGES OR ANY DAMAGES WHATSOEVER RESULTING FROM LOSS OF USE, DATA OR PROFITS, WHETHER IN AN ACTION OF CONTRACT, NEGLIGENCE OR OTHER TORTIOUS ACTION, ARISING OUT OF OR IN CONNECTION WITH THE USE OR PERFORMANCE OF INFORMATION AVAILABLE FROM THE SERVICES. THE DOCUMENTS AND RELATED GRAPHICS CONTAINED HEREIN COULD INCLUDE TECHNICAL INACCURACIES OR TYPOGRAPHICAL ERRORS. CHANGES ARE PERIODICALLY ADDED TO THE INFORMATION HEREIN. MICROSOFT AND/OR ITS RESPECTIVE SUPPLIERS MAY MAKE IMPROVEMENTS AND/OR CHANGES IN THE PRODUCT(S) AND/OR THE PROGRAM(S) DESCRIBED HEREIN AT ANY TIME. PARTIAL SCREEN SHOTS MAY BE VIEWED IN FULL WITHIN THE SOFTWARE VERSION SPECIFIED.

MICROSOFT®, WINDOWS®, AND MICROSOFT OFFICE® ARE REGISTERED TRADEMARKS OF THE MICROSOFT CORPORATION IN THE U.S.A. AND OTHER COUNTRIES. THIS BOOK IS NOT SPONSORED OR ENDORSED BY OR AFFILIATED WITH THE MICROSOFT CORPORATION.

Library of Congress Cataloging-in-Publication Data

De Veaux, Richard D.
   Stats : data and models / Richard D. De Veaux, Williams College, Paul F. Velleman, Cornell University, David E. Bock, Cornell University. —4th edition.
      pages cm
   Includes index.
   ISBN 978-0-321-98649-8
 1. Statistics—Textbooks. 2. Mathematical statistics—Textbooks. I. Velleman, Paul F., 1949- II. Bock, David E. III. Title.
   QA276.12.D417 2016
   519.5—dc23

                                       2014019094

1 2 3 4 5 6 7 8 9 10—RRD—18 17 16 15 14

ISBN 13: 978-0-321-99028-0 (Instructor's Edition)
ISBN 10:   0-321-99028-5 (Instructor's Edition)

ISBN 13: 978-0-321-98649-8 (Student Edition)
ISBN 10:   0-321-98649-0 (Student Edition)

www.pearsonhighered.com

*To Sylvia, who has helped me in more ways than she'll ever know,
and to Nicholas, Scyrine, Frederick, and Alexandra,
who make me so proud in everything that they are and do*
—*Dick*

*To my sons, David and Zev, from whom I've learned so much,
and to my wife, Sue, for taking a chance on me*
—*Paul*

*To Greg and Becca, great fun as kids and great friends as adults,
and especially to my wife and best friend, Joanna, for her
understanding, encouragement, and love*
—*Dave*

Meet the Authors

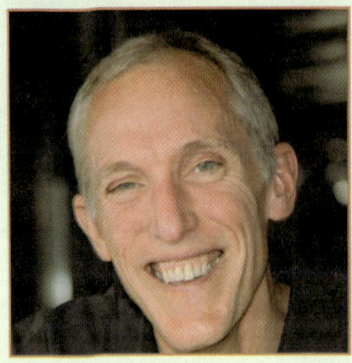

Richard D. De Veaux is an internationally known educator and consultant. He has taught at the Wharton School and the Princeton University School of Engineering, where he won a "Lifetime Award for Dedication and Excellence in Teaching." Since 1994, he has been Professor of Statistics at Williams College. Dick has won both the Wilcoxon and Shewell awards from the American Society for Quality. He is a fellow of the American Statistical Association (ASA) and an elected member of the International Statistical Institute (ISI). In 2008, he was named Statistician of the Year by the Boston Chapter of the ASA. Dick is also well known in industry, where for more than 25 years he has consulted for such Fortune 500 companies as American Express, Hewlett-Packard, Alcoa, DuPont, Pillsbury, General Electric, and Chemical Bank. Because he consulted with Mickey Hart on his book *Planet Drum*, he has also sometimes been called the "Official Statistician for the Grateful Dead." His real-world experiences and anecdotes illustrate many of this book's chapters.

Dick holds degrees from Princeton University in Civil Engineering (B.S.E.) and Mathematics (A.B.) and from Stanford University in Dance Education (M.A.) and Statistics (Ph.D.), where he studied dance with Inga Weiss and Statistics with Persi Diaconis. His research focuses on the analysis of large data sets and data mining in science and industry.

In his spare time, he is an avid cyclist and swimmer. He also is the founder of the "Diminished Faculty," an a cappella Doo-Wop quartet at Williams College and sings bass in the college concert choir. Dick is the father of four children.

Paul F. Velleman has an international reputation for innovative Statistics education. He is the author and designer of the multimedia Statistics program *ActivStats*, for which he was awarded the EDUCOM Medal for innovative uses of computers in teaching statistics, and the ICTCM Award for Innovation in Using Technology in College Mathematics. He also developed the award-winning statistics program, *Data Desk*, and the Internet site Data and Story Library (DASL) (lib.stat.cmu.edu/DASL/), which provides data sets for teaching Statistics. Paul's understanding of using and teaching with technology informs much of this book's approach.

Paul has taught Statistics at Cornell University since 1975. He holds an A.B. from Dartmouth College in Mathematics and Social Science, and M.S. and Ph.D. degrees in Statistics from Princeton University, where he studied with John Tukey. His research often deals with statistical graphics and data analysis methods. Paul co-authored (with David Hoaglin) *ABCs of Exploratory Data Analysis.* Paul is a Fellow of the American Statistical Association and of the American Association for the Advancement of Science. Paul is the father of two boys.

David E. Bock taught mathematics at Ithaca High School for 35 years. He has taught Statistics at Ithaca High School, Tompkins-Cortland Community College, Ithaca College, and Cornell University. Dave has won numerous teaching awards, including the MAA's Edyth May Sliffe Award for Distinguished High School Mathematics Teaching (twice), Cornell University's Outstanding Educator Award (three times), and has been a finalist for New York State Teacher of the Year.

Dave holds degrees from the University at Albany in Mathematics (B.A.) and Statistics/Education (M.S.). Dave has been a reader and table leader for the AP Statistics exam, serves as a Statistics consultant to the College Board, and leads workshops and institutes for AP Statistics teachers. He has recently served as K–12 Education and Outreach Coordinator and a senior lecturer for the Mathematics Department at Cornell University. His understanding of how students learn informs much of this book's approach.

Dave and his wife relax by biking or hiking, spending much of their free time in Canada, the Rockies, or the Blue Ridge Mountains. They have a son, a daughter, and four grandchildren.

Table of Contents

Preface ix

Part I Exploring and Understanding Data

1 Stats Starts Here   1
1.1 What Is Statistics? ■ **1.2** Data ■ **1.3** Variables

2 Displaying and Describing Categorical Data   16
2.1 Summarizing and Displaying a Single Categorical Variable ■ **2.2** Exploring the Relationship Between Two Categorical Variables

3 Displaying and Summarizing Quantitative Data   45
3.1 Displaying Quantitative Variables ■ **3.2** Shape ■ **3.3** Center ■ **3.4** Spread ■ **3.5** Boxplots and 5-Number Summaries ■ **3.6** The Center of Symmetric Distributions: The Mean ■ **3.7** The Spread of Symmetric Distributions: The Standard Deviation ■ **3.8** Summary—What to *Tell* About a Quantitative Variable

4 Understanding and Comparing Distributions   84
4.1 Comparing Groups with Histograms ■ **4.2** Comparing Groups with Boxplots ■ **4.3** Outliers ■ **4.4** Timeplots: Order, Please! ■ **4.5** Re-Expressing Data: A First Look

5 The Standard Deviation as a Ruler and the Normal Model   112
5.1 Standardizing with z-Scores ■ **5.2** Shifting and Scaling ■ **5.3** Normal Models ■ **5.4** Finding Normal Percentiles ■ **5.5** Normal Probability Plots

Part II Exploring Relationships Between Variables

6 Scatterplots, Association, and Correlation   151
6.1 Scatterplots ■ **6.2** Correlation ■ **6.3** Warning: Correlation ≠ Causation ■ **6.4** Straightening Scatterplots

7 Linear Regression   182
7.1 Least Squares: The Line of "Best Fit" ■ **7.2** The Linear Model ■ **7.3** Finding the Least Squares Line ■ **7.4** Regression to the Mean ■ **7.5** Examining the Residuals ■ **7.6** R^2—The Variation Accounted For by the Model ■ **7.7** Regression Assumptions and Conditions

8 Regression Wisdom   219
8.1 Examining Residuals ■ **8.2** Extrapolation: Reaching Beyond the Data ■ **8.3** Outliers, Leverage, and Influence ■ **8.4** Lurking Variables and Causation ■ **8.5** Working with Summary Values

9 Re-expressing Data: Get It Straight!   247
9.1 Straightening Scatterplots – The Four Goals ■ **9.2** Finding a Good Re-Expression

Part III Gathering Data

10 Understanding Randomness — 280
10.1 What Is Randomness? ■ 10.2 Simulating by Hand

11 Sample Surveys — 294
11.1 The Three Big Ideas of Sampling ■ 11.2 Populations and Parameters ■ 11.3 Simple Random Samples ■ 11.4 Other Sampling Designs ■ 11.5 From the Population to the Sample: You Can't Always Get What You Want ■ 11.6 The Valid Survey ■ 11.7 Common Sampling Mistakes, or How to Sample Badly

12 Experiments and Observational Studies — 318
12.1 Observational Studies ■ 12.2 Randomized, Comparative Experiments ■ 12.3 The Four Principles of Experimental Design ■ 12.4 Control Treatments ■ 12.5 Blocking ■ 12.6 Confounding

Part IV Randomness and Probability

13 From Randomness to Probability — 348
13.1 Random Phenomena ■ 13.2 Modeling Probability ■ 13.3 Formal Probability

14 Probability Rules! — 366
14.1 The General Addition Rule ■ 14.2 Conditional Probability and the General Multiplication Rule ■ 14.3 Independence ■ 14.4 Picturing Probability: Tables, Venn Diagrams, and Trees ■ 14.5 Reversing the Conditioning and Bayes' Rule

15 Random Variables — 389
15.1 Center: The Expected Value ■ 15.2 Spread: The Standard Deviation ■ 15.3 Shifting and Combining Random Variables ■ 15.4 Continuous Random Variables

16 Probability Models — 412
16.1 Bernoulli Trials ■ 16.2 The Geometric Model ■ 16.3 The Binomial Model ■ 16.4 Approximating the Binomial with a Normal Model ■ 16.5 The Continuity Correction ■ 16.6 The Poisson Model ■ 16.7 Other Continuous Random Variables: The Uniform and the Exponential

Part V From the Data at Hand to the World at Large

17 Sampling Distribution Models — 443
17.1 Sampling Distribution of a Proportion ■ 17.2 When Does the Normal Model Work? Assumptions and Conditions ■ 17.3 The Sampling Distribution of Other Statistics ■ 17.4 The Central Limit Theorem: The Fundamental Theorem of Statistics ■ 17.5 Sampling Distributions: A Summary

18 Confidence Intervals for Proportions — 472
18.1 A Confidence Interval ■ 18.2 Interpreting Confidence Intervals: What Does 95% Confidence Really Mean? ■ 18.3 Margin of Error: Certainty vs. Precision ■ 18.4 Assumptions and Conditions

19 Testing Hypotheses About Proportions — 494
19.1 Hypotheses ■ 19.2 P-Values ■ 19.3 The Reasoning of Hypothesis Testing ■ 19.4 Alternative Alternatives ■ 19.5 P-Values and Decisions: What to Tell About a Hypothesis Test

20 Inferences About Means — 518
20.1 Getting Started: The Central Limit Theorem (Again) ■ 20.2 Gosset's t ■ 20.3 Interpreting Confidence Intervals ■ 20.4 A Hypothesis Test for the Mean ■ 20.5 Choosing the Sample Size

21 More About Tests and Intervals — 548
21.1 Choosing Hypotheses ■ 21.2 How to Think About P-Values ■ 21.3 Alpha Levels ■ 21.4 Critical Values for Hypothesis Tests ■ 21.5 Errors

Part VI Accessing Associations Between Variables

22 Comparing Groups — 585
22.1 The Standard Deviation of a Difference ■ 22.2 Assumptions and Conditions for Comparing Proportions ■ 22.3 A Confidence Interval for the Difference Between Two Proportions ■ 22.4 The Two Sample z-Test: Testing for the Difference Between Proportions 22.5 A Confidence Interval for the Difference Between Two Means ■ 22.6 The Two-Sample t-Test: Testing for the Difference Between Two Means ■ 22.7 The Pooled t-Test: Everyone into the Pool?

23 Paired Samples and Blocks — 630
23.1 Paired Data ■ 23.2 Assumptions and Conditions ■ 23.3 Confidence Intervals for Matched Pairs ■ 23.4 Blocking

24 Comparing Counts — 655
24.1 Goodness-of-Fit Tests ■ 24.2 Chi-Square Test of Homogeneity ■ 24.3 Examining the Residuals ■ 24.4 Chi-Square Test of Independence

25 Inferences for Regression — 689
25.1 The Population and the Sample ■ 25.2 Assumptions and Conditions ■ 25.3 Intuition About Regression Inference ■ 25.4 Regression Inference ■ 25.5 Standard Errors for Predicted Values ■ 25.6 Confidence Intervals for Predicted Values ■ 25.7 Logistic Regression

Part VII Inference When Variables Are Related

*26 Analysis of Variance — 747
26.1 Testing Whether the Means of Several Groups Are Equal ■ 26.2 The ANOVA Table ■ 26.3 Assumptions and Conditions ■ 26.4 Comparing Means ■ 26.5 ANOVA on Observational Data

27 Multifactor Analysis of Variance — 782
27.1 A Two Factor ANOVA Model ■ 27.2 Assumptions and Conditions ■ 27.3 Interactions

28 Multiple Regression — 817
28.1 What Is Multiple Regression? ■ 28.2 Interpreting Multiple Regression Coefficients ■ 28.3 The Multiple Regression Model—Assumptions and Conditions ■ 28.4 Multiple Regression Inference ■ 28.5 Comparing Multiple Regression Models

29 Multiple Regression Wisdom (available on DVD and also online: pearsonhighered.com/dvb) — 859
29.1 Indicators ■ 29.2 Diagnosing Regression Models: Looking at the Cases ■ 29.3 Building Multiple Regression Models

Appendixes
A Answers **A-1** ■ B Photo Acknowledgments **A-51** ■ C Index **A-53** ■ D Tables and Selected Formulas **A-69**

---

*Optional Chapter

Preface

We are often asked why we write Statistics texts. After all, it takes a lot of work to find new and better examples, to keep datasets current, and to make a book an enjoyable and effective learning tool. So we thought we'd address that question first.

We do it because it's fun.

Of course, we care about teaching students to think statistically; we are teachers and professional statisticians. But Statistics can be a particularly challenging subject to teach. The student encounters many new concepts, many new methods, and many new terms. And we want to change the way students think about the world. From the start, our challenge has been to write a book that students would read, learn from, and enjoy. And we return to that goal with each new edition.

The book you hold is quicker to the point and, we hope, even more readable than previous editions. Of course, we've kept our conversational style and background anecdotes.[1] But we've tightened discussions and adjusted the order of some topics to move the story we tell about Statistics even more quickly to interesting real-world questions. We've focused even more on statistical thinking.

More and more high school math teachers are using examples from Statistics to provide intuitive examples of how a little bit of math can help us say a lot about the world. So students expect Statistics to be about real-world insights. This edition of *Stats: Data and Models* keeps your students engaged and interested because we show Statistics in action right from the start. Students will be solving problems of the kind they're likely to encounter in real life sooner. In Chapter 4, they will be comparing groups and in Chapter 6, they'll see relationships between two quantitative variables—and, of course, always with real, modern data. By emphasizing statistical thinking, these early chapters lay the foundation for the more advanced chapters on Analysis of Variance and Multiple Regression at the end of the book, telling a consistent story throughout.

There are few things more fun and useful for students than being empowered to discover something new about the world. And few things more fun for authors than helping students make those discoveries.

So, What's New in This Edition?

We've rewritten sections throughout the book to make them clearer and more interesting. We've introduced new up-to-the-minute motivating examples throughout. Many chapters lead with new examples—and we follow through with analyses of the data from those examples.

We've added a number of new features, each with the goal of making it even easier for students to put the concepts of Statistics together into a coherent whole.

1. **New and improved pedagogical tools:** A new section head list at the beginning of each chapter provides a road map. Section heads within each chapter are reorganized and reworded to be clear and specific. Chapter study materials now include Learning Objectives as well as terms. Students who understand the objectives and know the terms are well on their way to being ready for any test.

2. **Streamlined design:** Our goal has always been an accessible text. This edition sports an entirely new design that clarifies the purpose of each text element. The major theme of each chapter is more linear and easier to follow without distraction. Essential supporting material is clearly boxed and shaded, so students know where to focus their study efforts. Enriching—and often entertaining—side material is boxed, but not shaded.

3. **Streamlined content:** Our reorganization has shortened the book from 31 to 29 chapters (Chapter 29 is provided on DVD and also online at pearsonhighered.com/dvb). Each chapter is still a focused discussion, and most can be taught in one lesson. We've combined topics that are conceptually similar and reduced time spent on secondary topics. We've grouped important concepts, often in new presentation orders. The result is a more readable text.

4. **Content changes:** Here's how we've reorganized the content:

    a. Chapter 1 now gets down to business immediately rather than just providing an introduction to the book's features.

    b. The discussions of probability and random variables are reorganized to improve clarity and flow; they are tighter and more to the point.

    c. We've moved the discussion of inference for means earlier. We still lead the discussion of inference with inference for proportions (for reasons we explain in the *Test Bank and Resource Guide*), but now we turn immediately to inference for means so students can see the methods side by side. Students can then also see that the reasoning is really the same.

    d. When we discuss comparing groups, we now discuss both proportions and means, which helps students to see the parallels.

5. **Exercises:** We've updated most exercises that use real-world data, retired some that were getting old, and added new exercises. Each chapter's exercises now start with single-concept exercises for each

---
[1] And our footnotes.

section, labeled with the section number so students can find exercises to review for any topic they wish to check.

6. We've updated the choice of technologies supported in the **On the Computer** sections at the end of chapters. You'll now find advice on *StatCrunch* and *R* along with the other packages we have traditionally discussed.

7. Sections are numbered to help with navigation and reading assignments.

Our Approach

Statistics is practiced with technology. We think a modern statistics text should recognize that fact from the start. And so do our students. You won't find tedious calculations worked by hand. But you will find equation forms that favor intuition over calculation. You'll find extensive use of real data—even large data sets. And, most important, you'll find a focus on statistical thinking rather than calculation. The question that motivates each of our hundreds of examples is not "how do you *find* the answer?" but "how do you *think* about the answer?"

Textbooks are defined more by what they choose not to cover than by what they do cover. We've structured this text so that each new topic fits into a student's growing understanding. Several topic orders can support this goal. We explain our reasons for our topic order in the *Test Bank and Resource Guide*. We also describe some alternative orders supported by these materials.

GAISE Guidelines

The Guidelines for Assessment and Instruction in Statistics Education (GAISE) report adopted by the American Statistical Association urges that Statistics education should

1. emphasize Statistical literacy and develop Statistical thinking,
2. use real data,
3. stress conceptual understanding rather than mere knowledge of procedures,
4. foster active learning,
5. use technology for developing concepts and analyzing data, and
6. make assessment a part of the learning process.

We've designed our text and supplementary materials to support this approach to the introductory course. We urge you to think about these guidelines with each class meeting.

Our Goal: Read This Book!

The best text in the world is of little value if students don't read it. Here are some of the ways we have made this edition even more approachable:

- **Readability.** This book doesn't read like other Statistics texts. Our style is both colloquial and informative, engaging students to actually read the book to see what it says. We've tightened the discussions and removed digressions.
- **Humor.** We know that humor is the best way to promote learning. You will find quips and wry comments throughout the narrative, in margin notes, and in footnotes.
- **Informality.** Our informal diction doesn't mean that we treat the subject matter lightly or informally. We try to be precise and, wherever possible, we offer deeper (but not more mathematical) explanations and justifications than those found in most introductory texts.
- **Focused lessons.** The chapters are shorter than in most other texts so instructors and students can focus on one topic at a time.
- **Consistency.** We try to avoid the "do what we say, not what we do" trap. Having taught the importance of plotting data and checking assumptions and conditions, we model that behavior right through the rest of the book. (Check the exercises in Chapter 28. You'll see that we still require and demonstrate the plots and checks that were introduced in the early chapters.) This consistency helps reinforce these fundamental principles.
- **The need to read.** Students who just skim the book, or start from an exercise and look for a similar example in a box to copy, may find our presentation frustrating. The important concepts, definitions, and sample solutions don't sit in little boxes. Statistics is a consistent story about how to understand the world when we have data. The story can't be told piecemeal. This is a book that needs to be read, so we've tried to make the reading experience enjoyable.

Mathematics

Mathematics can

1. provide a concise, clear statement of important concepts.
2. describe calculations to be performed with data.
3. embody proofs of fundamental results.

Of these, we emphasize the first. Mathematics can make discussions of Statistics concepts, probability, and inference clear and concise. We don't shy away from using math where it can clarify without intimidating. But we know that some students are put off by equations, so we always provide a verbal description and a numerical example as well. Some theorems about Statistics are quite interesting, and many are important. Often, though, their proofs are not enlightening to introductory Statistics students and can distract the audience from the concepts we want them to understand. So we avoid them here.

Nor do we slide in the opposite direction and concentrate on calculation. Although statistics calculations are generally straightforward, they are also usually tedious. And, more to the point, they are often unnecessary. Today, virtually all statistics are calculated with technology, so there is little need for students to spend time summing squared deviations by hand. We have selected the equations that do appear for their focus on illuminating concepts and methods. Although these equations may be the best way to understand the concepts, they may not be optimal for hand calculation. When that happens, we give an alternative formula, better suited for hand calculation, for those who find following the process a better way to learn about the result.

Technology and Data

To experience the real world of Statistics, use modern technology to explore real data sets.

Technology. We assume that you are using some form of technology—a statistics package, a calculator, a spreadsheet, or some combination of these—in your Statistics course. We also assume that you'll put little emphasis on calculating answers by hand, even though we often show how. However, this is not a technology-heavy book. The role of technology in this book is to get the calculations out of the way so we can focus on statistical thinking. We discuss generic computer output, but we don't adopt any particular statistics software. We do offer guidance to help students get started on eight common software platforms: Excel®, Minitab®, *Data Desk*, JMP®, SPSS®, TI-83/84 Plus graphing calculators, StatCrunch®, and R®. The **On the Computer** section at the end of most chapters is specific to the methods learned in that chapter. All of these packages have inexpensive student options. But your students can have a choice of three of them at *no-cost: StatCrunch* (accessed through *MyStatLab*, available from Pearson with the text), *Data Desk* (found on the DVD and website that accompanies the book and free to move to their computers), and *R* (found online, but run on their computers.) The book's DVD includes the e-book *ActivStats*; the statistics package *Data Desk*; as well as versions of the data sets used in the book appropriate for a variety of packages.

Data. Because we use technology for computing, we don't limit ourselves to small, artificial data sets. You'll find some small data sets, but we also base examples and exercises on real data with a moderate number of cases—usually more than you would want to enter by hand into a program or calculator. Machine-readable versions of the data are included on the DVD and on the book's website, **www.pearsonhighered.com/dvb**.

Continuing Features

Enhancing Understanding

Where Are We Going? Each chapter starts with a paragraph that points out the kinds of questions students will learn how to answer in the chapter. A new chapter outline helps organize major topics for the students.

Each chapter ends with a **What Have We Learned?** summary, which includes new learning objectives and definitions of terms introduced in the chapter. Students can think of these as study guides.

In each chapter, our innovative **What Can Go Wrong?** sections highlight the most common errors that people make and the misconceptions they have about Statistics. One of our goals is to arm students with the tools to detect statistical errors and to offer practice in debunking misuses of statistics, whether intentional or not.

Margin and in-text boxed notes. Throughout each chapter, boxed margin and in-text notes enhance and enrich the text. Boxes with essential information are screened. Conversational notes that enhance the text and entertain the reader are unscreened.

By Hand. Even though we encourage the use of technology to calculate statistical quantities, we realize the pedagogical benefits of occasionally doing a calculation by hand. The By Hand boxes break apart the calculation of many simpler formulas to help the student through the calculation of a worked example.

Math Boxes. In many chapters, we present the mathematical underpinnings of the statistical methods and concepts. By setting these proofs, derivations, and justifications apart from the narrative, we allow the student to continue to follow the logical development of the topic at hand, yet also refer to the underlying mathematics for greater depth.

Reality Check. We regularly remind students that Statistics is about understanding the world with data. Results that make no sense are probably wrong, no matter how carefully we think we did the calculations. Mistakes are often easy to spot with a little thought, so we ask students to stop for a reality check before interpreting their result.

Notation Alert. Throughout this book, we emphasize the importance of clear communication, and proper notation is part of the vocabulary of Statistics. We've found that it helps students when we are clear about the letters and symbols statisticians use to mean very specific things, so we've included Notation Alerts whenever we introduce a special notation that students will see again.

Connections. Each chapter has a Connections feature to link key terms and concepts with previous discussions and to point out the continuing themes. Connections help students fit newly learned concepts into a growing understanding of Statistics.

Learning by Example

For Example. As we introduce each important concept, we provide a focused example applying it—usually with real up-to-the-minute data. Many For Examples carry the discussion through the chapter, picking up the story and moving it forward as students learn more about the topic.

Step-by-Step Examples: Think, Show, Tell. Step-by-Step examples repeat the mantra of Think, Show, and Tell in every chapter. These longer, worked examples guide students through the process of analyzing the problem with the general explanation on the left and the worked-out problem on the right. They emphasize the importance of thinking about a Statistics question (What do we know? What do we hope to learn? Are the assumptions and conditions satisfied?) and reporting our findings (the Tell step). The Show step contains the mechanics of calculating results and conveys our belief that it is only one part of the process. The result is a better understanding of the concept, not just number crunching. In the fourth edition, we've updated Think, Show, Tell Step-by-Step examples with new examples and data.

Testing Understanding

Just Checking. Just Checking questions are quick checks throughout the chapter; most involve very little calculation. These questions encourage students to pause and think about what they've just read. The Just Checking answers are at the end of the exercise sets in each chapter so students can easily check themselves.

Exercises. We've added section-specific single-concept exercises at the beginning of each exercise set so students can be sure they have a clear understanding of each important topic before they're asked to tie them all together in more comprehensive exercises. Exercises have been updated with the most recent data. Many come from news stories; some from recent research articles. Whenever possible, the data are on the bound-in DVD so students can explore them further.

Technology

***ActivStats* Pointers.** The DVD bound into new copies of the book includes *ActivStats*, so we've included occasional pointers to the *ActivStats* activities when they parallel discussions in the book. Many students choose to look at these first, before reading the chapter or attending a class on each subject.

Data Sources. Most of the data used in examples and exercises are from real-world sources, and whenever we can, we include references to the Internet data sources used, often in the form of URLs. The data we use are usually on the DVD. If you seek the data—or an updated version of the data—on the Internet, we try to direct you to a good starting point.

On the Computer. In the real world, Statistics is practiced with computers. We prefer not to choose a particular Statistics program. Instead, at the end of most chapters, we summarize what students can find in the most common packages, often with annotated output. We then offer specific guidance for several of the most common packages (*Data Desk*, Excel®, JMP®, Minitab®, R®, SPSS®, StatCrunch®, and TI-83/84 Plus[2]) to help students get started with the software of their choice.

On the DVD

The DVD accompanying new books holds a number of supporting materials, including *ActivStats,* the *Data Desk* statistics package, animations, all large data sets from the text formatted for the most popular technologies, and one additional chapter.

***ActivStats* (for *Data Desk*).** The award-winning *ActivStats* multimedia program supports learning chapter by chapter. It complements the book with videos of real-world stories, worked examples, animated expositions of each of the major Statistics topics, and tools for performing simulations, visualizing inference, and learning to use statistics software. *ActivStats* includes

- more than 1000 homework exercises, plus answers to the "odd numbered" exercises.
- 17 short video clips, 70 animated activities, 117 teaching applets, and more than 300 data sets.

Data Desk, a student version of the professional statistics package, gives students the ability to perform any analysis in the textbook. Its interactive dynamic displays help students visualize relationships and models.

Data. Data for exercises marked are available on the DVD, Companion website, and MyStatLab™ formatted for multiple statistics software applications.

Additional Chapter. An additional chapter covering more advanced topics in multiple regression (Chapter 29). This chapter discusses modern diagnostic and model building methods.

---

[2]For brevity, we will write TI-83/84 Plus for the TI-83 Plus and/or TI-84 Plus. Keystrokes and output remain the same for the TI-83 Plus and the TI-84 Plus, so instructions and examples serve for both calculators.

Supplements

For the Student

***Stats: Data and Models*, Fourth Edition,** for-sale student edition (ISBN-13: 978-0-321-98649-8; ISBN-10: 0-321-98649-0)

Student's Solutions Manual, by William Craine, provides detailed, worked-out solutions to odd-numbered exercises. (ISBN-13: 978-0-321-98997-0; ISBN-10: 0-321-98997-X)

Study card for the De Veaux/Velleman/Bock Statistics Series is a resource for students containing important formulas, definitions, and tables that correspond precisely to the De Veaux/Velleman/Bock Statistics series. This card can work as a reference for completing homework assignments or as an aid in studying. (ISBN-13: 978-0-321-82626-8; ISBN-10: 0-321-82626-4)

Videos for the De Veaux/Velleman/Bock Series, Fourth Edition, available to download from MyStatLab®. Concept Videos use humor to promote learning. Unique characters in fun situations explain the key concepts of statistics, covering important definitions and procedures for most chapters. Also available are videos of worked solutions for many of the Step-by-Step examples in the text.

For the Instructor

Instructor's Edition contains answers to all exercises. (ISBN-13: 978-0-321-99028-0; ISBN-10: 0-321-99028-5)

Instructor website contains the following resources at: **www.pearsonhighered.com/dvb**

- **Getting Started:** The De Veaux/Velleman/Bock Approach, Sample syllabi, Getting Started with Technology.
- **Preparing for Class:** Chapter and Lesson Support: ActivStats Pointers, Planning Your Lessons, Instructor's Supplements.

Instructor's Solutions Manual (download only), by William Craine, contains detailed solutions to all of the exercises. The Instructor's Solutions Manual is available to download from within MyStatLab® and in the Instructor Resource Center at **www.pearsonhighered.com/irc**.

Online Test Bank and Resource Guide (download only), by William Craine, includes chapter-by-chapter comments on the major concepts, tips on presenting topics (and what to avoid), extra teaching examples, a list of resources, chapter quizzes, part-level tests, and suggestions for projects. The Online Test Bank and Resource Guide is available to download from within MyStatLab® and in the Instructor Resource Center at **www.pearsonhighered.com/irc**.

Instructor's Podcasts (10 points in 10 minutes). These audio podcasts focus on key points in each chapter to help you with class preparation. They can be easily downloaded from MyStatLab and the Instructor Resource Center (**www.pearsonhighered.com/irc**).

Technology Resources

MyStatLab™ Online Course (access code required)

MyStatLab from Pearson is the world's leading online resource for teaching and learning statistics; integrating interactive homework, assessment, and media in a flexible, easy-to-use format. MyStatLab is a course management system that delivers **proven results** in helping individual students succeed.

- MyStatLab can be implemented successfully in any environment—lab-based, hybrid, fully online, traditional—and demonstrates the quantifiable difference that integrated usage has on student retention, subsequent success, and overall achievement.

- MyStatLab's comprehensive online gradebook automatically tracks students' results on tests, quizzes, homework, and in the study plan. Instructors can use the gradebook to provide positive feedback or intervene if students have trouble. Gradebook data can be easily exported to a variety of spreadsheet programs, such as Microsoft Excel.

MyStatLab provides **engaging experiences** that personalize, stimulate, and measure learning for each student. In addition to the resources below, each course includes a full interactive online version of the accompanying textbook.

- **Personalized Learning:** We now offer your course with an optional focus on adaptive learning, to allow your students to work on just what they need to learn when it makes the most sense, to maximize their potential for understanding and success.

- **Tutorial Exercises with Multimedia Learning Aids:** The homework and practice exercises in MyStatLab align with the exercises in the textbook, and most regenerate algorithmically to give students unlimited opportunity for practice and mastery. Exercises offer immediate helpful feedback, guided solutions, sample problems, animations, videos, statistical software tutorial videos and eText clips for extra help at point-of-use.

- **MyStatLab Accessibility:** MyStatLab is compatible with the JAWS screen reader, and enables multiple-choice and free-response problem-types to be read, and interacted with via keyboard controls and math notation input. MyStatLab also works with screen enlargers, including ZoomText, MAGic, and SuperNova. And all MyStatLab videos accompanying texts with copyright 2009 and later have closed captioning. More information on this functionality is available at http://mymathlab.com/accessibility.

- **StatTalk Videos:** Fun-loving statistician Andrew Vickers takes to the streets of Brooklyn, NY, to demonstrate important statistical concepts through interesting stories and real-life events. This series of 24 fun and engaging videos will help students actually understand statistical concepts. Available with an instructor's user guide and assessment questions.

- **Additional Question Libraries:** In addition to algorithmically regenerated questions that are aligned with your textbook, MyStatLab courses come with two additional question libraries.
  - **450 exercises** in **Getting Ready for Statistics** cover the developmental math topics students need for the course. These can be assigned as a prerequisite to other assignments, if desired.
  - **1000 exercises** in the **Conceptual Question Library** require students to apply their statistical understanding.
- **StatCrunch™:** MyStatLab integrates the web-based statistical software, StatCrunch, within the online assessment platform so that students can easily analyze data sets from exercises and the text. In addition, MyStatLab includes access to **www.StatCrunch.com**, a vibrant online community where users can access tens of thousands of shared data sets, create and conduct online surveys, perform complex analyses using the powerful statistical software, and generate compelling reports. Designed for today's students, StatCrunch works on any mobile device.
- **Statistical Software Support and Integration:** We make it easy to copy our data sets, both from the eText and the MyStatLab questions, into software such as StatCrunch, Minitab, Excel, and more. Students have access to a variety of support tools—Technology Tutorial Videos, Technology Study Cards, and Technology Manuals for select titles—to learn how to effectively use statistical software.

And, MyStatLab comes from an **experienced partner** with educational expertise and an eye on the future.
- Knowing that you are using a Pearson product means knowing that you are using quality content. That means that our eTexts are accurate and our assessment tools work. It means we are committed to making MyMathLab as accessible as possible.
- Whether you are just getting started with MyStatLab, or have a question along the way, we're here to help you learn about our technologies and how to incorporate them into your course.

To learn more about how MyStatLab combines proven learning applications with powerful assessment, visit **www.mystatlab.com** or contact your Pearson representative.

MyStatLab™ Ready to Go Course (access code required)

These new Ready to Go courses provide students with all the same great MyStatLab features, but make it easier for instructors to get started. Each course includes pre-assigned homework and quizzes to make creating a course even simpler. Ask your Pearson representative about the details for this particular course or to see a copy of this course.

MathXL® for Statistics Online Course (access code required)

MathXL® is the homework and assessment engine that runs MyStatLab. (MyStatLab is MathXL plus a learning management system.)

With MathXL for Statistics, instructors can:
- Create, edit, and assign online homework and tests using algorithmically generated exercises correlated at the objective level to the textbook.
- Create and assign their own online exercises and import TestGen tests for added flexibility.
- Maintain records of all student work, tracked in MathXL's online gradebook.

With MathXL for Statistics, students can:
- Take chapter tests in MathXL and receive personalized study plans and/or personalized homework assignments based on their test results.
- Use the study plan and/or the homework to link directly to tutorial exercises for the objectives they need to study.
- Students can also access supplemental animations and video clips directly from selected exercises.
- Knowing that students often use external statistical software, we make it easy to copy our data sets, both from the eText and the MyStatLab questions, into software like StatCrunch™, Minitab, Excel and more.

MathXL for Statistics is available to qualified adopters. For more information, visit our website at www.mathxl.com, or contact your Pearson representative.

StatCrunch™

StatCrunch is powerful web-based statistical software that allows users to perform complex analyses, share data sets, and generate compelling reports of their data. The vibrant online community offers tens of thousands of shared data sets for students to analyze.
- **Collect.** Users can upload their own data to StatCrunch or search a large library of publicly shared data sets, spanning almost any topic of interest. Also, an online survey tool allows users to quickly collect data via web-based surveys.
- **Crunch.** A full range of numerical and graphical methods allow users to analyze and gain insights from any data set. Interactive graphics help users understand statistical concepts, and are available for export to enrich reports with visual representations of data.
- **Communicate.** Reporting options help users create a wide variety of visually-appealing representations of their data.

Full access to StatCrunch is available with a MyStatLab kit, and StatCrunch is available by itself to qualified adopters. StatCrunch Mobile works on any mobile device and is now available, visit www.statcrunch.com from the browser on your smart phone or tablet. For more information, visit our website at www.StatCrunch.com, or contact your Pearson representative.

TestGen®

TestGen® (www.pearsoned.com/testgen) enables instructors to build, edit, print, and administer tests using a computerized bank of questions developed to cover all the objectives of the text. TestGen is algorithmically based, allowing instructors to create multiple but equivalent versions of the same question or test with the click of a button. Instructors can also modify test bank questions or add new questions. The software and testbank are available for download from Pearson Education's online catalog.

PowerPoint® Lecture Slides

PowerPoint® Lecture Slides provide an outline to use in a lecture setting, presenting definitions, key concepts, and figures from the text. These slides are available within MyStatLab and in the Instructor Resource Center at **www.pearsonhighered.com/irc**.

Active Learning Questions

Prepared in PowerPoint®, these questions are intended for use with classroom response systems. Several multiple-choice questions are available for each chapter of the book, allowing instructors to quickly assess mastery of material in class. The Active Learning Questions are available to download from within MyStatLab® and in the Instructor Resource Center at **www.pearsonhighered.com/irc**.

Acknowledgments

Many people have contributed to this book in all of its editions. This edition never would have seen the light of day without the assistance of the incredible team at Pearson. Our Editor-in-Chief, Deirdre Lynch, was central to the genesis, development, and realization of this project from day one. Shannon Steed, Project Manager, kept the cogs from getting into the wheels where they often wanted to wander with much needed humor and grace. Senior Marketing Manager Erin Kelly made sure the word got out. Chere Bemelmans, Program Manager, Salena Casha and Justin's, Editorial Assistants, were essential in managing all of the behind-the-scenes work that needed to be done. Stephanie Green, Media Producer, put together a top-notch media package for this book. Melissa Welch of Studio Montage and Barbara Atkinson are responsible for the wonderful way the book looks. Carol Melville, Procurement Specialist, worked miracles to get this book in your hands, and Greg Tobin, President, was supportive and good-humored throughout all aspects of the project.

A special thanks goes out to Nancy Kincade, Project Manager at Lumina Datamatics, for her close attention to detail.

We are grateful for expert help from William Craine who developed and wrote supplements to accompany this new edition.

We'd also like to thank our accuracy checkers whose monumental task was to make sure we said what we thought we were saying. They are Cathleen Zucco-Tevloff, Rider University; and Stanley Seltzer, Ithaca College. Special thanks to Kurt Mederer, University of Michigan, for his careful reading of the text.

We extend our sincere thanks for the suggestions and contributions made by the following reviewers of this edition:

Nazanin Azarnia
Santa Fe Community College

Patricia Humphrey
Georgia Southern University

Steve Marsden
Glendale College

Cathy Zucco-Teveloff
Rider University

Jay Xu
Williams College

Ken Grace
Anoka Ramsey Community College

Joseph Kupresanin
Cecil College

Jackie Miller
The Ohio State University

Dottie Walton
Cuyahoga Community College

We also extend our sincere thanks for the suggestions and contributions made by the following reviewers of the previous editions:

Mary Kay Abbey
Montgomery College

Froozan Pourboghnaf Afiat
Community College of Southern Nevada

Mehdi Afiat
Community College of Southern Nevada

Nazanin Azarnia
Santa Fe Community College

Sanjib Basu
Northern Illinois University

Carl D. Bodenschatz
University of Pittsburgh

Steven Bogart
Shoreline Community College

Ann Cannon
Cornell College

Robert L. Carson
Hagerstown Community College

Jerry Chen
Suffolk County CommunityCollege

Rick Denman
Southwestern University

Jeffrey Eldridge
Edmonds Community College

Karen Estes
St. Petersburg Junior College

Richard Friary

Kim (Robinson) Gilbert
Clayton College & State University

Ken Grace
Anoka-Ramsey Community College

Jonathan Graham
University of Montana

Nancy Heckman
University of British Columbia

James Helreich
Marist College

Susan Herring
Sonoma State University

Mary R. Hudachek-Buswell
Clayton State University

Patricia Humphrey
Georgia Southern University

Becky Hurley
Rockingham CommunityCollege

Debra Ingram
Arkansas State University

Kelly Jackson
Camden County College

Martin Jones
College of Charleston

Rebecka Jornsten
Rutgers University

Michael Kinter
Cuesta College

Kathleen Kone
Community College of Allegheny County

Michael Lichter
State University of New York–Buffalo

Susan Loch
University of Minnesota

Pamela Lockwood
Western Texas A & M University

Wei-Yin Loh
University of Wisconsin–Madison

Catherine Matos
Clayton College & State University

Elaine McDonald
Sonoma State University

Hari Mukerjee
Witchita State University

Helen Noble
San Diego State University

Monica Oabos
Santa Barbara City College

Linda Obeid
Reedley College

Charles C. Okeke
Community College of Southern Nevada

Pamela Omer
Western New England College

Mavis Pararai
Indiana University of Pennsylvania

Gina Reed
Gainesville College

Juana Sanchez
UCLA

Gerald Schoultz
Grand Valley State University

Jim Smart
Tallahassee Community College

Chamont Wang
The College of New Jersey

Edward Welsh
Westfield State College

Heydar Zahedani
California State University, San Marcos

Chapter 1: Stats Starts Here[1]

- 1.1 What Is Statistics?
- 1.2 Data
- 1.3 Variables

Where are we going?

Statistics gets no respect. People say things like "You can prove anything with Statistics." People will write off a claim based on data as "just a statistical trick." And Statistics courses don't have the reputation of being students' first choice for a fun elective.

But Statistics *is* fun. That's probably not what you heard on the street, but it's true. Statistics is about how to think clearly with data. A little practice thinking statistically is all it takes to start seeing the world more clearly and accurately.

This is a book about understanding the world by using data. So we'd better start by understanding data. There's more to that than you might have thought.

> "But where shall I begin?" asked Alice. "Begin at the beginning," the King said gravely, "and go on till you come to the end: then stop."
>
> —*Lewis Carroll, Alice's Adventures in Wonderland*

1.1 What Is Statistics?

People around the world have one thing in common—they all want to figure out what's going on. You'd think with the amount of information available to everyone today this would be an easy task, but actually, as the amount of information grows, so does our need to understand what it can tell us.

At the base of all this information, on the Internet and all around us, are data. We'll talk about data in more detail in the next section, but for now, think of **data** as any collection of numbers, characters, images, or other items that provide information about something. What sense can we make of all this data? You certainly can't make a coherent picture from random pieces of information. Whenever there are data and a need for understanding the world, you'll find Statistics.

This book will help you develop the skills you need to understand and communicate the knowledge that can be learned from data. By thinking clearly about the question you're trying to answer and learning the statistical tools to show what the data are saying, you'll acquire the skills to tell clearly what it all means. Our job is to help you make sense of the concepts and methods of Statistics and to turn it into a powerful, effective approach to understanding the world through data.

---

[1] We were thinking of calling this chapter "Introduction" but nobody reads the introduction, and we wanted you to read this. We feel safe admitting this down here in the footnotes because nobody reads footnotes either.

FRAZZ © 2003 Jef Mallett. Distributed by Universal Uclick. Reprinted with permission. All rights reserved.

"Data is king at Amazon. Clickstream and purchase data are the crown jewels at Amazon. They help us build features to personalize the Web site experience."

—Ronny Kohavi,
former Director of Data Mining and Personalization, Amazon.com

Q: What is Statistics?
A: Statistics is a way of reasoning, along with a collection of tools and methods, designed to help us understand the world.
Q: What are statistics?
A: Statistics (plural) are particular calculations made from data.
Q: So what is data?
A: You mean, "what *are* data?" Data is the plural form. The singular is datum.
Q: OK, OK, so what are data?
A: Data are values along with their context.

The ads say, "Don't drink and drive; you don't want to be a statistic." But you can't be a statistic.
 We say, "Don't be a datum."

Data vary. Ask different people the same question and you'll get a variety of answers. Statistics helps us to make sense of the world described by our data by seeing past the underlying variation to find patterns and relationships. This book will teach you skills to help with this task and ways of thinking about variation that are the foundation of sound reasoning about data.

Consider the following:

- If you have a Facebook account, you have probably noticed that the ads you see online tend to match your interests and activities. Coincidence? Hardly. According to the *Wall Street Journal* (10/18/2010),[2] much of your personal information has probably been sold to marketing or tracking companies. Why would Facebook give you a free account and let you upload as much as you want to its site? Because your data are valuable! Using your Facebook profile, a company might build a profile of your interests and activities: what movies and sports you like; your age, sex, education level, and hobbies; where you live; and, of course, who your friends are and what *they* like. From Facebook's point of view, your data are a potential gold mine. Gold ore in the ground is neither very useful nor pretty. But with skill, it can be turned into something both beautiful and valuable. What we're going to talk about in this book is how you can mine your own data and learn valuable insights about the world.

- In 2012, in the United States, wireless subscribers (about 90% of the U.S. population) sent a total of 2.19 trillion text (SMS) messages. That's over 182 billion a month, up from about 12 million a month just 10 years earlier.[3] Some of these messages are sent or read while the sender or the receiver is driving. How dangerous is texting while driving?

 How can we study the effect of texting while driving? One way is to measure reaction times of drivers faced with an unexpected event while driving and texting. Researchers at the University of Utah tested drivers on simulators that could present emergency situations. They compared reaction times of sober drivers, drunk drivers, and texting drivers.[4] The results were striking. The texting drivers actually responded more slowly and were more dangerous than those who were above the legal limit for alcohol.

In this book, you'll learn how to design and analyze experiments like this. You'll learn how to interpret data and to communicate the message you see to others. You'll also learn how to spot deficiencies and weaknesses in conclusions drawn by others that you see in newspapers and on the Internet every day. Statistics can help you become a more informed citizen by giving you the tools to understand, question, and interpret data.

---

[2] blogs.wsj.com/digits/2010/10/18/referers-how-facebook-apps-leak-user-ids/

[3] CTIA The International Association for the Wireless Telecommunications Industry (www.ctia.org/your-wireless-life/how-wireless-works/wireless-quick-facts).

[4] "Text Messaging During Simulated Driving," Drews, F. A. et al., Human Factors: hfs.sagepub.com/content/51/5/762

Statistics in a Word

It can be fun, and sometimes useful, to summarize a discipline in only a few words. So,

Economics is about . . . *Money (and why it is good).*
Psychology: *Why we think what we think (we think).*
Paleontology: *Previous Life.*
Biology: *Life.*
Religion: *After Life*
Anthropology: *Who?*
History: *What, where, and when?*
Philosophy: *Why?*
Engineering: *How?*
Accounting: *How much?*

In such a caricature, Statistics is about . . . **Variation.**

> **Statistics is about variation** Data vary because we don't see everything and because even what we do see and measure, we measure imperfectly.
> So, in a very basic way, Statistics is about the real, imperfect world in which we live.

1.2 Data

Amazon.com opened for business in July 1995, billing itself as "Earth's Biggest Bookstore." By 1997, Amazon had a catalog of more than 2.5 million book titles and had sold books to more than 1.5 million customers in 150 countries. In 2012, the company's sales reached $61 billion (a 27% increase from the previous year). Amazon has sold a wide variety of merchandise, including a $400,000 necklace, yak cheese from Tibet, and the largest book in the world. How did Amazon become so successful and how can it keep track of so many customers and such a wide variety of products? The answer to both questions is *data*.

But what are data? Think about it for a minute. What exactly *do* we mean by "data"? Do data have to be numbers? The amount of your last purchase in dollars is numerical data. But your name and address in Amazon's database are also data even though they are not numerical. What about your ZIP code? That's a number, but would Amazon care about, say, the *average* ZIP code of its customers?

Let's look at some hypothetical values that Amazon might collect:

105-2686834-3759466	Ohio	Nashville	Kansas	10.99	440	N	B0000015Y6	Katherine H.
105-9318443-4200264	Illinois	Orange County	Boston	16.99	312	Y	B000002BK9	Samuel P.
105-1872500-0198646	Massachusetts	Bad Blood	Chicago	15.98	413	N	B000068ZVQ	Chris G.
103-2628345-9238664	Canada	Let Go	Mammals	11.99	902	N	B0000010AA	Monique D.
002-1663369-6638649	Ohio	Best of Kansas	Kansas	10.99	440	N	B002MXA7Q0	Katherine H.

Try to guess what they represent. Why is that hard? Because there is no *context*. If we don't know what values are measured and what is measured about them, the values are meaningless. We can make the meaning clear if we organize the values into a **data table** such as this one:

Order Number	Name	State/Country	Price	Area Code	Previous Album Download	Gift?	ASIN	New Purchase Artist
105-2686834-3759466	Katherine H.	Ohio	10.99	440	Nashville	N	B0000015Y6	Kansas
105-9318443-4200264	Samuel R	Illinois	16.99	312	Orange County	Y	B000002BK9	Boston
105-1372500-0198646	Chris G.	Massachusetts	15.98	413	Bad Blood	N	B000068ZVQ	Chicago
103-2628345-9238664	Monique D.	Canada	11.99	902	Let Go	N	B0000010AA	Mammals
002-1663369-6638649	Katherine H.	Ohio	10.99	440	Best of Kansas	N	B002MXA7Q0	Kansas

Now we can see that these are purchase records for album download orders from Amazon. The column titles tell what has been recorded. Each row is about a particular purchase.

What information would provide a **context**? Newspaper journalists know that the lead paragraph of a good story should establish the "Five W's": *who, what, when, where,* and (if possible) *why*. Often, we add *how* to the list as well. The answers to the first two questions are essential. If we don't know *what* values are measured and *who* those values are measured on, the values are meaningless.

Who and What

In general, the rows of a data table correspond to individual **cases** about *whom* (or about which, if they're not people) we record some characteristics. Cases go by different names, depending on the situation.

- Individuals who answer a survey are called **respondents**.
- People on whom we experiment are **subjects** or (in an attempt to acknowledge the importance of their role in the experiment) **participants**.
- Animals, plants, websites, and other inanimate subjects are often called **experimental units**.
- Often we simply call cases what they are: for example, *customers, economic quarters,* or *companies*.
- In a database, rows are called **records**—in this example, purchase records. Perhaps the most generic term is *cases*, but in any event the rows represent the *who* of the data.

> **Data trumps intuition** Amazon monitors and evolves its website to better serve customers and maximize sales. To decide which changes to make, analysts experiment with new designs, offers, recommendations, and links. Statisticians want to know how long you'll spend browsing the site and whether you'll follow the links or purchase the suggested items. As Ronny Kohavi, former director of Data Mining and Personalization for Amazon, said, "Data trumps intuition. Instead of using our intuition, we experiment on the live site and let our customers tell us what works for them."

Look at all the columns to see exactly what each row refers to. Here the cases are different purchase records. You might have thought that each customer was a case, but notice that, for example, Katherine H. appears twice, both in the first and the last row. A common place to find out exactly what each row refers to is the leftmost column. That value often identifies the cases, in this example, it's the order number. If you collect the data yourself, you'll know what the cases are. But, often, you'll be looking at data that someone else collected and you'll have to ask or figure that out yourself.

Often the cases are a **sample** from some larger **population** that we'd like to understand. Amazon doesn't care about just these customers; it wants to understand the buying patterns of *all* its customers, and, generalizing further, it wants to know how to attract other Internet users who may not have made a purchase from Amazon's site. To be able to generalize from the sample of cases to the larger population, we'll want the sample to be *representative* of that population—a kind of snapshot image of the larger world.

We must know *who* and *what* to analyze data. Without knowing these two, we don't have enough information to start. Of course, we'd always like to know more. The more we know about the data, the more we'll understand about the world. If possible, we'd like to know the *when* and *where* of data as well. Values recorded in 1803 may mean something different than similar values recorded last year. Values measured in Tanzania may differ in meaning from similar measurements made in Mexico. And knowing *why* the data were collected can tell us much about its reliability and quality.

How the Data Are Collected

How the data are collected can make the difference between insight and nonsense. As we'll see later, data that come from a voluntary survey on the Internet are almost always worthless. One primary concern of Statistics, to be discussed in Part III, is the design of sound methods for collecting data. Throughout this book, whenever we introduce data, we'll provide a margin note listing the W's (and H) of the data. Identifying the W's is a habit we recommend.

The first step of any data analysis is to know what you are trying to accomplish and what you want to know. To help you use Statistics to understand the world and make decisions, we'll lead you through the entire process of *thinking* about the problem, *showing* what you've found, and *telling* others what you've learned. Every guided example in this book is broken into these three steps: *Think, Show,* and *Tell*. Identifying the problem and the *who* and *what* of the data is a key part of the *Think* step of any analysis. Make sure you know these before you proceed to *Show* or *Tell* anything about the data.

A | S **Activity: Collect data in an experiment on yourself.** With the computer, you can experiment on yourself and then save the data. Go on to the subsequent related activities to check your understanding.

For Example IDENTIFYING THE "WHO"

In December 2013, *Consumer Reports* published an evaluation of 46 tablets from a variety of manufacturers.

QUESTION: Describe the population of interest, the sample, and the *Who* of the study.

ANSWER: The magazine is interested in the performance of tablets currently offered for sale. It tested a sample of 46 tablets, which are the "Who" for these data. Each tablet selected represents all similar tablets offered by that manufacturer.

1.3 Variables

The characteristics recorded about each individual are called **variables**. They are usually found as the columns of a data table with a name in the header that identifies what has been recorded. In the Amazon data table we find the variables *Order Number, Name, State/Country, Price,* and so on.

Categorical Variables

Some variables just tell us what group or category each individual belongs to. Are you male or female? Pierced or not? We call variables like these **categorical**, or **qualitative, variables**. (You may also see them called **nominal variables** because they name categories.) Some variables are clearly categorical, like the variable *State/Country*. Its values are text and those values tell us what category the particular case falls into. But numerals are often used to label categories, so categorical variable values can also be numerals. For example, Amazon collects telephone area codes that *categorize* each phone number into a geographical region. So area code is considered a categorical variable even though it has numeric values. (But see the story in the following box.)

> **Area codes—numbers or categories?** The *What* and *Why* of area codes are not as simple as they may first seem. When area codes were first introduced, AT&T was still the source of all telephone equipment, and phones had dials.
>
> To reduce wear and tear on the dials, the area codes with the lowest digits (for which the dial would have to spin least) were assigned to the most populous regions—those with the most phone numbers and thus the area codes most likely to be dialed. New York City was assigned 212, Chicago 312, and Los Angeles 213, but rural upstate New York was given 607, Joliet was 815, and San Diego 619. For that reason, at one time the numerical value of an area code could be used to guess something about the population of its region. Since the advent of push-button phones, area codes have finally become just categories.

"Far too many scientists have only a shaky grasp of the statistical techniques they are using. They employ them as an amateur chef employs a cookbook, believing the recipes will work without understanding why. A more *cordon bleu* attitude . . . might lead to fewer statistical soufflés failing to rise."

—*The Economist, June 3, 2004, "Sloppy stats shame science"*

Descriptive responses to questions are often categories. For example, the responses to the questions "Who is your cell phone provider?" or "What is your marital status?" yield categorical values. When Amazon considers a special offer of free shipping to customers, it might first analyze how purchases have been shipped in the recent past. Amazon might start by counting the number of purchases shipped in each category: ground transportation, second-day air, and overnight air. Counting is a natural way to summarize a categorical variable like *Shipping Method*. Chapter 2 discusses summaries and displays of categorical variables more fully.

Quantitative Variables

When a variable contains measured numerical values with measurement *units*, we call it a **quantitative variable**. Quantitative variables typically record an amount or degree of something. For quantitative variables, its measurement **units** provide a meaning for the numbers. Even more important, units such as yen, cubits, carats, angstroms, nanoseconds, miles per hour, or degrees Celsius tell us the *scale* of measurement, so we know how far apart two values are. Without units, the values of a measured variable have no meaning. It does little good to be promised a raise of 5000 a year if you don't know whether it will be paid in Euros, dollars, pennies, yen, or Estonian krooni. Chapter 3 discusses quantitative variables. We'll see how to display and summarize them there.

Sometimes a variable with numeric values can be treated as either categorical or quantitative depending on what we want to know from it. Amazon could record your *Age* in years. That seems quantitative, and it would be if the company wanted to know the average age of those customers who visit their site after 3 AM. But suppose Amazon wants to decide which album to feature on its site when you visit. Then thinking of your age in one of the categories Child, Teen, Adult, or Senior might be more useful. So, sometimes whether a variable is treated as categorical or quantitative is more about the question we want to ask rather than an intrinsic property of the variable itself.

Identifiers

> **Privacy and the Internet** You have many Identifiers: a social security number, a student ID number, possibly a passport number, a health insurance number, and probably a Facebook account name. Privacy experts are worried that Internet thieves may match your identity in these different areas of your life, allowing, for example, your health, education, and financial records to be merged. Even online companies such as Facebook and Google are able to link your online behavior to some of these identifiers, which carries with it both advantages and dangers. The National Strategy for Trusted Identities in Cyberspace (www.wired.com/images_blogs/threatlevel/2011/04/NSTICstrategy_041511.pdf) proposes ways that we may address this challenge in the near future.

For a categorical variable like *Sex*, each individual is assigned one of two possible values, say *M* or *F*. But for a variable with ID numbers, such as a *student ID*, each individual receives a unique value. We call a variable like this, which has exactly as many values as cases, an **identifier variable**. Identifiers are useful, but not typically for analysis.

Amazon wants to know who you are when you sign in again and doesn't want to confuse you with some other customer. So it assigns you a unique identifier. Amazon also wants to send you the right product, so it assigns a unique Amazon Standard Identification Number (ASIN) to each item it carries. You'll want to recognize when a variable is playing the role of an identifier so you aren't tempted to analyze it.

Identifier variables themselves don't tell us anything useful about their categories because we know there is exactly one individual in each. However, they are crucial in this era of large data sets because by uniquely identifying the cases, they make it possible to combine data from different sources, protect (or violate) privacy, and provide unique labels. Many large databases are *relational* databases. In a relational database, different data tables link to one another by matching identifiers. In the Amazon example, the *Customer Number*, *ASIN*, and *Transaction Number* are all identifiers. The IP (Internet protocol) address of your computer is another identifier, needed so that the electronic messages sent to you can find you.

Ordinal Variables

A typical course evaluation survey asks, "How valuable do you think this course will be to you?" 1 = Worthless; 2 = Slightly; 3 = Middling; 4 = Reasonably; 5 = Invaluable. Is *Educational Value* categorical or quantitative? Often the best way to tell is to look to the *why* of the study. A teacher might just count the number of students who gave each response for her course, treating *Educational Value* as a categorical variable. When she wants to see whether the course is improving, she might treat the responses as the *amount* of perceived value—in effect, treating the variable as quantitative.

But what are the units? There is certainly an *order* of perceived worth: Higher numbers indicate higher perceived worth. A course that averages 4.5 seems more valuable than one that averages 2, but we should be careful about treating *Educational Value* as purely quantitative. To treat it as quantitative, she'll have to imagine that it has "educational value units" or some similar arbitrary construct. Because there are no natural units, she should be cautious. Variables that report order without natural units are often called **ordinal variables**. But saying "that's an ordinal variable" doesn't get you off the hook. You must still look to the *why* of your study and understand what you want to learn from the variable to decide whether to treat it as categorical or quantitative.

For Example IDENTIFYING "WHAT" AND "WHY" OF TABLETS

RECAP: A *Consumer Reports* article about 46 tablets lists each tablet's manufacturer, price, battery life (hrs.), the operating system (Android, iOS, or Windows), an overall quality score (0–100), and whether or not it has a memory card reader.

QUESTION: Are these variables categorical or quantitative? Include units where appropriate, and describe the "Why" of this investigation.

ANSWER: The variables are
- manufacturer (categorical)
- price (quantitative, $)
- battery life (quantitative, hrs.)
- operating system (categorical)
- performance score (quantitative, no units)
- memory card reader (categorical)

The magazine hopes to provide consumers with the information to choose a good tablet.

Just Checking

In the 2004 Tour de France, Lance Armstrong made history by winning the race for an unprecedented sixth time. In 2005, he became the only 7-time winner and set a new record for the fastest average speed—41.65 kilometers per hour—that stands to this day. In 2012, he was banned for life for doping offenses and stripped of all of his titles. You can find data on all the Tour de France races in the data set **Tour de France 2014**. Here are the first three and last eight lines of the data set. Keep in mind that the entire data set has over 100 entries.

1. List as many of the W's as you can for this data set.
2. Classify each variable as categorical or quantitative; if quantitative, identify the units.

Year	Winner	Country of Origin	Age	Team	Total Time (h/min/s)	Avg. Speed (km/h)	Stages	Total Distance Ridden (km)	Starting Riders	Finishing Riders
1903	Maurice Garin	France	32	La Française	94.33.00	25.7	6	2428	60	21
1904	Henri Cornet	France	20	Cycles JC	96.05.00	25.3	6	2428	88	23
1905	Louis Trousseller	France	24	Peugeot	112.18.09	27.1	11	2994	60	24
⋮										
2007	Alberto Contador	Spain	24	Discovery Channel	91.00.26	38.97	20	3547	189	141
2008	Carlos Sastre	Spain	33	CSC-Saxo Bank	87.52.52	40.50	21	3559	199	145
2009	Alberto Contador	Spain	26	Astana	85.48.35	40.32	21	3460	180	156
2010	Andy Schleck	Luxembourg	25	Saxo Bank	91.59.27	39.590	20	3642	180	170
2011	Cadel Evans	Australia	34	BMC	86.12.22	39.788	21	3430	198	167
2012	Bradley Wiggins	Great Britain	32	Sky	87.34.47	39.827	20	3488	198	153
2013	Christopher Froome	Great Britain	28	Sky	83.56.40	40.551	21	3404	198	169
2014	Vincenzo Nibali	Italy	29	Astana	89.56.06	40.735	21	3663.5	198	164

There's a world of data on the Internet These days, one of the richest sources of data is the Internet. With a bit of practice, you can learn to find data on almost any subject. Many of the data sets we use in this book were found in this way. The Internet has both advantages and disadvantages as a source of data. Among the advantages are the fact that often you'll be able to find even more current data than those we present. The disadvantage is that references to Internet addresses can "break" as sites evolve, move, and die.

Our solution to these challenges is to offer the best advice we can to help you search for the data, wherever they may be residing. We usually point you to a website. We'll sometimes suggest search terms and offer other guidance.

Some words of caution, though: Data found on Internet sites may not be formatted in the best way for use in statistics software. Although you may see a data table in standard form, an attempt to copy the data may leave you with a single column of values. You may have to work in your favorite statistics or spreadsheet program to reformat the data into variables. You will also probably want to remove commas from large numbers and extra symbols such as money indicators ($, ¥, £); few statistics packages can handle these.

WHAT CAN GO WRONG?

- **Don't label a variable as categorical or quantitative without thinking about the data and what they represent.** The same variable can sometimes take on different roles.
- **Don't assume that a variable is quantitative just because its values are numbers.** Categories are often given numerical labels. Don't let that fool you into thinking they have quantitative meaning. Look at the context.
- **Always be skeptical.** One reason to analyze data is to discover the truth. Even when you are told a context for the data, it may turn out that the truth is a bit (or even a lot) different. The context colors our interpretation of the data, so those who want to influence what you think may slant the context. A survey that seems to be about all students may in fact report just the opinions of those who visited a fan website. The question that respondents answered may be posed in a way that influences responses.

What Have We Learned?

Learning Objectives

Understand that data are values, whether numerical or labels, together with their context.

- *Who, what, why, where, when* (and *how*)—the W's—help nail down the context of the data.
- We must know *who, what,* and *why* to be able to say anything useful based on the data. The *who* are the cases. The *what* are the variables. A variable gives information about each of the cases. The *why* helps us decide which way to treat the variables.
- Stop and identify the W's whenever you have data, and be sure you can identify the cases and the variables.

Consider the source of your data and the reasons the data were collected. That can help you understand what you might be able to learn from the data.

Identify whether a variable is being used as categorical or quantitative.
- Categorical variables identify a category for each case. Usually we think about the counts of cases that fall in each category. (An exception is an identifier variable that just names each case.)
- Quantitative variables record measurements or amounts of something; they must have units.
- Sometimes we may treat the same variable as categorical or quantitative depending on what we want to learn from it, which means some variables can't be pigeonholed as one type or the other.

Review of Terms

The key terms are in chapter order so you can use this list to review the material in the chapter.

Data	Recorded values whether numbers or labels, together with their context (p. 1).
Data table	An arrangement of data in which each row represents a case and each column represents a variable (p. 3).
Context	The context ideally tells *who* was measured, *what* was measured, *how* the data were collected, *where* the data were collected, and *when* and *why* the study was performed (p. 4).
Case	An individual about whom or which we have data (p. 4).
Respondent	Someone who answers, or responds to, a survey (p. 4).
Subject	A human experimental unit. Also called a participant (p. 4).
Participant	A human experimental unit. Also called a subject (p. 4).
Experimental unit	An individual in a study for which or for whom data values are recorded. Human experimental units are usually called subjects or participants (p. 4).
Record	Information about an individual in a database (p. 4).
Sample	A subset of a population, examined in hope of learning about the population (p. 4).
Population	The entire group of individuals or instances about whom we hope to learn (p. 4).
Variable	A variable holds information about the same characteristic for many cases (p. 5).
Categorical (or qualitative) variable	A variable that names categories with words or numerals (p. 5).
Nominal variable	The term "nominal" can be applied to a variable whose values are used only to name categories (p. 5).
Quantitative variable	A variable in which the numbers are values of measured quantities with units (p. 5).
Units	A quantity or amount adopted as a standard of measurement, such as dollars, hours, or grams (p. 5).
Identifier variable	A categorical variable that records a unique value for each case, used to name or identify it (p. 6).
Ordinal variable	The term "ordinal" can be applied to a variable whose categorical values possess some kind of order (p. 6).

On the Computer: ENTERING DATA

Most often we find statistics on a computer using a program, or *package*, designed for that purpose. There are many different statistics packages, but they all do essentially the same things.

If you understand what the computer needs to know to do what you want and what it needs to show you in return, you can figure out the specific details of most packages pretty easily.

For example, to get your data into a computer statistics package, you need to tell the computer:

- Where to find the data. This usually means directing the computer to a file stored on your computer's disk or to data on a database. Or it might just mean that you have copied the data from a spreadsheet program or Internet site and it is currently on your computer's clipboard. Usually, the data should be in the form of a data table with cases in the rows and variables in the columns. Most computer statistics packages prefer the *delimiter* that marks the division between elements of a data table to be a tab character (comma is another common delimiter) and the delimiter that marks the end of a case to be a *return* character.
- Where to put the data. (Usually this is handled automatically.)
- What to call the variables. Some data tables have variable names as the first row of the data, and often statistics packages can take the variable names from the first row automatically.
- Excel is often used to help organize, manipulate and prepare data for other software packages. Many of the other packages take Excel files as inputs. Alternatively, you can copy a data table from Excel and Paste it into many packages, or export Excel spreadsheets as tab delimited (.txt) or comma delimited files (.csv), which can be easily shared and imported into other programs. All data files provided with this text are in these two formats.

EXCEL

To open a file containing data in Excel:

- Choose **File > Open**.
- Browse to find the file to open. Excel supports many file formats.
- Other programs can import data from a variety of file formats, but all take both tab delimited (.txt) and comma delimited (.csv) text files.
- You can also copy tables of data from other sources, such as Internet sites, and paste them into an Excel spreadsheet. Excel can recognize the format of many tables copied this way, but this method may not work for some tables.
- Excel may not recognize the format of the data. If data include dates or other special formats ($, €, ¥, etc.), identify the desired format. Select the cells or columns to reformat and choose **Format > Cell**. Often, the General format is the best option.

DATA DESK

To read data into Data Desk

- **File > Open**. The dialog lets you specify variable names (or take them from the first row of the data), the delimiter, or how to read formatted data.
- The **File > Import** command works in the same way, but instead of starting a new data file, it adds the data in the file to the current data file. Data Desk can work with multiple data tables in the same file.
- If the data are already in another program, such as, for example, a spreadsheet, **Copy** the data table (including the column headings). In Data Desk choose **Edit > Paste** Variables. There is no need to create variables first; Data Desk does that automatically. You'll see the same dialog as for Open and Import.

*Screenshot from Microsoft Excel®. Used by permission of Microsoft Corporation.

JMP

To import a text file, choose:

- **File > Open** and select the file from the dialog. At the bottom of the dialog screen you'll see **Open As:**—be sure to change to **Data (Using Preview)**. This will allow you to specify the delimiter and make sure the variable names are correct. (**JMP** also allows various formats to be imported directly, including .xls files.)

You can also paste a data set in directly (with or without variable names) by selecting:

- **File > New > New Data Table** and then **Edit > Paste (or Paste with Column Names** if you copied the names of the variables as well).

Finally, you can import a data set from a URL directly by selecting:

- **File > Internet Open** and pasting in the address of the web site. JMP will attempt to find data on the page. It may take a few tries and some edits to get the data set in correctly.

MINITAB

To import a text or Excel file, choose:

- **File > Open Worksheet**. From **Files of type,** choose **Text (*.txt)** or **Excel (*.xls; *xlsx)**.
- Browse to find and select the file.

- In the lower right corner of the dialog, choose **Open** to open the data file alone, or **Merge** to add the data to an existing worksheet.
- Click **Open**.

R

R can import many types of files, but text files (tab or comma delimited) are easiest. If the file is tab delimited and contains the variable names in the first row, then:

> mydata = read.delim(file.choose())

will give a dialog where you can pick the file you want to import. It will then be in a data frame called mydata. If the file is comma delimted, use:

>mydata=read.csv(file.choose())

COMMENTS

(RStudio provides an interactive dialog that may be easier to use). For other options including the case that the file does not contain variable names, consult **R** help.

SPSS

To import a text file, choose:

- **File > Open > Data**. Under "Files of type", choose **Text (*.txt,*.dat)**. Select the file you want to import. Click **Open.**

- A window will open called **Text Import Wizard**. Follow the steps, depending on the type of file you want to import.

STATCRUNCH

Statcrunch offers several ways to enter data. Click MyStatCrunch > My Data. Click a dataset to analyze the data or edit its properties.

MY DATA

Click a data set link to analyze the data or edit its properties. Want to load a new data set?

- Select a file on my computer.
- Enter the WWW address of a file.

- Paste data into a form.
- Type or paste data into a blank data table.

For the "select a file on my computer" option, Statcrunch offers a choice of whitespace, comma, tab or semicolon delimiters. You may also choose to use the first line as the names of the variables.

After making your choices, select the **Load File** button at the bottom of the screen.

TI 83/84 PLUS

Locate the TI List on the book's website pearsonhighered.com/deveaux or **ActivStats DVD**.
- Link your calculator to the computer using the USB cable.
- Right-click on the file name and choose "Send to TI-Device."

Exercises

Section 1.1

1. **Grocery shopping** Many grocery store chains offer customers a card they can scan when they check out and offer discounts to people who do so. To get the card, customers must give information, including a mailing address and e-mail address. The actual purpose is not to reward loyal customers but to gather data. What data do these cards allow stores to gather, and why would they want that data?

2. **Online shopping** Online retailers such as Amazon.com keep data on products that customers buy, and even products they look at. What does Amazon hope to gain from such information?

Section 1.2

3. **Super Bowl** Sports announcers love to quote statistics. During the Super Bowl, they particularly love to announce when a record has been broken. They might have a list of all Super Bowl games, along with the scores of each team, total scores for the two teams, margin of victory, passing yards for the quarterbacks, and many more bits of information. Identify the *who* in this list.

4. **Nobel laureates** The website www.nobelprize.org allows you to look up all the Nobel prizes awarded in any year. The data are not listed in a table. Rather you drag a slider to the year and see a list of the awardees for that year. Describe the *who* in this scenario.

Section 1.3

5. **Grade levels** A person's grade in school is generally identified by a number.
   a) Give an example of a *why* in which grade level is treated as categorical.
   b) Give an example of a *why* in which grade level is treated as quantitative.

6. **ZIP codes** The Postal Service uses five-digit ZIP codes to identify locations to assist in delivering mail.
   a) In what sense are ZIP codes categorical?
   b) Is there any ordinal sense to ZIP codes? In other words, does a larger ZIP code tell you anything about a location compared to a smaller ZIP code?

7. **Voters** A February 2010 Gallup Poll question asked, "In politics, as of today, do you consider yourself a Republican, a Democrat, or an Independent?" The possible responses were "Democrat," "Republican," "Independent," "Other," and "No Response." What kind of variable is the response?

8. **Job hunting** A June 2011 Gallup Poll asked Americans, "Thinking about the job situation in America today, would you say that it is now a good time or a bad time to find a quality job?" The choices were "Good time" or "Bad time." What kind of variable is the response?

9. **Medicine** A pharmaceutical company conducts an experiment in which a subject takes 100 mg of a substance orally. The researchers measure how many minutes it takes for half of the substance to exit the bloodstream. What kind of variable is the company studying?

10. **Stress** A medical researcher measures the increase in heart rate of patients under a stress test. What kind of variable is the researcher studying?

Chapter Exercises

11. **The news** Find a newspaper or magazine article in which some data are reported. For the data discussed in the article, identify as many of the W's as you can. Include a copy of the article with your report.

12. **The Internet** Find an Internet source that reports on a study and describes the data. Print out the description and identify as many of the W's as you can.

(*Exercises 13–20*) *For each description of data, identify Who and What were investigated and the Population of interest.*

13. **Gaydar** A study conducted by a team of American and Canadian researchers found that during ovulation, a woman can tell whether a man is gay or straight by looking at his face. To explore the subject, the authors conducted three investigations, the first of which involved 40 undergraduate women who were asked to guess the sexual orientation of 80 men based on photos of their face. Half of the men were gay, and the other half were straight. All held similar expressions in the photos or were deemed to be equally attractive. None of the women were using any contraceptive drugs at the time of the test. The result: the closer a woman was to her peak ovulation the more accurate her guess. (http://health.usnews.com/health-news/family-health/brain-and-behavior/articles/2011/06/27/ovulation-seems-to-aid-womens-gaydar)

14. **Hula-hoops** The hula-hoop, a popular children's toy in the 1950s, has gained popularity as an exercise in recent years. But does it work? To answer this question, the American Council on Exercise conducted a study to evaluate the cardio and calorie-burning benefits of "hooping." Researchers recorded heart rate and oxygen consumption of participants, as well as their individual ratings of perceived exertion, at regular intervals during a 30-minute workout. (www.acefitness.org/certifiednewsarticle/1094/)

15. **Bicycle safety** Ian Walker, a psychologist at the University of Bath, wondered whether drivers treat bicycle riders differently when they wear helmets. He rigged his bicycle with an ultrasonic sensor that could measure how close each car was that passed him. He then rode on alternating days with and without a helmet. Out of 2500 cars passing him, he found that when he wore his helmet, motorists passed 3.35 inches closer to him, on average, than when his head was bare. (Source: *NY Times*, Dec. 10, 2006)

16. **Investments** Some companies offer 401(k) retirement plans to employees, permitting them to shift part of their before-tax salaries into investments such as mutual funds. Employers typically match 50% of the employees' contribution up to about 6% of salary. One company, concerned with what it believed was a low employee participation rate in its 401(k) plan, sampled 30 other companies with similar plans and asked for their 401(k) participation rates.

17. **Honesty** Coffee stations in offices often just ask users to leave money in a tray to pay for their coffee, but many people cheat. Researchers at Newcastle University alternately taped two posters over the coffee station. During one week, it was a picture of flowers; during the other, it was a pair of staring eyes. They found that the average contribution was significantly higher when the eyes poster was up than when the flowers were there. Apparently, the mere feeling of being watched—even by eyes that were not real—was enough to encourage people to behave more honestly. (Source: *NY Times*, Dec. 10, 2006)

18. **Blindness** A study begun in 2011 examines the use of stem cells in treating two forms of blindness, Stargardt's disease and dry age-related macular degeneration. Each of the 24 patients entered one of two separate trials in which embryonic stem cells were to be used to treat the condition. (http://www.blindness.org/index.php?view=article&id=2514:stem-cell-clinical-trial-for-stargardt-disease-set-to-begin-&option=com_content&Itemid=122)

19. **Not-so-diet soda** A look at 474 participants in the San Antonio Longitudinal Study of Aging found that participants who drank two or more diet sodas a day "experienced waist size increases six times greater than those of people who didn't drink diet soda." (news.yahoo.com/diet-sodas-dont-help-dieting-175007737.html)

20. **Molten iron** The Cleveland Casting Plant is a large, highly automated producer of gray and nodular iron automotive castings for Ford Motor Company. The company is interested in keeping the pouring temperature of the molten iron (in degrees Fahrenheit) close to the specified value of 2550 degrees. Cleveland Casting measured the pouring temperature for 10 randomly selected crankshafts.

(*Exercises 21–34*) *For each description of data, identify the W's, name the variables, specify for each variable whether its use indicates that it should be treated as categorical or quantitative, and, for any quantitative variable, identify the units in which it was measured (or note that they were not provided).*

21. **Weighing bears** Because of the difficulty of weighing a bear in the woods, researchers caught and measured 54 bears, recording their weight, neck size, length, and sex. They hoped to find a way to estimate weight from the other, more easily determined quantities.

22. **Schools** The State Education Department requires local school districts to keep these records on all students: age, race or ethnicity, days absent, current grade level, standardized test scores in reading and mathematics, and any disabilities or special educational needs.

23. **Arby's menu** A listing posted by the Arby's restaurant chain gives, for each of the sandwiches it sells, the type of meat in the sandwich, the number of calories, and the serving size in ounces. The data might be used to assess the nutritional value of the different sandwiches.

24. **Age and party** The Gallup Poll conducted a representative telephone survey of 1180 American voters during the first quarter of 2007. Among the reported results were the voter's region (Northeast, South, etc.), age, party affiliation, and whether or not the person had voted in the 2006 midterm congressional election.

25. **Babies** Medical researchers at a large city hospital investigating the impact of prenatal care on newborn health collected data from 882 births during 1998–2000. They kept track of the mother's age, the number of weeks the pregnancy lasted, the type of birth (cesarean, induced, natural), the level of prenatal care the mother had (none, minimal, adequate), the birth weight and sex of the baby, and whether the baby exhibited health problems (none, minor, major).

26. **Flowers** In a study appearing in the journal *Science*, a research team reports that plants in southern England are flowering earlier in the spring. Records of the first flowering dates for 385 species over a period of 47 years show that flowering has advanced an average of 15 days per decade, an indication of climate warming, according to the authors.

27. **Herbal medicine** Scientists at a major pharmaceutical firm conducted an experiment to study the effectiveness of an herbal compound to treat the common cold. They exposed each patient to a cold virus, then gave them either the herbal compound or a sugar solution known to have no effect on colds. Several days later they assessed each patient's condition, using a cold severity scale ranging from 0 to 5. They found no evidence of benefits of the compound.

28. **Vineyards** Business analysts hoping to provide information helpful to American grape growers compiled these data about vineyards: size (acres), number of years in existence, state, varieties of grapes grown, average case price, gross sales, and percent profit.

29. **Streams** In performing research for an ecology class, students at a college in upstate New York collect data on streams each year. They record a number of biological, chemical, and physical variables, including the stream name, the substrate of the stream (limestone, shale, or mixed), the acidity of the water (pH), the temperature (°C), and the BCI (a numerical measure of biological diversity).

30. **Fuel economy** The Environmental Protection Agency (EPA) tracks fuel economy of automobiles based on information from the manufacturers (Ford, Toyota, etc.). Among the data the agency collects are the manufacturer, vehicle type (car, SUV, etc.), weight, horsepower, and gas mileage (mpg) for city and highway driving.

31. **Refrigerators** In 2013, *Consumer Reports* published an article evaluating refrigerators. It listed 353 models, giving the brand, cost, size (cu ft), type (such as top freezer), estimated annual energy cost, an overall rating (good, excellent, etc.), and the repair history for that brand (percentage requiring repairs over the past 5 years).

32. **Walking in circles** People who get lost in the desert, mountains, or woods often seem to wander in circles rather than walk in straight lines. To see whether people naturally walk in circles in the absence of visual clues, researcher Andrea Axtell tested 32 people on a football field. One at a time, they stood at the center of one goal line, were blindfolded, and then tried to walk to the other goal line. She recorded each individual's sex, height, handedness, the number of yards each was able to walk before going out of bounds, and whether each wandered off course to the left or the right. No one made it all the way to the far end of the field without crossing one of the sidelines. (Source: *STATS* No. 39, Winter 2004)

33. **Kentucky Derby 2014** The Kentucky Derby is a horse race that has been run every year since 1875 at Churchill Downs, Louisville, Kentucky. The race started as a 1.5-mile race, but in 1896, it was shortened to 1.25 miles because experts felt that 3-year-old horses shouldn't run such a long race that early in the season. (It has been run in May every year but one—1901—when it took place on April 29). Here are the data for the first four and several recent races.

Year	Winner	Jockey	Trainer	Owner	Time
2013	Orb	J. Rosario	S. McGaughey	Stuart Janney & Phipps Stable	2:02.89
2012	I'll Have Another	M. Gutierrez	D. O'Neill	Reddam Racing	2:01.83
2011	Animal Kingdom	J. Velazquez	H. G. Motion	Team Valor	2:02.04
2010	Super Saver	C. Borel	T. Pletcher	WinStar Farm	2:04.45
2009	Mine That Bird	C. Borel	B. Woolley	Double Eagle Ranch	2:02.66
2008	Big Brown	K. Desormeaux	R. Dutrow	IEAH Stables, Pompa et al	2:01.82
...					
1878	Day Star	J. Carter	L. Paul	T. J. Nichols	2:37.25
1877	Baden Baden	W. Walker	E. Brown	Daniel Swigert	2:38
1876	Vagrant	R. Swim	J. Williams	William Astor	2:38.25
1875	Aristides	O. Lewis	A. Williams	H. P. McGrath	2:37.75

Source: Excerpt from HorseHats.com. Published by Thoroughbred Promotions.

34. **Indianapolis 500** The 2.5-mile Indianapolis Motor Speedway has been the home to a race on Memorial Day nearly every year since 1911. Even during the first race, there were controversies. Ralph Mulford was given the checkered flag first but took three extra laps just to make sure he'd completed 500 miles. When he finished, another driver, Ray Harroun, was being presented with the winner's trophy, and Mulford's protests were ignored. Harroun averaged 74.6 mph for the 500 miles. In 2013, the winner, Tony Kanaan, averaged over 187 mph, beating the previous record by over 17 mph!

Here are the data for the first five races and five recent Indianapolis 500 races.

Year	Driver	Time (hr:min:sec)	Speed (mph)
1911	Ray Harroun	6:42:08	74.602
1912	Joe Dawson	6:21:06	78.719
1913	Jules Goux	6:35:05	75.933
1914	René Thomas	6:03:45	82.474
1915	Ralph DePalma	5:33:55.51	89.840
...			
2009	Hélio Castroneves	3:19:34.6427	150.318
2010	Dario Franchitti	3:05:37.0131	161.623
2011	Dan Wheldon	2:56:11.7267	170.265
2012	Dario Franchitti	2:58:51.2532	167.734
2013	Tony Kanaan	2:40:03.4181	187.433

Just Checking ANSWERS

1. *Who*—Tour de France races; *What*—year, winner, country of origin, age, team, total time, average speed, stages, total distance ridden, starting riders, finishing riders; *How*—official statistics at race; *Where*—France (for the most part); *When*—1903 to 2014; *Why*—not specified (To see progress in speeds of cycling racing?)

2.
Variable	Type	Units
Year	Quantitative or Identifier	Years
Winner	Categorical	
Country of Origin	Categorical	
Age	Quantitative	Years
Team	Categorical	
Total Time	Quantitative	Hours/minutes/seconds
Average Speed	Quantitative	Kilometers per hour
Stages	Quantitative	Counts (stages)
Total Distance	Quantitative	Kilometers
Starting Riders	Quantitative	Counts (riders)
Finishing Riders	Quantitative	Counts (riders)

chapter 2
Displaying and Describing Categorical Data

2.1 Summarizing and Displaying a Single Categorical Variable

2.2 Exploring the Relationship Between Two Categorical Variables

Where are we going?

What is your class: Freshman, Sophomore, Junior, or Senior? What is your blood type? Which candidate do you plan to vote for?

Data are not just numbers—the answers to all these questions are categories. We'll study categorical variables like these, but things get even more exciting when we look at how categorical variables work together. Are men or women more likely to be Democrats? Are people with blue eyes more likely to be left-handed?

You often see categorical data displayed in pie charts or bar charts, or summarized in tables, sometimes in confusing ways. But, with a little skill, it's not hard to do it right.

Table 2.1
Part of a data table showing four variables for nine people aboard the *Titanic*.

What happened on the *Titanic* at 11:40 on the night of April 14, 1912, is well known. Frederick Fleet's cry of "Iceberg, right ahead" and the three accompanying pulls of the crow's nest bell signaled the beginning of a nightmare that has become legend. By 2:15 a.m., the *Titanic*, thought by many to be unsinkable, had sunk, leaving more than 1500 passengers and crew members on board to meet their icy fate.

Below are some data about the passengers and crew aboard the *Titanic*. Each case (row) of the data table represents a person on board the ship. The variables are the person's *Survival* status (Dead or Alive), *Age* (Adult or Child), *Sex* (Male or Female), and ticket *Class* (First, Second, Third, or Crew). (Data in **Titanic**)

Survival	Age	Sex	Class
Dead	Adult	Male	Third
Dead	Adult	Male	Crew
Dead	Adult	Male	Third
Dead	Adult	Male	Crew
Dead	Adult	Male	Crew
Dead	Adult	Male	Crew
Alive	Adult	Female	First
Dead	Adult	Male	Third
Dead	Adult	Male	Crew

A S **Video: The Incident** tells the story of the *Titanic*, and includes rare film footage.

The problem with a data table like this—and in fact with all data tables—is that you can't *see* what's going on. And seeing is just what we want to do. We need ways to show the data so that we can see patterns, relationships, trends, and exceptions.

2.1 Summarizing and Displaying a Single Categorical Variable

Who	People on the *Titanic*
What	Survival status, age, sex, ticket class
When	April 14, 1912
Where	North Atlantic
How	A variety of sources and Internet sites
Why	Historical interest

So, what should we do with data like these? There are three things you should always do first with data.

The Three Rules of Data Analysis

1. **Make a picture.** A display of your data will reveal things you're not likely to see in a table of numbers and will help you to *Think* clearly about the patterns and relationships that may be hiding in your data.
2. **Make a picture.** A well-designed display will *Show* the important features and patterns in your data. It could also show you things you did not expect to see: extraordinary (possibly wrong) data values or unexpected patterns.
3. **Make a picture.** The best way to *Tell* others about your data is with a well-chosen picture.

These are the three rules of data analysis. There are pictures of data throughout the book, and new kinds keep showing up. These days, technology makes drawing pictures of data easy, so there is no reason not to follow the three rules.

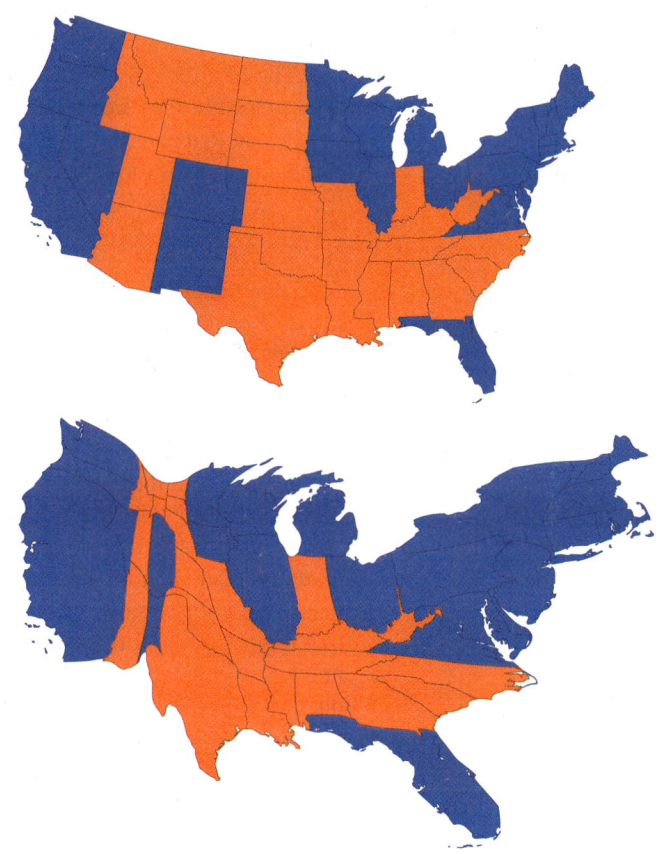

Figure 2.1

A picture to tell a story. In November 2012, Barack Obama was re-elected as President of the United States. News reports commonly showed the election results with maps like the one on top, coloring states won by Obama blue and those won by his opponent, Mitt Romney, red. Even though Romney lost, doesn't it look like there's more red than blue? That's because some of the larger states like Montana and Wyoming have far fewer voters than some of the smaller states like Maryland and Connecticut. The strange-looking map on the bottom cleverly distorts the states to resize them proportional to their populations. By sacrificing an accurate display of the land areas, we get a better impression of the votes cast, giving us a clear picture of Obama's victory.
(Source: "A Picture to tell a Story" by M. E. J. Newman from Maps of the 2008 US Presidential Election Results. Copyright © 2008 by Mark Newman. Used By Permission of Mark Newman.)

The Area Principle

We're ready to follow the three rules of data analysis and make a picture of the data. But a bad picture can distort our understanding rather than help it. On the next page is a graph of the *Titanic* data. What impression do you get about who was aboard the ship?

It looks like most of the people on the *Titanic* were crew members, with a few passengers along for the ride. That doesn't seem right. What's wrong? The lengths of the ships *do* match the number of people in each ticket class category. However, experience and psychological tests show that our eyes tend to be more impressed by the *area* than by other aspects of each ship image. So, even though the *length* of each ship matches up with one of the totals, it's the associated *area* in the image that we notice. There were about 3 times as many crew as second-class passengers, and the ship depicting the number of

Figure 2.2
How many people were in each class on the *Titanic*? From this display, it looks as though the service must have been great, since most aboard were crew members. Although the length of each ship here corresponds to the correct number, the impression is all wrong. In fact, only about 40% were crew.

Activity: Make and examine a table of counts. Even data on something as simple as hair color can reveal surprises when you organize it in a data table.

Class	Count
First	325
Second	285
Third	706
Crew	885

Table 2.2
A frequency table of the *Titanic* passengers.

Class	%
First	14.77
Second	12.95
Third	32.08
Crew	40.21

Table 2.3
A relative frequency table for the same data.

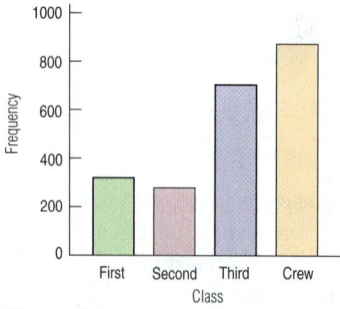

Figure 2.3
People on the Titanic by Ticket Class. With the area principle satisfied, we can see the true distribution more clearly.

crew is about 3 times longer than the ship depicting second-class passengers. The problem is that it occupies about 9 times the area. That just isn't a correct impression.

The best data displays observe a fundamental principle of graphing data called the **area principle**. The area principle says that the area occupied by a part of the graph should correspond to the magnitude of the value it represents. Violations of the area principle are a common way to lie (or, since most mistakes are unintentional, we should say err) with Statistics.

Frequency Tables

A display that distorts the number of cases can be misleading. But before we can follow the area principle to accurately display a categorical variable, we need to organize the number of cases associated with each category. Even when we have thousands of cases, a variable like ticket *Class,* with only a few categories, is easy to organize. We just count the number of cases corresponding to each category and put them in a table. A **frequency table** (Table 2.2) also records the totals and uses the category names to label each row. We use the names of the categories to label each row in the frequency table. For ticket *Class,* these are "First," "Second," "Third," and "Crew."

For a variable with dozens or hundreds of categories, a frequency table will be much harder to read. You might want to combine categories into larger headings. For example, instead of counting the number of students from each state, you might group the states into regions like "Northeast," "South," "Midwest," "Mountain States," and "West." If the number of cases in several categories is relatively small, you can put them together into one category labeled "Other."

Counts are useful, but sometimes we want to know the fraction or *proportion* of the data in each category, so we divide the counts by the total number of cases. Usually we multiply by 100 to express these proportions as percentages. A **relative frequency table** displays the *percentages,* rather than the counts, of the values in each category. Both types of tables show how the cases are distributed across the categories. In this way, they describe the **distribution** of a categorical variable because they name the possible categories and tell how frequently each occurs.

Bar Charts

Figure 2.3 shows a chart that obeys the area principle. It's not as visually entertaining as the ships, but it does give an *accurate* visual impression of the distribution. The height of each bar shows the count for its category. The bars are the same width, so their heights determine their areas, and the areas are proportional to the counts in each class. Now it's easy to see that the majority of people on board were *not* crew, as the ships picture led us

CHAPTER 2 Displaying and Describing Categorical Data

Activity: Bar Charts. Watch bar charts grow from data; then use your statistics package to create some bar charts for yourself.

to believe. We can also see that there were about 3 times as many crew as second-class passengers. And there were more than twice as many third-class passengers as either first- or second-class passengers, something you may have missed in the frequency table. Bar charts make these kinds of comparisons easy and natural.

A **bar chart** displays the distribution of a categorical variable, showing the counts for each category next to each other for easy comparison. Bar charts should have small spaces between the bars to indicate that these are freestanding bars that could be rearranged into any order. The bars are lined up along a common base.

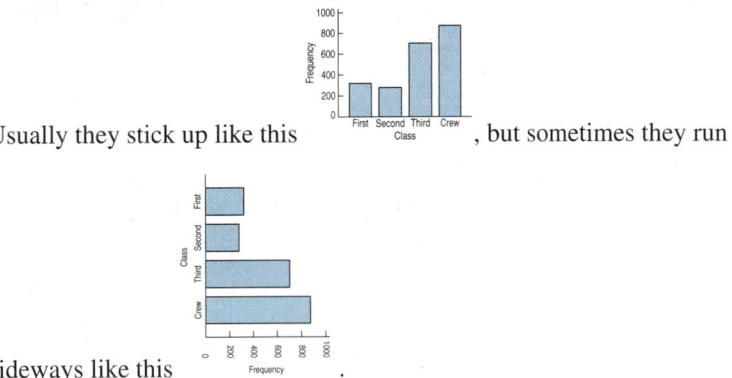

Usually they stick up like this [chart], but sometimes they run sideways like this [chart].

If we really want to draw attention to the relative proportion of passengers falling into each of these classes, we could replace the counts with percentages and use a **relative frequency bar chart**.

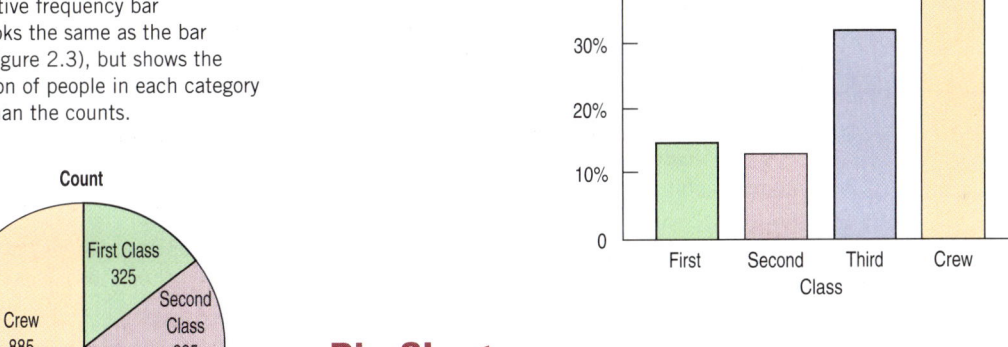

Figure 2.4
The relative frequency bar chart looks the same as the bar chart (Figure 2.3), but shows the proportion of people in each category rather than the counts.

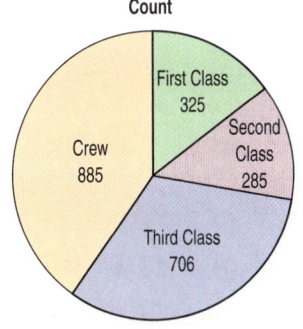

Figure 2.5
Number of *Titanic* passengers in each class.

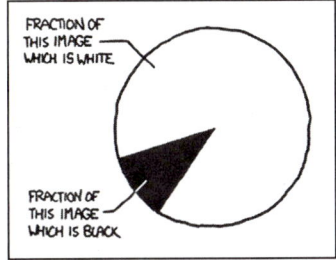

© 2013 Randall Munroe. Reprinted with permission. All rights reserved.

Pie Charts

Another common display that shows how a whole group breaks into several categories is a pie chart. **Pie charts** show the whole group of cases as a circle. They slice the circle into pieces whose size is proportional to the fraction of the whole in each category.

Pie charts give a quick impression of how a whole group is partitioned into smaller groups. Because we're used to cutting up pies into 2, 4, or 8 pieces, pie charts are good for seeing relative frequencies near 1/2, 1/4, or 1/8. For example, you may be able to tell that the pink slice, representing the second-class passengers, is very close to 1/8 of the total. It's harder to see that there were about twice as many third-class as first-class passengers. Which category had the most passengers? Were there more crew or more third-class passengers? Comparisons such as these are easier in a bar chart.

For example, Figure 2.6 shows three pie charts that look pretty much alike along with bar charts of the same data. The bar charts show three distinctly different patterns, but it is almost impossible to see those in the pie charts.

No matter what display you choose for your variable, make sure that the data are counts or percentages of cases. For example, a pie chart showing the price of each ticket class wouldn't make sense. Price is quantitative. So before making a bar chart or a pie chart, always check the **categorical data condition** that the data are counts or percentages of individuals in categories.

Figure 2.6
Patterns that are easy to see in the bar charts are often hard to see in the corresponding pie charts.

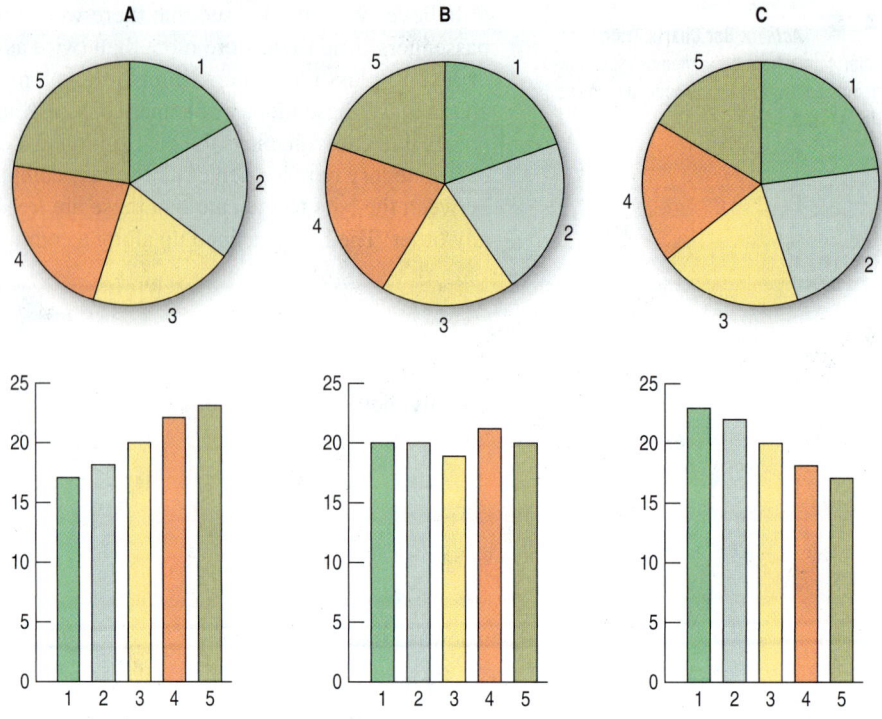

> **THINK BEFORE YOU DRAW**
> Our first rule of data analysis is *Make a picture*. But what kind of picture? We don't have a lot of options—yet. There's more to Statistics than pie charts and bar charts, and knowing when to use each type of graph is a critical first step in data analysis. That decision depends in part on what type of data we have.
> If you want to make a relative frequency bar chart or a pie chart, you'll need to also make sure that the categories don't overlap so that no individual is counted twice. If the categories do overlap, you can still make a bar chart, but the percentages won't add up to 100%. For the *Titanic* data, either kind of display is appropriate because the categories don't overlap.
> For any statistical analysis, a crucial decision is the choice of a proper method. Be sure to think about the situation at hand, and make sure to check that the type of analysis you plan is appropriate. The Categorical Data Condition is just the first of many such checks.

2.2 Exploring the Relationship Between Two Categorical Variables

Activity: Children at Risk. This activity looks at the fates of children aboard the *Titanic*; the subsequent activity shows how to make such tables on a computer.

Contingency Tables: Children and First-Class Ticket Holders First?

Only 32% of those aboard the *Titanic* survived. Was that survival rate the same for men and women? For children and adults? For all ticket classes? It's often more interesting to ask if one variable relates to another. For example, was there a relationship between the kind of ticket a passenger held and the passenger's chances of making it into a lifeboat?

To answer that question we can arrange the counts for the two categorical variables, *Survival* and ticket *Class*, in a table. Table 2.4 shows each person aboard the *Titanic*

Table 2.4

Contingency table of ticket **Class** *and* **Survival.** The bottom line of "Totals" is the same as the previous frequency table.

		Class				
		First	Second	Third	Crew	Total
Survival	Alive	203	118	178	212	711
	Dead	122	167	528	673	1490
	Total	325	285	706	885	2201

classified according to both their ticket *Class* and their *Survival*. Because the table shows how the individuals are distributed along each variable, contingent on the value of the other variable, such a table is called a **contingency table**.

Each **cell** of the table gives the count for a combination of values of the two variables. The margins of the table, both on the right and at the bottom, give totals. The bottom line of the table is just the frequency distribution of ticket *Class*. The right column of the table is the frequency distribution of the variable *Survival*. When presented like this, in the margins of a contingency table, the frequency distribution of one of the variables is called its **marginal distribution**. The marginal distribution can be expressed either as counts or percentages.

If you look down the column for second-class passengers to the first row, you'll find the cell containing the 118 second-class passengers who survived. Looking at the cell to its right we see that more third-class passengers (178) survived. But, does that mean that third-class passengers were more *likely* to survive? It's true that *more* third-class passengers survived, but there were many more third-class passengers on board the ship. To compare the two numbers fairly, we need to express them as percentages—but as a percentage of what?

For any cell, there are three choices of percentage. We could express the 118 second-class survivors as 5.4% of all the passengers on the *Titanic* (the *overall percent*), as 16.6% of all the survivors (the *row percent*), or as 41.4% of all second-class passengers (the *column percent*). Each of these percentages is potentially interesting.

Statistics programs offer all three. Unfortunately, they often put them all together in each cell of the table. The resulting table holds lots of information, but it can be hard to understand:

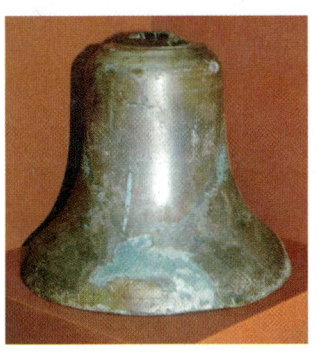

A bell-shaped artifact from the *Titanic*.

Table 2.5

Another contingency table of ticket **Class.** This time we see not only the counts for each combination of *Class* and *Survival* (in bold) but the percentages these counts represent. For each count, there are three choices for the percentage: by row, by column, and by table total. There's probably too much information here for this table to be useful.

			Class				
			First	Second	Third	Crew	Total
Survival	Alive	Count	203	118	178	212	711
		% of Row	28.6%	16.6%	25.0%	29.8%	100%
		% of Column	62.5%	41.4%	25.2%	24.0%	32.3%
		% of Overall Total	9.2%	5.4%	8.1%	9.6%	32.3%
	Dead	Count	122	167	528	673	1490
		% of Row	8.2%	11.2%	35.4%	45.2%	100%
		% of Column	37.5%	58.6%	74.8%	76.0%	67.7%
		% of Overall Total	5.5%	7.6%	24.0%	30.6%	67.7%
	Total	Count	325	285	706	885	2201
		% of Row	14.8%	12.9%	32.1%	40.2%	100%
		% of Column	100%	100%	100%	100%	100%
		% of Overall Total	14.8%	12.9%	32.1%	40.2%	100%

To simplify the table, let's first pull out the overall percent values:

Table 2.6

A contingency table of *Class* by *Survival* with only the overall percentages.

		Class				
		First	Second	Third	Crew	Total
Survival	Alive	9.2%	5.4%	8.1%	9.6%	32.3%
	Dead	5.5%	7.6%	24.0%	30.6%	67.7%
	Total	14.8%	12.9%	32.1%	40.2%	100%

These percentages tell us what percent of *all* passengers belong to each combination of column and row category. For example, we see that 8.1% of the people aboard the *Titanic* were surviving third-class ticket holders, but only 5.4% were surviving second-class ticket holders. Comparing these percentages, you might think that the chances of surviving were better in third class than in second. But be careful. There were many more third-class than second-class passengers on the *Titanic*, so there were more third-class survivors. Comparing the overall percentages, 5.4% and 8.1%, doesn't answer the question about whether second-class passengers were more or less likely to survive than third-class passengers. Overall percentages can be useful, but they don't answer questions involving comparisons of percentages of one variable across levels of the other.

PERCENT OF WHAT?

The English language can be tricky when we talk about percentages. If you're asked "What percent *of the survivors* were in second class?" it's pretty clear that we're interested only in survivors. So we should restrict our attention only to the survivors and look at the number of second-class passengers among all the survivors—in other words, the row percent.

But if you're asked "What percent were second-class passengers who survived?" you have a different question. Be careful; now we need to consider everyone on board, so 2201 should be the denominator, and the answer is the overall percent.

And if you're asked "What percent of the second-class passengers survived?" you have a third question. Now the denominator is the 285 second-class passengers, and the answer is the column percent.

Always be sure to ask "percent of what?" That will help you to know *Who* we're talking about and whether we want *row*, *column*, or *overall* percentages.

For Example FINDING MARGINAL DISTRIBUTIONS

A recent Gallup poll asked 1008 Americans age 18 and over whether they planned to watch the upcoming Super Bowl. The pollster also asked those who planned to watch whether they were looking forward more to seeing the football game or the commercials. The results are summarized in the table:

		Sex		
		Male	Female	Total
Response	Game	279	200	479
	Commercials	81	156	237
	Won't Watch	132	160	292
	Total	492	516	1008

QUESTION: What's the marginal distribution of the responses?

ANSWER: To determine the percentages for the three responses, divide the count for each response by the total number of people polled:

$$\frac{479}{1008} = 47.5\% \quad \frac{237}{1008} = 23.5\% \quad \frac{292}{1008} = 29.0\%.$$

According to the poll, 47.5% of American adults were looking forward to watching the Super Bowl game, 23.5% were looking forward to watching the commercials, and 29% didn't plan to watch at all.

Conditional Distributions

Rather than looking at the overall percentages, it's more interesting to ask whether the chance of surviving the *Titanic* sinking *depended* on ticket class. We can look at this question in two ways. First, we could ask how the distribution of ticket *Class* changes between survivors and nonsurvivors. To do that, we look at the *row percentages*:

Table 2.7
The conditional distribution of ticket *Class* conditioned on each value of *Survival: Alive* and *Dead*.

		Class				
		First	Second	Third	Crew	Total
Survival	Alive	203	118	178	212	711
		28.6%	16.6%	25.0%	29.8%	100%
	Dead	122	167	528	673	1490
		8.2%	11.2%	35.4%	45.2%	100%

By focusing on each row separately, we see the distribution of class for each *condition* of surviving or not. The sum of the percentages in each row is 100%, and we divide that up by ticket class. Here, we temporarily restrict our attention to survivors and make a pie chart for them. Then we focus on the nonsurvivors and make their pie chart. These pie charts show the distribution of ticket classes *for each row* of the table: survivors and nonsurvivors. The distributions we create this way are called **conditional distributions** because they show the distribution of one variable for just those cases that satisfy a condition on another variable.

Figure 2.7
Pie charts of the conditional distributions of ticket Class for the survivors and nonsurvivors, separately. Do the distributions appear to be the same? We're primarily concerned with percentages here, so pie charts are a reasonable choice.

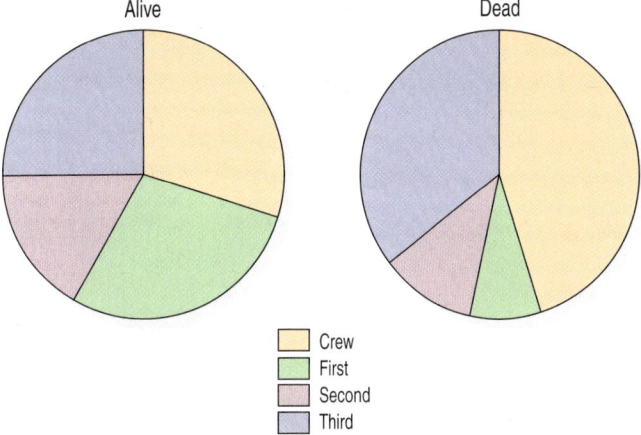

For Example   FINDING CONDITIONAL DISTRIBUTIONS

RECAP: The table shows results of a poll asking adults whether they were looking forward to the Super Bowl game, looking forward to the commercials, or didn't plan to watch.

		Sex		
		Male	Female	Total
Response	Game	279	200	479
	Commercials	81	156	237
	Won't Watch	132	160	292
	Total	492	516	1008

QUESTION: How do the conditional distributions of *Sex* differ among the responses to the question about commercials?

ANSWER: Look at the group of people who responded "Commercials" and determine what percent of them were male and female:

$$\frac{81}{237} = 34.2\% \quad \frac{156}{237} = 65.8\%.$$

Women make up a sizable majority of the adult Americans who look forward to seeing Super Bowl commercials more than the game itself. Nearly 66% of people who voiced a preference for the commercials were women, and only 34% were men.

But we still haven't answered our question. Instead, we can turn the conditioning around and look at the distribution of *Survival* for each category of ticket *Class*. The *column percentages* show us whether the chance of surviving was roughly the same *for each of the four classes.* Now the percentages in each column add to 100% because we've restricted our attention, in turn, to each of the four ticket classes:

		Class				
		First	Second	Third	Crew	Total
Survival	Alive Count	203	118	178	212	711
	% of Column	62.5%	41.4%	25.2%	24.0%	32.3%
	Dead Count	122	167	528	673	1490
	% of Column	37.5%	58.6%	74.8%	76.0%	67.7%
	Total Count	325	285	706	885	2201
		100%	100%	100%	100%	100%

Table 2.8

A contingency table of Class by Survival with only counts and column percentages. Each column represents the conditional distribution of *Survival* for a given category of ticket *Class*.

Looking at how the percentages change across each row, it sure looks like ticket class mattered in whether a passenger survived. To make it more vivid, we could display the percentages for surviving and not for each *Class* in a side-by-side bar chart:

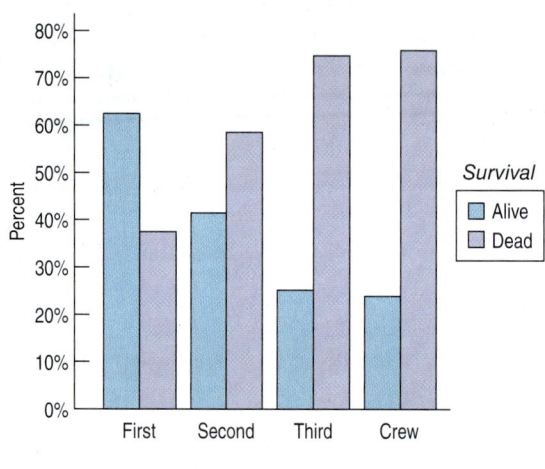

Figure 2.8

Side-by-side bar chart showing the conditional distribution of *Survival* for each category of ticket *Class*. The corresponding pie charts would have only two categories in each of four pies, so bar charts seem the better alternative.

These bar charts are particularly simple because, for the variable *Survival*, we have only two alternatives: Alive and Dead. When we have only two categories, we can simplify the display even more by dropping one category. Here are the percentages of dying *across the classes* displayed in one chart:

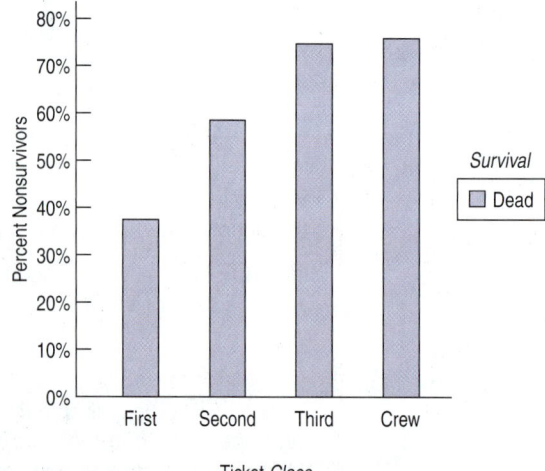

Figure 2.9

Bar chart showing just nonsurvivor percentages for each value of ticket *Class*. Because we have only two values, the second bar doesn't add any information. Compare this chart to the side-by-side bar chart shown earlier.

Now it's easy to compare the risks. Among first-class passengers, 37.5% perished, compared to 58.6% for second-class ticket holders, 74.8% for those in third class, and 76.0% for crew members.

If the risk had been about the same across the ticket classes, we would have said that survival was *independent* of class. But it's not. The differences we see among these conditional distributions suggest that survival may have depended on ticket class. You may find it useful to look at the conditional distributions of each variable in a contingency table to explore the dependence between them.

It is interesting to know that *Class* and *Survival* are associated. That's an important part of the *Titanic* story. And we know how important this is because the margins show us the actual numbers of people involved.

Variables can be associated in many ways and to different degrees. The best way to tell whether two variables are associated is to ask whether they are *not*.[1] In a contingency table, when the distribution of *one* variable is the same for all categories of another, we say that the variables are **independent**. That tells us there's no association between these variables. We'll see a way to check for independence formally later in the book. For now, we'll just compare the distributions.

For Example LOOKING FOR ASSOCIATIONS BETWEEN VARIABLES

RECAP: The table shows results of a poll asking adults whether they were looking forward to the Super Bowl game, looking forward to the commercials, or didn't plan to watch.

		Sex		
		Male	Female	Total
Response	Game	279	200	479
	Commercials	81	156	237
	Won't Watch	132	160	292
	Total	492	516	1008

QUESTION: Does it seem that there's an association between interest in Super Bowl TV coverage and a person's sex?

ANSWER: First find the distribution of the three responses for the men (the column percentages):

$$\frac{279}{492} = 56.7\% \quad \frac{81}{492} = 16.5\% \quad \frac{132}{492} = 26.8\%.$$

Then do the same for the women who were polled, and display the two distributions with a side-by-side bar chart.

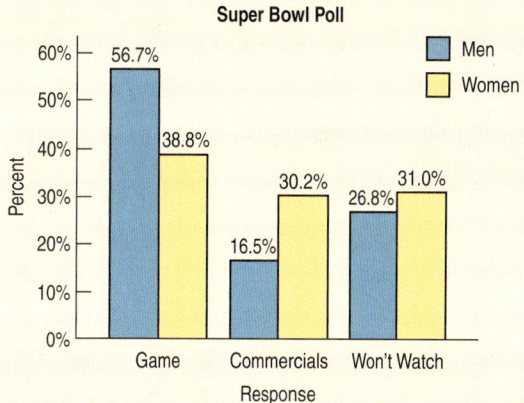

Based on this poll, it appears that women were only slightly more likely than men to say they won't watch the Super Bowl: 31% of the women said they didn't plan to watch, compared to just under 27% of men. Among those who planned to watch, however, there appears to be an association between the viewer's sex and what the

---

[1] This kind of "backward" reasoning shows up surprisingly often in science—and in Statistics. We'll see it again.

viewer is most looking forward to. While more women are interested in the game (38.8%) than the commercials (30.2%), the margin among men is much wider: 56.7% of men said they were looking forward to seeing the game, compared to only 16.5% who cited the commercials.

Just Checking

A Statistics class reports the following data on *Sex* and *Eye Color* for students in the class:

		Eye Color			
		Blue	Brown	Green/Hazel/Other	Total
Sex	Male	6	20	6	32
	Female	4	16	12	32
	Total	10	36	18	64

1. What percent of females are brown-eyed?
2. What percent of brown-eyed students are female?
3. What percent of students are brown-eyed females?
4. What's the distribution of *Eye Color*?
5. What's the conditional distribution of *Eye Color* for the males?
6. Compare the percent who are female among the blue-eyed students to the percent of all students who are female.
7. Does it seem that *Eye Color* and *Sex* are independent? Explain.

Segmented Bar Charts

We could display the *Titanic* information by dividing up bars rather than circles. The resulting **segmented bar chart** treats each bar as the "whole" and divides it proportionally into segments corresponding to the percentage in each group. We can clearly see that the distributions of ticket *Class* are different, indicating again that survival was not independent of ticket *Class*.

Figure 2.10

A segmented bar chart for **Class** *by* **Survival**. Notice that although the totals for survivors and nonsurvivors are quite different, the bars are the same height because we have converted the numbers to *percentages*. Compare this display with the side-by-side pie charts of the same data in Figure 2.7.

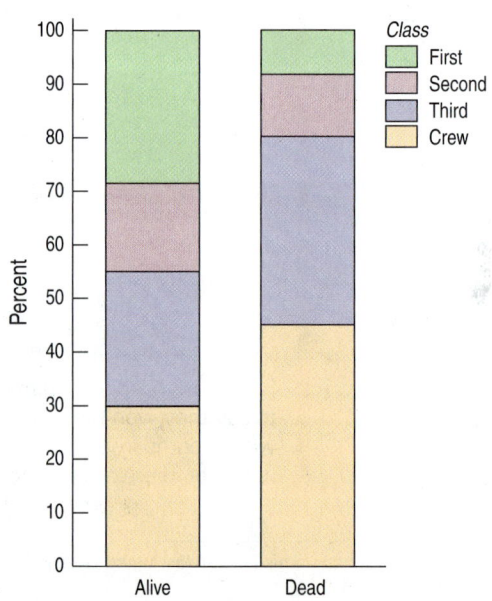

Step-by-Step Example EXAMINING CONTINGENCY TABLES

We asked for a picture of a man eating fish. This is what we got.

Medical researchers followed 6272 Swedish men for 30 years to see whether there was any association between the amount of fish in their diet and prostate cancer. Their results are summarized in this table:

		Prostate Cancer	
		No	Yes
Fish Consumption	Never/Seldom	110	14
	Small Part of Diet	2420	201
	Moderate Part	2769	209
	Large Part	507	42

Question: Is there an association between fish consumption and prostate cancer?

THINK ▶ Plan
State what the problem is about.

I want to know if there is an association between fish consumption and prostate cancer.

Variables
Identify the variables and report the W's.

The individuals are 6272 Swedish men followed by medical researchers for 30 years. The variables record their fish consumption and whether or not they were diagnosed with prostate cancer.

Be sure to check the appropriate condition.

✓ **Categorical Data Condition:** I have counts for both fish consumption and cancer diagnosis. The categories of diet do not overlap, and the diagnoses do not overlap. It's okay to draw pie charts or bar charts.

SHOW ▶ Mechanics
Check the marginal distributions first before looking at the two variables together.

		Prostate Cancer		
		No	Yes	Total
Fish Consumption	Never/Seldom	110	14	124 (2.0%)
	Small Part of Diet	2420	201	2621 (41.8%)
	Moderate Part	2769	209	2978 (47.5%)
	Large Part	507	42	549 (8.8%)
	Total	5806 (92.6%)	466 (7.4%)	6272 (100%)

Two categories of the diet are quite small, with only 2.0% Never/seldom eating fish and 8.8% in the "Large part" category. Overall, 7.4% of the men in this study had prostate cancer.

Then, make appropriate displays to see whether there is a difference in the relative proportions. These pie charts compare fish consumption for men who have prostate cancer to fish consumption for men who don't.

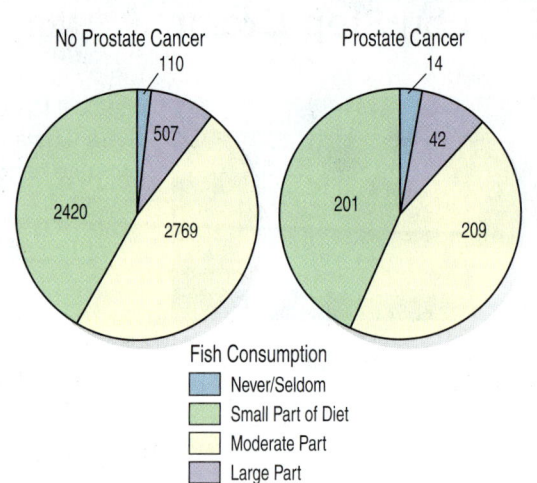

It's hard to see much difference in the pie charts. So, I made a display of the row percentages. Because there are only two alternatives, I chose to display the risk of prostate cancer for each group:

Both pie charts and bar charts can be used to compare conditional distributions. Here we compare prostate cancer rates based on differences in fish consumption.

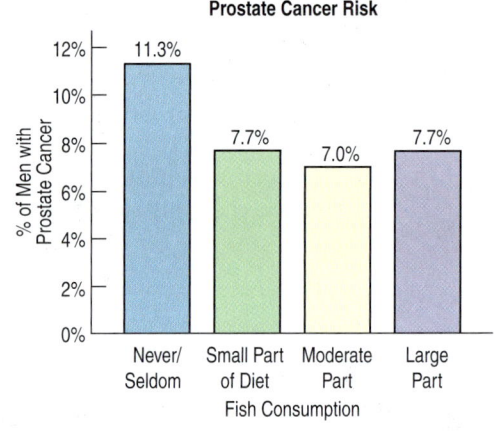

TELL ➡ Conclusion Interpret the patterns in the table and displays in context. If you can, discuss possible real-world consequences. Be careful not to overstate what you see. The results may not generalize to other situations.

Overall, there is a 7.4% rate of prostate cancer among men in this study. Most of the men (89.3%) ate fish either as a moderate or small part of their diet. From the pie charts, it's hard to see a difference in cancer rates among the groups. But in the bar chart, it looks like the cancer rate for those who never/seldom ate fish may be somewhat higher.

However, only 124 of the 6272 men in the study fell into this category, and only 14 of them developed prostate cancer. More study would probably be needed before we would recommend that men change their diets.[2]

---

[2]"Fatty fish consumption and risk of prostate cancer," *Lancet*, June 2001. The original study actually used pairs of twins, which enabled the researchers to discern that the risk of cancer for those who never ate fish actually *was* substantially greater. Using pairs is a special way of gathering data. We'll discuss such study design issues and how to analyze the data in the later chapters.

This study is an example of looking at a sample of data to learn something about a larger population, one of the main goals of this book. We care about more than these particular 6272 Swedish men. We hope that learning about their experiences will tell us something about the value of eating fish in general. That raises several questions. What population do we think this sample might represent? Do we hope to learn about all Swedish men? About all men? How do we know that other factors besides that amount of fish they ate weren't associated with prostate cancer? Perhaps men who eat fish often have other habits that distinguish them from the others and maybe those other habits are what actually kept their cancer rates lower.

Observational studies, like this one, often lead to contradictory results because we can't control all the other factors. In fact, a later paper, published in 2011, based on data from a cancer prevention trial on 3400 men from 1994 to 2003, showed that some fatty acids may actually increase the risk of prostate cancer.[3] We'll discuss the pros and cons of observational studies and experiments where we can control the factors in Chapter 12.

WHAT CAN GO WRONG?

- **Don't violate the area principle.** This is probably the most common mistake in a graphical display. It is often made in the cause of artistic presentation. Here, for example, are two displays of the pie chart of the *Titanic* passengers by class:

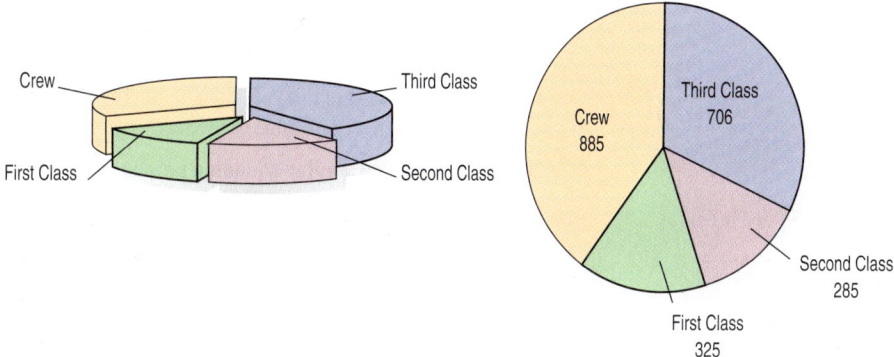

The one on the left looks pretty, doesn't it? But showing the pie on a slant violates the area principle and makes it much more difficult to compare fractions of the whole made up of each class—the principal feature that a pie chart ought to show.

- **Keep it honest.** Here's a pie chart that displays data on the percentage of high school students who engage in specified dangerous behaviors as reported by the Centers for Disease Control and Prevention. What's wrong with this plot?

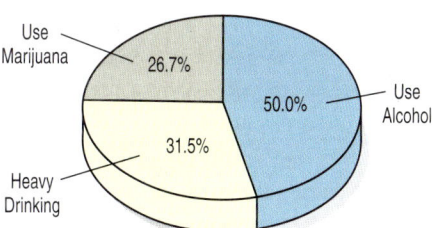

Try adding up the percentages. Or look at the 50% slice. Does it look right? Then think: What are these percentages of? Is there a "whole" that has been sliced up? In a pie chart, the proportions shown by each slice of the pie must add up to 100% and each individual must fall into only one category. Of course, showing the pie on a slant makes it even harder to detect the error.

---

[3]"Serum phospholipid fatty acids and prostate cancer risk: Results from the Prostate Cancer Prevention Trial," *American Journal of Epidemiology*, 2011.

Here's another one. The following chart shows the average number of texts in various time periods by American cell phone customers in the period 2011 to 2014.

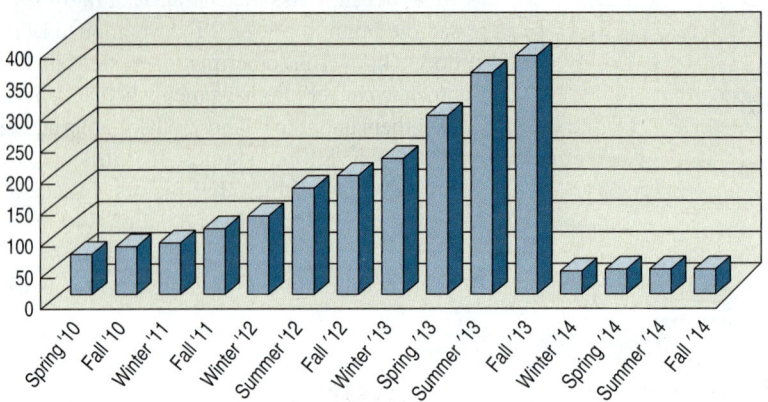

It may look as though text messaging decreased suddenly in 2014, which probably doesn't jibe with your experience. In fact, this chart has several problems. First, it's not a bar chart. Bar charts display counts of categories. This bar chart is a plot of a quantitative variable (average number of texts) against time—although to make it worse, some of the time periods are missing. Even though these flaws are already fatal, the worst mistake is one that can't be seen from the plot. In 2014, the company reporting the data switched from reporting the average number of texts per month (reported each quarter) to average number of texts per day. So, the numbers in the last four bars should be multiplied by 12 to make them comparable to the rest.

- **Don't confuse similar-sounding percentages.** These percentages sound similar but are different:

		Class				Total
		First	Second	Third	Crew	
Survival	Alive	203	118	178	212	711
	Dead	122	167	528	673	1490
	Total	325	285	706	885	2201

- The percentage of the passengers who were both in first class and survived: This would be 203/2201, or 9.2%.
- The percentage of the first-class passengers who survived: This is 203/325, or 62.5%.
- The percentage of the survivors who were in first class: This is 203/711, or 28.6%.

In each instance, pay attention to the *Who* implicitly defined by the phrase. Often there is a restriction to a smaller group (all aboard the *Titanic,* those in first class, and those who survived, respectively) before a percentage is found. Your discussion of results must make these differences clear.

- **Don't forget to look at the variables separately, too.** When you make a contingency table or display a conditional distribution, be sure you also examine the marginal distributions. It's important to know how many cases are in each category.

- **Be sure to use enough individuals.** When you consider percentages, take care that they are based on a large enough number of individuals. Take care not to make a report such as this one: *We found that 66.67% of the rats improved their performance with training. The other rat died.*

- **Don't overstate your case.** Independence is an important concept, but it is rare for two variables to be *entirely* independent. We can't conclude that one variable has no effect whatsoever on another. Usually, all we know is that little effect was observed in our study. Other studies of other groups under other circumstances could find different results.

Simpson's Paradox

■ **Be wary of looking only at overall percentages in a contingency table.** It's easy to make up an example showing that looking at overall proportions can be misleading. Suppose there are two pilots, Moe and Jill. Moe argues that he's the better pilot of the two, since he managed to land 83% of his last 120 flights on time compared with Jill's 78%. But let's look at the data a little more closely. Here are the results for each of their last 120 flights, broken down by the time of day they flew:

Table 2.10

On-time flights by Time of Day and Pilot. Look at the percentages within each *Time of Day* category. Who has a better on-time record during the day? At night? Who is better overall?

		Time of Day		
		Day	Night	Overall
Pilot	Moe	90 out of 100 90%	10 out of 20 50%	100 out of 120 83%
	Jill	19 out of 20 95%	75 out of 100 75%	94 out of 120 78%

Look at the daytime and nighttime flights separately. For day flights, Jill had a 95% on-time rate and Moe only a 90% rate. At night, Jill was on time 75% of the time and Moe only 50%. So Moe is better "overall," but Jill is better both during the day and at night. How can this be?

What's going on here is a problem known as **Simpson's paradox**, named for the statistician who described it in the 1960s. It comes up rarely in real life, but there have been several well-publicized cases. As we can see from the pilot example, the problem is combining percentages over different groups. Jill has mostly night flights, which are more difficult, so her *overall average* is heavily influenced by her nighttime average. Moe, on the other hand, benefits from flying mostly during the day, with its high on-time percentage. With their very different patterns of flying conditions, taking an overall percentage is misleading. It's not a fair comparison.

Don't use inappropriate averages. Sometimes they aren't as obvious as the Centerville sign.

The moral of Simpson's paradox is to be careful when you combine across different levels of a second variable. It's always better to compare percentages *within* each level of the other variable. The overall percentage may be misleading.

SIMPSON'S PARADOX

One famous example of Simpson's paradox arose during an investigation of admission rates for men and women at the University of California at Berkeley's graduate schools. As reported in an article in *Science*,[4] about 45% of male applicants were admitted, but only about 30% of female applicants got in. It looked like a clear case of discrimination. However, when the data were broken down by school (Engineering, Law, Medicine, etc.), it turned out that, within each school, the women were admitted at nearly the same or, in some cases, much *higher* rates than the men. How could this be? Women applied in large numbers to schools with very low admission rates (Law and Medicine, for example, admitted fewer than 10%). Men tended to apply to Engineering and Science. Those schools have admission rates above 50%. When the *average* was taken, the women had a much lower *overall* rate, but the average didn't really make sense.

All of the methods of this chapter work with *categorical variables*. Focusing on the total, row, or column percentages in a contingency table changes who you are interested in. Restricting the *who* of the study in different ways answers different questions.

---

[4] P. J. Bickel, E. A. Hammel, and J. W. O'Connell (1975). "Sex Bias in Graduate Admissions: Data From Berkeley." *Science* **187** (4175): 398–404. doi:10.1126/science.187.4175.398.

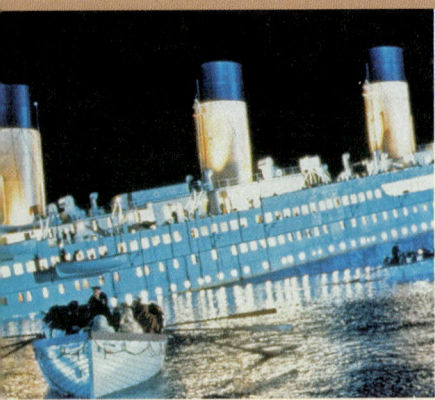

What Have We Learned?

Learning Objectives

Make and interpret a frequency table for a categorical variable.
- We can summarize categorical data by counting the number of cases in each category, sometimes expressing the resulting distribution as percentages.

Make and interpret a bar chart or pie chart.
- We display categorical data using the area principle in either a **bar chart** or a **pie chart**.

Make and interpret a contingency table.
- When we want to see how two categorical variables are related, we put the counts (and/or percentages) in a two-way table called a **contingency table**.

Make and interpret bar charts and pie charts of marginal distributions.
- We look at the **marginal distribution** of each variable (found in the margins of the table). We also look at the **conditional distribution** of a variable within each category of the other variable.
- Comparing conditional distributions of one variable across categories of another tells us about the association between variables. If the conditional distributions of one variable are (roughly) the same for every category of the other, the variables are **independent**.

Review of Terms

The key terms are in chapter order so you can use this list to review the material in the chapter.

Frequency table (Relative frequency table) A frequency table lists the categories in a categorical variable and gives the count (or percentage) of observations for each category (p. 18).

Distribution The distribution of a variable gives
- the possible values of the variable and
- the relative frequency of each value (p. 18).

Area principle In a statistical display, each data value should be represented by the same amount of area (p. 18).

Bar chart (Relative frequency bar chart) Bar charts show a bar whose area represents the count (or percentage) of observations for each category of a categorical variable (p. 19).

Pie chart Pie charts show how a "whole" divides into categories by showing a wedge of a circle whose area corresponds to the proportion in each category (p. 19).

Categorical data condition The methods in this chapter are appropriate for displaying and describing categorical data. Be careful not to use them with quantitative data (p. 19).

Contingency table A contingency table displays counts and, sometimes, percentages of individuals falling into named categories on two or more variables. The table categorizes the individuals on all variables at once to reveal possible patterns in one variable that may be contingent on the category of the other (p. 21).

Marginal distribution In a contingency table, the distribution of either variable alone is called the marginal distribution. The counts or percentages are the totals found in the margins (last row or column) of the table (p. 21).

Conditional distribution The distribution of a variable restricting the Who to consider only a smaller group of individuals is called a conditional distribution (p. 23).

Independence Variables are said to be independent if the conditional distribution of one variable is the same for each category of the other. We'll show how to check for independence in a later chapter (p. 25).

Segmented bar chart A segmented bar chart displays the conditional distribution of a categorical variable within each category of another variable (p. 26).

Simpson's paradox When averages are taken across different groups, they can appear to contradict the overall averages. This is known as "Simpson's paradox" (p. 31).

On the Computer DISPLAYING CATEGORICAL DATA

Although every package makes a slightly different bar chart, they all have similar features:

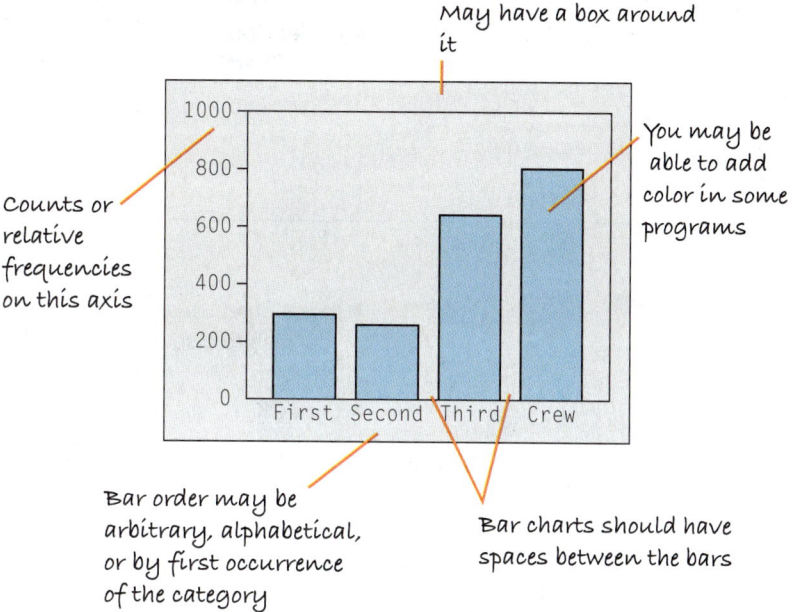

Sometimes the count or a percentage is printed above or on top of each bar to give some additional information. You may find that your statistics package sorts category names in annoying orders by default. For example, many packages sort categories alphabetically or by the order the categories are seen in the data set. Often, neither of these is the best choice.

DATA DESK

To make a bar chart or pie chart:
- Select the variable.
- In the **Plot** menu, choose **Bar Chart** or **Pie Chart**.

To make a frequency table:
- In the **Calc** menu choose **Frequency Table**.

Data Desk cannot make stacked bar charts.

COMMENTS

These commands treat data as categorical even if they are numerals. If you select a quantitative variable by mistake, you'll see an error message warning of too many categories. The **Replicate Y by X** command can generate variables from summary counts with values for each case that are thus suitable for these commands.

EXCEL

To make a bar chart:
- First make a pivot table (Excel's name for a frequency table). From the **Data** menu, choose **Pivot Table** and **Pivot Chart Report**.
- When you reach the Layout window, drag your variable to the row area and drag your variable again to the data area. This tells Excel to count the occurrences of each category. Once you have an Excel pivot table, you can construct bar charts and pie charts.
- Click inside the Pivot Table.
- Click the Pivot Table Chart Wizard button. Excel creates a bar chart.
- A longer path leads to a pie chart; see your Excel documentation.

COMMENTS

Excel uses the pivot table to specify the category names and find counts within each category. If you already have that information, you can proceed directly to the Chart Wizard.

JMP

JMP makes a bar chart and frequency table together:
- From the **Analyze** menu, choose **Distribution**.
- In the Distribution dialog, drag the name of the variable into the empty variable window beside the label "Y, Columns"; click **OK**.

To make a pie chart:
- Choose **Chart** from the **Graph** menu.
- In the Chart dialog, select the variable name from the Columns list.
- Click on the button labeled "Statistics," and select "N" from the drop-down menu.
- Click the "Categories, X, Levels" button to assign the same variable name to the *x*-axis.
- Under Options, click on the **second** button—labeled "**Bar Chart**"—and select "Pie" from the drop-down menu.

MINITAB

To make a bar chart:
- Choose **Bar Chart** from the **Graph** menu.
- Select "Counts of unique values" in the first menu, and select "Simple" for the type of graph. Click **OK**.
- In the Chart dialog, enter the name of the variable that you wish to display in the box labeled "Categorical variables."
- Click **OK**.

R

To make a bar chart or pie chart in R, you first need to create the frequency table for the desired variable:
- **table(X)** will give a frequency table for a single variable X.
- **barplot(table(X))** will give a bar chart for X.
- Similarly **pie(table(X))** will give a pie chart.

COMMENTS

Stacked bar charts of two variables, X and Y, can be made using **barplot(xtabs(~X+Y))** or directly from a two-way table of counts or percentages. Legends and other options are available for all charts using various functions.

SPSS

To make a bar chart:
- Open the **Chart Builder** from the **Graphs** menu.
- Click the **Gallery** tab.
- Choose **Bar Chart** from the list of chart types.
- Drag the appropriate bar chart onto the canvas.
- Drag a categorical variable onto the *x*-axis drop zone.
- Click **OK**.

COMMENT

A similar path makes a pie chart by choosing **Pie chart** from the list of chart types.

STATCRUNCH

To make a bar chart or pie chart:
- Click on **Graphics**.
- Choose the type of plot **» with data** or **» with summary**.
- Choose the variable name from the list of **Columns**; if using summaries, also, choose the counts.
- Click on **Next**.
- Choose **Frequency/Counts** or (usually) **Relative frequency/Percents**. Note that you may elect to group categories under a specified percentage as "Other."
- Click on **Create Graph**.

TI-83/84 PLUS

The TI-83 won't do displays for categorical variables.

Exercises

Section 2.1

1. **Automobile fatalities** The table gives the numbers of passenger car occupants killed in accidents in 2011 by car type.

Subcompact and Mini	1351
Compact	3789
Intermediate	4050
Full	2627
Unknown	164

    Source: www.nhtsa.gov

    Convert this table to a relative frequency table.

2. **Nonoccupant fatalities** The frequencies of traffic fatalities of nonoccupants of vehicles are shown in the following bar chart. Change this to a bar chart of relative frequencies.

    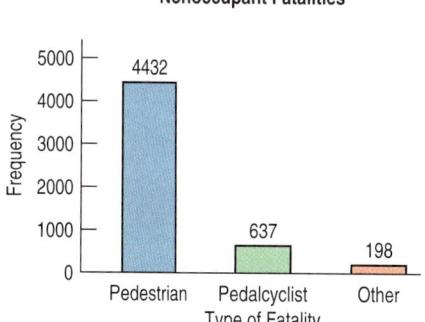

3. **Movie genres** The Motion Picture Association of America (MPAA) rates each film to designate the appropriate audience. The ratings are G, PG, PG-13, and R. The ratings of the 20 top-grossing movies in the years 1996, 2006, 2008 and 2012 are shown in the following bar charts. The pie charts show the same data, but are unlabeled. Match each pie chart with the correct year.

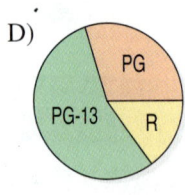

4. **Marriage in decline** Changing attitudes about marriage and families prompted Pew Research to ask how people felt about particular recent trends (pewresearch.org/pubs/1802/decline-marriage-rise-new-families). For each trend, participants were asked whether the trend "is a good thing," "is a bad thing," or "makes no difference." Some participants said they didn't know or chose not to respond. The following bar charts show the number of each response for each trend. The pie charts show the same data without the trends identified. Match each pie chart with the correct trend and bar chart.

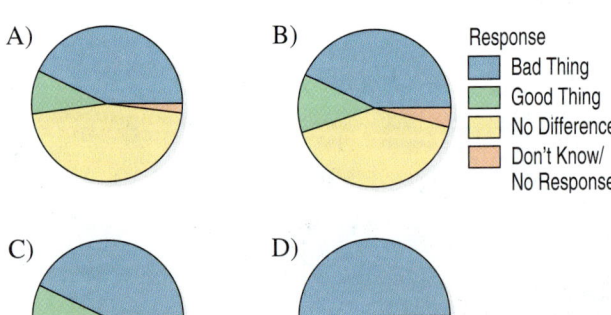

Section 2.2

5. Movies again The following table summarizes 348 films from 2010 that have been classified into a genre and have a MPAA rating.

	G	PG	PG-13	R	Total
Action/Adventure	3	18	18	14	53
Comedy	2	18	29	41	90
Documentary	4	8	8	9	29
Drama	0	16	36	65	117
Musical	0	2	2	0	4
Suspense/Thriller/Horror	0	1	13	41	55
Total	9	63	106	170	348

Source: www.the-numbers.com/

a) What percentage of these films were rated R?
b) What percentage of these films were R-rated comedies?
c) What percentage of R-rated films were comedies?
d) What percentage of comedies were rated R?

6. Labor force The following table shows the labor force status of people 16 years of age and older in the United States. Counts are in thousands of people.

Age	Employed	Unemployed	Not in Labor Force	Total
16 to 19 Years	4378	1528	10,995	16,901
20 to 24 Years	12,699	2329	6019	21,047
25 to 54 Years	94,082	8858	22,350	125,290
55 to 64 Years	21,636	1660	12,589	35,885
65 Years and Over	6268	449	31,988	38,705
Totals	139,063	14,824	83,941	237,828

a) What percentage of the population were unemployed?
b) What percentage of the population were unemployed and aged 25 to 54 years?
c) What percentage of 20 to 24 year olds were unemployed?
d) What percentage of employed people were aged 16 to 19 years?

Chapter Exercises

7. Graphs in the news Find a bar chart of categorical data from a newspaper, a magazine, or the Internet.
a) Is the graph clearly labeled?
b) Does it violate the area principle?
c) Does the accompanying article tell the W's of the variable?
d) Do you think the article correctly interprets the data? Explain.

8. Graphs in the news II Find a pie chart of categorical data from a newspaper, a magazine, or the Internet.
a) Is the graph clearly labeled?
b) Does it violate the area principle?
c) Does the accompanying article tell the W's of the variable?
d) Do you think the article correctly interprets the data? Explain.

9. Tables in the news Find a frequency table of categorical data from a newspaper, a magazine, or the Internet.
a) Is it clearly labeled?
b) Does it display percentages or counts?
c) Does the accompanying article tell the W's of the variable?
d) Do you think the article correctly interprets the data? Explain.

10. Tables in the news II Find a contingency table of categorical data from a newspaper, a magazine, or the Internet.
a) Is it clearly labeled?
b) Does it display percentages or counts?
c) Does the accompanying article tell the W's of the variables?
d) Do you think the article correctly interprets the data? Explain.

11. Movie genres II The pie chart summarizes the genres of 193 movies shown in a suburban multiplex theatre in 2012.

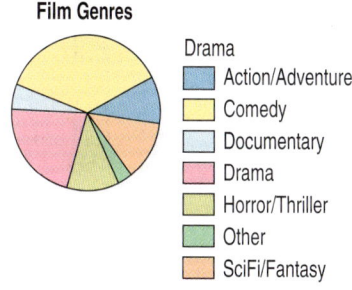

a) Is this an appropriate display for the genres? Why or why not?
b) Which genre was least common?

12. Movie ratings The pie chart shows the ratings assigned to the 20 top-grossing movies for each year from 2009 to 2013.

a) Is this an appropriate display for these data? Explain.
b) Which was the most common rating?

13. Genres again Here is a bar chart summarizing the 2012 movie genres from the suburban multiplex theatre, as seen in the pie chart in Exercise 11.

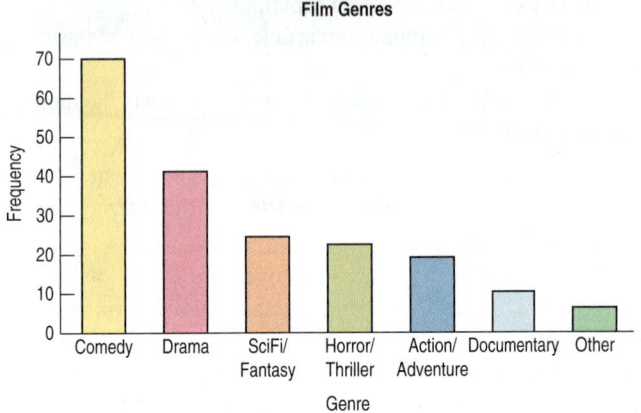

a) Were Action/Adventure or SciFi/Fantasy films more common?
b) Is it easier to see that in a pie chart or the bar chart? Explain.

14. Ratings again Here is a bar chart of the movie ratings from 2009 to 2013 seen in the pie chart in Exercise 12.

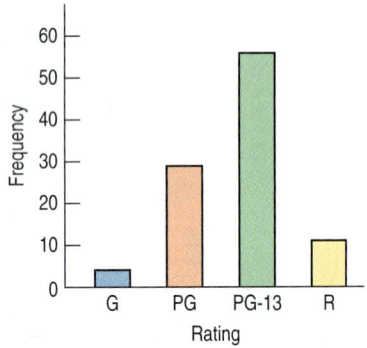

a) Which was the least common rating?
b) An editorial claimed that there's been a growth in PG-13 rated films that, according to the writer, "have too much sex and violence," at the expense of G-rated films that offer "good, clean fun." The writer offered the following bar chart as evidence to support his claim.

Does the bar chart support his claim? Explain.

15. Magnet schools An article in the Winter 2003 issue of *Chance* magazine reported on the Houston Independent School District's magnet schools programs. Of the 1755 qualified applicants, 931 were accepted, 298 were waitlisted, and 526 were turned away for lack of space. Find the relative frequency distribution of the decisions made, and write a sentence describing it.

16. Magnet schools again The *Chance* article about the Houston magnet schools program described in Exercise 15 also indicated that 517 applicants were black or Hispanic, 292 Asian, and 946 white. Summarize the relative frequency distribution of ethnicity with a sentence or two (in the proper context, of course).

17. Causes of death 2011 The Centers for Disease Control and Prevention lists causes of death in the United States during 2011:

Cause of Death	Percent
Heart Disease	23.7%
Cancer	22.9%
Lung Diseases	5.7%
Circulatory Diseases and Stroke	5.1%
Accidents	4.9%

Source: www.cdc.gov/nchs/fastats/deaths.htm

a) Is it reasonable to conclude that heart or lung diseases were the cause of approximately 29.4% of U.S. deaths in 2011?
b) What percent of deaths were from causes not listed here?
c) Create an appropriate display for these data.

18. Plane crashes An investigation compiled information about recent nonmilitary plane crashes. The causes, to the extent that they could be determined, are summarized in the table.

Causes of Fatal Accidents	
Cause	Percent
Pilot Error	46%
Other Human Error	8
Weather	9
Mechanical Failure	28
Sabotage	9
Other Causes	1

Source: www.planecrashinfo.com/cause.htm

a) Is it reasonable to conclude that the weather or mechanical failures caused only about 37% of recent plane crashes?
b) Why do the numbers in the table add to 101%?
c) Create an appropriate display for these data.

19. Oil spills 2013 Data from the International Tanker Owners Pollution Federation Limited give the cause of spillage for 459 oil tanker accidents resulting in spills of more than 700 tons of oil from 1970–2013 (www.itopf.com/information-services/data-and-statistics/statistics). On the next page are displays.

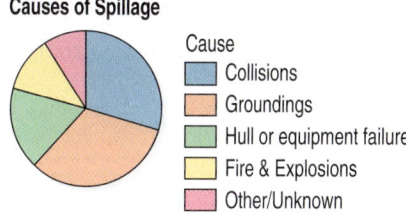

a) Write a brief report interpreting what the displays show.
b) Is a pie chart an appropriate display for these data? Why or why not?
c) Were there more spills due to Collisions or to Groundings? Is it easier to tell that from the bar chart or the pie chart?

T 20. Winter Olympics 2010 Twenty-six countries won medals in the 2010 Winter Olympics. The table lists them, along with the total number of medals each won:

Country	Medals	Country	Medals
United States	37	Poland	6
Germany	30	Italy	5
Canada	26	Japan	5
Norway	23	Finland	5
Austria	16	Australia	3
Russia	15	Belarus	3
South Korea	14	Slovakia	3
China	11	Croatia	3
Sweden	11	Slovenia	3
France	11	Latvia	2
Switzerland	9	Great Britain	1
Netherlands	8	Estonia	1
Czech Republic	6	Kazakhstan	1

a) Try to make a display of these data. What problems do you encounter?
b) Organize the data so that the graph is more successful.

21. Global warming The Pew Research Center for the People and the Press (people-press.org) has asked a representative sample of U.S. adults about global warming, repeating the question over time. In October 2010, the responses reflected a decreased belief that global warming is real and due to human activity. Here's a display of the percentages of respondents choosing each of the major alternatives offered:

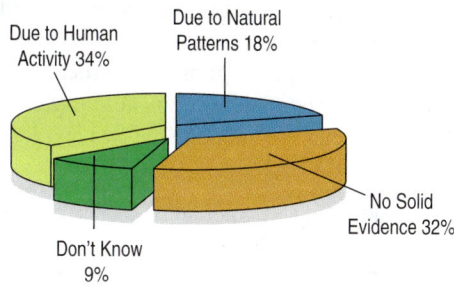

List the errors in this display.

22. Modalities A survey of athletic trainers (Source: Scott F. Nadler, Michael Prybicien, Gerard A. Malanga, and Dan Sicher, "Complications from Therapeutic Modalities: Results of a National Survey of Athletic Trainers." *Archives of Physical Medical Rehabilitation* 84 [June 2003]) asked what modalities (treatment methods such as ice, whirlpool, ultrasound, or exercise) they commonly use to treat injuries. Respondents were each asked to list three modalities. The article included the following figure reporting the modalities used:

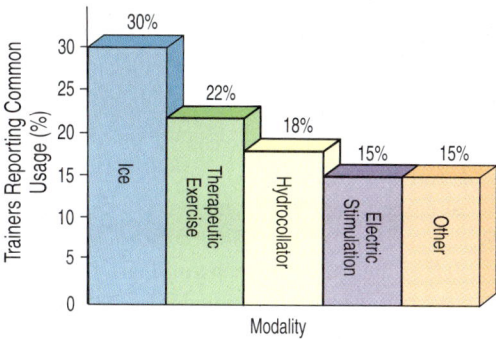

a) What problems do you see with the graph?
b) Consider the percentages for the named modalities. Do you see anything odd about them?

23. Teen smokers The organization Monitoring the Future (www.monitoringthefuture.org) asked 2048 eighth graders who said they smoked cigarettes what brands they preferred. The table below shows brand preferences for two regions of the country. Write a few sentences describing the similarities and differences in brand preferences among eighth graders in the two regions listed.

Brand Preference	South	West
Marlboro	58.4%	58.0%
Newport	22.5%	10.1%
Camel	3.3%	9.5%
Other (over 20 brands)	9.1%	9.5%
No Usual Brand	6.7%	12.9%

24. Handguns In an effort to reduce the number of gun-related homicides, some cities have run buyback programs in which the police offer cash (often $50) to anyone who turns in an operating handgun. *Chance* magazine looked at results from a four-year period in Milwaukee. The table below shows what types of guns were turned in and what types were used in homicides during a four-year period. Write a few sentences comparing the two distributions.

Caliber of Gun	Buyback	Homicide
Small (.22, .25, .32)	76.4%	20.3%
Medium (.357, .38, 9 mm)	19.3%	54.7%
Large (.40, .44, .45)	2.1%	10.8%
Other	2.2%	14.2%

25. Movies by genre and rating Here's a table that classifies movies released in 2010 by genre and MPAA rating:

	G	PG	PG-13	R	Total
Action/Adventure	33.33	28.57	16.98	8.24	15.23
Comedy	22.22	28.57	27.36	24.12	25.86
Documentary	44.44	12.70	7.55	5.29	8.33
Drama	0.00	25.40	33.96	38.24	33.62
Musical	0.00	3.17	1.89	0.00	1.15
Suspense/Horror	0.00	1.59	12.26	24.12	15.80
Total	100%	100%	100%	100%	100%

a) The table gives column percents. How could you tell that from the table itself?
b) What percentage of these movies were comedies?
c) What percentage of the PG-rated movies were comedies?
d) Which of the following can you learn from this table? Give the answer if you can find it from the table.
  i. The percentage of PG-13 movies that were comedies
  ii. The percentage of dramas that were R-rated
  iii. The percentage of dramas that were G-rated
  iv. The percentage of 2010 movies that were PG-rated comedies

26. The last picture show Here's another table showing information on 348 movies released in 2010. This table gives percentages of the overall total:

	G	PG	PG-13	R	Total
Action/Adventure	0.86%	5.17	5.17	4.02	15.23
Comedy	0.57	5.17	8.33	11.78	25.86
Documentary	1.15	2.30	2.30	2.59	8.33
Drama	0.00	4.60	10.34	18.68	33.62
Musical	0.00	0.57	0.57	0.00	1.15
Suspense/Horror	0.00	0.29	3.74	11.78	15.80
Total	2.59	18.10	30.46	48.85	100%

a) How can you tell that this table holds overall percentages (rather than row or column percentages)?
b) What was the most common genre/rating combination in 2010 movies?
c) How many of these movies were PG-rated comedies?
d) How many were G-rated?
e) An editorial about the movies noted, "More than three-quarters of the movies made today can be seen only by patrons 13 years old or older." Does this table support that assertion? Explain.

27. Seniors Prior to graduation, a high school class was surveyed about its plans. The following table displays the results for white and minority students (the "Minority" group included African-American, Asian, Hispanic, and Native American students):

		Seniors	
		White	Minority
Plans	4-Year College	198	44
	2-Year College	36	6
	Military	4	1
	Employment	14	3
	Other	16	3

a) What percent of the seniors are white?
b) What percent of the seniors are planning to attend a 2-year college?
c) What percent of the seniors are white and planning to attend a 2-year college?
d) What percent of the white seniors are planning to attend a 2-year college?
e) What percent of the seniors planning to attend a 2-year college are white?

28. Politics Students in an Intro Stats course were asked to describe their politics as "Liberal," "Moderate," or "Conservative." Here are the results:

		Politics			
		L	M	C	Total
Sex	Female	35	36	6	77
	Male	50	44	21	115
	Total	85	80	27	192

a) What percent of the class is male?
b) What percent of the class considers themselves to be "Conservative"?
c) What percent of the males in the class consider themselves to be "Conservative"?
d) What percent of all students in the class are males who consider themselves to be "Conservative"?

29. More about seniors Look again at the table of post-graduation plans for the senior class in Exercise 27.
  a) Find the conditional distributions (percentages) of plans for the white students.
  b) Find the conditional distributions (percentages) of plans for the minority students.
  c) Create a graph comparing the plans of white and minority students.
  d) Do you see any important differences in the post-graduation plans of white and minority students? Write a brief summary of what these data show, including comparisons of conditional distributions.

30. Politics revisited Look again at the table of political views for the Intro Stats students in Exercise 28.
  a) Find the conditional distributions (percentages) of political views for the females.
  b) Find the conditional distributions (percentages) of political views for the males.
  c) Make a graphical display that compares the two distributions.
  d) Do the variables *Politics* and *Sex* appear to be independent? Explain.

31. Magnet schools revisited The *Chance* magazine article described in Exercise 15 further examined the impact of an applicant's ethnicity on the likelihood of admission to the Houston Independent School District's magnet schools programs. Those data are summarized in the table below.

		Admission Decision			
		Accepted	Wait-Listed	Turned Away	Total
Ethnicity	Black/Hispanic	485	0	32	517
	Asian	110	49	133	292
	White	336	251	359	946
	Total	931	300	524	1755

  a) What percent of all applicants were Asian?
  b) What percent of the students accepted were Asian?
  c) What percent of Asians were accepted?
  d) What percent of all students were accepted?

32. More politics Look once more at the table summarizing the political views of Intro Stats students in Exercise 28.
  a) Produce a graphical display comparing the conditional distributions of males and females among the three categories of politics.
  b) Comment briefly on what you see from the display in part a.

33. Back to school Examine the table about ethnicity and acceptance for the Houston Independent School District's magnet schools program, shown in Exercise 31. Does it appear that the admissions decisions are made independent of the applicant's ethnicity? Explain.

34. Parking lots A survey of autos parked in student and staff lots at a large university classified the brands by country of origin, as seen in the table.

		Driver	
		Student	Staff
Origin	American	107	105
	European	33	12
	Asian	55	47

  a) What percent of all the cars surveyed were foreign?
  b) What percent of the American cars were owned by students?
  c) What percent of the students owned American cars?
  d) What is the marginal distribution of origin?
  e) What are the conditional distributions of origin by driver classification?
  f) Do you think that the origin of the car is independent of the type of driver? Explain.

T 35. Weather forecasts Just how accurate are the weather forecasts we hear every day? The following table compares the daily forecast with a city's actual weather for a year:

		Actual Weather	
		Rain	No Rain
Forecast	Rain	27	63
	No Rain	7	268

  a) On what percent of days did it actually rain?
  b) On what percent of days was rain predicted?
  c) What percent of the time was the forecast correct?
  d) Do you see evidence of an association between the type of weather and the ability of forecasters to make an accurate prediction? Write a brief explanation, including an appropriate graph.

T 36. Twin births In 2000, the *Journal of the American Medical Association (JAMA)* published a study that examined pregnancies that resulted in the birth of twins. Births were classified as preterm with intervention (induced labor or cesarean), preterm without procedures, or term/post-term. Researchers also classified the pregnancies by the level of prenatal medical care the mother received (inadequate, adequate, or intensive). The data, from the years 1995–1997, are summarized in the table below. Figures are in thousands of births. (Source: *JAMA* 284 [2000]:335–341)

		Twin Births 1995–1997 (in thousands)			
		Preterm (induced or cesarean)	Preterm (without procedures)	Term or Post-Term	Total
Level of Prenatal Care	Intensive	18	15	28	61
	Adequate	46	43	65	154
	Inadequate	12	13	38	63
	Total	76	71	131	278

a) What percent of these mothers received inadequate medical care during their pregnancies?
b) What percent of all twin births were preterm?
c) Among the mothers who received inadequate medical care, what percent of the twin births were preterm?
d) Create an appropriate graph comparing the outcomes of these pregnancies by the level of medical care the mother received.
e) Write a few sentences describing the association between these two variables.

37. **Blood pressure** A company held a blood pressure screening clinic for its employees. The results are summarized in the table below by age group and blood pressure level:

		Age		
		Under 30	30–49	Over 50
Blood Pressure	Low	27	37	31
	Normal	48	91	93
	High	23	51	73

a) Find the marginal distribution of blood pressure level.
b) Find the conditional distribution of blood pressure level within each age group.
c) Compare these distributions with a segmented bar graph.
d) Write a brief description of the association between age and blood pressure among these employees.
e) Does this prove that people's blood pressure increases they age? Explain.

38. **Obesity and exercise** The Centers for Disease Control and Prevention (CDC) has estimated that 19.8% of Americans over 15 years old are obese. The CDC conducts a survey on obesity and various behaviors. Here is a table on self-reported exercise classified by body mass index (BMI):

		Body Mass Index		
		Normal (%)	Overweight (%)	Obese (%)
Physical Activity	Inactive	23.8	26.0	35.6
	Irregularly Active	27.8	28.7	28.1
	Regular, Not Intense	31.6	31.1	27.2
	Regular, Intense	16.8	14.2	9.1

a) Are these percentages column percentages, row percentages, or table percentages?
b) Use graphical displays to show different percentages of physical activities for the three BMI groups.
c) Do these data prove that lack of exercise causes obesity? Explain.

39. **Anorexia** Hearing anecdotal reports that some patients undergoing treatment for the eating disorder anorexia seemed to be responding positively to the antidepressant Prozac, medical researchers conducted an experiment to investigate. They found 93 women being treated for anorexia who volunteered to participate. For one year, 49 randomly selected patients were treated with Prozac and the other 44 were given an inert substance called a placebo. At the end of the year, patients were diagnosed as healthy or relapsed, as summarized in the table:

	Prozac	Placebo	Total
Healthy	35	32	67
Relapse	14	12	26
Total	49	44	93

Do these results provide evidence that Prozac might be helpful in treating anorexia? Explain.

40. **Antidepressants and bone fractures** For a period of five years, physicians at McGill University Health Center followed more than 5000 adults over the age of 50. The researchers were investigating whether people taking a certain class of antidepressants (SSRIs) might be at greater risk of bone fractures. Their observations are summarized in the table:

	Taking SSRI	No SSRI	Total
Experienced Fractures	14	244	258
No Fractures	123	4627	4750
Total	137	4871	5008

Do these results suggest there's an association between taking SSRI antidepressants and experiencing bone fractures? Explain.

41. **Drivers' licenses 2011** The table on the next page shows the number of licensed U.S. drivers (in millions) by age and by sex (www.dot.gov).
a) What percent of total drivers are under 20?
b) What percent of total drivers are male?
c) Write a few sentences comparing the number of male and female licensed drivers in each age group.
d) Do a driver's age and sex appear to be independent? Explain.

Age	Male Drivers (millions)	Female Drivers (millions)	Total
19 and Under	5.1	4.9	10.0
20–24	8.7	8.6	17.3
25–29	9.2	9.2	18.4
30–34	8.9	8.9	17.8
35–39	9.7	9.6	19.3
40–44	9.9	9.8	19.7
45–49	10.6	10.7	21.3
50–54	10.1	10.2	20.3
55–59	8.7	8.9	17.6
60–64	7.2	7.3	14.5
65–69	5.3	5.4	10.7
70–74	3.8	4.0	7.8
75–79	2.9	3.2	6.1
80–84	2.0	2.4	4.4
85 and Over	1.4	1.7	3.1
Total	103.5	104.8	208.3

42. **Tattoos** A study by the University of Texas Southwestern Medical Center examined 626 people to see if an increased risk of contracting hepatitis C was associated with having a tattoo. If the subject had a tattoo, researchers asked whether it had been done in a commercial tattoo parlor or elsewhere. Write a brief description of the association between tattooing and hepatitis C, including an appropriate graphical display.

	Tattoo Done in Commercial Parlor	Tattoo Done Elsewhere	No Tattoo
Has Hepatitis C	17	8	18
No Hepatitis C	35	53	495

43. **Hospitals** Most patients who undergo surgery make routine recoveries and are discharged as planned. Others suffer excessive bleeding, infection, or other postsurgical complications and have their discharges from the hospital delayed. Suppose your city has a large hospital and a small hospital, each performing major and minor surgeries. You collect data to see how many surgical patients have their discharges delayed by postsurgical complications, and you find the results shown in the following table.

	Discharge Delayed	
	Large Hospital	Small Hospital
Major Surgery	120 of 800	10 of 50
Minor Surgery	10 of 200	20 of 250

a) Overall, for what percent of patients was discharge delayed?
b) Were the percentages different for major and minor surgery?
c) Overall, what were the discharge delay rates at each hospital?
d) What were the delay rates at each hospital for each kind of surgery?
e) The small hospital advertises that it has a lower rate of postsurgical complications. Do you agree?
f) Explain, in your own words, why this confusion occurs.

44. **Delivery service** A company must decide which of two delivery services it will contract with. During a recent trial period, the company shipped numerous packages with each service and kept track of how often deliveries did not arrive on time. Here are the data:

Delivery Service	Type of Service	Number of Deliveries	Number of Late Packages
Pack Rats	Regular	400	12
	Overnight	100	16
Boxes R Us	Regular	100	2
	Overnight	400	28

a) Compare the two services' overall percentage of late deliveries.
b) On the basis of the results in part a, the company has decided to hire Pack Rats. Do you agree that Pack Rats delivers on time more often? Explain.
c) The results here are an instance of what phenomenon?

45. **Graduate admissions** A 1975 article in the magazine *Science* examined the graduate admissions process at Berkeley for evidence of sex discrimination. The table below shows the number of applicants accepted to each of four graduate programs:

		Males Accepted (of applicants)	Females Accepted (of applicants)
Program	1	511 of 825	89 of 108
	2	352 of 560	17 of 25
	3	137 of 407	132 of 375
	4	22 of 373	24 of 341
	Total	1022 of 2165	262 of 849

a) What percent of total applicants were admitted?
b) Overall, was a higher percentage of males or females admitted?
c) Compare the percentage of males and females admitted in each program.
d) Which of the comparisons you made do you consider to be the most valid? Why?

46. Be a Simpson Can you design a Simpson's paradox? Two companies are vying for a city's "Best Local Employer" award, to be given to the company most committed to hiring local residents. Although both employers hired 300 new people in the past year, Company A brags that it deserves the award because 70% of its new jobs went to local residents, compared to only 60% for Company B. Company B concedes that those percentages are correct, but points out that most of its new jobs were full-time, while most of Company A's were part-time. Not only that, says Company B, but a higher percentage of its full-time jobs went to local residents than did Company A's, and the same was true for part-time jobs. Thus, Company B argues, it's a better local employer than Company A.

Show how it's possible for Company B to fill a higher percentage of both full-time and part-time jobs with local residents, even though Company A hired more local residents overall.

Just Checking ANSWERS

1. 50.0%
2. 44.4%
3. 25.0%
4. 15.6% Blue, 56.3% Brown, 28.1% Green/Hazel/Other
5. 18.8% Blue, 62.5% Brown, 18.8% Green/Hazel/Other
6. 40% of the blue-eyed students are female, while 50% of all students are female.
7. Since blue-eyed students appear less likely to be female, it seems that *Sex* and *Eye Color* may not be independent. (But the numbers are small.)

CHAPTER 3
Displaying and Summarizing Quantitative Data

3.1 Displaying Quantitative Variables

3.2 Shape

3.3 Center

3.4 Spread

3.5 Boxplots and 5-Number Summaries

3.6 The Center of Symmetric Distributions: The Mean

3.7 The Spread of Symmetric Distributions: The Standard Deviation

3.8 Summary—What to *Tell* About a Quantitative Variable

> *Where are we going?*
> If someone asked you to summarize a variable, what would you say? You might start by making a picture. For quantitative data, that first picture would probably be a histogram. We've all looked at histograms, but what should we look *for*? We'll describe the histogram, and we'll often do more—reporting numerical summaries of the center and spread as well. Spread measures are a bit less common, but in Statistics, they are even more important. This chapter is where we'll first encounter the single most important value in all of Statistics.

On March 11, 2011, the most powerful earthquake ever recorded in Japan created a wall of water that devastated the northeast coast of Japan and left 20,000 people dead or missing. Tsunamis like this are most often caused by earthquakes beneath the sea that shift the earth's crust, displacing a large mass of water. The 2011 tsunami in Japan was caused by a 9.0 magnitude earthquake. It was particularly noted for the damage it caused to the Fukushima Daiichi nuclear power plant, bringing it perilously close to a complete meltdown and causing international concern.

As disastrous as it was, the Japan tsunami was not nearly as deadly as the tsunami of December 26, 2004, off the west coast of Sumatra, which killed an estimated 297,248 people, making it the most lethal tsunami on record. The earthquake that caused it was magnitude 9.1—more than 25% more powerful than the Japanese earthquake.[1] Were these earthquakes truly extraordinary, or did they just happen at unlucky times and places? The U.S. National Geophysical Data Center[2] has data on more than 5000 earthquakes dating back to 2150 B.C.E., and we have estimates of the magnitude of the underlying earthquake for the 1168 that were known to cause tsunamis. What can we learn from these data?

3.1 Displaying Quantitative Variables

Let's start with a picture. For categorical variables, it is easy to draw the distribution because it's natural to count the cases in each category. But for quantitative variables, there are no categories. Instead, we usually slice up all the possible values into bins and

---
[1] Earthquake magnitudes are measured on a log scale.
[2] www.ngdc.noaa.gov.

then count the number of cases that fall into each bin. **The bins, together with these counts, give the distribution of the quantitative variable and provide the building blocks for the display of the distribution, called a histogram. By representing the counts as bars and plotting them against the bin values, the histogram displays the distribution at a glance.**

Histograms

Here is a histogram of the *Magnitudes* (on the Richter scale) of the 1168 earthquakes in the NGDC data:

Figure 3.1

A histogram shows the distribution of magnitudes (in Richter scale units) of tsunami-generating earthquakes. (Data in **Tsunami Earthquakes**)

Who	1168 earthquakes known to have caused tsunamis for which we have data or good estimates
What	Magnitude (Richter scale[3]), depth (m), date, location, and other variables
When	From 2000 B.C.E. to the present
Where	All over the earth

The height of each bar of the histogram shows the number of cases that fall in its bin. In this histogram of earthquake magnitudes, each bin has a width of 0.5 units, so, for example, the height of the tallest bar says that there were about 350 earthquakes with magnitudes between 7.0 and 7.5.

> **DESIGNING YOUR HISTOGRAM**
>
> Different features of the distribution may appear more obvious at different bin width choices. When you use technology, it's usually easy to vary the bin width interactively so you can make sure that a feature you think you see isn't a consequence of a certain bin width choice.

Stepping back from the histogram, you can see the entire distribution of earthquake magnitudes. Does this distribution look as you expected? It is often a good idea to *imagine* what the distribution might look like before you make the display. That way you'll be less likely to be fooled by errors in the data or when you accidentally graph the wrong variable. We can see that these earthquakes typically have magnitudes around 7. Most are between 5.5 and 8.5, but one is less than 4 and a few are 9 or bigger. Now we can answer the question about the Sumatra and Japan tsunamis. With values of 9.1 and 9.0 it's clear that these earthquakes were extraordinarily powerful—among the largest on record.

In a histogram, the bins slice up all the values of the quantitative variable, so that every value falls in a bin. Choosing the bin width used to be an important exercise in introductory statistics courses. Today, technology will choose a default width for you, but usually you can vary the bin width interactively. You'll probably want to have between 5 and 30 bins (unless your data set is very large) and you'll want the bin boundaries to be aesthetically pleasing (such as 7.0 and 7.5 instead of 7.098 and 7.464). You may find that some features of the distribution appear or disappear as you change the bin width. Here's a histogram with a smaller bin width of 0.2 for the tsunami data:

---

[3]Technically, Richter scale values are in units of log dyne-cm. But the Richter scale is so common now that usually the units are assumed. The U.S. Geological Survey gives the background details of Richter scale measurements on its website www.usgs.gov.

Figure 3.2
A histogram of earthquake magnitudes with a smaller bin size highlights the surprising number of earthquakes with values at 7.0.

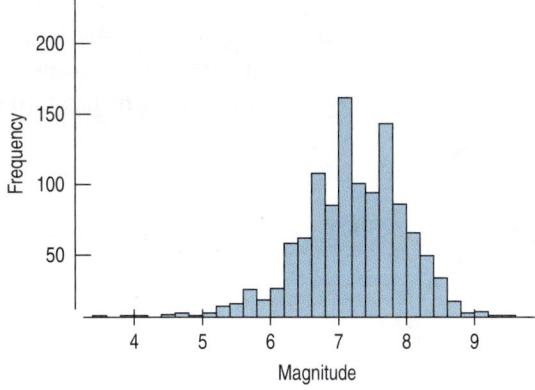

The overall distribution appears similar, as it should, but we're struck by two features that we didn't see before. There is a spike around magnitude 7.0. The 1168 magnitudes include historical values, some more than 2000 years old. Many of the values were estimated by experts and not measured by modern seismographs. Perhaps the experts rounded some of the values to 7, reasoning that that was a typical value for a moderately sized tsunami-causing earthquake. That might explain the overabundance of values of 7.0. There is also a gap in the histogram between 4.2 and 4.4. Unlike a bar chart that uses gaps just to separate categories, a **gap** in a histogram indicates that there is a bin (or bins) with no values. There were no earthquakes among the 1168 that had values between 4.2 and 4.4. Gaps often reveal interesting features about the distribution.

Sometimes it is useful to make a **relative frequency histogram**, replacing the counts on the vertical axis with the *percentage* of the total number of cases falling in each bin. Of course, the shape of the histogram is exactly the same; only the vertical scale is different.

Figure 3.3
A relative frequency histogram looks just like a frequency histogram except for the labels on the *y*-axis, which now show the percentage of earthquakes in each bin. Compare this one with Figure 3.1.

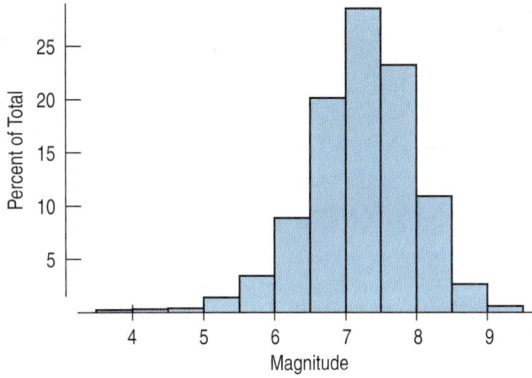

Stem-and-Leaf Displays

Histograms provide an easy-to-understand summary of the distribution of a quantitative variable, but they don't show the data values themselves. Here's a histogram of the pulse rates of 24 women, taken by a researcher at a health clinic:

Figure 3.4
The pulse rates of 24 women at a health clinic.

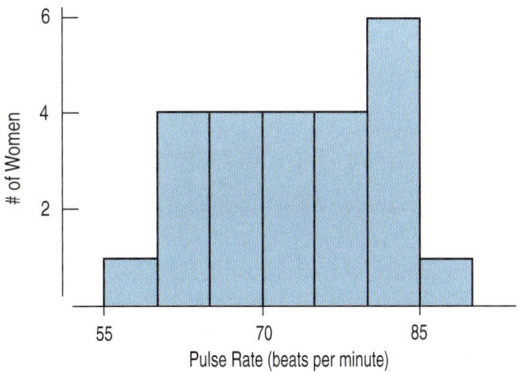

> **Stem-and-leaf or stemplot?** The stem-and-leaf display was devised by John W. Tukey, one of the greatest statisticians of the 20th century. It is called a "Stemplot" in some texts and computer programs, but we prefer Tukey's original name for it.

The story seems pretty clear. We can see the entire span of the data and can easily see what a typical pulse rate might be. But is that all there is to these data?

A **stem-and-leaf display** is like a histogram, but it shows the individual values. It's also easier to make by hand. Here's a stem-and-leaf display of the same data:

Turn the stem-and-leaf on its side (or turn your head to the right) and squint at it. It should look roughly like the histogram of the same data. Does it?[4]

What does the first line of the display, which says 5│6 mean? It stands for a pulse of 56 beats per minute (bpm). We've taken the tens place of the number and made that the "stem." Then we sliced off the ones place and made it a "leaf." The next line down is 6│0444. That shows that there was one pulse rate of 60 and three of 64 bpm.

Stem-and-leaf displays are especially useful when you make them by hand for batches of fewer than a few hundred data values. They are a quick way to display—and even to record—numbers. Because the leaves show the individual values, we can sometimes see even more in the data than the distribution's shape. Take another look at all the leaves of the pulse data. See anything unusual? At a glance you can see that they are all even. With a bit more thought you can see that they are all multiples of 4—something you couldn't possibly see from a histogram. How do you think the nurse took these pulses? Counting beats for a full minute or counting for only 15 seconds and multiplying by 4?

> **A S** *Activity:* **Stem-and-Leaf Displays.** As you might expect of something called "stem-and-leaf," these displays grow as you consider each data value.

By Hand MAKING A STEM-AND-LEAF

Stem-and-leaf displays work like histograms, but they show more information. They use part of the number itself (called the stem) to name the bins. To make the "bars," they use the next digit of the number. For example, if we had a test score of 76, we could write it 7│6, where 7 serves as the stem and 6 as the leaf. Then, to display the scores 83, 76, and 88 together, we would write

```
7│6
8│38
```

For the pulse data, we have

---

[4] You could make the stem-and-leaf with the higher values on the top. Putting the lower value at the top matches the histogram; putting the higher value at the top matches the way a vertical axis works in other displays.

This display is OK, but a little crowded. A histogram might split each line into two bars. With a stem-and-leaf, we can do the same by putting the leaves 0–4 on one line and 5–9 on another, as we saw earlier:

```
5 | 6
6 | 0444
6 | 8888
7 | 2222
7 | 6666
8 | 000044
8 | 8
```
Pulse Rate
(5|6 means 56 beats/min)

For numbers with three or more digits, you'll often decide to truncate (or round) the number to two places, using the first digit as the stem and the second as the leaf. So, if you had 432, 540, 571, and 638, you might display them as shown below with an indication that 6|3 means 630–639.

```
4 | 3
5 | 47
6 | 3
```

When you make a stem-and-leaf by hand, make sure to give each digit the same width to preserve the area principle. (That can lead to some fat 1's and thin 8's—but it makes the display honest.)

A|S *Activity:* **Dotplots.** Click on points to see their values and even drag them around.

Dotplots

A **dotplot** is a simple display. It just places a dot along an axis for each case in the data. It's like a stem-and-leaf display, but with dots instead of digits for all the leaves. Dotplots are a great way to display a small data set (especially if you forget how to write the digits from 0 to 9). Here's a dotplot of the time (in seconds) that the winning horse took to win the Kentucky Derby in each race between the first Derby in 1875 and the 2014 Derby.

Dotplots show basic facts about the distribution. We can find the slowest and quickest races by finding times for the topmost and bottommost dots. It's also clear that there are two clusters of points, one just below 160 seconds and the other at about 122 seconds. Something strange happened to the Derby times. Once we know to look for it, we can find out that in 1896 the distance of the Derby race was changed from 1.5 miles to the current 1.25 miles. That explains the two clusters of winning times.

Some dotplots stretch out horizontally, with the counts on the vertical axis, like a histogram. Others, such as the one shown here, run vertically, like a stem-and-leaf display. Some dotplots place points next to each other when they would otherwise overlap. Others just place them on top of one another. Newspapers sometimes offer dotplots using little pictures, called pictographs, instead of dots.

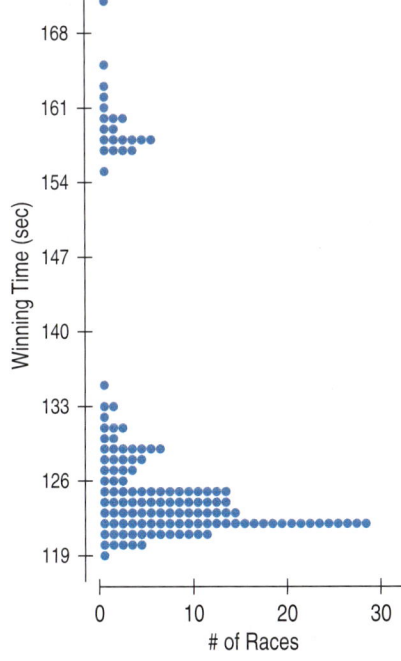

Figure 3.5
A dotplot of Kentucky Derby winning times plots each race as its own dot. We can see two distinct groups corresponding to the two different race distances (Data in **Kentucky Derby 2014**).

Think Before You Draw

Suddenly, we face a lot more options when it's time to invoke our first rule of data analysis and make a picture. *Think* carefully to decide which type of graph to make. In the previous chapter, you learned to check the Categorical Data Condition before making a pie chart or a bar chart. Now, before making a stem-and-leaf display, a histogram, or a dotplot, you need to check the

Quantitative Data Condition: The data are values of a quantitative variable whose units are known.

Although a bar chart and a histogram may look somewhat similar, they're not the same display. You can't display categorical data in a histogram or quantitative data in a bar chart. Always check the condition that confirms what type of data you have before proceeding with your display.

Step back from the histogram or stem-and-leaf display. What can you say about the distribution? Every histogram can tell its own story, but you need to develop a vocabulary to help you explain it to others. When you describe a distribution, you should always start by talking about three things: its *shape, center, and spread*.

3.2 Shape

> **WHAT IS THE MODE?**
> The mode is sometimes defined as the single value that appears most often. That definition is fine for categorical variables because all we need to do is count the number of cases for each category. For quantitative variables, the mode is more ambiguous. What is the mode of the Kentucky Derby times? Well, seven races were timed at 122.2 seconds—more than any other race time. Should that be the mode? Probably not. For quantitative data, it makes more sense to use the term "mode" in the more general sense of the peak of the histogram rather than as a single summary value. In this sense, the important feature of the Kentucky Derby races is that there are two distinct modes, representing the two different versions of the race and warning us to consider those two versions separately.

1. *Does the histogram have a single, central hump or several separated humps?* These humps are called **modes**.[5] The earthquake magnitudes have a single mode at just about 7. A histogram with one peak, such as the earthquake magnitudes, is dubbed **unimodal**; histograms with two peaks are **bimodal**, and those with three or more are called **multimodal**.[6] For example, here's a bimodal histogram.

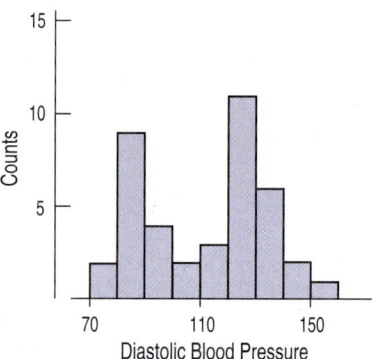

Figure 3.6
A bimodal histogram has two apparent peaks.

A histogram that doesn't appear to have any mode and in which all the bars are approximately the same height is called **uniform**.

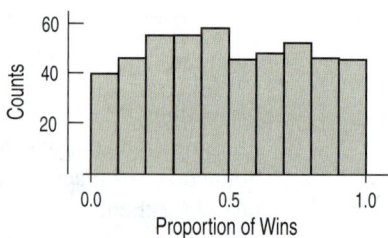

Figure 3.7
In this histogram, the bars are all about the same height. The histogram doesn't appear to have a mode and is called uniform.

---

[5]Well, technically, it's the value on the horizontal axis of the histogram that is the mode, but anyone asked to point to the mode would point to the hump.

[6]Apparently, statisticians don't like to count past two.

2. *Is the histogram **symmetric**?* Can you fold it along a vertical line through the middle and have the edges match pretty closely, or are more of the values on one side?

Figure 3.8
A symmetric histogram can fold in the middle so that the two sides almost match.

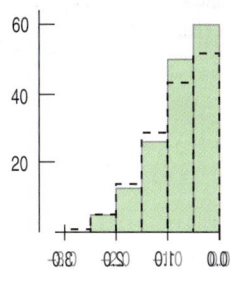

> **Pie à la mode?** You've heard of pie à la mode. Is there a connection between pie and the mode of a distribution? Actually, there is! The mode of a distribution is a *popular* value near which a lot of the data values gather. And "à la mode" means "in style"—*not* "with ice cream." That just happened to be a *popular* way to have pie in Paris around 1900.

The (usually) thinner ends of a distribution are called the **tails**. If one tail stretches out farther than the other, the histogram is said to be **skewed** to the side of the longer tail.

Figure 3.9
Two skewed histograms showing data on two variables for all female heart attack patients in New York State in one year. The blue one (age in years) is skewed to the left. The purple one (charges in $) is skewed to the right.

3. *Do any unusual features stick out?* Often such features tell us something interesting or exciting about the data. You should always mention any stragglers, or **outliers**, that stand away from the body of the distribution. If you're collecting data on nose lengths and Pinocchio is in the group, you'd probably notice him, and you'd certainly want to mention it.

Outliers can affect almost every method we discuss in this course. So we'll always be on the lookout for them. An outlier can be the most informative part of your data. Or it might just be an error. But don't throw it away without comment. Treat it specially and discuss it when you tell about your data. Or find the error and fix it if you can. Be sure to look for outliers. Always.

A bit later in this chapter, you'll learn a handy rule of thumb for deciding when a point might be considered an outlier.

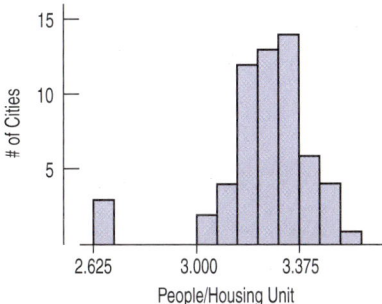

Figure 3.10
A histogram with outliers. There are three cities in the leftmost bar. We wonder why they might be different.

Activity: Attributes of Distribution Shape. This activity shows off aspects of distribution shape through animation and example, then lets you make and interpret histograms with your statistics package.

For Example DESCRIBING HISTOGRAMS

A credit card company wants to see how much customers in a particular segment of their market use their credit card. They have provided you with data[7] on the amount spent by 500 selected customers during a 3-month period and have asked you to summarize the expenditures. Of course, you begin by making a histogram.

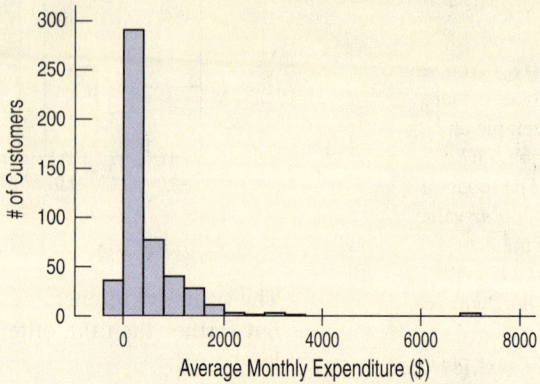

QUESTION: Describe the shape of this distribution. (Data in **Credit Card Exp**)

ANSWER: The distribution of expenditures is unimodal and skewed to the high end. There is an extraordinarily large value at about $7000, and some of the expenditures are negative.

Are there any gaps in the distribution? The Kentucky Derby data that we saw in the dotplot on page 49 has a large gap between two groups of times, one near 120 seconds and one near 160. Gaps help us see multiple modes and encourage us to notice when the data may come from different sources or contain more than one group.

Toto, I've a Feeling We're Not in Math Class Anymore

When Dorothy and her dog Toto land in Oz, everything is more vivid and colorful, but also more dangerous and exciting. Dorothy has new choices to make. She can't always rely on the old definitions, and the yellow brick road has many branches. You may be coming to a similar realization about Statistics.

When we summarize data, our goal is usually more than just developing a detailed knowledge of the data we have at hand. We want to know what the data say about the world, so we'd like to know whether the patterns we see in histograms and summary statistics generalize to other individuals and situations. Scientists generally don't care about the particular guinea pigs in their experiment, but rather about what how they react to different treatments says about how other animals (and, perhaps, humans) might respond.

Because we want to see broad patterns, rather than focus on the details of the data set we're looking at, many of the most important concepts in Statistics are not precisely defined. Whether a histogram is symmetric or skewed, whether it has one or more modes, whether a case is far enough from the rest of the data to be considered an outlier—these are all somewhat vague concepts. They all require judgment.

You may be used to finding a single correct and precise answer, but in Statistics, there may be more than one interpretation. That may make you a little uncomfortable at first, but soon you'll see that leaving room for judgment brings you both power and responsibility. It means that your own knowledge about the world and your judgment matter. You'll use them, along with the statistical evidence, to draw conclusions and make decisions about the world.

---

[7]These data are real, but cannot be further identified for obvious privacy reasons.

Just Checking

It's often a good idea to think about what the distribution of a data set might look like before we collect the data. What do you think the distribution of each of the following data sets will look like? Be sure to discuss its shape. Where do you think the center might be? How spread out do you think the values will be?

1. Number of miles run by Saturday morning joggers at a park.
2. Hours spent by U.S. adults watching football on Thanksgiving Day.
3. Amount of winnings of all people playing a particular state's lottery last week.
4. Ages of the faculty members at your school.
5. Last digit of phone numbers on your campus.

3.3 Center

Let's return to the tsunami earthquakes. But this time, let's look at just 25 years of data: 221 earthquakes that occurred from 1989 through 2013 (selected from **Tsunami Earthquakes**). See Figure 3.11. These should be more accurately measured than prehistoric quakes because seismographs were in wide use. Try to put your finger at the value on the histogram you think is typical. (Read the value from the horizontal axis and remember it.) When we think of a typical value, we usually look for the **center** of the distribution. Where do you think the center of this distribution is? For a unimodal, symmetric distribution such as the earthquake data, it's easy. We'd all agree on the center of symmetry, where we would fold the histogram to match the two sides. But when the distribution is skewed or possibly multimodal, it's not immediately clear what we even mean by the center.

The Median

One natural choice of typical value is the value that is literally in the middle, with half the values below it and half above it.

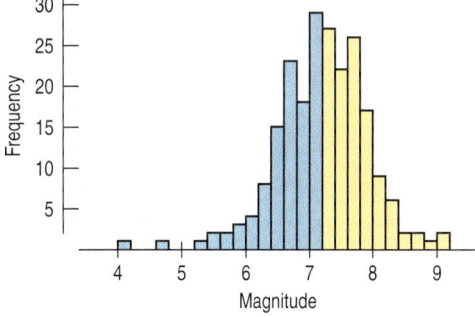

Figure 3.11

Tsunami-causing earthquakes (1989–2013). The median splits the histogram into two halves of equal area.

Histograms follow the area principle, and each half of the data has about 88 earthquakes, so each colored region has the same area in the display. The middle value that divides the histogram into two equal areas is called the **median**.

The median has the same units as the data. Be sure to include the units whenever you discuss the median.

For the recent tsunamis, there are 221 earthquakes, so the median is found at the $(221 + 1)/2 = 111$th place in the sorted data. The median earthquake magnitude is 7.2.

■ NOTATION ALERT

We always use n to indicate the number of values. Some people even say, "How big is the n?" when they mean the number of data values.

By Hand FINDING THE MEDIAN

Finding the median of a batch of n numbers is easy as long as you remember to order the values first. If n is odd, the median is the middle value. Counting in from the ends, we find this value in the $\frac{n+1}{2}$ position.

When n is even, there are two middle values. So, in this case, the median is the average of the two values in positions $\frac{n}{2}$ and $\frac{n}{2}+1$.

Here are two examples:
Suppose the batch has these values: 14.1, 3.2, 25.3, 2.8, −17.5, 13.9, 45.8.
First we order the values: −17.5, 2.8, 3.2, 13.9, 14.1, 25.3, 45.8.
Since there are 7 values, the median is the $(7 + 1)/2 = 4th$ value, counting from the top or bottom: 13.9. Notice that 3 values are lower, 3 higher.
Suppose we had the same batch with another value at 35.7. Then the ordered values are −17.5, 2.8, 3.2, 13.9, 14.1, 25.3, 35.7, 45.8.
The median is the average of the 8/2 or 4th, and the (8/2) + 1, or 5th, values. So the median is $(13.9 + 14.1)/2 = 14.0$. Four data values are lower, and four higher.

The median is one way to find the center of the data. But there are many others. We'll look at an even more important measure later in this chapter.

Knowing the median, we could say that in recent years, a typical tsunami-causing earthquake has been about 7.2 on the Richter scale. How much does that really say? How well does the median describe the data? After all, not every earthquake has a Richter scale value of 7.2. Whenever we find the center of data, the next step is always to ask how well it actually summarizes the data.

3.4 Spread

WHAT WE DON'T KNOW

Statistics pays close attention to what we *don't* know as well as what we do know. Understanding how spread out the data are is a first step in understanding what a summary *cannot* tell us about the data. It's the beginning of telling us what we don't know.

If every earthquake that caused a tsunami registered 7.2 on the Richter scale, then knowing the median would tell us everything about the distribution of earthquake magnitudes. The more the data vary, however, the less the median alone can tell us. So we need to measure the **spread**—or how much the data values vary around the center. In other words, how spread out are they? When we describe a distribution numerically, we always report a measure of its spread along with its center. Spread, like center and shape, is a general description of the distribution, not a precise quantity. There are several different measures of spread. Some are simple, some more complex, and, as we'll see, each has its own advantages and disadvantages.

The Range

How should we measure the spread? We could simply look at the extent of the data. How far apart are the two extremes? The **range** of the data is defined as the *difference* between the maximum and minimum values:

$$Range = max - min.$$

Notice that the range is a *single number, not* an interval of values, as you might think from its use in common speech. The maximum magnitude of these earthquakes is 9.1 and the minimum is 4.0, so the *range* is $9.1 - 4.0 = 5.1$.

The range has the disadvantage that a single extreme value can make it very large, giving a value that doesn't really represent the data overall.

The Interquartile Range

A better way to describe the spread of a variable might be to ignore the extremes and concentrate on the middle of the data. We could, for example, find the range of just the middle half of the data. What do we mean by the middle half? Divide the data in half at the median. Now divide both halves in half again, cutting the data into four quarters. We call these new dividing points **quartiles**. One quarter of the data lies below the **lower quartile**, and one quarter of the data lies above the **upper quartile**, so half the data lies between them. The quartiles border the middle half of the data. For any percentage of the data, there is a corresponding **percentile**—the value that leaves that percentage of the data below it. The lower and upper quartiles are also known as the 25th and 75th percentiles of the data, respectively, since 25% of the data falls below the lower quartile and 75% of the data falls below the upper quartile. The median is the 50th percentile.

By Hand FINDING QUARTILES

A simple way to find the quartiles is to start by splitting the batch into two halves at the median. (When n is odd, some statisticians include the median in both halves; others omit it.[8]) The lower quartile is the median of the lower half, and the upper quartile is the median of the upper half.

Here are our two examples again.

The ordered values of the first batch were −17.5, 2.8, 3.2, 13.9, 14.1, 25.3, and 45.8, with a median of 13.9. Notice that 7 is odd; we'll include the median in both halves to get −17.5, 2.8, 3.2, 13.9 and 13.9, 14.1, 25.3, 45.8.

Each half has 4 values, so the median of each is the average of its 2nd and 3rd values. So, the lower quartile is $(2.8 + 3.2)/2 = 3.0$ and the upper quartile is $(14.1 + 25.3)/2 = 19.7$.

The second batch of data had the ordered values −17.5, 2.8, 3.2, 13.9, 14.1, 25.3, 35.7, and 45.8. Here n is even, so the two halves of 4 values are −17.5, 2.8, 3.2, 13.9 and 14.1, 25.3, 35.7, 45.8. Now the lower quartile is $(2.8 + 3.2)/2 = 3.0$ and the upper quartile is $(25.3 + 35.7)/2 = 30.5$.

> **Why the quartiles?**
> Could we use other percentiles besides the quartiles to measure the spread? Sure, we could, but the IQR is the most commonly used percentile difference and the one you're most likely to see in practice.

The difference between the quartiles tells us how much territory the middle half of the data covers and is called the **interquartile range**. It's commonly abbreviated IQR (and pronounced "eye-cue-are"):

$$IQR = upper\ quartile - lower\ quartile.$$

For the earthquakes, there are 111 values below the median and 111 values above the median (including the median in each half). The midpoint of the lower half is the 56th value in the ordered data; that turns out to be 6.7. In the upper half, we find the 166th value has a magnitude of 7.6 as the third quartile. The *difference* between the quartiles gives the IQR:

$$IQR = 7.6 - 6.7 = 0.9.$$

Now we know that the middle half of the earthquake magnitudes extends across a (interquartile) range of 0.9 Richter scale units. This seems like a reasonable summary of the spread of the distribution, as we can see from this histogram:

Figure 3.12
The quartiles bound the middle 50% of the values of the distribution. This gives a visual indication of the spread of the data. Here we see that the IQR is 0.9 Richter scale units.

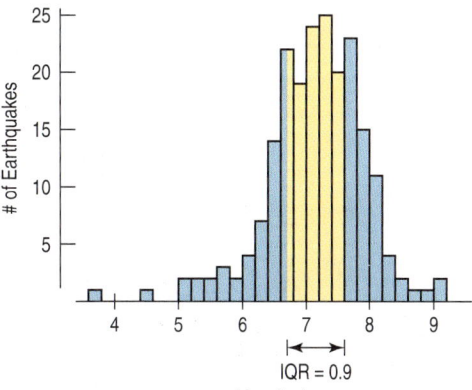

---
[8]Unless otherwise noted, we will use the method of including the median in both halves when calculating the quartiles in examples and answers. Your answer may vary slightly.

56 PART I Exploring and Understanding Data

The IQR is almost always a reasonable summary of the spread of a distribution. Even if the distribution itself is skewed or has some outliers, the IQR should provide useful information. The one exception is when the data are strongly bimodal. For example, remember the dotplot of winning times in the Kentucky Derby (page 49)? Because the race distance was changed, we really have data on two different races, and they shouldn't be summarized together.

> **SO, WHAT IS A QUARTILE ANYWAY?**
> Finding the quartiles sounds easy, but surprisingly, the quartiles are not well-defined. It's not always clear how to find a value such that exactly one quarter of the data lies above or below that value. We offered a simple rule for finding quartiles in the box on page 55: Find the median of each half of the data split by the median. When n is odd, we include the median with each of the halves. Some other texts omit the median from each half before finding the quartiles. Both methods are commonly used. If you are willing to do a bit more calculating, there are several other methods that locate a quartile somewhere between adjacent data values. We know of at least six different rules for finding quartiles. Remarkably, each one is in use in some software package or calculator.
>
> So don't worry too much about getting the "exact" value for a quartile. All of the methods agree pretty closely when the data set is large. When the data set is small, different rules will disagree more, but in that case there's little need to summarize the data anyway.
>
> Remember, Statistics is about understanding the world, not about calculating the right number. The "answer" to a statistical question is a sentence about the issue raised in the question.

3.5 Boxplots and 5-Number Summaries

The **5-number summary** of a distribution reports its median, quartiles, and extremes (maximum and minimum). The 5-number summary for the recent tsunami earthquake *Magnitudes* looks like this:

Max	9.1
Q3	7.6
Median	7.2
Q1	6.7
Min	4.0

■ **NOTATION ALERT**
We always use Q1 to label the lower (25%) quartile and Q3 to label the upper (75%) quartile. We skip the number 2 because the median would, by this system, naturally be labeled Q2—but we don't usually call it that.

It's a good idea to report the number of data values and the identity of the cases (the *Who*). Here there are 221 earthquakes.

The 5-number summary provides a good overview of the distribution of magnitudes of these tsunami-causing earthquakes. For a start, we can see that the median magnitude is 7.2. Because the IQR is only 7.6 − 6.7 = 0.9, we see that many quakes are close to the median magnitude. Indeed, the quartiles show us that the middle half of these earthquakes had magnitudes between 6.7 and 7.6. One quarter of the earthquakes had magnitudes above 7.6, although one tsunami was caused by a quake measuring only 4.0 on the Richter scale.

Boxplots

Once we have a 5-number summary of a (quantitative) variable, we can display that information in a **boxplot**. To make a boxplot of the earthquake magnitudes, follow these steps:

1. Draw a single vertical axis spanning the extent of the data.[9] Draw short horizontal lines at the lower and upper quartiles and at the median. Then connect them with vertical lines to form a box. The box can have any width that looks OK.[10]

---
[9] The axis could also run horizontally.
[10] Some computer programs draw wider boxes for larger data sets. That can be useful when comparing groups.

2.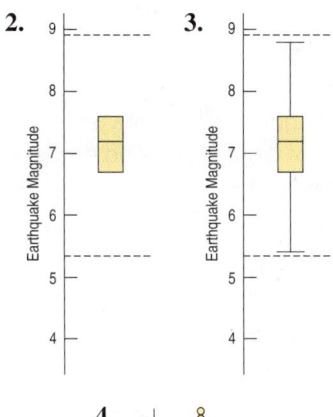

2. To construct the boxplot, erect "fences" around the main part of the data. Place the upper fence 1.5 IQRs above the upper quartile and the lower fence 1.5 IQRs below the lower quartile. For the earthquake magnitude data, we compute

$$Upper\ fence = Q3 + 1.5\ IQR = 7.6 + 1.5 \times 0.9 = 8.95$$

and

$$Lower\ fence = Q1 - 1.5\ IQR = 6.7 - 1.5 \times 0.9 = 5.35.$$

The fences are just for construction and are not part of the display. We show them here with dotted lines for illustration. You should never include them in your boxplot.

3. We use the fences to grow "whiskers." Draw lines from the ends of the box up and down to *the most extreme data values found within the fences*. If a data value falls outside one of the fences, we do *not* connect it with a whisker.

4. Finally, we add the outliers by displaying any data values beyond the fences with special symbols. (We often use a different symbol for "**far outliers**"—data values farther than 3 IQRs from the quartiles.)

A boxplot highlights several features of the distribution. The central box shows the middle half of the data, between the quartiles. The height of the box is equal to the IQR. If the median is roughly centered between the quartiles, then the middle half of the data is roughly symmetric. If the median is not centered, the distribution is skewed. The whiskers show skewness as well if they are not roughly the same length. Any outliers are displayed individually, both to keep them out of the way for judging skewness and to encourage you to give them special attention. They may be mistakes, or they may be the most interesting cases in your data.

For the recent tsunami earthquake data, the central box contains all the earthquakes whose magnitudes are between 6.7 and 7.6 on the Richter scale. From the shape of the box, it looks like the central part of the distribution of earthquakes is roughly symmetric, but the longer lower whisker indicates that the distribution stretches out slightly at the lower end. We also see a few very small magnitude earthquakes and the two large earthquakes we've discussed. Boxplots are particularly good at pointing out outliers.

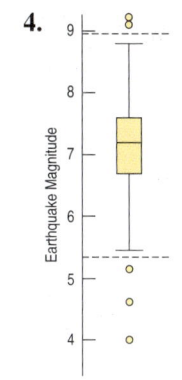

A S **Boxplots.** Watch a boxplot under construction.

Figure 3.13
By turning the boxplot and putting it on the same scale as the histogram, we can compare both displays of the earthquake magnitudes and see how each represents the distribution.

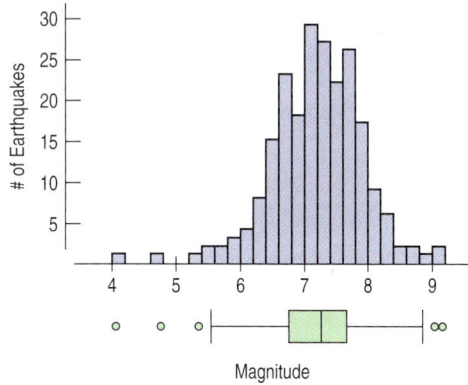

Why 1.5 IQRs? One of the authors asked the prominent statistician, John W. Tukey, the originator of the boxplot, why the outlier nomination rule cut at 1.5 IQRs beyond each quartile. He answered that the reason was that 1 IQR would be too small and 2 IQRs would be too large. That works for us.

Step-by-Step Example SHAPE, CENTER, AND SPREAD: FLIGHT CANCELLATIONS

The U.S. Bureau of Transportation Statistics (www.bts.gov) reports data on airline flights. Let's look at data giving the percentage of flights cancelled each month between 1994 and 2013. (Data in **Pct Cancelled 2013**)

Question: How often are flights cancelled?

Who	Months
What	Percentage of flights cancelled at U.S. airports
When	1994–2013
Where	United States

THINK ➡ Variable Identify the *variable*, and decide how you wish to display it.

To identify a variable, report the W's.

Select an appropriate display based on the nature of the data and what you want to know.

I want to learn about the monthly percentage of flight cancellations at U.S airports.

I have data from the U.S. Bureau of Transportation Statistics giving the percentage of flights cancelled at U.S. airports each month between 1994 and 2013.

✓ **Quantitative Data Condition:** Percentages are quantitative. A histogram and numerical summaries would be appropriate.

SHOW ➡ Mechanics We usually make histograms with a computer or graphing calculator.

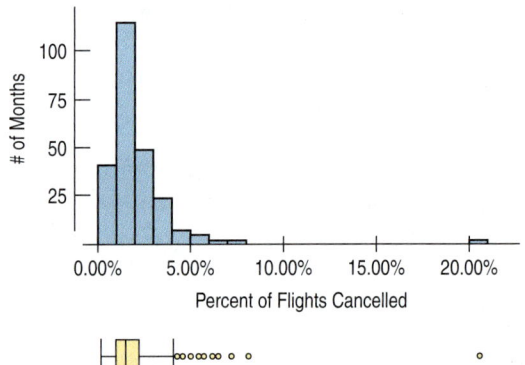

The histogram shows a distribution skewed to the high end and one extreme outlier, a month in which more than 20% of flights were cancelled.

REALITY CHECK It's always a good idea to think about what you expect to see so that you can check whether the histogram looks like what you expected.

In most months, fewer than 5% of flights are cancelled and usually only about 2% or 3%. That seems reasonable.

Count	238
Max	20.24
Q3	2.39
Median	1.68
Q1	1.16
Min	0.38
IQR	1.23

With 238 months, we probably have more data than you'd choose to work with by hand. The results given here are from technology.

TELL → Interpretation Describe the shape, center, and spread of the distribution. Report on the symmetry, number of modes, and any gaps or outliers. You should also mention any concerns you may have about the data.

The distribution of cancellations is skewed to the right, and this makes sense: The values can't fall below 0%, but can increase almost arbitrarily due to bad weather or other events.

The median is 1.68% and the IQR is 1.23%. The low IQR indicates that in most months the cancellation rate is close to the median. In fact, it's between 1.16 and 2.39% in the middle 50% of all months, and in only 1/4 of the months were more than 2.39% of flights cancelled.

There is one extraordinary value: 20.2%. Looking it up, I find that the extraordinary month was September 2001. The attacks of September 11 shut down air travel for several days, accounting for this outlier.

3.6 The Center of Symmetric Distributions: The Mean

■ **NOTATION ALERT**
In Algebra, you used letters to represent values in a problem, but it didn't matter what letter you picked. You could call the width of a rectangle X or you could call it w (or *Fred*, for that matter). But in Statistics, the notation is part of the vocabulary. For example, in Statistics n is always the number of data values. Always.

We have already begun to point out such special notation conventions: n, Q1, and Q3. Think of them as part of the terminology you need to learn in this course.

Here's another one: Whenever we put a bar over a symbol, it means "find the mean."

Medians do a good job of summarizing the center of a distribution, even when the shape is skewed or when there is an outlier, as with the flight cancellations. But when we have symmetric data, there's another alternative. You probably already know how to average values. In fact, to find the median when n is even, we said you should average the two middle values, and you didn't even flinch.

The earthquake magnitudes are pretty close to symmetric, so we can also summarize their center with a mean. The mean recent tsunami earthquake magnitude is 7.10—about what we might expect from the histogram. You already know how to average values, but this is a good place to introduce notation that we'll use throughout the book. We use the Greek capital letter sigma, Σ, to mean "sum" (sigma is "S" in Greek), and we'll write

$$\bar{y} = \frac{Total}{n} = \frac{\Sigma y}{n}.$$

The formula says that to find the **mean**, add up all the values of the variable and divide that sum by the number of data values, n—just as you've always done.[11]

Once we've averaged the data, you'd expect the result to be called the *average*, but that would be too easy. Informally, we speak of the "average person" but we don't add up

---
[11] You may also see the variable called x and the equation written $\bar{x} = \frac{Total}{n} = \frac{\Sigma x}{n}$. Don't let that throw you.

You are free to name the variable anything you want, but we'll generally use y for variables like this that we want to summarize, model, or predict. (Later we'll talk about variables that are used to explain, model, or predict y. We'll call them x.)

people and divide by the number of people. A median is also a kind of average. To make this distinction, the value we calculated is called the mean, $\bar{y}$, and pronounced "y-bar."

The mean feels like the center because it is the point where the histogram balances:

Figure 3.14
The mean is located at the *balancing point* of the histogram. For the recent earthquake magnitudes, $\bar{y} = 7.15$.

Average or mean?
In everyday language, sometimes "average" *does* mean what we want it to mean. We don't talk about your grade point mean or a baseball player's batting mean or the Dow Jones industrial mean. So we'll continue to say "average" when that seems most natural. When we do, though, you may assume that what we mean is the mean.

Mean or Median?

Using the center of balance makes sense when the data are symmetric. But data are not always this well behaved. If the distribution is skewed or has outliers, the center is not so well defined and the mean may not be what we want. For example, the mean of the flight cancellations doesn't give a very good idea of the typical percentage of cancellations.

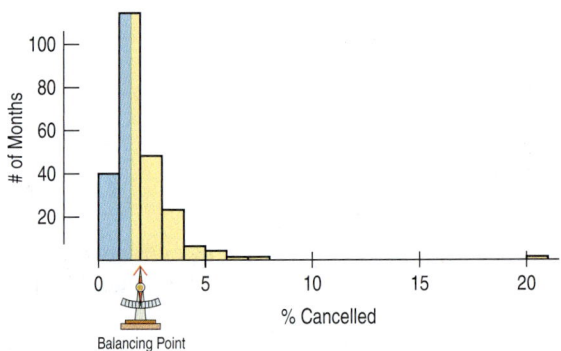

Figure 3.15
The median splits the area of the histogram in half at 1.68%. Because the distribution is skewed to the right, the mean (2.01%) is higher than the median. The points at the right have pulled the mean toward them away from the median.

The mean is 2.01%, but nearly two-thirds of months had cancellation rates below that, so the mean doesn't feel like a good overall summary. Why is the balancing point so high? The large outlying value pulls it to the right. For data like these, the median is a better summary of the center.

Because the median considers only the order of the values, it is **resistant** to values that are extraordinarily large or small; it simply notes that they are one of the "big ones" or the "small ones" and ignores their distance from the center.

For the 1168 tsunami earthquake magnitudes, it doesn't seem to make much difference—the mean is 7.05; the median is 7.0. When the data are symmetric, the mean and median will be close, but when the data are skewed, the median is likely to be a better choice. So, why not just use the median? Well, for one, the median can go overboard. It's not just resistant to occasional outliers, but can be unaffected by changes in up to half the data values. By contrast, the mean includes input from each data value and gives each one equal weight. It's also easier to work with, so when the distribution is unimodal and symmetric, we'll use the mean.

A S Activity: The Center of a Distribution. Compare measures of center by dragging points up and down and seeing the consequences. Another activity shows how to find summaries with your statistics package.

Of course, to choose between mean and median, we'll start by looking at the data. If the histogram is symmetric and there are no outliers, we'll prefer the mean. However, if the histogram is skewed or has outliers, we're usually better off with the median. If you're not sure, report both and discuss why they might differ.

For Example DESCRIBING CENTER

RECAP: You want to summarize the expenditures of 500 credit card company customers, and have looked at a histogram.

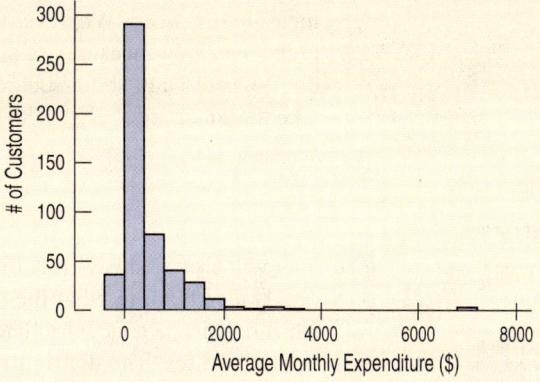

QUESTION: You have found the mean expenditure to be $478.19 and the median to be $216.28. Which is the more appropriate measure of center, and why?

ANSWER: Because the distribution of expenditures is skewed, the median is the more appropriate measure of center. Unlike the mean, it's not affected by the large outlying value or by the skewness. Half of these credit card customers had average monthly expenditures less than $216.48 and half more.

WHEN TO EXPECT SKEWNESS

Even without making a histogram, we can expect some variables to be skewed. When values of a quantitative variable are bounded on one side but not the other, the distribution may be skewed. For example, incomes and waiting times can't be less than zero, so they are often skewed to the right. Amounts of things (dollars, employees) are often skewed to the right for the same reason. If a test is too easy, the distribution will be skewed to the left because many scores will bump against 100%. And combinations of things are often skewed. In the case of the cancelled flights, flights are more likely to be cancelled in January (due to snowstorms) and in August (thunderstorms). Combining values across months leads to a skewed distribution.

3.7 The Spread of Symmetric Distributions: The Standard Deviation

The IQR is always a reasonable summary of spread, but because it uses only the two quartiles of the data, it ignores much of the information about how individual values vary. A more powerful approach uses the standard deviation, which takes into account how far *each* value is from the mean. Like the mean, the standard deviation is appropriate only for symmetric data.

One way to think about spread is to examine how far each data value is from the mean. This difference is called a *deviation*. We could just average the deviations, but the positive and negative differences always cancel each other out. So the average deviation is always zero—not very helpful.

To keep them from canceling out, we *square* each deviation. Squaring always gives a positive value, so the sum won't be zero. That's great. Squaring also emphasizes larger differences—a feature that turns out to be both good and bad.

When we add up these squared deviations and find their average (almost), we call the result the **variance**:

$$s^2 = \frac{\sum (y - \bar{y})^2}{n - 1}.$$

■ **NOTATION ALERT**

s^2 always means the variance of a set of data, and s always denotes the standard deviation.

Why almost? It *would* be a mean if we divided the sum by n. Instead, we divide by $n - 1$. Why? Well, you might think it's just to drive you crazy, but actually s^2 would be too small if we divided by n. Why $n - 1$ fixes it is an amazing fact, but showing it is beyond the scope of this course. Nevertheless, we'll say more in a later chapter.

The variance will play an important role later in this book, but it has a problem as a measure of spread. Whatever the units of the original data are, the variance is in *squared* units. We want measures of spread to have the same units as the data. And we probably don't want to talk about squared dollars or mpg^2. So, to get back to the original units, we take the square root of s^2. The result, s, is the **standard deviation**:

$$s = \sqrt{\frac{\sum(y - \bar{y})^2}{n - 1}}.$$

Who	52 adults
What	Resting heart rates
Units	Beats per minute

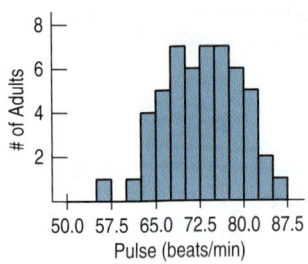

You will almost always rely on a calculator or computer to do the calculating.

Understanding what the standard deviation really means will take some time, and we'll revisit the concept in later chapters. For now, have a look at this histogram of resting pulse rates. The distribution is roughly symmetric, so it's okay to choose the mean and standard deviation as our summaries of center and spread. The mean pulse rate is 72.7 beats per minute, and we can see that's a typical heart rate. We also see that some heart rates are higher and some lower—but how much? Well, the standard deviation of 6.5 beats per minute indicates that, on average, we might expect people's heart rates to differ from the mean rate by about 6.5 beats per minute. Looking at the histogram, we can see that 6.5 beats above or below the mean appears to be a typical deviation.

A S *Activity:* **The Spread of a Distribution.** What happens to measures of spread when some of the data values change may not be quite what you expect.

 By Hand HOW TO CALCULATE A STANDARD DEVIATION

To find the standard deviation, start with the mean, $\bar{y}$. Then find the *deviations* by taking $\bar{y}$ from each value: $(y - \bar{y})$. Square each deviation: $(y - \bar{y})^2$.

Now just add up the squared deviations and divide by $n - 1$. That gives you the variance, s^2. To find the standard deviation, s, take the square root. Here we go:

Suppose the batch of values is 14, 13, 20, 22, 18, 19, and 13.
The mean is $\bar{y} = 17$. So the deviations are found by subtracting 17 from each value:

Original Values	Deviations	Squared Deviations
14	14 − 17 = −3	$(-3)^2 = 9$
13	13 − 17 = −4	$(-4)^2 = 16$
20	20 − 17 = 3	9
22	22 − 17 = 5	25
18	18 − 17 = 1	1
19	19 − 17 = 2	4
13	13 − 17 = −4	16

Add up the squared deviations: 9 + 16 + 9 + 25 + 1 + 4 + 16 = 80
Now divide by $n - 1$:   80/6 = 13.33
Finally, take the square root: $s = \sqrt{13.33} = 3.65$

Thinking About Variation

Statistics is about variation, so spread is an important fundamental concept in Statistics. Measures of spread help us to be precise about what we *don't* know. If many data values are scattered far from the center, the IQR and the standard deviation will be large. If the data values are close to the center, then these measures of spread will be small. If all our data values were exactly the same, we'd have no question about summarizing the center,

A S *Activity:* **Displaying Spread.** What does the standard deviation look like on a histogram? How about the IQR?

A S *Activity*: **Playing with Summaries.** See how different summary measures behave as you place and drag values, and see how sensitive some statistics are to individual data values.

and all measures of spread would be zero—and we wouldn't need Statistics. You might think this would be a big plus, but it would make for a boring world. Fortunately (at least for Statistics), data do vary.

Measures of spread tell how well other summaries describe the data. That's why we always (always!) report a spread along with any summary of the center.

> **For Example** DESCRIBING SPREAD
>
> **RECAP:** The histogram has shown that the distribution of credit card expenditures is skewed, and you have used the median to describe the center. The quartiles are $73.84 and $624.80.
>
> **QUESTION:** What is the IQR and why is it a suitable measure of spread?
>
> **ANSWER:** For these data, the interquartile range (IQR) is $624.80 − $73.84 = $550.96. Like the median, the IQR is not affected by the outlying value or by the skewness of the distribution, so it is an appropriate measure of spread for the given expenditures.

Just Checking

6. The U.S. Census Bureau reports the median family income in its summary of census data. Why do you suppose they use the median instead of the mean? What might be the disadvantages of reporting the mean?

7. You've just bought a new car that claims to get a highway fuel efficiency of 31 miles per gallon. Of course, your mileage will "vary." If you had to guess, would you expect the IQR of gas mileage attained by all cars like yours to be 30 mpg, 3 mpg, or 0.3 mpg? Why?

8. A company selling a new MP3 player advertises that the player has a mean lifetime of 5 years. If you were in charge of quality control at the factory, would you prefer that the standard deviation of life spans of the players you produce be 2 years or 2 months? Why?

3.8 Summary—What to *Tell* About a Quantitative Variable

A S *Activity*: **Playing with Summaries.** Here's a Statistics game about summaries that even some experienced statisticians find . . . well, challenging. Your intuition may be better. Give it a try!

What should you *Tell* about a quantitative variable?

- Start by making a histogram or stem-and-leaf display, and discuss the shape of the distribution.
- Next, discuss the center and spread.
  - Always pair the median with the IQR and the mean with the standard deviation. It's not useful to report one without the other. Reporting a center without a spread is dangerous. You may think you know more than you do about the distribution. Reporting only the spread leaves us wondering where we are.
  - If the shape is skewed, report the median and IQR. You may want to include the mean and standard deviation as well, but you should point out why the mean and median differ.
  - If the shape is symmetric, report the mean and standard deviation and possibly the median and IQR as well. For unimodal symmetric data, the IQR is usually a bit larger than the standard deviation. If that's not true of your data set, look again to make sure that the distribution isn't skewed and there are no outliers.

64 PART I Exploring and Understanding Data

- Also, discuss any unusual features.
  - If there are multiple modes, try to understand why. If you can identify a reason for separate modes (for example, women and men typically have heart attacks at different ages), it may be a good idea to split the data into separate groups.
  - If there are any clear outliers, you should point them out. If you are reporting the mean and standard deviation, report them with the outliers present and with the outliers omitted. The differences may be revealing. (Of course, the median and IQR won't be affected very much by the outliers.)

> **HOW "ACCURATE" SHOULD WE BE?**
> Don't think you should report means and standard deviations to a zillion decimal places; such implied accuracy is really meaningless. Although there is no ironclad rule, statisticians commonly report summary statistics to one or two decimal places more than the original data have.

For Example CHOOSING SUMMARY STATISTICS

RECAP: You have provided the credit card company's board of directors with a histogram of customer expenditures, and you have summarized the center and spread with the median and IQR. Knowing a little Statistics, the directors now insist on having the mean and standard deviation as summaries of the spending data.

QUESTION: Although you know that the mean is $478.19 and the standard deviation is $741.87, you need to explain to them why these are not suitable summary statistics for these expenditures data. What would you give as reasons?

ANSWER: The high outlier at $7000 pulls the mean up substantially and inflates the standard deviation. Locating the mean value on the histogram shows that it is not a typical value at all, and the standard deviation suggests that expenditures vary much more than they do. The median and IQR are more resistant to the presence of skewness and outliers, giving more realistic descriptions of center and spread.

Step-by-Step Example SUMMARIZING A DISTRIBUTION

One of the authors owned a Nissan Maxima for 8 years. Being a statistician, he recorded the car's fuel efficiency (in mpg) each time he filled the tank. He wanted to know what fuel efficiency to expect as "ordinary" for his car. (Hey, he's a statistician. What would you expect?[12]) Knowing this, he was able to predict when he'd need to fill the tank again and to notice if the fuel efficiency suddenly got worse, which could be a sign of trouble.

Question: How would you describe the distribution of *Fuel efficiency* for this car?

THINK ▶ Plan State what you want to find out.	I want to summarize the distribution of Nissan Maxima fuel efficiency.
Variable Identify the variable and report the W's.	The data are the fuel efficiency values in miles per gallon for the first 100 fill-ups of a Nissan Maxima.
Be sure to check the appropriate condition.	✓ **Quantitative Data Condition:** The fuel efficiencies are quantitative with units of miles per gallon. Histograms are an appropriate way to display the distribution. Numerical summaries are appropriate as well.

---

[12] He also recorded the time of day, temperature, price of gas, and phase of the moon. (OK, maybe not phase of the moon.) (Data in **Nissan**)

SHOW → Mechanics Make a histogram. Based on the shape, choose appropriate numerical summaries.

A histogram of the data shows a fairly symmetric distribution with a low outlier.

Count	100
Mean	22.4 mpg
StdDev	2.45
Q1	20.8
Median	22.0
Q3	24.0
IQR	3.2

The mean and median are close, so the outlier doesn't seem to be a problem. I can use the mean and standard deviation.

REALITY CHECK A value of 22 mpg seems reasonable for such a car. The spread is reasonable, although the range looks a bit large.

TELL → Conclusion Summarize and interpret your findings in context. Be sure to discuss the distribution's shape, center, spread, and unusual features (if any).

The distribution of mileage is unimodal and roughly symmetric with a mean of 22.4 mpg. There is a low outlier that should be investigated, but it does not influence the mean very much. The standard deviation suggests that from tankful to tankful, I can expect the car's fuel economy to differ from the mean by an average of about 2.45 mpg.

> **I GOT A DIFFERENT ANSWER: DID I MESS UP?**
> When you calculate a mean, the computation is clear: You sum all the values and divide by the sample size. You may round your answer less or more than someone else (we recommend one more decimal place than the data), but all books and technologies agree on how to find the mean. Some statistics, however, are more problematic. For example, we've already pointed out that methods of finding quartiles differ.
> Differences in numeric results can also arise from decisions in the middle of calculations. For example, if you round off your value for the mean before you calculate the sum of squared deviations, your standard deviation probably won't agree with a computer program that calculates using many decimal places. (We do recommend that you use as many digits as you can during the calculation and round only when you are done.)
> Don't be overly concerned with these discrepancies, especially if the differences are small. They don't mean that your answer is "wrong," and they usually won't change any conclusion you might draw about the data. Sometimes (in footnotes and in the answers in the back of the book) we'll note alternative results, but we could never list all the possible values, so we'll rely on your common sense to focus on the meaning rather than on the digits. Remember: Answers are sentences—not single numbers!

WHAT CAN GO WRONG?

A data display should tell a story about the data. To do that, it must speak in a clear language, making plain what variable is displayed, what the axes show, and what the values of the data are. And it must be consistent in those decisions.

A display of quantitative data can go wrong in many ways. The most common failures arise from only a few basic errors:

- **Don't make a histogram of a categorical variable.** Just because the variable contains numbers doesn't mean that it's quantitative. Here's a histogram of the insurance policy numbers of some workers.

Figure 3.16
It's not appropriate to display these data with a histogram.

It's not very informative because the policy numbers are just identifiers. Similarly, a histogram or stem-and-leaf display of a categorical variable makes no sense. A bar chart or pie chart would be more appropriate.

- **Don't look for shape, center, and spread of a bar chart.** A bar chart showing the sizes of the piles displays the distribution of a categorical variable, but the bars could be arranged in any order left to right. Concepts like symmetry, center, and spread make sense only for quantitative variables.

- **Don't use bars in every display—save them for histograms and bar charts.** In a bar chart, the bars indicate how many cases of a categorical variable are piled in each category. Bars in a histogram indicate the number of cases piled in each interval of a quantitative variable. In both bar charts and histograms, the bars represent counts of data values. Some people create other displays that use bars to represent individual data values. Beware: Such graphs are neither bar charts nor histograms. For example, a student was asked to make a histogram from data showing the number of juvenile bald eagles seen during each of the 13 weeks in the winter at a site in Rock Island, Illinois. Instead, he made this plot:

Figure 3.17
This isn't a histogram or a bar chart. It's an ill-conceived graph that uses bars to represent individual data values (number of eagles sighted) week by week.

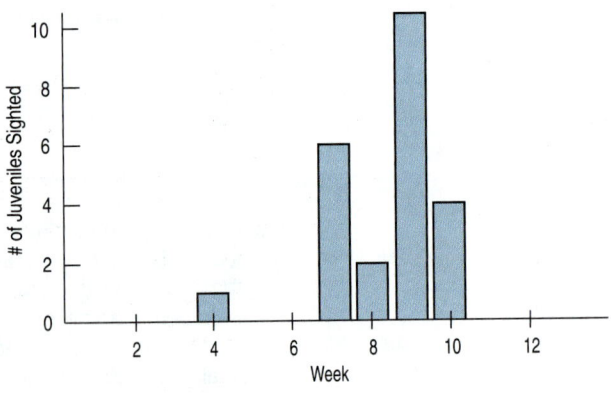

Look carefully. That's not a histogram. A histogram shows *What* we've measured along the horizontal axis and counts of the associated *Who*'s represented as bar heights. This student has it backwards: He used bars to show counts of birds for each week.[13] We need counts of weeks. A correct histogram should have a tall bar at "0" to show there were many weeks when no eagles were seen, like this:

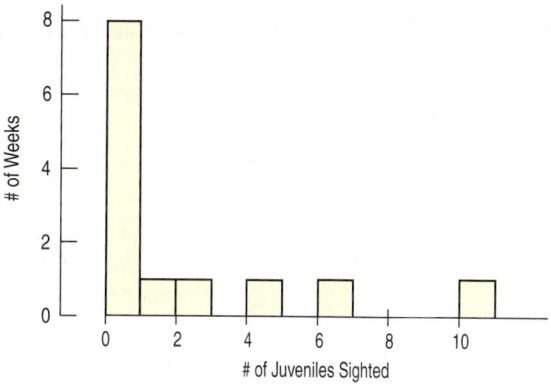

Figure 3.18
A histogram of the eagle-sighting data shows the number of weeks in which different counts of eagles occurred. This display shows the distribution of juvenile-eagle sightings.

- **Choose a bin width appropriate to the data.** Computer programs usually do a pretty good job of choosing histogram bin widths. Often there's an easy way to adjust the width, sometimes interactively. Here are the tsunami earthquakes with two (rather extreme) choices for the bin size:

For the usual amounts of data we display, between 5 and 30 bins show the features of a distribution best. Too few bins tend to hide too much and too many bins make unimportant features too prominent. The task of summarizing a quantitative variable is relatively simple, and there is a simple path to follow. However, you need to watch out for certain features of the data that make summarizing them with a number dangerous. Here's some advice:

- **Don't forget to do a reality check.** Don't let the computer or calculator do your thinking for you. Make sure the calculated summaries make sense. For example, does the mean look like it is in the center of the histogram? Think about the spread: An IQR of 50 mpg would clearly be wrong for gas mileage. And no measure of spread can be negative. The standard deviation can take the value 0, but only in the very unusual case that all the data values equal the same number. If you see an IQR or standard deviation equal to 0, it's probably a sign that something's wrong with the data.

- **Don't forget to sort the values before finding the median or percentiles.** It seems obvious, but when you work by hand, it's easy to forget to sort the data first before counting in to find medians, quartiles, or other percentiles. Don't report that the median of the five values 194, 5, 1, 17, and 893 is 1 just because 1 is the middle number.

---

[13]Edward Tufte, in his book *The Visual Display of Quantitative Information*, proposes that graphs should have a high data-to-ink ratio. That is, we shouldn't waste a lot of ink to display a single number when a dot would do the job.

- **Don't worry about small differences when using different methods.** Finding the 10th percentile or the lower quartile in a data set sounds easy enough. But it turns out that the definitions are not exactly clear. If you compare different statistics packages or calculators, you may find that they give slightly different answers for the same data. These differences, though, are unlikely to be important in interpreting the data, the quartiles, or the IQR, so don't let them worry you.

- **Don't compute numerical summaries of a categorical variable.** Neither the mean ZIP code nor the standard deviation of Social Security numbers is meaningful. If the variable is categorical, you should instead report summaries such as percentages of individuals in each category. It is easy to make this mistake when using technology to do the summaries for you. After all, the computer doesn't care what the numbers mean.

Gold Card Customers—Regions National Banks		
Month	April 2014	May 2014
Average ZIP Code	45,034.34	38,743.34

- **Don't report too many decimal places.** Statistics programs and calculators often report a ridiculous number of digits. A general rule for numerical summaries is to report one or two more digits than the number of digits in the data. For example, earlier we saw a dotplot of the Kentucky Derby race times. The mean and standard deviation of those times could be reported as

$$\bar{y} = 129.56779 \text{ sec} \qquad s = 13.05657 \text{ sec}$$

But we knew the race times only to the nearest quarter second, so the extra digits are meaningless.

- **Don't round in the middle of a calculation.** Don't *report* too many decimal places, but it's best not to do any rounding until the end of your calculations. Even though you might report the mean of the earthquake magnitudes as 7.05, it's really 7.05281. Use the more precise number in your calculations if you're finding the standard deviation by hand—or be prepared to see small differences in your final result.

- **Watch out for multiple modes.** The summaries of the Kentucky Derby times are meaningless for another reason. As we saw in the dotplot, the Derby was initially a longer race. It would make much more sense to report that the old 1.5 mile Derby had a mean time of 159.6 seconds, while the current Derby has a mean time of 124.3 seconds. If the distribution has multiple modes, consider separating the data into different groups and summarizing each group separately.

- **Beware of outliers.** The median and IQR are resistant to outliers, but the mean and standard deviation are not. To help spot outliers . . .

- **Don't forget to: Make a picture (make a picture, make a picture).** The sensitivity of the mean and standard deviation to outliers is one reason you should always make a picture of the data. Summarizing a variable with its mean and standard deviation when you have not looked at a histogram or dotplot to check for outliers or skewness invites disaster. You may find yourself drawing absurd or dangerously wrong conclusions about the data. And, of course, you should demand no less of others. Don't accept a mean and standard deviation blindly without some evidence that the variable they summarize is unimodal, symmetric, and free of outliers.

CONNECTIONS

Distributions of quantitative variables, like those of categorical variables, show the possible values and their relative frequencies. A histogram shows the distribution of values in a quantitative variable with adjacent bars. Don't confuse histograms with bar charts, which display categorical variables. For categorical data, the mode is the category with the biggest count. For quantitative data, modes are peaks in the histogram.

The shape of the distribution of a quantitative variable is an important concept in most of the subsequent chapters. We will be especially interested in distributions that are unimodal and symmetric.

In addition to their shape, we summarize distributions with center and spread, usually pairing a measure of center with a measure of spread: median with IQR and mean with standard deviation. We favor the mean and standard deviation when the shape is unimodal and symmetric, but choose the median and IQR for skewed distributions or when there are outliers we can't otherwise set aside.

What Have We Learned?

Learning Objectives

Make and interpret displays of the distribution of a variable.
- We display the distribution of a quantitative variable with a histogram, stem-and-leaf display, or dotplot.
- We understand distributions in terms of their shape, center, and spread.

Describe the shape of a distribution.
- A **symmetric** distribution has roughly the same shape reflected around the center.
- A **skewed** distribution extends farther on one side than on the other.
- A **unimodal** distribution has a single major hump or mode; a bimodal distribution has two; multimodal distributions have more.
- **Outliers** are values that lie far from the rest of the data.
- Report any other **unusual feature** of the distribution such as gaps.

Summarize the center of a distribution by computing the mean and median, and know when it is best to use each.
- The **mean** is the sum of the values divided by the count. It is a suitable summary for unimodal, symmetric distributions.
- The **median** is the middle value; half the values are above and half are below the median. It is a better summary when the distribution is skewed or has outliers.

Compute the standard deviation and interquartile range (IQR), and know when it is best to use each to summarize the spread.
- The **standard deviation** is roughly the square root of the average squared difference between each data value and the mean. It is the summary of choice for the spread of unimodal, symmetric variables.
- The **IQR** is the difference between the quartiles. It is often a better summary of spread for skewed distributions or data with outliers.
- We'll report the median and IQR when the distribution is skewed. If it's symmetric, we'll summarize the distribution with the mean and standard deviation (and possibly the median and IQR as well). Always pair the median with the IQR and the mean with the standard deviation.

Recognize that the methods of this chapter assume that the data are quantitative.
- The **Quantitative Data Condition** serves as a reminder to check that the data are quantitative. A good way to be sure is to report the measurement units.

Find a 5-number summary and, using it, make a boxplot. Use the boxplot's outlier nomination rule to identify cases that may deserve special attention.
- A **5-number summary** consists of the median, the quartiles, and the extremes of the data.
- A **boxplot** shows the quartiles as the upper and lower ends of a central box, the median as a line across the box, and "whiskers" that extend to the most extreme values that are not nominated as outliers.
- Boxplots display separately any case that is more than 1.5 IQRs beyond each quartile. These cases should be considered as possible outliers.

Review of Terms

Distribution The distribution of a quantitative variable slices up all the possible values of the variable into equal-width bins and gives the number of values (or counts) falling into each bin (p. 46).

Histogram (relative frequency histogram) A histogram uses adjacent bars to show the distribution of a quantitative variable. Each bar represents the frequency (or relative frequency) of values falling in each bin (p. 46).

Gap A region of the distribution where there are no values (p. 47).

Stem-and-leaf display	A display that shows quantitative data values in a way that sketches the distribution of the data. It's best described in detail by example (p. 48).
Dotplot	A dotplot graphs a dot for each case against a single axis (p. 49).
Shape	To describe the shape of a distribution, look for (p. 50) • single vs. multiple modes. • symmetry vs. skewness. • outliers and gaps.
Mode	A hump or local high point in the shape of the distribution of a variable. The apparent location of modes can change as the scale of a histogram is changed (p. 50).
Unimodal (Bimodal)	Having one mode. This is a useful term for describing the shape of a histogram when it's generally mound-shaped. Distributions with two modes are called **bimodal**. Those with more than two are **multimodal** (p. 50).
Uniform	A distribution that doesn't appear to have any mode and in which all the bars of its histogram are approximately the same height (p. 50).
Symmetric	A distribution is symmetric if the two halves on either side of the center look approximately like mirror images of each other (p. 51).
Tails	The parts of a distribution that typically trail off on either side. Distributions can be characterized as having long tails (if they straggle off for some distance) or short tails (if they don't) (p. 51).
Skewed	A distribution is skewed if it's not symmetric and one tail stretches out farther than the other. Distributions are said to be **skewed left** when the longer tail stretches to the left, and **skewed right** when it goes to the right (p. 51).
Outliers	Outliers are extreme values that don't appear to belong with the rest of the data. They may be unusual values that deserve further investigation, or they may be just mistakes; there's no obvious way to tell. Don't delete outliers automatically—you have to think about them. Outliers can affect many statistical analyses, so you should always be alert for them. Boxplots display points more than 1.5 IQR from either end of the box individually, but this is just a rule-of-thumb and not a definition of what is an outlier (p. 51).
Center	The place in the distribution of a variable that you'd point to if you wanted to attempt the impossible by summarizing the entire distribution with a single number. Measures of center include the mean and median (p. 53).
Median	The median is the middle value, with half of the data above and half below it. If n is even, it is the average of the two middle values. It is usually paired with the IQR (p. 53).
Spread	A numerical summary of how tightly the values are clustered around the center. Measures of spread include the IQR and standard deviation (p. 54).
Range	The difference between the lowest and highest values in a data set (p. 54). $$Range = max - min$$
Quartile	The lower quartile (Q1) is the value with a quarter of the data below it. The upper quartile (Q3) has three quarters of the data below it. The median and quartiles divide data into four parts with equal numbers of data values (p. 55).
Percentile	The ith percentile is the number that falls above i% of the data (p. 55).
Interquartile range (IQR)	The IQR is the difference between the first and third quartiles. IQR = Q3 − Q1. It is usually reported along with the median (p. 55).
5-number summary	The 5-number summary of a distribution reports the minimum value, Q1, the median, Q3, and the maximum value (p. 56).
Boxplot	A boxplot displays the 5-number summary as a central box with whiskers that extend to the nonoutlying data values. Boxplots are particularly effective for comparing groups and for displaying possible outliers (p. 56).
Mean	The mean is found by summing all the data values and dividing by the count: $$\bar{y} = \frac{Total}{n} = \frac{\sum y}{n}.$$ It is usually paired with the standard deviation (p. 59).

Resistant — A calculated summary is said to be resistant if outliers have only a small effect on it (p. 60).

Variance — The variance is the sum of squared deviations from the mean, divided by the count minus 1:

$$s^2 = \frac{\sum(y - \bar{y})^2}{n - 1}.$$

It is useful in calculations later in the book (p. 61).

Standard deviation — The standard deviation is the square root of the variance:

$$s = \sqrt{\frac{\sum(y - \bar{y})^2}{n - 1}}$$

It is usually reported along with the mean (p. 62).

On the Computer DISPLAYING AND SUMMARIZING QUANTITATIVE VARIABLES

Almost any program that displays data can make a histogram, but some will do a better job of determining where the bars should start and how they should partition the span of the data.

The vertical scale may be counts or proportions. Sometimes it isn't clear which. But the shape of the histogram is the same either way.

Most packages choose the number of bars for you automatically. Often you can adjust that choice.

The axis should be clearly labeled so you can tell what "pile" each bar represents. You should be able to tell the lower and upper bounds of each bar.

Many statistics packages offer a prepackaged collection of summary measures. The result might look like this:

```
Variable: Weight
N = 234
Mean = 143.3      Median = 139
St. Dev = 11.1    IQR = 14
```

Alternatively, a package might make a table for several variables and summary measures:

Variable	N	mean	median	stdev	IQR
Weight	234	143.3	139	11.1	14
Height	234	68.3	68.1	4.3	5
Score	234	86	88	9	5

A S *Case Study:* **Describing Distribution Shapes.** Who's safer in a crash—passengers or the driver? Investigate with your statistics package.

It is usually easy to read the results and identify each computed summary. You should be able to read the summary statistics produced by any computer package.

Packages often provide many more summary statistics than you need. Of course, some of these may not be appropriate when the data are skewed or have outliers. It is your responsibility to check a histogram or stem-and-leaf display and decide which summary statistics to use.

It is common for packages to report summary statistics to many decimal places of "accuracy." Of course, it is rare data that have such accuracy in the original measurements. The ability to calculate to six or seven digits beyond the decimal point doesn't mean that those digits have any meaning. Generally it's a good idea to round these values, allowing perhaps one more digit of precision than was given in the original data.

Displays and summaries of quantitative variables are among the simplest things you can do in most statistics packages.

DATA DESK

To make a histogram:
- Select the variable to display.
- In the **Plot** menu, choose **Histogram**.

To calculate summaries:
- In the **Calc** menu, open the **summaries** submenu. **Options** offer separate tables, a single unified table, and other formats.

EXCEL

In Excel, there is another way to find some of the standard summary statistics. For example, to compute the mean:
- Click on an empty cell.
- Go to the Formulas tab in the Ribbon. Click on the drop-down arrow next to AutoSum and choose **Average**.
- Enter the data range in the formula displayed in the empty box you selected earlier.
- Press **Enter**. This computes the mean for the values in that range.

To compute the standard deviation:
- Click on an empty cell.
- Go to the Formulas tab in the Ribbon and click the drop-down arrow next to AutoSum and select **More functions...**
- In the dialog window that opens, select **STDEV** from the list of functions and click **OK**. A new dialog window opens. Enter a range of fields into the text fields and click **OK**.

Excel computes the standard deviation for the values in that range and places it in the specified cell of the spreadsheet.

JMP

To make a histogram and find summary statistics:
- Choose **Distribution** from the **Analyze** menu.
- In the **Distribution** dialog, drag the name of the variable that you wish to analyze into the empty window beside the label "Y, Columns."
- Click **OK**. JMP computes standard summary statistics along with displays of the variables.

MINITAB

To make a histogram:
- Choose **Histogram** from the **Graph** menu.
- Select "Simple" for the type of graph and click **OK**.
- Enter the name of the quantitative variable you wish to display in the box labeled "Graph variables." Click **OK**.

To calculate summary statistics:
- Choose **Basic statistics** from the **Stat** menu. From the **Basic Statistics** submenu, choose **Display Descriptive Statistics**.
- Assign variables from the variable list box to the Variables box. MINITAB makes a Descriptive Statistics table.

R

For a quantitative variable X:
- **summary(X)** gives a 5-number summary and the mean
- **mean(X)** gives the mean and **sd(X)** gives the standard deviation
- **hist(X)** produces a histogram and **boxplot(X)** makes a boxplot

COMMENTS

Many other summaries are available, including **min( )**, **max( )**, **quantile(X,prob=p)** (where p is a probability between 0 and 1), and **median( )**. Both hist and boxplot have many options.

SPSS

To make a histogram in SPSS open the Chart Builder from the Graphs menu:
- Click the **Gallery** tab.
- Choose **Histogram** from the list of chart types.
- Drag the histogram onto the canvas.
- Drag a scale variable to the y-axis drop zone.
- Click **OK**.

To calculate summary statistics:
- Choose **Explore** from the **Descriptive Statistics** sub-menu of the **Analyze** menu. In the Explore dialog, assign one or more variables from the source list to the Dependent List and click the **OK** button.

STATCRUNCH

To make a histogram, dotplot, or stem-and-leaf plot:
- Click on **Graphics**.
- Choose the type of plot.
- Choose the variable name from the list of **Columns**.
- Click on **Next**.
- (For a histogram) Choose **Frequency** or (usually) **Relative frequency**, and (if desired) set the axis scale by entering the **Start** value and **Bin width**.
- Click on **Create Graph**.

To calculate summaries:
- Click on **Stat**.
- Choose **Summary Stats » Columns**.
- Choose the variable name from the list of **Columns**.
- Click on **Calculate**.

COMMENTS

- You may need to hold down the Ctrl or command key to choose more than one variable to summarize.
- Before calculating, click on **Next** to choose additional summary statistics.

TI-83/84 PLUS

To make a histogram:
- Turn a **STATPLOT** on.
- Choose the **histogram icon** and specify the List where the data are stored.
- **ZoomStat**, then adjust the **WINDOW** appropriately.

To calculate summary statistics:
- Choose **1-VarStats** from the **STAT CALC** menu and specify the List where the data are stored. You must scroll down to see the 5-number summary.

- To make a boxplot, set up a **STAT PLOT** using the boxplot icon.

COMMENTS

If the data are stored as a frequency table (say, with data values in L1 and frequencies in L2), set up the Plot with Xlist: L1 and Freq: L2.

Exercises

Section 3.1

1. Details Here are histograms for four manufactured sets of numbers. The histograms look rather different, but all four sets have the same 5-number summary, so the boxplots for all four sets are identical to the one shown. Using these plots as examples, explain some limitations in using boxplots to describe the shape of a distribution.

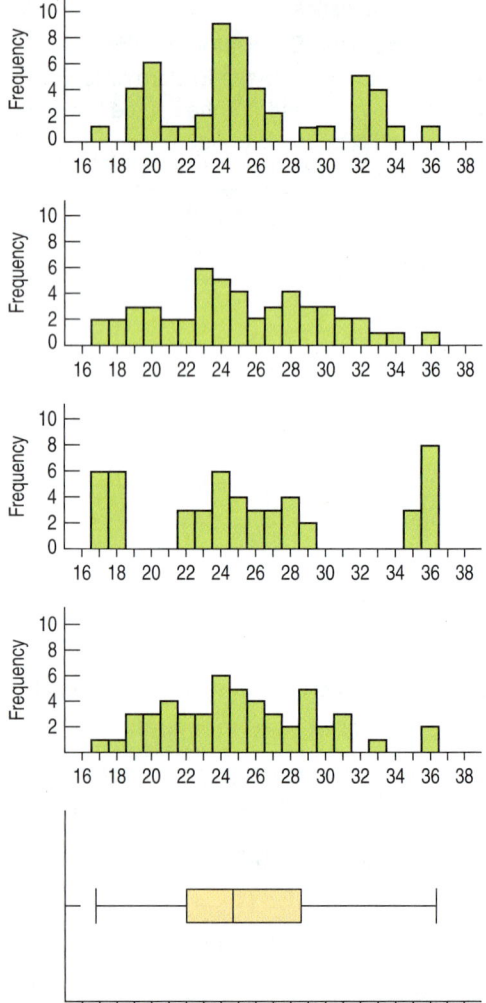

2. Opposites In a way, boxplots are the opposite of histograms. A histogram divides the number line into equal intervals and displays the number of data values in each interval. A boxplot divides the *data* into equal parts and displays the portion of the number line each part covers. These two plots display the number of incarcerated prisoners in each state as of 2012.

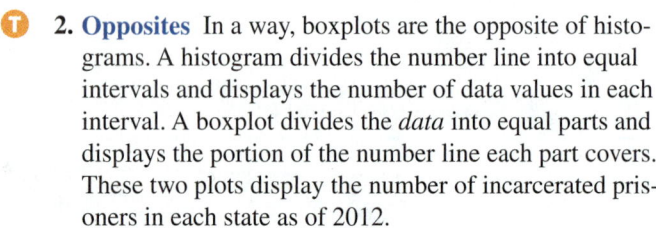

a) Explain how you could tell, by looking at a boxplot, where the tallest bars on the histogram would be located.
b) Explain how both the boxplot and the histogram can indicate a skewed distribution.
c) Identify one feature of the distribution that the histogram shows but a boxplot does not.
d) Identify one feature of the distribution that the boxplot shows but the histogram does not.

Section 3.2

3. Outliers The 5-number summary for the run times in minutes of the 150 highest grossing movies of 2010 looks like this:

Min	Q1	Med	Q3	Max
43	98	104.5	116	160

Are there any outliers in these data? How can you tell?

4. Adoptions The following table shows the mean and median ages in years of children adopted in different states in 2010. In all 50 states, the mean age is higher than the median. What might that imply about the shape of the distribution of ages of adopted children in each state?

State	Mean Age	Median Age
Alabama	7.3	6.5
Alaska	7.1	6.6
Arizona	6.4	5.1
Arkansas	6.0	4.5
California	5.9	4.6
Colorado	5.9	4.7

Section 3.3

5. Adoptions II Here is a histogram showing the total number of adoptions in each of the 50 states and the District of Columbia.

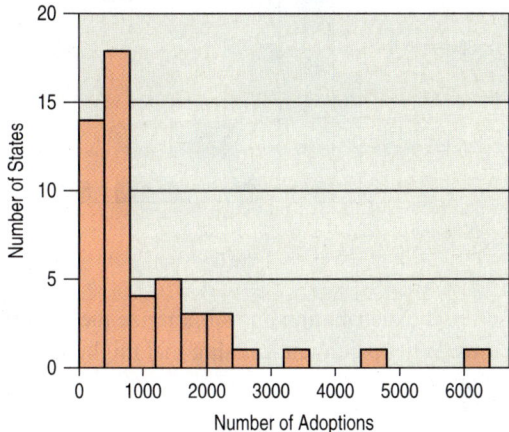

Would you expect the mean number of adoptions or the median number of adoptions to be higher? Why?

6. Test score centers After entering the test scores from her Statistics class of 25 students, the instructor calculated the mean and the median of the scores. Upon checking, she discovered that she had entered the top score as 46, but it should have been 56. When she corrects this score, how will the mean and median be affected?

Section 3.4

7. Test score spreads The professor in question 6 had also calculated the IQR and standard deviation of the test scores before realizing her mistake. (She had entered the top score as 46 but it was actually 56.) What effect will correcting the error have on the IQR and the standard deviation?

8. Fuel economy The boxplot shows the fuel economy ratings for 67 model year 2012 subcompact cars. Some summary statistics are also provided. The extreme outlier is the **Mitsubishi i-MiEV**, an electric car whose electricity usage is *equivalent* to 112 miles per gallon.

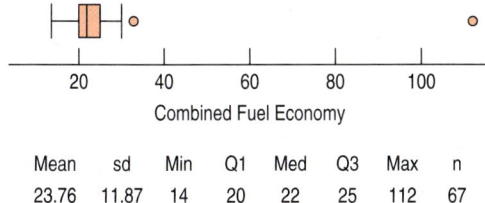

Mean	sd	Min	Q1	Med	Q3	Max	n
23.76	11.87	14	20	22	25	112	67

If that electric car is removed from the data set, how will the standard deviation be affected? The IQR?

Chapter Exercises

9. Histogram Find a histogram that shows the distribution of a variable in a newspaper, a magazine, or the Internet.
 a) Does the article identify the W's?
 b) Discuss whether the display is appropriate.
 c) Discuss what the display reveals about the variable and its distribution.
 d) Does the article accurately describe and interpret the data? Explain.

10. Not a histogram Find a graph other than a histogram that shows the distribution of a quantitative variable in a newspaper, a magazine, or the Internet.
 a) Does the article identify the W's?
 b) Discuss whether the display is appropriate for the data.
 c) Discuss what the display reveals about the variable and its distribution.
 d) Does the article accurately describe and interpret the data? Explain.

11. In the news Find an article in a newspaper, a magazine, or the Internet that discusses an "average."
 a) Does the article discuss the W's for the data?
 b) What are the units of the variable?
 c) Is the average used the median or the mean? How can you tell?
 d) Is the choice of median or mean appropriate for the situation? Explain.

12. In the news II Find an article in a newspaper, a magazine, or the Internet that discusses a measure of spread.
 a) Does the article discuss the W's for the data?
 b) What are the units of the variable?
 c) Does the article use the range, IQR, or standard deviation?
 d) Is the choice of measure of spread appropriate for the situation? Explain.

13. Thinking about shape Would you expect distributions of these variables to be uniform, unimodal, or bimodal? Symmetric or skewed? Explain why.
 a) The number of speeding tickets each student in the senior class of a college has ever had.
 b) Players' scores (number of strokes) at the U.S. Open golf tournament in a given year.
 c) Weights of female babies born in a particular hospital over the course of a year.
 d) The length of the average hair on the heads of students in a large class.

14. More shapes Would you expect distributions of these variables to be uniform, unimodal, or bimodal? Symmetric or skewed? Explain why.
 a) Ages of people at a Little League game.
 b) Number of siblings of people in your class.
 c) Pulse rates of college-age males.
 d) Number of times each face of a die shows in 100 tosses.

15. Sugar in cereals The histogram displays the sugar content (as a percent of weight) of 49 brands of breakfast cereals.

a) Describe this distribution.
b) What do you think might account for this shape?

16. Singers The display shows the heights of some of the singers in a chorus, collected so that the singers could be positioned on stage with shorter ones in front and taller ones in back.

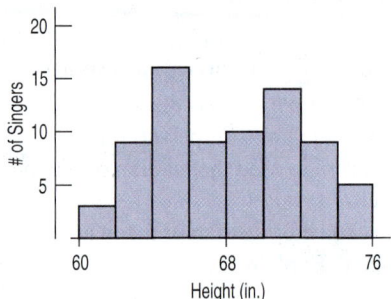

a) Describe the distribution.
b) Can you account for the features you see here?

17. Vineyards The histogram shows the sizes (in acres) of 36 vineyards in the Finger Lakes region of New York.

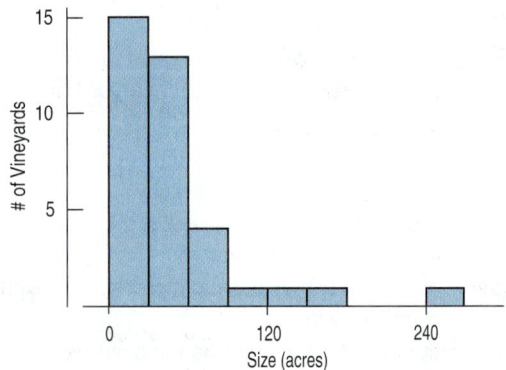

a) Approximately what percentage of these vineyards are under 60 acres?
b) Write a brief description of this distribution (shape, center, spread, unusual features).

18. Run times One of the authors collected the times (in minutes) it took him to run 4 miles on various courses during a 10-year period. Here is a histogram of the times.

Describe the distribution and summarize the important features. What is it about running that might account for the shape you see?

19. Heart attack stays The histogram shows the lengths of hospital stays (in days) for all the female patients admitted to hospitals in New York during one year with a primary diagnosis of acute myocardial infarction (heart attack).

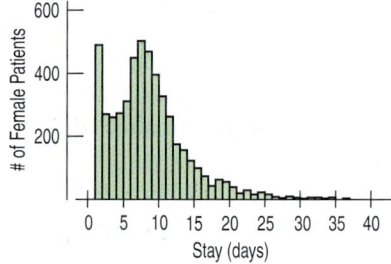

a) From the histogram, would you expect the mean or median to be larger? Explain.
b) Write a few sentences describing this distribution (shape, center, spread, unusual features).
c) Which summary statistics would you choose to summarize the center and spread in these data? Why?

20. E-mails A university teacher saved every e-mail received from students in a large Introductory Statistics class during an entire term. He then counted, for each student who had sent him at least one e-mail, how many e-mails each student had sent.

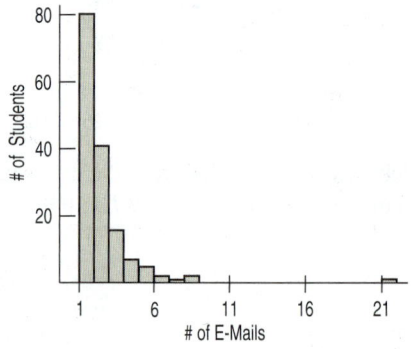

a) From the histogram, would you expect the mean or the median to be larger? Explain.
b) Write a few sentences describing this distribution (shape, center, spread, unusual features).
c) Which summary statistics would you choose to summarize the center and spread in these data? Why?

21. **Super Bowl points 2013** How many points do football teams score in the Super Bowl? Here are the total numbers of points scored by both teams in each of the first 48 Super Bowl games:

45, 47, 23, 30, 29, 27, 21, 31, 22, 38, 46, 37, 66, 50, 37, 47, 44, 47, 54, 56, 59, 52, 36, 65, 39, 61, 69, 43, 75, 44, 56, 55, 53, 39, 41, 37, 69, 61, 45, 31, 46, 31, 50, 48, 56, 31, 21, 34

a) Find the median.
b) Find the quartiles.
c) Write a description based on the 5-number summary.

22. **Super Bowl wins 2013** In the Super Bowl, by how many points does the winning team outscore the losers? Here are the winning margins for the first 48 Super Bowl games:

25, 19, 9, 16, 3, 21, 7, 17, 10, 4, 18, 17, 4, 12, 17, 5, 10, 29, 22, 36, 19, 32, 4, 45, 1, 13, 35, 17, 23, 10, 14, 7, 15, 7, 27, 3, 27, 3, 3, 11, 12, 3, 4, 14, 6, 6, 4, 3

a) Find the median.
b) Find the quartiles.
c) Write a description based on the 5-number summary.

23. **Summaries** Here are costs of nine compact refrigerators rated very good or excellent by *Consumer Reports* on their website www.consumerreports.org.

$ 150 150 160 180 150 140 120 130 120

Find these statistics:
a) mean
b) median and quartiles
c) range and IQR

24. **Tornadoes 2013** Here are the annual numbers of deaths from tornadoes in the United States from 1998 through 2013 (Source: NOAA):

130 94 40 40 555 54 35 38 67 81 125 21 45 553 70 54

Find these statistics:
a) mean
b) median and quartiles
c) range and IQR

25. **Mistake** A clerk entering salary data into a company spreadsheet accidentally put an extra "0" in the boss's salary, listing it as $2,000,000 instead of $200,000. Explain how this error will affect these summary statistics for the company payroll:

a) measures of center: median and mean.
b) measures of spread: range, IQR, and standard deviation.

26. **Sick days** During contract negotiations, a company seeks to change the number of sick days employees may take, saying that the annual "average" is 7 days of absence per employee. The union negotiators counter that the "average" employee misses only 3 days of work each year. Explain how both sides might be correct, identifying the measure of center you think each side is using and why the difference might exist.

27. **Standard deviation I** For each lettered part, a through c, examine the two given sets of numbers. Without doing any calculations, decide which set has the larger standard deviation and explain why. Then check by finding the standard deviations.

	Set 1	Set 2
a)	3, 5, 6, 7, 9	2, 4, 6, 8, 10
b)	10, 14, 15, 16, 20	10, 11, 15, 19, 20
c)	2, 6, 6, 9, 11, 14	82, 86, 86, 89, 91, 94

28. **Standard deviation II** For each lettered part, a through c, examine the two given sets of numbers. Without doing any calculations, decide which set has the larger standard deviation and explain why. Then check by finding the standard deviations.

	Set 1	Set 2
a)	4, 7, 7, 7, 10	4, 6, 7, 8, 10
b)	100, 140, 150, 160, 200	10, 50, 60, 70, 110
c)	10, 16, 18, 20, 22, 28	48, 56, 58, 60, 62, 70

29. **Pizza prices** The histogram shows the distribution of the prices of plain pizza slices (in $) for 156 weeks in Dallas, TX.

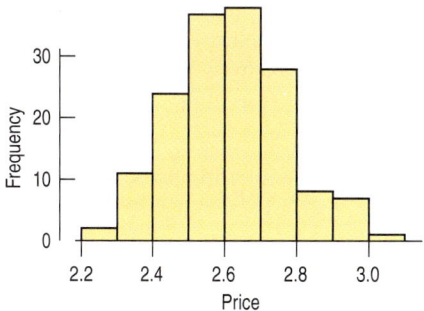

Which summary statistics would you choose to summarize the center and spread in these data? Why?

30. **Neck size** The histogram shows the neck sizes (in inches) of 250 men recruited for a health study in Utah.

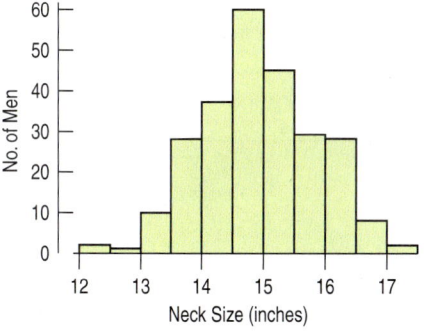

Which summary statistics would you choose to summarize the center and spread in these data? Why?

31. Pizza prices again Look again at the histogram of the pizza prices in Exercise 29.
 a) Is the mean closer to $2.40, $2.60, or $2.80? Why?
 b) Is the standard deviation closer to $0.15, $0.50, or $1.00? Explain.

32. Neck sizes again Look again at the histogram of men's neck sizes in Exercise 30.
 a) Is the mean closer to 14, 15, or 16 inches? Why?
 b) Is the standard deviation closer to 1 inch, 3 inches, or 5 inches? Explain.

33. Movie lengths 2010 The histogram shows the running times in minutes of the 150 top-grossing feature films released in 2010.

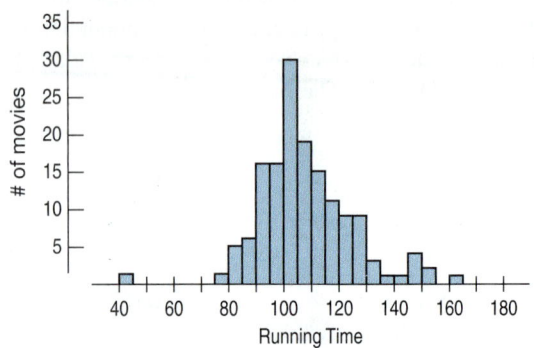

 a) You plan to see a movie this weekend. Based on these movies, how long do you expect a typical movie to run?
 b) Would you be surprised to find that your movie ran for $2\frac{1}{2}$ hours (150 minutes)?
 c) Which would you expect to be higher: the mean or the median run time for all movies? Why?

34. Golf drives 2013 The display shows the average drive distance (in yards) for 155 professional golfers during a week on the men's PGA tour in 2013.

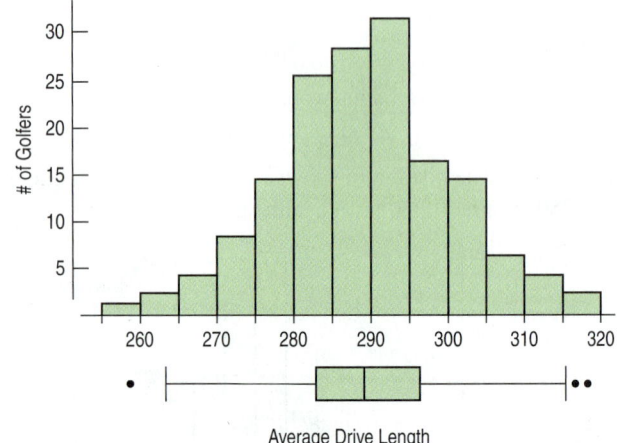

 a) Describe this distribution.
 b) Approximately what proportion of professional male golfers drive, on average, 280 yards or less?
 c) Estimate the mean by examining the histogram.
 d) Do you expect the mean to be smaller than, approximately equal to, or larger than the median? Why?

35. Movie lengths II 2010 Exercise 33 looked at the running times of movies released in 2010. The standard deviation of these running times is 16.6 minutes, and the quartiles are $Q_1 = 98$ minutes and $Q_3 = 116$ minutes.
 a) Write a sentence or two describing the spread in running times based on
  i) the quartiles.
  ii) the standard deviation.
 b) Do you have any concerns about using either of these descriptions of spread? Explain.

36. Golf drives II 2013 Exercise 34 looked at distances PGA golfers can hit the ball. The standard deviation of these average drive distances is 11.2 yards, and the quartiles are $Q_1 = 282.5$ yards and $Q_3 = 295.6$ yards.
 a) Write a sentence or two describing the spread in distances based on
  i) the quartiles.
  ii) the standard deviation.
 b) Do you have any concerns about using either of these descriptions of spread? Explain.

37. Movie earnings 2013 The histogram shows total gross earnings (in millions of dollars) of the top 200 major release movies in 2013.

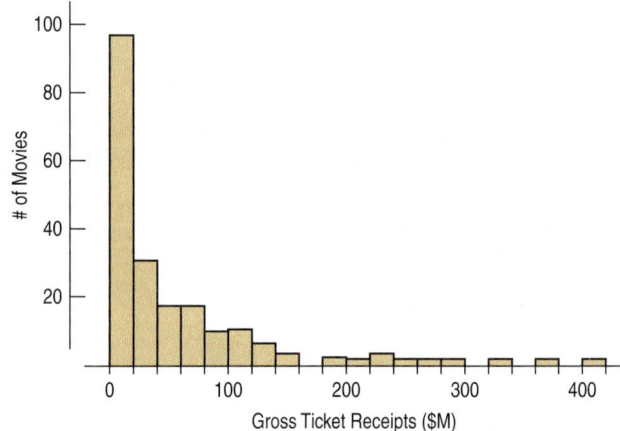

An industry publication reports that the typical movie makes 29.6 million, but a watchdog group concerned with rising ticket prices says that the average earnings are closer to 49.3 million. What statistic do you think each group is using? Explain.

38. Cold weather A meteorologist preparing a talk about global warming compiled a list of weekly low temperatures (in degrees Fahrenheit) he observed at his southern Florida home last year. The coldest temperature for any week was 36°F, but he inadvertently recorded the Celsius value of 2°. Assuming that he correctly listed all the other temperatures, explain how this error will affect these summary statistics:

a) measures of center: mean and median.
b) measures of spread: range, IQR, and standard deviation.

39. **Payroll** A small warehouse employs a supervisor at $1200 a week, an inventory manager at $700 a week, six stock boys at $400 a week, and four drivers at $500 a week.

a) Find the mean and median wage.
b) How many employees earn more than the mean wage?
c) Which measure of center best describes a typical wage at this company: the mean or the median?
d) Which measure of spread would best describe the payroll: the range, the IQR, or the standard deviation? Why?

T 40. **Singers full choir** The frequency table shows the heights (in inches) of 130 members of a choir.

Height	Count	Height	Count
60	2	69	5
61	6	70	11
62	9	71	8
63	7	72	9
64	5	73	4
65	20	74	2
66	18	75	4
67	7	76	1
68	12		

a) Find the median and IQR.
b) Find the mean and standard deviation.
c) Display these data with a histogram.
d) Write a few sentences describing the distribution.

41. **Gasoline 2014** In January 2014, 16 gas stations in eastern Wisconsin posted these prices for a gallon of regular gasoline:

3.23	3.26	3.23	3.39
3.45	3.43	3.42	3.45
3.46	3.11	3.15	3.22
3.21	3.26	3.27	3.28

a) Make a stem-and-leaf display of these gas prices. Use split stems; for example, use two 3.2 stems—one for prices between $3.20 and $3.24 and the other for prices from $3.25 to $3.29.
b) Describe the shape, center, and spread of this distribution.
c) What unusual feature do you see?

42. **The Great One** During his 20 seasons in the NHL, Wayne Gretzky scored 50% more points than anyone who ever played professional hockey. He accomplished this amazing feat while playing in 280 fewer games than Gordie Howe, the previous record holder. Here are the number of games Gretzky played during each season:

79, 80, 80, 80, 74, 80, 80, 79, 64, 78, 73, 78, 74, 45, 81, 48, 80, 82, 82, 70

a) Create a stem-and-leaf display for these data, using split stems.
b) Describe the shape of the distribution.
c) Describe the center and spread of this distribution.
d) What unusual feature do you see? What might explain this?

43. **States** The stem-and-leaf display shows populations of the 50 states, in millions of people, according to the 2010 census.

```
0 | 111111111112222333333 4444
0 | 555556666667789
1 | 000233
1 | 99
2 |
2 | 5
3 |
3 | 7
(2|5 means 25)
```

a) What measures of center and spread are most appropriate?
b) Without doing any calculations, which must be larger: the median or the mean? Explain how you know.
c) From the stem-and-leaf display, find the median and the interquartile range.
d) Write a few sentences describing this distribution.

44. **Wayne Gretzky** In Exercise 42, you examined the number of games played by hockey great Wayne Gretzky during his 20-year career in the NHL.

a) Would you use the median or the mean to describe the center of this distribution? Why?
b) Find the median.
c) Without actually finding the mean, would you expect it to be higher or lower than the median? Explain.

45. **A-Rod 2013** Alex Rodriguez (known to fans as A-Rod) was the youngest player ever to hit 500 home runs. Here is a stem-and-leaf display of the number of home runs hit by A-Rod during the 1994–2013 seasons. (www.baseball-reference.com/players/r/rodrial01.shtml.) Describe the distribution, mentioning its shape and any unusual features.

```
0 | 057
1 | 68
2 | 3
3 | 005566
4 | 12278
5 | 247
(5|2 means 52)
```

46. Bird species 2013 The Cornell Lab of Ornithology holds an annual Christmas Bird Count (www.birdsource.org), in which bird watchers at various locations around the country see how many different species of birds they can spot. Here are the number of species counted from the 20 sites with the most species in 2013:

184	98	101	126	150
166	82	136	124	118
133	83	86	101	105
97	88	131	128	106

a) Create a stem-and-leaf display of these data.
b) Write a brief description of the distribution. Be sure to discuss the overall shape as well as any unusual features.

47. Major Hurricanes 2013 The following data give the number of hurricanes classified as major hurricanes in the Atlantic Ocean each year from 1944 through 2013, as reported by NOAA (www.nhc.noaa.gov):
3, 3, 1, 2, 4, 3, 8, 5, 3, 4, 2, 6, 2, 2, 5, 2, 2, 7, 1, 2, 6, 1, 3, 1, 0, 5, 2, 1, 0, 1, 2, 3, 2, 1, 2, 2, 2, 3, 1, 1, 1, 3, 0, 1, 3, 2, 1, 2, 1, 1, 0, 5, 6, 1, 3, 5, 3, 4, 2, 3, 6, 7, 2, 2, 5, 2, 5, 4, 2, 0

a) Create a dotplot of these data.
b) Describe the distribution.

48. Horsepower Create a stem-and-leaf display for these horsepowers of autos reviewed by *Consumer Reports*, and describe the distribution:

155	103	130	80	65
142	125	129	71	69
125	115	138	68	78
150	133	135	90	97
68	105	88	115	110
95	85	109	115	71
97	110	65	90	
75	120	80	70	

49. A-Rod again 2013 Students were asked to make a histogram of the number of home runs hit by Alex Rodriguez from 1994 to 2013 (see Exercise 45). One student submitted the following display (he noted that A Rod had hit no home runs in 1994):

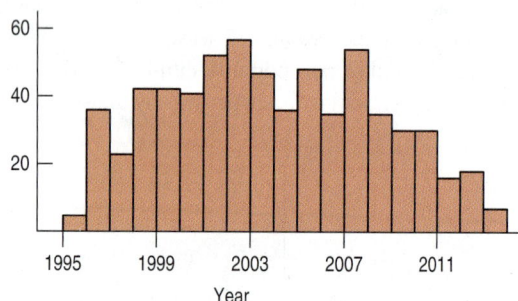

a) Comment on this graph.
b) Create your own histogram of the data.

50. Return of the birds 2013 Students were given the assignment to make a histogram of the data on bird counts reported in Exercise 46. One student submitted the following display:

a) Comment on this graph.
b) Create your own histogram of the data.

51. Acid rain Two researchers measured the pH (a scale on which a value of 7 is neutral and values below 7 are acidic) of water collected from rain and snow over a 6-month period in Allegheny County, Pennsylvania. Describe their data with a graph and a few sentences:
4.57 5.62 4.12 5.29 4.64 4.31 4.30 4.39 4.45 5.67
4.39 4.52 4.26 4.26 4.40 5.78 4.73 4.56 5.08 4.41
4.12 5.51 4.82 4.63 4.29 4.60

52. Marijuana 2007 In 2007 the Council of Europe published a report entitled *The European School Survey Project on Alcohol and Other Drugs* (www.espad.org). Among other issues, the survey investigated the percentages of 16-year-olds who had used marijuana. Shown here are the results for 34 European countries. Create an appropriate graph of these data, and describe the distribution.

Country	Percentage	Country	Percentage
Armenia	3	Italy	23
Austria	17	Latvia	18
Belgium	24	Lithuania	18
Bulgaria	22	Malta	13
Croatia	18	Monaco	28
Cyprus	5	Netherlands	28
Czech Republic	45	Norway	6
Estonia	26	Poland	16
Faroe Islands	6	Portugal	13
Finland	8	Romania	4
France	31	Russia	19
Germany	20	Slovak Republic	32
Greece	6	Slovenia	22
Hungary	13	Sweden	7
Iceland	9	Switzerland	33
Ireland	20	Ukraine	14
Isle of Man	34	United Kingdom	29

53. Final grades A professor (of something other than Statistics!) distributed the following histogram to show the distribution of grades on his 200-point final exam. Comment on the display.

54. Final grades revisited After receiving many complaints about his final-grade histogram from students currently taking a Statistics course, the professor from Exercise 53 distributed the following revised histogram:

a) Comment on this display.
b) Describe the distribution of grades.

55. ZIP codes Holes-R-Us, an Internet company that sells piercing jewelry, keeps transaction records on its sales. At a recent sales meeting, one of the staff presented a histogram of the ZIP codes of the last 500 customers, so that the staff might understand where sales are coming from. Comment on the usefulness and appropriateness of the display.

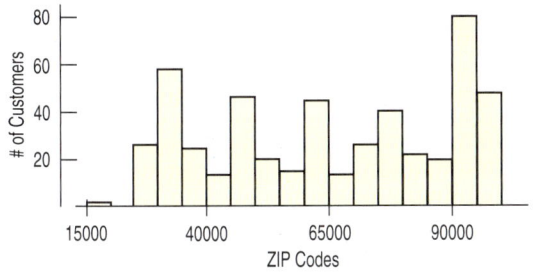

56. ZIP codes revisited Here are some summary statistics to go with the histogram of the ZIP codes of 500 customers from the Holes-R-Us Internet Jewelry Salon that we saw in Exercise 55:

Count	500
Mean	64,970.0
StdDev	23,523.0
Median	64,871
IQR	44,183
Q1	46,050
Q3	90,233

What can these statistics tell you about the company's sales?

57. Math scores 2013 The National Center for Education Statistics (nces.ed.gov/nationsreportcard/) reports average mathematics achievement scores for eighth graders in all 50 states:

State	Score	State	Score
Alabama	269	Montana	289
Alaska	282	Nebraska	285
Arizona	280	Nevada	278
Arkansas	278	New Hampshire	296
California	276	New Jersey	296
Colorado	290	New Mexico	273
Connecticut	285	New York	282
Delaware	282	North Carolina	286
Florida	281	North Dakota	291
Georgia	279	Ohio	290
Hawaii	281	Oklahoma	276
Idaho	286	Oregon	284
Illinois	285	Pennsylvania	290
Indiana	288	Rhode Island	284
Iowa	285	South Carolina	280
Kansas	290	South Dakota	287
Kentucky	281	Tennessee	278
Louisiana	273	Texas	288
Maine	289	Utah	284
Maryland	287	Vermont	295
Massachusetts	301	Virginia	288
Michigan	280	Washington	290
Minnesota	295	West Virginia	274
Mississippi	271	Wisconsin	289
Missouri	283	Wyoming	288

a) Find the median, the IQR, the mean, and the standard deviation of these state averages.
b) Which summary statistics would you report for these data? Why?
c) Write a brief summary of the performance of eighth graders nationwide.

58. Boomtowns 2011 In 2011, the website NewGeography.com listed its ranking of the best cities for job growth in the United States. Here is the magazine's top 20 larger cites, along with their weighted job rating index.

1	Austin-Round Rock-San Marcos, TX	89.1
2	New Orleans-Metairie-Kenner, LA	88.9
3	Houston-Sugar Land-Baytown, TX	82.1
4	San Antonio-New Braunfels, TX	80.7
5	Dallas-Plano-Irving, TX Metropolitan Division	80.6
6	Washington-Arlington-Alexandria, DC-VA-MD-WV Metropolitan Division	74.1
7	Northern Virginia, VA	73.8
8	Nashville-Davidson–Murfreesboro–Franklin, TN	73.1
9	New York City, NY	72.6
10	Philadelphia City, PA	71.9
11	Pittsburgh, PA	71.7
12	Bethesda-Rockville-Frederick, MD Metropolitan Division	70.7
13	Boston-Cambridge-Quincy, MA NECTA Division	70.4
14	Raleigh-Cary, NC	67.8
15	Fort Worth-Arlington, TX Metropolitan Division	67.8
16	Rochester, NY	66.7
17	Nassau-Suffolk, NY Metropolitan Division	65.5
18	Buffalo-Niagara Falls, NY	64.7
19	Columbus, OH	64.3
20	Salt Lake City, UT	64.2

a) Make a suitable display of the weighted growth indices.
b) Summarize the typical growth index among these cities with a median and mean. Why do they differ?
c) Given what you know about the distribution, which of the measures in part b does the better job of summarizing the growth indices? Why?
d) Summarize the spread of the growth index distribution with a standard deviation and with an IQR.
e) Given what you know about the distribution, which of the measures in part d does the better job of summarizing the growth rates? Why?
f) Suppose we subtract from each of the preceding growth rates the average U.S. large-city growth index of 49.23%, so that we can look at how much these indices exceed the U.S. rate. How would this change the values of the summary statistics you calculated above? (*Hint:* You need not recompute any of the summary statistics from scratch.)
g) If we were to omit Austin-Round Rock-San Marcos, TX from the data, how would you expect the mean, median, standard deviation, and IQR to change? Explain your expectations for each.
h) Write a brief report about all of these growth indices.

59. Population growth 2010 The following data show the percentage change in population for the 50 states and the District of Columbia from the 2000 census to the 2010 census. Using appropriate graphical displays and summary statistics, write a report on the percentage change in population by state.

Alabama	7.5
Alaska	13.3
Arizona	24.6
Arkansas	9.1
California	10.0
Colorado	16.9
Connecticut	4.9
Delaware	14.6
District of Columbia	5.2
Florida	17.6
Georgia	18.3
Hawaii	12.3
Idaho	21.1
Illinois	3.3
Indiana	6.6
Iowa	4.1
Kansas	6.1
Kentucky	7.4
Louisiana	1.4
Maine	4.2
Maryland	9.0
Massachusetts	3.1
Michigan	−0.6
Minnesota	7.8
Mississippi	4.3
Missouri	7.0
Montana	9.7
Nebraska	6.7
Nevada	35.1
New Hampshire	6.5
New Jersey	4.5
New Mexico	13.2
New York	2.1
North Carolina	18.5
North Dakota	4.7
Ohio	1.6
Oklahoma	8.7
Oregon	12.0
Pennsylvania	3.4
Rhode Island	0.4
South Carolina	15.3
South Dakota	7.9
Tennessee	11.5
Texas	20.6
Utah	23.8
Vermont	2.8
Virginia	13.0
Washington	14.1
West Virginia	2.5
Wisconsin	6.0
Wyoming	14.1

Source: www.census.gov/compendia/statab/rankings.html

T 60. Prisons 2013 A report from the U.S. Department of Justice (www.ojp.usdoj.gov/bjs/) reported the percent changes in federal prison populations in 20 northeastern and midwestern states from 2000 to 2012. Using appropriate graphical displays and summary statistics, write a report on the changes in prison populations.

State	Percent Change	State	Percent Change
Connecticut	−4.5	New Hampshire	23.6
Delaware	−0.1	New Jersey	−22.0
Indiana	43.3	New York	−22.8
Iowa	9.8	Ohio	11.0
Kentucky	48.2	Pennsylvania	38.7
Maine	25.6	Rhode Island	−1.7
Maryland	−8.6	Vermont	19.9
Massachusetts	5.5	Virginia	22.8
Michigan	−8.6	West Virginia	83.4
Minnesota	59.3	Wisconsin	8.9

Just Checking ANSWERS

(Thoughts will vary.)

1. Slightly skewed to the right. Center around 3 miles? Few over 10 miles.

2. Bimodal. Center between 1 and 2 hours? Many people watch no football; others watch most of one or more games. Probably only a few values over 5 hours.

3. Strongly skewed to the right, with almost everyone at $0; a few small prizes, with the winner an outlier.

4. Fairly symmetric, somewhat uniform, perhaps slightly skewed to the right. Center in the 40s? Few ages below 25 or above 70.

5. Uniform, symmetric. Center near 5. Roughly equal counts for each digit 0–9.

6. Incomes are probably skewed to the right and not symmetric, making the median the more appropriate measure of center. The mean will be influenced by the high end of family incomes and not reflect the "typical" family income as well as the median would. It will give the impression that the typical income is higher than it is.

7. An IQR of 30 mpg would mean that only 50% of the cars get gas mileages in an interval 30 mpg wide. Fuel economy doesn't vary that much. 3 mpg is reasonable. It seems plausible that 50% of the cars will be within about 3 mpg of each other. An IQR of 0.3 mpg would mean that the gas mileage of half the cars varies little from the estimate. It's unlikely that cars, drivers, and driving conditions are that consistent.

8. We'd prefer a standard deviation of 2 months. Making a consistent product is important for quality. Customers want to be able to count on the MP3 player lasting somewhere close to 5 years, and a standard deviation of 2 years would mean that life spans were highly variable.

chapter 4
Understanding and Comparing Distributions

- 4.1 Comparing Groups with Histograms
- 4.2 Comparing Groups with Boxplots
- 4.3 Outliers
- 4.4 Timeplots: Order, Please!
- 4.5 Re-Expressing Data: A First Look

Where are we going?

Are heart attack rates the same for men and women? Is that expensive beverage container really worth the price? Are wind patterns the same throughout the year? We can answer much more interesting questions about variables when we compare distributions for different groups. These are the kinds of questions where Statistics can really help. Some simple graphical displays and summaries can start us thinking about patterns, trends, and models—something we'll do throughout the rest of this book.

Who	Days during 2011
What	Average daily wind speed (mph), Average barometric pressure (mb), Average daily temperature (deg Celsius)
When	2011
Where	Hopkins Forest, in western Massachusetts
Why	Long-term observations to study ecology and climate (Data in **Hopkins Forest 2011**)

The Hopkins Memorial Forest is a 2500-acre reserve in Massachusetts, New York, and Vermont managed by the Williams College Center for Environmental Studies (CES). As part of their mission, CES monitors forest resources and conditions over the long term.

One of the variables measured in the forest is wind speed. For each day, remote sensors record the average, minimum, and maximum wind speeds. Wind is caused as air flows from areas of high pressure to areas of low pressure. Centers of low pressure often accompany storms, so both high winds and low pressure are associated with some of the fiercest storms. Wind speeds can vary greatly both during a day and from day to day, but if we step back a bit farther, we can see patterns. By modeling these patterns, we can understand things about *Average Wind Speed* and the insights it provides about weather that we may not have known.

In Chapter 2, we looked at the association between two categorical variables using contingency tables and displays. Here, we'll look at different ways to examine the relationship between two variables when one is quantitative, and the other indicates groups to compare. We are given wind speed averages for each day of the year. But we can gather the days together into groups and compare the wind speeds among them. If we partition *Time* in this way, we'll gain enormous flexibility. When we change the size of the groups, from the entire year, to seasons, to months, and finally to days, different patterns emerge that together increase our understanding of the entire pattern.

Let's start with the "big picture." Here's a histogram and 5-number summary of the *Average Wind Speed* for every day in 2011. Because of the skewness, we'll report the median and IQR. We can see that the distribution of *Average Wind Speed* is unimodal and skewed to the right. Median daily wind speed is about 1.12 mph, and on half of the days, the average wind speed is between 0.46 and 2.28 mph. We also see several possible outliers with a rather windy 6.73-mph day. Were these unusual weather events or just the windiest days of the year? To answer that, we'll need to work with the summaries a bit more.

Figure 4.1

A histogram of daily *Average Wind Speed* for 2011. It is unimodal and skewed to the right. The boxplot below the histogram suggests several possible outliers that may deserve our attention.

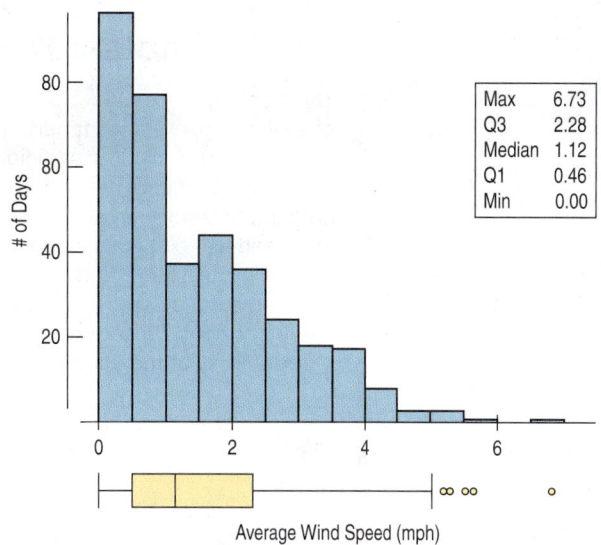

4.1 Comparing Groups with Histograms

It is almost always more interesting to compare groups. Is it windier in the winter or the summer? Are any months particularly windy? Are weekends a special problem? Let's split the year into two groups: April through September (Summer) and October through March (Winter). To compare the groups, we create two histograms, being careful to use the same scale. Here are displays of the average daily wind speed for Summer (on the left) and Winter (on the right):

Figure 4.2

Histograms of *Average Wind Speed* for days in Summer (left) and Winter (right) show very different patterns.

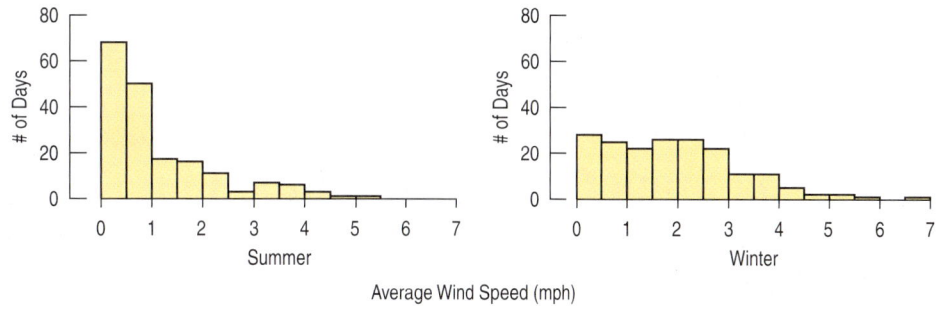

The distribution of wind speeds for the summer months is unimodal and skewed to the right. By contrast, the distribution for the winter months is less strongly skewed and more nearly uniform. During the summer, a typical day has a mean wind speed of less than 1 mph. In winter, the typical wind speed is higher and there is at least one noticeably high value. There is also more variability in the winter (IQR = 1.82 mph) than in summer (IQR = 1.27 mph).

Summaries for *Average Wind Speed* by Season				
Season	Mean	StdDev	Median	IntQRange
Summer	1.11	1.10	0.71	1.27
Winter	1.90	1.29	1.72	1.82

For Example   COMPARING GROUPS WITH STEM-AND-LEAF

The Nest Egg Index, devised by the investment firm of A.G. Edwards, is a measure of saving and investment performance for each of the 50 states, based on 12 economic factors, including participation in retirement savings plans, personal debt levels, and home ownership. The average index is 100 and the numbers indicate the percentage above or below the average. There are only 50 values, so a back-to-back stem-and-leaf plot is an effective display. Here's one comparing the nest egg index in the Northeast and Midwest states to those in the South and West. In this display, the stems run down the middle of the plot, with the leaves for the two regions to the left or right. Be careful when you read the values on the left: 5|8 means nest egg index of 85% of the southern or western states.

```
    South            Northeast
  and West           and Midwest

       5778 |  8 | 8
      12344 |  9 | 03
 6667778899 |  9 | 67
      02334 | 10 | 012233334
         56 | 10 | 6779
            | 11 | 122444
```

(4|9|3 means 94% for a Northeast/Midwest
state and 93% for a South/West state)

QUESTION: How do nest egg indices compare for these regions?

ANSWER: Nest egg indices were generally higher for states in the Northeast and Midwest than in the South and West. The distribution for the northeastern and midwestern states has most of its values above 100, varying from a low of 88% to a high of 114%. Nine northeastern and midwestern states had higher indices than any states in the South and West.

4.2 Comparing Groups with Boxplots

A S   Can Diet Prolong Life? Here's a subject that's been in the news: Can you live longer by eating less? (Or would it just seem longer?) Look at the data in subsequent activities, and you'll find that you can learn a lot by comparing two groups with boxplots.

Are some months windier than others? Even residents may not have a good idea which parts of the year are the windiest. (Do you know for your hometown?) And what about the spreads? Are wind speeds equally variable from month to month, or do some months show more variation?

Histograms or stem-and-leaf displays are a fine way to look at one distribution or two. But it would be hard to see patterns by comparing 12 histograms. Boxplots offer an ideal balance of information and simplicity, hiding the details while displaying the overall summary information. So we often plot them side by side for groups or categories we wish to compare.

By placing boxplots side by side, we can easily see which groups have higher medians, which have the greater IQRs, where the central 50% of the data is located in each group, and which have the greater overall range. And, when the boxes are in an order, we can get a general idea of patterns in both the centers and the spreads. Equally important, we can see past any outliers in making these comparisons because boxplots display them separately.

At the top of the next page are boxplots of the *Average Daily Wind Speed* by month.

Figure 4.3
Boxplots of the *Average Daily Wind Speed* plotted for each *Month* show seasonal patterns in both the centers and spreads. New outliers appear because they are now judged relative to the *Month* in which they occurred.

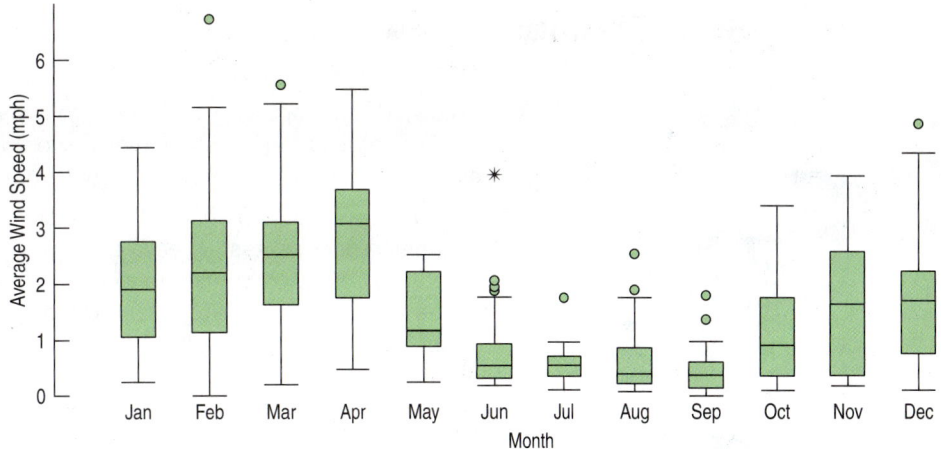

Here we see that wind speeds tend to decrease in the summer. The months in which the winds are both strongest and most variable are October through April.

When we looked at a boxplot of wind speeds for the entire year (see Figure 4.1), there were only 5 outliers. But the monthly boxplots show different outliers than before because some days that seemed ordinary when placed against the entire year's data look like outliers for the month that they're in. Outliers are context dependent. Those windy days in August certainly wouldn't stand out in November or December, but for August, they were remarkable—as we'll soon see.

For Example COMPARING GROUPS

Roller coaster[1] riders want a coaster that goes fast. There are two main types of roller coasters: those with wooden tracks and those with steel tracks. Do they typically run at different speeds? Here are boxplots (Data in **Coasters 2014**):

QUESTION: Compare the speeds of wood and steel roller coasters.

ANSWER: The median speed for the wood track coasters is a bit higher than for the steel coasters but overall, their typical speeds are about the same. However, the speeds of the steel track coasters vary much more and include both high and low outliers.

---
[1] See the Roller Coaster Data Base at www.rcdb.com.

88 PART I Exploring and Understanding Data

Step-by-Step Example COMPARING GROUPS

Most scientific studies compare two or more groups. It is almost always a good idea to start an analysis of data from such studies by comparing boxplots for the groups. Here's an example:

For her class project, a student compared the efficiency of various coffee cups. She compared four different cups, testing each of them eight different times. Each time, she heated water to 180°F, poured it into a cup, and sealed it. After 30 minutes, she measured the temperature again and recorded the difference in temperature. Because these are temperature differences, smaller differences mean that the liquid stayed hot—just what we would want in a coffee cup. (Data in **Cups**)

Question: What can we say about the effectiveness of these four cups?

THINK ▶ Plan State what you want to find out.

Variables Identify the *variables* and report the W's.

Be sure to check the appropriate condition.

I want to compare the effectiveness of four different brands of cups in maintaining temperature. I have eight measurements of Temperature Change for a single example of each of the brands.

✓ **Quantitative Data Condition:** The Temperature Changes are quantitative with units of °F. Boxplots and summary statistics are appropriate for comparing the groups.

SHOW ▶ Mechanics Report the 5-number summaries of the four groups. Include the IQR as a check.

	Min	Q1	Median	Q3	Max	IQR
CUPPS	6°F	6	8.25	14.25	18.50	8.25
Nissan	0	1	2	4.50	7	3.50
SIGG	9	11.50	14.25	21.75	24.50	10.25
Starbucks	6	6.50	8.50	14.25	17.50	7.75

Make a picture. Because we want to compare the distributions for four groups, boxplots are an appropriate choice.

TELL ➡ Conclusion Interpret what the boxplots and summaries say about the ability of these cups to retain heat. Compare the shapes, centers, and spreads, and note any outliers.

The individual distributions of temperature changes are all slightly skewed to the high end. The Nissan cup does the best job of keeping liquids hot, with a median loss of only 2°F, and the SIGG cup does the worst, typically losing 14°F. The difference is large enough to be important: A coffee drinker would be likely to notice a 14° drop in temperature. And the cups are clearly different: 75% of the Nissan tests showed less heat loss than any of the other cups in the study. The IQR of results for the Nissan cup is also the smallest of these test cups, indicating that it is a consistent performer.

Just Checking

The Bureau of Transportation Statistics of the U.S. Department of Transportation collects and publishes statistics on airline travel (www.transtats.bts.gov). Here are three displays of the % of flights arriving on time each month from 1994 through 2013 (Data in **Flight Ontime 2013**)

1. Describe what the histogram says about on-time arrivals.
2. What does the boxplot of on-time arrivals suggest that you can't see in the histogram?
3. Describe the patterns shown in the boxplots by month. At what time of year are flights most likely to be on time? Can you suggest reasons for this pattern?

4.3 Outliers

In the boxplots for the *Average Wind Speed* by *Month*, several days are nominated by the boxplots as possible outliers. Cases that stand out from the rest of the data almost always deserve attention. An outlier is a value that doesn't fit with the rest of the data, but exactly how different it should be to receive special treatment is a judgment call. Boxplots provide a rule of thumb to highlight these unusual cases, but that rule is only a guide, and it doesn't tell you what to do with them. So, what *should* we do with outliers?

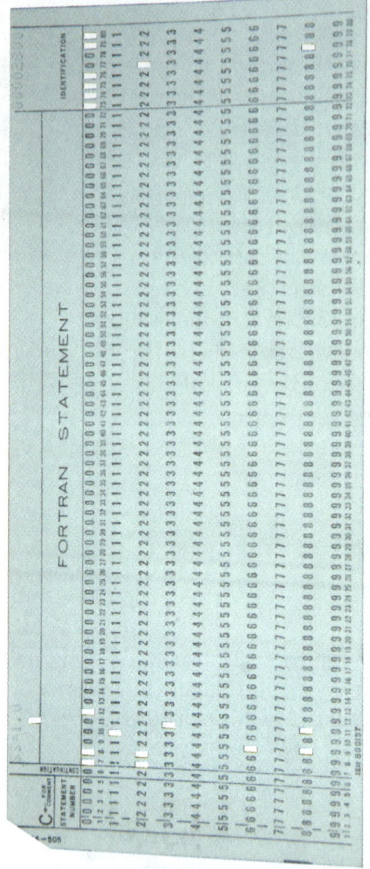

Outliers arise for many reasons. They *may* be the most important values in the data set, pointing out an exceptional case or illuminating a pattern by being the exception to the rule. They may be values that just happen to lie above the limits suggested by the box plot rule. Or they may be errors. A decimal point may have been misplaced, digits transposed, or digits repeated or omitted. Some outliers are obviously wrong. For example, if a class survey includes a student who claims to be 170 inches tall (about 14 feet, or 4.3 meters), you can be sure that's an error. But, maybe the units were wrong. (Did the student mean 170 centimeters—about 65 inches?). There are many ways errors can creep into data sets and such errors occur remarkably often. If you can identify the correct value, then you should certainly use it. One important reason to look into outliers is to correct errors in your data.

> **14-year-old widowers?** Two researchers, Ansley Coale and Fred Stephan, looking at data from the 1950 census noticed that the number of widowed 14-year-old boys had increased from 85 in 1940 to a whopping 1600 in 1950. The number of divorced 14-year-old boys had increased, too, from 85 to 1240. Oddly, the number of teenaged widowers and divorcees *decreased* for every age group after 14, from 15 to 19. When Coale and Stephan also noticed a large increase in the number of young Native Americans in the Northeast United States, they began to look for the cause of these problems. Data in the 1950 census were recorded on computer cards. Cards are hard to read and mistakes are easy to make. It turned out that data punches had been shifted to the right by one column on hundreds of cards. Because each card column meant something different, the shift turned 43-year-old widowed males into 14-year-olds, 42-year-old divorcees into 14-year-olds, and children of white parents into Native Americans. Not all outliers have such a colorful (or famous) story, but it is always worthwhile to investigate them. And, as in this case, the explanation is often surprising. (Source: A. Coale and F. Stephan, "The case of the Indians and the teen-age widows," *J. Am. Stat. Assoc.* 57 [Jun 1962]: 338–347.)

Values that are large (or small) in one context may not be remarkable in another. The boxplots of *Average Wind Speed* by *Month* show possible outliers in several of the months. The windiest day in February was an outlier not only in that windy month, but for the entire year as well. The windiest day in June was a "far" outlier—lying more than 3 IQRs from the upper quartile for that month, but it wouldn't have been unusual for a winter day. And, the windiest day in August seemed much windier than the days around it even though it was relatively calm.

Many outliers are not wrong; they're just different. And most repay the effort to understand them. You can sometimes learn more from the extraordinary cases than from summaries of the entire data set.

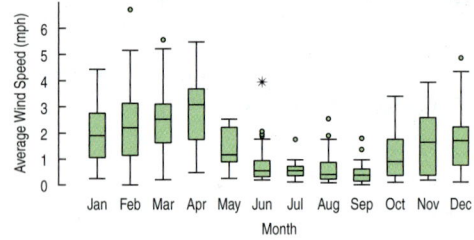

What about those outliers in the Hopkins wind data? That blustery day in February turned out to be a blast that brought four days of subzero temperatures ($-17°F$) to the region. John Quinlan of the National Weather Service predicted at the time: "The Berkshires have definitely had some of the highest gusts with this wind event. This is definitely the worst we'll be seeing."

What about that June day? It was June 2. A search for weather events on the Internet for June 2, 2011, finds a rare tornado in western Massachusetts. The Berkshire Eagle reported: "It probably never occurred to most Berkshire residents horrified by the photos and footage of recent tornado damage in Alabama and Missouri that tornadoes would soon strike close to home. They did, of course, on Wednesday, sowing destruction and in some cases costing lives in Springfield, West Springfield, Monson and other towns near enough to us that the destruction had a sobering effect."

Finally, let's look at the more extreme outlier in August. It wouldn't be remarkable in most other months. It turns out that that was Hurricane Irene, whose eye passed right over the Hopkins Forest. According to *The New York Times*, "it was probably the greatest number of people ever threatened by a single storm in the United States." In fact, the average wind speed that day, 2.53 mph, was much lower than the predictions for Irene had warned.

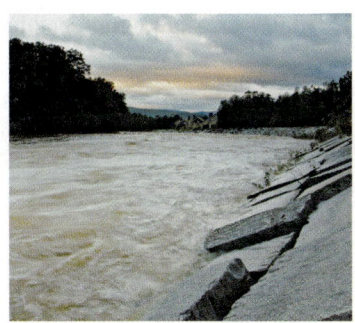

In the aftermath of Hurricane Irene, the Hoosic River in western Massachusetts rose more than 10 feet over its banks, swallowing portions of Williams College, including the soccer and baseball fields.

A | S *Case Study:* **Are passengers or drivers safer in a crash?** Practice the skills of this chapter by comparing these two groups.

Not all outliers are as dramatic as Hurricane Irene, but all deserve attention. If you can correct an error, you should do so (and note the correction). If you can't correct it, or if you confirm that it is correct, you can simply note its existence and leave it in the data set. But the safest path is to report summaries and analyses with *and* without the outlier so that a reader can judge the influence of the outlier for him- or herself.

There are two things you should *never* do with outliers. You should not leave an outlier in place and proceed as if nothing were unusual. Analyses of data with outliers are very likely to be wrong. The other is to omit an outlier from the analysis without comment. If you want to exclude an outlier, you must announce your decision and, to the extent you can, justify it. But a case lying just over the fence suggested by the boxplot may just be the largest (or smallest) value at the end of a stretched-out tail. A histogram is often a better way to see more detail about how the outlier fits in (or doesn't) with the rest of the data, and how large the gap is between it and the rest of the data.

For Example CHECKING OUT THE OUTLIERS

RECAP: We've looked at the speeds of roller coasters and found a difference between steel- and wooden-track coasters. We also noticed an extraordinarily high value.

QUESTION: The fastest coaster in this collection turns out to be "Formula Rossa" at Ferrari world in Abu Dhabi. What might make this roller coaster unusual? You'll have to do some research, but that's often what happens with outliers.

ANSWER: Formula Rossa is easy to find in an Internet search. We learn that it is a "hydraulic launch" coaster. That is, it doesn't get its remarkable speed just from gravity, but rather from a kick-start by a hydraulic piston. That could make it different from the other roller coasters. It accelerates to 149 miles per hour (240 km/h) in 4.8 seconds. Riders must wear safety goggles.

4.4 Timeplots: Order, Please!

The Hopkins Forest wind speeds are reported as daily averages.[2] Previously, we grouped the days into months or seasons, but we could look at the wind speed values day by day. Whenever we have data measured over time, it is a good idea to look for

---

[2] Here's a another warning that the word "average" is used to mean different things. Many "daily average windspeeds" are computed as the midrange—the average of the highest and lowest speeds seen during the day. But the Hopkins Forest values are actually measured continuously all day and averaged over all those observations. When you see the word "average" it is wise to check for the details.

patterns by plotting the data in time order. Here are the daily average wind speeds plotted over time:

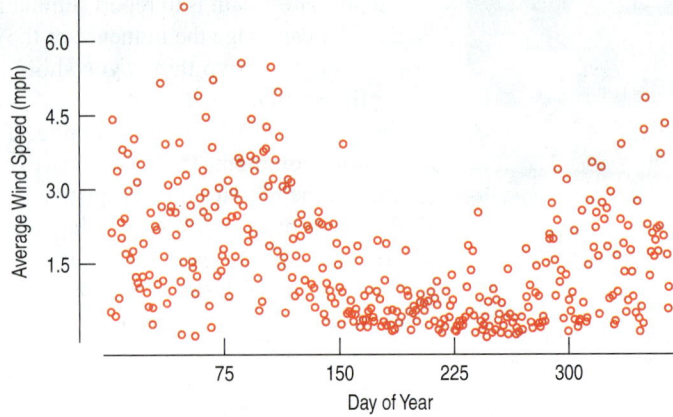

Figure 4.4
A timeplot of *Average Wind Speed* shows the overall pattern and changes in variation.

A display of values against time is sometimes called a **timeplot**. This timeplot reflects the pattern that we saw when we plotted the wind speeds by month. But without the arbitrary divisions between months, we can see a calm period during the summer, starting around day 150 (the beginning of June), when the wind is relatively mild and doesn't vary greatly from day to day. We can also see that the wind becomes both more variable and stronger during the late fall and winter months.

*Smoothing Timeplots

Timeplots often show a great deal of point-to-point variation, as this one does. We usually want to see past this variation to understand any underlying smooth trends, and then also think about how the values vary around that trend—the timeplot version of center and spread. You could take your pencil, squint at the plot a bit to blur things, and sketch a smooth curve. And you'd probably do a pretty good job. But we can also ask a computer to do it for us.

There are many ways for computers to run a smooth trace through a timeplot. Some provide an equation that gives a typical value for any given time point, others just offer a smooth trace. For our purposes here, it doesn't matter all that much. You can try the methods offered by your favorite program and pick the one that looks best for the timeplot you are looking at and the questions you'd like to answer.

You'll often see timeplots drawn with all the points connected, especially in financial publications. This trace through the data falls at one extreme, following every bump. For the wind speeds, the daily fluctuations are so large that connecting the points doesn't help us a great deal:

Figure 4.5
The *Average Wind Speeds* of Figure 4.4, drawn by connecting all the points. Sometimes this can help in seeing the underlying pattern, but here there is too much daily variation for this to be very useful.

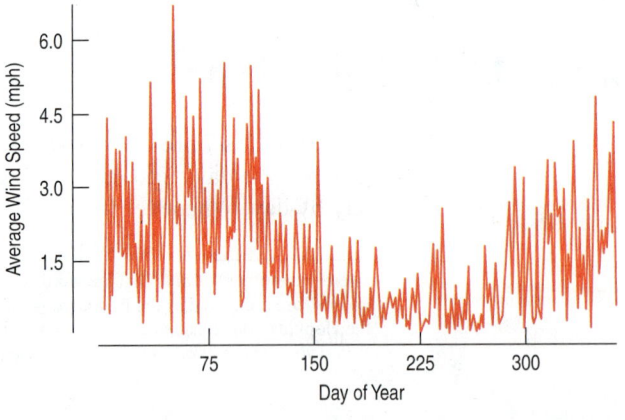

A smooth trace can highlight long-term patterns and help us see them through the more local variation. Figure 4.6 shows the daily average wind speed values with a smooth trace found by a method called *lowess*, available in many statistics programs.

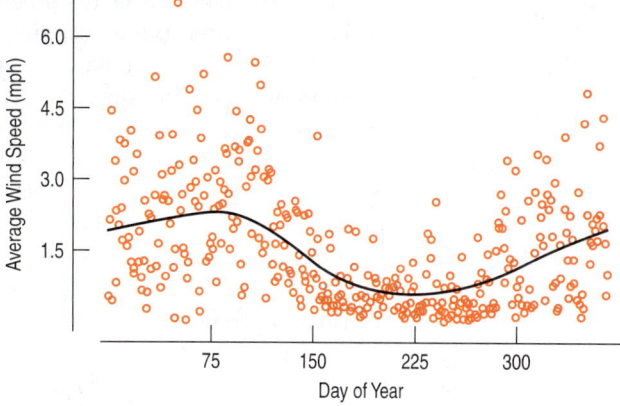

Figure 4.6
The *Average Wind Speeds* of Figure 4.4, with a smooth trace added to help your eye see the long-term pattern.

With the smooth trace, it's a bit easier to see a pattern. The trace helps our eye follow the main trend and alerts us to points that don't fit the overall pattern.

For Example TIMEPLOTS AND SMOOTHING

RECAP: We have looked at the world's fastest roller coasters. Have coaster tracks been getting shorter? Our data include coasters going back to 1909. Here's a time-plot with a *lowess* smooth trace:

QUESTION: What does the timeplot say about the trend in roller coaster track lengths?

ANSWER: Roller coaster tracks for these coasters seem to be getting shorter, although there are still some with very long tracks. The trend to shorter tracks seems to have begun in the mid 1970s and is continuing to this day.

> **How do smoothers work?** If you planned to go camping in the Hopkins Forest around July 15 and wanted to know how windy it was likely to be, you'd probably look at a typical value for mid-July. Maybe you'd consider all the wind speeds for July, or maybe only those from July 10th to July 20th. That's just the kind of thing we do when we smooth a timeplot.
>   One simple way to smooth is with a **moving average**. To find a smooth value for a particular time, we average the values around that point in an interval called the "window." To find the value for the next point, we *move* the window by one point in time and take the new *average*. The size of the window you choose affects how smooth the resulting trace will be. For the Hopkins Forest winds, we might use a 5-day moving average. Stock analysts often use a 50- or 200-day moving average to help them (attempt) to see the underlying pattern in stock price movements.
>   Can we use smoothing to predict the future? We have only the values of a stock in the past, but (unfortunately) none of the future values. We could use the recent past as our window and take the simple average of those values. A more sophisticated method, **exponential smoothing**, gives more weight to the recent past values and less and less weight to values as they recede into the past.

Looking Into the Future

People have always wanted to peer into the future, and we predict that they will continue to do so. Sometimes that makes sense. Most likely, the Hopkins Forest climate follows regular seasonal patterns. It's probably safe to predict a less windy June next year and a windier December. But we certainly wouldn't predict another hurricane on August 28.

Other patterns are riskier to extend into the future. If a stock has been rising, will it continue to go up? No stock has ever increased in value indefinitely, and no stock analyst has consistently been able to forecast when a stock's value will turn around. Stock prices, unemployment rates, and other economic, social, or psychological concepts are much harder to predict than physical quantities. The path a ball will follow when thrown from a certain height at a given speed and direction is well understood. Even some dogs can predict the flight of a frisbee well enough to run to where it will be within reach and catch it. The path interest rates will take is much less clear. Unless you have strong (nonstatistical) reasons for doing otherwise, you should resist the temptation to think that any trend you see will continue, even into the near future.

Statistical models often tempt those who use them to think beyond the data. We'll pay close attention later in this book to understanding when, how, and how much we can justify doing that.

4.5 Re-Expressing Data: A First Look

Re-Expressing to Improve Symmetry

When data are skewed, it can be hard to summarize them simply with a center and spread, and hard to decide whether the most extreme values are outliers or just part of the stretched-out tail. How can we say anything useful about such data? The secret is to *re-express* the data by applying a simple function to each value.

Many relationships and "laws" in the sciences and social sciences include functions such as logarithms, square roots, and reciprocals. Similar relationships often show up in data.

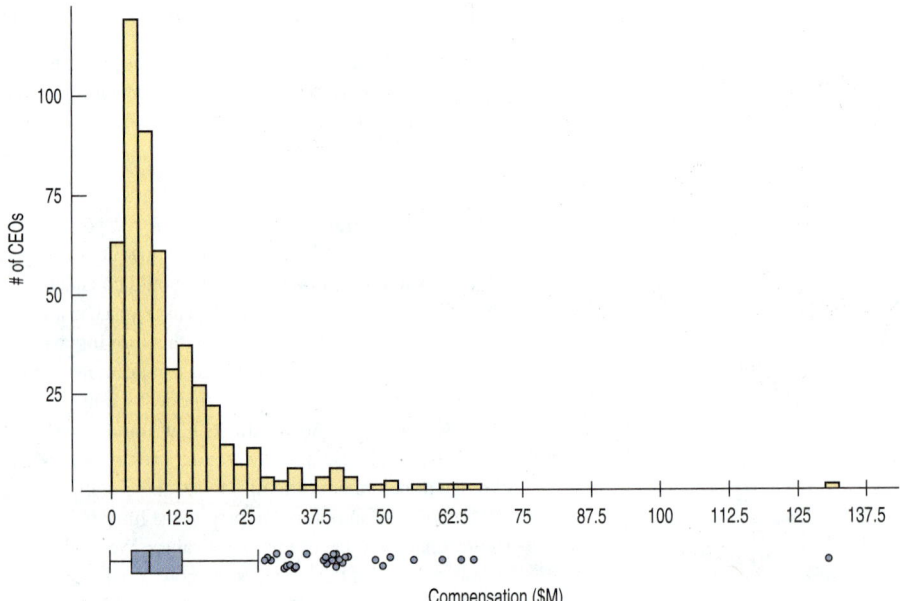

Figure 4.7

CEOs' compensation for the Forbes 500 companies in 2012.

Consider this example: In 1980, large companies' chief executive officers (CEOs) made, on average, about 42 times what workers earned. In the next two decades, CEO compensation soared compared to the average worker. By 2008, that multiple had jumped[3] to 344. What does the distribution of the compensation of Forbes 500 companies' CEOs look like? Here's a histogram and boxplot for 2012 compensation:

We have 500 CEOs and nearly 60 possible histogram bins, about half of which are empty—and many of which contain only one or two CEOs. The boxplot indicates that some CEOs received extraordinarily high pay, while the majority received relatively "little." But look at the values of the bins. The first bin, with about a quarter of the CEOs, covers compensations of $0 to $2,500,000. Imagine receiving a salary survey with these categories:

What is your income?
a) $0 to $2,500,000
b) $2,500,001 to $5,000,000
c) $5,000,001 to $7,500,000
d) More than $7,500,000

What we *can* see from this histogram and boxplot is that this distribution is highly skewed to the right.

It can be hard to decide what we mean by the "center" of a skewed distribution, so it's hard to pick a typical value to summarize the distribution. What would you say was a typical CEO total compensation? The mean value is $10,475,830, while the median is "only" $6,967,500. Each tells us something different about the data.

One approach is to **re-express**, or **transform**, the data by applying a simple function to make the skewed distribution more symmetric. For example, we could take the square root or logarithm of each pay value. Taking logs works pretty well for the CEO compensations, as you can see:[4]

Figure 4.8

The logarithms of 2012 CEO compensations are much more nearly symmetric.

The histogram of the logs of the total CEO compensations is much more nearly symmetric, so we can see that the log compensations are between 5.2, which corresponds to $160,000, and 8.1, corresponding to about $125,000,000. And it's easier to talk about a typical value for the logs. Both the mean and median log compensations are about 6.84. (That's about $6,900,000 but who's counting?)

Against the background of a generally symmetric main body of data, it's easier to decide whether the largest or smallest pay values are outliers. In fact, the three most highly compensated CEOs are identified as outliers by the boxplot rule of thumb even after this re-expression. It's impressive to be an outlier CEO in annual compensation. It's even more impressive to be an outlier in the log scale!

Variables that are skewed to the right often benefit from a re-expression by square roots, logs, or reciprocals. Those skewed to the left may benefit from squaring the data. Because computers and calculators can do the calculating, re-expressing data is quite easy. Consider re-expression as a helpful tool whenever you have skewed data.

---

[3]www.forbes.com/lists/2012/12/ceo-compensation-12_land.html (Data in **CEO salaries 2012**)

[4]Three CEOs, including Larry Page of Google—accepted no pay. When we took logs, their pay dropped out because we couldn't take log(0).

> **DEALING WITH LOGARITHMS**
>
> You have probably learned about logs in math courses and seen them in psychology or science classes. In this book, we use them only for making data behave better. Base 10 logs are the easiest to understand, but natural logs are often used as well. (Either one is fine.) You can think of base 10 logs as roughly one less than the number of digits you need to write the number. So 100, which is the smallest number to require three digits, has a $\log_{10}$ of 2. And 1000 has a $\log_{10}$ of 3. The $\log_{10}$ of 500 is between 2 and 3, but you'd need a calculator to find that it's approximately 2.7. All salaries of "six figures" have $\log_{10}$ between 5 and 6. Logs are incredibly useful for making skewed data more symmetric. But don't worry—nobody does logs without technology and neither should you. Often, remaking a histogram or other display of the data is as easy as pushing another button.

Re-Expressing to Equalize Spread Across Groups

Researchers measured the concentration (nanograms per milliliter) of cotinine in the blood of three groups of people: nonsmokers who have not been exposed to smoke, nonsmokers who have been exposed to smoke (ETS), and passive smokers. Cotinine is left in the blood when the body metabolizes nicotine, so this measure gives a direct measurement of the effect of passive smoke exposure. The boxplots of the cotinine levels of the three groups tell us that the smokers have higher cotinine levels, but if we want to compare the levels of the passive smokers to those of the nonsmokers, we're in trouble, because on this scale, the cotinine levels for both nonsmoking groups are too low to be seen.

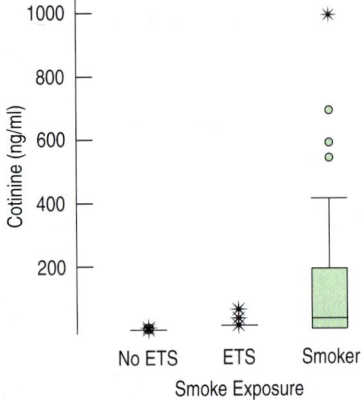

Figure 4.9

Cotinine levels (nanograms per milliliter) for three groups with different exposures to tobacco smoke. Can you compare the ETS (exposed to smoke) and No ETS groups?

Re-expressing can help alleviate the problem of comparing groups that have very different spreads. For measurements like the cotinine data, whose values can't be negative and whose distributions are skewed to the high end, a good first guess at a re-expression is the logarithm.

After taking logs, we can compare the groups and see that the nonsmokers exposed to environmental smoke (the ETS group) do show increased levels of (log) cotinine, although not the high levels found in the blood of smokers.

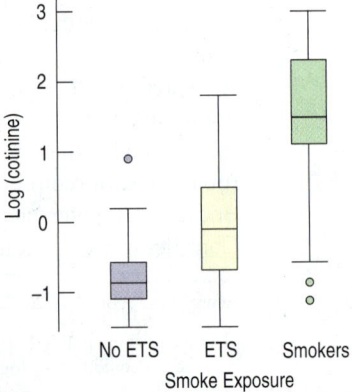

Figure 4.10

Blood cotinine levels after taking logs. What a difference a log makes!

Notice that the same re-expression has also improved the symmetry of the cotinine distribution for smokers and pulled in most of the apparent outliers in all of the groups. It is not unusual for a re-expression that improves one aspect of data to improve others as well. We'll talk about other ways to re-express data as the need arises throughout the book.

WHAT CAN GO WRONG?

- **Avoid inconsistent scales.** Parts of displays should be mutually consistent—no fair changing scales in the middle or plotting two variables on different scales but on the same display. When comparing two groups, be sure to compare them on the same scale.

- **Label clearly.** Variables should be identified clearly and axes labeled so a reader knows what the plot displays.

  Here's a remarkable example of a plot gone wrong. It illustrated a news story about rising college costs. It uses timeplots, but it gives a misleading impression. First think about the story you're being told by this display. Then try to figure out what has gone wrong.

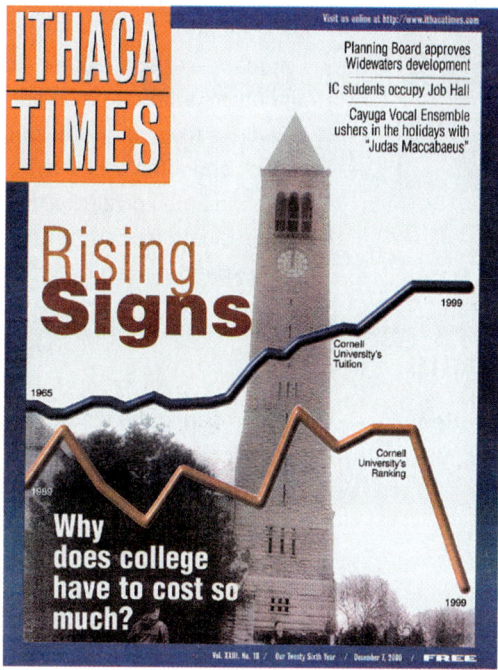

  What's wrong? Just about everything.

  - The horizontal scales are inconsistent. Both lines show trends over time, but exactly for what years? The tuition sequence starts in 1965, but rankings are graphed from 1989. Plotting them on the same (invisible) scale makes it seem that they're for the same years.
  - The vertical axis isn't labeled. That hides the fact that it's inconsistent. Does it graph dollars (of tuition) or ranking (of Cornell University)?

  This display violates three of the rules. And it's even worse than that: It violates a rule that we didn't even bother to mention.

  - The two inconsistent scales for the vertical axis don't point in the same direction! The line for Cornell's rank shows that it has "plummeted" from 15th place to 6th place in academic rank. Most of us think that's an *improvement*, but that's not the message of this graph.

- **Beware of outliers.** If the data have outliers and you can correct them, you should do so. If they are clearly wrong or impossible, you should remove them and report on them. Otherwise, consider summarizing the data both with and without the outliers.

CONNECTIONS

We discussed the value of summarizing a distribution with shape, center, and spread in Chapter 3, and we developed several ways to measure these attributes. Now we've seen the value of comparing distributions for different groups and of looking at patterns in a quantitative variable measured over time. Although it can be interesting to summarize a single variable for a single group, it is almost always more interesting to compare groups and look for patterns across several groups and over time. We'll continue to make comparisons like these throughout the rest of our work.

What Have We Learned?

Learning Objectives

Choose the right tool for comparing distributions.
- Compare the distributions of two or three groups with histograms.
- Compare several groups with boxplots, which make it easy to compare centers and spreads and spot outliers, but hide much of the detail of distribution shape.

Treat outliers with attention and care.
- Outliers may be "local" or global, especially in a time series.
- Track down the background for outliers—it may be informative.

Timeplots graph individual values over time.

Re-express data to make it easier to work with.
- Re-expression can make skewed distributions more nearly symmetric.
- Re-expression can make the spreads of different groups more nearly comparable.

Review of Terms

Comparing distributions When comparing the distributions of several groups using histograms or stem-and-leaf displays, consider their:
- Shape
- Center
- Spread (p. 85).

Comparing boxplots When comparing groups with boxplots:
- Compare the shapes. Do the boxes look symmetric or skewed? Are there differences between groups?
- Compare the medians. Which group has the higher center? Is there any pattern to the medians?
- Compare the IQRs. Which group is more spread out? Is there any pattern to how the IQRs change?
- Using the IQRs as a background measure of variation, do the medians seem to be different, or do they just vary in the way that you'd expect from the overall variation?
- Check for possible outliers. Identify them if you can and discuss why they might be unusual. Of course, correct them if you find that they are errors (p. 86).

Outliers Values that are large or small in one context may not be remarkable in another. Be sure to consider outliers in their appropriate context (p. 89).

Timeplot A timeplot displays data that change over time. Often, successive values are connected with lines to show trends more clearly. Sometimes a smooth curve is added to the plot to help show long-term patterns and trends (p. 92).

On the Computer COMPARING DISTRIBUTIONS

Most programs for displaying and analyzing data can display plots to compare the distributions of different groups. Typically, these are boxplots displayed side by side.

Side-by-side boxplots should be on the same y-axis scale so they can be compared.

Some programs offer a graphical way to assess how much the medians differ by drawing a band around the median or by "notching" the boxes.

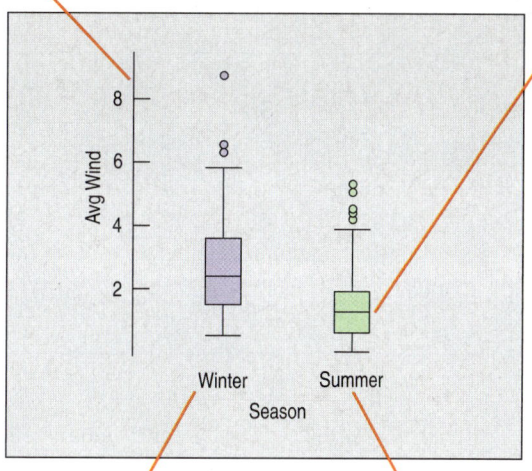

Boxes are typically labeled with a group name. Often they are placed in alphabetical order by group name—not the most useful order.

There are two ways to organize data when we want to compare groups. Each group can be in its own variable (or list, on a calculator). In this form, the experiment comparing cups would have four variables, one for each type of cup:

CUPPS	SIGG	Nissan	Starbucks
6	2	12	13
6	1.5	16	7
6	2	9	7
18.5	3	23	17.5
10	0	11	10
17.5	7	20.5	15.5
11	0.5	12.5	6
6.5	6	24.5	6

But there's another way to think about and organize the data. What is the variable of interest (the *What*) in this experiment? It's the number of degrees lost by the water in each cup. And the *Who* is each time she tested a cup. We could gather all of the temperature values into one variable and put the names of the cups in a second variable listing the individual results, one on each row. Now, the *Who* is clearer—it's an experimental run, one row of the table. Most statistics packages prefer data on groups organized in this way.

Cup	Temperature Difference	Cup	Temperature Difference
CUPPS	6	SIGG	12
CUPPS	6	SIGG	16
CUPPS	6	SIGG	9
CUPPS	18.5	SIGG	23
CUPPS	10	SIGG	11
CUPPS	17.5	SIGG	20.5
CUPPS	11	SIGG	12.5
CUPPS	6.5	SIGG	24.5
Nissan	2	Starbucks	13
Nissan	1.5	Starbucks	7
Nissan	2	Starbucks	7
Nissan	3	Starbucks	17.5
Nissan	0	Starbucks	10
Nissan	7	Starbucks	15.5
Nissan	0.5	Starbucks	6
Nissan	6	Starbucks	6

That's actually the way we've thought about the wind speed data in this chapter, treating wind speeds as one variable and the groups (whether seasons, months, or days) as a second variable.

DATA DESK

If the data are in separate variables:
- Select the variables and choose **Boxplot side by side** from the **Plot** menu. The boxes will appear in the order in which the variables were selected.

If the data are a single quantitative variable and a second variable holding group names:
- Select the quantitative variable as Y and the group variable as X.
- Then choose **Boxplot y by x** from the **Plot** menu. The boxes will appear in alphabetical order by group name.

Data Desk offers options for assessing whether any pair of medians differ.

EXCEL

Excel cannot make boxplots.

JMP

- Choose **Fit y by x**.
- Assign a continuous response variable to **Y, Response** and a nominal group variable holding the group names to **X, Factor**, and click **OK**. JMP will offer (among other things) dotplots of the data.
- Click the red triangle and, under **Display Options**, select Boxplots.

Note: if the variables are of the wrong type, the display options might not offer boxplots.

MINITAB

- Choose **Boxplot...** from the **Graph** menu.

If your data are in the form of one quantitative variable and one group variable:
- Choose **One Y** and **with Groups**.

If your data are in separate columns of the worksheet:
- Choose **Multiple Y's**.

R

If the data for each group are in separate variables, X1, X2, X3,… then:
- boxplot(X1,X2,X3...) will produce side-by-side boxplots.

If the quantitative values are in variable X and the grouping values are in categorical variable Y:
- boxplot(X~Y) will produce side-by-side boxplots—one for each level of Y.

SPSS

To make a boxplot in SPSS, open the **Chart Builder** from the Graphs menu:
- Click the **Gallery** tab.
- Choose **Boxplot** from the list of chart types.
- Drag a single or 2-D (side-by-side) boxplot onto the canvas.

- Drag a scale variable to the *y*-axis drop zone.
- To make side-by-side boxplots, drag a categorical variable to the *x*-axis drop zone.
- Click **OK**.

STATCRUNCH

To make a boxplot:
- Click on **Graphics**.
- Choose **Boxplot**.

- Choose the variable name from the list of **Columns**.
- Click on **Next**.

TI-83/84 PLUS

To make a boxplot, set up a **STAT PLOT** by using the boxplot icon.

To compare groups with boxplots, enter the data in lists, for example in L1 and L2.

Set up STATPLOT's Plot1 to make a boxplot of the L1 data:
- Turn the plot On.
- Choose the first boxplot icon (so the plot will indicate outliers).

- Specify Xlist:L1 and Freq:1, and select the Mark you want the calculator to use for displaying any outliers.
- Use ZoomStat to display the boxplot for L1. You can now TRACE to see the statistics in the 5-number summary.
- Then set up Plot2 to display the L2 data. This time when you use ZoomStat with both plots turned on, the display shows the boxplots in parallel.

Exercises

Section 4.1

1. **Load factors, 2013** The Research and Innovative Technology Administration of the Bureau of Transportation Statistics (www.TranStats.bts.gov/Data_Elements.aspx?Data=2) reports load factors (passenger-miles as a percentage of available seat-miles) for commercial airlines for every month from 2000 through September 2013. Here are histograms comparing the domestic and international load factors for this time period:

 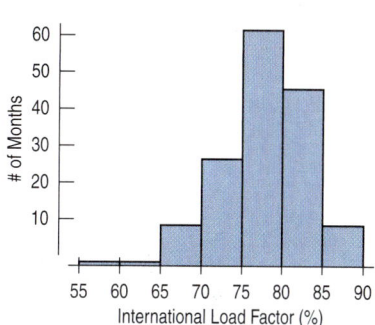

	Mean	Median	Std Dev	IQR	n	Missing
Domestic	77.36715	78.02	6.45481	9.20	165	0
International	77.75976	77.98	5.22681	6.78	165	0

Compare and contrast the distributions.

2. **Load factors, 2013 by season** We can also compare domestic load factors for September through March versus those for April through August:

Compare and contrast what these histograms show.

Section 4.2

3. **Load factors 2013 by month** Here is a display of the international load factors by month for the period from 2000 to 2013.

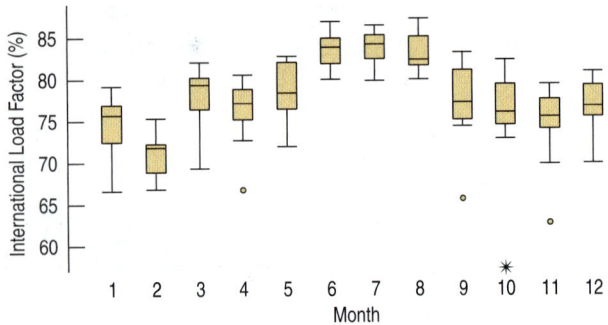

Discuss the patterns you see in this display.

4. **Load factors 2013 by year** Here is a display of the domestic load factors by year.

Discuss the patterns you see in this display.

Section 4.3

5. **Extraordinary months** Look at the boxplots by month in Exercise 3. The three outliers in September, October, and November are all from the year 2001 and reflect the terrorist attacks of 9/11. Do you think the data for these months should be set aside in any overall analysis of load factor patterns by month? Explain.

6. **Extraordinary months again** Here's a boxplot of all of the domestic load factors. It shows a single outlier. The boxplots of Exercise 4 show the same data, but have no outliers. Why do you think this is?

Section 4.4

7. **Load factors 2013 over time** Here is a timeplot of each monthly load factor for domestic flights for 2000 to 2013 along with a lowess smooth.

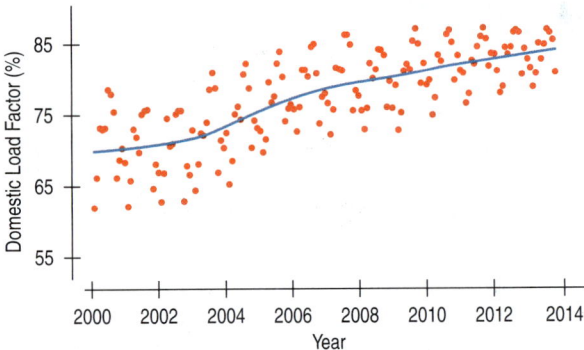

a) Describe the patterns you see in this plot.
b) Do you expect the overall pattern to continue for another decade? Why or why not?

8. **Load factors 2013 over time, a second look** Here is a timeplot of the same domestic load factors as in Exercise 7, but this time with a different smoother (called a median smoother) applied.

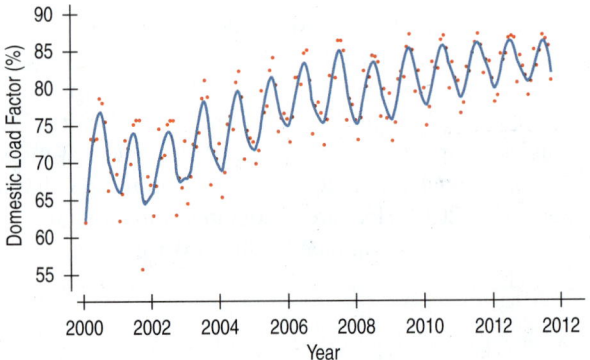

a) What patterns do you see in this plot?
b) Do you expect the patterns you see here to continue for the next decade? Why or why not?

Section 4.5

9. **Exoplanets** Discoveries of planets beyond our solar system have grown rapidly. Here is a histogram showing the distance (in light-years) from earth to stars having known planets (as of 2014).

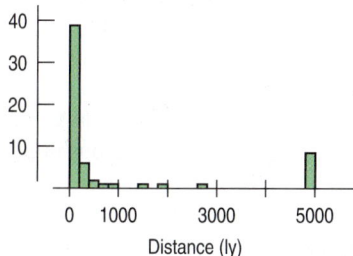

Explain why it might be beneficial to re-express these distances.

10. **Exoplanets re-expressed** Here are the exoplanet distances of Exercise 9, re-expressed to the log scale.

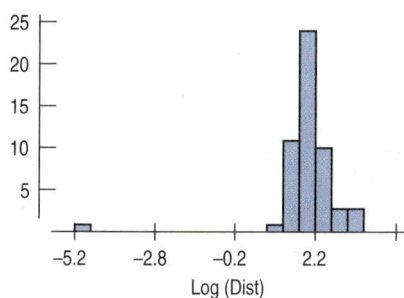

a) Is this a better scale to understand these distances?
b) The low outlier is "sol"– that is, it is the distance of the sun from the earth. Do you think it belongs with these data?

Chapter Exercises

11. **In the news** Find an article in a newspaper, a magazine, or the Internet that compares two or more groups of data.
    a) Does the article discuss the W's?
    b) Is the chosen display appropriate? Explain.
    c) Discuss what the display reveals about the groups.
    d) Does the article accurately describe and interpret the data? Explain.

12. **In the news again** Find an article in a newspaper, a magazine, or the Internet that shows a timeplot.
    a) Does the article discuss the W's?
    b) Is the timeplot appropriate for the data? Explain.
    c) Discuss what the timeplot reveals about the variable.
    d) Does the article accurately describe and interpret the data? Explain.

13. **Time on the Internet** Find data on the Internet (or elsewhere) that give results recorded over time. Make an appropriate display and discuss what it shows.

14. **Groups on the Internet** Find data on the Internet (or elsewhere) for two or more groups. Make appropriate displays to compare the groups, and interpret what you find.

15. **Pizza prices by city** A company that sells frozen pizza to stores in four markets in the United States (Denver, Baltimore, Dallas, and Chicago) wants to examine the prices that the stores charge for pizza slices. Here are boxplots comparing data from a sample of stores in each market:

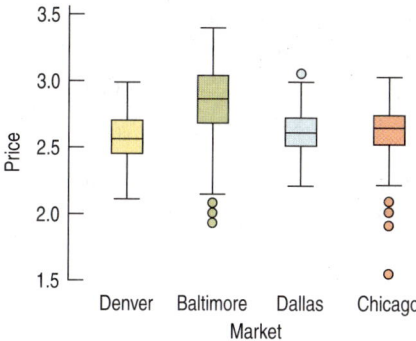

a) Do prices appear to be the same in the four markets? Explain.
b) Does the presence of any outliers affect your overall conclusions about prices in the four markets?

16. **Costs** To help travelers know what to expect, researchers collected the prices of commodities in 16 cities throughout the world. Here are boxplots comparing the prices of a ride on public transportation, a newspaper, and a cup of coffee in the 16 cities (prices are all in US$).

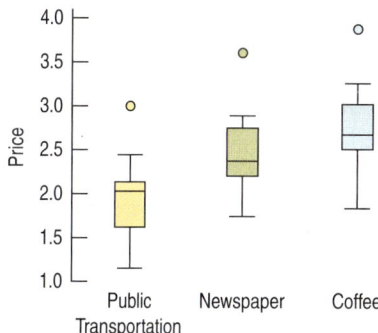

a) On average, which commodity is the most expensive?
b) Is a newspaper always more expensive than a ride on public transportation? Explain.
c) Does the presence of outliers affect your conclusions in part a or b?

17. **Rock concert accidents** Crowd Management Strategies (www.crowdsafe.com) monitors accidents at rock concerts. In their database, they list the names and other variables of victims whose deaths were attributed to "crowd crush" at rock concerts. Here are the histogram and boxplot of the victims' ages for data from a recent one-year period.

a) What features of the distribution can you see in both the histogram and the boxplot?
b) What features of the distribution can you see in the histogram that you could not see in the boxplot?
c) What summary statistic would you choose to summarize the center of this distribution? Why?
d) What summary statistic would you choose to summarize the spread of this distribution? Why?

18. Slalom times 2010 The Men's Giant Slalom skiing event consists of two runs whose times are added together for a final score. Two displays of the giant slalom times in the 2010 Winter Olympics are shown below.

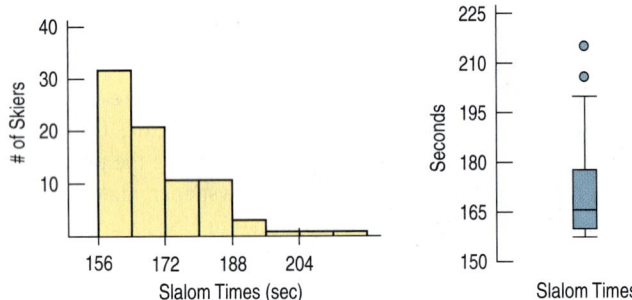

a) What features of the distribution can you see in both the histogram and the boxplot?
b) What summary statistic would you choose to summarize the center of this distribution? Why?
c) What summary statistic would you choose to summarize the spread of this distribution? Why?

19. Cereals Sugar is a major ingredient in many breakfast cereals. The histogram displays the sugar content as a percentage of weight for 49 brands of cereal. The boxplots compare sugar content for adult and children's cereals.

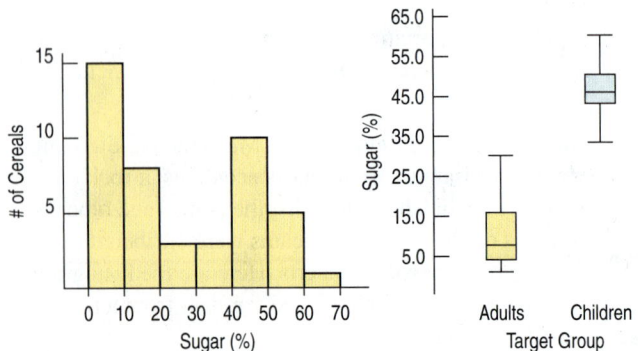

a) What is the range of the sugar contents of these cereals?
b) Describe the shape of the distribution.
c) What aspect of breakfast cereals might account for this shape?
d) Are all children's cereals higher in sugar than adult cereals?
e) Which group of cereals varies more in sugar content? Explain.

20. Tendon transfers People with spinal cord injuries may lose function in some, but not all, of their muscles. The ability to push oneself up is particularly important for shifting position when seated and for transferring into and out of wheelchairs. Surgeons compared two operations to restore the ability to push up in children. The histogram shows scores rating pushing strength two years after surgery and the boxplots compare results for the two surgical methods. (Source: Mulcahey, Lutz, Kozen, Betz, "Prospective Evaluation of Biceps to Triceps and Deltoid to Triceps for Elbow Extension in Tetraplegia," *Journal of Hand Surgery*, 28, 6, 2003)

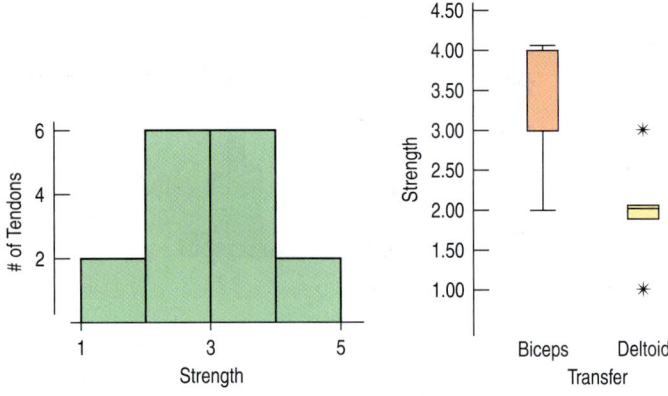

a) Describe the shape of this distribution.
b) What is the range of the strength scores?
c) What fact about results of the two procedures is hidden in the histogram?
d) Which method had the higher (better) median score?
e) Was that method always best?
f) Which method produced the most consistent results? Explain.

21. Population growth 2010 by region On the next page is a "back-to-back" stem-and-leaf display that shows two data sets at once—one going to the left, one to the right. The display compares the percent change in population for two regions of the United States (based on census figures for 2000 and 2010). The fastest growing state was Nevada at 35%. To show the distributions better, this display breaks each stem into two lines, putting leaves 0–4 on one stem and leaves 5–9 on the other.

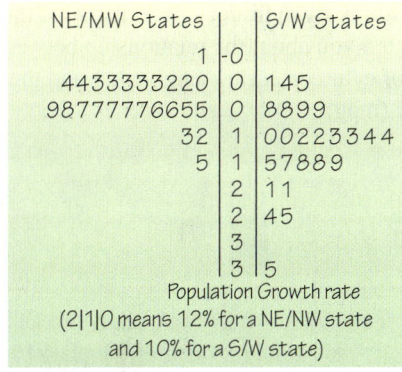

```
NE/MW States  |   | S/W States
            1 |-0 |
   4433333220 | 0 | 145
   98777776655| 0 | 8899
           32 | 1 | 00223344
            5 | 1 | 57889
              | 2 | 11
              | 2 | 45
              | 3 |
              | 3 | 5
```
Population Growth rate
(2|1|0 means 12% for a NE/NW state
and 10% for a S/W state)

a) Use the data displayed in the stem-and-leaf display to construct comparative boxplots.
b) Write a few sentences describing the difference in growth rates for the two regions of the United States.

22. **Camp sites** Shown below are the histogram and summary statistics for the number of camp sites at public parks in Vermont.

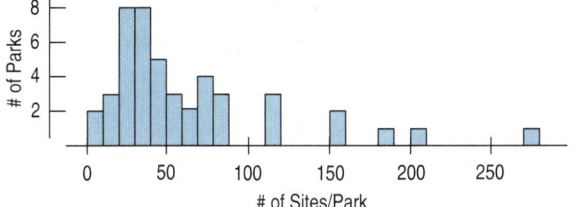

Count	46
Mean	62.8 sites
Median	43.5
StdDev	56.2
Min	0
Max	275
Q1	28
Q3	78

a) Which statistics would you use to identify the center and spread of this distribution? Why?
b) How many parks would you classify as outliers? Explain.
c) Create a boxplot for these data.
d) Write a few sentences describing the distribution.

23. **Hospital stays** The U.S. National Center for Health Statistics compiles data on the length of stay by patients in short-term hospitals and publishes its findings in *Vital and Health Statistics*. Data from a sample of 39 male patients and 35 female patients on length of stay (in days) are displayed in the following histograms.

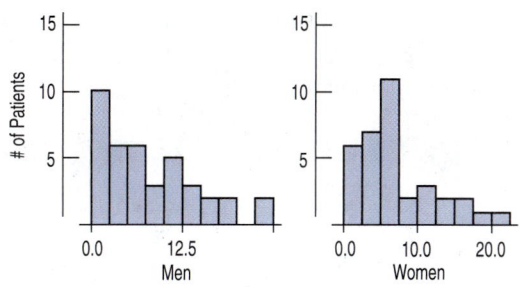

a) What would you change about these histograms to make them easier to compare?
b) Describe these distributions by writing a few sentences comparing the duration of hospitalization for men and women.
c) Can you suggest a reason for the peak in women's length of stay?

T 24. **Deaths 2011** A National Vital Statistics Report (www.cdc.gov/nchs/) indicated that nearly 290,000 black Americans died in 2011, compared with just over 2 million white Americans. Below are histograms displaying the distributions of their ages at death.

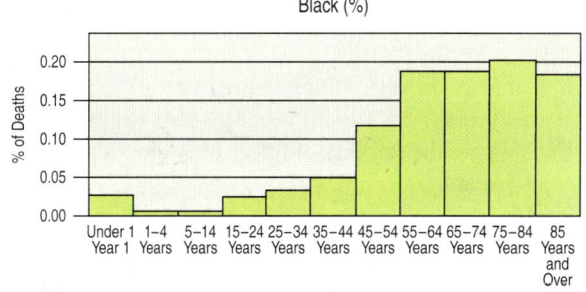

a) Describe the overall shapes of these distributions.
b) How do the distributions differ?
c) Look carefully at the bar definitions. Where do these plots violate the rules for statistical graphs?

25. Women's basketball Here are boxplots of the points scored during the first 10 games of the season for both Scyrine and Alexandra:

a) Summarize the similarities and differences in their performance so far.
b) The coach can take only one player to the state championship. Which one should she take? Why?

26. Gas prices 2013 Here are boxplots of weekly gas prices for regular gas in the United States as reported by the U.S. Energy Information Administration for 2009 through 2013.

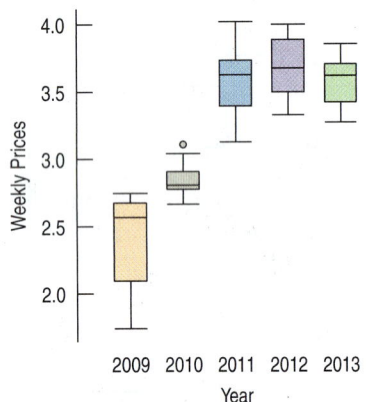

a) Compare the distribution of prices over the five years.
b) Compare the stability of prices over the five years.

27. Marriage age In 1975, did men and women marry at the same age? Here are boxplots of the age at first marriage for a sample of U.S. citizens then. Write a brief report discussing what these data show.

28. Fuel economy by cylinders Describe what these boxplots tell you about the relationship between the number of cylinders a car's engine has and the car's fuel economy (mpg).

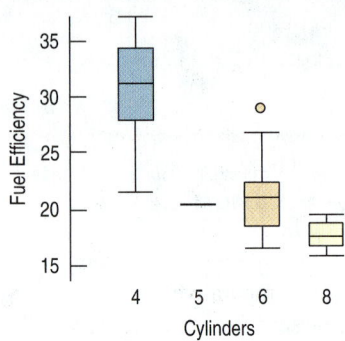

29. Fuel economy 2012 The Environmental Protection Agency provides fuel economy and pollution information on over 2000 car models. Here is a boxplot of *Combined Fuel Economy* (using an average of driving conditions) in *miles per gallon* by vehicle *Type* (midsize car, standard pickup truck, or SUV) for 2012 model vehicles. Summarize what you see about the fuel economies of the three vehicle types.

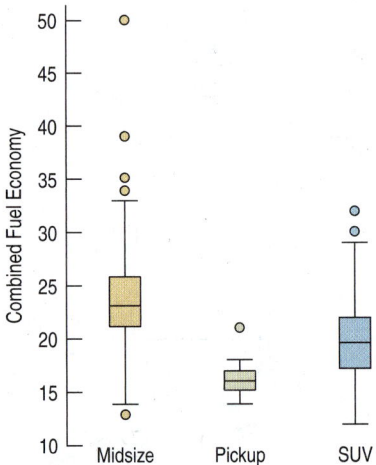

30. Ozone Ozone levels (in parts per billion, ppb) were recorded at sites in New Jersey monthly between 1926 and 1971. Here are boxplots of the data for each month (over the 46 years), lined up in order (January = 1):

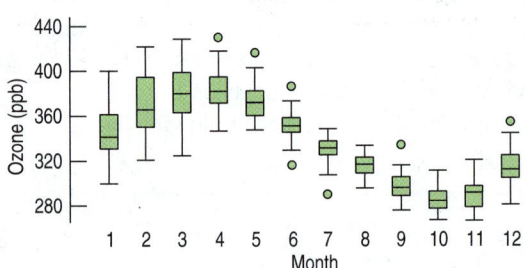

a) In what month was the highest ozone level ever recorded?
b) Which month has the largest IQR?
c) Which month has the smallest range?
d) Write a brief comparison of the ozone levels in January and June.
e) Write a report on the annual patterns you see in the ozone levels.

31. **Test scores** Three Statistics classes all took the same test. Histograms and boxplots of the scores for each class are shown below. Match each class with the corresponding boxplot.

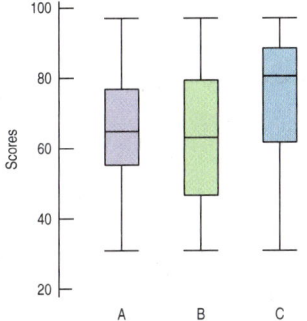

32. **Eye and hair color** A survey of 1021 school-age children was conducted by randomly selecting children from several large urban elementary schools. Two of the questions concerned eye and hair color. In the survey, the following codes were used:

Hair Color	Eye Color
1 = Blond	1 = Blue
2 = Brown	2 = Green
3 = Black	3 = Brown
4 = Red	4 = Grey
5 = Other	5 = Other

The Statistics students analyzing the data were asked to study the relationship between eye and hair color. They produced this plot:

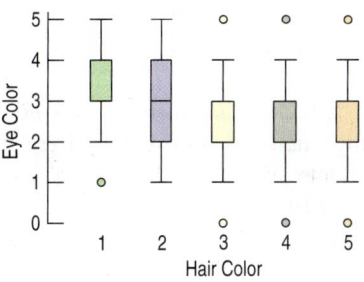

Is their graph appropriate? If so, summarize the findings. If not, explain why not.

33. **Graduation?** A survey of major universities asked what percentage of incoming freshmen usually graduate "on time" in 4 years. Use the summary statistics given to answer the questions that follow.

	% on Time
Count	48
Mean	68.35
Median	69.90
StdDev	10.20
Min	43.20
Max	87.40
Range	44.20
25th %tile	59.15
75th %tile	74.75

a) Would you describe this distribution as symmetric or skewed? Explain.
b) Are there any outliers? Explain.
c) Create a boxplot of these data.
d) Write a few sentences about the graduation rates.

34. **Vineyards revisted** Here are summary statistics for the sizes (in acres) of the Finger Lakes vineyards (of Exercise 3.17):

Count	36
Mean	46.50 acres
StdDev	47.76
Median	33.50
IQR	36.50
Min	6
Q1	18.50
Q3	55
Max	250

Suppose you didn't have access to the data. Answer the following questions from the summary statistics alone:

a) Would you describe this distribution as symmetric or skewed? Explain.
b) Are there any outliers? Explain.
c) Create a boxplot of these data.
d) Write a few sentences about the sizes of the vineyards.

35. Caffeine A student study of the effects of caffeine asked volunteers to take a memory test 2 hours after drinking soda. Some drank caffeine-free cola, some drank regular cola (with caffeine), and others drank a mixture of the two (getting a half-dose of caffeine). Here are the 5-number summaries for each group's scores (number of items recalled correctly) on the memory test:

	n	Min	Q1	Median	Q3	Max
No Caffeine	15	16	20	21	24	26
Low Caffeine	15	16	18	21	24	27
High Caffeine	15	12	17	19	22	24

a) Describe the W's for these data.
b) Name the variables and classify each as categorical or quantitative.
c) Create side-by-side boxplots to display these results as best you can with this information.
d) Write a few sentences comparing the performances of the three groups.

36. SAT scores Here are the summary statistics for Verbal SAT scores for a high school graduating class:

	n	Mean	Median	SD	Min	Max	Q1	Q3
Male	80	590	600	97.2	310	800	515	650
Female	82	602	625	102.0	360	770	530	680

a) Create side-by-side boxplots comparing the scores of boys and girls as best you can from the information given.
b) Write a brief report on these results. Be sure to discuss the shape, center, and spread of the scores.

37. Derby speeds 2014 (Data in **Kentucky Derby 2014**) How fast do horses run? Kentucky Derby winners top 30 miles per hour, as shown in this graph. The graph shows the percentage of Derby winners that have run *slower* than each given speed. Note that few have won running less than 33 miles per hour, but about 86% of the winning horses have run less than 37 miles per hour. (A cumulative frequency graph like this is called an "ogive.")

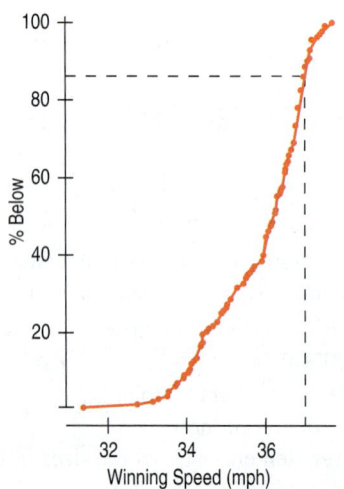

Suppose you had only the ogive (and no access to the data):
a) Estimate the median winning speed.
b) Estimate the quartiles.
c) Estimate the range and the IQR.
d) Create a boxplot of these speeds.
e) Write a few sentences about the speeds of the Kentucky Derby winners.

38. Cholesterol The Framingham Heart Study recorded the cholesterol levels of more than 1400 men. Here is an ogive of the distribution of these cholesterol measures. (An ogive shows the percentage of cases at or below a certain value.) Construct a boxplot for these data, and write a few sentences describing the distribution.

39. Reading scores A class of fourth graders takes a diagnostic reading test, and the scores are reported by reading grade level. The 5-number summaries for the 14 boys and 11 girls are shown:

Boys	2.0	3.9	4.3	4.9	6.0
Girls	2.8	3.8	4.5	5.2	5.9

a) Which group had the highest score?
b) Which group had the greater range?
c) Which group had the greater interquartile range?
d) Which group generally did better on the test? Explain.
e) If the mean reading level for boys is 4.2 and for girls is 4.6, what is the overall mean for the class?

40. Cloud seeding In an experiment to determine whether seeding clouds with silver iodide increases rainfall, 52 clouds were randomly assigned to be seeded or not. The amount of rain they generated was then measured (in acre-feet). Here are the summary statistics:

	n	Mean	Median	SD	IQR	Q1	Q3
Unseeded	26	164.59	44.20	278.43	138.60	24.40	163
Seeded	26	441.98	221.60	650.79	337.60	92.40	430

a) Which of the summary statistics are most appropriate for describing these distributions. Why?
b) Do you see any evidence that seeding clouds may be effective? Explain.

41. Industrial experiment Engineers at a computer production plant tested two methods for accuracy in drilling holes into a PC board. They tested how fast they could set the drilling machine by running 10 boards at each of two different speeds. To assess the results, they measured the distance (in inches) from the center of a target on the board to the center of the hole. The data and summary statistics are shown in the table:

	Distance (in.)	Speed	Distance (in.)	Speed
	0.000101	Fast	0.000098	Slow
	0.000102	Fast	0.000096	Slow
	0.000100	Fast	0.000097	Slow
	0.000102	Fast	0.000095	Slow
	0.000101	Fast	0.000094	Slow
	0.000103	Fast	0.000098	Slow
	0.000104	Fast	0.000096	Slow
	0.000102	Fast	0.975600	Slow
	0.000102	Fast	0.000097	Slow
	0.000100	Fast	0.000096	Slow
Mean	0.000102		0.097647	
StdDev	0.000001		0.308481	

Write a report summarizing the findings of the experiment. Include appropriate visual and verbal displays of the distributions, and make a recommendation to the engineers if they are most interested in the accuracy of the method.

42. Cholesterol and Smoking A study examining the health risks of smoking measured the cholesterol levels of people who had smoked for at least 25 years and people of similar ages who had smoked for no more than 5 years and then stopped. Create appropriate graphical displays for both groups, and write a brief report comparing their cholesterol levels. Here are the data:

Smokers				Ex-Smokers		
225	211	209	284	250	134	300
258	216	196	288	249	213	310
250	200	209	280	175	174	328
225	256	243	200	160	188	321
213	246	225	237	213	257	292
232	267	232	216	200	271	227
216	243	200	155	238	163	263
216	271	230	309	192	242	249
183	280	217	305	242	267	243
287	217	246	351	217	267	218
200	280	209		217	183	228

43. MPG A consumer organization wants to compare gas mileage figures for several models of cars made in the United States with autos manufactured in other countries. The data for a random sample of cars classified as "midsize" are shown in the table.
a) Create graphical displays for these two groups.
b) Write a few sentences comparing the distributions.

Gas Mileage (mpg)	Country	Gas Mileage (mpg)	Country
22	U.S.	17	Other
39	U.S.	26	Other
39	U.S.	18	Other
22	U.S.	20	Other
22	U.S.	24	Other
21	U.S.	22	Other
29	U.S.	28	Other
21	U.S.	23	Other
21	U.S.	30	Other
24	U.S.	19	Other
23	U.S.	27	Other
17	U.S.	21	Other
30	U.S.	22	Other
19	U.S.	29	Other
23	U.S.	29	Other
21	U.S.	28	Other
24	U.S.	26	Other
50	Other	21	Other
24	Other	20	Other
35	Other	21	Other

44. Baseball 2013 American League baseball teams play their games with the designated hitter rule, meaning that pitchers do not bat. The League believes that replacing the pitcher, typically a weak hitter, with another player in the batting order produces more runs and generates more interest among fans. Following are the average number of runs scored by each team in the 2013 season.

American	Average Runs/Game	National	Runs/Game
BOS	5.27	STL	4.83
DET	4.91	COL	4.36
OAK	4.73	CIN	4.31
BAL	4.60	ATL	4.25
CLE	4.60	ARI	4.23
LAA	4.52	WSN	4.05
TEX	4.48	LAD	4.01
TOR	4.40	MIL	3.95
TBR	4.29	PIT	3.91
NYY	4.01	SFG	3.88
KCR	4.00	NYM	3.82
SEA	3.85	SDP	3.81
MIN	3.79	PHI	3.77
HOU	3.77	CHC	3.72
CHW	3.69	MIA	3.17

Source: www.baseball-reference.com

a) Create an appropriate graphical display of these data.
b) Write a few sentences comparing the average number of runs scored per game in the two leagues. (Remember: shape, center, spread, unusual features!)
c) The World Series teams, the Boston Red Sox and St. Louis Cardinals, both were at the top of their respective leagues. Did either team score an unusually large number of runs per game? Explain.

T 45. **Fruit flies** Researchers tracked a population of 1,203,646 fruit flies, counting how many died each day for 171 days. Here are three timeplots offering different views of these data. One shows the number of flies alive on each day, one the number who died that day, and the third the mortality rate—the fraction of the number alive who died. On the last day studied, the last 2 flies died, for a mortality rate of 1.0.

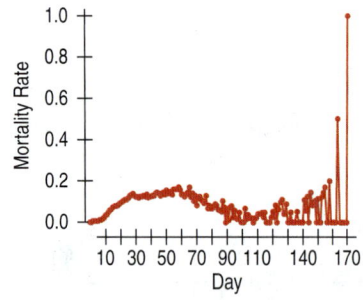

a) On approximately what day did the most flies die?
b) On what day during the first 100 days did the largest *proportion* of flies die?
c) When did the number of fruit flies alive stop changing very much from day to day?

T 46. **Drunk driving 2012** Accidents involving drunk drivers account for about 40% of all deaths on the nation's highways. The table in the next column tracks the number of alcohol-related fatalities for 26 years. (www.madd.org/drunk-driving/about/history.html)

Year	Deaths (thousands)	Year	Deaths (thousands)
1982	26.2	1998	16.7
1983	24.6	1999	16.6
1984	24.8	2000	17.4
1985	23.2	2001	17.4
1986	25.0	2002	17.5
1987	24.1	2003	17.1
1988	23.8	2004	16.9
1989	22.4	2005	16.9
1990	22.6	2006	15.8
1991	20.2	2007	15.4
1992	18.3	2008	13.8
1993	17.9	2009	10.8
1994	17.3	2010	10.2
1995	17.7	2011	9.9
1996	17.7	2012	10.3
1997	16.7		

a) Create a stem-and-leaf display or a histogram of these data.
b) Create a timeplot.
c) Using features apparent in the stem-and-leaf display (or histogram) and the timeplot, write a few sentences about deaths caused by drunk driving.

47. **Assets** Here is a histogram of the assets (in millions of dollars) of 79 companies chosen from the *Forbes* list of the nation's top corporations:

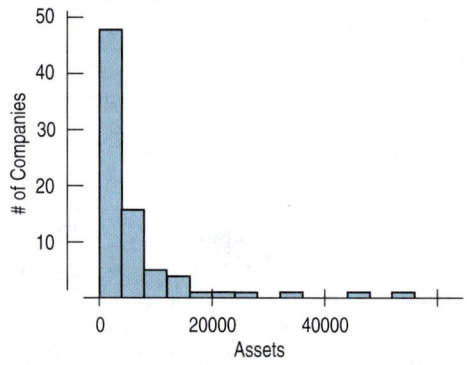

a) What aspect of this distribution makes it difficult to summarize, or to discuss, center and spread?
b) What would you suggest doing with these data if we want to understand them better?

T 48. **Music library** Students were asked how many songs they had in their digital music libraries. Here's a display of the responses:

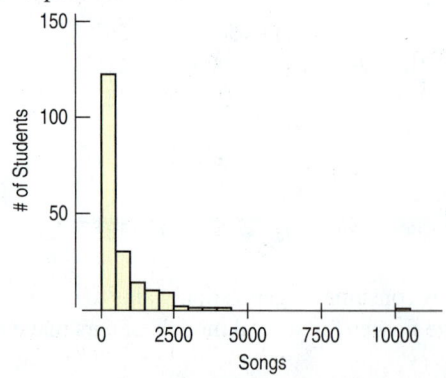

a) What aspect of this distribution makes it difficult to summarize, or to discuss, center and spread?
b) What would you suggest doing with these data if we want to understand them better?

49. Assets again Here are the same data you saw in Exercise 47 after re-expressions as the square root of assets (in $M) and the logarithm of assets (in $M):

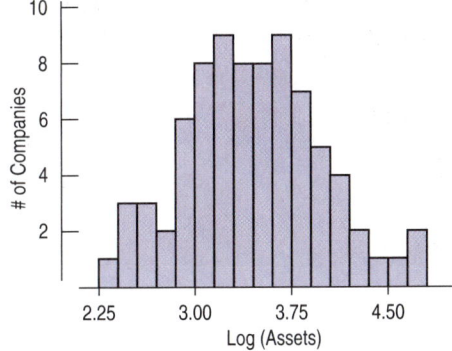

a) Which re-expression do you prefer? Why?
b) In the square root re-expression, what does the value 50 actually indicate about the company's assets?
c) In the logarithm re-expression, what does the value 3 actually indicate about the company's assets?

T 50. Rainmakers The table lists the amount of rainfall (in acre-feet) from the 26 clouds seeded with silver iodide discussed in Exercise 40:

2745	703	302	242	119	40	7
1697	489	274	200	118	32	4
1656	430	274	198	115	31	
978	334	255	129	92	17	

a) Why is acre-feet a good way to measure the amount of precipitation produced by cloud seeding?
b) Plot these data, and describe the distribution.
c) Create a re-expression of these data that produces a more advantageous distribution.
d) Explain what your re-expressed scale means.

T 51. Stereograms Stereograms appear to be composed entirely of random dots. However, they contain separate images that a viewer can "fuse" into a three-dimensional (3D) image by staring at the dots while defocusing the eyes. An experiment was performed to determine whether knowledge of the embedded image affected the time required for subjects to fuse the images. One group of subjects (group NV) received no information or just verbal information about the shape of the embedded object. A second group (group VV) received both verbal information and visual information (specifically, a drawing of the object). The experimenters measured how many seconds it took for the subject to report that he or she saw the 3D image.

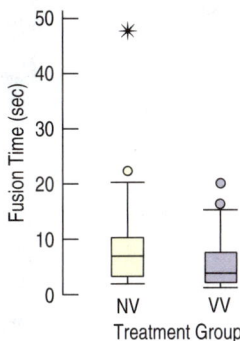

a) What two variables are discussed in this description?
b) For each variable, is it quantitative or categorical? If quantitative, what are the units?
c) The boxplots compare the fusion times for the two treatment groups. Write a few sentences comparing these distributions. What does the experiment show?

T 52. Stereograms, revisited Because of the skewness of the distributions of fusion times described in Exercise 51, we might consider a re-expression. Here are the boxplots of the *log* of fusion times. Is it better to analyze the original fusion times or the log fusion times? Explain.

 Just Checking ANSWERS

1. The % on-time arrivals have a unimodal, symmetric distribution centered at about 79%. In most months, between 70% and 86% of the flights arrived on time.

2. The boxplot of % on-time arrivals nominates several months as low outliers.

3. The boxplots by month show a strong seasonal pattern. Flights are more likely to be on time in the spring and fall and less likely to be on time in the winter and summer. One possible reason for the pattern is snowstorms in the winter and thunderstorms in the summer.

chapter 5
The Standard Deviation as a Ruler and the Normal Model

5.1 Standardizing with z-Scores
5.2 Shifting and Scaling
5.3 Normal Models
5.4 Finding Normal Percentiles
5.5 Normal Probability Plots

> **Where are we going?**
> A college admissions officer is looking at the files of two candidates, one with a total SAT score of 1500, another with an ACT score of 21. Which candidate scored better? How do we compare things when they're measured on different scales?
>
> To answer a question like this, we need to standardize the results. To do that, we need to know two things. First, we need a base value for comparison—we'll often use the mean for that. Next, we need to know how far away we are from the mean. So, we'll need some sort of ruler. Fortunately, the standard deviation is just the thing we need.
>
> The idea of measuring distances from means by counting standard deviations shows up throughout Statistics, and it starts right here.

The women's heptathlon in the Olympics consists of seven track-and-field events: the 200 m and 800 m runs, 100 m high hurdles, shot put, javelin, high jump, and long jump. In the 2012 Olympics, Tatanya Chernova of Russia posted a long jump of 6.54 meters—about half a meter farther than the mean distance for all contestants.

Jessica Ennis of Great Britian won the 200 m run with a time of 22.83 seconds—more than two seconds faster than the average. Each contestant is awarded points for each event based on her performance. So, which performance deserves more points? It's not clear how to compare them. They aren't measured in the same units, or even in the same *direction* (longer jumps are better but shorter times are better.)

We want to compare individual values, not whole groups as we did earlier. To see which value is more extraordinary, we need a way to judge them against the background of data they come from. The tool we'll use is one you've already seen; the standard deviation. We'll judge how unusual a value is by how far, in standard deviation units, it lies from the mean. Statisticians use the standard deviation as a ruler throughout Statistics, and we'll do that for the rest of the book.

We'll start with stem-and-leaf displays because they show individual values, and because it is easy to orient them either high-to-low or low-to-high so the best performances can be at the top of both displays.

Which of the two winning scores is the better one? Chernova's 6.54 m long jump was 0.63 m longer than the mean jump of 5.91 m. The standard deviation for this event was 0.56 m, so her jump was a little more than one SD better (longer) than the mean. More precisely, it was $(6.54 - 5.91)/0.56 = 1.1$ standard deviations better than the mean, so we could give her 1.1 points. For the 200 m run, Ennis's winning time of

Figure 5.1

Stem-and-leaf displays for the 200 m race and the long jump in the 2012 Olympic Heptathlon. Jessica Ennis (green scores) won the 200 m, and Tatanya Chernova (red scores) won the long jump. The stems for the 200 m race run from faster to slower and the stems for the long jump from longer to shorter so that the best scores are at the top of each display. (Data in **Womens Heptathalon 2012**)

```
200 m Run              Long Jump
Stem  Leaf          Stem  Leaf
 22  88              6  55
 23  3               6  0000111111122233334
 23  6677779       6 5  77888889999
 24  12333444        5  4
 24  556778          4
 25  012334444     4  1
 25  679             3  7
22|8 = 22.8 seconds   6|5 = 6.5 meters
```

	Long Jump	200 m
Mean (all contestants)	5.91 m	24.48 s
SD	0.56 m	0.80 s
n	35	36
Chernova	6.54 m	23.67 s
Ennis	6.48 m	22.83 s

22.83 was more impressive because it was just over 2 SD's better (faster) than the mean. She would get 2.1 points for that.

We could turn each result into a score like this, and simply add them up over the 7 events to determine the winner. If we did this, we'd find that Ennis got a total of 8.4 (and won the gold). Chernova got a 4.3 to take the bronze. Olympic judges actually use a point system based on similar calculations (but based on performances from many more competitions). In those points, Ennis got 6955 and Chernova 6628. However, the correlation between our system and the Olympic point system is 0.99!

5.1 Standardizing with z-Scores

Expressing a distance from the mean in standard deviations *standardizes* the performances. To **standardize** a value, we subtract the mean and then divide this difference by the standard deviation:

$$z = \frac{y - \bar{y}}{s}.$$

The values are called **standardized values**, and are commonly denoted with the letter z. Usually we just call them **z-scores**.

■ **NOTATION ALERT**

We always use the letter z to denote values that have been standardized with the mean and standard deviation.

z-scores measure the distance of a value from the mean in standard deviations. A z-score of 2 says that a data value is 2 standard deviations above the mean. It doesn't matter whether the original variable was measured in fathoms, dollars, or carats; those units don't apply to z-scores. Data values below the mean have negative z-scores, so a z-score of -1.6 means that the data value was 1.6 standard deviations below the mean. Of course, regardless of the direction, the farther a data value is from the mean, the more unusual it is, so a z-score of -1.3 is more extraordinary than a z-score of 1.2.

For Example STANDARDIZING SKIING TIMES

The men's super combined skiing event debuted in the 2010 Winter Olympics in Vancouver. It consists of two races: a downhill and a slalom. Times for the two events are added together, and the skier with the lowest total time wins. At Vancouver, the mean slalom time was 52.67 seconds with a standard deviation of 1.614 seconds. The mean downhill time was 116.26 seconds with a standard deviation of 1.914 seconds. Bode Miller of the United States, who won the gold medal with a combined time of 164.92 seconds, skied the slalom in 51.01 seconds and the downhill in 113.91 seconds.

QUESTION: On which race did he do better compared to the competition?

ANSWER: $z_{slalom} = \dfrac{y - \bar{y}}{s} = \dfrac{51.01 - 52.67}{1.614} = -1.03$

$z_{downhill} = \dfrac{113.91 - 116.26}{1.914} = -1.23$

Keeping in mind that faster times are *below* the mean, Miller's downhill time of 1.23 SDs below the mean is even more remarkable than his slalom time, which was 1.03 SDs below the mean.

By using the standard deviation as a ruler to measure statistical distance from the mean, we can compare values from the same data set, or even two values measured on different variables or with different scales. However, the comparison must still make sense. The z-score simply tells us how far a value is from its mean in terms of standard deviations.

We can even combine individual z-scores for the heptathlon to get an overall score. In the table below we combine the two events, the long jump and the 200 m run. Notice that for the running events we would naturally reverse the sign of the z-score for those events because a time *below* the mean is better.

		Event	
		Long Jump	200 m Run
	Mean	5.91 m	24.48 s
	SD	0.56 m	0.80 s
Chernova	Performance	6.54 m	23.67 s
	z-score	$(6.54 - 5.91)/0.56 = 1.13$	$(23.67 - 24.48)/0.80 = -1.01$
	Total for two events	$1.13 + 1.01 = 2.14$	
Ennis	Performance	6.48 m	22.83 s
	z-score	$(6.48 - 5.91)/0.56 = 1.02$	$(22.83 - 24.48)/0.80 = -2.06$
	Total for two events	$1.02 + 2.06 = 3.08$	

For Example COMBINING z-SCORES

At the 2010 Vancouver Winter Olympics, Bode Miller had earlier earned a bronze medal in the downhill and a silver medal in the super-G slalom before winning the gold in the men's super combined. Ted Ligety, the winner of the gold medal at the 2006 games, had super combined times in 2010 of 50.76 seconds in the slalom (fastest of the 35 finalists) and 115.06 seconds in the downhill for a total of 165.82 seconds, almost a second behind Miller. But he finished in fifth place in 2010. He beat Miller in the slalom, but Miller beat him in the downhill. The downhill is longer than the slalom and so, counts for more in the total.

QUESTION: Would the placement have changed if each event had been treated equally by standardizing each and adding the standardized scores?

ANSWER: We've seen that Miller's z-scores for slalom and downhill were -1.03 and -1.23, respectively. That's a total of -2.26. Ligety's z-scores were

$$z_{slalom} = \dfrac{y - \bar{y}}{s} = \dfrac{50.76 - 52.67}{1.614} = -1.18$$

$$z_{downhill} = \dfrac{115.06 - 116.26}{1.914} = -0.63$$

So his total z-score was -1.81. That's not as good as Miller's total of -2.26. So, although Ligety beat Miller in the slalom, Miller beat him by *more* in the downhill. Using the standardized scores would not have put Ligety ahead of Miller.

CHAPTER 5 The Standard Deviation as a Ruler and the Normal Model

Just Checking

1. Your Statistics teacher has announced that the lower of your two tests will be dropped. You got a 90 on test 1 and an 80 on test 2. You're all set to drop the 80 until she announces that she grades "on a curve." She standardized the scores in order to decide which is the lower one. If the mean on the first test was 88 with a standard deviation of 4 and the mean on the second was 75 with a standard deviation of 5,

   a) Which one will be dropped?

   b) Does this seem "fair"?

5.2 Shifting and Scaling

Who	80 male participants of the NHANES survey between the ages of 19 and 24 who measured between 68 and 70 inches tall
What	Their weights
Unit	Kilograms
When	2001–2002
Where	United States
Why	To study nutrition, and health issues and trends
How	National survey

There are two steps to finding a *z*-score. First, the data are *shifted* by subtracting the mean. Then, they are *rescaled* by dividing by the standard deviation. How do these operations work?

Shifting to Adjust the Center

Since the 1960s, the Centers for Disease Control and Prevention's National Center for Health Statistics has been collecting health and nutritional information on people of all ages and backgrounds. The most recent survey, the National Health Interview Survey (NHIS), interviewed nearly 75,000 children about their health. The previous survey, the National Health and Nutrition Examination Survey (NHANES) 2001–2002,[1] measured a wide variety of variables, including body measurements, cardiovascular fitness, blood chemistry, and demographic information on more than 11,000 individuals.

Included in this group were 80 men between 19 and 24 years old of average height (between 5'8" and 5'10" tall). Here are a histogram and boxplot of their weights (Data in **NHANES**)

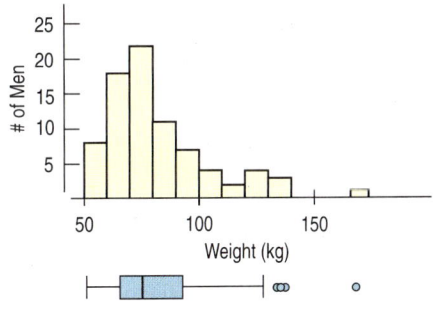

Figure 5.2
Histogram and boxplot for the men's weights. The shape is skewed to the right with several high outliers.

Their mean weight is 82.36 kg. For this age and height group, the National Institutes of Health recommends a maximum healthy weight of 74 kg, but we can see that some of the men are heavier than the recommended weight. To compare their weights to the

---
[1] www.cdc.gov/nchs/nhanes.htm

recommended maximum, we could subtract 74 kg from each of their weights. What would that do to the center, shape, and spread of the histogram? Here's the picture:

Figure 5.3
Subtracting 74 kilograms shifts the entire histogram down but leaves the spread and the shape exactly the same.

 Activity: Changing the Baseline. What happens when we shift data? Do measures of center and spread change?

> **Shifting ideal heights**
> Doctors' height and weight charts sometimes give ideal weights for various heights that include 2-inch heels. If the mean height of adult women is 66 inches including 2-inch heels, what is the mean height of women without shoes? Each woman is shorter by 2 inches when barefoot, so the mean is decreased by 2 inches, to 64 inches.

On average, they weigh 82.36 kg, so on average they're 8.36 kg overweight. And, after subtracting 74 from each weight, the mean of the new distribution is $82.36 - 74 = 8.36$ kg. In fact, when we **shift** the data by adding (or subtracting) a constant to each value, all measures of position (center, percentiles, min, max) will increase (or decrease) by the same constant.

What about the spread? What does adding or subtracting a constant value do to the spread of the distribution? Look at the two histograms again. Adding or subtracting a constant changes each data value equally, so the entire distribution just shifts. Its shape doesn't change and neither does the spread. None of the measures of spread we've discussed—not the range, not the IQR, not the standard deviation—changes.

Adding (or subtracting) a constant to every data value adds (or subtracts) the same constant to measures of position, but leaves measures of spread unchanged.

Rescaling to Adjust the Scale

Not everyone thinks naturally in metric units. Suppose we want to look at the weights in pounds instead. We'd have to rescale the data. Because there are about 2.2 pounds in every kilogram, we'd convert the weights by multiplying each value by 2.2. Multiplying or dividing each value by a constant—**rescaling** the data—changes the measurement units. Here are histograms of the two weight distributions, plotted on the same scale, so you can see the effect of multiplying:

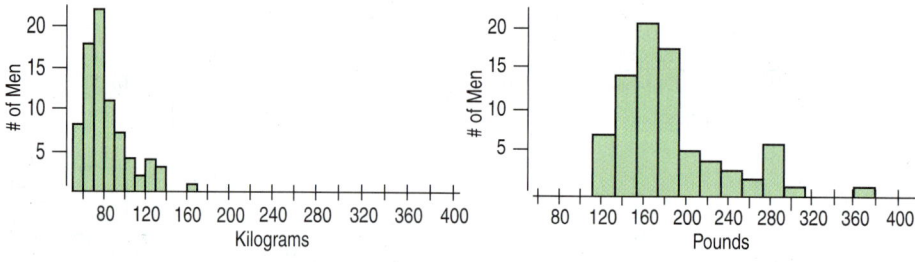

Figure 5.4
Men's weights in both kilograms and pounds. How do the distributions and numerical summaries change?

 Simulation: Changing the Units. Change the center and spread values for a distribution and watch the summaries change (or not, as the case may be).

What happens to the shape of the distribution? Although the histograms don't look exactly alike, we see that the shape really hasn't changed: Both are unimodal and skewed to the right.

What happens to the mean? Not too surprisingly, it gets multiplied by 2.2 as well. The men weigh 82.36 kg on average, which is 181.19 pounds. As the boxplots and 5-number summaries show, all measures of position act the same way. They all get multiplied by this same constant.

What happens to the spread? Take a look at the boxplots. The spread in pounds (on the right) is larger. How much larger? If you guessed 2.2 times, you've figured out how measures of spread get rescaled.

Figure 5.5
The boxplots (drawn on the same scale) show the weights measured in kilograms (on the left) and pounds (on the right). Because 1 kg is about 2.2 lb, all the points in the right box are 2.2 times larger than the corresponding points in the left box. So each measure of position and spread is 2.2 times as large when measured in pounds rather than kilograms.

	Weight (kg)	Weight (lb)
Min	54.3	119.46
Q1	67.3	148.06
Median	76.85	169.07
Q3	92.3	203.06
Max	161.5	355.30
IQR	25	55
SD	22.27	48.99

When we multiply (or divide) all the data values by any constant, all measures of position (such as the mean, median, and percentiles) and measures of spread (such as the range, the IQR, and the standard deviation) are multiplied (or divided) by that same constant.

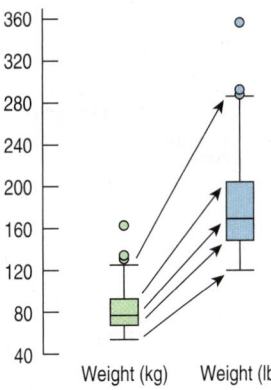

For Example RESCALING THE MEN'S COMBINED TIMES

RECAP: The times in the men's combined event at the winter Olympics are reported in minutes and seconds. The mean and standard deviation of the 34 final super combined times at the 2010 Olympics were 168.93 seconds and 2.90 seconds, respectively.

QUESTION: Suppose instead that we had reported the times in minutes—that is, that each individual time was divided by 60. What would the resulting mean and standard deviation be?

ANSWER: Dividing all the times by 60 would divide both the mean and the standard deviation by 60:

Mean = 168.93/60 = 2.816 minutes;   SD = 2.90/60 = 0.048 minute.

✓ Just Checking

2. In 1995, the Educational Testing Service (ETS) adjusted the scores of SAT tests. Before ETS recentered the SAT Verbal test, the mean of all test scores was 450.

   a) How would adding 50 points to each score affect the mean?

   b) The standard deviation was 100 points. What would the standard deviation be after adding 50 points?

   c) Suppose we drew boxplots of test takers' scores a year before and a year after the recentering. How would the boxplots of the two years differ?

3. A company manufactures wheels for in-line skates. The diameters of the wheels have a mean of 3 inches and a standard deviation of 0.1 inch. Because so many of their customers use the metric system, the company decided to report their production statistics in millimeters (1 inch = 25.4 mm). They report that the standard deviation is now 2.54 mm. A corporate executive is worried about this increase in variation. Should he be concerned? Explain.

 Activity: Standardizing. What if we both shift and rescale? The result is so nice that we give it a name.

Shifting, Scaling, and z-Scores

Standardizing data into z-scores is just shifting them by the mean and rescaling them by the standard deviation. Now we can see how standardizing affects the distribution. When we subtract the mean of the data from every data value, we shift the mean to zero. As we have seen, such a shift doesn't change the standard deviation.

When we *divide* each of these shifted values by *s*, however, the standard deviation should be divided by *s* as well. Since the standard deviation was *s* to start with, the new standard deviation becomes 1.

How, then, does standardizing affect the distribution of a variable? Let's consider the three aspects of a distribution: the shape, center, and spread.

> z-scores have mean 0 and standard deviation 1.

- *Standardizing into z-scores does not change the **shape** of the distribution of a variable.*
- *Standardizing into z-scores changes the **center** by making the mean 0.*
- *Standardizing into z-scores changes the **spread** by making the standard deviation 1.*

Step-by-Step Example WORKING WITH STANDARDIZED VARIABLES

Many colleges and universities require applicants to submit scores on standardized tests such as the SAT Writing, Math, and Critical Reading (Verbal) tests. The college your little sister wants to apply to says that while there is no minimum score required, the middle 50% of their students have combined SAT scores between 1530 and 1850. You'd feel confident if you knew her score was in their top 25%, but unfortunately she took the ACT test, an alternative standardized test.

Question: How high does her ACT need to be to make it into the top quarter of equivalent SAT scores?

To answer that question you'll have to standardize all the scores, so you'll need to know the mean and standard deviations of scores for some group on both tests. The college doesn't report the mean or standard deviation for their applicants on either test, so we'll use the group of all test takers nationally. For college-bound seniors, the average combined SAT score is about 1500 and the standard deviation is about 250 points. For the same group, the ACT average is 20.8 with a standard deviation of 4.8.

THINK ▶ Plan State what you want to find out.

Variables Identify the variables and report the W's (if known).

Check the appropriate conditions.

I want to know what ACT score corresponds to the upper-quartile SAT score. I know the mean and standard deviation for both the SAT and ACT scores based on all test takers, but I have no individual data values.

✓ **Quantitative Data Condition:** Scores for both tests are quantitative but have no meaningful units other than points.

SHOW ▶ Mechanics Standardize the variables.

The middle 50% of SAT scores at this college fall between 1530 and 1850 points. To be in the top quarter, my sister would have to have a score of at least 1850. That's a z-score of

$$z = \frac{(1850 - 1500)}{250} = 1.40$$

So an SAT score of 1850 is 1.40 standard deviations above the mean of all test takers.

The ACT value we seek is 1.40 standard deviations above the mean.

For the ACT, 1.40 standard deviations above the mean is $20.8 + 1.40(4.8) = 27.52$.

TELL ▶ Conclusion Interpret your results in context.

To be in the top quarter of applicants in terms of combined SAT score, she'd need to have an ACT score of at least 27.52.

5.3 Normal Models

> **Is the Normal normal?**
> Don't be misled. The name "Normal" doesn't mean that these are the *usual* shapes for histograms. The name follows a tradition of positive thinking in Mathematics and Statistics in which functions, equations, and relationships that are easy to work with or have other nice properties are called "normal," "common," "regular," "natural," or similar terms. It's as if by calling them ordinary, we could make them actually occur more often and simplify our lives.

A z-score gives an indication of how unusual a value is because it tells how far it is from the mean. If the data value sits right at the mean, it's not very far at all and its z-score is 0. A z-score of 1 tells us that the data value is 1 standard deviation above the mean, while a z-score of −1 tells us that the value is 1 standard deviation below the mean. How far from 0 does a z-score have to be to be interesting or unusual? We know that 50% of the data lie between the quartiles. For symmetric data, the standard deviation is usually a bit smaller than the IQR, and it's not uncommon for at least half of the data to have z-scores between −1 and 1. But no matter what the shape of the distribution, a z-score of 3 (plus or minus) or more is rare, and a z-score of 6 or 7 shouts out for attention.

To say more about how big we expect a z-score to be, we need to *model* the data's distribution. A model will let us say much more precisely how often we'd be likely to see z-scores of different sizes. Of course, like all models of the real world, the model will be wrong—in the sense that it can't match reality exactly. But it can still be useful. Like a physical model, it's something we can look at and manipulate to learn more about the real world.

Models help our understanding in many ways. Just as a model of an airplane in a wind tunnel can give insights even though it doesn't show every rivet,[2] models of data give us summaries that we can learn from and use, even though they don't fit each data value exactly. Without models, what we can learn about the world at large is limited to only what we can say about the data we have at hand.

> "All models are wrong—but some are useful."
> —George Box, famous statistician

There is no universal standard for z-scores, but there is a model that shows up over and over in Statistics. You may have heard of "bell-shaped curves." Statisticians call them Normal models. **Normal models** are appropriate for distributions whose shapes are unimodal and roughly symmetric. For these distributions, they provide a measure of how extreme a z-score is. Fortunately, there is a Normal model for every possible combination of mean and standard deviation. We write $N(\mu, \sigma)$ to represent a Normal model with a mean of μ and a standard deviation of σ. Why the Greek? Well, *this* mean and standard deviation are not numerical summaries of data. They are part of the model. They don't come from the data. Rather, they are numbers that we choose to help specify the model. Such numbers are called **parameters** of the model.

> ■ **NOTATION ALERT**
> $N(\mu, \sigma)$ always denotes a Normal model. The μ, pronounced "mew," is the Greek letter for "m" and always represents the mean in a model. The σ, sigma, is the lowercase Greek letter for "s" and always represents the standard deviation in a model.

We don't want to confuse the parameters with summaries of the data such as $\bar{y}$ and s, so we use special symbols. In Statistics, we almost always use Greek letters for parameters. By contrast, summaries of data are called **statistics** and are usually written with Latin letters.

If we model data with a Normal model and standardize them using the corresponding μ and σ, we still call the standardized value a **z-score**, and we write

$$z = \frac{y - \mu}{\sigma}.$$

> **IS THE STANDARD NORMAL A STANDARD?**
> Yes. We call it the "Standard Normal" because it models standardized values. It is also a "standard" because this is the particular Normal model that we almost always use.

Usually it's easier to standardize data first (using its mean and standard deviation). Then we need only the model $N(0, 1)$. The Normal model with mean 0 and standard deviation 1 is called the **standard Normal model** (or the **standard Normal distribution**).

But be careful. You shouldn't use a Normal model for just any data set. Remember that standardizing won't change the shape of the distribution. If the distribution is not unimodal and symmetric to begin with, standardizing won't make it Normal.

When we use the Normal model, we assume that the distribution of the data is, well, Normal. Practically speaking, there's no way to check whether this **Normality Assumption** is

---

[2] In fact, the model is useful *because* it doesn't have every rivet. It is because models offer a simpler view of reality that they are so useful as we try to understand reality.

true. In fact, it almost certainly is not true. Real data don't behave like mathematical models. Models are idealized; real data are real. The good news, however, is that to use a Normal model, it's sufficient to check the following condition:

> **Nearly Normal Condition.** The shape of the data's distribution is unimodal and symmetric. Check this by making a histogram (or a Normal probability plot, which we'll explain later).

Don't model data with a Normal model without checking whether the condition is satisfied.

All models make **assumptions**. Whenever we model—and we'll do that often—we'll be careful to point out the assumptions that we're making. And, what's even more important, we'll check the associated **conditions** in the data to make sure that those assumptions are reasonable.

Activity: Working with Normal Models. Learn more about the Normal model and see what data drawn at random from a Normal model might look like.

The 68–95–99.7 Rule

Normal models give us an idea of how extreme a value is by telling us how likely it is to find one that far from the mean. We'll soon show how to find these numbers precisely—but one simple rule is usually all we need.

It turns out that in a Normal model, about 68% of the values fall within 1 standard deviation of the mean, about 95% of the values fall within 2 standard deviations of the mean, and about 99.7%—almost all—of the values fall within 3 standard deviations of the mean. These facts are summarized in a rule that we call (let's see . . .) the **68–95–99.7 Rule**.[3]

Figure 5.6
Reaching out one, two, and three standard deviations on a Normal model gives the 68–95–99.7 Rule, seen as proportions of the area under the curve.

> **One in a million** These magic 68, 95, 99.7 values come from the Normal model. As a model, it can give us corresponding values for any z-score. For example, it tells us that fewer than 1 out of a million values have z-scores smaller than -5.0 or larger than $+5.0$. So if someone tells you you're "one in a million," they must really admire your z-score.

For Example USING THE 68–95–99.7 RULE

QUESTION: In the 2010 Winter Olympics men's slalom, Li Lei of China skied in a total time of 120.86 sec for two runs—about 1 standard deviation slower than the mean. If a Normal model is useful in describing slalom times, about how many of the 48 skiers finishing the event would you expect skied the slalom *slower* than Li Lei?

ANSWER: From the 68–95–99.7 Rule, we expect 68% of the skiers to be within one standard deviation of the mean. Of the remaining 32%, we expect half on the high end and half on the low end. 16% of 48 is 7.7, so, conservatively, we'd expect about 7 skiers to do worse than Li Lei.

---

[3] This rule is also called the "Empirical Rule" because it originally came from observation. The rule was first published by Abraham de Moivre in 1733, 75 years before the Normal model was discovered. Maybe it should be called "de Moivre's Rule," but those names wouldn't help us remember the important numbers, 68, 95, and 99.7.

Just Checking

4. As a group, the Dutch are among the tallest people in the world. The average Dutch man is 184 cm tall—just over 6 feet (and the average Dutch woman is 170.8 cm tall—just over 5′7″). If a Normal model is appropriate and the standard deviation for men is about 8 cm, what percentage of all Dutch men will be over 2 meters (6′6″) tall?

5. Suppose it takes 20 minutes, on average, to drive to school, with a standard deviation of 2 minutes. Suppose a Normal model is appropriate for the distributions of driving times.

   a) How often will you arrive at school in less than 22 minutes?
   b) How often will it take you more than 24 minutes?
   c) Do you think the distribution of your driving times is unimodal and symmetric?
   d) What does this say about the accuracy of your predictions? Explain.

Activity: Working with Normal Models. Well, actually playing with them. This interactive tool lets you do what this chapter's figures can't do, move them!

The First Three Rules for Working with Normal Models

1. Make a picture.
2. Make a picture.
3. Make a picture.

Although we're thinking about models, not histograms of data, the three rules don't change. To help you think clearly, a simple hand-drawn sketch is all you need. Even experienced statisticians sketch pictures to help them think about Normal models. You should too.

Of course, when we have data, we'll also need to make a histogram to check the **Nearly Normal Condition** to be sure we can use the Normal model to model the data's distribution. Other times, we may be told that a Normal model is appropriate based on prior knowledge of the situation or on theoretical considerations.

Activity: Normal Models. Normal models have several interesting properties—see them here.

> **How to sketch a normal curve that looks normal** To sketch a good Normal curve, you need to remember only three things:
>
> - The Normal curve is bell-shaped and symmetric around its mean. Start at the middle, and sketch to the right and left from there.
> - Even though the Normal model extends forever on either side, you need to draw it only for 3 standard deviations. After that, there's so little left that it isn't worth sketching.
> - The place where the bell shape changes from curving downward to curving back up—the *inflection point*–is exactly one standard deviation away from the mean.

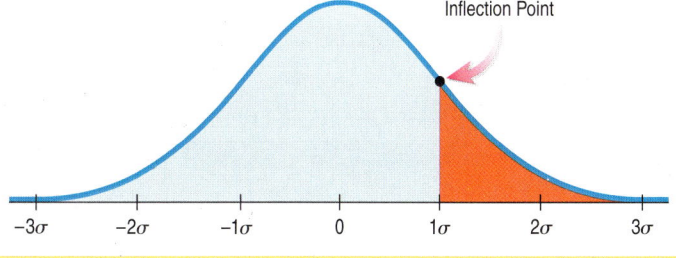

Step-by-Step Example: WORKING WITH THE 68–95–99.7 RULE

Each part of the SAT Reasoning Test has a distribution that is roughly unimodal and symmetric and is designed to have an overall mean of about 500 and a standard deviation of 100 for all test takers. In any one year, the mean and standard deviation may differ from these target values by a small amount, but they are a good overall approximation.

Question: Suppose you earned a 600 on one part of your SAT. Where do you stand among all students who took that test?

You could calculate your z-score and find out that it's $z = (600 - 500)/100 = 1.0$, but what does that tell you about your percentile? You'll need the Normal model and the 68–95–99.7 Rule to answer that question.

THINK → Plan State what you want to know.

I want to see how my SAT score compares with the scores of all other students. To do that, I'll need to model the distribution.

Variables Identify the variable and report the W's.
Be sure to check the appropriate conditions.

Let y = my SAT score. Scores are quantitative but have no meaningful units other than points.

✓ **Nearly Normal Condition:** If I had data, I would check the histogram. I have no data, but I am told that the SAT scores are roughly unimodal and symmetric.

Specify the parameters of your model.

I will model SAT score with a $N(500, 100)$ model.

SHOW → Mechanics Make a picture of this Normal model. (A simple sketch is all you need.)

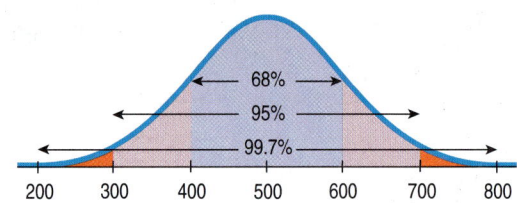

Locate your score.

My score of 600 is 1 standard deviation above the mean. That corresponds to one of the points of the 68–95–99.7 Rule.

TELL → Conclusion Interpret your result in context.

About 68% of those who took the test had scores that fell no more than 1 standard deviation from the mean, so 100% − 68% = 32% of all students had scores more than 1 standard deviation away. Only half of those were on the high side, so about 16% (half of 32%) of the test scores were better than mine. My score of 600 is higher than about 84% of all scores on this test.

The bounds of SAT scoring at 200 and 800 can also be explained by the 68–95–99.7 Rule. Because 200 and 800 are three standard deviations from 500, it hardly pays to extend the scoring any farther on either side. We'd get more information only on $100 - 99.7 = 0.3\%$ of students.

5.4 Finding Normal Percentiles

A|S *Activity: Your Pulse z-Score.* Is your pulse rate high or low? Find its z-score with the Normal Model Tool.

A|S *Activity: The Normal Table.* Table Z just sits there, but this version of the Normal table changes so it always Makes a Picture that fits.

An SAT score of 600 is easy to assess, because we can think of it as one standard deviation above the mean. If your score was 680, though, where do you stand among the rest of the people tested? Your z-score is 1.80, so you're somewhere between 1 and 2 standard deviations above the mean. We figured out that no more than 16% of people score better than 600. By the same logic, no more than 2.5% of people score better than 700. Can we be more specific than "between 16% and 2.5%"?

When the value doesn't fall exactly 1, 2, or 3 standard deviations from the mean, we can look it up in a table of **Normal percentiles** or use technology.[4] Either way, we first convert our data to z-scores before using the table. Your SAT score of 680 has a z-score of $(680 - 500)/100 = 1.80$.

Figure 5.7
A table of Normal percentiles (Table Z in Appendix D) lets us find the percentage of individuals in a Standard Normal distribution falling below any specified z-score value.

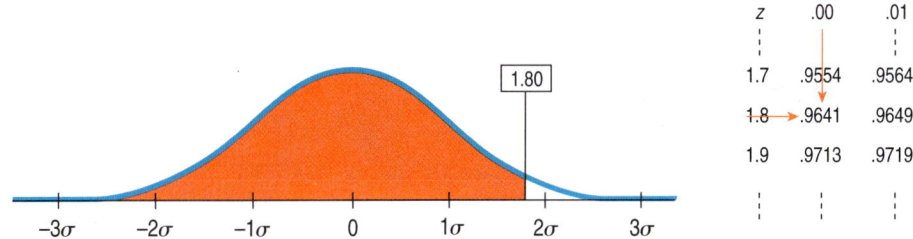

In the piece of the table shown, we find your z-score by looking down the left column for the first two digits, 1.8, and across the top row for the third digit, 0. The table gives the percentile as 0.9641. That means that 96.4% of the z-scores are less than 1.80. Only 3.6% of people, then, scored better than 680 on the SAT.

A|S *Working with Normal Models.* Well, actually playing with them. This interactive tool lets you do what this chapter's figures can't do, because they don't move when you push on them!

A|S *Normal Models.* Normal models have several interesting properties—see them here.

These days, finding percentiles from a Normal probability table is a "desert island" method—something we might do if we desperately needed a Normal percentile and were stranded miles from the mainland with only a Normal probability table. (Of course, you might feel just that way during a Statistics exam, so it's a good idea to know how to do it.) Fortunately, most of the time, we can use a calculator, a computer, a smart phone or tablet app, or the Internet. Most technology tools that find Normal percentiles offer to draw the picture as well. With modern technology tools, it's easy to answer a question like what percentage of all test takers scores between 950 and 1080 on the combined SAT.

Other Models

Of course, the Normal is not the only model for data. There are models for skewed data (watch out for χ^2 later in the book) and we'll see models for variables that can take on only a few values (Binomial, for example). But the Normal will return in an important and surprising way.

---

[4] See Table Z in Appendix D, if you're curious. But many statistics computer packages do this, too—and more easily!

Step-by-Step Example WORKING WITH NORMAL MODELS PART I

The Normal model is our first model for data. It's the first in a series of modeling situations where we step away from the data at hand to make more general statements about the world. We'll become more practiced in thinking about and learning the details of models as we progress through the book. To give you some practice in thinking about the Normal model, here are several problems that ask you to find percentiles in detail.

Question: What proportion of SAT scores fall between 450 and 600?

THINK ➡ Plan State the problem.

I want to know the proportion of SAT scores between 450 and 600.

Variables Name the variable.

Let y = SAT score.

Check the appropriate conditions and specify which Normal model to use.

✓ **Nearly Normal Condition:** We are told that SAT scores are nearly Normal.

I'll model SAT scores with a $N(500, 100)$ model, using the mean and standard deviation specified for them.

SHOW ➡ Mechanics Make a picture of this Normal model. Locate the desired values and shade the region of interest.

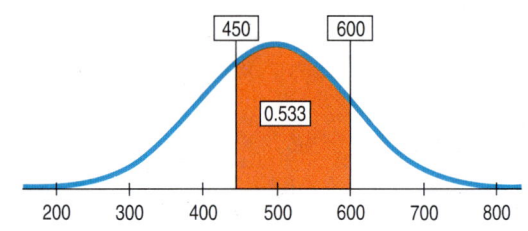

Find z-scores for the cut points 450 and 600. Use technology to find the desired proportions, represented by the area under the curve.

Standardizing the two scores, I find that

$$z = \frac{(y - \mu)}{\sigma} = \frac{(600 - 500)}{100} = 1.00$$

and

$$z = \frac{(450 - 500)}{100} = -0.50.$$

So,
Area$(450 < y < 600)$ = Area$(-0.5 < z < 1.0)$
$= 0.5328.$

(**OR:** From Table Z, Area$(z < 1.0) = 0.8413$ and Area$(z < -0.5) = 0.3085$, so the proportion of z-scores between them is $0.8413 - 0.3085 = 0.5328$, or 53.28%.)

(If you use a table, then you need to subtract the two areas to find the area *between* the cut points.)

TELL ➡ Conclusion Interpret your result in context.

The normal model estimates that about 53.3% of SAT scores fall between 450 and 600.

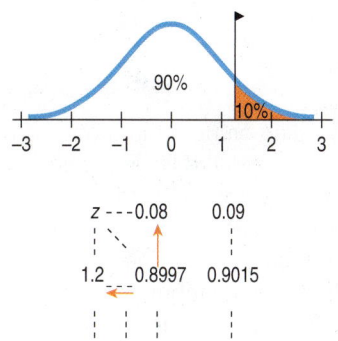

From Percentiles to Scores: z in Reverse

Finding areas from z-scores is the simplest way to work with the Normal model. But sometimes we start with areas and are asked to work backward to find the corresponding z-score or even the original data value. For instance, what z-score cuts off the top 10% in a Normal model?

Make a picture like the one shown, shading the rightmost 10% of the area. Notice that this is the 90th percentile. Look in Table Z for an area of 0.900. The exact area is not there, but 0.8997 is pretty close. That shows up in the table with 1.2 in the left margin and 0.08 in the top margin. The z-score for the 90th percentile, then, is approximately $z = 1.28$.

Computers and calculators will determine the cut point more precisely (and more easily).

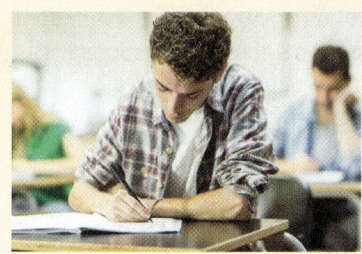

Step-by-Step Example WORKING WITH NORMAL MODELS PART II

Question: Suppose a college says it admits only people with SAT Verbal test scores among the top 10%. How high a score does it take to be eligible?

THINK ▶ Plan State the problem.

How high an SAT Verbal score do I need to be in the top 10% of all test takers?

Variable Define the variable.

Let y = my SAT score.

Check to see if a Normal model is appropriate, and specify which Normal model to use.

✓ **Nearly Normal Condition:** I am told that SAT scores are nearly Normal. I'll model them with $N(500, 100)$.

SHOW ▶ Mechanics Make a picture of this Normal model. Locate the desired percentile approximately by shading the rightmost 10% of the area.

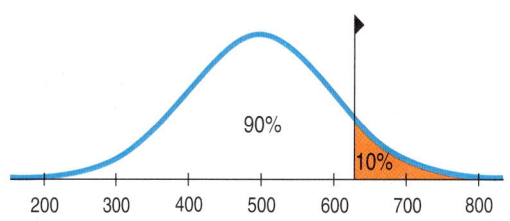

The college takes the top 10%, so its cutoff score is the 90th percentile. Find the corresponding z-score using Table Z as shown previously.

The cut point is $z = 1.28$.

Convert the z-score back to the original units.

A z-score of 1.28 is 1.28 standard deviations above the mean. Since the SD is 100, that's 128 SAT points. The cutoff is 128 points above the mean of 500, or 628.

TELL ▶ Conclusion Interpret your results in the proper context.

Because the school wants SAT Verbal scores in the top 10%, the cutoff is 628. (Actually, since SAT scores are reported only in multiples of 10, I'd have to score at least a 630.)

Step-by-Step Example: MORE WORKING WITH NORMAL MODELS

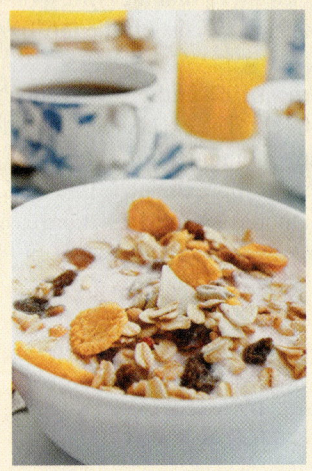

Working with Normal percentiles can be a little tricky, depending on how the problem is stated. Here are a few more worked examples of the kind you're likely to see.

A cereal manufacturer has a machine that fills the boxes. Boxes are labeled "16 ounces," so the company wants to have that much cereal in each box, but since no packaging process is perfect, there will be minor variations. If the machine is set at exactly 16 ounces and the Normal model applies (or at least the distribution is roughly symmetric), then about half of the boxes will be underweight, making consumers unhappy and exposing the company to bad publicity and possible lawsuits. To prevent underweight boxes, the manufacturer has to set the mean a little higher than 16.0 ounces.

Based on their experience with the packaging machine, the company believes that the amount of cereal in the boxes fits a Normal model with a standard deviation of 0.2 ounce. The manufacturer decides to set the machine to put an average of 16.3 ounces in each box. Let's use that model to answer a series of questions about these cereal boxes.

Question 1: What fraction of the boxes will be underweight?

THINK ▶ Plan State the problem.

What proportion of boxes weigh less than 16 ounces?

Variable Name the variable.

Let y = weight of cereal in a box.

Check to see if a Normal model is appropriate.

✓ **Nearly Normal Condition:** I have no data, so I cannot make a histogram, but I am told that the company believes the distribution of weights from the machine is Normal.

Specify which Normal model to use.

I'll use a $N(16.3, 0.2)$ model.

SHOW ▶ Mechanics Make a picture of this Normal model. Locate the value you're interested in on the picture, label it, and shade the appropriate region.

REALITY CHECK Estimate from the picture the percentage of boxes that are underweight. (This will be useful later to check that your answer makes sense.) It looks like a low percentage. Less than 20% for sure.

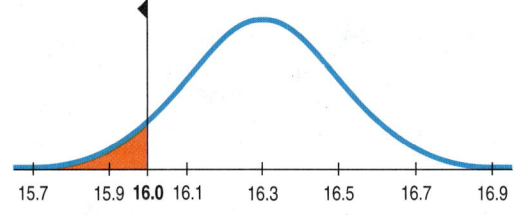

I want to know what fraction of the boxes will weigh less than 16 ounces.

Convert your cutoff value into a z-score.

$$z = \frac{y - \mu}{\sigma} = \frac{16 - 16.3}{0.2} = -1.50$$

Find the area with your calculator (or use the Normal table).

Area$(y < 16)$ = Area$(z < -1.50)$ = 0.0668

TELL ▶ Conclusion State your conclusion, and check that it's consistent with your earlier guess. It's below 20%—seems okay.

I estimate that approximately 6.7% of the boxes will contain less than 16 ounces of cereal.

Question 2: The company's lawyers say that 6.7% is too high. They insist that no more than 4% of the boxes can be underweight. So the company needs to set the machine to put a little more cereal in each box. What mean setting do they need?

THINK ➡ Plan State the problem.

What mean weight will reduce the proportion of underweight boxes to 4%?

Variable Name the variable.

Check to see if a Normal model is appropriate.

Let y = weight of cereal in a box.

✓ **Nearly Normal Condition:** I am told that a Normal model applies.

Specify which Normal model to use. This time you are not given a value for the mean!

I don't know μ, the mean amount of cereal. The standard deviation for this machine is 0.2 ounce. The model is $N(\mu, 0.2)$.

REALITY CHECK We found out earlier that setting the machine to $\mu = 16.3$ ounces made 6.7% of the boxes too light. We'll need to raise the mean a bit to reduce this fraction.

No more than 4% of the boxes can be below 16 ounces.

SHOW ➡ Mechanics Make a picture of this Normal model. Center it at μ (since you don't know the mean), and shade the region below 16 ounces.

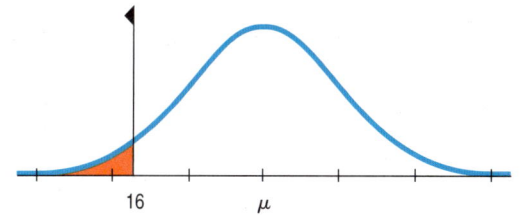

Using your calculator (or the Normal table), find the z-score that cuts off the lowest 4%.

The z-score that has 0.04 area to the left of it is $z = -1.75$.

Use this information to find μ. It's located 1.75 standard deviations to the right of 16. Since σ is 0.2, that's 1.75×0.2, or 0.35 ounce more than 16.

For 16 to be 1.75 standard deviations below the mean, the mean must be

$$16 + 1.75(0.2) = 16.35 \text{ ounces.}$$

TELL ➡ Conclusion Interpret your result in context.

(This makes sense; we knew it would have to be just a bit higher than 16.3.)

The company must set the machine to average 16.35 ounces of cereal per box.

Question 3: The company president vetoes that plan, saying the company should give away less free cereal, not more. Her goal is to set the machine no higher than 16.2 ounces and still have only 4% underweight boxes. The only way to accomplish this is to reduce the standard deviation. What standard deviation must the company achieve, and what does that mean about the machine?

THINK ➡ Plan State the problem.

What standard deviation will allow the mean to be 16.2 ounces and still have only 4% of boxes underweight?

Variable Name the variable.

Let y = weight of cereal in a box.

Check conditions to be sure that a Normal model is appropriate.

✓ **Nearly Normal Condition:** The company believes that the weights are described by a Normal model.

Specify which Normal model to use. This time you don't know σ.

REALITY CHECK We know the new standard deviation must be less than 0.2 ounce.

I know the mean, but not the standard deviation, so my model is $N(16.2, \sigma)$.

SHOW ➡ Mechanics Make a picture of this Normal model. Center it at 16.2, and shade the area you're interested in. We want 4% of the area to the left of 16 ounces.

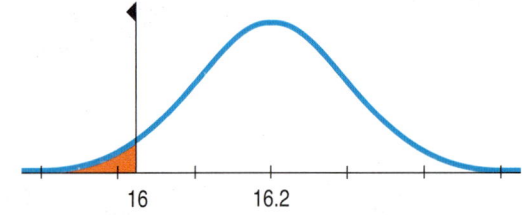

Find the z-score that cuts off the lowest 4%.

I know that the z-score with 4% below it is $z = -1.75$.

Solve for σ. (We need 16 to be 1.75 σ's below 16.2, so 1.75 σ must be 0.2 ounce. You could just start with that equation.)

$$z = \frac{y - \mu}{\sigma}$$

$$-1.75 = \frac{16 - 16.2}{\sigma}$$

$$1.75\sigma = 0.2$$

$$\sigma = 0.114$$

TELL ➡ Conclusion Interpret your result in context.

As we expected, the standard deviation is lower than before—actually, quite a bit lower.

The company must get the machine to box cereal with a standard deviation of only 0.114 ounce. This means the machine must be more consistent (by nearly a factor of 2) in filling the boxes.

5.5 Normal Probability Plots

In the examples we've worked through, we've assumed that the underlying data distribution was roughly unimodal and symmetric, so that using a Normal model makes sense. When you have data, you must *check* to see whether a Normal model is reasonable. How? Make a picture, of course! Drawing a histogram of the data and looking at the shape is one good way to see whether a Normal model might work.

There's a more specialized graphical display that can help you to decide whether the Normal model is appropriate: the **Normal probability plot**. If the distribution of the data is roughly Normal, the plot is roughly a diagonal straight line. Deviations from a straight line indicate that the distribution is not Normal. This plot is usually able to show deviations from Normality more clearly than the corresponding histogram, but it's usually easier to understand *how* a distribution fails to be Normal by looking at its histogram.

Some data on a car's fuel efficiency provide an example of data that are nearly Normal. The overall pattern of the Normal probability plot is straight. The two trailing low values correspond to the values in the histogram that trail off the low end. They're not quite in line with the rest of the data set. The Normal probability plot shows us that they're a bit lower than we'd expect of the lowest two values in a Normal model.

Figure 5.8
Histogram and Normal probability plot for gas mileage (mpg) recorded by one of the authors over the 8 years he owned a Nissan Maxima. The vertical axes are the same, so each dot on the probability plot would fall into the bar on the histogram immediately to its left.

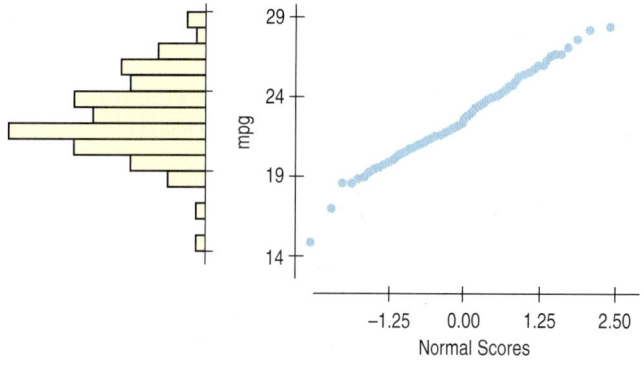

By contrast, the Normal probability plot of the men's *Weight*s from the NHANES Study is far from straight. The weights are skewed to the high end, and the plot is curved. We'd conclude from these pictures that approximations using the 68–95–99.7 Rule for these data would not be very accurate.

Figure 5.9
Histogram and Normal probability plot for men's weights. Note how a skewed distribution corresponds to a bent probability plot.

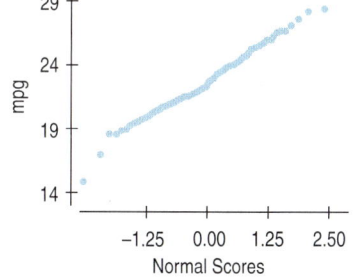

How Does a Normal Probability Plot Work?

Why does the Normal probability plot work like that? We looked at 100 fuel efficiency measures for the author's car. The smallest of these has a z-score of −3.16. The Normal model can tell us what value to expect for the smallest z-score in a batch of 100 if a Normal model were appropriate. That turns out to be −2.58. So our first data value is smaller than we would expect from the Normal.

We can continue this and ask a similar question for each value. For example, the 14th-smallest fuel efficiency has a z-score of almost exactly −1, and that's just what we should expect (well, −1.1 to be exact). A Normal probability plot takes each data value and plots it against the z-score you'd expect that point to have if the distribution were perfectly Normal.[5]

When the values match up well, the line is straight. If one or two points are surprising from the Normal's point of view, they don't line up. When the entire distribution is skewed or different from the Normal in some other way, the values don't match up very well at all and the plot bends.

It turns out to be tricky to find the values we expect. They're called *Normal scores*, but you can't easily look them up in the tables. That's why probability plots are best made with technology and not by hand.

The best advice on using Normal probability plots is to see whether they are straight. If so, then your data look like data from a Normal model. If not, make a histogram to understand how they differ from the model.

Activity: Assessing Normality.
This activity guides you through the process of checking the Nearly Normal Condition using your statistics package.

WHAT CAN GO WRONG?

- **Don't use a Normal model when the distribution is not unimodal and symmetric.** Normal models are so easy and useful that it is tempting to use them even when they don't describe the data very well. That can lead to wrong conclusions. Don't use a Normal model without first checking the **Nearly Normal Condition**. Look at a picture of the data to check that it is unimodal and symmetric. A histogram, or a Normal probability plot, can help you tell whether a Normal model is appropriate.

  The CEOs (p. 94) had a mean total compensation of $10.48M and a standard deviation of $11.46M. Using the Normal model rule, we should expect about 68% of the CEOs to have compensations between −$12.44M and $33.40M. In fact, more than 95% of the CEOs have annual compensations in this range. What went wrong? The distribution is skewed, not symmetric. Using the 68–95–99.7 Rule for data like these will lead to silly results.

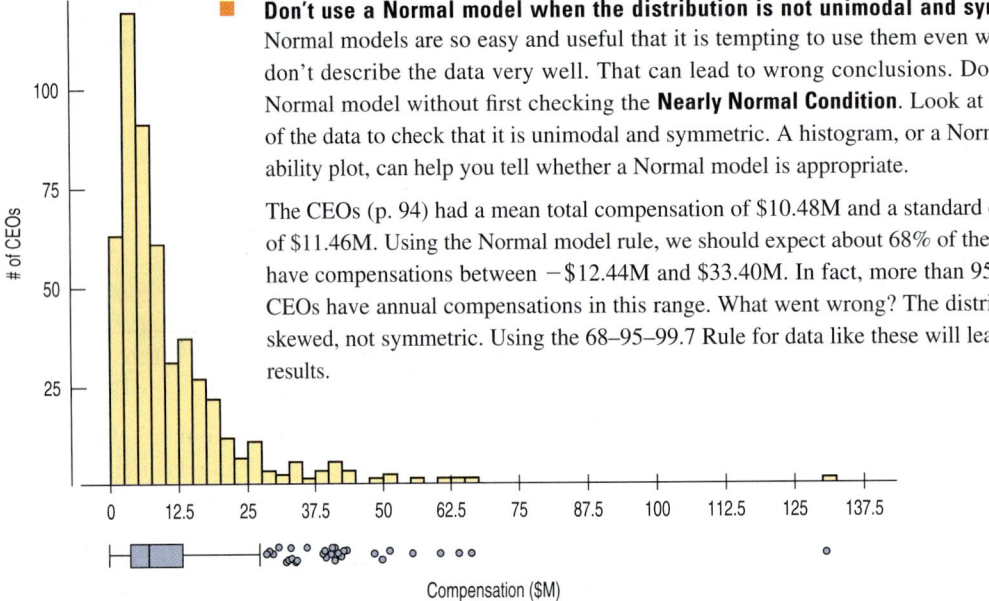

- **Don't use the mean and standard deviation when outliers are present.** Both means and standard deviations can be distorted by outliers, and no model based on distorted values will do a good job. A z-score calculated from a distribution with outliers may be misleading. It's always a good idea to check for outliers. How? Make a picture.

- **Don't round your results in the middle of a calculation.** We reported the mean of the heptathletes' 200 m run as 24.48 seconds. More precisely, it was 24.4844444444444456 seconds.

  You should use all the precision available in the data for all the intermediate steps of a calculation. Using the more precise value for the mean (and also carrying 15 digits for the SD), the z-score calculation for Ennis's run comes out to

  $$z = \frac{22.83 - 24.4844444444444456}{0.80034437032511907} = -2.06716571739284216.$$

  We'd likely report that as −2.067, as opposed to the rounded-off value of −2.1 we got earlier from the table.

---

[5] Sometimes, a Normal probability plot is drawn with the two axes switched, putting the data on the *x*-axis and the *z*-scores on the *y*-axis.

- **Do what we say, not what we do.** When we showed the z-score calculations for Fountain, we rounded the mean to 24.48 s and the SD to 0.80 s. Then to make the story clearer we used *those values* in the displayed calculation.

  We'll continue to show simplified calculations in the book to make the story simpler. When you calculate with full precision, your results may differ slightly from ours. So, we also advise . . .

- **Don't worry about minor differences in results.** Because various calculators and programs may carry different precision in calculations, your answers may differ slightly from those in the text and in the Step-By-Steps, or even from the values given in the answers in the back of the book. Those differences aren't anything to worry about. They're not the main story Statistics tries to tell.

CONNECTIONS

Changing the center and spread of a variable is equivalent to changing its *units*. Indeed, the only part of the data's context changed by standardizing is the units. All other aspects of the context do not depend on the choice or modification of measurement units. This fact points out an important distinction between the numbers the data provide for calculation and the meaning of the variables and the relationships among them. Standardizing can make the numbers easier to work with, but it does not alter the meaning.

Another way to look at this is to note that standardizing may change the center and spread values, but it does not affect the *shape* of a distribution. A histogram or boxplot of standardized values looks just the same as the histogram or boxplot of the original values except, perhaps, for the numbers on the axes.

When we summarized *shape, center,* and *spread* for histograms, we compared them to unimodal, symmetric shapes. You couldn't ask for a nicer example than the Normal model. And if the shape *is* like a Normal, we'll use the mean and standard deviation to standardize the values.

What Have We Learned?

Learning Objectives

Understand how z-scores facilitate comparisons by standardizing variables to have zero mean and unit standard deviation. Recognize normally distributed data by making a histogram and checking whether it is unimodal, symmetric, and bell-shaped, or by making a normal probability plot using technology and checking whether the plot is roughly a straight line.

- The Normal model is a distribution that will be important for much of the rest of this course.
- Before using a Normal model, we should check that our data are plausibly from a normally distributed population.
- A Normal probability plot provides evidence that the data are Normally distributed if it is linear.

Understand how to use the Normal model to judge whether a value is extreme.

- Standardize values to make z-scores and obtain a standard scale. Then refer to a standard Normal distribution.
- Use the 68–95–99.7 Rule as a rule-of-thumb to judge whether a value is extreme.

Know how to refer to tables or technology to find the probability of a value randomly selected from a Normal model falling in any interval.

- Know how to perform calculations about Normally distributed values and probabilities.

Review of Terms

Standardize We standardize to eliminate units. Standardized values can be compared and combined even if the original variables had different units and magnitudes (p. 113).

Standardized value A value found by subtracting the mean and dividing by the standard deviation (p. 113).

Shifting Adding a constant to each data value adds the same constant to the mean, the median, and the quartiles, but does not change the standard deviation or IQR (p. 116).

Rescaling Multiplying each data value by a constant multiplies both the measures of position (mean, median, and quartiles) and the measures of spread (standard deviation and IQR) by that constant (p. 116).

Normal model A useful family of models for unimodal, symmetric distributions (p. 119).

Parameter A numerically valued attribute of a model. For example, the values of μ and σ in a $N(\mu, \sigma)$ model are parameters (p. 119).

Statistic A value calculated from data to summarize aspects of the data. For example, the mean, $\bar{y}$, and standard deviation, s, are statistics (p. 119).

z-score A z-score tells how many standard deviations a value is from the mean; z-scores have a mean of 0 and a standard deviation of 1 (pp. 113, 119).

When working with data, use the statistics $\bar{y}$ and s:

$$z = \frac{y - \bar{y}}{s}.$$

When working with models, use the parameters μ and σ:

$$z = \frac{y - \mu}{\sigma}.$$

Standard Normal model A Normal model, $N(\mu, \sigma)$ with mean $\mu = 0$ and standard deviation $\sigma = 1$. Also called the standard Normal distribution (p. 119).

Nearly Normal Condition A distribution is nearly Normal if it is unimodal and symmetric. We can check by looking at a histogram or a Normal probability plot (p. 120).

68–95–99.7 Rule In a Normal model, about 68% of values fall within 1 standard deviation of the mean, about 95% fall within 2 standard deviations of the mean, and about 99.7% fall within 3 standard deviations of the mean (p. 120).

Normal percentile The Normal percentile corresponding to a z-score gives the percentage of values in a standard Normal distribution found at that z-score or below (p. 123).

Normal probability plot A display to help assess whether a distribution of data is approximately Normal. If the plot is nearly straight, the data satisfy the Nearly Normal Condition (p. 129).

On the Computer NORMAL PROBABILITY PLOTS

The best way to tell whether your data can be modeled well by a Normal model is to make a picture or two. We've already talked about making histograms. Normal probability plots are almost never made by hand because the values of the Normal scores are tricky to find. But most statistics software make Normal plots, though various packages call the same plot by different names and array the information differently.

DATA DESK

To make a "Normal Probability Plot" in Data Desk:
- Select the Variable.
- Choose **Normal Prob Plot** from the **Plot** menu.

COMMENTS

Data Desk places the ordered data values on the vertical axis and the Normal scores on the horizontal axis.

EXCEL

Excel offers a "Normal probability plot" as part of the Regression command in the Data Analysis extension, but (as of this writing) it is not a correct Normal probability plot and should not be used.

JMP

To make a "Normal Quantile Plot" in JMP:
- Make a histogram using **Distributions** from the **Analyze** menu.
- Click on the drop-down menu next to the variable name.
- Choose **Normal Quantile Plot** from the drop-down menu.
- JMP opens the plot next to the histogram.

COMMENTS

JMP places the ordered data on the vertical axis and the Normal scores on the horizontal axis. The vertical axis aligns with the histogram's axis, a useful feature.

MINITAB

To make a "Normal Probability Plot" in MINITAB:
- Choose **Probability Plot** from the **Graph** menu.
- Select "Single" for the type of plot. Click **OK**.
- Enter the name of the variable in the "Graph variables" box. Click **OK**.

COMMENTS

MINITAB places the ordered data on the horizontal axis and the Normal scores on the vertical axis.

R

To make a Normal probability (Q-Q) plot for X:
- qqnorm(X) will produce the plot.

To standardize a variable X:
- Z = (X − mean(X))/sd(X) will create a standardized variable Z.

COMMENTS

By default, R places the ordered data on the vertical axis and the Normal scores on the horizontal axis, but that can be reversed by setting **datax = TRUE** inside qqnorm.

SPSS

To make a Normal "P-P plot" in SPSS:
- Choose **P-P** from the **Graphs** menu.
- Select the variable to be displayed in the source list.
- Click the arrow button to move the variable into the target list.
- Click the **OK** button.

COMMENTS

SPSS places the ordered data on the horizontal axis and the Normal scores on the vertical axis. You may safely ignore the options in the P-P dialog.

STATCRUNCH

To make a Normal probability plot:
- Click on **Graphics**.
- Choose **QQ Plot**.
- Choose the variable name from the list of **Columns**.
- Click on **Create Graph**.

To work with Normal percentiles:
- Click on **Stat**.
- Choose **Calculators » Normal**.
- Choose a lower tail ($\leq$) or upper tail ($\geq$) region.
- Enter the z-score cutoff, and then click on **Compute** to find the probability.

OR

Enter the desired probability, and then click on **Compute** to find the z-score cutoff.

TI-83/84 PLUS

To create a "Normal Percentile Plot" on the TI-83:
- Set up a **STAT PLOT** using the last of the Types.
- Specify your datalist, and the axis you choose to represent the data.
- Although most people wouldn't open a statistics package just to find a Normal model value they could find in a table, you *would* use a calculator for that function.
- So . . . to find what percent of a Normal model lies between two z-scores, choose **normalcdf** from the **DISTR**ibutions menu and enter the command **normalcdf(zLeft, zRight)**.
- To find the z-score that corresponds to a given percentile in a Normal model, choose **invNorm** from the **DISTR**ibutions menu and enter the command **invNorm(percentile)**.

COMMENTS

We often want to find Normal percentages from a certain z-score to infinity. On the calculator, indicate "infinity" as a very large z-score, say, 99. For example, the percentage of a Normal model over 2 standard deviations above the mean can be evaluated with **normalcdf(2, 99)**.

To make a Normal Probability plot:
- Turn a STATPLOT On.
- Tell it to make a Normal probability plot by choosing the last of the icons.
- Specify your datalist and which axis you want the data on. (Use Y to make the plot look like those here.)
- Specify the Mark you want the plot to use.
- Now ZoomStat does the rest.

Exercises

Section 5.1

1. **Stats test** The mean score on the Stats exam was 75 points with a standard deviation of 5 points, and Gregor's z-score was -2. How many points did he score?

2. **Mensa** People with z-scores above 2.5 on an IQ test are sometimes classified as geniuses. If IQ scores have a mean of 100 and a standard deviation of 16 points, what IQ score do you need to be considered a genius?

3. **Temperatures** A town's January high temperatures average 36°F with a standard deviation of 10°, while in July the mean high temperature is 74° and the standard deviation is 8°. In which month is it more unusual to have a day with a high temperature of 55°? Explain.

4. **Placement exams** An incoming freshman took her college's placement exams in French and mathematics. In French, she scored 82 and in math 86. The overall results on the French exam had a mean of 72 and a standard deviation of 8, while the mean math score was 68, with a standard deviation of 12. On which exam did she do better compared with the other freshmen?

Section 5.2

5. **Shipments** A company selling clothing on the Internet reports that the packages it ships have a median weight of 68 ounces and an IQR of 40 ounces.

   a) The company plans to include a sales flyer weighing 4 ounces in each package. What will the new median and IQR be?

b) If the company recorded the shipping weights of these new packages in pounds instead of ounces, what would the median and IQR be? (1 lb. = 16 oz.)

6. **Hotline** A company's customer service hotline handles many calls relating to orders, refunds, and other issues. The company's records indicate that the median length of calls to the hotline is 4.4 minutes with an IQR of 2.3 minutes.

   a) If the company were to describe the duration of these calls in seconds instead of minutes, what would the median and IQR be?
   b) In an effort to speed up the customer service process, the company decides to streamline the series of pushbutton menus customers must navigate, cutting the time by 24 seconds. What will the median and IQR of the length of hotline calls become?

Section 5.3

7. **Guzzlers?** Environmental Protection Agency (EPA) fuel economy estimates for automobile models tested recently predicted a mean of 24.8 mpg and a standard deviation of 6.2 mpg for highway driving. Assume that a Normal model can be applied.

   a) Draw the model for auto fuel economy. Clearly label it, showing what the 68–95–99.7 Rule predicts.
   b) In what interval would you expect the central 68% of autos to be found?
   c) About what percent of autos should get more than 31 mpg?
   d) About what percent of cars should get between 31 and 37.2 mpg?
   e) Describe the gas mileage of the worst 2.5% of all cars.

8. **IQ** Some IQ tests are standardized to a Normal model, with a mean of 100 and a standard deviation of 16.

   a) Draw the model for these IQ scores. Clearly label it, showing what the 68–95–99.7 Rule predicts.
   b) In what interval would you expect the central 95% of IQ scores to be found?
   c) About what percent of people should have IQ scores above 116?
   d) About what percent of people should have IQ scores between 68 and 84?
   e) About what percent of people should have IQ scores above 132?

Section 5.4

9. **Normal cattle** The Virginia Cooperative Extension reports that the mean weight of yearling Angus steers is 1152 pounds. Suppose that weights of all such animals can be described by a Normal model with a standard deviation of 84 pounds. What percent of steers weigh

   a) over 1250 pounds?
   b) under 1200 pounds?
   c) between 1000 and 1100 pounds?

10. **IQs revisited** Based on the Normal model $N(100, 16)$ describing IQ scores, what percent of people's IQs would you expect to be

    a) over 80?
    b) under 90?
    c) between 112 and 132?

Section 5.5

11. **Music library** Corey has 4929 songs in his computer's music library. The lengths of the songs have a mean of 242.4 seconds and standard deviation of 114.51 seconds. A Normal probability plot of the song lengths looks like this:

    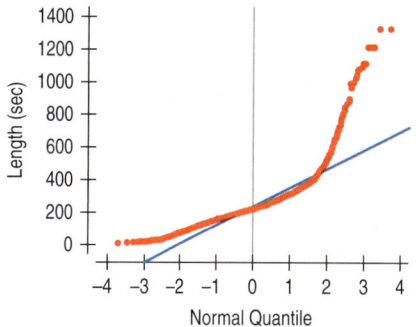

    a) Do you think the distribution is Normal? Explain.
    b) If it isn't Normal, how does it differ from a Normal model?

12. **Wisconsin ACT math** The histogram shows the distribution of mean ACT mathematics scores for all Wisconsin public schools in 2011. The vertical lines show the mean and one standard deviation above and below the mean. 78.8% of the data points are between the two outer lines.

    a) Give two reasons that a Normal model is not appropriate for these data.

    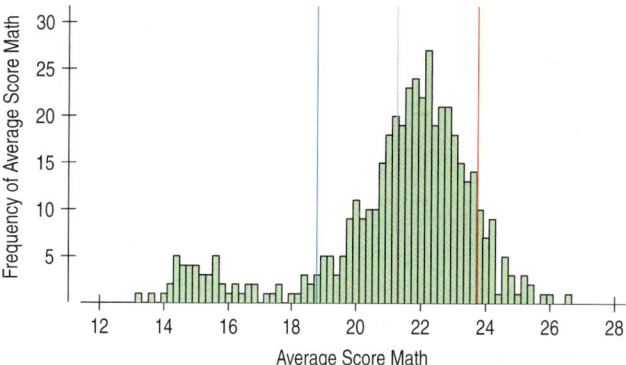

    b) The Normal probability plot on the left shows the distribution of these scores. The plot on the right shows the same data with the Milwaukee area schools (mostly in the low mode) removed. What do these plots tell you about the shape of the distributions?

Chapter Exercises

13. **Payroll** Here are the summary statistics for the weekly payroll of a small company: lowest salary = $300, mean salary = $700, median = $500, range = $1200, IQR = $600, first quartile = $350, standard deviation = $400.
    a) Do you think the distribution of salaries is symmetric, skewed to the left, or skewed to the right? Explain why.
    b) Between what two values are the middle 50% of the salaries found?
    c) Suppose business has been good and the company gives every employee a $50 raise. Tell the new value of each of the summary statistics.
    d) Instead, suppose the company gives each employee a 10% raise. Tell the new value of each of the summary statistics.

14. **Hams** A specialty foods company sells "gourmet hams" by mail order. The hams vary in size from 4.15 to 7.45 pounds, with a mean weight of 6 pounds and standard deviation of 0.65 pounds. The quartiles and median weights are 5.6, 6.2, and 6.55 pounds.
    a) Find the range and the IQR of the weights.
    b) Do you think the distribution of the weights is symmetric or skewed? If skewed, which way? Why?
    c) If these weights were expressed in ounces (1 pound = 16 ounces) what would the mean, standard deviation, quartiles, median, IQR, and range be?
    d) When the company ships these hams, the box and packing materials add 30 ounces. What are the mean, standard deviation, quartiles, median, IQR, and range of weights of boxes shipped (in ounces)?
    e) One customer made a special order of a 10-pound ham. Which of the summary statistics of part d might *not* change if that data value were added to the distribution?

15. **SAT or ACT?** Each year thousands of high school students take either the SAT or the ACT, standardized tests used in the college admissions process. Combined SAT Math and Verbal scores go as high as 1600, while the maximum ACT composite score is 36. Since the two exams use very different scales, comparisons of performance are difficult. A convenient rule of thumb is $SAT = 40 \times ACT + 150$; that is, multiply an ACT score by 40 and add 150 points to estimate the equivalent SAT score. An admissions officer reported the following statistics about the ACT scores of 2355 students who applied to her college one year. Find the summaries of equivalent SAT scores.

    Lowest score = 19 Mean = 27 Standard deviation = 3
    Q3 = 30 Median = 28 IQR = 6

16. **Cold U?** A high school senior uses the Internet to get information on February temperatures in the town where he'll be going to college. He finds a website with some statistics, but they are given in degrees Celsius. The conversion formula is $°F = 9/5°C + 32$. Determine the Fahrenheit equivalents for the summary information below.

    Maximum temperature = 11°C Range = 33°
    Mean = 1° Standard deviation = 7°
    Median = 2° IQR = 16°

17. **Stats test, part II** Suppose your Statistics professor reports test grades as z-scores, and you got a score of 2.20 on an exam. Write a sentence explaining what that means.

18. **Checkup** One of the authors has an adopted grandson whose birth family members are very short. After examining him at his 2-year checkup, the boy's pediatrician said that the z-score for his height relative to American 2-year-olds was −1.88. Write a sentence explaining what that means.

19. **Music library again** Corey has 4929 songs in his computer's music library. The songs have a mean duration of 242.4 seconds with a standard deviation of 114.51 seconds. *On the Nickel*, by Tom Waits, is 380 seconds long. What is its z-score?

20. **Windy** In the last chapter, we looked at three outliers arising from a plot of *Average Wind Speed* by *Month* in the Hopkins Forest. Each was associated with an unusually strong storm, but which was the most remarkable for its month?
    Here are the summary statistics for each of those three months:

	February	June	August
Mean	2.324	0.857	0.63
SD	1.577	0.795	0.597

    The outliers had values of 6.73 mph, 3.93 mph, and 2.53 mph, respectively.
    a) What are their z-scores?
    b) Which was the most extraordinary wind event?

21. **Combining test scores** The first Stats exam had a mean of 65 and a standard deviation of 10 points; the second had a mean of 80 and a standard deviation of 5 points. Derrick scored an 80 on both tests. Julie scored a 70 on the first test and a 90 on the second. They both totaled 160 points on the two exams, but Julie claims that her total is better. Explain.

22. Combining scores again The first Stat exam had a mean of 80 and a standard deviation of 4 points; the second had a mean of 70 and a standard deviation of 15 points. Reginald scored an 80 on the first test and an 85 on the second. Sara scored an 88 on the first but only a 65 on the second. Although Reginald's total score is higher, Sara feels she should get the higher grade. Explain her point of view.

23. Final exams Anna, a language major, took final exams in both French and Spanish and scored 83 on each. Her roommate Megan, also taking both courses, scored 77 on the French exam and 95 on the Spanish exam. Overall, student scores on the French exam had a mean of 81 and a standard deviation of 5, and the Spanish scores had a mean of 74 and a standard deviation of 15.
   a) To qualify for language honors, a major must maintain at least an 85 average for all language courses taken. So far, which student qualifies?
   b) Which student's overall performance was better?

24. MP3s Two companies market new batteries targeted at owners of personal music players. DuraTunes claims a mean battery life of 11 hours, while RockReady advertises 12 hours.
   a) Explain why you would also like to know the standard deviations of the battery lifespans before deciding which brand to buy.
   b) Suppose those standard deviations are 2 hours for DuraTunes and 1.5 hours for RockReady. You are headed for 8 hours at the beach. Which battery is most likely to last all day? Explain.
   c) If your beach trip is all weekend, and you probably will have the music on for 16 hours, which battery is most likely to last? Explain.

25. Cattle Using $N(1152, 84)$, the Normal model for weights of Angus steers in Exercise 9,
   a) How many standard deviations from the mean would a steer weighing 1000 pounds be?
   b) Which would be more unusual, a steer weighing 1000 pounds or one weighing 1250 pounds?

T 26. Car speeds John Beale of Stanford, California, recorded the speeds of cars driving past his house, where the speed limit read 20 mph. The mean of 100 readings was 23.84 mph, with a standard deviation of 3.56 mph. (He actually recorded every car for a two-month period. These are 100 representative readings.)
   a) How many standard deviations from the mean would a car going under the speed limit be?
   b) Which would be more unusual, a car traveling 34 mph or one going 10 mph?

27. More cattle Recall that the beef cattle described in Exercise 25 had a mean weight of 1152 pounds, with a standard deviation of 84 pounds.
   a) Cattle buyers hope that yearling Angus steers will weigh at least 1000 pounds. To see how much over (or under) that goal the cattle are, we could subtract 1000 pounds from all the weights. What would the new mean and standard deviation be?
   b) Suppose such cattle sell at auction for 40 cents a pound. Find the mean and standard deviation of the sale prices (in dollars) for all the steers.

T 28. Car speeds again For the car speed data in Exercise 26, recall that the mean speed recorded was 23.84 mph, with a standard deviation of 3.56 mph. To see how many cars are speeding, John subtracts 20 mph from all speeds.
   a) What is the mean speed now? What is the new standard deviation?
   b) His friend in Berlin wants to study the speeds, so John converts all the original miles-per-hour readings to kilometers per hour by multiplying all speeds by 1.609 (km per mile). What is the mean now? What is the new standard deviation?

29. Cattle, part III Suppose the auctioneer in Exercise 27 sold a herd of cattle whose minimum weight was 980 pounds, median was 1140 pounds, standard deviation 84 pounds, and IQR 102 pounds. They sold for 40 cents a pound, and the auctioneer took a $20 commission on each animal. Then, for example, a steer weighing 1100 pounds would net the owner $0.40(1100) - 20 = \$420$. Find the minimum, median, standard deviation, and IQR of the net sale prices.

30. Caught speeding Suppose police set up radar surveillance on the Stanford street described in Exercise 26. They handed out a large number of tickets to speeders going a mean of 28 mph, with a standard deviation of 2.4 mph, a maximum of 33 mph, and an IQR of 3.2 mph. Local law prescribes fines of $100, plus $10 per mile per hour over the 20 mph speed limit. For example, a driver convicted of going 25 mph would be fined $100 + 10(5) = \$150$. Find the mean, maximum, standard deviation, and IQR of all the potential fines.

31. Professors A friend tells you about a recent study dealing with the number of years of teaching experience among current college professors. He remembers the mean but can't recall whether the standard deviation was 6 months, 6 years, or 16 years. Tell him which one it must have been, and why.

32. Rock concerts A popular band on tour played a series of concerts in large venues. They always drew a large crowd, averaging 21,359 fans. While the band did not announce (and probably never calculated) the standard deviation, which of these values do you think is most likely to be correct: 20, 200, 2000, or 20,000 fans? Explain your choice.

33. Small steer In Exercise 25, we suggested the model $N(1152, 84)$ for weights in pounds of yearling Angus steers. What weight would you consider to be unusually low for such an animal? Explain.

34. High IQ Exercise 8, proposes modeling IQ scores with $N(100, 16)$. What IQ would you consider to be unusually high? Explain.

35. Trees A forester measured 27 of the trees in a large woods that is up for sale. He found a mean diameter of 10.4 inches and a standard deviation of 4.7 inches. Suppose that these trees provide an accurate description of the whole forest and that a Normal model applies.
a) Draw the Normal model for tree diameters.
b) What size would you expect the central 95% of all trees to be?
c) About what percent of the trees should be less than an inch in diameter?
d) About what percent of the trees should be between 5.8 and 10.4 inches in diameter?
e) About what percent of the trees should be over 15 inches in diameter?

36. Rivets A company that manufactures rivets believes the shear strength (in pounds) is modeled by $N(800, 50)$.
a) Draw and label the Normal model.
b) Would it be safe to use these rivets in a situation requiring a shear strength of 750 pounds? Explain.
c) About what percent of these rivets would you expect to fall below 900 pounds?
d) Rivets are used in a variety of applications with varying shear strength requirements. What is the maximum shear strength for which you would feel comfortable approving this company's rivets? Explain your reasoning.

37. Trees, part II Later on, the forester in Exercise 35 shows you a histogram of the tree diameters he used in analyzing the woods that was for sale. Do you think he was justified in using a Normal model? Explain, citing some specific concerns.

38. Car speeds, the picture For the car speed data in Exercise 26, here is the histogram, boxplot, and Normal probability plot of the 100 readings. Do you think it is appropriate to apply a Normal model here? Explain.

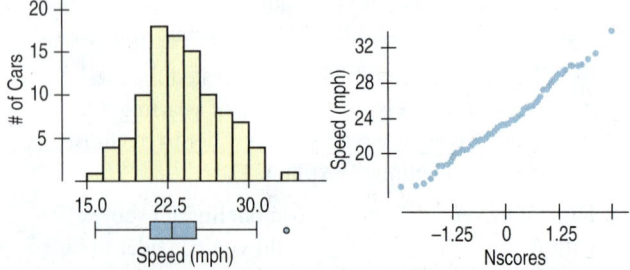

39. Winter Olympics 2010 downhill Fifty-nine men completed the men's alpine downhill race in Vancouver. The gold medal winner finished in 114.3 seconds. Here are the times (in seconds) for all competitors (espn.go.com/olympics/winter/2010/results/_/sport/1/event/2):

114.3	115.0	115.7	116.7	118.6	119.8
114.4	115.2	115.8	116.7	118.7	120.0
114.4	115.2	116.0	117.0	118.8	120.1
114.5	115.3	116.1	117.2	118.9	120.6
114.6	115.3	116.2	117.2	119.2	121.7
114.7	115.4	116.2	117.4	119.5	121.7
114.8	115.4	116.3	117.7	119.6	122.6
114.8	115.5	116.4	117.9	119.7	123.4
114.9	115.6	116.6	118.1	119.8	124.4
114.9	115.7	116.7	118.4	119.8	

a) The mean time was 117.34 seconds, with a standard deviation of 2.465 seconds. If the Normal model is appropriate, what percent of times will be less than 114.875 seconds?
b) What is the actual percent of times less than 114.875 seconds?
c) Why do you think the two percentages don't agree?
d) Make a histogram of these times. What do you see?

40. Check the model The mean of the 100 car speeds in Exercise 26 was 23.84 mph, with a standard deviation of 3.56 mph.
a) Using a Normal model, what values should border the middle 95% of all car speeds?
b) Here are some summary statistics.

Percentile		Speed
100%	Max	34.060
97.5%		30.976
90.0%		28.978
75.0%	Q3	25.785
50.0%	Median	23.525
25.0%	Q1	21.547
10.0%		19.163
2.5%		16.638
0.0%	Min	16.270

From your answer in part a, how well does the model do in predicting those percentiles? Are you surprised? Explain.

41. Receivers 2013 NFL data from the 2013 football season reported the number of yards gained by each of the league's 181 wide receivers:

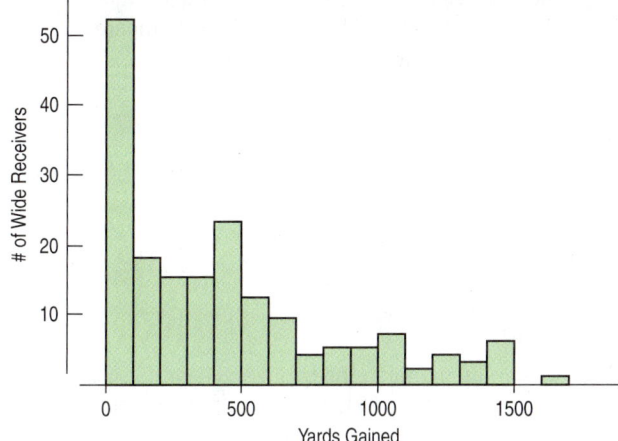

The mean is 426.98 yards, with a standard deviation of 408.34 yards.
a) According to the Normal model, what percent of receivers would you expect to gain more yards than 2 standard deviations above the mean number of yards?
b) For these data, what does that mean?
c) Explain the problem in using a Normal model here.

42. Customer database A large philanthropic organization keeps records on the people who have contributed. In addition to keeping records of past giving, the organization buys demographic data on neighborhoods from the U.S. Census Bureau. Eighteen of these variables concern ethnicity. Here are a histogram and summary statistics for the percentage of whites in the neighborhoods of 500 donors:

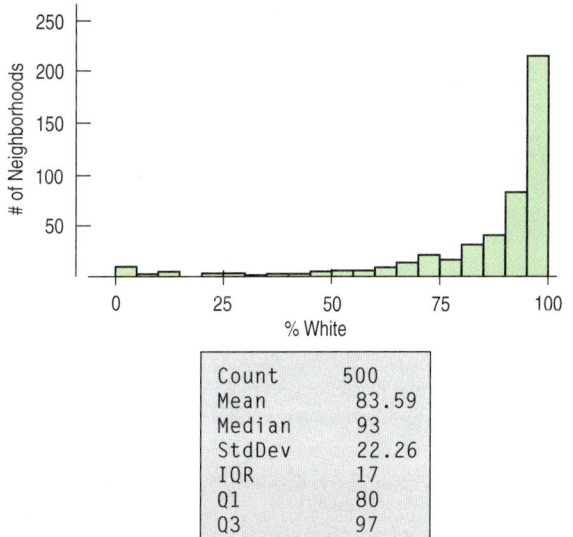

```
Count     500
Mean      83.59
Median    93
StdDev    22.26
IQR       17
Q1        80
Q3        97
```

a) Which is a better summary of the percentage of white residents in the neighborhoods, the mean or the median? Explain.
b) Which is a better summary of the spread, the IQR or the standard deviation? Explain.

c) From a Normal model, about what percentage of neighborhoods should have a percent white within one standard deviation of the mean?
d) What percentage of neighborhoods actually have a percent white within one standard deviation of the mean?
e) Explain the discrepancy between parts c and d.

43. More cattle Based on the model $N(1152, 84)$ describing Angus steer weights from Exercise 25, what are the cutoff values for
a) the highest 10% of the weights?
b) the lowest 20% of the weights?
c) the middle 40% of the weights?

44. More IQs In the Normal model $N(100, 16)$ from Exercise 2, what cutoff value bounds
a) the highest 5% of all IQs?
b) the lowest 30% of the IQs?
c) the middle 80% of the IQs?

45. Cattle, finis Consider the Angus weights model $N(1152, 84)$ one last time.
a) What weight represents the 40th percentile?
b) What weight represents the 99th percentile?
c) What's the IQR of the weights of these Angus steers?

46. IQ, finis Consider the IQ model $N(100, 16)$ one last time.
a) What IQ represents the 15th percentile?
b) What IQ represents the 98th percentile?
c) What's the IQR of the IQs?

47. Cholesterol Assume the cholesterol levels of adult American women can be described by a Normal model with a mean of 188 mg/dL and a standard deviation of 24.
a) Draw and label the Normal model.
b) What percent of adult women do you expect to have cholesterol levels over 200 mg/dL?
c) What percent of adult women do you expect to have cholesterol levels between 150 and 170 mg/dL?
d) Estimate the IQR of the cholesterol levels.
e) Above what value are the highest 15% of women's cholesterol levels?

48. Tires A tire manufacturer believes that the treadlife of its snow tires can be described by a Normal model with a mean of 32,000 miles and standard deviation of 2500 miles.
a) If you buy one of these tires, would it be reasonable for you to hope it will last 40,000 miles? Explain.
b) Approximately what fraction of these tires can be expected to last less than 30,000 miles?
c) Approximately what fraction of these tires can be expected to last between 30,000 and 35,000 miles?
d) Estimate the IQR of the treadlives.
e) In planning a marketing strategy, a local tire dealer wants to offer a refund to any customer whose tires fail to last a certain number of miles. However, the dealer does not want to take too big a risk. If the dealer is willing to give refunds to no more than 1 of every 25 customers, for what mileage can he guarantee these tires to last?

49. Kindergarten Companies that design furniture for elementary school classrooms produce a variety of sizes for kids of different ages. Suppose the heights of kindergarten children can be described by a Normal model with a mean of 38.2 inches and standard deviation of 1.8 inches.
   a) What fraction of kindergarten kids should the company expect to be less than 3 feet tall?
   b) In what height interval should the company expect to find the middle 80% of kindergarteners?
   c) At least how tall are the biggest 10% of kindergarteners?

50. Body temperatures Most people think that the "normal" adult body temperature is 98.6°F. That figure, based on a 19th-century study, has recently been challenged. In a 1992 article in the *Journal of the American Medical Association*, researchers reported that a more accurate figure may be 98.2°F. Furthermore, the standard deviation appeared to be around 0.7°F. Assume that a Normal model is appropriate.
   a) In what interval would you expect most people's body temperatures to be? Explain.
   b) What fraction of people would be expected to have body temperatures above 98.6°F?
   c) Below what body temperature are the coolest 20% of all people?

51. Eggs Hens usually begin laying eggs when they are about 6 months old. Young hens tend to lay smaller eggs, often weighing less than the desired minimum weight of 54 grams.
   a) The average weight of the eggs produced by the young hens is 50.9 grams, and only 28% of their eggs exceed the desired minimum weight. If a Normal model is appropriate, what would the standard deviation of the egg weights be?
   b) By the time these hens have reached the age of 1 year, the eggs they produce average 67.1 grams, and 98% of them are above the minimum weight. What is the standard deviation for the appropriate Normal model for these older hens?
   c) Are egg sizes more consistent for the younger hens or the older ones? Explain.

52. Tomatoes Agricultural scientists are working on developing an improved variety of Roma tomatoes. Marketing research indicates that customers are likely to bypass Romas that weigh less than 70 grams. The current variety of Roma plants produces fruit that averages 74 grams, but 11% of the tomatoes are too small. It is reasonable to assume that a Normal model applies.
   a) What is the standard deviation of the weights of Romas now being grown?
   b) Scientists hope to reduce the frequency of undersized tomatoes to no more than 4%. One way to accomplish this is to raise the average size of the fruit. If the standard deviation remains the same, what target mean should they have as a goal?
   c) The researchers produce a new variety with a mean weight of 75 grams, which meets the 4% goal. What is the standard deviation of the weights of these new Romas?
   d) Based on their standard deviations, compare the tomatoes produced by the two varieties.

Just Checking
ANSWERS

1. a) On the first test, the mean is 88 and the SD is 4, so $z = (90 - 88)/4 = 0.5$. On the second test, the mean is 75 and the SD is 5, so $z = (80 - 75)/5 = 1.0$. The first test has the lower z-score, so it is the one that will be dropped.
   b) No. The second test is 1 standard deviation above the mean, farther away than the first test, so it's the better score relative to the class.

2. a) The mean would increase to 500.
   b) The standard deviation is still 100 points.
   c) The two boxplots would look nearly identical (the shape of the distribution would remain the same), but the later one would be shifted 50 points higher.

3. The standard deviation is now 2.54 millimeters, which is the same as 0.1 inch. Nothing has changed. The standard deviation has "increased" only because we're reporting it in millimeters now, not inches.

4. The mean is 184 centimeters, with a standard deviation of 8 centimeters. 2 meters is 200 centimeters, which is 2 standard deviations above the mean. We expect 5% of the men to be more than 2 standard deviations below or above the mean, so half of those, 2.5%, are likely to be above 2 meters.

5. a) We know that 68% of the time we'll be within 1 standard deviation (2 min) of 20. So 32% of the time we'll arrive in less than 18 or more than 22 minutes. Half of those times (16%) will be greater than 22 minutes, so 84% will be less than 22 minutes.
   b) 24 minutes is 2 standard deviations above the mean. Because of the 95% rule, we know 2.5% of the times will be more than 24 minutes.
   c) Traffic incidents may occasionally increase the time it takes to get to school, so the driving times may be skewed to the right, and there may be outliers.
   d) If so, the Normal model would not be appropriate and the percentages we predict would not be accurate.

Review of Part I

Exploring and Understanding Data

Quick Review

It's time to put it all together. Real data don't come tagged with instructions for use. So let's step back and look at how the key concepts and skills we've seen work together. This brief list and the review exercises that follow should help you check your understanding of Statistics so far.

- We treat data two ways: as categorical and as quantitative.

- To describe categorical data:
  - Make a picture. Bar graphs work well for comparing counts in categories.
  - Summarize the distribution with a table of counts or relative frequencies (percents) in each category.
  - Pie charts and segmented bar charts display divisions of a whole.
  - Compare distributions with plots side by side.
  - Look for associations between variables by comparing marginal and conditional distributions.

- To describe quantitative data:
  - Make a picture. Use histograms, boxplots, stem-and-leaf displays, or dotplots. Stem-and-leafs are great when working by hand and good for small data sets. Histograms are a good way to see the distribution. Boxplots are best for comparing several distributions.
  - Describe distributions in terms of their shape, center, and spread, and note any unusual features such as gaps or outliers.
  - The shape of most distributions you'll see will likely be uniform, unimodal, or bimodal. It may be multimodal. If it is unimodal, then it may be symmetric or skewed.
  - A 5-number summary makes a good numerical description of a distribution: min, Q1, median, Q3, and max.

- If the distribution is skewed, be sure to include the median and interquartile range (IQR) when you describe its center and spread.

- A distribution that is severely skewed may benefit from re-expressing the data. If it is skewed to the high end, taking logs often works well.

- If the distribution is unimodal and symmetric, describe its center and spread with the mean and standard deviation.

- Use the standard deviation as a ruler to tell how unusual an observed value may be, or to compare or combine measurements made on different scales.

- Shifting a distribution by adding or subtracting a constant affects measures of position but not measures of spread. Rescaling by multiplying or dividing by a constant affects both.

- When a distribution is roughly unimodal and symmetric, a Normal model may be useful. For Normal models, the 68–95–99.7 Rule is a good rule of thumb.

- If the Normal model fits well (check a histogram or Normal probability plot), then Normal percentile tables or functions found in most statistics technology can provide more detailed values.

Need more help with some of this? It never hurts to reread sections of the chapters! And in the following pages we offer you more opportunities[1] to review these concepts and skills.

The exercises that follow use the concepts and skills you've learned in the first five chapters. To be more realistic and more useful for your review, they don't tell you which of the concepts or methods you need. But neither will the exam.

---

[1] If you doubted that we are teachers, this should convince you. Only a teacher would call additional homework exercises "opportunities."

Review Exercises

1. **Bananas** Here are the prices (in cents per pound) of bananas reported from 15 markets surveyed by the U.S. Department of Agriculture.

51	52	45
48	53	52
50	49	52
48	43	46
45	42	50

   a) Display these data with an appropriate graph.
   b) Report appropriate summary statistics.
   c) Write a few sentences about this distribution.

2. **Prenatal care** Results of a 1996 American Medical Association report about the infant mortality rate for twins carried for the full term of a normal pregnancy are shown on the next page, broken down by the level of prenatal care the mother had received.

Full-Term Pregnancies, Level of Prenatal Care	Infant Mortality Rate Among Twins (deaths per thousand live births)
Intensive	5.4
Adequate	3.9
Inadequate	6.1
Overall	5.1

a) Is the overall rate the average of the other three rates? Should it be? Explain.
b) Do these results indicate that adequate prenatal care is important for pregnant women? Explain.
c) Do these results suggest that a woman pregnant with twins should be wary of seeking too much medical care? Explain.

3. **Singers by parts** The boxplots shown display the heights (in inches) of 130 members of the choir we saw in Exercise 3.40. Now we have information about what part they sing.

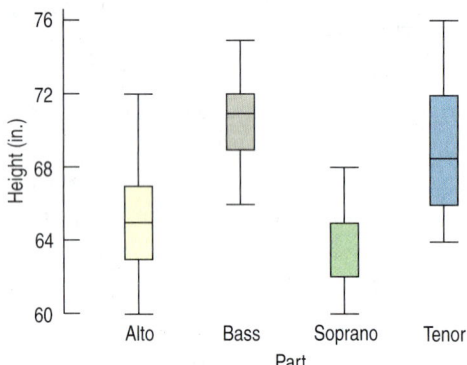

a) It appears that the median height for sopranos is missing, but actually the median and the upper quartile are equal. How could that happen?
b) Write a few sentences describing what you see.

4. **Dialysis** In a study of dialysis, researchers found that "of the three patients who were currently on dialysis, 67% had developed blindness and 33% had their toes amputated." What kind of display might be appropriate for these data? Explain.

5. **Beanstalks** Beanstalk Clubs are social clubs for very tall people. To join, a man must be over 6′2″ tall, and a woman over 5′10″. The National Health Survey suggests that heights of adults may be Normally distributed, with mean heights of 69.1″ for men and 64.0″ for women. The respective standard deviations are 2.8″ and 2.5″.
a) You are probably not surprised to learn that men are generally taller than women, but what does the greater standard deviation for men's heights indicate?

b) Who are more likely to qualify for Beanstalk membership, men or women? Explain.

6. **Bread** Clarksburg Bakery is trying to predict how many loaves to bake. In the past 100 days, they have sold between 95 and 140 loaves per day. Here is a histogram of the number of loaves they sold for the past 100 days.

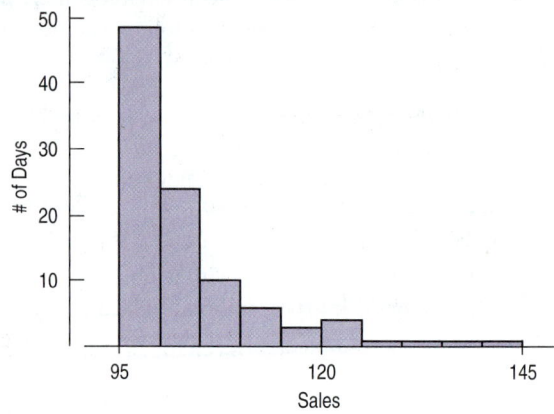

a) Describe the distribution.
b) Which should be larger, the mean number of sales or the median? Explain.
c) Here are the summary statistics for Clarksburg Bakery's bread sales. Use these statistics and the histogram above to create a boxplot. You may approximate the values of any outliers.

Summary of Sales	
Median	100
Min	95
Max	140
25th %tile	97
75th %tile	105.5

d) For these data, the mean was 103 loaves sold per day, with a standard deviation of 9 loaves. Do these statistics suggest that Clarksburg Bakery should expect to sell between 94 and 112 loaves on about 68% of the days? Explain.

7. **State University** Public relations staff at State U. phoned 850 local residents. After identifying themselves, the callers asked the survey participants their ages, whether they had attended college, and whether they had a favorable opinion of the university. The official report to the university's directors claimed that, in general, people had very favorable opinions about the university.
a) Identify the W's of these data.

b) Identify the variables, classify each as categorical or quantitative, and specify units if relevant.
c) Are you confident about the report's conclusion? Explain.

8. **Acid rain** Based on long-term investigation, researchers have suggested that the acidity (pH) of rainfall in the Shenandoah Mountains can be described by the Normal model $N(4.9, 0.6)$.
   a) Draw and carefully label the model.
   b) What percent of storms produce rainfall with pH over 6?
   c) What percent of storms produce rainfall with pH under 4?
   d) The lower the pH, the more acidic the rain. What is the pH level for the most acidic 20% of all storms?
   e) What is the pH level for the least acidic 5% of all storms?
   f) What is the IQR for the pH of rainfall?

9. **Fraud detection** A credit card bank is investigating the incidence of fraudulent card use. The bank suspects that the type of product bought may provide clues to the fraud. To examine this situation, the bank looks at the Standard Industrial Code (SIC) of the business related to the transaction. This is a code that was used by the U.S. Census Bureau and Statistics Canada to identify the type of every registered business in North America.[2] For example, 1011 designates Meat and Meat Products (except Poultry), 1012 is Poultry Products, 1021 is Fish Products, 1031 is Canned and Preserved Fruits and Vegetables, and 1032 is Frozen Fruits and Vegetables.

    A company intern produces the following histogram of the SIC codes for 1536 transactions:

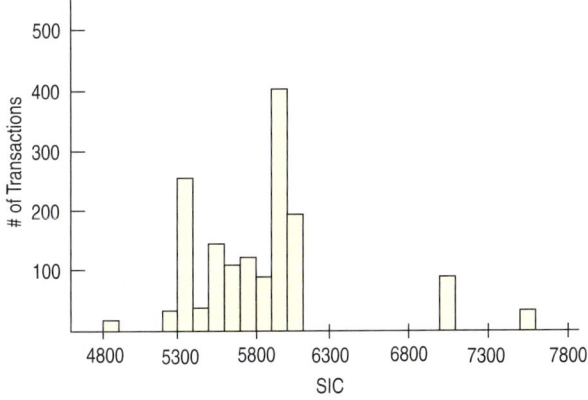

He also reports that the mean SIC is 5823.13 with a standard deviation of 488.17.

---
[2]Since 1997, the SIC has been replaced by the NAICS, a code of six letters.

a) Comment on any problems you see with the use of the mean and standard deviation as summary statistics.
b) How well do you think the Normal model will work on these data? Explain.

10. **Streams** As part of the course work, a class at an upstate NY college collects data on streams each year. Students record a number of biological, chemical, and physical variables, including the stream name, the substrate of the stream (*limestone (L)*, *shale (S)*, or *mixed (M)*), the pH, the temperature (°C), and the BCI, a measure of biological diversity.

Group	Count	%
Limestone	77	44.8
Mixed	26	15.1
Shale	69	40.1

a) Name each variable, indicating whether it is categorical or quantitative, and give the units if available.
b) These streams have been classified according to their substrate—the composition of soil and rock over which they flow—as summarized in the table. What kind of graph might be used to display these data?

11. **Cramming** One Thursday, researchers gave students enrolled in a section of basic Spanish a set of 50 new vocabulary words to memorize. On Friday, the students took a vocabulary test. When they returned to class the following Monday, they were retested—without advance warning. Both sets of test scores for the 28 students are shown below.

Fri	Mon	Fri	Mon
42	36	50	47
44	44	34	34
45	46	38	31
48	38	43	40
44	40	39	41
43	38	46	32
41	37	37	36
35	31	40	31
43	32	41	32
48	37	48	39
43	41	37	31
45	32	36	41
47	44		

a) Create a graphical display to compare the two distributions of scores.

b) Write a few sentences about the scores reported on Friday and Monday.
c) Create a graphical display showing the distribution of the *changes* in student scores.
d) Describe the distribution of changes.

12. **e-Books** A study by the Pew Internet & American Life Project found that 78% of U.S residents over 16 years old read a book in the past 12 months. They also found that 21% had read an e-book using a reader or computer during that period. A newspaper reporting on these findings concluded that 99% of U.S. adult residents had read a book in some fashion in the past year. (http://libraries.pewinternet.org/2012/04/04/the-rise-of-e-reading/) Do you agree? Explain.

13. **Let's play cards** You pick a card from a deck (see description in Chapter 9) and record its denomination (7, say) and its suit (maybe spades).
    a) Is the variable *suit* categorical or quantitative?
    b) Name a game you might be playing for which you would consider the variable *denomination* to be categorical. Explain.
    c) Name a game you might be playing for which you would consider the variable *denomination* to be quantitative. Explain.

T 14. **Accidents** Progressive Insurance asked customers who had been involved in auto accidents how far they were from home when the accident happened. The data are summarized in the table.

Miles from Home	% of Accidents
Less than 1	23
1 to 5	29
6 to 10	17
11 to 15	8
16 to 20	6
Over 20	17

a) Create an appropriate graph of these data.
b) Do these data indicate that driving near home is particularly dangerous? Explain.

T 15. **Hard water** In an investigation of environmental causes of disease, data were collected on the annual mortality rate (deaths per 100,000) for males in 61 large towns in England and Wales. In addition, the water hardness was recorded as the calcium concentration (parts per million, ppm) in the drinking water.
    a) What are the variables in this study? For each, indicate whether it is quantitative or categorical and what the units are.
    b) Here are histograms of calcium concentration and mortality. Describe the distributions of the two variables.

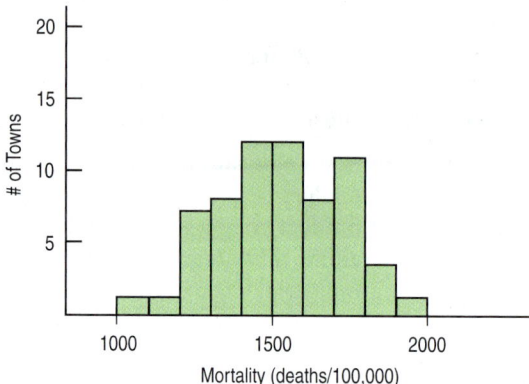

T 16. **Hard water II** The data set from England and Wales also notes for each town whether it was south or north of Derby. Here are some summary statistics and a comparative boxplot for the two regions.

Summary of Mortality				
Group	Count	Mean	Median	StdDev
North	34	1631.59	1631	138.470
South	27	1388.85	1369	151.114

a) What is the overall mean mortality rate for the two regions?
b) Do you see evidence of a difference in mortality rates? Explain.

17. **Seasons** Average daily temperatures in January and July for 60 large U.S. cities are graphed in the histograms below.

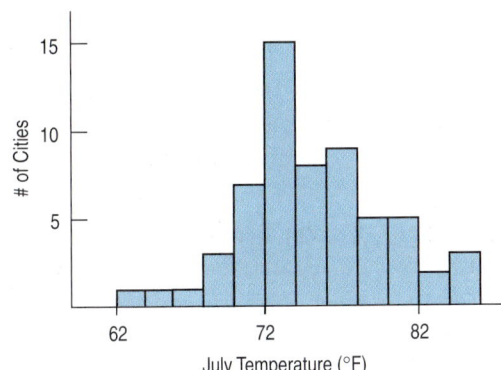

a) What aspect of these histograms makes it difficult to compare the distributions?
b) What differences do you see between the distributions of January and July average temperatures?

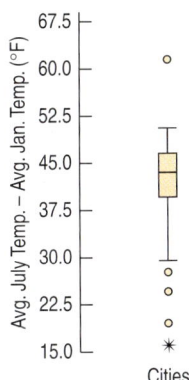

c) Differences in temperatures (July–January) for each of the cities are displayed in the boxplot above. Write a few sentences describing what you see.

18. **Old Faithful** It is a common belief that Yellowstone's most famous geyser erupts once an hour at very predictable intervals. The histogram below shows the time gaps (in minutes) between 222 successive eruptions. Describe this distribution.

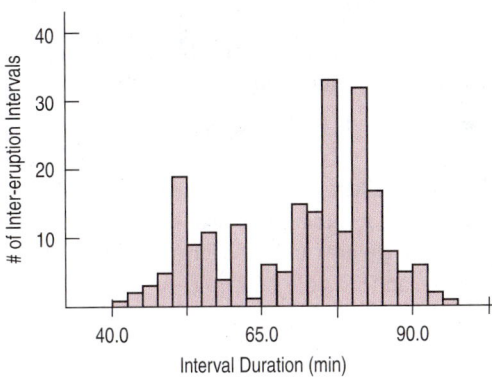

19. **Old Faithful?** Does the duration of an eruption have an effect on the length of time that elapses before the next eruption?

a) The histogram below shows the duration (in minutes) of those 222 eruptions. Describe this distribution.

b) Explain why it is not appropriate to find summary statistics for this distribution.
c) Let's classify the eruptions as "long" or "short," depending upon whether or not they last at least 3 minutes. Describe what you see in the comparative boxplots.

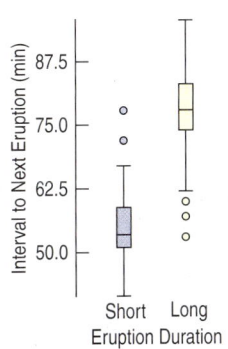

20. Teen drivers 2013 The National Highway Traffic Safety Administration reported that there were 3206 fatal accidents involving drivers between the ages of 15 and 19 years old the previous year, of which 65.5% involved male drivers. Of the male drivers, 18.4% involved drinking, while of the female drivers, 10.8% involved drinking. Assuming roughly equal numbers of male and female drivers, use these statistics to explain the concept of independence.

T 21. Liberty's nose Is the Statue of Liberty's nose too long? Her nose measures 4′6″, but she is a large statue, after all. Her arm is 42 feet long. That means her arm is $42/45 = 9.3$ times as long as her nose. Is that a reasonable ratio? Shown in the table are arm and nose lengths of 18 girls in a Statistics class, and the ratio of arm-to-nose length for each.

Arm (cm)	Nose (cm)	Arm/Nose Ratio
73.8	5.0	14.8
74.0	4.5	16.4
69.5	4.5	15.4
62.5	4.7	13.3
68.6	4.4	15.6
64.5	4.8	13.4
68.2	4.8	14.2
63.5	4.4	14.4
63.5	5.4	11.8
67.0	4.6	14.6
67.4	4.4	15.3
70.7	4.3	16.4
69.4	4.1	16.9
71.7	4.5	15.9
69.0	4.4	15.7
69.8	4.5	15.5
71.0	4.8	14.8
71.3	4.7	15.2

a) Make an appropriate plot and describe the distribution of the ratios.
b) Summarize the ratios numerically, choosing appropriate measures of center and spread.
c) Is the ratio of 9.3 for the Statue of Liberty unrealistically low? Explain.

T 22. Winter Olympics 2010 speed skating The top 36 women's 500-m speed skating times are listed in the table in the next column.
a) The mean finishing time was 40.72 seconds, with a standard deviation of 9.82 seconds. If the Normal model is appropriate, what percent of the times should be within 5 second of 40.72?
b) What percent of the times actually fall within this interval?
c) Explain the discrepancy between parts a and b.

Nation	Athlete	Result
Korea	Sang-Hwa Lee	38.249
Germany	Jenny Wolf	38.307
China	Beixing Wang	38.487
Netherlands	Margot Boer	38.511
China	Shuang Zhang	38.530
Japan	Sayuri Yoshii	38.566
Russian Federation	Yulia Nemaya	38.594
China	Peiyu Jin	38.686
United States	Heather Richardson	38.698
Germany	Monique Angermuller	38.761
China	Aihua Xing	38.792
Japan	Nao Kodaira	38.835
Canada	Christine Nesbitt	38.881
Netherlands	Thijsje Oenema	38.892
DPR Korea	Hyon-Suk Ko	38.893
Japan	Shihomi Shinya	38.964
Japan	Tomomi Okazaki	38.971
United States	Elli Ochowicz	39.002
Kazakhstan	Yekaterina Aydova	39.024
United States	Jennifer Rodriguez	39.182
Netherlands	Laurine van Riessen	39.302
Canada	Shannon Rempel	39.351
Germany	Judith Hesse	39.357
Russian Federation	Olga Fatkulina	39.359
Czech Republic	Karolina Erbanova	39.365
Korea	Bo-Ra Lee	39.396
Russian Federation	Svetlana Kaykan	39.422
Italy	Chiara Simionato	39.480
United States	Lauren Cholewinski	39.514
Korea	Jee-Min Ahn	39.595
Australia	Sophie Muir	39.649
Russian Federation	Yekaterina Malysheva	39.782
Korea	Min-Jee Oh	39.816
Canada	Anastasia Bucsis	39.879
Belarus	Svetlana Radkevich	39.899
Netherlands	Annette Gerritsen	97.952

23. Sample A study in South Africa focusing on the impact of health insurance identified 1590 children at birth and then sought to conduct follow-up health studies 5 years later. Only 416 of the original group participated in the 5-year follow-up study. This made researchers concerned that the follow-up group might not accurately resemble the total group in terms of health insurance. The following table summarizes the two groups by race and by presence of medical insurance when the child was born. Carefully explain how this study demonstrates Simpson's paradox. (Source: *Birth to Ten Study*, Medical Research Council, South Africa)

	Number (%) Insured	
Race	Follow-Up	Not Traced
Black	36 of 404 (8.9%)	91 of 1048 (8.7%)
White	10 of 12 (83.3%)	104 of 126 (82.5%)
Overall	46 of 416 (11.1%)	195 of 1174 (16.6%)

24. **Sluggers** Babe Ruth was the first great "slugger" in baseball. His record of 60 home runs in one season held for 34 years until Roger Maris hit 61 in 1961. Mark McGwire (with the aid of steroids) set a new standard of 70 in 1998. Listed below are the home run totals for each season McGwire played. Also listed are Babe Ruth's home run totals.

    McGwire: 3*, 49, 32, 33, 39, 22, 42, 9*, 9*, 39, 52, 58, 70, 65, 32*, 29*

    Ruth: 54, 59, 35, 41, 46, 25, 47, 60, 54, 46, 49, 46, 41, 34, 22

    a) Find the 5-number summary for McGwire's career.
    b) Do any of his seasons appear to be outliers? Explain.
    c) McGwire played in only 18 games at the end of his first big league season, and missed major portions of some other seasons because of injuries to his back and knees. Those seasons might not be representative of his abilities. They are marked with asterisks in the list above. Omit these values and make parallel boxplots comparing McGwire's career to Babe Ruth's.
    d) Write a few sentences comparing the two sluggers.
    e) Create a side-by-side stem-and-leaf display comparing the careers of the two players.
    f) What aspects of the distributions are apparent in the stem-and-leaf displays that did not clearly show in the boxplots?

25. **Be quick!** Avoiding an accident when driving can depend on reaction time. That time, measured from the moment the driver first sees the danger until he or she steps on the brake pedal, is thought to follow a Normal model with a mean of 1.5 seconds and a standard deviation of 0.18 second.
    a) Use the 68–95–99.7 Rule to draw the Normal model.
    b) Write a few sentences describing driver reaction times.
    c) What percent of drivers have a reaction time less than 1.25 seconds?
    d) What percent of drivers have reaction times between 1.6 and 1.8 seconds?
    e) What is the interquartile range of reaction times?
    f) Describe the reaction times of the slowest 1/3 of all drivers.

26. **Music and memory** Is it a good idea to listen to music when studying for a big test? In a study conducted by some Statistics students, 62 people were randomly assigned to listen to rap music, Mozart, or no music while attempting to memorize objects pictured on a page. They were then asked to list all the objects they could remember. Here are the 5-number summaries for each group:

	n	Min	Q1	Median	Q3	Max
Rap	29	5	8	10	12	25
Mozart	20	4	7	10	12	27
None	13	8	9.5	13	17	24

    a) Describe the W's for these data: *Who, What, Where, Why, When, How*.
    b) Name the variables and classify each as categorical or quantitative.
    c) Create parallel boxplots as best you can from these summary statistics to display these results.
    d) Write a few sentences comparing the performances of the three groups.

27. **Mail** Here are the number of pieces of mail received at a school office for 36 days.

123	70	90	151	115	97
80	78	72	100	128	130
52	103	138	66	135	76
112	92	93	143	100	88
118	118	106	110	75	60
95	131	59	115	105	85

    a) Plot these data.
    b) Find appropriate summary statistics.
    c) Write a brief description of the school's mail deliveries.
    d) What percent of the days actually lie within one standard deviation of the mean? Comment.

28. **Birth order** Is your birth order related to your choice of major? A Statistics professor at a large university polled his students to find out what their majors were and what position they held in the family birth order. The results are summarized in the table.
    a) What percent of these students are oldest or only children?
    b) What percent of Humanities majors are oldest children?
    c) What percent of oldest children are Humanities students?

d) What percent of the students are oldest children majoring in the Humanities?

	Birth Order*				
Major		1	2	3	4+	Total
Math/Science	34	14	6	3	57
Agriculture	52	27	5	9	93
Humanities	15	17	8	3	43
Other	12	11	1	6	30
Total	113	69	20	21	223

*1 = oldest or only child

29. **Herbal medicine** Researchers for the Herbal Medicine Council collected information on people's experiences with a new herbal remedy for colds. They went to a store that sold natural health products. There they asked 100 customers whether they had taken the cold remedy and, if so, to rate its effectiveness (on a scale from 1 to 10) in curing their symptoms. The Council concluded that this product was highly effective in treating the common cold.
   a) Identify the W's of these data.
   b) Identify the variables, classify each as categorical or quantitative, and specify units if relevant.
   c) Are you confident about the Council's conclusion? Explain.

30. **Birth order revisited** Consider again the data on birth order and college majors in Exercise 28.
   a) What is the marginal distribution of majors?
   b) What is the conditional distribution of majors for the oldest children?
   c) What is the conditional distribution of majors for the children born second?
   d) Do you think that college major appears to be independent of birth order? Explain.

31. **Engines** One measure of the size of an automobile engine is its "displacement," the total volume (in liters or cubic inches) of its cylinders. Summary statistics for several models of new cars are shown. These displacements were measured in cubic inches.

Summary of Displacement	
Count	38
Mean	177.29
Median	148.5
StdDev	88.88
Range	275
25th %tile	105
75th %tile	231

a) How many cars were measured?
b) Why might the mean be so much larger than the median?
c) Describe the center and spread of this distribution with appropriate statistics.
d) Your neighbor is bragging about the 227-cubic-inch engine he bought in his new car. Is that engine unusually large? Explain.
e) Are there any engines in this data set that you would consider to be outliers? Explain.
f) Is it reasonable to expect that about 68% of car engines measure between 88 and 266 cubic inches? (That's 177.289 ± 88.8767.) Explain.
g) We can convert all the data from cubic inches to cubic centimeters (cc) by multiplying by 16.4. For example, a 200-cubic-inch engine has a displacement of 3280 cc. How would such a conversion affect each of the summary statistics?

32. **Engines, again** Horsepower is another measure commonly used to describe auto engines. Here are the summary statistics and histogram displaying horsepowers of the same group of 38 cars discussed in Exercise 31.

Summary of Horsepower	
Count	38
Mean	101.7
Median	100
StdDev	26.4
Range	90
25th %tile	78
75th %tile	125

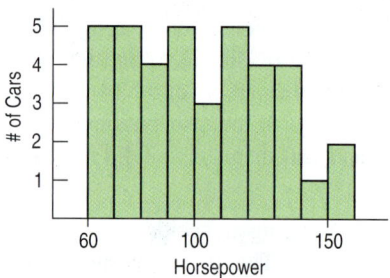

a) Describe the shape, center, and spread of this distribution.
b) What is the interquartile range?
c) Are any of these engines outliers in terms of horsepower? Explain.
d) Do you think the 68–95–99.7 Rule applies to the horsepower of auto engines? Explain.

e) From the histogram, make a rough estimate of the percentage of these engines whose horsepower is within one standard deviation of the mean.
f) A fuel additive boasts in its advertising that it can "add 10 horsepower to any car." Assuming that is true, what would happen to each of these summary statistics if this additive were used in all the cars?

33. **Age and party 2011** The Pew Research Center conducts surveys regularly asking respondents which political party they identify with or lean toward. Among their results is the following table relating preferred political party and age. (people-press.org)

	Party			
Age	Republican/ Lean Rep.	Democrat/ Lean Dem.	Neither	Total
18–29	318	424	73	815
30–49	991	1058	203	2252
50–64	1260	1407	264	2931
65 +	1136	1087	193	2416
Total	3705	3976	733	8414

a) What percent of people surveyed were Republicans or leaned Republican?
b) Do you think this might be a reasonable estimate of the percentage of all voters who are Republicans or lean Republicans? Explain.
c) What percent of people surveyed were under 30 or over 65?
d) What percent of people were classified as "Neither" and under the age of 30?
e) What percent of the people classified as "Neither" were under 30?
f) What percent of people under 30 were classified as "Neither"?

34. **Pay** According to the Bureau of Labor Statistics, the mean hourly wage for Chief Executives in 2009 was $80.43 and the median hourly wage was $77.27. By contrast, for General and Operations Managers, the mean hourly wage was $53.15 and the median was $44.55. Are these wage distributions likely to be symmetric, skewed left, or skewed right? Explain.

35. **Age and party 2011 II** Consider again the Pew Research Center results on age and political party in Exercise 33.
a) What is the marginal distribution of party affiliation?
b) Create segmented bar graphs displaying the conditional distribution of party affiliation for each age group.
c) Summarize these poll results in a few sentences that might appear in a newspaper article about party affiliation in the United States.
d) Do you think party affiliation is independent of the voter's age? Explain.

36. **Bike safety 2012** The Bicycle Helmet Safety Institute website includes a report on the number of bicycle fatalities per year in the United States. The table below shows the counts for the years 1994–2012.

Year	Bicycle Fatalities
1994	802
1995	833
1996	765
1997	814
1998	760
1999	754
2000	693
2001	732
2002	665
2003	629
2004	727
2005	784
2006	769
2007	699
2008	716
2009	628
2010	616
2011	675
2012	726

a) What are the W's for these data?
b) Display the data in a stem-and-leaf display.
c) Display the data in a timeplot.
d) What is apparent in the stem-and-leaf display that is hard to see in the timeplot?
e) What is apparent in the timeplot that is hard to see in the stem-and-leaf display?
f) Write a few sentences about bicycle fatalities in the United States.

37. **Some assembly required** A company that markets build-it-yourself furniture sells a computer desk that is advertised with the claim "less than an hour to assemble." However, through postpurchase surveys the company has learned that only 25% of its customers succeeded in building the desk in under an hour. The mean time was 1.29 hours. The company assumes that consumer assembly time follows a Normal model.
a) Find the standard deviation of the assembly time model.

b) One way the company could solve this problem would be to change the advertising claim. What assembly time should the company quote in order that 60% of customers succeed in finishing the desk by then?
c) Wishing to maintain the "less than an hour" claim, the company hopes that revising the instructions and labeling the parts more clearly can improve the 1-hour success rate to 60%. If the standard deviation stays the same, what new lower mean time does the company need to achieve?
d) Months later, another postpurchase survey shows that new instructions and part labeling did lower the mean assembly time, but only to 55 minutes. Nonetheless, the company did achieve the 60%-in-an-hour goal, too. How was that possible?

T 38. Profits Here is a stem-and-leaf display showing profits as a percent of sales for 29 of the *Forbes* 500 largest U.S. corporations. The stems are split; each stem represents a span of 5%, from a loss of 9% to a profit of 25%.

```
-0 | 99
-0 | 1234
 0 | 111123444
 0 | 5555679
 1 | 00113
 1 |
 2 | 2
 2 | 5
```
Profits (% of sales)
(−0|3 means a loss of 3%)

a) Find the 5-number summary.
b) Draw a boxplot for these data.
c) Find the mean and standard deviation.
d) Describe the distribution of profits for these corporations.

chapter 6
Scatterplots, Association, and Correlation

6.1 Scatterplots
6.2 Correlation
6.3 Warning: Correlation ≠ Causation
*6.4 Straightening Scatterplots

Where are we going?

Is the price of sneakers related to how long they last? Is your alertness in class related to how much (or little) sleep you got the night before?

In this chapter, we'll look at relationships between two quantitative variables. We'll start by looking at scatterplots and describing the essence of what we see—the direction, form, and strength of the association. Then, as we did for histograms, we'll find a quantitative summary of what we learned from the display. We'll use the correlation to measure the strength of the association we see in the scatterplot.

Who	Years 1970–2012
What	Year and Mean error in the position of Atlantic hurricanes as predicted 72 hours ahead by the NHC
Units	Years and Nautical miles
When	1970–2012
Where	Atlantic Ocean, the Gulf of Mexico and the Carribean
Why	NHC wants to improve prediction models

The typhoon that hit the Philippines in late 2013 was the strongest cyclone ever recorded both in terms of wind speed at landfall and low pressure. It devastated the coast of that island country and killed over 6000 people. As of early 2014, there are still bodies being found. In the Atlantic Ocean, such storms are known as hurricanes. In 2011, Hurricane Irene was smaller and less powerful, but it still caused 7 billion dollars in damage and killed 33 people. More than 2.5 million people had to evacuate and 4 million lost electricity. Irene's impact was high for a smaller storm because of its track, along the heavily populated East Coast, directly over New York City, and into New England.

Where will a hurricane go? The impact of a hurricane depends on both its path and strength, so the National Hurricane Center (NHC) of the National Oceanic and Atmospheric Administration (NOAA) tries to predict the path each hurricane will take. But hurricanes tend to wander around aimlessly and are pushed by fronts and other weather phenomena in their area, so they are notoriously difficult to predict. Even relatively small changes in a hurricane's track can make big differences in the damage it causes.

To improve hurricane prediction, NOAA[1] relies on sophisticated computer models and has been working for decades to improve them. How well are they doing? Have predictions improved in recent years? Has the improvement been consistent? Here's a timeplot of the mean error in nautical miles of the NHC's 72-hour predictions of Atlantic hurricanes since 1970. NOAA refers to these as the Forecast error or the Prediction error and reports annual results.

---
[1] www.nhc.noaa.gov (Data in **Tracking Hurricanes 2012**)

PART II Exploring Relationships Between Variables

Figure 6.1
A scatterplot of the average tracking error in nautical miles of the predicted position of Atlantic hurricanes for predictions made by the National Hurricane Center of NOAA, plotted against the *Year* in which the predictions were made.

> **WHERE DID THE ORIGIN GO?**
> Scatterplots usually don't—and shouldn't—show the origin, because often neither variable has values near 0. The display should focus on the part of the coordinate plane that actually contains the data. In our example about hurricanes, none of the prediction errors or years were anywhere near 0, so the computer drew the scatterplot with axes that don't meet.

A S *Activity:* **Heights of Husbands and Wives.** Husbands are usually taller than their wives. Or are they?

Clearly, predictions have improved. The plot shows a fairly steady decline in the average tracking error, from almost 500 nautical miles in the late 1970s to less than 100 nautical miles in 2012. We can also see a few years when predictions were unusually good and that 1972 was a really bad year for predicting hurricane tracks.

This timeplot is an example of a more general kind of display called a **scatterplot**. Scatterplots may be the most common displays for data. By just looking at them, you can see patterns, trends, relationships, and even the occasional extraordinary value sitting apart from the others. As the great philosopher Yogi Berra[2] once said, "You can observe a lot just by watching."[3] Scatterplots are the best way to start observing the relationship between two *quantitative* variables.

Relationships between variables are often at the heart of what we'd like to learn from data:

- Are grades higher now than they used to be?
- Do people tend to reach puberty at a younger age than in previous generations?
- Does applying magnets to parts of the body relieve pain? If so, are stronger magnets more effective?
- Do students learn better when they get more sleep?

Questions such as these relate two quantitative variables and ask whether there is an **association** between them. Scatterplots are the ideal way to *picture* such associations.

6.1 Scatterplots

A S *Activity:* **Making and Understanding Scatterplots.** See the best way to make scatterplots—using a computer.

How would you describe the association of hurricane *Prediction Error* and *Year*? Everyone looks at scatterplots. But, if asked, many people would find it hard to say what to look for in a scatterplot. What do *you* see? Try to describe the scatterplot of *Prediction Error* against *Year*.

You might say that the **direction** of the association is important. Over time, the NHC's prediction errors have decreased. A pattern like this that runs from the upper left to the lower right is said to be **negative**. A pattern running the other way is called **positive**.

> Look for **Direction**: What's my sign—positive, negative, or neither?

The second thing to look for in a scatterplot is its **form**. A plot that appears as a cloud or swarm of points stretched out in a generally consistent, straight form is called **linear**. For example, the scatterplot of *Prediction Error* vs. *Year* has such an underlying linear form, although some points stray away from it.

---
[2]Hall of Fame catcher, outfielder, and manager of the New York Mets and Yankees.
[3]But then he also said "I really didn't say everything I said." So we can't really be sure.

> Look for **Form**: straight, curved, something exotic, or no pattern at all?

If the relationship isn't straight, but curves gently while still increasing or decreasing steadily, we can often find ways to make it more nearly straight. But if it curves sharply up and down, for example like this: , there is much less we can say about it with the methods of this book.

The third feature to look for in a scatterplot is the **strength** of the relationship. At one extreme, do the points appear tightly clustered in a single stream (whether straight, curved, or bending all over the place)? Or, at the other extreme, does the swarm of points seem to form a vague cloud through which we can barely discern any trend or pattern?

> Look for **Strength**: how much scatter?

The *Prediction error* vs. *Year* plot shows moderate scatter around a generally straight form. This indicates that the linear trend of improving prediction is pretty consistent and moderately strong.

Finally, always look for the unexpected. Often the most interesting thing to see in a scatterplot is something you never thought to look for. One example of such a surprise is an **outlier** standing away from the overall pattern of the scatterplot. Such a point is almost always interesting and always deserves special attention. In the scatterplot of prediction errors, the year 1972 stands out as a year with very high prediction errors. An Internet search shows that it was a relatively quiet hurricane season. However, it included the very unusual—and deadly—Hurricane Agnes, which combined with another low-pressure center to ravage the northeastern United States, killing 122 and causing 1.3 billion 1972 dollars in damage. Possibly, Agnes was also unusually difficult to predict.

> Look for **Unusual Features**: are there outliers or subgroups?

You should also look for clusters or subgroups that stand away from the rest of the plot or that show a trend in a different direction. Deviating groups should raise questions about why they are different. They may be a clue that you should split the data into subgroups instead of looking at them all together.

For Example COMPARING PRICES WORLDWIDE

If you travel overseas, you know that what's really important is not the amount in your wallet but the amount it can buy. UBS (one of the largest banks in the world) prepared a report comparing prices, wages, and other economic conditions in cities around the world for their international clients. Some of the variables they measured in 73 cities are *Cost of Living, Food Costs, Average Hourly Wage*, average number of *Working Hours* per Year, average number of *Vacation Days*, hours of work (at the average wage) needed to buy an *iPhone*, minutes of work needed to buy a *Big Mac*, and *Women's Clothing Costs*.[4] For your burger fix, you might want to live in Tokyo where it takes only about 9 minutes of work to afford a Big Mac. In Nairobi, you'd have to work almost an hour and a half.

Of course, these variables are associated, but do they consistently reflect costs of living? Plotting pairs of variables can reveal how and even if they are associated. The variety of these associations illustrate different directions and kinds of association patterns you might see in other scatterplots.

---

[4]Detail of the methodology can be found in the report *Prices and Earning: A comparison of purchasing power around the globe/2012 edition*. www.economist.com/node/14288808?story_id=14288808 (Data in **Prices and Earnings 2012**)

QUESTION: Describe the patterns shown by each of these plots.

ANSWER: In Plot A, the association between *Food Costs* and *Cost of Women's Clothes* is positive and straight with a few high outliers, and moderately strong. In Plot B, the association between *Hours to Earn an iPhone4S* and *Average Hourly Wage* is negative and strong, but the form is not straight. In Plot C, the association between the *Clothes Index* and *Working Hours* is weak, negative, and generally straight with one or two possible high outliers. In Plot D, there does not appear to be any association between *Food Costs* and *Vacation Days*.

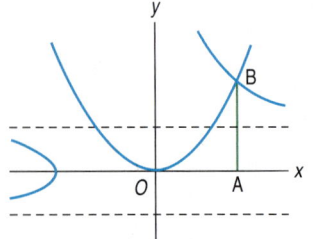

A figure from Newton's *Enumeration of lines of the third order*[5] labels the axes in the modern way.

> **Scatterplot history** The idea of using two axes at right angles to define a field on which to display values can be traced back to the philosopher, scientist, and mathematician René Descartes (1596–1650). The playing field he defined in this way is formally called a Cartesian plane in his honor. Sir Isaac Newton (1642–1727) may have been the first to use the now-standard convention of denoting the vertical axis y and the horizontal axis x.
>
> Both Descartes and Newton used the Cartesian plane to plot functions, as you have probably done yourself. Plotting individual data values works in a similar way. Each case is displayed as a dot or symbol on a scatterplot at a position that corresponds to its values on two variables—one measured along the x-axis and the other along the y-axis. When you work with data you should label the axes with the names of the variables rather than just using y and x.

Roles for Variables

Which variable should go on the x-axis and which on the y-axis? What we want to know about the relationship can tell us how to make the plot. We often have questions such as

- Do baseball teams that score more runs sell more tickets to their games?
- Do older houses sell for less than newer ones of comparable size and quality?
- Do students who score higher on their SAT tests have higher grade point averages in college?
- Can we estimate a person's percent body fat accurately by measuring their waist or wrist size?

---

[5] Sir Isaac Newton's *Enumeration of lines of the third order, generation of curves by shadows, organic description of curves, and construction of equations by curves*. p 86. Translated from the Latin. With notes and examples, by C.R.M. Talbot. Available at quod.lib.umich.edu/u/umhistmath/ABQ9451.0001.001/81?rgn=full+text;view=pdf

CHAPTER 6 Scatterplots, Association, and Correlation

■ **NOTATION ALERT**

In Statistics, the assignment of variables to the *x*- and *y*-axes (and the choice of notation for them in formulas) often conveys information about their roles as predictor or response variable. So *x* and *y* are reserved letters as well, but not just for labeling the axes of a scatterplot.

In these examples, the two variables play different roles. We'll call the variable of interest the **response variable** and the other the **explanatory** or **predictor variable**.[6] We'll continue our practice of naming the variable of interest *y*. Naturally we'll plot it on the *y*-axis and place the explanatory variable on the *x*-axis. Sometimes, we'll call them the ***x*- and *y*-variables**. For example, we wanted to know how accurate hurricane prediction is given the year, so prediction error was on the *y*-axis and year on the *x*-axis. When you make a scatterplot, you can assume that those who view it will think this way, so choose which variables to assign to which axes carefully.

The roles that we choose for variables are more about how we *think* about them than about the variables themselves. Just placing a variable on the *x*-axis doesn't necessarily

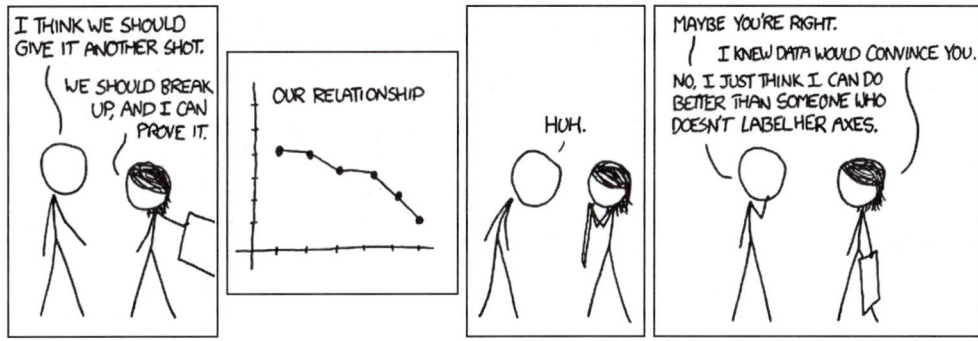

© 2013 Randall Munroe. Reprinted with permission. All rights reserved.

A/S *Self-Test:* Scatterplot Check.
Can you identify a scatterplot's direction, form, and strength?

mean that it explains or predicts *anything*. And the variable on the *y*-axis may not respond to it in any way. We plotted prediction error on the *y*-axis against year on the *x*-axis because the National Hurricane Center is interested in how their predictions have changed over time. Could we have plotted them the other way? In this case, it's hard to imagine reversing the roles—knowing the prediction error and wanting to guess in what year it happened. But for some scatterplots, it can make sense to use either choice, so you have to think about how the choice of role helps to answer the question you have.

6.2 Correlation

Data collected from students in Statistics classes included their *Height* (in inches) and *Weight* (in pounds). It's no great surprise that there is a positive association between the two. As you might suspect, taller students trend to weigh more. (If we had reversed the roles and chosen height as the explanatory variable, we might say that heavier students trend to be taller.)[7] And the form of the scatterplot is fairly straight as well, although there appears to be a high outlier.

Figure 6.2
Weight (lbs) versus *Height* (in) from a statistics class. (Data in **Heights and Weights**)

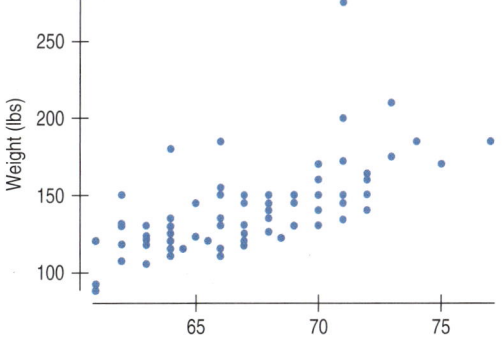

Who	Students
What	Height (inches), weight (pounds)
Where	Ithaca, NY
Why	Data for class
How	Survey

[6]The *x*- and *y*-variables have sometimes been referred to as the *independent* and *dependent* variables, respectively. The idea was that the *y*-variable depended on the *x*-variable and the *x*-variable acted independently to make *y* respond. These names, however, conflict with other uses of the same terms in Statistics.

[7]The young son of one of the authors, when told (as he often was) that he was tall for his age, used to point out that, actually, he was young for his height.

The pattern in the scatterplots looks straight, and the positive association is clear, but how strong is that association? If you had to put a number (say, between 0 and 1) on the strength, what would you say? The strength of the association shouldn't depend on the units we choose for *x* or *y*. For example, if we had measured weight in *kg* instead of *lbs* and height in *cm* instead of *inches*, the plot (Figure 6.3) would look the same.

Figure 6.3
Changing the units (from pounds to kilograms for *Weight* and from inches to centimeters for *Height*) did not change the direction, form, or strength of the association between the variables.

If we subtract the mean from each variable, that shouldn't change the strength of the association either. We would just move the means of both variables to zero. (See Figure 6.4.) This makes it easier to see what each point contributes to the strength of the association because we can tell which points are above and which are below the means of each variable. Figure 6.4 colors them to make this even clearer. The green points in the upper right (first quadrant) and lower left (third quadrant) are consistent with a positive association while the red points in the other two quadrants are consistent with a negative association. Points lying on the *x*- or *y*-axis don't really add any information; they are colored blue.

A S **Activity: Correlation.** Here's a good example of how correlation works to summarize the strength of a linear relationship and disregard scaling.

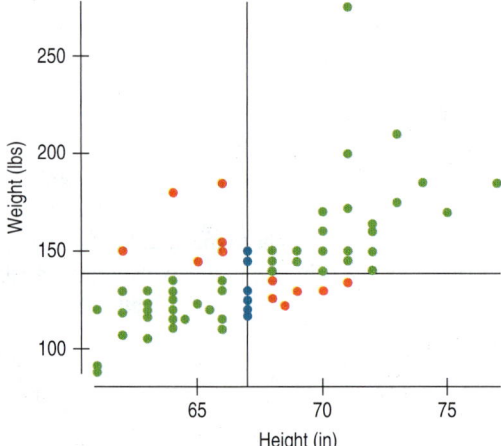

Figure 6.4
In this scatterplot, points are colored according to how they affect the association: green for positive, red for negative, and blue for neutral.

We said that the units shouldn't affect our measure of strength. Now that we've subtracted means, a natural way to remove the units is to standardize each variable and work instead with the *z*-scores. Recall that to find *z*-scores for any variable, we subtract its mean from each value and divide by its standard deviation. So instead of plotting (x, y), we plot:

$$(z_x, z_y) = \left(\frac{x - \bar{x}}{s_x}, \frac{y - \bar{y}}{s_y} \right).$$

z-Scores do just what we want. They center the plotted points around the origin and they remove the original units, replacing them with standard deviations. As we've seen, measuring differences in standard deviations is a fundamental idea in Statistics.

Figure 6.5
Standardized *Weight* and *Height*. We use the standard deviation to judge the distance of each point from the center of the plot and the influence it should have.

For the green points in Figure 6.5, both z-scores are positive, or both are negative. In either case, their product, $z_x z_y$, is positive. But the red points have opposite signs, so their products are negative. We can measure the strength of the association by adding up these products: $\sum z_x z_y$. That summarizes both the direction and strength of the association. Points farther from the origin have larger z-scores, so they'll contribute more to the sum. But the sum can keep growing as we consider more points. To keep that from happening, the natural (for Statisticians, anyway) thing to do is to divide the sum by $n - 1$.[8]

The result is the famous **correlation coefficient**:

$$r = \frac{\sum z_x z_y}{n - 1}.$$

Dividing the sum by $n - 1$ serves two purposes. It adjusts the strength for the number of points and it makes the correlation lie between values of -1 and $+1$. For the students' heights and weights, the correlation is 0.644. Because it is based on z-scores, which have no units, the correlation has no units either. It will stay the same if you change from inches to centimeters, fathoms, or Ångstroms.

There are a number of alternative formulas for the correlation coefficient, using x and y in their original units, that you may encounter. Here are two of the most common:

$$r = \frac{\sum (x - \bar{x})(y - \bar{y})}{\sqrt{\sum (x - \bar{x})^2 (y - \bar{y})^2}} = \frac{\sum (x - \bar{x})(y - \bar{y})}{(n - 1)s_x s_y}.$$

These formulas are convenient for computing correlation by hand, but the z-score form is best for understanding what correlation means.

■ **NOTATION ALERT**

The letter r is always used for correlation, so you can't use it for anything else in Statistics. Whenever you see an r, it's safe to assume it's a correlation.

By Hand FINDING THE CORRELATION COEFFICIENT

To find the correlation coefficient by hand, start with the summary statistics for both variables: $\bar{x}, \bar{y}, s_x,$ and s_y. For each data pair, find the deviations from the means for both variables: $(x - \bar{x})$ and $(y - \bar{y})$, and multiply these deviations together: $(x - \bar{x})(y - \bar{y})$. Add up these products for all data pairs.

To get the correlation, r, divide this sum by the product $(n - 1)s_x s_y$.

Here's an example:

Suppose the data pairs are

x	6	10	14	19	21
y	5	3	7	8	12

---

[8] Yes, the same $n - 1$ as in the standard deviation calculation. And we offer the same promise to explain it later.

Then $\bar{x} = 14, \bar{y} = 7, s_x = 6.20$, and $s_y = 3.39$

Deviations from $\bar{x}$	Deviations from $\bar{y}$	Product $(x - \bar{x})(y - \bar{y})$
6 − 14 = −8	5 − 7 = −2	−8 × −2 = 16
10 − 14 = −4	3 − 7 = −4	16
14 − 14 = 0	7 − 7 = 0	0
19 − 14 = 5	8 − 7 = 1	5
21 − 14 = 7	12 − 7 = 5	35

Add the products: 16 + 16 + 0 + 5 + 35 = 72
Finally, divide by $(n − 1) \times s_x \times s_y = (5 − 1) \times 6.20 \times 3.39 = 84.07$
The ratio is the correlation coefficient:

$$r = 72/84.07 = 0.856$$

Assumptions and Conditions for Correlation

Correlation measures the strength of *linear* association between two *quantitative* variables. To interpret a correlation, we must assume that there is a true underlying linear relationship. Of course, in general, we can't know that. But we *can* check whether that assumption is plausible by looking at the data we do have. To distinguish between what we *assume* about variables from what we can check by looking at the data, we'll call the things we check *conditions*. Checking the conditions almost always means either thinking about how the data were collected or looking at a graph—but if you follow the first three rules, you'll have already made the plot, so that takes no extra effort.

There are three conditions to check before you use a correlation:

- **Quantitative Variables Condition** Don't make the common error of calling an association involving a categorical variable a correlation. Correlation is only about quantitative variables.
- **Straight Enough Condition** The best check for the assumption that the variables are truly linearly related is to look at the scatterplot to see whether it looks reasonably straight. That's a judgment call, but not a difficult one.
- **No Outliers Condition** Outliers can distort the correlation dramatically, making a weak association look strong or a strong one look weak. Outliers can even change the sign of the correlation. But it's easy to see outliers in the scatterplot, so to check this condition, just look.

Correlations are often reported without supporting data or plots. Nevertheless, you can still think about the conditions. Be cautious in interpreting (or accepting others' interpretations of) a correlation when you can't check the conditions.

For Example CORRELATIONS FOR SCATTERPLOT PATTERNS

Look back at the scatterplots of the economic variables in cities around the world (p. 152). The correlations for those plots are (a to d) 0.614, −0.791, −0.388, and −0.040, respectively.

QUESTION: Check the conditions for using correlation. If you feel they are satisfied, interpret the correlation.

A S *Activity:* **Correlation and Relationship Strength.** What does a correlation of 0.8 look like? How about 0.3?

> **ANSWER:** All of the variables examined are quantitative and none of the plots shows an outlier. However, the relationship between *Hours to Earn an iPhone* and *Average Wage* is not straight, so the correlation coefficient isn't an appropriate summary. For the others:
>
> A correlation of 0.614 between *Food Costs* and *Women's Clothing Costs* indicates a moderately strong positive association.
>
> A correlation of −0.388 between *Clothes Index* and *Working Hours* indicates a moderately weak negative association.
>
> The small correlation value of −0.040 between *Food Costs* and *Vacation Days* suggests that there may be no linear association between them.

Just Checking

Your Statistics instructor tells you that the correlation between the scores (points out of 50) on Exam 1 and Exam 2 was 0.75.

1. Before answering any questions about the correlation, what would you like to see? Why?
2. If she adds 10 points to each Exam 1 score, how will this change the correlation?
3. If she standardizes scores on each exam, how will this affect the correlation?
4. In general, if someone did poorly on Exam 1, are they likely to have done poorly or well on Exam 2? Explain.
5. If someone did poorly on Exam 1, can you be sure that they did poorly on Exam 2 as well? Explain.

Step-by-Step Example LOOKING AT ASSOCIATION

When blood pressure is measured, it is reported as two values: systolic blood pressure and diastolic blood pressure.

Questions: How are these variables related to each other? Do they tend to be both high or both low? How strongly associated are they?

THINK ➡ Plan State what you are trying to investigate.

Variables Identify the two quantitative variables whose relationship you wish to examine. Report the W's, and be sure both variables are recorded for the same individuals.

I'll examine the relationship between two measures of blood pressure. (Data in **Framingham**)

The variables are systolic and diastolic blood pressure (*SBP* and *DBP*), recorded in millimeters of mercury (mm Hg) for each of 1406 participants in the Framingham Heart Study, a long-running health study in Framingham, Massachusetts.[9]

---

[9] www.nhlbi.nih.gov/about/framingham

Plot Make the scatterplot. Use a computer program or graphing calculator if possible.	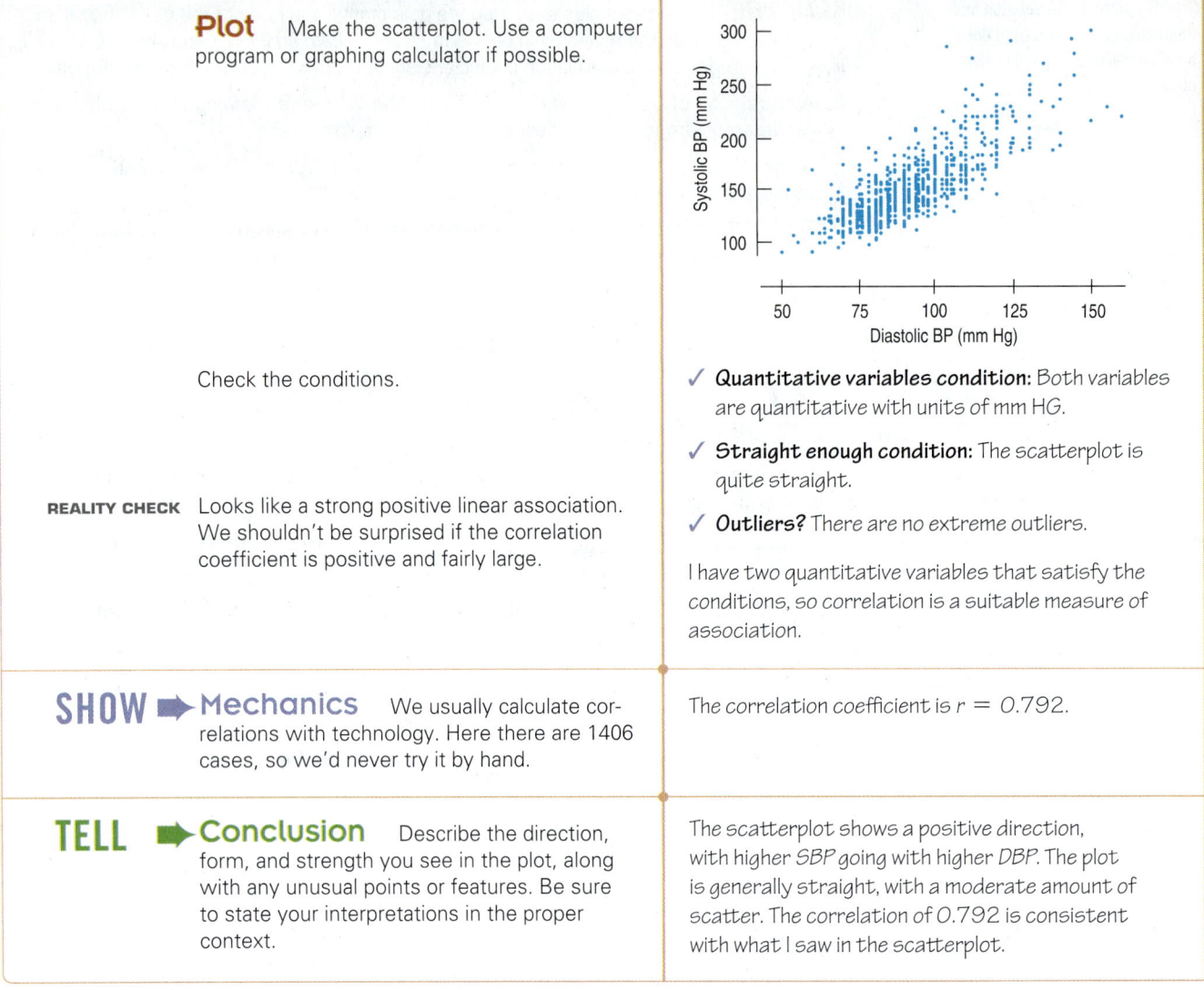
Check the conditions.	✓ **Quantitative variables condition:** Both variables are quantitative with units of mm HG. ✓ **Straight enough condition:** The scatterplot is quite straight. ✓ **Outliers?** There are no extreme outliers. I have two quantitative variables that satisfy the conditions, so correlation is a suitable measure of association.
REALITY CHECK Looks like a strong positive linear association. We shouldn't be surprised if the correlation coefficient is positive and fairly large.	
SHOW ➡ **Mechanics** We usually calculate correlations with technology. Here there are 1406 cases, so we'd never try it by hand.	The correlation coefficient is $r = 0.792$.
TELL ➡ **Conclusion** Describe the direction, form, and strength you see in the plot, along with any unusual points or features. Be sure to state your interpretations in the proper context.	The scatterplot shows a positive direction, with higher SBP going with higher DBP. The plot is generally straight, with a moderate amount of scatter. The correlation of 0.792 is consistent with what I saw in the scatterplot.

A S *Activity:* **Construct Scatterplots with a Given Correlation.** Try to make a scatterplot that has a given correlation. How close can you get?

Correlation Properties

Here's a useful list of facts about the correlation coefficient:

- The sign of a correlation coefficient gives the direction of the association.
- Correlation is always between -1 and $+1$. Correlation *can* be exactly equal to -1.0 or $+1.0$, but these values are unusual in real data because they mean that all the data points fall *exactly* on a single straight line.
- Correlation treats x and y symmetrically. The correlation of x with y is the same as the correlation of y with x.
- Correlation has no units. This fact can be especially appropriate when the data's units are somewhat vague to begin with (IQ score, personality index, socialization, and so on). Correlation is sometimes given as a percentage, but you probably shouldn't do that because it suggests a percentage of *something*—and correlation, lacking units, has no "something" of which to be a percentage.
- Correlation is not affected by changes in the center or scale of either variable. Changing the units or baseline of either variable has no effect on the correlation coefficient. Correlation depends only on the z-scores, and they are unaffected by changes in center or scale.

- Correlation measures the strength of the *linear* association between the two variables. Variables can be strongly associated but still have a small correlation if the association isn't linear.
- Correlation is sensitive to outliers. A single outlying value can make a small correlation large or make a large one small.

> **How strong is strong?** You'll often see correlations characterized as "weak," "moderate," or "strong," but be careful. There's no agreement on what those terms mean. The same numerical correlation might be strong in one context and weak in another. You might be thrilled to discover a correlation of 0.7 between the new summary of the economy you've come up with and stock market prices, but you'd consider it a design failure if you found a correlation of "only" 0.7 between two tests intended to measure the same skill. Deliberately vague terms like "weak," "moderate," or "strong" that describe a linear association can be useful additions to the numerical summary that correlation provides. But be sure to include the correlation and show a scatterplot so others can judge for themselves.

Sir Maurice Kendall (1907–1983) was a prominent statistician, writer, and teacher. He made significant contributions to the study of randomness (developing tests for randomness that are still used), the theory of statistics, and rank-based statistics. He also served as director of the World Fertility Study, for which the United Nations awarded him their Peace Medal.

For Example CHANGING SCALES

RECAP: Several measures of prices and wages in cities around the world show a variety of relationships, some of which we can summarize with correlations.

QUESTION: Suppose that, instead of measuring prices in U.S. dollars and recording work time in hours, we had used Euros and minutes. How would those changes affect the conditions, the values of correlation, or our interpretation of the relationships involving those variables?

ANSWER: Not at all. Correlation is based on standardized values (*z*-scores), so the conditions, value of *r*, and interpretation are all unaffected by changes in units.

PRONUNCIATION NOTE:
Many people pronounce Likert's name with a long "i" as in "like." But he actually pronounced it with a short "i" as in "lick."

■ NOTATION ALERT
The Greek letter tau is most often used for Kendall's tau, but it is sometimes used for other things in Statistics. The context is usually clear.

*Measuring Trend: Kendall's Tau

Survey researchers often ask people to rank their opinions or attitudes on a scale that lacks units. (Common scales run from 1 to 5, 1 to 7, or 1 to 10.) For example, we might ask:

"How would you assess the pace of your Statistics course so far?"

1 = *Way too slow*   2 = *A little too slow*   3 = *About right*
4 = *A little too fast*   5 = *Way too fast*

Scales of this sort that attempt to measure attitudes numerically are called Likert scales, after their developer, Rensis Likert. Likert scales have order. The higher the number, the faster you think the course has gone, but a 4 doesn't necessarily mean *twice* as fast as 2. So we may want to be careful when applying methods such as correlation that require the Straight Enough Condition. That may be too strong a requirement when all we may want to ask of the data is whether there is a consistent trend.

Suppose we collected data from lots of courses and tried to relate perceived pace to the number of students in the courses. We might wonder if students perceive the course as being faster in larger courses, but it isn't necessarily clear what it would mean for *pace* to be related to *class size* in a linear way. So the correlation coefficient might not be the appropriate measure. Still, we might want to know if the speed of the course and its size are associated. To answer this question, we can use an alternative measure of association, Kendall's tau.

A MONOTONOUS RELATIONSHIP?
You probably think of monotonous as *boring*. A monotone relationship shares the same root, but it's only the *sameness* that's important. A monotone relationship is one that increases or decreases consistently (but it doesn't have to be boring).

Figure 6.7
For each pair of points, Kendall's tau records whether the slope between them is positive (red), negative (blue), or zero.

Kendall's tau is a statistic designed to assess how close the relationship between two variables is to being *monotone*. A monotone relationship is one that consistently increases or decreases, but not necessarily in a linear fashion. That is often all we are interested in. And, of course, if we only care about a consistent trend, we don't have to worry that one or both of our variables might not be on a perfectly quantitative scale.

Kendall's tau measures monotonicity directly. For each pair of points in a scatterplot, it records only whether the slope of a line between those two points is positive, negative, or zero. (If the points have the same *x*-value, the slope between them is ignored.) In a monotone plot, these pairwise slopes would all be positive or all be negative. In a non-monotone plot, the counts of positive and negative slopes will tend to balance out. Tau (often written with the Greek letter τ) is the difference between the number of positive slopes and the number of negative slopes divided by the total number of slopes between pairs.

Obviously, tau can take values between -1.0 and $+1.0$. If every pairwise slope is negative, then tau will equal -1.0, and the plot will show a single stream of points trending consistently downward. If every pairwise slope is positive, then tau will equal $+1.0$, and the plot will show a single stream of points trending consistently upward. In a generally horizontal plot or one that goes up and then down, the counts of positive and negative slopes will balance out and tau will be near zero.

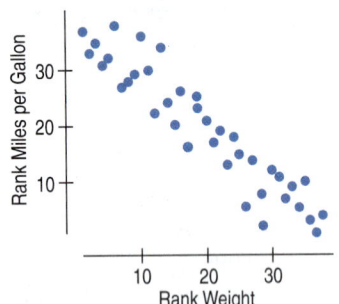

Figure 6.8
Spearman's rho finds the correlation of the *ranks* of the two variables.

*Nonparametric Association: Spearman's Rho

One of the problems we have seen with the correlation coefficient is that it is very sensitive to violations of the Straight Enough Condition. Both outliers and bends in the data make it impossible to interpret correlation. **Spearman's rho** (often denoted by the Greek letter ρ) can deal with both of these problems. Rho replaces the original data values with their *ranks* within each variable. That is, it replaces the lowest value in *x* by the number 1, the next lowest value by 2, and so on, until the highest value is assigned the value *n*. The same ranking method is applied to the *y*-variable. A scatterplot of the *y*-ranks against the *x*-ranks shows the same general *trend* as the original data, going up whenever the data went up and down whenever the data went down. If the original scatterplot shows a consistent but bent trend, however, the scatterplot of the ranks is likely to be more nearly linear. And if either variable has an extreme outlying value, the ranking process keeps it from being extreme by just counting it as the highest or lowest value, ignoring how extreme it really is.

Spearman's rho is the correlation of the two rank variables. Because this is a correlation coefficient, it must be between -1.0 and 1.0.

Both Kendall's tau and Spearman's rho have advantages over the correlation coefficient *r*. For one, they can be used even when we know only the ranks. Also, they measure the consistency of the trend between the variables without insisting that the trend be linear.

■ NOTATION ALERT
The Greek letter rho has some other uses in Statistics, but the context usually makes the meaning clear.

Charles Edward Spearman, FRS (1863–1945) was an English psychologist known for work in Statistics, as a pioneer of factor analysis, and for Spearman's rank correlation coefficient. He also did fundamental work on measures of human intelligence.

They have the added advantage that they are not much affected by outliers. Spearman's rho limits the outlier to the value of its rank, and Kendall's tau cares only about the sign of each slope between points, not how *steep* it might be. Neither tau nor rho is changed at all by re-expressing variables by functions that don't alter the order of values (that is, by just the sort of functions we might use to straighten a scatterplot). But, unlike the correlation coefficient, neither statistic can be used as a base for more advanced or complex methods, so they tend to be specialized methods used when we care primarily about consistent trend between two variables.

Both of these measures are examples of what are called *nonparametric* or *distribution-free* methods. We first discussed parameters in Chapter 5. They are the constants in a model. The correlation coefficient attempts to estimate a particular parameter in the Normal model for two quantitative variables. Kendall's tau and Spearman's rho are less specific. They measure association, but there is no parameter that they are tied to and no specific model they require.

We'll run into more examples of nonparametric methods throughout the book. Because they tend to be somewhat specialized methods, we'll continue to put them in starred sections in the chapters where they appear.

6.3 Warning: Correlation ≠ Causation

Whenever we have a strong correlation, it's tempting to try to explain it by imagining that the predictor variable has *caused* the response to change. Humans are like that; we tend to see causes and effects in everything.

Sometimes this tendency can be amusing. A scatterplot of the human population (y) of Oldenburg, Germany, in the beginning of the 1930s plotted against the number of storks nesting in the town (x) shows a tempting pattern.

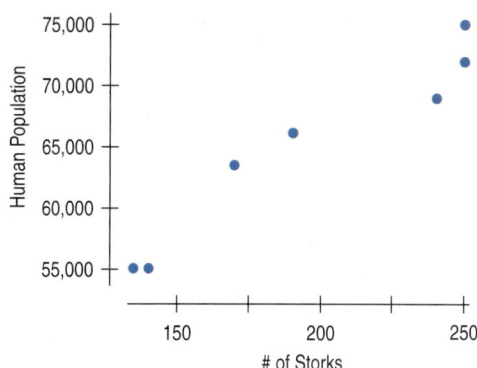

Figure 6.9
The number of storks in Oldenburg, Germany, plotted against the population of the town for 7 years in the 1930s. The association is clear. How about the causation?
(Source: *Statistics for Experimenters*, Box, Hunter and Hunter. Originally from *Ornithologische Monatsberichte*, 44, no. 2) (Data in **Storks**)

Anyone who has seen the beginning of the movie *Dumbo* remembers Mrs. Jumbo anxiously waiting for the stork to bring her new baby. Even though you know it's silly, you can't help but think for a minute that this plot shows that storks are the culprits. The two variables are obviously related to each other (the correlation is 0.97!), but that doesn't prove that storks bring babies.

It turns out that storks nest on house chimneys. More people means more babies, but also more houses, more nesting sites, and so more storks. The causation is actually in the *opposite* direction, but you can't tell from the scatterplot or correlation. You need additional information—not just the data—to determine the real mechanism.

A scatterplot of the damage (in dollars) caused to houses by fires would show a strong correlation with the number of firefighters at the scene. Surely the damage doesn't cause firefighters. And firefighters do seem to cause damage, spraying water all around and chopping holes. Does that mean we shouldn't call the fire department? Of course not. There is an underlying variable that leads to both more damage and more firefighters: the size of the blaze.

We call a hidden variable, like the fire size, that stands behind a relationship and determines it by simultaneously affecting the other two variables a **lurking variable**. You can often debunk claims made about data by finding a lurking variable behind the scenes.

However, whether you find a lurking variable behind a scatterplot or not, remember: Scatterplots and correlation coefficients *never* prove causation. That's one reason it took so long for the U.S. Surgeon General to get warning labels on cigarettes. Although there was plenty of evidence that increased smoking was *associated* with increased levels of lung cancer, it took years to provide evidence that smoking actually *causes* lung cancer.

© 2013 Randall Munroe. Reprinted with permission. All rights reserved.

> **Does cancer cause smoking?** Even if the correlation of two variables is due to a causal relationship, the correlation itself cannot tell us what causes what.
>
> Sir Ronald Aylmer Fisher (1890–1962) was one of the greatest statisticians of the 20th century. Fisher testified in court (in testimony paid for by the tobacco companies) that a causal relationship might underlie the correlation of smoking and cancer:
>
> > "Is it possible, then, that lung cancer . . . is one of the causes of smoking cigarettes? I don't think it can be excluded . . . the pre-cancerous condition is one involving a certain amount of slight chronic inflammation. . . .
> > A slight cause of irritation . . . is commonly accompanied by pulling out a cigarette, and getting a little compensation for life's minor ills in that way. And . . . is not unlikely to be associated with smoking more frequently."
>
> Ironically, the proof that smoking indeed is the cause of many cancers came from experiments conducted following the principles of experiment design and analysis that Fisher himself developed—and that we'll see in Chapter 12.

Correlation Tables

It is common in some fields to compute the correlations between every pair of variables in a collection of variables and arrange these correlations in a table. The row and column headings name the variables, and the cells hold the correlations.

Correlation tables are compact and give a lot of summary information at a glance. They can be an efficient way to start to look at a large data set, but a dangerous one. By presenting all of these correlations without any checks for linearity and outliers, the correlation table risks showing truly small correlations that have been inflated by outliers, truly large correlations that are hidden by outliers, and correlations of any size that may be meaningless because the underlying form is not linear.

The diagonal cells of a correlation table always show correlations of exactly 1. (Can you see why?) Correlation tables are commonly offered by statistics packages on computers. These same packages often offer simple ways to make all the scatterplots that go with these correlations.

	Assets	Sales	Market Value	Profits	Cash Flow	Employees
Assets	1.000					
Sales	0.746	1.000				
Market Value	0.682	0.879	1.000			
Profits	0.602	0.814	0.968	1.000		
Cash Flow	0.641	0.855	0.970	0.989	1.000	
Employees	0.594	0.924	0.818	0.762	0.787	1.000

Table 6.1

A correlation table of data reported by *Forbes* magazine on several financial measures for large companies. From this table, can you be sure that the variables are linearly associated and free from outliers?

*6.4 Straightening Scatterplots

Correlation is a suitable measure of strength for straight relationships only. When a scatterplot shows a bent form that has an increasing or decreasing trend, we can often straighten the form of the plot by re-expressing one or both variables.

Some camera lenses have an adjustable aperture, the hole that lets the light in. The size of the aperture is expressed in a mysterious number called the f/stop. Each increase of one f/stop number corresponds to a halving of the light that is allowed to come through. The f/stops of one digital camera are

f/stop: 2.8 4 5.6 8 11 16 22 32.

When you halve the shutter speed, you cut down the light, so you have to open the aperture one notch. We could experiment to find the best f/stop value for each shutter speed. A table of recommended shutter speeds and f/stops for a camera lists the relationship like this:

Shutter speed:	1/1000	1/500	1/250	1/125	1/60	1/30	1/15	1/8
f/stop:	2.8	4	5.6	8	11	16	22	32

The correlation of these shutter speeds and f/stops is 0.979. That sounds pretty high. But a high correlation doesn't necessarily mean a strong linear relationship. When we check the scatterplot (we *always* check the scatterplot), it shows that something is not quite right:

Figure 6.10
A scatterplot of *f/stop* vs. *Shutter Speed* shows a bent relationship.

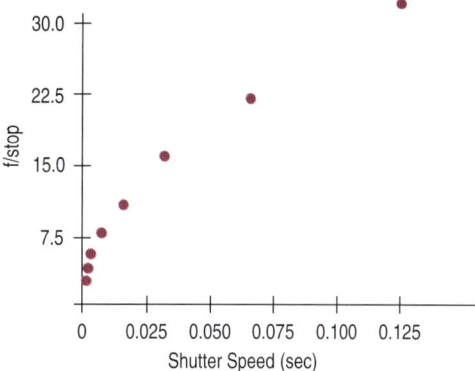

We can see that the f/stop is not *linearly* related to the shutter speed. Can we find a transformation of *f/stop* that straightens out the line?

In Chapter 4, we saw that **re-expressing** data by raising each value to a power or taking the logarithm could make the shape of the histogram of the data more nearly symmetric and could make the spreads of different groups more nearly equal. Now we hope to make a curved relationship straighter. Remarkably, these three goals are closely related, so we can use the same methods to achieve all of them.

> **Re-expressions in scientific laws** Scientific laws often include simple re-expressions. For example, in Psychology, Fechner's Law states that sensation increases as the logarithm of stimulus intensity ($S = k \log R$).

The Ladder of Powers

The secret is to select a re-expression from a simple collection of functions—the powers and the logarithm. We can raise each data value in a quantitative variable to the same power. It is important to recall a few facts about powers:

- The ½ power is the same as taking the square root.
- The -1 power is the reciprocal: $y^{-1} = 1/y$
- Putting those together, $y^{-1/2} = 1/\sqrt{y}$

In addition, we'll use the logarithm in place of the "0" power. Although it doesn't matter for this purpose, the base-10 logarithm is usually easier to think about.[10]

Where to start? It turns out that certain kinds of data are more likely to be helped by particular re-expressions. Knowing that gives you a good place to start your search for a re-expression.

Power	Name	Comment
2	The square of the data values, y^2.	Try this for unimodal distributions that are skewed to the left or to re-express y in a scatterplot that bends downward.
1	The raw data—no change at all. This is "home base." The farther you step from here up or down the ladder, the greater the effect.	Data that can take on both positive and negative values with no bounds are less likely to benefit from re-expression.
1/2	The square root of the data values, $\sqrt{y}$.	Counts often benefit from a square root re-expression. For counted data, start here.
"0"	Although mathematicians define the "0-th" power differently,[11] for us the place is held by the logarithm. You may feel uneasy about logarithms. Don't worry; the computer or calculator does the work.[12]	Measurements that cannot be negative, and especially values that grow by percentage increases such as salaries or populations, often benefit from a log re-expression. When in doubt, start here. If your data have zeros, try adding a small constant to each value before finding the logs.
−1/2	The (negative) reciprocal square root, $-1/\sqrt{y}$.	An uncommon re-expression, but sometimes useful. Changing the sign to take the *negative* of the reciprocal square root preserves the direction of relationships, making things a bit simpler.
−1	The (negative) reciprocal, $-1/y$.	Ratios of two quantities (miles per hour, for example) often benefit from a reciprocal. (You have about a 50-50 chance that the original ratio was taken in the "wrong" order for simple statistical analysis and would benefit from re-expression.) Often, the reciprocal will have simple units (hours per mile). Change the sign if you want to preserve the direction of relationships. If your data have zeros, try adding a small constant to all values before finding the reciprocal.

The powers and logarithm have an important advantage as re-expressions: They are *invertible*. That is, you can get back to the original values easily. For example, if we re-express y by the square root, we can get back to the original values by squaring the re-expressed values. To undo a logarithm, raise 10 to the power of each value.[13]

The positive powers and the logarithm preserve the order of values—the larger of two values is still the larger one after re-expression. Negative powers reverse the order ($4 > 2$ but $¼ < ½$) so we sometimes negate the values ($-¼ > -½$) to preserve the order.

Most important, the *effect* that any of these re-expressions has on straightening a scatterplot is ordered. Choose a power. If it doesn't straighten the plot by enough, then try the next power. If it overshoots so the plot bends the other way, then step back. That insight makes it easy to find an effective re-expression, and it leads to referring to these functions as the **ladder of powers**.

But don't forget—when you take a negative power, the direction of the relationship will change. That's OK. You can always change the sign of the response variable if you want to keep the same direction. With modern technology, finding a suitable re-expression is no harder than the push of a button.

---

[10]You may have learned facts about calculating with logarithms. Don't worry; you won't need them here.

[11]You may remember that for any nonzero number y, $y^0 = 1$. This is not a very exciting re-expression for data; every data value would be the same. We use the logarithm in its place.

[12]Your calculator or software package probably gives you a choice between "base 10" logarithms and "natural (base e)" logarithms. Don't worry about that. It doesn't matter at all which you use; they have exactly the same effect on the data. If you want to choose, base-10 logarithms can be a bit easier to interpret.

[13]Your computer or calculator knows how.

f/Stops Again

Following the advice in the table, we can try taking the logarithm of the f/stop values.

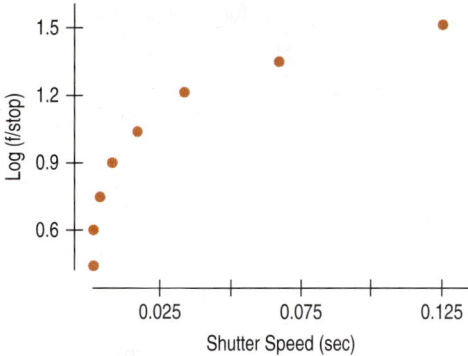

Figure 6.11
Log(f/stop) values vs. *Shutter Speed.*

But, as Figure 6.12 shows, the logarithm of f/stop values is even *more* curved vs. *Shutter Speed.* The Ladder of Powers says that we'd do better if we move in the opposite direction. That would take us to the square.

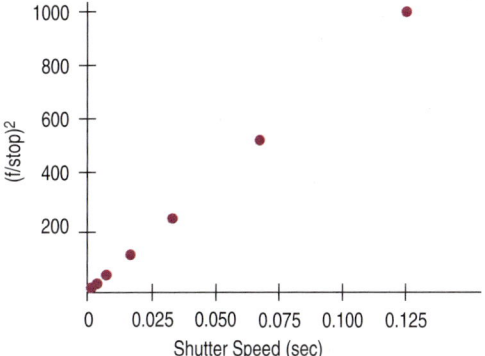

Figure 6.12
(f/stop)^2 vs. Shutter Speed.

That scatterplot looks much straighter.[14] It's appropriate to summarize the strength of this relationship with a correlation, which turns out to be 0.998.[15] You can find more discussion of finding appropriate re-expressions in Chapter 9.

WHAT CAN GO WRONG?

- **Don't say "correlation" when you mean "association."** How often have you heard the word "correlation"? Chances are pretty good that when you've heard the term, it's been misused. It's one of the most widely misused Statistics terms, and given how often statistics are misused, that's saying a lot. One of the problems is that many people use the specific term *correlation* when they really mean the more general term *association*. "Association" is a deliberately vague term describing the relationship between two variables.

  "Correlation" is a precise term that measures the strength and direction of the linear relationship between quantitative variables.

- **Don't correlate categorical variables.** Did you know that there's a strong correlation between playing an instrument and drinking coffee? No? One reason might be that the statement

---

[14]Sometimes, we can do a "reality check" on our choice of re-expression. In this case, a bit of research reveals that f/stops are related to the diameter of the open shutter. Since the amount of light that enters is determined by the *area* of the open shutter, which is related to the diameter by squaring, the square re-expression seems reasonable. Not all re-expressions have such nice explanations, but it's a good idea to think about them.

[15]That's a high correlation, but it does not confirm that we have the best re-expression. After all, the correlation of the original relationship was 0.979—high enough to satisfy almost anyone. To judge a re-expression look at the scatterplot. Your eye is a better judge of straightness than any statistic.

doesn't make sense. People who misuse the term "correlation" to mean "association" often fail to notice whether the variables they discuss are quantitative. But correlation is only valid for quantitative variables. Be sure to check that the variables are quantitative.

- **Don't confuse correlation with causation.** One of the most common mistakes people make in interpreting statistics occurs when they observe a high correlation between two variables and jump to the perhaps tempting conclusion that one thing must be causing the other. Scatterplots and correlations *never* demonstrate causation. At best, these statistical tools can only reveal an association between variables, and that's a far cry from establishing cause and effect.

- **Make sure the association is linear.** Not all associations between quantitative variables are linear. Correlation can miss even a strong nonlinear association. A student project evaluating the quality of brownies baked at different temperatures reports a correlation of −0.05 between judges' scores and baking temperature. That seems to say there is no relationship—until we look at the scatterplot.

The relationship between brownie taste *Score* and *Baking Temperature* is strong, but not at all linear.

There is a strong association, but the relationship is not linear. Don't forget to check the Straight Enough Condition.

- **Don't assume the relationship is linear just because the correlation coefficient is high.** The correlation of f/stops and shutter speeds is 0.979 and yet the relationship is clearly not straight. A high correlation is no guarantee of straightness. Nor is it safe to use correlation to judge the best re-expression. It's always important to look at the scatterplot.

A scatterplot of *f/stop* vs. *Shutter Speed* shows a bent relationship even though the correlation is $r = 0.979$.

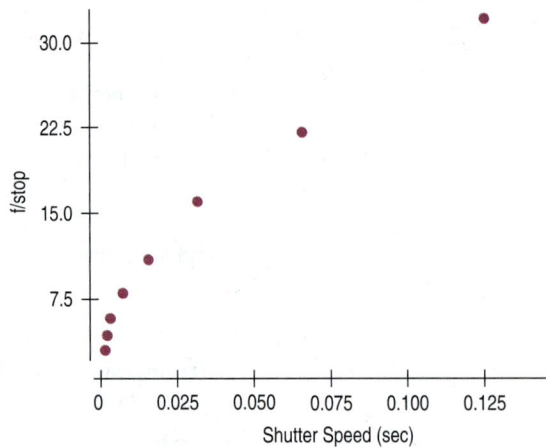

- **Beware of outliers.** You can't interpret a correlation coefficient safely without a background check for outliers. Here's a silly example:

The relationship between IQ and shoe size among comedians shows a surprisingly strong positive correlation of 0.50. To check assumptions, we look at the scatterplot.

What is the relationship? The correlation is 0.50, but it is all due to the outlier (the green x) in the upper right corner. Who does that point represent?

The outlier is Bozo the Clown, known for his large shoes, and widely acknowledged to be a comic "genius." Without Bozo, the correlation is near zero.

Even a single outlier can dominate the correlation value because its z-scores will be large and their product even larger. That's why you need to check the scatterplot for outliers.

CONNECTIONS

Scatterplots are the basic tool for examining the relationship between two quantitative variables. We start with a picture when we want to understand the distribution of a single variable, and we always make a scatterplot to begin to understand the relationship between two quantitative variables. Always.

We used z-scores as a way to measure the statistical distance of data values from their means. Now we've seen the z-scores of x and y working together to build the correlation coefficient. Correlation is a summary statistic like the mean and standard deviation—only it summarizes the strength of a linear relationship. And we interpret it as we did z-scores, using the standard deviations as our rulers in both x and y.

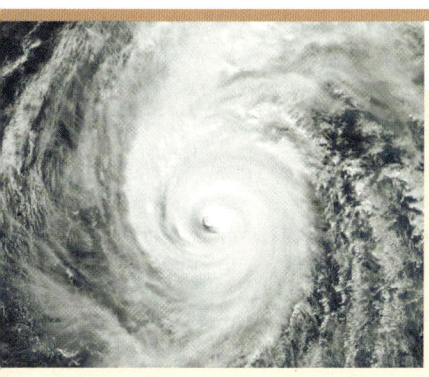

A S *Simulation*: **Correlation, Center, and Scale.** If you have any lingering doubts that shifting and rescaling the data won't change the correlation, watch nothing happen right before your eyes!

What Have We Learned?
Learning Objectives

Make a scatterplot to display the relationship between two quantitative variables.
- Look at the direction, form, and strength of the relationship, and any outliers that stand away from the overall pattern.

Provided the form of the relationship is straight, summarize its strength with a correlation, r.
- The sign of the correlation gives the direction of the relationship.
- $-1 \leq r \leq 1$; A correlation of 1 or -1 is a perfect linear relationship. A correlation of 0 indicates that there is no linear relationship.
- Correlation has no units, so shifting or scaling the data, standardizing, or even swapping the variables has no effect on the numerical value.

A large correlation is not a sign of a causal relationship

Recognize nonlinearity, and straighten plots.
- If a scatterplot of y vs. x isn't straight, correlation isn't appropriate.
- Re-expressing one or both variables can often improve the straightness of the relationship.
- The powers, roots, and the logarithm provide an ordered collection of re-expressions so you can search up and down the "ladder of powers" to find an appropriate one.

Review of Terms

Scatterplots A scatterplot shows the relationship between two quantitative variables measured on the same cases (p. 152).

Association
- **Direction:** A positive direction or association means that, in general, as one variable increases, so does the other. When increases in one variable generally correspond to decreases in the other, the association is negative (p. 152).
- **Form:** The form we care about most is straight, but you should certainly describe other patterns you see in scatterplots (p. 152).
- **Strength:** A scatterplot is said to show a strong association if there is little scatter around the underlying relationship (p. 153).

Outlier A point that does not fit the overall pattern seen in the scatterplot (p. 153).

Response variable, Explanatory variable, x-variable, y-variable In a scatterplot, you must choose a role for each variable. Assign to the y-axis the response variable that you hope to predict or explain. Assign to the x-axis the explanatory or predictor variable that accounts for, explains, predicts, or is otherwise responsible for the y-variable (p. 155).

Correlation Coefficient The correlation coefficient is a numerical measure of the direction and strength of a linear association (p. 157).

$$r = \frac{\sum z_x z_y}{n-1}$$

Kendall's tau A nonparametric measure of association that measures monotonicity by finding the slopes between all pairs of points in the data. It is computed as

$$\frac{\text{positive slopes} - \text{negative slopes}}{\frac{1}{2}n(n-1)}$$

Spearman's rho A nonparametric measure of association found by first converting both variable to ranks and then finding the correlation between the ranks.

Lurking variable A variable other than x and y that simultaneously affects both variables, accounting for the correlation between the two (p. 163).

Re-expression We re-express data by taking the logarithm, the square root, the reciprocal, or some other mathematical operation of all values of a variable (p. 165).

Ladder of Powers The Ladder of Powers places in order the effects that many re-expressions have on the data (p. 166).

On the Computer SCATTERPLOTS AND CORRELATION

Statistics packages generally make it easy to look at a scatterplot to check whether the correlation is an appropriate summary of the relationship. Some packages make this easier than others.

Many packages allow you to modify or enhance a scatterplot, altering the axis labels, the axis numbering, the plot symbols, or the colors used. Some options, such as color and symbol choice, can be used to display additional information on the scatterplot.

DATA DESK

To make a scatterplot of two variables:
- Click to select one variable as Y.
- Shift-click to select the other as X.
- Choose **Scatterplot** from the **Plot** menu.
- Then find the correlation by choosing **Correlation** from the scatterplot's HyperView menu.

Alternatively, select the two variables and choose **Pearson Product-Moment** from the **Correlations** submenu of the **Calc** menu.

COMMENTS

We prefer that you look at the scatterplot first and then find the correlation. But if you've found the correlation first, click on the correlation value to drop down a menu that offers to make the scatterplot.

EXCEL

To make a scatterplot in Excel:
- Select the columns of data to use in the scatterplot. You can select more than one column by holding down the control key while clicking.
- In the Insert tab, click on the **Scatter** button and select the **Scatter with only Markers** chart from the menu.

Unfortunately, the plot this creates is often statistically useless. To make the plot useful, we need to change the display:
- With the chart selected click on the **Gridlines** button in the Layout tab to cause the Chart Tools tab to appear.
- Within Primary Horizontal Gridlines, select **None**. This will remove the gridlines from the scatterplot.
- To change the axis scaling, click on the numbers of each axis of the chart, and click on the **Format Selection** button in the Layout tab.
- Select the **Fixed** option instead of the Auto option, and type a value more suited for the scatterplot. You can use the pop-up dialog window as a straightedge to approximate the appropriate values.

Excel automatically places the leftmost of the two columns you select on the x-axis, and the rightmost one on the y-axis. If that's not what you'd prefer for your plot, you'll want to switch them.

To switch the X- and Y-variables:
- Click the chart to access the **Chart Tools** tabs.
- Click on the **Select Data** button in the Design tab.
- In the pop-up window's Legend Entries box, click on **Edit**.
- Highlight and delete everything in the Series X Values line, and select new data from the spreadsheet. (Note that selecting the column would inadvertently select the title of the column, which would not work well here.)
- Do the same with the Series Y Values line.
- Press **OK**, then press **OK** again.

JMP

To make a scatterplot and compute correlation:
- Choose **Fit Y by X** from the **Analyze** menu.
- In the Fit Y by X dialog, drag the Y variable into the "Y, Response" box, and drag the X variable into the "X, Factor" box.
- Click the **OK** button.

Once JMP has made the scatterplot, click on the red triangle next to the plot title to reveal a menu of options:
- Select **Density Ellipse** and select 0.95. JMP draws an ellipse around the data and reveals the **Correlation** tab.
- Click the blue triangle next to Correlation to reveal a table containing the correlation coefficient.

MINITAB

To make a scatterplot:
- Choose **Scatterplot** from the **Graph** menu.
- Choose "Simple" for the type of graph. Click **OK**.
- Enter variable names for the Y variable and X variable into the table. Click **OK**.

To compute a correlation coefficient:
- Choose **Basic Statistics** from the **Stat** menu.
- From the Basic Statistics submenu, choose **Correlation**. Specify the names of at least two quantitative variables in the "Variables" box.
- Click **OK** to compute the correlation table.

R

To make a scatterplot of two variables, X and Y:
- **plot(X,Y)** or equivalently **plot(Y~X)** will produce the plot.
- **cor(X,Y)** finds the correlation.

COMMENTS

Your variables X and Y may be variables in a data frame. If so, and DATA is the name of the data frame, then you will need to attach the data frame, or use

with(DATA,plot(Y ~ X)) or plot (Y ~ X,data = DATA).

SPSS

To make a scatterplot in SPSS:
- Open the Chart Builder from the **Graphs** menu. Then
- Click the **Gallery** tab.
- Choose Scatterplot from the list of chart types.
- Drag the scatterplot onto the canvas.
- Drag a scale variable you want as the response variable to the y-axis drop zone.
- Drag a scale variable you want as the factor or predictor to the x-axis drop zone.
- Click **OK**.

To compute a correlation coefficient:
- Choose **Correlate** from the **Analyze** menu.
- From the Correlate submenu, choose **Bivariate**.
- In the Bivariate Correlations dialog, use the arrow button to move variables between the source and target lists.

Make sure the **Pearson** option is selected in the Correlation Coefficients field.

STATCRUNCH

To make a scatterplot:
- Click on **Graphics**.
- Choose **Scatter Plot**.
- Choose X and Y variable names from the list of **Columns**.
- Click on **Create Graph**.

To find a correlation:
- Click on **Stat**.
- Choose **Summary Stats » Correlation**.
- Choose two variable names from the list of **Columns**. (You may need to hold down the ctrl or command key to choose the second one.)
- Click on **Calculate**.

TI-83/84 PLUS

To create a scatterplot:
- Set up the **STAT PLOT** by choosing the **scatterplot** icon (the first option). Specify the lists where the data are stored as Xlist and Ylist.
- Set the graphing **WINDOW** to the appropriate scale and **GRAPH** (or take the easy way out and just ZoomStat!).

To find the correlation:
- Go to **STAT CALC** menu and select **8: LinReg(a + bx)**.
- Then specify the lists where the data are stored. The final command you will enter should look like **LinReg(a + bx) L1, L2**.

COMMENTS

Notice that if you **TRACE** the scatterplot, the calculator will tell you the x- and y-value at each point.

If the calculator does not tell you the correlation after you enter a LinReg command, try this: Hit **2nd CATALOG**. You now see a list of everything the calculator knows how to do. Scroll down until you find **DiagnosticOn**. Hit **ENTER** twice. (It should say Done.) Now and forevermore (or until you change batteries), you can find a correlation by using your calculator.

Exercises

Section 6.1

1. **Association** Suppose you were to collect data for each pair of variables. You want to make a scatterplot. Which variable would you use as the explanatory variable and which as the response variable? Why? What would you expect to see in the scatterplot? Discuss the likely direction, form, and strength.
   a) Apples: weight in grams, weight in ounces
   b) Apples: circumference (inches), weight (ounces)
   c) College freshmen: shoe size, grade point average
   d) Gasoline: number of miles you drove since filling up, gallons remaining in your tank

2. **Association II** Suppose you were to collect data for each pair of variables. You want to make a scatterplot. Which variable would you use as the explanatory variable and which as the response variable? Why? What would you expect to see in the scatterplot? Discuss the likely direction, form, and strength.
   a) T-shirts at a store: price each, number sold
   b) Scuba diving: depth, water pressure
   c) Scuba diving: depth, visibility
   d) All elementary school students: weight, score on a reading test

3. **Bookstore sales** Consider the following data from a small bookstore.

Number of Sales People Working	Sales (in $1000)
2	10
3	11
7	13
9	14
10	18
10	20
12	20
15	22
16	22
20	26
$\bar{x} = 10.4$	$\bar{y} = 17.6$
$SD(x) = 5.64$	$SD(y) = 5.34$

   a) Prepare a scatterplot of *Sales* against *Number of sales people* working.
   b) What can you say about the direction of the association?
   c) What can you say about the form of the relationship?
   d) What can you say about the strength of the relationship?
   e) Does the scatterplot show any outliers?

4. **Disk drives 2014** Disk drives have been getting larger. Their capacity is now often given in *terabytes* (TB) where 1 TB = 1000 gigabytes, or about a trillion bytes. A search of prices for external disk drives on Amazon.com in early 2014 found the following data:

Capacity (in TB)	Price (in $)
0.50	59.99
1.0	79.99
2.0	111.97
3.0	109.99
4.0	149.99
6.0	423.34
8.0	596.11
12.0	1079.99
32.0	4461.00

   a) Prepare a scatterplot of *Price* against *Capacity*.
   b) What can you say about the direction of the association?
   c) What can you say about the form of the relationship?
   d) What can you say about the strength of the relationship?
   e) Does the scatterplot show any outliers?

Section 6.2

5. **Correlation facts** If we assume that the conditions for correlation are met, which of the following are true? If false, explain briefly.
   a) A correlation of -0.98 indicates a strong, negative association.
   b) Multiplying every value of x by 2 will double the correlation.
   c) The units of the correlation are the same as the units of y.

6. **Correlation facts II** If we assume that the conditions for correlation are met, which of the following are true? If false, explain briefly.
   a) A correlation of 0.02 indicates a strong positive association.
   b) Standardizing the variables will make the correlation 0.
   c) Adding an outlier can dramatically change the correlation.

Section 6.3

7. **Bookstore sales again** A larger firm is considering acquiring the bookstore of Exercise 3. An analyst for the firm, noting the relationship seen in Exercise 3, suggests that when they acquire the store they should hire more people because that will drive higher sales. Is his conclusion justified? What alternative explanations can you offer? Use appropriate statistics terminology.

8. Blizzards A study finds that during blizzards, online sales are highly associated with the number of snow plows on the road; the more plows, the more online purchases. The director of an association of online merchants suggests that the organization should encourage municipalities to send out more plows whenever it snows because, he says, that will increase business. Comment.

Section 6.4

*9. **Salaries and logs** For an analysis of the salaries of your company, you plot the salaries of all employees against the number of years they have worked for the company. You find that plotting the base-10 logarithm of salary makes the plot much straighter. A part-time shipping clerk, who has worked at the company for one year earns $10,000. A manager earns $100,000 after 15 years with the firm. The CEO, who founded the company 30 years ago, receives $1,000,000. What are the values you will plot? Will the plot of these three points be straight enough?

*10. **Dexterity scores** Scores on a test of dexterity are recorded by timing a subject who is inverting pegs by picking them up with one hand, manipulating them to turn them over, and then placing them back in a frame. A typical 4-year-old needs about 2.5 seconds to invert a peg, but a 9-year-old takes 2 seconds and a 12-year-old can do it in 1.5 seconds. Plot these three points. Now try the reciprocal re-expression $(1/y)$ vs. age. Which version is straighter?

Chapter Exercises

11. Association III Suppose you were to collect data for each pair of variables. You want to make a scatterplot. Which variable would you use as the explanatory variable and which as the response variable? Why? What would you expect to see in the scatterplot? Discuss the likely direction, form, and strength.
   a) When climbing mountains: altitude, temperature
   b) For each week: ice cream cone sales, air-conditioner sales
   c) People: age, grip strength
   d) Drivers: blood alcohol level, reaction time

12. Association IV Suppose you were to collect data for each pair of variables. You want to make a scatterplot. Which variable would you use as the explanatory variable and which as the response variable? Why? What would you expect to see in the scatterplot? Discuss the likely direction, form, and strength.
   a) Legal consultation time, cost
   b) Lightning strikes: distance from lightning, time delay of the thunder
   c) A streetlight: its apparent brightness, your distance from it
   d) Cars: weight of car, age of owner

13. Scatterplots Which of these scatterplots show
   a) little or no association?
   b) a negative association?
   c) a linear association?
   d) a moderately strong association?
   e) a very strong association?

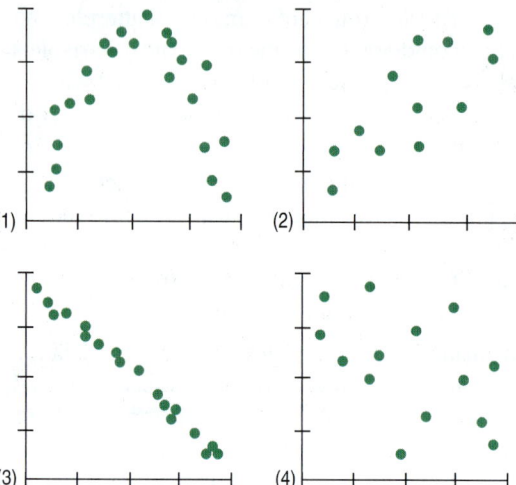

14. Scatterplots II Which of these scatterplots show
   a) little or no association?
   b) a negative association?
   c) a linear association?
   d) a moderately strong association?
   e) a very strong association?

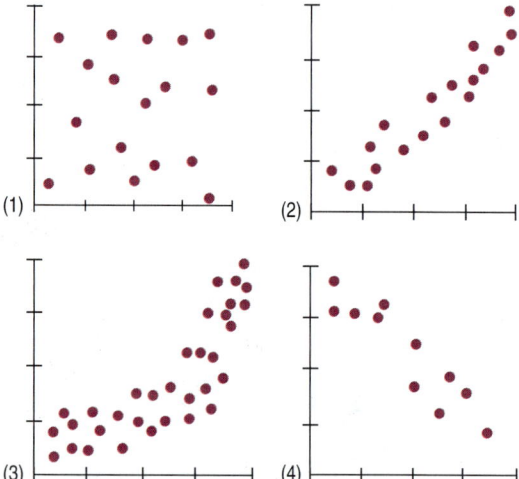

T 15. Performance IQ scores vs. brain size A study examined brain size (measured as pixels counted in a digitized magnetic resonance image [MRI] of a cross section of the brain) and IQ (4 Performance scales of the Weschler IQ test) for college students. The scatterplot shows the *Performance IQ* scores vs. *Brain Size*. Comment on the association between *Brain Size* and *Performance IQ* as seen in the scatterplot.

T 16. Kentucky Derby 2014 The fastest horse in Kentucky Derby history was Secretariat in 1973. The scatterplot shows speed (in miles per hour) of the winning horses each year.

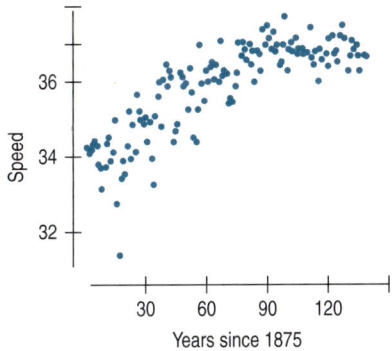

What do you see? In most sporting events, performances have improved and continue to improve, so surely we anticipate a positive direction. But what of the form? Has the performance increased at the same rate throughout the past 140 years?

17. Firing pottery A ceramics factory can fire eight large batches of pottery a day. Sometimes a few of the pieces break in the process. In order to understand the problem better, the factory records the number of broken pieces in each batch for 3 days and then creates the scatterplot shown.

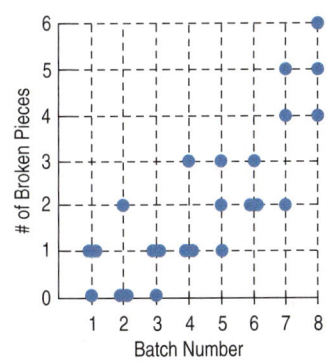

a) Make a histogram showing the distribution of the number of broken pieces in the 24 batches of pottery examined.
b) Describe the distribution as shown in the histogram. What feature of the problem is more apparent in the histogram than in the scatterplot?
c) What aspect of the company's problem is more apparent in the scatterplot?

18. Coffee sales Owners of a new coffee shop tracked sales for the first 20 days and displayed the data in a scatterplot (by day).

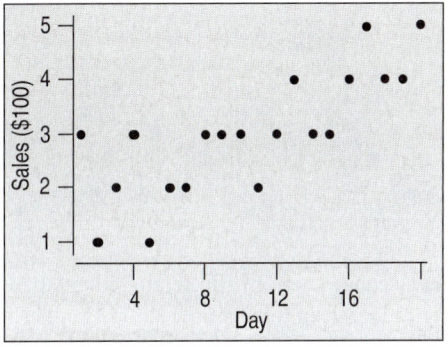

a) Make a histogram of the daily sales since the shop has been in business.
b) State one fact that is obvious from the scatterplot, but not from the histogram.
c) State one fact that is obvious from the histogram, but not from the scatterplot.

19. Matching Here are several scatterplots. The calculated correlations are −0.923, −0.487, 0.006, and 0.777. Which is which?

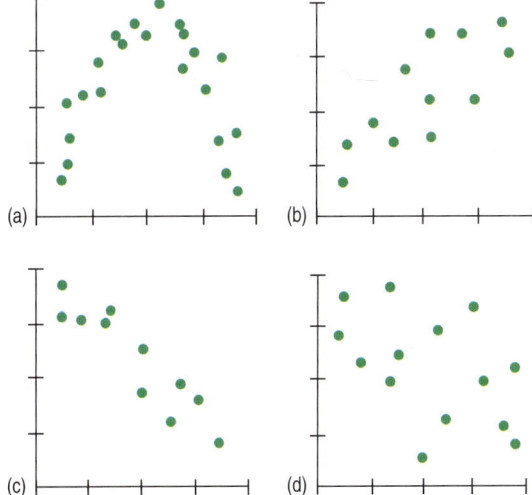

20. **Matching II** Here are several scatterplots. The calculated correlations are −0.977, −0.021, 0.736, and 0.951. Which is which?

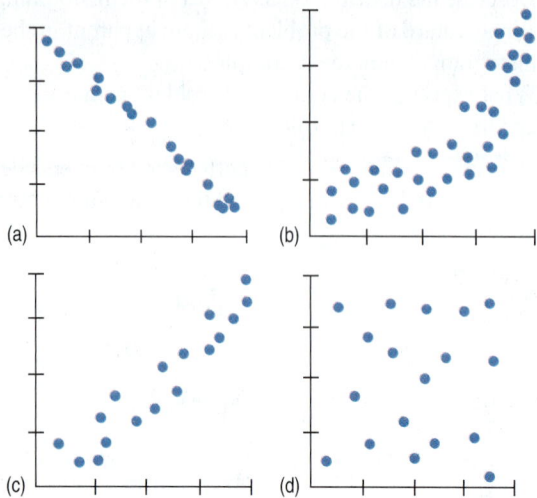

21. **Politics** A candidate for office claims that "there is a correlation between television watching and crime." Criticize this statement on statistical grounds.

22. **Car thefts** The National Insurance Crime Bureau reports that Honda Accords, Honda Civics, and Toyota Camrys are the cars most frequently reported stolen, while Ford Tauruses, Pontiac Vibes, and Buick LeSabres are stolen least often. Is it reasonable to say that there's a correlation between the type of car you own and the risk that it will be stolen?

23. **Roller coasters 2014** Most roller coasters get their speed by dropping down a steep initial incline, so it makes sense that the height of that drop might be related to the speed of the coaster. Here's a scatterplot of top *Speed* and largest *Drop* for 107 roller coasters around the world.

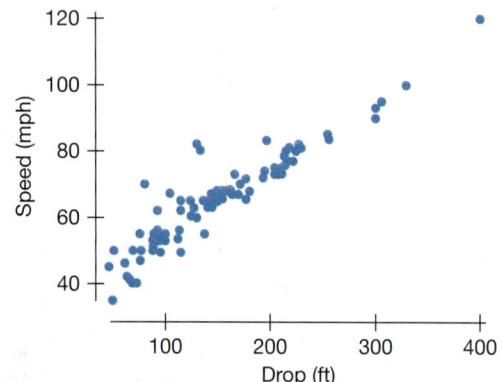

a) Does the scatterplot indicate that it is appropriate to calculate the correlation? Explain.
b) In fact, the correlation of *Speed* and *Drop* is 0.937. Describe the association.

24. **Antidepressants** A study compared the effectiveness of several antidepressants by examining the experiments in which they had passed the FDA requirements. Each of those experiments compared the active drug with a placebo, an inert pill given to some of the subjects. In each experiment some patients treated with the placebo had improved, a phenomenon called the *placebo effect*. Patients' depression levels were evaluated on the Hamilton Depression Rating Scale, where larger numbers indicate greater improvement. (The Hamilton scale is a widely accepted standard that was used in each of the independently run studies.) The following scatterplot compares mean improvement levels for the antidepressants and placebos for several experiments.

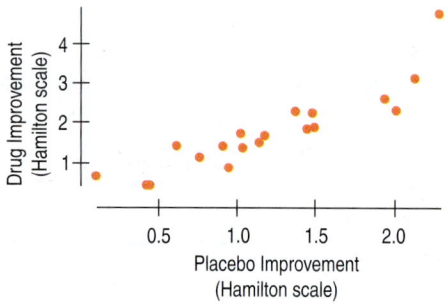

a) Is it appropriate to calculate the correlation? Explain.
b) The correlation is 0.898. Explain what we have learned about the results of these experiments.

25. **Streams and hard water** In a study of streams in the Adirondack Mountains, the following relationship was found between the water's pH and its hardness (measured in grains):

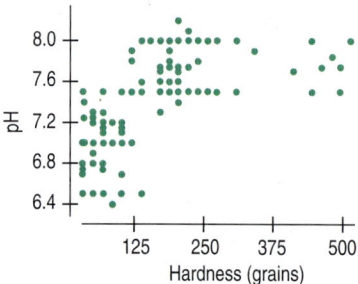

Is it appropriate to summarize the strength of association with a correlation? Explain.

26. **Traffic headaches** A study of traffic delays in 68 U.S. cities found the following relationship between *Total Delays* (in total hours lost) and *Mean Highway Speed*:

Is it appropriate to summarize the strength of association with a correlation? Explain.

27. Cold nights Is there an association between time of year and the nighttime temperature in North Dakota? A researcher assigned the numbers 1–365 to the days January 1–December 31 and recorded the temperature at 2:00 a.m. for each. What might you expect the correlation between *DayNumber* and *Temperature* to be? Explain.

28. Association V A researcher investigating the association between two variables collected some data and was surprised when he calculated the correlation. He had expected to find a fairly strong association, yet the correlation was near 0. Discouraged, he didn't bother making a scatterplot. Explain to him how the scatterplot could still reveal the strong association he anticipated.

29. Prediction units The errors in predicting hurricane tracks (examined in this chapter) were given in nautical miles. A statutory mile is 0.86898 nautical mile. Most people living on the Gulf Coast of the United States would prefer to know the prediction errors in statutory miles rather than nautical miles. Explain why converting the errors to miles would not change the correlation between *Prediction Error* and *Year*.

30. More predictions Hurricane Katrina's hurricane force winds extended 120 miles from its center. Katrina was a big storm, and that affects how we think about the prediction errors. Suppose we add 120 miles to each error to get an idea of how far from the predicted track we might still find damaging winds. Explain what would happen to the correlation between *Prediction Error* and *Year*, and why.

31. Correlation errors Your Economics instructor assigns your class to investigate factors associated with the gross domestic product (*GDP*) of nations. Each student examines a different factor (such as *Life Expectancy, Literacy Rate*, etc.) for a few countries and reports to the class. Apparently, some of your classmates do not understand Statistics very well because you know several of their conclusions are incorrect. Explain the mistakes in their statements:
a) "My very low correlation of −0.772 shows that there is almost no association between *GDP* and *Infant Mortality Rate*."
b) "There was a correlation of 0.44 between *GDP* and *Continent*."

32. More correlation errors Students in the Economics class discussed in Exercise 31 also wrote these conclusions. Explain the mistakes they made.
a) "There was a very strong correlation of 1.22 between *Life Expectancy* and *GDP*."
b) "The correlation between *Literacy Rate* and *GDP* was 0.83. This shows that countries wanting to increase their standard of living should invest heavily in education."

33. Height and reading A researcher studies children in elementary school and finds a strong positive linear association between height and reading scores.
a) Does this mean that taller children are generally better readers?
b) What might explain the strong correlation?

34. Smart phones and life expectancy A survey of the world's nations in 2014 shows a strong positive correlation between percentage of the country using smart phones and life expectancy in years at birth.
a) Does this mean that smart phones are good for your health?
b) What might explain the strong correlation?

35. Correlation conclusions I The correlation between *Age* and *Income* as measured on 100 people is $r = 0.75$. Explain whether or not each of these possible conclusions is justified:
a) When *Age* increases, *Income* increases as well.
b) The form of the relationship between *Age* and *Income* is straight.
c) There are no outliers in the scatterplot of *Income* vs. *Age*.
d) Whether we measure *Age* in years or months, the correlation will still be 0.75.

36. Correlation conclusions II The correlation between *Fuel Efficiency* (as measured by miles per gallon) and *Price* of 150 cars at a large dealership is $r = -0.34$. Explain whether or not each of these possible conclusions is justified:
a) The more you pay, the lower the fuel efficiency of your car will be.
b) The form of the relationship between *Fuel Efficiency* and *Price* is moderately straight.
c) There are several outliers that explain the low correlation.
d) If we measure *Fuel Efficiency* in kilometers per liter instead of miles per gallon, the correlation will increase.

37. Baldness and heart disease Medical researchers followed 1435 middle-aged men for a period of 5 years, measuring the amount of *Baldness* present (none = 1, little = 2, some = 3, much = 4, extreme = 5) and presence of *Heart Disease* (No = 0, Yes = 1). They found a correlation of 0.089 between the two variables. Comment on their conclusion that this shows that baldness is not a possible cause of heart disease.

38. Sample survey A polling organization is checking its database to see if the two data sources it used sampled the same ZIP codes. The variable *Datasource* = 1 if the data source is MetroMedia, 2 if the data source is DataQwest, and 3 if it's RollingPoll. The organization finds that the correlation between five-digit ZIP code and *Datasource* is −0.0229. It concludes that the correlation is low enough to state that there is no dependency between *ZIP Code* and *Source of Data*. Comment.

T 39. Income and housing The Office of Federal Housing Enterprise Oversight (www.fhfa.gov) collects data on various aspects of housing costs around the United States. Here is a scatterplot of the *Housing Cost Index* versus the *Median Family Income* for each of the 50 states. The correlation is 0.65.

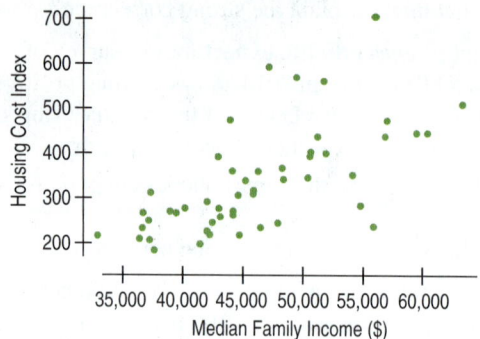

a) Describe the relationship between the *Housing Cost Index* and the *Median Family Income* by state.
b) If we standardized both variables, what would the correlation coefficient between the standardized variables be?
c) If we had measured *Median Family Income* in thousands of dollars instead of dollars, how would the correlation change?
d) Washington, DC, has a housing cost index of 548 and a median income of about $45,000. If we were to include DC in the data set, how would that affect the correlation coefficient?
e) Do these data provide proof that by raising the median family income in a state, the housing cost index will rise as a result? Explain.
*f) For these data Kendall's tau is 0.51. Does that provide proof that by raising the median income in a state, the Housing Cost Index will rise as a result? Explain what Kendall's tau says and does not say.

40. **Interest rates and mortgages 2013** Since 1985, average mortgage interest rates have fluctuated from a low of nearly 3% to a high of over 14%. Is there a relationship between the amount of money people borrow and the interest rate that's offered? Here is a scatterplot of *Mortgage Loan Amount* in the United States (in trillions of 2013 dollars) versus yearly *Interest Rate* since 1985. The correlation is −0.80.

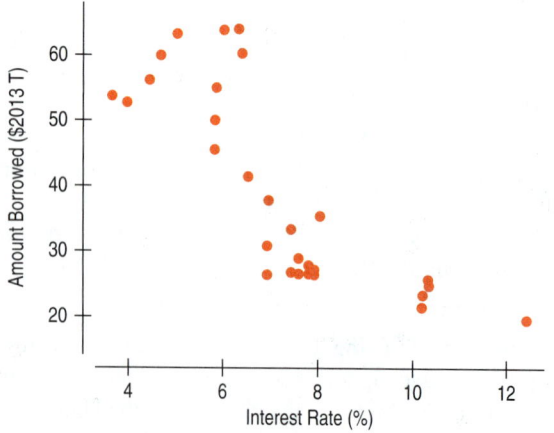

a) Describe the relationship between *Mortgage Loan Amount* and *Interest Rate*.
b) If we standardized both variables, what would the correlation coefficient between the standardized variables be?
c) If we were to measure *Mortgage Loan Amount* in billions of dollars instead of trillions of dollars, how would the correlation coefficient change?
d) Suppose that next year, interest rates were 11% and mortgages totaled $60 trillion. How would including that year with these data affect the correlation coefficient?
e) Do these data provide proof that if mortgage rates are lowered, people will take out larger mortgages? Explain.
*f) For these data Kendall's tau is −0.65. Does that provide proof that if mortgage rates are lowered, people will take out more mortgages? Explain what Kendall's tau says and does not say.

41. **Fuel economy 2014** Here are advertised engine size (in liters) and gas mileage (estimated combined city and highway) for several 2014 vehicles. (www.kbb.com)

Manufacturer	Model	Eng Displ	Comb MPG
Audi	A4 quattro	2	26
LEXUS	IS 250	2.5	24
Mercedes-Benz	CL 63	5.5	18
Chrysler	Mini Cooper Paceman	1.6	31
BMW	ActiveHybrid 5	3	26
Nissan	SENTRA	1.8	30
Ford	TAURUS AWD FFV	3.5	21
Porsche	Panamera GTS	4.8	19
Subaru	IMPREZA WAGON AWD	2	28
GMC	TERRAIN FWD	2.4	26
Jeep	Cherokee FWD	2.4	25
BMW	X3 xDrive35i	3	21
Chevrolet	EQUINOX AWD	3.6	19
Volvo	XC60 AWD	3.2	20
Chevrolet	C10 TAHOE 2WD	5.3	17
Porsche	Boxster S	3.4	23
BMW	650i Coupe	4.4	20
Chevrolet	SONIC	1.4	31
Volkswagen	BEETLE	2	26
Chevrolet	SONIC 5	1.4	33
Kia	OPTIMA	2.4	27
Mercedes-Benz	E 350	3.5	24
Toyota	AVALON HYBRID	2.5	40
Chevrolet	IMPALA	2.4	29

Manufacturer	Model	Eng Displ	Comb MPG
Chrysler	Charger AWD	3.6	21
Porsche	Panamera Turbo Executive	4.8	18
Volkswagen	JETTA SPORTWAGEN	2	34
Mercedes-Benz	E 350 4Matic (Wagon)	3.5	21
Nissan	TITAN 4WD	5.6	14
GMC	H1500 SAVANA AWD PASS	5.3	14
Nissan	QX60 AWD	3.5	21
Ford	MKT AWD	3.5	18
Nissan	QX60 AWD	2.5	26
Jeep	Grand Cherokee SRT8	6.4	15
Toyota	GX 460	4.6	17

a) Make a scatterplot for these data.
b) Describe the direction, form, and strength of the plot.
c) Find the correlation between engine size and miles per gallon.
d) Write a few sentences telling what the plot says about fuel economy.

42. Drug abuse A survey was conducted in the United States and 10 countries of Western Europe to determine the percentage of teenagers who had used marijuana and other drugs. The results are summarized in the following table.

Country	Percent Who Have Used	
	Marijuana	Other Drugs
Czech Rep.	22	4
Denmark	17	3
England	40	21
Finland	5	1
Ireland	37	16
Italy	19	8
No. Ireland	23	14
Norway	6	3
Portugal	7	3
Scotland	53	31
United States	34	24

a) Create a scatterplot.
b) What is the correlation between the percent of teens who have used marijuana and the percent who have used other drugs?
c) Write a brief description of the association.
d) Do these results confirm that marijuana is a "gateway drug," that is, that marijuana use leads to the use of other drugs? Explain.

43. Burgers Fast food is often considered unhealthy because much of it is high in both fat and sodium. But are the two related? Here are the fat and sodium contents of several brands of burgers.

Fat (g)	19	31	34	35	39	39	43
Sodium (mg)	920	1500	1310	860	1180	940	1260

a) Analyze the association between fat content and sodium using correlation and scatterplots.
*b) Find Spearman's rho for these data. Compare it with the Pearson correlation. Comment.

44. Burgers II In the previous exercise you analyzed the association between the amounts of fat and sodium in fast food hamburgers. What about fat and calories? Here are data for the same burgers:

Fat (g)	19	31	34	35	39	39	43
Calories	410	580	590	570	640	680	660

a) Analyze the association between fat content and calories using correlation and scatterplots.
*b) Repeat your analysis using Spearman's rho. Explain any differences you see.

45. Attendance 2013 American League baseball games are played under the designated hitter rule, meaning that pitchers, often weak hitters, do not come to bat. Baseball owners believe that the designated hitter rule means more runs scored, which in turn means higher attendance. Is there evidence that more fans attend games if the teams score more runs? Data collected from American League games during the 2013 season indicate a correlation of 0.384 between runs scored and the number of people at the game. (espn.go.com/mlb/attendance)

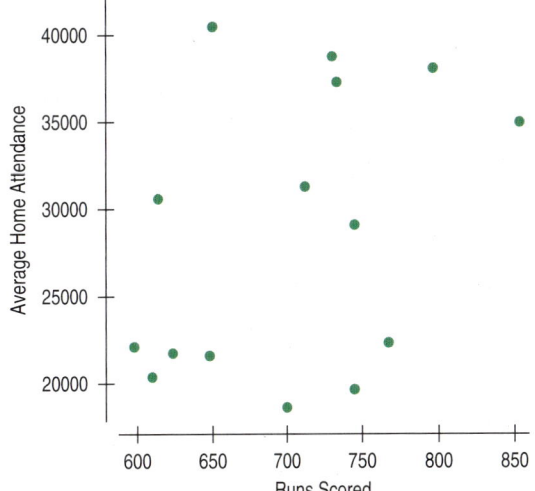

a) Does the scatterplot indicate that it's appropriate to calculate a correlation? Explain.
b) Describe the association between attendance and runs scored.
c) Does this association prove that the owners are right that more fans will come to games if the teams score more runs?

T 46. Second inning 2013 Perhaps fans are just more interested in teams that win. The displays below are based on American League teams for the 2013 season. (espn.go.com) (Use the data set Attendance_2013).

	Wins	Runs	Attendance
Wins	1.000		
Runs	0.793	1.000	
Attendance	0.254	0.384	1.000

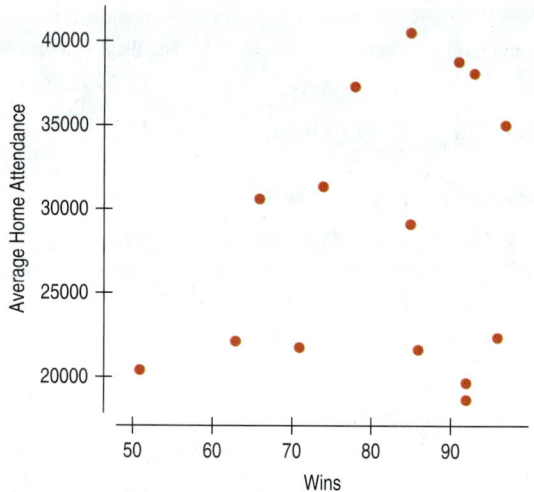

a) Do winning teams generally enjoy greater attendance at their home games? Describe the association.
b) Is attendance more strongly associated with winning or scoring runs? Explain.
c) How strongly is scoring runs associated with winning games?

T 47. Thrills Since 1994, the Best Roller Coaster Poll (www.ushsho.com/bestrollercoasterpoll.htm) has been ranking the world's best roller coasters. In 2011, Bizarro earned the top steel coaster rank for the sixth straight year. Here are data on the top 10 steel coasters from this poll:

Rank	Roller Coaster	Park	Location	Initial Drop (ft.)	Duration (sec)	Height (ft.)	Max Speed (mph)	Max Vert Angle (degrees)	Length (ft.)
1	Bizarro	Six Flags New England	MA	221	155	208	77	72	5400
2	Expedition GeForce	Holiday Park	DE	184	75	188	74.6	82	4003
3	Intimidator 305	Kings Dominion	VA	300	98	305	93	85	5100
4	Kawasemi	Tobu Zoo	JP		60	108	54	67.4	2454
5	Nemesis	Alton Towers	UK	104	80	43	50	40	2349
6	Piraten	Djurs Sommerland	DK	100	61	105	56	70	2477
7	Goliath	Walibi World	NL	152	92	155	66.9	70	3984
8	Millennium Force	Cedar Point	OH	300	120	310	93	80	6595
9	Katun	Mirabilandia	IT	148	142	169	65		3937
10	iSpeed	Mirabilandia	IT	181	60	180	74.6	90	3281

What do these data indicate about the *Length* of the track and the *Duration* of the ride you can expect?

T 48. Thrills II For the roller coaster data in Exercise 47:
a) Examine the relationship between *Initial Drop* and *Speed*.
b) Examine the relationship between *Initial Drop* and *Height*.
c) What conclusions can you safely draw about the initial drop of a roller coaster? Is *Initial Drop* strongly correlated with other variables as well?

T 49. Thrills III For the roller coaster data in Exercise 47:
a) Explain why in looking for a variable that explains rank, you will be hoping for a negative correlation.
b) Do any of the provided variables provide a strong predictor for roller coaster rank?
c) What other (unaccounted) for variables might help explain the rank?

T 50. Vehicle weights The Minnesota Department of Transportation hoped that they could measure the weights of big trucks without actually stopping the vehicles by using a newly developed "weight-in-motion" scale. To see if the new device was accurate, they conducted a calibration test. They weighed several stopped trucks (*Static Weight*) and assumed that this weight was correct.

* Then they weighed the trucks again while they were moving to see how well the new scale could estimate the actual weight. Their data are given in the table at the top of the next column.

Weights (1000s of lbs)	
Weight-in-Motion	Static Weight
26.0	27.9
29.9	29.1
39.5	38.0
25.1	27.0
31.6	30.3
36.2	34.5
25.1	27.8
31.0	29.6
35.6	33.1
40.2	35.5

a) Make a scatterplot for these data.
b) Describe the direction, form, and strength of the plot.
c) Write a few sentences telling what the plot says about the data. (*Note:* The sentences should be about weighing trucks, not about scatterplots.)
d) Find the correlation.
e) If the trucks were weighed in kilograms, how would this change the correlation? (1 kilogram = 22 pounds)
f) Do any points deviate from the overall pattern? What does the plot say about a possible recalibration of the weight-in-motion scale?

51. **Planets (more or less)** On August 24, 2006, the International Astronomical Union voted that Pluto is not a planet. Some members of the public have been reluctant to accept that decision. Let's look at some of the data. Is there any pattern to the locations of the planets? The table shows the average distance of each of the traditional nine planets from the sun.

Planet	Position Number	Distance from Sun (million miles)
Mercury	1	36
Venus	2	67
Earth	3	93
Mars	4	142
Jupiter	5	484
Saturn	6	887
Uranus	7	1784
Neptune	8	2796
Pluto	9	3666

a) Make a scatterplot and describe the association. (Remember: direction, form, and strength!)

b) Why would you not want to talk about the correlation between a planet's *Position Number* and *Distance* from the sun?
c) Make a scatterplot showing the logarithm of *Distance* vs. *Position Number*. What is better about this scatterplot?
d) Looking only at the scatterplot, what is Kendall's tau for these data? Explain.

*52. **Flights 2012** Here are the number of domestic flights flown in each year from 2000 to 2012 (www.TranStats.bts.gov/Data_Elements.aspx?Data=2).

Year	Flights
2000	7,905,617
2001	7,626,312
2002	8,085,083
2003	9,458,818
2004	9,968,047
2005	10,038,373
2006	9,712,750
2007	9,839,578
2008	9,378,227
2009	8,768,938
2010	8,702,365
2011	8,649,087
2012	8,446,201

a) Find the correlation of *Flights* with *Year*.
b) Make a scatterplot and describe the trend.
c) Why is the correlation you found in part a not a suitable summary of the strength of the association?
d) Find Kendall's tau for these data. Would that be an appropriate measure for summarizing their relationship? Explain.

Just Checking ANSWERS

1. We know the scores are quantitative. We should check to see whether the Straight Enough Condition and the Outlier Condition are satisfied by looking at a scatterplot of the two scores.
2. It won't change.
3. It won't change.
4. They are likely to have done poorly. The positive correlation means that low scores on Exam 1 are associated with low scores on Exam 2 (and similarly for high scores).
5. No. The general association is positive, but individual performances may vary.

Chapter 7: Linear Regression

- 7.1 Least Squares: The Line of "Best Fit"
- 7.2 The Linear Model
- 7.3 Finding the Least Squares Line
- 7.4 Regression to the Mean
- 7.5 Examining the Residuals
- 7.6 R^2—The Variation Accounted For by the Model
- 7.7 Regression Assumptions and Conditions

Where are we going?

It's often hard to measure wind speeds of a hurricane directly. But lower pressure in storms is associated with higher wind speeds. In fact, for hurricanes, the correlation is -0.88. If we know that a hurricane has a central pressure of 860 mb, how high would we expect the wind speed to be? The correlation alone won't tell us the answer.

We need a model to be able to use one variable to predict another. Using linear regression, we can understand the relationship between two quantitative variables better and make predictions.

Who	Items on the Burger King menu
What	Protein content and total fat content
Units	Grams of protein Grams of fat
How	Supplied by BK on request or at their website

Video: Manatees and Motorboats. Are motorboats killing more manatees in Florida? Here's the story on video.

The Whopper™ has been Burger King's signature sandwich since 1957. One Triple Whopper with cheese provides 71 grams of protein—all the protein you need in a day. It also supplies 1230 calories and 82 grams of fat. The Daily Value (based on a 2000-calorie diet) for fat is 65 grams. So after a Triple Whopper you'll want the rest of your calories that day to be fat-free.[1]

Of course, the Whopper isn't the only item Burger King sells. How are fat and protein related on the entire BK menu? The scatterplot of the *Fat* (in grams) vs. the *Protein* (in grams) for foods sold at Burger King shows a positive, moderately strong, linear relationship.

Figure 7.1

Total *Fat* vs. *Protein* for 122 items on the BK menu. The Triple Whopper is in the upper right corner. It's extreme, but is it out of line? (Data in **Burger King Items**)

---
[1] Sorry about the fries.

7.1 Least Squares: The Line of "Best Fit"

A|S *Activity:* **Linear Equations.** For a quick review of linear equations, view this activity and play with the interactive tool.

> "Statisticians, like artists, have the bad habit of falling in love with their models."
>
> —George Box, famous statistician

A|S *Activity:* **Residuals.** Residuals are the basis for fitting lines to scatterplots. See how they work.

If you want 25 grams of protein in your lunch, how much fat should you expect to consume at Burger King? The correlation between *Fat* and *Protein* is 0.76, which tells us that the linear association seen in the scatterplot is fairly strong. But *strength* of the relationship is only part of the picture. The correlation says, "The linear association between these two variables is fairly strong," but it doesn't tell us *what the line is*.

Of course, the points in the scatterplot don't all line up. But anyone looking at the plot can see that the points move from the lower left to the upper right in a consistent way. You could probably sketch a straight line that summarizes that relationship pretty well. We'll take that one step further and find a **linear model** that gives an equation of a straight line through the data. This model can predict the fat content of any Burger King food, given its protein.

Of course, no line can go through all the points, but a linear model can summarize the general pattern with only a couple of parameters. Like all **models** of the real world, the line will be wrong—wrong in the sense that it can't match reality *exactly*. But it can help us understand how the variables are associated. Without models, our understanding of the world is limited to only the data we have at hand.

Not only can't we draw a line through all the points, the best line might not even hit *any* of the points. Then how can it be the "best" line? We want to find the line that somehow comes *closer* to all the points than any other line. Some of the points will be above the line and some below. For example, the line might suggest that a BK Tendercrisp® chicken sandwich (without mayonnaise) with 31 grams of protein[2] should have 36.7 grams of fat when, in fact, it actually has only 22. We call the estimate made from a model the **predicted value**, and write it as $\hat{y}$ (called *y-hat*) to distinguish it from the observed value, *y*. The difference between the observed value and its associated predicted value is called the **residual**. The residual value tells us how far off the model's prediction is at that point. The BK Tendercrisp chicken residual would be $y - \hat{y} = 22 - 36.6 = -14.6g$ of fat.

Figure 7.2
The residual is the difference between the actual *y*-value and the predicted $\hat{y}$-value. So it is a difference in the *y* direction.

To find the residuals, we always subtract the predicted value from the observed one.

Residual = Observed value − Predicted value

The negative residual tells us that the actual fat content of the BK Tendercrisp chicken is about 14.6 grams *less* than the model predicts for a typical Burger King menu item with 31 grams of protein. We often use residuals in this way. The line tells us the average fat content we'd expect for a given amount of protein. Items with negative residuals have less fat than we'd expect, while those with positive residuals have more.

Our challenge now is how to find the right line.

RESIDUALS
A **positive** residual means the actual value is larger than the predicted value—an underestimate. A **negative** residual shows that the model made an overestimate.

---

[2] The sandwich comes with mayo unless you ask for it without. That adds an extra 24 grams of fat, which is more than the original sandwich contained.

184 PART II Exploring Relationships Between Variables

A | S *Activity:* **The Least Squares Criterion.** Does your sense of "best fit" look like the least squares line?

The size of the residuals tells us how well our line fits the data. A line that fits well will have small residuals. But we can't just add up all the residuals to see how well we've done, because some of the residuals are negative. We faced the same issue when we calculated a standard deviation to measure spread. And we deal with it the same way here: by squaring the residuals. Now that they're all positive, adding them up makes sense. Squaring also emphasizes the large residuals. After all, points near the line are consistent with the model, but we're more concerned about points far from the line. When we add all the squared residuals together, that sum indicates how well the line we drew fits the data—the smaller the sum, the better the fit. A different line will produce a different sum, maybe bigger, maybe smaller. The **line of best fit** is the line for which the sum of the squared residuals is smallest, the **least squares** line.

You might think that finding this line would be pretty hard. Surprisingly, it's not, although it was an exciting mathematical discovery when Legendre published it in 1805.

> **Who's on first** In 1805, the French mathematician Adrien-Marie Legendre was the first to publish the "least squares" solution to the problem of fitting a line to data when the points don't all fall exactly on the line. The main challenge was how to distribute the errors "fairly." After considerable thought, he decided to minimize the sum of the squares of what we now call the residuals. When Legendre published his paper, though, the German mathematician Carl Friedrich Gauss claimed he had been using the method since 1795. Gauss later referred to the "least squares" solution as "*our*" method" (*principium nostrum*), which certainly didn't help his relationship with Legendre.

7.2 The Linear Model

■ **NOTATION ALERT**
"Putting a hat on it" is standard Statistics notation to indicate that something has been predicted by a model. Whenever you see a hat over a variable name or symbol, you can assume it is the predicted version of that variable or symbol (and look around for the model).

In a linear model, we use b_1 for the slope and b_0 for the y-intercept.

You may remember from Algebra that a straight line can be written as

$$y = mx + b.$$

We'll use this form for our linear model, but in Statistics we use slightly different notation:

$$\hat{y} = b_0 + b_1 x.$$

We write $\hat{y}$ (y-hat) to emphasize that the points that satisfy this equation are just our *predicted* values, not the actual data values (which scatter around the line). If the model is a good one, the data values will scatter closely around it.

We write b_1 and b_0 for the slope and intercept of the line. The b's are called the *coefficients* of the linear model. The coefficient b_1 is the **slope**, which tells how rapidly $\hat{y}$ changes with respect to x. The coefficient b_0 is the **intercept**, which tells where the line hits (intercepts) the y-axis.[3]

For the Burger King menu items, the line of best fit is

$$\widehat{Fat} = 8.4 + 0.91\, Protein.$$

What does this mean? The slope, 0.91, says that a Burger King item with one more gram of protein can be expected, on average, to have 0.91 more grams of fat. Less formally, we might say that, on average, Burger King foods pack about 0.91 grams of fat per gram of protein. Slopes are always expressed in y-units per x-unit. They tell how the y-variable changes (in its units) for a one-unit change in the x-variable. When you see a phrase like "students per teacher" or "kilobytes per second," think slope.

What about the intercept, 8.4? Algebraically, that's the value the line takes when x is zero. Here, our model predicts that even a BK item with no protein would have,

A | S *Simulation:* **Interpreting Equations.** This activity demonstrates how to use and interpret linear equations.

---

[3] We change from $mx + b$ to $b_0 + b_1 x$ for a reason: Eventually, we'll want to add more x's to the model to make it more realistic, and we don't want to use up the entire alphabet. What would we use for the next coefficient after m? The next letter is n and that one's already taken. o? See our point? Sometimes subscripts are the best approach.

on average, about 8.4 grams of fat. Is that reasonable? Well, there are two items on the menu with 0 grams of protein (apple fries and applesauce). Neither has any fat either (they are essentially pure carbohydrate), but we could imagine a protein-less item with 6.8 grams of fat. But often 0 is not a plausible value for *x* (the year 0, a baby born weighing 0 grams, . . .). Then the intercept serves only as a starting value for our predictions, and we don't interpret it as a meaningful predicted value.

For Example A LINEAR MODEL FOR HURRICANES

The barometric pressure at the center of a hurricane is often used to measure the strength of the hurricane because it can predict the maximum wind speed of the storm. A scatterplot shows that the relationship is straight, strong, and negative. It has a correlation of −0.896. (Data in **Hurricanes 2014**)

Using technology to fit the straight line, we find

$$\widehat{MaxWindSpeed} = 1024.464 - 0.968\, CentralPressure.$$

QUESTION: Interpret this model. What does the slope mean in this context? Does the intercept have a meaningful interpretation?

ANSWER: The negative slope says that lower *CentralPressure* goes with higher *MaxWindSpeed*. This is consistent with the way hurricanes work: Low central pressure pulls in moist air, driving the rotation and the resulting destructive winds. The slope's value says that, on average, the maximum wind speed is about 0.968 knots higher for every 1-millibar drop in central pressure.

It's not meaningful, however, to interpret the intercept as the wind speed predicted for a central pressure of 0—that would be a vacuum. Instead, it is merely a starting value for the model.

7.3 Finding the Least Squares Line

> **SLOPE**
>
> $$b_1 = r\frac{s_y}{s_x}$$

How do we find the actual values of slope and intercept for the least squares line? The formulas are built from the summary statistics you already know: the correlation (to tell us the strength of the linear association), the standard deviations (to give us the units), and the means (to tell us where to put the line). Remarkably, that's all you need to find the value of the slope:

$$b_1 = r\frac{s_y}{s_x}.$$

UNITS OF y PER UNIT OF x

Get into the habit of identifying the units by writing down "y-units per x-unit," with the unit names put in place. You'll find it'll really help you to Tell about the line in context.

We've already seen that the correlation tells us the *sign* as well as the strength of the relationship, so it should be no surprise that the slope inherits this sign as well.

Correlations don't have units, but slopes do. How x and y are measured—what units they have—doesn't affect their correlation, but can change their slope. If children grow an average of 3 inches per year, that's 0.21 millimeters per day. For the slope, it matters whether you express age in days or years and whether you measure height in inches or millimeters. Changing the units of x and y affects their standard deviations directly. And that's how the slope gets its units—from the ratio of the two standard deviations. So, the units of the slope are a ratio too. The units of the slope are always the units of y per unit of x.

What about the intercept? If you had to predict the y-value for a data point whose x-value was the average, $\bar{x}$, you'd probably pick $\bar{y}$, and that's just what the least squares line does. Putting that into our equation, we find

$$\bar{y} = b_0 + b_1\bar{x},$$

or, rearranging the terms,

$$b_0 = \bar{y} - b_1\bar{x}.$$

For Example FINDING THE REGRESSION EQUATION

Let's try out the slope and intercept formulas on the "BK data" and see how to find the line. We checked the conditions when we calculated the correlation. The summary statistics are as follows:

Protein	Fat
$\bar{x} = 18.0$ g	$\bar{y} = 24.8$ g
$s_x = 13.5$ g	$s_y = 16.2$ g
$r = 0.76$	

So, the slope is

$$b_1 = r\frac{s_y}{s_x} = 0.76 \times \frac{16.2 \text{ g fat}}{13.5 \text{ g protein}} = 0.91 \text{ grams of fat per gram of protein}.$$

The intercept is

$$b_0 = \bar{y} - b_1\bar{x} = 24.8 \text{ g fat} - 0.91\frac{\text{g fat}}{\text{g protein}} \times 18.0 \text{ g protein} = 8.4 \text{ g fat}.$$

Putting these results back into the equation gives

$$\widehat{Fat} = 8.4 \text{ g fat} + 0.91\frac{\text{g fat}}{\text{g protein}} Protein,$$

or more simply,

$$\widehat{Fat} = 8.4 + 0.91 \, Protein.$$

In summary, the predicted fat of a Burger King food item in grams is $8.4 + 0.91 \times$ the protein content (in grams).

With an estimated linear model, it's easy to predict fat content for any menu item we want. For example, for the BK Tendercrisp chicken sandwich with 31 grams of protein, we can plug in 31 grams for the amount of protein and see that the predicted fat content is $8.4 + 0.91(31) = 36.6$ grams of fat. (Actually, rounding has caught up with us. Using full precision not shown here, the predicted value to one decimal place should be 36.7.

Always use full precision until you report your results). Because the BK Tendercrisp chicken sandwich actually has 22 grams of fat, its residual is

$$Fat - \widehat{Fat} = 22 - 36.7 = -14.7 \text{ g.}$$

Least squares lines are commonly called regression lines. In a few pages, we'll see that this name is an accident of history. For now, you just need to know that "regression" almost always means "the linear model fit by least squares."

Figure 7.3
Burger King menu items with the regression line. The regression model lets us make predictions. The predicted fat value for an item with 31 grams of protein is 36.6 g.

A S *Activity:* **Find a Regression Equation.** Now that we've done it by hand, try it with technology using the statistics package paired with your version of *ActivStats*.

To use a regression model, we should check the same **conditions** for regression as we did for correlation, making sure that the variables are **Quantitative**, that the relationship is **Straight Enough**, and that there are **No Outliers**.

Step-by-Step Example CALCULATING A REGRESSION EQUATION

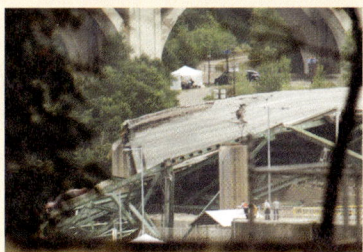

During the evening rush hour of August 1, 2007, an eight-lane steel truss bridge spanning the Mississippi River in Minneapolis, Minnesota, collapsed without warning, sending cars plummeting into the river, killing 13 and injuring 145. Although similar events had brought attention to our aging infrastructure, this disaster put the spotlight on the problem and raised the awareness of the general public.

How can we tell which bridges are safe?

Most states conduct regular safety checks, giving a bridge a structural deficiency score on various scales. The New York State Department of Transportation uses a scale that runs from 1 to 7, with a score of 5 or less indicating "deficient."

Many factors contribute to the deterioration of a bridge, including amount of traffic, material used in the bridge, weather, and bridge design.

New York has more than 17,000 bridges. We have available data on the 194 bridges of Tompkins County.[4] (Data in **Bridges Tompkins County 2014**)

One natural concern is the age of a bridge. A model that relates age to safety score might help the DOT to focus inspectors' efforts where they are most needed.

Question: Is there a relationship between the age of a bridge and its safety rating?

THINK ➡ Plan State the problem.	I want to know whether there is a relationship between the age of a bridge in Tompkins County, New York, and its safety rating.
Variables Identify the variables and report the W's.	I have data giving the Safety Score and Age for 194 bridges constructed or replaced since 1886.

---

[4] Coincidentally, the home county of two of the authors.

Just as we did for correlation, check the conditions for a regression by making a picture. Never fit a regression without looking at the scatterplot first.

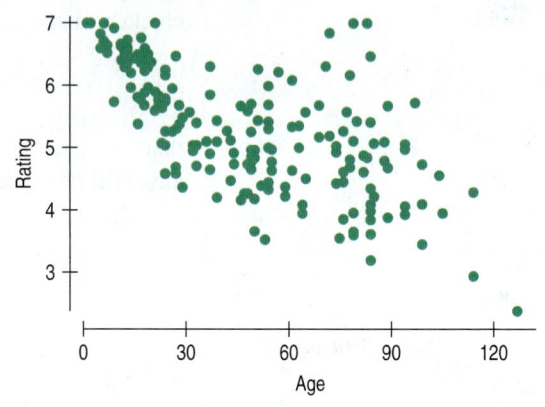

Conditions:

✓ **Quantitative Variables:** Yes, although Condition rating has no units. Age is in years.

✓ **Straight Enough:** Scatterplot looks straight.

✓ **No Outliers:** None are evident.

It is OK to use a linear regression to model this relationship.

SHOW → Mechanics
Find the equation of the regression line. Summary statistics give the building blocks of the calculation.

(We generally report summary statistics to one more digit of accuracy than the data. We do the same for intercept and predicted values, but for slopes we usually report an additional digit. Remember, though, not to round until you finish computing an answer.)[5]

Find the slope, b_1.

Find the intercept, b_0.

Write the equation of the model, using meaningful variable names.

Age

$\bar{x} = 48.52$

$s_x = 28.84$

Safety Score (points from 1 to 7)

$\bar{y} = 5.27$

$s_y = 0.959$

Correlation

$r = -0.659$

$b_1 = r \dfrac{s_y}{s_x}$

$\quad = -0.659 \dfrac{0.959}{28.84}$

$\quad = -0.0219$ points per year

$b_0 = \bar{y} - b_1 \bar{x} = 5.27 - (-0.0219)48.52$

$\quad = 6.33$

The least squares line is

$\hat{y} = 6.33 - 0.0219x$

or

$\widehat{Rating} = 6.33 - 0.0219\,Age$

---

[5] We warned you that we'll round in the intermediate steps of a calculation to show the steps more clearly, and we've done that here. If you repeat these calculations yourself on a calculator or statistics program, you may get somewhat different results.

TELL ➡ **Conclusion** Interpret what you have found in the context of the question. Discuss in terms of the variables and their units.

The condition of the bridges in Tompkins County, New York, decreases with the age of the bridges at the time of inspection. Bridge condition declines by about 0.022 points on the scale from 1 to 7 for each year of age. The model uses a base of 6.3, which is quite reasonable because a new bridge (0 years of age) should have a safety score near the maximum of 7.

Because I have only data from one county, I can't tell from these data whether this model would apply to bridges in other counties of New York or in other locations.

7.4 Regression to the Mean

Suppose you were told that a new male student was about to join the class, and you were asked to guess his height in inches. What would be your guess? Since you have no other information, a good guess would be the mean height of all male students. (If you thought about this in z-scores, you'd say $\hat{z}_{Height} = 0$). Now suppose you are also told that this student has a grade point average (*GPA*) of 3.9—about 2 SDs above the mean *GPA*. Would that change your guess? Probably not. The correlation between *GPA* and *Height* is near 0, knowing something about *GPA* wouldn't tell you anything about *Height*.

On the other hand, if you were told that, measured in centimeters, the student's *Height* was 2 SDs above the mean, you'd know his *Height* in inches exactly. There's a perfect correlation between *Height* in Inches and *Height* in Centimeters, so you'd be able to predict perfectly. In fact, you'd predict that he's 2 SDs above the mean *Height* in inches as well. Or, writing this as z-scores, you could say: $\hat{z}_{Htin} = z_{Htcm}$.

What if you're told that the student is 2 SDs above the mean in *Shoe Size*? Would you still guess that he's of average *Height*? You might guess that he's taller than average, since there's a positive correlation between *Height* and *Shoe Size*. But would you guess that he's 2 SDs above the mean? When there was no correlation, we didn't move away from the mean at all. With a perfect correlation, we moved our guess the full 2 SDs. Any correlation between these extremes should lead us to guess somewhere between 0 and 2 SDs above the mean. As we'll see in a minute, if we think in z-scores, the relationship is remarkably simple. When the correlation between *x* and *y* is *r*, and *x* has a z-score of z_x, the predicted value of *y* is just $\hat{z}_y = rz_x$. For example, if our student was 2 SDs above the mean *Shoe Size*, the formula tells us to guess $r \times 2$ standard deviations above the mean for his *Height*.

How did we get that? We know from the least squares line that $\hat{y} = b_0 + b_1 x$. We also know that $b_0 = \bar{y} - b_1 \bar{x}$. If we standardize *x* and *y*, both their means will be 0, and so for standardized variables, $b_0 = 0$. And the slope $b_1 = r\frac{s_y}{s_x}$. For standardized variables, both s_y and s_x are 1, so $b_1 = r$. In other words, $\hat{z}_y = rz_x$.

Even more important than deriving the formula, is understanding what it says about regression. While we rarely standardize variables to do a regression, we can think about what the z-score regression means. It says that when the correlation is *r*, and an *x*-value lies *k* standard deviations from its mean, our prediction for *y* is $r \times k$ standard deviations from its mean as we can see in the following plot.

The z-score equation has another profound consequence. It says that for linear relationships, you can never predict that *y* will be *farther* from its mean than *x* was from its mean. That's because *r* is always between -1 and 1. So, each predicted *y* tends to be closer to its mean (in standard deviations) than its corresponding *x* was. Sir Francis Galton discovered this property when trying to predict heights of offspring

Why is correlation "r"? In his original paper, Galton standardized the variables and so the slope was the same as the correlation. He used *r* to stand for this coefficient from the (standardized) regression.

Figure 7.4

Standardized Fat vs. *Standardized Protein* with the regression line. Each one-standard-deviation change in *Protein* results in a predicted change of *r* standard deviations in *Fat*.

Just Checking

A scatterplot of house *Price* (in dollars) vs. house *Size* (in square feet) for houses sold recently in Saratoga, New York, shows a relationship that is straight, with only moderate scatter and no outliers. The correlation between house *Price* and house *Size* is 0.76. The least squares line is

$$\widehat{Price} = 8400 + 88.67\ Size.$$

1. You go to an open house and find that the house is 1 SD above the mean *Size*. How many SDs above the mean *Price* would you predict it to cost?

2. You read an ad for a house that's 2 SDs below the mean *Size*. What would you guess about its price?

3. A friend tells you about a house whose *Size* in square meters (he's European) is 1.5 standard deviations above the mean. What would you guess about its *Size* in square feet?

4. Suppose the standard deviation of the house *Prices* is $77,000. How much more expensive than the mean *Price* would you predict for the house in Part 1 (that's one SD above the mean in *Size*)?

5. Suppose the mean *Price* of all these houses is $168,000. What is the predicted *Price* for the house in Part 4 that's one SD above the mean in *Size*?

6. The mean *Size* of the houses is 1800 sq ft. and the standard deviation is 660 sq ft. Predict the *Price* of a 2460 sq ft. house (1 SD above the mean) using the least squares equation in the original units to verify that you get the same prediction as in Part 5.

from their parents. What he found was that even though tall parents tend to have tall children, their children are closer to the mean of all offspring than their parents were. He called this phenomenon "regression to mediocrity". He had hoped that tall parents would have children whose heights were even more extraordinary, but that's not what he found. Often when people relate essentially the same variable in two different groups, or at two different times, they see this same phenomenon—the tendency of the response variable to be closer to the mean than the predicted value. Unfortunately, people try to interpret this by thinking that the performance of those far from the mean is deteriorating, but it's just a mathematical fact about the correlation. So, today we try to be less judgmental about this phenomenon and we call it **regression to the mean**. We managed to get rid of the term "mediocrity," but the name regression stuck as a name for the whole least squares fitting procedure—and that's where we get the term **regression line**.

MATH BOX

Where does the equation of the line of best fit come from? And what does it really mean?

We can write the equation of any line if we know a point on the line and the slope. Finding a point on the line isn't hard. For the BK sandwiches, it's logical to predict that a sandwich with average protein will have average fat. In fact, it's fairly easy to show that the best fit line *always* passes through the point $(\bar{x}, \bar{y})$.

To find the slope, we'll again convert both x and y to z-scores. We need to keep a few facts in mind:

1. The mean of any set of z-scores is 0. Since we know the line passes through $(\bar{x}, \bar{y})$ this tells us that the line of best fit for z-scores passes through the origin (0, 0).

2. The standard deviation of a set of z-scores is 1, and so, of course, are the variances. This means, in particular, that

$$\frac{\sum(z_y - \bar{z}_y)^2}{(n-1)} = \frac{\sum(z_y - 0)^2}{(n-1)} = \frac{\sum z_y^2}{(n-1)} = 1,$$

a fact that will be important soon.

3. Remember from Chapter 6, the correlation is $r = \dfrac{\sum z_x z_y}{(n-1)}$, also important soon.

We want to find the slope of the best fit line. Because it passes through the origin, its equation will be of the simple form $\hat{z}_y = mz_x$. We want to find the value of m that minimizes the sum of the squared residuals $\sum(z_y - \hat{z}_y)^2$. Because it actually makes things simpler, we'll divide that sum by $(n-1)$ and minimize this quantity instead.

Minimize: $\dfrac{\sum(z_y - \hat{z}_y)^2}{(n-1)}$

Since $\hat{z}_y = mz_x$, we are minimizing $\dfrac{\sum(z_y - mz_x)^2}{(n-1)}$

We can expand this and write it as: $\dfrac{\sum z_y^2}{(n-1)} - 2m\dfrac{\sum z_x z_y}{(n-1)} + m^2 \dfrac{\sum z_x^2}{(n-1)}$

4. Notice that the first term is 1 (because it's the variance of the z_y), the second term is $-2mr$ and the last term is m^2 (because the variance of z_x is 1 as well). So to minimize the sum of squared residuals we only have to minimize: $1 - 2mr + m^2$. This last expression is quadratic, which, as you may remember, describes a parabola. And a parabola of the form $y = ax^2 + bx + c$ reaches its minimum at its turning point, which occurs when $x = -\dfrac{b}{2a}$. So, for this particular quadratic, we can minimize the sum of the squared residuals by choosing $m = -\dfrac{-(2r)}{2(1)} = r$.

This is really an amazing fact and explains to some extent the importance of the correlation as a summary of the strength of the linear relationship between x and y. To say it again: The slope of the best fit line for the z-scores (the standardized values) of any two variables x and y is their correlation, r.

We've converted to z-scores to make the mathematics easier, but what does this result mean for any two variables x and y?

- A slope of r for the best fit line for z-scores means that if we move 1 standard deviation in z_x, we expect to move r standard deviations in z_y. So a BK item that is one standard deviation above the mean in protein would be expected to be r standard deviations above the mean in fat. And a sandwich that is 2 standard deviations higher than the mean in protein would be expected to be 2 × r standard deviations higher than the mean in fat. If the correlation were 1, we'd expect it to be a full 2 standard deviations higher in fat. On the other hand, if the correlation were 0, we'd predict it to have the mean fat content. For the Burger King items, the correlation actually is 0.83, so for a sandwich 2 standard deviations higher than the mean in protein, we'd predict its fat content to be 1.66 standard deviations higher than the average fat content.

Rather than just saying that we'd predict a BK sandwich to be r standard deviations above the mean in fat for each standard deviation above the mean in protein, we can write the relationship in the original units by multiplying by the ratio of the standard deviations. The slope of the regression line is then $b = r\dfrac{s_y}{s_x}$.

- We know that choosing $m = r$ minimizes the sum of the squared residuals, but how small does that sum get? Equation 4 told us that the sum of the squared residuals divided by $(n - 1)$ is $1 - 2mr + m^2$. When $m = r$, this quantity is $1 - 2r \times r + r^2 = 1 - r^2$. Because it's a sum of squares it has to be positive, so $1 - r^2 \geq 0$. That means that $r^2 \leq 1$, and therefore, $-1 \leq r \leq 1$. In other words, it shows that the correlation is always between -1 and $+1$.

The quantity r^2 is also important in its own right. Look back at the expression $\dfrac{\sum(z_y - \hat{z}_y)^2}{(n - 1)}$. Compare it to the variance of the z_y: $Var(z_y) = \dfrac{\sum(z_y - \bar{z}_y)^2}{(n - 1)}$. Almost the same, aren't they? The variance of z_y measures its variation from the mean. The sum of the residuals measures variation from the predicted values, from the regression line. The variance of z_y is 1. Now that we've minimized the variation around the regression line it's only $1 - r^2$. In other words, the regression line has removed (or accounted for) r^2 of the original variance in y. Knowing how much variance the regression removes helps us assess the strength of our model. We'll use r^2 to help us do just that.

7.5 Examining the Residuals

■ **NOTATION ALERT**
Galton used r for correlation, so we use e for "error." But it's not a mistake. Statisticians often refer to variability not explained by a model as error.

The residuals are defined as

$$\text{Residual} = \text{Observed value} - \text{Predicted value},$$

or in symbols,

$$e = y - \hat{y}.$$

When we want to know how well the model fits, we can ask instead what the model missed. No model is perfect, so it's important to know how and where it fails. To see that, we look at the residuals.

For Example KATRINA'S RESIDUAL

RECAP: The linear model relating hurricanes' wind speeds to their central pressures was

$$\widehat{MaxWindSpeed} = 1024.464 - 0.968\, CentralPressure.$$

Let's use this model to make predictions and see how those predictions do.

QUESTION: Hurricane Katrina had a central pressure measured at 920 millibars. What does our regression model predict for her maximum wind speed? How good is that prediction, given that Katrina's actual wind speed was measured at 110 knots?

ANSWER: Substituting 920 for the central pressure in the regression model equation gives

$$\widehat{MaxWindSpeed} = 1024.464 - 0.968(920) = 133.90.$$

The regression model predicts a maximum wind speed of 133.90 knots for Hurricane Katrina.

> The residual for this prediction is the observed value minus the predicted value:
>
> 150 − 133.90 = 16.1 kts.
>
> In the case of Hurricane Katrina, the model predicts a wind speed 16.1 knots lower than was actually observed.

Figure 7.5
The residuals for the BK menu regression look appropriately boring. There are no obvious patterns, although there are a few points with large residuals. The negative ones turn out to be grilled chicken items, which are among the lowest fat (per protein) items on the menu. The two high outliers contain hash browns, which are high fat per protein items.

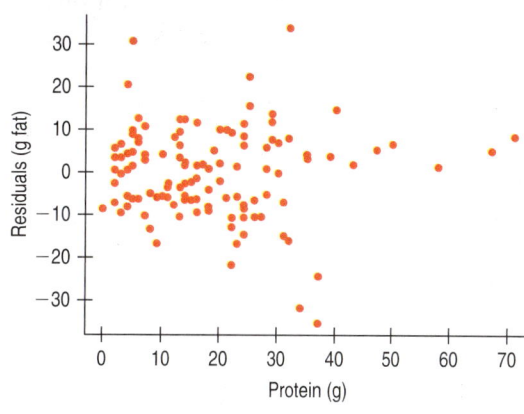

Residuals help us see whether the model makes sense. When a regression model is appropriate, it should model the underlying relationship. Nothing interesting should be left behind. So after we fit a regression model, we usually plot the residuals in the hope of finding . . . nothing interesting.

A scatterplot of the residuals vs. the *x*-values should be the most boring scatterplot you've ever seen. It shouldn't have any interesting features, like a direction or shape. It should stretch horizontally, with about the same amount of scatter throughout. It should show no bends, and it should have no outliers. If you see any of these features, investigate them if possible, and find out what the regression model missed.

Most computer statistics packages plot the residuals against the predicted values $\hat{y}$, rather than against *x*.[6] When the slope is negative, the two versions are mirror images. When the slope is positive, they're virtually identical except for the axis labels. Since all we care about is the patterns (or, better, lack of patterns) in the plot, it really doesn't matter which way we plot the residuals.

The Residual Standard Deviation

If the residuals show no interesting pattern when we plot them against *x*, we can look at how big they are. After all, we're trying to make them as small as possible. Since their mean is always zero, though, it's only sensible to look at how much they vary. The standard deviation of the residuals, s_e, gives us a measure of how much the points spread around the regression line. Of course, for this summary to make sense, the residuals should all share the same underlying spread. That's why we check to make sure that the residual plot has about the same amount of scatter throughout.

This gives us a new assumption: the **Equal Variance Assumption**. The associated condition to check is the **Does the Plot Thicken? Condition**. We ask if the plot thickens because the usual violation to the assumption is that the spread increases as *x* or the predicted values increase. However, to be thorough, we check to make sure that the spread is about the same throughout. We can check that either in the original scatterplot of *y* against *x* or in the scatterplot of residuals.

We estimate the **standard deviation of the residuals** in almost the way you'd expect:

$$s_e = \sqrt{\frac{\sum e^2}{n-2}}.$$

WHY n − 2?
Why n − 2 rather than n − 1? We used n − 1 for s when we estimated the mean. Now we're estimating both a slope and an intercept. Looks like a pattern—and it is. We subtract one more for each parameter we estimate.

---
[6]They have a good reason for this choice. When regression models use more than one *x*-variable, there can be many *x*'s to plot against, but there's still only one $\hat{y}$.

We don't need to subtract the mean because the mean of the residuals $\bar{e} = 0$. The $n - 2$, like the $n - 1$ in the denominator of the variance, is an adjustment we'll discuss in a later chapter.

For the Burger King foods, the standard deviation of the residuals is 10.6 grams of fat. That looks about right in the scatterplot of residuals. The residual for the BK Tendercrisp chicken was -14.7 grams, just under 1.5 standard deviations.

It's a good idea to make a histogram of the residuals. If we see a unimodal, symmetric histogram, then we can apply the 68–95–99.7 Rule to see how well the regression model describes the data. In particular, we know that 95% of the residuals should be no farther from 0 than $2s_e$. The Burger King residuals look like this:

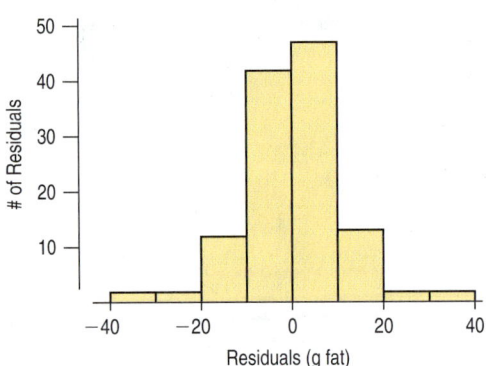

Figure 7.6
The histogram of residuals is symmetric and unimodal, centered at 0, with a standard deviation of 10.6. Only a few values lie outside of 2 standard deviations. The low ones are the chicken items mentioned before. The high ones contain hash browns.

Sure enough, most are within 2(10.6), or 21.2, grams of fat from 0.

7.6 R^2—The Variation Accounted For by the Model

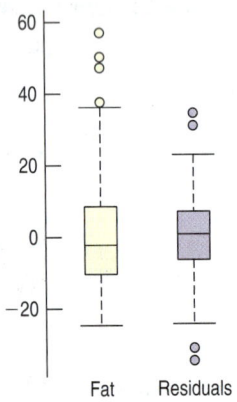

Figure 7.7
Compare the variability of total Fat with the residuals from the regression. The means have been subtracted to make it easier to compare spreads. The variation left in the residuals is unaccounted for by the model, but it's less than the variation in the original data.

A S *Simulation:* **Interpreting Equations.** This demonstrates how to use and interpret linear equations.

The variation in the residuals is the key to assessing how well the model fits the data. Let's compare the variation of the response variable with the variation of the residuals. The total *Fat* has a standard deviation of 16.2 grams. The standard deviation of the residuals is 10.6 grams. If the correlation were 1.0 and the model predicted the *Fat* values perfectly, the residuals would all be zero and have no variation. We couldn't possibly do any better than that.

On the other hand, if the correlation were zero, the model would simply predict 24.8 grams of *Fat* (the mean) for all menu items. The residuals from that prediction would just be the observed *Fat* values minus their mean. These residuals would have the same variability as the original data because, as we know, just subtracting the mean doesn't change the spread.

How well does the BK regression model do? Look at the boxplots. The variation in the residuals is smaller than in the data, but certainly bigger than zero. That's nice to know, but how much of the variation is still left in the residuals? If you had to put a number between 0% and 100% on the fraction of the variation left in the residuals, what would you say?

All regression models fall somewhere between the two extremes of zero correlation and perfect correlation. We'd like to gauge where our model falls. Can we use the correlation to do that? Well, a regression model with correlation -0.5 is doing as well as one with correlation $+0.5$. They just have different directions. But if we *square* the correlation coefficient, we'll get a value between 0 and 1, and the direction won't matter. All regression analyses include this statistic, although by tradition, it is written with a capital letter, R^2, and pronounced "R-squared." The squared correlation, r^2, gives the fraction of the data's variation accounted for by the model, and $1 - r^2$ is the fraction of the original variation left in the residuals. For the Burger King model, $r^2 = 0.76^2 = 0.58$, and $1 - r^2$ is 0.42, so 42% of the variability in total *Fat* has been left in the residuals. How close was that to your guess?

An R^2 of 0 means that none of the variance in the data is in the model; all of it is still in the residuals. It would be hard to imagine using that model for anything. Because R^2 is a fraction of a whole, it is often given as a percentage.[7] For the Burger King data, R^2 is 58%.

When interpreting a regression model, you need to be able to explain what R^2 means. According to our linear model, 58% of the variability in the fat content of Burger King sandwiches is accounted for by variation in the protein content.

---

[7] By contrast, we usually give correlation coefficients as decimal values between -1.0 and 1.0.

> **TWICE AS STRONG?**
> Is a correlation of 0.80 twice as strong as a correlation of 0.40? Not if you think in terms of R^2. A correlation of 0.80 means an R^2 of $0.80^2 = 64\%$. A correlation of 0.40 means an R^2 of $0.40^2 = 16\%$ — only a quarter as much of the variability accounted for. A correlation of 0.80 gives an R^2 *four* times as strong as a correlation of 0.40 and accounts for four times as much of the variability.

> **HOW CAN WE SEE THAT R^2 IS REALLY THE FRACTION OF VARIANCE ACCOUNTED FOR BY THE MODEL?**
> It's a simple calculation. The variance of the fat content of the Burger King foods is $16.2^2 = 262.44$. If we treat the residuals as data, the variance of the residuals is 110.53.[8] As a fraction, that's $110.53/262.44 = 0.42$ or 42%. That's the fraction of the variance that is *not* accounted for by the model. The fraction that is accounted for is $100\% - 42\% = 58\%$, just the value we got for R^2.

For Example INTERPRETING R^2

RECAP: Our regression model that predicts maximum wind speed in hurricanes based on the storm's central pressure has $R^2 = 80.3\%$.

QUESTION: What does that say about our regression model?

ANSWER: An R^2 of 80.3% indicates that 80.3% of the variation in maximum wind speed can be accounted for by the hurricane's central pressure. Other factors, such as temperature and whether the storm is over water or land, may account for some of the remaining variation.

Just Checking

Back to our regression of house *Price* ($) on house *Size* (square feet):

$$\widehat{Price} = 8400 + 88.67\, Size.$$

The R^2 value is reported as 57.8%, and the standard deviation of the residuals is $53,790.

7. What does the R^2 value mean about the relationship of price and size?
8. Is the correlation of price and size positive or negative? How do you know?
9. You find that your house is worth $50,000 more than the regression model predicts. You are undoubtedly pleased, but is this actually a surprisingly large residual?

> **Some extreme tales**
> One major company developed a method to differentiate between proteins. To do so, they had to distinguish between regressions with R^2 of 99.99% and 99.98%. For this application, 99.98% was not high enough.
> On the other hand, the head of a hedge fund reports that although his regressions give R^2 below 2%, they are highly successful because those used by his competition are even lower.

How Big Should R^2 Be?

R^2 is always between 0% and 100%. But what's a "good" R^2 value? The answer depends on the kind of data you are analyzing and on what you want to do with it. Just as with correlation, there is no value for R^2 that automatically determines that the regression is "good." Data from scientific experiments often have R^2 in the 80% to 90% range and even higher. Data from observational studies and surveys, though, often show relatively weak associations because it's so difficult to measure responses reliably. An R^2 of 50% to 30% or even lower might be taken as evidence of a useful regression. The standard deviation of the residuals can give us more information about the usefulness of the regression by telling us how much scatter there is around the line.

As we've seen, an R^2 of 100% is a perfect fit, with no scatter around the line. The s_e would be zero. All of the variance is accounted for by the model and none is left in the residuals at all. This sounds great, but it's too good to be true for real data.[9]

---
[8] This isn't quite the same as squaring the s_e that we discussed on the previous pages, but it's very close. We'll deal with the distinction in Chapter 23.
[9] If you see an R^2 of 100%, it's a good idea to figure out what happened. You may have discovered a new law of Physics, but it's much more likely that you accidentally regressed two variables that measure the same thing.

Along with the slope and intercept for a regression, you should always report R^2 so that readers can judge for themselves how successful the regression is at fitting the data. Statistics is about variation, and R^2 measures the success of the regression model in terms of the fraction of the variation of y accounted for by the regression. R^2 is the first part of a regression that many people look at because, along with the scatterplot, it tells whether the regression model is even worth thinking about.

Predicting in the Other Direction—A Tale of Two Regressions

Regression slopes may not behave exactly the way you'd expect at first. Our regression model for the Burger King sandwiches was $\widehat{Fat} = 8.4 + 0.917\ Protein$. That equation allowed us to estimate that a sandwich with 31 grams of protein would have 36.7 grams of fat. Suppose, though, that we knew the fat content and wanted to predict the amount of protein. It might seem natural to think that by solving the equation for *Protein* we'd get a model for predicting *Protein* from *Fat*. But that doesn't work.

Our original model is $\hat{y} = b_0 + b_1 x$, but the new one needs to evaluate an $\hat{x}$ based on a value of y. We don't have y in our original model, only $\hat{y}$, and that makes all the difference. Our model doesn't fit the BK data values perfectly, and the least squares criterion focuses on the *vertical* (y) errors the model makes in using x to model y—not on *horizontal* errors related to x.

A quick look at the equations reveals why. Simply solving for x would give a new line whose slope is the reciprocal of ours. To model y in terms of x, our slope is $b_1 = r\dfrac{s_y}{s_x}$. To model x in terms of y, we'd need to use the slope $b_1 = r\dfrac{s_x}{s_y}$. That's *not* the reciprocal of ours.

Sure, if the correlation, r, were 1.0 or -1.0 the slopes *would* be reciprocals, but that would happen only if we had a perfect fit. Real data don't follow perfect straight lines, so in the real world y and $\hat{y}$ aren't the same, r is a fraction, and the slopes of the two models are not simple reciprocals of one another. Also, if the standard deviations were equal—for example, if we standardize both variables—the two slopes would be *the same*. Far from being reciprocals, both would be equal to the correlation—but we already knew that the correlation of x with y is the same as the correlation of y with x.

Otherwise, slopes of the two lines will not be reciprocals, so we can't derive one equation from the other. If we want to predict x from y, we need to create that model from the data. For example, to predict *Protein* from *Fat* we can't just invert the model we found before. Instead we find, the slope, $b_1 = 0.76\ \dfrac{13.5}{16.2} = 0.63$ grams of protein per gram of fat. The regression model is then $\widehat{Protein} = 2.29 + 0.63\ Fat$.

Moral of the story: Decide which variable you want to use (x) to predict values for the other (y). Then find the model that does that. If, later, you want to make predictions in the other direction, start over and create the other model from scratch.

Protein	Fat
$\bar{x}$ = 18.0 g	$\bar{y}$ = 4.8 g
s_x = 13.5 g	s_y = 16.2 g
$r = 0.76$	

7.7 Regression Assumptions and Conditions

The linear regression model may be the most widely used model in all of Statistics. It has everything we could want in a model: two easily estimated parameters, a meaningful measure of how well the model fits the data, and the ability to predict new values. It even provides an easy way to see violations of conditions in plots of the residuals.

Like all models, though, linear models are only appropriate if some assumptions are true. We can't confirm assumptions, but we often can check related conditions.

First, be sure that both variables are quantitative. It makes no sense to perform a regression on categorical variables. After all, what could the slope possibly mean? Always check the **Quantitative Variables Condition**.

The linear model only makes sense if the relationship is linear. It is easy to check the associated **Straight Enough Condition**. Just look at the scatterplot of y vs. x. You don't need a *perfectly* straight plot, but it must be straight enough for the linear model to make

> **MAKE A PICTURE (OR TWO)**
> You can't check the conditions just by checking boxes. You need to examine both the original scatterplot of y against x before you fit the model, and the plot of residuals afterward. These plots can save you from making embarrassing errors and losing points on the exam.

sense. If you try to model a curved relationship with a straight line, you'll usually get just what you deserve. If the scatterplot is not straight enough, stop here. You can't use a linear model for *any* two variables, even if they are related. They must have a *linear* association, or the model won't mean a thing.

For the standard deviation of the residuals to summarize the scatter of all the residuals, the residuals must share the same spread. That's an assumption. But if the scatterplot of *y* vs. *x* looks equally spread out everywhere and (often more vividly) if the *residual plot* of residuals *vs.* predicted values also has a consistent spread, then the assumption is reasonable. The most common violation of that equal variance assumption is for the residuals to spread out more for *larger* values of *x*, so a good mnemonic for this check is the **Does the Plot Thicken? Condition**.

Outlying points can dramatically change a regression model. They can even change the sign of the slope, which would give a very different impression of the relationship between the variables if you only look at the regression model. So check the **Outlier Condition**. Check both the scatterplot of *y* against *x*, and the residual plot to be sure there are no outliers. The residual plot often shows violations more clearly and may reveal other unexpected patterns or interesting quirks in the data. Of course, any outliers are likely to be interesting and informative, so be sure to look into why they are unusual.

To summarize:

Before starting, be sure to check the

- **Quantitative Variable Condition** If either *y* or *x* is categorical, you can't make a scatterplot and you can't perform a regression. Stop.

From the scatterplot of *y* against *x*, check the

- **Straight Enough Condition** Is the relationship between *y* and *x* straight enough to proceed with a linear regression model?
- **Outlier Condition** Are there any outliers that might dramatically influence the fit of the least squares line?
- **Does the Plot Thicken? Condition** Does the spread of the data around the generally straight relationship seem to be consistent for all values of *x*?

After fitting the regression model, make a plot of residuals against the predicted values and look for

- Any bends that would violate the **Straight Enough Condition**,
- Any outliers that weren't clear before, and
- Any change in the spread of the residuals from one part of the plot to another.

Step-by-Step Example REGRESSION

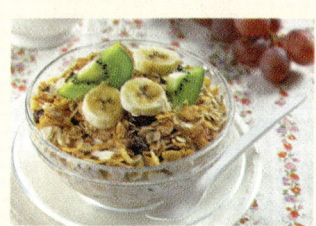

If you plan to hit the fast-food joints for lunch, you should have a good breakfast. Nutritionists, concerned about "empty calories" in breakfast cereals, recorded facts about 77 cereals, including their *Calories* per serving and *Sugar* content (in grams). (Data in **Cereals**)

Question: How are calories and sugar content related in breakfast cereals?

THINK ➡ Plan State the problem.

I am interested in the relationship between sugar content and calories in cereals.

Variables Name the variables and report the W's.

I have two quantitative variables, *Calories* and *Sugar* content per serving, measured on 77 breakfast cereals. The units of measurement are calories and grams of sugar, respectively.

✓ **Quantitative Variables:** Both variables are quantitative. The units of measurement for *Calories* and *Sugar* are calories and grams respectively.

I'll check the remaining conditions from the scatterplot:

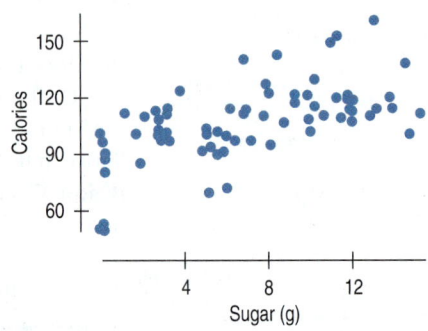

Check the conditions for a regression by making a picture. Never fit a regression without looking at the scatterplot first. Calories are reported to the nearest 10 calories and sugar content is reported to the nearest gram, so many cereals have the same values for both variables. In order to show them better, we have added a little random noise to both variables in the scatterplot, a process called *jittering*. Of course, we use the actual values for calculation.

✓ **Outlier Condition**: There are no obvious outliers or groups.

✓ **Straight Enough:** The scatterplot looks straight to me.

✓ **Does the Plot Thicken?** The spread around the line looks about the same throughout, but I'll check it again in the residuals.

I will fit a regression model to these data.

SHOW ➡ Mechanics If there are no clear violations of the conditions, fit a straight line model of the form $\hat{y} = b_0 + b_1 x$ to the data. Summary statistics give the building blocks of the calculation.

Calories
$$\bar{y} = 106.88 \text{ calories}$$
$$s_y = 19.48 \text{ calories}$$

Sugar
$$\bar{x} = 6.94 \text{ grams}$$
$$s_x = 4.42 \text{ grams}$$

Correlation
$$r = 0.564$$

Find the slope.
$$b_1 = r\frac{s_y}{s_x} = 0.564 \frac{19.48}{4.42}$$
$$= 2.49 \text{ calories per gram of sugar}$$

Find the intercept.
$$b_0 = \bar{y} - b_1 \bar{x} = 106.88 - 2.49(6.94)$$
$$= 89.6 \text{ calories}$$

So the least squares line is

Write the equation, using meaningful variable names.
$$\hat{y} = 89.6 + 2.49\, x,$$
or $\widehat{Calories} = 89.6 + 2.49\, Sugar.$

State the value of R^2.
Squaring the correlation gives
$$R^2 = 0.564^2 = 0.318 \text{ or } 31.8\%.$$

TELL → Conclusion Describe what the model says in words and numbers. Be sure to use the names of the variables and their units.

The key to interpreting a regression model is to start with the phrase "b_1 y-units per x-unit," substituting the estimated value of the slope for b_1 and the names of the respective units.

The intercept is then a starting or base value. It may (as in this example) be meaningful, or (when $x = 0$ is not realistic) it may just be a starting value.

R^2 gives the fraction of the variability of y accounted for by the linear regression model.

Find the standard deviation of the residuals, s_e, and compare it to the original, s_y.

The scatterplot shows a positive, linear relationship and no outliers. The least squares regression line fit through these data has the equation

$$\widehat{Calories} = 89.6 + 2.49 \, Sugar.$$

The slope says that cereals have about 2.49 calories per gram of sugar.

The intercept predicts that a serving of a sugar-free cereal would average about 89.6 calories.

The R^2 says that 31.8% of the variability in Calories is accounted for by variation in Sugar content.

$s_e = 16.2$ calories. That's smaller than the original SD of 19.5, but still fairly large.

THINK AGAIN → Check Again Even though we looked at the scatterplot *before* fitting a regression model, a plot of the residuals is essential to any regression analysis because it is the best check for additional patterns and interesting quirks in the data. (As we did earlier, the points have been jittered to see the pattern more clearly.)

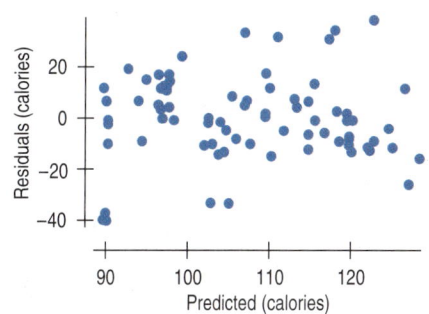

The residuals show a horizontal direction, a shapeless form, and roughly equal scatter for all predicted values. The linear model appears to be appropriate.

Reality Check: Is the Regression Reasonable?

> **Regression: adjective, noun, or verb?** You may see the term *regression* used in different ways. There are many ways to fit a line to data, but the term "regression line" or "regression" without any other qualifiers always means least squares. People also use *regression* as a verb when they speak of *regressing* a *y*-variable on an *x*-variable to mean fitting a linear model.

Statistics don't come out of nowhere. They are based on data. The results of a statistical analysis should reinforce your common sense, not fly in its face. If the results are surprising, then either you've learned something new about the world or your analysis is wrong.

Whenever you perform a regression, think about the coefficients and ask whether they make sense. Is a slope of 2.5 calories per gram of sugar reasonable? That's hard to say right off. We know from the summary statistics that a typical cereal has about 100 calories and 7 grams of sugar per serving. A gram of sugar contributes some calories (actually, 4, but you don't need to know that), so calories should go up with increasing sugar. The direction of the slope seems right.

To see if the *size* of the slope is reasonable, a useful trick is to consider its order of magnitude. Start by asking if deflating the slope by a factor of 10 seems reasonable. Is 0.25 calories per gram of sugar enough? The 7 grams of sugar found in the average cereal would contribute less than 2 calories. That seems too small.

Then try inflating the slope by a factor of 10. Is 25 calories per gram reasonable? The average cereal would have 175 calories from sugar alone. The average cereal has only 100 calories per serving, though, so that slope seems too big.

We have tried inflating the slope by a factor of 10 and deflating it by 10 and found both to be unreasonable. So, like Goldilocks, we're left with the value in the middle that's just right. And an increase of 2.5 calories per gram of sugar is certainly *plausible*.

The small effort of asking yourself whether the regression equation is plausible is repaid whenever you catch errors or avoid saying something silly or absurd about the data. It's too easy to take something that comes out of a computer at face value and assume that it makes sense.

Always be skeptical and ask yourself if the answer is reasonable.

For Example CAUSATION AND REGRESSION

RECAP: Our regression model predicting hurricane wind speeds from the central pressure was reasonably successful. The negative slope indicates that, in general, storms with lower central pressures have stronger winds.

QUESTION: Can we conclude that lower central barometric pressure *causes* the higher wind speeds in hurricanes?

ANSWER: No. While it may be true that lower pressure causes higher winds, a regression model for observed data such as these cannot demonstrate causation. Perhaps higher wind speeds reduce the barometric pressure, or perhaps both pressure and wind speed are driven by some other variable we have not observed.

(As it happens, in hurricanes it is reasonable to say that the low central pressure at the eye is responsible for the high winds because it draws moist, warm air into the center of the storm, where it swirls around, generating the winds. But as is often the case, things aren't quite that simple. The winds themselves contribute to lowering the pressure at the center of the storm as it becomes organized into a hurricane. The lesson is that to understand causation in hurricanes, we must do more than just model the relationship of two variables; we must study the mechanism itself.)

WHAT CAN GO WRONG?

There are many ways in which data that appear at first to be good candidates for regression analysis may be unsuitable. And there are ways that people use regression that can lead them astray. Here's an overview of the most common problems. We'll discuss them at length in the next chapter.

- **Don't fit a straight line to a nonlinear relationship.** Linear regression is suited only to relationships that are, well, *linear*. Fortunately, we can often improve the linearity easily by using re-expression. We'll come back to that topic in Chapter 8.

- **Don't ignore outliers.** Outliers can have a serious impact on the fitted model. You should identify them and think about why they are extraordinary. If they turn out not to be obvious errors, read the next chapter for advice.

- **Don't invert the regression.** The BK regression model was $\widehat{Fat} = 8.4 + 0.91\ Protein$. Knowing protein content, we can predict the amount of fat. But that doesn't let us switch the regression around. We can't use this model to predict protein values from fat values. To model y from x, the least squares slope is $b_1 = r\dfrac{s_y}{s_x}$. To model x in terms of y, we'd find $b_1 = r\dfrac{s_x}{s_y}$. That's not the reciprocal of the first slope (unless the correlation is 1.0). To swap the predictor–response roles of the variables in a regression (which can sometimes make sense), we must fit a new regression equation (it turns out to be $\widehat{Protein} = 2.29 + 0.63\ Fat$).

CONNECTIONS

The linear model is one of the most important models in Statistics. Chapter 5 talked about the assignment of variables to the y- and x-axes. That didn't matter to correlation, but it does matter to regression because y is predicted by x in the regression model.

The connection of R^2 to correlation is obvious, although it may not be immediately clear that just by squaring the correlation we can learn the fraction of the variability of y accounted for by a regression on x. We'll return to this in subsequent chapters.

We made a big fuss about knowing the units of your quantitative variables. We didn't need units for correlation, but without the units we can't define the slope of a regression. A regression makes no sense if you don't know the *Who*, the *What*, and the *Units* of both your variables.

We've summed squared deviations before when we computed the standard deviation and variance. That's not coincidental. They are closely connected to regression.

What Have We Learned?

Learning Objectives

Model linear relationships with a linear model.
- Examine residuals to assess how well a model fits the data.

The best fit model is the one that minimizes the sum of the squared residuals—the Least Squares Model.

Find the coefficients of the least squares line from the correlation and the summary statistics of each variable: $b_1 = r\dfrac{s_y}{s_x}$.
$$b_0 = \bar{y} - b_1\bar{x}.$$

Understand the correlation coefficient as the number of standard deviations by which one variable is expected to change for a one standard deviation change in the other.
- Recognize the sometimes-surprising phenomenon of regression to the mean that can occur when two variables measuring the same, or similar, things are correlated.

Always examine the residuals to check for violations of assumptions and conditions and to identify any outliers.

Always report and interpret R^2, which reports the fraction of the variation of y accounted for by the linear model.
- R^2 is the square of the correlation coefficient.

Be sure to check the conditions for regression before reporting or interpreting a regression model. Check that:
- The relationship is Straight Enough.
- A scatterplot does not "thicken" in any region.
- There are no outliers.

Interpret a regression slope as y-units per x-unit.

Review of Terms

Linear model An equation of the form
$$\hat{y} = b_0 + b_1 x.$$
To interpret a linear model, we need to know the variables (along with their W's) and their units (p. 181).

Model An equation or formula that simplifies and represents reality (p. 183).

Predicted value The value of $\hat{y}$ found for a given x-value in the data. A predicted value is found by substituting the x-value in the regression equation. The predicted values are the values on the fitted line; the points $(x, \hat{y})$ all lie exactly on the fitted line (p. 183).

	The predicted values are found from the linear model that we fit: $$\hat{y} = b_0 + b_1 x.$$
Residuals	The differences between data values and the corresponding values predicted by the regression model—or, more generally, values predicted by any model (p. 183). $$\text{Residual} = \text{Observed value} - \text{Predicted value} = y - \hat{y}$$
Regression line (Line of best fit)	The particular linear equation $$\hat{y} = b_0 + b_1 x$$ that satisfies the least squares criterion is called the least squares regression line. Casually, we often just call it the regression line, or the line of best fit (p. 184).
Least squares	The least squares criterion specifies the unique line that minimizes the variance of the residuals or, equivalently, the sum of the squared residuals (p. 184).
Slope	The slope, b_1, gives a value in "y-units *per* x-unit." Changes of one unit in x are associated with changes of b_1 units in predicted values of y (p. 184). The slope can be found by $$b_1 = r \frac{s_y}{s_x}.$$
Intercept	The intercept, b_0, gives a starting value in y-units. It's the $\hat{y}$-value when x is 0. You can find the intercept from $b_0 = \bar{y} - b_1 \bar{x}$ (p. 184).
Regression to the mean	Because the correlation is always less than 1.0 in magnitude, each predicted $\hat{y}$ tends to be fewer standard deviations from its mean than its corresponding x was from its mean (p. 190).
Standard deviation of the residuals (s_e)	The standard deviation of the residuals is found by $s_e = \sqrt{\frac{\sum e^2}{n-2}}$. When the assumptions and conditions are met, the residuals can be well described by using this standard deviation and the 68–95–99.7 Rule (p. 193).
R^2	The square of the correlation between y and x (p. 194).

- R^2 gives the fraction of the variability of y accounted for by the least squares linear regression on x.

- R^2 is an overall measure of how successful the regression is in linearly relating y to x.

On the Computer REGRESSION

All statistics packages make a table of results for a regression. These tables may differ slightly from one package to another, but all are essentially the same—and all include much more than we need to know for now. Every computer regression table includes a section that looks something like this:

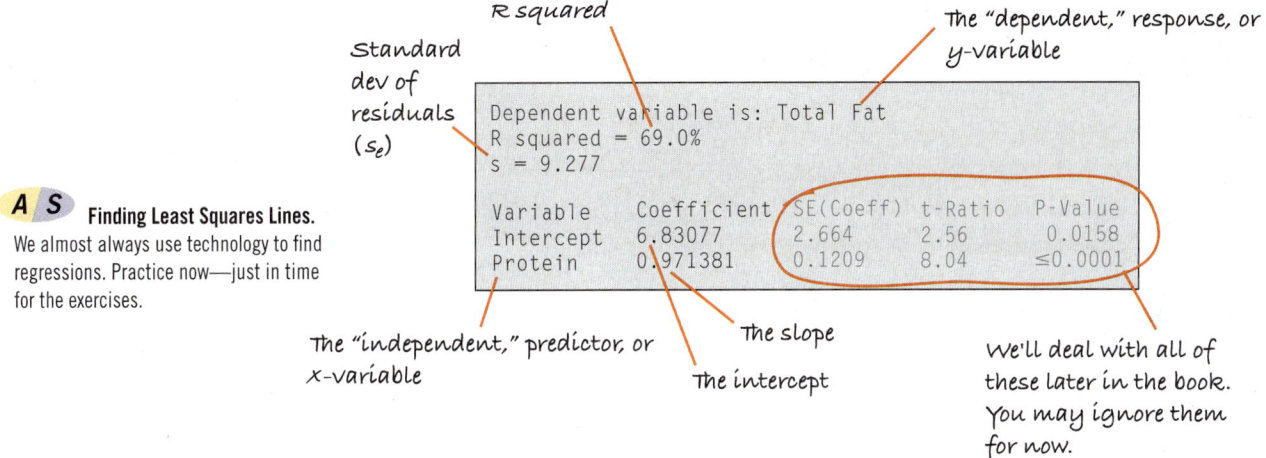

Finding Least Squares Lines. We almost always use technology to find regressions. Practice now—just in time for the exercises.

The slope and intercept coefficient are given in a table such as this one. Usually the slope is labeled with the name of the x-variable, and the intercept is labeled "Intercept" or "Constant." So the regression equation shown here is

$$\widehat{Fat} = 6.83077 + 0.97138 \, Protein.$$

It is not unusual for statistics packages to give many more digits of the estimated slope and intercept than could possibly be estimated from the data. (The original data were reported to the nearest gram.) Ordinarily, you should round most of the reported numbers to one digit more than the precision of the data, and the slope to two. We will learn about the other numbers in the regression table later in the book. For now, all you need to be able to do is find the coefficients, the s_e, and the R^2 value.

DATA DESK

- Select the *y*-variable and the *x*-variable.
- In the **Plot** menu, choose **Scatterplot**.
- From the scatterplot HyperView menu, choose **Add Regression Line** to display the line.
- From the HyperView menu, choose **Regression** to compute the regression.
- To plot the residuals, click on the **HyperView** menu on the **Regression** output table.
- A menu drops down that offers scatterplots of residuals against predicted values (as well as other options).

COMMENTS

Alternatively, find the regression first with the **Regression** command in the **Calc** menu. Click on the *x*-variable's name to open a menu that offers the scatterplot.

EXCEL

- Click on a blank cell in the spreadsheet.
- Go to the **Formulas** tab in the Ribbon and click **More Functions → Statistical**.

COMMENTS

The correlation is computed in the selected cell. Correlations computed this way will update if any of the data values are changed.

- Choose the **CORREL** function from the drop-down menu of functions.
- In the dialog that pops up, enter the range of one of the variables in the space provided.
- Enter the range of the other variable in the space provided.
- Click **OK**.

Before you interpret a correlation coefficient, always make a scatterplot to check for nonlinearity and outliers. If the variables are not linearly related, the correlation coefficient cannot be interpreted.

JMP

- Choose **Fit Y by X** from the **Analyze** menu.
- Specify the y-variable in the Select Columns box and click the **y, Response** button.
- Specify the x-variable and click the **X, Factor** button.
- Click **OK** to make a scatterplot.

- In the scatterplot window, click on the red triangle beside the heading labeled "Bivariate Fit ..." and choose **Fit Line**. JMP draws the least squares regression line on the scatterplot and displays the results of the regression in tables below the plot.

MINITAB

- Choose **Regression** from the **Stat** menu.
- From the Regression submenu, choose **Fitted Line Plot**.
- In the Fitted Line Plot dialog, click in the **Response Y** box, and assign the y-variable from the variable list.

- Click in the **Predictor X** box, and assign the x-variable from the Variable list. Make sure that the Type of Regression Model is set to Linear. Click the **OK** button.

R

- lm(Y~X) produces the linear model.
- summary(lm(Y~X)) produces more information about the model.

COMMENTS
Typically, your variables X and Y will be in a data frame. If DATA is the name of the data frame, then
- lm(Y~X, data = DATA) is the preferred syntax.

SPSS

- Choose **Interactive** from the **Graphs** menu.
- From the interactive Graphs submenu, choose **Scatterplot**.
- In the Create Scatterplot dialog, drag the y-variable into the **y-axis target**, and the x-variable into the **x-axis target**.

- Click on the **Fit** tab.
- Choose **Regression** from the **Method** popup menu. Click the **OK** button.

STATCRUNCH

- Click on **Stat**.
- Choose **Regression » Simple Linear**.
- Choose X and Y variable names from the list of columns.
- Click on **Next** (twice) to **Plot the fitted line** on the scatterplot.
- Click on **Calculate** to see the regression analysis.
- Click on **Next** to see the scatterplot.

COMMENTS

Remember to check the scatterplot to be sure a linear model is appropriate.

Note that before you **Calculate**, clicking on **Next** also allows you to:
- enter an X-value for which you want to find the predicted Y-value;
- save all the fitted values;
- save the residuals;
- ask for a residuals plot.

TI-83/84 PLUS

To find the equation of the regression line (add the line to a scatterplot), choose **LinReg(a+bx)**, tell it the list names, and then add a comma to specify a function name (from **VARS Y-Vars 1:Function**). The final command looks like

LinReg(a+bx)L1, L2, Y1.

- To make a residuals plot, set up a **STATPLOT** as a scatterplot.
- Specify your explanatory data list as Xlist.
- For Ylist, import the name RESID from the **LIST NAMES** menu. **ZoomStat** will now create the residuals plot.

COMMENTS

Each time you execute a **LinReg** command, the calculator automatically computes the residuals and stores them in a data list named RESID. If you want to see them, go to **STAT EDIT**. Space through the names of the lists until you find a blank. Import RESID from the LIST NAMES menu. Now every time you have the calculator compute a regression analysis, it will show you the residuals.

Exercises

Section 7.1

1. **True or false** If false, explain briefly.
   a) We choose the linear model that passes through the most data points on the scatterplot.
   b) The residuals are the observed y-values minus the y-values predicted by the linear model.
   c) Least squares means that the square of the largest residual is as small as it could possibly be.

2. **True or false II** If false, explain briefly.
   a) Some of the residuals from a least squares linear model will be positive and some will be negative.
   b) Least Squares means that some of the squares of the residuals are minimized.
   c) We write $\hat{y}$ to denote the predicted values and y to denote the observed values.

Section 7.2

3. **Least squares interpretations** A least squares regression line was calculated to relate the length (cm) of newborn boys to their weight in kg. The line is $\widehat{weight} = -5.94 + 0.1875\, length$. Explain in words what this model means. Should new parents (who tend to worry) be concerned if their newborn's length and weight don't fit this equation?

4. **Residual interpretations** The newborn grandson of one of the authors was 48 cm long and weighed 3 kg. According to the regression model of Exercise 3, what was his residual? What does that say about him?

Section 7.3

5. Bookstore sales revisited Recall the data we saw in Chapter 6, exercise 3 for a bookstore. The manager wants to predict *Sales* from *Number of Sales People Working*.

Number of Sales People Working	Sales (in $1000)
2	10
3	11
7	13
9	14
10	18
10	20
12	20
15	22
16	22
20	26
$\bar{x} = 10.4$	$\bar{y} = 17.6$
$SD(x) = 5.64$	$SD(y) = 5.34$
$r = 0.965$	

a) Find the slope estimate, b_1.
b) What does it mean, in this context?
c) Find the intercept, b_0.
d) What does it mean, in this context? Is it meaningful?
e) Write down the equation that predicts *Sales* from *Number of Sales People Working*.
f) If 18 people are working, what *Sales* do you predict?
g) If sales are actually $25,000, what is the value of the residual?
h) Have we overestimated or underestimated the sales?

6. Disk drives 2014 again Recall the data on disk drives we saw in chapter 6 exercise 4. Suppose we want to predict *Price* from *Capacity*.

Capacity (in TB)	Price (in $)
0.50	59.99
1.0	79.99
2.0	111.97
3.0	109.99
4.0	149.99
6.0	423.34
8.0	596.11
12.0	1079.99
32.0	4461.00
$\bar{x} = 7.6111$	$\bar{y} = 785.819$
$SD(x) = 9.855$	$SD(y) = 1418.67$
$r = 0.9876$	

a) Find the slope estimate, b_1.
b) What does it mean, in this context?
c) Find the intercept, b_0.
d) What does it mean, in this context? Is it meaningful?
e) Write down the equation that predicts *Price* from *Capacity*.
f) What would you predict for the price of a 20 TB drive?
g) A 20 TB drive on Amazon.com was listed at $2017.86. According to the model, does this seem like a good buy? How much would you save compared to what you expected to pay?
h) Does the model overestimate or underestimate the price?
i) The correlation is very high. Does this mean that the model is accurate? Explain. (Hint: Revisit the scatterplot from chapter 6 exercise 4.)

Section 7.4

7. Sophomore slump? A CEO complains that the winners of his "rookie junior executive of the year" award often turn out to have less impressive performance the following year. He wonders whether the award actually encourages them to slack off. Can you offer a better explanation?

8. Sophomore slump again? An online investment blogger advises investing in mutual funds that have performed badly the past year because "regression to the mean tells us that they will do well next year." Is he correct?

Section 7.5

9. Bookstore sales once more Here are the residuals for a regression of *Sales* on *Number of Sales People Working* for the bookstore of exercise 5:

Number of Sales People Working	Residual
2	0.07
3	0.16
7	−1.49
9	−2.32
10	0.77
10	2.77
12	0.94
15	0.20
16	−0.72
20	−0.37

a) What are the units of the residuals?
b) Which residual contributes the most to the sum that was minimized according to the Least Squares Criterion to find this regression?
c) Which residual contributes least to that sum?

10. Disk drives 2014, residuals Here are the residuals for a regression of *Price* on *Capacity* for the hard drives of Exercise 6. (based on the hand-computed coefficients.)

Capacity	Residual
0.50	285.16
1.0	234.07
2.0	123.88
3.0	−20.27
4.0	−122.44
6.0	−133.43
8.0	−245.00
12.0	−329.80
32.0	207.81

a) Which residual contributes the most to the sum that is minimized by the Least Squares criterion?
b) Five of the residuals are negative. What does that mean about those drives? Be specific and use the correct units.

Section 7.6

11. Bookstore sales last time For the regression model for the bookstore of exercise 5, what is the value of R^2 and what does it mean?

12. Disk drives encore For the hard drive data of Exercise 6, find and interpret the value of R^2.

Section 7.7

13. Residual plots Here are residual plots (residuals plotted against predicted values) for three linear regression models. Indicate which condition appears to be violated (linearity, outlier or equal spread) in each case.

a)

b)

c)
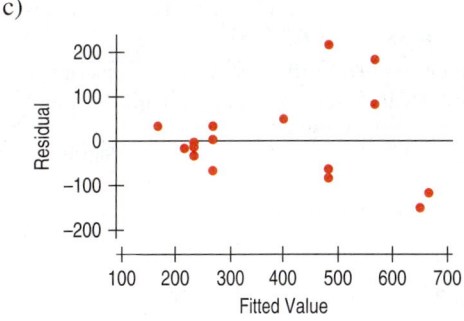

14. Disk drives 2014, residuals again Here is a scatterplot of the residuals from the regression of the hard drive prices on their sizes from Exercise 6.

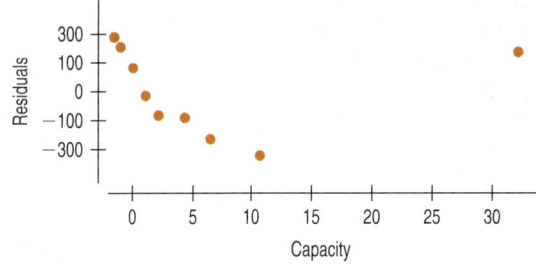

a) Are any assumptions or conditions violated? If so, which ones?
b) What would you recommend about this regression?

Chapter Exercises

15. Cereals For many people, breakfast cereal is an important source of fiber in their diets. Cereals also contain potassium, a mineral shown to be associated with maintaining a healthy blood pressure. An analysis of the amount of fiber (in grams) and the potassium content (in milligrams) in servings of 77 breakfast cereals produced the regression model $\widehat{Potassium} = 38 + 27\ Fiber$. If your cereal provides 9 grams of fiber per serving, how much potassium does the model estimate you will get?

16. Engine size In Chapter 6, Exercise 41 we examined the relationship between the fuel economy (mpg) and *Engine Size* for 35 models of cars (Data in **Fuel Economy 2014**). Further analysis produces the regression model $\widehat{mpg} = 36.25 - 3.867\ Engine\ size$. If the car you are thinking of buying has a 4 liter engine, what does this model suggest your gas mileage would be?

17. More cereal Exercise 15 describes a regression model that estimates a cereal's potassium content from the amount of fiber it contains. In this context, what does it mean to say that a cereal has a negative residual?

18. Engine size again Exercise 16 describes a regression model that uses a car's engine size to estimate its fuel economy. In this context, what does it mean to say that a certain car has a positive residual?

19. Another bowl In Exercise 15, the regression model $\widehat{Potassium} = 38 + 27\, Fiber$ relates fiber (in grams) and potassium content (in milligrams) in servings of breakfast cereals. Explain what the slope means.

20. More engine size In Exercise 16, the regression model $\widehat{mpg} = 36.25 - 3.867\, Engine\ Size$ relates cars' engine size to their fuel economy (in mpg). Explain what the slope means.

21. Cereal again The correlation between a cereal's fiber and potassium contents is $r = 0.903$. What fraction of the variability in potassium is accounted for by the amount of fiber that servings contain?

22. Another car The correlation between a car's engine size and its fuel economy (in mpg) is $r = -0.8476$. What fraction of the variability in fuel economy is accounted for by the engine size?

23. Last bowl! For Exercise 15's regression model predicting potassium content (in milligrams) from the amount of fiber (in grams) in breakfast cereals, $s_e = 30.77$. Explain in this context what that means.

24. Last tank! For Exercise 16's regression model predicting fuel economy (in mpg) from the car's engine size, $s_e = 3.265$. Explain in this context what that means.

25. Regression equations Fill in the missing information in the following table.

	$\bar{x}$	s_x	$\bar{y}$	s_y	r	$\hat{y} = b_0 + b_1 x$
a)	10	2	20	3	0.5	
b)	2	0.06	7.2	1.2	-0.4	
c)	12	6			-0.8	$\hat{y} = 200 - 4x$
d)	2.5	1.2		100		$\hat{y} = -100 + 50x$

26. More regression equations Fill in the missing information in the following table.

	$\bar{x}$	s_x	$\bar{y}$	s_y	r	$\hat{y} = b_0 + b_1 x$
a)	30	4	18	6	-0.2	
b)	100	18	60	10	0.9	
c)		0.8	50	15		$\hat{y} = -10 + 15x$
d)			18	4	-0.6	$\hat{y} = 30 - 2x$

27. Residuals Tell what each of the residual plots below indicates about the appropriateness of the linear model that was fit to the data.

28. Residuals Tell what each of the residual plots below indicates about the appropriateness of the linear model that was fit to the data.

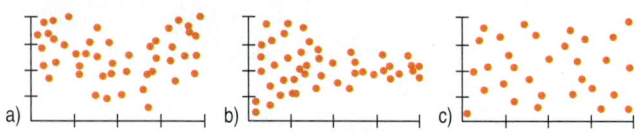

29. Real estate A random sample of records of home sales from Feb. 15 to Apr. 30, 1993, from the files maintained by the Albuquerque Board of Realtors gives the *Price* and *Size* (in square feet) of 117 homes. A regression to predict *Price* (in thousands of dollars) from *Size* has an R^2 of 71.4%. The residuals plot indicated that a linear model is appropriate.

a) What are the variables and units in this regression?
b) What units does the slope have?
c) Do you think the slope is positive or negative? Explain.

30. Coasters 2014, revisited The data set on roller coasters from Chapter 6, Exercise 23 lists the *Duration* of the ride in seconds in addition to the *Drop* height in feet. One coaster (the "Tower of Terror") is unusual for having large drop but a short ride. After setting it aside, a regression to predict *Duration* from *Drop* for the remaining 85 coasters has $R^2 = 32.0\%$.

a) What are the variables and units in this regression?
b) What units does the slope have?
c) Do you think the slope is positive or negative? Explain.

31. What slope? If you create a regression model for predicting the *Weight* of a car (in pounds) from its *Length* (in feet), is the slope most likely to be 3, 30, 300, or 3000? Explain.

32. What slope again? If you create a regression model for estimating the *Height* of a pine tree (in feet) based on the *Circumference* of its trunk (in inches), is the slope most likely to be 0.1, 1, 10, or 100? Explain.

33. Real estate again The regression of *Price* on *Size* of homes in Albuquerque had $R^2 = 71.4\%$, as described in Exercise 29. Write a sentence (in context, of course) summarizing what the R^2 says about this regression.

34. Coasters 2014 again Exercise 30 examined the association between the *Duration* of a roller coaster ride and the height of its initial *Drop*, reporting that $R^2 = 32.0\%$. Write a sentence (in context, of course) summarizing what the R^2 says about this regression.

35. Misinterpretations A Biology student who created a regression model to use a bird's *Height* when perched for predicting its *Wingspan* made these two statements. Assuming the calculations were done correctly, explain what is wrong with each interpretation.

a) My R^2 of 93% shows that this linear model is appropriate.
b) A bird 10 inches tall will have a wingspan of 17 inches.

36. More misinterpretations A Sociology student investigated the association between a country's *Literacy Rate* and *Life Expectancy*, and then drew the conclusions listed below. Explain why each statement is incorrect. (Assume that all the calculations were done properly.)
a) The R^2 of 64% means that the *Literacy Rate* determines 64% of the *Life Expectancy* for a country.
b) The slope of the line shows that an increase of 5% in *Literacy Rate* will produce a 2-year improvement in *Life Expectancy*.

37. Real estate redux The regression of *Price* on *Size* of homes in Albuquerque had $R^2 = 71.4\%$, as described in Exercise 29.
a) What is the correlation between *Size* and *Price*?
b) What would you predict about the *Price* of a home 1 SD above average in *Size*?
c) What would you predict about the *Price* of a home 2 SDs below average in *Size*?

38. Another ride The regression of *Duration* of a roller coaster ride on the height of its initial *Drop*, described in Exercise 30, had $R^2 = 32.0\%$.
a) What is the correlation between *Drop* and *Duration*?
b) What would you predict about the *Duration* of the ride on a coaster whose initial *Drop* was 1 standard deviation below the mean *Drop*?
c) What would you predict about the *Duration* of the ride on a coaster whose initial *Drop* was 3 standard deviations above the mean *Drop*?

39. ESP People who claim to "have ESP" participate in a screening test in which they have to guess which of several images someone is thinking of. You and a friend both took the test. You scored 2 standard deviations above the mean, and your friend scored 1 standard deviation below the mean. The researchers offer everyone the opportunity to take a retest.
a) Should you choose to take this retest? Explain.
b) Now explain to your friend what his decision should be and why.

40. SI jinx Players in any sport who are having great seasons, turning in performances that are much better than anyone might have anticipated, often are pictured on the cover of *Sports Illustrated*. Frequently, their performances then falter somewhat, leading some athletes to believe in a "*Sports Illustrated* jinx." Similarly, it is common for phenomenal rookies to have less stellar second seasons—the so-called "sophomore slump." While fans, athletes, and analysts have proposed many theories about what leads to such declines, a statistician might offer a simpler (statistical) explanation. Explain.

41. More real estate Consider the Albuquerque home sales from Exercise 29 again. The regression analysis gives the model $\widehat{Price} = 47.82 + 0.061\ Size$.
a) Explain what the slope of the line says about housing prices and house size.
b) What price would you predict for a 3000-square-foot house in this market?
c) A real estate agent shows a potential buyer a 1200-square-foot home, saying that the asking price is $6000 less than what one would expect to pay for a house of this size. What is the asking price, and what is the $6000 called?

42. Last ride Consider the roller coasters (with the outlier removed) described in Exercise 30 again. The regression analysis gives the model $\widehat{Duration} = 85.83 + 0.397\ Drop$.
a) Explain what the slope of the line says about how long a roller coaster ride may last and the height of the coaster.
b) A new roller coaster advertises an initial drop of 200 feet. How long would you predict the rides last?
c) Another coaster with a 150-foot initial drop advertises a 2 and a half-minute ride. Is this longer or shorter than you'd expect? By how much? What's that called?

43. Cigarettes Is the nicotine content of a cigarette related to the "tar"? A collection of data (in milligrams) on 29 cigarettes produced the scatterplot, residuals plot, and regression analysis shown:

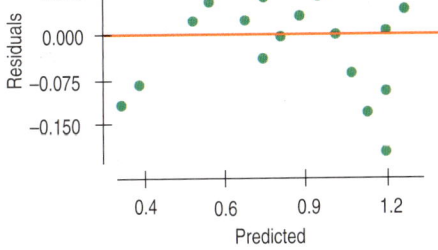

Dependent variable is Nicotine
R-squared = 92.4%

Variable	Coefficient
Constant	0.154030
Tar	0.065052

a) Do you think a linear model is appropriate here? Explain.
b) Explain the meaning of R^2 in this context.

44. Attendance 2013, revisited In the previous chapter, you looked at the relationship between the number of wins by American League baseball teams and the average attendance at their home games for the 2013 season. Here are the scatterplot, the residuals plot, and part of the regression analysis:

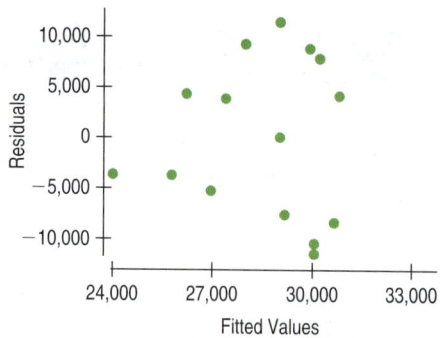

Dependent variable is Home Attendance
R-squared = 6.45%

Variable	Coefficient
Intercept	16484.0
Wins	147.4

a) Do you think a linear model is appropriate here? Explain.
b) Interpret the meaning of R^2 in this context.
c) Do the residuals show any pattern worth remarking on?
d) The point at the top of the plots is the New York Yankees. What can you say about the residual for the Yankees?

45. Another cigarette Consider again the regression of *Nicotine* content on *Tar* (both in milligrams) for the cigarettes examined in Exercise 43.

a) What is the correlation between *Tar* and *Nicotine*?
b) What would you predict about the average *Nicotine* content of cigarettes that are 2 standard deviations below average in *Tar* content?
c) If a cigarette is 1 standard deviation above average in *Nicotine* content, what do you suspect is true about its *Tar* content?

46. Attendance 2013, revisited Consider again the regression of *Average Attendance* on *Wins* for the baseball teams examined in Exercise 44.

a) What is the correlation between *Wins* and *Average Attendance*?

b) What would you predict about the *Average Attendance* for a team that is 2 standard deviations above average in *Wins*?
c) If a team is 1 standard deviation below average in attendance, what would you predict about the number of games the team has won?

47. Last cigarette Take another look at the regression analysis of tar and nicotine content of the cigarettes in Exercise 43.

a) Write the equation of the regression line.
b) Estimate the *Nicotine* content of cigarettes with 4 milligrams of *Tar*.
c) Interpret the meaning of the slope of the regression line in this context.
d) What does the *y*-intercept mean?
e) If a new brand of cigarette contains 7 milligrams of tar and a nicotine level whose residual is −0.5 mg, what is the nicotine content?

48. Attendance 2013, last inning Refer again to the regression analysis for average attendance and games won by American League baseball teams, seen in Exercise 44.

a) Write the equation of the regression line.
b) Estimate the *Average Attendance* for a team with 50 *Wins*.
c) Interpret the meaning of the slope of the regression line in this context.
d) In general, what would a negative residual mean in this context?
e) The St. Louis Cardinals, the 2013 NL Champions, are not included in these data because they are a National League team. During the 2013 regular season, the Cardinals won 97 games and averaged 41,602 fans at their home games. Calculate the residual for this team, and explain what it means.

49. Income and housing revisited In Chapter 6, Exercise 39, we learned that the Office of Federal Housing Enterprise Oversight (OFHEO) collects data on various aspects of housing costs around the United States. Here's a scatterplot (by state) of the *Housing Cost Index* (HCI) vs. the *Median Family Income* (MFI) for the 50 states. The correlation is $r = 0.65$. The mean HCI is 338.2, with a standard deviation of 116.55. The mean MFI is $46,234, with a standard deviation of $7072.47.

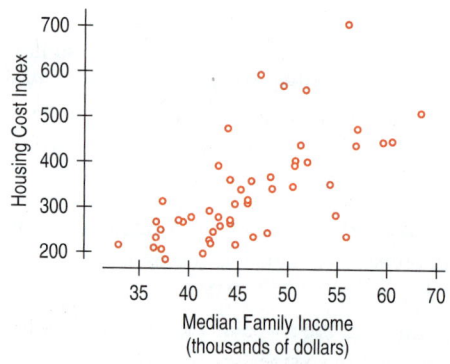

a) Is a regression analysis appropriate? Explain.
b) What is the equation that predicts Housing Cost Index from median family income?
c) For a state with MFI = $44,993, what would be the predicted HCI?
d) Washington, DC, has an MFI of $44,993 and an HCI of 548.02. How far off is the prediction in part b from the actual HCI?
e) If we standardized both variables, what would be the regression equation that predicts standardized HCI from standardized MFI?
f) If we standardized both variables, what would be the regression equation that predicts standardized MFI from standardized HCI?

50. Interest rates and mortgages 2013 again In Chapter 6, Exercise 40, we saw a plot of mortgages in the United States (in trillions of 2013 dollars) vs. the interest rate at various times over the past 25 years. The correlation is $r = -0.80$. The mean mortgage amount is $3.926 T and the mean interest rate is 7.194%. The standard deviations are $1.515 T for mortgage amounts and 2.055% for the interest rates.

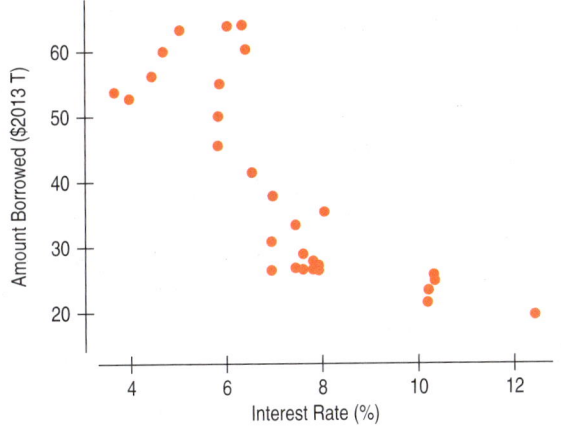

a) Is a regression model appropriate for predicting mortgage amount from interest rates? Explain.
b) Regardless of your answer to a), find the equation that predicts mortgage amount from interest rates?
c) What would you predict the mortgage amount would be if the interest rates climbed to 13%?
d) Do you have any reservations about your prediction in part c? Explain.
e) If we standardized both variables, what would be the regression equation that predicts standardized mortgage amount from standardized interest rates?
f) If we standardized both variables, what would be the regression equation that predicts standardized interest rates from standardized mortgage amount?

51. Online clothes An online clothing retailer keeps track of its customers' purchases. For those customers who signed up for the company's credit card, the company also has information on the customer's *Age* and *Income*. A random sample of 500 of these customers shows the following scatterplot of *Total Yearly Purchases* by *Age*:

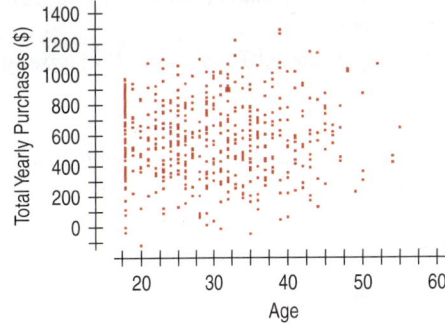

The correlation between *Total Yearly Purchases* and *Age* is $r = 0.037$. Summary statistics for the two variables are:

	Mean	SD
Age	29.67 yrs	8.51 yrs
Total Yearly Purchase	$572.52	$253.62

a) What is the linear regression equation for predicting *Total Yearly Purchase* from *Age*?
b) Do the assumptions and conditions for regression appear to be met?
c) What is the predicted *Total Yearly Purchase* for an 18-year-old? For a 50-year-old?
d) What percent of the variability in *Total Yearly Purchases* is accounted for by this model?
e) Do you think the regression might be a useful one for the company? Explain.

52. Online clothes II For the online clothing retailer discussed in the previous problem, the scatterplot of *Total Yearly Purchases* by *Income* looks like this:

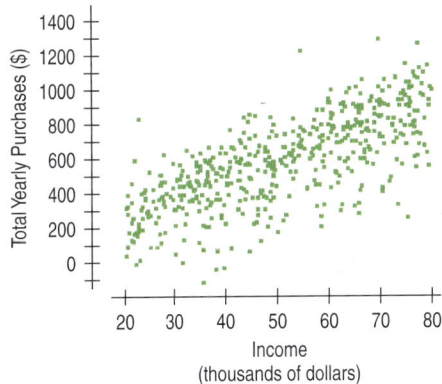

The correlation between *Total Yearly Purchases* and *Income* is 0.722. Summary statistics for the two variables are:

	Mean	SD
Income	$50,343.40	$16,952.50
Total Yearly Purchase	$572.52	$253.62

a) What is the linear regression equation for predicting *Total Yearly Purchase* from *Income*?
b) Do the assumptions and conditions for regression appear to be met?
c) What is the predicted *Total Yearly Purchase* for someone with a yearly *Income* of $20,000? For someone with an annual *Income* of $80,000?
d) What percent of the variability in *Total Yearly Purchases* is accounted for by this model?
e) Do you think the regression might be a useful one for the company? Comment.

53. SAT scores The SAT is a test often used as part of an application to college. SAT scores are between 200 and 800, but have no units. Tests are given in both Math and Verbal areas. SAT-Math problems require the ability to read and understand the questions, but can a person's verbal score be used to predict the math score? Verbal and math SAT scores of a high school graduating class are displayed in the scatterplot, with the regression line added.

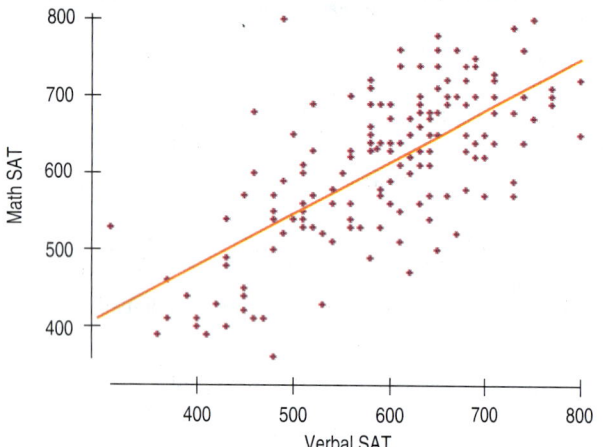

a) Describe the relationship.
b) Are there any students whose scores do not seem to fit the overall pattern?
c) For these data, $r = 0.685$. Interpret this statistic.
d) These verbal scores averaged 596.3, with a standard deviation of 99.5, and the math scores averaged 612.2, with a standard deviation of 96.1. Write the equation of the regression line.
e) Interpret the slope of this line.
f) Predict the math score of a student with a verbal score of 500.
g) Every year, some students score a perfect 1600. Based on this model, what would such a student's residual be for her math score?

54. Success in college Colleges use SAT scores in the admissions process because they believe these scores provide some insight into how a high school student will perform at the college level. Suppose the entering freshmen at a certain college have mean combined *SAT Scores* of 1222, with a standard deviation of 123. In the first semester, these students attained a mean *GPA* of 2.66, with a standard deviation of 0.56. A scatterplot showed the association to be reasonably linear, and the correlation between *SAT* score and *GPA* was 0.47.

a) Write the equation of the regression line.
b) Explain what the *y*-intercept of the regression line indicates.
c) Interpret the slope of the regression line.
d) Predict the GPA of a freshman who scored a combined 1400.
e) Based upon these statistics, how effective do you think SAT scores would be in predicting academic success during the first semester of the freshman year at this college? Explain.
f) As a student, would you rather have a positive or a negative residual in this context? Explain.

55. SAT, take 2 Suppose we wanted to use SAT math scores to estimate verbal scores based on the information in Exercise 53.

a) What is the correlation?
b) Write the equation of the line of regression predicting verbal scores from math scores.
c) In general, what would a positive residual mean in this context?
d) A person tells you her math score was 500. Predict her verbal score.
e) Using that predicted verbal score and the equation you created in Exercise 53, predict her math score.
f) Why doesn't the result in part e come out to 500?

56. Success, part 2 Based on the statistics for college freshmen given in Exercise 54, what SAT score would you predict for a freshmen who attained a first-semester GPA of 3.0?

57. Wildfires 2012 The National Interagency Fire Center (www.nifc.gov) reports statistics about wildfires. Here's an analysis of the number of wildfires between 1985 and 2012.

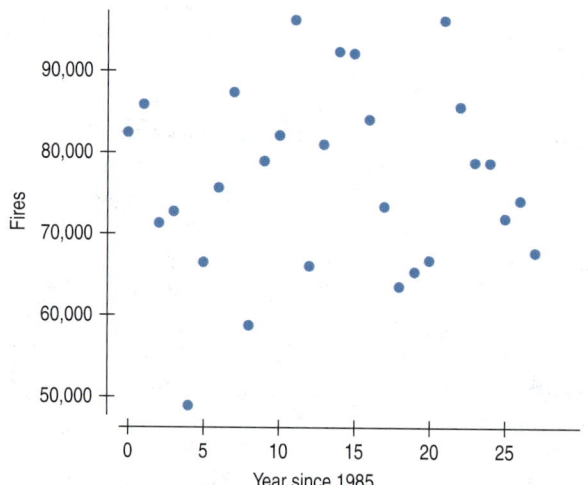

Dependent variable is Fires
R-squared = 0.032%
s = 11690

Variable	Coefficient
Intercept	75609.0
Years since 1985	78.5

a) Is a linear model appropriate for these data? Explain.
b) Interpret the slope in this context.
c) Can we interpret the intercept? Why or why not?
d) What does the value of s_e say about the size of the residuals? What does it say about the effectiveness of the model?
e) What does R^2 mean in this context?

58. Wildfires 2012—sizes We saw in Exercise 57 that the number of fires was nearly constant. But has the damage they cause remained constant as well? Here's a regression that examines the trend in *Acres per Fire*, (in hundreds of thousands of acres) together with some supporting plots:

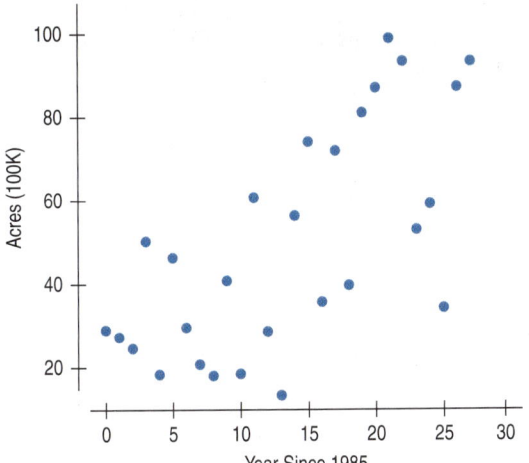

Dependent variable is Acres/fire
R-squared = 46.25%
s = 20.08

Variable	Coefficient
Intercept	19.62
Years since 1985	2.22

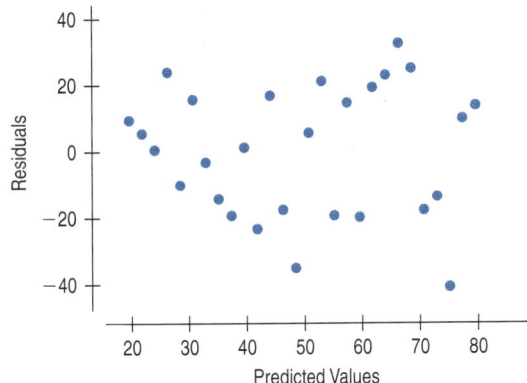

a) Is the regression model appropriate for these data? Explain.
b) Interpret the slope in this context.
c) Can we interpret the intercept? Why or why not?
d) What interpretation (if any) can you give for the R^2 in the regression table?

59. Used cars 2014 Carmax.com lists numerous Toyota Corollas for sale within a 250 mile radius of Redlands, CA. Listed at the top of the next column are the ages of the cars and the advertised prices.

a) Make a scatterplot for these data.
b) Describe the association between *Age* and *Price* of a used Corolla.
c) Do you think a linear model is appropriate?
d) Computer software says that $R^2 = 75.2\%$. What is the correlation between *Age* and *Price*?
e) Explain the meaning of R^2 in this context.
f) Why doesn't this model explain 100% of the variability in the price of a used Corolla?

Age (yr)	Price Advertised ($)
9	11,599
4	14,998
4	12,998
7	10,998
3	15,998
5	14,599
5	11,559
8	9,998
9	9,998
1	15,998
5	12,599
3	16,998
5	13,998

60. Drug abuse revisited In the exercises of the last chapter, you examined results of a survey conducted in the United States and 10 countries of Western Europe to determine the percentage of teenagers who had used marijuana and other drugs. Below is the scatterplot. Summary statistics showed that the mean percent that had used marijuana was 23.9%, with a standard deviation of 15.6%. An average of 11.6% of teens had used other drugs, with a standard deviation of 10.2%.

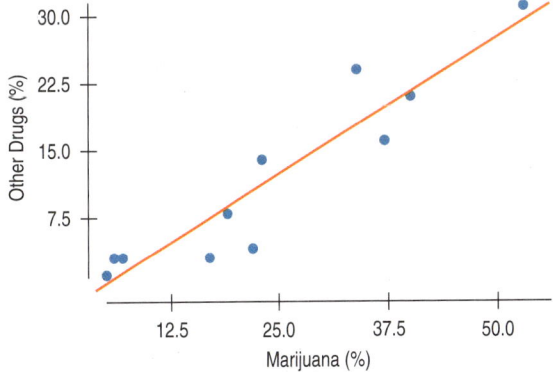

a) Do you think a linear model is appropriate? Explain.
b) For this regression, R^2 is 87.3%. Interpret this statistic in this context.

c) Write the equation you would use to estimate the percentage of teens who use other drugs from the percentage who have used marijuana.
d) Explain in context what the slope of this line means.
e) Do these results confirm that marijuana is a "gateway drug," that is, that marijuana use leads to the use of other drugs?

61. **More used cars 2014** Use the advertised prices for Toyota Corollas given in Exercise 59 to create a linear model for the relationship between a car's *Age* and its *Price*.
   a) Find the equation of the regression line.
   b) Explain the meaning of the slope of the line.
   c) Explain the meaning of the y-intercept of the line.
   d) If you want to sell a 7-year-old Corolla, what price seems appropriate?
   e) You have a chance to buy one of two cars. They are about the same age and appear to be in equally good condition. Would you rather buy the one with a positive residual or the one with a negative residual? Explain.
   f) You see a "For Sale" sign on a 10-year-old Corolla stating the asking price as $8,500. What is the residual?
   g) Would this regression model be useful in establishing a fair price for a 25-year-old car? Explain.

62. **Veggie burgers** Burger King introduced a meat-free burger in 2002. The nutrition label is shown here:

Nutrition Facts

Calories	330
Fat	10g*
Sodium	760g
Sugars	5g
Protein	14g
Carbohydrates	43g
Dietary Fiber	4g
Cholesterol	0

* (2 grams of saturated fat)

RECOMMENDED DAILY VALUES
(based on a 2,000-calorie/day diet)

Iron	20%
Vitamin A	10%
Vitamin C	10%
Calcium	6%

   a) Use the regression model created in this chapter, $\widehat{Fat} = 6.8 + 0.97\,Protein$, to predict the fat content of this burger from its protein content.
   b) What is its residual? How would you explain the residual?
   c) Write a brief report about the *Fat* and *Protein* content of this menu item. Be sure to talk about the variables by name and in the correct units.

63. **Burgers revisited** In the last chapter, you examined the association between the amounts of *Fat* and *Calories* in fast-food hamburgers. Here are the data:

Fat (g)	19	31	34	35	39	39	43
Calories	410	580	590	570	640	680	660

   a) Create a scatterplot of *Calories* vs. *Fat*.
   b) Interpret the value of R^2 in this context.
   c) Write the equation of the line of regression.
   d) Use the residuals plot to explain whether your linear model is appropriate.
   e) Explain the meaning of the y-intercept of the line.
   f) Explain the meaning of the slope of the line.
   g) A new burger containing 28 grams of fat is introduced. According to this model, its residual for calories is +33. How many calories does the burger have?

64. **Chicken** Chicken sandwiches are often advertised as a healthier alternative to beef because many are lower in fat. Tests on 11 brands of fast-food chicken sandwiches produced the following summary statistics and scatterplot from a graphing calculator:

	Fat (g)	Calories
Mean	20.6	472.7
St. Dev.	9.8	144.2
Correlation	0.947	

   a) Do you think a linear model is appropriate in this situation?
   b) Describe the strength of this association.
   c) Write the equation of the regression line.
   d) Explain the meaning of the slope.
   e) Explain the meaning of the y–intercept.
   f) What does it mean if a certain sandwich has a negative residual?
   g) If a chicken sandwich and a burger each advertised 35 grams of fat, which would you expect to have more calories (see Exercise 63)?
   h) McDonald's Filet-O-Fish sandwich has 26 grams of fat and 470 calories. Does the fat–calorie relationship in this sandwich appear to be very different from that found in chicken sandwiches or in burgers (see Exercise 63)? Explain.

65. **A second helping of burgers** In Exercise 63, you created a model that can estimate the number of *Calories* in a burger when the *Fat* content is known.
   a) Explain why you cannot use that model to estimate the fat content of a burger with 600 calories.
   b) Using an appropriate model, estimate the fat content of a burger with 600 calories.

66. **Cost of living 2013** Numbeo.com lists the cost of living (COL) for many cities around the world. These rankings

scale New York City as 100, and express the cost of living in other cities as a percentage of the New York cost. For example, the table below shows 25 of the most expensive cities in the world, and indicates that the cost of living in Paris was 30.3% higher than New York in 2010, but only 25.0% higher in 2013.

a) Using the scatterplot below, describe the association between costs of living in 2010 and 2013.
b) The correlation is 0.714. Find and interpret the value of R^2.
c) The regression equation predicting the 2013 cost of living from the 2010 figure is $\widehat{COL\ 2013} = -38.292 + 1.319\ COL\ 2010$. Use this equation to find the residual for Oslo.
d) Explain what the residual means.

City	COL 2013	COL 2010
Oslo	170.04	152.85
Copenhagen	124.9	138.91
Zurich	166.51	132.03
Paris	125.02	130.3
Geneva	156.17	122.69
Milan	106.12	122.58
Dublin	121.5	120.79
Brussels	111.55	120
London	116.06	118.47
Berlin	86.65	115.24
Lyon	104.08	115.04
Helsinki	111.76	113.91
Tampere	109.04	111.96
Amsterdam	107.61	111.3
Stockholm	126.96	111.02
Oxford	109.15	110.15
Tampa	81.32	109.64
Munich	105.01	107.96
Sydney	139.25	104.83
Vienna	99.5	103.88
Madrid	88.27	103.67
Atlanta	77.77	100.99
Edmonton	105.24	100.36
Seattle	91.79	100.33
New York	100	100

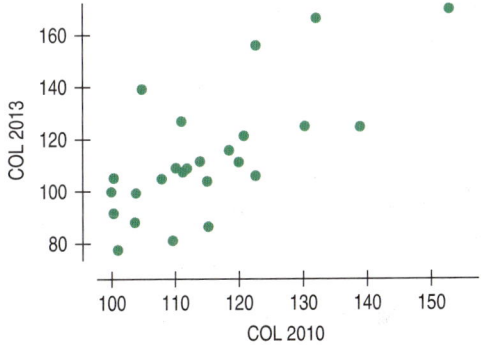

67. New York bridges We saw in this chapter that in Tompkins County, New York, older bridges were in worse condition than newer ones. Tompkins is a rural area. Is this relationship true in New York City as well? Here are data on the *Condition* (as measured by the state Department of Transportation Condition Index) and *Age at Inspection* for bridges in New York City.

Dependent variable is Condition
R-squared = 2.6%
s = 0.6708

Variable	Coefficient
Intercept	4.95147
Age@Inspection	−0.00481

a) New York State defines any bridge with a condition score less than 5 as *deficient*. What does this model predict for the condition scores of New York City bridges?
b) Our earlier model found that the condition of bridges in Tompkins County was decreasing at about 0.025 per year. What does this model say about New York City bridges?
c) How much faith would you place in this model? Explain.

68. Birthrates 2010 The table shows the number of live births per 1000 population in the United States, starting in 1965. (National Center for Health Statistics, www.cdc.gov/nchs/)

Year	1965	1970	1975	1980	1985	1990	1995	2000	2005	2010
Rate	19.4	18.4	14.8	15.9	15.6	16.4	14.8	14.4	14.0	13.0

a) Make a scatterplot and describe the general trend in *Birthrates*. (Enter *Year* as years since 1900: 65, 70, 75, etc.)
b) Find the equation of the regression line.
c) Check to see if the line is an appropriate model. Explain.
d) Interpret the slope of the line.
e) The table gives rates only at 5-year intervals. Estimate what the rate was in 1978.
f) In 1978, the birthrate was actually 15.0. How close did your model come?
g) Predict what the *Birthrate* will be in 2010. Comment on your faith in this prediction.
h) Predict the *Birthrate* for 2050. Comment on your faith in this prediction.

69. Climate change 2013 The earth's climate is getting warmer. The most common theory attributes the increase to an increase in atmospheric levels of carbon dioxide (CO_2), a greenhouse gas. Here is a scatterplot showing the mean annual CO_2 concentration in the atmosphere, measured in parts per million (ppm) at the top of Mauna Loa in Hawaii, and the mean annual air temperature over both land and sea across the globe, in degrees Celsius (°C) for the years 1970 to 2013.

A regression predicting *Temperature* from CO_2 produces the following output table (in part):

Dependent variable is Global Temperature (°C)
R-squared = 73.6%

Variable	Coefficient
Intercept	10.64
CO_2 (ppm)	0.0103

a) What is the correlation between CO_2 and *Temperature*?
b) Explain the meaning of *R*-squared in this context.
c) Give the regression equation.
d) What is the meaning of the slope in this equation?
e) What is the meaning of the *y*-intercept of this equation?
f) Here is a scatterplot of the residuals vs. predicted values. Does this plot show evidence of the violation of any assumptions behind the regression? If so, which ones?

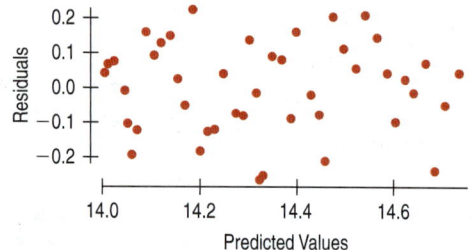

g) CO_2 levels will probably reach 400 ppm by 2020. What mean *Temperature* does the regression predict for that concentration of CO_2?
h) Does the answer in g) mean that when the CO_2 level hits 400 ppm, the Temperature will reach the predicted level? Explain briefly.

70. Climate change 2013, revisited In Exercise 69, we saw the relationship between CO_2 measured at Mauna Loa and average global temperatures from 1970 to 2013. Here is a plot of average global temperatures plotted against the yearly average of the Dow Jones Industrial Average for the same time period

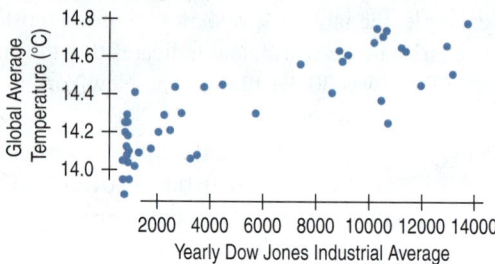

A regression produces the following output table (in part):

Dependent variable is Global Temperature (°C)
R-squared = 68.6%

Variable	Coefficient
Intercept	14.09
DJIA	0.0000459

a) What is the correlation between the *DJIA* and *Temperature*?
b) Explain the meaning of R-squared in this context.
c) Give the regression equation.
d) What is the meaning of the slope in this equation?
e) What is the meaning of the *y*-intercept of this equation?
f) Here is a scatterplot of the residuals vs. predicted values. Does this plot show evidence of the violation of any assumptions behind the regression? If so, which ones?

g) Suppose the Dow hits 20,000 in 2020. What mean Temperature does the regression predict for that level?
h) Does the answer in g) mean that when the Dow hits 20,000, the Temperature will reach the predicted level? Explain briefly.

71. Body fat It is difficult to determine a person's body fat percentage accurately without immersing him or her in water. Researchers hoping to find ways to make a good estimate immersed 20 male subjects, then measured their waists and recorded their weights shown in the table at the top of the next column.

a) Create a model to predict %*Body Fat* from *Weight*.
b) Do you think a linear model is appropriate? Explain.
c) Interpret the slope of your model.
d) Is your model likely to make reliable estimates? Explain.
e) What is the residual for a person who weighs 190 pounds and has 21% body fat?

Waist (in.)	Weight (lb)	Body Fat (%)	Waist (in.)	Weight (lb)	Body Fat (%)
32	175	6	33	188	10
36	181	21	40	240	20
38	200	15	36	175	22
33	159	6	32	168	9
39	196	22	44	246	38
40	192	31	33	160	10
41	205	32	41	215	27
35	173	21	34	159	12
38	187	25	34	146	10
38	188	30	44	219	28

T 72. Body fat again Would a model that uses the person's *Waist* size be able to predict the *%Body Fat* more accurately than one that uses *Weight*? Using the data in Exercise 71, create and analyze that model.

T 73. Heptathlon revisited We discussed the women's 2012 Olympic heptathlon in Chapter 6. Here are the results from the high jump, 800-meter run, and long jump for the 26 women who successfully completed all three events in the 2004 Olympics (www.espn.com):

Name	Country	High Jump (m)	800-m (sec)	Long Jump (m)
Carolina Klüft	SWE	1.91	134.15	6.51
Austra Skujyte	LIT	1.76	135.92	6.30
Kelly Sotherton	GBR	1.85	132.27	6.51
Shelia Burrell	USA	1.70	135.32	6.25
Yelena Prokhorova	RUS	1.79	131.31	6.21
Sonja Kesselschlaeger	GER	1.76	135.21	6.42
Marie Collonville	FRA	1.85	133.62	6.19
Natalya Dobrynska	UKR	1.82	137.01	6.23
Margaret Simpson	GHA	1.79	137.72	6.02
Svetlana Sokolova	RUS	1.70	133.23	5.84
J. J. Shobha	IND	1.67	137.28	6.36
Claudia Tonn	GER	1.82	130.77	6.35
Naide Gomes	POR	1.85	140.05	6.10
Michelle Perry	USA	1.70	133.69	6.02
Aryiro Strataki	GRE	1.79	137.90	5.97
Karin Ruckstuhl	NED	1.85	133.95	5.90
Karin Ertl	GER	1.73	138.68	6.03
Kylie Wheeler	AUS	1.79	137.65	6.36
Janice Josephs	RSA	1.70	138.47	6.21
Tiffany Lott Hogan	USA	1.67	145.10	6.15
Magdalena Szczepanska	POL	1.76	133.08	5.98
Irina Naumenko	KAZ	1.79	134.57	6.16
Yuliya Akulenko	UKR	1.73	142.58	6.02
Soma Biswas	IND	1.70	132.27	5.92
Marsha Mark-Baird	TRI	1.70	141.21	6.22
Michaela Hejnova	CZE	1.70	145.68	5.70

Let's examine the association among these events. Perform a regression to predict high-jump performance from the 800-meter results.

a) What is the regression equation? What does the slope mean?
b) What percent of the variability in high jumps can be accounted for by differences in 800-m times?
c) Do good high jumpers tend to be fast runners? (Be careful—low times are good for running events and high distances are good for jumps.)
d) What does the residuals plot reveal about the model?
e) Do you think this is a useful model? Would you use it to predict high-jump performance? (Compare the residual standard deviation to the standard deviation of the high jumps.)

T 74. Heptathlon revisited again We saw the data for the women's 2004 Olympic heptathlon in Exercise 73. Are the two jumping events associated? Perform a regression of the long-jump results on the high-jump results.

a) What is the regression equation? What does the slope mean?
b) What percentage of the variability in long jumps can be accounted for by high-jump performances?
c) Do good high jumpers tend to be good long jumpers?
d) What does the residuals plot reveal about the model?
e) Do you think this is a useful model? Would you use it to predict long-jump performance? (Compare the residual standard deviation to the standard deviation of the long jumps.)

T 75. Hard water revisited In an investigation of environmental causes of disease, data were collected on the annual mortality rate (deaths per 100,000) for males in 61 large towns in England and Wales. In addition, the water hardness was recorded as the calcium concentration (parts per million, ppm) in the drinking water. The following display shows the relationship between *Mortality* and *Calcium* concentration for these towns:

a) Describe what you see in this scatterplot, in context.
b) Here is the regression analysis of *Mortality* and *Calcium* concentration. What is the regression equation?

Dependent variable is Mortality
R-squared = 43%
s = 143.0

Variable	Coefficient	SE(Coeff)	t-Ratio	P-Value
Intercept	1676	29.30	57.2	<0.0001
Calcium	−3.23	0.48	26.66	<0.0001

c) Interpret the slope and *y*-intercept of the line, in context.
d) The largest residual, with a value of −348.6, is for the town of Exeter. Explain what this value means.
e) The hardness of Derby's municipal water is about 100 ppm of calcium. Use this equation to predict the mortality rate in Derby.
f) Explain the meaning of *R*-squared in this situation.

T 76. Gators Wildlife researchers monitor many wildlife populations by taking aerial photographs. Can they estimate the weights of alligators accurately from the air? Here is a regression analysis of the *Weight* of alligators (in pounds) and their *Length* (in inches) based on data collected about captured alligators.

Dependent variable is Weight
R-squared = 83.6%
s = 54.01

Variable	Coefficient	SE(Coeff)	t-Ratio	P-Value
Intercept	−393.3	47.53	−8.27	<0.0001
Length	5.9	0.5448	10.8	<0.0001

a) Did they choose the correct variable to use as the dependent variable and the predictor? Explain.
b) What is the correlation between an alligator's length and weight?
c) Write the regression equation.
d) Interpret the slope of the equation in this context.
e) Do you think this equation will allow the scientists to make accurate predictions about alligators? What part of the regression analysis indicates this? What additional concerns do you have?

77. Least squares Consider the four points $(10,10)$, $(20,50)$, $(40,20)$, and $(50,80)$. The least squares line is $\hat{y} = 7.0 + 1.1x$. Explain what "least squares" means, using these data as a specific example.

78. Least squares Consider the four points $(200,1950)$, $(400,1650)$, $(600,1800)$, and $(800,1600)$. The least squares line is $\hat{y} = 1975 - 0.45x$. Explain what "least squares" means, using these data as a specific example.

Just Checking ANSWERS

1. You should expect the price to be 0.76 standard deviations above the mean.
2. You should expect the size to be $2(0.76) = 1.52$ standard deviations below the mean.
3. The home is 1.5 standard deviations above the mean in size no matter how size is measured.
4. $0.76 \times \$77{,}000 = \$58{,}520$
5. $\$168{,}000 + 58{,}520 = \$226{,}520$
6. $8400 + 88.67 \times 2460 = \$226{,}528$. The difference is due to rounding error when the regression coefficients are kept to only 2 decimal places.
7. Differences in the size of houses account for about 57.8% of the variation in the house prices.
8. It's positive. The correlation and the slope have the same sign.
9. No, the standard deviation of the residuals is $53,790. We shouldn't be surprised by any residual smaller in magnitude than 2 standard deviations, and a residual of $50.000 is less than 1 standard deviation from 0.

chapter 8
Regression Wisdom

8.1 Examining Residuals
8.2 Extrapolation: Reaching Beyond the Data
8.3 Outliers, Leverage, and Influence
8.4 Lurking Variables and Causation
8.5 Working with Summary Values

> **Where are we going?**
> What happens when we fit a regression model to data that aren't straight? How bad will the predictions be? How can we tell if the model is appropriate? Questions like these are as important as fitting the model itself. In this chapter, we'll see how to tell whether a regression model is sensible and what to do if it isn't.

A S *Activity:* **Construct a Plot with a Given Slope.** How's your feel for regression lines? Can you make a scatterplot that has a specified slope?

Regression is used every day throughout the world to predict customer loyalty, numbers of admissions at hospitals, sales of automobiles, and many other things. Because regression is so widely used, it's also widely abused and misinterpreted. This chapter presents examples of regressions in which things are not quite as simple as they may have seemed at first and shows how you can still use regression to discover what the data have to say.

8.1 Examining Residuals

No regression analysis is complete without a display of the residuals to check that the linear model is reasonable. Because the residuals are what is "left over" after the model describes the relationship, they often reveal subtleties that were not clear from a plot of the original data. Sometimes these are additional details that help confirm or refine our understanding. Sometimes they reveal violations of the regression conditions that require our attention.

For example, the fundamental assumption in working with a linear model is that the relationship you are modeling is, in fact, linear. That sounds obvious, but you can't take it for granted. It may be hard to detect nonlinearity from the scatterplot you looked at before you fit the regression model. Sometimes you can see important features such as nonlinearity more readily when you plot the residuals.

> **STRAIGHT ENOUGH CONDITION**
> We can't *know* whether the **Linearity Assumption** is true, but we can see if it's *plausible* by checking the **Straight Enough Condition**.

Getting the "Bends": When the Residuals Aren't Straight

Jessica Meir and Paul Ponganis study emperor penguins at the Scripps Institution of Oceanography's Center for Marine Biotechnology and Biomedicine at the University of California at San Diego. Says Jessica:

> *Emperor penguins are the most accomplished divers among birds, making routine dives of 5–12 minutes, with the longest recorded dive over 27 minutes. These birds can also dive to depths of over 500 meters! Since air-breathing animals like penguins must hold their breath while submerged, the duration of any given dive depends on how much oxygen is in the bird's body at the beginning of the dive, how quickly that oxygen gets used, and the lowest level of oxygen the bird can tolerate. The rate of oxygen depletion is primarily determined by the penguin's heart rate. Consequently, studies of heart rates during dives can help us understand how these animals regulate their oxygen consumption in order to make such impressive dives.* *

The researchers equip emperor penguins with devices that record their heart rates during dives. Here's a scatterplot of the *Dive Heart Rate* (beats per minute) and the *Duration* (minutes) of dives by these high-tech penguins. (Data in **Penguins**)

Figure 8.1
The scatterplot of *Dive Heart Rate* in beats per minute (bpm) vs. *Duration* (minutes) shows a strong, roughly linear, negative association.

The scatterplot has a consistent trend from the upper left to the lower right and a moderately strong negative association ($R^2 = 71.5\%$). The linear regression equation

$$\widehat{DiveHeartRate} = 96.9 - 5.47\,Duration$$

says that for longer dives, the average *Dive Heart Rate* is lower by about 5.47 beats per dive minute, starting from a value of 96.9 beats per minute.

Figure 8.2
Plotting the residuals against *Duration* reveals a bend. It was also in the original scatterplot, but here it's easier to see.

The scatterplot of the residuals against *Duration* makes things clearer. The Linearity Assumption says we should see no pattern here, but the residual plot shows a strong bend, starting high on the left, dropping down in the middle of the plot, and rising again at the right. Graphs of residuals often reveal patterns such as this that were not readily apparent in the original scatterplot.

---

*Excerpt from Research Note on Emperor Penguins, Scripps Institution of Oceanography's Center for Marine Biotechnology and Biomedicine at the University of California at San Diego by Jessica Meir. Published by Meir, Jessica.

Now looking back at the original scatterplot, we can see that the scatter of points isn't really straight. There's a slight bend to that plot, but the bend is much easier to see in the residuals. Even though it means rechecking the Straight Enough Condition *after* you find the regression, it's always a good idea to check your scatterplot of the residuals for bends that you might have overlooked in the original scatterplot. We probably want to consider re-expressing one of the variables (as discussed in section 6.4) to help straighten out the relationship.

Sifting Residuals for Groups

In the Step-By-Step analysis in Chapter 7, to predict *Calories* from *Sugar* content in breakfast cereals, we examined a scatterplot of the residuals (Data in **Cereals**). Our first impression was that it had no particular structure. But let's look again.

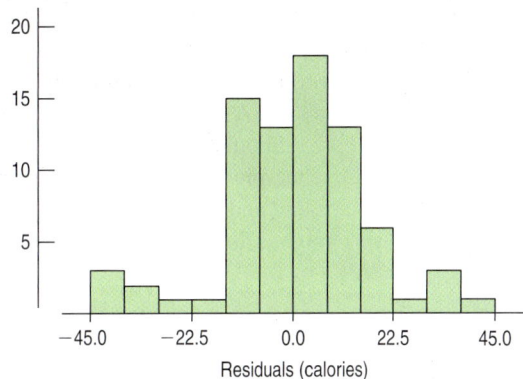

Figure 8.3
A histogram of the breakfast cereal regression residuals from Chapter 7 shows small modes both above and below the central large mode. These may be worth a second look.

Look at the histogram of the residuals. How would you describe its shape? It looks like there might be small modes on both sides of the central body of the data. One group of cereals seems to stand out (on the left) as having large negative residuals, with fewer calories than our regression model predicts, and another stands out (on the right) with large positive residuals. Whenever we suspect multiple modes, we ask whether there might be different subgroups in the data.

Let's look more carefully at those points in the residual plot (see Figure 8.4). Now we can see that those two groups do stand away from the central pattern in the scatterplot. The high-residual cereals are Just Right Fruit & Nut; Muesli Raisins, Dates & Almonds; Peaches & Pecans; Mueslix Crispy Blend; and Nutri-Grain Almond Raisin. Do these cereals seem to have something in common? They all present themselves as "healthy." This might be surprising, but in fact, "healthy" cereals often contain more fat, and therefore more calories, than we might expect from looking at their sugar content alone.

Figure 8.4
A scatterplot of the breakfast cereal regression residuals vs. predicted values (points have been jittered). The green "**x**" points are cereals whose calorie content is higher than the linear model predicts. The red "**–**" points show cereals with fewer calories than the model predicts. Is there something special about these cereals?

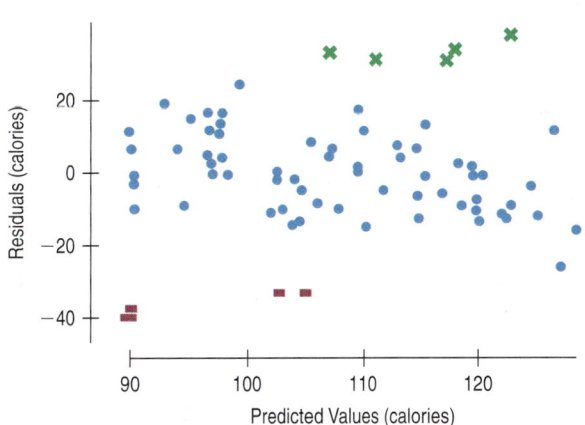

The low-residual cereals are Puffed Rice, Puffed Wheat, three bran cereals, and Golden Crisp. You might not have grouped these cereals together before. What they have in common is a low calorie count *relative to their sugar content*—even though their sugar contents are quite different.

These observations may not lead us to question the overall linear model, but they do help us understand that other factors may be part of the story. An examination of residuals often leads us to discover groups of observations that are different from the rest.

When we discover that there is more than one group in a regression, we may decide to analyze the groups separately, using a different model for each group. Or we can stick with the original model and simply note that there are groups that are a little different. Either way, the model will be wrong, but useful, so it will improve our understanding of the data.

Subsets

Cereal manufacturers aim cereals at different segments of the market. Supermarkets and cereal manufacturers try to attract different customers by placing different types of cereals on certain shelves. Cereals for kids tend to be on the "kid's shelf," at their eye level. Toddlers wouldn't be likely to grab a box from this shelf and beg, "Mom, can we please get this All-Bran with Extra Fiber?"

Should we take this extra information into account in our analysis? Figure 8.5 shows a scatterplot of *Calories* and *Sugar*, colored according to the shelf on which the cereals were found and with a separate regression line fit for each. The top shelf is clearly different. We might want to report two regressions, one for the top shelf and one for the bottom two shelves.[1]

> Here's an important unstated condition for fitting models: **All the data must come from the same population.**

Figure 8.5
Calories and *Sugar* (y values jittered) colored according to the shelf on which the cereal was found in a supermarket, with regression lines fit for each shelf individually. Do these data appear homogeneous? That is, do all the cereals seem to be from the same population of cereals? Or are there different kinds of cereals that we might want to consider separately?

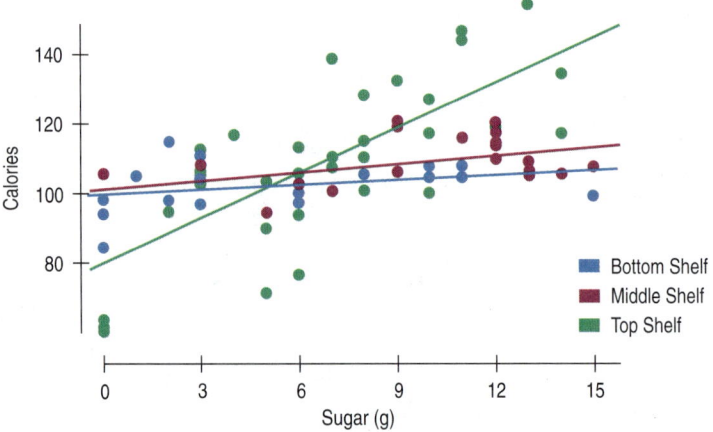

8.2 Extrapolation: Reaching Beyond the Data

Case Study: **Predicting Manatee Kills.** Can we use regression to predict the number of manatees that will be killed by power boats this year?

"Prediction is difficult, especially about the future."
—Niels Bohr, Danish physicist

Linear models give a predicted value for each case in the data. Put a new *x*-value into the equation, and it gives a predicted value, $\hat{y}$, to go with it. But when the new *x*-value lies far from the data we used to build the regression, how trustworthy is the prediction?

The simple answer is that the farther the new *x*-value is from $\bar{x}$, the less trust we should place in the predicted value. Once we venture into new *x* territory, such a prediction is called an **extrapolation**. Extrapolations are dubious because they require the very questionable assumption that nothing about the relationship between *x* and *y* changes even at extreme values of *x* and beyond.

Extrapolation is a good way to see just where the limits of our model may be. But it requires caution. When the *x*-variable is *Time*, extrapolation becomes an attempt to peer

---

[1] More complex models can take into account both sugar content and shelf information. This kind of *multiple regression* model is a natural extension of the model we're using here. We will study these models in Chapter 28.

into the future. People have always wanted to see into the future, and it doesn't take a crystal ball to foresee that they always will. In the past, seers, oracles, and wizards were called on to predict the future. Today, mediums, fortune-tellers, astrologers, and Tarot card readers still find many customers.

Those with a more scientific outlook may use a linear model as their digital crystal ball. Some physical phenomena do exhibit a kind of "inertia" that allows us to guess that current systematic behavior will continue, but be careful in counting on that kind of regularity in phenomena such as stock prices, sales figures, hurricane tracks, or public opinion.

Extrapolating from current trends is so tempting that even professional forecasters sometimes expect too much from their models—and sometimes the errors are striking. In the mid-1970s, oil prices surged and long lines at gas stations were common. In 1970, oil cost about $17 a barrel (in 2014 dollars)—about what it had cost for 20 years or so. But then, within just a few years, the price surged to over $40. In 1975, a survey of 15 top econometric forecasting models (built by groups that included Nobel prize–winning economists) found predictions for 1985 oil prices that ranged from $300 to over $700 a barrel (in 2014 dollars). How close were these forecasts?

Here's a scatterplot (with regression line) of oil prices from 1971 to 1982 (in 2014 dollars; Data in **Historical Oil Prices 2014**).

> **WHEN THE DATA ARE YEARS . . .**
>
> . . . we usually don't enter them as four-digit numbers. Here we used 0 for 1970, 10 for 1980, and so on. Or we may simply enter two digits, using 82 for 1982, for instance. Rescaling years like this often makes calculations easier and equations simpler. We recommend you do it, too. But be careful: if 1982 is 82, then 2004 is 104 (not 4), right?

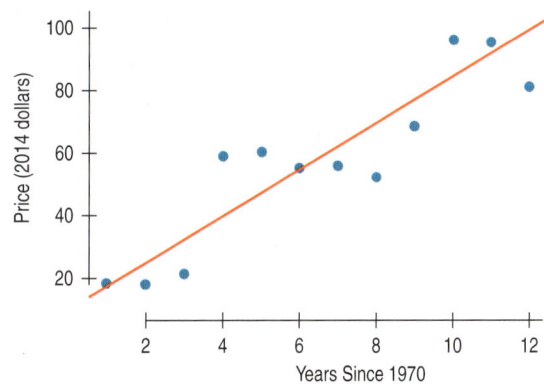

Figure 8.6

The scatterplot shows an average *increase* in the price of a barrel of oil of about $6.75 per year from 1971 to 1982.

The regression model

$$\widehat{Price} = 12.79 + 6.75 \, Years\ Since\ 1970$$

says that prices had been going up 6.75 dollars per year, or about $67.50 in 10 years. If you assume that they would *keep going up*, it's not hard to imagine almost any price you want. That would result in a forecast of about $316 for the year 2015.

Figure 8.7

This scatterplot of oil prices from 1981 to 1998 shows an average *decrease* of nearly $5 per year.

So, how did the forecasters do? Well, in the period from 1982 to 1998 oil prices didn't exactly continue that steady increase. In fact, they went down so much that by 1998, prices (adjusted for inflation) were the lowest they'd been since before World War II.

Not one of the experts' models predicted that.

Of course, these decreases clearly couldn't continue, or oil would have been free by the year 2000. The Energy Information Administration (EIA) has provided the U.S. Congress with both short- and long-term oil price forecasts every year since 1979. In that year they predicted that 1995 prices would be $87 (in today's $). As we've just seen, they were closer to $25. So, 20 years later, what did the EIA learn about forecasting? Let's see how well their 1998 and 1999 forecasts for the present have done. Here's a timeplot of the EIA's predictions and the actual prices (in 2014 dollars).

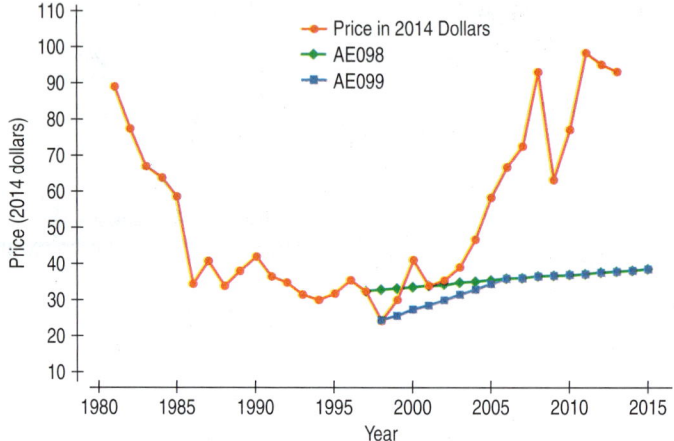

Figure 8.8

Here are the EIA forecasts with the actual prices from 1981 to 2013. Neither forecast predicted the sharp run-up in the past few years.

Oops! They seemed to have missed the sharp run-up in oil prices leading up to the financial crisis of 2008, and then the sharp drop in prices at the beginning of 2009 back to about $64 per barrel and the bounce back up to $106 after that (and the subsequent drop, and . . .).

Where do you think oil prices will go in the next decade? *Your* guess may be as good as anyone's!

Of course, knowing that extrapolation requires thought and caution doesn't stop people. The temptation to see into the future is hard to resist. So our more realistic advice is this:

If you extrapolate into the future, at least don't believe blindly that the prediction will come true.

For Example EXTRAPOLATION: REACHING BEYOND THE DATA

The U.S. Census Bureau (www.census.gov) reports the median age at first marriage for men and women (Data in **Marriage age 2011**). Here's a regression of median *Age* (at first marriage) for men against *Year* (since 1890) at every census from 1890 to 1940:

R-squared = 92.6%
s = 0.2417

Variable	Coefficient
Intercept	26.07
Year	−0.04

The regression equation is

$$\widehat{Age} = 26.07 - 0.04\, Year.$$

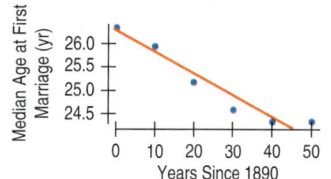

The median age at which men first married fell at the rate of about a year every 25 years from 1890 to 1940.

> **QUESTION:** What would this model predict as the age at first marriage for men in the year 2000?
>
> **ANSWER:** When *Year* counts from 0 in 1890, the year 2000 is "110." Substituting 110 for *Year*, we find that the model predicts a first marriage *Age* of $26.07 - 0.04 \times 110 = 21.7$ years old.
>
> **QUESTION:** In the year 2011, the median *Age* at first marriage for men was 28.7 years. What's gone wrong?
>
> **ANSWER:** It is never safe to extrapolate beyond the data very far. The regression was calculated for years up to 1940. To see how absurd a prediction from that period can be when extrapolated into the present look at a scatterplot of the median *Age* at first marriage for men for all the data from 1890 to 2011:
>
>
>
> Median *Age* at first marriage (years of age) for men in the United States vs. *Year*. The regression model above was fit only to the first 50 years of the data (shown in blue), which looked nicely linear. But the linear pattern could not have continued, and in fact it changed in direction, steepness, and strength.
>
> Now we can see why the extrapolation failed. Although the trend in *Age* at first marriage was linear and negative for the first part of the century, after World War II, it leveled off for about 30 years. Since 1980 or so, it has risen steadily. To characterize age at first marriage, we should probably treat these three time periods separately.

Predicting Changes

Not only is it incorrect and dangerous to interpret association with causation, but when using regression there's a more subtle danger. Never interpret a regression slope coefficient as predicting how *y* is likely to change if its *x* value in the data were changed. Here's an example: In Chapter 7, we found a regression model relating calories in breakfast cereals to their sugar content as

$$\widehat{Calories} = 89.6 + 2.49\, Sugar.$$

It might be tempting to interpret this slope as implying that adding 1 gram of sugar is expected to *lead to* a change of 2.49 calories. Predicting a change of b_1 in *y* for a 1 unit change in *x* might be true in some instances, but it isn't a consequence of the regression. In fact, adding a gram of sugar to a cereal increases its calorie content by 3.90 calories—that's the calorie content of a gram of sugar.

A safer interpretation of the slope is that cereals that have a gram more sugar in them tend to have about 2.49 more calories per serving. That is, the regression model describes how the cereals differ, but does not tell us how they might change if circumstances were different.

We've warned against using a regression model to predict far from the *x*-values in the data. You should also be cautious about using a regression model to predict what might happen to cases in the regression if they were changed. Regression models describe the data as they are, not as they might be under other circumstances.

8.3 Outliers, Leverage, and Influence

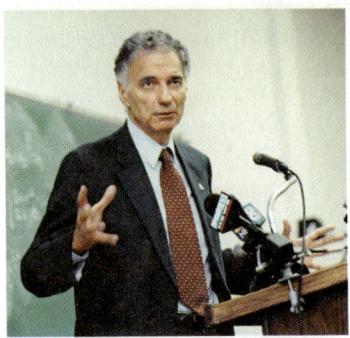

The outcome of the 2000 U.S. presidential election was determined in Florida amid much controversy. Even years later, historians continue to debate who really received the most votes. The main race was between George W. Bush and Al Gore, but two minor candidates played a significant role. To the political right of the major party candidates was Pat Buchanan, while to the political left was Ralph Nader. Generally, Nader earned more votes than Buchanan throughout the state. We would expect counties with larger vote totals to give more votes to each candidate. Here's a regression relating *Buchanan*'s vote totals by county in the state of Florida to *Nader*'s (Data in **Election 2000**):

Dependent variable is Buchanan
R-squared = 42.8%

Variable	Coefficient
Intercept	50.3
Nader	0.14

The regression model:

$$\widehat{Buchanan} = 50.3 + 0.14\, Nader$$

says that, in each county, Buchanan received about 0.14 times (or 14% of) the vote Nader received, starting from a base of 50.3 votes.

This seems like a reasonable regression, with an R^2 of almost 43%. But we've violated all three Rules of Data Analysis by going straight to the regression table without making a picture.

Here's a scatterplot that shows the vote for Buchanan in each county of Florida plotted against the vote for Nader. The striking outlier is Palm Beach County.

"Nature is nowhere accustomed more openly to display her secret mysteries than in cases where she shows traces of her workings apart from the beaten path."

—William Harvey
(1657)

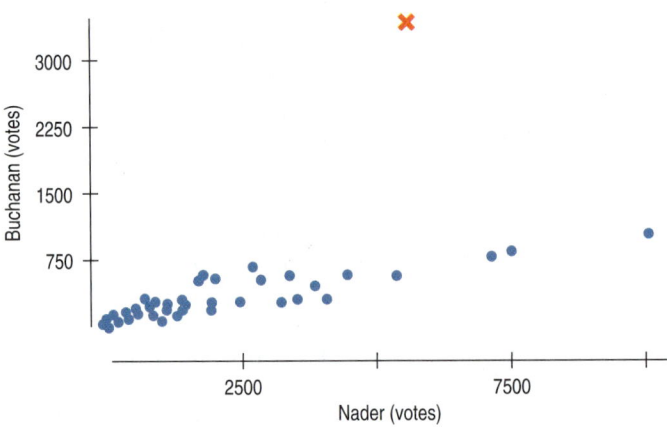

Figure 8.9

Votes received by Buchanan against votes for Nader in all Florida counties in the presidential election of 2000. The red "x" point is Palm Beach County, home of the "butterfly ballot."

The so-called "butterfly ballot," used only in Palm Beach County, was a source of controversy. It has been claimed that the format of this ballot confused voters so that some who intended to vote for the Democrat, Al Gore, punched the wrong hole next to his name and, as a result, voted for Buchanan.

Figure 8.10
The red line shows the effect that one unusual point can have on a regression.

"Give me a place to stand and I will move the Earth."

—*Archimedes*
(287–211 B.C.E.)

A **S**   **Activity: Leverage.** You may be surprised to see how sensitive to a single influential point a regression line is.

"For whoever knows the ways of Nature will more easily notice her deviations; and, on the other hand, whoever knows her deviations will more accurately describe her ways."

—*Francis Bacon*
(1561–1626)

The scatterplot shows a strong, positive, linear association, and one striking point. With Palm Beach removed from the regression, the R^2 jumps from 42.8% to 82.1% and the slope of the line changes to 0.1, suggesting that Buchanan received only about 10% of the vote that Nader received. Palm Beach County now stands out, not as a Buchanan stronghold, but rather as a clear violation of the model that begs for explanation.

One of the great values of models is that they help us to see when and how data values are unusual. In regression, a point can stand out in two different ways. First, a data value can have a large residual, as Palm Beach County does in this example. Because they seem to be different from the other cases, points whose residuals are large always deserve special attention.

A data point can also be unusual if its *x*-value is far from the mean of the *x*-values. Such a point is said to have high **leverage**. The physical image of a lever is exactly right. We know the line must pass through $(\bar{x}, \bar{y})$, so you can picture that point as the fulcrum of the lever. Just as sitting farther from the hinge on a see-saw gives you more leverage to pull it your way, points with values far from $\bar{x}$ pull more strongly on the regression line.

A point with high leverage has the potential to change the regression line. But it doesn't always use that potential. If the point lines up with the pattern of the other points, then including it doesn't change our estimate of the line. How can you tell if a high-leverage point actually changes the model? Just fit the linear model twice, both with and without the point in question. We say that a point is **influential** if omitting it from the analysis changes the model enough to make a meaningful difference.[2]

Influence depends on both leverage and residual; a case with high leverage whose *y*-value sits right on the line fit to the rest of the data is not influential. Removing that case won't change the slope, even if it does affect R^2. A case with modest leverage but a very large residual (such as Palm Beach County) can be influential. Of course, if a point has enough leverage, it can pull the line right to it. Then it's highly influential, but its residual is small. The only way to be sure is to fit both regressions.

Unusual points in a regression often tell us a great deal about the data and the model. We face a challenge: The best way to identify unusual points is against the background of a model, but a model dominated by a single case is unlikely to be useful for identifying unusual cases. (That insight's at least 400 years old. See the quote in the margin.) You can set aside cases and discuss what the model looks like with and without them, but arbitrarily deleting cases can give a false sense of how well the model fits the data. Your goal should be understanding the data, not making R^2 as big as you can.

In 2000, George W. Bush won Florida (and thus the presidency) by only a few hundred votes, so Palm Beach County's residual is big enough to be meaningful. It's the rare unusual point that determines a presidency, but all are worth examining and trying to understand.

---

[2]How big a change would make a difference depends on the circumstances and is a judgment call. Note that some textbooks use the term *influential point* for any observation that influences the slope, intercept, or R^2. We'll reserve the term for points that influence the slope.

Figure 8.11
Bozo's extraordinarily large shoes give his data point high leverage in the regression. Wherever Bozo's *IQ* falls, the regression line will follow.

Figure 8.12
If Bozo's IQ were low, the regression slope would change from positive to negative. A single influential point can change a regression model drastically.

A point with so much influence that it pulls the regression line close to it can make its residual deceptively small. Influential points like that can have a shocking effect on the regression. Here's a plot of *IQ* against *Shoe Size*, again from the fanciful study of intelligence and foot size in comedians we saw in Chapter 6. The linear regression output shows

Dependent variable is IQ
R-squared = 24.8%

Variable	Coefficient
Intercept	93.3265
Shoe size	2.08318

Although this is a silly example, it illustrates an important and common potential problem: Almost all of the variance accounted for ($R^2 = 24.8\%$) is due to *one* point, namely, Bozo. Without Bozo, there is almost no correlation between *Shoe Size* and *IQ*. Look what happens to the regression when we take him out:

Dependent variable is IQ
R-squared 0.7%

Variable	Coefficient
Intercept	105.458
Shoe size	−0.460194

The R^2 value is now 0.7%—a very weak linear relationship (as one might expect!). One single point exhibits a great influence on the regression analysis.

What would have happened if Bozo hadn't shown his comic genius on IQ tests? Suppose his measured *IQ* had been only 50. The slope of the line would then drop from 0.96 IQ points/shoe size to −0.69 IQ points/shoe size. No matter where Bozo's *IQ* is, the line follows it because his *Shoe Size*, being so far from the mean *Shoe Size*, makes Bozo a high-leverage point.

This example is far-fetched, but similar situations occur in real life. For example, a regression of sales against floor space for hardware stores that looked primarily at small-town businesses could be dominated in a similar way if Lowe's was included.

> **WARNING**
> Influential points can hide in plots of residuals. Points with high leverage pull the line close to them, so they often have small residuals. You'll see influential points more easily in scatterplots of the original data or by finding a regression model with and without the points.

✓ Just Checking

Each of these scatterplots shows an unusual point. For each, tell whether the point is a high-leverage point, would have a large residual, or is influential.

8.4 Lurking Variables and Causation

CAUSAL INTERPRETATIONS

One common way to interpret a regression slope is to say that "a change of 1 unit in x results in a change of b_1 units in y." This way of saying things encourages causal thinking. Beware.

In Chapter 6, we tried to make it clear that no matter how strong the correlation is between two variables, it doesn't show that one variable causes the other. Putting a regression line through a cloud of points just increases the temptation to think and to say that the *x*-variable *causes* the *y*-variable. Just to make sure, let's repeat the point again: No matter how strong the association, no matter how large the R^2 value, no matter how straight the line, you cannot conclude from a regression alone that one variable *causes* the other. There's always the possibility that some third variable is driving both of the variables you have observed. With observational data, as opposed to data from a designed experiment, there is no way to be sure that such a **lurking variable** is not responsible for the apparent association.[3]

Here's an example: The scatterplot shows the *Life Expectancy* (average of men and women, in years) for each of 40 countries of the world, plotted against the square root of the number of *Doctors per person* in the country. (The square root is here to make the relationship satisfy the Straight Enough Condition, as we saw back in Chapter 6; Data in **Doctors and Life Expectancy**).

Figure 8.13

The relationship between *Life Expectancy* (years) and availability of *Doctors* (measured as √Doctors/Person) for countries of the world is strong, positive, and linear.

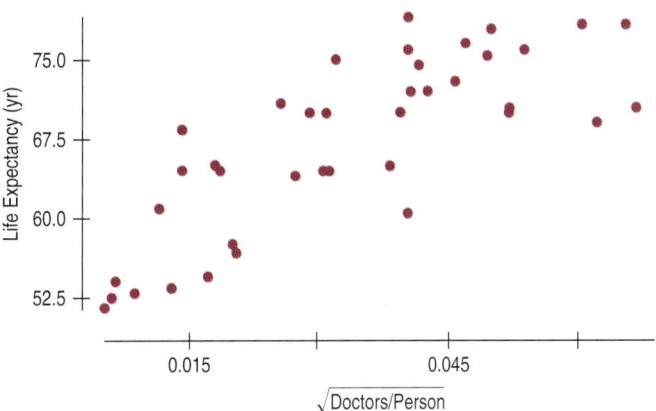

The strong positive association ($R^2 = 62.9\%$) seems to confirm our expectation that more *Doctors per person* improves health care, leading to longer lifetimes and a greater *Life Expectancy*. The strength of the association would *seem* to argue that we should send more doctors to developing countries to increase life expectancy.

That conclusion is about the consequences of a change. Would sending more doctors increase life expectancy? Specifically, do doctors *cause* greater life expectancy? Perhaps, but these are observed data, so there may be another explanation for the association.

In Figure 8.14, the similar-looking scatterplot's *x*-variable is the square root of the number of *Televisions per person* in each country. The positive association in this scatterplot is even *stronger* than the association in the previous plot ($R^2 = 72.5\%$). We can fit the linear model, and quite possibly use the number of TVs as a way to predict life expectancy. Should we conclude that increasing the number of TVs actually extends lifetimes? If so, we should send TVs instead of doctors to developing countries. Not only is the correlation with life expectancy higher, but TVs are much cheaper than doctors and possibly more fun.

---

[3]Chapter 12 discusses observational data and experiments at greater length.

Figure 8.14
To increase life expectancy, don't send doctors, send TVs; they're cheaper and more fun. Or maybe that's not the right interpretation of this scatterplot of *life expectancy* against availability of TVs (as $\sqrt{TVs/Person}$).

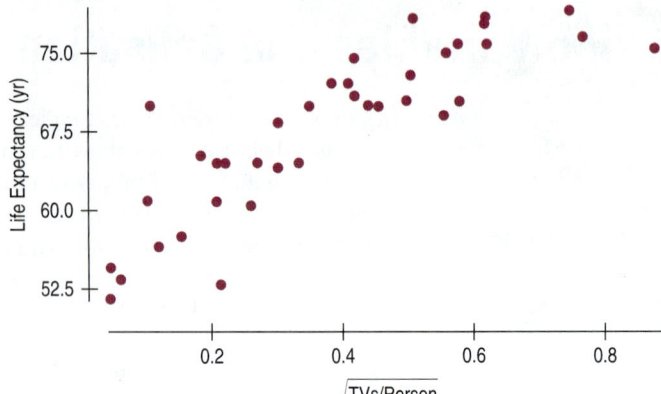

What's wrong with this reasoning? Maybe we were a bit hasty earlier when we concluded that doctors *cause* longer lives. Maybe there's a lurking variable here. Countries with higher standards of living have both longer life expectancies *and* more doctors (and more TVs). Could higher living standards cause changes in the other variables? If so, then improving living standards might be expected to prolong lives, increase the number of doctors, and increase the number of TVs.

From this example, you can see how easy it is to fall into the trap of mistakenly inferring causality from a regression. For all we know, doctors (or TVs!) *do* increase life expectancy. But we can't tell that from data like these, no matter how much we'd like to. Resist the temptation to conclude that *x* causes *y* from a regression, no matter how obvious that conclusion seems to you.

8.5 Working with Summary Values

Scatterplots of statistics summarized over groups tend to show less variability than we would see if we measured the same variable on individuals. This is because the summary statistics themselves vary less than the data on the individuals do—a fact we will make more specific in coming chapters.

In Chapter 6, we looked at the heights and weights of individual students (Data in **Heights and Weights**). There, we saw a correlation of 0.644, so R^2 is 41.5%.

Figure 8.15
Weight (lb) against *Height* (in.) for a sample of men. There's a strong, positive, linear association.

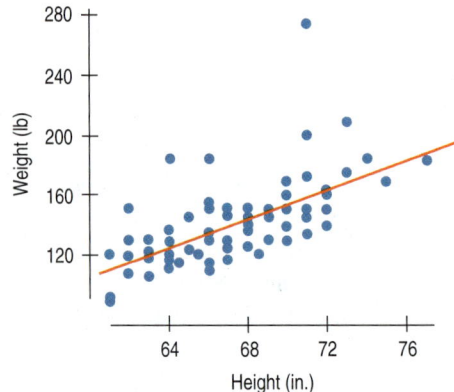

Suppose, instead of data on individuals, we knew only the mean weight for each height value. The scatterplot of mean weight by height would show less scatter. And R^2 would increase to 80.1%.

Figure 8.16
Mean Weight (lb) shows a stronger linear association with *Height* than do the weights of individuals. Means vary less than individual values.

Scatterplots of summary statistics show less scatter than the baseline data on individuals and can give a false impression of how well a line summarizes the data. There's no simple correction for this phenomenon. Once we're given summary data, there's no simple way to get the original values back.

In the life expectancy and TVs example, we have no good measure of exposure to doctors or to TV on an individual basis. But if we did, we should expect the scatterplot to show more variability and the corresponding R^2 to be smaller. The bottom line is that you should be a bit suspicious of conclusions based on regressions of summary data. They may look better than they really are.

To think about lurking variables you must think "outside the box." What variables are *not* in your data that ought to be?[4]

For Example USING SEVERAL OF THESE METHODS TOGETHER

Motorcycles designed to run off-road, often known as dirt bikes, are specialized vehicles.

We have data on 114 dirt bikes (Data in **Dirt Bikes 2014**). Some cost as little as $1399, while others are substantially more expensive. Let's investigate how the size and type of engine contribute to the cost of a dirt bike. As always, we start with a scatterplot.

Here's a scatterplot of the manufacturer's suggested retail price (*MSRP*) in dollars against the engine *Displacement* (in cubic centimeters), along with a regression analysis:

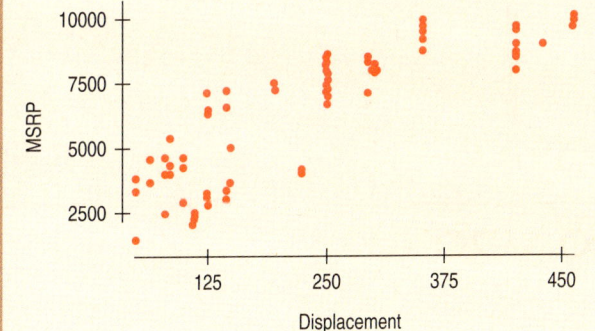

Dependent variable is MSRP
R-squared = 70.4%   s = 1335

Variable	Coefficient
Intercept	2856.84
Displacement	15.2567

QUESTION: What do you see in the scatterplot?

ANSWER: There is a strong positive association between the engine displacement of dirt bikes and the manufacturer's suggested retail price. There is some curvature and there seem to be a group of points with small displacement whose prices are substantially below the rest but show a similar slope.

---

[4]If you wonder about the graphic: this is one of the boxes you should think outside of. There are many others. Stay alert for them.

Here's a scatterplot of the residuals:

QUESTION: What do you see in the residuals plot?

ANSWER: The points at the far right and the bottom left don't fit well with the other dirt bikes. Overall, there appears to be a bend in the relationship, so a linear model may not be appropriate.

Let's try a re-expression. Here's a scatterplot showing *MSRP* against the cube root of *Displacement* to make the relationship closer to straight. (Since displacement is measured in cubic centimeters, its cube root has the simple units of centimeters.) In addition, we've colored the plot according to the cooling method used in the bike's engine: liquid or air. Each group is shown with its own regression line, as we did for the cereals on different shelves.

QUESTION: What does this plot say about dirt bikes?

ANSWER: There appears to be a positive, linear relationship between *MSRP* and the cube root of *Displacement*. In general, the larger the engine a bike has, the higher the suggested price. Liquid-cooled dirt bikes, however, typically cost more than air-cooled bikes with comparable displacement.

WHAT CAN GO WRONG?

This entire chapter has held warnings about things that can go wrong in a regression analysis. So let's just recap. When you make a linear model:

- **Make sure the relationship is straight.** Check the Straight Enough Condition. Always examine the residuals for evidence that the Linearity Assumption has failed. It's often easier to see deviations from a straight line in the residuals plot than in the scatterplot of the original data. Pay special attention to the most extreme residuals because they may have something to add to the story told by the linear model.

- **Be on guard for different groups in your regression.** Check for evidence that the data consist of separate subsets. If you find subsets that behave differently, consider fitting a different linear model to each subset.

- **Beware of extrapolating.** Beware of extrapolation beyond the *x*-values that were used to fit the model. Although it's common to use linear models to extrapolate, the practice is dangerous.

- **Beware especially of extrapolating into the future!** Be especially cautious about extrapolating into the future with linear models. To predict the future, you must assume that future changes will continue at the same rate you've observed in the past. Predicting the future is particularly tempting and particularly dangerous.

- **Look for unusual points.** Unusual points always deserve attention and may well reveal more about your data than the rest of the points combined. Always look for them and try to understand why they stand apart. A scatterplot of the data is a good way to see high-leverage and influential points. A scatterplot of the residuals against the predicted values is a good tool for finding points with large residuals.

- **Beware of high-leverage points and especially of those that are influential.** Influential points can alter the regression model a great deal. The resulting model may say more about one or two points than about the overall relationship.

- **Consider comparing two regressions.** To see the impact of outliers on a regression, it's often wise to run two regressions, one with and one without the extraordinary points, and then to discuss the differences.

- **Treat unusual points honestly.** If you remove enough carefully selected points, you will eventually get a regression with a high R^2, but it won't give you much understanding. Some variables are not related in a way that's simple enough for a linear model to fit very well. When that happens, report the failure and stop.

- **Beware of lurking variables.** Think about lurking variables before interpreting a linear model. It's particularly tempting to explain a strong regression by thinking that the *x*-variable *causes* the *y*-variable. A linear model alone can never demonstrate such causation, in part because it cannot eliminate the chance that a lurking variable has caused the variation in both *x* and *y*.

- **Watch out when dealing with data that are summaries.** Be cautious in working with data values that are themselves summaries, such as means or medians. Such statistics are less variable than the data on which they are based, so they tend to inflate the impression of the strength of a relationship.

CONNECTIONS

We should always be alert to things that could go wrong if we were to use statistics without thinking carefully. Regression opens new vistas of potential problems. But each one relates to issues we've thought about before.

It is always important that our data be from a single homogeneous group and not made up of disparate groups. We looked for multiple modes in single variables. Now we check scatterplots for evidence of subgroups in our data. As with modes, it's often best to split the data and analyze the groups separately.

Our concern with unusual points and their potential influence also harkens back to our earlier concern with outliers in histograms and boxplots—and for many of the same reasons. As we've seen here, regression offers such points new scope for mischief.

The risks of interpreting linear models as causal or predictive arose in Chapters 6 and 7. And they're important enough to mention again in later chapters.

What Have We Learned?

Learning Objectives

Be skeptical of regression models. Always plot and examine the residuals for unexpected behavior. Be alert to a variety of possible violations of the standard regression assumptions and know what to do when you find them.

Be alert for subgroups in the data.

- Often these will turn out to be separate groups that should not be analyzed together in a single analysis.
- Often identifying subgroups can help us understand what is going on in the data.

Be especially cautious about extrapolating beyond the data.

Look out for unusual and extraordinary observations.

- Cases that are extreme in x have high leverage and can affect a regression model strongly.
- Cases that are extreme in y have large residuals and are called outliers.
- Cases that have both high leverage and large residuals are influential. Setting them aside will change the regression model, so you should consider whether their influence on the model is appropriate or desirable.

Interpret regression models appropriately. Don't infer causation from a regression model.

Notice when you are working with summary values.

- Summaries vary less than the data they summarize, so they may give the impression of greater certainty than your model deserves.

Diagnose and treat nonlinearity.

- If a scatterplot of y vs. x isn't straight, a linear regression model isn't appropriate.
- Re-expressing one or both variables can often improve the straightness of the relationship.
- The powers, roots, and the logarithm provide an ordered collection of re-expressions so you can search up and down the "ladder of powers" to find an appropriate one.

Review of Terms

Extrapolation Although linear models provide an easy way to predict values of y for a given value of x, it is unsafe to predict for values of x far from the ones used to find the linear model equation. Such extrapolation may pretend to see into the future, but the predictions should not be trusted (p. 222).

Outlier Any data point that stands away from the others can be called an outlier. In regression, cases can be extraordinary in two ways: by having a large residual—being an outlier—or by having high leverage (p. 226).

Leverage Data points whose x-values are far from the mean of x are said to exert leverage on a linear model. High-leverage points pull the line close to them, and so they can have a large effect on the line, sometimes completely determining the slope and intercept. With high enough leverage, their residuals can be deceptively small (p. 227).

Influential point A point that, if omitted from the data, results in a very different regression model (p. 227).

Lurking variable A variable that is not explicitly part of a model but affects the way the variables in the model appear to be related. Because we can never be certain that observational data are not hiding a lurking variable that influences both x and y, it is never safe to conclude that a linear model demonstrates a causal relationship, no matter how strong the linear association (p. 229).

On the Computer REGRESSION DIAGNOSIS

Most statistics technology offers simple ways to check whether your data satisfy the conditions for regression. We have already seen that these programs can make a simple scatterplot. They can also help check the conditions by plotting residuals.

DATA DESK

- Click on the **HyperView** menu on the **Regression** output table. A menu drops down to offer scatterplots of residuals against predicted values, Normal probability plots of residuals, or just the ability to save the residuals and predicted values.
- Click on the name of a predictor in the regression table to be offered a scatterplot of the residuals against that predictor.

COMMENTS
If you change any of the variables in the regression analysis, Data Desk will offer to update the plots of residuals.

EXCEL

The Data Analysis add-in for Excel includes a Regression command. The dialog box it shows offers to make plots of residuals.

COMMENTS
Do not use the Normal probability plot offered in the regression dialog. It is not what it claims to be and is wrong.

JMP

- From the **Analyze** menu, choose **Fit Y by X**. Select **Fit Line**.
- Under Linear Fit, select **Plot Residuals**. You can also choose to **Save Residuals**.

- Subsequently, from the **Distribution** menu, choose **Normal quantile plot** or **histogram** for the residuals.

MINITAB

- From the **Stat** menu, choose **Regression**.
- From the **Regression** submenu, select **Regression** again.
- In the Regression dialog, enter the response variable name in the "Response" box and the predictor variable name in the "Predictor" box.
- To specify saved results, in the Regression dialog, click **Storage**.

- Check "Residuals" and "Fits." Click **OK**.
- To specify displays, in the Regression dialog, click **Graphs**.
- Under "Residual Plots," select "Individual plots" and check "Residuals versus fits."
- Click **OK**. Now back in the Regression dialog, click **OK**. Minitab computes the regression and the requested saved values and graphs.

R

Save the regression model object by giving it a name, such as "myreg":

- myreg = lm(Y~X) or myreg = lm(Y~X, data = DATA) where DATA is the name of the data frame.
- plot(residuals(myreg)~predict(myreg)) plots the residuals against the predicted values.

- qqnorm(residuals(myreg)) gives a normal probability plot of the residuals.
- plot(myreg) gives similar plots (but not exactly the same).

SPSS

- From the **Analyze** menu, choose **Regression**.
- From the Regression submenu, choose **Linear**.
- After assigning variables to their roles in the regression, click the "**Plots ...**" button.

In the Plots dialog, you can specify a Normal probability plot of residuals and scatterplots of various versions of standardized residuals and predicted values.

COMMENTS

A plot of ***ZRESID** against ***PRED** will look most like the residual plots we've discussed. SPSS standardizes the residuals by dividing by their standard deviation. (There's no need to subtract their mean; it must be zero.) The standardization doesn't affect the scatterplot.

STATCRUNCH

To create a residuals plot:
- Click on **Stat**.
- Choose **Regression » Simple Linear** and choose X and Y.
- Click on **Next** and click on **Next** again.
- Indicate which type of residuals plot you want.
- Click on **Calculate**.

COMMENTS

Note that before you click on **Next** for the second time you may indicate that you want to save the values of the residuals. Residuals becomes a new column, and you may use that variable to create a histogram or residuals plot.

TI-83/84 PLUS

- To make a residuals plot, set up a **STATPLOT** as a scatterplot.
- Specify your explanatory data list as **Xlist**.
- For **Ylist**, import the name **RESID** from the **LIST NAMES** menu. **ZoomStat** will now create the residuals plot.

COMMENTS

Each time you execute a **LinReg** command, the calculator automatically computes the residuals and stores them in a data list named **RESID**. If you want to see them, go to **STAT EDIT**. Space through the names of the lists until you find a blank. Import **RESID** from the **LIST NAMES** menu. Now every time you have the calculator compute a regression analysis, it will show you the residuals.

Exercises

Section 8.1

1. **Credit card spending** An analysis of spending by a sample of credit card bank cardholders shows that spending by cardholders in January (*Jan*) is related to their spending in December (*Dec*):

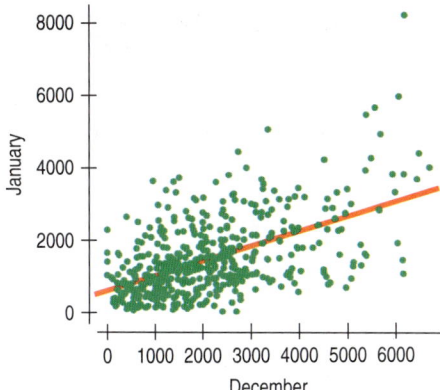

The assumptions and conditions of the linear regression seemed to be satisfied and an analyst was about to predict January spending using the model

$$\widehat{Jan} = \$612.07 + 0.403\ Dec.$$

Another analyst worried that different types of cardholders might behave differently. She examined the spending patterns of the cardholders and placed them into five market *Segments*. When she plotted the data using different colors and symbols for the five different segments, she found the following:

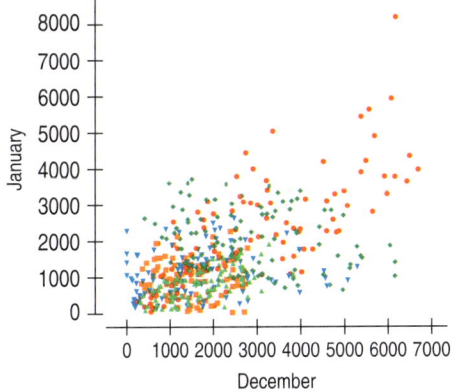

Look at this plot carefully and discuss why she might be worried about the predictions from the model $\widehat{Jan} = \$612.07 + 0.403\ Dec.$

2. **Revenue and talent cost** A concert production company examined its records. The manager made the following scatterplot. The company places concerts in two venues, a smaller, more intimate theater (plotted with blue circles) and a larger auditorium-style venue (red x's).

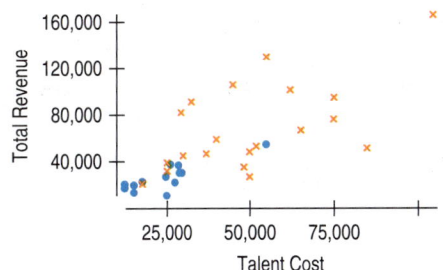

a) Describe the relationship between *Talent Cost* and *Total Revenue*. (Remember: direction, form, strength, outliers.)
b) How are the results for the two venues similar?
c) How are they different?

3. **Market segments** The analyst in Exercise 1 tried fitting the regression line to each market segment separately and found the following:

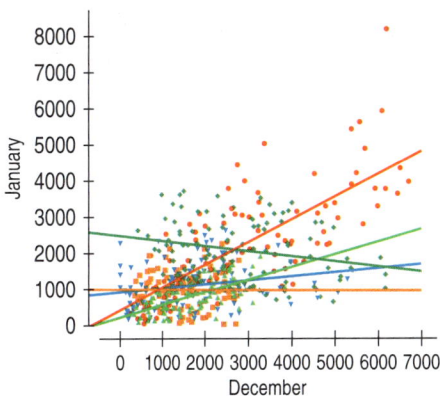

What does this say about her concern in Exercise 1? Was she justified in worrying that the overall model $\widehat{Jan} = \$612.07 + 0.403\ Dec$ might not accurately summarize the relationship? Explain briefly.

4. **Revenue and ticket sales** The concert production company of Exercise 2 made a second scatterplot, this time relating *Total Revenue* to *Ticket Sales*.

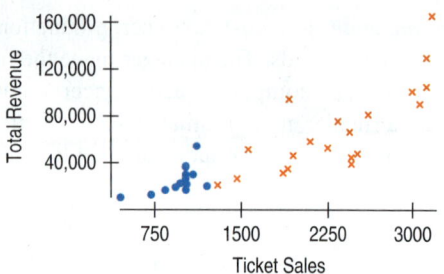

a) Describe the relationship between *Ticket Sales* and *Total Revenue*.
b) How are the results for the two venues similar?
c) How are they different?

Section 8.2

5. **Cell phone costs** Noting a recent study predicting the increase in cell phone costs, a friend remarks that by the time he's a grandfather, no one will be able to afford a cell phone. Explain where his thinking went awry.

6. **Stopping times** Using data from 20 compact cars, a consumer group develops a model that predicts the stopping time for a vehicle by using its weight. You consider using this model to predict the stopping time for your large SUV. Explain why this is not advisable.

7. **Revenue and large venues** A regression of *Total Revenue* on *Ticket Sales* by the concert production company of Exercises 2 and 4 finds the model

$$\widehat{Revenue} = -14{,}228 + 36.87\, TicketSales.$$

a) Management is considering adding a stadium-style venue that would seat 10,000. What does this model predict that revenue would be if the new venue were to sell out?
b) Why would it be unwise to assume that this model accurately predicts revenue for this situation?

8. **Revenue and advanced sales** The production company of Exercise 7 offers advanced sales to "Frequent Buyers" through its website. Here's a relevant scatterplot:

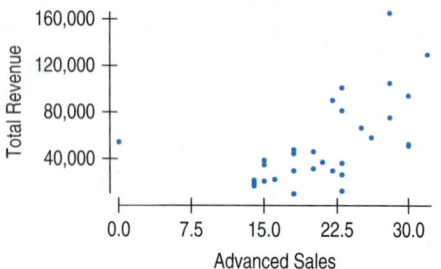

One performer refused to permit advanced sales. What effect has that point had on the regression to model *Total Revenue* from *Advanced Sales*?

Section 8.3

9. **Abalone** Abalones are edible sea snails that include over 100 species. A researcher is working with a model that uses the number of rings in an Abalone's shell to predict its age. He finds an observation that he believes has been miscalculated. After deleting this outlier, he redoes the calculation. Does it appear that this outlier was exerting very much influence?

Before:
Dependent variable is Age
R-squared = 67.5%

Variable	Coefficient
Intercept	1.736
Rings	0.45

After:
Dependent variable is Age
R-squared = 83.9%

Variable	Coefficient
Intercept	1.56
Rings	1.13

10. **Abalone again** The researcher in Exercise 9 is content with the second regression. But he has found a number of shells that have large residuals and is considering removing all them. Is this good practice?

Section 8.4

11. **Skinned knees** There is a strong correlation between the temperature and the number of skinned knees on playgrounds. Does this tell us that warm weather causes children to trip?

12. **Cell phones and life expectancy** The correlation between cell phone usage and life expectancy is very high. Should we buy cell phones to help people live longer?

Section 8.5

13. **Grading** A team of Calculus teachers is analyzing student scores on a final exam compared to the midterm scores. One teacher proposes that they already have every teacher's class averages and they should just work with those averages. Explain why this is problematic.

14. **Average GPA** An athletic director proudly states that he has used the average GPAs of the university's sports teams and is predicting a high graduation rate for the teams. Why is this method unsafe?

Chapter Exercises

15. Marriage age 2011 Is there evidence that the age at which women get married has changed over the past 100 years? The scatterplot shows the trend in age at first marriage for American women (www.census.gov).

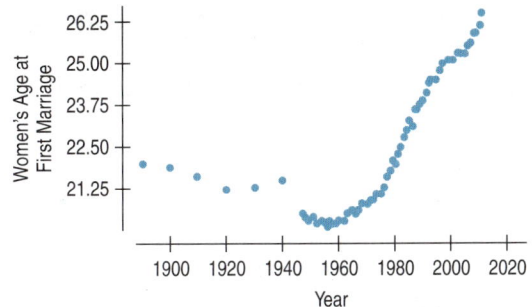

a) Is there a clear pattern? Describe the trend.
b) Is the association strong?
c) Is the correlation high? Explain.
d) Is a linear model appropriate? Explain.

16. Smoking 2011 The Centers for Disease Control and Prevention track cigarette smoking in the United States. How has the percentage of people who smoke changed since the danger became clear during the last half of the 20th century? The scatterplot shows percentages of smokers among men 18–24 years of age, as estimated by surveys, from 1965 through 2011 (http://www.cdc.gov/nchs/).

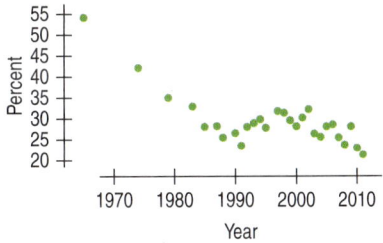

a) Is there a clear pattern? Describe the trend.
b) Is the association strong?
c) Is a linear model appropriate? Explain.

17. Human Development Index 2012 The United Nations Development Programme (UNDP) uses the Human Development Index (HDI) in an attempt to summarize in one number the progress in health, education, and economics of a country. In 2012, the HDI was as high as 0.955 for Norway and as low as 0.304 for Niger. The gross domestic product per capita (GDPPC), by contrast, is often used to summarize the *overall* economic strength of a country. Is the HDI related to the GDPPC? At the top of the next column is a scatterplot of *HDI* against *GDPPC*.

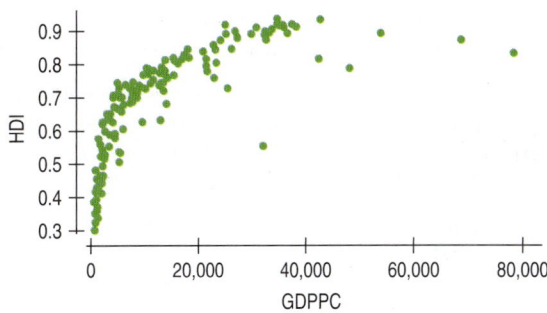

a) Explain why fitting a linear model to these data might be misleading.
b) If you fit a linear model to the data, what do you think a scatterplot of residuals versus predicted *HDI* will look like?

18. HDI 2012 revisited The United Nations Development Programme (UNDP) uses the Human Development Index (HDI) in an attempt to summarize in one number the progress in health, education, and economics of a country. The number of Internet users per 100 people is positively associated with economic progress in a country. Can the number of Internet users be used to predict the HDI? Here is a scatterplot of HDI against Internet users:

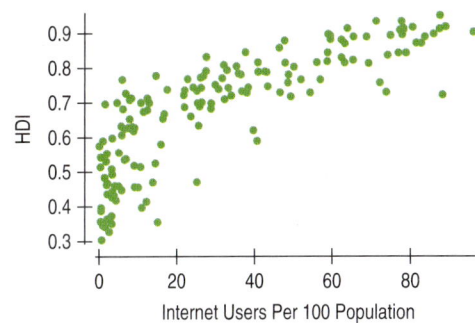

a) Explain why fitting a linear model to these data might be misleading.
b) If you fit a linear model to the data, what do you think a scatterplot of residuals vs. predicted *HDI* will look like?

19. Good model? In justifying his choice of a model, a student wrote, "I know this is the correct model because $R^2 = 99.4\%$."

a) Is this reasoning correct? Explain.
b) Does this model allow the student to make accurate predictions? Explain.

20. Bad model? A student who has created a linear model is disappointed to find that her R^2 value is a very low 13%.
   a) Does this mean that a linear model is not appropriate? Explain.
   b) Does this model allow the student to make accurate predictions? Explain.

21. Movie dramas Here's a scatterplot of the production budgets (in millions of dollars) vs. the running time (in minutes) for major release movies in 2005. Dramas are plotted as red x's and all other genres are plotted as blue dots. (The re-make of *King Kong* is plotted as a black "-". At the time it was the most expensive movie ever made, and not typical of any genre.) A separate least squares regression line has been fitted to each group. For the following questions, just examine the plot:

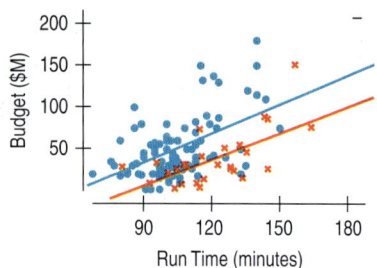

   a) What are the units for the slopes of these lines?
   b) In what way are dramas and other movies similar with respect to this relationship?
   c) In what way are dramas different from other genres of movies with respect to this relationship?

22. Smoking 2011, women and men In Exercise 16, we examined the percentage of men aged 18–24 who smoked from 1965 to 2011 according to the Centers for Disease Control and Prevention. How about women? Here's a scatterplot showing the corresponding percentages for both men and women:

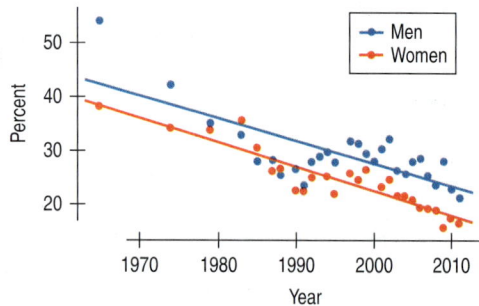

   a) In what ways are the trends in smoking behavior similar for men and women?
   b) How do the smoking rates for women differ from those for men?
   c) Viewed alone, the trend for men may have seemed to violate the Linearity Condition. How about the trend for women? Does the consistency of the two patterns encourage you to think that a linear model for the trend in men might be appropriate? (Note: there is no correct answer to this question; it is raised for you to think about.)

23. Oakland passengers 2013 The scatterplot below shows the number of passengers at Oakland (CA) airport month by month since 1997. (www.oaklandairport.com)

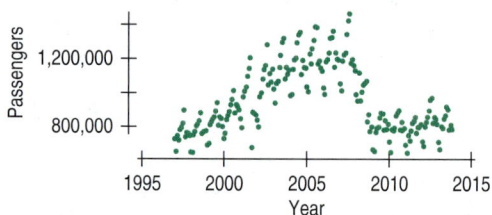

   a) Describe the patterns in passengers at Oakland airport that you see in this time plot.
   b) Until 2009, analysts got fairly good predictions using a linear model. Why might that not be the case now?
   c) If they considered only the data from 2009 to the present, might they get reasonable predictions into the future?

24. Tracking hurricanes 2012 In a previous chapter, we saw data on the errors (in nautical miles) made by the National Hurricane Center in predicting the path of hurricanes. The scatterplot at the top of the next page shows the trend in the 24-hour tracking errors since 1970 (www.nhc.noaa.gov).

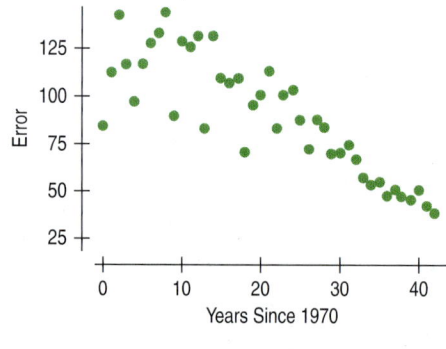

Dependent variable is Error
R-squared = 72.5% s_e = 16.13

Variable	Coefficient
Intercept	133.024
Years - 1970	−2.0602

   a) Interpret the slope and intercept of the model.
   b) Interpret s_e in this context.
   c) The Center would like to achieve an average tracking error of 45 nautical miles by 2015. Will they make it? Defend your response.
   d) What if their goal were an average tracking error of 25 nautical miles?
   e) What cautions would you state about your conclusion?

25. **Unusual points** Each of these four scatterplots shows a cluster of points and one "stray" point. For each, answer these questions:
   1) In what way is the point unusual? Does it have high leverage, a large residual, or both?
   2) Do you think that point is an influential point?
   3) If that point were removed, would the correlation become stronger or weaker? Explain.
   4) If that point were removed, would the slope of the regression line increase or decrease? Explain.

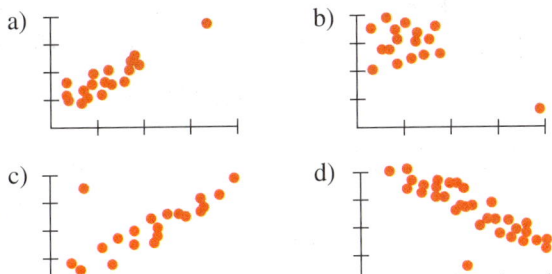

26. **More unusual points** Each of the following scatterplots shows a cluster of points and one "stray" point. For each, answer these questions:
   1) In what way is the point unusual? Does it have high leverage, a large residual, or both?
   2) Do you think that point is an influential point?
   3) If that point were removed, would the correlation become stronger or weaker? Explain.
   4) If that point were removed, would the slope of the regression line increase or decrease? Explain.

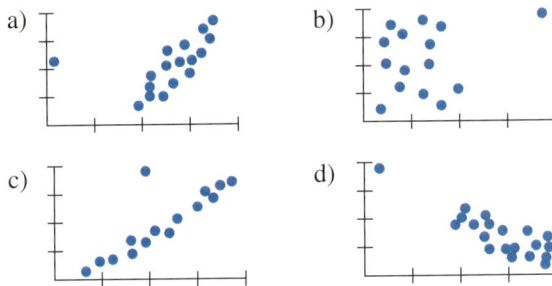

27. **The extra point** The scatterplot shows five blue data points at the left. Not surprisingly, the correlation for these points is $r = 0$. Suppose *one* additional data point is added at one of the five positions suggested below in green. Match each point (a–e) with the correct new correlation from the list given.
   1) −0.90   4) 0.05
   2) −0.40   5) 0.75
   3) 0.00

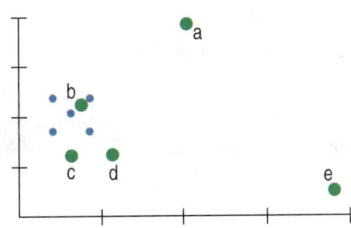

28. **The extra point revisited** The original five points in Exercise 27 produce a regression line with slope 0. Match each of the green points (a–e) with the slope of the line after that one point is added:
   1) −0.45   4) 0.05
   2) −0.30   5) 0.85
   3) 0.00

29. **What's the cause?** Suppose a researcher studying health issues measures blood pressure and the percentage of body fat for several adult males and finds a strong positive association. Describe three different possible cause-and-effect relationships that might be present.

30. **What's the effect?** A researcher studying violent behavior in elementary school children asks the children's parents how much time each child spends playing computer games and has their teachers rate each child on the level of aggressiveness they display while playing with other children. Suppose that the researcher finds a moderately strong positive correlation. Describe three different possible cause-and-effect explanations for this relationship.

31. **Reading** To measure progress in reading ability, students at an elementary school take a reading comprehension test every year. Scores are measured in "grade-level" units; that is, a score of 4.2 means that a student is reading at slightly above the expected level for a fourth grader. The school principal prepares a report to parents that includes a graph showing the mean reading score for each grade. In his comments, he points out that the strong positive trend demonstrates the success of the school's reading program.

   a) Does this graph indicate that students are making satisfactory progress in reading? Explain.

b) What would you estimate the correlation between *Grade* and *Average Reading Level* to be?
c) If, instead of this plot showing average reading levels, the principal had produced a scatterplot of the reading levels of all the individual students, would you expect the correlation to be the same, higher, or lower? Explain.
d) Although the principal did not do a regression analysis, someone as statistically astute as you might do that. (But don't bother.) What value of the slope of that line would you view as demonstrating acceptable progress in reading comprehension? Explain.

32. **Grades** A college admissions officer, defending the college's use of SAT scores in the admissions process, produced the following graph. It shows the mean GPAs for last year's freshmen, grouped by SAT scores. How strong is the evidence that *SAT Score* is a good predictor of *GPA*? What concerns you about the graph, the statistical methodology or the conclusions reached?

33. **Heating** After keeping track of his heating expenses for several winters, a homeowner believes he can estimate the monthly cost from the average daily Fahrenheit temperature by using the model $\widehat{Cost} = 133 - 2.13\ Temp$. Here is the residuals plot for his data:

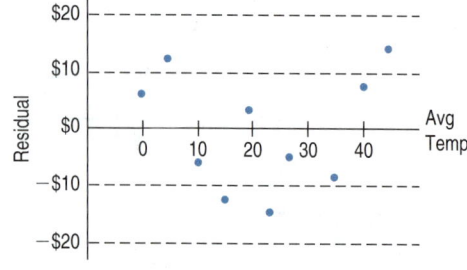

a) Interpret the slope of the line in this context.
b) Interpret the *y*-intercept of the line in this context.
c) During months when the temperature stays around freezing, would you expect cost predictions based on this model to be accurate, too low, or too high? Explain.

d) What heating cost does the model predict for a month that averages 10°?
e) During one of the months on which the model was based, the temperature did average 10°. What were the actual heating costs for that month?
f) Should the homeowner use this model? Explain.
g) Would this model be more successful if the temperature were expressed in degrees Celsius? Explain.

34. **Speed** How does the speed at which you drive affect your fuel economy? To find out, researchers drove a compact car for 200 miles at speeds ranging from 35 to 75 miles per hour. From their data, they created the model $\widehat{Fuel\ Efficiency} = 32 - 0.1\ Speed$ and created this residual plot:

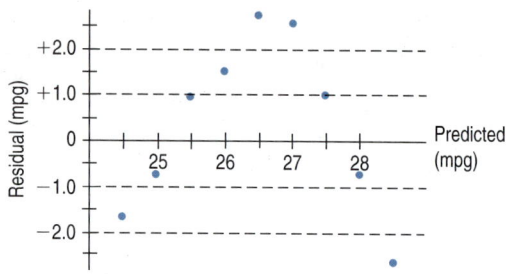

a) Interpret the slope of this line in context.
b) Explain why it's silly to attach any meaning to the *y*-intercept.
c) When this model predicts high *Fuel Efficiency*, what can you say about those predictions?
d) What *Fuel Efficiency* does the model predict when the car is driven at 50 mph?
e) What was the actual *Fuel Efficiency* when the car was driven at 45 mph?
f) Do you think there appears to be a strong association between *Speed* and *Fuel Efficiency*? Explain.
g) Do you think this is the appropriate model for that association? Explain.

35. **Interest rates 2014** Here's a plot showing the federal rate on 3-month Treasury bills from 1950 to 1980, and a regression model fit to the relationship between the *Rate* (in %) and *Years Since 1950* (www.gpoaccess.gov/eop/).

a) What is the correlation between *Rate* and *Year*?
b) Interpret the slope and intercept.
c) What does this model predict for the interest rate in the year 2000?
d) Would you expect this prediction to have been accurate? Explain.

(continued)

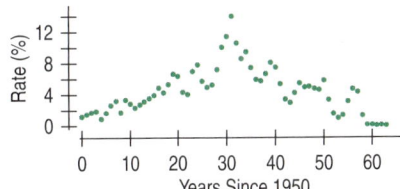

Dependent variable is Rate
R-squared = 77.6%   s = 1.232

Variable	Coefficient
Intercept	0.61149
Year - 1950	0.24788

36. Marriage age, 2011 The graph shows the ages of both men and women at first marriage (www.census.gov).

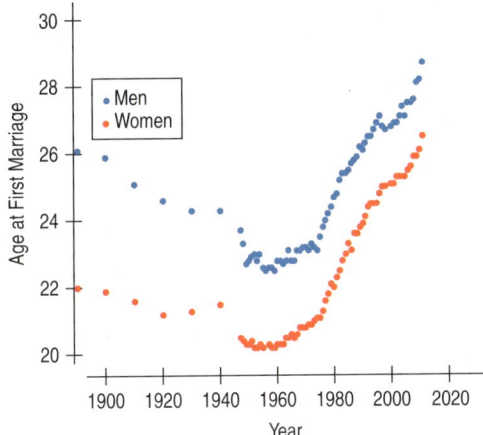

Clearly, the patterns for men and women are similar. But are the two lines getting closer together?

Here's a timeplot showing the *difference* in average age (men's age − women's age) at first marriage, the regression analysis, and the associated residuals plot.

Dependent variable is Age Difference
R-squared = 74.9%   s = 0.2346

Variable	Coefficient
Intercept	32.968
Year	−0.01549

a) What is the correlation between *Age Difference* and *Year*?
b) Interpret the slope of this line.
c) Predict the average age difference in 2015.
d) Describe reasons why you might not place much faith in that prediction.

37. Interest rates 2014 revisited In Exercise 35, you investigated the federal rate on 3-month Treasury bills between 1950 and 1980. The scatterplot below shows that the trend changed dramatically after 1980, so we computed a new regression model for the years 1981 to 2013.

Here's the model for the data from 1981 to 2013:

Dependent variable is Rate
R-squared = 77.0%   s = 1.626

Variable	Coefficient
Intercept	18.87316
Year - 1950	−0.30317

a) How does this model compare to the one in Exercise 35?
b) What does this model estimate the interest rate was in 2000? How does this compare to the rate you predicted in Exercise 35?
c) Do you trust this newer predicted value? Explain.
d) What does each of the two models predict the interest rate on 3-month Treasury bills will be in 2020?
e) Comment on your predictions in d)

38. Marriage age 2011 again Has the trend of decreasing difference in age at first marriage seen in Exercise 36 gotten stronger recently? The scatterplot and residual plot for the data from 1980 through 2011, along with a regression for just those years, are below.

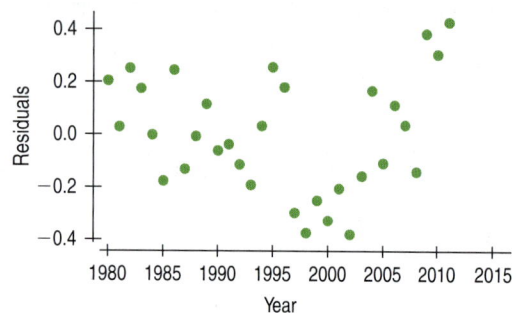

Dependent variable is Men − Women
R-squared = 49.4%   s = 0.225

Variable	Coefficient
Intercept	48.727
Year	−0.0234

a) Is this linear model appropriate for the post-1980 data? Explain.
b) What does the slope say about marriage ages?
c) Explain why it's not reasonable to interpret the y-intercept.

39. Gestation For humans, pregnancy lasts about 280 days. In other species of animals, the length of time from conception to birth varies. Is there any evidence that the gestation period is related to the animal's life span? The first scatterplot shows *Gestation Period* (in days) vs. *Life Expectancy* (in years) for 18 species of mammals. The highlighted point at the far right represents humans.

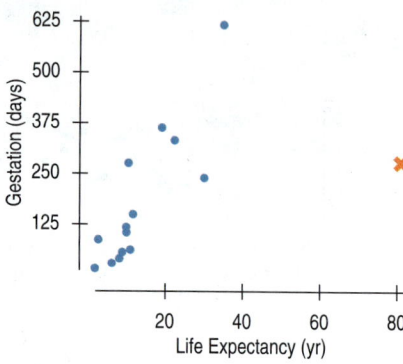

a) For these data, $r = 0.54$, not a very strong relationship. Do you think the association would be stronger or weaker if humans were removed? Explain.
b) Is there reasonable justification for removing humans from the data set? Explain.
c) Here are the scatterplot and regression analysis for the 17 nonhuman species. Comment on the strength of the association.

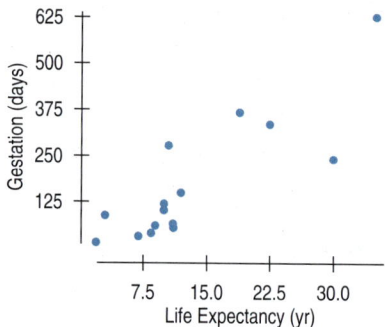

Dependent variable is Gestation
R-squared = 72.2%

Variable	Coefficient
Constant	−39.5172
LifExp	15.4980

d) Interpret the slope of the line.
e) Some species of monkeys have a life expectancy of about 20 years. Estimate the expected gestation period of one of these monkeys.

40. Swim the lake 2013 People swam across Lake Ontario 52 times between 1974 and 2013 (www.soloswims.com). We might be interested in whether they are getting any faster or slower. Here are the regression of the crossing

(continued)

Times (minutes) against the *Year* since 1974 of the crossing and the residuals plot:

Dependent variable is Time
R-squared = 1.2%   s = 439.1

Variable	Coefficient
Intercept	1213.423
Year since 1974	3.986

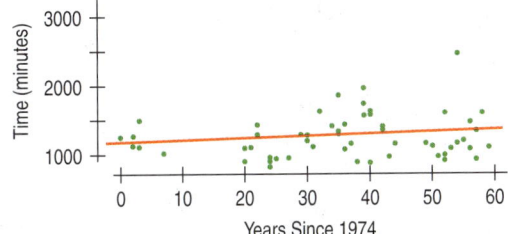

a) What does the R^2 mean for this regression?
b) Are the swimmers getting faster or slower? Explain.
c) The outlier seen in the residuals plot is a crossing by Vicki Keith in 1987 in which she swam a round trip, north to south, and then back again. Clearly, this swim doesn't belong with the others. Would removing it change the model a lot? Explain.

41. Elephants and hippos We removed humans from the scatterplot in Exercise 39 because our species was an outlier in life expectancy. The resulting scatterplot (below) shows two points that now may be of concern. The point in the upper right corner of this scatterplot is for elephants, and the other point at the far right is for hippos.

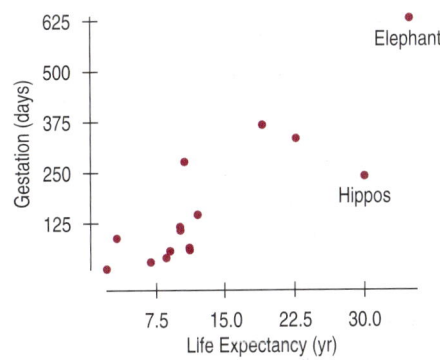

a) By removing one of these points, we could make the association appear to be stronger. Which point? Explain.
b) Would the slope of the line increase or decrease?
c) Should we just keep removing animals to increase the strength of the model? Explain.
d) If we remove elephants from the scatterplot, the slope of the regression line becomes 11.6 days per year. Do you think elephants were an influential point? Explain.

42. Another swim 2013 In Exercise 40, we saw that Vicki Keith's round-trip swim of Lake Ontario was an obvious outlier among the other one-way times. Here is the new regression after this unusual point is removed:

Dependent variable is Time
R-Squared = 4.5%   s = 322.7

Variable	Coefficient
Intercept	1139.794
Year since 1974	5.691

a) In this new model, the value of s_e is smaller. Explain what that means in this context.
b) Now would you be willing to say that the Lake Ontario swimmers are getting faster (or slower)?

43. Marriage age 2010 revisited Suppose you wanted to predict the trend in marriage age for American women into the early part of this century.

a) How could you use the data graphed in Exercise 15 to get a good prediction? Marriage ages in selected years starting in 1900 are listed below. Use all or part of these data to create an appropriate model for predicting the average age at which women will first marry in 2020.

1900–1950 (10-yr intervals):	21.9, 21.6, 21.2, 21.3, 21.5, 20.3
1955–2010 (5-yr intervals):	20.3, 20.3, 20.6, 20.8, 21.1, 22.0, 23.3, 23.9, 24.5, 25.1, 25.3, 26.1

b) How much faith do you place in this prediction? Explain.
c) Do you think your model would produce an accurate prediction about your grandchildren, say, 50 years from now? Explain.

44. Bridges covered In Chapter 7, (Data in **Tompkins County Bridges 2014**) we found a relationship between the age of a bridge in Tompkins County, New York, and its condition as found by inspection. But we considered only bridges built or replaced since 1886. Tompkins County is the home of the oldest covered bridge in daily use in New York State. Built in 1853, it was recently judged to have a condition of 4.57.

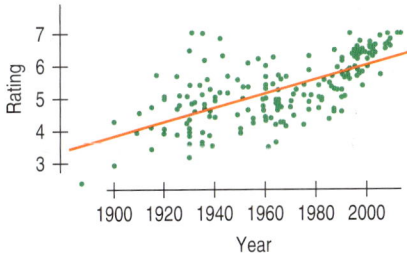

Dependent variable is Condition
R-squared = 43.4%   s = 0.7236

Variable	Coefficient
Intercept	−37.7775
Year	0.02190

a) If we use this regression to predict the condition of the covered bridge, what would its residual be?
b) If we add the covered bridge to the data, what would you expect to happen to the regression slope? Explain.

c) If we add the covered bridge to the data, what would you expect to happen to the R^2? Explain.
d) The bridge was extensively restored in 1972. If we use that date instead, do you find the condition of the bridge remarkable?

45. Life expectancy 2013 Data for 26 Western Hemisphere countries can be used to examine the association between life expectancy and the birth rate (number of births per 1000 population).

Country	Birth Rate (births/1000 population)	Life Expectancy
Argentina	17.12	76.1
Barbados	12.1	77
Belize	25.58	76.3
Canada	10.28	81.1
Chile	14.12	79.3
Colombia	16.98	73.9
Costa Rica	16.25	79.4
Cuba	9.92	79.3
Dominican Republic	19.21	73.6
Ecuador	19.23	75.8
El Salvador	17.12	72.4
Guatemala	25.99	71.4
Guyana	16.31	70.2
Haiti	23.35	62.4
Honduras	24.16	73.4
Jamaica	18.65	73.3
Mexico	18.61	77.1
Nicaragua	18.77	74.3
Panama	18.91	76.3
Paraguay	16.95	72.7
Saint Kitts and Nevis	13.79	73.3
Saint Lucia	14.19	74.8
Saint Vincent and the Grenadines	14.12	72.5
Trinidad and Tobago	14.07	70.3
Uruguay	13.28	77.2

a) Create a scatterplot relating *Life Expectancy* to *Birth Rate* and describe the association. Are the regression assumptions satisfied?
b) Is there an outlier? If so, set it aside and identify it before computing the regression.
c) Find the equation of the regression line with the outlier removed.
d) Make a plot of the residuals and comment.
e) Is the line an appropriate model?.
f) Interpret the value of R^2.
g) If government leaders want to increase life expectancy, should they encourage women to have fewer children? Explain.

46. Tour de France 2014 We met the Tour de France data set in Chapter 1 (in Just Checking). One hundred years ago, the fastest rider finished the course at an average speed of about 25.3 kph (around 15.8 mph). By the 21st century, riders were averaging over 40 kph (nearly 25 mph).

a) Make a scatterplot of *Avg Speed* against *Year*. Describe the relationship of *Avg Speed* by *Year*, being careful to point out any unusual features in the plot.
b) Find the regression equation of *Avg Speed* on *Year*.
c) Are the conditions for regression met? Comment.

47. Inflation 2011 The Consumer Price Index (CPI) tracks the prices of consumer goods in the United States, as shown in the following table. The CPI is reported monthly, but we can look at selected values. The table shows the January CPI at five-year intervals. It indicates, for example, that the average item costing $17.90 in 1926 cost $220.22 in the year 2011.

Year	JanCPI	Year	JanCPI
1916	10.4	1966	31.8
1921	19.0	1971	39.8
1926	17.9	1976	55.6
1931	15.9	1981	87.0
1936	13.8	1986	109.6
1941	14.1	1991	134.6
1946	18.2	1996	154.4
1951	25.4	2001	175.1
1956	26.8	2006	198.3
1961	29.8	2011	220.223

a) Make a scatterplot showing the trend in consumer prices. Describe what you see.
b) Be an economic forecaster: Project increases in the cost of living over the next decade. Justify decisions you make in creating your model.

48. Second stage 2014 Look once more at the data from the Tour de France. In Exercise 46, we looked at the whole history of the race, but now let's consider just the post-World War II era.

a) Find the regression of *Avg Speed* by *Year* only for years from 1947 to the present. Are the conditions for regression met?
b) Interpret the slope.
c) In 1981, Bernard Hinault averaged 39.86 kph, while in 2013, Christopher Froome averaged 40.55 kph. Which was the more impressive performance? Why?

Just Checking ANSWERS

1. Not high leverage, not influential, large residual
2. High leverage, not influential, small residual
3. High leverage, influential, not large residual

Chapter 9
Re-expressing Data: Get It Straight!

9.1 Straightening Scatterplots—The Four Goals

9.2 Finding a Good Re-expression

Where are we going?
What should we do when the data don't satisfy one of the conditions we're checking? What can we do if they are really skewed, or fail to satisfy the Straight Enough Condition?
Re-expressing the data—replacing the original by a simple function of it such as a logarithm or reciprocal—can often improve things. That makes the methods in this book far more useful because they can be applied to many more kinds of data.
And technology makes it all pretty easy.

A S Activity: Re-expressing Data. Should you re-express data? Actually, you already do.

Scan through any Physics book. Most equations have powers, reciprocals, or logs.

How fast can you go on a bicycle? If you measure your speed, you probably do it in miles per hour or kilometers per hour. In a 12-mile-long time trial in the 2005 Tour de France, Dave Zabriskie *averaged* nearly 35 mph (54.7 kph), beating Lance Armstrong by 2 seconds. You probably realize that's a tough act to follow. It's fast. You can tell that at a glance because you have no trouble thinking in terms of distance covered per time.

OK, then, if you averaged 12.5 mph (20.1 kph) for a mile *run,* would *that* be fast? Would it be fast for a 100 m dash? Even if you run the mile often, you probably have to stop and calculate. Running a mile in under 5 minutes (12 mph) is fast. A mile at 16 mph would be a world record (that's a 3-minute, 45-second mile). When Usain Bolt set the world record for the 100m at 9.58 sec, he averaged 23.35 mph for the 100 meters. At his peak, between 60 and 80 meters, he ran nearly 28 mph, a speed no other human has ever matched. There's no single *natural* way to measure speed. Sometimes we use time over distance; other times we use the *reciprocal*, distance over time. Neither one is *correct*. We're just used to thinking that way in each case.

So, how does this insight help us understand data? All quantitative data come to us measured in some way, with units specified. But maybe those units aren't the best choice. It's not that meters are better (or worse) than fathoms or leagues. What we're talking about is re-expressing the data another way by applying a function, such as a square root, log, or reciprocal. You already use some of them, even though you may not know it. For example, the Richter scale of earthquake strength (logs), the decibel scale for sound intensity (logs), the f/stop scale for camera aperture openings (squares), and the gauges of shotguns (square roots) all include simple functions of this sort.

Why bother? As with speeds, some expressions of the data may be easier to think about. And some may be much easier to analyze with statistical methods. We've seen that symmetric distributions are easier to summarize and straight scatterplots are easier to model with regressions. We often look to re-express our data if doing so makes them more suitable for our methods.

9.1 Straightening Scatterplots – The Four Goals

We introduced the ladder of powers in Chapter 6 as a guide to finding a good re-expression. In this chapter we'll expand the discussion and give examples to show you how to extend your ability to analyze data by using re-expressions.

We know from common sense and from Physics that heavier cars need more fuel, but exactly how does a car's weight affect its fuel efficiency? Here are the scatterplot of *Weight* (in pounds) and *Fuel Efficiency* (in miles per gallon) for 38 cars, and the residuals plot (Data in **Fuel Efficiency**):

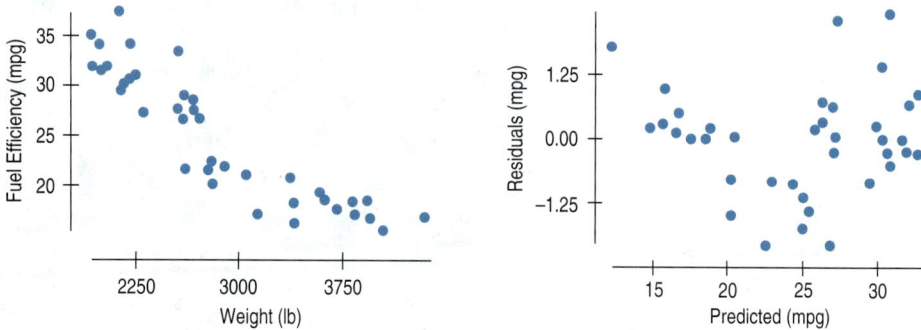

Figure 9.1

Fuel Efficiency (mpg) vs. *Weight* for 38 cars as reported by *Consumer Reports*. The scatterplot shows a negative direction, roughly linear shape, and strong relationship. However, the residuals from a regression of *Fuel Efficiency* on *Weight* reveal a bent shape when plotted against the predicted values. Looking back at the original scatterplot, you may be able to see the bend.

Hmm…. Even though R^2 is 81.6%, the residuals don't show the random scatter we were hoping for. The shape is clearly bent. Looking back at the first scatterplot, you can probably see the slight bending. Think about the regression line through the points. How heavy would a car have to be to have a predicted gas mileage of 0? It looks like the *Fuel Efficiency* would go negative at about 6000 pounds. The Maybach 62 was a luxury car produced by Daimler AG to compete with Rolls-Royce and Bentley. It was the heaviest "production car" in the world at 6184 lbs. Although not known for its fuel efficiency (rated at 10 mpg city), it did get more than the *minus* 2.6 mpg predicted from the model. Sadly after selling only 2100 of them in 10 years, Daimler stopped production in 2013. The 2012 model sold (without extra features) for $427,700. Extrapolation always requires caution, but it can go dangerously wrong when your model is wrong, because wrong models tend to do even worse the farther you get from the middle of the data.

Figure 9.2

Extrapolating the regression line gives an absurd answer for vehicles that weigh as little as 6000 pounds.

The bend in the relationship between *Fuel Efficiency* and *Weight* is the kind of failure to satisfy the conditions for an analysis that we can repair by re-expressing the data. Instead of looking at miles per gallon, we could take the reciprocal and work with gallons per hundred miles.[1]

> **"Gallons per hundred miles—what an absurd way to measure fuel efficiency! Who would ever do it that way?"** Not all re-expressions are easy to understand, but in this case the answer is "Everyone except U.S. drivers." Most of the world measures fuel efficiency in liters per 100 kilometers (L/100 km). This is the same reciprocal form (fuel amount per distance driven) and differs from gallons per 100 miles only by a constant multiple of about 2.38. It has been suggested that most of the world thinks, "I've got to go 100 km; how much gas do I need?" But Americans think, "I've got 10 gallons in the tank. How far can I drive?" In much the same way, re-expressions "think" about the data differently but don't change what they mean.

The direction of the association is positive now, since we're measuring gas consumption and heavier cars consume more gas per mile. The relationship is much straighter, as we can see from a scatterplot of the regression residuals.

---

[1] Multiplying by 100 to get gallons per 100 miles simply makes the numbers easier to think about: You might have a good idea of how many gallons your car needs to drive 100 miles, but probably a much poorer sense of how much gas you need to go just 1 mile.

Figure 9.3

The reciprocal (1/y) is measured in gallons per mile. Gallons per 100 miles gives more meaningful numbers. The reciprocal is more nearly linear against *Weight* than the original variable, but the re-expression changes the direction of the relationship. The residuals from the regression of *Fuel Consumption* (gal/100 mi) on *Weight* show less of a pattern than before.

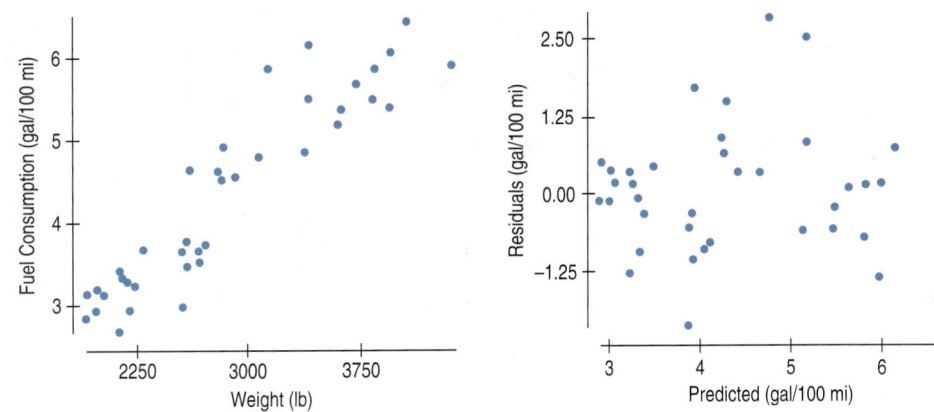

Who	262 large companies
What	Assets, sales, and market sector
Units	$
How	Public records
When	2014
Why	By *Forbes* magazine in reporting on the Forbes Global 2000 (Data in **Forbes Global 2000**) for that year

This is more the kind of boring residuals plot (no direction, no particular shape, no outliers, no bends) that we hope to see, so we have reason to think that the Straight Enough Condition is now satisfied. Now here's the payoff: What does the reciprocal model say about the Maybach? The regression line fit to *Fuel Consumption* vs. *Weight* predicts somewhere near 9.26 for a car weighing 6184 pounds. What does this mean? It means the car is predicted to use 9.26 gallons for every 100 miles, or in other words,

$$\frac{100 \text{ miles}}{9.26 \text{ gallons}} = 10.8 \text{ mpg}.$$

That's a much more reasonable prediction and very close to the reported value of 10 miles per gallon (of course, *your* mileage may vary ...).

Goals of Re-expression

We re-express data for several reasons. Each of these goals helps make the data more suitable for analysis by our methods.

Goal 1

Make the distribution of a variable (as seen in its histogram, for example) more symmetric. It's easier to summarize the center of a symmetric distribution, and for nearly symmetric distributions, we can use the mean and standard deviation. If the distribution is unimodal, then the resulting distribution may be closer to the Normal model, allowing us to use the 68–95–99.7 Rule.

Here are a histogram, quite skewed, showing the *Assets* of 262 companies selected from the Forbes Global 2000 (Data in **Forbes Global 2000**) list (in $B) and the more

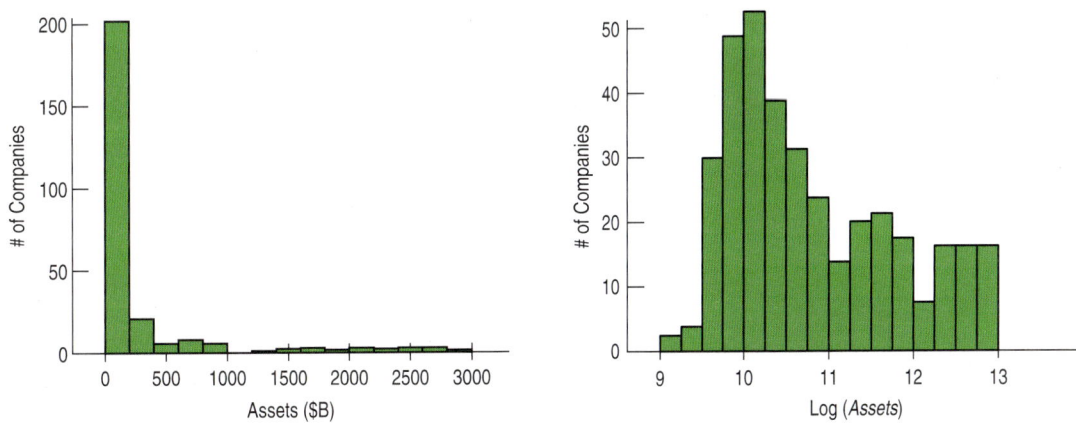

A | S Simulation: Re-expression in Action. Slide the re-expression power and watch the histogram change.

Figure 9.4

The distribution of the *Assets* of large companies is skewed to the right. Data on wealth often look like this. Taking logs makes the distribution more nearly symmetric. (We display the *Assets* in $B, but when we take the log, we use the original units [$]. Since the average *Assets* is near 10 $B, the average of *Log Assets* is around 10.)

symmetric histogram after taking logs.[1] The distribution of *Log Assets* is still skewed to the right, but much less so. And now there is the suggestion of a second, smaller, mode on the right that we couldn't see before.

Goal 2

Make the spread of several groups (as seen in side-by-side boxplots) more alike, even if their centers differ. Groups that share a common spread are easier to compare. We'll see methods later in the book that can be applied only to groups with a common standard deviation. We saw an example of re-expression for comparing groups with boxplots in Chapter 4.

Here are the *Assets* of these companies by *Market Sector:*

Figure 9.5

Assets of large companies by *Market Sector.* It's hard to see much of anything because the assets of the companies in the Finance sector are so large compared to the other sectors.

Taking logs makes the individual boxplots more symmetric and gives them spreads that are more nearly equal. When a re-expression simplifies the structure in several groups at once, it seems to be a natural choice for those data.

Figure 9.6

After re-expressing by logs, it's much easier to compare many market sectors. The boxplots are more nearly symmetric, most have similar spreads, and companies that seemed to be outliers before are no longer extraordinary.

Doing this makes it possible to compare assets across market sectors. It can also reveal other features in the data. Some companies that looked like outliers on the high end turned out to be more typical.

Goal 3

Make the form of a scatterplot more nearly linear. Scatterplots with a straight form are easier to model. We saw an example of scatterplot straightening in Chapter 6. The greater value of re-expression to straighten a relationship is that we can fit a linear model once the relationship is straight.

Because the assets of large companies vary so much by market sector and size, taking logs can help us see the differences more clearly. It makes sense that companies with more assets will have more sales as well. Here's a plot of *Sales* against *Log Assets*. We've already seen that companies in the financial sector are different, so we'll plot them with a different color.

---

[1]Logarithms can be "base 10" or "natural" (using the base *e*.) For the purposes of data re-expression, it really doesn't matter which one you use; they have the same effect on the shape of a histogram or scatterplot. But base-10 logs are a bit easier to relate to the original data because they count how many digits were in the original values. The base-10 log of 100 is 2. The log of 1000 is 3. Assets with a base-10 log of 10 are about $10,000,000,000. You may assume that the logs in this book are base-10 logs.

Plotting the log *Sales* against log *Assets* straightens out the relationship. Now the two groups are clear and the relationship between the variables is linear. Even more compelling, the slopes of the two groups are almost identical.

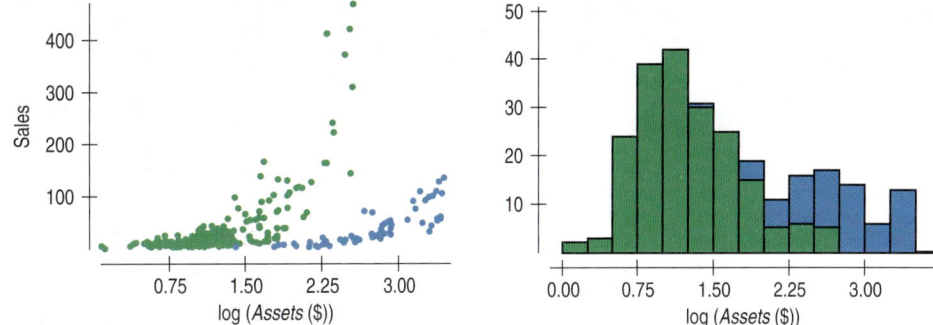

Figure 9.7
A plot of *Sales* against *Log Assets*. There is a clear relationship between the variables, and the differences between financial sector companies and others are evident. Looking back at the histogram of *Log Assets*, we can see that the second mode was indeed the Financial sector companies. But, the scatterplot is clearly not linear, so we are not done yet.

Goal 4

Make the scatter in a scatterplot spread out evenly rather than thickening at one end. Having an even scatter is a condition of many methods of Statistics, as we'll see in later chapters. This goal is closely related to Goal 2, but it often comes along with Goal 3. Indeed, a glance back at the scatterplot (Figure 9.7) shows that the plot for *Sales* vs. *Log*

For Example RECOGNIZING WHEN A RE-EXPRESSION CAN HELP

RECAP: In Chapter 8, we saw the awesome ability of emperor penguins to slow their heart rates while diving. Here are three displays relating to the diving heart rates:

(The boxplots show the diving heart rates for each of the 9 penguins whose dives were tracked. The names are those given by the researchers; EP = emperor penguin.)

QUESTION: What features of each of these displays suggest that a re-expression might be helpful?

ANSWER: The scatterplot shows a curved relationship, concave upward, between the duration of the dives and penguins' heart rates. Re-expressing either variable may help to straighten the pattern.

The histogram of heart rates is skewed to the high end. Re-expression often helps to make skewed distributions more nearly symmetric.

The boxplots each show skewness to the high end as well. The medians are low in the boxes, and several show high outliers.

Figure 9.8
The plot of log *Sales* vs. log *Assets* shows a clear linear relationship. The slopes for the two groups appear to be about the same, but the intercept for financial sector companies (in blue) is larger.

Assets is much more spread out on the right than on the left, while the plot for log *Sales* vs. *Log Assets* (Figure 9.8) has roughly the same variation in *Log Sales* for each value of *Log Assets*.

9.2 Finding a Good Re-Expression

Activity: Re-expression in Action. Here's the animated version of the Ladder of Powers. Slide the power and watch the change.

Symmetric distributions are easier to summarize and understand. It is even difficult to say just where the center of a skewed distribution is. And linear relationships are simpler and are easier to model than curved relationships. But just haphazardly trying various re-expressions to seek a useful transformation can be frustrating. We introduced the Ladder of Powers in Chapter 6. By systematically going up or down the ladder, you can often find a re-expression that will make your analysis simpler. The comments next to each power in the ladder suggest where to start looking for a re-expression for a single variable depending on your situation. Here is the Ladder of Powers again, for convenience. (See Table 9.1 next page.)

But, for scatterplots, which direction on the ladder should you go? A simple rule due to John Tukey can help guide you. Look at Figure 9.9. Now look at your scatterplot and see which quarter of the circle the curve of your plot most resembles. For example, the plot of *Fuel Efficiency* vs. *Weight* in Figure 9.1 looks like quadrant 3 in the lower left. Tukey's rule says to take y up the ladder if the curve looks like either quadrant 1 or 2 (the top half of the circle) and down the ladder if it most resembles quadrant 3 or 4. It's also possible to re-express x, but we will concentrate on just looking at the response. If you think of the curve of the circle as a pointer, it points in the directions for re-expression.

One note of caution–if your curve looks like two quadrants at the same time or curves in other ways, you won't be able to straighten the relationship with the methods in this chapter. You can also use the correlation between the variables (or equivalently the R^2 value from the regression) as a guide to help you decide among re-expressions. But this only works for comparing alternative re-expressions. A high R^2–even a *very* high R^2 –does not guarantee that the relationship is straight. And beware of outliers, which can influence the correlation, either increasing or decreasing it. It's best to look at the scatterplots and choose the re-expression that makes the relationship straightest, the spread most constant, and the histogram of the response variable the most symmetric. It's often easy to find a single re-expression that does a good job on all three criteria. But no re-expression will be perfect. You often will need to use your judgment to choose among the best two.

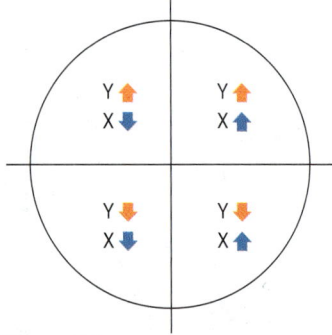

Figure 9.9
Tukey's circle of transformations shows what direction to move on the ladder for each variable depending on what quadrant the scatterplot resembles. For example, if the scatterplot is curved upward like quadrant 3 (bottom left) then either y or x should move down the ladder.

Table 9.1
The Ladder of Powers with comments to help guide in an initial choice of re-expression for a single variable.

Power	Name	Comment
2	The square of the data values, y^2.	Try this for unimodal distributions that are skewed to the left.
1	The raw data—no change at all. This is "home base." The farther you step from here up or down the ladder, the greater the effect.	Data that can take on both positive and negative values with no bounds are less likely to benefit from re-expression.
1/2	The square root of the data values, $\sqrt{y}$.	Counts often benefit from a square root re-expression. For counted data, start here.
"0"	The logarithm of y. It doesn't matter whether you take the log base 10, the natural log, (or any other base).	Measurements that cannot be negative, and especially values that grow by percentage increases such as salaries or populations, often benefit from a log re-expression. When in doubt, start here. If your data have zeros, try adding a small constant to all values before finding the logs.
−1/2	The (negative) reciprocal square root, $-1/\sqrt{y}$.	An uncommon re-expression, but sometimes useful. Changing the sign to take the *negative* of the reciprocal square root preserves the direction of relationships, making things a bit simpler.
−1	The (negative) reciprocal, $-1/y$.	Ratios of two quantities (miles per hour, for example) often benefit from a reciprocal. (You have about a 50–50 chance that the original ratio was taken in the "wrong" order for simple statistical analysis and would benefit from re-expression.) Often, the reciprocal will have simple units (hours per mile). Change the sign if you want to preserve the direction of relationships. If your data have zeros, try adding a small constant to all values before finding the reciprocal.

Just Checking

1. You want to model the relationship between the number of birds counted at a nesting site and the temperature (in degrees Celsius). The scatterplot of counts vs. temperature shows an upwardly curving pattern, with more birds spotted at higher temperatures. What transformation (if any) of the bird counts might you start with?

2. You want to model the relationship between prices for various items in Paris and in Hong Kong. The scatterplot of Hong Kong prices vs. Parisian prices shows a generally straight pattern with a small amount of scatter. What transformation (if any) of the Hong Kong prices might you start with?

3. You want to model the population growth of the United States over the past 200 years. The scatterplot shows a strongly upwardly curved pattern. What transformation (if any) of the population might you start with?

Step-by-Step Example RE-EXPRESSING TO STRAIGHTEN A SCATTERPLOT

We've seen curvature in the relationship between emperor penguins' diving heart rates and the duration of the dive. In addition, the histogram of heart rates was skewed and the boxplots of heart rates by individual birds showed skewness to the high end.

Questions: What re-expression might straighten out the relationship?

Think ➔ Plan State the problem.

Variables Identify the variables and report the W's.

Plot Check that even if there is a curve, the overall pattern does not reach a minimum or maximum and then turn around and go back. An up-and-down curve can't be fixed by re-expression.

I want to fit a linear model for the *Diving Heart Rate (DHR)* and *Duration* of a penguin dive.

I have the *DHR* in beats per minute and the *Duration* in minutes.

The plot shows a negative direction and an association that is curved like the third quadrant of Tukey's circle. The correlation is −0.846.

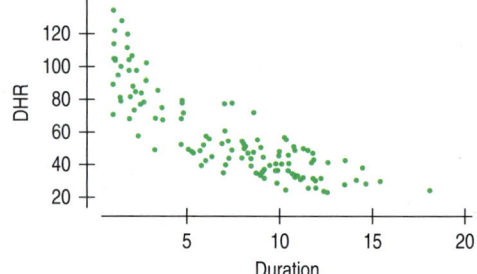

The histogram of heart rates shows right skewness.

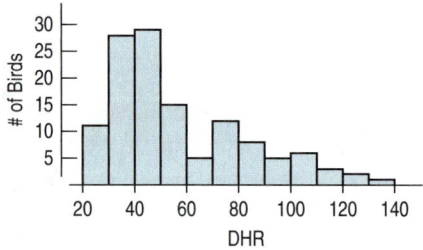

SHOW ▶ Mechanics

Try a re-expression. See what quadrant the curve most resembles and try moving a re-expression of *y* in that direction. The lesson of the Ladder of Powers is that if we're moving in the right direction but have not had sufficient effect, we should go farther along the ladder.

Tukey's circle suggests going down the ladder. We'll try the square root of DHR. Here's a plot of the square root of DHR against Duration:

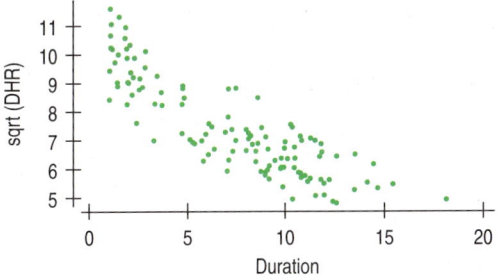

This example shows improvement, but is still not straight. We'll go further down the ladder and try logarithm.

The plot is less bent, but still not straight. Stepping from the 1/2 power to the "0" power, we try the logarithm of DHR against Duration:

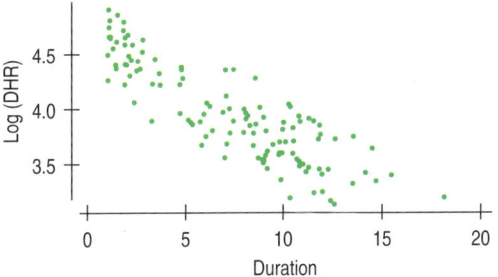

Often a re-expression that improves straightness improves the symmetry of the response variable as well. That's the case here.

Let's look at the histogram of log(DHR) as well:

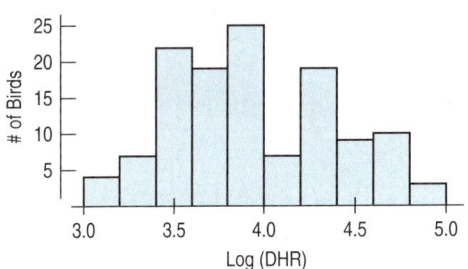

There is no harm in going too far down the ladder as we can always "back up". That way we know we've gone far enough. It looks here like we've gone too far. If anything the curve is now curving downward instead of upward. We used the negative reciprocal to keep the direction the same (negative in this case).

The histogram is more symmetric. This reexpression might work. The correlation is now −0.869. However, I'll go a little farther to make sure I've gone far enough. Here's a plot of the (negative) reciprocal of DHR against Duration:

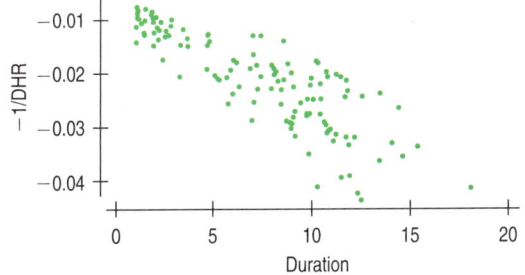

The histogram also reflects that we've gone too far. Now the skewness is in the opposite direction.

Now the plot seems to curve slightly downward, it spreads out more on the right, the correlation is −0.845, and my histogram now shows skewness in the opposite direction than it did before.

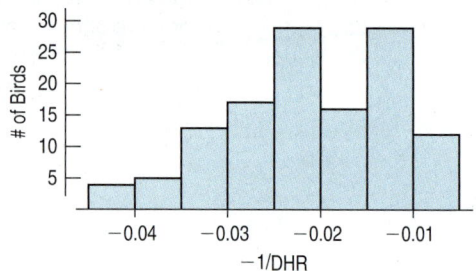

It appears that I've gone too far. I'll back up one "rung" and try -1/sqrt(DHR):

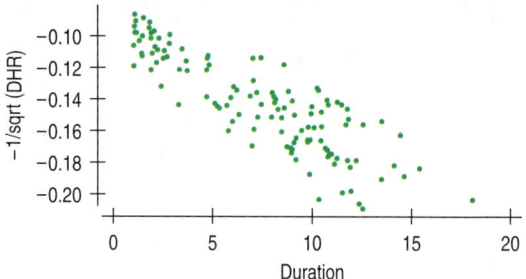

We may have to choose between two adjacent re-expressions. For most data analyses, it really doesn't matter which we choose.

This looks pretty good and the correlation is −0.863. Let's check the histogram:

The histogram is symmetric as well.

Tell ➡ **Conclusion** Specify your choice of re-expression. If there's some natural interpretation (as for gallons per 100 miles), give that.

It is hard to choose between this re-expression and the log, but I'll stay with this one.

Now that the re-expressed data satisfy the Straight Enough Condition, we can fit a linear model by least squares. We find that

$$\frac{-1}{\sqrt{\widehat{DHR}}} = -0.0955 - 0.00642 \, Duration.$$

We can use this model to predict the heart beats resulting from a dive of, say, 10 minutes:

$$\frac{-1}{\sqrt{\widehat{DHR}}} = -0.0955 - 0.00642 \times 10 = -0.1597$$

We could leave the result in these units ($-1/\sqrt{beats}$). Sometimes the new units may be as meaningful as the original, but here we want to transform the predicted value back into beats. Fortunately, each of the re-expressions in the Ladder of Powers can be reversed.

To reverse the process, we first take the reciprocal: $\sqrt{\widehat{DHR}} = -1/(-0.1597) = 6.2617$. Then squaring gets us back to the original units:

$$\widehat{DHR} = 6.2617^2 = 39.2 \, beats.$$

This may be the most painful part of the re-expression. Getting back to the original units can sometimes be a little work. Nevertheless, it's worth the effort to always consider re-expression. Re-expressions extend the reach of all of your Statistics tools by helping more data to satisfy the conditions they require. Just think how much more useful this course just became!

For Example COMPARING RE-EXPRESSIONS

RECAP: We've concluded that it was a toss-up between the reciprocal square root and the logarithm of heart rate. We chose the square root, but let's look at the logarithm. Here are the resulting displays:

QUESTIONS: Comment on these displays. Now that we've looked at the original data (rung 1 on the Ladder), the reciprocal square root (rung −1/2), and the logarithm (rung 0), which re-expression of *Diving Heart Rate* would you choose?

ANSWER: The scatterplot is linear and the histogram is symmetric. The boxplots are still a bit skewed to the high end, but less so than for the original Diving Heart Rate values. We don't expect real data to cooperate perfectly, and the logarithm also seems like a good compromise re-expression, improving several different aspects of the data.

Plan B: Attack of the Logarithms

The Ladder of Powers is often successful at finding an effective re-expression. Sometimes, though, the curvature is more stubborn, and we're not satisfied with the residual plots. What then?

When none of the data values is zero or negative, logarithms can be a helpful ally in the search for a useful model. Try taking the logs of both the x- and y-variables. Then re-express the data using some combination of x or $\log(x)$ vs. y or $\log(y)$. You may find that one of these works pretty well. This can be especially useful if the ladder is telling you to try large powers of y. It may be better to re-express x.

Model Name	x-axis	y-axis	Comment
Exponential	x	$\log(y)$	This model is the "0" power in the ladder approach, useful for values that grow by percentage increases.
Logarithmic	$\log(x)$	y	A wide range of x-values, or a scatterplot descending rapidly at the left but leveling off toward the right, may benefit from trying this model.
Power	$\log(x)$	$\log(y)$	The Goldilocks model: When one of the ladder's powers is too big and the next is too small, this one may be just right.

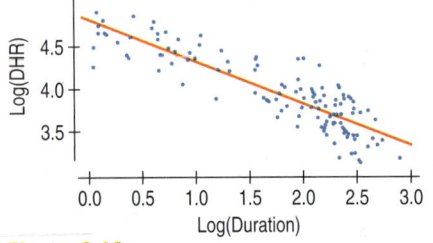

Figure 9.10
Plotting log (*Length*) against log (*Strength*) gives a straighter shape.

When we tried to model the relationship between the *DHR* and the dive *Duration*, we were torn between the reciprocal square root, the "−1/2" power, and the logarithm, the "0" power.

The scatterplot shows what happens when we graph the logarithm of *DHR* against the logarithm of *Duration*. Technology reveals that the equation of our log–log model is

$$\widehat{\log(DHR)} = 4.81 - 0.491 \log(Duration)$$

This approach suggests we re-express *Duration* with the reciprocal square root.

A warning, though! Don't expect to be able to straighten every curved scatterplot you find. It may be that there just isn't a very effective re-expression to be had. You'll certainly encounter situations when nothing seems to work the way you wish it would. Don't set your sights too high—you won't find a perfect model. Keep in mind: We seek a *useful* model, not perfection (or even "the best").

Multiple Benefits

We often choose a re-expression for one reason and then discover that it has helped other aspects of an analysis. For example, we might re-express a variable to make its histogram more nearly symmetric. It wouldn't be surprising to find that the same re-expression also straightens scatterplots or makes spreads more nearly equal. This phenomenon is one reason we encourage that data re-expression be used more often. Sometimes there's an obvious "best" or "right" re-expression for a variable. For example, it makes sense that things that tend to grow by a constant percentage, so that larger values increase faster (populations, bacteria counts, wealth), grow exponentially. Logarithms straighten out the exponential trend and pull in the long right tail in the histogram. When several groups are made more symmetric or a relationship seen in different groups is made more linear with the same slope, the argument for that re-expression is even stronger.

We saw just this phenomenon in the companies example earlier. Each of our goals was met by re-expressing *Assets* with logarithms. That single re-expression improved all

> **OCCAM'S RAZOR**
>
> If you think that simpler explanations and simpler models are more likely to give a true picture of the way things work, then you should look for opportunities to re-express your data and simplify your analyses.
>
> The general principle that simpler explanations are likely to be the better ones is known as Occam's Razor, after the English philosopher and theologian William of Occam (1284–1347).

four of the goals at the same time. And the plot of *Log Sales* against *Log Assets* showed two parallel straight relationships. Multiple successes like this turn out not to be all that unusual.

Measurement errors are often larger when measuring larger quantities than when measuring smaller ones. (The error in your height may be only a centimeter or two, but the error in the height of a tree could be ten times that much.) Again, logarithms are likely to help. Measurements of rates (e.g., time to complete a task) are often plagued by infinities for those who never finish. The reciprocal of "minutes per task" is "tasks per minute"—a speed measure. And the unfinished tasks that once took "infinite" time now simply rate a zero speed. Here once more the re-expression seems natural.

In other cases, the only evidence we have to favor re-expression is that it seems to work well and that it leads to simpler models. Often we can find a re-expression for a variable that simplifies our analysis in several ways at once, making the distribution symmetric, making the relationship linear in terms of other variables of interest, or stabilizing its variance. If so, re-expressing the variable certainly simplifies our efforts to analyze and understand it.

Why Not Just Use a Curve?

When a clearly curved pattern shows up in the scatterplot, why not just fit a curve to the data? We saw earlier that the association between the *Weight* of a car and its *Fuel Efficiency* was not a straight line. Instead of trying to find a way to straighten the plot, why not find a curve that seems to describe the pattern well?

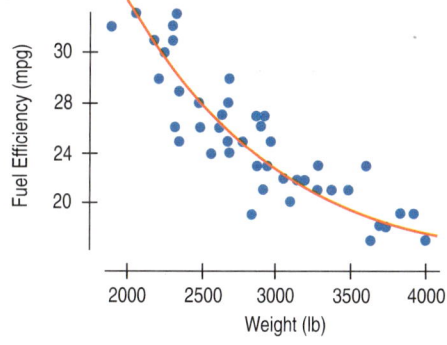

We can find "curves of best fit" using essentially the same approach that led us to linear models. You won't be surprised, though, to learn that the mathematics and the calculations are considerably more difficult for curved models. Many calculators and computer packages do have the ability to fit curves to data, but this approach has many drawbacks.

Straight lines are easy to understand. We know how to think about the slope and the y-intercept, for example. We often want some of the other benefits mentioned earlier, such as making the spread around the model more nearly the same everywhere. In later chapters you will learn more advanced statistical methods for analyzing linear associations.

We give all of that up when we fit a model that is not linear. For many reasons, then, it is usually better to re-express the data to straighten the plot.

WHAT CAN GO WRONG?

- **Don't expect your model to be perfect.** In Chapter 5 we quoted statistician George Box: "All models are wrong, but some are useful." Be aware that the real world is a messy place and data can be uncooperative. Don't expect to find one elusive re-expression that magically irons out every kink in your scatterplot and produces perfect residuals. You aren't looking for the Right Model, because that mythical creature doesn't exist. Find a useful model and use it wisely.

- **Don't stray too far from the ladder.** It's wise not to stray too far from the powers that we suggest in the Ladder of Powers. Taking the y-values to an extremely high power may artificially inflate R^2, but it won't give a useful or meaningful model, so it doesn't really simplify anything. It's better to stick to powers between 2 and −2. Even in that range, you should prefer the simpler powers in the ladder to those in the cracks. A square root is easier to understand than the 0.413 power. That simplicity may compensate for a slightly less straight relationship.

- **Don't choose a model based on R^2 alone.** You've tried re-expressing your data to straighten a curved relationship and found a model with a high R^2. Beware: That doesn't mean the pattern is straight now. On the next page is a plot of a relationship with an R^2 of 98.3%.

  The R^2 is about as high as we could ask for, but if you look closely, you'll see that there's a consistent bend. Plotting the residuals from the least squares line makes the bend much easier to see.

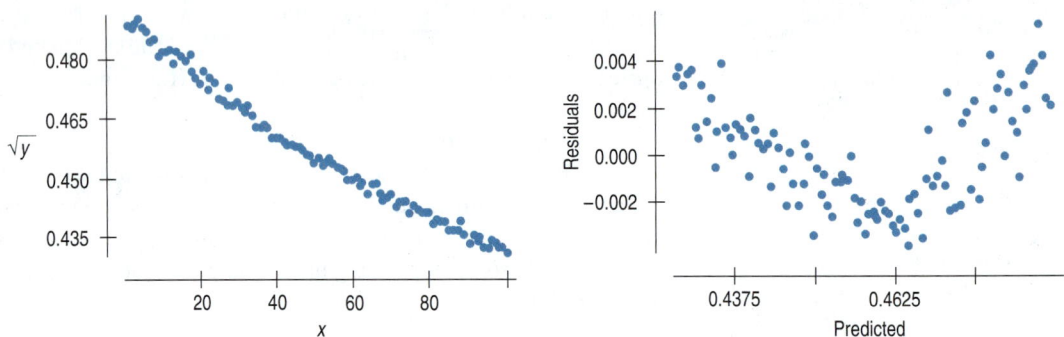

  Remember the basic rule of data analysis: *Make a picture*. Before you fit a line, always look at the pattern in the scatterplot. After you fit the line, check for linearity again by plotting the residuals.

- **Be alert for multiple modes.** Re-expression can often make a skewed unimodal histogram more nearly symmetric, but it cannot pull separate modes together. A suitable re-expression may, however, make the separation of the modes clearer, simplifying their interpretation and making it easier to separate them to analyze individually.

- **Watch out for scatterplots that turn around.** Re-expression can straighten many bent relationships but not those that go up and then down or down and then up. You should refuse to analyze such data with methods that require a linear form.

Figure 9.11
The shape of the scatterplot of *Birth Rates* (births per 100,000 women) in the United States shows an oscillation that cannot be straightened by re-expressing the data.

- **Watch out for negative data values.** It's impossible to re-express negative values by any power that is not a whole number on the Ladder of Powers or to re-express values that are zero for negative powers. Most statistics programs will just mark the result of trying to re-express such values "missing" if they can't be re-expressed. But that might mean that when you try a re-expression, you inadvertently lose a bunch of data values. The effect of that loss may be surprising and may substantially change your analysis. Because you are likely to be working with a computer package or calculator, take special care that you do not lose otherwise good data values when you choose a re-expression.

  One possible cure for zeros and small negative values is to add a constant ($\frac{1}{2}$ and $\frac{1}{6}$ are often used) to bring all the data values above zero.

- **Watch for data far from 1.** Data values that are all very far from 1 may not be much affected by re-expression unless the range is very large. Re-expressing numbers between 1 and 100 will have a much greater effect than re-expressing numbers between 100,001 and 100,100. When all your data values are large (for example, working with years), consider subtracting a constant to bring them back near 1. (For example, consider "years since 1950" as an alternative variable for re-expression. Unless your data start at 1950, then avoid creating a zero by using "years since 1949.")

CONNECTIONS

We have seen several ways to model or summarize data. Each requires that the data have a particular simple structure. We seek symmetry for summaries of center and spread and to use a Normal model. We seek equal variation across groups when we compare groups with boxplots or want to compare their centers. We seek linear shape in a scatterplot so that we can use correlation to summarize the scatter and regression to fit a linear model.

Data do often satisfy the requirements to use Statistics methods. But often they do not. Our choice is to stop with just displays, to use much more complex methods, or to re-express the data so that we can use the simpler methods we have developed.

In this fundamental sense, this chapter connects to everything we have done thus far and to all of the methods we will introduce throughout the rest of the book. Re-expression greatly extends the reach and applicability of all of these methods.

What Have We Learned?

Learning Objectives

- Understand that models aren't perfect, but recognize when a well-chosen re-expression can help you improve and simplify your analysis.
- Understand the value of re-expressing data to improve symmetry, to make the scatter around a line more constant, or to make a scatterplot more linear.
- Recognize when the pattern of the data indicates that no re-expression can improve the structure of the data.
- Know how to re-express data with powers and how to find an effective re-expression for your data using your statistics software or calculator. Understand how to use the Ladder of Powers or the log-log approach to find a good re-expression.
- Be able to reverse any of the common re-expressions to put a predicted value or residual back into the original units.
- Be able to describe a summary or display of a re-expressed variable, making clear how it was re-expressed and giving its re-expressed units.
- Be able to describe a regression model fit to re-expressed data in terms of the reexpressed variables.

Review of Terms

Re-expression We re-express data by taking the logarithm, the square root, the reciprocal, or some other mathematical operation of all values of a variable (p. 249).

Ladder of Powers The Ladder of Powers places in order the effects that many re-expressions have on the data (p. 252).

On the Computer RE-EXPRESSION

Computers and calculators make it easy to re-express data. Most statistics packages offer a way to re-express and compute with variables. Some packages permit you to specify the power of a re-expression with a slider or other moveable control, possibly while watching the consequences of the re-expression on a plot or analysis. This, of course, is a very effective way to find a good re-expression.

DATA DESK

To re-express a variable in Data Desk, select the variable and Choose the function to re-express it from the **Manip > Transform** menu. Square root, log, reciprocal, and reciprocal root are immediately available. For others, make a derived variable and type the function. Data Desk makes a new derived variable that holds the re-expressed values. Any value changed in the original variable will immediately be re-expressed in the derived variable.

COMMENTS

Or choose **Manip > Transform > Dynamic > Box-Cox** to generate a continuously changeable variable and a slider that specifies the power. Set plots to **Automatic Update** in their HyperView menus and watch them change dynamically as you drag the slider.

EXCEL

To re-express a variable in Excel, use Excel's built-in functions as you would for any calculation. Changing a value in the original column will change the re-expressed value.

JMP

To re-express a variable in JMP, double-click to the right of the last column of data to create a new column. Name the new column and select it. Choose **Formula** from the **Cols** menu. In the Formula dialog, choose the transformation and variable that you wish to assign to the new column. Click the **OK** button. JMP places the re-expressed data in the new column.

As of version 11, JMP allows re-expression "on the fly" in most dialogs. For example, in the **Fit Y by X** platform, right click on either variable to bring up a **Transform** option that offers most re-expressions.

COMMENTS

The log and square root re-expressions are found in the **Transcendental** menu of functions in the formula dialog.

MINITAB

To re-express a variable in MINITAB, choose **Calculator** from the **Calc** menu. In the Calculator dialog, specify a name for the new re-expressed variable. Use the **Functions List**, the calculator buttons, and the **Variables list** box to build the expression. Click **OK**.

R

To re-express a variable in R, either create a new variable, as, for example:
> Log.y = log(y) #natural log

or re-express inside a formula. For example:
> lm(log(y) ~ x)

CHAPTER 9 Re-expressing Data: Get It Straight!

SPSS

To re-express a variable in SPSS, choose **Compute** from the **Transform** menu. Enter a name in the Target Variable field. Use the calculator and Function List to build the expression.

Move a variable to be re-expressed from the source list to the Numeric Expression field. Click the **OK** button.

STATCRUNCH

From the **Data** menu, select **Compute Expression**

In the Compute Expression box, type the expression and name the new variable

Click the **Compute!** button to make the new variable.

TI-83/84 PLUS

To re-express data stored in a list, perform the re-expression on the whole list and store it in another list. For example, to use the logarithms of the data in L1, enter the command **log(L1) STO L2**.

Exercises

Section 9.1

1. **Residuals** Suppose you have fit a linear model to some data and now take a look at the residuals. For each of the following possible residuals plots, tell whether you would try a re-expression and, if so, why.

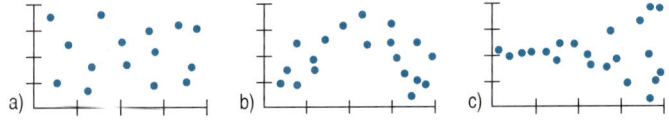

2. **More Residuals** Suppose you have fit a linear model to some data and now take a look at the residuals. For each of the following possible residuals plots, tell whether you would try a re-expression and, if so, why.

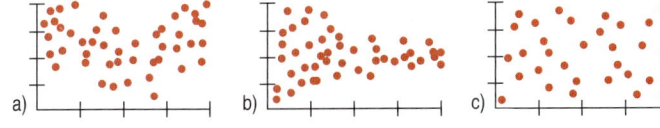

3. **BK Protein** Recall the data about the Burger King menu items. Here are boxplots of protein content comparing items that contain meat with those that do not. The plot on the right graphs log(*Protein*). Which of the goals of re-expression does this illustrate?

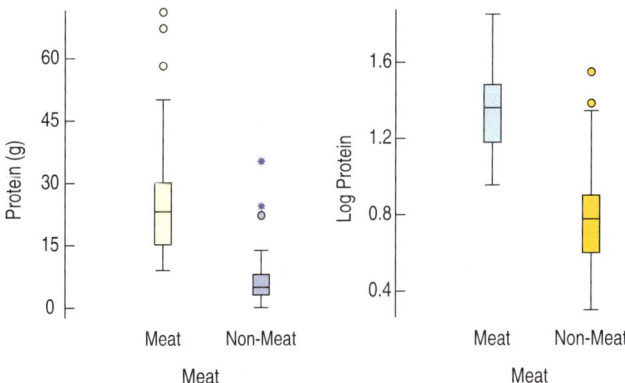

4. Life Expectancy and TV Recall the example of life expectancy *vs* TVs per person in chapter 6. In that example, we use the square root of TVs per person. Here are the original data and the square root. Which of the goals of re-expression does this illustrate?

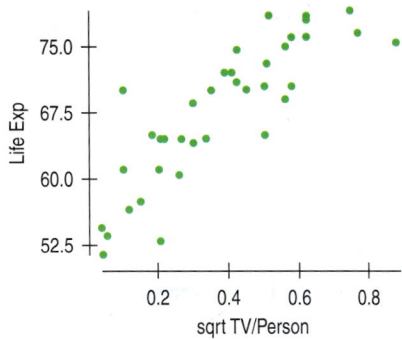

Section 9.2

5. BK Protein again Exercise 3 looked at the distribution of protein in the Burger King menu items, comparing meat and non-mean items. That exercise offered the logarithm as a re-expression of Protein. Here are two other alternatives, the square root and the reciprocal. Would you still prefer the log? Explain why

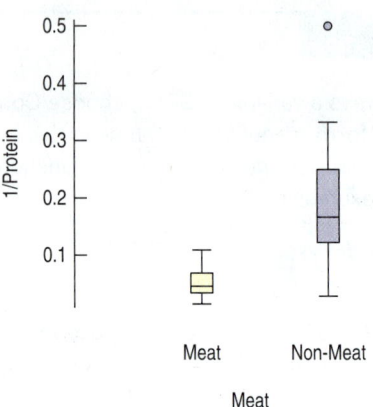

6. Life Expectancy and TV again Exercise 4 revisited the relationship between life expectancy and TVs per capita and saw that re-expression to the square root of TVs per capita made the plot more nearly straight. But was that the best choice of re-expression. Here is a scatterplot of lige expectancy *vs* the logarithm of TV's per capita. How can you tell that this re-expression is too far along the Ladder of Powers?

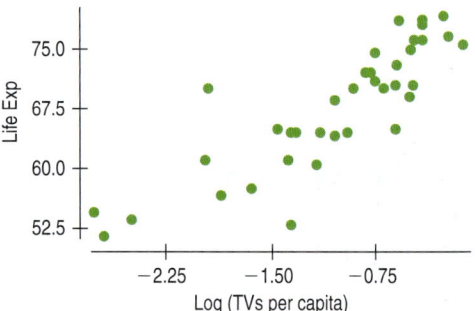

Chapter Exercises

7. Oakland passengers 2013 revisited In Chapter 8, Exercise 9, we created a linear model describing the trend in the number of passengers departing from the Oakland (CA) airport each month since the start of 1997. Here's the residual plot, but with lines added to show the order of the values in time:

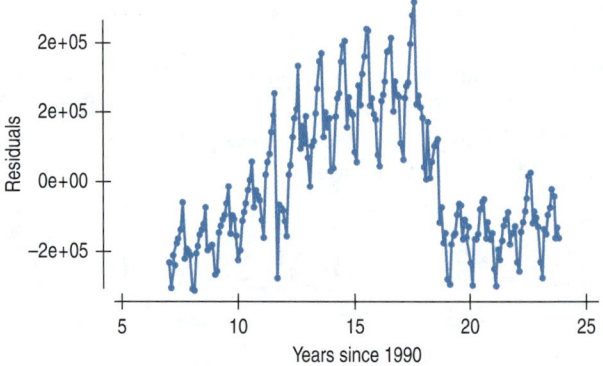

a) Can you account for the pattern shown here?
b) Would a re-expression help us deal with this pattern? Explain.

8. **Hopkins winds, revisited** In Chapter 4, we examined the wind speeds in the Hopkins forest over the course of a year. Here's the scatterplot we saw then (Data in **Hopkins Forest 2011**):

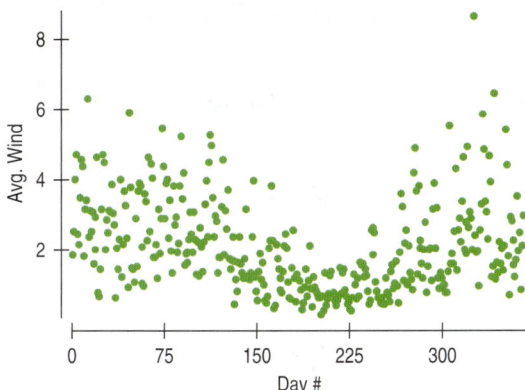

a) Describe the pattern you see here.
b) Should we try re-expressing either variable to make this plot straighter? Explain.

9. **Models** For each of the models listed below, predict y when $x = 2$.

a) $\ln \hat{y} = 1.2 + 0.8x$ d) $\hat{y} = 1.2 + 0.8 \ln x$
b) $\sqrt{\hat{y}} = 1.2 + 0.8x$ e) $\log \hat{y} = 1.2 + 0.8 \log x$
c) $\dfrac{1}{\hat{y}} = 1.2 + 0.8x$

10. **More models** For each of the models listed below, predict y when $x = 2$.

a) $\hat{y} = 1.2 + 0.8 \log x$ d) $\hat{y}^2 = 1.2 + 0.8x$
b) $\log \hat{y} = 1.2 + 0.8x$ e) $\dfrac{1}{\sqrt{\hat{y}}} = 1.2 + 0.8x$
c) $\ln \hat{y} = 1.2 + 0.8 \ln x$

11. **Gas mileage** As the example in the chapter indicates, one of the important factors determining a car's *Fuel Efficiency* is its *Weight*. Let's examine this relationship again, for 11 cars.

a) Describe the association between these variables shown in the scatterplot on the next page.

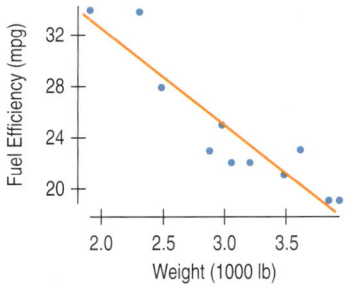

b) Here is the regression analysis for the linear model. What does the slope of the line say about this relationship?

Dependent variable is: Fuel Efficiency
R-squared = 85.9%

Variable	Coefficient
Intercept	47.9636
Weight	−7.65184

c) Do you think this linear model is appropriate? Use the residuals plot to explain your decision.

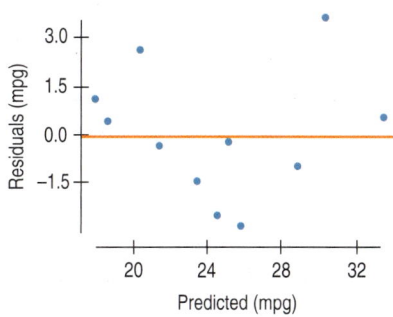

12. **Crowdedness** In a *Chance* magazine article (Summer 2005), Danielle Vasilescu and Howard Wainer used data from the United Nations Center for Human Settlements to investigate aspects of living conditions for several countries. Among the variables they looked at were the country's per capita gross domestic product (*GDP*, in $) and *Crowdedness,* defined as the average number of persons per room living in homes there. This scatterplot displays these data for 56 countries:

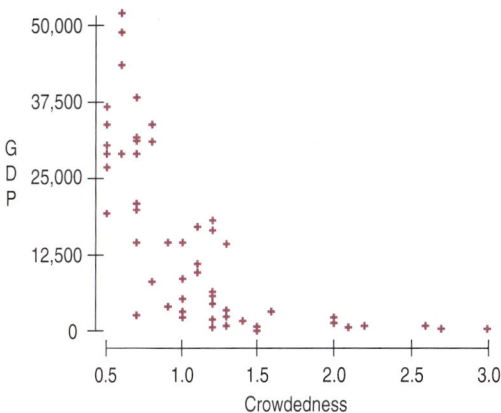

a) Explain why you should re-express these data before trying to fit a model.
b) What re-expression of *GDP* would you try as a starting point?

13. **Gas mileage, revisited** Let's try the re-expressed variable *Fuel Consumption* (gal/100 mi) to examine the fuel efficiency of the 11 cars in Exercise 11. Here are the revised regression analysis and residuals plot:

Dependent variable is: Fuel Consumption
R-squared = 89.2%

Variable	Coefficient
Intercept	0.624932
Weight	1.17791

a) Explain why this model appears to be better than the linear model.
b) Using the regression analysis above, write an equation of this model.
c) Interpret the slope of this line.
d) Based on this model, how many miles per gallon would you expect a 3500-pound car to get?

14. Crowdedness again In Exercise 12 we looked at United Nations data about a country's *GDP* and the average number of people per room (*Crowdedness*) in housing there. For a re-expression, a student tried the reciprocal $-10000/GDP$, representing the number of people per $10,000 of gross domestic product. Here are the results, plotted against *Crowdedness*:

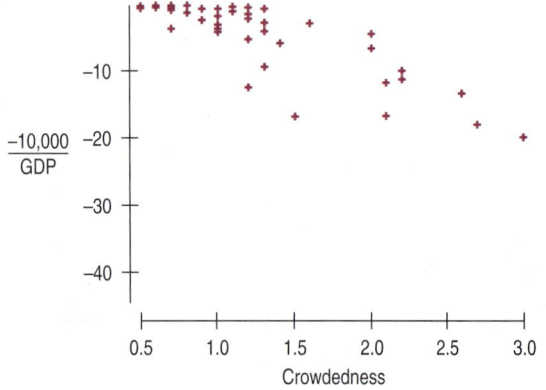

a) Is this a useful re-expression? Explain.
b) What re-expression would you suggest this student try next?

15. GDP 2013 The scatterplot shows the gross domestic product (GDP) of the United States in billions of (2013) dollars plotted against years since 1950.

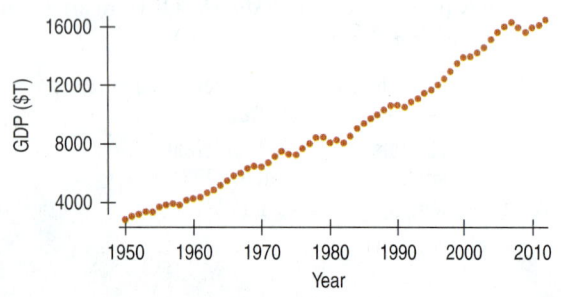

A linear model fit to the relationship looks like this:

Dependent variable is: GDP
R-squared = 98.3%   s = 553.2

Variable	Coefficient
Intercept	2058.981
Year-1950	227.176

a) Does the value 98.3% suggest that this is a good model? Explain.
b) Here's a scatterplot of the residuals. Now do you think this is a good model for these data? Explain?

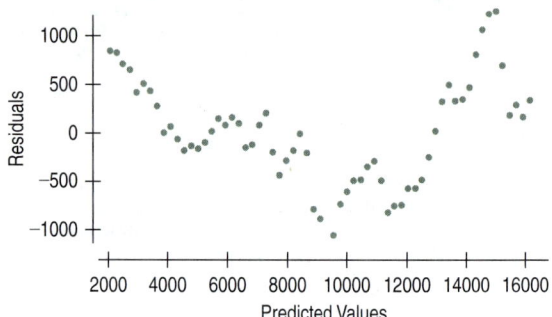

16. Interest rates 2014, once more The 3-month Treasury bill interest rate is watched by investors and economists. Here's a scatterplot of the 3-month Treasury bill rate since 1950 (Data in **Interest Rates 2014**):

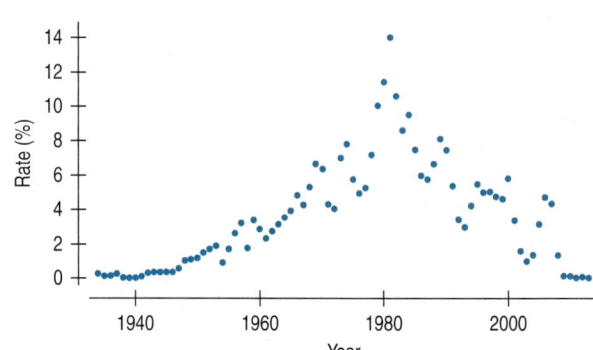

Clearly, the relationship is not linear. Can it be made nearly linear with a re-expression? If so, which one would you suggest? If not, why not?

17. Better GDP model? Consider again the post-1950 trend in U.S. GDP we examined in Exercise 15. Here are a regression and residual plot when we use the log of GDP in the model. Is this a better model for GDP? Explain.

Dependent variable is: LogGDP
R-squared = 99.1%   s = 0.0554

Variable	Coefficient
Intercept	0.79993
Year-1950	0.03139

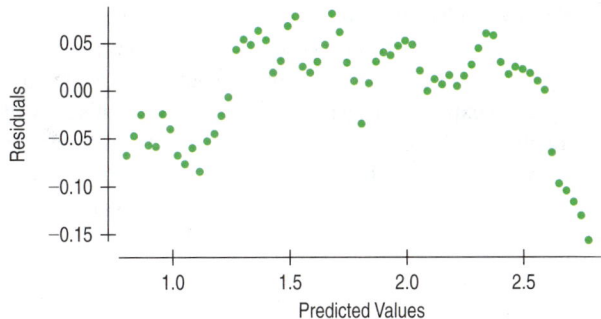

18. Pressure Scientist Robert Boyle examined the relationship between the volume in which a gas is contained and the pressure in its container. He used a cylindrical container with a moveable top that could be raised or lowered to change the volume. He measured the *Height* in inches by counting equally spaced marks on the cylinder, and measured the *Pressure* in inches of mercury (as in a barometer). Some of his data are listed in the table. Create an appropriate model.

Height	48	44	40	36	32	28
Pressure	29.1	31.9	35.3	39.3	44.2	50.3
Height	24	20	18	16	14	12
Pressure	58.8	70.7	77.9	87.9	100.4	117.6

19. Brakes The following table shows stopping distances in feet for a car tested 3 times at each of 5 speeds. We hope to create a model that predicts *Stopping Distance* from the *Speed* of the car.

Speed (mph)	Stopping Distances (ft)
20	64, 62, 59
30	114, 118, 105
40	153, 171, 165
50	231, 203, 238
60	317, 321, 276

a) Explain why a linear model is not appropriate.
b) Re-express the data to straighten the scatterplot.
c) Create an appropriate model.
d) Estimate the stopping distance for a car traveling 55 mph.
e) Estimate the stopping distance for a car traveling 70 mph.
f) How much confidence do you place in these predictions? Why?

20. Pendulum A student experimenting with a pendulum counted the number of full swings the pendulum made in 20 seconds for various lengths of string. Her data are shown below.

Length (in.)	6.5	9	11.5	14.5	18	21	24	27	30	37.5
Number of Swings	22	20	17	16	14	13	13	12	11	10

a) Explain why a linear model is not appropriate for using the *Length* of a pendulum to predict the *Number of Swings* in 20 seconds.
b) Re-express the data to straighten the scatterplot.
c) Create an appropriate model.
d) Estimate the number of swings for a pendulum with a 4-inch string.
e) Estimate the number of swings for a pendulum with a 48-inch string.
f) How much confidence do you place in these predictions? Why?

21. Baseball salaries 2013 Ballplayers have been signing ever larger contracts. The highest salaries (in millions of dollars per season) for each year since 1874 is in the file on the CD. (Data in **Baseball salaries 2013**). Here are recent entries:

Player	Year	Salary (million $)
Nolan Ryan	1980	1.0
George Foster	1982	2.0
Kirby Puckett	1990	3.0
Jose Canseco	1990	4.7
Roger Clemens	1991	5.3
Ken Griffey, Jr.	1986	8.5
Albert Belle	1997	11.0
Pedro Martinez	1998	12.5
Mike Piazza	1999	12.5
Mo Vaughn	1999	13.3
Kevin Brown	1999	15.0
Carlos Delgado	2001	17.0
Alex Rodriguez	2001	22.0
Manny Ramirez	2004	22.5
Alex Rodriguez	2005	26.0
Derek Jeter	2006	20.6
Jason Giambi	2007	23.4
Alex Rodriguez	2009	33.0
Alex Rodriguez	2010	33.0
Alex Rodriguez	2011	32.0
Alex Rodriguez	2012	30.0
Alex Rodriguez	2013	29.0

a) Examine a scatterplot of the data. Does it look straight?
b) Part of the problem is that the salaries have not been adjusted for inflation. Plot the *Adjusted* (for inflation) *Salary* against *Year*. Is it straight now?
c) Find the best re-expression that you can to straighten out the scatterplot in b) and plot that against *Year*.

d) Fit the regression for the re-expression you found in c)
e) What is the R^2 of the regression in d)?
f) Plot the residuals of the model in d) against *Year*
g) Comment on the accuracy of the model you found given the residual plot in f).

22. **Planets, distances and years** At a meeting of the International Astronomical Union (IAU) in Prague in 2006, Pluto was determined not to be a planet, but rather the largest member of the Kuiper belt of icy objects. Let's examine some facts. Here is a table of the 9 sun-orbiting objects formerly known as planets:

Planet	Position Number	Distance from Sun (million miles)	Length of Year (Earth years)
Mercury	1	36	0.24
Venus	2	67	0.61
Earth	3	93	1.00
Mars	4	142	1.88
Jupiter	5	484	11.86
Saturn	6	887	29.46
Uranus	7	1784	84.07
Neptune	8	2796	164.82
Pluto	9	3707	247.68

a) Plot the *Length* of the year against the *Distance* from the sun. Describe the shape of your plot.
b) Re-express one or both variables to straighten the plot. Use the re-expressed data to create a model describing the length of a planet's year based on its distance from the sun.
c) Comment on how well your model fits the data.

23. **Planets, distances and order** Let's look again at the pattern in the locations of the planets in our solar system seen in the table in Exercise 22.
a) Re-express the distances to create a model for the *Distance* from the sun based on the planet's *Position*.
b) Based on this model, would you agree with the International Astronomical Union that Pluto is not a planet? Explain.

24. **Planets, and asteroids** The asteroid belt between Mars and Jupiter may be the remnants of a failed planet. If so, then Jupiter is really in position 6, Saturn is in 7, and so on. Repeat Exercise 23, using this revised method of numbering the positions. Which method seems to work better?

25. **Planets, and Eris** In July 2005, astronomers Mike Brown, Chad Trujillo, and David Rabinowitz announced the discovery of a sun-orbiting object, since named Eris,[2] that is 5% larger than Pluto. Eris orbits the sun once every 560 earth years at an average distance of about 6300 million miles from the sun. Based on its *Position*, how does Eris's *Distance* from the sun (re-expressed to logs) compare with the prediction made by your model of Exercise 23?

26. **Planets, models, and laws** The model you found in Exercise 22 is a relationship noted in the 17th century by Kepler as his Third Law of Planetary Motion. It was subsequently explained as a consequence of Newton's Law of Gravitation. The models for Exercises 23–25 relate to what is sometimes called the Titius-Bode "law," a pattern noticed in the 18th century but lacking any scientific explanation.

Compare how well the re-expressed data are described by their respective linear models. What aspect of the model of Exercise 22 suggests that we have found a physical law? In the future, we may learn enough about a planetary system around another star to tell whether the Titius-Bode pattern applies there. If you discovered that another planetary system followed the same pattern, how would it change your opinion about whether this is a real natural "law"? What would you think if the next system we find does not follow this pattern?

27. **Logs (not logarithms)** The value of a log is based on the number of board feet of lumber the log may contain. (A board foot is the equivalent of a piece of wood 1 inch thick, 12 inches wide, and 1 foot long. For example, a $2'' \times 4''$ piece that is 12 feet long contains 8 board feet.) To estimate the amount of lumber in a log, buyers measure the diameter inside the bark at the smaller end. Then they look in a table based on the Doyle Log Scale. The table below shows the estimates for logs 16 feet long.

Diameter of Log	8"	12"	16"	20"	24"	28"
Board Feet	16	64	144	256	400	576

a) What model does this scale use?
b) How much lumber would you estimate that a log 10 inches in diameter contains?
c) What does this model suggest about logs 36 inches in diameter?

28. **Weightlifting 2014** Listed below are the Olympic record men's weight-lifting performances as of 2014.

Weight Class	Record Holder	Country	Total Weight
56	Halil Mutlu	Turkey	305
62	Kim Un-Guk	North Korea	327
69	Galain Boevski	Bulgaria	357
77	Lu Xiaojun	China	379
85	Andrei Rybakou	Belarus	394
94	Ilya Ilyin	Kazakhstan	418
105	Andrei Aramnau	Belarus	436

---

[2]Eris is the Greek goddess of warfare and strife who caused a quarrel among the other goddesses that led to the Trojan war. In the astronomical world, Eris stirred up trouble when the question of its proper designation led to the raucous meeting of the IAU in Prague where IAU members voted to demote Pluto and Eris to dwarf-planet status www.gps.caltech.edu/~mbrown/planetlila/#paper.

a) Create a linear model for the *Weight Lifted* in each *Weight Class*.
b) Check the residuals plot. Is your linear model appropriate?
c) Create a better model by re-expression *Total Weight Lifted* and explain how you found it.
d) Explain why you think your new model is better.
e) The record holder of the Unlimited weight class weighs 152 kg. Predict how much he can lift with the models from a) and c).

29. Life expectancy history The data in the next column list the *Life Expectancy* for white males in the United States every decade during the past 110 years (1 = 1900 to 1910, 2 = 1911 to 1920, etc.). Create a model to predict future increases in life expectancy. (National Vital Statistics Report) Hint: Try "Plan B".

Decade	1	2	3	4	5	6	7	8	9	10	11	12
Life exp.	46.6	48.6	54.4	59.7	62.1	66.5	67.4	68.0	70.7	72.7	74.9	76.5

30. Lifting more weight In Exercise 28 you examined the record weight-lifting performances for the Olympics. You found a re-expression of *Total Weight Lifted*.
a) Find a model for Total Weight Lifted by re-expressing Weight Class instead of Total Weight Lifted.
b) Compare this model to the one you found in Exercise 24.
c) Predict the Total Weight Lifted by the 152 kg record holder in Exercise 24 e).
d) Which prediction do you think is better? Explain.
e) The record holder is Hossein Rezazadeh, who lifted 472 kg at the 2000 Sydney Olympics. Which model predicted it better?

31. Slower is cheaper? Researchers studying how a car's *Fuel Efficiency* varies with its *Speed* drove a compact car 200 miles at various speeds on a test track. Their data are shown in the table.

Speed (mph)	35	40	45	50	55	60	65	70	75
Fuel Eff. (mpg)	25.9	27.7	28.5	29.5	29.2	27.4	26.4	24.2	22.8

Create a linear model for this relationship and report any concerns you may have about the model.

32. Orange production The table below shows that as the number of oranges on a tree increases, the fruit tends to get smaller. Create a model for this relationship, and express any concerns you may have.

Number of Oranges/Tree	Average Weight/Fruit (lb)
50	0.60
100	0.58
150	0.56
200	0.55
250	0.53
300	0.52
350	0.50
400	0.49
450	0.48
500	0.46
600	0.44
700	0.42
800	0.40
900	0.38

33. Years to live, 2013 Insurance companies and other organizations use actuarial tables to estimate the remaining lifespans of their customers. On the next page are the estimated additional years of life for black males in the United States, according to a 2013 National Vital Statistics Report. (www.cdc.gov/nchs/deaths.htm)

Age	10	20	30	40	50	60	70	80	90	100
Years Left	63.2	53.6	44.5	35.5	26.8	19.3	13.1	7.9	4.4	2.5

a) Find a re-expression to create an appropriate model.
b) Predict the remaining lifespan of an 18-year-old black man.
c) Are you satisfied that your model has accounted for the relationship between *Years Left* and *Age*? Explain.

34. Tree growth A 1996 study examined the growth of grapefruit trees in Texas, determining the average trunk *Diameter* (in inches) for trees of varying *Ages*:

Age (yr)	2	4	6	8	10	12	14	16	18	20
Diameter (in.)	2.1	3.9	5.2	6.2	6.9	7.6	8.3	9.1	10.0	11.4

a) Fit a linear model to these data. What concerns do you have about the model?
b) If data had been given for individual trees instead of averages, would you expect the fit to be stronger, less strong, or about the same? Explain.

Just Checking ANSWERS

1. Counts are often best transformed by using the square root.
2. None. The relationship is already straight.
3. Even though, technically, the population values are counts, you should probably try a stronger transformation like log(population) because populations grow in proportion to their size.

Review of Part II

Exploring Relationships Between Variables

Quick Review

You have now survived your second major unit of Statistics. Here's a brief summary of the key concepts and skills:

- We treat data two ways: as categorical and as quantitative.
- To explore relationships in categorical data, review Chapter 2.
- To explore relationships in quantitative data:
  - Make a picture. Use a scatterplot. Put the explanatory variable on the *x*-axis and the response variable on the *y*-axis.
  - Describe the association between two quantitative variables in terms of direction, form, and strength.
  - The amount of scatter determines the strength of the association.
  - If, as one variable increases so does the other, the association is positive. If one increases as the other decreases, it's negative.
  - If the form of the association is linear, calculate a correlation to measure its strength numerically, and do a regression analysis to model it.
  - Correlations closer to −1 or +1 indicate stronger linear associations. Correlations near 0 indicate weak linear relationships, but other forms of association may still be present.
  - The line of best fit is also called the least squares regression line because it minimizes the sum of the squared residuals.
- The regression line predicts values of the response variable from values of the explanatory variable.
- A residual is the difference between the true value of the response variable and the value predicted by the regression model.
- The slope of the line is a rate of change, best described in "*y*-units" per "*x*-unit."
- R^2 gives the fraction of the variation in the response variable that is accounted for by the model.
- The standard deviation of the residuals measures the amount of scatter around the line.
- Outliers and influential points can distort any of our models.
- If you see a pattern (a curve) in the residuals plot, your chosen model is not appropriate; use a different model. You may, for example, straighten the relationship by re-expressing one of the variables.
- To straighten bent relationships, re-express the data using logarithms or a power (squares, square roots, reciprocals, etc.).
- Always remember that an association is not necessarily an indication that one of the variables causes the other.

Need more help with some of this? Try rereading some sections of Chapters 6 through 9. And see below for more opportunities to review these concepts and skills.

> "One must learn by doing the thing; though you think you know it, you have no certainty until you try."
>
> —Sophocles (495–406 B.C.E.)

Review Exercises

1. **College** Every year, *US News and World Report* publishes a special issue on many U.S. colleges and universities. The scatterplots below have *Student/Faculty Ratio* (number of students per faculty member) for the colleges and universities on the *y*-axes plotted against 4 other variables. The correct correlations for these scatterplots appear in this list. Match them.

   −0.98   −0.71   −0.51   0.09   0.23   0.69

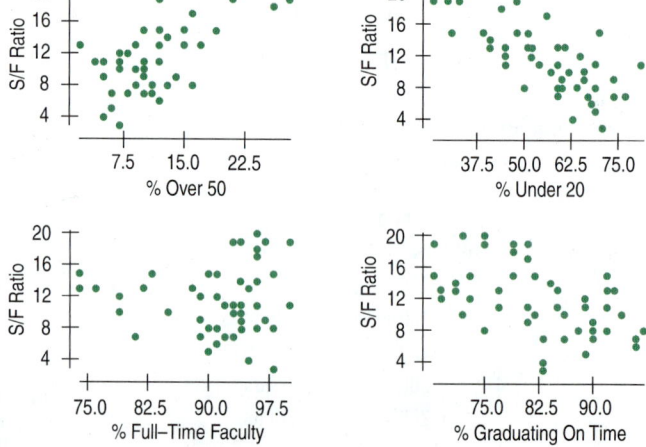

2. **Togetherness** Are good grades in high school associated with family togetherness? A random sample of 142 high school students was asked how many meals per week their families ate together. Their responses produced a mean of 3.78 meals per week, with a standard deviation of 2.2. Researchers then matched these responses against the students' grade point averages (GPAs). The scatterplot appeared to be reasonably linear, so they created a line of regression. No apparent pattern emerged in the residuals plot. The equation of the line was $\widehat{GPA} = 2.73 + 0.11\, Meals$.

a) Interpret the y-intercept in this context.
b) Interpret the slope in this context.
c) What was the mean GPA for these students?
d) If a student in this study had a negative residual, what did that mean?
e) Upon hearing of this study, a counselor recommended that parents who want to improve the grades their children get should get the family to eat together more often. Do you agree with this interpretation? Explain.

3. **Vineyards, more information** Here are the scatterplot and regression analysis for *Case Prices* of 36 wines from vineyards in the Finger Lakes region of New York State and the *Ages* of the vineyards. (Data in **Vineyards full**):

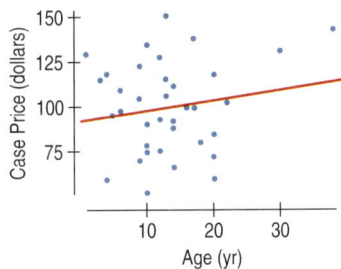

Dependent variable is Case Price
R-squared = 2.7%

Variable	Coefficient
Constant	92.7650
Age	0.567284

a) Does it appear that vineyards in business longer get higher prices for their wines? Explain.
b) What does this analysis tell us about vineyards in the rest of the world?
c) Write the regression equation.
d) Explain why that equation is essentially useless.

4. **Vineyards again** Instead of *Age*, perhaps the *Size* of the vineyard (in acres) is associated with the price of the wines. Look at the scatterplot (Data in **Vineyards full**):

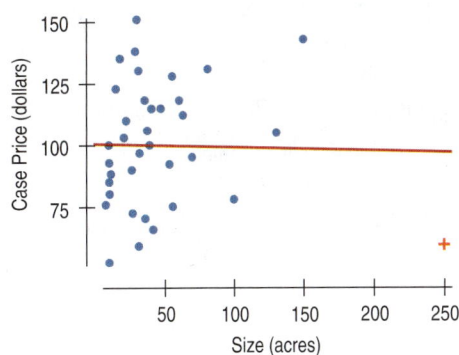

a) Do you see any evidence of an association?
b) What concern do you have about this scatterplot?
c) If the red "+" data point is removed, would the correlation become stronger or weaker? Explain.
d) If the red "+" data point is removed, would the slope of the line increase or decrease? Explain.

5. **More twins 2014?** In January 2012, the *New York Times* published a story called *Twin Births in the U.S., Like Never Before*, in which they reported a 76 percent increase in the rate of twin births from 1980 to 2009. Here are the number of twins born each year (per 1000 live births). (www.cdc.gov/nchs/births.htm)

Year	Twin Birth (per 1000 live births)	Year	Twin Birth (per 1000 live births)
1980	18.92	1997	26.83
1981	19.21	1998	28.08
1982	19.46	1999	28.87
1983	19.86	2000	29.30
1984	19.88	2001	30.12
1985	20.50	2002	31.11
1986	21.30	2003	31.46
1987	21.36	2004	32.15
1988	21.80	2005	32.17
1989	22.41	2006	32.11
1990	22.46	2007	32.19
1991	23.05	2008	32.62
1992	23.35	2009	33.22
1993	23.88	2010	33.15
1994	24.39	2011	33.20
1995	24.86	2012	33.15
1996	25.84		

a) Using the data only up to 2009 (as in the *NYTimes* article), find the equation of the regression line for predicting the number of twin births by *Years since 1980*.
b) Explain in this context what the slope means.
c) Predict the number of twin births in the United States for the year 2014. Comment on your faith in that prediction.
d) Comment on the residuals plot and the appropriateness of the model. Look at the past three years' rates.

6. Dow Jones 2014 The Dow Jones stock index measures the performance of the stocks of America's largest companies (finance.yahoo.com). A regression of the Dow prices on years 1972–2014 looks like this:

Dependent variable is Dow Index
R-squared = 88.3%   s = 1604

Variable	Coefficient
Intercept	−2485.66
Year since 1970	354.24

a) What is the correlation between *Dow Index* and *Year*?
b) Write the regression equation.
c) Explain in this context what the equation says.
d) Here's a scatterplot of the residuals. Comment on the appropriateness of the model.

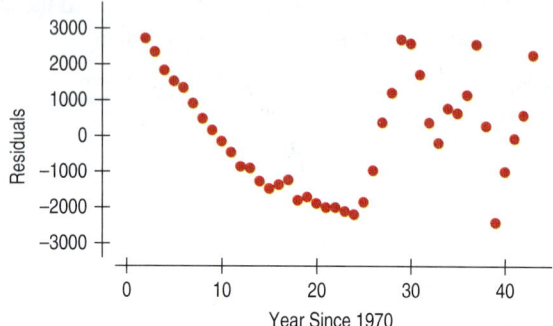

7. Acid rain Biologists studying the effects of acid rain on wildlife collected data from 163 streams in the Adirondack Mountains. They recorded the *pH* (acidity) of the water and the *BCI*, a measure of biological diversity, and they calculated $R^2 = 27\%$. Here's a scatterplot of *BCI* against *pH*:

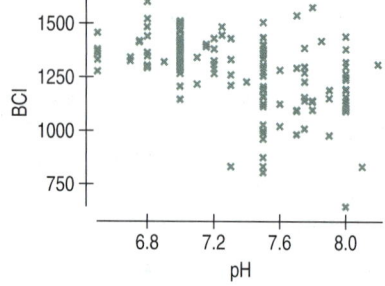

a) What is the correlation between *pH* and *BCI*?
b) Describe the association between these two variables.
c) If a stream has average *pH*, what would you predict about the *BCI*?
d) In a stream where the *pH* is 3 standard deviations above average, what would you predict about the *BCI*?

8. Manatees 2013 Marine biologists warn that the growing number of powerboats registered in Florida threatens the existence of manatees. The data below come from the Florida Fish and Wildlife Conservation Commission (myfwc.com/research/manatee/) and the National Marine Manufacturers Association (www.nmma.org/).

Year	Manatees Killed	Powerboat Registrations (in 1000s)
1982	13	447
1983	21	460
1984	24	481
1985	16	498
1986	24	513
1987	20	512
1988	15	527
1989	34	559
1990	33	585
1992	33	614
1993	39	646
1994	43	675
1995	50	711
1996	47	719
1997	53	716
1998	38	716
1999	35	716
2000	49	735
2001	81	860
2002	95	923
2003	73	940
2004	69	946
2005	79	974
2006	92	988
2007	73	992
2008	90	932
2009	97	949
2010	83	914
2011	88	890
2012	81	870
2013	72	865

a) In this context, which is the explanatory variable?
b) Make a scatterplot of these data and describe the association you see.
c) Find the correlation between *Boat Registrations* and *Manatee Deaths*.
d) Interpret the value of R^2.
e) Does your analysis prove that powerboats are killing manatees?

9. A manatee model 2013 Continue your analysis of the manatee situation from the previous exercise.
a) Create a linear model of the association between *Manatee Deaths* and *Powerboat Registrations*.
b) Interpret the slope of your model.
c) Interpret the *y*-intercept of your model.
d) How accurately did your model predict the high number of manatee deaths in 2009?
e) Which is better for the manatees, positive residuals or negative residuals? Explain.
f) What does your model suggest about the future for the manatee?

10. **Grades** A Statistics instructor created a linear regression equation to predict students' final exam scores from their midterm exam scores. The regression equation was $\widehat{Fin} = 10 + 0.9\,Mid$.
   a) If Susan scored a 70 on the midterm, what did the instructor predict for her score on the final?
   b) Susan got an 80 on the final. How big is her residual?
   c) If the standard deviation of the final was 12 points and the standard deviation of the midterm was 10 points, what is the correlation between the two tests?
   d) How many points would someone need to score on the midterm to have a predicted final score of 100?
   e) Suppose someone scored 100 on the final. Explain why you can't estimate this student's midterm score from the information given.
   f) One of the students in the class scored 100 on the midterm but got overconfident, slacked off, and scored only 15 on the final exam. What is the residual for this student?
   g) No other student in the class "achieved" such a dramatic turnaround. If the instructor decides not to include this student's scores when constructing a new regression model, will the R^2 value of the regression increase, decrease, or remain the same? Explain.
   h) Will the slope of the new line increase or decrease?

11. **Traffic** Highway planners investigated the relationship between traffic *Density* (number of automobiles per mile) and the average *Speed* of the traffic on a moderately large city thoroughfare. The data were collected at the same location at 10 different times over a span of 3 months. They found a mean traffic *Density* of 68.6 cars per mile (cpm) with standard deviation of 27.07 cpm. Overall, the cars' average *Speed* was 26.38 mph, with standard deviation of 9.68 mph. These researchers found the regression line for these data to be $\widehat{Speed} = 50.55 - 0.352\,Density$.
   a) What is the value of the correlation coefficient between *Speed* and *Density*?
   b) What percent of the variation in average *Speed* is explained by traffic *Density*?
   c) Predict the average *Speed* of traffic on the thoroughfare when the traffic *Density* is 50 cpm.
   d) What is the value of the residual for a traffic *Density* of 56 cpm with an observed *Speed* of 32.5 mph?
   e) The data set initially included the point *Density* = 125 cpm, *Speed* = 55 mph. This point was considered an outlier and was not included in the analysis. Will the slope increase, decrease, or remain the same if we redo the analysis and include this point?
   f) Will the correlation become stronger, weaker, or remain the same if we redo the analysis and include this point $(125, 55)$?
   g) A European member of the research team measured the *Speed* of the cars in kilometers per hour (1 km $\approx$ 0.62 miles) and the traffic *Density* in cars per kilometer. Find the value of his calculated correlation between speed and density.

12. **Cramming** One Thursday, researchers gave students enrolled in a section of basic Spanish a set of 50 new vocabulary words to memorize. On Friday, the students took a vocabulary test. When they returned to class the following Monday, they were retested—without advance warning. Here are the test scores for the 25 students.

Fri.	Mon.	Fri.	Mon.	Fri.	Mon.
42	36	48	37	39	41
44	44	43	41	46	32
45	46	45	32	37	36
48	38	47	44	40	31
44	40	50	47	41	32
43	38	34	34	48	39
41	37	38	31	37	31
35	31	43	40	36	41
43	32				

   a) What is the correlation between *Friday* and *Monday* scores?
   b) What does a scatterplot show about the association between the scores?
   c) What does it mean for a student to have a positive residual?
   d) What would you predict about a student whose *Friday* score was one standard deviation below average?
   e) Write the equation of the regression line.
   f) Predict the *Monday* score of a student who earned a 40 on Friday.

13. **Cars, correlations** What factor most explains differences in *Fuel Efficiency* among cars? Below is a correlation matrix exploring that relationship for the car's *Weight*, *Horsepower*, engine *size* (*Displacement*), and number of *Cylinders*. (Data in **Cars**)

	MPG	Weight	Horse-power	Displacement	Cylinders
MPG	1.000				
Weight	−0.903	1.000			
Horsepower	−0.871	0.917	1.000		
Displacement	−0.786	0.951	0.872	1.000	
Cylinders	−0.806	0.917	0.864	0.940	1.000

   a) Which factor seems most strongly associated with *Fuel Efficiency*?
   b) What does the negative correlation indicate?
   c) Explain the meaning of R^2 for that relationship.

14. **Cars, revisited** Look again at the correlation table for cars in the previous exercise.
   a) Which two variables in the table exhibit the strongest association?
   b) Is that strong association necessarily cause-and-effect? Offer at least two explanations why that association might be so strong.
   c) Engine displacements for U.S.-made cars are often measured in cubic inches. For many foreign cars, the

units are either cubic centimeters or liters. How would changing from cubic inches to liters affect the calculated correlations involving *Displacement*?
d) What would you predict about the *Fuel Efficiency* of a car whose engine *Displacement* is one standard deviation above the mean?

15. **Cars, one more time!** Can we predict the *Horsepower* of the engine that manufacturers will put in a car by knowing the *Weight* of the car? Here are the regression analysis and residuals plot:

Dependent variable is Horsepower
R-squared = 84.1%

Variable	Coefficient
Intercept	3.49834
Weight	34.3144

a) Write the equation of the regression line.
b) Do you think the car's *Weight* is measured in pounds or thousands of pounds? Explain.
c) Do you think this linear model is appropriate? Explain.
d) The highest point in the residuals plot, representing a residual of 22.5 horsepower, is for a Chevy weighing 2595 pounds. How much horsepower does this car have?

16. **Colorblind** Although some women are colorblind, this condition is found primarily in men. Why is it wrong to say there's a strong correlation between *Sex* and *Colorblindness*?

17. **Old Faithful again** There is evidence that eruptions of Old Faithful can best be predicted by knowing the duration of the previous eruption.
a) Describe what you see in the scatterplot of *Intervals* between eruptions vs. *Duration* of the previous eruption.

b) Write the equation of the line of best fit. Here's the regression analysis:

Dependent variable is Interval
R-squared = 77.0%   s = 6.16 min

Variable	Coefficient
Intercept	33.9668
Duration	10.3582

c) Carefully explain what the slope of the line means in this context.
d) How accurate do you expect predictions based on this model to be? Cite statistical evidence.
e) If you just witnessed an eruption that lasted 4 minutes, how long do you predict you'll have to wait to see the next eruption?
f) So you waited, and the next eruption came in 79 minutes. Use this as an example to define a residual.

18. **Crocodile lengths** The ranges inhabited by the Indian gharial crocodile and the Australian saltwater crocodile overlap in Bangladesh. Suppose a very large crocodile skeleton is found there, and we wish to determine the species of the animal. Wildlife scientists have measured the lengths of the heads and the complete bodies of several crocs (in centimeters) of each species, creating the regression analyses below:

Indian Crocodile
Dependent variable is IBody
R-squared = 97.2%

Variable	Coefficient
Intercept	−69.3693
IHead	7.40004

Australian Crocodile
Dependent variable is ABody
R-squared = 98.1%

Variable	Coefficient
Intercept	−21.3429
AHead	7.82761

a) Do the associations between the sizes of the heads and bodies of the two species appear to be strong? Explain.
b) In what ways are the two relationships similar? Explain.
c) What is different about the two models? What does that mean?
d) The crocodile skeleton found had a head length of 62 cm and a body length of 380 cm. Which species do you think it was? Explain why.

19. **How old is that tree?** One can determine how old a tree is by counting its rings, but that requires either cutting the tree down or extracting a sample from the tree's core. Can we estimate the tree's age simply from its diameter? A forester measured 27 trees of the same species that had been cut down, and counted the rings to determine the ages of the trees.

Diameter (in.)	Age (yr)	Diameter (in.)	Age (yr)
1.8	4	10.3	23
1.8	5	14.3	25
2.2	8	13.2	28
4.4	8	9.9	29
6.6	8	13.2	30
4.4	10	15.4	30
7.7	10	17.6	33
10.8	12	14.3	34
7.7	13	15.4	35
5.5	14	11.0	38
9.9	16	15.4	38
10.1	18	16.5	40
12.1	20	16.5	42
12.8	22		

a) Find the correlation between *Diameter* and *Age*. Does this suggest that a linear model may be appropriate? Explain.
b) Create a scatterplot and describe the association.
c) Create the linear model.
d) Check the residuals. Explain why a linear model is probably not appropriate.
e) If you used this model, would it generally overestimate or underestimate the ages of very large trees? Explain.

20. Improving trees In the last exercise, you saw that the linear model had some deficiencies. Let's create a better model.
  a) Perhaps the cross-sectional area of a tree would be a better predictor of its age. Since area is measured in square units, try re-expressing the data by squaring the diameters. Does the scatterplot look better?
  b) Create a model that predicts *Age* from the square of the *Diameter*.
  c) Check the residuals plot for this new model. Is this model more appropriate? Why?
  d) Estimate the age of a tree 18 inches in diameter.

21. Big screen An electronics website collects data on the size of new HD flat panel televisions (measuring the diagonal of the screen in inches) to predict the cost (in hundreds of dollars). Which of these is most likely to be the slope of the regression line: 0.03, 0.3, 3, 30? Explain.

22. Smoking and pregnancy 2011 The Child Trends Data Bank monitors issues related to children. The table shows a 50-state average of the percent of expectant mothers who smoked cigarettes during their pregnancies.

Year	% Smoking While Pregnant	Year	% Smoking While Pregnant
1990	19.2	2001	13.8
1991	18.7	2002	13.3
1992	17.9	2003	12.7
1993	16.8	2004	10.9
1994	16.0	2005	10.1
1995	15.4	2006	10.0
1996	15.3	2007	10.4
1997	14.9	2008	9.7
1998	14.8	2009	9.3
1999	14.1	2010	9.2
2000	14.0	2011	9.0

a) Create a scatterplot and describe the trend you see.
b) Find the correlation.
c) How is the value of the correlation affected by the fact that the data are averages rather than percentages for each of the 50 states?
d) Write a linear model and interpret the slope in context.

23. No smoking? The downward trend in smoking you saw in the last exercise is good news for the health of babies, but will it ever stop?
  a) Explain why you can't use the linear model you created in Exercise 22 to see when smoking during pregnancy will cease altogether.
  b) Create a model that could estimate the year in which the level of smoking would be 0%.
  c) Comment on the reliability of such a prediction.

24. Tips It's commonly believed that people use tips to reward good service. A researcher for the hospitality industry examined tips and ratings of service quality from 2645 dining parties at 21 different restaurants. The correlation between ratings of service and tip percentages was 0.11. (Source: M. Lynn and M. McCall, "Gratitude and Gratuity." *Journal of Socio-Economics* 29: 203–214)
  a) Describe the relationship between *Quality of Service* and *Tip Size*.
  b) Find and interpret the value of R^2 in this context.

25. U.S. cities Data from 50 large U.S. cities show the mean *January Temperature* and the *Latitude*. Describe what you see in the scatterplot.

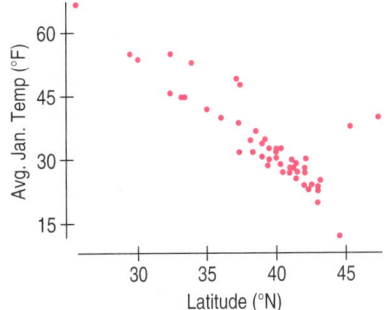

26. Correlations The study of U.S. cities in Exercise 25 found the mean *January Temperature* (degrees Fahrenheit), *Altitude* (feet above sea level), and *Latitude* (degrees north of the equator) for 55 cities. Here's the correlation matrix:

	Jan. Temp	Latitude	Altitude
Jan. Temp	1.000		
Latitude	−0.848	1.000	
Altitude	−0.369	0.184	1.000

a) Which seems to be more useful in predicting *January Temperature: Altitude* or *Latitude*? Explain.
b) If the *Temperature* were measured in degrees Celsius, what would be the correlation between Temperature and Latitude?
c) If the *Temperature* were measured in degrees Celsius and the *Altitude* in meters, what would be the correlation? Explain.
d) What would you predict about the January *Temperature* in a city whose *Altitude* is two standard deviations higher than the average *Altitude*?

27. Winter in the city Summary statistics for the data relating the latitude and average January temperature for 55 large U.S. cities are given below.

Variable	Mean	StdDev
Latitude	39.02	5.42
JanTemp	26.44	13.49

Correlation = −0.848

a) What percent of the variation in January *Temperature* can be explained by variation in *Latitude*?
b) What is indicated by the fact that the correlation is negative?
c) Write the equation of the line of regression for predicting average January *Temperature* from *Latitude*.
d) Explain what the slope of the line means.
e) Do you think the *y*-intercept is meaningful? Explain.
f) The latitude of Denver is 40°N. Predict the mean January temperature there.
g) What does it mean if the residual for a city is positive?

28. Depression The September 1998 issue of the *American Psychologist* published an article by Kraut et al. that reported on an experiment examining "the social and psychological impact of the Internet on 169 people in 73 households during their first 1 to 2 years online." In the experiment, 73 households were offered free Internet access for 1 or 2 years in return for allowing their time and activity online to be tracked. The members of the households who participated in the study were also given a battery of tests at the beginning and again at the end of the study. The conclusion of the study made news headlines: Those who spent more time online tended to be more depressed at the end of the experiment. Although the paper reports a more complex model, the basic result can be summarized in the following regression of *Depression* (at the end of the study, in "depression scale units") vs. *Internet Use* (in mean hours per week):

Dependent variable is Depression
R-squared = 4.6%   s = 0.4563

Variable	Coefficient
Intercept	0.5655
Internet use	0.0199

The news reports about this study clearly concluded that using the Internet causes depression. Discuss whether such a conclusion can be drawn from this regression. If so, discuss the supporting evidence. If not, say why not.

29. Jumps 2012 How are Olympic performances in various events related? The plot shows winning long-jump and high-jump distances, in meters, for the Summer Olympics from 1912 through 2012.

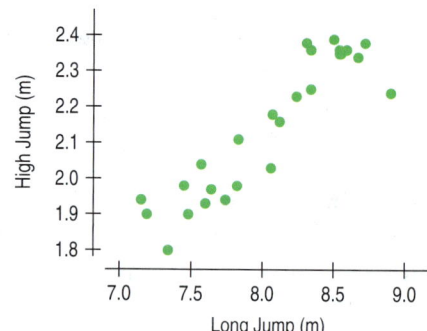

a) Describe the association.
b) Do long-jump performances somehow influence the high-jumpers? How do you account for the relationship you see?
c) The correlation for the given scatterplot is 0.912. If we converted the jump lengths to centimeters by multiplying by 100, would that make the actual correlation higher or lower?
d) What would you predict about the long jump in a year when the high-jumper jumped one standard deviation better than the average high jump?

30. Modeling jumps 2012 Here are the summary statistics for the Olympic long jumps and high jumps displayed in the previous exercise.

Event	Mean	StdDev
High Jump	2.148077	0.1938663
Long Jump	8.050000	0.5135913

Correlation = 0.91246

a) Write the equation of the line of regression for estimating *High Jump* from *Long Jump*.
b) Interpret the slope of the line.
c) In a year when the long jump is 8.9 m, what high jump would you predict?

d) Why can't you use this line to estimate the long jump for a year when you know the high jump was 2.25 m?
e) Write the equation of the line you need to make that prediction.

31. French Consider the association between a student's score on a French vocabulary test and the weight of the student. What direction and strength of correlation would you expect in each of the following situations? Explain.
a) The students are all in third grade.
b) The students are in third through twelfth grades in the same school district.
c) The students are in tenth grade in France.
d) The students are in third through twelfth grades in France.

32. Twins Twins are often born at less than 9 months gestation. The graph from the *Journal of the American Medical Association (JAMA)* shows the rate of preterm twin births in the United States over the past 20 years. In this study, *JAMA* categorized mothers by the level of prenatal medical care they received: inadequate, adequate, or intensive.
a) Describe the overall trend in preterm twin births.
b) Describe any differences you see in this trend, depending on the level of prenatal medical care the mother received.
c) Should expectant mothers be advised to cut back on the level of medical care they seek in the hope of avoiding preterm births? Explain.

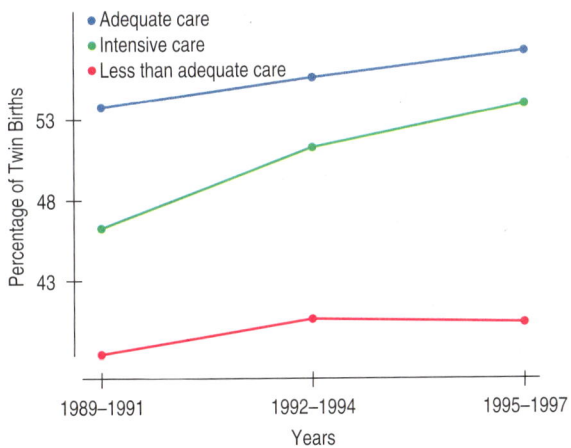

Preterm Birth Rate per 100 live twin births among U.S. twins by intensive, adequate, and less than adequate prenatal care utilization, 1981–1997. (Source: *JAMA* 284[2000]: 335–341)

33. Lunchtime Does how long toddlers sit at the lunch table help predict how much they eat? The table and graph show the number of minutes the kids stayed at the table and the number of calories they consumed. Create and interpret a model for these data.

Calories	Time	Calories	Time
472	21.4	450	42.4
498	30.8	410	43.1
465	37.7	504	29.2
456	33.5	437	31.3
423	32.8	489	28.6
437	39.5	436	32.9
508	22.8	480	30.6
431	34.1	439	35.1
479	33.9	444	33.0
454	43.8	408	43.7

34. Gasoline Since clean-air regulations have dictated the use of unleaded gasoline, the supply of leaded gas in New York state has diminished. The following table was given on the August 2001 New York State Math B exam, a statewide achievement test for high school students.

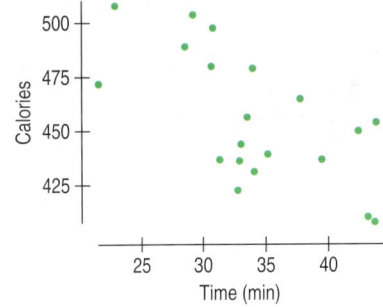

Year	1984	1988	1992	1996	2000
Gallons (1000's)	150	124	104	76	50

a) Create a linear model and predict the number of gallons that will be available in 2015. Comment.
b) The exam then asked students to estimate the year when leaded gasoline will first become unavailable, expecting them to use the model from part a to answer the question. Explain why that method is incorrect.
c) Create a model that *would* be appropriate for that task, and make the estimate.
d) The "wrong" answer from the other model is fairly accurate in this case. *Why?*

35. Tobacco and alcohol Are people who use tobacco products more likely to consume alcohol? Here are data on household spending (in pounds) taken by the British government on 11 regions in Great Britain. Do tobacco and alcohol spending appear to be related? What questions do you have about these data? What conclusions can you draw?

Region	Alcohol	Tobacco
North	6.47	4.03
Yorkshire	6.13	3.76
Northeast	6.19	3.77
East Midlands	4.89	3.34
West Midlands	5.63	3.47
East Anglia	4.52	2.92
Southeast	5.89	3.20
Southwest	4.79	2.71
Wales	5.27	3.53
Scotland	6.08	4.51
Northern Ireland	4.02	4.56

36. Football weights The Sears Cup was established in 1993 to honor institutions that maintain a broad-based athletic program, achieving success in many sports, both men's and women's. Since its Division III inception in 1995, the cup has been won by Williams College 15 of 17 years. Their football team has an 85.3% winning record under their current coach. Why does the football team win so much? Is it because they're heavier than their opponents? The table shows the average team weights for selected years from 1973 to 1993.

Year	Weight (lb)	Year	Weight (lb)
1973	185.5	1983	192.0
1975	182.4	1987	196.9
1977	182.1	1989	202.9
1979	191.1	1991	206.0
1981	189.4	1993	198.7

a) Fit a straight line to the relationship between *Weight* and *Year*.
b) Does a straight line seem reasonable?
c) Predict the average weight of the team for the year 2015. Does this seem reasonable?
d) What about the prediction for the year 2103? Explain.
e) What about the prediction for the year 3003? Explain.

37. Models Find the predicted value of y, using each model for $x = 10$.
a) $\hat{y} = 2 + 0.8 \ln x$
b) $\log \hat{y} = 5 - 0.23x$
c) $\dfrac{1}{\sqrt{\hat{y}}} = 17.1 - 1.66x$

38. Williams vs. Texas Here are the average weights of the football team for the University of Texas for various years in the 20th century.

Year	1905	1919	1932	1945	1955	1965
Weight (lb)	164	163	181	192	195	199

a) Fit a straight line to the relationship of *Weight* by *Year* for Texas football players.
b) According to these models, in what year will the predicted weight of the Williams College team from Exercise 36 first be more than the weight of the University of Texas team?
c) Do you believe this? Explain.

39. Vehicle weights The Minnesota Department of Transportation hoped that they could measure the weights of big trucks without actually stopping the vehicles by using a newly developed "weigh-in-motion" scale. After installation of the scale, a study was conducted to find out whether the scale's readings correspond to the true weights of the trucks being monitored. In Chapter 6, Exercise 50, you examined the scatterplot for the data they collected, finding the association to be approximately linear with $R^2 = 93\%$. Their regression equation is $\widehat{Wt} = 10.85 + 0.64\, Scale$, where both the scale reading and the predicted weight of the truck are measured in thousands of pounds.

a) Estimate the weight of a truck if this scale read 31,200 pounds.
b) If that truck actually weighed 32,120 pounds, what was the residual?
c) If the scale reads 35,590 pounds, and the truck has a residual of -2440 pounds, how much does it actually weigh?
d) In general, do you expect estimates made using this equation to be reasonably accurate? Explain.
e) If the police plan to use this scale to issue tickets to trucks that appear to be overloaded, will negative or positive residuals be a greater problem? Explain.

40. Profit How are a company's profits related to its sales? Let's examine data from 71 large U.S. corporations. All amounts are in millions of dollars.

a) Histograms of *Profits* and *Sales* and histograms of the logarithms of *Profits* and *Sales* are seen below. Why are the re-expressed data better for regression?

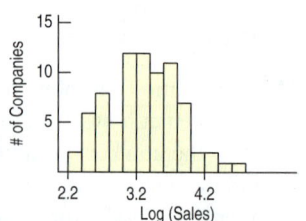

b) Here are the scatterplot and residuals plot for the regression of logarithm of *Profits* vs. log of *Sales*. Do you think this model is appropriate? Explain.

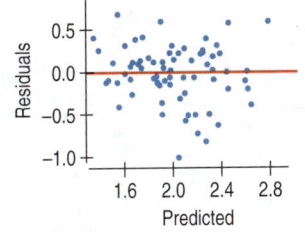

c) Here's the regression analysis. Write the equation.

Dependent variable is Log Profit
R-squared = 48.1%

Variable	Coefficient
Intercept	−0.106259
LogSales	0.647798

d) Use your equation to estimate profits earned by a company with sales of 2.5 billion dollars. (That's 2500 million.)

T 41. Down the drain Most water tanks have a drain plug so that the tank may be emptied when it's to be moved or repaired. How long it takes a certain size of tank to drain depends on the size of the plug, as shown in the table. Create a model.

Plug Dia (in.)	$\frac{3}{8}$	$\frac{1}{2}$	$\frac{3}{4}$	1	$1\frac{1}{4}$	$1\frac{1}{2}$	2
Drain Time (min.)	140	80	35	20	13	10	5

T 42. Chips A start-up company has developed an improved electronic chip for use in laboratory equipment. The company needs to project the manufacturing cost, so it develops a spreadsheet model that takes into account the purchase of production equipment, overhead, raw materials, depreciation, maintenance, and other business costs. The spreadsheet estimates the cost of producing 10,000 to 200,000 chips per year, as seen in the table. Develop a regression model to predict *Costs* based on the *Level* of production.

Chips Produced (1000s)	Cost per Chip ($)	Chips Produced (1000s)	Cost per Chip ($)
10	146.10	90	47.22
20	105.80	100	44.31
30	85.75	120	42.88
40	77.02	140	39.05
50	66.10	160	37.47
60	63.92	180	35.09
70	58.80	200	34.04
80	50.91		

chapter 10 Understanding Randomness

10.1 What Is Randomness?
10.2 Simulating by Hand

> *Where are we going?*
> Few things in life are certain. But that doesn't mean that we all understand randomness. We'll be formal about it in a few chapters, but for now, we'll simulate what we want to understand to see the patterns that emerge.

"The most decisive conceptual event of twentieth century physics has been the discovery that the world is not deterministic. . . . A space was cleared for chance."

— Ian Hocking,
The Taming of Chance

We all know what it means for something to be random. Or do we? Many children's games rely on chance outcomes. Rolling dice, spinning spinners, and shuffling cards all select at random. Adult games use randomness as well, from card games to lotteries to bingo. What's the most important aspect of the randomness in these games? It must be fair.

What is it about **random** selection that makes it seem fair? It's really two things. First, nobody can guess the outcome before it happens. Second, when we want things to be fair, usually some underlying set of outcomes will be equally likely (although in many games, some combinations of outcomes are more likely than others).

Randomness is not always what we might think of as "at random." Random outcomes have a lot of structure, especially when viewed in the long run. You can't predict how a fair coin will land on any single toss, but you're pretty confident that if you flipped it thousands of times you'd see about 50% heads. As we will see, randomness is an essential tool of Statistics. Statisticians don't think of randomness as the annoying tendency of things to be unpredictable or haphazard. Statisticians use randomness as a tool. In fact, without deliberately applying randomness, we couldn't do most of Statistics, and this book would stop right about here.[1]

But truly random values are surprisingly hard to get. Just to see how fair humans are at selecting, pick a number at random from the top of the next page. Go ahead. Turn the page, look at the numbers quickly, and pick a number at random.

Ready?
Go.

---
[1] Don't get your hopes up.

1 2 3 4

10.1 What Is Randomness?

> "The generation of random numbers is too important to be left to chance."
>
> —Robert R. Coveyou, Oak Ridge National Laboratory

> "Anyone who considers arithmetical methods of producing random digits is, of course, in a state of sin."
>
> —John Von Neumann (famous mathematician and developer of the digital computer)

Activity: Random Behavior. ActivStats' Random Experiment Tool lets you experiment with truly random outcomes. We'll use it a lot in the coming chapters.

Did you pick 3? If so, you've got company. Almost 75% of all people pick the number 3. About 20% pick either 2 or 4. If you picked 1, well, consider yourself a little different. Only about 5% choose 1. Psychologists have proposed reasons for this phenomenon, but for us, it simply serves as a lesson that we've got to find a better way to choose things at random.

So how should we **generate random numbers**? As we'll see, there are many different types of randomness, but they all share the property that they are unpredictable. It's surprisingly difficult to get random values even when they're equally likely. Computers, calculators, and even smartphones have become popular ways to generate random numbers, but they can't generate truly random numbers either. Start a program from the same place, and it will always follow exactly the same path. So numbers generated by a program are not truly random. Technically, "random" numbers generated this way are *pseudorandom* numbers. Fortunately, pseudorandom values are virtually indistinguishable from truly random numbers even though they aren't really random. For most purposes, that's good enough.

DILBERT © 2001 Scott Adams. Distributed by Universal Uclick. Reprinted with permission. All rights reserved.

Activity: Truly Random Values on the Internet. This activity will take you to an Internet site (www.random.org) that generates all the truly random numbers you could want.

There *are* ways to generate random numbers so that they are both equally likely and truly random. And, along with almost anything else, you can find truly random numbers on the Internet. Several sites are available that use methods like timing the decay of a radioactive element to generate truly random digits.[2]

A string of random digits might look like this:

```
2217726304387410092537086270581997622725849795907032825001108963
3217535822643800292254644943760642389043766557204107354186024508
8906427308645681412198226653885873285801699027843110380420067664
8740522639824530519902027044464984322000946238678577902639002954
8887003319933147508331265192321413908608674496383528968974910533
6944182713168919406022181281304751019321546303870481407676636740
6070204916508913632855351361361043794293428486909462881431793360
7706356513310563210508993624272872250535395513645991015328128202
```

---
[2] For example, www.random.org or www.randomnumbers.info.

The best ways we know to generate data that give a fair and accurate picture of the world rely on randomness, and the ways in which we draw conclusions from those data depend on the randomness, too.

Playing with a full deck? An ordinary deck of playing cards, like the ones used in bridge and many other card games, consists of 52 cards. There are numbered cards (2 through 10), and face cards (Jack, Queen, King, Ace) whose value depends on the game you are playing. Each card is also marked by one of four suits (clubs, diamonds, hearts, or spades) whose significance is also game-specific.

How much do you have to shuffle? Even something as common as card shuffling may not be as random as you might think. If you shuffle cards by the usual method in which you split the deck in half and try to let cards fall roughly alternately from each half, you're doing a "riffle shuffle."

How many times should you shuffle cards in this way to make the deck random? A surprising fact was discovered by statisticians Persi Diaconis, Ronald Graham, and W. M. Kantor. It takes seven riffle shuffles. Fewer than seven leaves order in the deck, but after that, more shuffling does little good. Most people, though, don't shuffle that many times.

When computers were first used to generate hands in bridge tournaments, some professional bridge players complained that the computer was making too many "weird" hands—hands with 10 cards of one suit, for example. Suddenly these hands were appearing more often than players were used to when cards were shuffled by hand. The players assumed that the computer was doing something wrong. But it turns out that it's humans who hadn't been shuffling enough to make the decks really random and have those "weird" hands appear as often as they should.

10.2 Simulating by Hand

Suppose a cereal manufacturer puts pictures of famous athletes on cards in boxes of cereal in the hope of boosting sales. The manufacturer announces that 20% of the boxes contain a picture of Hope Solo, 30% a picture of Danica Patrick, and the rest a picture of Blake Griffin. You want all three pictures. How many boxes of cereal do you expect to have to buy in order to get the complete set?

How can we answer questions like this? Well, one way is to buy hundreds of boxes of cereal to see what might happen. But let's not. Instead, we'll consider using a random model. Why random? When we pick a box of cereal off the shelf, we don't know what picture is inside. We'll assume that the pictures are randomly placed in the boxes and that the boxes are distributed randomly to stores around the country. Why a model? Because we won't actually buy the cereal boxes. We can't afford all those boxes and we don't want to waste food. So we need an imitation of the real process that we can manipulate and control. In short, we're going to simulate reality. A **simulation** mimics reality by using random numbers to represent the outcomes of real events. Just as pilots use flight simulators to learn about and practice real situations, we can learn a great deal about the real events by carefully modeling the randomness and analyzing the simulation results.

The question we've asked is how many boxes do you expect to buy to get a complete card collection. But we can't answer our question by completing a card collection just once. We want to understand the *typical* number of boxes to open, how that number

> **Randomness is fundamental** Modern physics has shown that randomness is not just a mathematical game; it is fundamentally the way the universe works.
>
> *Regardless of improvements in data collection or in computer power, the best we can ever do, according to quantum mechanics . . . is predict the probability that an electron, or a proton, or a neutron, or any other of nature's constituents, will be found here or there. Probability reigns supreme in the microcosmos.*
>
> Excerpt from "*The Fabric of the Cosmos: Space, Time, and the Texture of Reality*" by Brian Greene. Published by Knopf Doubleday Publishing Group, © 2004.

varies, and, often, the shape of the distribution. So we'll have to do this over and over. We call each time we obtain a simulated answer to our question a **trial**.

For the sports cards, a trial's outcome is the number of boxes. We'll need at least 3 boxes to get one of each card, but with really bad luck, you could empty the shelves of several supermarkets before finding the card you need to get all 3. So, the possible outcomes of a trial are 3, 4, 5, or lots more. But we can't simply pick one of those numbers at random, because they're not equally likely. We'd be surprised if we only needed 3 boxes to get all the cards, but we'd probably be even more surprised to find that it took exactly 7,359 boxes. In fact, the reason we're doing the simulation is that it's hard to guess how many boxes we'd expect to open.

Building a Simulation

The relative frequencies of the cards are: 20% Hope Solo, 30% Danica Patrick, and 50% Blake Griffin. So, to use the equally likely random digits we get from the Internet, we can interpret the digits 0 and 1 as finding Hope; 2, 3, and 4 as finding Danica; and 5 through 9 as finding Blake to simulate opening one box. Opening one box is the basic building block, called a **component** of our simulation. But the component's outcome isn't the result we want. We need to observe a sequence of components until our card collection is complete. The *trial's* outcome is called the **response variable**; for this simulation that's the *number* of components (boxes) in the sequence.

Let's look at the steps for making a simulation:

Specify how to model a component outcome using equally likely random digits:

1. **Identify the component to be repeated.** In this case, our component is the opening of a box of cereal.
2. **Explain how you will model the component's outcome.** We'll assign outcomes to the equally likely random digits as follows:

    0, 1 Hope Solo   2, 3, 4 Danica Patrick   5, 6, 7, 8, 9 Blake Griffin.

Specify how to simulate trials:

3. **Explain how you will combine the components to model a trial.** We pretend to open boxes (repeat components) until our collection is complete. We do this by looking at each random digit and indicating what picture it represents. We continue until we've found all three.
4. **State clearly what the response variable is.** We want to find out the number of boxes it might take to get all three pictures.

Put it all together to run the simulation:

5. **Run several trials.** For example, consider the third line of random digits shown earlier (p. 282):

    8906427308645681412198226653885873285801699027843110380420067664.

    Let's see what happened.

    The first random digit, 8, means you get Blake's picture. So the first component's outcome is Blake. The second digit, 9, means Blake's picture is also in the next box. Continuing to interpret the random digits, we get Hope's picture (0) in the third, Blake's (6) again in the fourth, and finally Danica (4) on the fifth box. Since we've now found all three pictures, we've finished one trial of our simulation. This trial's outcome is 5 boxes.

    Now we keep going, running more trials by looking at the rest of our line of random digits:

    89064 2730 8645681 41219 822665388587328580 169902 78431 1038 042006 7664.

It's best to create a chart to keep track of what happens:

Trial Number	Component Outcomes	Trial Outcomes: y = Number of Boxes
1	89064 = Blake, Blake, Hope, Blake, Danica	5
2	2730 = Danica, Blake, Danica, Hope	4
3	8645681 = Blake, Blake, Danica, ..., Hope	7
4	41219 = Danica, Hope, Danica, Hope, Blake	5
5	822665388587328580 = Blake, Danica, ..., Hope	18
6	169902 = Hope, Blake, Blake, Blake, Hope, Danica	6
7	78431 = Blake, Blake, Danica, Danica, Hope	5
8	1038 = Hope, Hope, Danica, Blake	4
9	042006 = Hope, Danica, Danica, Hope, Hope, Blake	6
10	7664... = Blake, Blake, Blake, Danica...	?

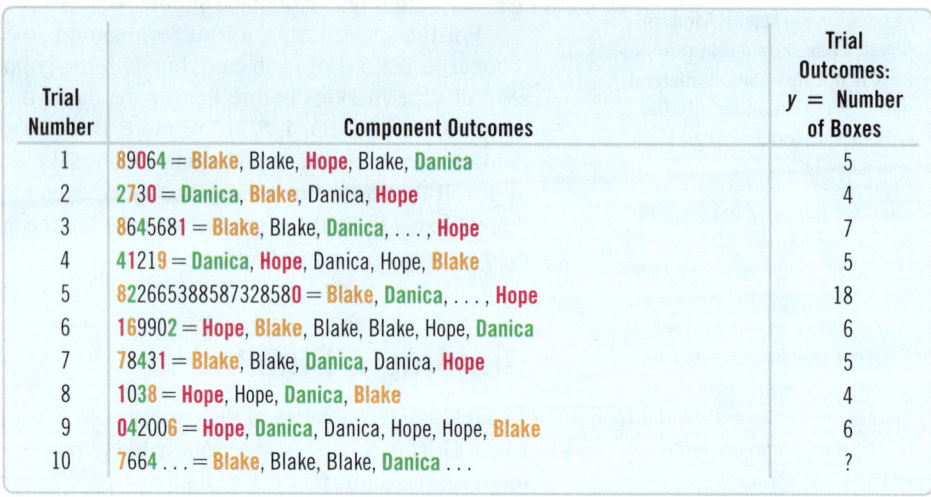

Analyze the response variable:

6. **Collect and summarize the results of all the trials.** You know how to summarize and display a response variable. You'll certainly want to report the shape, center, and spread, and depending on the question asked, you may want to include more.

7. **State your conclusion**, as always, in the context of the question you wanted to answer. Based on this simulation, we estimate that customers hoping to complete their card collection will need to open a median of 5 boxes, but it could take a lot more.

If you fear that these may not be accurate estimates because we ran only nine trials, you are absolutely correct. The more trials the better, and nine is woefully inadequate. Twenty trials is probably a reasonable minimum if you are doing this by hand. Even better, use a computer and run a few hundred trials.

 Activity: Bigger Samples Are Better. The random simulation tool can generate lots of outcomes with a single click, so you can see more of the long run with less effort.

For Example SIMULATING A DICE GAME

The game of 21 can be played with an ordinary 6-sided die. Competitors each roll the die repeatedly, trying to get the highest total less than or equal to 21. If your total exceeds 21, you lose.

Suppose your opponent has rolled an 18. Your task is to try to beat him by getting more than 18 points without going over 21. How many rolls do you expect to make, and what are your chances of winning?

QUESTION: How will you simulate the components?

ANSWER: A component is one roll of the die. I'll simulate each roll by looking at a random digit from a table or an Internet site. The digits 1 through 6 will represent the results on the die; I'll ignore digits 7–9 and 0.

QUESTION: How will you combine components to model a trial? What's the response variable?

ANSWER: I'll add components until my total is greater than 18, counting the number of rolls. If my total is greater than 21, it is a loss; if not, it is a win. There are two response variables. I'll count the number of times I roll the die, and I'll keep track of whether I win or lose.

QUESTION: How would you use these random digits to run trials? Show your method clearly for two trials.

91129 58757 69274 92380 82464 33089

ANSWER: I've marked the discarded digits in color.

Trial #1:	9	1	1	2	9	5	8	7	5	7	6	
Total:		1	2	4		9			14		20	Outcomes: 6 rolls, won
Trial #2:	9	2	7	4	9	2	3	8	0	8	2	4	6
Total:		2		6		8	11				13	17	23	Outcomes: 7 rolls, lost

QUESTION: Suppose you run 30 trials, getting the outcomes tallied here. What is your conclusion?

ANSWER: Based on my simulation, when competing against an opponent who has a score of 18, I expect my turn to usually last 5 or 6 rolls, and I should win about 70% of the time.

Number of Rolls		Result	
4	///	Won	ᚎ ᚎ ᚎ ᚎ /
5	ᚎ ᚎ	Lost	ᚎ ////
6	ᚎ ᚎ /		
7	ᚎ		
8	/		

Just Checking

The baseball World Series consists of up to seven games. The first team to win four games wins the series. The first two are played at one team's home ballpark, the next three at the other team's park, and the final two (if needed) are played back at the first park. Records over the past century show that there is a home field advantage; the home team has about a 55% chance of winning. Does the current system of alternating ballparks even out the home field advantage? How often will the team that begins at home win the series?

Let's set up the simulation:

1. What is the component to be repeated?
2. How will you model each component from equally likely random digits?
3. How will you model a trial by combining components?
4. What is the response variable?
5. How will you analyze the response variable?

Step-by-Step Example SIMULATION

Fifty-seven students participated in a lottery for a particularly desirable dorm room—a large triple close to the student center. Twenty of the participants were members of the same varsity team. When all three winners were members of the team, the other students cried foul.

Question: Could an all-team outcome reasonably be expected to happen if everyone had a fair shot at the room?

THINK ▶ Plan
State the problem. Identify the important parts of your simulation.

I'll use a simulation to investigate whether it's unlikely that three varsity athletes would get the great room in the dorm if the lottery were fair.

Components
Identify the components.

A component is the selection of a student.

Outcomes
State how you will model each component using equally likely random digits. You can't just use the digits from 0 to 9 because the outcomes you are simulating are not multiples of 10%.

I'll look at two-digit random numbers.

Let 00–19 represent the 20 varsity applicants.

Let 20–56 represent the other 37 applicants.

Skip 57–99. If I get a number in this range, I'll throw it away and go back for another two-digit random number.

There are 20 and 37 students in the two groups. This time you must use *pairs* of random digits (and ignore some of them) to represent the 57 students.

Trial
Explain how you will combine the components to simulate a trial. In each of these trials, you can't choose the same student twice, so you'll need to ignore a random number if it comes up a second or third time. Be sure to mention this in describing your simulation.

Each trial consists of identifying pairs of digits as V (varsity) or N (nonvarsity) until 3 people are chosen, ignoring out-of-range or repeated numbers (X)—I can't put the same person in the room twice.

Response Variable
Define your response variable.

The response variable is whether or not all three selected students are on the varsity team.

SHOW ▶ Mechanics
Run several trials. Carefully record the random numbers, indicating

1) the corresponding component outcomes (Varsity, Nonvarsity, or ignored number) and
2) the value of the response variable.

Trial Number	Component Outcomes	All Varsity?
1	74 02 94 39 02 77 55 X V X N X X N	No
2	18 63 33 25 V X N N	No
3	05 45 88 91 56 V N X X N	No
4	39 09 07 N V V	No
5	65 39 45 95 43 X N N X N	No
6	98 95 11 68 77 12 17 X X V X X V V	Yes
7	26 19 89 93 77 27 N V X X X N	No
8	23 52 37 N N N	No
9	16 50 83 44 V N X N	No
10	74 17 46 85 09 X V N X V	No

Analyze
Summarize the results across all trials to answer the initial question.

"All varsity" occurred once, or 10% of the time.

> **TELL ➡ Conclusion** Describe what the simulation shows, and interpret your results in the context of the real world.
>
> In my simulation of "fair" room draws, the three people chosen were all varsity team members only 10% of the time. While this result could happen by chance, it is not particularly likely. I'm suspicious, but I'd need many more trials and a smaller frequency of the all-varsity outcome before I would make an accusation of unfairness.

WHAT CAN GO WRONG?

Activity: Estimating Summaries from Random Outcomes. See how well you can estimate something you can't know just by generating random outcomes.

- **Don't overstate your case.** Let's face it: In some sense, a simulation is *always* wrong. After all, it's not the real thing. We didn't actually buy any cereal or run a room draw. So beware of confusing what *really* happens with what a simulation suggests *might* happen. Never forget that future results will not match your simulated results exactly.

- **Model outcome chances accurately.** A common mistake in constructing a simulation is to adopt a strategy that may appear to produce the right kind of results, but that does not accurately model the situation. For example, in our room draw, we could have gotten 0, 1, 2, or 3 team members. Why not just see how often these digits occur in random digits from 0 to 9, ignoring the digits 4 and up?

  3 2 1 7 9 0 0 5 9 7 3 7 9 2 5 2 4 1 3 8
  3 2 1 x x 0 0 x x x 3 x x 2 x 2 x 1 3 x

  This "simulation" makes it seem fairly likely that three team members would be chosen. There's a big problem with this approach, though: The digits 0, 1, 2, and 3 occur with equal frequency among random digits, making each outcome appear to happen 25% of the time. In fact, the selection of 0, 1, 2, or all 3 team members are not all equally likely outcomes. In our correct simulation, we estimated that all 3 would be chosen only about 10% of the time. If your simulation overlooks important aspects of the real situation, your model will not be accurate.

- **Run enough trials.** Simulation is cheap and fairly easy to do. Don't try to draw conclusions based on 5 or 10 trials (even though we did for illustration purposes here). We'll see how many trials to use in later chapters. For now, err on the side of large numbers of trials.

CONNECTIONS

Simulations often generate many outcomes of a response variable, and we are often interested in the distribution of these responses. The tools we use to display and summarize the distribution of any real variable are appropriate for displaying and summarizing randomly generated responses as well.

Make histograms, boxplots, and Normal probability plots of the response variables from simulations, and summarize them with measures of center and spread. Be especially careful to report the variation of your response variable.

Don't forget to think about your analyses. Simulations can hide subtle errors. A careful analysis of the responses can save you from erroneous conclusions based on a faulty simulation.

You may be less likely to find an outlier in simulated responses, but if you find one, you should certainly determine how it happened.

What Have We Learned?

Learning Objectives

- Know how to obtain random digits from computer programs or the Internet.
- Be able to design and execute, by hand, a simulation of a random phenomenon, specifying the design of the simulation to properly reflect the relative frequencies of random events.
- Understand that although individual random outcomes cannot be predicted, in the long run, simulated values tend to settle down. The goal of simulations is to obtain and understand those values.

Review of Terms

Random — An outcome is random if we know the possible values it can have, but not which particular value it takes (p. 280).

Generating random numbers — Random numbers are hard to generate. Nevertheless, several Internet sites offer an unlimited supply of equally likely random values (p. 281).

Simulation — A simulation models a real-world situation by using random-digit outcomes to mimic the uncertainty of a response variable of interest (p. 282).

Trial — The sequence of several components representing events that we are pretending will take place (p. 283).

Component — A component (of a simulation) uses equally likely random digits to model simple random occurrences whose outcomes may not be equally likely (p. 283).

Response variable — Values of the response variable record the results of each trial with respect to what we were interested in (p. 283).

On the Computer SIMULATION

Simulations are best done with the help of technology simply because more trials make a better simulation, and computers are fast. There are special computer programs designed for simulation, but most statistics packages can generate random numbers and support a simulation.

All technology-generated random numbers are *pseudorandom*. The random numbers available on the Internet may technically be better, but the differences won't matter for any simulation of modest size. Pseudorandom numbers generate the next random value from the previous one by a specified algorithm. But they have to start somewhere. This starting point is called the "seed." Most programs let you set the seed. There's usually little reason to do this, but if you wish to, go ahead. If you reset the seed to the same value, the programs will generate the same sequence of "random" numbers.

 Activity: Creating Random Values. Learn to use your statistics package to generate random outcomes.

DATA DESK

Generate random numbers in Data Desk with the **Generate Random Numbers . . .** command in the **Manip** menu. A dialog guides you in specifying the number of variables to fill, the number of cases, and details about the values. For most simulations, generate random uniform values.

EXCEL

The **RAND** function generates a random value between 0 and 1. You can multiply to scale it up to any range you like and use the INT function to turn the result into an integer.

COMMENTS

Published tests of Excel's random-number generation have declared it to be inadequate. However, for simple simulations, it should be OK. Don't trust it for important large simulations.

JMP

- In a new column, in the **Cols** menu choose **Column Info**....
- In the dialog, click the **New Property** button, and choose **Formula** from the drop-down menu.
- Click the **Edit Formula** button, and in the **Functions(grouped)** window click on **Random**. **Random Integer (10)**, for example, will generate a random integer between 1 and 10.

MINITAB

- In the **Calc** menu, choose **Random Data**....
- In the Random Data submenu, choose **Uniform**....

A dialog guides you in specifying details of range and number of columns to generate.

R

R has many built-in random number generators.
To generate k random integers between 1 and n:
- **Sample(1:n,k,replace = T)** where k and n are integers.
- **Sample(1:n,k)** will ensure that there are no repeats. Here k ≤ n.

COMMENTS

To generate k random digits, use **sample(0:9,k,replace = T)** where k is an integer.

SPSS

The **RV.UNIFORM(min, max)** function returns a random value that is equally likely between the min and max limits.

STATCRUNCH

To generate a list of random numbers:
- Click on **Data**.
- Choose **Simulate data » Uniform**.
  Enter the number of rows and columns of random numbers you want. (Often you'll specify the desired number of random values as **Rows** and just 1 **Column**.)
- Enter the interval of possible values for the random numbers (a ≤ x < b).
- Click on **Simulate**. The random numbers will appear in the data table.

COMMENTS

To get random *integers* from 0 to 99, set a = 0 and b = 100, and then simply ignore the decimal places in the numbers generated.

OR
- Under **Data** choose **Computer expression**.
- As Y, choose the **Column** of random numbers.
- Choose **Floor(Y)** from the functions.
- Click on **Set Expression**.
- Click on **Compute**.

TI-83/84 PLUS

To generate random numbers, use **5:RandInt** from the **Math PRB** menu. This command will produce any number of random integers in a specified range.

COMMENTS

Some examples:

RandInt(0,1) randomly chooses a 0 or a 1. This is an effective simulation of a coin toss.

RandInt(1,6,2) randomly returns two integers between 1 and 6. This is a good way to simulate rolling two dice. To run several trials, just hit ENTER repeatedly.

RandInt(0,57,3) produces three random integers between 0 and 57-a nice way to simulate the chapter's dorm room lottery.

Exercises

Section 10.1

1. **Random outcomes** For each of the following scenarios, decide if the outcome is random.
   a) Flip a coin to decide who takes out the trash. Is who takes out the trash random?
   b) A friend asks you to quickly name a professional sports team. Is the sports team named random?
   c) Names are selected out of a hat to decide roommates in a dormitory. Is your roommate for the year random?

2. **More random outcomes** For each of the following scenarios, decide if the outcome is random.
   a) You enter a contest in which the winning ticket is selected from a large drum of entries. Was the winner of the contest random?
   b) When playing a board game, the number of spaces you move is decided by rolling a six-sided die. Is the number of spaces you move random?
   c) Before flipping a coin, your friend asks you to "call it." Is your choice (heads or tails) random?

Section 10.2

3. **Components** Playing a board game, you are stuck in jail until you roll "doubles" (same numbers) on a pair of dice. You will simulate using random digits to estimate how many turns you expect to remain in jail. Identify the component if you were to run a simulation.

4. **More components** Your local burger joint gives away a game piece with each purchase. Game pieces are labeled *Burger*, *Fries*, and *Shake* and are equally likely to be received. You will simulate to estimate how many purchases you will need to collect the entire set to win a free meal. Identify the component if you were to run a simulation.

5. **Response variable** For the board game described in Exercise 3, how will you combine components to model a trial? What is the response variable?

6. **Response variable, take two** For the burger joint contest described in Exercise 4, how will you combine components to model a trial? What is the response variable?

Chapter Exercises

7. **The lottery** Many states run lotteries, giving away millions of dollars if you match a certain set of winning numbers. How are those numbers determined? Do you think this method guarantees randomness? Explain.

8. **Games** Many kinds of games people play rely on randomness. Cite three different methods commonly used in the attempt to achieve this randomness, and discuss the effectiveness of each.

9. **Birth defects** The American College of Obstetricians and Gynecologists says that out of every 100 babies born in the United States, 3 have some kind of major birth defect. How would you assign random numbers to conduct a simulation based on this proportion?

10. **Colorblind** By some estimates, about 10% of all males have some color perception defect, most commonly red–green colorblindness. How would you assign random numbers to conduct a simulation based on this percentage?

11. **Geography** An elementary school teacher with 25 students plans to have each of them make a poster about two different states. The teacher first numbers the states (in alphabetical order, from 1-Alabama to 50-Wyoming), then uses a random number table to decide which states each student gets. Here are the random digits:

    45921 01710 22892 37076

    a) Which two state numbers does the first student get?
    b) Which two state numbers go to the second student?

12. **Get rich** Your state's BigBucks Lottery prize has reached $100,000,000, and you decide to play. You have to pick five numbers between 1 and 60, and you'll win if your numbers match those drawn by the state. You decide to pick your "lucky" numbers using a random number table. Which numbers do you play, based on these random digits?

    43680 98750 13092 76561 58712

13. **Play the lottery** Some people play state-run lotteries by always playing the same favorite "lucky" numbers. Assuming that the lottery is truly random, is this strategy better, worse, or the same as choosing different numbers for each play? Explain.

14. **Play it again, Sam** In Exercise 12, you imagined playing the lottery by using random digits to decide what numbers to play. Is this a particularly good or bad strategy? Explain.

15. **Bad simulations** Explain why each of the following simulations fails to model the real situation properly:
    a) Use a random integer from 0 through 9 to represent the number of heads when 9 coins are tossed.
    b) A basketball player takes a foul shot. Look at a random digit, using an odd digit to represent a good shot and an even digit to represent a miss.
    c) Use random integers from 1 through 13 to represent the denominations of the cards in a five-card poker hand.

16. **More bad simulations** Explain why each of the following simulations fails to model the real situation:
    a) Use random numbers 2 through 12 to represent the sum of the faces when two dice are rolled.
    b) Use a random integer from 0 through 5 to represent the number of boys in a family of 5 children.
    c) Simulate a baseball player's performance at bat by letting 0 = an out, 1 = a single, 2 = a double, 3 = a triple, and 4 = a home run.

17. **Wrong conclusion** A Statistics student properly simulated the length of checkout lines in a grocery store and then reported, "The average length of the line will be 3.2 people." What's wrong with this conclusion?

18. **Another wrong conclusion** After simulating the spread of a disease, a researcher wrote, "24% of the people contracted the disease." What should the correct conclusion be?

19. **Election** You're pretty sure that your candidate for class president has about 55% of the votes in the entire school. But you're worried that only 100 students will show up to vote. How often will the underdog (the one with 45% support) win? To find out, you set up a simulation.
    a) Describe how you will simulate a component.
    b) Describe how you will simulate a trial.
    c) Describe the response variable.

20. **Two pair or three of a kind?** When drawing five cards randomly from a deck, which is more likely, two pairs or three of a kind? A pair is exactly two of the same denomination. Three of a kind is exactly 3 of the same denomination. (Don't count three 8's as a pair—that's 3 of a kind. And don't count 4 of the same kind as two pair—that's 4 of a kind, a very special hand.) How could you simulate 5-card hands? Be careful; once you've picked the 8 of spades, you can't get it again in that hand.
    a) Describe how you will simulate a component.
    b) Describe how you will simulate a trial.
    c) Describe the response variable.

21. **Cereal** In the chapter's example, 20% of the cereal boxes contained a picture of Hope Solo, 30% Danica Patrick, and the rest Blake Griffin. Suppose you buy five boxes of cereal. Estimate the probability that you end up with a complete set of the pictures. Your simulation should have at least 20 runs.

22. **Cereal, again** Suppose you really want the Hope Solo picture. How many boxes of cereal do you need to buy to be pretty sure of getting at least one? First, define what "pretty sure" means to you. Then run a simulation with at least 10 trials.

23. **Multiple choice** You take a quiz with 6 multiple choice questions. After you studied, you estimated that you would have about an 80% chance of getting any individual question right. What are your chances of getting them all right? Use at least 20 trials.

24. **Lucky guessing?** A friend of yours who took the multiple choice quiz in Exercise 23 got all 6 questions right, but now claims to have guessed blindly on every question. If each question offered 4 possible answers, do you believe her? Explain, basing your argument on a simulation involving at least 10 trials.

25. **Beat the lottery** Many states run lotteries to raise money. A website advertises that it knows "how to increase YOUR chances of Winning the Lottery." They offer several systems and criticize others as foolish. One system is called *Lucky Numbers*. People who play the *Lucky Numbers* system just pick a "lucky" number to play, but

maybe some numbers are luckier than others. Let's use a simulation to see how well this system works.

To make the situation manageable, simulate a simple lottery in which a single digit from 0 to 9 is selected as the winning number. Pick a single value to bet, such as 1, and keep playing it over and over. You'll want to run at least 100 trials. (If you can program the simulations on a computer, run several hundred. Or generalize the questions to a lottery that chooses two- or three-digit numbers—for which you'll need thousands of trials.)

a) What proportion of the time do you expect to win?
b) Would you expect better results if you picked a "luckier" number, such as 7? (Try it if you don't know.) Explain.

26. **Random is as random does** The "beat the lottery" website discussed in Exercise 25 suggests that because lottery numbers are random, it is better to select your bet randomly. For the same simple lottery in Exercise 25 (random values from 0 to 9), generate each bet by choosing a separate random value between 0 and 9. Play many games. What proportion of the time do you win?

27. **It evens out in the end** The "beat the lottery" website of Exercise 25 notes that in the long run we expect each value to turn up about the same number of times. That leads to their recommended strategy. First, watch the lottery for a while, recording the winners. Then bet the value that has turned up the least, because it will need to turn up more often to even things out. If there is more than one "rarest" value, just take the lowest one (since it doesn't matter). Simulating the simplified lottery described in Exercise 25, play many games with this system. What proportion of the time do you win?

28. **Play the winner?** Another strategy for beating the lottery is the reverse of the system described in Exercise 27. Simulate the simplified lottery described in Exercise 25. Each time, bet the number that just turned up. The website suggests that this method should do worse. Does it? Play many games and see.

29. **Driving test** You are about to take the road test for your driver's license. You hear that only 34% of candidates pass the test the first time, but the percentage rises to 72% on subsequent retests. Estimate the average number of tests drivers take in order to get a license. Your simulation should use at least 20 runs.

30. **Still learning?** As in Exercise 29, assume that your chance of passing the driver's test is 34% the first time and 72% for subsequent retests. Estimate the percentage of those tested who still do not have a driver's license after two attempts.

31. **Basketball strategy** Late in NCAA basketball games, foul shots are often "one and one," meaning a player gets a second shot only if he makes the first shot. Suppose a player has made 72% of his foul shots this season. Estimate the number of points he will score in a one-and-one situation.

32. **Blood donors** A person with type O-positive blood can receive blood only from other type O donors. About 44% of the U.S. population has type O blood. At a blood drive, how many potential donors do you expect to examine in order to get three units of type O blood?

33. **Free groceries** To attract shoppers, a supermarket runs a weekly contest that involves "scratch-off" cards. With each purchase, customers get a card with a black spot obscuring a message. When the spot is scratched away, most of the cards simply say, "Sorry—please try again." But during the week, 100 customers will get cards that make them eligible for a drawing for free groceries. Ten of the cards say they may be worth $200, 10 others say $100, 20 may be worth $50, and the rest could be worth $20. To register those cards, customers write their names on them and put them in a barrel at the front of the store. At the end of the week, the store manager draws cards at random, awarding the lucky customers free groceries in the amount specified on their card. The drawings continue until the store has given away more than $500 of free groceries. Estimate the average number of winners each week with a simulation.

34. **Find the ace** A technology store holds a contest to attract shoppers. Once an hour, someone at checkout is chosen at random to play in the contest. Here's how it works: An ace and four other cards are shuffled and placed face down on a table. The customer gets to turn over cards one at a time, looking for the ace. The person wins $100 of store credit if the ace is the first card, $50 if it is the second card, and $20, $10, or $5 if it is the third, fourth, or last card chosen. What is the average dollar amount of store credit given away in the contest? Estimate with a simulation.

35. **The family** Many couples want to have both a boy and a girl. If they decide to continue to have children until they have one child of each sex, what would the average family size be? Assume that boys and girls are equally likely.

36. **A bigger family** Suppose a couple will continue having children until they have at least two children of each sex (two boys *and* two girls). How many children might they expect to have?

37. **Dice game** You are playing a children's game in which the number of spaces you get to move is determined by the rolling of a die. You must land exactly on the final space to win. If you are 10 spaces away, how many turns might it take you to win?

38. **Parcheesi** You are three spaces from a win in Parcheesi. On each turn, you will roll two dice. To win, you must roll a total of 3 or roll a 3 on one of the dice. How many turns might you expect this to take?

39. **The hot hand** A basketball player with a 65% shooting percentage has just made 6 shots in a row. The announcer says this player "is hot tonight! She's in the zone!" Assume the player takes about 20 shots per game. Is it unusual for her to make 6 or more shots in a row during a game?

40. **The World Series** The World Series ends when a team wins 4 games. Suppose that sports analysts consider one team a bit stronger, with a 55% chance to win any individual game. Estimate the likelihood that the underdog wins the series.

41. **Teammates** Four couples at a dinner party play a board game after the meal. They decide to play as teams of two and to select the teams randomly. All eight people write their names on slips of paper. The slips are thoroughly mixed, then drawn two at a time. How likely is it that every person will be teamed with someone other than the person he or she came to the party with?

42. **Second team** Suppose the couples in Exercise 41 choose the teams by having one member of each couple write their names on the cards and the other people each pick a card at random. How likely is it that every person will be teamed with someone other than the person he or she came with?

43. **Job discrimination?** A company with a large sales staff announces openings for three positions as regional managers. Twenty-two of the current salespersons apply, 12 men and 10 women. After the interviews, when the company announces the newly appointed managers, all three positions go to women. The men complain of job discrimination. Do they have a case? Simulate a random selection of three people from the applicant pool, and make a decision about the likelihood that a fair process would result in hiring all women.

44. **Smartphones** A proud legislator claims that your state's new law banning texting and hand-held phones while driving reduced occurrences of infractions to less than 10% of all drivers. While on a long drive home from your college, you notice a few people seemingly texting. You decide to count everyone using their smartphones illegally who pass you on the expressway for the next 20 minutes. It turns out that 5 out of the 20 drivers were actually using their phones illegally. Does this cast doubt on the legislator's figure of 10%? Use a simulation to estimate the likelihood of seeing at least 5 out of 20 drivers using their phones illegally if the actual usage rate is only 10%. Explain your conclusion clearly.

Just Checking ANSWERS

1. The component is one game.
2. I'll generate random numbers and assign numbers from 00 to 54 to the home team's winning and from 55 to 99 to the visitors' winning.
3. I'll generate components until one team wins 4 games. I'll record which team wins the series.
4. The response is who wins the series.
5. I'll calculate the proportion of wins by the team that starts at home.

Chapter 11 Sample Surveys

11.1 The Three Big Ideas of Sampling

11.2 Populations and Parameters

11.3 Simple Random Samples

11.4 Other Sampling Designs

11.5 From the Population to the Sample: You Can't Always Get What You Want

11.6 The Valid Survey

11.7 Common Sampling Mistakes, or How to Sample Badly

> *Where are we going?*
> We see surveys all the time. How can asking just a thousand people tell us much about a national election? And how *do* they select the respondents? It turns out that there are many ways to select a good sample. But there are just three main ideas to understand.

In May and June 2012, Pew Research conducted a survey to assess where Americans got their news. Pew sampled 3003 U.S. adults, and asked, "Where did you get news yesterday?". They report that 55% of respondents say their main source for news is television. But the most striking change is in the percentage who report getting their news online or on a mobile device, which now exceeds the percentage who rely on newspapers or radio. Pew claimed that these estimates were close to the true percentages that they would have found if they had asked all U.S. adults. That step from a small sample to the entire population is impossible without understanding Statistics. To make business decisions, to do science, to choose wise investments, or to understand how voters think they'll vote in the next election, we need to stretch beyond the data at hand to the world at large.

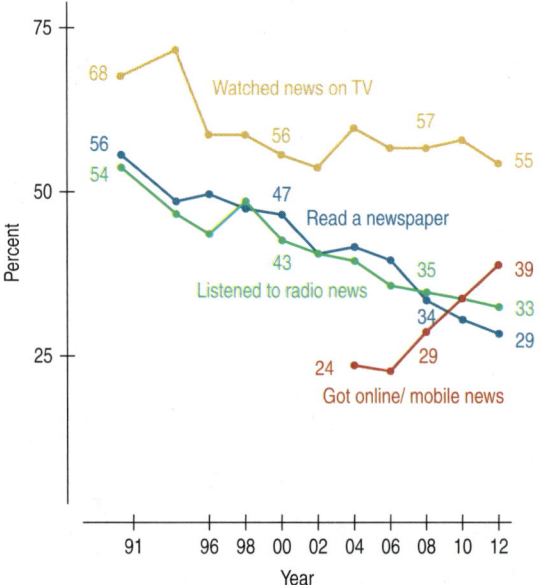

Figure 11.1
A timeplot showing the responses of 18–29 year olds to Pew polls asking them where they get most of their news about national and international news. Respondents could name up to 2 sources. (Source: "A Time plot showing the Digital NEWS" from In Changing News Landscape, Even Television is Vulnerable, Trends in News Consumption Survey: 1991–2012 by Pew Research Center for the People & the Press. Copyright © 2012. Used by Permission of Pew Research Center.)

11.1 The Three Big Ideas of Sampling

To make that stretch, we need three ideas. You'll find the first one natural. The second may be more surprising. The third is one of the strange but true facts that often confuse those who don't know Statistics.

Idea 1: Examine a Part of the Whole

A S Activity: **Populations and Samples.** Explore the differences between populations and samples.

The first idea is to draw a sample. We'd like to know about an entire group of individuals—a **population**—but examining all of them is usually impractical, if not impossible. So we settle for examining a smaller group of individuals—a **sample**—selected from the population.

You do this every day. For example, suppose you wonder how the vegetable soup you're cooking for dinner tonight is going to go over with your friends. To decide whether it meets your standards, you only need to try a small amount. You might taste just a spoonful or two. You certainly don't have to consume the whole pot. You trust that the taste will *represent* the flavor of the entire pot. The idea behind your tasting is that a small sample, if selected properly, can represent the entire population.

It's hard to go a day without hearing about the latest opinion poll. These polls are examples of **sample surveys**, designed to ask questions of a small group of people in the hope of learning something about the entire population. Most likely, you've never been selected to be part of one of these national opinion polls. That's true of most people. So how can the pollsters claim that a sample is representative of the entire population? The answer is that professional pollsters work quite hard to ensure that the "taste"—the sample that they take—represents the population.

THE W'S AND SAMPLING

The population we are interested in is usually determined by *Why* we are doing the study. The individuals on whom we collect data are the sample.

Bias Selecting a sample to represent the population fairly is more difficult than it sounds. Polls or surveys most often fail because they use a sampling method that tends to over- or underrepresent parts of the population. The method may overlook subgroups that are harder to find (such as the homeless) or favor others (such as Internet users who like to respond to online surveys). Sampling methods that, by their nature, tend to over- or underemphasize some characteristics of the population are said to be **biased**. Bias is the bane of sampling—the one thing above all to avoid. Conclusions based on samples drawn with biased methods are inherently flawed. There is usually no way to fix bias after the sample is drawn and no way to salvage useful information from it.

A S Video: **The *Literary Digest* Poll and the Election of 1936.** Hear the story of one of the most famous polling failures in history.

Here's a famous example of a really dismal failure. By the beginning of the 20th century, it was common for newspapers to ask readers to return "straw" ballots on a variety of topics. (Today's Internet surveys are the same idea, gone electronic.) The earliest known example of such a straw vote in the United States dates back to 1824.

During the period from 1916 to 1936, the magazine *Literary Digest* regularly surveyed public opinion and forecast election results correctly. During the 1936 presidential campaign between Alf Landon and Franklin Delano Roosevelt, it mailed more than 10 million ballots and got back an astonishing 2.4 million. (Polls were still a relatively novel idea, and many people thought it was important to send back their opinions.) The results were clear: Alf Landon would be the next president by a landslide, 57% to 43%. You remember President Landon? No? In fact, Landon carried only two states. Roosevelt won, 62% to 37%, and, perhaps coincidentally, the *Digest* went bankrupt soon afterward.

What went wrong? One problem was that the *Digest*'s sample wasn't representative. Where would *you* find 10 million names and addresses to sample? The *Digest* used the phone book, as many surveys do.[1] But in 1936, at the height of the Great Depression,

---

[1] Today, phone numbers are computer-generated to make sure that unlisted numbers are included. But special methods must be used for the growing part of the population that do not have "land-line" access.

In 1936, a young pollster named George Gallup used a subsample of only 3000 of the 2.4 million responses that the *Literary Digest* received to reproduce the wrong prediction of Landon's victory over Roosevelt. He then used an entirely different sample of 50,000 and predicted that Roosevelt would get 56% of the vote to Landon's 44%. His sample was apparently much more *representative* of the actual voting populace. The Gallup Organization went on to become one of the leading polling companies.

telephones were a luxury, so the *Digest* under-sampled poor voters. The campaign of 1936 focused on the economy, and those who were less well off were more likely to vote for the Democrat. So the *Digest*'s sample was hopelessly biased.

How do modern polls get their samples to *represent* the entire population? You might think that they'd handpick individuals to sample with care and precision. But in fact, they do something quite different: They select individuals to sample *at random.*

Idea 2: Randomize

Think back to the soup sample. Suppose you add some salt to the pot. If you sample it from the top before stirring, you'll get the misleading idea that the whole pot is salty. If you sample from the bottom, you'll get an equally misleading idea that the whole pot is bland. By stirring, you *randomize* the amount of salt throughout the pot, making each taste more typical of the whole pot.

Not only does randomization protect you against factors that you know are in the data, it can also help protect against factors that you didn't know were there. Suppose, while you weren't looking, a friend added a handful of peas to the soup. If they're down at the bottom of the pot, and you don't randomize the soup by stirring, your test spoonful won't have any peas. By stirring in the salt, you *also* randomize the peas throughout the pot, making your sample taste more typical of the overall pot *even though you didn't know the peas were there*. So randomizing protects us even in this case.

How do we "stir" people in a survey? We select them at random. **Randomizing** protects us from the influences of *all* the features of our population by making sure that, *on average*, the sample looks like the rest of the population. The importance of deliberately using randomness is one of the great insights of Statistics.

> **WHY NOT MATCH THE SAMPLE TO THE POPULATION?**
> Rather than randomizing, we could try to design our sample so that the people we choose are typical in terms of every characteristic we can think of. We might want the income levels of those we sample to match the population. How about age? Political affiliation? Marital status? Having children? Living in the suburbs? We can't possibly think of all the things that might be important. Even if we could, we wouldn't be able to match our sample to the population for all these characteristics.

A S *Activity:* Sampling from Some Real Populations. Draw random samples to see how closely they resemble each other and the population.

For Example IS A RANDOM SAMPLE REPRESENTATIVE?

Here are summary statistics comparing two samples of 8000 drawn at random from a company's database of 3.5 million customers:

	Mean Age (yr)	White (%)	Female (%)	Mean # of Children	Mean Income Bracket (1–7)	Mean Wealth Bracket (1–9)	Homeowner? (% Yes)
Sample 1	61.4	85.12	56.2	1.54	3.91	5.29	71.36
Sample 2	61.2	84.44	56.4	1.51	3.88	5.33	72.30

QUESTION: Do you think these samples are representative of the population of the company's customers? Explain.

ANSWER: The two samples look very similar with respect to these seven variables. It appears that randomizing has automatically matched them pretty closely. We can reasonably assume that since the two samples don't differ too much from each other, they don't differ much from the rest of the population either.

Idea 3: It's the Sample Size

How large a random sample do we need for the sample to be reasonably representative of the population? Most people think that we need a large percentage, or *fraction,* of the population, but it turns out that all that matters is the *number of individuals in the sample.* A random sample of 100 students in a college represents the student body just about as well as a random sample of 100 voters represents the entire electorate of the United States. This is the *third* idea and probably the most surprising one in designing surveys.

How can it be that only the size of the sample, and not the population, matters? Well, let's return one last time to that pot of soup. If you're cooking for a banquet rather than just for a few people, your pot will be bigger, but do you need a bigger spoon to decide how the soup tastes? Of course not. The same-size spoonful is probably enough to make a decision about the entire pot, no matter how large the pot. The *fraction* of the population that you've sampled doesn't matter.[2] It's the **sample size**—or the number of individuals in the sample—that's important.

How big a sample do you need? That depends on what you're estimating. To get an idea of what's really in the soup, you'll need a large enough taste to get a *representative* sample from the pot. For a survey that tries to find the proportion of the population falling into a category, you'll usually need several hundred respondents to say anything precise enough to be useful.[3]

A S *Activity:* **Does the Population Size Matter?** Here's the narrated version of this important idea about sampling.

> **Larger samples?** A friend who knows that you are taking Statistics asks your advice on her study. What can you possibly say that will be helpful? Just say, "If you could just get a larger sample, it would probably improve your study." Even though a larger sample might not be worth the cost, it will almost always make the results more precise.

> **What do the pollsters do?** How do professional polling agencies do their work? The most common polling method today is to contact respondents by telephone. Computers generate random telephone numbers, so pollsters can even call some people with unlisted phone numbers. The interviewer may then ask to speak with the adult who is home who had the most recent birthday or use some other essentially random way to select among those available. In phrasing questions, pollsters often list alternative responses (such as candidates' names) in different orders to avoid biases that might favor the first name on the list.
>
> Do these methods work? The Pew Research Center for the People and the Press, reporting on one survey, says that
>
> > Across five days of interviewing, surveys today are able to make some kind of contact with the vast majority of households (76%). But because of busy schedules, skepticism and outright refusals, interviews were completed in just 38% of households that were reached using standard polling procedures.
>
> Nevertheless, studies indicate that those actually sampled can give a good snapshot of larger populations from which the surveyed households were drawn.

A S *Video:* **Frito-Lay Sampling for Quality.** How does a potato chip manufacturer make sure to cook only the best potatoes?

Does a Census Make Sense? Why bother determining the right sample size? Wouldn't it be better to just include everyone and "sample" the entire population? Such a special sample is called a **census**. Although a census would appear to provide the best possible information about the population, there are a number of reasons why it might not.

First, it can be difficult to complete a census. Some individuals in the population will be hard (and expensive) to locate. Or a census might just be impractical. If you were a taste tester for the Hostess™ Company, you probably wouldn't want to census *all* the Twinkies on the production line. Not only might this be life-endangering, but you wouldn't have any left to sell.

---

[2] Well, that's not exactly true. If the population is small enough and the sample is more than 10% of the whole population, it *can* matter. It doesn't matter whenever, as usual, our sample is a very small fraction of the population.

[3] Chapter 17 gives the details behind this statement and shows how to decide on a sample size for a survey.

Activity: Can a Large Sample Protect Against Bias? Explore how we can learn about the population from large or repeated samples.

Second, populations rarely stand still. In populations of people, babies are born and folks die or leave the country. In opinion surveys, events may cause a shift in opinion during the survey. A census takes longer to complete and the population changes while you work. A sample surveyed in just a few days may give more accurate information.

Third, taking a census can be more complex than sampling. For example, the U.S. Census records too many college students. Many are counted once with their families and are then counted a second time in a report filed by their schools.

> **The undercount** It's particularly difficult to compile a complete census of a population as large, complex, and spread out as the U.S. population. The U.S. Census is known to miss some residents. On occasion, the undercount has been striking. For example, there have been blocks in inner cities in which the number of residents recorded by the Census was smaller than the number of electric meters for which bills were being paid. What makes the problem particularly important is that some groups have a higher probability of being missed than others—undocumented immigrants, the homeless, the poor. The Census Bureau proposed the use of random sampling to estimate the number of residents missed by the ordinary census. Unfortunately, the resulting debate has become more political than statistical.

11.2 Populations and Parameters

STATISTICS AND PARAMETERS I
Any quantity that we calculate from data could be called a "statistic." But in practice, we usually use a statistic to estimate a population parameter.

Activity: Statistics and Parameters. Explore the difference between statistics and parameters.

STATISTICS AND PARAMETERS II
Remember: Population parameters are not just unknown—they are almost always *unknowable*. The best we can do is to estimate them with sample statistics.

A study found that teens were less likely to "buckle up." The National Center for Chronic Disease Prevention and Health Promotion reports that 21.7% of U.S. teens never or rarely wear seat belts. We're sure they didn't take a census, so what *does* the 21.7% mean? We can't know what percentage of teenagers wear seat belts. Reality is just too complex. But we can simplify the question by building a model.

Models use mathematics to represent reality. Parameters are the key numbers in those models. A parameter used in a model for a population is sometimes called (redundantly) a **population parameter**.

But let's not forget about the data. We use summaries of the data to estimate the population parameters. As we know, any summary found from the data is a **statistic**. Sometimes you'll see the (also redundant) term **sample statistic**.[4]

We've already met two parameters in Chapter 5: the mean, μ, and the standard deviation, σ. We'll try to keep denoting population model parameters with Greek letters and the corresponding statistics with Latin letters. Usually, but not always, the letter used for the statistic and the parameter correspond in a natural way. So the standard deviation of the data is s, and the corresponding parameter is σ (Greek for s). In Chapter 6, we used r to denote the sample correlation. The corresponding correlation in a model for the population would be called ρ (rho). In Chapters 7 and 8, b_1 represented the slope of a linear regression estimated from the data. But when we think about a (linear) *model* for the population, we denote the slope parameter β_1 (beta).

Get the pattern? Good. Now it breaks down. We denote the mean of a population model with μ (because μ is the Greek letter for m). It might make sense to denote the sample mean with m, but long-standing convention is to put a bar over anything when we average it, so we write $\bar{y}$. What about proportions? Suppose we want to talk about the proportion of teens who don't wear seat belts. If we use p to denote the proportion from the data, what is the corresponding model parameter? By all rights, it should be π. But statements like $\pi = 0.25$ might be confusing because π has been equal to 3.1415926... for so long, and it's worked so *well*. So, once again we violate the rule. We'll use p for the population model parameter and $\hat{p}$ for the proportion from the data (since, like $\hat{y}$ in regression, it's an estimated value).

---
[4]Where else besides a sample *could* a statistic come from?

■ NOTATION ALERT

This entire table is a notation alert.

Here's a table summarizing the notation:

Name	Statistic	Parameter
Mean	$\bar{y}$	μ (mu, pronounced "meeoo," not "moo")
Standard Deviation	s	σ (sigma)
Correlation	r	ρ (rho, pronounced like "row")
Regression Coefficient	b	β (beta, pronounced "baytah"[5])
Proportion	$\hat{p}$	p (pronounced "pee"[6])

We draw samples because we can't work with the entire population, but we want the statistics we compute from a sample to reflect the corresponding parameters accurately. A sample that does this is said to be **representative**. A biased sampling methodology tends to over- or underestimate the parameter of interest.

Just Checking

1. Various claims are often made for surveys. Why is each of the following claims not correct?

   a) It is always better to take a census than to draw a sample.

   b) Stopping students on their way out of the cafeteria is a good way to sample if we want to know about the quality of the food there.

   c) We drew a sample of 100 from the 3000 students in a school. To get the same level of precision for a town of 30,000 residents, we'll need a sample of 1000.

   d) A poll taken at a statistics support website garnered 12,357 responses. The majority said they enjoy doing statistics homework. With a sample size that large, we can be pretty sure that most Statistics students feel this way too.

   e) The true percentage of all Statistics students who enjoy the homework is called a "population statistic."

11.3 Simple Random Samples

How would you select a representative sample? Most people would say that every individual in the population should have an equal chance to be selected, and certainly that seems fair. But it's not sufficient. There are many ways to give everyone an equal chance that still wouldn't give a representative sample. Consider, for example, a school that has equal numbers of males and females. We could sample like this: Flip a coin. If it comes up heads, select 100 female students at random. If it comes up tails, select 100 males at random. Everyone has an equal chance of selection, but every sample is of only a single sex—hardly representative.

We need to do better. Suppose we insist that every possible *sample* of the size we plan to draw has an equal chance to be selected. This ensures that situations like the one just described are not likely to occur and still guarantees that each person has an equal chance of being selected. What's different is that with this method, each *combination* of people has an equal chance of being selected as well. A sample drawn in this way is called a **Simple Random Sample**, usually abbreviated **SRS**. An SRS is the standard against which

---
[5]If you're from the United States. If you're British or Canadian, it's "beetah."
[6]Just in case you weren't sure.

we measure other sampling methods, and the sampling method on which the theory of working with sampled data is based.

To select a sample at random, we first need to define where the sample will come from. The **sampling frame** is a list of individuals from which the sample is drawn. For example, to draw a random sample of students at a college, we might obtain a list of all registered full-time students and sample from that list. In defining the sampling frame, we must deal with the details of defining the population. Are part-time students included? How about those who are attending school elsewhere and transferring credits back to the college?

Once we have a sampling frame, the easiest way to choose an SRS is to assign a random number to each individual in the sampling frame. We then select only those whose random numbers satisfy some rule.[7] Let's look at some ways to do this.

- We can be more efficient when we're choosing a larger sample from a sampling frame stored in a data file. First we assign a random number with several digits (say, from 0 to 10,000) to each individual. Then we sort the random numbers into numerical order, keeping each name with its number. Choosing the first n names from this re-arranged list will give us a random sample of that size.
- Often the sampling frame is so large that it would be too tedious to number everyone consecutively. If our intended sample size is approximately 10% of the sampling frame, we can assign each individual a single random digit 0 to 9. Then we select only those with a specific random digit, say, 5.

Samples drawn at random generally differ one from another. Each draw of random numbers selects *different* people for our sample. These differences lead to different values for the variables we measure. We call these sample-to-sample differences **sampling variability**. Surprisingly, sampling variability isn't a problem; it's an opportunity. In future chapters we'll investigate what the variation in a sample can tell us about its population.

> **ERROR OKAY; BIAS NO WAY!**
>
> Sampling variability is sometimes referred to as *sampling error*, making it sound like it's some kind of mistake. It's not. We understand that samples will vary, so "sampling error" is to be expected. It's *bias* we must strive to avoid. Bias means our sampling method distorts our view of the population, and that will surely lead to mistakes.

> **For Example** USING RANDOM NUMBERS TO GET AN SRS
>
> There are 80 students enrolled in an introductory Statistics class; you are to select a sample of 5.
>
> **QUESTION:** How can you select an SRS of 5 students using these random digits found on the Internet: 05166 29305 77482?
>
> **ANSWER:** First, I'll number the students from 00 to 79. Taking the random numbers two digits at a time gives me 05, 16, 62, 93, 05, 77, and 48. I'll ignore 93 because the students were numbered only up to 79. And, so as not to pick the same person twice, I'll skip the repeated number 05. My simple random sample consists of students with the numbers 05, 16, 62, 77, and 48.

11.4 Other Sampling Designs

Simple random sampling is not the only fair way to sample. More complicated designs may save time or money or help avoid potential bias and improve sample accuracy. All statistical sampling designs have in common the idea that chance, rather than human choice, is used to select the sample.

Stratified Sampling

Designs that are used to sample from large populations—especially populations residing across large areas—are often more complicated than simple random samples. Sometimes the population is first sliced into homogeneous groups, called **strata**, before the sample is

---
[7] Chapter 10 presented ways of finding and working with random numbers.

selected. Then simple random sampling is used within each stratum before the results are combined. This common sampling design is called **stratified random sampling**.

Why would we want to complicate things? Here's an example. Suppose we want to learn how students feel about funding for the football team at a large university. The campus is 60% men and 40% women, and we suspect that men and women have different views on the funding. If we use simple random sampling to select 100 people for the survey, we could end up with 70 men and 30 women or 35 men and 65 women. Our resulting estimates of the level of support for the football funding could vary widely. To help reduce this sampling variability, we can decide to force a representative balance, selecting 60 men at random and 40 women at random. This would guarantee that the proportions of men and women within our sample match the proportions in the population, and that should make such samples more accurate in representing population opinion.

You can imagine the importance of stratifying by race, income, age, and other characteristics, depending on the questions in the survey. Samples taken within a stratum vary less, so our estimates can be more precise. This reduced sampling variability is the most important benefit of stratifying.

Stratified sampling can also help us notice important differences among groups. But be careful. As we saw in Chapter 2, if we unthinkingly combine grouped data, we risk reaching the wrong conclusion, becoming victims of Simpson's paradox.

For Example STRATIFYING THE SAMPLE

RECAP: You're trying to find out what freshmen think of the food served on campus. Food Services believes that men and women typically have different opinions about the importance of the salad bar.

QUESTION: How should you adjust your sampling strategy to allow for this difference?

ANSWER: I will stratify my sample by drawing an SRS of men and a separate SRS of women—assuming that the data from the registrar include information about each person's sex.

Cluster and Multistage Sampling

Suppose we wanted to assess the reading level of this textbook based on the length of the sentences. Simple random sampling could be awkward; we'd have to number each sentence, then find, for example, the 576th sentence or the 2482nd sentence, and so on.

It would be much easier to pick a few *pages* at random and count the lengths of the sentences on those pages. That works if we believe that each page is representative of the entire book in terms of reading level. Splitting the population into representative **clusters** can make sampling more practical. Then we could simply select one or a few clusters at random and perform a census within each of them. This sampling design is called **cluster sampling**. Clusters are generally selected for reasons of efficiency, practicality, or cost. Ideally, if each cluster represents the full population fairly, cluster sampling will be unbiased. But often we just hope that by choosing a sample of clusters, we can obtain a representative sample of the entire population.

For Example CLUSTER SAMPLING

RECAP: In trying to find out what freshmen think about the food served on campus, you've considered both an SRS and a stratified sample. Now you have run into a problem: It's simply too difficult and time consuming to track down the individuals whose names were chosen for your sample. Fortunately, freshmen at your school are all housed in 10 freshman dorms.

QUESTIONS: How could you use this fact to draw a cluster sample? How might that alleviate the problem? What concerns do you have?

ANSWER: To draw a cluster sample, I would select a few freshman dorms at random and then try to contact everyone in each selected dorm. I could save time by simply knocking on doors on a given evening and interviewing people. I'd have to assume that freshmen were assigned to dorms pretty much at random and that the people I'm able to contact are representative of everyone in the dorm.

What's the difference between cluster sampling and stratified sampling? We stratify to ensure that our sample represents different groups in the population, and we sample randomly within each stratum. Strata are internally homogeneous, but differ from one another. By contrast, we select clusters to make sampling more practical or affordable. Clusters can be heterogeneous; we want our sample of clusters to provide a representative sample.

STRATIFIED VS. CLUSTER SAMPLING

Boston cream pie consists of a layer of yellow cake, a layer of custard, another cake layer, and then a chocolate frosting. Suppose you are a professional taster (yes, there really are such people) whose job is to check your company's pies for quality. You'd need to eat small samples of randomly selected pies, tasting all three components: the cake, the custard, and the frosting.

One approach is to cut a thin vertical slice out of the pie. Such a slice will be a lot like the entire pie, so by eating that slice, you'll learn about the whole pie. This vertical slice containing all the different ingredients in the pie would be a *cluster* sample.

Another approach is to sample in *strata*: Select some tastes of the cake at random, some tastes of custard at random, and some bits of frosting at random. You'll end up with a reliable judgment of the pie's quality.

Many populations you might want to learn about are like this Boston cream pie. You can think of the subpopulations of interest as horizontal strata, like the layers of pie. Cluster samples slice vertically across the layers to obtain clusters, each of which may have parts of the entire population. Stratified samples represent the population by drawing some from each layer, reducing variability in the results that could arise because of the differences among the layers.

Boston cream pie is the official dessert of Massachusetts. It is, of course, a cake and not a pie.

Sometimes we use a variety of sampling methods together. Sampling schemes that combine several methods are called **multistage samples**. In trying to assess the reading level of this book, we might worry that it starts out easy and then gets harder as the concepts become more difficult. If so, we'd want to avoid samples that selected heavily from early or from late chapters. To guarantee a fair mix of chapters, we could randomly choose one chapter from each of the six parts of the book and then randomly select a few pages from each of those chapters. If, altogether, that made too many sentences, we might select a few sentences at random from each of the chosen pages. So, what is our sampling strategy? First, we stratify by the part of the book and randomly choose a chapter to represent each stratum. Within each selected chapter, we choose pages as clusters. Finally, we consider an SRS of sentences within each cluster. Most surveys conducted by professional polling organizations use some combination of stratified and cluster sampling as well as simple random samples.

For Example   MULTISTAGE SAMPLING

RECAP: Learning that freshmen are housed in separate dorms allowed you to sample their attitudes about the campus food by going to dorms chosen at random, but you're still concerned about possible differences in opinions between men and women. It turns out that these freshman dorms house men and women on alternate floors.

> **QUESTION:** How can you design a sampling plan that uses this fact to your advantage?
>
> **ANSWER:** Now I can stratify my sample by sex. I would first choose one or two dorms at random and then select the same number of dorm floors at random from among those that house men and, separately, from among those that house women. I could then draw an SRS of residents on the selected floors.

Systematic Samples

Some samples select individuals systematically. For example, you might survey every 10th person on an alphabetical list of students. To make it random, you still must start the systematic selection from a randomly selected individual. When the order of the list is not associated in any way with the responses sought, **systematic sampling** can give a representative sample.

Think about the reading-level sampling example again. Suppose we have chosen a chapter of the book at random, then three pages at random from that chapter, and now we want to select a sample of 10 sentences from the 73 sentences found on those pages. Instead of numbering each sentence so we can pick a simple random sample, it would be easier to sample systematically. A quick calculation shows $73/10 = 7.3$, so we can get our sample by just picking every seventh sentence on the page. But where should you start? At random, of course. We've accounted for $10 \times 7 = 70$ of the sentences, so we'll throw the extra 3 into the starting group and choose a sentence at random from the first 10. Then we pick every seventh sentence after that and record its length.

Systematic sampling can be much less expensive than true random sampling. When you use a systematic sample, you should justify the assumption that the systematic method is not associated with any of the measured variables. For example, if you decided to sample students in their dorm rooms by knocking on every other door, a dorm in which male and female rooms were alternated could result in a sample that was all male or all female.

 Just Checking

2. We need to survey a random sample of the 300 passengers on a flight from San Francisco to Tokyo. Name each sampling method described below.

   a) Pick every 10th passenger as people board the plane.

   b) From the boarding list, randomly choose 5 people flying first class and 25 of the other passengers.

   c) Randomly generate 30 seat numbers and survey the passengers who sit there.

   d) Randomly select a seat position (right window, right center, right aisle, etc.) and survey all the passengers sitting in those seats.

Step-by-Step Example SAMPLING

The assignment says, "Conduct your own sample survey to find out how many hours per week students at your school spend watching TV during the school year." Let's see how we might do this step by step. (Remember, though—actually collecting the data from your sample can be difficult and time consuming.)

Question: How would you design this survey?

THINK ➡ Plan
State what you want to know.

Population and Parameter
Identify the W's of the study. The *Why* determines the population and the associated sampling frame. The *What* identifies the parameter of interest and the variables measured. The *Who* is the sample we actually draw. The *How*, *When*, and *Where* are given by the sampling plan.

Often, thinking about the *Why* will help us see whether the sampling frame and plan are adequate to learn about the population.

Sampling Plan
Specify the sampling method and the sample size, n. Specify how the sample was actually drawn. What is the sampling frame? How was the randomization performed?

A good description should be complete enough to allow someone to replicate the procedure, drawing another sample from the same population in the same manner.

I wanted to design a study to find out how many hours of TV students at my school watch.

The population studied was students at our school. I obtained a list of all students currently enrolled and used it as the sampling frame. The parameter of interest was the number of TV hours watched per week during the school year, which I attempted to measure by asking students how much TV they watched during the previous week.

I decided against stratifying by class or sex because I didn't think TV watching would differ much between males and females or across classes. I selected a simple random sample of students from the list. I obtained an alphabetical list of students, assigned each a random digit between 0 and 9, and then selected all students who were assigned a "4." This method generated a sample of 212 students from the population of 2133 students.

SHOW ➡ Sampling Practice
Specify *When*, *Where*, and *How* the sampling was performed. Specify any other details of your survey, such as how respondents were contacted, what incentives were offered to encourage them to respond, how nonrespondents were treated, and so on.

The survey was taken over the period Oct. 15 to Oct. 25. Surveys were sent to selected students by e-mail, with the request that they respond by e-mail as well. Students who could not be reached by e-mail were handed the survey in person.

TELL ➡ Summary and Conclusion
This report should include a discussion of all the elements. In addition, it's good practice to discuss any special circumstances. Professional polling organizations report the *When* of their samples but will also note, for example, any important news that might have changed respondents' opinions during the sampling process. In this survey, perhaps, a major news story or sporting event might change students' TV viewing behavior.

The question you ask also matters. It's better to be specific ("How many hours did you watch TV last week?") than to ask a general question ("How many hours of TV do you usually watch in a week?").

During the period Oct. 15 to Oct. 25, 212 students were randomly selected, using a simple random sample from a list of all students currently enrolled. The survey they received asked the following question: "How many hours did you spend watching television last week?"

Of the 212 students surveyed, 110 responded. It's possible that the nonrespondents differ in the number of TV hours watched from those who responded, but I was unable to follow up on them due to limited time and funds. The 110 respondents reported an average 3.62 hours of TV watching per week. The median was only 2 hours per week. A histogram of the data shows that the distribution is highly right-skewed, indicating that the median might be a more appropriate summary of the typical TV watching of the students.

The report should show a display of the data, provide and interpret the statistics from the sample, and state the conclusions that you reached about the population.

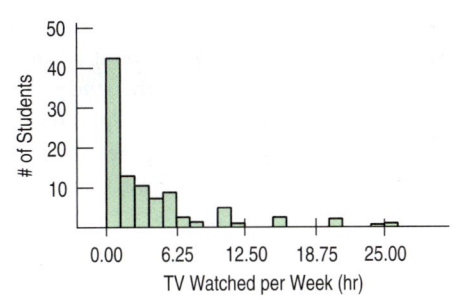

Most of the students (90%) watch between 0 and 10 hours per week, while 30% reported watching less than 1 hour per week. A few watch much more. About 3% reported watching more than 20 hours per week.

11.5 From the Population to the Sample: You Can't Always Get What You Want

WHAT'S THE SAMPLE?

The population we want to study is determined by asking *why*. When we design a survey, we use the term "sample" to refer to the individuals selected, from whom we hope to obtain responses. Unfortunately, the real sample is just those we can reach to obtain responses—the *who* of the study. These are slightly different uses of the same term *sample*. The context usually makes clear which we mean, but it's important to realize that the difference between the two samples could undermine even a well-designed study.

Before you start a survey, think first about the population you want to study. You may find that it's not the well-defined group you thought it was. Who, exactly, is a student, for example? Even if the population seems well defined, it may not be a practical group from which to draw a sample. For example, election polls want to sample from all those who will vote in the next election—a population that is impossible to identify before Election Day

Next, you must specify the sampling frame—the individuals who are actually available to be sampled. (Do you have a list of students to sample from? How about a list of registered voters?) Usually, the sampling frame is not the group you *really* want to know about. (All those registered to vote are not equally likely to show up.) The sampling frame limits what your survey can find out.

Then there's your target sample. These are the individuals for whom you *intend* to measure responses. You're not likely to get responses from all of them. ("I know it's dinnertime, but I'm sure you wouldn't mind answering a few questions. It'll only take 20 minutes or so. Oh, you're busy?") Nonresponse is a problem in many surveys.

Finally, there's your sample—the actual respondents. These are the individuals about whom you *do* get data and can draw conclusions. Unfortunately, they might not be representative of the sampling frame or the population.

CALVIN AND HOBBES © 1993 Watterson. Distributed by Universal Uclick. Reprinted with permission. All rights reserved.

At each step, the group we can study may be constrained further. The *Who* keeps changing, and each constraint can introduce biases. Whenever you work with or learn about a sample, pay attention not only to who was sampled but to who was *left out* of the study. A careful study should address the question of how well each group matches the population of interest. One of the main benefits of simple random sampling is that it never loses its sense of who's *Who*. The *Who* in an SRS is the population from which we've drawn a representative sample. That's not always true for other kinds of samples.

11.6 The Valid Survey

It isn't sufficient to just draw a sample and start asking questions. We want our survey to be *valid*. A valid survey yields the information we are seeking about the population we are interested in. Before setting out to survey, ask yourself

- What do I want to know?
- Am I asking the right respondents?
- Am I asking the right questions?
- What would I do with the answers if I had them; would they address the things I want to know?

These questions may sound obvious, but there are a number of pitfalls to avoid.

Know what you want to know. Before considering a survey, understand what you hope to learn and about whom you hope to learn it.

Use the right sampling frame. A valid survey obtains responses from the appropriate respondents. Be sure you have a suitable *sampling frame*. Have you identified the population of interest and sampled from it appropriately? A company might survey customers who returned warranty registration cards—a readily available sampling frame. But if the company wants to make their product more attractive, the most important population is the customers who rejected their product in favor of one from a competitor.

Tune your instrument. You may be tempted to ask questions you don't really need, but longer survey instruments yield fewer responses and thus a greater chance of non-response bias.

Ask specific rather than general questions. People are not very good at estimating their typical behavior, so it is better to ask "How many hours did you sleep last night?" than "How much do you usually sleep?" Sure, some responses will include some unusual events (My dog was sick; I was up all night.), but overall you'll get better data.

Ask for quantitative results when possible. "How many magazines did you read last week?" is better than "How much do you read: A lot, A moderate amount, A little, or None at all?"

Be careful in phrasing questions. A respondent may not understand the question—or may understand the question differently than the researcher intended it. ("Does anyone in your family ride a motorcycle?" Do you mean just me, my spouse, and my children? Or does "family" include my father, my siblings, and my second cousin once removed? And does a motor scooter count?) Respondents are unlikely (or may not have the opportunity) to ask for clarification. A question like "Do you approve of the recent actions of the Ambassador to Mexico?" is likely not to measure what you want if many respondents don't know who the Ambassador is or what he or she recently did.

Respondents may even lie or shade their responses if they feel embarrassed by the question ("Did you have too much to drink last night?"), are intimidated or insulted by the question ("Could you understand our new *Instructions for Dummies* manual, or was it too difficult for you?"), or if they want to avoid offending the interviewer ("Would

> **A short survey** Given that the *New York Times* reports[8] that statisticians can earn $125,000 at top companies their first year on the job, do you think this course will be valuable to *you*?

> **Pilot testing a survey question** A researcher distributed a survey to an organization before some economizing changes were made. She asked how people felt about a proposed cutback in secretarial and administrative support on a seven-point scale from Very Happy to Very Unhappy.
>
> But virtually all respondents were very unhappy about the cutbacks, so the results weren't particularly useful. If she had pretested the question, she might have chosen a scale that ran from unhappy to outraged.

you hire a man with a tattoo?" asked by a tattooed interviewer). Also, be careful to avoid phrases that have double or regional meanings. "How often do you go to town?" might be interpreted differently by different people and cultures.

Even subtle differences in phrasing can make a difference. In January 2006, the *New York Times* asked half of the 1229 U.S. adults in their sample the following question:

> *After 9/11, President Bush authorized government wiretaps on some phone calls in the U.S. without getting court warrants, saying this was necessary to reduce the threat of terrorism. Do you approve or disapprove of this?*[9]

They found that 53% of respondents approved. But when they asked the other half of their sample a question with different phrasing,

> *After 9/11, George W. Bush authorized government wiretaps on some phone calls in the U.S. without getting court warrants. Do you approve or disapprove of this?*[10]

only 46% approved.

Be careful in phrasing answers. It's often a good idea to offer choices rather than inviting a free response. Open-ended answers can be difficult to analyze. "How did you like the movie?" may start an interesting debate, but it may be better to give a range of possible responses. Be sure to phrase answers in a neutral way. When asking "Do you support higher school taxes?" positive responses could be worded "Yes," "Yes, it is important for our children," or "Yes, our future depends on it." But those are not equivalent answers.

The best way to protect a survey from unanticipated measurement errors is to perform a pilot survey. A **pilot** is a trial run of the survey you eventually plan to give to a larger group, using a draft of your survey questions administered to a small sample drawn from the same sampling frame you intend to use. By analyzing the results from this smaller survey, you can often discover ways to improve your survey instrument.

11.7 Common Sampling Mistakes, or How to Sample Badly

Bad sample designs yield worthless data. Many of the most convenient forms of sampling can be seriously biased. And there is no way to correct for the bias from a bad sample. So it's wise to pay attention to sample design—and to beware of reports based on poor samples.

Mistake 1: Sample Volunteers

One of the most common dangerous sampling methods is a voluntary response sample. In a **voluntary response sample**, a large group of individuals is invited to respond, and those who choose to respond are counted. The respondents, rather than the researcher, decide who will be in the sample. This method is used by call-in shows, 900 numbers, Internet polls, and letters written to members of Congress. Voluntary response samples are almost always biased, so conclusions drawn from them are almost always wrong.

It's often hard to define the sampling frame of a voluntary response study. Practically, the frames are groups such as Internet users who frequent a particular website or those who happen to be watching a particular TV show at the moment. But those sampling frames don't correspond to the population of interest.

Even within the sampling frame, voluntary response samples are often biased toward those with strong opinions or those who are strongly motivated. People with very negative opinions tend to respond more often than those with equally strong positive opinions. The sample is not representative, even though every individual in the population may have been offered the chance to respond. The resulting **voluntary response bias** invalidates the survey.

---
[8]www.nytimes.com/2009/08/06/technology/06stats.html
[9]Excerpt from "The New York Times/CBS News Poll, January 20–25, 2006". Published in The New York Times, © 2006.
[10]ibid

Activity: Sources of Sampling Bias. Here's a narrated exploration of sampling bias.

> **If you had it to do over again, would you have children?** Ann Landers, the advice columnist, asked parents this question. The overwhelming majority–70% of the more than 10,000 people who wrote in–said no, kids weren't worth it. A more carefully designed survey later showed that about 90% of parents actually are happy with their decision to have children. What accounts for the striking difference in these two results? What parents do you think are most likely to respond to the original question?

For Example VOLUNTARY RESPONSE SAMPLE

RECAP: You're trying to find out what freshmen think of the food served on campus, and have thought of a variety of sampling methods, all time consuming. A friend suggests that you set up a "Tell Us What You Think" website and invite freshmen to visit the site to complete a questionnaire.

QUESTION: What's wrong with this idea?

ANSWER: Letting each freshman decide whether to participate makes this a voluntary response survey. Students who were dissatisfied might be more likely to go to the website to record their complaints, and this could give me a biased view of the opinions of all freshmen.

Mistake 2: Sample Conveniently

Another sampling method that doesn't work is convenience sampling. As the name suggests, in **convenience sampling** we simply include the individuals who are convenient for us to sample. Unfortunately, this group may not be representative of the population. Here's an amusing example. Back in 2001, when computer use in the home was not as common as it is today, a survey of 437 potential home buyers in Orange County, California, reached the surprising conclusion that

> All but 2 percent of the buyers have at least one computer at home, and 62 percent have two or more. Of those with a computer, 99 percent are connected to the Internet (Source: Jennifer Hieger, "Portrait of Homebuyer Household: 2 Kids and a PC," Orange County Register, 27 July 2001).[11]

How was the survey conducted? On the Internet!

Many surveys conducted at shopping malls suffer from the same problem. People in shopping malls are not necessarily representative of the population of interest. Mall shoppers tend to be more affluent and include a larger percentage of teenagers and retirees than the population at large. To make matters worse, survey interviewers tend to select individuals who look "safe," or easy to interview.

> Do you use the Internet?
> Click here ○ for yes
> Click here ○ for no

INTERNET SURVEYS
Internet convenience surveys are worthless. As voluntary response surveys, they have no well-defined sampling frame (all those who use the Internet and visit their site?) and thus report no useful information. Do not believe them.

For Example CONVENIENCE SAMPLE

RECAP: To try to gauge freshman opinion about the food served on campus, Food Services suggests that you just stand outside a school cafeteria at lunchtime and stop people to ask them questions.

QUESTION: What's wrong with this sampling strategy?

ANSWER: This would be a convenience sample, and it's likely to be biased. I would miss people who use the cafeteria for dinner, but not for lunch, and I'd never hear from anyone who hates the food so much that they have stopped coming to the school cafeteria.

---

[11] Excerpt from "Portrait of Home buyer Household: 2 Kids and a PC" by Jennifer Hieger. Published by Orange County Register, © 27 July 2001.

Mistake 3: Use a Bad Sampling Frame

An SRS from an incomplete sampling frame introduces bias because the individuals included may differ from the ones not in the frame. People in prison, homeless people, students, and long-term travelers are all likely to be missed. Professional polling companies now need special procedures to be sure they include in their sampling frame people who can be reached only by cell phone.

Mistake 4: Undercoverage

Many survey designs suffer from **undercoverage**, in which some portion of the population is not sampled at all or has a smaller representation in the sample than it has in the population. Undercoverage can arise for a number of reasons, but it's always a potential source of bias.

Telephone surveys are usually conducted when you are likely to be home, such as dinnertime. If you eat out often, you may be less likely to be surveyed, a possible source of undercoverage.

Nonresponse Bias

A common and serious potential source of bias for most surveys is **nonresponse bias**. No survey succeeds in getting responses from everyone. The problem is that those who don't respond may differ from those who do. And they may differ on just the variables we care about. The lack of response will bias the results. Rather than sending out a large number of surveys for which the response rate will be low, it is often better to design a smaller randomized survey for which you have the resources to ensure a high response rate. One of the problems with nonresponse bias is that it's usually impossible to tell what the nonrespondents might have said.

It turns out that the *Literary Digest* survey was wrong on two counts. First, their list of 10 million people was not representative. There was a selection bias in their sampling frame. There was also a nonresponse bias. We know this because the *Digest* also surveyed a *systematic* sample in Chicago, sending the same question used in the larger survey to every third registered voter. They *still* got a result in favor of Landon, even though Chicago voted overwhelmingly for Roosevelt in the election. This suggests that the Roosevelt supporters were less likely to respond to the *Digest* survey. There's a modern version of this problem: It's been suggested that those who screen their calls with caller ID or an answering machine, and so might not talk to a pollster, may differ in wealth or political views from those who just answer the phone.

Video: **Biased Question Wording**. Watch a hapless interviewer make every mistake in the book.

Response Bias

Response bias[12] refers to anything in the survey design that influences the responses. Response biases include the tendency of respondents to tailor their responses to try to please the interviewer, the natural unwillingness of respondents to reveal personal facts or admit to illegal or unapproved behavior, and the ways in which the wording of the questions can influence responses.

How to Think About Biases

- **Look for biases in any survey you encounter.** If you design one of your own, ask someone else to help look for biases that may not be obvious to you. And do this *before* you collect your data. There's no way to recover from a biased sampling method or a survey that asks biased questions.

    Sorry, it just can't be done.

---

[12]Response bias is not the opposite of nonresponse bias. (We don't make these terms up; we just try to explain them.)

A bigger sample size for a biased study just gives you a bigger useless study.

A really big sample gives you a really big useless study. (Think of the 2.4 million *Literary Digest* responses.)

- **Spend your time and resources reducing biases.** No other use of resources is as worthwhile as reducing the biases.
- **Think about the members of the population who could have been excluded from your study.** Be careful not to claim that you have learned anything about them.
- **If you can, pilot-test your survey.** Administer the survey in the exact form that you intend to use it to a small sample drawn from the population you intend to sample. Look for misunderstandings, misinterpretation, confusion, or other possible biases. Then refine your survey instrument.
- **Always report your sampling methods in detail.** Others may be able to detect biases where you did not expect to find them.

WHAT CAN GO WRONG?

- The principal thing that can go wrong in sampling is that the sample can fail to represent the population. Unfortunately, this can happen in many different ways and for many different reasons. We've considered many of them in the chapter.

- It is also an error to draw too small a sample for your needs regardless of how well you draw the sample. However, generally it is more worthwhile to devote effort on improving the quality of your sample than on expanding its size.

CONNECTIONS

With this chapter, we take our first formal steps to relate our sample data to a larger population. Some of these ideas have been lurking in the background as we sought patterns and summaries for data. Even when we only worked with the data at hand, we often thought about implications for a larger population of individuals.

Notice the ongoing central importance of models. We've seen models in several ways in previous chapters. Here, we recognize the value of a model for a population. The parameters of such a model are values we will often want to estimate using statistics such as those we've been calculating. The connections to summary statistics for center, spread, correlation, and slope are obvious.

We now have a specific application for random numbers. The idea of applying randomness deliberately showed up in Chapter 10 for simulation. Now we need randomization to get good-quality data from the real world.

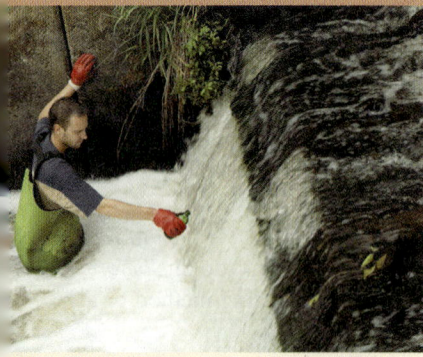

What Have We Learned?

Learning Objectives

Know the three ideas of sampling.
- Examine a part of the whole: A sample can give information about the population.
- Randomize to make the sample representative.
- The sample size is what matters. It's the size of the sample—and not its fraction of the larger population—that determines the precision of the statistics it yields.

Be able to draw a Simple Random Sample (SRS) using a table of random digits or a list of random numbers from technology or an Internet site.
- In a **simple random sample** (SRS), every possible group of n individuals has an equal chance of being our sample.

Understand the advantages and disadvantages of other sampling methods:
- **Stratified samples** can reduce sampling variability by identifying homogeneous subgroups and then randomly sampling within each.
- **Cluster samples** randomly select among heterogeneous subgroups, making our sampling tasks more manageable.
- **Systematic samples** can work in some situations and are often the least expensive method of sampling. But we still want to start them randomly.
- **Multistage samples** combine several random sampling methods.

Identify and avoid causes of bias.
- **Voluntary response samples** are almost always biased and should be avoided and distrusted.
- **Convenience samples** are likely to be flawed for similar reasons.
- **Bad sampling frames** can lead to samples that don't represent the population of interest.
- **Undercoverage** occurs when individuals from a subgroup of the population are selected less often than they should be.
- **Nonresponse bias** can arise when sampled individuals will not or cannot respond.
- **Response bias** arises when respondents' answers might be affected by external influences, such as question wording or interviewer behavior.
- Use best practices when designing a sample survey to improve the chance that your results will be valid.

Review of Terms

Population — The entire group of individuals or instances about whom we hope to learn (p. 295).

Sample — A (representative) subset of a population, examined in the hope of learning about the population (p. 295).

Sample survey — A study that asks questions of a sample drawn from some population in the hope of learning something about the entire population. Polls taken to assess voter preferences are common sample surveys (p. 295).

Bias — Any systematic failure of a sampling method to represent its population is bias. Biased sampling methods tend to over- or underestimate parameters. It is almost impossible to recover from bias, so efforts to avoid it are well spent (p. 295).

Common errors include
- relying on voluntary response.
- undercoverage of the population.
- nonresponse bias.
- response bias.

Randomization — The best defense against bias is randomization, in which each individual is given a fair, random chance of selection (p. 296).

Sample size — The number of individuals in a sample. The sample size determines how well the sample represents the population, not the fraction of the population sampled (p. 297).

Census — A sample that consists of the entire population (p. 297).

Population parameter — A numerically valued attribute of a model for a population. We rarely expect to know the true value of a population parameter, but we do hope to estimate it from sampled data. For example, the mean income of all employed people in the country is a population parameter (p. 298).

Statistic, sample statistic — Statistics are values calculated for sampled data. Those that correspond to, and thus estimate, a population parameter, are of particular interest. For example, the mean income of all employed people in a representative sample can provide a good estimate of the corresponding population parameter. The term "sample statistic" is sometimes used, usually to parallel the corresponding term "population parameter" (p. 298).

Representative	A sample is said to be representative if the statistics computed from it accurately reflect the corresponding population parameters (p. 299).
Simple Random Sample (SRS)	A Simple Random Sample of sample size n is a sample in which each set of n elements in the population has an equal chance of selection (p. 299).
Sampling frame	A list of individuals from whom the sample is drawn is called the sampling frame. Individuals who may be in the population of interest, but who are not in the sampling frame, cannot be included in any sample (p. 300).
Sampling variability	The natural tendency of randomly drawn samples to differ, one from another. Sometimes, unfortunately, called *sampling error*, sampling variability is no error at all, but just the natural result of random sampling (p. 300).
Stratified random sample	A sampling design in which the population is divided into several subpopulations, or **strata**, and random samples are then drawn from each stratum. If the strata are homogeneous, but are different from each other, a stratified sample may yield more consistent results than an SRS (pp. 300, 301).
Cluster sample	A sampling design in which entire groups, or **clusters**, are chosen at random. Cluster sampling is usually selected as a matter of convenience, practicality, or cost. Clusters are heterogeneous. A sample of clusters should be representative of the population (p. 301).
Multistage sample	Sampling schemes that combine several sampling methods. For example, a national polling service may stratify the country by geographical regions, select a random sample of cities from each region, and then interview a cluster of residents in each city (p. 302).
Systematic sample	A sample drawn by selecting individuals systematically from a sampling frame. When there is no relationship between the order of the sampling frame and the variables of interest, a systematic sample can be representative (p. 303).
Pilot study	A small trial run of a survey to check whether questions are clear. A pilot study can reduce errors due to ambiguous questions (p. 307).
Voluntary response sample	A sample in which individuals can choose on their own whether to participate. Samples based on voluntary response are always invalid and cannot be recovered, no matter how large the sample size (p. 307).
Voluntary response bias	Bias introduced to a sample when individuals can choose on their own whether to participate in the sample. Samples based on voluntary response are always invalid and cannot be recovered, no matter how large the sample size (p. 307).
Convenience sample	A convenience sample consists of the individuals who are conveniently available. Convenience samples often fail to be representative because every individual in the population is not equally convenient to sample (p. 308).
Undercoverage	A sampling scheme that biases the sample in a way that gives a part of the population less representation in the sample than it has in the population suffers from undercoverage (p. 309).
Nonresponse bias	Bias introduced when a large fraction of those sampled fails to respond. Those who do respond are likely to not represent the entire population. Voluntary response bias is a form of nonresponse bias, but nonresponse may occur for other reasons. For example, those who are at work during the day won't respond to a telephone survey conducted only during working hours (p. 309).
Response bias	Anything in a survey design that influences responses falls under the heading of response bias. One typical response bias arises from the wording of questions, which may suggest a favored response. Voters, for example, are more likely to express support of "the president" than support of the particular person holding that office at the moment (p. 309).

On the Computer SAMPLING

Computer-generated pseudorandom numbers are usually good enough for drawing random samples. But there is little reason not to use the truly random values available on the Internet.

Here's a convenient way to draw an SRS of a specified size using a computer-based sampling frame. The sampling frame can be a list of names or of identification numbers arrayed, for example, as a column in a spreadsheet, statistics program, or database:

1. Generate random numbers of enough digits so that each exceeds the size of the sampling frame list by several digits. This makes duplication unlikely.
2. Assign the random numbers arbitrarily to individuals in the sampling frame list. For example, put them in an adjacent column.
3. Sort the list of random numbers, carrying along the sampling frame list.
4. Now the first n values in the sorted sampling frame column are an SRS of n values from the entire sampling frame.

Exercises

Section 11.1

1. **Texas A&M** Administrators at Texas A&M University were interested in estimating the percentage of students who are the first in their family to go to college. The A&M student body has about 46,000 members. How might the administrators answer their question by applying the three Big Ideas?

2. **Satisfied workers** The managers of a large company wished to know the percentage of employees who feel "extremely satisfied" to work there. The company has roughly 24,000 employees. They contacted a random sample of employees and asked them about their job satisfaction, obtaining 437 completed responses. How does their study deal with the three Big Ideas of sampling?

Section 11.2

3. **A&M again** The president of the university plans a speech to an alumni group. He plans to talk about the proportion of students who responded in the survey that they are the first in their family to attend college, but the first draft of his speech treats that proportion as the actual proportion of current A&M students who are the first in their families to attend college. Explain to the president the difference between the proportion of respondents who are first attenders and the proportion of the entire student body that are first attenders. Use appropriate statistics terminology.

4. **Satisfied respondents** The company's annual report states, "Our survey shows that 87.34% of our employees are 'very happy' working here." Comment on that claim. Use appropriate statistics terminology.

Section 11.3

5. **Sampling students** A professor teaching a large lecture class of 350 students samples her class by rolling a die. Then, starting with the row number on the die (1 to 6), she passes out a survey to every fourth row of the large lecture hall. She says that this is a Simple Random Sample because everyone had an equal opportunity to sit in any seat and because she randomized the choice of rows. What do you think? Be specific.

6. **Sampling satisfaction** A company hoping to assess employee satisfaction surveys employees by assigning computer-generated random numbers to each employee on a list of all employees and then contacting all those whose assigned random number is divisible by 7. Is this a simple random sample?

Section 11.4

7. **Sampling A&M students** For each scenario, identify the kind of sample used by the university administrators from Exercise 1:
   a) Select several dormitories at random and contact everyone living in the selected dorms.
   b) Using a computer-based list of registered students, contact 200 freshmen, 200 sophomores, 200 juniors, and 200 seniors selected at random from each class.
   c) Using a computer-based alphabetical list of registered students, select one of the first 25 on the list by random and then contact the student whose name is 50 names later, and then every 50 names beyond that.

Section 11.5

8. **Satisfactory satisfaction samples** For each scenario, determine the sampling method used by the managers from Exercise 2.
   a) Use the company e-mail directory to contact 150 employees from among those employed for less than 5 years, 150 from among those employed for 5–10 years, and 150 from among those employed for more than 10 years.
   b) Use the company e-mail directory to contact every 50th employee on the list.
   c) Select several divisions of the company at random. Within each division, draw an SRS of employees to contact.

Section 11.6

9. **Survey students** What problems do you see with asking the following question of students? "Are you the first member of your family to seek higher education?"

10. **Happy employees** The company plans to have the head of each corporate division hold a meeting of their employees to ask whether they are happy on their jobs. They will ask people to raise their hands to indicate whether they are happy. What problems do you see with this plan?

Section 11.7

11. **Student samples** The university administration of Exercise 1 is considering a variety of ways to sample students for a survey. For each of these proposed survey designs, identify the problem.
    a) Publish an advertisement inviting students to visit a website and answer questions.
    b) Set up a table in the student union and ask students to stop and answer a survey.

12. **Surveying employees** The company of Exercise 2 is considering ways to survey their employees. For each of these proposed designs, identify the problem.
    a) Leave a stack of surveys out in the employee cafeteria so people can pick them up and return them.
    b) Stuff a questionnaire in the mailbox of each employee with the request that they fill it out and return it.

Chapter Exercises

13. **Roper** Through their *Roper Reports Worldwide*, GfK Roper conducts a global consumer survey to help multinational companies understand different consumer attitudes throughout the world. Within 30 countries, the researchers interview 1000 people aged 13–65. Their samples are designed so that they get 500 males and 500 females in each country. (www.gfkamerica.com)
    a) Are they using a simple random sample? Explain.
    b) What kind of design do you think they are using?

14. **Student center survey** For their class project, a group of Statistics students decide to survey the student body to assess opinions about the proposed new student center. Their sample of 200 contained 50 first-year students, 50 sophomores, 50 juniors, and 50 seniors.
    a) Do you think the group was using an SRS? Why?
    b) What sampling design do you think they used?

15. **Drug tests** Major League Baseball tests players to see whether they are using performance-enhancing drugs. Officials select a team at random, and a drug-testing crew shows up unannounced to test all 40 players on the team. Each testing day can be considered a study of drug use in Major League Baseball.
    a) What kind of sample is this?
    b) Is that choice appropriate?

16. **Gallup** At its website (www.gallup.com), the Gallup Poll publishes results of a new survey each day. Scroll down to the end, and you'll find a statement that includes words such as these:

    Results are based on telephone interviews with 1016 national adults, aged 18 and older, conducted March 7–10, 2014. . . . In addition to sampling error, question wording and practical difficulties in conducting surveys can introduce error or bias into the findings of public opinion polls.

    a) For this survey, identify the population of interest.
    b) Gallup performs its surveys by phoning numbers generated at random by a computer program. What is the sampling frame?
    c) What problems, if any, would you be concerned about in matching the sampling frame with the population?

In exercises 17 to 23, for the following reports about statistical studies, identify the following items (if possible). If you can't tell, then say so—this often happens when we read about a survey.

a) The population
b) The population parameter of interest
c) The sampling frame
d) The sample
e) The sampling method, including whether or not randomization was employed
f) Who (if anyone) was left out of the study.
g) Any potential sources of bias you can detect and any problems you see in generalizing to the population of interest

17. **Medical treatments** Consumers Union asked all subscribers whether they had used alternative medical treatments and, if so, whether they had benefited from them. For almost all of the treatments, approximately 20% of those responding reported cures or substantial improvement in their condition.

18. **Social life** A question posted on the gamefaqs.com website on August 1, 2011, asked visitors to the site, "Do you have an active social life outside the Internet?" 22% of the 55,581 respondents said "No" or "Not really, most of my personal contact is online."

19. Mayoral race Hoping to learn what issues may resonate with voters in the coming election, the campaign director for a mayoral candidate selects one block from each of the city's election districts. Staff members go there and interview all the adult residents they can find.

20. Soil samples The Environmental Protection Agency took soil samples at 16 locations near a former industrial waste dump and checked each for evidence of toxic chemicals. They found no elevated levels of any harmful substances.

21. Roadblock State police set up a roadblock to estimate the percentage of cars with up-to-date registration, insurance, and safety inspection stickers. They usually find problems with about 10% of the cars they stop.

22. Snack foods A company packaging snack foods maintains quality control by randomly selecting 10 cases from each day's production and weighing the bags. Then they open one bag from each case and inspect the contents.

23. Milk samples Dairy inspectors visit farms unannounced and take samples of the milk to test for contamination. If the milk is found to contain dirt, antibiotics, or other foreign matter, the milk will be destroyed and the farm reinspected until purity is restored.

24. Mistaken poll A local TV station conducted a "Pulse-Poll" about the upcoming mayoral election. Evening news viewers were invited to text in their votes, with the results to be announced on the late-night news. Based on the texts, the station predicted that Amabo would win the election with 52% of the vote. They were wrong: Amabo lost, getting only 46% of the vote. Do you think the station's faulty prediction is more likely to be a result of bias or sampling error? Explain.

25. Another mistaken poll Prior to the mayoral election discussed in Exercise 24, the newspaper also conducted a poll. The paper surveyed a random sample of registered voters stratified by political party, age, sex, and area of residence. This poll predicted that Amabo would win the election with 52% of the vote. The newspaper was wrong: Amabo lost, getting only 46% of the vote. Do you think the newspaper's faulty prediction is more likely to be a result of bias or sampling error? Explain.

26. Parent opinion, part 1 In a large city school system with 20 elementary schools, the school board is considering the adoption of a new policy that would require elementary students to pass a test in order to be promoted to the next grade. The PTA wants to find out whether parents agree with this plan. Listed below are some of the ideas proposed for gathering data. For each, indicate what kind of sampling strategy is involved and what (if any) biases might result.
   a) Put a big ad in the newspaper asking people to log their opinions on the PTA website.
   b) Randomly select one of the elementary schools and contact every parent by phone.
   c) Send a survey home with every student, and ask parents to fill it out and return it the next day.
   d) Randomly select 20 parents from each elementary school. Send them a survey, and follow up with a phone call if they do not return the survey within a week.

27. Parent opinion, part 2 Let's revisit the school system described in Exercise 26. Four new sampling strategies have been proposed to help the PTA determine whether parents favor requiring elementary students to pass a test in order to be promoted to the next grade. For each, indicate what kind of sampling strategy is involved and what (if any) biases might result.
   a) Run a poll on the local TV news, asking people to text in whether they favor or oppose the plan.
   b) Hold a PTA meeting at each of the 20 elementary schools, and tally the opinions expressed by those who attend the meetings.
   c) Randomly select one class at each elementary school and contact each of those parents.
   d) Go through the district's enrollment records, selecting every 40th parent. PTA volunteers will go to those homes to interview the people chosen.

28. Churches For your political science class, you'd like to take a survey from a sample of all the Catholic Church members in your city. A list of churches shows 17 Catholic churches within the city limits. Rather than try to obtain a list of all members of all these churches, you decide to pick 3 churches at random. For those churches, you'll ask to get a list of all current members and contact 100 members at random.
   a) What kind of design have you used?
   b) What could go wrong with your design?

29. Playground Some people have been complaining that the children's playground at a municipal park is too small and is in need of repair. Managers of the park decide to survey city residents to see if they believe the playground should be rebuilt. They hand out questionnaires to parents who bring children to the park. Describe possible biases in this sample.

30. Roller coasters An amusement park has opened a new roller coaster. It is so popular that people are waiting for up to 3 hours for a 2-minute ride. Concerned about how patrons (who paid a large amount to enter the park and ride on the rides) feel about this, they survey every 10th person on the line for the roller coaster, starting from a randomly selected individual.
   a) What kind of sample is this?
   b) What is the sampling frame?
   c) Is it likely to be representative?
   d) What members of the population of interest are omitted?

31. Playground, act two The survey described in Exercise 29 asked: *Many people believe this playground is too small and in need of repair. Do you think the playground should be repaired and expanded even if that means raising the entrance fee to the park?*

Describe two ways this question may lead to response bias.

32. Wording the survey Two members of the PTA committee in Exercises 26 and 27 have proposed different questions to ask in seeking parents' opinions.

Question 1: Should elementary school–age children have to pass high-stakes tests in order to remain with their classmates?
Question 2: Should schools and students be held accountable for meeting yearly learning goals by testing students before they advance to the next grade?

a) Do you think responses to these two questions might differ? How? What kind of bias is this?
b) Propose a question with more neutral wording that might better assess parental opinion.

33. Banning ephedra An online poll on a website asked:

A nationwide ban of the diet supplement ephedra went into effect recently. The herbal stimulant has been linked to 155 deaths and many more heart attacks and strokes. Ephedra manufacturer NVE Pharmaceuticals, claiming that the FDA lacked proof that ephedra is dangerous if used as directed, was denied a temporary restraining order on the ban by a federal judge. Do you think that ephedra should continue to be banned nationwide?

65% of 17,303 respondents said "yes." Comment on each of the following statements about this poll:

a) With a sample size that large, we can be pretty certain we know the true proportion of Americans who think ephedra should be banned.
b) The wording of the question is clearly very biased.
c) The sampling frame is all Internet users.
d) Results of this voluntary response survey can't be reliably generalized to any population of interest.

34. Survey questions Examine each of the following questions for possible bias. If you think the question is biased, indicate how and propose a better question.

a) Should companies that pollute the environment be compelled to pay the costs of cleanup?
b) Given that 18-year-olds are old enough to vote and to serve in the military, is it fair to set the drinking age at 21?

35. More survey questions Examine each of the following questions for possible bias. If you think the question is biased, indicate how and propose a better question.

a) Do you think high school students should be required to wear uniforms?
b) Given humanity's great tradition of exploration, do you favor continued funding for space flights?

36. Phone surveys Any time we conduct a survey, we must take care to avoid undercoverage. Suppose we plan to select 500 names from the city phone book, call their homes between noon and 4 P.M., and interview whoever answers, anticipating contacts with at least 200 people.

a) Why is it difficult to use a simple random sample here?
b) Describe a more convenient, but still random, sampling strategy.
c) What kinds of households are likely to be included in the eventual sample of opinion? Excluded?
d) Suppose, instead, that we continue calling each number, perhaps in the morning or evening, until an adult is contacted and interviewed. How does this improve the sampling design?
e) Random-digit dialing machines can generate the phone calls for us. How would this improve our design? Is anyone still excluded?

37. Cell phone survey What about drawing a random sample only from cell phone exchanges? Discuss the advantages and disadvantages of such a sampling method compared with surveying randomly generated telephone numbers from non–cell phone exchanges. Do you think these advantages and disadvantages have changed over time? How do you expect they'll change in the future?

38. Arm length How long is your arm compared with your hand size? Put your right thumb at your left shoulder bone, stretch your hand open wide, and extend your hand down your arm. Put your thumb at the place where your little finger is, and extend down the arm again. Repeat this a third time. Now your little finger will probably have reached the back of your left hand. If your arm is less than four hand widths, turn your hand sideways and count finger widths until you reach the end of your middle finger.

a) How many hand and finger widths is your arm?
b) Suppose you repeat your measurement 10 times and average your results. What parameter would this average estimate? What is the population?
c) Suppose you now collect arm lengths measured in this way from 9 friends and average these 10 measurements. What is the population now? What parameter would this average estimate?
d) Do you think these 10 arm lengths are likely to be representative of the population of arm lengths in your community? In the country? Why or why not?

39. Fuel economy Occasionally, when I fill my car with gas, I figure out how many miles per gallon my car got. I wrote down those results after six fill-ups in the past few months. Overall, it appears my car gets 28.8 miles per gallon.

a) What statistic have I calculated?
b) What is the parameter I'm trying to estimate?
c) How might my results be biased?
d) When the Environmental Protection Agency (EPA) checks a car like mine to predict its fuel economy, what parameter is it trying to estimate?

40. Accounting Between quarterly audits, a company likes to check on its accounting procedures to address any problems before they become serious. The accounting staff processes payments on about 120 orders each day. The next day, the supervisor rechecks 10 of the transactions to be sure they were processed properly.
  a) Propose a sampling strategy for the supervisor.
  b) How would you modify that strategy if the company makes both wholesale and retail sales, requiring different bookkeeping procedures?

41. Happy workers? A manufacturing company employs 14 project managers, 48 supervisors, and 377 laborers. In an effort to keep informed about any possible sources of employee discontent, management wants to conduct job satisfaction interviews with a sample of employees every month.
  a) Do you see any potential danger in the company's plan? Explain.
  b) Propose a sampling strategy that uses a simple random sample.
  c) Why do you think a simple random sample might not provide the representative opinion the company seeks?
  d) Propose a better sampling strategy.
  e) Listed below are the last names of the project managers. Use random numbers to select two people to be interviewed. Explain your method carefully.

Barrett	Bowman	Chen
DeLara	DeRoos	Grigorov
Maceli	Mulvaney	Pagliarulo
Rosica	Smithson	Tadros
Williams	Yamamoto	

42. Quality control Sammy's Salsa, a small local company, produces 20 cases of salsa a day. Each case contains 12 jars and is imprinted with a code indicating the date and batch number. To help maintain consistency, at the end of each day, Sammy selects three jars of salsa, weighs the contents, and tastes the product. Help Sammy select the sample jars. Today's cases are coded 07N61 through 07N80.
  a) Carefully explain your sampling strategy.
  b) Show how to use random numbers to pick 3 jars.
  c) Did you use a simple random sample? Explain.

43. A fish story Concerned about reports of discolored scales on fish caught downstream from a newly sited chemical plant, scientists set up a field station in a shoreline public park. For one week, they asked fishermen there to bring any fish they caught to the field station for a brief inspection. At the end of the week, the scientists said that 18% of the 234 fish that were submitted for inspection displayed the discoloration. From this information, can the researchers estimate what proportion of fish in the river have discolored scales? Explain.

44. Sampling methods Consider each of these situations. Do you think the proposed sampling method is appropriate? Explain.
  a) We want to know what percentage of local doctors accept Medicaid patients. We call the offices of 50 doctors randomly selected from local Yellow Pages listings.
  b) We want to know what percentage of local businesses anticipate hiring additional employees in the upcoming month. We randomly select a page in the Yellow Pages and call every business listed there.

45. More sampling methods Consider each of these situations. Do you think the proposed sampling method is appropriate? Explain.
  a) We want to know if there is neighborhood support to turn a vacant lot into a playground. We spend a Saturday afternoon going door-to-door in the neighborhood, asking people to sign a petition.
  b) We want to know if students at our college are satisfied with the selection of food available on campus. We go to the largest cafeteria and interview every 10th person in line.

Just Checking ANSWERS

1. a) It can be hard to reach all members of a population, and it can take so long that circumstances change, affecting the responses. A well-designed sample is often a better choice.
   b) This sample is probably biased—students who didn't like the food at the cafeteria might not choose to eat there.
   c) No, only the sample size matters, not the fraction of the overall population.
   d) Students who frequent this website might be more enthusiastic about Statistics than the overall population of Statistics students. A large sample cannot compensate for bias.
   e) It's the population "parameter." "Statistics" describe samples.

2. a) systematic
   b) stratified
   c) simple
   d) cluster

chapter 12 Experiments and Observational Studies

- 12.1 Observational Studies
- 12.2 Randomized, Comparative Experiments
- 12.3 The Four Principles of Experimental Design
- 12.4 Control Treatments
- 12.5 Blocking
- 12.6 Confounding

> **Where are we going?**
> Experiments are the "Gold Standard" of data collection. No drug comes to market without at least one FDA-approved experiment to demonstrate its safety and effectiveness. Much of what we know in science and social science comes from carefully designed experiments.
> The Four Principles of Experimental Design (Control what you can, Randomize for the rest, Replicate the trials, and, when appropriate, Block to remove identifiable variation) describe what makes a sound experiment and how to understand the results.

Who gets good grades? And, more important, why? Is there something schools and parents could do to help weaker students improve their grades? Some people think they have an answer: music! No, not your iPod, but an instrument. In a study conducted at Mission Viejo High School, in California, researchers compared the scholastic performance of music students with that of non-music students. Guess what? The music students had a much higher overall grade point average than the non-music students, 3.59 to 2.91. Not only that: A whopping 16% of the music students had all A's compared with only 5% of the non-music students.

As a result of this study and others, many parent groups and educators pressed for expanded music programs in the nation's schools. They argued that the work ethic, discipline, and feeling of accomplishment fostered by learning to play an instrument also enhance a person's ability to succeed in school. They thought that involving more students in music would raise academic performance. What do you think? Does this study provide solid evidence? Or are there other possible explanations for the difference in grades? Is there any way to really prove such a conjecture?

12.1 Observational Studies

This research tried to show an association between music education and grades. But it wasn't a survey. Nor did it assign students to get music education. Instead, it simply observed students "in the wild," recording the choices they made and the outcome. Such studies are called **observational studies**. In observational studies, researchers don't *assign* choices; they simply observe them. In addition, this was a **retrospective study**, because researchers first identified subjects who studied music and then collected data on their past grades.

What's wrong with concluding that music education causes good grades? The claim that music study *caused* higher grades depends on there being *no other differences* between the groups that could account for the differences in grades. But there are lots of lurking variables that might cause the groups to perform differently. Students who study music may have better work habits to start with, making them successful in both music and course work. Music students may have more parental support (someone had to pay for all those lessons), and that support may have enhanced their academic performance, too. Maybe they came from wealthier homes and had other advantages. Or it could be that smarter kids just like to play musical instruments.

Observational studies are valuable for discovering trends and possible relationships. They are used widely in public health and marketing. Observational studies that try to discover variables related to rare outcomes, such as specific diseases, are often retrospective. They first identify people with the disease and then look into their history and heritage in search of things that may be related to their condition. But retrospective studies have a restricted view of the world because they are usually limited to a small part of the entire population. And because retrospective records are based on historical data and memories, they can have errors. (Do you recall *exactly* what you ate even yesterday? How about last Wednesday?)

A more reliable approach is to observe individuals over time, recording the variables of interest and ultimately seeing how things turn out. For example, we might start by selecting young students who have not begun music lessons. We could then track their academic performance over several years, comparing those who later choose to study music with those who do not. If we identify subjects in advance and collect data as events unfold, we have a **prospective study**.

Although an observational study may identify important variables related to an outcome of interest, there is no guarantee that we have found the right or the most important related variables. Students who choose to study an instrument might still differ from the others in some important way that we failed to observe. It may be this difference—whether we know what it is or not—rather than music itself that leads to better grades. It's just not possible for observational studies, whether prospective or retrospective, to demonstrate a causal relationship.

> **Retrospective studies can give valuable clues** For rare illnesses, it's not practical to draw a large enough sample to see many ill respondents, so the only option remaining is to develop retrospective data. For example, researchers can interview those who have become ill. The likely causes of both Legionnaires' disease and HIV were identified from such retrospective studies of the small populations who were first infected. But to confirm the causes, researchers needed laboratory-based experiments.

For Example   DESIGNING AN OBSERVATIONAL STUDY

In early 2007, a larger-than-usual number of cats and dogs developed kidney failure; many died. Initially, researchers didn't know why, so they used an observational study to investigate.

QUESTION: Suppose you were called on to plan a study seeking the cause of this problem. Would your design be retrospective or prospective? Explain why.

ANSWER: I would use a retrospective observational study. Even though the incidence of disease was higher than usual, it was still rare. Surveying all pets would have been impractical. Instead, it makes sense to locate some who were sick and ask about their diets, exposure to toxins, and other possible causes.

12.2 Randomized, Comparative Experiments

> "He that leaves nothing to chance will do few things ill, but he will do very few things."
>
> —Lord Halifax (1633–1695)

Is it *ever* possible to get convincing evidence of a cause-and-effect relationship? Well, yes it is, but we have to take a different approach. We could take a group of third graders, randomly assign half to take music lessons, and forbid the other half to do so. Then we could compare their grades several years later. This kind of study design is an experiment.

An experiment requires a **random assignment** of subjects to treatments. Only an experiment can justify a claim like "Music lessons cause higher grades." Questions such as "Does taking vitamin C reduce the chance of getting a cold?" and "Does working with computers improve performance in Statistics class?" and "Is this drug a safe and effective treatment for that disease?" require a designed experiment to establish cause and effect.

Experimental design was advanced in the 19th century by work in psychophysics by Gustav Fechner (1801–1887), the founder of experimental psychology. Fechner designed ingenious experiments that exhibited many of the features of modern designed experiments. Fechner was careful to control for the effects of factors that might affect his results. For example, in his 1860 book *Elemente der Psychophysik* he cautioned readers to group experiment trials together to minimize the possible effects of time of day and fatigue.

Experiments study the relationship between two or more variables. An experimenter must identify at least one explanatory variable, called a **factor**, to manipulate and at least one **response variable** to measure. What distinguishes an **experiment** from other types of investigation is that the experimenter actively and deliberately manipulates the factors to control the details of the possible treatments, and assigns the subjects to those treatments *at random*. The experimenter then observes the response variable and *compares* responses for different groups of subjects who have been treated differently. For example, we might design an experiment to see whether the amount of sleep and exercise you get affects your grades.

The individuals on whom or which we experiment are known by a variety of terms. When they are human, we usually call them **subjects** or **participants**. Other individuals (rats, days, petri dishes of bacteria) are commonly referred to by the more generic term **experimental unit**. When we recruit subjects for our sleep deprivation experiment, we'll probably have better luck inviting participants than advertising for experimental units.

The specific values that the experimenter chooses for a factor are called the **levels** of the factor. We might assign our participants to sleep for 4, 6, or 8 hours. Often there are several factors at a variety of levels. (Our participants will also be assigned to a treadmill for 0 or 30 minutes.) The combination of specific levels from all the factors that an experimental unit receives is known as its **treatment**. (Our participants could have any one of six different treatments—three sleep levels, each at two exercise levels.)

How should we assign our participants to these treatments? Some students prefer 4 hours of sleep, while others need 8. Some exercise regularly; others are couch potatoes. Should we let the students choose the treatments they'd prefer? No. That would not be a good idea. Nor should we assign them according to what we feel might be best. To have any hope of drawing a fair conclusion, we must assign our participants to their treatments *at random*.

It may be obvious to you that we shouldn't let the students choose the treatment they'd prefer, but the need for random assignment is a lesson that was once hard for some to accept. For example, physicians naturally prefer to assign patients to the therapy that they think best rather than have a random element such as a coin flip determine the treatment.[1] But we've known for more than a century that for the results of an experiment to be valid, we must use deliberate randomization.

> **IN SUMMARY—AN EXPERIMENT**
>
> *Manipulates* the factor levels to create treatments. *Randomly assigns* subjects to these treatment levels. *Compares* the responses of the subject groups across treatment levels.

> **The Women's Health Initiative** In 2010, the National Center for Biotechnology Information released data from the Women's Health Initiative, a major 15-year research program funded by the National Institutes of Health to address the most common causes of death, disability, and poor quality of life in older women.[2] It consists of both an observational study with more than 93,000 participants and several randomized comparative experiments. The goals of this study included
>
> - giving reliable estimates of the extent to which known risk factors predict heart disease, cancers, and fractures;
> - identifying "new" risk factors for these and other diseases in women;
> - comparing risk factors, presence of disease at the start of the study, and new occurrences of disease during the study across all study components; and
> - creating a future resource to identify biological indicators of disease, especially substances and factors found in blood.
>
> That is, the study sought to identify possible risk factors and assess how serious they might be and to build up data that might be checked retrospectively as the women in the study continue to be followed. There would be no way to find out these things with an experiment because the task includes identifying new risk factors. If we don't know those risk factors, we could never control them as factors in an experiment.
>
> By contrast, one of the clinical trials (randomized experiments) that received much press attention randomly assigned postmenopausal women to take either hormone replacement therapy or an inactive pill. The results concluded that hormone replacement with estrogen carried increased risks of stroke.

---

[1] If we (or the physician) knew which was the better treatment, we wouldn't be performing the experiment.
[2] www.nhlbi.nih.gov/whi/background.htm

> **The FDA** No drug can be sold in the United States without first showing, in a suitably designed experiment approved by the Food and Drug Administration (FDA), that it's safe and effective. The small print on the booklet that comes with many prescription drugs usually describes the outcomes of that experiment.

For Example DETERMINING THE TREATMENTS AND RESPONSE VARIABLE

RECAP: In 2007, deaths of a large number of pet dogs and cats were ultimately traced to contamination of some brands of pet food. The manufacturer now claims that the food is safe, but before it can be released, it must be tested.

QUESTION: In an experiment to test whether the food is now safe for dogs to eat,[3] what would be the treatments and what would be the response variable?

ANSWER: The treatments would be ordinary-size portions of two dog foods: the new one from the company (the *test food*) and one that I was certain was safe (perhaps prepared in my kitchen or laboratory). The response would be a veterinarian's assessment of the health of the test animals.

12.3 The Four Principles of Experimental Design

> **Video: An Industrial Experiment.** Manufacturers often use designed experiments to help them perfect new products. Watch this video about one such experiment.

1. **Control.** We control sources of variation other than the factors we are testing by making conditions as similar as possible for all treatment groups. For human subjects, we try to treat them alike. However, there is always a question of degree and practicality. Controlling extraneous sources of variation reduces the variability of the responses, making it easier to detect differences among the treatment groups.

   Of course, when a controlled factor has only particular levels it is hard to generalize to levels that haven't been tested. For example, if we test two laundry detergents and carefully control the water temperature at 180°F, we would reduce the variation in our results due to water temperature, but we couldn't say much about the detergents' performance in cold water.

   Although we control both experimental factors and other sources of variation, we think of them and deal with them very differently. We control a factor by assigning subjects to different factor levels so we can see how the response changes at those different levels. But we control other sources of variation to *prevent* them from changing and affecting the response.

2. **Randomize.** As in sample surveys, randomization allows us to equalize the effects of unknown or uncontrollable sources of variation. It does not eliminate the effects of these sources, but by distributing them equally (on average) across the treatment levels, it makes comparisons among the treatments fair. Assigning experimental units to treatments at random allows us to use the powerful methods of Statistics to draw conclusions from an experiment. When we assign subjects to treatments at random, we reduce bias due to uncontrolled sources of variation.

   Experimenters often control factors that are easy, or inexpensive to control. They randomize to protect against the effects of other factors, even factors they haven't thought about. How to choose between the two strategies is best summed up in the old adage that says "control what you can, and randomize the rest."

3. **Replicate.** Drawing conclusions about the world is impossible unless we repeat, or **replicate**, our results. Two kinds of replication show up in comparative experiments. First, we should apply each treatment to a number of subjects. Only with such replication can we estimate the variability of responses. If we have not assessed the variation, the experiment is not complete. The outcome of an experiment on a single subject is an anecdote, not data.

The deep insight that experiments should use random assignment is quite an old one. It can be attributed to the American philosopher and scientist C. S. Peirce in his experiments with J. Jastrow, published in 1885.

---

[3] It may disturb you (as it does us) to think of deliberately putting dogs at risk in this experiment, but in fact that is what is done. The risk is borne by a small number of dogs so that the far larger population of dogs can be kept safe.

DILBERT © 2002 Scott Adams. Distributed by Universal Uclick. Reprinted with permission. All rights reserved

Activity: The Three Rules of Experimental Design. Watch an animated discussion of three rules of design.

Activity: Perform an Experiment. How well can you read pie charts and bar charts? Find out as you serve as the subject in your own experiment.

A second kind of **replication** shows up when the entire experiment is repeated on a different population of experimental units. We may believe that what is true of the students in Psych 101 who volunteered for the sleep experiment is true of all humans, but we'll feel more confident if our results for the experiment are *replicated* in another part of the country, with people of different ages, and at different times of the year. Replication of an entire experiment with the controlled sources of variation at different levels is an essential step in science.

4. **Block.** Some of the variation in our data may be due to differences among our participants that we can allow for in our experiment design. For example, suppose the participants available for a study of balance include two members of the varsity girls gymnastics team and 10 other students with no gymnastics experience. Randomizing may place both gymnasts in the same treatment group. In the long run, if we could perform the experiment over and over, it would all equalize. But wouldn't it be better to assign one gymnast to each group (at random) and five of the other students to each group (at random)? By doing this, we would improve fairness in the short run. This approach recognizes the variation due to practice and experience and allocates the participants at random *within* each experience level. When we do this, the variable "Experience" is a blocking variable, and the levels of "Experience" are called blocks.

Sometimes, attributes of the experimental units that we are not studying and that we can't control may nevertheless affect the outcomes of an experiment. If we group similar individuals together and then randomize within each of these **blocks**, we can account for much of the variability due to the difference among the blocks so it won't obscure our comparison of the treatment groups. Blocking is an important compromise between randomization and control. However, unlike the first three principles, blocking is not *required* in an experimental design.

For Example CONTROL, RANDOMIZE, AND REPLICATE

RECAP: We're planning an experiment to see whether the new pet food is safe for dogs to eat. We'll feed some animals the new food and others a food known to be safe, comparing their health after a period of time.

QUESTION: In this experiment, how will you implement the principles of control, randomization, and replication?

ANSWER: I'd control the portion sizes eaten by the dogs. To reduce possible variability from factors other than the food, I'd standardize other aspects of their environments—housing the dogs in similar pens and ensuring that each got the same amount of water, exercise, play, and sleep time, for example. I might restrict the experiment to a single breed of dog and to adult dogs to further minimize variation.

To equalize traits, pre-existing conditions, and other unknown influences, I would assign dogs to the two feed treatments randomly.

I would replicate by assigning more than one dog to each treatment to allow for variability among individual dogs. If I had the time and funding, I might replicate the entire experiment using, for example, a different breed of dog.

Diagrams

An experiment is carried out over time with specific actions occurring in a specified order. A diagram of the procedure can help in thinking about experiments.[4]

The diagram emphasizes the random allocation of subjects to treatment groups, the separate treatments applied to these groups, and the ultimate comparison of results. It's best to specify the responses that will be compared. A good way to start comparing results for the treatment groups is with boxplots.

Step-by-Step Example: DESIGNING AN EXPERIMENT

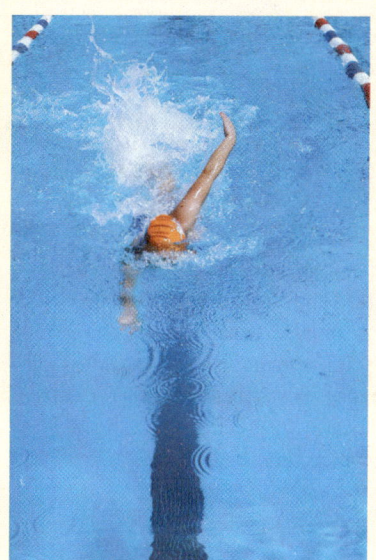

A student, preparing for a triathlon, suspected that the 45 minutes each day she spent training in a chlorinated pool was damaging her nail polish. She wished to investigate whether the color of the nail polish might make a difference. How could she design an experiment to investigate?

Let's work through the design, step by step. We'll design the simplest kind of experiment, a completely randomized experiment in one factor. The statements in the right column are the kinds of things you would need to say in proposing an experiment. You'd include them in the "methods" section of a report once the experiment is run.

> A **completely randomized experiment** is the ideal simple design, just as a *simple random sample* is the ideal simple sample—and for many of the same reasons.

Question: How would you design an experiment to test whether two nail colors resisted chipping differently when challenged by soaks in chlorine water?

THINK → Plan State what you want to know.

I want to know whether two colors of nail polish resist chipping differently when challenged by soaks in chlorine water.

Response Specify the response variable.

I can measure the percentage of the nail polish chipped away by examining scanned images of the nails with a computer program.

Treatments Specify the factor levels and the treatments.

The factor is nail polish *Color*. The factor levels are *Red* and *Nude*, two common nail polish colors.

Experimental Units Specify the experimental units.

To control the surface colored by the polish, reducing variability that might be found with natural nails, I'll obtain 30 acrylic nails and glue each one to a chopstick.

---
[4] Diagrams of this sort were introduced by David Moore in his textbooks and are still widely used.

Experiment Design Observe the principles of design:

Control any sources of variability you know of and can control.

I will use the same brand of acrylic nails and the same brand of nail polish. All nails will be painted at the same time and soaked at the same time in the same chlorine solution for the same amount of time.

Replicate results by placing more than one nail in each treatment group.

I'll use 15 nails in each group.

Randomly assign experimental units to treatments to equalize the effects of unknown or uncontrollable sources of variation.

I will randomly divide the nails into two groups for painting with each color.

Describe how the randomization will be accomplished.

I will number the nails from 1 to 30. I will obtain 30 random numbers from an Internet site and sort them, carrying along the nail ID numbers. Then I'll assign the first 15 nails to one color and the remaining ones to the other.

Make a Picture A diagram of your design can help you think about it clearly.

Specify any other experiment details. Give enough details so that another experimenter could exactly replicate your experiment. It's generally better to include details that might seem irrelevant than to leave out matters that could turn out to make a difference.

All nails will be soaked together for 45 minutes in a chlorine solution that matches typical swimming pool levels of chlorine. After the nails dry, I will tap each one 100 times on a computer keyboard to simulate daily stresses. I will repeat this treatment for 5 days and measure the result at the end of that time.

Specify how to measure the response.

I will scan each nail and use a digitized image of the nail to measure the fraction of the polish that has chipped off by comparing the before and after areas of nail polish using Adobe Illustrator.

SHOW ➡ Once you collect the data, you'll need to display them and compare the results for the treatment groups.

I will display the results with side-by-side boxplots to compare the treatment groups.

I will then compare the means of the groups.

TELL ➡ To answer the initial question, we ask whether the differences observed in the means of the groups are meaningful.

Because this is a randomized experiment, we can attribute significant differences to the treatments. To do this properly, we'll need methods from "statistical inference," the subject of the rest of this book.

If the difference in mean percent chipped between the groups is greater than I would expect from the variation I observe among the nails within each treatment group, I may be able to conclude that these difference can be attributed to the differences in nail polish color.

Does the Difference Make a Difference?

If the differences among the treatment groups are big enough, we'll attribute the differences to the treatments, but how can we decide whether the differences are big enough?

We wouldn't expect the group means to be identical. Even if the treatment made no difference at all there would still be some variation. We assigned the nails to treatments at random. But a different random assignment would have led to different results. Even a repeat of the *same* treatment on different randomly assigned sets of nails would lead to a different mean. The real question is whether the differences we observed between treatment groups are about as big as we might get just from the randomization alone, or whether they're bigger than that. If we decide that they're bigger, we'll attribute the differences to the treatments and say the differences are **statistically significant**.

Activity: **Graph the Data.** Do you think there's a significant difference in your perception of pie charts and bar charts? Explore the data from your plot perception experiment.

How can we decide whether the treatment means differ enough to be considered statistically significant? Later chapters will offer methods to help answer that question, but to get some intuition, think about deciding whether a coin is fair. If we flip a fair coin 100 times, we expect, *on average*, to get 50 heads. Suppose we get 54 heads out of 100. That doesn't seem very surprising. It's well within the bounds of ordinary random fluctuations. What if we'd seen 94 heads? That's clearly outside the bounds. We'd be pretty sure that the coin flips were not random. But what about 74 heads? Is that far enough from 50% to arouse our suspicions? That's the sort of question we need to ask of our experiment results.

In Statistics terminology, 94 heads would be a statistically significant difference from 50, and 54 heads would not. Whether 74 is *statistically significant* or not would depend on the chance of getting at least 74 heads in 100 flips of a fair coin and on our tolerance for believing that rare events can happen to us.

In the nails experiment, we ask whether the differences we see among the treatment groups are the kind of differences we'd expect from randomization. A good way to get a feeling for that is to look at how much our results vary among nails that get the *same* treatment. Boxplots of our results by treatment group can give us a general idea.

For example, Figure 12.1 shows two pairs of boxplots whose centers differ by exactly the same amount. In the upper set, that difference appears to be larger than we'd expect just by chance. Why? Because the variation is quite small *within* treatment groups, so the larger difference *between* the groups is unlikely to be just from the randomization. In the bottom pair, that same difference between the centers looks less impressive. There, the variation *within* each group swamps the difference *between* the two medians. We'd say the difference is statistically significant in the upper pair and not statistically significant in the lower pair.

In later chapters, we'll see statistical tests that quantify this intuition. For now, the important point is that a difference is statistically significant if we don't believe that it's likely to have occurred only by chance.

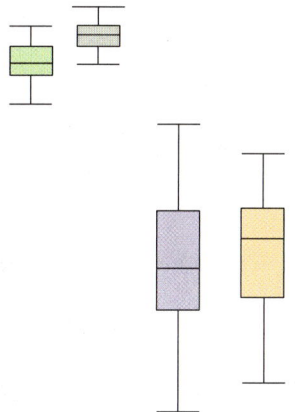

Figure 12.1
The boxplots in both pairs have centers the same distance apart, but when the spreads are large, the observed difference may be just from random fluctuation.

Just Checking

1. At one time, a method called "gastric freezing" was used to treat people with peptic ulcers. An inflatable bladder was inserted down the esophagus and into the stomach, and then a cold liquid was pumped into the bladder. Now you can find the following notice on the Internet site of a major insurance company:

    [Our company] does not cover gastric freezing (intragastric hypothermia) for chronic peptic ulcer disease. . . .

    Gastric freezing for chronic peptic ulcer disease is a non-surgical treatment which was popular about 20 years ago but now is seldom performed. It has been abandoned due to a high complication rate, only temporary improvement experienced by patients, and a lack of effectiveness when tested by double-blind, controlled clinical trials.

What did that "controlled clinical trial" (experiment) probably look like? (Don't worry about "double-blind"; we'll get to that soon.)

a) What was the factor in this experiment?

b) What was the response variable?

c) What were the treatments?

d) How did researchers decide which subjects received which treatment?

e) Were the results statistically significant?

Experiments and Samples

Both experiments and sample surveys use randomization to get unbiased data. But they do so in different ways and for different purposes. Sample surveys try to estimate population parameters, so the sample needs to be as representative of the population as possible. By contrast, experiments try to assess the effects of treatments. Experimental units are not always drawn randomly from the population. For example, a medical experiment may deal only with local patients who have the disease under study. The randomization is in the assignment of their therapy. We want a sample to exhibit the diversity and variability of the population, but for an experiment the more homogeneous the subjects the more easily we'll spot differences in the effects of the treatments.

Unless the experimental units are chosen from the population at random, you should be cautious about generalizing experiment results to larger populations until the experiment has been repeated under different circumstances. Results become more persuasive if they remain the same in completely different settings, such as in a different season, in a different country, or for a different species.

Nevertheless, experiments can draw stronger conclusions than surveys. By looking only at the differences across treatment groups, experiments cancel out many sources of bias. For example, the entire pool of subjects may be biased and not representative of the population. (College students need more sleep, on average, than the general population.) But when we assign subjects randomly to treatment groups, all the groups are biased, *in the same way*. When we consider the differences in their responses, these biases cancel out, allowing us to see the *differences* due to treatment effects more clearly.

A RANDOM SAMPLE? PROBABLY NOT

Experiments are rarely performed on random samples from a population. Don't describe the subjects in an experiment as a random sample unless they really are. More likely, the randomization was in assigning subjects to treatments.

12.4 Control Treatments

Activity: Control Groups in Experiments. Is a control group really necessary?

Suppose you wanted to test a $300 piece of software designed to shorten download times. You could just try it on several files and record the download times, but you probably want to *compare* the speed with what would happen *without* the software installed. Such a baseline measurement is called a **control treatment**, and the experimental units assigned to it are called a **control group**.

This is a use of the word "control" in an entirely different context. Previously, we controlled extraneous sources of variation by keeping them constant. Here, we use a control treatment as another *level* of the factor included so we can compare the treatment results to a situation in which "nothing happens."

Blinding

Humans are notoriously susceptible to errors in judgment.[5] All of us. When we know what treatment was assigned, it's difficult not to let that knowledge influence our assessment of the response, even when we try to be careful.

---

[5] For example, here we are in Chapter 12 and you're still reading the footnotes.

> **Blinding by misleading** Social science experiments can sometimes blind subjects by misleading them about the purpose of a study. One of the authors participated as an undergraduate in a (now infamous) psychology experiment using such a blinding method. The subjects were told that the experiment was about three-dimensional spatial perception and were assigned to draw a horse. While they were drawing, they heard a loud noise and groaning from the next room. The *real* purpose of the experiment was to see whether the social pressure of being in groups made people react to disaster differently. Subjects had been randomly assigned to draw either in groups or alone; that was the treatment. The experimenter was not interested in the drawings, but the subjects were blinded to the treatment because they were misled.

Suppose you were trying to investigate which brand of cola students really prefer. You set up an experiment to see which of the three competing brands students prefer (or whether they can tell the difference at all). But people have brand loyalties. You probably prefer one brand already. If you knew which brand you were tasting, it might influence your rating. To avoid this problem, it would be better to disguise the brands as much as possible. This strategy is called **blinding** the participants to the treatment.[6]

But it isn't just the subjects who should be blind. Experimenters themselves often subconsciously behave in ways that favor what they believe. Even technicians may treat plants or test animals differently if, for example, they expect them to die. An animal that starts doing a little better than others by showing an increased appetite may get fed a bit more than the experimental protocol specifies.

People are so good at picking up subtle cues about treatments that the best (in fact, the *only*) defense against such biases in experiments on human subjects is to keep *anyone* who could affect the outcome or the measurement of the response from knowing which subjects have been assigned to which treatments. So, not only should your cola-tasting subjects be blinded, but also *you*, as the experimenter, shouldn't know which drink is which, either—at least until you're ready to analyze the results.

There are two main classes of individuals who can affect the outcome of the experiment:

- those who could influence the results (the subjects, treatment administrators, or technicians)
- those who evaluate the results (judges, treating physicians, etc.)

When all the individuals in either one of these classes are blinded, an experiment is said to be **single-blind**. When everyone in *both* classes is blinded, we call the experiment **double-blind**. Even if several individuals in one class are blinded—for example, both the patients and the technicians who administer the treatment—the study would still be just single-blind. If only some of the individuals in a class are blind—for example, if subjects are not told of their treatment, but the administering technician is not blind—there is a substantial risk that subjects can discern their treatment from subtle cues in the technician's behavior or that the technician might inadvertently treat subjects differently. Such experiments cannot be considered truly blind.

To improve the nail polish experiment design, the experimenter could have the amount of chipping measured by an independent judge who was shown only monochrome images and didn't know how the nails had been treated. That would make the experiment single-blind. To make the experiment double-blind, we could put colored glasses on the person painting the nails, soaking them in the chlorine water, and tapping them on the keyboard so she couldn't see the colors of the nails.

For Example BLINDING

RECAP: In our experiment to see if the new pet food is now safe, we're feeding one group of dogs the new food and another group a food we know to be safe. Our response variable is the health of the animals as assessed by a veterinarian.

QUESTIONS: Should the vet be blinded? Why or why not? How would you do this? (Extra credit: Can this experiment be double-blind? Would that mean that the test animals wouldn't know what they were eating?)

ANSWER: Whenever the response variable involves judgment, it is a good idea to blind the evaluator to the treatments. The veterinarian should not be told which dogs ate which foods.

---

[6] C. S. Peirce, in the same 1885 work in which he introduced randomization, also recommended blinding.

Activity: Blinded Experiments. This narrated account of blinding isn't a placebo!

Extra credit: There is a need for double-blinding. In this case, the workers who care for and feed the animals should not be aware of which dogs are receiving which food. We'll need to make the "safe" food look as much like the "test" food as possible.

Placebos

Active placebos The placebo effect is stronger when placebo treatments are administered with authority or by a figure who appears to be an authority. "Doctors" in white coats generate a stronger effect than salespeople in polyester suits. But the placebo effect is not reduced much even when subjects know that the effect exists. People often suspect that they've gotten the placebo if nothing at all happens. So, recently, drug manufacturers have gone so far in making placebos realistic that they cause the same side effects as the drug being tested! Such "active placebos" usually induce a stronger placebo effect. When those side effects include loss of appetite or hair, the practice may raise ethical questions.

Often, simply applying *any* treatment can induce an improvement. Every parent knows the medicinal value of a kiss to make a toddler's scrape or bump stop hurting. Some of the improvement seen with a treatment—even an effective treatment—can be due simply to the act of treating. In addition, we know that regression to the mean may account for some patients' improvement. To separate the effect of the treatment from these other effects we can use a control treatment that mimics the treatment itself.

A "fake" treatment that looks just like the treatments being tested is called a **placebo**. Placebos are the best way to blind human participants from knowing whether they are receiving the treatment or not. One common version of a placebo in drug testing is a "sugar pill." The fact is that subjects treated with a placebo sometimes improve. It's not unusual for 20% or more of subjects given a placebo to report reduction in pain, improved movement, or greater alertness, or even to demonstrate improved health or performance. This **placebo effect** highlights both the importance of effective blinding and the importance of comparing treatments with a control. Placebo controls are so effective that you should use them as an essential tool for blinding whenever possible.

In summary, the best experiments are usually

- randomized.
- comparative.
- double-blind.
- placebo-controlled.

Real world example: Does ginkgo biloba improve memory?
Researchers investigated the purported memory-enhancing effect of ginkgo biloba tree extract.[7] In a randomized, comparative, double-blind, placebo-controlled study, they administered treatments to 230 elderly community members. One group received Ginkoba™ according to the manufacturer's instructions. The other received a similar-looking placebo. Thirteen different tests of memory were administered before and after treatment. The placebo group showed greater improvement on 7 of the tests, the treatment group on the other 6. None showed any significant differences. In the margin are boxplots of one measure.

"The test results are in for our new drug. 9 out of 10 doctors recommended the placebo."
Chris Wildt/CartoonStock

---
[7] P. R. Solomon, F. Adams, A. Silver, J. Zimmer, R. De Veaux, "Ginkgo for Memory Enhancement. A Randomized Controlled Trial." *JAMA* 288 [2002]: 835–840.

Just Checking

2. Recall the experiment about gastric freezing, an old method for treating peptic ulcers that you read about in the first Just Checking. Doctors would insert an inflatable bladder down the patient's esophagus and into the stomach and then pump in a cold liquid. A major insurance company now states that it doesn't cover this treatment because "double-blind, controlled clinical trials" failed to demonstrate that gastric freezing was effective.

   a) What does it mean that the experiment was double-blind?

   b) Why would you recommend a placebo control?

   c) Suppose that researchers suspected that the effectiveness of the gastric freezing treatment might depend on whether a patient had recently developed the peptic ulcer or had been suffering from the condition for a long time. How might the researchers have designed the experiment?

12.5 Blocking

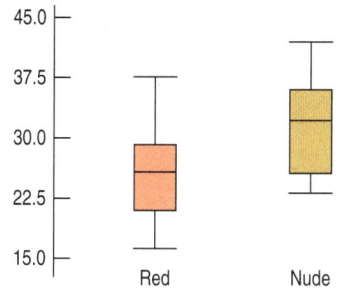

The nail polish experiment did show a difference in the durability of the two colors (see the boxplots in the margin). But the student wondered whether her artificial treatments in fact reflected the experience of swimming and dealing with the everyday stresses on her nails. So she designed a follow-up study. Her plan was to paint her own nails and record chipping during several days as she swam each day and went about her usual work schedule. In this experiment, her experimental units would be her own fingernails. One possible design would have been to paint her left hand with one color and her right hand with the other. But she is right-handed and thought that the stresses on her nails might be different between the two hands.

How could she design the experiment so that the differences in hands don't mess up her attempts to see differences in the nail polish colors? When groups of experimental units are similar, it's often a good idea to gather them together into **blocks** and then randomly assign the experimental units within each block to the treatments. By blocking, we isolate the variability attributable to the differences between the blocks, so that we can see the differences caused by the treatments more clearly.

Here, she could define each hand to be a block. She could then randomly assign half the fingers of a hand to one color and half to the other and randomly do the same for the other hand. Each treatment (color) is still assigned to the same number of fingers. The randomization occurs only within the blocks. The design is a **randomized block design**.

In effect, she can conduct a separate experiment for the fingers of each hand and then combine the results. If her right hand is harder on nail polish because she is right-handed, she can still compare the two polish colors within that hand. The picture depicts the design.

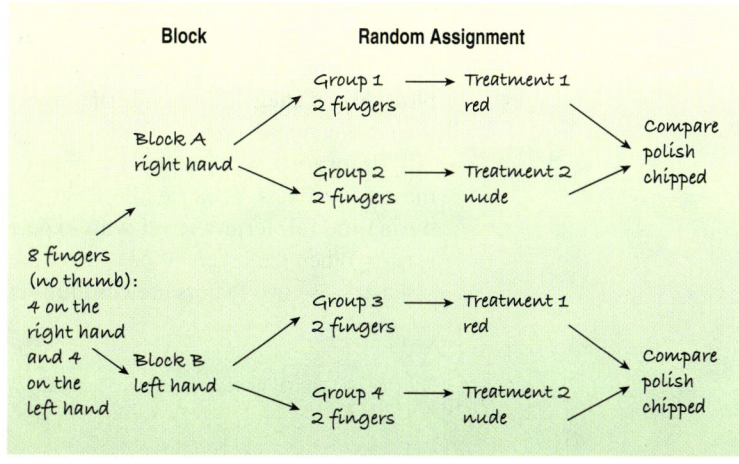

In an observational study, we sometimes pair participants that are similar in ways *not* under study. By **matching** participants in this way, we can reduce variation in much the same way as blocking. For example, a retrospective study of music education and grades might match each student who studies an instrument with someone of the same sex who is similar in family income but didn't study an instrument. We could then compare grades of music students with those of non-music students. The matching would reduce the variation due to income and sex differences.

Blocking for experiments is the same idea as stratifying is for sampling. Both methods group together subjects that are similar and randomize within those groups as a way to remove unwanted variation. (But be careful to keep the terms straight. Don't say that we "stratify" an experiment or "block" a sample.) We use blocks to reduce variability so we can see the effects of the factors; we're not usually interested in studying the effects of the blocks themselves.

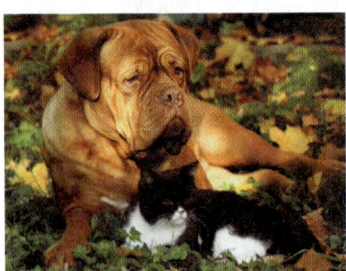

For Example BLOCKING

RECAP: In 2007, pet food contamination put cats at risk, as well as dogs. Our experiment should probably test the safety of the new food on both animals.

QUESTIONS: Why shouldn't we randomly assign a mix of cats and dogs to the two treatment groups? What would you recommend instead?

ANSWER: Dogs and cats might respond differently to the foods, and that variability could obscure my results. Blocking by species can remove that superfluous variation. I'd randomize cats to the two treatments (test food and safe food) separately from the dogs. I'd measure their responses separately and look at the results afterward.

12.6 Confounding

Professor Stephen Ceci of Cornell University performed an experiment to investigate the effect of a teacher's classroom style on student evaluations. He taught a class in developmental psychology during two successive terms to a total of 472 students in two very similar classes. He kept everything about his teaching identical (same text, same syllabus, same office hours, etc.) and modified only his style in class. During the fall term, he maintained a subdued demeanor. During the spring term, he used expansive gestures and lectured with more enthusiasm, varying his vocal pitch and using more hand gestures. He administered a standard student evaluation form at the end of each term.

The students in the fall term class rated him only an average teacher. Those in the spring term class rated him an excellent teacher, praising his knowledge and accessibility, and even the quality of the textbook. On the question "How much did you learn in the course?," the average response changed from 2.93 to 4.05 on a 5-point scale.[8]

How much of the difference he observed was due to his difference in manner, and how much might have been due to the season of the year? Fall term in Ithaca, New York (home of Cornell University), starts out colorful and pleasantly warm but ends cold and bleak. Spring term starts out bitter and snowy and ends with blooming flowers and singing birds. Might students' overall happiness have been affected by the season and reflected in their evaluations?

Unfortunately, there's no way to tell. Nothing in the data enables us to tease apart these two effects, because all the students who experienced the subdued manner did so during the fall term and all who experienced the expansive manner did so during the spring. When the levels of one factor are associated with the levels of another factor, we say that these two factors are **confounded**.

---

[8] But the two classes performed almost identically well on the final exam.

In some experiments, such as this one, it's just not possible to avoid some confounding. Professor Ceci could have randomly assigned students to one of two classes during the same term, but then we might question whether mornings or afternoons were better, or whether he really delivered the same class the second time (after practicing on the first class). Or he could have had another professor deliver the second class, but that would have raised more serious issues about differences in the two professors and concern over more serious confounding.

For Example CONFOUNDING

RECAP: After many dogs and cats suffered health problems caused by contaminated foods, we're trying to find out whether a newly formulated pet food is safe. Our experiment will feed some animals the new food and others a food known to be safe, and a veterinarian will check the response.

QUESTION: Why would it be a bad design to feed the test food to some dogs and the safe food to cats?

ANSWER: This would create confounding. We would not be able to tell whether any differences in animals' health were attributable to the food they had eaten or to differences in how the two species responded.

A TWO-FACTOR EXAMPLE

Confounding can also arise from a badly designed multifactor experiment. Here's a classic. A credit card bank wanted to test the sensitivity of the market to two factors: the annual fee charged for a card and the annual percentage rate charged. Not wanting to scrimp on sample size, the bank selected 100,000 people at random from a mailing list. It sent out 50,000 offers with a low rate and no fee and 50,000 offers with a higher rate and a $50 annual fee. Guess what happened? That's right—people preferred the low-rate, no-fee card. No surprise. In fact, they signed up for that card at over twice the rate as the other offer. And because of the large sample size, the bank was able to estimate the difference precisely. But the question the bank really wanted to answer was "how much of the change was due to the rate, and how much was due to the fee?" Unfortunately, there's simply no way to separate out the two effects. If the bank had sent out all four possible different treatments—low rate with no fee, low rate with $50 fee, high rate with no fee, and high rate with $50 fee—each to 25,000 people, it could have learned about both factors and could have also seen what happens when the two factors occur in combination.

Lurking or Confounding?

Confounding may remind you of the problem of lurking variables we discussed back in Chapters 6 and 8. Confounding variables and lurking variables are alike in that they interfere with our ability to interpret our analyses simply. Each can mislead us, but there are important differences in both how and where the confusion may arise.

A lurking variable creates an association between two other variables that tempts us to think that one may cause the other. This can happen in a regression analysis or an observational study when a lurking variable influences both the explanatory and response variables. Recall that countries with more TV sets per capita tend to have longer life expectancies. We shouldn't conclude it's the TVs "causing" longer life. We suspect instead that a generally higher standard of living may mean that people can afford more TVs and get better health care, too. Our data revealed an association between TVs and life expectancy, but economic conditions were a likely lurking variable. ==A lurking variable, then, is usually thought of as a variable associated with both y and x that makes it appear that x may be causing y.==

Confounding can arise in experiments when some other variable associated with a factor has an effect on the response variable. However, in a designed experiment, the experimenter *assigns* treatments (at random) to participants rather than just observing them. A confounding variable can't be thought of as causing that assignment. Professor Ceci's choice of teaching styles was not caused by the weather, but because he used one style in the fall and the other in the spring, he was unable to tell how much of his students' reactions were attributable to his teaching and how much to the weather. A confounding variable, then, is associated in a noncausal way with a factor and affects the response. Because of the confounding, we find that we can't tell whether any effect we see was caused by our factor or by the confounding variable—or even by both working together.

Both confounding and lurking variables are outside influences that make it harder to understand the relationship we are modeling. However, the nature of the causation is different in the two situations. In regression and observational studies, we can only observe associations between variables. Although we can't demonstrate a causal relationship, we often imagine whether *x* could cause *y*. We can be misled by a lurking variable that influences both. In a designed experiment, we often hope to show that the factor causes a response. Here we can be misled by a confounding variable that's associated with the factor and causes or contributes to the differences we observe in the response.

It's worth noting that the role of blinding in an experiment is to combat a possible source of confounding. There's a risk that knowledge about the treatments could lead the subjects or those interacting with them to behave differently or could influence judgments made by the people evaluating the responses. That means we won't know whether the treatments really do produce different results or if we're being fooled by these confounding influences.

WHAT CAN GO WRONG?

- **Don't give up just because you can't run an experiment.** Sometimes we can't run an experiment because we can't identify or control the factors. Sometimes it would simply be unethical to run the experiment. (Consider randomly assigning students to take—and be graded in—a Statistics course deliberately taught to be boring and difficult or one that had an unlimited budget to use multimedia, real-world examples, and field trips to make the subject more interesting.) If we can't perform an experiment, often an observational study is a good choice.

- **Beware of confounding.** Use randomization whenever possible to ensure that the factors not in your experiment are not confounded with your treatment levels. Be alert to confounding that cannot be avoided, and report it along with your results.

- **Bad things can happen even to good experiments.** Protect yourself by recording additional information. An experiment in which the air conditioning failed for 2 weeks, affecting the results, was saved by recording the temperature (although that was not originally one of the factors) and estimating the effect the higher temperature had on the response.[9]

  It's generally good practice to collect as much information as possible about your experimental units and the circumstances of the experiment. For example, in the nail polish experiment, it would be wise to record details (temperature, humidity) that might affect the durability of the polish on the acrylic nails. Sometimes we can use this extra information during the analysis to reduce biases.

- **Don't spend your entire budget on the first run.** Just as it's a good idea to pretest a survey, it's always wise to try a small pilot experiment before running the full-scale experiment. You may learn, for example, how to choose factor levels more effectively, about effects you forgot to control, and about unanticipated confoundings.

---

[9]R. D. De Veaux and M. Szelewski, "Optimizing Automatic Splitless Injection Parameters for Gas Chromatographic Environmental Analysis." *Journal of Chromatographic Science* 27, no. 9 (1989): 513–518.

CONNECTIONS

The fundamental role of randomization in experiments clearly points back to our discussions of randomization, to our experiments with simulations, and to our use of randomization in sampling. The similarities and differences between experiments and samples are important to keep in mind and can make each concept clearer.

If you think that blocking in an experiment resembles stratifying in a sample, you're quite right. Both are ways of removing variation we can identify to help us see past the variation in the data.

Experiments compare groups of subjects that have been treated differently. Graphics such as boxplots that help us compare groups are closely related to these ideas. Think about what we look for in a boxplot to tell whether two groups look really different, and you'll be thinking about the same issues as experiment designers.

Generally, we're going to consider how different the mean responses are for different treatment groups. And we're going to judge whether those differences are large by using standard deviations as rulers. (That's why we needed to replicate results for each treatment; we need to be able to estimate those standard deviations.) The discussion in Chapter 5 introduced this fundamental statistical thought, and it's going to keep coming back over and over again. Statistics is about variation.

We'll see a number of ways to analyze results from experiments in subsequent chapters.

What Have We Learned?

Learning Objectives

Recognize observational studies.
- A *retrospective* study looks at an outcome in the present and looks for facts in the past that relate to it.
- A *prospective* study selects subjects and follows them as events unfold.

Know the elements of a designed randomized experiment.
- *Experimental units* (sometimes called *subjects* or *participants*) are assigned at random to *treatments*.
  - The experimenter manipulates *factors*, setting them to specified *levels* to establish the treatments.
- A quantitative *response variable* is measured or observed for each experimental unit.
- We can attribute differences in the response to the differences among the treatments.

State and apply the Four Principles of Experimental Design.
- *Control* sources of variation other than the factors being tested. Make the conditions as similar as possible for all treatment groups except for differences among the treatments.
- *Randomize* the assignment of participants to treatments.
- *Replicate* by applying each treatment to more than one participant.
- *Block* the experiment by grouping together participants who are similar in important ways that you cannot control.

Work with *blinding* and *control groups*.
- A *single-blind* study is one in which *either* all those who can affect the results or all those who evaluate the results are kept ignorant of which subjects receive which treatments.
- A *double-blind* study is one in which *both* all who can affect the results *and* all who evaluate the results are ignorant of which subjects receive which treatments.
- A *control group* is assigned to a null treatment or to the best available alternative treatment.
  - Control participants are often administered a *placebo* or null treatment that mimics the treatment being studied but is known to be inactive. This is one way to blind participants.

- Understand the differences between experiments and surveys.
  - Surveys try to estimate facts (parameter) about a population, so they require a representative random sample from that population.
  - Experiments try to estimate the differences in the effects of treatments. They randomize a group of experimental units to treatments, but there is no need for the experimental units to be a representative sample from the population.
- Be alert for possible confounding due to a variable that is not under control affecting the responses differentially.

Review of Terms

Observational study — A study based on data in which no manipulation of factors has been employed (p. 318).

Retrospective study — An observational study in which subjects are selected and then their previous conditions or behaviors are determined. Retrospective studies need not be based on random samples and they usually focus on estimating differences between groups or associations between variables (p. 318).

Prospective study — An observational study in which subjects are followed to observe future outcomes. Because no treatments are deliberately applied, a prospective study is not an experiment. Nevertheless, prospective studies typically focus on estimating differences among groups that might appear as the groups are followed during the course of the study (p. 319).

Random assignment — To be valid, an experiment must assign experimental units to treatment groups using some form of randomization (p. 319).

Factor — A variable whose levels are manipulated by the experimenter. Experiments attempt to discover the effects that differences in factor levels may have on the responses of the experimental units (p. 320).

Response variable — A variable whose values are compared across different treatments. In a randomized experiment, large response differences can be attributed to the effect of differences in treatment level (p. 320).

Experiment — An experiment *manipulates* factor levels to create treatments, *randomly assigns* subjects to these treatment levels, and then *compares* the responses of the subject groups across treatment levels (p. 320).

Subjects or participants — The individuals who participate in an experiment, especially when they are human. A more general term is experimental unit (p. 320).

Experimental units — Individuals on whom an experiment is performed. Usually called **subjects** or **participants** when they are human (p. 320).

Levels — The specific values that the experimenter chooses for a factor (p. 320).

Treatment — The process, intervention, or other controlled circumstance applied to randomly assigned experimental units. Treatments are the different levels of a single factor or are made up of combinations of levels of two or more factors (p. 320).

Principles of experimental design
- **Control** aspects of the experiment that we know may have an effect on the response, but that are not the factors being studied (p. 321).
- **Randomize** subjects to treatments to even out effects that we cannot control (p. 321).
- **Replicate** over as many subjects as possible. Results for a single subject are just anecdotes. If, as often happens, the subjects of the experiment are not a representative sample from the population of interest, replicate the entire study with a different group of subjects, preferably from a different part of the population (p. 321, 322).
- **Block** to reduce the effects of identifiable attributes of the subjects that cannot be controlled (p. 322).

Completely randomized design	In a **completely randomized design**, all experimental units have an equal chance of receiving any treatment (p. 323).
Statistically significant	When an observed difference is too large for us to believe that it is likely to have occurred naturally, we consider the difference to be statistically significant. Subsequent chapters will show specific calculations and give rules, but the principle remains the same (p. 325).
Control group	The experimental units assigned to a baseline treatment level (called the "control treatment") typically either the default treatment, which is well understood, or a null, placebo treatment. Their responses provide a basis for comparison (p. 326).
Blinding	Any individual associated with an experiment who is not aware of how subjects have been allocated to treatment groups is said to be blinded (p. 327).
Single-blind and double-blind	There are two main classes of individuals who can affect the outcome of an experiment: ■ Those who could *influence the results* (the subjects, treatment administrators, or technicians). ■ Those who *evaluate the results* (judges, treating physicians, etc.). When every individual in *either* of these classes is blinded, an experiment is said to be single-blind (p. 327). When everyone in *both* classes is blinded, we call the experiment double-blind (p. 327).
Placebo	A treatment known to have no effect, administered so that all groups experience the same conditions. Many subjects respond to such a treatment (a response known as a placebo effect). Only by comparing with a placebo can we be sure that the observed effect of a treatment is not due simply to the placebo effect (p. 328).
Placebo effect	The tendency of many human subjects (often 20% or more of experiment subjects) to show a response even when administered a placebo (p. 328).
Block	When groups of experimental units are similar in a way that is not a factor under study, it is often a good idea to gather them together into blocks and then randomize the assignment of treatments within each block. By blocking in this way, we isolate the variability attributable to the differences between the blocks so that we can see the differences caused by the treatments more clearly (p. 329).
Randomized block design	An experiment design in which participants are randomly assigned to treatments within each block (p. 329).
Matching	In a retrospective or prospective study, participants who are similar in ways not under study may be matched and then compared with each other on the variables of interest. Matching, like blocking, reduces unwanted variation (p. 330).
Confounding	When the levels of one factor are associated with the levels of another factor in such a way that their effects cannot be separated, we say that these two factors are confounded (p. 330).

On the Computer EXPERIMENTS

Most experiments are analyzed with a statistics package. You should almost always display the results of a comparative experiment with side-by-side boxplots. You may also want to display the means and standard deviations of the treatment groups in a table.

The analyses offered by statistics packages for comparative randomized experiments fall under the general heading of Analysis of Variance, usually abbreviated ANOVA.

Exercises

Section 12.1

1. **Steroids** The 1990s and early 2000s could be considered the steroids era in Major League Baseball, as many players have admitted to using the drug to increase performance on the field. If a sports writer wanted to compare home run totals from the steroids era to an earlier decade, say the 1960s, explain why this would be an observational study. Could the writer conclude that it was the steroids that caused the increase in home runs? Why or why not?

2. **E-commerce** A business student conjectures that the Internet caused companies to become more profitable, since many transactions previously handled "face-to-face" could now be completed online. The student compares earnings from a sample of companies from the 1980s to a sample from the 2000s. Explain why this is an observational study. If indeed profitability increased, can she conclude the Internet was the cause? Why or why not?

Section 12.2

3. **Tips** A pizza delivery driver, always trying to increase tips, runs an experiment on his next 40 deliveries. He flips a coin to decide whether or not to call a customer from his mobile phone when he is five minutes away, hoping this slight bump in customer service will lead to a slight bump in tips. After 40 deliveries, he will compare the average tip percentage between the customers he called and those he did not. What are the experimental units and how did he randomize treatments?

4. **Tomatoes** You want to compare the tastiness and juiciness of tomatoes grown with three amounts of a new fertilizer: none, half the recommended amount, and the full recommended amount. You allocate 6 tomato plants to receive each amount of fertilizer, assigning them at random. What are the experimental units? What is the response variable?

5. **Tips II** For the experiment described in Exercise 3, list the factor, the levels, and the response variable.

6. **Tomatoes II** For the experiment described in Exercise 4, name the factor and its levels. How might the response be measured?

Section 12.3

7. **Tips again** For the experiment of Exercise 3, name some variables the driver did or should have controlled. Was the experiment randomized and replicated?

8. **Tomatoes again** For the experiment of Exercise 4, discuss variables that could be controlled or that could not be controlled. Is the experiment randomized and replicated?

Section 12.4

9. **More tips** Is the experiment of Exercise 3 blinded? Can it be double-blinded? Explain.

10. **More tomatoes** If the tomato taster doesn't know how the tomatoes have been treated, is the experiment single or double-blind? How might the blinding be improved further?

Section 12.5

11. **Block that tip** The driver of Exercise 3 wants to know about tipping in general. So he recruits several other drivers to participate in the experiment. Each driver randomly decides whether to phone customers before delivery and records the tip percentage. Is this experiment blocked? Is that a good idea?

12. **Blocking tomatoes** To obtain enough plants for the tomato experiment of Exercise 4, experimenters have to purchase plants from two different garden centers. They then randomly assign the plants from each garden center to all three fertilizer treatments. Is the experiment blocked? Is that a good idea?

Section 12.6

13. **Confounded tips** For the experiment of Exercise 3, name some confounding variables that might influence the experiment's results?

14. **Tomatoes finis** What factors might confound the results of the experiment in Exercise 4?

Chapter Exercises

15. **Standardized test scores** For his Statistics class experiment, researcher J. Gilbert decided to study how parents' income affects children's performance on standardized tests like the SAT. He proposed to collect information from a random sample of test takers and examine the relationship between parental income and SAT score.
    a) Is this an experiment? If not, what kind of study is it?
    b) If there is a relationship between parental income and SAT score, why can't we conclude that differences in score are caused by differences in parental income?

16. **Heart attacks and height** Researchers who examined health records of thousands of males found that men who died of myocardial infarction (heart attack) tended to be shorter than men who did not.
    a) Is this an experiment? If not, what kind of study is it?
    b) Is it correct to conclude that shorter men are at higher risk of dying from a heart attack? Explain.

17. **MS and vitamin D** Multiple sclerosis (MS) is an autoimmune disease that strikes more often the farther people live

from the equator. Could vitamin D—which most people get from the sun's ultraviolet rays—be a factor? Researchers compared vitamin D levels in blood samples from 150 U.S. military personnel who have developed MS with blood samples of nearly 300 who have not. The samples were taken, on average, five years before the disease was diagnosed. Those with the highest blood vitamin D levels had a 62% lower risk of MS than those with the lowest levels. (The link was only in whites, not in blacks or Hispanics.)

a) What kind of study was this?
b) Is that an appropriate choice for investigating this problem? Explain.
c) Who were the subjects?
d) What were the variables?

18. **Super Bowl commercials** When spending large amounts to purchase advertising time, companies want to know what audience they'll reach. In January 2011, a poll asked 1008 American adults whether they planned to watch the upcoming Super Bowl. Men and women were asked separately whether they were looking forward more to the football game or to watching the commercials. Among the men, 16% were planning to watch and were looking forward primarily to the commercials. Among women, 30% were looking forward primarily to the commercials.

a) Was this a stratified sample or a blocked experiment? Explain.
b) Was the design of the study appropriate for the advertisers' questions?

19. **Menopause** Researchers studied the herb black cohosh as a treatment for hot flashes caused by menopause. They randomly assigned 351 women aged 45 to 55 who reported at least two hot flashes a day to one of five groups: (1) black cohosh, (2) a multiherb supplement with black cohosh, (3) the multiherb supplement plus advice to consume more soy foods, (4) estrogen replacement therapy, or (5) a placebo. After a year, only the women given estrogen replacement therapy had symptoms different from those of the placebo group. (Source: *Annals of Internal Medicine* 145:12, 869–897, 2006)

a) What kind of study was this?
b) Is that an appropriate choice for this problem?
c) Who were the subjects?
d) Identify the treatment and response variables.

20. **Honesty** Coffee stations in offices often just ask users to leave money in a tray to pay for their coffee, but many people cheat. Researchers at Newcastle University replaced the picture of flowers on the wall behind the coffee station with a picture of staring eyes. They found that the average contribution increased significantly above the well-established standard when people felt they were being watched, even though the eyes were patently not real. (Source: *The New York Times* 12/10/06)

a) Was this a survey, an observational study, or an experiment? How can we tell?
b) Identify the variables.
c) What does "increased significantly" mean in a statistical sense?

21–34. What's the design? *Read each brief report of statistical research, and identify*

a) whether it was an observational study or an experiment.

If it was an observational study, identify (if possible)

b) whether it was retrospective or prospective.
c) the subjects studied and how they were selected.
d) the parameter of interest.
e) the nature and scope of the conclusion the study can reach.

If it was an experiment, identify (if possible)

b) the subjects studied.
c) the factor(s) in the experiment and the number of levels for each.
d) the number of treatments.
e) the response variable measured.
f) the design (completely randomized, blocked, or matched).
g) whether it was blind (or double-blind).
h) the nature and scope of the conclusion the experiment can reach.

21. Over a 4-month period, among 30 people with bipolar disorder, patients who were given a high dose (10 g/day) of omega-3 fats from fish oil improved more than those given a placebo. (Source: *Archives of General Psychiatry* 56 [1999]: 407)

22. Among a group of disabled women aged 65 and older who were tracked for several years, those who had a vitamin B_{12} deficiency were twice as likely to suffer severe depression as those who did not. (Source: *American Journal of Psychiatry* 157 [2000]: 715)

23. In a test of roughly 200 men and women, those with moderately high blood pressure (averaging 164/89 mm Hg) did worse on tests of memory and reaction time than those with normal blood pressure. (Source: *Hypertension* 36 [2000]: 1079)

24. Is diet or exercise effective in combating insomnia? Some believe that cutting out desserts can help alleviate the problem, while others recommend exercise. Forty volunteers suffering from insomnia agreed to participate in a month-long test. Half were randomly assigned to a special no-desserts diet; the others continued desserts as usual. Half of the people in each of these groups were randomly assigned to an exercise program, while the others did not exercise. Those who ate no desserts and engaged in exercise showed the most improvement.

25. After menopause, some women take supplemental estrogen. There is some concern that if these women also drink alcohol, their estrogen levels will rise too high. Twelve volunteers who were receiving supplemental estrogen were randomly divided into two groups, as were

12 other volunteers not on estrogen. In each case, one group drank an alcoholic beverage, the other a nonalcoholic beverage. An hour later, everyone's estrogen level was checked. Only those on supplemental estrogen who drank alcohol showed a marked increase.

26. Researchers have linked an increase in the incidence of breast cancer in Italy to dioxin released by an industrial accident in 1976. The study identified 981 women who lived near the site of the accident and were under age 40 at the time. Fifteen of the women had developed breast cancer at an unusually young average age of 45. Medical records showed that they had heightened concentrations of dioxin in their blood and that each tenfold increase in dioxin level was associated with a doubling of the risk of breast cancer. (Source: *Science News*, Aug. 3, 2002)

27. In 2002, the journal *Science* reported that a study of women in Finland indicated that having sons shortened the life spans of mothers by about 34 weeks per son, but that daughters helped to lengthen the mothers' lives. The data came from church records from the period 1640 to 1870.

28. Scientists at a major pharmaceutical firm investigated the effectiveness of an herbal compound to treat the common cold. They exposed each subject to a cold virus, then gave him or her either the herbal compound or a sugar solution known to have no effect on colds. Several days later, they assessed the patient's condition, using a cold severity scale ranging from 0 to 5. They found no evidence of benefits associated with the compound.

29. The May 4, 2000, issue of *Science News* reported that, contrary to popular belief, depressed individuals cry no more often in response to sad situations than nondepressed people. Researchers studied 23 men and 48 women with major depression and 9 men and 24 women with no depression. They showed the subjects a sad film about a boy whose father has died, noting whether or not the subjects cried. Women cried more often than men, but there were no significant differences between the depressed and nondepressed groups.

30. Some people who race greyhounds give the dogs large doses of vitamin C in the belief that the dogs will run faster. Investigators at the University of Florida tried three different diets in random order on each of five racing greyhounds. They were surprised to find that when the dogs ate high amounts of vitamin C they ran more slowly. (Source: *Science News*, July 20, 2002)

31. Some people claim they can get relief from migraine headache pain by drinking a large glass of ice water. Researchers plan to enlist several people who suffer from migraines in a test. When a participant experiences a migraine headache, he or she will take a pill that may be a standard pain reliever or a placebo. Half of each group will also drink ice water. Participants will then report the level of pain relief they experience.

32. A dog food company wants to compare a new lower-calorie food with their standard dog food to see if it's effective in helping inactive dogs maintain a healthy weight. They have found several dog owners willing to participate in the trial. The dogs have been classified as small, medium, or large breeds, and the company will supply some owners of each size of dog with one of the two foods. The owners have agreed not to feed their dogs anything else for a period of 6 months, after which the dogs' weights will be checked.

33. Athletes who had suffered hamstring injuries were randomly assigned to one of two exercise programs. Those who engaged in static stretching returned to sports activity in a mean of 15.2 days faster than those assigned to a program of agility and trunk stabilization exercises. (Source: *Journal of Orthopaedic & Sports Physical Therapy* 34 [March 2004]: 3)

34. Pew Research compared respondents to an ordinary 5-day telephone survey with respondents to a 4-month-long rigorous survey designed to generate the highest possible response rate. They were especially interested in identifying any variables for which those who responded to the ordinary survey were different from those who could be reached only by the rigorous survey.

35. Omega-3 Exercise 21 describes an experiment that showed that high doses of omega-3 fats might be of benefit to people with bipolar disorder. The experiment involved a control group of subjects who received a placebo. Why didn't the experimenters just give everyone the omega-3 fats to see if they improved?

36. Insomnia Exercise 24 describes an experiment showing that exercise helped people sleep better. The experiment involved other groups of subjects who didn't exercise. Why didn't the experimenters just have everyone exercise and see if their ability to sleep improved?

37. Omega-3, revisited Exercises 21 and 35 describe an experiment investigating a dietary approach to treating bipolar disorder. Researchers randomly assigned 30 subjects to two treatment groups, one group taking a high dose of omega-3 fats and the other a placebo.
  a) Why was it important to randomize in assigning the subjects to the two groups?
  b) What would be the advantages and disadvantages of using 100 subjects instead of 30?

38. Insomnia, again Exercises 24 and 36 describe an experiment investigating the effectiveness of exercise in combating insomnia. Researchers randomly assigned half of the 40 volunteers to an exercise program.
  a) Why was it important to randomize in deciding who would exercise?
  b) What would be the advantages and disadvantages of using 100 subjects instead of 40?

39. Omega-3, finis Exercises 21, 35, and 37 describe an experiment investigating the effectiveness of omega-3 fats in treating bipolar disorder. Suppose some of the 30 subjects were very active people who walked a lot or got vigorous exercise several times a week, while others tended to be more sedentary, working office jobs and watching a lot of TV. Why might researchers choose to block the subjects by activity level before randomly assigning them to the omega-3 and placebo groups?

40. **Insomnia, at last** Exercises 24, 36, and 38 describe an experiment investigating the effectiveness of exercise in combating insomnia. Suppose some of the 40 subjects had maintained a healthy weight, but others were quite overweight. Why might researchers choose to block the subjects by weight level before randomly assigning some of each group to the exercise program?

41. **Injuries** Exercise 33 describes an experiment that studies hamstring injuries. Describe a strategy to randomly assign injured athletes to the two exercise programs. What should be done if one of the athletes says he'd prefer the other program?

42. **Tomatoes II** Describe a strategy to randomly split 24 tomato plants into the three groups for the completely randomized single factor experiment of Exercise 4.

43. **Shoes** A running-shoe manufacturer wants to test the effect of its new sprinting shoe on 100-meter dash times. The company sponsors 5 athletes who are running the 100-meter dash in the 2012 Summer Olympic games. To test the shoe, it has all 5 runners run the 100-meter dash with a competitor's shoe and then again with their new shoe. The company uses the difference in times as the response variable.
    a) Suggest some improvements to the design.
    b) Why might the shoe manufacturer not be able to generalize the results they find to all runners?

44. **Swimsuits** A swimsuit manufacturer wants to test the speed of its newly designed suit. The company designs an experiment by having 6 randomly selected Olympic swimmers swim as fast as they can with their old swimsuit first and then swim the same event again with the new, expensive swimsuit. The company will use the difference in times as the response variable. Criticize the experiment and point out some of the problems with generalizing the results.

45. **Hamstrings** Exercise 33 discussed an experiment to see if the time it took athletes with hamstring injuries to be able to return to sports was different depending on which of two exercise programs they engaged in.
    a) Explain why it was important to assign the athletes to the two different treatments randomly.
    b) There was no control group consisting of athletes who did not participate in a special exercise program. Explain the advantage of including such a group.
    c) How might blinding have been used?
    d) One group returned to sports activity in a mean of 37.4 days ($SD = 27.6$ days) and the other in a mean of 22.2 days ($SD = 8.3$ days). Do you think this difference is statistically significant? Explain.

46. **Diet and blood pressure** An experiment showed that subjects fed the DASH diet were able to lower their blood pressure by an average of 6.7 points compared to a group fed a "control diet." All meals were prepared by dieticians.
    a) Why were the subjects randomly assigned to the diets instead of letting people pick what they wanted to eat?
    b) Why were the meals prepared by dieticians?
    c) Why did the researchers need the control group? If the DASH diet group's blood pressure was lower at the end of the experiment than at the beginning, wouldn't that prove the effectiveness of that diet?
    d) What additional information would you want to know in order to decide whether an average reduction in blood pressure of 6.7 points was statistically significant?

47. **Mozart** Will listening to a Mozart piano sonata make you smarter? In a 1995 study published in the journal *Psychological Science*, Rauscher, Shaw, and Ky reported that when students were given a spatial reasoning section of a standard IQ test, those who listened to Mozart for 10 minutes improved their scores more than those who simply sat quietly.
    a) These researchers said the differences were statistically significant. Explain what that means in context.
    b) Steele, Bass, and Crook tried to replicate the original study. In their study, also published in *Psychological Science* (1999), the subjects were 125 college students who participated in the experiment for course credit. Subjects first took the test. Then they were assigned to one of three groups: listening to a Mozart piano sonata, listening to music by Philip Glass, and sitting for 10 minutes in silence. Three days after the treatments, they were retested. Draw a diagram displaying the design of this experiment.
    c) These boxplots show the differences in score before and after treatment for the three groups. Did the Mozart group show improvement?

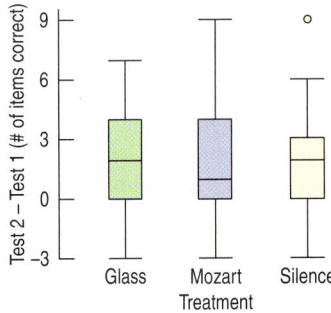

    d) Do you think the results prove that listening to Mozart is beneficial? Explain.

48. **Contrast Baths** Contrast bath treatments use the immersion of an injured limb alternately in water of two contrasting temperatures. Those who use the method claim that it can reduce swelling. Researchers compared three treatments: (1) contrast baths and exercise, (2) contrast baths alone, and (3) exercise alone. (Source: R. G. Janssen,

D. A. Schwartz, and P.F. Velleman "A Randomized Controlled Study of Contrast Baths on Patients with Carpal Tunnel Syndrome." *Journal of Hand Therapy*, 2009). They report the following boxplots comparing the change in hand volume after treatment:

a) The researchers conclude that the differences were not statistically significant. Explain what that means in context.
b) The title says that the study was randomized and controlled. Explain what that probably means for this study.
c) The study did not use a placebo treatment. What was done instead? Do you think that was an appropriate choice? Explain.

49. **Wine** A 2001 Danish study published in the *Archives of Internal Medicine* casts significant doubt on suggestions that adults who drink wine have higher levels of "good" cholesterol and fewer heart attacks. These researchers followed a group of individuals born at a Copenhagen hospital between 1959 and 1961 for 40 years. Their study found that in this group the adults who drank wine were richer and better educated than those who did not.
a) What kind of study was this?
b) It is generally true that people with high levels of education and high socioeconomic status are healthier than others. How does this call into question the supposed health benefits of wine?
c) Can studies such as these prove causation (that wine helps prevent heart attacks, that drinking wine makes one richer, that being rich helps prevent heart attacks, etc.)? Explain.

50. **Swimming** Recently, a group of adults who swim regularly for exercise were evaluated for depression. It turned out that these swimmers were less likely to be depressed than the general population. The researchers said the difference was statistically significant.
a) What does "statistically significant" mean in this context?
b) Is this an experiment or an observational study? Explain.
c) News reports claimed this study proved that swimming can prevent depression. Explain why this conclusion is not justified by the study. Include an example of a possible lurking variable.
d) But perhaps it is true. We wonder if exercise can ward off depression, and whether anaerobic exercise (like weight training) is as effective as aerobic exercise (like swimming). We find 120 volunteers not currently engaged in a regular program of exercise. Design an appropriate experiment.

51. **Dowsing** Before drilling for water, many rural homeowners hire a dowser (a person who claims to be able to sense the presence of underground water using a forked stick.) Suppose we wish to set up an experiment to test one dowser's ability. We get 20 identical containers, fill some with water, and ask him to tell which ones they are.
a) How will we randomize this procedure?
b) The dowser correctly identifies the contents of 12 out of 20 containers. Do you think this level of success is statistically significant? Explain.
c) How many correct identifications (out of 20) would the dowser have to make to convince you that the forked-stick trick works? Explain.

52. **Healing** A medical researcher suspects that giving postsurgical patients large doses of vitamin E will speed their recovery times by helping their incisions heal more quickly. Design an experiment to test this conjecture. Be sure to identify the factors, levels, treatments, response variable, and the role of randomization.

53. **Reading** Some schools teach reading using phonics (the sounds made by letters) and others using whole language (word recognition). Suppose a school district wants to know which method works better. Suggest a design for an appropriate experiment.

54. **Gas mileage** Do cars get better gas mileage with premium instead of regular unleaded gasoline? It might be possible to test some engines in a laboratory, but we'd rather use real cars and real drivers in real day-to-day driving, so we get 20 volunteers. Design the experiment.

55. **Weekend deaths** A study published in the *New England Journal of Medicine* (Aug. 2001) suggests that it's dangerous to enter a hospital on a weekend. During a 10-year period, researchers tracked over 4 million emergency admissions to hospitals in Ontario, Canada. Their findings revealed that patients admitted on weekends had a much higher risk of death than those who went on weekdays.
a) The researchers said the difference in death rates was "statistically significant." Explain in this context what that means.
b) What kind of study was this? Explain.
c) If you think you're quite ill on a Saturday, should you wait until Monday to seek medical help? Explain.
d) Suggest some possible explanations for this troubling finding.

56. **Shingles** A research doctor has discovered a new ointment that she believes will be more effective than the current medication in the treatment of shingles (a painful skin rash). Eight patients have volunteered to participate in the initial trials of this ointment. You are the statistician hired as a consultant to help design an experiment.

a) Describe how you will conduct this experiment.
b) Suppose the eight patients' last names start with the letters A to H. Using the random numbers listed below, show which patients you will assign to each treatment. Explain your randomization procedure clearly.

   41098   18329   78458   31685   55259

c) Can you make this experiment double-blind? How?
d) The initial experiment revealed that males and females may respond differently to the ointment. Further testing of the drug's effectiveness is now planned, and many patients have volunteered. What changes in your first design, if any, would you make for this second stage of testing?

57. **Beetles** Hoping to learn how to control crop damage by a certain species of beetle, a researcher plans to test two different pesticides in small plots of corn. A few days after application of the chemicals, he'll check the number of beetle larvae found on each plant. The researcher wants to know whether either pesticide works and whether there is a significant difference in effectiveness between them. Design an appropriate experiment.

58. **SAT prep** Can special study courses actually help raise SAT scores? One organization says that the 30 students they tutored achieved an average gain of 60 points when they retook the test.
a) Explain why this does not necessarily prove that the special course caused the scores to go up.
b) Propose a design for an experiment that could test the effectiveness of the tutorial course.
c) Suppose you suspect that the tutorial course might be more helpful for students whose initial scores were particularly low. How would this affect your proposed design?

59. **Safety switch** An industrial machine requires an emergency shutoff switch that must be designed so that it can be easily operated with either hand. Design an experiment to find out whether workers will be able to deactivate the machine as quickly with their left hands as with their right hands. Be sure to explain the role of randomization in your design.

60. **Washing clothes** A consumer group wants to test the effectiveness of a new "organic" laundry detergent and make recommendations to customers about how to best use the product. They intentionally stain 30 white T-shirts with grass in order to see how well the detergent will clean them. They want to try the detergent in cold water and in hot water on both the "regular" and "delicates" wash cycles. Design an appropriate experiment, indicating the number of factors, levels, and treatments. Explain the role of randomization in your experiment.

61. **Skydiving, anyone?** A humor piece published in the *British Medical Journal* ("Parachute use to prevent death and major trauma related to gravitational challenge: Systematic review of randomized control trials," Gordon, Smith, and Pell, *BMJ*, 2003:327) notes that we can't tell for sure whether parachutes are safe and effective because there has never been a properly randomized, double-blind, placebo-controlled study of parachute effectiveness in skydiving. (Yes, this is the sort of thing statisticians find funny. . . .) Suppose you were designing such a study:
a) What is the factor in this experiment?
b) What experimental units would you propose?[10]
c) What would serve as a placebo for this study?
d) What would the treatments be?
e) What would the response variable be?
f) What sources of variability would you control?
g) How would you randomize this "experiment"?
h) How would you make the experiment double-blind?

Just Checking ANSWERS

1. a) The factor was type of treatment for peptic ulcer.
   b) The response variable could be a measure of relief from gastric ulcer pain or an evaluation by a physician of the state of the disease.
   c) Treatments would be gastric freezing and some alternative control treatment.
   d) Treatments should be assigned randomly.
   e) No. The Web site reports "lack of effectiveness," indicating that no large differences in patient healing were noted.

2. a) Neither the patients who received the treatment nor the doctor who evaluated them to see if they had improved knew what treatment they had received.
   b) The placebo is needed to accomplish blinding. The best alternative would be using body-temperature liquid rather than the freezing liquid.
   c) The researchers should block the subjects by the length of time they had had the ulcer, then randomly assign subjects in each block to the freezing and placebo groups.

---
[10] Don't include your Statistics instructor!

Review of Part III

Gathering Data

Quick Review

Before you can make a boxplot, calculate a mean, describe a distribution, or fit a line, you must have meaningful data to work with. Getting good data is essential to any investigation. No amount of clever analysis can make up for badly collected data. Here's a brief summary of the key concepts and skills:

- The way you gather data depends both on what you want to discover and on what is practical.

- To get some insight into what might happen in a real situation, model it with a **simulation** using random numbers.

- To answer questions about a target population, collect information from a sample with a **survey** or poll.
  - Choose the sample randomly. Random sampling designs include simple, stratified, systematic, cluster, and multistage.
  - A simple random sample draws without restriction from the entire target population.
  - When there are subgroups within the population that may respond differently, use a stratified sample.
  - Avoid bias, a systematic distortion of the results. Sample designs that allow undercoverage or response bias and designs such as voluntary response or convenience samples don't faithfully represent the population.
  - Samples will naturally vary one from another. This sample-to-sample variation is called sampling error. Each sample only approximates the target population.

- **Observational studies** collect information from a sample drawn from a target population.
  - Retrospective studies examine existing data. Prospective studies identify subjects in advance, then follow them to collect data as the data are created, perhaps over many years.
  - Observational studies can spot associations between variables but cannot establish cause and effect. It's impossible to eliminate the possibility of lurking or confounding variables.

- To see how different treatments influence a response variable, design an **experiment**.
  - Assign subjects to treatments randomly. If you don't assign treatments randomly, your experiment is not likely to yield valid results.
  - Control known sources of variation as much as possible. Reduce variation that cannot be controlled by using blocking, if possible.
  - Replicate the experiment, assigning several subjects to each treatment level.
  - If possible, replicate the entire experiment with an entirely different collection of subjects.
  - A well-designed experiment can provide evidence that changes in the factors cause changes in the response variable.

Now for more opportunities to review these concepts and skills . . .

Review Exercises

1–18. What design? *Analyze the design of each research example reported. Is it a sample survey, an observational study, or an experiment? If a sample, what are the population, the parameter of interest, and the sampling procedure? If an observational study, was it retrospective or prospective? If an experiment, describe the factors, treatments, randomization, response variable, and any blocking, matching, or blinding that may be present. In each, what kind of conclusions can be reached?*

1. Researchers identified 242 children in the Cleveland area who had been born prematurely (at about 29 weeks). They examined these children at age 8 and again at age 20, comparing them to another group of 233 children not born prematurely. Their report, published in the *New England Journal of Medicine,* said the "preemies" engaged in significantly less risky behavior than the others. Differences showed up in the use of alcohol and marijuana, conviction of crimes, and teenage pregnancy.

2. The journal *Circulation* reported that among 1900 people who had heart attacks, those who drank an average of 19 cups of tea a week were 44% more likely than non-drinkers to survive at least 3 years after the attack.

3. Researchers at the Purina Pet Institute studied Labrador retrievers for evidence of a relationship between diet and longevity. At 8 weeks of age, 2 puppies of the same sex and weight were randomly assigned to one of two groups—a total of 48 dogs in all. One group was allowed to eat all they wanted, while the other group was fed a diet about 25% lower in calories. The median life span of dogs

fed the restricted diet was 22 months longer than that of other dogs. (Source: *Science News* 161, no. 19)

4. The radioactive gas radon, found in some homes, poses a health risk to residents. To assess the level of contamination in their area, a county health department wants to test a few homes. If the risk seems high, they will publicize the results to emphasize the need for home testing. Officials plan to use the local property tax list to randomly choose 25 homes from various areas of the county.

5. Data were collected over a decade from 1021 men and women with a recent history of precancerous colon polyps. Participants were randomly assigned to receive folic acid (a B vitamin) or a placebo, and the study concluded that those receiving the folic acid may actually increase their risk of developing additional precancerous growths. Previous studies suggested that taking folic acid may help to prevent colorectal cancer. (Source: *JAMA* 2007, 297)

6. In the journal *Science,* a research team reported that plants in southern England are flowering earlier in the spring. Records of the first flowering dates for 385 species over a period of 47 years indicate that flowering has advanced an average of 15 days per decade, an indication of climate warming, according to the authors.

7. Fireworks manufacturers face a dilemma. They must be sure that the rockets work properly, but test-firing a rocket essentially destroys it. On the other hand, not testing the product leaves open the danger that they sell a bunch of duds, leading to unhappy customers and loss of future sales. The solution, of course, is to test a few of the rockets produced each day, assuming that if those tested work properly, the others are ready for sale.

8. People who read the last page of a mystery novel first generally like stories better. Researchers recruited 819 college students to read short stories, and for one story, they were given a spoiler paragraph beforehand. On the second and third story, the spoiler was incorporated as the opening paragraph or not given at all. Overall, participants liked the stories best after first reading spoilers. (Source: *Psychological Science*, August 12, 2011)

9. Does keeping a child's lunch in an insulated bag, even with ice packs, protect the food from warming to temperatures where germs can proliferate? Researchers used an electric temperature gun on 235 lunches at preschools 90 minutes before they were to be eaten. Of the lunches with ice packs, over 90% of them were at unsafe temperatures. The study was of particular interest because preschoolers develop up to four times as many foodborne infections as do adults. (Source: *Science News*, August 9, 2011)

10. Some doctors have expressed concern that men who have vasectomies seemed more likely to develop prostate cancer. Medical researchers used a national cancer registry to identify 923 men who had had prostate cancer and 1224 men of similar ages who had not. Roughly one quarter of the men in each group had undergone a vasectomy, many more than 25 years before the study. The study's authors concluded that there is strong evidence that having the operation presents no long-term risk for developing prostate cancer. (Source: *Science News*, July 20, 2002)

11. Widely used antidepressants may reduce ominous brain plaques associated with Alzheimer's disease. In the study, mice genetically engineered to have large amounts of brain plaque were given a class of antidepressants that boost serotonin in the brain. After a single dose, the plaque levels dropped, and after four months, the mice had about half the brain plaques as the mice that didn't take the drug. (Source: *Proceedings of the National Academy of Sciences*, August 22, 2011)

12. An artisan wants to create pottery that has the appearance of age. He prepares several samples of clay with four different glazes and test fires them in a kiln at three different temperature settings.

13. Tests of gene therapy on laboratory rats have raised hopes of stopping the degeneration of tissue that characterizes chronic heart failure. Researchers at the University of California, San Diego, used hamsters with cardiac disease, randomly assigning 30 to receive the gene therapy and leaving the other 28 untreated. Five weeks after treatment the gene therapy group's heart muscles stabilized, while those of the untreated hamsters continued to weaken. (Source: *Science News*, July 27, 2002)

14. People aged 50 to 71 were initially contacted in the mid-1990s to participate in a study about smoking and bladder cancer. Data were collected from more than 280,000 men and 186,000 women from eight states who answered questions about their health, smoking history, alcohol intake, diet, physical activity, and other lifestyle factors. When the study ended in 2006, about half the bladder cancer cases in adults age 50 and older were traceable to smoking. (Source: *Journal of the American Medical Association*, August 17, 2011)

15. An orange-juice processing plant will accept a shipment of fruit only after several hundred oranges selected from various locations within the truck are carefully inspected. If too many show signs of unsuitability for juice (bruised, rotten, unripe, etc.), the whole truckload is rejected.

16. A soft-drink manufacturer must be sure the bottle caps on the soda are fully sealed and will not come off easily. Inspectors pull a few bottles off the production line at regular intervals and test the caps. If they detect any problems, they will stop the bottling process to adjust or repair the machine that caps the bottles.

17. Older Americans with a college education are significantly more likely to be emotionally well-off than are people in this age group with less education. Among those aged 65 and older, 35% scored 90 or above on the Emotional Health Index, but for those with a college degree, the percentage rose to 43% (post-graduate degree, 46%). The results are based on phone interviews conducted between January 2010 and July 2011. (Source: gallup.com, August 19, 2011)

18. Does the use of computer software in Introductory Statistics classes lead to better understanding of the concepts? A professor teaching two sections of Statistics decides to investigate. She teaches both sections using the same lectures and assignments, but gives one class statistics software to help them with their homework. The classes take the same final exam, and graders do not know which students used computers during the semester. The professor is also concerned that students who have had calculus may perform differently from those who have not, so she plans to compare software vs. no-software scores separately for these two groups of students.

19. **Point spread** When taking bets on sporting events, bookmakers often include a "point spread" that awards the weaker team extra points. In theory, this makes the outcome of the bet a toss-up. Suppose a gambler places a $10 bet and picks the winners of five games. If he's right about fewer than three of the games, he loses. If he gets three, four, or all five correct, he's paid $10, $20, and $50, respectively. Estimate the amount such a bettor might expect to lose over many weeks of gambling.

20. **The lottery** Many people spend a lot of money trying to win huge jackpots in state lotteries. Let's play a simplified version using only the numbers from 1 to 20. You bet on three numbers. The state picks five winning numbers. If your three are all among the winners, you are rich!
    a) Simulate repeated plays. How long did it take you to win?
    b) In real lotteries, there are many more choices (often 54) and you must match all five winning numbers. Explain how these changes affect your chances of hitting the jackpot.

21. **Everyday randomness** Aside from casinos, lotteries, and games, there are other situations you encounter in which something is described as "random" in some way. Give three different examples. Describe how randomness is (or is not) achieved in each.

22. **Cell phone risks** Researchers at the Washington University School of Medicine randomly placed 480 rats into one of three chambers containing radio antennas. One group was exposed to digital cell phone radio waves, the second to analog cell phone waves, and the third group to no radio waves. Two years later, the rats were examined for signs of brain tumors. In June 2002, the scientists said that differences among the three groups were not statistically significant.
    a) Is this a study or an experiment? Explain.
    b) Explain in this context what "not statistically significant" means.
    c) Comment on the fact that this research was funded by Motorola, a manufacturer of cell phones.

23. **Tips** In restaurants, servers rely on tips as a major source of income. Does serving candy after the meal produce larger tips? To find out, two waiters determined randomly whether or not to give candy to 92 dining parties. They recorded the sizes of the tips and reported that guests getting candy tipped an average of 17.8% of the bill, compared with an average tip of only 15.1% from those who got no candy. (Source: "Sweetening the Till: The Use of Candy to Increase Restaurant Tipping." *Journal of Applied Social Psychology* 32, no. 2 [2002]: 300–309)
    a) Was this an experiment or an observational study? Explain.
    b) Is it reasonable to conclude that the candy caused guests to tip more? Explain.
    c) The researchers said the difference was statistically significant. Explain in this context what that means.

24. **Tips, take 2** In another experiment to see if getting candy after a meal would induce customers to leave a bigger tip, a waitress randomly decided what to do with 80 dining parties. Some parties received no candy, some just one piece, and some two pieces. Others initially got just one piece of candy, and then the waitress suggested that they take another piece. She recorded the tips received, finding that, in general, the more candy, the higher the tip, but the highest tips (23%) came from the parties who got one piece and then were offered more. (Source: "Sweetening the Till: The Use of Candy to Increase Restaurant Tipping." *Journal of Applied Social Psychology* 32, no. 2 [2002]: 300–309)
    a) Diagram this experiment.
    b) How many factors are there? How many levels?
    c) How many treatments are there?
    d) What is the response variable?
    e) Did this experiment involve blinding? Double-blinding?
    f) In what way might the waitress, perhaps unintentionally, have biased the results?

25. **Timing** In August 2011, a Sodahead.com voluntary response poll asked site visitors, "Obama is on Vacation Again: Does He Have the Worst Timing Ever?" 56% of the 629 votes were for "Yes." During the week of the poll, a 5.8 earthquake struck near Washington, D.C., and Hurricane Irene made its way up the East coast. What types of bias may be present in the results of the poll?

26. **Laundry** An experiment to test a new laundry detergent, SparkleKleen, is being conducted by a consumer advocate group. They would like to compare its performance

with that of a laboratory standard detergent they have used in previous experiments. They can stain 16 swatches of cloth with 2 tsp of a common staining compound and then use a well-calibrated optical scanner to detect the amount of the stain left after washing. To save time in the experiment, several suggestions have been made. Comment on the possible merits and drawbacks of each one.

a) Since data for the laboratory standard detergent are already available from previous experiments, for this experiment wash all 16 swatches with SparkleKleen, and compare the results with the previous data.

b) Use both detergents with eight separate runs each, but to save time, use only a 10-second wash time with very hot water.

c) To ease bookkeeping, first run all of the standard detergent washes on eight swatches, then run all of the SparkleKleen washes on the other eight swatches.

d) Rather than run the experiment, use data from the company that produced SparkleKleen, and compare them with past data from the standard detergent.

27. When to stop? You play a game that involves rolling a die. You can roll as many times as you want, and your score is the total for all the rolls. But . . . if you roll a 6 your score is 0 and your turn is over. What might be a good strategy for a game like this?

a) One of your opponents decides to roll 4 times, then stop (hoping not to get the dreaded 6 before then). Use a simulation to estimate his average score.

b) Another opponent decides to roll until she gets at least 12 points, then stop. Use a simulation to estimate her average score.

c) Propose another strategy that you would use to play this game. Using your strategy, simulate several turns. Do you think you would beat the two opponents?

28. Rivets A company that manufactures rivets believes the shear strength of the rivets they manufacture follows a Normal model with a mean breaking strength of 950 pounds and a standard deviation of 40 pounds.

a) What percentage of rivets selected at random will break when tested under a 900-pound load?

b) You're trying to improve the rivets and want to examine some that fail. Use a simulation to estimate how many rivets you might need to test in order to find three that fail at 900 pounds (or below).

29. Homecoming A college Statistics class conducted a survey concerning community attitudes about the college's large homecoming celebration. That survey drew its sample in the following manner: Telephone numbers were generated at random by selecting one of the local telephone exchanges (first three digits) at random and then generating a random four-digit number to follow the exchange. If a person answered the phone and the call was to a residence, then that person was taken to be the subject for interview. (Undergraduate students and those under voting age were excluded, as was anyone who could not speak English.) Calls were placed until a sample of 200 eligible respondents had been reached.

a) Did every telephone number that could occur in that community have an equal chance of being generated?

b) Did this method of generating telephone numbers result in a simple random sample (SRS) of local residences? Explain.

c) Did this method generate an SRS of local voters? Explain.

d) Is this method unbiased in generating samples of households? Explain.

30. Youthful appearance *Readers' Digest* (April 2002, p. 152) reported results of several surveys that asked graduate students to examine photographs of men and women and try to guess their ages. Researchers compared these guesses with the number of times the people in the pictures reported having sexual intercourse. It turned out that those who had been more sexually active were judged as looking younger, and that the difference was described as "statistically significant." Psychologist David Weeks, who compiled the research, speculated that lovemaking boosts hormones that "reduce fatty tissue and increase lean muscle, giving a more youthful appearance."

a) What does "statistically significant" mean in this context?

b) Explain in statistical terms why you might be skeptical about Dr. Weeks's conclusion. Propose an alternative explanation for these results.

31. Smoking and Alzheimer's Medical studies indicate that smokers are less likely to develop Alzheimer's disease than people who never smoked.

a) Does this prove that smoking may offer some protection against Alzheimer's? Explain.

b) Offer an alternative explanation for this association.

c) How would you conduct a study to investigate this?

32. Antacids A researcher wants to compare the performance of three types of antacid in volunteers suffering from acid reflux disease. Because men and women may react differently to this medication, the subjects are split into two groups, by sex. Subjects in each group are randomly assigned to take one of the antacids or to take a sugar pill made to look the same. The subjects will rate their level of discomfort 30 minutes after eating.

a) What kind of design is this?

b) The experiment uses volunteers rather than a random sample of all people suffering from acid reflux disease. Does this make the results invalid? Explain.

c) How may the use of the placebo confound this experiment? Explain.

33. Sex and violence Does the content of a television program affect viewers' memory of the products advertised in commercials? Design an experiment to compare the ability of viewers to recall brand names of items featured in commercials during programs with violent content, sexual content, or neutral content.

34. **Pubs** In England, a Leeds University researcher said that the local watering hole's welcoming atmosphere helps men get rid of the stresses of modern life and is vital for their psychological well-being. Author of the report, Dr. Colin Gill, said rather than complain, women should encourage men to "pop out for a swift half." "Pub-time allows men to bond with friends and colleagues," he said. "Men need break-out time as much as women and are mentally healthier for it." Gill added that men might feel unfulfilled or empty if they had not been to the pub for a week. The report, commissioned by alcohol-free beer brand Kaliber, surveyed 900 men on their reasons for going to the pub. More than 40% said they went for the conversation, with relaxation and a friendly atmosphere being the other most common reasons. Only 1 in 10 listed alcohol as the overriding reason.

Let's examine this news story from a statistical perspective.
a) What are the W's: *Who, What, When, Where, Why, How?*
b) What population does the researcher think the study applies to?
c) What is the most important thing about the selection process that the article does *not* tell us?
d) How do *you* think the 900 respondents were selected? (Name a method of drawing a sample that is likely to have been used.)
e) Do you think the report that only 10% of respondents listed alcohol as an important reason for going to the pub might be a biased result? Why?

35. **Age and party 2008** The Pew Research Center collected data from national exits polls conducted by *NBC News* after the 2008 presidential election. The following table shows information regarding voter age and party preference:

	Republican	Democrat	Other	Total
18–29	260	390	351	1001
30–44	320	379	300	999
45–64	329	369	300	998
65+	361	392	251	1004
Total	1270	1530	1202	4002

a) What sampling strategy do you think the pollsters used? Explain.
b) What percentage of the people surveyed were Democrats?
c) Do you think this is a good estimate of the percentage of voters in the United States who are registered Democrats? Why or why not?
d) In creating this sample design, what question do you think the pollsters were trying to answer?

36. **Bias?** Political analyst Michael Barone has written that "conservatives are more likely than others to refuse to respond to polls, particularly those polls taken by media outlets that conservatives consider biased" (Source: *The Weekly Standard*, March 10, 1997). The Pew Research Foundation tested this assertion by asking the same questions in a national survey run by standard methods and in a more rigorous survey that was a true SRS with careful follow-up to encourage participation. The response rate in the "standard survey" was 42%. The response rate in the "rigorous survey" was 71%.
a) What kind of bias does Barone claim may exist in polls?
b) What is the population for these surveys?
c) On the question of political position, the Pew researchers report the following table:

	Standard Survey	Rigorous Survey
Conservative	37%	35%
Moderate	40%	41%
Liberal	19%	20%

What makes you think these results are incomplete?
d) The Pew researchers report that differences between opinions expressed on the two surveys were not statistically significant. Explain what "not statistically significant" means in this context.

37. **Save the grapes.** Vineyard owners have problems with birds that like to eat the ripening grapes. Some vineyards use scarecrows to try to keep birds away. Others use netting that covers the plants. Owners really would like to know if either method works and, if so, which one is better. One owner has offered to let you use his vineyard this year for an experiment. Propose a design. Carefully indicate how you would set up the experiment, specifying the factor(s) and response variable.

38. **Bats** It's generally believed that baseball players can hit the ball farther with aluminum bats than with the traditional wooden ones. Is that true? And, if so, how much farther? Players on your local high school baseball team have agreed to help you find out. Design an appropriate experiment.

39. **Acupuncture** Research reported in 2008 brings to light the effectiveness of treating chronic lower back pain with different methods. One-third of nearly 1200 volunteers were administered conventional treatment (drugs, physical therapy, and exercise). The remaining patients got 30-minute acupuncture sessions. Half of these patients were punctured at sites suspected of being useful and half received needles at other spots on their bodies. Comparable shares of each acupuncture group, roughly 45%, reported decreased back pain for at least six months after their sessions ended. This was almost twice

as high as those receiving the conventional therapy, leading the researchers to conclude that results were statistically significant.

a) Why did the researchers feel it was necessary to have some of the patients undergo a "fake" acupuncture?
b) Because patients had to consent to participate in this experiment, the subjects were essentially self-selected—a kind of voluntary response group. Explain why that does not invalidate the findings of the experiment.
c) What does "statistically significant" mean in the context of this experiment?

40. **NBA draft lottery** Professional basketball teams hold a "draft" each year in which they get to pick the best available college and high school players. In an effort to promote competition, teams with the worst records get to pick first, theoretically allowing them to add better players. To combat the fear that teams with no chance to make the playoffs might try to get better draft picks by intentionally losing late-season games, the NBA's Board of Governors adopted a weighted lottery system in 1990. Under this system, the 11 teams that did not make the playoffs were eligible for the lottery. The NBA prepared 66 cards, each naming one of the teams. The team with the worst win-loss record was named on 11 of the cards, the second-worst team on 10 cards, and so on, with the team having the best record among the nonplayoff clubs getting only one chance at having the first pick. The cards were mixed, then drawn randomly to determine the order in which the teams could draft players. Suppose there are two exceptional players available in this year's draft and your favorite team had the third-worst record. Use a simulation to find out how likely it is that your team gets to pick first or second. Describe your simulation carefully.

41. **Security** There are 20 first-class passengers and 120 coach passengers scheduled on a flight. In addition to the usual security screening, 10% of the passengers will be subjected to a more complete search.
a) Describe a sampling strategy to randomly select those to be searched.
b) Here is the first-class passenger list and a set of random digits. Select two passengers to be searched, carefully demonstrating your process.

65436 71127 04879 41516 20451 02227 94769 23593

Bergman	Cox	Fontana	Perl
Bowman	DeLara	Forester	Rabkin
Burkhauser	Delli-Bovi	Frongillo	Roufaiel
Castillo	Dugan	Furnas	Swafford
Clancy	Febo	LePage	Testut

c) Explain how you would use a random number table to select the coach passengers to be searched.

42. **Profiling?** Among the 20 first-class passengers on the flight described in Exercise 41, there were four businessmen from the Middle East. Two of them were the two passengers selected to be searched. They complained of profiling, but the airline claims that the selection was random. What do you think? Support your conclusion with a simulation.

43. **Par 4** In theory, a golfer playing a par-4 hole tees off, hitting the ball in the fairway, then hits an approach shot onto the green. The first putt (usually long) probably won't go in, but the second putt (usually much shorter) should. Sounds simple enough, but how many strokes might it really take? Use a simulation to estimate a pretty good golfer's score based on these assumptions:
- The tee shot hits the fairway 70% of the time.
- A first approach shot lands on the green 80% of the time from the fairway, but only 40% of the time otherwise.
- Subsequent approach shots land on the green 90% of the time.
- The first putt goes in 20% of the time, and subsequent putts go in 90% of the time.

44. **The back nine** Use simulations to estimate more golf scores, similar to the procedure in Exercise 43.
a) On a par 3, the golfer hopes the tee shot lands on the green. Assume that the tee shot behaves like the first approach shot described in Exercise 43.
b) On a par 5, the second shot will reach the green 10% of the time and hit the fairway 60% of the time. If it does not hit the green, the golfer must play an approach shot as described in Exercise 43.
c) Create a list of assumptions that describe your golfing ability, and then simulate your score on a few holes. Explain your simulation clearly.

chapter 13 From Randomness to Probability

13.1 Random Phenomena
13.2 Modeling Probability
13.3 Formal Probability

Where are we going?

Flip a coin. Can you predict the outcome? It's hard to guess the outcome of just one flip because the outcome is random. If you flip it over and over, though, you can predict the *proportion* of heads you're likely to see in the long run.

It's this long-term predictability of randomness that we'll use throughout the rest of the book. To do that, we'll need to talk about the probability of different outcomes and learn some rules for dealing with them.

Early humans saw a world filled with random events. To help them make sense of the chaos around them, they sought out seers, consulted oracles, and read tea leaves. As science developed, we learned to recognize some events as predictable. We can now forecast the change of seasons, tell when eclipses will occur precisely, and even make a reasonably good guess at how warm it will be tomorrow. But many other events are still essentially random. Will the stock market go up or down today? When will the next car pass this corner?

But we have also learned to understand randomness. The surprising fact is that in the long run, many truly random phenomena settle down in a way that's consistent and predictable. It's this property of random phenomena that makes the next steps we're about to take in Statistics possible. The previous three chapters showed that randomness plays a critical role in gathering data. That fact alone makes it important to understand how random events behave. From here on, randomness will be fundamental to how we think about data.

13.1 Random Phenomena

Every day you drive through the intersection at College and Main. Even though it may seem that the light is never green when you get there, you know this can't really be true. In fact, if you try really hard, you can recall just sailing through the green light once in a while.

What's random here? The light itself is governed by a timer. Its pattern isn't haphazard. In fact, the light may even be red at precisely the same times each day. It's the time you arrive at the light that is *random*. It isn't that your driving is erratic. Even if you try to leave your house at exactly the same time every day, whether the light is red or green as *you* reach the intersection is a random phenomenon.[1] For us, a **random phenomenon** is a situation

---

[1] If you somehow managed to leave your house at *precisely* the same time every day and there was *no* variation in the time it took you to get to the light, then there wouldn't be any randomness, but that's not very realistic.

in which we know what outcomes can possibly occur, but we don't know which particular outcome will happen. Even though the color of the light is random[2] as you approach it, some *fraction* of the time, the light will be green. How can you figure out what that fraction is?

You might record what happens at the intersection each day and graph the *accumulated percentage* of green lights like this:

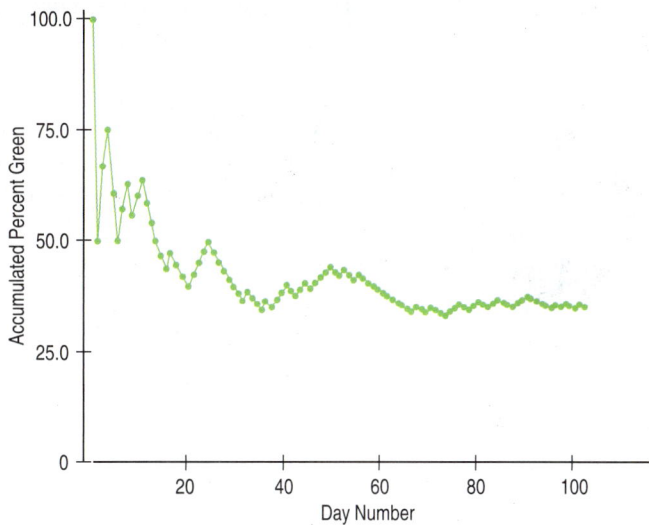

Figure 13.1
The accumulated percentage of times the light is green settles down as you see more outcomes.

Day	Light	% Green
1	Green	100
2	Red	50
3	Green	66.7
4	Green	75
5	Red	60
6	Red	50
⋮	⋮	⋮

TRIALS, OUTCOMES, AND EVENTS

A phenomenon consists of trials. Each trial has an outcome. Outcomes combine to make events.

THE SAMPLE SPACE

For a random phenomenon, the sample space, **S**, is the set of all possible outcomes of each trial.

The first day you recorded the light, it was green. Then on the next five days, it was red, then green again, then green, red, and red. When you plot the percentage of green lights against days, the graph starts at 100% (because the first time, the light was green, so 1 out of 1, for 100%). Then the next day it was red, so the accumulated percentage drops to 50% (1 out of 2). The third day it was green again (2 out of 3, or 67% green), then green (3 out of 4, or 75%), then red twice in a row (3 out of 5, for 60% green, and then 3 out of 6, for 50%), and so on. As you collect a new data value for each day, each new outcome becomes a smaller and smaller fraction of the accumulated experience, so, in the long run, the graph settles down. As it settles down, you can see that, in fact, the light is green about 35% of the time.

In general, each occasion upon which we observe a random phenomenon is called a **trial**. At each trial, we note the value of the random phenomenon, and call that the trial's **outcome**. (If this language reminds you of Chapter 10, that's *not* unintentional.)

For the traffic light, there are really three possible outcomes: red, yellow, or green. Often we're more interested in a combination of outcomes rather than in the individual ones. When you see the light turn yellow, what do *you* do? If you race through the intersection, then you treat the yellow more like a green light. If you step on the brakes, you treat it more like a red light. Either way, you might want to group the yellow with one or the other. When we combine outcomes like that, the resulting combination is an **event**.[3] We call the collection of *all possible outcomes* the **sample space**.[4] We'll denote the sample space **S**. (Some books are even fancier and use the Greek letter Ω.) For the traffic light, **S** = {red, green, yellow}. If you flip a coin once, the sample space is very simple: **S** = {H, T}. If you flip two coins, it's more complicated because now there are four outcomes, so S = {HH, HT, TH, TT}. If *ABC News* takes a sample of 1023 randomly chosen U.S. adults for a poll, the sample space is incomprehensibly enormous because it would list every combination of 1023 adults you could take from the approximately 250 million adults in the United States.

---

[2] Even though the randomness here comes from the uncertainty in our arrival time, we can think of the light itself as showing a color at random.

[3] Each individual outcome is also an event.

[4] Mathematicians like to use the term "space" as a fancy name for a set. Sort of like referring to that closet colleges call a dorm room as "living space." But it's really just the set of all outcomes.

> "For even the most stupid of men . . . is convinced that the more observations have been made, the less danger there is of wandering from one's goal."
>
> —*Jacob Bernoulli, 1713, discoverer of the LLN*

Although he could have said it much more gently, Bernoulli is trying to tell us how intuitive the LLN actually is, even though it wasn't formally proved until the 18th century. Most of us would guess that the law is true from our everyday experiences.

EMPIRICAL PROBABILITY

For any event **A**,

$$P(A) = \frac{\#\text{ times } A \text{ occurs}}{\text{total \# of trials}}$$

in the long run.

THE LAW OF AVERAGES

Don't let yourself think that there's a Law of Averages that promises short-term compensation for recent deviations from expected behavior. A belief in such a "Law" can lead to money lost in gambling and to poor business decisions.

The Law of Large Numbers

What's the *probability* of a green light at College and Main? Based on the graph, it looks like the relative frequency of green lights settles down to about 35%, so saying that the probability is about 0.35 seems like a reasonable answer. But do random phenomena always behave well enough for this to make sense? Might the relative frequency of an event bounce back and forth between two values forever, never settling on just one number?

Fortunately, a principle called the **Law of Large Numbers** (LLN) gives us the guarantee we need. The LLN says that as we repeat a random process over and over, the proportion of times that an event occurs does settle down to one number. We call this number the **probability** of the event. But the law of large numbers requires two key assumptions. First, the random phenomenon we're studying must not change—the outcomes must have the same probabilities for each trial. And, the events must be **independent**.[5] Informally, independence means that the outcome of one trial doesn't affect the outcomes of the others. (We'll see a formal definition of independent events in the next chapter.) The LLN says that as the number of independent trials increases, the long-run *relative frequency* of repeated events gets closer and closer to a single value.

Because the LLN guarantees that relative frequencies settle down in the long run, we can give a name to the value that they approach. We call it the probability of the event. If the relative frequency of green lights at that intersection settles down to 35% in the long run, we say that the probability of encountering a green light is 0.35, and we write $P(\text{green}) = 0.35$. Because this definition is based on repeatedly observing the event's outcome, this definition of probability is often called **empirical probability**.

The Nonexistent Law of Averages

Even though the LLN seems natural, it is often misunderstood because the idea of the *long run* is hard to grasp. Many people believe, for example, that an outcome of a random event that hasn't occurred in many trials is "due" to occur. Many gamblers bet on numbers that haven't been seen for a while, mistakenly believing that they're likely to come up sooner. A common term for this is the "Law of Averages." After all, we know that in the long run, the relative frequency will settle down to the probability of that outcome, so now we have some "catching up" to do, right?

Wrong. The Law of Large Numbers says nothing about short-run behavior. Relative frequencies even out *only in the long run*. And, according to the LLN, the long run is *really* long (*infinitely* long, in fact).

The so-called Law of Averages doesn't exist at all. But you'll hear people talk about it as if it does. Is a good hitter in baseball who has struck out the last six times *due* for a hit his next time up? If you've been doing particularly well in weekly quizzes in Statistics class, are you *due* for a bad grade? No. This isn't the way random phenomena work. There is *no* Law of Averages for short runs.

For Example COINS AND THE LAW OF AVERAGES

You've just flipped a fair coin and seen six heads in a row.

QUESTION: Does the coin "owe" you some tails? Suppose you spend that coin and your friend gets it in change. When she starts flipping the coin, should she expect a run of tails?

ANSWER: Of course not. Each flip is a new event. The coin can't "remember" what it did in the past, so it can't "owe" any particular outcomes in the future.

---

[5] There are stronger forms of the Law that don't require independence, but for our purposes, this form is general enough.

The law of averages in everyday life
"Dear Abby: My husband and I just had our eighth child. Another girl, and I am really one disappointed woman. I suppose I should thank God she was healthy, but, Abby, this one was supposed to have been a boy. Even the doctor told me that the law of averages was in our favor 100 to one." (Excerpt from "The Law of Averages" by Abigail Van Buren from The Nature of Mathematics, by Karl J. Smith. Published by Brooks/Cole, Cengage Learning, © 1991.)

Just to see how this works in practice, the authors ran a simulation of 100,000 flips of a fair coin. We collected 100,000 random numbers, letting the numbers 0 to 4 represent heads and the numbers 5 to 9 represent tails. In our 100,000 "flips," there were 2981 streaks of at least 5 heads. The "Law of Averages" suggests that the next flip after a run of 5 heads should be tails more often to even things out. Actually, the next flip was heads more often than tails: 1550 times to 1431 times. That's 52% heads. You can perform a similar simulation easily on a computer. Try it!

Of course, sometimes an apparent drift from what we expect means that the probabilities are, in fact, *not* what we thought. If you get 10 heads in a row, maybe the coin has heads on both sides!

The lesson of the LLN is that sequences of random events don't compensate in the *short* run and don't need to do so to get back to the right long-run probability. If the probability of an outcome doesn't change and the events are independent, the probability of any outcome in another trial is *always* what it was, no matter what has happened in other trials.

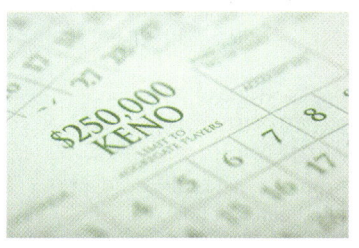

"Slump? I ain't in no slump. I just ain't hittin'."

—*Yogi Berra*

Beat the casino Keno is a simple casino game in which numbers from 1 to 80 are chosen. The numbers, as in most lottery games, are supposed to be equally likely. Payoffs are made depending on how many of those numbers you match on your card. A group of graduate students from a Statistics department decided to take a field trip to Reno. They (*very* discreetly) wrote down the outcomes of the games for a couple of days, then drove back to test whether the numbers were, in fact, equally likely. It turned out that some numbers were *more likely* to come up than others. Rather than bet on the Law of Averages and put their money on the numbers that were "due," the students put their faith in the LLN—and all their (and their friends') money on the numbers that had come up before. After they pocketed more than $50,000, they were escorted off the premises and invited never to show their faces in that casino again.

 Just Checking

1. One common proposal for beating the lottery is to note which numbers have come up lately, eliminate those from consideration, and bet on numbers that have not come up for a long time. Proponents of this method argue that in the long run, every number should be selected equally often, so those that haven't come up are due. Explain why this is faulty reasoning.

13.2 Modeling Probability

A S Activity: What Is Probability? The best way to get a feel for probabilities is to experiment with them. We'll use this random-outcomes tool many more times.

Probability was first studied extensively by a group of French mathematicians who were interested in games of chance.[6] Rather than *experiment* with the games (and risk losing their money), they developed mathematical models. When the probability comes from a mathematical model and not from observation, it is called **theoretical probability**. To make things simple (as we usually do when we build models), they started by looking at games in which the different outcomes were equally likely. Fortunately, many games of chance are like that. Any of 52 cards is equally likely to be the next one dealt from a well-shuffled deck. Each face of a die is equally likely to land up (or at least it *should be*).

---
[6] OK, gambling.

It's easy to find probabilities for events that are made up of several *equally likely* outcomes. We just count all the outcomes that the event contains. The probability of the event is the number of outcomes in the event divided by the total number of possible outcomes. We can write

$$P(A) = \frac{\#\text{ outcomes in } A}{\#\text{ of possible outcomes}}.$$

For example, the probability of drawing a face card (JQK) from a deck is

$$P(\text{face card}) = \frac{\#\text{ face cards}}{\#\text{ cards}} = \frac{12}{52} = \frac{3}{13}.$$

■ **NOTATION ALERT**

We often use capital letters—and usually from the beginning of the alphabet—to denote events. We *always* use P to denote probability. So,

$$P(A) = 0.35$$

means "the probability of the event A is 0.35."

When being formal, use decimals (or fractions) for the probability values, but sometimes, especially when talking more informally, it's easier to use percentages.

> **How hard can counting be?** Finding the probability of any event when the outcomes are equally likely is straightforward, but not necessarily easy. It gets hard when the number of outcomes in the event (and in the sample space) gets big. Think about flipping two coins. The sample space is $S = \{HH, HT, TH, TT\}$ and each outcome is equally likely. So, what's the probability of getting exactly one head and one tail? Let's call that event A. Well, there are two outcomes in the event $A = \{HT, TH\}$ out of the 4 possible equally likely ones in S, so $P(A) = \frac{2}{4}$, or $\frac{1}{2}$.
>
> OK, now flip 100 coins. What's the probability of exactly 67 heads? Well, first, how many outcomes are in the sample space? $S = \{HHHHHHHHHHH \ldots H, HH \ldots T, \ldots \}$ Um . . . . a lot. In fact, there are 1,267,650,600,228,229,401,496,703, 205,376 different outcomes possible when flipping 100 coins. And that's just the denominator of the probability! To answer the question, we'd still have to figure out how many ways there are to get 67 heads. We'll see how in Chapter 16; stay tuned!

A S *Activity:* **Multiple Discrete Outcomes.** The world isn't all heads or tails. Experiment with an event with 4 random alternative outcomes.

Don't get trapped into thinking that random events are always equally likely. The chance of winning a lottery—especially lotteries with very large payoffs—is small. Regardless, people continue to buy tickets. In an attempt to understand why, an interviewer asked someone who had just purchased a lottery ticket, "What do you think your chances are of winning the lottery?" The reply was, "Oh, about 50–50." The shocked interviewer asked, "How do you get that?" to which the response was, "Well, the way I figure it, either I win or I don't!"

The moral of this story is that events are *not* always equally likely.

Personal Probability

What's the probability that your grade in this Statistics course will be an A? You may be able to come up with a number that seems reasonable. Of course, no matter how confident or depressed you feel about your chance of success, your probability should be between 0 and 1. How did you come up with this probability? It can't be an empirical probability. For that, you'd have to take the course over and over (and over . . .), and forget everything after each time so the probability of getting an A would stay the same. But people use probability in a third sense as well.

We use the language of probability in everyday speech to express a degree of uncertainty *without* basing it on long-run relative frequencies or mathematical models. Your personal assessment of your chances of getting an A expresses your uncertainty about the outcome. That uncertainty may be based on how comfortable you're feeling in the course or on your midterm grade, but it can't be based on long-run behavior. We call this third kind of probability a **subjective** or **personal probability**.

Although personal probabilities may be based on experience, they're not based either on long-run relative frequencies or on equally likely events. So they don't display the kind of consistency that we'll need probabilities to have. For that reason, we'll stick to formally defined probabilities. You should be alert to the difference.

> **Which kind of probability?** The line between personal probability and the other two probabilities can be a fuzzy one. When a weather forecaster predicts a 40% probability of rain, is this a personal probability or a relative frequency probability? The claim may be that 40% of the time, when the map looks like this, it has rained (over some period of time). Or the forecaster may be stating a personal opinion that is based on years of experience and reflects a sense of what has happened in the past in similar situations. When you hear a probability stated, try to ascertain what kind of probability is intended.

The First Three Rules for Working with Probability

1. Make a picture.
2. Make a picture.
3. Make a picture.

We're dealing with probabilities now, not data, but the three rules don't change. The most common kind of picture to make is called a Venn diagram. We'll use Venn diagrams throughout the rest of this chapter. Even experienced statisticians make Venn diagrams to help them think about probabilities of compound and overlapping events. You should, too.

John Venn (1834–1923) created the Venn diagram. His book on probability, *The Logic of Chance*, was "strikingly original and considerably influenced the development of the theory of Statistics," according to John Maynard Keynes, one of the luminaries of Economics.

13.3 Formal Probability

For some people, the phrase "50/50" means something vague like "I don't know" or "whatever." But when we discuss probabilities of outcomes, it takes on the precise meaning of *equally likely*. Speaking vaguely about probabilities will get us into trouble, so whenever we talk about probabilities, we'll need to be precise.[7] And to do that, we'll need to develop some formal rules[8] about how probability works.

Rule 1. If the probability is 0, the event never occurs, and likewise if it has probability 1, it *always* occurs. Even if you think an event is very unlikely, its probability can't be negative, and even if you're sure it will happen, its probability can't be greater than 1. (Think about relative frequencies.) So we require that

> **A probability is a number between 0 and 1.**
> **For any event A, $0 \leq P(A) \leq 1$.**

> **Surprising probabilities** We've been careful to discuss probabilities only for situations in which the outcomes were finite, or even countably infinite. But if the outcomes can take on *any* numerical value at all (we say they are *continuous*), things can get surprising. For example, what is the probability that a randomly selected child will be *exactly* 3 feet tall? Well, if we mean 3.00000 . . . feet, the answer is zero. No randomly selected child—even one whose height would be recorded as 3 feet, will be *exactly* 3 feet tall (to an infinite number of decimal places). But, if you've grown taller than 3 feet, there must have been a time in your life when you actually *were* exactly 3 feet tall, even if only for a second. So this is an outcome with probability 0 that not only has happened—it has happened to *you*.

---

[7]And to be precise, we will be talking only about sample spaces where we can enumerate all the outcomes. Mathematicians call this a countable number of outcomes.

[8]Actually, in mathematical terms, these are axioms—statements that we assume to be true of probability. We'll derive other rules from these in the next chapter.

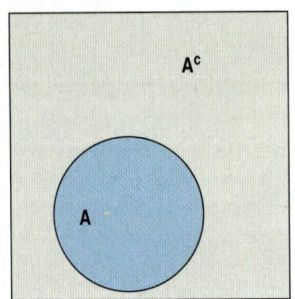

The set **A** and its complement **A**ᶜ. Together, they make up the entire sample space **S**.

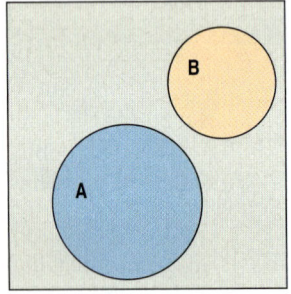

A simple Venn diagram of two disjoint sets, **A** and **B**.

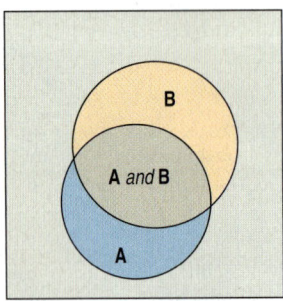

Two sets **A** and **B** that are not disjoint. The event (**A** and **B**) is their intersection.

Rule 2. If a random phenomenon has only one possible outcome, it's not very interesting (or very random). So we need to distribute the probabilities among all the outcomes a trial can have. How can we do that so that it makes sense? For example, consider what you're doing as you read this book. The possible outcomes might be

A: You read to the end of this chapter before stopping.
B: You finish this section but stop reading before the end of the chapter.
C: You bail out before the end of this section.

When we assign probabilities to these outcomes, the first thing to be sure of is that we distribute all of the available probability. If we look at all the events in the entire sample space, the probability of that collection of events has to be 1. So the probability of the entire sample space is 1.

Making this more formal gives the **Probability Assignment Rule**:

The set of all possible outcomes of a trial must have probability 1.

$$P(S) = 1$$

Rule 3. Suppose the probability that you get to class on time is 0.8. What's the probability that you don't get to class on time? Yes, it's 0.2. The set of outcomes that are *not* in the event **A** is called the complement of **A**, and is denoted $\mathbf{A}^C$. This leads to the **Complement Rule**:

The probability of an event not occurring is 1 minus the probability that it does occur.

$$P(\mathbf{A}^C) = 1 - P(\mathbf{A})$$

For Example APPLYING THE COMPLEMENT RULE

RECAP: We opened the chapter by looking at the traffic light at the corner of College and Main, observing that when we arrive at that intersection, the light is green about 35% of the time.

QUESTION: If $P(\text{green}) = 0.35$, what's the probability the light isn't green when you get to College and Main?

ANSWER: "Not green" is the complement of "green," so $P(\text{not green}) = 1 - P(\text{green}) = 1 - 0.35 = 0.65$

There's a 65% chance I won't have a green light.

Rule 4. Suppose the probability that a randomly selected student is a sophomore (**A**) is 0.20, and the probability that he or she is a junior (**B**) is 0.30. What is the probability that the student is *either* a sophomore *or* a junior, written $P(\mathbf{A} \text{ or } \mathbf{B})$? If you guessed 0.50, you've deduced the **Addition Rule**, which says that you can add the probabilities of events that are disjoint.[9] To see whether two events are disjoint, we take them apart into their component outcomes and check whether they have any outcomes in common. **Disjoint** (or mutually exclusive) events have no outcomes in common. The **Addition Rule** states,

For two disjoint events A and B, the probability that one *or* the other occurs is the sum of the probabilities of the two events.

$$P(\mathbf{A} \text{ or } \mathbf{B}) = P(\mathbf{A}) + P(\mathbf{B}), \text{ provided that A and B are disjoint.}$$

---

[9] You may see $P(\mathbf{A} \text{ or } \mathbf{B})$ written as $P(\mathbf{A} \cup \mathbf{B})$. The symbol ∪ means "union," representing outcomes that are in event **A** or event **B** (or both). The symbol ∩ means "intersection," representing outcomes that are in both event **A** and event **B**. You may sometimes see $P(\mathbf{A} \text{ and } \mathbf{B})$ written as $P(\mathbf{A} \cap \mathbf{B})$.

For Example   APPLYING THE ADDITION RULE

RECAP: When you get to the light at College and Main, it's red, green, or yellow. We know that $P(\text{green}) = 0.35$.

QUESTION: Suppose we find out that $P(\text{yellow})$ is about 0.04. What's the probability the light is red?

ANSWER: The light must be red, green, or yellow, so if we can figure out the probability that the light is green or yellow, we can use the Complement Rule to find the probability that it is red. To find the probability that the light is green or yellow, I can use the Addition Rule because these are disjoint events: The light can't be both green and yellow at the same time.

$$P(\text{green OR yellow}) = 0.35 + 0.04 = 0.39$$

Red is the only remaining alternative, and the probabilities must add up to 1, so

$$\begin{aligned} P(\text{red}) &= P(\text{not green OR yellow}) \\ &= 1 - P(\text{green OR yellow}) \\ &= 1 - 0.39 = 0.61 \end{aligned}$$

> "Baseball is 90% mental. The other half is physical."
>
> —*Yogi Berra*

The Addition Rule can be extended to any number of disjoint events, and that's helpful for checking probability assignments. Because individual sample space outcomes are always disjoint, we have an easy way to check whether the probabilities we've assigned to the possible outcomes are legitimate. The Probability Assignment Rule tells us that to be a **legitimate assignment of probabilities**, the sum of the probabilities of all possible outcomes must be exactly 1. No more, no less. For example, if we were told that the probabilities of selecting at random a freshman, sophomore, junior, or senior from all the undergraduates at a school were 0.25, 0.23, 0.22, and 0.20, respectively, we would know that something was wrong. These "probabilities" sum to only 0.90, so this is not a legitimate probability assignment. Either a value is wrong, or we just missed some possible outcomes, like "pre-freshman" or "postgraduate" categories that soak up the remaining 0.10. Similarly, a claim that the probabilities were 0.26, 0.27, 0.29, and 0.30 would be wrong because these "probabilities" sum to more than 1.

But be careful: The Addition Rule doesn't work for events that aren't disjoint. If the probability of owning a smartphone is 0.50 and the probability of owning a computer is 0.90, the probability of owning either a smartphone or a computer may be pretty high, but it is *not* 1.40! Why can't you add probabilities like this? Because these events are not disjoint. You *can* own both. In the next chapter, we'll see how to add probabilities for events like these, but we'll need another rule.

Rule 5. Suppose your job requires you to fly from Atlanta to Houston every Monday morning. The airline's website reports that this flight is on time 85% of the time. What's the chance that it will be on time two weeks in a row? That's the same as asking for the probability that your flight is on time this week *and* it's on time again next week. For independent events, the answer is very simple. Remember that independence means that the outcome of one event doesn't influence the outcome of the other. What happens with your flight this week doesn't influence whether it will be on time next week, so it's reasonable to assume that those events are independent. The **Multiplication Rule** says that for independent events, to find the probability that both events occur, we just multiply the probabilities together. Formally,

> For two independent events A and B, the probability that both A *and* B occur is the product of the probabilities of the two events.
>
> $P(\text{A and B}) = P(\text{A}) \times P(\text{B})$, provided that A and B are independent.

A S *Activity:* **Multiplication Rule for Independent Events.** Experiment with independent random events to explore the Multiplication Rule.

This rule can be extended to more than two independent events. What's the chance of your flight being on time for a month—four Mondays in a row? We can multiply the probabilities of it happening each week:

$$0.85 \times 0.85 \times 0.85 \times 0.85 = 0.522$$

A S *Activity:* **Probabilities of Compound Events.** The Random tool also lets you experiment with Compound random events to see if they are independent.

or just over 50–50. Of course, to calculate this probability, we have used the assumption that the four events are independent.

Many Statistics methods require an **Independence Assumption**, but *assuming* independence doesn't make it true. Always *Think* about whether that assumption is reasonable before using the Multiplication Rule.

For Example APPLYING THE MULTIPLICATION RULE (AND OTHERS)

RECAP: We've determined that the probability that we encounter a green light at the corner of College and Main is 0.35, a yellow light 0.04, and a red light 0.61. Let's think about how many times during your morning commute in the week ahead you might hit a red light there.

QUESTION: What's the probability you find the light red both Monday and Tuesday?

ANSWER: Because the color of the light I see on Monday doesn't influence the color I'll see on Tuesday, these are independent events; I can use the Multiplication Rule:

$$P(\text{red Monday AND red Tuesday}) = P(\text{red}) \times P(\text{red})$$
$$= (0.61)(0.61)$$
$$= 0.3721$$

There's about a 37% chance I'll hit red lights both Monday and Tuesday mornings.

QUESTION: What's the probability you don't encounter a red light until Wednesday?

ANSWER: For that to happen, I'd have to see green or yellow on Monday, green or yellow on Tuesday, and then red on Wednesday. I can simplify this by thinking of it as not red on Monday, not red on Tuesday, and then red on Wednesday.

$$P(\text{not red}) = 1 - P(\text{red}) = 1 - 0.61 = 0.39, \text{ so}$$
$$P(\text{not red Monday AND not red Tuesday AND red Wednesday})$$
$$= P(\text{not red}) \times P(\text{not red}) \times P(\text{red})$$
$$= (0.39)(0.39)(0.61)$$
$$= 0.092781$$

There's about a 9% chance that this week I'll hit my first red light there on Wednesday morning.

QUESTION: What's the probability that you'll have to stop *at least once* during the week?

ANSWER: Having to stop at least once means that I have to stop for the light 1, 2, 3, 4, or 5 times next week. It's easier to think about the complement: never having to stop at a red light. Having to stop at least once means that I didn't make it through the week with no red lights.

$P(\text{having to stop at the light at least once in 5 days})$
$= 1 - P(\text{no red lights for 5 days in a row})$
$= 1 - P(\text{not red AND not red AND not red AND not red AND not red})$
$= 1 - (0.39)(0.39)(0.39)(0.39)(0.39)$
$= 1 - (0.39)^5$
$= 1 - 0.0090$
$= 0.991$

I'm not likely to make it through the intersection of College and Main without having to stop sometime this week. There's over a 99% chance I'll hit at least one red light there.

> **AT LEAST**
> Note that the phrase "at least" is often a tip-off to think about the complement. Something that happens *at least once* does happen. Happening at least once is the complement of not happening at all, and that's easier to find.

Just Checking

2. Opinion polling organizations contact their respondents by telephone. Random telephone numbers are generated, and interviewers try to contact those households. According to the Pew Research Center for the People and the Press, in 2012, the contact rate was about 62%. We can reasonably assume each household's response to be independent of the others. What is the probability that

   a) the interviewer successfully contacts the next household on her list?

   b) the interviewer successfully contacts both of the next two households on her list?

   c) the interviewer's first successful contact is the third household on the list?

   d) the interviewer makes at least one successful contact among the next five households on the list?

> **SOME = AT LEAST ONE**
>
> In informal English, you may see "some" used to mean "at least one." "What's the probability that some of the eggs in that carton are broken?" means at least one.

Step-by-Step Example PROBABILITY

The five rules we've seen can be used in a number of different combinations to answer a surprising number of questions. Let's try one to see how we might go about it.

M&M's® Milk Chocolate candies now come in 7 colors, but they've changed over time. In 1995, Americans voted to change tan M&M's (which had replaced violet in 1949) to blue. In 2002, Mars™, the parent company of M&M's, used the Internet to solicit global opinion for a seventh color. To decide which color to add, Mars surveyed kids in nearly every country of the world and asked them to vote among purple, pink, and teal. The global winner was purple! In the United States, 42% of those who voted said purple, 37% said teal, and only 19% said pink. But in Japan, the percentages were 38% pink, 36% teal, and only 16% purple. Let's use Japan's percentages to ask some questions:

1. What's the probability that a Japanese M&M's survey respondent selected at random chose either pink or teal?
2. If we pick two respondents at random, what's the probability that both chose purple?
3. If we pick three respondents at random, what's the probability that *at least one* chose purple?

THINK ➡ The probability of an event is its long-term relative frequency. It can be determined in several ways: by looking at many replications of an event, by deducing it from equally likely events, or by using some other information. Here, we are told the relative frequencies of the three responses.

Make sure the probabilities are legitimate. Here, they're not. Either there was a mistake, or the other voters must have chosen a color other than the three given. A check of the reports from other countries shows a similar deficit, so probably we're missing those who had no preference or who wrote in another color.

According to the M&M's website:

$$P(\text{pink}) = 0.38$$
$$P(\text{teal}) = 0.36$$
$$P(\text{purple}) = 0.16$$

Each is between 0 and 1, but they don't add up to 1. The remaining 10% must have not expressed a preference or written in another color. I'll put them together into "other" and add $P(\text{other}) = 0.10$.

Now I have a legitimate assignment of probabilities.

Question 1. What's the probability that a Japanese M&M's survey respondent selected at random chose either pink or teal?

THINK → Plan Decide which rules to use and check the conditions they require.

The events "pink" and "teal" are disjoint (because one respondent can't choose both). I can apply the Addition Rule.

SHOW → Mechanics Show your work.

$$P(\text{pink or teal}) = P(\text{pink}) + P(\text{teal})$$
$$= 0.38 + 0.36 = 0.74$$

TELL → Conclusion Interpret your results in the proper context.

The probability that the respondent chose pink or teal is 0.74.

Question 2. If we pick two respondents at random, what's the probability that they both said purple?

THINK → Plan The word "both" suggests we want $P(A \text{ and } B)$, which calls for the Multiplication Rule. Think about the assumption.

✓ **Independence Assumption:** The choice made by one respondent does not affect the choice of the other, so the events are independent. I can use the Multiplication Rule.

SHOW → Mechanics Show your work. For both respondents to choose purple, each one has to choose purple.

$$P(\text{both purple})$$
$$= P(\text{first purple and second purple})$$
$$= P(\text{first purple}) \times P(\text{second purple})$$
$$= 0.16 \times 0.16 = 0.0256$$

TELL → Conclusion Interpret your results in the proper context.

The probability that both chose purple is 0.0256.

Question 3. If we pick three respondents at random, what's the probability that at least one chose purple?

THINK → Plan The phrase "at least . . ." often flags a question best answered by looking at the complement, and that's the best approach here. The complement of "At least one preferred purple" is "None of them preferred purple."

Think about the assumption.

$P(\text{at least one purple})$
$= P(\{\text{none purple}\}^C)$
$= 1 - P(\text{none purple})$.
$P(\text{none purple}) = P(\text{not purple and not purple and not purple})$.

✓ **Independence Assumption:** These are independent events because they are choices by three random respondents. I can use the Multiplication Rule.

SHOW ▶ Mechanics We calculate P(none purple) by using the Multiplication Rule.	$P(\text{none purple}) = P(\text{first not purple}) \times P(\text{second not purple}) \times P(\text{third not purple})$ $= [P(\text{not purple})]^3.$ $P(\text{not purple}) = 1 - P(\text{purple})$ $= 1 - 0.16 = 0.84.$ So $P(\text{none purple}) = (0.84)^3 = 0.5927.$
Then we can use the Complement Rule to get the probability we want.	$P(\text{at least 1 purple}) = 1 - P(\text{none purple})$ $= 1 - 0.5927 = 0.4073.$
TELL ▶ Conclusion Interpret your results in the proper context.	There's about a 40.7% chance that at least one of the respondents chose purple.

WHAT CAN GO WRONG?

- **Beware of probabilities that don't add up to 1.** To be a legitimate probability assignment, the sum of the probabilities for all possible outcomes must total 1. If the sum is less than 1, you may need to add another category ("other") and assign the remaining probability to that outcome. If the sum is more than 1, check that the outcomes are disjoint. If they're not, then you can't assign probabilities by just counting relative frequencies.

- **Don't add probabilities of events if they're not disjoint.** Events must be disjoint to use the Addition Rule. The probability of being younger than 80 *or* a female is not the probability of being younger than 80 *plus* the probability of being female. That sum may be more than 1.

- **Don't multiply probabilities of events if they're not independent.** The probability of selecting a student at random who is over 6'10" tall *and* on the basketball team is *not* the probability the student is over 6'10" tall *times* the probability he's on the basketball team. Knowing that the student is over 6'10" changes the probability of his being on the basketball team. You can't multiply these probabilities. The multiplication of probabilities of events that are not independent is one of the most common errors people make in dealing with probabilities.

- **Don't confuse disjoint and independent.** Disjoint events *can't* be independent. If **A** = {you get an A in this class} and **B** = {you get a B in this class}, **A** and **B** are disjoint. Are they independent? If you find out that **A** is true, does that change the probability of **B**? You bet it does! So they can't be independent. We'll return to this issue in the next chapter.

CONNECTIONS

We began thinking about independence back in Chapter 2 when we looked at contingency tables and asked whether the distribution of one variable was the same for each category of another. Then, in Chapter 11, we saw that independence was fundamental to drawing a Simple Random Sample. For computing compound probabilities, we again ask about independence. And we'll continue to think about independence throughout the rest of the book.

Our interest in probability extends back to the start of the book. We've talked about "relative frequencies" often. But—let's be honest—that's just a casual term for probability. For example, you can now rephrase the 68–95–99.7 Rule we saw in Chapter 5 to talk about the *probability* that a random value selected from a Normal model will fall within 1, 2, or 3 standard deviations of the mean.

Why not just say "probability" from the start? Well, we didn't need any of the formal rules of this chapter (or the next one), so there was no point to weighing down the discussion with those rules. And "relative frequency" is the right intuitive way to think about probability in this course, so you've been thinking right all along.

Keep it up.

What Have We Learned?

Learning Objectives

Apply the facts about probability to determine whether an assignment of probabilities is legitimate.
- Probability is long-run relative frequency.
- Individual probabilities must be between 0 and 1.
- The sum of probabilities assigned to all outcomes must be 1.

Understand the Law of Large Numbers and that the common understanding of the "Law of Averages" is false.

Know the rules of probability and how to apply them.
- **Rule 1:** A probability is a number between 0 and 1.
$$\text{For any event } A, 0 \leq P(A) \leq 1.$$
- **Rule 2: Probability Assignment Rule:**
$$P(S) = 1.$$
- **Rule 3:** The **Complement Rule** says that $P(\text{not } A) = P(A^C) = 1 - P(A)$.
- **Rule 4:** The **Addition Rule** for disjoint events says that $P(A \text{ or } B) = P(A) + P(B)$ provided events A and B are disjoint.
- **Rule 5:** The **Multiplication Rule** for independent events says that $P(A \text{ and } B) = P(A) \times P(B)$ provided events A and B are independent.

Review of Terms

Random phenomenon A phenomenon is random if we know what outcomes could happen, but not which particular values will happen (p. 348).

Trial A single attempt or realization of a random phenomenon (p. 349).

Outcome The value measured, observed, or reported for an individual instance of trial (p. 349).

Event A collection of outcomes. Usually, we identify events so that we can attach probabilities to them. We denote events with bold capital letters such as **A**, **B**, or **C** (p. 349).

Sample space The collection of all possible outcome values. The collection of values in the sample space has a probability of 1. We denote the sample space with a boldface capital **S** (p. 349).

Law of Large Numbers This law states that the long-run *relative frequency* of an event's occurrence gets closer and closer to the *true* relative frequency as the number of trials increases (p. 350).

Independence (informally) Two events are *independent* if learning that one event occurs does not change the probability that the other event occurs (p. 350).

Probability The probability of an event is a number between 0 and 1 that reports the likelihood of that event's occurrence. We write $P(A)$ for the probability of the event **A** (p. 350).

Empirical probability When the probability comes from the long-run relative frequency of the event's occurrence, it is an **empirical probability** (p. 350).

Theoretical probability When the probability comes from a model (such as equally likely outcomes), it is a **theoretical probability** (p. 351).

Personal (or subjective) probability When the probability is subjective and represents your personal degree of belief, it is a **personal probability** (p. 352).

Probability Assignment Rule	The probability of the entire sample space must be 1. $P(S) = 1$ (p. 354).
Complement Rule	The probability of an event not occurring is 1 minus the probability that it does occur (p. 354). $$P(A^C) = 1 - P(A)$$
Addition Rule	If A and B are disjoint events, then the probability of A *or* B is (p. 354) $$P(A \text{ or } B) = P(A) + P(B).$$
Disjoint (mutually exclusive)	Two events are disjoint if they share no outcomes in common. If A and B are disjoint, then knowing that A occurs tells us that B cannot occur. Disjoint events are also called "mutually exclusive" (p. 354).
Legitimate assignment of probabilities	An assignment of probabilities to outcomes is legitimate if (p. 355) ■ each probability is between 0 and 1 (inclusive). ■ the sum of the probabilities is 1.
Multiplication Rule	If A and B are independent events, then the probability of A *and* B is (p. 355) $$P(A \text{ and } B) = P(A) \times P(B).$$

Exercises

Section 13.1

1. **Flipping a coin** Flipping a fair coin is said to randomly generate heads and tails with equal probability. Explain what random means in this context.

2. **Dice** Rolling a fair six-sided die is supposed to randomly generate the numbers 1 through 6. Explain what random means in this context.

3. **Flipping a coin II** Your friend says: "I flipped five heads in a row! The next one has to be tails!" Explain why this thinking is incorrect.

4. **Dice II** After rolling doubles on a pair of dice three times in a row, your friend exclaims, "I can't get doubles four times in a row!" Explain why this thinking is incorrect.

Section 13.2

5. **Wardrobe** In your dresser are five blue shirts, three red shirts, and two black shirts.
   a) What is the probability of randomly selecting a red shirt?
   b) What is the probability that a randomly selected shirt is not black?

6. **Playlists** Your list of favorite songs contains 10 rock songs, 7 rap songs, and 3 country songs.
   a) What is the probability that a randomly played song is a rap song?
   b) What is the probability that a randomly played song is not country?

Section 13.3

7. **Cell phones and surveys** A 2010 study conducted by the National Center for Health Statistics found that 25% of U.S. households had no landline service. This raises concerns about the accuracy of certain surveys, as they depend on random-digit dialing to households via landlines. We are going to pick five U.S. households at random:
   a) What is the probability that all five of them have a landline?
   b) What is the probability that at least one of them does not have a landline?
   c) What is the probability that at least one of them does have a landline?

8. **Cell phones and surveys II** The survey by the National Center for Health Statistics further found that 49% of adults ages 25–29 had only a cell phone and no landline. We randomly select four 25–29-year-olds:
   a) What is the probability that all of these adults have a only a cell phone and no landline?
   b) What is the probability that none of these adults have only a cell phone and no landline?
   c) What is the probability that at least one of these adults has only a cell phone and no landline?

Chapter Exercises

9. **Sample spaces** For each of the following, list the sample space and tell whether you think the events are equally likely:
   a) Toss 2 coins; record the order of heads and tails.
   b) A family has 3 children; record the number of boys.
   c) Flip a coin until you get a head or 3 consecutive tails; record each flip.
   d) Roll two dice; record the larger number (or simply the number in case of a tie).

10. **Sample spaces** For each of the following, list the sample space and tell whether you think the events are equally likely:
    a) Roll two dice; record the sum of the numbers.
    b) A family has 3 children; record each child's sex in order of birth.
    c) Toss four coins; record the number of tails.
    d) Toss a coin 10 times; record the length of the longest run of heads.

11. **Roulette** A casino claims that its roulette wheel is truly random. What should that claim mean?

12. **Rain** The weather reporter on TV makes predictions such as a 25% chance of rain. What do you think is the meaning of such a phrase?

13. **Winter** Comment on the following quotation:
    "What I think is our best determination is it will be a colder than normal winter," said Pamela Naber Knox, a Wisconsin state climatologist. "I'm basing that on a couple of different things. First, in looking at the past few winters, there has been a lack of really cold weather. Even though we are not supposed to use the law of averages, we are due." (Source: Excerpt from "*The Law of Averages*" by Naber Knox from Activity-Based Statistics: Student Guide by Richard L. Scheaffer, Jeffrey Witmer, Ann Watkins, Mrudulla Gnanadesikan. Published by Springer/Sci-Tech/Trade, © 1996.)

14. **Snow** After an unusually dry autumn, a radio announcer is heard to say, "Watch out! We'll pay for these sunny days later on this winter." Explain what he's trying to say, and comment on the validity of his reasoning.

15. **Cold streak** A basketball player missed 8 of 11 shots in the final portion of a game. When talking to reporters afterward, he says that he's not worried about his next game because he's due to make a lot of shots. Comment on his reasoning.

16. **Crash** Commercial airplanes have an excellent safety record. Nevertheless, there are crashes occasionally, with the loss of many lives. In the weeks following a crash, airlines often report a drop in the number of passengers, probably because people are afraid to risk flying.
    a) A travel agent suggests that since the law of averages makes it highly unlikely to have two plane crashes within a few weeks of each other, flying soon after a crash is the safest time. What do you think?
    b) If the airline industry proudly announces that it has set a new record for the longest period of safe flights, would you be reluctant to fly? Are the airlines due to have a crash?

17. **Auto insurance** Insurance companies collect annual payments from drivers in exchange for paying for the cost of accidents.
    a) Why should you be reluctant to accept a $1500 payment from your neighbor to cover his automobile accidents in the next year?
    b) Why can the insurance company make that offer?

18. **Jackpot** On February 11, 2009, the AP news wire released the following story:

    (LAS VEGAS, Nev.)—A man in town to watch the NCAA basketball tournament hit a $38.7 million jackpot on Friday, the biggest slot machine payout ever. The 25-year-old software engineer from Los Angeles, whose name was not released at his request, won after putting three $1 coins in a machine at the Excalibur hotel-casino, said Rick Sorensen, a spokesman for slot machine maker International Game Technology.

    a) How can the Excalibur afford to give away millions of dollars on a $3 bet?
    b) Why was the maker willing to make a statement? Wouldn't most businesses want to keep such a huge loss quiet?

19. **Spinner** The plastic arrow on a spinner for a child's game stops rotating to point at a color that will determine what happens next. Which of the following probability assignments are legitimate?

	Probabilities of ...			
	Red	Yellow	Green	Blue
a)	0.25	0.25	0.25	0.25
b)	0.10	0.20	0.30	0.40
c)	0.20	0.30	0.40	0.50
d)	0	0	1.00	0
e)	0.10	0.20	1.20	−1.50

20. **Scratch off** Many stores run "secret sales": Shoppers receive cards that determine how large a discount they get, but the percentage is revealed by scratching off that black stuff (what *is* that?) only after the purchase has been totaled at the cash register. The store is required to reveal (in the fine print) the distribution of discounts available. Which of these probability assignments are legitimate?

	Probabilities of ...			
	10% Off	20% Off	30% Off	50% Off
a)	0.20	0.20	0.20	0.20
b)	0.50	0.30	0.20	0.10
c)	0.80	0.10	0.05	0.05
d)	0.75	0.25	0.25	−0.25
e)	1.00	0	0	0

21. **Electronics** Suppose that 46% of families living in a certain county own a computer and 18% own an HDTV. The Addition Rule might suggest, then, that 64% of families own either a computer or an HDTV. What's wrong with that reasoning?

22. **Homes** Funding for many schools comes from taxes based on assessed values of local properties. People's homes are assessed higher if they have extra features such as garages and swimming pools. Assessment records in a certain school district indicate that 37% of the homes have garages and 3% have swimming pools. The Addition Rule might suggest, then, that 40% of residences have a garage or a pool. What's wrong with that reasoning?

23. **Speeders** Traffic checks on a certain section of highway suggest that 60% of drivers are speeding there. Since $0.6 \times 0.6 = 0.36$, the Multiplication Rule might suggest that there's a 36% chance that two vehicles in a row are both speeding. What's wrong with that reasoning?

24. **Lefties** Although it's hard to be definitive in classifying people as right- or left-handed, some studies suggest that about 14% of people are left-handed. Since $0.14 \times 0.14 = 0.0196$, the Multiplication Rule might suggest that there's about a 2% chance that a brother and a sister are both lefties. What's wrong with that reasoning?

25. **College admissions** For high school students graduating in 2013, college admissions to the nation's most selective schools were the most competitive in memory. Harvard accepted about 5.8% of its applicants, Dartmouth 10%, and Penn 12.1%. Jorge has applied to all three. Assuming that he's a typical applicant, he figures that his chances of getting into both Harvard and Dartmouth must be about 0.58%.
    a) How has he arrived at this conclusion?
    b) What additional assumption is he making?
    c) Do you agree with his conclusion?

26. **College admissions II** In Exercise 25, we saw that in 2013 Harvard accepted about 5.8% of its applicants, Dartmouth 10%, and Penn 12.1%. Jorge has applied to all three. He figures that his chances of getting into at least one of the three must be about 27.9%.
    a) How has he arrived at this conclusion?
    b) What assumption is he making?
    c) Do you agree with his conclusion?

27. **Car repairs** A consumer organization estimates that over a 1-year period 17% of cars will need to be repaired only once, 7% will need repairs exactly twice, and 4% will require three or more repairs. What is the probability that a car chosen at random will need
    a) no repairs?
    b) no more than one repair?
    c) some repairs?

28. **Stats projects** In a large Introductory Statistics lecture hall, the professor reports that 55% of the students enrolled have never taken a Calculus course, 32% have taken only one semester of Calculus, and the rest have taken two or more semesters of Calculus. The professor randomly assigns students to groups of three to work on a project for the course. What is the probability that the first groupmate you meet has studied
    a) two or more semesters of Calculus?
    b) some Calculus?
    c) no more than one semester of Calculus?

29. **More repairs** Consider again the auto repair rates described in Exercise 27. If you own two cars, what is the probability that
    a) neither will need repair?
    b) both will need repair?
    c) at least one car will need repair?

30. **Another project** You are assigned to be part of a group of three students from the Intro Stats class described in Exercise 28. What is the probability that of your other two groupmates,
    a) neither has studied Calculus?
    b) both have studied at least one semester of Calculus?
    c) at least one has had more than one semester of Calculus?

31. **Repairs again** You used the Multiplication Rule to calculate repair probabilities for your cars in Exercise 29.
    a) What must be true about your cars in order to make that approach valid?
    b) Do you think this assumption is reasonable? Explain.

32. **Final project** You used the Multiplication Rule to calculate probabilities about the Calculus background of your Statistics groupmates in Exercise 30.
    a) What must be true about the groups in order to make that approach valid?
    b) Do you think this assumption is reasonable? Explain.

33. **Energy 2011** A Gallup Poll (www.gallup.com/poll/161729/americans-divided-energy-environment-trade-off.aspx) in March 2013 asked 529 U.S. adults whether increasing oil, gas, and coal production or developing alternative energy sources such as wind and solar power should be given a higher priority. Here are the results:

Response	Number
Oil, gas, coal	164
Wind and solar	312
Equally Important	37
No Opinion	16
Total	529

    If we select a person at random from this sample of 529 adults,
    a) what is the probability that the person responded "Increase oil, gas, and coal"?
    b) what is the probability that the person responded "Equally important" or had no opinion?

34. **Failing fathers?** A Pew Research poll in 2011 asked 2005 U.S. adults whether being a father today is harder than it was a generation ago. Here's how they responded:

Response	Number
Easier	501
Same	802
Harder	682
No Opinion	20
Total	2005

If we select a respondent at random from this sample of 2005 adults,

a) what is the probability that the selected person responded "Harder"?
b) what is the probability that the person responded the "Same" or "Easier"?

35. **More energy** Exercise 33 shows the results of a Gallup Poll about energy. Suppose we select three people at random from this sample.
   a) What is the probability that all three responded "Develop wind and solar"?
   b) What is the probability that none responded "Equally important"?
   c) What assumption did you make in computing these probabilities?
   d) Explain why you think that assumption is reasonable.

36. **Fathers, revisited** Consider again the results of the poll about fathering discussed in Exercise 34. If we select two people at random from this sample,
   a) what is the probability that both think that being a father is easier today?
   b) what is the probability that neither thinks being a father is easier today?
   c) what is the probability that the first person thinks being a father is easier today and the second one doesn't?
   d) what assumption did you make in computing these probabilities?
   e) explain why you think that assumption is reasonable.

37. **Polling** As mentioned in the chapter, opinion-polling organizations contact their respondents by sampling random telephone numbers. Although interviewers can reach about 62% of U.S. households, the percentage of those contacted who agree to cooperate with the survey fell from 43% in 1997 to 14% in 2012 (Source: Pew Research Center for the People and the Press). Each household, of course, is independent of the others. Using the cooperation rate from 2012,
   a) what is the probability that the next household on the list will be contacted but will refuse to cooperate?
   b) what is the probability of failing to contact a household or of contacting the household but not getting them to agree to the interview?
   c) show another way to calculate the probability in part b.

38. **Polling, part II** According to Pew Research, the contact rate (probability of contacting a selected household) was 90% in 1997 and 62% in 2012. However, the cooperation rate (probability of someone at the contacted household agreeing to be interviewed) was 43% in 1997 and dropped to 14% in 2012.
   a) What is the probability (in 2012) of obtaining an interview with the next household on the sample list? (To obtain an interview, an interviewer must both contact the household and then get agreement for the interview.)
   b) Was it more likely to obtain an interview from a randomly selected household in 1997 or in 2012?

39. **M&M's** The Mars company says that before the introduction of purple, yellow candies made up 20% of their plain M&M's, red another 20%, and orange, blue, and green each made up 10%. The rest were brown.
   a) If you pick an M&M at random, what is the probability that
      1. it is brown?
      2. it is yellow or orange?
      3. it is not green?
      4. it is striped?
   b) If you pick three M&M's in a row, what is the probability that
      1. they are all brown?
      2. the third one is the first one that's red?
      3. none are yellow?
      4. at least one is green?

40. **Blood** The American Red Cross says that about 45% of the U.S. population has Type O blood, 40% Type A, 11% Type B, and the rest Type AB.
   a) Someone volunteers to give blood. What is the probability that this donor
      1. has Type AB blood?
      2. has Type A or Type B?
      3. is not Type O?
   b) Among four potential donors, what is the probability that
      1. all are Type O?
      2. no one is Type AB?
      3. they are not all Type A?
      4. at least one person is Type B?

41. **Disjoint or independent?** In Exercise 39, you calculated probabilities of getting various M&M's. Some of your answers depended on the assumption that the outcomes described were *disjoint*; that is, they could not both happen at the same time. Other answers depended on the assumption that the events were *independent*; that is, the occurrence of one of them doesn't affect the probability of the other. Do you understand the difference between disjoint and independent?
   a) If you draw one M&M, are the events of getting a red one and getting an orange one disjoint, independent, or neither?
   b) If you draw two M&M's one after the other, are the events of getting a red on the first and a red on the second disjoint, independent, or neither?
   c) Can disjoint events ever be independent? Explain.

42. **Disjoint or independent?** In Exercise 40, you calculated probabilities involving various blood types. Some of your answers depended on the assumption that the outcomes described were *disjoint;* that is, they could not both happen at the same time. Other answers depended on the assumption that the events were *independent*; that is, the occurrence of one of them doesn't affect the probability

of the other. Do you understand the difference between disjoint and independent?

a) If you examine one person, are the events that the person is Type A and that the same person is Type B disjoint, independent, or neither?

b) If you examine two people, are the events that the first is Type A and the second Type B disjoint, independent, or neither?

c) Can disjoint events ever be independent? Explain.

43. **Dice** You roll a fair die three times. What is the probability that

a) you roll all 6's?
b) you roll all odd numbers?
c) none of your rolls gets a number divisible by 3?
d) you roll at least one 5?
e) the numbers you roll are not all 5's?

44. **Slot machine** A slot machine has three wheels that spin independently. Each has 10 equally likely symbols: 4 bars, 3 lemons, 2 cherries, and a bell. If you play, what is the probability that

a) you get 3 lemons?
b) you get no fruit symbols?
c) you get 3 bells (the jackpot)?
d) you get no bells?
e) you get at least one bar (an automatic loser)?

45. **Champion bowler** A certain bowler can bowl a strike 70% of the time. If the bowls are independent, what's the probability that she

a) goes three consecutive frames without a strike?
b) makes her first strike in the third frame?
c) has at least one strike in the first three frames?
d) bowls a perfect game (12 consecutive strikes)?

46. **The train** To get to work, a commuter must cross train tracks. The time the train arrives varies slightly from day to day, but the commuter estimates he'll get stopped on about 15% of work days. During a certain 5-day work week, what is the probability that he

a) gets stopped on Monday and again on Tuesday?
b) gets stopped for the first time on Thursday?
c) gets stopped every day?
d) gets stopped at least once during the week?

47. **Voters** Suppose that in your city 37% of the voters are registered as Democrats, 29% as Republicans, and 11% as members of other parties (Liberal, Right to Life, Green, etc.). Voters not aligned with any official party are termed "Independent." You are conducting a poll by calling registered voters at random. In your first three calls, what is the probability you talk to

a) all Republicans?
b) no Democrats?
c) at least one Independent?

48. **Religion** Census reports for a city indicate that 62% of residents classify themselves as Christian, 12% as Jewish, and 16% as members of other religions (Muslims, Buddhists, etc.). The remaining residents classify themselves as nonreligious. A polling organization seeking information about public opinions wants to be sure to talk with people holding a variety of religious views, and makes random phone calls. Among the first four people they call, what is the probability they reach

a) all Christians?
b) no Jews?
c) at least one person who is nonreligious?

49. **Lights** You purchased a five-pack of new light bulbs that were recalled because 6% of the lights did not work. What is the probability that at least one of your lights is defective?

50. **Pepsi** For a sales promotion, the manufacturer places winning symbols under the caps of 10% of all Pepsi bottles. You buy a six-pack. What is the probability that you win something?

51. **9/11?** On September 11, 2002, the first anniversary of the terrorist attack on the World Trade Center, the New York State Lottery's daily number came up 9–1–1. An interesting coincidence or a cosmic sign?

a) What is the probability that the winning three numbers match the date on any given day?
b) What is the probability that a whole year passes without this happening?
c) What is the probability that the date and winning lottery number match at least once during any year?
d) If every one of the 50 states has a three-digit lottery, what is the probability that at least one of them will come up 9–1–1 on September 11?

52. **Red cards** You shuffle a deck of cards and then start turning them over one at a time. The first one is red. So is the second. And the third. In fact, you are surprised to get 10 red cards in a row. You start thinking, "The next one is due to be black!"

a) Are you correct in thinking that there's a higher probability that the next card will be black than red? Explain.
b) Is this an example of the Law of Large Numbers? Explain.

Just Checking ANSWERS

1. The LLN works only in the long run, not in the short run. The random methods for selecting lottery numbers have no memory of previous picks, so there is no change in the probability that a certain number will come up.

2. a) 0.62
   b) $(0.62)(0.62) = 0.3844$
   c) $(1 - 0.62)^2(0.62) = 0.089528$
   d) $1 - (1 - 0.62)^5 = 0.992076$

chapter 14
Probability Rules!

- **14.1** The General Addition Rule
- **14.2** Conditional Probability and the General Multiplication Rule
- **14.3** Independence
- **14.4** Picturing Probability: Tables, Venn Diagrams, and Trees
- **14.5** Reversing the Conditioning and Bayes' Rule

> *Where are we going?*
> Is the probability of a car accident the same for all age groups? Insurance companies want to know so they can set their rates. Is everyone at equal risk for getting the flu next winter? Medical researchers use factors such as age, sex, lifestyle, and family history to estimate the probability for each individual. These are examples of **conditional** probabilities. We'll see how the probability of an event can change when we learn more about the situation at hand. Sometimes the results are surprising.

Pull a bill from your wallet or pocket without looking at it. An outcome of this trial is the bill you select. The sample space is all the bills in circulation: $S = \{\$1 \text{ bill}, \$2 \text{ bill}, \$5 \text{ bill}, \$10 \text{ bill}, \$20 \text{ bill}, \$50 \text{ bill}, \$100 \text{ bill}\}$.[1] These are *all* the possible outcomes. (In spite of what you may have seen in bank robbery movies, there are no $500 or $1000 bills.)

We can combine the outcomes in different ways to make many different events. For example, the event $A = \{\$1, \$5, \$10\}$ represents selecting a $1, $5, or $10 bill. The event $B = \{\text{a bill that does not have a president on it}\}$ is the collection of outcomes (Don't look! Can you name them?): $\{\$10 \text{ (Hamilton)}, \$100 \text{ (Franklin)}\}$. The event $C = \{\text{enough money to pay for a \$12 meal with one bill}\}$ is the set of outcomes $\{\$20, \$50, \$100\}$.

Notice that these outcomes are not equally likely. You'd no doubt be more surprised (and pleased) to pull out a $100 bill than a $1 bill—it's not as likely, though. You probably carry many more $1 than $100 bills, but without information about the probability of each outcome, we can't calculate the probability of an event.

The probability of the event **C** (getting a bill worth more than $12) is *not* 3/7. There are 7 possible outcomes, and 3 of them exceed $12, but they are *not equally* likely.

14.1 The General Addition Rule

Now look at the bill in your hand. There are images of famous buildings in the center of the backs of all but two bills in circulation. The $1 bill has the word ONE in the center, and the $2 bill shows the signing of the Declaration of Independence.

---

[1] Well, technically, the sample space is all the bills in your pocket. You may be quite sure there isn't a $100 bill in there, but *we* don't know that, so humor us that it's at least *possible* that any legal bill could be there.

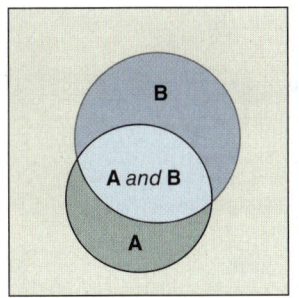

Events **A** and **B** and their intersection.

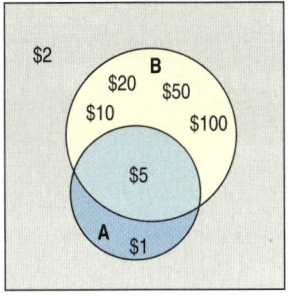

Denominations of bills that are odd (**A**) or that have a building on the reverse side (**B**). The two sets both include the $5 bill, and both exclude the $2 bill.

What's the probability of randomly selecting **A** = {a bill with an odd-numbered value} or **B** = {a bill with a building on the reverse}? We know **A** = {$1, $5} and **B** = {$5, $10, $20, $50, $100}. But $P(A \text{ or } B)$ is not simply the sum $P(A) + P(B)$, because the events **A** and **B** are not disjoint (the $5 bill is in both sets). So what can we do? We'll need a new probability rule.

As the diagrams show, we can't use the Addition Rule and add the two probabilities because the events are not disjoint; they overlap. There's an outcome (the $5 bill) in the *intersection* of **A** and **B**. The Venn diagram represents the sample space. Notice that the $2 bill has neither a building nor an odd denomination, so it sits outside both circles.

The $5 bill plays a crucial role here because it is both odd *and* has a building on the reverse. It's in both **A** and **B**, which places it in the *intersection* of the two circles. The reason we can't simply add the probabilities of **A** and **B** is that we'd count the $5 bill twice.

If we did add the two probabilities, we could compensate by *subtracting* the probability of that $5 bill. So,

$P(\text{odd number value or building})$
$= P(\text{odd number value}) + P(\text{building}) - P(\text{odd number value and building})$
$= P(\$1, \$5) + P(\$5, \$10, \$20, \$50, \$100) - P(\$5).$

This method works in general. We add the probabilities of two events and then subtract out the probability of their intersection. This approach gives us the **General Addition Rule**, which does not require disjoint events:

$$P(A \text{ or } B) = P(A) + P(B) - P(A \text{ and } B).$$

Drawing a picture using Venn diagrams often makes the rule easier to see and understand.

For Example USING THE GENERAL ADDITION RULE

A survey of college students asked the questions: "Are you currently in a relationship?" and "Are you involved in intercollegiate or club sports?" The survey found that 33% were currently in a relationship, and 25% were involved in sports. Eleven percent responded "yes" to both.

QUESTION: What's the probability that a randomly selected student either is in a relationship or is involved in athletics?

ANSWER: Let **R** = {student in a relationship} and **S** = {student involved in sports}.

$P(\text{a student either is in a relationship or involved in sports}) = P(R \text{ or } S)$
$= P(R) + P(S) - P(R \text{ and } S)$
$= 0.33 + 0.25 - 0.11$
$= 0.47$

There's a 47% chance that a randomly selected college student either is in a relationship or is involved in sports. Notice that this use of the word "or" includes "or both." That's the usual meaning of *or* in English. If we wanted to exclude those students who were involved in sports and in a relationship we'd have to ask, "What's the probability that a selected student either is in a relationship or is involved in sports, but not both?"

> **Would you like dessert or coffee?** Natural language can be ambiguous. In this question, what are your choices? Can you have only one of the two, or could you simply answer "yes" and have both dessert and coffee?
>
> In Statistics, when we say "or" we always mean the inclusive version. In other words, the probability of **A** or **B** means the probability of either **A** or **B** or both. But in everyday language,

"or" is usually the exclusive version meaning one or the other but not both. For example, if you are asked "Would you like the steak or the vegetarian entrée?" you can't get both. This difference can be confusing for students seeing the statistical language for the first time.

The General Addition Rule subtracts the probability of the outcomes in **A** *and* **B** because we've counted those outcomes *twice*. But they're still there. On the other hand, if we really want the probability of **A** or **B**, but NOT both, we have to get rid of the outcomes in $\{A \text{ and } B\}$. So $P(A \text{ or } B, \text{ but } not \text{ both}) = P(A \text{ or } B) - P(A \text{ and } B) = P(A) + P(B) - 2 \times P(A \text{ and } B)$. Now we've subtracted $P(A \text{ and } B)$ twice—once because we don't want to double-count these events and a second time because we really didn't want to count them at all. Confused? *Make a picture.*

For Example USING VENN DIAGRAMS

RECAP: We return to our survey of college students: 33% are in a relationship, 25% are involved in sports, and 11% are in both.

QUESTIONS: Based on a Venn diagram, what is the probability that a randomly selected student

a. is not in a relationship and is not involved in sports?
b. is in a relationship but not involved in sports?
c. either is in a relationship or is involved in sports, but not both?

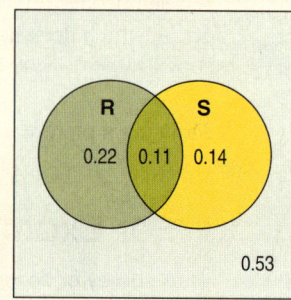

ANSWER: Let $R = \{\text{student is in a relationship}\}$ and $S = \{\text{student is involved in sports}\}$. In the Venn diagram, the intersection of the circles is $P(R \text{ and } S) = 0.11$. Since $P(R) = 0.33$, $P(R \text{ and } S^C) = 0.33 - 0.11 = 0.22$. Also, $P(R^C \text{ and } S) = 0.25 - 0.11 = 0.14$. Now, $0.22 + 0.11 + 0.14 = 0.47$, leaving $1 - 0.47 = 0.53$ for the region outside both circles.

Now, $P(\text{is not in a relationship and is not involved in sports})$
$= P(R^C \text{ and } S^C) = 0.53$

$P(\text{is in a relationship but not involved in sports}) = P(R \text{ and } S^C) = 0.22$

$P(\text{is in a relationship or involved in sports, but not both})$
$= P(R \text{ and } S^C) + P(R^C \text{ and } S) = 0.22 + 0.14 = 0.36.$

Or

$P(R) + P(S) - 2P(R \text{ and } S) = 0.33 + 0.25 - 2(0.11) = 0.36.$

Just Checking

1. Back in Chapter 1, we suggested that you sample some pages of this book at random to see whether they held some sort of data display or an equation. We actually did just that. We drew a representative sample and found the following:

   48% of pages had some kind of data display,
   27% of pages had an equation, and
   7% of pages had both a data display and an equation.

a) Display these results in a Venn diagram.

b) What is the probability that a randomly selected sample page had neither a data display nor an equation?

c) What is the probability that a randomly selected sample page had a data display but no equation?

Step-by-Step Example USING THE GENERAL ADDITION RULE

You almost certainly qualify as an "online adult"—an adult who uses the Internet. Pew Research tracks the behavior of online adults. In 2013 they found that 71% were Facebook users, 18% used Twitter, and 15% used both. (http://www.pewinternet.org/Commentary/2012/March/Pew-Internet-Social-Networking-full-detail.aspx) (updated in September 2013)

Questions: For a randomly selected online adult, what is the probability that the person

1. uses either Facebook or Twitter?
2. uses either Facebook or Twitter, but not both?
3. doesn't use Facebook or Twitter?

THINK ▶ Plan Define the events we're interested in. There are no conditions to check; the General Addition Rule works for any events!

Plot Make a picture, and use the given probabilities to find the probability for each region.

The blue region represents **A** but not **B**. The green intersection region represents **A** and **B**. Note that since $P(\mathbf{A}) = 0.71$ and $P(\mathbf{A} \text{ and } \mathbf{B}) = 0.15$, the probability of **A** but not **B** must be $0.71 - 0.15 = 0.56$.

The yellow region is **B** but not **A**.

The gray region outside both circles represents the outcome neither **A** nor **B**. All the probabilities must total 1, so you can determine the probability of that region by subtraction.

Now, figure out what you want to know. The probabilities can come from the diagram or a formula. Sometimes translating the words to equations is the trickiest step.

Let $\mathbf{A} = \{\text{respondent uses Facebook}\}$.
Let $\mathbf{B} = \{\text{respondent uses Twitter}\}$.

I know that
$$P(\mathbf{A}) = 0.71$$
$$P(\mathbf{B}) = 0.18$$
$$P(\mathbf{A} \text{ and } \mathbf{B}) = 0.15$$

So
$$P(\mathbf{A} \text{ and } \mathbf{B}^C) = 0.71 - 0.15 = 0.56$$
$$P(\mathbf{B} \text{ and } \mathbf{A}^C) = 0.18 - 0.15 = 0.03$$
$$P(\mathbf{A}^C \text{ and } \mathbf{B}^C) = 1 - (0.56 + 0.15 + 0.03)$$
$$= 0.26$$

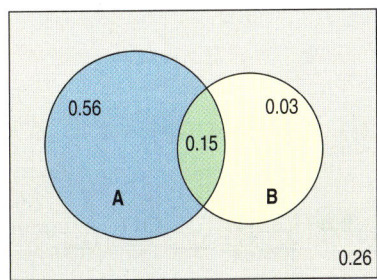

Question 1. What is the probability that the person uses either Facebook or Twitter?

SHOW → Mechanics The probability that the respondent uses Facebook or Twitter is $P(\mathbf{A}\text{ or }\mathbf{B})$. We can use the General Addition Rule, or we can add the probabilities seen in the diagram.

$$P(\mathbf{A}\text{ or }\mathbf{B}) = P(\mathbf{A}) + P(\mathbf{B}) - P(\mathbf{A}\text{ and }\mathbf{B})$$
$$= 0.71 + 0.18 - 0.15$$
$$= 0.74$$

OR

$$P(\mathbf{A}\text{ or }\mathbf{B}) = 0.56 + 0.15 + 0.03 = 0.74$$

TELL → Conclusion Don't forget to interpret your result in context.

76% of online adults use Facebook or Twitter.

Question 2. What is the probability that the person uses either Facebook or Twitter, but not both?

SHOW → Mechanics We can use the rule, or just add the appropriate probabilities seen in the Venn diagram.

$$P(\mathbf{A}\text{ or }\mathbf{B}\text{ but NOT both}) = P(\mathbf{A}\text{ or }\mathbf{B}) - P(\mathbf{A}\text{ and }\mathbf{B})$$
$$= 0.74 - 0.15 = 0.59$$

OR

$$P(\mathbf{A}\text{ or }\mathbf{B}\text{ but NOT both})$$
$$= P(\mathbf{A}\text{ and }\mathbf{B}^C) + P(\mathbf{B}\text{ and }\mathbf{A}^C)$$
$$= 0.56 + 0.03 = 0.59$$

TELL → Conclusion Interpret your result in context.

59% of online adults use Facebook or Twitter, but not both.

Question 3. What is the probability that the person doesn't use Facebook or Twitter?

SHOW → Mechanics Not using Facebook or Twitter is the complement of using one or the other. Use the Complement Rule or just notice that "not using Facebook or Twitter" is represented by the region outside both circles.

$$P(\text{not Facebook or Twitter}) =$$
$$1 - P(\text{either Facebook or Twitter})$$
$$= 1 - P(\mathbf{A}\text{ or }\mathbf{B})$$
$$= 1 - 0.74 = 0.26$$

OR

$$P(\mathbf{A}^C\text{ and }\mathbf{B}^C) = 0.26$$

TELL → Conclusion Interpret your result in context.

Only 26% of online adults use neither Facebook nor Twitter.

14.2 Conditional Probability and the General Multiplication Rule

Two psychologists surveyed 478 children in grades 4, 5, and 6 in elementary schools in Michigan. They stratified their sample, drawing roughly 1/3 from rural, 1/3 from suburban, and 1/3 from urban schools. Among other questions, they asked the students whether their primary goal was to get good grades, to be popular, or to be good at sports. One question of interest was whether boys and girls at this age had similar goals.

Here's a *contingency table* giving counts of the students by their goals and sex:

Table 14.1
The distribution of goals for boys and girls.

		Goals			
		Grades	Popular	Sports	Total
Sex	Boy	117	50	60	227
	Girl	130	91	30	251
	Total	247	141	90	478

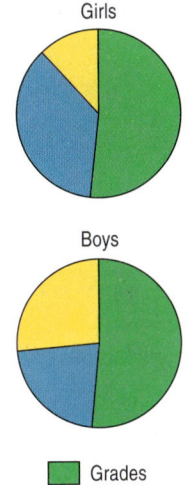

Figure 14.1
The distribution of goals for boys and girls.

We looked at contingency tables and graphed *conditional distributions* back in Chapter 2. The pie charts show the *relative frequencies* with which boys and girls named the three goals. It's only a short step from these relative frequencies to probabilities.

Let's focus on this study and make the sample space just the set of these 478 students. If we select a student at random from this study, the probability we select a girl is just the corresponding relative frequency (since we're equally likely to select any of the 478 students). There are 251 girls in the data out of a total of 478, giving a probability of

$$P(\text{girl}) = 251/478 = 0.525$$

The same method works for more complicated events like intersections. For example, what's the probability of selecting a girl whose goal is to be popular? Well, 91 girls named popularity as their goal, so the probability is

$$P(\text{girl } and \text{ popular}) = 91/478 = 0.190$$

The probability of selecting a student whose goal is to excel at sports is

$$P(\text{sports}) = 90/478 = 0.188$$

What if we are given the information that the selected student is a girl? Would that change the probability that the selected student's goal is sports? You bet it would! The pie charts show that girls are much less likely to say their goal is to excel at sports than are boys. When we restrict our focus to girls, we look only at the girls' row of the table. Of the 251 girls, only 30 of them said their goal was to excel at sports.

We write the probability that a selected student wants to excel at sports *given that we have selected a girl* as

$$P(\text{sports}|\text{girl}) = 30/251 = 0.120$$

For boys, we look at the conditional distribution of goals given "boy" shown in the top row of the table. There, of the 227 boys, 60 said their goal was to excel at sports. So, $P(\text{sports}|\text{boy}) = 60/227 = 0.264$, more than twice the girls' probability.

In general, when we want the probability of an event from a *conditional* distribution, we write $P(\mathbf{B}|\mathbf{A})$ and pronounce it "the probability of **B** *given* **A**." A probability that takes into account a given *condition* such as this is called a **conditional probability**.

Activity: Birth Weights and Smoking. Does smoking increase the chance of having a baby with low birth weight?

Let's look at what we did. We worked with the counts, but we could work with the probabilities just as well. There were 30 students who both were girls and had sports as their goal, and there are 251 girls. So we found the probability to be 30/251. To find the probability of the event **B** *given* the event **A**, we restrict our attention to the

NOTATION ALERT

$P(\mathbf{B}|\mathbf{A})$ is the conditional probability of $\mathbf{B}$ given $\mathbf{A}$.

outcomes in **A**. We then find in what fraction of *those* outcomes **B** also occurred. Formally, we write

$$P(\mathbf{B}|\mathbf{A}) = \frac{P(\mathbf{A} \text{ and } \mathbf{B})}{P(\mathbf{A})}.$$

Thinking this through, we can see that it's just what we've been doing, but now with probabilities rather than with counts. Look back at the girls for whom sports was the goal. How did we calculate $P(\text{sports}|\text{girl})$?

The rule says to use probabilities. It says to find $P(\mathbf{A} \text{ and } \mathbf{B})/P(\mathbf{A})$. The result is the same whether we use counts or probabilities because the total number in the sample cancels out:

$$\frac{P(\text{sports and girl})}{P(\text{girl})} = \frac{30/478}{251/478} = \frac{30}{251} = 0.12.$$

Activity: Conditional Probability. Simulation is great for seeing conditional probabilities at work.

To use the formula for conditional probability, we're supposed to insist on one restriction. The formula doesn't work if $P(\mathbf{A})$ is 0. After all, we can't be "given" the fact that **A** was true if the probability of **A** is 0!

Let's take our rule out for a spin. What's the probability that we have selected a girl *given* that the selected student's goal is popularity? Applying the rule, we get

$$P(\text{girl}|\text{popular}) = \frac{P(\text{girl and popular})}{P(\text{popular})}$$

$$= \frac{91/478}{141/478} = \frac{91}{141} = 0.65.$$

For Example FINDING A CONDITIONAL PROBABILITY

RECAP: Our survey found that 33% of college students are in a relationship, 25% are involved in sports, and 11% are both.

QUESTION: While dining in a campus facility open only to students, you meet a student who is on a sports team. What is the probability that your new acquaintance is in a relationship?

ANSWER: Let **R** = {student in a relationship} and **S** = {student is involved in sports}.

$P(\text{student is in a relationship given that the student is involved in sports}) = P(\mathbf{R}|\mathbf{S})$

$$= \frac{P(\mathbf{R} \text{ and } \mathbf{S})}{P(\mathbf{S})}$$

$$= \frac{0.11}{0.25}$$

$$\approx 0.44$$

There's a probability of about 0.44 that a student involved with sports is in a relationship. Notice that this is higher than the probability for all students.

The General Multiplication Rule

Remember the Multiplication Rule for the probability of **A** and **B**? It said

$P(\mathbf{A} \text{ and } \mathbf{B}) = P(\mathbf{A}) \times P(\mathbf{B})$ when **A** and **B** are independent.

Now we can write a more general rule that doesn't require independence. In fact, we've *already* written it down. We just need to rearrange the equation a bit.

The equation in the definition for conditional probability contains the probability of **A** *and* **B**. Rewriting the equation gives

$$P(\mathbf{A} \text{ and } \mathbf{B}) = P(\mathbf{A}) \times P(\mathbf{B}|\mathbf{A}).$$

A S *Activity:* **The General Multiplication Rule.** The best way to understand the General Multiplication Rule is with an experiment.

This is a **General Multiplication Rule** for compound events that does not require the events to be independent. Better than that, it even makes sense. The probability that two events, **A** and **B**, *both* occur is the probability that event **A** occurs multiplied by the probability that event **B** *also* occurs—that is, by the probability that event **B** occurs *given* that event **A** occurs.

Of course, there's nothing special about which set we call **A** and which one we call **B**. We should be able to state this the other way around. And indeed we can. It is equally true that

$$P(\mathbf{A} \text{ and } \mathbf{B}) = P(\mathbf{B}) \times P(\mathbf{A}|\mathbf{B}).$$

14.3 Independence

> **INDEPENDENCE**
> If we had to pick one idea in this chapter that you should understand and remember, it's the definition and meaning of independence. We'll need this idea in every one of the chapters that follow.

A S *Activity:* **Independence.** *Are Smoking* and *Low Birth Weight* independent?

Let's return to the question of just what it means for events to be independent. We've said informally that what we mean by independence is that the outcome of one event does not influence the probability of the other. With our new notation for conditional probabilities, we can write a formal definition: Events **A** and **B** are **independent** if (and *only* if)

$$P(\mathbf{B}|\mathbf{A}) = P(\mathbf{B}).$$

From now on, use the formal rule whenever you can. The Multiplication Rule for independent events that we saw in Chapter 13 is just a special case of the General Multiplication Rule. The general rule says

$$P(\mathbf{A} \text{ and } \mathbf{B}) = P(\mathbf{A}) \times P(\mathbf{B}|\mathbf{A}).$$

whether the events are independent or not. But when events **A** and **B** are independent, we can write $P(\mathbf{B})$ for $P(\mathbf{B}|\mathbf{A})$ and we get back our simple rule:

$$P(\mathbf{A} \text{ and } \mathbf{B}) = P(\mathbf{A}) \times P(\mathbf{B}).$$

Sometimes people use this statement as the definition of independent events, but we find the other definition more intuitive. Either way, the idea is that for independent events, the probability of one doesn't change when the other occurs.

Is the probability of having good grades as a goal independent of the sex of the responding student? Looks like it might be. We need to check whether

$$P(\text{grades}|\text{girl}) = P(\text{grades})$$

$$\frac{130}{251} = 0.52 \stackrel{?}{=} \frac{247}{478} = 0.52$$

To two decimal place accuracy, it looks like we can consider choosing good grades as a goal to be independent of sex.

On the other hand, $P(\text{sports})$ is 90/478, or about 18.8%, but $P(\text{sports}|\text{boy})$ is $60/227 = 26.4\%$. Because these probabilities aren't equal, we can be pretty sure that choosing success in sports as a goal is not independent of the student's sex.

For Example CHECKING FOR INDEPENDENCE

RECAP: Our survey told us that 33% of college students are in a relationship, 25% are involved in sports, and 11% are both.

QUESTION: Are being in a relationship and being involved in sports independent? Are they disjoint?

> **ANSWER:** Let **R** = {student is in a relationship} and **S** = {student is involved in sports}. If these events are independent, then knowing that a student does sports doesn't affect the probability that he or she is in a relationship.
>
> But we saw that $P(R|S) = P(R \text{ and } S)/P(S) = 0.44$ and we know that $P(R) = 0.33$, so the events are not independent. Students who are involved in sports are more likely to be in a relationship on this campus.
>
> These events are not disjoint either. In fact, 11% of all students on this campus are both involved in sports and in a relationship.

Independent ≠ Disjoint

Are disjoint events independent? Both disjoint and independent seem to imply separation and distinctness, but in fact disjoint events *cannot* be independent.[2] Let's see why. Consider the two disjoint events {you get an A in this course} and {you get a B in this course}. They're disjoint because they have no outcomes in common. Suppose you learn that you *did* get an A in the course. Now what is the probability that you got a B? You can't get both grades, so it must be 0.

Think about what that means. Knowing that the first event (getting an A) occurred changed your probability for the second event (down to 0). So these events aren't independent.

Mutually exclusive events can't be independent. They have no outcomes in common, so if one occurs, the other doesn't. A common error is to treat disjoint events as if they were independent, and apply the Multiplication Rule for independent events. Don't make that mistake.

Figure 14.2
Because events **A** and **B** are mutually exclusive, learning that **A** happened tells us that **B** didn't. The probability of **B** has changed from whatever it was to zero. So disjoint events **A** and **B** are not independent.

Just Checking

2. The American Association for Public Opinion Research (AAPOR) is an association of about 1600 individuals who share an interest in public opinion and survey research. They report that typically as few as 10% of random phone calls result in a completed interview. Reasons are varied, but some of the most common include no answer, refusal to cooperate, and failure to complete the call.
   Which of the following events are independent, which are disjoint, and which are neither independent nor disjoint?

   a) **A** = Your telephone number is randomly selected. **B** = You're not at home at dinnertime when they call.

   b) **A** = As a selected subject, you complete the interview. **B** = As a selected subject, you refuse to cooperate.

   c) **A** = You are not at home when they call at 11 a.m. **B** = You are employed full-time.

 Video: Is There a Hot Hand in Basketball? Most coaches and fans believe that basketball players sometimes get "hot" and make more of their shots. What do the conditional probabilities say?

Depending on Independence

It's much easier to think about independent events than to deal with conditional probabilities, but treating all events as independent can lead to wrong, and possibly costly, decisions.

---
[2] Well, technically two disjoint events *can* be independent, but only if the probability of one of the events is 0. For practical purposes, though, we can ignore this case. After all, as statisticians we don't anticipate having data about things that never happen.

Activity: Hot Hand Simulation. Can you tell the difference between real and simulated sequences of basketball shot hits and misses?

For example, insurance companies that insure against natural disasters like hurricanes or earthquakes might want to avoid insuring too many buildings in the same city. For an event like the Japan earthquake, the probability of two buildings suffering damage is not independent if they're in the same city. Even though the probability of a building being destroyed by an earthquake may be small, the probability that two neighboring buildings are both destroyed by an earthquake is not the product of the two. In fact, it may be close to the probability that a single one is destroyed. The probabilities can't be multiplied unless the events are independent. For that reason, it's safer for a company to hold policies on buildings that are geographically spread out because then the events are closer to being independent.

Whenever you see probabilities multiplied together, stop and ask whether you think they are really independent.

14.4 Picturing Probability: Tables, Venn Diagrams, and Trees

In the Facebook–Twitter example, we found that 71% of online adults use Facebook, 18% use Twitter, and 15% use both. That may not look like enough information to answer all the questions you[3] might have about the problem. But often, the right picture can help. Surprisingly there are three very different ways to do this. We've seen two of them already: Venn diagrams and probability tables. We'll meet a third in a minute.

Let's try to put what we know into a table. Translating percentages to probabilities, what we know looks like this:

		Use Facebook		
		Yes	No	Total
Use Twitter	Yes	0.15		0.18
	No			
	Total	0.71		1.00

Notice that the 0.71 and 0.18 are *marginal* probabilities and so they go into the *margins*. The 0.15 is the probability of using both sites—Facebook and Twitter—so that's a *joint* probability. Those belong in the interior of the table.

Because the cells of the table show disjoint events, the probabilities always add to the marginal totals going across rows or down columns. So, filling in the rest of the table is quick:

		Use Facebook		
		Yes	No	Total
Use Twitter	Yes	0.15	0.03	0.18
	No	0.56	0.26	0.82
	Total	0.71	0.29	1.00

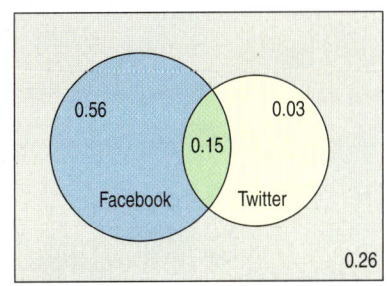

Compare this with the Venn diagram. Notice which entries in the table match up with the sets in this diagram. Whether a Venn diagram or a table is better to use will depend on what you are given and the questions you're being asked. Try both.

---

[3] or your instructor.

Step-by-Step Example: ARE THE EVENTS DISJOINT? INDEPENDENT?

Let's take another look at the social networking survey. Researchers report that 71% of online adults use Facebook, 18% use Twitter, and 15% use both.

Questions: 1. Are using Facebook and using Twitter mutually exclusive?
2. Are Facebooking and tweeting independent?

THINK → Plan Define the events we're interested in. State the given probabilities.

Let A = {respondent uses Facebook}.
Let B = {respondent uses Twitter}.

I know that $P(A) = 0.71$
$P(B) = 0.18$
$P(A \text{ and } B) = 0.15$

Question 1. Are using Facebook and using Twitter mutually exclusive?

SHOW → Mechanics Disjoint events cannot *both* happen at the same time, so check to see if $P(A \text{ and } B) = 0$.

$P(A \text{ and } B) = 0.15$. Since some adults use both Facebook and Twitter, $P(A \text{ and } B) \neq 0$. The events are not mutually exclusive.

TELL → Conclusion State your conclusion in context.

15% of online adults use both Facebook and Twitter, so these are not disjoint events.

Question 2. Are Facebooking and tweeting independent?

THINK → Plan Make a table.

		Use Facebook		
		Yes	No	Total
Use Twitter	Yes	0.15	0.03	0.18
	No	0.56	0.26	0.82
	Total	0.71	0.29	1.00

SHOW → Mechanics Does using Facebook change the probability of using twitter? That is, does $P(B|A) = P(B)$?

Because the two probabilities are *not* the same, the events are not independent.

$P(B|A) = \dfrac{P(A \text{ and } B)}{P(A)} = \dfrac{0.15}{0.71} \approx 0.21$

$P(B) = 0.62$

$P(B|A) \neq P(B)$

TELL → Conclusion Interpret your results in context.

Overall, 18% of online adults use Twitter, but 21% of Facebook users use Twitter. Since respondents who use Facebook are more likely to tweet, the two events are not independent.

Just Checking

3. Remember our sample of pages in this book from the earlier Just Checking...?

   48% of pages had a data display,
   27% of pages had an equation, and
   7% of pages had both a data display and an equation.

   a) Make a contingency table for the variables *display* and *equation*.
   b) What is the probability that a randomly selected sample page with an equation also had a data display?
   c) Are having an equation and having a data display disjoint events?
   d) Are having an equation and having a data display independent events?

"Why," said the Dodo, "the best way to explain it is to do it."

—Lewis Carroll

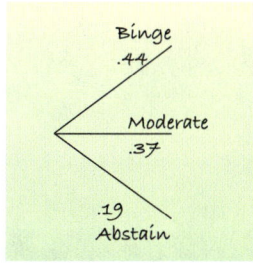

Figure 14.3
We can diagram the three outcomes of drinking and indicate their respective probabilities with a simple tree diagram.

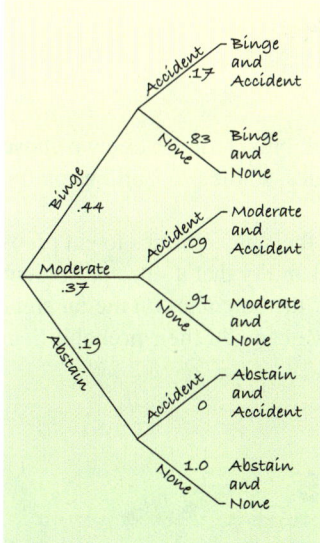

Figure 14.4
Extending the tree diagram, we can show both drinking and accident outcomes. The accident probabilities are conditional on the drinking outcomes, and they change depending on which branch we follow. Because we are concerned only with *alcohol-related* accidents, the conditional probability $P(\text{accident} | \text{abstinence})$ must be 0.

Tree Diagrams

For men, binge drinking is defined as having five or more drinks in a row, and for women as having four or more drinks in a row. (The difference is because of the average difference in weight.) According to a study by the Harvard School of Public Health, 44% of college students engage in binge drinking, 37% drink moderately, and 19% abstain entirely. Another study, published in the *American Journal of Health Behavior*, finds that among binge drinkers aged 21 to 34, 17% have been involved in an alcohol-related automobile accident, while among non-bingers of the same age, only 9% have been involved in such accidents.

What's the probability that a randomly selected college student will be a binge drinker and has had an alcohol-related car accident?

To start, we see that the probability of selecting a binge drinker is about 44%. To find the probability of selecting someone who is both a binge drinker and a driver with an alcohol-related accident, we would need to pull out the General Multiplication Rule and multiply the probability of one of the events by the conditional probability of the other given the first.

Or we *could* make a picture. Which would you prefer?

We thought so.

Neither tables nor Venn diagrams can handle conditional probabilities. The kind of picture that helps us look at conditional probabilities is called a **tree diagram** because it shows sequences of events as paths that look like branches of a tree. It is a good idea to make a tree diagram almost any time you plan to use the General Multiplication Rule. The number of different paths we can take can get large, so we usually draw the tree starting from the left and growing vine-like across the page, although sometimes you'll see them drawn from the bottom up or top down.

The first branch of our tree (Figure 14.3) separates students according to their drinking habits. We label each branch of the tree with a possible outcome and its corresponding probability.

Notice that we cover all possible outcomes with the branches. The probabilities add up to one. But we're also interested in car accidents. The probability of having an alcohol-related accident *depends* on one's drinking behavior. Because the probabilities are *conditional*, we draw the alternatives separately on each branch of the tree.

On each of the second set of branches, we write the possible outcomes associated with having an alcohol-related car accident (having an accident or not) and the associated probability (Figure 14.4). These probabilities are different because they are *conditional* depending on the student's drinking behavior. (It shouldn't be too surprising that those who binge drink have a higher probability of alcohol-related accidents.) The probabilities add up to one, because given the outcome on the first branch, these outcomes cover all the possibilities. Looking back at the General Multiplication Rule, we can see how the tree depicts the calculation. To find the probability that a randomly selected student will be a binge drinker who has had an alcohol-related car accident, we follow the top branches. The probability of selecting a binger is 0.44. The conditional probability of an accident *given* binge drinking is 0.17. The General Multiplication Rule tells us that to find the *joint* probability of being a binge drinker and having an accident, we multiply these two probabilities together:

$$P(\text{binge and accident}) = P(\text{binge}) \times P(\text{accident}|\text{binge})$$
$$= 0.44 \times 0.17 = 0.075$$

And we can do the same for each combination of outcomes:

Figure 14.5
We can find the probabilities of compound events by multiplying the probabilities along the branch of the tree that leads to the event, just the way the General Multiplication Rule specifies. The probability of abstaining and having an alcohol-related accident is, of course, zero.

All the outcomes at the far right are disjoint because at each branch of the tree we chose between disjoint alternatives. And they are *all* the possibilities, so the probabilities on the far right must add up to one.

Because the final outcomes are disjoint, we can add up their probabilities to get probabilities for compound events. For example, what's the probability that a selected student has had an alcohol-related car accident? We simply find *all* the outcomes on the far right in which an accident has happened. There are three and we can add their probabilities: $0.075 + 0.033 + 0 = 0.108$—almost an 11% chance.

14.5 Reversing the Conditioning and Bayes' Rule

If we know a student has had an alcohol-related accident, what's the probability that the student is a binge drinker? That's an interesting question, but we can't just read it from the tree. The tree gives us $P(\text{accident}|\text{binge})$, but we want $P(\text{binge}|\text{accident})$—conditioning in the other direction. The two probabilities are definitely *not* the same. We have reversed the conditioning.

We may not have the conditional probability we want, but we do know everything we need to know to find it. To find a conditional probability, we need the probability that both events happen divided by the probability that the given event occurs. We have already found the probability of an alcohol-related accident: $0.075 + 0.033 + 0 = 0.108$.

The joint probability that a student is both a binge drinker and someone who's had an alcohol-related accident is found at the top branch: 0.075. We've restricted the *Who* of the problem to the students with alcohol-related accidents, so we divide the two to find the conditional probability:

$$P(\text{binge} \mid \text{accident}) = \frac{P(\text{binge and accident})}{P(\text{accident})}$$

$$= \frac{0.075}{0.108} = 0.694$$

The chance that a student who has an alcohol-related car accident is a binge drinker is more than 69%! As we said, reversing the conditioning is rarely intuitive, but tree diagrams help us keep track of the calculation when there aren't too many alternatives to consider.

Step-by-Step Example REVERSING THE CONDITIONING

When the authors were in college, there were only three requirements for graduation that were the same for all students: You had to be able to tread water for 2 minutes, you had to learn a foreign language, and you had to be free of tuberculosis. For the last requirement, all freshmen had to take a TB screening test that consisted of a nurse jabbing what looked like a corncob holder into your forearm. You were then expected to report back in 48 hours to have it checked. If you were healthy and TB-free, your arm was supposed to look as though you'd never had the test.

Sometime during the 48 hours, one of us had a reaction. When he finally saw the nurse, his arm was about 50% bigger than normal and a very unhealthy red. Did he have TB? The nurse had said that the test was about 99% effective, so it seemed that the chances must be pretty high that he had TB. How high do you think the chances were? Go ahead and guess. Guess low.

We'll call **TB** the event of actually having TB and **+** the event of testing positive. To start a tree, we need to know $P(\textbf{TB})$, the probability of having TB.[4] We also need to know the conditional probabilities $P(+ \mid \textbf{TB})$ and $P(+ \mid \textbf{TB}^C)$. Diagnostic tests can make two kinds of errors. They can give a positive result for a healthy person (a *false positive*) or a negative result for a sick person (a *false negative*). Being 99% accurate usually means a false-positive rate of 1%. That is, someone who doesn't have the disease has a 1% chance of testing positive anyway. We can write $P(+ \mid \textbf{TB}^C) = 0.01$.

Since a false negative is more serious (because a sick person might not get treatment), tests are usually constructed to have a lower false-negative rate. We don't know exactly, but let's assume a 0.1% false-negative rate. So only 0.1% of sick people test negative. We can write $P(- \mid \textbf{TB}) = 0.001$.

THINK ▶ Plan Define the events we're interested in and their probabilities.

Let **TB** = {having TB} and **TB**^{C} = {no TB}
+ = {testing positive} and
− = {testing negative}

I know that $P(+ \mid \textbf{TB}^C) = 0.01$ and $P(- \mid \textbf{TB}) = 0.001$. I also know that $P(\textbf{TB}) = 0.00005$.

Figure out what you want to know in terms of the events. Use the notation of conditional probability to write the event whose probability you want to find.

I'm interested in the probability that the author had TB given that he tested positive: $P(\textbf{TB} \mid +)$.

---
[4]This isn't given, so we looked it up. Although TB is a matter of serious concern to public health officials, it is a fairly uncommon disease, with an incidence of about 5 cases per 100,000 in the United States (see www.cdc.gov/tb/default.htm).

SHOW ➡ Plot Draw the tree diagram. When probabilities are very small like these are, be careful to keep all the significant digits.

To finish the tree we need $P(\mathbf{TB}^C)$, $P(-|\mathbf{TB}^C)$, and $P(+|\mathbf{TB})$. We can find each of these from the Complement Rule:

$$P(\mathbf{TB}^C) = 1 - P(\mathbf{TB}) = 0.99995$$
$$P(-|\mathbf{TB}^C) = 1 - P(+|\mathbf{TB}^C)$$
$$= 1 - 0.01 = 0.99 \text{ and}$$
$$P(+|\mathbf{TB}) = 1 - P(-|\mathbf{TB})$$
$$= 1 - 0.001 = 0.999$$

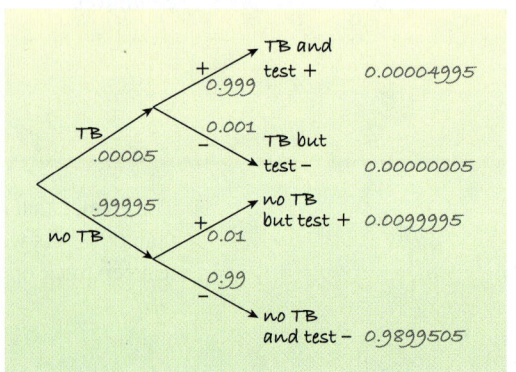

Mechanics Multiply along the branches to find the probabilities of the four possible outcomes. Check your work by seeing if they total 1.

Add up the probabilities corresponding to the condition of interest—in this case, testing positive. We can add the probabilities from the tree twigs that correspond to testing positive because the tree shows disjoint events.

Divide the probability of both events occurring (here, having TB and a positive test) by the probability of satisfying the condition (testing positive).

(Check: $0.00004995 + 0.00000005 + 0.0099995 + 0.98995050 = 1$)

$$P(+) = P(\mathbf{TB} \text{ and } +) + P(\mathbf{TB}^C \text{ and } +)$$
$$P(+) = 0.00004995 + 0.0099995$$
$$= 0.01004945$$

$$P(\mathbf{TB}|+) = \frac{P(\mathbf{TB} \text{ and } +)}{P(+)}$$
$$= \frac{0.00004995}{0.01004945}$$
$$= 0.00497$$

TELL ➡ Conclusion Interpret your result in context.

The chance of having TB after a positive test is less than 0.5%.

When we reverse the order of conditioning, we change the *Who* we are concerned with. With events of low probability, the result can be surprising. That's the reason patients who test positive for HIV, for example, are always told to seek medical counseling. They may have only a small chance of actually being infected. That's why global drug or disease testing can have unexpected consequences if people interpret *testing* positive as *being* positive.

The Reverend Thomas Bayes is credited posthumously with the rule that is the foundation of Bayesian Statistics.

Bayes' Rule

When we have $P(\mathbf{A}|\mathbf{B})$ but want the *reverse* probability $P(\mathbf{B}|\mathbf{A})$, we need to find $P(\mathbf{A} \text{ and } \mathbf{B})$ and $P(\mathbf{A})$. A tree is often a convenient way of finding these probabilities. It can work even when we have more than two possible events, as we saw in the binge-drinking example. Instead of using the tree, we *could* write the calculation algebraically, showing exactly how we found the quantities that we needed: $P(\mathbf{A} \text{ and } \mathbf{B})$ and $P(\mathbf{A})$. The result is a formula known as Bayes' Rule, after the Reverend Thomas Bayes (1702?–1761), who was credited with the rule after his death, when he could no longer defend himself. Bayes' Rule is quite important in Statistics and is the foundation of an approach to Statistical analysis known as Bayesian Statistics. Although the simple rule deals with two alternative outcomes, the rule can be extended to the situation in which there are more than two branches to the first split of the tree. The principle remains the same (although the math gets more difficult). Bayes' Rule is just a formula for reversing the probability from the conditional probability that you're originally given. Bayes' Rule for two events says that $P(\mathbf{B}|\mathbf{A}) = \dfrac{P(\mathbf{A}|\mathbf{B})P(\mathbf{B})}{P(\mathbf{A}|\mathbf{B})P(\mathbf{B}) + P(\mathbf{A}|\mathbf{B}^C)P(\mathbf{B}^C)}$. Masochists may wish to try it with the TB testing probabilities. (It's easier to just draw the tree, isn't it?)

> ### For Example REVERSING THE CONDITIONING
>
> A recent Maryland highway safety study found that in 77% of all accidents the driver was wearing a seatbelt. Accident reports indicated that 92% of those drivers escaped serious injury (defined as hospitalization or death), but only 63% of the nonbelted drivers were so fortunate.
>
> **QUESTION:** What's the probability that a driver who was seriously injured wasn't wearing a seatbelt?
>
> **ANSWER:** Let B = the driver was wearing a seatbelt, and NB = no belt.
> Let I = serious injury or death, and OK = not seriously injured.
> I know $P(B) = 0.77$, so $P(NB) = 1 - 0.77 = 0.23$.
> Also, $P(OK|B) = 0.92$, so $P(I|B) = 0.08$
> and $P(OK|NB) = 0.63$, so $P(I|NB) = 0.37$
>
>
>
> $$P(NB|I) = \dfrac{P(NB \text{ and } I)}{P(I)} = \dfrac{0.0851}{0.0616 + 0.0851} = 0.58$$
>
> Even though only 23% of drivers weren't wearing seatbelts, they accounted for 58% of all the deaths and serious injuries.
>
> Just some advice from your friends, the authors: *Please buckle up!* (We want you to finish this course.)

Nicholas Saunderson (1682–1739) was a blind English mathematician who invented a tactile board to help other blind people do mathematics. And he may have been the true originator of "Bayes' Rule."

> ### Who discovered Bayes' Rule?
> Stigler's "Law of Eponymy" states that discoveries named for someone (eponyms) are usually named for the wrong person. Steven Stigler, who admits he didn't originate the law, is an expert on the history of Statistics, and he suspected that the law might apply to Bayes' Rule. He looked at the possibility that another candidate–one Nicholas Saunderson–was the real discoverer, not the Reverend Bayes. He assembled historical evidence and compared probabilities that the historical events would have happened *given* that Bayes was the discoverer of the rule, with the corresponding probabilities *given* that Saunderson was the discoverer. Of course, what he really wanted to know were the probabilities that Bayes or Saunderson was the discoverer *given* the historical events. How did he *reverse* the conditional probabilities? He used Bayes' Rule and concluded that, actually, it's more likely that Saunderson is the real originator of the rule.
>
> But that doesn't change our tradition of naming the rule for Bayes and calling the branch of Statistics arising from this approach Bayesian Statistics.

WHAT CAN GO WRONG?

- **Don't use a simple probability rule where a general rule is appropriate.** Don't assume independence without reason to believe it. Don't assume that outcomes are disjoint without checking that they are. Remember that the general rules always apply, even when outcomes are in fact independent or disjoint.

- **Don't reverse conditioning naïvely.** As we have seen, the probability of **A** *given* **B** may not, and, in general does not, resemble the probability of **B** *given* **A**. The true probability may be counterintuitive.

- **Don't confuse "disjoint" with "independent."** Disjoint events *cannot* happen at the same time. When one happens, you know the other did not, so $P(\mathbf{B}|\mathbf{A}) = 0$. Independent events *must* be able to happen at the same time. When one happens, you know it has no effect on the other, so $P(\mathbf{B}|\mathbf{A}) = P(\mathbf{B})$.

CONNECTIONS

This chapter shows the unintuitive side of probability. If you've been thinking, "My mind doesn't work this way," you're probably right. Humans don't seem to find conditional and compound probabilities natural and often have trouble with them. But, by being careful to follow the rules outlined in this chapter you can avoid the common mistakes.

Our central connection is to the guiding principle that Statistics is about understanding the world. The events discussed in this chapter are close to the kinds of real-world situations in which understanding probabilities matters. The methods and concepts of this chapter are the tools you need to understand the part of the real world that deals with the outcomes of complex, uncertain events.

What Have We Learned?

Learning Objectives

Know the general rules of probability and how to apply them
- The General Addition Rule says that $P(\mathbf{A} \text{ or } \mathbf{B}) = P(\mathbf{A}) + P(\mathbf{B}) - P(\mathbf{A} \text{ and } \mathbf{B})$
- The General Multiplication Rule says that $P(\mathbf{A} \text{ and } \mathbf{B}) = P(\mathbf{A}) \times P(\mathbf{B}|\mathbf{A})$

Know that the conditional probability of an event **B** given the event **A** is $P(\mathbf{B}|\mathbf{A}) = P(\mathbf{A} \text{ and } \mathbf{B})/P(\mathbf{A})$

Know how to define and use independence.
- Events **A** and **B** are independent if $P(\mathbf{A}|\mathbf{B}) = P(\mathbf{A})$

Know how to make and use tables, Venn diagrams, and tree diagrams to solve problems involving probability

Know how to use tree diagrams and Bayes' Rule to reverse the conditioning

Review of Terms

General Addition Rule For any two events, **A** and **B**, the probability of **A** *or* **B** is (p. 367)

$$P(\mathbf{A} \text{ or } \mathbf{B}) = P(\mathbf{A}) + P(\mathbf{B}) - P(\mathbf{A} \text{ and } \mathbf{B}).$$

Conditional probability The conditional probability of the event **B** given the event **A** has occurred is:

$$P(\mathbf{B}|\mathbf{A}) = \frac{P(\mathbf{A} \text{ and } \mathbf{B})}{P(\mathbf{A})}.$$

$P(\mathbf{B}|\mathbf{A})$ is read "the probability of **B** *given* **A**" (p. 371).

General Multiplication Rule   For any two events, **A** and **B**, the probability of **A** and **B** is (p. 373)

$$P(A \text{ and } B) = P(A) \times P(B|A).$$

Independence (used formally)   Events **A** and **B** are independent when $P(B|A) = P(B)$ (p. 373).

Tree diagram   A display of conditional events or probabilities that is helpful in thinking through conditioning (p. 377).

Exercises

Section 14.1

1. **Pet ownership** Suppose that 25% of people have a dog, 29% of people have a cat, and 12% of people own both. What is the probability that someone owns a dog or a cat?

2. **Cooking and shopping** Forty-five percent of Americans like to cook and 59% of Americans like to shop, while 23% enjoy both activities. What is the probability that a randomly selected American either enjoys cooking or shopping or both?

Section 14.2

3. **Sports** What is the probability that a person likes to watch football, given that she also likes to watch basketball?

	Football	No Football
Basketball	27	13
No Basketball	38	22

4. **Sports again** From Exercise 3, if someone doesn't like to watch basketball, what is the probability that she will be a football fan?

5. **Late to the train** A student figures that he has a 30% chance of being let out of class late. If he leaves class late, there is a 45% chance that he will miss his train. What is the probability that it will cause him to miss the train?

6. **Field goals** A nervous kicker usually makes 70% of his first field goal attempts. If he makes his first attempt, his success rate rises to 90%. What is the probability that he makes his first two kicks?

Section 14.3

7. **Titanic** On the *Titanic*, the probability of survival was 0.323. Among first class passengers, it was 0.625. Were *survival* and *ticket class* independent? Explain.

8. **Births** If the sex of a child is independent of all other births, is the probability of a woman giving birth to a girl after having four boys greater than it was on her first birth? Explain.

Section 14.4

9. **Facebook** Facebook reports that 70% of its users are from outside the United States and that 50% of its users log on to Facebook every day. Suppose that 20% of its users are U.S. users who log on every day. Make a probability table. Why is a table better than a tree here?

10. **Online banking** A national survey indicated that 30% of adults conduct their banking online. It also found that 40% are younger than 50, and that 25% are younger than 50 and conduct their banking online. Make a probability table. Why is a table better than a tree here?

11. **Facebook again** Suppose that the information in Exercise 9 had been presented in the following way. Facebook reports that 70% of its users are from outside the United States. Of the U.S. users, two-thirds log on every day. Of the non-U.S. users, three-sevenths log on every day. Draw a tree for this situation. Why is the tree better than the table in this case? Where are the joint probabilities found?

12. **Online banking again** Suppose that the information in Exercise 10 had been presented in the following way. A national survey of bank customers finds that 40% are younger than 50. Of those younger than 50, 5 of 8 conduct their banking online. Of those older than 50, only 1 of 12 banks online. Draw a tree for this situation. Why is the tree better than the table in this case? Where are the joint probabilities found?

Section 14.5

13. **Facebook final** Given the probabilities in Exercise 11, what is the probability that a person is from the United States given that he logs on to Facebook every day? Has the probability that he is from the United States increased or decreased with the additional information?

14. **Online banking last time** Given the probabilities in Exercise 12, what is the probability that a person is younger than 50 given that she uses online banking? Has the probability that she is younger than 50 increased or decreased with the additional information?

Chapter Exercises

15. Phones Recent research suggests that 73% of Americans have a home phone, 83% have a cell phone, and 58% of people have both. What is the probability that an American has
a) a home or cell phone?
b) neither a home phone nor a cell phone?
c) a cell phone but no home phone?

16. Travel Suppose the probability that a U.S. resident has traveled to Canada is 0.18, to Mexico is 0.09, and to both countries is 0.04. What's the probability that an American chosen at random has
a) traveled to Canada but not Mexico?
b) traveled to either Canada or Mexico?
c) not traveled to either country?

17. Amenities A check of dorm rooms on a large college campus revealed that 38% had refrigerators, 52% had TVs, and 21% had both a TV and a refrigerator. What's the probability that a randomly selected dorm room has
a) a TV but no refrigerator?
b) a TV or a refrigerator, but not both?
c) neither a TV nor a refrigerator?

18. Workers Employment data at a large company reveal that 72% of the workers are married, that 44% are college graduates, and that half of the college grads are married. What's the probability that a randomly chosen worker
a) is neither married nor a college graduate?
b) is married but not a college graduate?
c) is married or a college graduate?

19. Global survey The marketing research organization GfK Roper conducts a yearly survey on consumer attitudes worldwide. They collect demographic information on the roughly 1500 respondents from each country that they survey. Here is a table showing the number of people with various levels of education in five countries:

Educational Level by Country

	Post-Graduate	College	Some High School	Primary or Less	No Answer	Total
China	7	315	671	506	3	1502
France	69	388	766	309	7	1539
India	161	514	622	227	11	1535
U.K.	58	207	1240	32	20	1557
U.S.	84	486	896	87	4	1557
Total	379	1910	4195	1161	45	7690

If we select someone at random from this survey,
a) what is the probability that the person is from the United States?
b) what is the probability that the person completed his or her education before college?
c) what is the probability that the person is from France *or* did some post-graduate study?
d) what is the probability that the person is from France *and* finished only primary school or less?

20. Birth order A survey of students in a large Introductory Statistics class asked about their birth order (1 = oldest or only child) and which college of the university they were enrolled in. Here are the data:

		Birth Order		
		1 or Only	2 or More	Total
College	Arts & Sciences	34	23	57
	Agriculture	52	41	93
	Human Ecology	15	28	43
	Other	12	18	30
	Total	113	110	223

Suppose we select a student at random from this class. What is the probability that the person is
a) a Human Ecology student?
b) a firstborn student?
c) firstborn *and* a Human Ecology student?
d) firstborn *or* a Human Ecology student?

21. Cards You draw a card at random from a standard deck of 52 cards. Find each of the following conditional probabilities:
a) The card is a heart, given that it is red.
b) The card is red, given that it is a heart.
c) The card is an ace, given that it is red.
d) The card is a queen, given that it is a face card.

22. Pets In its monthly report, the local animal shelter states that it currently has 24 dogs and 18 cats available for adoption. Eight of the dogs and 6 of the cats are male. Find each of the following conditional probabilities if an animal is selected at random:
a) The pet is male, given that it is a cat.
b) The pet is a cat, given that it is female.
c) The pet is female, given that it is a dog.

23. Health The probabilities that an adult American man has high blood pressure and/or high cholesterol are shown in the table.

		Blood Pressure	
		High	OK
Cholesterol	High	0.11	0.21
	OK	0.16	0.52

What's the probability that
a) a man has both conditions?
b) a man has high blood pressure?
c) a man with high blood pressure has high cholesterol?
d) a man has high blood pressure if it's known that he has high cholesterol?

24. **Immigration** The table shows the political affiliations of U.S. voters and their positions on supporting stronger immigration enforcement.

		Stronger Immigration Enforcement		
		Favor	Oppose	No Opinion
Party	Republican	0.30	0.04	0.03
	Democrat	0.22	0.11	0.02
	Other	0.16	0.07	0.05

   a) What's the probability that
      i) a randomly chosen voter favors stronger immigration enforcement?
      ii) a Republican favors stronger enforcement?
      iii) a voter who favors stronger enforcement is a Democrat?
   b) A candidate thinks she has a good chance of gaining the votes of anyone who is a Republican or in favor of stronger enforcement of immigration policy. What proportion of voters is that?

25. **Global survey, take 2** Look again at the table summarizing the Roper survey in Exercise 19.
   a) If we select a respondent at random, what's the probability we choose a person from the United States who has done post-graduate study?
   b) Among the respondents who have done post-graduate study, what's the probability the person is from the United States?
   c) What's the probability that a respondent from the United States has done post-graduate study?
   d) What's the probability that a respondent from China has only a primary-level education?
   e) What's the probability that a respondent with only a primary-level education is from China?

26. **Birth order, take 2** Look again at the data about birth order of Intro Stats students and their choices of colleges shown in Exercise 20.
   a) If we select a student at random, what's the probability the person is an Arts and Sciences student who is a second child (or more)?
   b) Among the Arts and Sciences students, what's the probability a student was a second child (or more)?
   c) Among second children (or more), what's the probability the student is enrolled in Arts and Sciences?
   d) What's the probability that a first or only child is enrolled in the Agriculture College?
   e) What is the probability that an Agriculture student is a first or only child?

27. **Sick kids** Seventy percent of kids who visit a doctor have a fever, and 30% of kids with a fever also have sore throats. What's the probability that a kid who goes to the doctor has a fever and a sore throat?

28. **Sick cars** Twenty percent of cars that are inspected have faulty pollution control systems. The cost of repairing a pollution control system exceeds $100 about 40% of the time. When a driver takes her car in for inspection, what's the probability that she will end up paying more than $100 to repair the pollution control system?

29. **Cards** You are dealt a hand of three cards, one at a time. Find the probability of each of the following.
   a) The first heart you get is the third card dealt.
   b) Your cards are all red (that is, all diamonds or hearts).
   c) You get no spades.
   d) You have at least one ace.

30. **Another hand** You pick three cards at random from a deck. Find the probability of each event described below.
   a) You get no aces.
   b) You get all hearts.
   c) The third card is your first red card.
   d) You have at least one diamond.

31. **Batteries** A junk box in your room contains a dozen old batteries, five of which are totally dead. You start picking batteries one at a time and testing them. Find the probability of each outcome.
   a) The first two you choose are both good.
   b) At least one of the first three works.
   c) The first four you pick all work.
   d) You have to pick five batteries to find one that works.

32. **Shirts** The soccer team's shirts have arrived in a big box, and people just start grabbing them, looking for the right size. The box contains 4 medium, 10 large, and 6 extra-large shirts. You want a medium for you and one for your sister. Find the probability of each event described.
   a) The first two you grab are the wrong sizes.
   b) The first medium shirt you find is the third one you check.
   c) The first four shirts you pick are all extra-large.
   d) At least one of the first four shirts you check is a medium.

33. **Eligibility** A university requires its biology majors to take a course called BioResearch. The prerequisite for this course is that students must have taken either a Statistics course or a computer course. By the time they are juniors, 52% of the Biology majors have taken Statistics, 23% have had a computer course, and 7% have done both.
   a) What percent of the junior Biology majors are ineligible for BioResearch?
   b) What's the probability that a junior Biology major who has taken Statistics has also taken a computer course?
   c) Are taking these two courses disjoint events? Explain.
   d) Are taking these two courses independent events? Explain.

34. Benefits Fifty-six percent of all American workers have a workplace retirement plan, 68% have health insurance, and 49% have both benefits. We select a worker at random.
  a) What's the probability he has neither employer-sponsored health insurance nor a retirement plan?
  b) What's the probability he has health insurance if he has a retirement plan?
  c) Are having health insurance and a retirement plan independent events? Explain.
  d) Are having these two benefits mutually exclusive? Explain.

35. Cell phones in the home A survey found that 73% of Americans have a home phone, 83% have a cell phone and 58% of people have both.
  a) If a person has a home phone, what's the probability that they have a cell phone also?
  b) Are having a home phone and a cell phone independent events? Explain.
  c) Are having a home phone and a cell phone mutually exclusive? Explain.

36. On the road again According to Exercise 16, the probability that a U.S. resident has traveled to Canada is 0.18, to Mexico is 0.09, and to both countries is 0.04.
  a) What's the probability that someone who has traveled to Mexico has visited Canada too?
  b) Are traveling to Mexico and to Canada disjoint events? Explain.
  c) Are traveling to Mexico and to Canada independent events? Explain.

37. Cards If you draw a card at random from a well-shuffled deck, is getting an ace independent of the suit? Explain.

38. Pets, again The local animal shelter in Exercise 22 reported that it currently has 24 dogs and 18 cats available for adoption; 8 of the dogs and 6 of the cats are male. Are the species and sex of the animals independent? Explain.

39. Unsafe food Early in 2010, *Consumer Reports* published the results of an extensive investigation of broiler chickens purchased from food stores in 23 states. Tests for bacteria in the meat showed that 62% of the chickens were contaminated with campylobacter, 14% with salmonella, and 9% with both.
  a) What's the probability that a tested chicken was not contaminated with either kind of bacteria?
  b) Are contamination with the two kinds of bacteria disjoint? Explain.
  c) Are contamination with the two kinds of bacteria independent? Explain.

40. Birth order, finis In Exercises 20 and 26, we looked at the birth orders and college choices of some Intro Stats students. For these students:
  a) Are enrolling in Agriculture and Human Ecology disjoint? Explain.
  b) Are enrolling in Agriculture and Human Ecology independent? Explain.
  c) Are being firstborn and enrolling in Human Ecology disjoint? Explain.
  d) Are being firstborn and enrolling in Human Ecology independent? Explain.

41. Men's health again Given the table of probabilities from Exercise 23, are high blood pressure and high cholesterol independent? Explain.

		Blood Pressure	
		High	OK
Cholesterol	High	0.11	0.21
	OK	0.16	0.52

42. Politics Given the table of probabilities from Exercise 24, are party affiliation and position on immigration independent? Explain.

		Stronger Immigration Enforcement		
		Favor	Oppose	No Opinion
Party	Republican	0.30	0.04	0.03
	Democrat	0.22	0.11	0.02
	Other	0.16	0.07	0.05

43. Phone service According to estimates from the federal government's 2010 National Health Interview Survey, based on face-to-face interviews in 16,676 households, approximately 63.6% of U.S. adults have both a landline in their residence and a cell phone, 25.4% have only cell phone service but no landline, and 1.8% have no telephone service at all.
  a) Polling agencies won't phone cell phone numbers because customers object to paying for such calls. What proportion of U.S. households can be reached by a landline call?
  b) Are having a cell phone and having a landline independent? Explain.

44. Snoring After surveying 995 adults, 81.5% of whom were over 30, the National Sleep Foundation reported that 36.8% of all the adults snored. 32% of the respondents were snorers over the age of 30.
  a) What percent of the respondents were 30 or less and did not snore?
  b) Is snoring independent of age? Explain.

45. Gender A poll conducted by Gallup classified respondents by sex and political party, as shown in the table. Is party affiliation independent of the respondents' sex? Explain.

	Democrat	Republican	Independent
Male	32	28	34
Female	41	25	26

46. Cars A random survey of autos parked in student and staff lots at a large university classified the brands by country of origin, as seen in the table. Is country of origin independent of type of driver?

	Driver	
Origin	Student	Staff
American	107	105
European	33	12
Asian	55	47

47. Luggage Leah is flying from Boston to Denver with a connection in Chicago. The probability her first flight leaves on time is 0.15. If the flight is on time, the probability that her luggage will make the connecting flight in Chicago is 0.95, but if the first flight is delayed, the probability that the luggage will make it is only 0.65.
 a) Are the first flight leaving on time and the luggage making the connection independent events? Explain.
 b) What is the probability that her luggage arrives in Denver with her?

48. Graduation A private college report contains these statistics:
 70% of incoming freshmen attended public schools.
 75% of public school students who enroll as freshmen eventually graduate.
 90% of other freshmen eventually graduate.
 a) Is there any evidence that a freshman's chances to graduate may depend upon what kind of high school the student attended? Explain.
 b) What percent of freshmen eventually graduate?

49. Late luggage Remember Leah (Exercise 47)? Suppose you pick her up at the Denver airport, and her luggage is not there. What is the probability that Leah's first flight was delayed?

50. Graduation, part II What percent of students who graduate from the college in Exercise 48 attended a public high school?

51. Absenteeism A company's records indicate that on any given day about 1% of their day-shift employees and 2% of the night-shift employees will miss work. Sixty percent of the employees work the day shift.
 a) Is absenteeism independent of shift worked? Explain.
 b) What percent of employees are absent on any given day?

52. E-readers PewInternet reported in January of 2014 that 32% of U.S. adults own at least one e-reader, and that 28% of U.S. adults read at least one e-book in the previous year (and thus, presumably, owned an e-reader). Overall, 76% of U.S. adults read at least one book (electronic or otherwise) in the previous year. (http://pewinternet.org/Reports/2014/E-Reading-Update/A-Snapshot-of-Reading-in-America-in-2013/Whos-reading-and-how-a-demographic-portrait.aspx).
 a) Explain how these statistics indicate that owning an e-reader and reading at least one book in the previous year are not independent.
 b) What's the probability that a randomly selected U.S. adult has an e-reader but didn't use it to read an e-book in the previous year?

53. Absenteeism, part II At the company described in Exercise 51, what percent of the absent employees are on the night shift?

54. E-readers II Given the e-reader data presented in Exercise 52.
 a) If a randomly selected U.S. adult has an e-reader, what is the probability that he or she hasn't read an e-book in the past year?
 b) Is it more or less likely that a randomly selected U.S. adult who does not own an e-reader would have read no books in the past year?

55. Drunks Police often set up sobriety checkpoints—roadblocks where drivers are asked a few brief questions to allow the officer to judge whether or not the person may have been drinking. If the officer does not suspect a problem, drivers are released to go on their way. Otherwise, drivers are detained for a Breathalyzer test that will determine whether or not they will be arrested. The police say that based on the brief initial stop, trained officers can make the right decision 80% of the time. Suppose the police operate a sobriety checkpoint after 9:00 p.m. on a Saturday night, a time when national traffic safety experts suspect that about 12% of drivers have been drinking.
 a) You are stopped at the checkpoint and, of course, have not been drinking. What's the probability that you are detained for further testing?
 b) What's the probability that any given driver will be detained?
 c) What's the probability that a driver who is detained has actually been drinking?
 d) What's the probability that a driver who was released had actually been drinking?

56. No-shows An airline offers discounted "advance-purchase" fares to customers who buy tickets more than 30 days before travel and charges "regular" fares for tickets purchased during those last 30 days. The company has noticed that 60% of its customers take advantage of the advance-purchase fares. The "no-show" rate among people who paid regular fares is 30%, but only 5% of customers with advance-purchase tickets are no-shows.
 a) What percent of all ticket holders are no-shows?
 b) What's the probability that a customer who didn't show had an advance-purchase ticket?
 c) Is being a no-show independent of the type of ticket a passenger holds? Explain.

57. Dishwashers Dan's Diner employs three dishwashers. Al washes 40% of the dishes and breaks only 1% of those he handles. Betty and Chuck each wash 30% of the dishes, and Betty breaks only 1% of hers, but Chuck breaks

3% of the dishes he washes. (He, of course, will need a new job soon. . . .) You go to Dan's for supper one night and hear a dish break at the sink. What's the probability that Chuck is on the job?

58. **Parts** A company manufacturing electronic components for home entertainment systems buys electrical connectors from three suppliers. The company prefers to use supplier A because only 1% of those connectors prove to be defective, but supplier A can deliver only 70% of the connectors needed. The company must also purchase connectors from two other suppliers, 20% from supplier B and the rest from supplier C. The rates of defective connectors from B and C are 2% and 4%, respectively. You buy one of these components, and when you try to use it you find that the connector is defective. What's the probability that your component came from supplier A?

59. **HIV testing** In July 2005, the journal *Annals of Internal Medicine* published a report on the reliability of HIV testing. Results of a large study suggested that among people with HIV, 99.7% of tests conducted were (correctly) positive, while for people without HIV 98.5% of the tests were (correctly) negative. A clinic serving an at-risk population offers free HIV testing, believing that 15% of the patients may actually carry HIV. What's the probability that a patient testing negative is truly free of HIV?

60. **Polygraphs** Lie detectors are controversial instruments, barred from use as evidence in many courts. Nonetheless, many employers use lie detector screening as part of their hiring process in the hope that they can avoid hiring people who might be dishonest. There has been some research, but no agreement, about the reliability of polygraph tests. Based on this research, suppose that a polygraph can detect 65% of lies, but incorrectly identifies 15% of true statements as lies.

A certain company believes that 95% of its job applicants are trustworthy. The company gives everyone a polygraph test, asking, "Have you ever stolen anything from your place of work?" Naturally, all the applicants answer "No," but the polygraph identifies some of those answers as lies, making the person ineligible for a job. What's the probability that a job applicant rejected under suspicion of dishonesty was actually trustworthy?

Just Checking ANSWERS

1. a)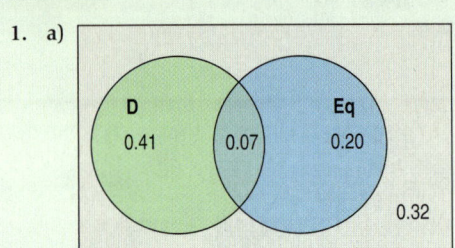

   b) 0.32
   c) 0.41

2. a) Independent
   b) Disjoint
   c) Neither

3. a)

	Equation		
	Yes	No	Total
Display Yes	0.07	0.41	0.48
Display No	0.20	0.32	0.52
Total	0.27	0.73	1.00

   b) $P(D|Eq) = P(D \text{ and } Eq)/P(Eq) = 0.07/0.27 = 0.259$
   c) No, pages can (and 7% do) have both.
   d) To be independent, we'd need $P(D|Eq) = P(D)$. $P(D/Eq) = 0.259$, but $P(D) = 0.48$. Overall, 48% of pages have data displays, but only about 26% of pages with equations do. They do not appear to be independent.

Chapter 15 Random Variables

15.1 Center: The Expected Value

15.2 Spread: The Standard Deviation

15.3 Shifting and Combining Random Variables

15.4 Continuous Random Variables

> **Where are we going?**
> How long do products last? Should you expect your computer to die just after the warranty runs out? How can you reduce your risk for developing hepatitis? Businesses, medical researchers, and other scientists all use probability to determine risk factors to help answer questions like these. To do that, they model the probability of outcomes using a special kind of variable— a *random* variable. Using random variables can help us talk about and predict random behavior.

Insurance companies make bets. They bet that you're going to live a long life. You bet that you're going to die sooner. Both you and the insurance company want the company to stay in business, so it's important to find a "fair price" for your bet. Of course, the right price for *you* depends on many factors, and nobody can predict exactly how long you'll live. But when the company averages over enough customers, it can make reasonably accurate estimates of the amount it can expect to collect on a policy before it has to pay its benefit.

Here's a simple example. An insurance company offers a "death and disability" policy that pays $10,000 when you die or $5000 if you are permanently disabled. It charges a premium of only $50 a year for this benefit. Is the company likely to make a profit selling such a plan? To answer this question, the company needs to know the *probability* that its clients will die or be disabled in any year. From actuarial information like this, the company can calculate the expected value of this policy.

> **What is an actuary?** Actuaries are the daring people who put a price on risk, estimating the likelihood and costs of rare events, so they can be insured. That takes financial, statistical, and business skills. It also makes them invaluable to many businesses. Actuaries are rather rare themselves; only about 19,000 work in North America. Perhaps because of this, they are well paid. If you're enjoying this course, you may want to look into a career as an actuary. Contact the Society of Actuaries or the Casualty Actuarial Society (who, despite what you may think, did not pay for this blurb).

15.1 Center: The Expected Value

We'll want to build a probability model in order to answer the questions about the insurance company's risk. First we need to define a few terms. The amount the company pays out on an individual policy is called a **random variable** because its numeric value is based on the

■ NOTATION ALERT
The most common letters for random variables are X, Y, and Z. But be cautious. If you see any capital letter, it just might denote a random variable.

 Activity: Random Variables. Learn more about random variables from this animated tour.

outcome of a random event. We use a capital letter, like X, to denote a random variable. We'll denote a particular value that it can have by the corresponding lowercase letter, in this case x. For the insurance company, x can be $10,000 (if you die that year), $5000 (if you are disabled), or $0 (if neither occurs). Because we can list all the outcomes, we might formally call this random variable a **discrete** random variable. Otherwise, we'd call it a **continuous** random variable. (We will say more about the properties of continuous random variables later in this chapter.) The collection of all the possible values and the probabilities that they occur is called the **probability model** for a discrete random variable.

Suppose, for example, that the death rate in any year is 1 out of every 1000 people, and that another 2 out of 1000 suffer some kind of disability. Then we can display the probability model for this insurance policy in a table like this:

Policyholder Outcome	Payout x	Probability $P(X = x)$
Death	10,000	$\frac{1}{1000}$
Disability	5000	$\frac{2}{1000}$
Neither	0	$\frac{997}{1000}$

To see what the insurance company can expect, imagine that it insures exactly 1000 people. Further imagine that, in perfect accordance with the probabilities, 1 of the policyholders dies, 2 are disabled, and the remaining 997 survive the year unscathed. The company would pay $10,000 to one client and $5000 to each of 2 clients. That's a total of $20,000, or an average of $20000/1000 = \$20$ per policy. Since it is charging people $50 for the policy, the company expects to make a profit of $30 per customer. Not bad!

We can't predict what *will* happen during any given year, but we can say what we *expect* to happen. To do this, we (or, rather, the insurance company) need the probability model. The expected value of a policy is a parameter of this model. In fact, it's the mean. We'll signify this with the notation μ (for population mean) or $E(X)$ for expected value. This isn't an average of some data values, so we won't estimate it. Instead, we assume that the probabilities are known and simply calculate the expected value from them.

■ NOTATION ALERT
The expected value (or mean) of a random variable is written $E(X)$ or μ.

How did we come up with $20 as the expected value of a policy payout? Here's the calculation. As we've seen, it often simplifies probability calculations to think about some (convenient) number of outcomes. For example, we could imagine that we have exactly 1000 clients. Of those, exactly 1 died and 2 were disabled, corresponding to what the probabilities would say.

$$\mu = E(X) = \frac{10,000(1) + 5000(2) + 0(997)}{1000}$$

So our total payout comes to $20,000, or $20 per policy.

Instead of writing the expected value as one big fraction, we can rewrite it as separate terms with a common denominator of 1000.

$$\mu = E(X)$$
$$= \$10,000\left(\frac{1}{1000}\right) + \$5000\left(\frac{2}{1000}\right) + \$0\left(\frac{997}{1000}\right)$$
$$= \$20.$$

How convenient! See the probabilities? For each policy, there's a 1/1000 chance that we'll have to pay $10,000 for a death and a 2/1000 chance that we'll have to pay $5000 for a disability. Of course, there's a 997/1000 chance that we won't have to pay anything.

Take a good look at the expression now. It's easy to calculate the **expected value** of a (discrete) random variable—just multiply each possible value by the probability that it occurs, and find the sum:

$$\mu = E(X) = \sum xP(x).$$

Be sure that every possible outcome is included in the sum. And verify that you have a valid probability model to start with—the probabilities should each be between 0 and 1 and should sum to one.

For Example LOVE AND EXPECTED VALUES

On Valentine's Day the *Quiet Nook* restaurant offers a *Lucky Lovers Special* that could save couples money on their romantic dinners. When the waiter brings the check, he'll also bring the four aces from a deck of cards. He'll shuffle them and lay them out face down on the table. The couple will then get to turn one card over. If it's a black ace, they'll owe the full amount, but if it's the ace of hearts, the waiter will give them a $20 Lucky Lovers discount. If they first turn over the ace of diamonds (hey—at least it's red!), they'll then get to turn over one of the remaining cards, earning a $10 discount for finding the ace of hearts this time.

QUESTION: Based on a probability model for the size of the Lucky Lovers discounts the restaurant will award, what's the expected discount for a couple?

ANSWER: Let X = the Lucky Lovers discount. The probabilities of the three outcomes are:

$$P(X = 20) = P(A\heartsuit) = \frac{1}{4}$$

$$P(X = 10) = P(A\diamondsuit, \text{then } A\heartsuit) = P(A\diamondsuit) \times P(A\heartsuit | A\diamondsuit) = \frac{1}{4} \times \frac{1}{3} = \frac{1}{12}$$

$$P(X = 0) = P(X \neq 20 \text{ or } 10) = 1 - \left(\frac{1}{4} + \frac{1}{12}\right) = \frac{2}{3}.$$

My probability model is:

Outcome	A♥	A♦, then A♥	Black Ace
x	20	10	0
$P(X = x)$	$\frac{1}{4}$	$\frac{1}{12}$	$\frac{2}{3}$

$$E(X) = 20 \times \frac{1}{4} + 10 \times \frac{1}{12} + 0 \times \frac{2}{3} = \frac{70}{12} \approx 5.83$$

Couples dining at the Quiet Nook can expect an average discount of $5.83.

✓ Just Checking

1. One of the authors took his minivan in for repair recently because the air conditioner was cutting out intermittently. The mechanic identified the problem as dirt in a control unit. He said that in about 75% of such cases, drawing down and then recharging the coolant a couple of times cleans up the problem—and costs only $60. If that fails, then the control unit must be replaced at an additional cost of $100 for parts and $40 for labor.

   a) Define the random variable and construct the probability model.
   b) What is the expected value of the cost of this repair?
   c) What does that mean in this context?

   Oh—in case you were wondering—the $60 fix worked!

15.2 Spread: The Standard Deviation

Of course, this expected value (or mean) is not what actually happens to any *particular* policyholder. No individual policy actually costs the company $20. We are dealing with random events, so some policyholders receive big payouts, others nothing. Because the insurance company must anticipate this variability, it needs to know the *standard deviation* of the random variable.

For data, we calculated the **standard deviation** by first computing the deviation from the mean and squaring it. We do that with (discrete) random variables as well. First, we find the deviation of each payout from the mean (expected value):

Policyholder Outcome	Payout x	Probability $P(X = x)$	Deviation $(x - \mu)$
Death	10,000	$\frac{1}{1000}$	$(10{,}000 - 20) = 9980$
Disability	5000	$\frac{2}{1000}$	$(5000 - 20) = 4980$
Neither	0	$\frac{997}{1000}$	$(0 - 20) = -20$

Next, we square each deviation. The **variance** is the expected value of those squared deviations, so we multiply each by the appropriate probability and sum those products. That gives us the variance of X. Here's what it looks like:

$$Var(X) = 9980^2\left(\frac{1}{1000}\right) + 4980^2\left(\frac{2}{1000}\right) + (-20)^2\left(\frac{997}{1000}\right) = 149{,}600.$$

Finally, we take the square root to get the standard deviation:

$$SD(X) = \sqrt{149{,}600} \approx \$386.78.$$

The insurance company can expect an average payout of $20 per policy, with a standard deviation of $386.78.

Think about that. The company charges $50 for each policy and expects to pay out $20 per policy. Sounds like an easy way to make $30. In fact, most of the time (probability 997/1000) the company pockets the entire $50. But would you consider selling your roommate such a policy? The problem is that occasionally the company loses big. With probability 1/1000, it will pay out $10,000, and with probability 2/1000, it will pay out $5000. That may be more risk than you're willing to take on. The standard deviation of $386.78 gives an indication that it's no sure thing. That's a pretty big spread (and risk) for an average profit of $30.

Here are the formulas for what we just did. Because these are parameters of our probability model, the variance and standard deviation can also be written as σ^2 and σ. You should recognize both kinds of notation.

$$\sigma^2 = Var(X) = \sum(x - \mu)^2 P(x)$$
$$\sigma = SD(X) = \sqrt{Var(X)}$$

For Example FINDING THE STANDARD DEVIATION

RECAP: The probability model for the Lucky Lovers restaurant discount is

Outcome	A♥	A♦, then A♥	Black Ace
x	20	10	0
$P(X = x)$	$\frac{1}{4}$	$\frac{1}{12}$	$\frac{2}{3}$

We found that couples can expect an average discount of $\mu = \$5.83$.

QUESTION: What's the standard deviation of the discounts?

ANSWER: First find the variance: $Var(X) = \sum (x - \mu)^2 \times P(x)$

$$= (20 - 5.83)^2 \times \frac{1}{4} + (10 - 5.83)^2 \times \frac{1}{12}$$

$$+ (0 - 5.83)^2 \times \frac{2}{3}$$

$$\approx 74.306.$$

So, $SD(X) = \sqrt{74.306} \approx \$8.62$

Couples can expect the Lucky Lovers discounts to average $5.83, with a standard deviation of $8.62.

Step-by-Step Example: EXPECTED VALUES AND STANDARD DEVIATIONS FOR DISCRETE RANDOM VARIABLES

As the head of inventory for Knowway Computer Company, you were thrilled that you had managed to ship 2 computers to your biggest client the day the order arrived. You are horrified, though, to find out that someone had restocked refurbished computers in with the new computers in your storeroom. The shipped computers were selected randomly from the 15 computers in stock, but 4 of those were actually refurbished.

If your client gets 2 new computers, things are fine. If the client gets one refurbished computer, it will be sent back at your expense—$100—and you can replace it. However, if both computers are refurbished, the client will cancel the order this month and you'll lose a total of $1000.

Question: What's the expected value and the standard deviation of the company's loss?

THINK ➡ Plan State the problem.

I want to find the company's expected loss for shipping refurbished computers and the standard deviation.

Variable Define the random variable.

Let X = amount of loss.

Plot Make a picture. This is another job for tree diagrams.

If you prefer calculation to drawing, find $P(NN)$ and $P(RR)$, then use the Complement Rule to find $P(NR \text{ or } RN)$.

Model List the possible values of the random variable, and determine the probability model.

Outcome	x	$P(X = x)$
Two Refurbs	1000	$P(RR) = 0.057$
One Refurb	100	$P(NR \text{ or } RN) = 0.2095 + 0.2095 = 0.419$
New/New	0	$P(NN) = 0.524$

SHOW ➡ **Mechanics** Find the expected value.		$E(X) = 0(0.524) + 100(0.419)$ $+ 1000(0.057)$ $= \$98.90$
Find the variance.		$Var(X) = (0 - 98.90)^2(0.524)$ $+ (100 - 98.90)^2(0.419)$ $+ (1000 - 98.90)^2(0.057)$ $= 51,408.79$
Find the standard deviation.		$SD(X) = \sqrt{51,408.79} = \$226.735$
TELL ➡ **Conclusion** Interpret your results in context. **REALITY CHECK** Both numbers seem reasonable. The expected value of $98.90 is between the extremes of $0 and $1000, and there's great variability in the outcome values.		I expect this mistake to cost the firm $98.90, with a standard deviation of $226.74. The large standard deviation reflects the fact that there's a pretty large range of possible losses.

15.3 Shifting and Combining Random Variables

Our insurance company expected to pay out an average of $20 per policy, with a standard deviation of about $387. If we take the $50 premium into account, we see the company makes a profit of $50 - 20 = \$30$ per policy. Suppose the company lowers the premium by $5 to $45. It's pretty clear that the expected profit also drops an average of $5 per policy, to $45 - 20 = \$25$.

What about the standard deviation? We know that adding or subtracting a constant from data shifts the mean but doesn't change the variance or standard deviation. The same is true of random variables.[1]

$$E(X \pm c) = E(X) \pm c \qquad Var(X \pm c) = Var(X).$$

For Example ADDING A CONSTANT

RECAP: We've determined that couples dining at the *Quiet Nook* can expect Lucky Lovers discounts averaging $5.83 with a standard deviation of $8.62. Suppose that for several weeks the restaurant has also been distributing coupons worth $5 off any one meal (one discount per table).

QUESTION: If every couple dining there on Valentine's Day brings a coupon, what will be the mean and standard deviation of the total discounts they'll receive?

ANSWER: Let D = total discount (Lucky Lovers plus the coupon); then $D = X + 5$.

$$E(D) = E(X + 5) = E(X) + 5 = 5.83 + 5 = \$10.83$$
$$Var(D) = Var(X + 5) = Var(X) = 8.62^2$$
$$SD(D) = \sqrt{Var(X)} = \$8.62$$

Couples with the coupon can expect total discounts averaging $10.83. The standard deviation is still $8.62.

---

[1] The rules in this section are true for both discrete *and* continuous random variables.

Back to insurance... What if the company decides to double all the payouts—that is, pay $20,000 for death and $10,000 for disability? This would double the average payout per policy and also increase the variability in payouts. We have seen that multiplying or dividing all data values by a constant changes both the mean and the standard deviation by the same factor. Variance, being the square of standard deviation, changes by the square of the constant. The same is true of random variables. In general, multiplying each value of a random variable by a constant multiplies the mean by that constant and the variance by the *square* of the constant.

$$E(aX) = aE(X) \qquad Var(aX) = a^2 Var(X)$$

For Example   DOUBLE THE LOVE

RECAP: On Valentine's Day at the *Quiet Nook*, couples may get a Lucky Lovers discount averaging $5.83 with a standard deviation of $8.62. When two couples dine together on a single check, the restaurant doubles the discount offer—$40 for the ace of hearts on the first card and $20 on the second.

QUESTION: What are the mean and standard deviation of discounts for such foursomes?

$$E(2X) = 2E(X) = 2(5.83) = \$11.66$$
$$Var(2X) = 2^2 Var(X) = 2^2 \cdot 8.62^2 = 297.2176$$
$$SD(2X) = \sqrt{297.2176} = \$17.24$$

ANSWER: If the restaurant doubles the discount offer, two couples dining together can expect to save an average of $11.66 with a standard deviation of $17.24.

This insurance company sells policies to more than just one person. How can we figure means and variances for a collection of customers? For example, how can the company find the total expected value (and standard deviation) of policies taken over all policyholders? Consider a simple case: just two customers, Mr. Ecks and Ms. Wye. With an expected payout of $20 on each policy, we might predict a total of $20 + $20 = $40 to be paid out on the two policies. Nothing surprising there. The expected value of the sum is the sum of the expected values.

$$E(X + Y) = E(X) + E(Y).$$

The variability is another matter. Is the risk of insuring two people the same as the risk of insuring one person for twice as much? We wouldn't expect both clients to die or become disabled in the same year. Because we've spread the risk, the standard deviation should be smaller. Indeed, this is the fundamental principle behind insurance. By spreading the risk among many policies, a company can keep the standard deviation quite small and predict costs more accurately.

But how much smaller is the standard deviation of the sum? It turns out that, if the random variables are independent, there is a simple Addition Rule for variances: *The variance of the sum of two independent random variables is the sum of their individual variances.*

For Mr. Ecks and Ms. Wye, the insurance company can expect their outcomes to be independent, so (using X for Mr. Ecks's payout and Y for Ms. Wye's)

$$\begin{aligned} Var(X + Y) &= Var(X) + Var(Y) \\ &= 149{,}600 + 149{,}600 \\ &= 299{,}200. \end{aligned}$$

If they had insured only Mr. Ecks for twice as much, there would only be one outcome rather than two *independent* outcomes, so the variance would have been

$$Var(2X) = 2^2 Var(X) = 4 \times 149{,}600 = 598{,}400,\text{ or}$$

twice as big as with two independent policies.

Of course, variances are in squared units. The company would prefer to know standard deviations, which are in dollars. The standard deviation of the payout for two independent policies is $\sqrt{299{,}200} = \$546.99$. But the standard deviation of the payout for a single policy of twice the size is $\sqrt{598{,}400} = \$773.56$, or about 40% more.

If the company has two customers, then, it will have an expected annual total payout of $40 with a standard deviation of about $547.

> ### For Example ADDING THE DISCOUNTS
>
> **RECAP:** The Valentine's Day Lucky Lovers discount for couples averages $5.83 with a standard deviation of $8.62. We've seen that if the restaurant doubles the discount offer for two couples dining together on a single check, they can expect to save $11.66 with a standard deviation of $17.24. Some couples decide instead to get separate checks and pool their two discounts.
>
> **QUESTION:** You and your amour go to this restaurant with another couple and agree to share any benefit from this promotion. Does it matter whether you pay separately or together?
>
> **ANSWER:** Let X_1 and X_2 represent the two separate discounts, and T the total; then $T = X_1 + X_2$.
>
> $$E(T) = E(X_1 + X_2) = E(X_1) + E(X_2) = 5.83 + 5.83 = \$11.66,$$
>
> so the expected saving is the same either way.
>
> The cards are reshuffled for each couple's turn, so the discounts couples receive are independent. It's okay to add the variances:
>
> $$Var(T) = Var(X_1 + X_2) = Var(X_1) + Var(X_2) = 8.62^2 + 8.62^2 = 148.6088$$
> $$SD(T) = \sqrt{148.6088} = \$12.19$$
>
> When two couples get separate checks, there's less variation in their total discount. The standard deviation is $12.19, compared to $17.24 for couples who play for the double discount on a single check.

In general,

- The mean of the sum of two random variables is the sum of the means.
- The mean of the difference of two random variables is the difference of the means.
- If the random variables are independent, the variance of their sum or difference is always the sum of the variances.

$$E(X \pm Y) = E(X) \pm E(Y) \qquad Var(X \pm Y) = Var(X) + Var(Y)$$

Wait a minute! Is that third part correct? Do we always *add* variances? Yes. Think about the two insurance policies. Suppose we want to know the mean and standard deviation of the *difference* in payouts to the two clients. Since each policy has an expected payout of $20, the expected difference is $20 - 20 = \$0$. If we also subtract variances, we get $0, too, and that surely doesn't make sense. Note that if the outcomes for the two clients are independent, the difference in payouts could range from $\$10{,}000 - \$0 = \$10{,}000$ to $\$0 - \$10{,}000 = -\$10{,}000$, a spread of $20,000. The variability in differences increases as much as the variability in sums. If the company has two customers, the difference in payouts has a mean of $0 and a standard deviation of about $547 (again).

PYTHAGOREAN THEOREM OF STATISTICS

We often use the standard deviation to measure variability, but when we add independent random variables, we use their variances. Think of the Pythagorean Theorem. In a right triangle (only), the *square* of the length of the hypotenuse is the sum of the *squares* of the lengths of the other two sides:

$$c^2 = a^2 + b^2.$$

For independent random variables (only), the *square* of the standard deviation of their sum is the sum of the *squares* of their standard deviations:

$$SD^2(X+Y) = SD^2(X) + SD^2(Y).$$

It's simpler to write this with *variances*:

For independent random variables, X and Y,

$$Var(X+Y) = Var(X) + Var(Y).$$

For Example WORKING WITH DIFFERENCES

RECAP: The Lucky Lovers discount at the *Quiet Nook* averages $5.83 with a standard deviation of $8.62. Just up the street, the *Wise Fool* restaurant has a competing Lottery of Love promotion. There a couple can select a specially prepared chocolate from a large bowl and unwrap it to learn the size of their discount. The restaurant's manager says the discounts vary with an average of $10.00 and a standard deviation of $15.00.

QUESTION: How much more can you expect to save at the *Wise Fool*? With what standard deviation?

ANSWER: Let W = discount at the *Wise Fool*, X = the discount at the *Quiet Nook*, and D = the difference: $D = W - X$. These are different promotions at separate restaurants, so the outcomes are independent.

$$E(W - X) = E(W) - E(X) = 10.00 - 5.83 = \$4.17$$

$$\begin{aligned} SD(W - X) &= \sqrt{Var(W - X)} \\ &= \sqrt{Var(W) + Var(X)} \\ &= \sqrt{15^2 + 8.62^2} \\ &\approx \$17.30 \end{aligned}$$

Discounts at the *Wise Fool* will average $4.17 more than at the *Quiet Nook*, with a standard deviation of the difference of $17.30.

FOR RANDOM VARIABLES, DOES $X + X + X = 3X$?

Maybe, but be careful. As we've just seen, insuring one person for $30,000 is not the same risk as insuring three people for $10,000 each. When each instance represents a different outcome for the same random variable, it's easy to fall into the trap of writing all of them with the same symbol. Don't make this common mistake. Make sure you write each instance as a *different* random variable. Just because each random variable describes a similar situation doesn't mean that each random outcome will be the same.

These are *random* variables, not the variables you saw in Algebra. Being random, they take on different values each time they're evaluated. So what you really mean is $X_1 + X_2 + X_3$. Written this way, it's clear that the sum shouldn't necessarily equal 3 times *anything*.

For Example SUMMING A SERIES OF OUTCOMES

RECAP: The *Quiet Nook*'s Lucky Lovers promotion offers couples discounts averaging $5.83 with a standard deviation of $8.62. The restaurant owner is planning on serving 40 couples on Valentine's Day.

QUESTION: What's the expected total of the discounts the owner will give? With what standard deviation?

ANSWER: Let $X_1, X_2, X_3, \ldots, X_{40}$ represent the discounts to the 40 couples, and T the total of all the discounts. Then:

$$\begin{aligned} T &= X_1 + X_2 + X_3 + \cdots + X_{40} \\ E(T) &= E(X_1 + X_2 + X_3 + \cdots + X_{40}) \\ &= E(X_1) + E(X_2) + E(X_3) + \cdots + E(X_{40}) \\ &= 5.83 + 5.83 + 5.83 + \cdots + 5.83 \\ &= \$233.20 \end{aligned}$$

Reshuffling cards between couples makes the discounts independent, so:

$$SD(T) = \sqrt{Var(X_1 + X_2 + X_3 + \cdots + X_{40})}$$
$$= \sqrt{Var(X_1) + Var(X_2) + Var(X_3) + \cdots + Var(X_{40})}$$
$$= \sqrt{8.62^2 + 8.62^2 + 8.62^2 + \cdots + 8.62^2}$$
$$\approx \$54.52$$

The restaurant owner can expect the 40 couples to win discounts totaling $233.20, with a standard deviation of $54.52.

Just Checking

2. Suppose the time it takes a customer to get and pay for seats at the ticket window of a baseball park is a random variable with a mean of 100 seconds and a standard deviation of 50 seconds. When you get there, you find only two people in line in front of you.

   a) How long do you expect to wait for your turn to get tickets?

   b) What's the standard deviation of your wait time?

   c) What assumption did you make about the two customers in finding the standard deviation?

Step-by-Step Example HITTING THE ROAD: COMBINING MEANS AND VARIANCES

You're planning to spend next year wandering through the mountains of Kyrgyzstan. You plan to sell your used SUV so you can purchase an off-road Honda motor scooter when you get there. Used SUVs of the year and mileage of yours are selling for a mean of $6940 with a standard deviation of $250. Your research shows that scooters in Kyrgyzstan are going for about 65,000 Kyrgyzstan som with a standard deviation of 500 som. One U.S. dollar is worth about 54.5 Kyrgyzstan som (54 som and 50 tylyn).

Question: How much cash can you expect to pocket after you sell your SUV and buy the scooter?

THINK ▶ Plan State the problem.

I want to model how much money I'd have (in som) after selling my SUV and buying the scooter.

Variables Define the random variables.

Let A = sale price of my SUV (in dollars),
B = price of a scooter (in som), and
D = profit (in som)

Write an appropriate equation. Think about the assumptions.

$D = 54.5A - B$

✓ **Independence Assumption:** The prices are independent.

SHOW ▶ Mechanics Find the expected value, using the appropriate rules.		$E(D) = E(54.5A - B)$ $ = 54.5E(A) - E(B)$ $ = 54.5(6,940) - (65,000)$ $E(D) = 313,230$ som
Find the variance, using the appropriate rules. Be sure to check the assumptions first!		Since sale and purchase prices are independent, $Var(D) = Var(54.5A - B)$ $ = Var(54.5A) + Var(B)$ $ = (54.5)^2 Var(A) + Var(B)$ $ = 2970.25(250)^2 + (500)^2$ $Var(D) = 185,890,625$
Find the standard deviation.		$SD(D) = \sqrt{185,890,625} = 13,634.17$ som
TELL ▶ Conclusion Interpret your results in context. (Here that means talking about dollars.)		I can expect to clear about 313,230 som ($5747) with a standard deviation of 13,634.17 som ($250).
REALITY CHECK Given the initial cost estimates, the mean and standard deviation seem reasonable.		

*Correlation and Covariance

In Chapter 6 we saw that the association of two variables could be measured with their correlation. What about random variables? We can talk about the correlation between random variables, too. But it's easier to start with a related concept called covariance.

If X is a random variable with expected value $E(X) = \mu$ and Y is a random variable with expected value $E(Y) = \nu$, then **the covariance of X and Y is defined as**

$$Cov(X, Y) = E((X - \mu)(Y - \nu)).$$

The covariance, like the correlation, measures how X and Y vary together ($co = together$). Think about a scatterplot of two positively correlated variables. When X is above its mean, Y tends to be, too. And when X is below its mean, Y tends to be below its mean. So the product in the covariance formula is positive.

Covariance has a few properties worth remembering:

- $Cov(X, Y) = Cov(Y, X)$
- $Cov(X, X) = Var(X)$
- $Cov(cX, dY) = cd Cov(X, Y)$, for any constants c and d
- $Cov(X, Y) = E(XY) - \mu\nu$
- If X and Y are independent, then $Cov(X, Y) = 0$—but be careful not to assume that the converse is true; it is possible for two variables to have zero covariance yet not be independent.

The covariance gives us the extra information we need to find the variance of the sum or difference of two random variables when they are not independent:

$$Var(X \pm Y) = Var(X) + Var(Y) \pm 2Cov(X, Y).$$

When X and Y are independent, their covariance is zero, so we have the Pythagorean theorem of statistics, as we saw earlier.[2]

Covariance, unlike correlation, doesn't have to be between -1 and 1. If X and Y have large values, the covariance will be large as well. And that's its problem. If we know the covariance between X and Y is -64.5, it doesn't give us a real sense of how negatively related they are. To fix the "problem," we can divide the covariance by each of the standard deviations to get the **correlation**:

$$Corr(X, Y) = \frac{Cov(X, Y)}{\sigma_X \sigma_Y}.$$

Correlation of random variables is usually denoted as ρ. Correlation of random variables has many of the properties we saw in Chapter 6 for the correlation coefficient, r.

- The sign of a correlation coefficient gives the direction of the association between random variables.
- Correlation is always between -1 and $+1$.
- Correlation treats X and Y symmetrically. The correlation of X with Y is the same as the correlation of Y with X.
- Unlike covariance, correlation has no units.
- Correlation is not affected by changes in the center or scale of either variable.

15.4 Continuous Random Variables

A Normal random variable differs from the discrete random variables that we have been discussing because a Normal random variable can take on any value. It's an example of a continuous random variable. You remember from Chapter 5 that for a Normal, we find the probability that it lies between two values by finding the area under the Normal curve. The curve we use to work with the Normal distribution is called the Normal model.[3]

For any continuous random variable, this curve (or model) doesn't give the probability directly as the probability models for discrete random variables did. Instead, to find the probability that the random variable is in any interval, you find the area below the curve over that interval as we did for the Normal model in Chapter 5. For the standard Normal,

Figure 15.1
The standard Normal curve (with mean 0 and standard deviation 1). The probability of finding a z-score in any interval is the area over that interval under the curve. For example, the probability that the z-score falls between -1 and 1 is about 68%, which can be seen approximately from the density function or found more precisely from a table or technology.

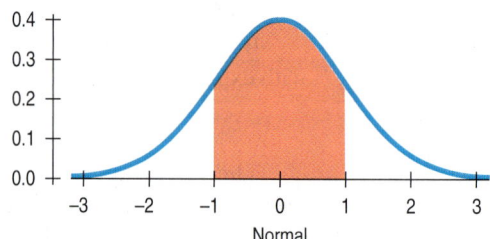

shown in Figure 15.1, we can see that the area below the curve between -1 and 1 is about 68%, which is where the 68–95–99.7 Rule comes from.

It's important to remember that, for a continuous random variable, X, $P(X = x)$ is 0 for every value of x! That may seem strange at first, but since the probability is the area under the curve over an interval, as the interval gets smaller and smaller, the probability does too. Finally, when the interval is just a point, there is no area—and no probability. (See the box on next page.)

---

[2] If you want to follow the geometry one step further, the more general form for correlated variables is related to the Law of Cosines.

[3] For continuous random variables, the curve is also called the probability density function (pdf) of the random variable.

> ### HOW CAN *EVERY* VALUE HAVE PROBABILITY 0?
>
> We can find a probability for any interval of z-scores. But the probability for a single z-score is zero. How can that be? Let's look at the standard Normal random variable, Z. We could find (from a table, website, or computer program) that the probability that Z lies between 0 and 1 is 0.3413.
>
>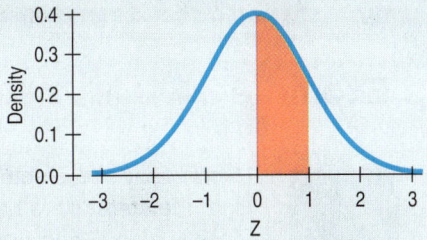
>
> That's the area under the Normal pdf (in red) between the values 0 and 1. So, what's the probability that Z is between 0 and 1/10?
>
>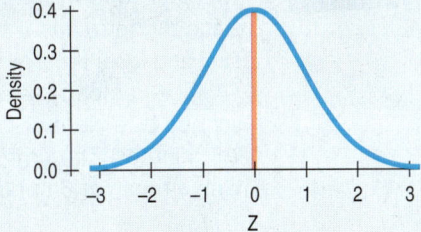
>
> That area is only 0.0398. What is the chance then that Z will fall between 0 and 1/100? There's not much area—the probability is only 0.0040. If we kept going, the probability would keep getting smaller. The probability that Z is between 0 and 1/100,000 is less than 0.00001.
>
>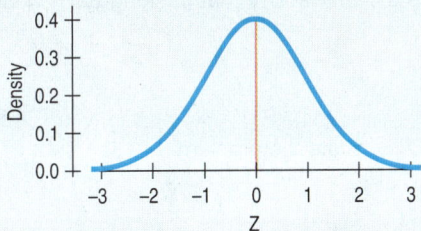
>
> So, what's the probability that Z is *exactly* 0? Well, there's *no* area under the curve right at $x = 0$, so the probability is 0. It's only intervals that have positive probability, but that's OK. In real life we never mean exactly 0.0000000000 or any other value. If you say "exactly 164 pounds," you might really mean between 163.5 and 164.5 pounds or even between 163.99 and 164.01 pounds, but realistically not 164.000000000 . . . pounds.

Activity: Means of Random Variables. Experiment with continuous random variables to learn how their expected values behave.

Some continuous random variables have Normal models; others may be skewed, uniform, or bimodal. Regardless of shape, all continuous random variables have means (which we also call *expected values*) and variances. In this book we won't worry about how to calculate them, but we can still work with models for continuous random variables when we're given these parameters. Chapter 16 gives some examples of other continuous random variables.

We know that the expected value of the sum of *any* two random variables is the sum of their expected values. We know how to compute the standard deviation of the sum as well. But, we also know that when we add two random variables together, the sum generally has a different probability model than the one we started with. Amazingly, independent Normal

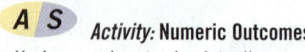 **Activity: Numeric Outcomes.** You've seen how to simulate discrete random outcomes. There's a tool for simulating continuous outcomes, too.

random variables behave better. The probability model for the sum of any number of independent Normal random variables is still Normal.

This is just one of the things that is special about the Normal. However, if this mathematical good behavior alone doesn't impress you, here's a real-world example of how we can use this fact.

Step-by-Step Example PACKAGING STEREOS

A company manufactures small stereo systems. At the end of the production line, the stereos are packaged and prepared for shipping. Stage 1 of this process is called "packing." Workers must collect all the system components (a main unit, two speakers, a power cord, an antenna, and some wires), put each in plastic bags, and then place everything inside a protective styrofoam form. The packed form then moves on to Stage 2, called "boxing." There, workers place the form and a packet of instructions in a cardboard box, close it, then seal and label the box for shipping.

The company says that times required for the packing stage can be described by a Normal model with a mean of 9 minutes and standard deviation of 1.5 minutes. The times for the boxing stage can also be modeled as Normal, with a mean of 6 minutes and standard deviation of 1 minute.

Questions:
1. What is the probability that packing two consecutive systems takes over 20 minutes?
2. What percentage of the stereo systems take longer to pack than to box?

Question 1: What is the probability that packing two consecutive systems takes over 20 minutes?

THINK ➡ Plan State the problem.

I want to estimate the probability that packing two consecutive systems takes over 20 minutes.

Let P_1 = time for packing the first system
P_2 = time for packing the second
T = total time to pack two systems

Variables Define your random variables.

Write an appropriate equation.

$T = P_1 + P_2$

Think about the assumptions. Sums of independent Normal random variables follow a Normal model. Such simplicity isn't true in general.

✓ **Normal Model Assumption:** We are told that both random variables follow Normal models.

✓ **Independence Assumption:** We can reasonably assume that the two packing times are independent.

SHOW ▶ Mechanics   Find the expected value.

$$E(T) = E(P_1 + P_2)$$
$$= E(P_1) + E(P_2)$$
$$= 9 + 9 = 18 \text{ minutes}$$

For sums of independent random variables, variances add. (We don't need the variables to be Normal for this to be true—just independent.)

Since the times are independent,
$$Var(T) = Var(P_1 + P_2)$$
$$= Var(P_1) + Var(P_2)$$
$$= 1.5^2 + 1.5^2$$
$$Var(T) = 4.50$$

Find the standard deviation.

$$SD(T) = \sqrt{4.50} \approx 2.12 \text{ minutes}$$

Now we use the fact that both random variables follow Normal models to say that their sum is also Normal.

I'll model T with $N(18, 2.12)$.

Sketch a picture of the Normal model for the total time, shading the region representing over 20 minutes.

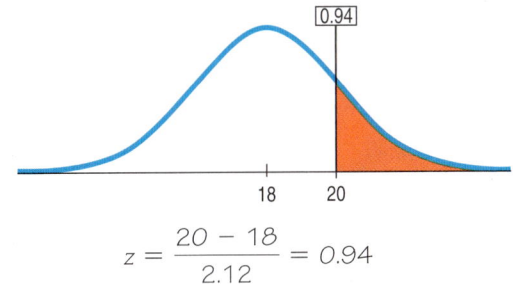

Find the z-score for 20 minutes.

Use technology or Table Z to find the probability.

$$z = \frac{20 - 18}{2.12} = 0.94$$

$$P(T > 20) = P(z > 0.94) = 0.1736$$

TELL ▶ Conclusion   Interpret your result in context.

There's a little more than a 17% chance that it will take a total of over 20 minutes to pack two consecutive stereo systems.

Question 2: What percentage of the stereo systems take longer to pack than to box?

THINK ▶ Plan   State the question.

I want to estimate the percentage of the stereo systems that take longer to pack than to box.

Variables   Define your random variables.

Let P = time for packing a system
B = time for boxing a system
D = difference in times to pack and box a system

Write an appropriate equation.

$$D = P - B$$

What are we trying to find? Notice that we can tell which of two quantities is greater by subtracting and asking whether the difference is positive or negative.

The probability that it takes longer to pack than to box a system is the probability that the difference $P - B$ is greater than zero.

Don't forget to think about the assumptions.

✓ **Normal Model Assumption:** We are told that both random variables follow Normal models.

✓ **Independence Assumption:** We can assume that the times it takes to pack and to box a system are independent.

SHOW ▶ Mechanics

Find the expected value.

$$E(D) = E(P - B)$$
$$= E(P) - E(B)$$
$$= 9 - 6 = 3 \text{ minutes}$$

For the difference of independent random variables, variances add.

Since the times are independent,
$$Var(D) = Var(P - B)$$
$$= Var(P) + Var(B)$$
$$= 1.5^2 + 1^2$$
$$Var(D) = 3.25$$

Find the standard deviation.

$$SD(D) = \sqrt{3.25} \approx 1.80 \text{ minutes}$$

State what model you will use.

I'll model D with $N(3, 1.80)$

Sketch a picture of the Normal model for the difference in times, and shade the region representing a difference greater than zero.

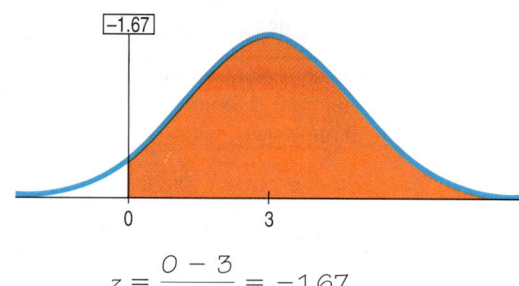

Find the z-score for 0 minutes, then use Table Z or technology to find the probability.

$$z = \frac{0 - 3}{1.80} = -1.67$$

$$P(D > 0) = P(z > -1.67) = 0.9525$$

TELL ▶ Conclusion

Interpret your result in context.

About 95% of all the stereo systems will require more time for packing than for boxing.

WHAT CAN GO WRONG?

- **Probability models are still just models.** Models can be useful, but they are not reality. Think about the assumptions behind your models. Are your dice really perfectly fair? (They are probably pretty close.) But when you hear that the probability of a nuclear accident is 1/10,000,000 per year, is that likely to be a precise value? Question probabilities as you would data.

- **If the model is wrong, so is everything else.** Before you try to find the mean or standard deviation of a random variable, check to make sure the probability model is reasonable. As a start, the probabilities in your model should add up to 1. If not, you may have calculated a probability incorrectly or left out a value of the random variable. For instance, in the insurance example, the description mentions only death and disability. Good health is by far the most likely outcome, not to mention the best for both you and the insurance company (who gets to keep your money). Don't overlook that.

- **Don't assume everything's Normal.** Just because a random variable is continuous or you happen to know a mean and standard deviation doesn't mean that a Normal model will be useful. You must *Think* about whether the **Normality Assumption** is justified. Using a Normal model when it really does not apply will lead to wrong answers and misleading conclusions.

To find the expected value of the sum or difference of random variables, we simply add or subtract means. Center is easy; spread is trickier. Watch out for some common traps.

- **Watch out for variables that aren't independent.** You can add expected values of *any* two random variables, but you can only add variances of independent random variables. Suppose a survey includes questions about the number of hours of sleep people get each night and also the number of hours they are awake each day. From their answers, we find the mean and standard deviation of hours asleep and hours awake. The expected total must be 24 hours; after all, people are either asleep or awake.[4] The means still add just fine. Since all the totals are exactly 24 hours, however, the standard deviation of the total will be 0. We can't add variances here because the number of hours you're awake depends on the number of hours you're asleep. Be sure to check for independence before adding variances.
- **Don't forget: Variances of independent random variables add. Standard deviations don't.**
- **Don't forget: Variances of independent random variables add, even when you're looking at the difference between them.**
- **Don't write independent instances of a random variable with notation that looks like they are the same variables.** Make sure you write each instance as a different random variable. Just because each random variable describes a similar situation doesn't mean that each random outcome will be the same. These are *random* variables, not the variables you saw in Algebra. Write $X_1 + X_2 + X_3$ rather than $X + X + X$.

We've seen means, variances, and standard deviations of data. We know that they estimate parameters of models for these data. Now we're looking at the probability models directly. We have only parameters because there are no data to summarize.

It should be no surprise that expected values and standard deviations adjust to shifts and changes of units in the same way as the corresponding data summaries. The fact that we can add variances of independent random quantities is fundamental and will explain why a number of statistical methods work the way they do.

What Have We Learned?
Learning Objectives

We've learned to work with random variables. We can use the probability model for a discrete random variable to find its expected value and its standard deviation.

We've learned that the mean of the sum or difference of two random variables, discrete or continuous, is just the sum or difference of their means. And we've learned the Pythagorean Theorem of Statistics: *For independent random variables,* the variance of their sum or difference is always the *sum* of their variances.

Finally, we've learned that Normal models are once again special. Sums or differences of Normally distributed random variables also follow Normal models.

Review of Terms

Random variable A random variable assumes any of several different numeric values as a result of some random event. Random variables are denoted by a capital letter such as X (p. 389).

Discrete random variable A random variable that can take one of a finite number[5] of distinct outcomes is called a discrete random variable (p. 390).

---

[4]Although some students do manage to attain a state of consciousness somewhere between sleeping and wakefulness during Statistics class.

[5]Technically, there could be an infinite number of outcomes, as long as they're *countable*. Essentially that means we can imagine listing them all in order, like the counting numbers 1, 2, 3, 4, 5, …

Continuous random variable	A random variable that can take any numeric value within a range of values is called a continuous random variable. The range may be infinite or bounded at either or both ends (p. 390).
Probability model	The probability model is a function that associates a probability P with each value of a discrete random variable X, denoted $P(X = x)$, or with any interval of values of a continuous random variable (pp. 390, 400).
Expected value	The expected value of a random variable is its theoretical long-run average value, the center of its model. Denoted μ or $E(X)$, it is found (if the random variable is discrete) by summing the products of variable values and probabilities (p. 390): $$\mu = E(X) = \sum x P(x).$$
Variance	The variance of a random variable is the expected value of the squared deviation from the mean. For discrete random variables, it can be calculated as (p. 392): $$\sigma^2 = Var(X) = \sum (x - \mu)^2 P(x).$$
Standard deviation	The standard deviation of a random variable describes the spread in the model, and is the square root of the variance (p. 392): $$\sigma = SD(X) = \sqrt{Var(X)}.$$
Changing a random variable by a constant	$E(X \pm c) = E(X) \pm c$ $\quad Var(X \pm c) = Var(X)$ (p. 394) $E(aX) = aE(X)$ $\quad Var(aX) = a^2 Var(X)$ (p. 395)
Adding or subtracting random variables	$E(X \pm Y) = E(X) \pm E(Y)$ (p. 395) $Var(X \pm Y) = Var(X) + Var(Y)$ If X and Y are independent, (p. 396) (The Pythagorean Theorem of Statistics) (p. 396)
*Covariance	The covariance of random variables X and Y is $Cov(X, Y) = E((X - \mu)(Y - \nu))$ where $\mu = E(X)$ and $\nu = E(Y)$. In general (no need to assume independence) $Var(X \pm Y) = Var(X) + Var(Y) \pm 2Cov(X, Y)$ (p. 397).

On the Computer RANDOM VARIABLES

Statistics packages deal with data, not with random variables. Nevertheless, the calculations needed to find means and standard deviations of random variables are little more than weighted means. Most packages can manage that, but then they are just being overblown calculators. For technological assistance with these calculations, we recommend you pull out your calculator.

TI-83/84 PLUS

To calculate the mean and standard deviation of a discrete random variable, enter the probability model in two lists:

- In one list (say, L1) enter the x-values of the variable.
- In a second list (say, L2) enter the associated probabilities $P(X = x)$.
- From the **STAT CALC** menu select **1-VarStats** and specify the two lists. You'll see the mean and standard deviation.

COMMENTS

You can enter the probabilities as fractions; the calculator will change them to decimals for you.

Notice that the calculator knows enough to call the standard deviation σ, but mistakenly uses $\bar{x}$ when it should say μ. Make sure you don't make that mistake!

Exercises

Section 15.1

1. **Oranges** A citrus farmer has observed the following distribution for the number of oranges per tree. How many oranges does he expect on average?

Oranges	25	30	35	40
Probability	0.10	0.40	0.30	0.20

2. **Caffeinated** A coffee shop tracks sales and has observed the distribution in the following table. What is the average daily sales that it can expect?

# of Sales	145	150	155	160	170
Probability	0.15	0.22	0.37	0.19	0.07

Section 15.2

3. **Oranges again** What is the standard deviation for Exercise 1?

4. **Caffeinated again** What is the standard deviation for Exercise 2?

Section 15.3

5. **Salary** An employer pays a mean salary for a 5-day workweek of $1250 with a standard deviation of $129. On the weekends, his salary expenses have a mean of $450 with a standard deviation of $57. What is the mean and standard deviation of his total weekly salaries?

6. **Golf scores** A golfer keeps track of his score for playing nine holes of golf (half a normal golf round). His mean score is 85 with a standard deviation of 11. Assuming that the second 9 has the same mean and standard deviation, what is the mean and standard deviation of his total score if he plays a full 18 holes?

Section 15.4

7. **Battery** The life span of a calculator battery is normally distributed with a mean of 45 hours and a standard deviation of 5 hours. What is the probability that a battery lasts more than 53 hours?

8. **Ketchup** An automatic filling machine in a factory fills bottles of ketchup with a mean of 16.1 oz and a standard deviation of 0.05 oz with a distribution that can be well modeled by a Normal model. What is the probability that your bottle of ketchup contains less than 16 oz?

Chapter Exercises

9. **Expected value** Find the expected value of each random variable:

   a)
   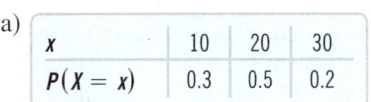

X	10	20	30
$P(X = x)$	0.3	0.5	0.2

   b)

X	2	4	6	8
$P(X = x)$	0.3	0.4	0.2	0.1

10. **Expected value** Find the expected value of each random variable:

    a)

X	0	1	2
$P(X = x)$	0.2	0.4	0.4

    b)

X	100	200	300	400
$P(X = x)$	0.1	0.2	0.5	0.2

11. **Pick a card, any card** You draw a card from a deck. If you get a red card, you win nothing. If you get a spade, you win $5. For any club, you win $10 plus an extra $20 for the ace of clubs.
    a) Create a probability model for the amount you win.
    b) Find the expected amount you'll win.
    c) What would you be willing to pay to play this game?

12. **You bet!** You roll a die. If it comes up a 6, you win $100. If not, you get to roll again. If you get a 6 the second time, you win $50. If not, you lose.
    a) Create a probability model for the amount you win.
    b) Find the expected amount you'll win.
    c) What would you be willing to pay to play this game?

13. **Kids** A couple plans to have children until they get a girl, but they agree that they will not have more than three children even if all are boys. (Assume boys and girls are equally likely.)
    a) Create a probability model for the number of children they might have.
    b) Find the expected number of children.
    c) Find the expected number of boys they'll have.

14. **Carnival** A carnival game offers a $100 cash prize for anyone who can break a balloon by throwing a dart at it. It costs $5 to play, and you're willing to spend up to $20 trying to win. You estimate that you have about a 10% chance of hitting the balloon on any throw.
    a) Create a probability model for this carnival game.
    b) Find the expected number of darts you'll throw.
    c) Find your expected winnings.

15. Software A small software company bids on two contracts and knows it can only get one of them. It anticipates a profit of $50,000 if it gets the larger contract and a profit of $20,000 on the smaller contract. The company estimates there's a 30% chance it will get the larger contract and a 60% chance it will get the smaller contract. Assuming the contracts will be awarded independently, what's the expected profit?

16. Racehorse A man buys a racehorse for $20,000 and enters it in two races. He plans to sell the horse afterward, hoping to make a profit. If the horse wins both races, its value will jump to $100,000. If it wins one of the races, it will be worth $50,000. If it loses both races, it will be worth only $10,000. The man believes there's a 20% chance that the horse will win the first race and a 30% chance it will win the second one. Assuming that the two races are independent events, find the man's expected profit.

17. Variation 1 Find the standard deviations of the random variables in Exercise 9.

18. Variation 2 Find the standard deviations of the random variables in Exercise 10.

19. Pick another card Find the standard deviation of the amount you might win drawing a card in Exercise 11.

20. The die Find the standard deviation of the amount you might win rolling a die in Exercise 12.

21. Kids, again Find the standard deviation of the number of children the couple in Exercise 13 may have.

22. Darts Find the standard deviation of your winnings throwing darts in Exercise 14.

23. Repairs The probability model below describes the number of repair calls that an appliance repair shop may receive during an hour.

Repair Calls	0	1	2	3
Probability	0.1	0.3	0.4	0.2

a) How many calls should the shop expect per hour?
b) What is the standard deviation?

24. Red lights A commuter must pass through five traffic lights on her way to work and will have to stop at each one that is red. She estimates the probability model for the number of red lights she hits, as shown below.

X = # of red	0	1	2	3	4	5
P(X = x)	0.05	0.25	0.35	0.15	0.15	0.05

a) How many red lights should she expect to hit each day?
b) What's the standard deviation?

25. Defects A consumer organization inspecting new cars found that many had appearance defects (dents, scratches, paint chips, etc.). While none had more than three of these defects, 7% had three, 11% two, and 21% one defect. Find the expected number of appearance defects in a new car and the standard deviation.

26. Insurance An insurance policy costs $100 and will pay policyholders $10,000 if they suffer a major injury (resulting in hospitalization) or $3000 if they suffer a minor injury (resulting in lost time from work). The company estimates that each year 1 in every 2000 policyholders may have a major injury, and 1 in 500 a minor injury only.

a) Create a probability model for the profit on a policy.
b) What's the company's expected profit on this policy?
c) What's the standard deviation?

27. Cancelled flights Mary is deciding whether to book the cheaper flight home from college after her final exams, but she's unsure when her last exam will be. She thinks there is only a 20% chance that the exam will be scheduled after the last day she can get a seat on the cheaper flight. If it is and she has to cancel the flight, she will lose $150. If she can take the cheaper flight, she will save $100.

a) If she books the cheaper flight, what can she expect to gain, on average?
b) What is the standard deviation?

28. Day trading An option to buy a stock is priced at $200. If the stock closes above 30 on May 15, the option will be worth $1000. If it closes below 20, the option will be worth nothing, and if it closes between 20 and 30 (inclusively), the option will be worth $200. A trader thinks there is a 50% chance that the stock will close in the 20–30 range, a 20% chance that it will close above 30, and a 30% chance that it will fall below 20 on May 15.

a) Should she buy the stock option?
b) How much does she expect to gain?
c) What is the standard deviation of her gain?

29. Contest You play two games against the same opponent. The probability you win the first game is 0.4. If you win the first game, the probability you also win the second is 0.2. If you lose the first game, the probability that you win the second is 0.3.

a) Are the two games independent? Explain.
b) What's the probability you lose both games?
c) What's the probability you win both games?
d) Let random variable X be the number of games you win. Find the probability model for X.
e) What are the expected value and standard deviation?

30. Contracts Your company bids for two contracts. You believe the probability you get contract #1 is 0.8. If you get contract #1, the probability you also get contract #2 will be 0.2, and if you do not get #1, the probability you get #2 will be 0.3.

a) Are the two contracts independent? Explain.
b) Find the probability you get both contracts.
c) Find the probability you get no contract.
d) Let X be the number of contracts you get. Find the probability model for X.
e) Find the expected value and standard deviation.

31. Batteries In a group of 10 batteries, 3 are dead. You choose 2 batteries at random.
   a) Create a probability model for the number of good batteries you get.
   b) What's the expected number of good ones you get?
   c) What's the standard deviation?

32. Kittens In a litter of seven kittens, three are female. You pick two kittens at random.
   a) Create a probability model for the number of male kittens you get.
   b) What's the expected number of males?
   c) What's the standard deviation?

33. Random variables Given independent random variables with means and standard deviations as shown, find the mean and standard deviation of:
   a) $3X$
   b) $Y + 6$
   c) $X + Y$
   d) $X - Y$
   e) $X_1 + X_2$

	Mean	SD
X	10	2
Y	20	5

34. Random variables Given independent random variables with means and standard deviations as shown, find the mean and standard deviation of:
   a) $X - 20$
   b) $0.5Y$
   c) $X + Y$
   d) $X - Y$
   e) $Y_1 + Y_2$

	Mean	SD
X	80	12
Y	12	3

35. Random variables Given independent random variables with means and standard deviations as shown, find the mean and standard deviation of:
   a) $0.8Y$
   b) $2X - 100$
   c) $X + 2Y$
   d) $3X - Y$
   e) $Y_1 + Y_2$

	Mean	SD
X	120	12
Y	300	16

36. Random variables Given independent random variables with means and standard deviations as shown, find the mean and standard deviation of:
   a) $2Y + 20$
   b) $3X$
   c) $0.25X + Y$
   d) $X - 5Y$
   e) $X_1 + X_2 + X_3$

	Mean	SD
X	80	12
Y	12	3

37. Eggs A grocery supplier believes that in a dozen eggs, the mean number of broken ones is 0.6 with a standard deviation of 0.5 eggs. You buy 3 dozen eggs without checking them.
   a) How many broken eggs do you expect to get?
   b) What's the standard deviation?
   c) What assumptions did you have to make about the eggs in order to answer this question?

38. Garden A company selling vegetable seeds in packets of 20 estimates that the mean number of seeds that will actually grow is 18, with a standard deviation of 1.2 seeds. You buy 5 different seed packets.
   a) How many bad (non-growing) seeds do you expect to get?
   b) What's the standard deviation?
   c) What assumptions did you make about the seeds? Do you think that assumption is warranted? Explain.

39. Repair calls Find the mean and standard deviation of the number of repair calls the appliance shop in Exercise 23 should expect during an 8-hour day.

40. Stop! Find the mean and standard deviation of the number of red lights the commuter in Exercise 24 should expect to hit on her way to work during a 5-day work week.

41. Tickets A delivery company's trucks occasionally get parking tickets, and based on past experience, the company plans that each truck will average 1.3 tickets a month, with a standard deviation of 0.7 tickets.
   a) If they have 18 trucks, what are the mean and standard deviation of the total number of parking tickets the company will have to pay this month?
   b) What assumption did you make in answering?

42. Donations Organizers of a televised fundraiser know from past experience that most people donate small amounts ($10–$25), some donate larger amounts ($50–$100), and a few people make very generous donations of $250, $500, or more. Historically, pledges average about $32 with a standard deviation of $54.
   a) If 120 people call in pledges, what are the mean and standard deviation of the total amount raised?
   b) What assumption did you make in answering this question?

43. Fire! An insurance company estimates that it should make an annual profit of $150 on each homeowner's policy written, with a standard deviation of $6000.
   a) Why is the standard deviation so large?
   b) If it writes only two of these policies, what are the mean and standard deviation of the annual profit?
   c) If it writes 10,000 of these policies, what are the mean and standard deviation of the annual profit?
   d) Is the company likely to be profitable? Explain.
   e) What assumptions underlie your analysis? Can you think of circumstances under which those assumptions might be violated? Explain.

44. Casino A casino knows that people play the slot machines in hopes of hitting the jackpot but that most of them lose their dollar. Suppose a certain machine pays out an average of $0.92, with a standard deviation of $120.
a) Why is the standard deviation so large?
b) If you play 5 times, what are the mean and standard deviation of the casino's profit?
c) If gamblers play this machine 1000 times in a day, what are the mean and standard deviation of the casino's profit?
d) Is the casino likely to be profitable? Explain.

45. Cereal The amount of cereal that can be poured into a small bowl varies with a mean of 1.5 ounces and a standard deviation of 0.3 ounces. A large bowl holds a mean of 2.5 ounces with a standard deviation of 0.4 ounces. You open a new box of cereal and pour one large and one small bowl.
a) How much more cereal do you expect to be in the large bowl?
b) What's the standard deviation of this difference?
c) If the difference follows a Normal model, what's the probability the small bowl contains more cereal than the large one?
d) What are the mean and standard deviation of the total amount of cereal in the two bowls?
e) If the total follows a Normal model, what's the probability you poured out more than 4.5 ounces of cereal in the two bowls together?
f) The amount of cereal the manufacturer puts in the boxes is a random variable with a mean of 16.3 ounces and a standard deviation of 0.2 ounces. Find the expected amount of cereal left in the box and the standard deviation.

46. Pets The American Veterinary Association claims that the annual cost of medical care for dogs averages $100, with a standard deviation of $30, and for cats averages $120, with a standard deviation of $35.
a) What's the expected difference in the cost of medical care for dogs and cats?
b) What's the standard deviation of that difference?
c) If the costs can be described by Normal models, what's the probability that medical expenses are higher for someone's dog than for her cat?
d) What concerns do you have?

47. More cereal In Exercise 45 we poured a large and a small bowl of cereal from a box. Suppose the amount of cereal that the manufacturer puts in the boxes is a random variable with mean 16.2 ounces and standard deviation 0.1 ounces.
a) Find the expected amount of cereal left in the box.
b) What's the standard deviation?
c) If the weight of the remaining cereal can be described by a Normal model, what's the probability that the box still contains more than 13 ounces?

48. More pets You're thinking about getting two dogs and a cat. Assume that annual veterinary expenses are independent and have a Normal model with the means and standard deviations described in Exercise 46.
a) Define appropriate variables and express the total annual veterinary costs you may have.
b) Describe the model for this total cost. Be sure to specify its name, expected value, and standard deviation.
c) What's the probability that your total expenses will exceed $400?

49. Medley In the 4 × 100 medley relay event, four swimmers swim 100 yards, each using a different stroke. A college team preparing for the conference championship looks at the times their swimmers have posted and creates a model based on the following assumptions:
- The swimmers' performances are independent.
- Each swimmer's times follow a Normal model.
- The means and standard deviations of the times (in seconds) are as shown:

Swimmer	Mean	SD
1 (backstroke)	50.72	0.24
2 (breaststroke)	55.51	0.22
3 (butterfly)	49.43	0.25
4 (freestyle)	44.91	0.21

a) What are the mean and standard deviation for the relay team's total time in this event?
b) The team's best time so far this season was 3:19.48. (That's 199.48 seconds.) Do you think the team is likely to swim faster than this at the conference championship? Explain.

50. Bikes Bicycles arrive at a bike shop in boxes. Before they can be sold, they must be unpacked, assembled, and tuned (lubricated, adjusted, etc.). Based on past experience, the shop manager makes the following assumptions about how long this may take:
- The times for each setup phase are independent.
- The times for each phase follow a Normal model.
- The means and standard deviations of the times (in minutes) are as shown:

Phase	Mean	SD
Unpacking	3.5	0.7
Assembly	21.8	2.4
Tuning	12.3	2.7

a) What are the mean and standard deviation for the total bicycle setup time?
b) A customer decides to buy a bike like one of the display models but wants a different color. The shop has one, still in the box. The manager says they can have it

ready in half an hour. Do you think the bike will be set up and ready to go as promised? Explain.

51. **Farmers' market** A farmer has 100 lb of apples and 50 lb of potatoes for sale. The market price for apples (per pound) each day is a random variable with a mean of 0.5 dollars and a standard deviation of 0.2 dollars. Similarly, for a pound of potatoes, the mean price is 0.3 dollars and the standard deviation is 0.1 dollars. It also costs him 2 dollars to bring all the apples and potatoes to the market. The market is busy with eager shoppers, so we can assume that he'll be able to sell all of each type of produce at that day's price.
    a) Define your random variables, and use them to express the farmer's net income.
    b) Find the mean.
    c) Find the standard deviation of the net income.
    d) Do you need to make any assumptions in calculating the mean? How about the standard deviation?

52. **Bike sale** The bicycle shop in Exercise 50 will be offering 2 specially priced children's models at a sidewalk sale. The basic model will sell for $120 and the deluxe model for $150. Past experience indicates that sales of the basic model will have a mean of 5.4 bikes with a standard deviation of 1.2, and sales of the deluxe model will have a mean of 3.2 bikes with a standard deviation of 0.8 bikes. The cost of setting up for the sidewalk sale is $200.
    a) Define random variables and use them to express the bicycle shop's net income.
    b) What's the mean of the net income?
    c) What's the standard deviation of the net income?
    d) Do you need to make any assumptions in calculating the mean? How about the standard deviation?

53. **Coffee and doughnuts** At a certain coffee shop, all the customers buy a cup of coffee; some also buy a doughnut. The shop owner believes that the number of cups he sells each day is normally distributed with a mean of 320 cups and a standard deviation of 20 cups. He also believes that the number of doughnuts he sells each day is independent of the coffee sales and is normally distributed with a mean of 150 doughnuts and a standard deviation of 12.
    a) The shop is open every day but Sunday. Assuming day-to-day sales are independent, what's the probability he'll sell more than 2000 cups of coffee in a week?
    b) If he makes a profit of 50 cents on each cup of coffee and 40 cents on each doughnut, can he reasonably expect to have a day's profit of over $300? Explain.
    c) What's the probability that on any given day he'll sell a doughnut to more than half of his coffee customers?

54. **Weightlifting** The Atlas BodyBuilding Company (ABC) sells "starter sets" of barbells that consist of one bar, two 20-pound weights, and four 5-pound weights. The bars weigh an average of 10 pounds with a standard deviation of 0.25 pounds. The weights average the specified amounts, but the standard deviations are 0.2 pounds for the 20-pounders and 0.1 pounds for the 5-pounders. We can assume that all the weights are normally distributed.
    a) ABC ships these starter sets to customers in two boxes: The bar goes in one box and the six weights go in another. What's the probability that the total weight in that second box exceeds 60.5 pounds? Define your variables clearly and state any assumptions you make.
    b) It costs ABC $0.40 per pound to ship the box containing the weights. Because it's an odd-shaped package, though, shipping the bar costs $0.50 a pound plus a $6.00 surcharge. Find the mean and standard deviation of the company's total cost for shipping a starter set.
    c) Suppose a customer puts a 20-pound weight at one end of the bar and the four 5-pound weights at the other end. Although he expects the two ends to weigh the same, they might differ slightly. What's the probability the difference is more than a quarter of a pound?

Just Checking ANSWERS

1. a)

Outcome	X = Cost	Probability
Recharging works	$60	0.75
Replace control unit	$200	0.25

   b) $60(0.75) + 200(0.25) = \$95$

   c) Car owners with this problem will spend an average of $95 to get it fixed.

2. a) $100 + 100 = 200$ seconds

   b) $\sqrt{50^2 + 50^2} = 70.7$ seconds

   c) The times for the two customers are independent.

Chapter 16: Probability Models

16.1 Bernoulli Trials
16.2 The Geometric Model
16.3 The Binomial Model
16.4 Approximating the Binomial with a Normal Model
*16.5 The Continuity Correction
16.6 The Poisson Model
16.7 Other Continuous Random Variables: The Uniform and the Exponential

Where are we going?

How many frogs do you have to kiss before you meet Prince Charming? Is finding 10 leukemia cases in a neighborhood adjacent to an EPA superfund site grounds for a lawsuit? We could simulate to answer almost any question involving probabilities. But in some situations, we can do better with a simple probability model. We'll meet a few of the most common models in this chapter.

Suppose a cereal manufacturer puts pictures of famous athletes on cards in boxes of cereal, in the hope of increasing sales. The manufacturer announces that 20% of the boxes contain a picture of Hope Solo, 30% a picture of Danica Patrick, and the rest a picture of Blake Griffin. Sound familiar? In Chapter 10 we simulated to find the number of boxes we'd need to open to get one of each card. That's a fairly complex question and one well-suited for simulation. But many important questions can be answered more directly by using simple probability models.

16.1 Bernoulli Trials

You're a huge Hope Solo fan. You don't care about completing the whole sports card collection, but you've just *got* to have the Hope Solo picture. How many boxes do you expect you'll have to open before you find her? This isn't the same question that we asked before, but this situation is simple enough for a probability model.

We'll keep the assumption that pictures are distributed at random and we'll trust the manufacturer's claim that 20% of the cards are Hope. So, when you open the box, the probability that you succeed in finding Hope is 0.20. Now we'll call the act of opening *each* box a trial, and note that:

- There are only two possible outcomes (called *success* and *failure*) on each trial. Either you get Hope's picture (success), or you don't (failure).
- The probability of success, denoted p, is the same on every trial. Here $p = 0.20$.
- The trials are independent. Finding Hope in the first box does not change what might happen when you reach for the next box.

Situations like this occur often and are called **Bernoulli trials**. Common examples of Bernoulli trials include tossing a coin, looking for defective products rolling off an

Daniel Bernoulli (1700–1782) was the nephew of Jacob, whom you saw in chapter 14. He was the first to work out the mathematics for what we now call Bernoulli trials.

A S **Activity: Bernoulli Trials.** Guess what! We've been generating Bernoulli trials all along. Look at the Random Simulation Tool in a new way.

CALVIN AND HOBBES © 1993 Watterson. Reprinted with permission of UNIVERSAL PRESS SYNDICATE. All rights reserved.

assembly line, or even shooting free throws in a basketball game. Just as we found equally likely random digits to be the building blocks for our simulation, we can use Bernoulli trials to build a wide variety of useful probability models.

Back to Hope. We want to know how many boxes we'll need to open to find her card. Let's call this random variable $Y = $ # boxes and build a probability model for it. What's the probability you find her picture in the first box of cereal? It's 20%, of course. We could write $P(Y = 1) = 0.2$.

How about the probability that you don't find Hope until the second box? Well, that means you fail on the first trial and then succeed on the second. With the probability of success 0.2, the probability of failure, denoted q, is $1 - 0.2 = 0.8$. Since the trials are independent, the probability of getting your first success on the second trial is $P(Y = 2) = (0.8)(0.2) = 0.16$.

Of course, you could have a run of bad luck. Maybe you won't find Hope until the fifth box of cereal. What are the chances of that? You'd have to fail 4 straight times and then succeed, so $P(Y = 5) = (0.8)^4(0.2) = 0.08192$.

How many boxes might you expect to have to open? We could reason that since Hope's picture is in 20% of the boxes, or 1 in 5, we expect to find her picture, on average, in the fifth box; that is, $E(Y) = \frac{1}{0.2} = 5$ boxes. That's correct, but not easy to prove.

16.2 The Geometric Model

We want to model how long it will take to achieve the first success in a series of Bernoulli trials. The model that tells us this probability is called the **Geometric probability model.** Geometric models are completely specified by one parameter, p, the probability of success, and are denoted Geom(p). Since achieving the first success on trial number x requires first experiencing $x - 1$ failures, the probabilities are easily expressed by a formula.

■ **NOTATION ALERT**

Now we have two more reserved letters. Whenever we deal with Bernoulli trials, p represents the probability of success and q the probability of failure. (Of course, $q = 1 - p$.)

GEOMETRIC PROBABILITY MODEL FOR BERNOULLI TRIALS: GEOM(p)

$p = $ probability of success (and $q = 1 - p = $ probability of failure)
$X = $ number of trials until the first success occurs

$$P(X = x) = q^{x-1} p$$

Expected value: $E(X) = \mu = \dfrac{1}{p}$

Standard deviation: $\sigma = \sqrt{\dfrac{q}{p^2}}$

For Example SPAM AND THE GEOMETRIC MODEL

Postini is a global company specializing in communications security. The company monitors over 1 billion Internet messages per day and recently reported that 91% of e-mails are spam!

Let's assume that your e-mail is typical—91% spam. We'll also assume you aren't using a spam filter, so every message gets dumped in your inbox. And, since spam comes from many different sources, we'll consider your messages to be independent.

QUESTIONS: Overnight your inbox collects e-mail. When you first check your e-mail in the morning, about how many spam e-mails should you expect to have to wade through and discard before you find a real message? What's the probability that the fourth message in your inbox is the first one that isn't spam?

ANSWER: There are two outcomes: a real message (success) and spam (failure). Since 91% of e-mails are spam, the probability of success $p = 1 - 0.91 = 0.09$.

Let X = the number of e-mails I'll check until I find a real message. I assume that the messages arrive independently and in a random order. I can use the model Geom(0.09).

$$E(X) = \frac{1}{p} = \frac{1}{0.09} = 11.1$$

$$P(X = 4) = (0.91)^3(0.09) = 0.0678$$

On average, I expect to have to check just over 11 e-mails before I find a real message. There's slightly less than a 7% chance that my first real message will be the fourth one I check.

MATH BOX

We want to find the mean (expected value) of random variable X, using a geometric model with probability of success p.

First, write the probabilities:

x	1	2	3	4	$\cdots$
$P(X = x)$	p	qp	q^2p	q^3p	$\cdots$

The expected value is: $E(X) = 1p + 2qp + 3q^2p + 4q^3p + \cdots$

Let $p = 1 - q$: $\quad = (1-q) + 2q(1-q) + 3q^2(1-q) + 4q^3(1-q) + \cdots$

Simplify: $\quad = 1 - q + 2q - 2q^2 + 3q^2 - 3q^3 + 4q^3 - 4q^4 + \cdots$

That's an infinite geometric series, with first term 1 and common ratio q:

$$= 1 + q + q^2 + q^3 + \cdots$$
$$= \frac{1}{1-q}$$

So, finally... $E(X) = \dfrac{1}{p}$.

Independence

One of the important requirements for Bernoulli trials is that the trials be independent. Sometimes that's a reasonable assumption—when tossing a coin or rolling a die, for example. But that becomes a problem when (often!) we're looking at situations involving samples chosen without replacement. We said that whether we find a Hope Solo card in one box has no effect on the probabilities in other boxes. This is *almost* true. Technically, if exactly 20% of the boxes have Hope Solo cards, then when you find one, you've reduced the number of remaining Hope Solo cards. With a few million boxes of cereal, though, the difference is hardly worth mentioning.

But if you knew there were 2 Hope Solo cards hiding in the 10 boxes of cereal on the market shelf, then finding one in the first box you try would clearly change your chances of finding Hope in the next box.

If we had an infinite number of boxes, there wouldn't be a problem. It's selecting from a finite population that causes the probabilities to change, making the trials not independent. Obviously, taking 2 out of 10 boxes changes the probability. Taking even a few hundred out of millions, though, makes very little difference. Fortunately, we have a rule of thumb for the in-between cases. It turns out that if we look at less than 10% of the population, we can pretend that the trials are independent and still calculate probabilities that are quite accurate.

> **The 10% Condition:** Bernoulli trials must be independent. If that assumption is violated, it is still okay to proceed as long as the sample is smaller than 10% of the population.

Step-By-Step Example WORKING WITH A GEOMETRIC MODEL

People with O-negative blood are called "universal donors" because O-negative blood can be given to anyone else, regardless of the recipient's blood type. Only about 6% of people have O-negative blood.

Questions:
1. If donors line up at random for a blood drive, how many do you expect to examine before you find someone who has O-negative blood?
2. What's the probability that the first O-negative donor found is one of the first four people in line?

THINK ➡ Plan   State the questions.

I want to estimate how many people I'll need to check to find an O-negative donor, and the probability that 1 of the first 4 people is O-negative.

Check to see that these are Bernoulli trials.

✓ There are two outcomes:
   success = O-negative
   failure = other blood types

✓ The probability of success for each person is $p = 0.06$, because they lined up randomly.

✓ **10% Condition:** Trials aren't independent because the population is finite, but the donors lined up are fewer than 10% of all possible donors.

Variable   Define the random variable.

Let X = number of donors until one is O-negative.

Model   Specify the model.

I can model X with Geom(0.06).

SHOW → Mechanics Find the mean.	$E(X) = \dfrac{1}{0.06} \approx 16.7$
Calculate the probability of success on one of the first four trials. That's the probability that $X = 1, 2, 3,$ or 4.	$P(X \le 4) = P(X = 1) + P(X = 2) + P(X = 3) + P(X = 4)$ $= (0.06) + (0.94)(0.06) + (0.94)^2(0.06) + (0.94)^3(0.06)$ ≈ 0.2193
TELL → Conclusion Interpret your results in context.	Blood drives such as this one expect to examine an average of 16.7 people to find a universal donor. About 22% of the time there will be one within the first 4 people in line.

16.3 The Binomial Model

We can use the Bernoulli trials to answer other questions. Suppose you buy 5 boxes of cereal. What's the probability you get *exactly* 2 pictures of Hope Solo? Before, we asked how long it would take until our first success. Now we want to find the probability of getting 2 successes among the 5 trials. We are still talking about Bernoulli trials, but we're asking a different question.

This time we're interested in the *number of successes* in the 5 trials, so we'll call it X = number of successes. We want to find $P(X = 2)$. This is an example of a **Binomial probability.** It takes two parameters to define this **Binomial model:** the number of trials, n, and the probability of success, p. We denote this model Binom(n, p). Here, $n = 5$ trials, and $p = 0.2$, the probability of finding a Hope Solo card in any trial.

Exactly 2 successes in 5 trials means 2 successes and 3 failures. It seems logical that the probability should be $(0.2)^2(0.8)^3$. Too bad! It's not that easy. That calculation would give you the probability of finding Hope in the first 2 boxes and not in the next 3—*in that order.* But you could find Hope in the third and fifth boxes and still have 2 successes. The probability of those outcomes in that particular order is $(0.8)(0.8)(0.2)(0.8)(0.2)$. That's also $(0.2)^2(0.8)^3$. In fact, the probability will always be the same, no matter what order the successes and failures occur in. Anytime we get 2 successes in 5 trials, no matter what the order, the probability will be $(0.2)^2(0.8)^3$. We just need to take account of all the possible orders in which the outcomes can occur.

Fortunately, these possible orders are *disjoint.* (For example, if your two successes came on the first two trials, they couldn't come on the last two.) So we can use the Addition Rule and add up the probabilities for all the possible orderings. Since the probabilities are all the same, we only need to know how many orders are possible. For small numbers, we can just make a tree diagram and count the branches. For larger numbers this isn't practical, so we let the computer or calculator do the work.

Each different order in which we can have k successes in n trials is called a "combination." The total number of ways that can happen is written $\binom{n}{k}$ or $_nC_k$ and pronounced "n choose k."

$$_nC_k = \frac{n!}{k!(n-k)!}$$ where $n!$ (pronounced "n factorial") $= n \times (n-1) \times \cdots \times 1$

A S *Activity:* **The Binomial Distribution.** It's more interesting to combine Bernoulli trials. Simulate this with the Random Tool to get a sense of how Binomial models behave.

■ **NOTATION ALERT**

Now punctuation! Throughout mathematics $n!$, pronounced "n factorial," is the product of all the integers from 1 to n. For example, $4! = 4 \cdot 3 \cdot 2 \cdot 1 = 24$.

For 2 successes in 5 trials,

$$_5C_2 = \frac{5!}{2!(5-2)!} = \frac{5 \times 4 \times 3 \times 2 \times 1}{2 \times 1 \times 3 \times 2 \times 1} = \frac{5 \times 4}{2 \times 1} = 10.$$

So there are 10 ways to get 2 Hope pictures in 5 boxes, and the probability of each is $(0.2)^2(0.8)^3$. Now we can find what we wanted:

$$P(\#\text{success} = 2) = 10(0.2)^2(0.8)^3 = 0.2048$$

In general, the probability of exactly k successes in n trials is $_nC_k p^k q^{n-k}$.

Using this formula, we could find the expected value by adding up $xP(X = x)$ for all values, but it would be a long, hard way to get an answer that you already know intuitively. What's the expected value? If we have 5 boxes, and Hope's picture is in 20% of them, then we would expect to have $5(0.2) = 1$ success. If we had 100 trials with probability of success 0.2, how many successes would you expect? Can you think of any reason not to say 20? It seems so simple that most people wouldn't even stop to think about it. You just multiply the probability of success by n. In other words, $E(X) = np$.

The standard deviation is less obvious; you can't just rely on your intuition. Fortunately, the formula for the standard deviation also boils down to something simple: $SD(X) = \sqrt{npq}$. (If you're curious about where that comes from, it's in the Math Box, too!) In 100 boxes of cereal, we expect to find 20 Hope Solo cards, with a standard deviation of $\sqrt{100 \times 0.8 \times 0.2} = 4$ boxes.

Time to summarize. A Binomial probability model describes the number of successes in a specified number of trials. It takes two parameters to specify this model: the number of trials n and the probability of success p.

BINOMIAL PROBABILITY MODEL FOR BERNOULLI TRIALS: BINOM(n, p)

n = number of trials
p = probability of success (and $q = 1 - p$ = probability of failure)
X = number of successes in n trials

$$P(X = x) = {_nC_x}\, p^x q^{n-x}, \text{ where } {_nC_x} = \frac{n!}{x!(n-x)!}$$

Mean: $\mu = np$
Standard Deviation: $\sigma = \sqrt{npq}$

MATH BOX

To derive the formulas for the mean and standard deviation of a Binomial model we start with the most basic situation.

Consider a single Bernoulli trial with probability of success p. Let's find the mean and variance of the number of successes.

Here's the probability model for the number of successes:

x	0	1
$P(X = x)$	q	p

Find the expected value:

$$E(X) = 0q + 1p$$
$$E(X) = p$$

And now the variance:

$$Var(X) = (0 - p)^2 q + (1 - p)^2 p$$
$$= p^2 q + q^2 p$$
$$= pq(p + q)$$
$$= pq(1)$$
$$Var(X) = pq$$

What happens when there is more than one trial, though? A Binomial model simply counts the number of successes in a series of n independent Bernoulli trials. That makes it easy to find the mean and standard deviation of a binomial random variable, Y.

$$\text{Let } Y = X_1 + X_2 + X_3 + \cdots + X_n$$
$$E(Y) = E(X_1 + X_2 + X_3 + \cdots + X_n)$$
$$= E(X_1) + E(X_2) + E(X_3) + \cdots + E(X_n)$$
$$= p + p + p + \cdots + p \text{ (There are } n \text{ terms.)}$$

So, as we thought, the mean is $E(Y) = np$.

And since the trials are independent, the variances add:
$$Var(Y) = Var(X_1 + X_2 + X_3 + \cdots + X_n)$$
$$= Var(X_1) + Var(X_2) + Var(X_3) + \cdots + Var(X_n)$$
$$= pq + pq + pq + \cdots + pq \text{ (Again, } n \text{ terms.)}$$
$$Var(Y) = npq$$

Voilà! The standard deviation is $SD(Y) = \sqrt{npq}$.

For Example SPAM AND THE BINOMIAL MODEL

RECAP: The communications monitoring company *Postini* has reported that 91% of e-mail messages are spam. Suppose your inbox contains 25 messages.

QUESTIONS: What are the mean and standard deviation of the number of real messages you should expect to find in your inbox? What's the probability that you'll find only 1 or 2 real messages?

ANSWER: I assume that messages arrive independently and at random, with the probability of success (a real message) $p = 1 - 0.91 = 0.09$. Let $X =$ the number of real messages among 25. I can use the model Binom(25, 0.09).

$$E(X) = np = 25(0.09) = 2.25$$
$$SD(X) = \sqrt{npq} = \sqrt{25(0.09)(0.91)} = 1.43$$
$$P(X = 1 \text{ or } 2) = P(X = 1) + P(X = 2)$$
$$= {}_{25}C_1(0.09)^1(0.91)^{24} + {}_{25}C_2(0.09)^2(0.91)^{23}$$
$$= 0.2340 + 0.2777$$
$$= 0.5117$$

Among 25 e-mail messages, I expect to find an average of 2.25 that aren't spam, with a standard deviation of 1.43 messages. There's just over a 50% chance that 1 or 2 of my 25 e-mails will be real messages.

Step-By-Step Example WORKING WITH A BINOMIAL MODEL

Suppose 20 donors come to a blood drive. Recall that 6% of people are "universal donors."

Questions:
1. What are the mean and standard deviation of the number of universal donors among them?
2. What is the probability that there are 2 or 3 universal donors?

THINK ▶ Plan State the question.	I want to know the mean and standard deviation of the number of universal donors among 20 people, and the probability that there are 2 or 3 of them.

	Check to see that these are Bernoulli trials.	✓ There are two outcomes: success = O-negative failure = other blood types ✓ $p = 0.06$, because people have lined up at random. ✓ **10% Condition:** Trials are not independent, because the population is finite, but fewer than 10% of all possible donors are lined up.
Variable	Define the random variable.	Let X = number of O-negative donors among $n = 20$ people.
Model	Specify the model.	I can model X with Binom(20, 0.06).
SHOW ➡ **Mechanics**	Find the expected value and standard deviation.	$E(X) = np = 20(0.06) = 1.2$ $SD(X) = \sqrt{npq} = \sqrt{20(0.06)(0.94)} \approx 1.06$ $P(X = 2 \text{ or } 3) = P(X = 2) + P(X = 3)$ $\phantom{P(X = 2 \text{ or } 3)} = {}_{20}C_2(0.06)^2(0.94)^{18}$ $\phantom{P(X = 2 \text{ or } 3) =} + {}_{20}C_3(0.06)^3(0.94)^{17}$ $\phantom{P(X = 2 \text{ or } 3)} \approx 0.2246 + 0.0860$ $\phantom{P(X = 2 \text{ or } 3)} = 0.3106$
TELL ➡ **Conclusion**	Interpret your results in context.	In groups of 20 randomly selected blood donors, I expect to find an average of 1.2 universal donors, with a standard deviation of 1.06. About 31% of the time, I'd find 2 or 3 universal donors among the 20 people.

16.4 Approximating the Binomial with a Normal Model

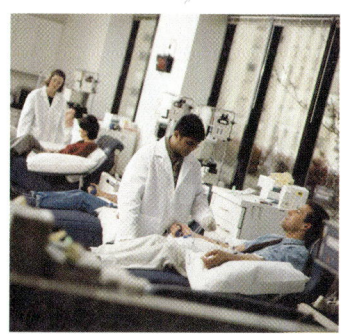

Activity: Normal Approximation. Binomial probabilities can be hard to calculate. With the Simulation Tool you'll see how well the Normal model can approximate the Binomial—a much easier method.

Suppose the Tennessee Red Cross anticipates the need for at least 1850 units of O-negative blood this year. It estimates that it will collect blood from 32,000 donors. How great is the risk that the Tennessee Red Cross will fall short of meeting its need? We've just learned how to calculate such probabilities. We can use the Binomial model with $n = 32,000$ and $p = 0.06$. The probability of getting *exactly* 1850 units of O-negative blood from 32,000 donors is ${}_{32000}C_{1850} \times 0.06^{1850} \times 0.94^{30150}$. No calculator on earth can calculate that first term (it has more than 100,000 digits).[1] And that's just the beginning. The problem said *at least* 1850, so we have to do it again for 1851, for 1852, and all the way up to 32,000. No thanks.

When we're dealing with a large number of trials like this, making direct calculations of the probabilities becomes tedious (or outright impossible). Here an old friend—the Normal model—comes to the rescue.

The Binomial model has mean $np = 1920$ and standard deviation $\sqrt{npq} \approx 42.48$. We could try approximating its distribution with a Normal model, using the same mean and

---

[1] If your calculator *can* find Binom(32000, 0.06), then it's smart enough to use an approximation. Read on to see how you can, too.

standard deviation. Remarkably enough, that turns out to be a very good approximation. (We'll see why in the next chapter.) With that approximation, we can find the *probability:*

$$P(X < 1850) = P\left(z < \frac{1850 - 1920}{42.48}\right) \approx P(z < -1.65) \approx 0.05$$

There seems to be about a 5% chance that this Red Cross chapter will run short of O-negative blood.

Can we always use a Normal model to make estimates of Binomial probabilities? No. Consider the Hope Solo situation—pictures in 20% of the cereal boxes. If we buy five boxes, the actual Binomial probabilities that we get 0, 1, 2, 3, 4, or 5 pictures of Hope are 33%, 41%, 20%, 5%, 1%, and 0.03%, respectively. The first histogram shows that this probability model is skewed. That makes it clear that we should not try to estimate these probabilities by using a Normal model.

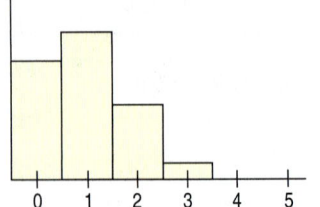

Now suppose we open 50 boxes of this cereal and count the number of Hope Solo pictures we find. The second histogram shows this probability model. It is centered at $np = 50(0.2) = 10$ pictures, as expected, and it appears to be fairly symmetric around that center. Let's have a closer look.

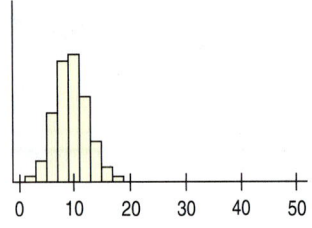

The third histogram again shows Binom(50, 0.2), this time magnified somewhat and centered at the expected value of 10 pictures of Hope. It looks close to Normal, for sure. With this larger sample size, it appears that a Normal model might be a useful approximation.

A Normal model, then, is a close enough approximation only for a large enough number of trials. And what we mean by "large enough" depends on the probability of success. We'd need a larger sample if the probability of success were very low (or very high). It turns out that a Normal model works pretty well if we expect to see at least 10 successes and 10 failures. That is, we check the **Success/Failure Condition.**

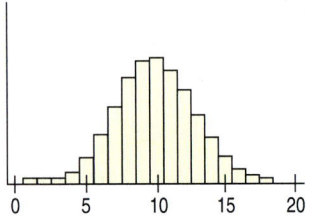

The Success/Failure Condition: A Binomial model is approximately Normal if we expect at least 10 successes and 10 failures:

$$np \geq 10 \text{ and } nq \geq 10.$$

MATH BOX

It's easy to see where the magic number 10 comes from. You just need to remember how Normal models work. The problem is that a Normal model extends infinitely in both directions. But a Binomial model must have between 0 and n successes, so if we use a Normal to approximate a Binomial, we have to cut off its tails. That's not very important if the center of the Normal model is so far from 0 and n that the lost tails have only a negligible area. More than three standard deviations should do it, because a Normal model has little probability past that.

So the mean needs to be at least 3 standard deviations from 0 and at least 3 standard deviations from n. Let's look at the 0 end.

We require:	$\mu - 3\sigma > 0$
Or in other words:	$\mu > 3\sigma$
For a Binomial, that's:	$np > 3\sqrt{npq}$
Squaring yields:	$n^2p^2 > 9npq$
Now simplify:	$np > 9q$
Since $q \leq 1$, we can require:	$np > 9$

For simplicity, we usually require that np (and nq for the other tail) be at least 10 to use the Normal approximation, the Success/Failure Condition.[2]

---

[2] Looking at the final step, we see that we need $np > 9$ in the worst case, when q (or p) is near 1, making the Binomial model quite skewed. When q and p are near 0.5—say between 0.4 and 0.6—the Binomial model is nearly symmetric and $np > 5$ ought to be safe enough. Although we'll always check for 10 expected successes and failures, keep in mind that for values of p near 0.5, we can be somewhat more forgiving.

CHAPTER 16 Probability Models

For Example: SPAM AND THE NORMAL APPROXIMATION TO THE BINOMIAL

RECAP: The communications monitoring company *Postini* has reported that 91% of e-mail messages are spam. Recently, you installed a spam filter. You observe that over the past week it okayed only 151 of 1422 e-mails you received, classifying the rest as junk. Should you worry that the filtering is too aggressive?

QUESTION: What's the probability that no more than 151 of 1422 e-mails is a real message?

ANSWER: I assume that messages arrive randomly and independently, with a probability of success (a real message) $p = 0.09$. The model Binom(1422, 0.09) applies, but will be hard to work with. Checking conditions for the Normal approximation, I see that:

✓ These messages represent less than 10% of all e-mail traffic.
✓ I expect $np = (1422)(0.09) = 127.98$ real messages and $nq = (1422)(0.91) = 1294.02$ spam messages, both far greater than 10.

It's okay to approximate this binomial probability by using a Normal model.

$$\mu = np = 1422(0.09) = 127.98$$
$$\sigma = \sqrt{npq} = \sqrt{1422(0.09)(0.91)} \approx 10.79$$
$$P(x \leq 151) = P\left(z \leq \frac{151 - 127.98}{10.79}\right)$$
$$= P(z \leq 2.13)$$
$$= 0.9834$$

Among my 1422 e-mails, there's over a 98% chance that no more than 151 of them were real messages, so the filter may be working properly.

*16.5 The Continuity Correction

When we used the Normal model to approximate the Binomial probabilites in the last section, we actually pulled a fast one on you. The Binomial is discrete, giving probabilities for specific counts, but the Normal is a **continuous random variable** that can take on *any value*. As we've seen, the probability that a Normal takes on any *specific* value is 0. So, for a Binomial with $n = 50$ and $p = 0.2$, the probability of exactly 10 successes is $P(X = 10) = 0.1398$. But for the Normal, $P(X = 10) = 0$. So there's a problem. The box below describes a solution, called the **continuity correction**.

> ***THE CONTINUITY CORRECTION**
>
> When we use a continuous model to model a set of discrete events, we may need to make an adjustment called the **continuity correction**. We approximated the Binomial distribution (50, 0.2) with a Normal distribution. But what does the Normal distribution say about the probability that $X = 10$? Every specific value in the Normal probability model has probability 0. That's not the answer we want.

Because X is really discrete, it takes on the exact values 0, 1, 2, . . . , 50, each with positive probability. The histogram holds the secret to the correction. Look at the bin corresponding to $X = 10$ in the histogram. It goes from 9.5 to 10.5. What we really want is to find the area under the Normal curve *between* 9.5 and 10.5. So when we use the Normal distribution to approximate discrete events, we go halfway to the next value on the left and/or the right. We approximate $P(X = 10)$ by finding $P(9.5 \leq X \leq 10.5)$. For a Binomial (50, 0.2), $\mu = 10$ and $\sigma = 2.83$.

$$\text{So } P(9.5 \leq X \leq 10.5) \approx P\left(\frac{9.5 - 10}{2.83} \leq z \leq \frac{10.5 - 10}{2.83}\right)$$
$$= P(-0.177 \leq z \leq 0.177)$$
$$= 0.1405$$

By comparison, the *exact* Binomial probability is 0.1398.

For Example USING THE NORMAL DISTRIBUTION

Some LCD panels have stuck or "dead" pixels that have defective transistors and are permanently unlit. If a panel has too many dead pixels, it must be rejected. A manufacturer knows that, when the production line is working correctly, the probability of rejecting a panel is 0.07.

QUESTIONS:

1. How many screens do they expect to reject in a day's production run of 500 screens? What is the standard deviation?
2. If they reject 40 screens today, is that a large enough number that they should be concerned that something may have gone wrong with the production line?
3. In the past week of 5 days of production, they've rejected 200 screens—an average of 40 per day. Should that raise concerns?

ANSWERS:

1. $\mu = 0.07 \times 500 = 35$ is the expected number of rejects
   $\sigma = \sqrt{npq} = \sqrt{500 \times 0.07 \times 0.93} = 5.7$
2. $P(X \geq 40) = P\left(z \geq \frac{40 - 35}{5.7}\right) = P(z \geq 0.877) \approx 0.19$, not an extraordinarily large number of rejects
3. Using the Normal approximation:
   $\mu = 0.07 \times 2500 = 175$
   $\sigma = \sqrt{2500 \times 0.07 \times 0.93} = 12.757$
   $P(X \geq 200) = P\left(z \geq \frac{200 - 175}{12.757}\right) = P(z \geq 1.96) \approx 0.025$

Yes, this seems to be a number of rejects that would occur by chance rarely if nothing were wrong.

16.6 The Poisson Model

In the early 1990s, a leukemia cluster was identified in the Massachusetts town of Woburn. Many more cases of leukemia, a cancer that originates in a cell in the marrow of bone, appeared in this small town than would be predicted. Was it evidence of a problem in the town, or was it chance? That question led to a famous trial in which the families of eight leukemia victims sued, and became grist for the book and movie *A Civil Action*. Following an 80-day trial, the judge called for a retrial after dismissing the jury's contradictory and confusing findings. Shortly thereafter, the chemical companies and the families settled.

When rare events occur together or in clusters, people often want to know if that happened just by chance or whether something else is going on. If we assume that the events occur independently, we can use a Binomial model to find the probability that a cluster of events like this occurs. For rare events, p will be quite small, and when n is large, it may be difficult to compute the exact probability that a certain size cluster occurs. To see why, let's try to compute the probability that a cluster of cases of size x occurs in Woburn. We'll use the *national average* of leukemia incidence to get a value for p, and the population of Woburn as our value for n. In the United States in the early 1990s, there were about 30,800 new cases of leukemia each year and about 280,000,000 people, giving a value for p of about 0.00011. The population of Woburn was about $n = 35,000$. We'd expect $np = 3.85$ new cases of leukemia in Woburn. How unlikely would 8 or more cases be? To answer that, we'll need to calculate the complement, adding the probabilities of no cases, exactly 1 case, etc., up to 7 cases. Each of those probabilities would have the form $_{35000}C_x p^x q^{35000-x}$. To find $_{35000}C_x$, we'll need 35,000!—a daunting number even for computers. In the 18th century, when people first were interested in computing such probabilities, it was impossible. We can easily go beyond the capabilities of today's computers by making n large enough and p small enough. We could use the Normal model to help us approximate the probabilities as long as np is at least 10, but for rare events such as leukemia in Woburn, when np falls much below 10, the Normal probabilities won't be accurate.

Simeon Denis Poisson[3] was a French mathematician interested in events with very small probability. He originally derived his model to approximate the Binomial model when the probability of a success, p, is very small and the number of trials, n, is very large. Poisson's contribution was providing a simple approximation to find that probability. When you see the formula, however, you won't necessarily see the connection to the Binomial. The Poisson's parameter is often denoted by λ, the mean of the distribution. To use the **Poisson model** to approximate the Binomial, we'll make their means match, so we set $\lambda = np$.

■ **NOTATION ALERT**
We almost always use λ (the Greek letter lambda) for the mean of a Poisson model.

> **POISSON PROBABILITY MODEL FOR SUCCESSES: *Poisson* (λ)**
> λ = mean number of successes.
> X = number of successes.
>
> $$P(X = x) = \frac{e^{-\lambda} \lambda^x}{x!}$$
>
> Expected value: $E(X) = \lambda$
> Standard deviation: $SD(X) = \sqrt{\lambda}$
>
> The Poisson model is a reasonably good approximation of the Binomial when $n \geq 20$ with $p \leq 0.05$ or $n \geq 100$ with $p \leq 0.10$.

---

[3] *Poisson* is a French name (meaning "fish"), properly pronounced "pwa sohn" and not with an "oy" sound as in poison.

Using Poisson's model, we can easily find the probabilities of a given size cluster of leukemia cases. Using $\lambda = np = 35000 \times 0.00011 = 3.85$ new cases a year, we find that the probability of seeing exactly x cases in a year is $P(X = x) = \dfrac{e^{-3.85}3.85^x}{x!}$.

By adding up these probabilities for $x = 0, 1, \ldots 7$, we'd find that the probability of 8 or more cases with $\lambda = 3.85$ is about 0.043. That's small but not terribly unusual.

> **Where does e come from?** You might know that $e = 2.7182818\ldots$ (to 7 decimal places), but does it have any real meaning? Yes! In fact, one of the places e originally turned up was in calculating how much more money you'd earn if you could get interest compounded more often. Suppose you receive an offer to earn 100% per year simple interest. (Take it!) At the end of the year, you'd have twice as much money as when you started. But suppose that you have the option of having the interest calculated and paid at the end of every month instead. Then each month you'll earn $(100/12)\%$ interest. At the end of the year, you'll wind up with $(1 + 1/12)^{12} = 2.613$ times as much instead of 2. If you could get the interest paid every day, you'd get $(1 + 1/365)^{365} = 2.715$ times as much. If you could convince the bank to compound and pay the interest every second, you'd get $(1 + 1/31536000)^{31536000} = 2.7182818$ times as much at year's end. This is where e shows up. In the limit, if you could get the interest compounded continually, you'd get e times as much. In other words, as n gets large, the limit of $(1 + 1/n)^n = e$. Who discovered this fact? Jacob Bernoulli, in 1683!

In spite of its origins as an approximation to the Binomial, the Poisson model is also used directly to model the probability of the occurrence of events for a variety of phenomena. It's a good model to consider whenever your data consist of counts of occurrences. It requires only that the events be independent and that the mean number of occurrences stay constant for the duration of the data collection (and beyond, if we hope to make predictions).

One nice feature of the Poisson model is that it scales according to the sample size.[4] For example, if we know that the average number of occurrences in a town the size of Woburn, with 35,000 people, is 3.85, we know that the average number of occurrences in a town of only 3500 residents is 0.385. We can use that new value of λ to calculate the probabilities for the smaller town. Using the Poisson formula, we'd find that the probability of 0 occurrences for a town of 3500 is about 0.68.

One of the consequences of the Poisson model is that, as long as the mean rate of occurrences stays constant, the occurrence of past events doesn't change the probability of future events. That is, even though events that occur according to the Poisson model *appear* to cluster, the probability of *another* event occurring is still the same. This is counterintuitive and may be one reason why so many people believe in the form of the "Law of Averages." When the Poisson model was used to model bombs dropping over London in World War II, the model said that even though several bombs had hit a particular sector, the probability of *another* hit was still the same. So, there was no point in moving people from one sector to another, even after several hits. You can imagine how difficult it must have been trying to convince people of that! This same phenomenon leads many people to think they see patterns in which an athlete, slot machine, or financial market gets "hot." But the occurrence of several successes in a row is not unusual and is rarely evidence that the mean has changed. Careful studies have been made of "hot streaks," but none has ever demonstrated that they actually exist.

> **Poisson in History**
> In his book *Gravity's Rainbow*, Thomas Pynchon describes using the Poisson to model the bombs dropping over London during World War II. "The rockets *are* distributing about London just as Poisson's equation in the textbook predicts. As the data keep coming in, Roger looks more and more like a prophet . . . all I'm doing is plugging numbers into a well-known equation, you can look it up in the book and do it yourself. . . ."

---
[4]Because *Poisson* means "fish" in French, you can remember this fact by the bilingual pun "*Poisson* scales"!

Just Checking

As we noted a few chapters ago, the Pew Research Center (www.pewresearch.org) reports that they are actually able to contact only 76% of the randomly selected households drawn for a telephone survey.

1. Explain why these phone calls can be considered Bernoulli trials.
2. Which of the models of this chapter (Geometric, Binomial, Normal) would you use to model the number of successful contacts from a list of 1000 sampled households? Explain.
3. Pew further reports that even after they contacted a household, only 38% agree to be interviewed, so the probability of getting a completed interview for a randomly selected household is only 0.29. Which of the models of this chapter would you use to model the number of households Pew has to call before they get the first completed interview?
4. Suppose that in the course of a year, legitimate survey organizations (not folks pretending to take a survey but actually trying to sell you something or change your vote) sample 70,000 of the approximately 107,000,000 households in the United States. You wonder if people from your school are represented in these polls. Counting all the living alumni from your school, there are 10,000 households. How would you model the number of these alumni households you'd expect will be contacted next year by a legitimate poll?

16.7 Other Continuous Random Variables: The Uniform and the Exponential

Many phenomena in business can be modeled by continuous random variables. The Normal model is important, but it is only one of many different models. Entire courses are devoted to studying which models work well in different situations, but we'll introduce just two others that are commonly used: the uniform and the exponential.

The Uniform Distribution

We've already seen the discrete version of the uniform probability model. It has the property that all events are equally likely: with n possible outcomes, $P(X = x) = 1/n$ for all possible values x. For the continuous uniform, we'd want all *intervals* of the same length to have the same probability. So, not surprisingly the probability model for the continuous uniform random variable looks flat. It can be defined by the formula

$$f(x) = \begin{cases} \dfrac{1}{b-a} & \text{if } a \leq x \leq b \\ 0 & \text{otherwise} \end{cases}$$

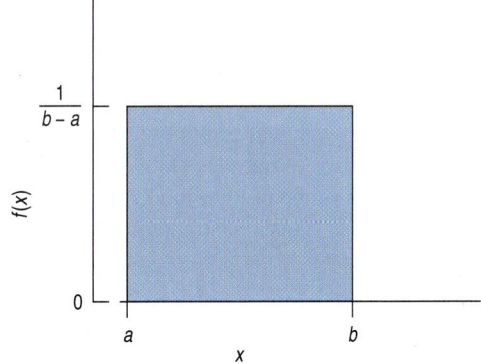

Figure 16.2
The probability model of a continuous uniform random variable on the interval from *a* to *b*.

From Figure 16.2, it's easy to see that the probability that X lies in any interval between a and b is the same as any other interval of the same length. In fact, the probability is just the ratio of the length of the interval to the total length: $b - a$. In other words:

For values c and d $(c \leq d)$ both within the interval $[a, b]$:

$$P(c \leq X \leq d) = \frac{(d - c)}{(b - a)}$$

As an example, suppose you arrive at a bus stop and want to model how long you'll wait for the next bus. The sign says that busses arrive about every 20 minutes, but no other information is given. You might assume that the arrival is equally likely to be anywhere in the next 20 minutes, and so the probability model would be

$$f(x) = \begin{cases} \frac{1}{20} & \text{if } 0 \leq x \leq 20 \\ 0 & \text{otherwise} \end{cases}$$

and would look as shown in Figure 16.3.

Figure 16.3

The probability model for a continuous uniform random variable on the interval [0,20]. Notice that the mean (the balancing point) of the distribution is at 10 minutes and that the area of the box is 1.

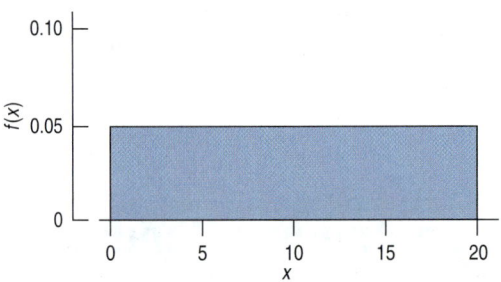

Just as the mean of a data distribution is the balancing point of a histogram, the mean of any continuous random variable is the balancing point of the density function. Looking at Figure 16.3, we can see that the balancing point is halfway between the end points at 10 minutes. In general, the expected value is:

$$E(X) = \frac{a + b}{2}$$

for a uniform distribution on the interval (a, b). With $a = 0$ and $b = 20$, the expected value would be 10 minutes.

The variance and standard deviation are less intuitive:

$$\text{Var}(X) = \frac{(b - a)^2}{12}; \quad SD(X) = \sqrt{\frac{(b - a)^2}{12}}.$$

Using these formulas, our bus wait will have an expected value of 10 minutes with a standard deviation of $\sqrt{\frac{(20 - 0)^2}{12}} = 5.77$ minutes.

The Exponential Model

We've seen that the Poisson model is a good model for the arrival, or occurrence, of events. Suppose, for example, that we use the Poisson to model the probability of x visits to our website within the next minute. Then, the exponential model with parameter λ can be used to model the time *between* those events. Because *time* is a continuous quantity, it has a continuous model with the form:

$$f(x) = \lambda e^{-\lambda x} \quad \text{for } x \geq 0 \text{ and } \lambda > 0$$

The use of the parameter λ again is not coincidental. It highlights the relationship between the exponential and the Poisson.

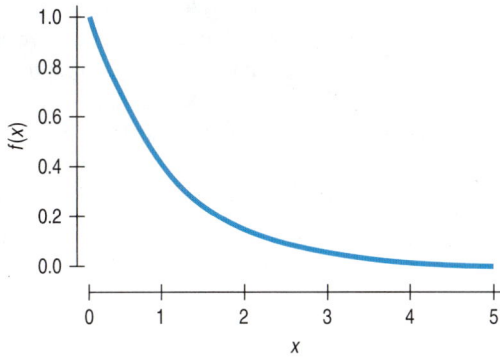

The exponential probability model (with $\lambda = 1$). The probability that x lies between any two values corresponds to the area under the curve between the two values.

If a discrete random variable can be modeled by a Poisson model with rate λ, then the times between events can be modeled by the continuous exponential model with the same parameter λ. The mean of the Poisson random variable is λ and the mean of the exponential is $1/\lambda$. The inverse relationship between the two means makes intuitive sense. If λ increases and we expect *more* hits per minute, then the expected time between hits should go down. The standard deviation of the exponential random variable is $1/\lambda$.

Probabilities of an exponential random variable can be found through the probability model. Fortunately, the probability that x lies between any two values, s and t ($s \leq t$), has a particularly easy form:

$$P(s \leq X \leq t) = e^{-\lambda s} - e^{-\lambda t}.$$

In particular, by setting s to be 0, we can find the probability that the waiting time will be less than t:[5]

$$P(X \leq t) = P(0 \leq X \leq t) = e^{-\lambda 0} - e^{-\lambda t} = 1 - e^{-\lambda t}.$$

If arrivals of hits to our website can be well modeled by a Poisson with $\lambda = 4$/minute, then the probability that we'll have to wait less than 20 seconds (1/3 of a minute) is $F(1/3) = P(0 \leq X \leq 1/3) = 1 - e^{-4/3} = 0.736$. That seems about right. Arrivals are coming about every 15 seconds on average, so we shouldn't be surprised that nearly 75% of the time we won't have to wait more than 20 seconds for the next hit.

WHAT CAN GO WRONG?

- **Be sure you have Bernoulli trials.** Be sure to check the requirements first: two possible outcomes per trial ("success" and "failure"), a constant probability of success, and independence. Remember to check the 10% Condition when sampling without replacement.

- **Don't confuse Geometric and Binomial models.** Both involve Bernoulli trials, but the issues are different. If you are repeating trials until your first success, that's a Geometric probability. You don't know in advance how many trials you'll need—theoretically, it could take forever. If you are counting the number of successes in a specified number of trials, that's a Binomial probability.

- **Don't use the Normal approximation with small n.** To use a Normal approximation in place of a Binomial model, there must be at least 10 expected successes and 10 expected failures.

---

[5]The function $P(X \leq t) = F(t)$ is called the cumulative distribution function (cdf) of the random variable X. The probability model is formally called the density.

- **Don't assume everything's Normal.** Just because a random variable is continuous or you happen to know a mean and standard deviation doesn't mean that a Normal model will be useful. You must think about whether the **Normality Assumption** is justified. Using a Normal model when it really does not apply will lead to wrong answers and misleading conclusions.

  A sample of CEOs has a mean total compensation of $10,307,311.87 with a standard deviation of $17,964,615.16. Using the Normal model rule, we should expect about 68% of the CEOs to have compensations between −$7,657,303.29 and $28,271,927.03. In fact, more than 90% of the CEOs have annual compensations in this range. What went wrong? The distribution is skewed, not symmetric. Using the 68–95–99.7 Rule for data like these will lead to silly results.

This chapter builds on what we know about random variables. We now have more probability models to join the Normal model.

There are a number of "forward" connections from this chapter. We'll see the **10% Condition** and the **Success/Failure Condition** often. And the facts about the Binomial distribution can help explain how proportions behave as we'll see in the next chapter.

What Have We Learned?

Learning Objectives

We've learned that Bernoulli trials show up in lots of places. Depending on the random variable of interest, we can use one of three models to estimate probabilities for Bernoulli trials:

- a Geometric model when we're interested in the number of Bernoulli trials until the next success;
- a Binomial model when we're interested in the number of successes in a certain number of Bernoulli trials;
- a Normal model to approximate a Binomial model when we expect at least 10 successes and 10 failures;

We can use a Poisson model to approximate a Binomial model when the probability of success, p, is very small and the number of trials, n, is very large;

- an Exponential model can model the time between two random events.

Review of Terms

Bernoulli trials, if . . .
1. there are two possible outcomes (p. 412).
2. the probability of success is constant.
3. the trials are independent.

Geometric probability model	A Geometric model is appropriate for a random variable that counts the number of Bernoulli trials until the first success (p. 413).
Binomial probability model	A Binomial model is appropriate for a random variable that counts the number of successes in a fixed number of Bernoulli trials (p. 417).
Success/Failure Condition	For a Normal model to be a good approximation of a Binomial model, we must expect at least 10 successes and 10 failures. That is, $np \geq 10$ and $nq \geq 10$ (p. 420).
Continuous random variable	A continuous random variable is one that can take on any value in an interval of values (p. 421).
Poisson probability model	A Poisson model can be used to approximate a Binomial model when p is small and n is large. Use $\lambda = np$ as the mean for the Poisson. The Poisson can also be used to model counts of a wide variety of phenomena (p. 423).
Uniform probability model	A Uniform model gives equal probability to intervals of the same size (p. 425).
Exponential probability model	An Exponential model can be used to estimate the time between two random events that occur following a Poisson probability model (p. 427).

On the Computer PROBABILITY MODELS

Most statistics packages offer functions that compute Binomial probabilities, and many offer functions for Geometric and Poisson probabilities as well. Some technology solutions automatically use the Normal approximation for the Binomial when the exact calculations become unmanageable. And many websites offer Normal probability calculations.

The only important differences among these functions are in what they are named and the order of their arguments. In these functions, pdf stands for "probability density function"—what we've been calling a probability model. For a discrete random variables (e.g., Geometric, Binomial, Poisson) the pdf(x) gives the probability $P(X = x)$. For a continuous random variable (e.g., Normal, Exponential), pdf(x) gives the value of the curve under which the probability lies. To find the probability, use the cdf(x), the "cumulative distribution function", that accumulates the probability and gives $P(X \leq x)$.

Generically, the functions are as follows:

Geometric pdf(*prob, x*)	Finds the individual geometric probability of getting the first success on trial *x* when the probability of success is *prob*. Note: Some Geometric models model the number of failures before the first success.	For example, the probability of finding the first Hope Solo picture in the fifth cereal box is Geometric pdf(0.2, 5)
Geometric cdf(*prob, x*)	Finds the cumulative probability of getting the first success on or before trial *x*, when the probability of success is *prob*.	For example, the total probability of finding Hope's picture in one of the first 4 boxes is Geometric cdf(0.2, 4)
Binomial pdf(*n, prob, x*)	Finds the probability of getting *x* successes in *n* trials when the probability of success is *prob*.	For example, Binomial pdf(5, 0.2, 2) is the probability of finding Hope's picture exactly twice among 5 boxes of cereal.
Binomial cdf(*n, prob, x*)	The cdf of the Binomial	
Poisson pdf(λ, *x*)	Finds the probability of *x* successes when the mean number of successes is $\lambda = np$.	Best for large *n* and small probability of success.
Poisson cdf(λ, *x*)	The cdf of the Poisson	
Norm (x, mu, sigma)	The pdf or cdf of the Normal.	
Exp(*x*, λ)	The pdf or cdf of the Exponential	

DATA DESK

BinomDistr(*x, n, prob*) (pdf)
CumBinomDistr(*x, n, prob*) (cdf)
PoisDistr(*x, mean*) (pdf)
CumPoisDistr(*x, mean*) (cdf)
ZDistr(*x*) (pdf)
CumZDistr*(x)* (cdf)

COMMENTS
Data Desk does not compute Geometric probabilities.
These functions work in derived variables or in scratchpads.

EXCEL

Binomdist(*x, n, prob, cumulative*)
Norm.dist(*x, mean, standard_dev, cumulative*)

COMMENTS
Set cumulative = *true* for cdf, *false* for pdf.
Excel's function fails when *x* or *n* is large.
Possibly, it does not use the Normal approximation.
Excel does not compute Geometric or Poisson probabilities.

JMP

Binomial Probability (*prob, n, x*) (pdf)
Binomial Distribution (*prob, n, x*) (cdf)
Poisson Probability (*mean, k*) (pdf)
Poisson Distribution (*mean, k*) (cdf)
Normal Distribution(*x, mean = 0, sd = 1*) (cdf)

COMMENTS
JMP does not compute Geometric probabilities.

MINITAB

Choose **Probability Distributions** from the **Calc** menu.
- dgeom(x,p) Gives $P(X = x)$ for geometric with probability of success p
- pgeom(x,p) Gives the cumulative probability $P(X \leq x)$ for the geometric

Choose **Binomial** from the Probability Distributions submenu.
To calculate the probability of getting *x* successes in *n* trials, choose **Probability**.

To calculate the probability of getting *x* or fewer successes among *n* trials, choose **Cumulative Probability**.
For Poisson, choose **Poisson** from the Probability Distribution submenu.
For Geometric, choose **Geometric** from the Probability Distribution submenu.

R

The standard library stats contains pdf's and cdf's of many common models:
- dgeom(x,p) # Gives $P(X = x)$ for the Geometric with probability *p*
- pgeom(x,p) # Gives the cdf $P(X \leq x)$ for the Geometric with probability *p*
- dbinom(x,n,p) # Gives $P(X = x)$ for the Binomial with *n* and *p*
- pbinom(x,n,p) # Gives the cdf $P(X \leq x)$ for the Binomial with *n* and *p*
- dpois(x,lambda) # Gives $P(X = x)$ for the Poisson with mean lambda
- ppois(x,lambda) # Gives the cdf $P(X \leq x)$ for the Poisson with mean lambda
- dnorm(x,mu,sig) # Gives the pdf for the Normal model
- pnorm(x,mu,sig) # Gives the cdf for the Normal model
- dexp(x,rate) # Gives the pdf for the Exponential with rate = λ
- pexp(x,rate) # Gives the cdf for the Exponential with rate = λ

SPSS

PDF.GEOM(x, prob)
CDF.GEOM(x, prob)
PDF.Poisson(x, mean)
CDF.Poisson(x, mean)

PDF.BINOM(x, n, prob)
CDF.BINOM(x, n, prob)
CDF.NORMAL(quant, mean, stddev)
PDF.NORMAL(quant, mean, stddev)

TI-83/84 PLUS

geometpdf(prob, x)
geometcdf(prob, x)
binompdf(n, prob, x)
binomcdf(n, prob, x)
poissonpdf(mean, x)
poissoncdf(mean, x)

COMMENTS
Find these commands in the **2nd DISTR** menu (the calculator refers to models as "distributions").

Exercises

Section 16.1

1. **Bernoulli** Do these situations involve Bernoulli trials? Explain.
   a) We roll 50 dice to find the distribution of the number of spots on the faces.
   b) How likely is it that in a group of 120 the majority may have Type A blood, given that Type A is found in 43% of the population?
   c) We deal 7 cards from a deck and get all hearts. How likely is that?
   d) We wish to predict the outcome of a vote on the school budget, and poll 500 of the 3000 likely voters to see how many favor the proposed budget.
   e) A company realizes that about 10% of its packages are not being sealed properly. In a case of 24, is it likely that more than 3 are unsealed?

2. **Bernoulli 2** Do these situations involve Bernoulli trials? Explain.
   a) You are rolling 5 dice and need to get at least two 6's to win the game.
   b) We record the distribution of eye colors found in a group of 500 people.
   c) A manufacturer recalls a doll because about 3% have buttons that are not properly attached. Customers return 37 of these dolls to the local toy store. Is the manufacturer likely to find any dangerous buttons?
   d) A city council of 11 Republicans and 8 Democrats picks a committee of 4 at random. What's the probability they choose all Democrats?
   e) A 2002 Rutgers University study found that 74% of high school students have cheated on a test at least once. Your local high-school principal conducts a survey in homerooms and gets responses that admit to cheating from 322 of the 481 students.

Section 16.2

3. **Toasters** A manufacturer ships toasters in cartons of 20. In each carton, they estimate a 5% chance that one of the toasters will need to be sent back for minor repairs. What is the probability that in a carton, there will be exactly 3 toasters that need repair?

4. **Soccer** A soccer team estimates that they will score on 8% of the corner kicks. In next week's game, the team hopes to kick 15 corner kicks. What are the chances that they will score on 2 of those opportunities?

Section 16.3

5. **Toasters again** In a batch of 10,000 toasters, what are the chances that fewer than 450 need to be returned?

6. **Soccer again** If this team has 200 corner kicks over the season, what are the chances that they score more than 22 times?

Section 16.4

7. **Sell!** A car dealership sells an average of 5 cars in a day. Using the Poisson model, what is the probability that the dealer sells 3 cars tomorrow?

8. **Passing on** A large hospital has an average of 7 fatalities in a week. Using the Poisson model, what is the probability that this week it has 10 fatalities?

Section 16.7

9. **Telephone numbers** A cable provider wants to contact customers in a particular telephone exchange to see how satisfied they are with the new digital TV service the company has provided. All numbers are in the 452 exchange, so there are 10,000 possible numbers from 452-0000 to 452-9999. If they select the numbers with equal probability:
   a) What distribution would they use to model the selection?
   b) The new business "incubator" was assigned the 200 numbers between 452-2500 and 452-2699, but these businesses don't subscribe to digital TV. What is the probability that the randomly selected number will be for an incubator business?
   c) Numbers above 9000 were only released for domestic use last year, so they went to newly constructed residences. What is the probability that a randomly selected number will be one of these?

10. **Serial numbers** In an effort to check the quality of their cell phones, a manufacturing manager decides to take a random sample of 10 cell phones from yesterday's production run, which produced cell phones with serial numbers ranging (according to when they were produced) from 43005000 to 43005999. If each of the 1000 phones is equally likely to be selected:
    a) What distribution would they use to model the selection?
    b) What is the probability that a randomly selected cell phone will be one of the last 100 to be produced?
    c) What is the probability that the first cell phone selected is either from the last 200 to be produced or from the first 50 to be produced?

11. **Component lifetimes** Lifetimes of electronic components can often be modeled by an exponential model. Suppose quality control engineers want to model the lifetime of a hard drive to have a mean lifetime of 3 years.
    a) What value of λ should they use?
    b) With this model, what would the probability be that a hard drive lasts 5 years or less?

12. **Website sales** Suppose occurrences of sales on a small company's website are well modeled by a Poisson model with $\lambda = 5$/hour.
    a) If a sale just occurred, what is the expected waiting time until the next sale?
    b) What is the probability that the next sale will happen in the next 6 minutes?

Chapter Exercises

13. **Simulating the model** Think about the Hope Solo picture search again. You are opening boxes of cereal one at a time looking for her picture, which is in 20% of the boxes. You want to know how many boxes you might have to open in order to find Hope.
    a) Describe how you would simulate the search for Hope using random numbers.
    b) Run at least 30 trials.
    c) Based on your simulation, estimate the probabilities that you might find your first picture of Hope in the first box, the second, etc.
    d) Calculate the actual probability model.
    e) Compare the distribution of outcomes in your simulation to the probability model.

14. **Simulation II** You are one space short of winning a child's board game and must roll a 1 on a die to claim victory. You want to know how many rolls it might take.
    a) Describe how you would simulate rolling the die until you get a 1.
    b) Run at least 30 trials.
    c) Based on your simulation, estimate the probabilities that you might win on the first roll, the second, the third, etc.
    d) Calculate the actual probability model.
    e) Compare the distribution of outcomes in your simulation to the probability model.

15. **Hope, again** Let's take one last look at the Hope Solo picture search. You know her picture is in 20% of the cereal boxes. You buy five boxes to see how many pictures of Hope you might get.
    a) Describe how you would simulate the number of pictures of Hope you might find in five boxes of cereal.
    b) Run at least 30 trials.
    c) Based on your simulation, estimate the probabilities that you get no pictures of Hope, 1 picture, 2 pictures, etc.
    d) Find the actual probability model.
    e) Compare the distribution of outcomes in your simulation to the probability model.

16. **Seatbelts** Suppose 75% of all drivers always wear their seatbelts. Let's investigate how many of the drivers might be belted among five cars waiting at a traffic light.
    a) Describe how you would simulate the number of seatbelt-wearing drivers among the five cars.
    b) Run at least 30 trials.
    c) Based on your simulation, estimate the probabilities there are no belted drivers, exactly one, two, etc.
    d) Find the actual probability model.
    e) Compare the distribution of outcomes in your simulation to the probability model.

17. **On time** A Department of Transportation report about air travel found that, nationwide, 76% of all flights are on time. Suppose you are at the airport and your flight is one of 50 scheduled to take off in the next two hours. Can you consider these departures to be Bernoulli trials? Explain.

18. **Lost luggage** A Department of Transportation report about air travel found that airlines misplace about 5 bags

per 1000 passengers. Suppose you are traveling with a group of people who have checked 22 pieces of luggage on your flight. Can you consider the fate of these bags to be Bernoulli trials? Explain.

19. **Hoops** A basketball player has made 80% of his foul shots during the season. Assuming the shots are independent, find the probability that in tonight's game he
    a) misses for the first time on his fifth attempt.
    b) makes his first basket on his fourth shot.
    c) makes his first basket on one of his first 3 shots.

20. **Chips** Suppose a computer chip manufacturer rejects 2% of the chips produced because they fail presale testing.
    a) What's the probability that the fifth chip you test is the first bad one you find?
    b) What's the probability you find a bad one within the first 10 you examine?

21. **More hoops** For the basketball player in Exercise 19, what's the expected number of shots until he misses?

22. **Chips ahoy** For the computer chips described in Exercise 20, how many do you expect to test before finding a bad one?

23. **Customer center operator** Raaj works at the customer service call center of a major credit card bank. Cardholders call for a variety of reasons, but regardless of their reason for calling, if they hold a platinum card, Raaj is instructed to offer them a double-miles promotion. About 10% of all cardholders hold platinum cards, and about 50% of those will take the double-miles promotion. On average, how many calls will Raaj have to take before finding the first cardholder to take the double-miles promotion?

24. **Cold calls** Justine works for an organization committed to raising money for Alzheimer's research. From past experience, the organization knows that about 20% of all potential donors will agree to give something if contacted by phone. They also know that of all people donating, about 5% will give $100 or more. On average, how many potential donors will she have to contact until she gets her first $100 donor?

25. **Blood** Only 4% of people have Type AB blood.
    a) On average, how many donors must be checked to find someone with Type AB blood?
    b) What's the probability that there is a Type AB donor among the first 5 people checked?
    c) What's the probability that the first Type AB donor will be found among the first 6 people?
    d) What's the probability that we won't find a Type AB donor before the 10th person?

26. **Color blindness** About 8% of males are color-blind. A researcher needs some color-blind subjects for an experiment and begins checking potential subjects.
    a) On average, how many men should the researcher expect to check to find one who is color-blind?
    b) What's the probability that she won't find anyone color-blind among the first 4 men she checks?
    c) What's the probability that the first color-blind man found will be the sixth person checked?
    d) What's the probability that she finds someone who is color-blind before checking the 10th man?

27. **Coins and intuition** If you flip a fair coin 100 times,
    a) Intuitively, how many heads do you expect?
    b) Use the formula for expected value to verify your intuition.

28. **Roulette and intuition** An American roulette wheel has 38 slots, of which 18 are red, 18 are black, and 2 are green (0 and 00). If you spin the wheel 38 times,
    a) Intuitively, how many times would you expect the ball to wind up in a green slot?
    b) Use the formula for expected value to verify your intuition.

29. **Lefties** Assume that 13% of people are left-handed. If we select 5 people at random, find the probability of each outcome.
    a) The first lefty is the fifth person chosen.
    b) There are some lefties among the 5 people.
    c) The first lefty is the second or third person.
    d) There are exactly 3 lefties in the group.
    e) There are at least 3 lefties in the group.
    f) There are no more than 3 lefties in the group.

30. **Arrows** An Olympic archer is able to hit the bull's-eye 80% of the time. Assume each shot is independent of the others. If she shoots 6 arrows, what's the probability of each of the following results?
    a) Her first bull's-eye comes on the third arrow.
    b) She misses the bull's-eye at least once.
    c) Her first bull's-eye comes on the fourth or fifth arrow.
    d) She gets exactly 4 bull's-eyes.
    e) She gets at least 4 bull's-eyes.
    f) She gets at most 4 bull's-eyes.

31. **Lefties, redux** Consider our group of 5 people from Exercise 29.
    a) How many lefties do you expect?
    b) With what standard deviation?
    c) If we keep picking people until we find a lefty, how long do you expect it will take?

32. **More arrows** Consider our archer from Exercise 30.
    a) How many bull's-eyes do you expect her to get?
    b) With what standard deviation?
    c) If she keeps shooting arrows until she hits the bull's-eye, how long do you expect it will take?

33. **Still more lefties** Suppose we choose 12 people instead of the 5 chosen in Exercise 29.
    a) Find the mean and standard deviation of the number of right-handers in the group.
    b) What's the probability that they're not all right-handed?

c) What's the probability that there are no more than 10 righties?
d) What's the probability that there are exactly 6 of each?
e) What's the probability that the majority is right-handed?

34. **Still more arrows** Suppose the archer from Exercise 30 shoots 10 arrows.
   a) Find the mean and standard deviation of the number of bull's-eyes she may get.
   b) What's the probability that she never misses?
   c) What's the probability that there are no more than 8 bull's-eyes?
   d) What's the probability that there are exactly 8 bull's-eyes?
   e) What's the probability that she hits the bull's-eye more often than she misses?

35. **Vision** It is generally believed that nearsightedness affects about 12% of all children. A school district tests the vision of 169 incoming kindergarten children. How many would you expect to be nearsighted? With what standard deviation?

36. **International students** At a certain college, 6% of all students come from outside the United States. Incoming students there are assigned at random to freshman dorms, where students live in residential clusters of 40 freshmen sharing a common lounge area. How many international students would you expect to find in a typical cluster? With what standard deviation?

37. **Tennis, anyone?** A certain tennis player makes a successful first serve 70% of the time. Assume that each serve is independent of the others. If she serves 6 times, what's the probability she gets
   a) all 6 serves in?
   b) exactly 4 serves in?
   c) at least 4 serves in?
   d) no more than 4 serves in?

38. **Frogs** A wildlife biologist examines frogs for a genetic trait he suspects may be linked to sensitivity to industrial toxins in the environment. Previous research had established that this trait is usually found in 1 of every 8 frogs. He collects and examines a dozen frogs. If the frequency of the trait has not changed, what's the probability he finds the trait in
   a) none of the 12 frogs?
   b) at least 2 frogs?
   c) 3 or 4 frogs?
   d) no more than 4 frogs?

39. **And more tennis** Suppose the tennis player in Exercise 37 serves 80 times in a match.
   a) What are the mean and standard deviation of the number of good first serves expected?
   b) Verify that you can use a Normal model to approximate the distribution of the number of good first serves.
   c) Use the 68–95–99.7 Rule to describe this distribution.
   d) What's the probability she makes at least 65 first serves?

40. **More arrows** The archer in Exercise 30 will be shooting 200 arrows in a large competition.
   a) What are the mean and standard deviation of the number of bull's-eyes she might get?
   b) Is a Normal model appropriate here? Explain.
   c) Use the 68–95–99.7 Rule to describe the distribution of the number of bull's-eyes she may get.
   d) Would you be surprised if she made only 140 bull's-eyes? Explain.

41. **Apples** An orchard owner knows that he'll have to use about 6% of the apples he harvests for cider because they will have bruises or blemishes. He expects a tree to produce about 300 apples.
   a) Describe an appropriate model for the number of cider apples that may come from that tree. Justify your model.
   b) Find the probability there will be no more than a dozen cider apples.
   c) Is it likely there will be more than 50 cider apples? Explain.

42. **Frogs, part II** Based on concerns raised by his preliminary research, the biologist in Exercise 38 decides to collect and examine 150 frogs.
   a) Assuming the frequency of the trait is still 1 in 8, determine the mean and standard deviation of the number of frogs with the trait he should expect to find in his sample.
   b) Verify that he can use a Normal model to approximate the distribution of the number of frogs with the trait.
   c) He found the trait in 22 of his frogs. Do you think this proves that the trait has become more common? Explain.

43. **Lefties, again** A lecture hall has 200 seats with folding arm tablets, 30 of which are designed for left-handers. The average size of classes that meet there is 188, and we can assume that about 13% of students are left-handed. What's the probability that a right-handed student in one of these classes is forced to use a lefty arm tablet?

44. **No-shows** An airline, believing that 5% of passengers fail to show up for flights, overbooks (sells more tickets than there are seats). Suppose a plane will hold 265 passengers, and the airline sells 275 tickets. What's the probability the airline will not have enough seats, so someone gets bumped?

45. **Annoying phone calls** A newly hired telemarketer is told he will probably make a sale on about 12% of his phone calls. The first week he called 200 people, but only made 10 sales. Should he suspect he was misled about the true success rate? Explain.

46. **The euro** Shortly after the introduction of the euro coin in Belgium, newspapers around the world published articles

claiming the coin is biased. The stories were based on reports that someone had spun the coin 250 times and gotten 140 heads—that's 56% heads. Do you think this is evidence that spinning a euro is unfair? Explain.

47. **Hurricanes, redux** We first looked at the occurrences of hurricanes in Chapter 3 (Exercise 47). Suppose we find and the arrivals can be modeled by a Poisson distribution with mean 2.45.
   a) What's the probability of no hurricanes next year?
   b) What's the probability that during the next two years, there's exactly 1 hurricane?

48. **Bank tellers** I am the only bank teller on duty at my local bank. I need to run out for 10 minutes, but I don't want to miss any customers. Suppose the arrival of customers can be modeled by a Poisson distribution with mean 2 customers per hour.
   a) What's the probability that no one will arrive in the next 10 minutes?
   b) What's the probability that 2 or more people arrive in the next 10 minutes?
   c) You've just served 2 customers who came in one after the other. Is this a better time to run out?

49. **TB, again** In Chapter 14 we saw that the probability of contracting TB is small, with p about 0.0005 for a new case in a given year. In a town of 8000 people:
   a) What's the expected number of new cases?
   b) Use the Poisson model to approximate the probability that there will be at least one new case of TB next year.

50. **Earthquakes** Suppose the probability of a major earthquake on a given day is 1 out of 10,000.
   a) What's the expected number of major earthquakes in the next 1000 days?
   b) Use the Poisson model to approximate the probability that there will be at least one major earthquake in the next 1000 days.

51. **Seatbelts II** Police estimate that 80% of drivers now wear their seatbelts. They set up a safety roadblock, stopping cars to check for seatbelt use.
   a) How many cars do they expect to stop before finding a driver whose seatbelt is not buckled?
   b) What's the probability that the first unbelted driver is in the sixth car stopped?
   c) What's the probability that the first 10 drivers are all wearing their seatbelts?
   d) If they stop 30 cars during the first hour, find the mean and standard deviation of the number of drivers expected to be wearing seatbelts.
   e) If they stop 120 cars during this safety check, what's the probability they find at least 20 drivers not wearing their seatbelts?

52. **Rickets** Vitamin D is essential for strong, healthy bones. Our bodies produce vitamin D naturally when sunlight falls upon the skin, or it can be taken as a dietary supplement. Although the bone disease rickets was largely eliminated in England during the 1950s, some people there are concerned that this generation of children is at increased risk because they are more likely to watch TV or play computer games than spend time outdoors. Recent research indicated that about 20% of British children are deficient in vitamin D. Suppose doctors test a group of elementary school children.
   a) What's the probability that the first vitamin D–deficient child is the eighth one tested?
   b) What's the probability that the first 10 children tested are all okay?
   c) How many kids do they expect to test before finding one who has this vitamin deficiency?
   d) They will test 50 students at the third-grade level. Find the mean and standard deviation of the number who may be deficient in vitamin D.
   e) If they test 320 children at this school, what's the probability that no more than 50 of them have the vitamin deficiency?

53. **ESP** Scientists wish to test the mind-reading ability of a person who claims to have ESP. They use five cards with different and distinctive symbols (square, circle, triangle, line, squiggle). Someone picks a card at random and thinks about the symbol. The "mind reader" must correctly identify which symbol was on the card. If the test consists of 100 trials, how many would this person need to get right in order to convince you that ESP may actually exist? Explain.

54. **True–false** A true–false test consists of 50 questions. How many does a student have to get right to convince you that he is not merely guessing? Explain.

55. **Hot hand** A basketball player who ordinarily makes about 55% of his free throw shots has made 4 in a row. Is this evidence that he has a "hot hand" tonight? That is, is this streak so unusual that it means the probability he makes a shot must have changed? Explain.

56. **New bow** The archer in Exercise 30 purchases a new bow, hoping that it will improve her success rate to more than 80% bull's-eyes. She is delighted when she first tests her new bow and hits 6 consecutive bull's-eyes. Do you think this is compelling evidence that the new bow is better? In other words, is a streak like this unusual for her? Explain.

57. **Hotter hand** The basketball player in Exercise 55 has new sneakers, which he thinks improve his game. Over his past 40 shots, he's made 32—much better than the 55% he usually shoots. Do you think his chances of making a shot really increased? In other words, is making at least 32 of 40 shots really unusual for him? (Do you think it's his sneakers?)

58. New bow, again The archer in Exercise 56 continues shooting arrows, ending up with 45 bull's-eyes in 50 shots. Now are you convinced that the new bow is better? Explain.

59. Web visitors A website manager has noticed that during the evening hours, about 3 people per minute check out from their shopping cart and make an online purchase. She believes that each purchase is independent of the others and wants to model the number of purchases per minute.
  a) What model might you suggest to model the number of purchases per minute?
  b) What is the probability that in any one minute at least one purchase is made?
  c) What is the probability that no one makes a purchase in the next two minutes?

60. Quality control In an effort to improve the quality of their cell phones, a manufacturing manager records the number of faulty phones in each day's production run. The manager notices that the number of faulty cell phones in a production run of cell phones is usually small and that the quality of one day's run seems to have no bearing on the next day.
  a) What model might you use to model the number of faulty cell phones produced in one day?
  b) If the mean number of faulty cell phones is 2 per day, what is the probability that no faulty cell phones will be produced tomorrow?
  c) If the mean number of faulty cell phones is 2 per day, what is the probability that 3 or more faulty cell phones were produced in today's run?

*61. Web visitors, part 2 The website manager in Exercise 59 wants to model the time between purchases. Recall that the mean number of purchases in the evening is 3 per minute.
  a) What model would you use to model the time between events?
  b) What is the mean time between purchases?
  c) What is the probability that the time to the next purchase will be between 1 and 2 minutes?

*62. Quality control, part 2 The cell phone manufacturer in Exercise 60 wants to model the time between events. The mean number of defective cell phones is 2 per day.
  a) What model would you use to model the time between events?
  b) What would the probability be that the time to the next failure is 1 day or less?
  c) What is the mean time between failures?

Just Checking ANSWERS

1. There are two outcomes (contact, no contact), the probability of contact is 0.76, and random calls should be independent.

2. Binomial, with $n = 1000$ and $p = 0.76$. For actual calculations, we could approximate using a Normal model with $\mu = np = 1000(0.76) = 760$ and $\sigma = \sqrt{npq} = \sqrt{1000(0.76)(0.24)} \approx 13.5$.

3. Geometric, with $p = 0.29$.

4. Poisson, with $\lambda = np = 10000\left(\dfrac{70000}{107000000}\right) = 6.54$.

Review of Part IV

Randomness and Probability

Quick Review

Here's a brief summary of the key concepts and skills in probability and probability modeling:

- The Law of Large Numbers says that the more times we try something, the closer the results will come to theoretical perfection.
    - Don't mistakenly misinterpret the Law of Large Numbers as the "Law of Averages." There's no such thing.
- Basic rules of probability can handle most situations:
    - To find the probability that an event OR another event happens, add their probabilities and subtract the probability that both happen.
    - To find the probability that an event AND another independent event both happen, multiply probabilities.
    - Conditional probabilities tell you how likely one event is to happen, knowing that another event has happened.
    - Mutually exclusive events (also called "disjoint") cannot both happen at the same time.
    - Two events are independent if the occurrence of one doesn't change the probability that the other happens.
- A probability model for a random variable describes the theoretical distribution of outcomes.
    - The mean of a random variable is its expected value.
    - For sums or differences of independent random variables, variances add.
    - To estimate probabilities involving quantitative variables, you may be able to use a Normal model—but only if the distribution of the variable is unimodal and symmetric.
    - To estimate the probability you'll get your first success on a certain trial, use a Geometric model.
    - To estimate the probability you'll get a certain number of successes in a specified number of independent trials, use a Binomial model.
    - To estimate the probability of the number of occurrences of a relatively rare phenomenon, consider using a Poisson model.
    - To estimate the interval between two occurrences of a random phenomenon, use the Exponential model.

Ready? Here are some opportunities to check your understanding of these ideas.

Review Exercises

1. **Quality control** A consumer organization estimates that 29% of new cars have a cosmetic defect, such as a scratch or a dent, when they are delivered to car dealers. This same organization believes that 7% have a functional defect—something that does not work properly—and that 2% of new cars have both kinds of problems.
   a) If you buy a new car, what's the probability that it has some kind of defect?
   b) What's the probability it has a cosmetic defect but no functional defect?
   c) If you notice a dent on a new car, what's the probability it has a functional defect?
   d) Are the two kinds of defects disjoint events? Explain.
   e) Do you think the two kinds of defects are independent events? Explain.

2. **Workers** A company's human resources officer reports a breakdown of employees by job type and sex shown in the table.

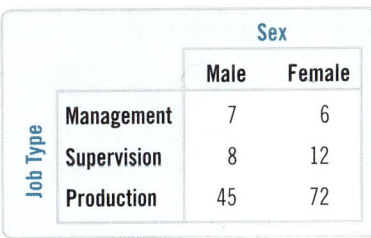

		Sex	
		Male	Female
Job Type	Management	7	6
	Supervision	8	12
	Production	45	72

   a) What's the probability that a worker selected at random is
      i) female?
      ii) female or a production worker?
      iii) female, if the person works in production?
      iv) a production worker, if the person is female?
   b) Do these data suggest that job type is independent of being male or female? Explain.

437

3. **Airfares** Each year a company must send 3 officials to a meeting in China and 5 officials to a meeting in France. Airline ticket prices vary from time to time, but the company purchases all tickets for a country at the same price. Past experience has shown that tickets to China have a mean price of $1000, with a standard deviation of $150, while the mean airfare to France is $500, with a standard deviation of $100.
   a) Define random variables and use them to express the total amount the company will have to spend to send these delegations to the two meetings.
   b) Find the mean and standard deviation of this total cost.
   c) Find the mean and standard deviation of the difference in price of a ticket to China and a ticket to France.
   d) Do you need to make any assumptions in calculating these means? How about the standard deviations?

4. **Bipolar** Psychiatrists estimate that about 1 in 100 adults suffers from bipolar disorder. What's the probability that in a city of 10,000 there are more than 200 people with this condition? Be sure to verify that a Normal model can be used here.

5. **A game** To play a game, you must pay $5 for each play. There is a 10% chance you will win $5, a 40% chance you will win $7, and a 50% chance you will win only $3.
   a) What are the mean and standard deviation of your net winnings?
   b) You play twice. Assuming the plays are independent events, what are the mean and standard deviation of your total winnings?

6. **Emergency switch** Safety engineers must determine whether industrial workers can operate a machine's emergency shutoff device. Among a group of test subjects, 66% were successful with their left hands, 82% with their right hands, and 51% with both hands.
   a) What percent of these workers could not operate the switch with either hand?
   b) Are success with right and left hands independent events? Explain.
   c) Are success with right and left hands mutually exclusive? Explain.

7. **Twins** In the United States, the probability of having twins (usually about 1 in 90 births) rises to about 1 in 10 for women who have been taking the fertility drug Clomid. Among a group of 10 pregnant women, what's the probability that
   a) at least one will have twins if none were taking a fertility drug?
   b) at least one will have twins if all were taking Clomid?
   c) at least one will have twins if half were taking Clomid?

8. **Deductible** A car owner may buy insurance that will pay the full price of repairing the car after an at-fault accident, or save $12 a year by getting a policy with a $500 deductible. Her insurance company says that about 0.5% of drivers in her area have an at-fault auto accident during any given year. Based on this information, should she buy the policy with the deductible or not? How does the value of her car influence this decision?

9. **More twins** A group of 5 women became pregnant while undergoing fertility treatments with the drug Clomid, discussed in Exercise 7. What's the probability that
   a) none will have twins?
   b) exactly 1 will have twins?
   c) at least 3 will have twins?

10. **At fault** The car insurance company in Exercise 8 believes that about 0.5% of drivers have an at-fault accident during a given year. Suppose the company insures 1355 drivers in that city.
    a) What are the mean and standard deviation of the number who may have at-fault accidents?
    b) Can you describe the distribution of these accidents with a Normal model? Explain.

11. **Twins, part III** At a large fertility clinic, 152 women became pregnant while taking Clomid. (See Exercise 7.)
    a) What are the mean and standard deviation of the number of twin births we might expect?
    b) Can we use a Normal model in this situation? Explain.
    c) What's the probability that no more than 10 of the women have twins?

12. **Child's play** In a board game you determine the number of spaces you may move by spinning a spinner and rolling a die. The spinner has three regions: Half of the spinner is marked "5," and the other half is equally divided between "10" and "20." The six faces of the die show 0, 0, 1, 2, 3, and 4 spots. When it's your turn, you spin and roll, adding the numbers together to determine how far you may move.
    a) Create a probability model for the outcome on the spinner.
    b) Find the mean and standard deviation of the spinner results.
    c) Create a probability model for the outcome on the die.
    d) Find the mean and standard deviation of the die results.
    e) Find the mean and standard deviation of the number of spaces you get to move.

13. **Language** Neurological research has shown that in about 80% of people, language abilities reside in the brain's left side. Another 10% display right-brain language centers, and the remaining 10% have two-sided language control. (The latter two groups are mainly left-handers; *Science News*, 161 no. 24 [2002].)

a) Assume that a freshman composition class contains 25 randomly selected people. What's the probability that no more than 15 of them have left-brain language control?
b) In a randomly assigned group of 5 of these students, what's the probability that no one has two-sided language control?
c) In the entire freshman class of 1200 students, how many would you expect to find of each type?
d) What are the mean and standard deviation of the number of these freshmen who might be right-brained in language abilities?
e) If an assumption of Normality is justified, use the 68–95–99.7 Rule to describe how many students in the freshman class might have right-brain language control.

14. **Play, again** If you land in a "penalty zone" on the game board described in Exercise 12, your move will be determined by subtracting the roll of the die from the result on the spinner. Now what are the mean and standard deviation of the number of spots you may move?

15. **Beanstalks** In some cities tall people who want to meet and socialize with other tall people can join Beanstalk Clubs. To qualify, a man must be over 6′2″ tall, and a woman over 5′10″. According to the National Health Survey, heights of adults may have a Normal model with mean heights of 69.1″ for men and 64.0″ for women. The respective standard deviations are 2.8″ and 2.5″.
    a) You're probably not surprised to learn that men are generally taller than women, but what does the greater standard deviation for men's heights indicate?
    b) Are men or women more likely to qualify for Beanstalk membership?
    c) Beanstalk members believe that height is an important factor when people select their spouses. To investigate, we select at random a married man and, independently, a married woman. Define two random variables, and use them to express how many inches taller the man is than the woman.
    d) What's the mean of this difference?
    e) What's the standard deviation of this difference?
    f) What's the probability that the man is taller than the woman (that the difference in heights is greater than 0)?
    g) Suppose a survey of married couples reveals that 92% of the husbands were taller than their wives. Based on your answer to part f, do you believe that people's choice of spouses is independent of height? Explain.

16. **Stocks** Since the stock market began in 1872, stock prices have risen in about 73% of the years. Assuming that market performance is independent from year to year, what's the probability that
    a) the market will rise for 3 consecutive years?
    b) the market will rise 3 years out of the next 5?
    c) the market will fall during at least 1 of the next 5 years?
    d) the market will rise during a majority of years over the next decade?

17. **Multiple choice** A multiple choice test has 50 questions, with 4 answer choices each. You must get at least 30 correct to pass the test, and the questions are very difficult.
    a) Are you likely to be able to pass by guessing on every question? Explain.
    b) Suppose, after studying for a while, you believe you have raised your chances of getting each question right to 70%. How likely are you to pass now?
    c) Assuming you are operating at the 70% level and the instructor arranges questions randomly, what's the probability that the third question is the first one you get right?

18. **Stock strategy** Many investment advisors argue that after stocks have declined in value for 2 consecutive years, people should invest heavily because the market rarely declines 3 years in a row.
    a) Since the stock market began in 1872, there have been two consecutive losing years eight times. In six of those cases, the market rose during the following year. Does this confirm the advice?
    b) Overall, stocks have risen in value during 95 of the 130 years since the market began in 1872. How is this fact relevant in assessing the statistical reasoning of the advisors?

19. **Insurance** A 65-year-old woman takes out a $100,000 term life insurance policy. The company charges an annual premium of $520. Estimate the company's expected profit on such policies if mortality tables indicate that only 2.6% of women age 65 die within a year.

20. **Teen smoking** The Centers for Disease Control and Prevention say that about 18% of high-school students smoke tobacco (down from a high of 38% in 1997). Suppose you randomly select high-school students to survey them on their attitudes toward scenes of smoking in the movies. What's the probability that
    a) none of the first 4 students you interview is a smoker?
    b) the first smoker is the sixth person you choose?
    c) there are no more than 2 smokers among 10 people you choose?

21. **Passing stats** Molly's college offers two sections of Statistics 101. From what she has heard about the two professors listed, Molly estimates that her chances of passing the course are 0.80 if she gets Professor Scedastic and 0.60 if she gets Professor Kurtosis. The registrar uses a lottery to randomly assign the 120 enrolled students based on the number of available seats in each class. There are 70 seats in Professor Scedastic's class and 50 in Professor Kurtosis's class.
    a) What's the probability that Molly will pass Statistics?
    b) At the end of the semester, we find out that Molly failed. What's the probability that she got Professor Kurtosis?

22. Teen smoking II Suppose that, as reported by the Centers for Disease Control, about 18% of high school students smoke tobacco. You randomly select 120 high school students to survey them on their attitudes toward scenes of smoking in the movies.
 a) What's the expected number of smokers?
 b) What's the standard deviation of the number of smokers?
 c) The number of smokers among 120 randomly selected students will vary from group to group. Explain why that number can be described with a Normal model.
 d) Using the 68–95–99.7 Rule, create and interpret a model for the number of smokers among your group of 120 students.

23. Random variables Given independent random variables with means and standard deviations as shown, find the mean and standard deviation of each of these variables:
 a) $X + 50$
 b) $10Y$
 c) $X + 0.5Y$
 d) $X - Y$
 e) $X + Y$

	Mean	SD
X	50	8
Y	100	6

24. Merger Explain why the facts you know about variances of independent random variables might encourage two small insurance companies to merge. (*Hint:* Think about the expected amount and potential variability in payouts for the separate and the merged companies.)

25. Youth survey According to a recent Gallup survey, 93% of teens use the Internet, but there are differences in how teen boys and girls say they use computers. The telephone poll found that 77% of boys had played computer games in the past week, compared with 65% of girls. On the other hand, 76% of girls said they had e-mailed friends in the past week, compared with only 65% of boys.
 a) For boys, the cited percentages are 77% playing computer games and 65% using e-mail. That total is 142%, so there is obviously a mistake in the report. No? Explain.
 b) Based on these results, do you think playing games and using e-mail are mutually exclusive? Explain.
 c) Do you think whether a child e-mails friends is independent of being a boy or a girl? Explain.
 d) Suppose that in fact 93% of the teens in your area do use the Internet. You want to interview a few who do not, so you start contacting teenagers at random. What is the probability that it takes you 5 interviews until you find the first person who does not use the Internet?

26. Meals A college student on a seven-day meal plan reports that the amount of money he spends daily on food varies with a mean of $13.50 and a standard deviation of $7.
 a) What are the mean and standard deviation of the amount he might spend in two consecutive days?
 b) What assumption did you make in order to find that standard deviation? Are there any reasons you might question that assumption?
 c) Estimate his average weekly food costs, and the standard deviation.
 d) Do you think it likely he might spend less than $50 in a week? Explain, including any assumptions you make in your analysis.

27. Travel to Kyrgyzstan Your pocket copy of *Kyrgyzstan on* 4237 ± 360 *Som a Day* claims that you can expect to spend about 4237 som each day with a standard deviation of 360 som. How well can you estimate your expenses for the trip?
 a) Your budget allows you to spend 90,000 som. To the nearest day, how long can you afford to stay in Kyrgyzstan, on average?
 b) What's the standard deviation of your expenses for a trip of that duration?
 c) You doubt that your total expenses will exceed your expectations by more than two standard deviations. How much extra money should you bring? On average, how much of a "cushion" will you have per day?

28. Picking melons Two stores sell watermelons. At the first store the melons weigh an average of 22 pounds, with a standard deviation of 2.5 pounds. At the second store the melons are smaller, with a mean of 18 pounds and a standard deviation of 2 pounds. You select a melon at random at each store.
 a) What's the mean difference in weights of the melons?
 b) What's the standard deviation of the difference in weights?
 c) If a Normal model can be used to describe the difference in weights, what's the probability that the melon you got at the first store is heavier?

29. Home, sweet home According to the 2000 Census, 66% of U.S. households own the home they live in. A mayoral candidate conducts a survey of 820 randomly selected homes in your city and finds only 523 owned by the current residents. The candidate then attacks the incumbent mayor, saying that there is an unusually low level of homeownership in the city. Do you agree? Explain.

30. Buying melons The first store in Exercise 28 sells watermelons for 32 cents a pound. The second store is having a sale on watermelons—only 25 cents a pound. Find the mean and standard deviation of the difference in the price you may pay for melons randomly selected at each store.

31. Who's the boss? The 2013 State of Women-Owned Businesses Report commissioned by American Express (http://www.womenable.com/userfiles/downloads/2013 _State_of_Women-Owned_Businesses_Report_FINAL .pdf) says that, excluding large, publicly traded firms, women-owned firms make up 30% of the privately held firm population, a number that has increased by 59% since 1997. You call some firms doing business locally, assuming that the national percentage is true in your area.
 a) What's the probability that the first 3 you call are all owned by women?

b) What's the probability that none of your first 4 calls finds a firm that is owned by a woman?

c) Suppose none of your first 5 calls found a firm owned by a woman. What's the probability that your next call does?

32. Jerseys A Statistics professor comes home to find that all four of his children got white team shirts from soccer camp this year. He concludes that this year, unlike other years, the camp must not be using a variety of colors. But then he finds out that in each child's age group there are 4 teams, only 1 of which wears white shirts. Each child just happened to get on the white team at random.

a) Why was he so surprised? If each age group uses the same 4 colors, what's the probability that all four kids would get the same-color shirt?

b) What's the probability that all 4 would get white shirts?

c) We lied. Actually, in the oldest child's group there are 6 teams instead of the 4 teams in each of the other three groups. How does this change the probability you calculated in part b?

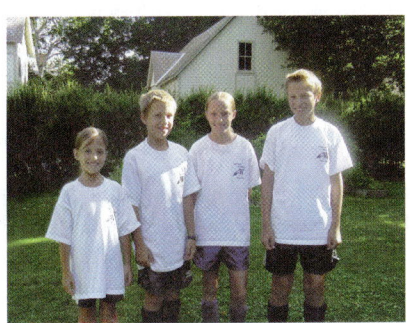

33. When to stop In Exercise 27 of the Review Exercises for Part III, we posed this question:

You play a game that involves rolling a die. You can roll as many times as you want, and your score is the total for all the rolls. But ... if you roll a 6, your score is 0 and your turn is over. What might be a good strategy for a game like this?

You attempted to devise a good strategy by simulating several plays to see what might happen. Let's try calculating a strategy.

a) On what roll would you expect to get a 6 for the first time?

b) So, roll *one time less* than that. Assuming all those rolls were not 6's, what's your expected score?

c) What's the probability that you can roll that many times without getting a 6?

34. Plan B Here's another attempt at developing a good strategy for the dice game in Exercise 33. Instead of stopping after a certain number of rolls, you could decide to stop when your score reaches a certain number of points.

a) How many points would you expect a roll to *add* to your score?

b) In terms of your current score, how many points would you expect a roll to *subtract* from your score?

c) Based on your answers in parts a and b, at what score will another roll "break even"?

d) Describe the strategy this result suggests.

35. Technology on campus Every 5 years the Conference Board of the Mathematical Sciences surveys college math departments. In 2000 the board reported that 51% of all undergraduates taking Calculus I were in classes that used graphing calculators and 31% were in classes that used computer assignments. Suppose that 16% used both calculators and computers.

a) What percent used neither kind of technology?

b) What percent used calculators but not computers?

c) What percent of the calculator users had computer assignments?

d) Based on this survey, do calculator and computer use appear to be independent events? Explain.

36. Dogs A census by the county dog control officer found that 18% of homes kept one dog as a pet, 4% had two dogs, and 1% had three or more. If a salesman visits two homes selected at random, what's the probability he encounters

a) no dogs?

b) some dogs?

c) dogs in each home?

d) more than one dog in each home?

37. O-rings Failures of O-rings on the space shuttle are fairly rare, but often disastrous, events. If we are testing O-rings, suppose that the probability of a failure of any one O-ring is 0.01. Let X be the number of failures in the next 10 O-rings tested.

a) What model might you use to model X?

b) What is the mean number of failures in the next 10 O-rings?

c) What is the probability that there is exactly one failure in the next 10 O-rings?

d) What is the probability that there is at least one failure in the next 10 O-rings?

38. Volcanoes Almost every year, there is some incidence of volcanic activity on the island of Japan. In 2005 there were 5 volcanic episodes, defined as either eruptions or sizable seismic activity. Suppose the mean number of episodes is 2.4 per year. Let X be the number of episodes in the 2-year period 2010–2011.

a) What model might you use to model X?

b) What is the mean number of episodes in this period?

c) What is the probability that there will be no episodes in this period?

d) What is the probability that there are more than three episodes in this period?

39. Socks In your sock drawer you have 4 blue socks, 5 gray socks, and 3 black ones. Half asleep one morning, you grab 2 socks at random and put them on. Find the probability you end up wearing

a) 2 blue socks.

b) no gray socks.

c) at least 1 black sock.
d) a green sock.
e) matching socks.

40. **Coins** A coin is to be tossed 36 times.
    a) What are the mean and standard deviation of the number of heads?
    b) Suppose the resulting number of heads is unusual, two standard deviations above the mean. How many "extra" heads were observed?
    c) If the coin were tossed 100 times, would you still consider the same number of extra heads unusual? Explain.
    d) In the 100 tosses, how many extra heads would you need to observe in order to say the results were unusual?
    e) Explain how these results refute the "Law of Averages" but confirm the Law of Large Numbers.

41. **The Drake equation** In 1961 astronomer Frank Drake developed an equation to try to estimate the number of extraterrestrial civilizations in our galaxy that might be able to communicate with us via radio transmissions. Now largely accepted by the scientific community, the Drake equation has helped spur efforts by radio astronomers to search for extraterrestrial intelligence.

    Here is the equation:

    $$N_C = N \cdot f_p \cdot n_e \cdot f_l \cdot f_i \cdot f_c \cdot f_L$$

    OK, it looks a little messy, but here's what it means:

Factor	What It Represents	Possible Value
N	Number of stars in the Milky Way Galaxy	200–400 billion
f_p	Probability that a star has planets	20%–50%
n_e	Number of planets in a solar system capable of sustaining earth-type life	1? 2?
f_l	Probability that life develops on a planet with a suitable environment	1%–100%
f_i	Probability that life evolves intelligence	50%?
f_c	Probability that intelligent life develops radio communication	10%–20%
f_L	Fraction of the planet's life for which the civilization survives	$\frac{1}{1,000,000}$?
N_C	Number of extraterrestrial civilizations in our galaxy with which we could communicate	?

    So, how many ETs are out there? That depends; values chosen for the many factors in the equation depend on ever-evolving scientific knowledge and one's personal guesses. But now, some questions.
    a) What quantity is calculated by the first product, $N \cdot f_p$?
    b) What quantity is calculated by the product, $N \cdot f_p \cdot n_e \cdot f_l$?
    c) What probability is calculated by the product $f_l \cdot f_i$?
    d) Which of the factors in the formula are conditional probabilities? Restate each in a way that makes the condition clear.

    Note: A quick Internet search will find you a site where you can play with the Drake equation yourself.

42. **Recalls** In a car rental company's fleet, 70% of the cars are American brands, 20% are Japanese, and the rest are German. The company notes that manufacturers' recalls seem to affect 2% of the American cars, but only 1% of the others.
    a) What's the probability that a randomly chosen car is recalled?
    b) What's the probability that a recalled car is American?

43. **Pregnant?** Suppose that 70% of the women who suspect they may be pregnant and purchase an in-home pregnancy test are actually pregnant. Further suppose that the test is 98% accurate. What's the probability that a woman whose test indicates that she is pregnant actually is?

44. **Door prize** You are among 100 people attending a charity fundraiser at which a large-screen TV will be given away as a door prize. To determine who wins, 99 white balls and 1 red ball have been placed in a box and thoroughly mixed. The guests will line up and, one at a time, pick a ball from the box. Whoever gets the red ball wins the TV, but if the ball is white, it is returned to the box. If none of the 100 guests gets the red ball, the TV will be auctioned off for additional benefit of the charity.
    a) What's the probability that the first person in line wins the TV?
    b) You are the third person in line. What's the probability that you win the TV?
    c) What's the probability that the charity gets to sell the TV because no one wins?
    d) Suppose you get to pick your spot in line. Where would you want to be in order to maximize your chances of winning?
    e) After hearing some protest about the plan, the organizers decide to award the prize by not returning the white balls to the box, thus ensuring that 1 of the 100 people will draw the red ball and win the TV. Now what position in line would you choose in order to maximize your chances?

Chapter 17

Sampling Distribution Models

17.1 Sampling Distribution of a Proportion

17.2 When Does the Normal Model Work? Assumptions and Conditions

17.3 The Sampling Distribution of Other Statistics

17.4 The Central Limit Theorem: The Fundamental Theorem of Statistics

17.5 Sampling Distributions: A Summary

Where are we going?

A poll based on a random sample of U.S. adults reports that 29% of those asked admit that they don't always drive as safely as they should because they are on their cell phone. How much can we trust this statistic? After all, they didn't ask *everyone*. Maybe the true proportion is 33% or 25%? How reliable *are* statistics based on random samples? We'll see that we can be surprisingly precise about how much we expect statistics from random samples to vary. This will enable us to start generalizing from samples we have at hand to the population at large.

Who	U.S. adults
What	Belief in global warming
When	March 2013
Where	United States
Why	Public attitudes

In March 2013 the Gallup Poll surveyed a random sample of 1022 adults. They report that 57% believe that "increases in the Earth's temperature over the last century are due more to human activity than to natural causes." At almost the same time in April 2013 the Yale School of Forestry found that 49% of respondents believe that global warming is caused mostly by human activities. Should we be surprised to find that we can get different values from properly selected random samples drawn from the same population?

17.1 Sampling Distribution of a Proportion

If we (or Gallup) had surveyed every possible sample of 1022 U.S. adults, we could find the proportion $\hat{p}$ of each sample who believe that global climate change is caused by human activity. Then we could see how much those sample proportions varied. Of course, we can't do that, but we can imagine drawing those samples. In fact, we can do better than that. We can simulate what the distribution will look like. For a simulation, we first have to pick a particular value to be the "true" population proportion. Recall that we denote the true population proportion with the letter p. We can pretend for now that Gallup got it exactly right and the true proportion is 57%. Then we could draw 1022 values that have a 43% chance of saying "no, climate change is not primarily caused by human activity" and a 57% chance of saying "Yes, it is."

443

444 PART V From the Data at Hand to the World at Large

The first three times we drew the 1022 values, we got 580, 612, and 604 simulated respondents (for $\hat{p} = 0.568, 0.599,$ and 0.591 respectively) thinking climate change is human-caused. (The proper term is *anthropogenic*.) When we ran the simulation 2000 times, a histogram of all the sample proportions looked like this:

Figure 17.1
The distribution of sample proportions for 2000 simulated samples of 1022 adults drawn from a population with proportion $p = 0.57$ of respondents who think climate change is anthropogenic.

It should be no surprise that we don't get the same proportion for each sample we draw, even though the underlying true value is the same for the population. Each $\hat{p}$ comes from a different simulated sample. The histogram above is a simulation of what we'd get if we could see *all the proportions from all possible samples*. That distribution has a special name. It is called the **sampling distribution** of the proportions.[1]

The histogram is unimodal, symmetric, and centered at p. That's probably not surprising. But, it's an amazing and fortunate fact that a Normal model is just the right one for the histogram of sample proportions. And, in fact, this was proved mathematically over 300 years ago.[2]

There is no reason you should guess that the Normal model would be the one we need here,[3] and, indeed, the importance of De Moivre's result was not immediately understood by his contemporaries. But (unlike De Moivre's contemporaries in 1718) we know how useful the Normal model can be.

Modeling how sample statistics, such as proportions or means, vary from sample to sample is one of the most powerful ideas we'll see in this course. A **sampling distribution model** for how a statistic from a sample varies from sample to sample allows us to quantify that variation and to make statements about where we think the corresponding population parameter is.

A S *Activity:* **Sampling Distribution of a Proportion.** You don't have to imagine—you can simulate.

NOTATION ALERT

The letter p is our choice for the *parameter* of the model for proportions. It violates our "Greek letters for parameters" rule, but if we stuck to that, our natural choice would be π. We could use π to be perfectly consistent, but then we'd have to write statements like $\pi = 0.46$. That just seems a bit weird to us. After all, we've known that $\pi = 3.1415926\ldots$ since the Greeks, and it's a hard habit to break.

So, we'll use p for the model parameter (the probability of a success) and $\hat{p}$ for the observed proportion in a sample. We'll also use q for the probability of a failure ($q = 1 - p$) and $\hat{q}$ for its observed value.

But be careful. We've already used capital P for a general probability. And we'll soon see another use of P in the next chapter! There are a lot of p's in this course; you'll need to think clearly about the context to keep them straight.

For Example USING THE SAMPLING DISTRIBUTION MODEL FOR A PROPORTION

According to the Centers for Disease Control and Prevention, about 18% of U.S. adults still smoke.

QUESTION: How much would we expect the proportion of smokers in a sample of size 1000 to vary from sample to sample?

ANSWER: We'll simulate by taking random integers between 1 and 100. We'll assign values between 1 and 18 to represent a smoker, and the rest to represent a nonsmoker. Then we'll repeat the process, drawing samples of size 1000 over and

---

[1] A word of caution. Until now, we've been plotting the *distribution of the sample*, a display of the actual data that were collected in that one sample. But now we've plotted the *sampling distribution*; a display of summary statistics ($\hat{p}$'s, for example) for many different samples. "Sample distribution" and "sampling distribution" sound a lot alike, but they refer to very different things. (Sorry about that—we didn't make up the terms.) And the distinction is critical. Whenever you read or write something about one of these, choose your terms carefully.

[2] By the mathematician De Moivre in 1718, long before the first computer was built.

[3] Well, the fact that we spent much of Chapter 5 on the Normal model might have been a hint.

over and recording the sample proportion of smokers for each sample. If we did this 10,000 times, we might get a histogram of sample proportions that looks like this:

The mean of this distribution is located at the population proportion, 0.18. The standard deviation of these 10,000 sample proportions is 0.0122 or 1.22%. Because the shape is symmetric and roughly Normal, the 68–95–99.7% Rule should work quite well. For example, we would expect about 95% of these samples to have proportions within 2 SDs (2.44%) of 18%. In fact, 9541 or 95.41% of them did.

Looking at lots of sample proportions is a big shift in our point of view. Up to now, you've thought of the sample proportion as a fixed value that comes from one particular sample. Given the number of successes and the size of your sample, the proportion was a number, a value that could be right or wrong on your homework. But now we're trying to understand how proportions from different random samples might behave. Each random sample has its own proportion and each of those proportions may be different, so we really have to think of proportions as random quantities. That's why simulating helps; it lets us see how sample proportions vary from sample to sample. We know (since 1718) that the Normal distribution provides a good model for the sampling distribution, but simulating different situations and seeing it can help you to believe it.[4]

Which Normal?

To use a Normal model, we need to specify two parameters: its mean and standard deviation. The center of the histogram is naturally at p, so that's what we'll use for the mean of the Normal model.

What about the standard deviation? For quantitative data, the mean gives us no information about the standard deviation. If we told you that a batch of bike helmets had a mean diameter of 26 centimeters and asked what the standard deviation was, you'd probably have no idea. And there's no reason why you should. There's no information about σ from knowing the value of μ for quantitative data.

But there's a special fact about proportions that makes the standard deviation easy to find. Once we know the mean, p, (and write $q = 1 - p$) we automatically also know the standard deviation of $\hat{p}$. We saw in Chapter 16 that for a Binomial model, the standard deviation of the *number* of success is $\sqrt{npq}$. Now we want the standard deviation of the *proportion* of successes, $\hat{p}$. The proportion is just the number of successes divided by the number of trials, n, so the standard deviation is also divided by n:

$$\sigma(\hat{p}) = SD(\hat{p}) = \frac{\sqrt{npq}}{n} = \sqrt{\frac{pq}{n}}.$$

When we draw simple random samples of n individuals, the proportions we find will vary from sample to sample. As long as n is reasonably large,[5] we can model the distribution of these sample proportions with a probability model that is

$$N\left(p, \sqrt{\frac{pq}{n}}\right).$$

Abraham de Moivre (1667–1754)

A S *Simulation:* **Simulating Sampling Distributions.** Watch the Normal model appear from random proportions.

---

[4] The saying isn't "being told is believing," but "seeing is believing" and there's a good reason for that.
[5] For smaller n, we could use a Binomial model (see Chapter 16).

Figure 17.2

A Normal model centered at p with a standard deviation of $\sqrt{\frac{pq}{n}}$ is a good model for a collection of proportions found for many random samples of size n from a population with success probability p.

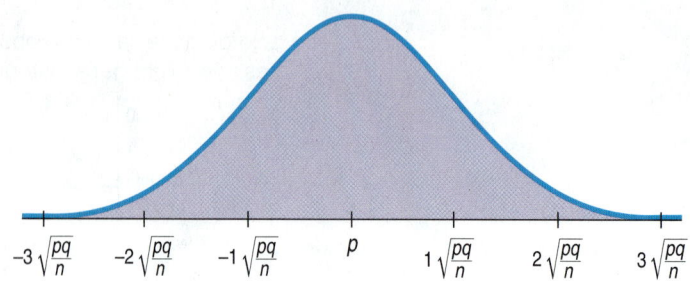

■ **NOTATION ALERT**

In Chapter 7, we introduced $\hat{y}$ as the predicted value for y. The "hat" here plays a similar role. It indicates that $\hat{p}$—the observed proportion in our data—is our *estimate* of the parameter p.

Although we'll never know the true proportion of adults who think global climate change is anthropogenic, we're supposing it to be 57%. Once we put the center at $p = 0.57$, the standard deviation for the Gallup poll is

$$SD(\hat{p}) = \sqrt{\frac{pq}{n}} = \sqrt{\frac{(0.57)(0.43)}{1022}} = 0.0155, \text{ or } 1.55\%.$$

Here's a picture of the Normal model for our simulation histogram:

Figure 17.3

Using 0.57 for p gives this Normal model for Figure 17.1's histogram of the sample proportions of adults who think climate change is anthropogenic ($n = 1022$).

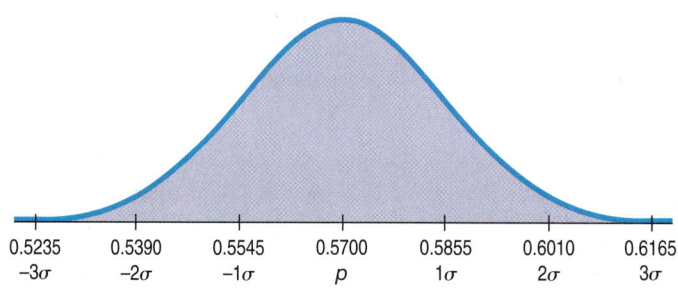

Let's see how well the theory works. In our simulation of smokers (page 444), the SD of $\hat{p}$ was 0.0122. The formula says it actually should be[6]

$$SD(\hat{p}) = \sqrt{\frac{pq}{n}} = \sqrt{\frac{(0.18)(0.82)}{1000}} = 0.0121.$$

That works.

Because we have a Normal model, we can use the 68–95–99.7 Rule or look up exact probabilities using a table or technology. For example, we know that 95% of Normally distributed values are within roughly two standard deviations of the mean, so we should not be surprised if various polls that may appear to ask the same question report a variety of results. Such sample-to-sample variation is sometimes called **sampling error**. It's not really an *error* at all, but just *variability* you'd expect to see from one sample to another. A better term would be **sampling variability**.[7]

Simulation: **The Standard Deviation of a Proportion.** Do you believe this formula for standard deviation? Don't just take our word for it—convince yourself with an experiment.

---

[6]The standard deviation is 0.012149 or 1.21%. Remember that the standard deviation always has the same units as the data. Here our units are %, or "percentage points." The standard deviation isn't 1.21% of anything, it is just 1.21 percentage points. If that's confusing, try writing the units out as "percentage points" instead of using the symbol %. Many polling agencies now do that too.

[7]Of course, polls differ for many reasons in addition to sampling variability. These can include minor differences in question wording (Gallup's question asked if the "increases in Earth's temperature" were due "more to human activity than to natural causes" but Yale asked if "global warming is caused mostly by human activities." Those may sound like the same questions, but the differences in wording could make a difference, for example, because "global warming" may be heard as a term with special or political connotations or because Gallup offered two alternatives.) Polls can also differ in the specific ways that the random samples are selected. But polling agencies report their "margin of error" based only on sampling variability. We'll see how they find that value in Chapter 18.

17.2 When Does the Normal Model Work? Assumptions and Conditions

De Moivre claimed that the sampling distribution can be modeled well by a Normal model. But, does it always work? Well, no. For example, if we drew samples of size 2, the only possible proportion values would be 0, 0.5, and 1. There's no way a histogram consisting only of those values could look like a Normal model.

Figure 17.4
Proportions from samples of size 2 can take on only three possible values. A Normal model does not work well.

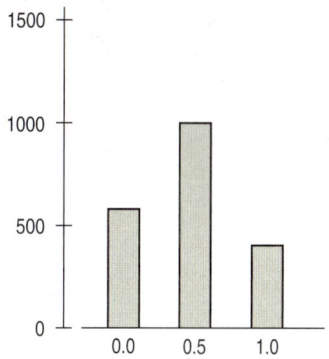

In fact, De Moivre's claim is only approximately true. (But that's OK. After all, models are only supposed to be approximately true.) The Normal model becomes a better and better representation of the sampling distribution as the size of the samples gets bigger.[8] Samples of size 1 or 2 won't work at all, but the distributions of sample proportions of larger samples are modeled well by the Normal model as long as p isn't too close to 0 or 1.

Populations with a true proportion, p, close to 0 or 1 can be a problem. Suppose a basketball coach surveys students to see how many male high school seniors are over 6'6". What will the proportions of samples of size 1000 look like? If the true proportion of students that tall is 0.001, then the coach is likely to get only a few seniors over 6'6" in any random sample of 1000. Most samples will have proportions of 0/1000, 1/1000, 2/1000, 3/1000, 4/1000 with only a very few samples having a higher proportion. Here's a simulation of 2000 surveys of size 1000 with $p = 0.001$:

Figure 17.5
The distribution of sample proportions for 2000 samples of size 1000 with $p = 0.001$. Because the true proportion is so small, the sampling distribution is skewed to the right and the Normal model won't work well.

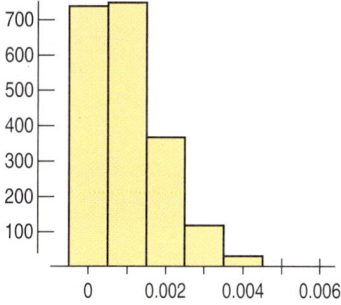

The problem is that the distribution is skewed to the right because p is so close to 0. (Had p been very close to 1, it would have been skewed to the left.) So, even though n is large, p is too small, and so the Normal model still won't work well.

When does the Normal model with mean p and standard deviation $\sqrt{\dfrac{pq}{n}}$ work well as a model for the sampling distribution of a sample proportion? We can summarize those cases by checking the following assumptions and conditions:

The Independence Assumption: The individuals in the samples (whose proportion we are finding) must be independent of each other. If not, the observations will

---
[8] Formally, we say the claim is true in the limit as n grows.

resemble each other too much and distort the standard deviation. You can't know if an assumption is true or not, but you *can* check to see whether the data were collected in a way that makes this assumption plausible.

If your data come from an experiment in which subjects were randomly assigned to treatments or from a survey based on a simple random sample, then the data satisfy the **Randomization Condition**. If not, you'll need to think carefully about whether it is honest to consider your cases independent with respect to the variables of interest and whether the cases you have are representative of the population you care about. If they aren't, then your inferences may not be valid.

10% Condition: It can also be a problem to sample too much of the population. Once you've sampled more than about 10% of the population, the remaining individuals are no longer really independent of each other. This **10%** is usually checked more casually than other assumptions and conditions. There are special formulas that you can use to adjust for sampling a large part of a population but, as the saying goes, they are "beyond the scope of this course."

We've seen that n has to be large enough and that p (and $q = 1 - p$ as well) can't be too close to 0 or 1. We put those two concerns together and check for at least 10 "successes" and 10 "failures." That is, make sure that $n\hat{p}$ and $n\hat{q}$ are both at least 10.

Success/Failure Condition: You should have at least 10 successes and 10 failures in your data.

The Success/Failure condition encourages a large sample size, but the 10% condition warns against sampling too much of a limited population. Nevertheless, this apparent conflict is rarely a problem. Most populations are safely large enough. Usually you should devote your effort and expense to gathering a larger sample or more experimental units.

In summary, if the assumptions and conditions are met, then you can use the Normal model to model the distribution of the sample proportion. In other words:

> **THE SAMPLING DISTRIBUTION MODEL FOR A PROPORTION**
> Provided that the sampled values are independent and the sample size is large enough, the sampling distribution of $\hat{p}$ is modeled by a Normal model with mean p and standard deviation $SD(\hat{p}) = \sqrt{\dfrac{pq}{n}}$.

Without sampling distribution models, the rest of Statistics just wouldn't exist. Sampling models are what make Statistics work. They inform us about the amount of variation we should expect when we sample. With a sampling distribution model, we can make informed decisions about how precise our estimate of the true value of a parameter might be. That's exactly what we'll be doing for the rest of this book.

Sampling distribution models enable us to say something about the population when all we have is data from a sample. By imagining what *might* happen if we were to draw many, many samples from the same population, we can learn a lot about how close the statistics computed from our one particular sample may be to the corresponding population parameters they estimate. That's the path to the *margin of error* you hear about in polls and surveys. We'll see how to determine that in the next chapter.

> **SUCCESSES AND FAILURES**
> The terms "success" and "failure" for the outcomes that have probability p and q are common in Statistics. But they are completely arbitrary labels. When we say that a disease occurs with probability p, we certainly don't mean that getting sick is a "success" in the ordinary sense of the word.

For Example USING THE SAMPLING DISTRIBUTION MODEL FOR PROPORTIONS

The Centers for Disease Control and Prevention report that 22% of 18-year-old women in the United States have a body mass index (BMI)[9] of 30 or more—a value considered by the National Heart, Lung, and Blood Institute to be associated with increased health risk.

---
[9] BMI = weight in kg/(height in m)^2.

As part of a routine health check at a large college, the physical education department usually requires students to come in to be measured and weighed. This year, the department decided to try out a self-report system. It asked 200 randomly selected female students to report their heights and weights (from which their BMIs could be calculated). Only 31 of these students had BMIs greater than 30.

QUESTION: Is this proportion of high-BMI students unusually small?

ANSWER: First, check:

✓ Randomization Condition: The department drew a random sample, so the respondents should be independent.

✓ 10% Condition: If the college class was only a few hundred students, we'd need those special formulas. For a "large college," we're probably OK.

✓ Success/Failure Condition: There were 31 "successes"—students with BMIs greater than 30 and 169 "failures." Both are at least 10.

It's OK to use a Normal model to describe the sampling distribution of the proportion of respondents with BMIs above 30.

The phys ed department observed $\hat{p} = \frac{31}{200} = 0.155$.

The department expected $p = 0.22$, with $SD(\hat{p}) = \sqrt{\frac{pq}{n}} = \sqrt{\frac{(0.22)(0.78)}{200}} = 0.029$,

so $z = \frac{\hat{p} - p}{SD(\hat{p})} = \frac{0.155 - 0.22}{0.029} = -2.24$.

By the 68–95–99.7 Rule, I know that values more than 2 standard deviations below the mean of a Normal model show up less than 2.5% of the time. Perhaps women at this college differ from the general population, or self-reporting may not provide accurate heights and weights.

Just Checking

1. You want to poll a random sample of 100 students on campus to see if they are in favor of the proposed location for the new student center. Of course, you'll get just one number, your sample proportion, $\hat{p}$. But if you imagined all the possible samples of 100 students you could draw and imagined the histogram of all the sample proportions from these samples, what shape would it have?

2. Where would the center of that histogram be?

3. If you think that about half the students are in favor of the plan, what would the standard deviation of the sample proportions be?

Step-by-Step Example WORKING WITH SAMPLING DISTRIBUTION MODELS FOR PROPORTIONS

Suppose that about 13% of the population is left-handed.[10] A 200-seat school auditorium has been built with 15 "lefty seats," seats that have the built-in desk on the left rather than the right arm of the chair. (For the right-handed readers among you, have you ever tried to take notes in a chair with the desk on the left side?)

Question: In a class of 90 students, what's the probability that there will not be enough seats for the left-handed students?

---

[10] Actually, there is little agreement among experts about the proportion of lefties in the population—or even how to define "left-handedness." Estimates range from 8% to 15%.

THINK ➡ Plan   State what we want to know.

I want to find the probability that in a group of 90 students, more than 15 will be left-handed. Since 15 out of 90 is 16.7%, I need the probability of finding more than 16.7% left-handed students out of a sample of 90 if the proportion of lefties is 13%.

Model   Think about the assumptions and check the conditions.

You might be able to think of cases where the **Independence Assumption** is not plausible—for example, if the students are all related, or if they were selected for being left- or right-handed. But for students in a class, the assumption of independence with respect to handedness seems reasonable.

✓ **Independence Assumption:** The 90 students in the class are not a random sample, but it is reasonable to assume that the probability that one student is left-handed is not changed by the fact that another student is right- or left-handed. It is also reasonable to assume that the students in the class are representative of students in general with respect to handedness.

✓ **10% Condition:** 90 is surely less than 10% of the population of all students so there is no need for special formulas. (Even if the school itself is small, I'm thinking of the population of all *possible* students who could have gone to the school.)

✓ **Success/Failure Condition:**

$$np = 90(0.13) = 11.7 \geq 10$$
$$nq = 90(0.87) = 78.3 \geq 10$$

State the parameters and the sampling distribution model.

The population proportion is $p = 0.13$. The conditions are satisfied, so I'll model the sampling distribution of $\hat{p}$ with a Normal model with mean 0.13 and a standard deviation of

$$SD(\hat{p}) = \sqrt{\frac{pq}{n}} = \sqrt{\frac{(0.13)(0.87)}{90}} \approx 0.035$$

My model for $\hat{p}$ is $N(0.13, 0.035)$.

SHOW ➡ Plot   Make a picture. Sketch the model and shade the area we're interested in, in this case the area to the right of 16.7%.

Mechanics   Use the standard deviation as a ruler to find the z-score of the cutoff proportion. We see that 16.7% lefties would be just over one standard deviation above the mean.

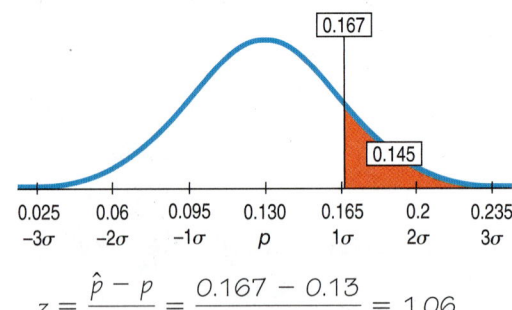

$$z = \frac{\hat{p} - p}{SD(\hat{p})} = \frac{0.167 - 0.13}{0.035} = 1.06$$

Find the resulting probability from a table of Normal probabilities, a computer program, or a calculator.

$$P(\hat{p} > 0.167) = P(z > 1.06) = 0.1446$$

TELL ➡ Conclusion   Interpret the probability in the context of the question.

There is about a 14.5% chance that there will not be enough seats for the left-handed students in the class.

17.3 The Sampling Distribution of Other Statistics

The sampling distribution for the proportion is especially useful because the Normal model provides such a good approximation. But, it might be useful to know the sampling distribution for *any* statistic that we can calculate, not just the sample proportion. Is the Normal model a good model for all statistics? Would you expect that the sampling distribution of the minimum or the maximum or the variance of a sample to be Normally distributed? What about the median?

SIMULATING THE SAMPLING DISTRIBUTIONS OF OTHER STATISTICS

In Chapter 7, we looked at a study of body measurements of 250 men (full data are in **Body fat complete**). The median of all 250 men's weights is 176.125 lbs and the variance is 730.9 lbs^2. We can treat these 250 men as a population, draw repeated random samples of 10, and compute the median, the variance, and the minimum for each sample.

Each of these histograms depicts the sampling distribution of its respective statistic. And it is easy to see that they aren't all the same. The sampling distribution of the medians is unimodal and symmetric. The sampling distribution of the variances is skewed to the right. And the sampling distribution of the minimums is, well, messy.

We can simulate to find the sampling distribution of *any* statistic we like: the maximum, the IQR, the 37th percentile, anything. But, both the proportion and the mean (as we'll see in the next section) have sampling distributions that are well approximated by a Normal model. And that observation hints at a fundamental fact about the sampling distribution models for the two summary statistics that we use most often.

Figure 17.6

Figure 17.7

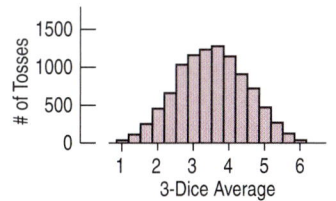

Figure 17.8

Simulating the Sampling Distribution of a Mean

Here's a simple simulation. Let's start with one fair die. If we toss this die 10,000 times, what should the histogram of the numbers on the face of the die look like? Figure 17.6 shows the results of a simulated 10,000 tosses.

Now let's toss a *pair* of dice and record the average of the two. If we repeat this (or at least simulate repeating it) 10,000 times, recording the average of each pair, what will the histogram of these 10,000 averages look like? Before you look, think a minute. Is getting an average of 1 on *two* dice as likely as getting an average of 3 or 3.5?

Let's see.

We're much more likely to get an average near 3.5 than we are to get one near 1 or 6. Without calculating those probabilities exactly, it's fairly easy to see that the *only* way to get an average of 1 is to get two 1's. To get a total of 7 (for an average of 3.5), though, there are many more possibilities. This distribution even has a name: the *triangular* distribution (see Figure 17.7).

What if we average 3 dice? We'll simulate 10,000 tosses of 3 dice and take their average. Figure 17.8 shows the result.

Figure 17.9

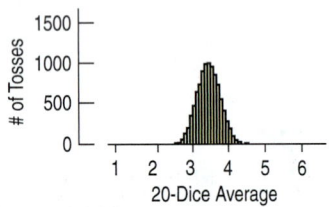

Figure 17.10

What's happening? First, it's getting harder to have averages near the ends. Getting an average of 1 or 6 with 3 dice requires all three to come up 1 or 6, respectively. That's less likely than for 2 dice to come up both 1 or both 6. The distribution is being pushed toward the middle. But what's happening to the shape? (This distribution doesn't have a name, as far as we know.)

Let's continue this simulation to see what happens with larger samples. Figure 17.9 shows a histogram of the averages for 10,000 tosses of 5 dice.

The pattern is becoming clearer. Two things continue to happen. The first fact we knew already from the Law of Large Numbers. It says that as the sample size (number of dice) gets larger, each sample average is more likely to be closer to the population mean. So, we see the means continuing to tighten around 3.5. More important, the shape of the distribution is approaching the Normal model.

Let's skip ahead and try 20 dice. The histogram of averages for 10,000 throws of 20 dice is in Figure 17.10.

Now we see the Normal shape clearly (and notice how much smaller the spread is). But can we count on this happening for situations other than dice throws?

17.4 The Central Limit Theorem: The Fundamental Theorem of Statistics

A S **Activity: The Sampling Distribution Model for Means.** Don't just sit there reading about the simulation—do it yourself.

"The theory of probabilities is at bottom nothing but common sense reduced to calculus."

—Laplace, in *Théorie analytique des probabilités,* 1812

What we saw with dice is true for means of repeated samples for almost every situation. When we looked at the sampling distribution of a proportion, we had to check only a few conditions. For means, the result is even more remarkable. *There are almost no conditions at all.*

Let's say that again: The sampling distribution of *any* mean becomes more nearly Normal as the sample size grows. All we need is for the observations to be independent and collected with randomization. We don't even care about the shape of the population distribution![11] This surprising fact was proved by Laplace in a fairly general form in 1810.[12] At the time, Laplace's theorem caused quite a stir (at least in mathematics circles) because the fact that it works no matter how the data are distributed seems remarkable. Laplace's result is called the **Central Limit Theorem**[13] (CLT).

Why should the Normal model show up again for the sampling distribution of means as well as proportions? We're not going to try to persuade you that it is obvious, clear, simple, or straightforward. In fact, the CLT is surprising and a bit weird. Not only does the distribution of means of repeated random samples get closer and closer to a Normal model as the sample size grows, *this is true regardless of the shape of the population distribution!* Even if we sample from a skewed or bimodal population, the Central Limit Theorem tells us that means of repeated random samples will tend to follow a Normal model as the sample size grows. If the sample size is small, it will work better the closer the population distribution is to a Normal model. If the data come from a population that's exactly Normal to start with, then the observations themselves are Normal. If we take samples of size 1, their "means" are just the observations—so, of course, they have a Normal sampling distribution.

But what if we start with a distribution that's strongly skewed, like the CEO compensations we saw back in Chapter 4?

---

[11]OK, one technical condition. The data must come from a population with a finite variance. You probably can't imagine a population with an infinite variance, but statisticians can construct such things, so we have to discuss them in footnotes like this. It really makes no difference in how you think about the important stuff, so you can just forget we mentioned it.

[12]Stigler, Stephen M. *The History of Statistics,* pp. 136–137.

[13]The word "central" in the name of the theorem means "fundamental." It doesn't refer to the center of a distribution.

> **Pierre-Simon Laplace** Laplace was one of the greatest scientists and mathematicians of his time. In addition to his contributions to probability and statistics, he published many new results in mathematics, physics, and astronomy (where his nebular theory was one of the first to describe the formation of the solar system in much the way it is understood today). He also played a leading role in establishing the metric system of measurement.
>
> His brilliance, though, sometimes got him into trouble. A visitor to the Académie des Sciences in Paris reported that Laplace let it be widely known that he considered himself the best mathematician in France. The effect of this on his colleagues was not eased by the fact that Laplace was right.

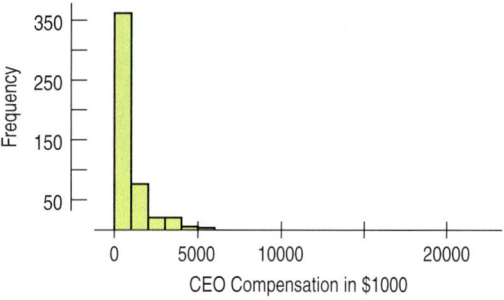

Figure 17.11
The distribution of CEO compensations is highly right skewed.

If we sample from this distribution, there's a good chance that we'll get an exceptionally large value. So some samples will have sample means much larger than others. Here is the sampling distribution of the means from 1000 samples of size 10:

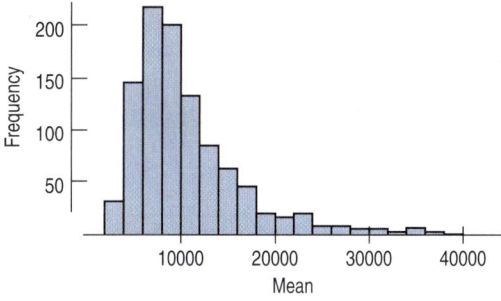

Figure 17.12
Samples of size 10 from the CEOs have means whose distribution is still right skewed, but less so than the distribution of the individual values shown in Figure 17.11.

This distribution is not *as* skewed as the original distribution, but still strongly right skewed. We wouldn't ordinarily find a mean of such a skewed distribution, but bear with us for a few steps. The mean of these values is around 10,000 (in $1000s), which actually matches the mean of all 500 CEOs in our population. What happens if we take a larger sample? Here is the sampling distribution of means from samples of size 50:

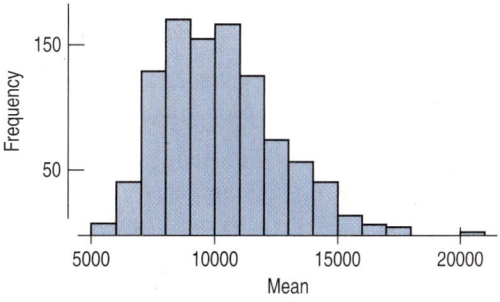

Figure 17.13
Means of samples of size 50 have a distribution that is only moderately right skewed.

This distribution is less skewed than the corresponding distribution from smaller samples and its mean is again near 10,000. Will this continue as we increase the sample size? Let's try samples of size 100:

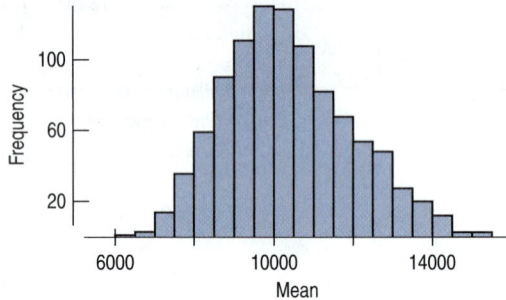

Figure 17.14
The means of samples of size 100 are nearly symmetric.

As we take larger samples, the distribution of means becomes more and more symmetric. By the time we get to samples of size 200, the distribution is quite symmetrical and, of course, has a mean quite close to 10,000.

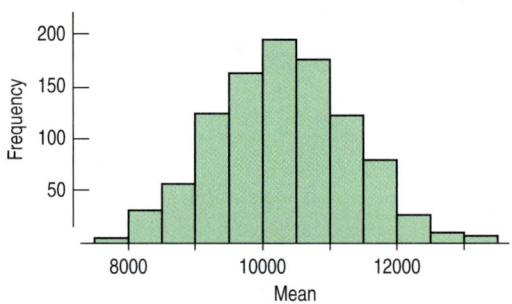

Figure 17.15
When the sample size is 200, even though we started with a highly skewed distribution, the sampling distribution of means is now almost perfectly symmetrical.

What about a really bimodal population one that consists of only 0's and 1's? The Central Limit Theorem says that even means of samples from this population will follow a Normal sampling distribution model. But wait. Suppose we have a categorical variable and we assign a 1 to each individual in the category and a 0 to each individual not in the category. And then we find the mean of these 0's and 1's. That's the same as counting the number of individuals who are in the category and dividing by n. That mean will be . . . the *sample proportion*, $\hat{p}$, of individuals who are in the category (a "success"). So maybe it wasn't so surprising after all that proportions, like means, have Normal sampling distribution models; they are actually just a special case of Laplace's remarkable theorem. Of course, for such an extremely bimodal population, we'll need a reasonably large sample size—and that's where the special conditions for proportions come in.

> **THE CENTRAL LIMIT THEOREM (CLT)**
> The mean of a random sample is a random variable whose sampling distribution can be approximated by a Normal model. The larger the sample, the better the approximation will be.

Assumptions and Conditions

The Central Limit Theorem requires essentially the same assumptions as we saw for modeling proportions:

Independence Assumption: The sampled values must be independent of each other. If they satisfy the **Randomization Condition** you may safely assume independence. Otherwise you must think carefully about whether they can be considered independent with respect to the variable of interest and whether the cases are representative of the population you care about.

Activity: **The Central Limit Theorem.** Simulate sampling distributions to see the CLT at work for yourself. Does it really work for samples from non-Normal populations?

The sample size, *n*, should be no more than 10% of the population.

> **Sample Size Condition:** Although the CLT tells us that a Normal model is useful in thinking about the behavior of sample means when the sample size is large enough, it doesn't tell us how large a sample we need. The truth is, it depends; there's no one-size-fits-all rule. If the population is unimodal and symmetric, even a fairly small sample is OK. If the population is strongly skewed, like the compensation for CEOs, we saw that it can take a pretty large sample for a Normal model to describe the distribution of sample means. For proportions, we have the success/failure conditions to guide us. For now you'll just need to think about your sample size in the context of what you know about the population, and then tell whether you believe the **Sample Size Condition** has been met.

But Which Normal?

The Central Limit Theorem says that the sampling distribution of any mean or proportion is approximately Normal. But which Normal model? We know that any Normal is specified by its mean and standard deviation. For proportions, the sampling distribution is centered at the population proportion. For means, it's centered at the population mean.

What about the standard deviations? We noticed in our dice simulation that the histograms got narrower as we averaged more and more dice together. This shouldn't be surprising. Means vary less than individual observations. Think about it for a minute. Which would be more surprising, having *one* person in your Statistics class who is over 6'9" tall or having the *mean* of all the students taking the course be over 6'9"? The first event is fairly rare.[14] You may have seen somebody this tall in one of your classes sometime. But a class of 100 whose mean height is over 6'9" tall just won't happen. Why? Because *means have smaller standard deviations than individuals*.

How much smaller? Well, we have good news and bad news. The good news is that the standard deviation of *y* falls as the sample size grows. The bad news is that it doesn't drop as fast as we might like. It only goes down by the *square root* of the sample size. That is, the sampling distribution of the mean has a standard deviation equal to

$$SD(\bar{y}) = \frac{\sigma}{\sqrt{n}}$$

where σ is the standard deviation of the population. To emphasize that this is a standard deviation *parameter* of the sampling distribution model for the sample mean, $\bar{y}$, we write $SD(\bar{y})$ or $\sigma(\bar{y})$.

> **THE SAMPLING DISTRIBUTION MODEL FOR A MEAN**
>
> When a random sample is drawn from any population with mean μ and standard deviation σ, its sample mean, $\bar{y}$, has a sampling distribution with the same *mean* μ but whose *standard deviation* is $\frac{\sigma}{\sqrt{n}}$ (we write $\sigma(\bar{y}) = SD(\bar{y}) = \frac{\sigma}{\sqrt{n}}$). No matter what population the random sample comes from, the *shape* of the sampling distribution is approximately Normal as long as the sample size is large enough. The larger the sample used, the more closely the Normal approximates the sampling distribution for the mean.

---

[14] If students are a random sample of adults, fewer than 1 out of 10,000 should be taller than 6'9". Why might college students not really be a random sample with respect to height? Even if they're not a perfectly random sample, a college student over 6'9" tall is still rare.

We now have two closely related sampling distribution models that we can use when the appropriate assumptions and conditions are met. Which one we use depends on which kind of data we have:

- When we have categorical data, we calculate a sample proportion, $\hat{p}$; the sampling distribution is roughly Normal[15] with a mean at the true proportion p and a standard deviation of $SD(\hat{p}) = \sqrt{\dfrac{pq}{n}} = \dfrac{\sqrt{pq}}{\sqrt{n}}$.

- When we have quantitative data, we calculate a sample mean, $\bar{y}$; the sampling distribution is roughly Normal[16] with a mean at the true mean, μ, and a standard deviation of $SD(\bar{y}) = \dfrac{\sigma}{\sqrt{n}}$.

The means of these models are easy to remember, so all you need to be careful about are the standard deviations. Remember that these are standard deviations of the *statistics* $\hat{p}$ and $\bar{y}$. They both have a square root of n in the denominator. That tells us that the larger the sample, the less either statistic will vary. The only difference is in the numerator. If you just start by writing $SD(\bar{y})$ for quantitative data and $SD(\hat{p})$ for categorical data, you'll be able to remember which formula to use.

For Example USING THE CENTRAL LIMIT THEOREM FOR MEANS

RECAP: A college physical education department asked a random sample of 200 female students to self-report their heights and weights, but the percentage of students with body mass indexes over 30 seemed suspiciously low. One possible explanation may be that the respondents "shaded" their weights down a bit. The CDC reports that the mean weight of 18-year-old women is 143.74 lb, with a standard deviation of 51.54 lb, but these 200 randomly selected women reported a mean weight of only 140 lb.

QUESTION: Based on the Central Limit Theorem and the 68–95–99.7 Rule, does the mean weight in this sample seem exceptionally low, or might this just be random sample-to-sample variation?

ANSWER: The conditions are satisfied:

✓ Randomization Condition: The women were a random sample and their weights can be assumed to be independent.

✓ Sample Size Condition: The distribution of college women's weights is likely to be unimodal and reasonably symmetric, so the CLT applies to means of even small samples; 200 values is plenty.

The sampling model for sample means is approximately Normal with $\mu(\bar{y}) = 143.7$ and $SD(\bar{y}) = \dfrac{\sigma}{\sqrt{n}} = \dfrac{51.54}{\sqrt{200}} = 3.64$. The expected distribution of sample means is:

The 68–95–99.7 Rule suggests that although the reported mean weight of 140 pounds is somewhat lower than expected, it does not appear to be unusual. Such variability is not all that extraordinary for samples of this size.

---

[15] Technically, we should say "the sampling distribution of the sample proportion can be well modeled by a Normal model," but no one would say that. So informally, we say that the distribution is roughly or approximately Normal.

[16] Technically, we should say "the sampling distribution of this random variable."

CHAPTER 17 Sampling Distribution Models

Step-by-Step Example: WORKING WITH THE SAMPLING DISTRIBUTION MODEL FOR THE MEAN

The Centers for Disease Control and Prevention reports that the mean weight of adult men in the United States is 190 lb with a standard deviation of 59 lb.[17]

Question: An elevator in our building has a weight limit of 10 persons or 2500 lb. What's the probability that if 10 men get on the elevator, they will overload its weight limit?

THINK → Plan State what we want to know.

Asking the probability that the total weight of a sample of 10 men exceeds 2500 pounds is equivalent to asking the probability that their mean weight is greater than 250 pounds.

Model Think about the assumptions and check the conditions.

✓ **Independence Assumption:** We don't have a random sample, but it's reasonable to think that the weights of 10 men in the elevator will be independent of each other and representative of the elevator-user population. (There could be exceptions—for example, if they were all from the same family or if the elevator were in a building with a diet clinic!)

If the sample were larger we'd be less concerned about the shape of the distribution of all weights.

✓ **Sample Size Condition:** I don't have data, but I suspect the distribution of population weights is roughly unimodal and symmetric, so a sample of 10 men seems large enough to use the CLT.

State the parameters and the sampling model.

The mean for all weights is $\mu = 190$ and the standard deviation is $\sigma = 59$ pounds. Since the conditions are satisfied, the CLT says that the sampling distribution of $\bar{y}$ has a Normal model with mean 190 and standard deviation

$$SD(\bar{y}) = \frac{\sigma}{\sqrt{n}} = \frac{59}{\sqrt{10}} \approx 18.66$$

SHOW → Plot Make a picture. Sketch the model and shade the area we're interested in. Here the mean weight of 250 pounds appears to be far out on the right tail of the curve.

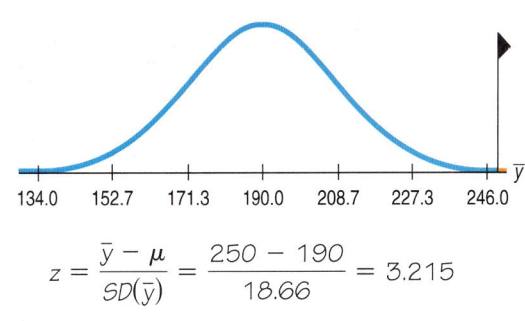

Mechanics Use the standard deviation as a ruler to find the z-score of the cutoff mean weight. We see that an average of 250 pounds is more than 3 standard deviations above the mean.

$$z = \frac{\bar{y} - \mu}{SD(\bar{y})} = \frac{250 - 190}{18.66} = 3.215$$

---

[17] Cynthia L. Ogden, Cheryl D. Fryar, Margaret D. Carroll, and Katherine M. Flegal, *Mean Body Weight, Height, and Body Mass Index, United States 1960–2002, Advance Data from Vital and Health Statistics Number 347*, Oct. 27, 2004. www.cdc.gov/nchs

Find the resulting probability from a table of Normal probabilities such as Table Z, a computer program, or a calculator.	$P(\bar{y} > 250) = P(z > 3.215) = 0.0007$
TELL ➡ Conclusion Interpret your result in the proper context, being careful to relate it to the original question.	The chance that a random collection of 10 men will exceed the elevator's weight limit is only 0.0007. So, unless my assumption of independence is false, it is quite unlikely that 10 people will exceed the total weight allowed on the elevator.

17.5 Sampling Distributions: A Summary

"The n's justify the means."

—Apocryphal statistical saying

Means vary less than individual data values. That makes sense. If the same test is given to many sections of a large course and the class average is, say, 80%, some students may score 95% because individual scores vary a lot. But we'd be shocked (and pleased!) if the *average* score of the students in any section was 95%. Averages are much less variable. Not only do group averages vary less than individual values, but common sense suggests that averages should be more consistent for larger groups. The fact that both $SD(\bar{y}) = \frac{\sigma}{\sqrt{n}}$ and $SD(\hat{p}) = \sqrt{(pq/n)}$ confirms this hunch because n is in the denominator in both cases. This shows that the variability of sample means decreases as the sample size increases. But, the standard deviation of the sampling distribution declines only with the square root of the sample size and not, for example, with $1/n$.

The mean of a random sample of 4 has half $\left(\frac{1}{\sqrt{4}} = \frac{1}{2}\right)$ the standard deviation of an individual data value. To cut the standard deviation in half again, we'd need a sample of 16, and a sample of 64 to halve it once more.

> **A billion-dollar misunderstanding?** In the late 1990s, the Bill and Melinda Gates Foundation began funding an effort to encourage the breakup of large schools into smaller schools. Why? It had been noticed that smaller schools were more common among the best-performing schools than one would expect. In time, the Annenberg Foundation, the Carnegie Corporation, the Center for Collaborative Education, the Center for School Change, Harvard's Change Leadership Group, the Open Society Institute, Pew Charitable Trusts, and the U.S. Department of Education's Smaller Learning Communities Program all supported the effort. Well over a billion dollars was spent to make schools smaller.
> But was it all based on a misunderstanding of sampling distributions? Statisticians Howard Wainer and Harris Zwerling[18] looked at the mean test scores of schools in Pennsylvania. They found that indeed 12% of the top-scoring 50 schools were from the smallest 3% of Pennsylvania schools—substantially more than the 3% we'd naively expect. But then they looked at the *bottom* 50. There they found that 18% were small schools! The explanation? Mean test scores behave like, well, means. We are looking at a rough real-world simulation in which each school is a trial. Even if all Pennsylvania schools were equivalent, we'd expect their mean scores to vary.

---

[18] Wainer, H. and Zwerling, H., "Legal and empirical evidence that smaller schools do not improve student achievement," *The Phi Delta Kappan* 2006, 87:300–303. Discussed in Howard Wainer, "The Most Dangerous Equation," *American Scientist*, May–June 2007, pp. 249–256; also at www.Americanscientist.org.

How much? The CLT tells us that means of test scores vary according to $\frac{\sigma}{\sqrt{n}}$. Smaller schools have (by definition) smaller *n*'s, so the sampling distributions of their mean scores naturally have larger standard deviations. It's natural, then, that small schools have both higher and lower mean scores.

On October 26, 2005, *The Seattle Times* reported:

> [T]he Gates Foundation announced last week it is moving away from its emphasis on converting large high schools into smaller ones and instead giving grants to specially selected school districts with a track record of academic improvement and effective leadership. Education leaders at the Foundation said they concluded that improving classroom instruction and mobilizing the resources of an entire district were more important first steps to improving high schools than breaking down the size.[19]

If only we had a much larger sample, we could get the standard deviation of the sampling distribution *really* under control so that the sample mean could tell us still more about the unknown population mean, but larger samples cost more and take longer to survey. And there are often practical limits to sample sizes. While we're gathering all that extra data, the population itself may change, or a news story may alter opinions. As we shall see, that nasty square root limits how much we can make a sample tell about the population. This is an example of the Law of Diminishing Returns.

The Real World and the Model World

Be careful. We have been slipping smoothly between the real world, in which we draw random samples of data, and a magical mathematical model world, in which we describe how the sample means and proportions we observe in the real world behave as random variables in all the random samples that we *might* have drawn. Now we have *two* distributions to deal with. The first is the real-world distribution of the sample, which we might display with a histogram (for quantitative data) or with a bar chart or table (for categorical data). The second is the math world *sampling distribution model* of the statistic, a Normal model based on the Central Limit Theorem. Don't confuse the two.

For example, don't mistakenly think the CLT says that the *data* are Normally distributed as long as the sample is large enough. In fact, as samples get larger, we expect the distribution of the data to look more and more like the population from which they are drawn—skewed, bimodal, whatever—but not necessarily Normal. You can collect a sample of CEO salaries for the next 1000 years,[20] but the histogram will never look Normal. It will be skewed to the right. The Central Limit Theorem doesn't talk about the distribution of the data from the sample. It talks about the distribution of sample *means* and sample *proportions* of many different random samples drawn from the same population. Of course, the CLT does require that the sample be big enough when the population shape is not unimodal and symmetric, but the fact that, even then, a Normal model is useful is still a surprising and powerful result.

Just Checking

4. Human gestation times have a mean of about 266 days, with a standard deviation of about 10 days.[21] If we record the gestation times of a sample of 100 women, do we know that a histogram of the times will be well modeled by a Normal model?

---

[19] Excerpt from "School size: Is smaller really better?" by Lynn Thompson. Published by The Seattle Times Company, © 2005.
[20] Don't forget to adjust for inflation.
[21] *Ultrasound Obstet. Gynecol.* 6 (1995) 353–357.

5. Suppose we look at the *average* gestation times for a sample of 100 women. If we imagined all the possible random samples of 100 women we could take and looked at the histogram of all the sample means, what shape would it have?
6. Where would the center of that histogram be?
7. What would be the standard deviation of that histogram?

Sampling Distribution Models—A Summary

Let's summarize what we've learned about sampling distributions. At the heart is the idea that *the statistic itself is a random variable.* We can't know what our statistic will be because it comes from a random sample. It's just one instance of something that happened for our particular random sample. A different random sample would have given a different result. This sample-to-sample variability is what generates the sampling distribution. The sampling distribution shows us the distribution of possible values that the statistic could have had.

We could simulate that distribution by pretending to take lots of samples. Fortunately, for the mean and the proportion, the CLT tells us that we can model their sampling distribution directly with a Normal model.

The two basic truths about sampling distributions are

1. Sampling distributions arise because samples vary. Each random sample will contain different cases and, so, a different value of the statistic.
2. Although we can always simulate a sampling distribution, the Central Limit Theorem saves us the trouble for means and proportions.

Here's a picture showing the process going into the sampling distribution model:

A S *Simulation: The CLT for Real Data.* Why settle for a picture when you can see it in action?

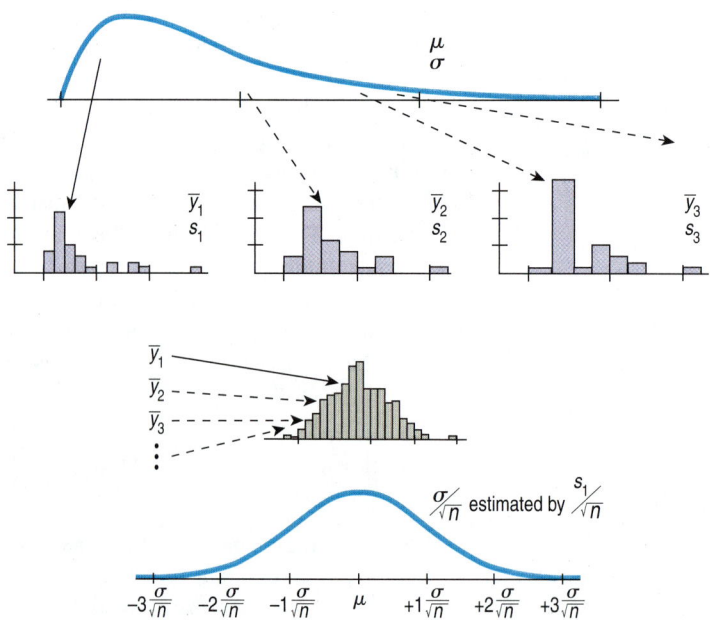

Figure 17.16
We start with a population model, which can have any shape. It can even be bimodal or skewed (as this one is). We label the mean of this model μ and its standard deviation, σ.

We draw one real sample (solid line) of size n and show its histogram and summary statistics. We imagine (or simulate) drawing many other samples (dotted lines), which have their own histograms and summary statistics.

We (imagine) gathering all the means into a histogram.

The CLT tells us we can model the shape of this histogram with a Normal model. The mean of this Normal is μ, and the standard deviation is $SD(\bar{y}) = \dfrac{\sigma}{\sqrt{n}}$.

WHAT CAN GO WRONG?

- **Don't confuse the sampling distribution with the distribution of the sample.** When you take a sample, you always look at the distribution of the values, usually with a histogram, and you may calculate summary statistics. Examining the distribution of the sample data is wise. But that's not the sampling distribution. The sampling distribution is an imaginary collection of all the values that a statistic *might* have taken for all possible random samples—the one you got and the ones that you didn't get. We use the sampling distribution model to make statements about how the statistic varies.

- **Beware of observations that are not independent.** The CLT depends crucially on the assumption of independence. If our elevator riders are related, are all from the same school (for example, an elementary school), or in some other way aren't a random sample, then the statements we try to make about the mean are going to be wrong. Unfortunately, this isn't something you can check in your data. You have to think about how the data were gathered. Good sampling practice and well-designed randomized experiments ensure independence.

- **Watch out for small samples from skewed or bimodal populations.** The CLT assures us that the sampling distribution model is Normal if n is large enough. If the population is nearly Normal, even small samples (like our 10 elevator riders) work. If the population is skewed or bimodal, then n will have to be large before the Normal model will work well. If we sampled 15 or even 20 CEOs and used $\bar{y}$ to make a statement about the mean of all CEOs' compensation, we'd likely get into trouble because the underlying data distribution is so skewed. Unfortunately, there's no good rule to follow.[22] It just depends on how skewed the data distribution is. Always plot the data to check.

CONNECTIONS

The concept of a sampling distribution connects to almost everything we have done. The fundamental connection is to the deliberate application of randomness in random sampling and randomized comparative experiments. If we didn't employ randomness to generate unbiased data, then repeating the data collection would just get the same data values again (with perhaps a few new measurement or recording errors). The distribution of statistic values arises directly because different random samples and randomized experiments would generate different statistic values.

The connection to the Normal distribution is obvious. We first introduced the Normal model before because it was "nice." As a unimodal, symmetric distribution with 99.7% of its area within three standard deviations of the mean, the Normal model is easy to work with. Now we see that the Normal model holds a special place among distributions because we can use it to model the sampling distributions of the mean and the proportion.

We use simulation to understand sampling distributions. In fact, some important sampling distributions were discovered first by simulation.

---

[22] For proportions, of course, there is a rule: the **Success/Failure Condition**. That works for proportions because the standard deviation of a proportion is linked to its mean.

What Have We Learned?

Learning Objectives

Model the variation in statistics from sample to sample with a sampling distribution.
- The Central Limit Theorem tells us that the sampling distribution of both the sample proportion and the sample mean are Normal.

Understand that, usually, the mean of a sampling distribution is the value of the parameter estimated.
- For the sampling distribution of $\hat{p}$, the mean is p.
- For the sampling distribution of $\bar{y}$, the mean is μ.

Interpret the standard deviation of a sampling distribution.
- The standard deviation of a sampling model is the most important information about it.
- The standard deviation of the sampling distribution of a proportion is $\sqrt{\frac{pq}{n}}$, where $q = 1 - p$.
- The standard deviation of the sampling distribution of a mean is $\frac{\sigma}{\sqrt{n}}$, where σ is the population standard deviation.

Understand that the Central Limit Theorem is a *limit* theorem.
- The sampling distribution of the mean is Normal, *no matter what the underlying distribution of the data is*.
- The CLT says that this happens in the limit, as the sample size grows. The Normal model applies sooner when sampling from a unimodal, symmetric population and more gradually when the population is very non-Normal.

Review of Terms

Sampling distribution Different random samples give different values for a statistic. The distribution of the statistics over all possible samples is called the sampling distribution. The sampling distribution model shows the behavior of the statistic over all the possible samples for the same size n (p. 444).

Sampling distribution model Because we can never see all possible samples, we often use a model as a practical way of describing the theoretical sampling distribution (p. 444).

Sampling variability (sampling error) The variability we expect to see from one random sample to another. It is sometimes called sampling error, but sampling variability is the better term (p. 446).

Sampling distribution model for a proportion If assumptions of independence and random sampling are met, and we expect at least 10 successes and 10 failures, then the sampling distribution of a proportion is modeled by a Normal model with a mean equal to the true proportion value, p, and a standard deviation equal to $\sqrt{\frac{pq}{n}}$ (p. 448).

Central Limit Theorem The Central Limit Theorem (CLT) states that the sampling distribution model of the sample mean (and proportion) from a random sample is approximately Normal for large n, *regardless of the distribution of the population, as long as the observations are independent* (p. 452).

Sampling distribution model for a mean If assumptions of independence and random sampling are met, and the sample size is large enough, the sampling distribution of the sample mean is modeled by a Normal model with a mean equal to the population mean, μ, and a standard deviation equal to $\frac{\sigma}{\sqrt{n}}$ (p. 455).

Exercises

Section 17.1

1. **Website** An investment website can tell what devices are used to access the site. The site managers wonder whether they should enhance the facilities for trading via smartphones so they want to estimate the proportion of users who access the site that way (even if they also use their computers sometimes). They draw a random sample of 200 investors from their customers. Suppose that the true proportion of smartphone users is 36%.
   a) What would you expect the shape of the sampling distribution for the sample proportion to be?
   b) What would be the mean of this sampling distribution?
   c) What would be the standard deviation of the sampling distribution?

2. **Marketing** The proportion of adult women in the United States is approximately 51%. A marketing survey telephones 400 people at random.
   a) What proportion of the sample of 400 would you expect to be women?
   b) What would the standard deviation of the sampling distribution be?
   c) How many women, on average, would you expect to find in a sample of that size?

3. **Send money** When they send out their fundraising letter, a philanthropic organization typically gets a return from about 5% of the people on their mailing list. To see what the response rate might be for future appeals, they did a simulation using samples of size 20, 50, 100, and 200. For each sample size, they simulated 1000 mailings with success rate $p = 0.05$ and constructed the histogram of the 1000 sample proportions, shown below. Explain how these histograms demonstrate what the Central Limit Theorem says about the sampling distribution model for sample proportions. Be sure to talk about shape, center, and spread.

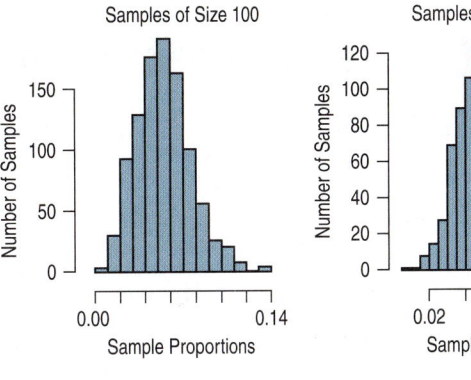

4. **Character recognition** An automatic character recognition device can successfully read about 85% of handwritten credit card applications. To estimate what might happen when this device reads a stack of applications, the company did a simulation using samples of size 20, 50, 75, and 100. For each sample size, they simulated 1000 samples with success rate $p = 0.85$ and constructed the histogram of the 1000 sample proportions, shown here. Explain how these histograms demonstrate what the Central Limit Theorem says about the sampling distribution model for sample proportions. Be sure to talk about shape, center, and spread.

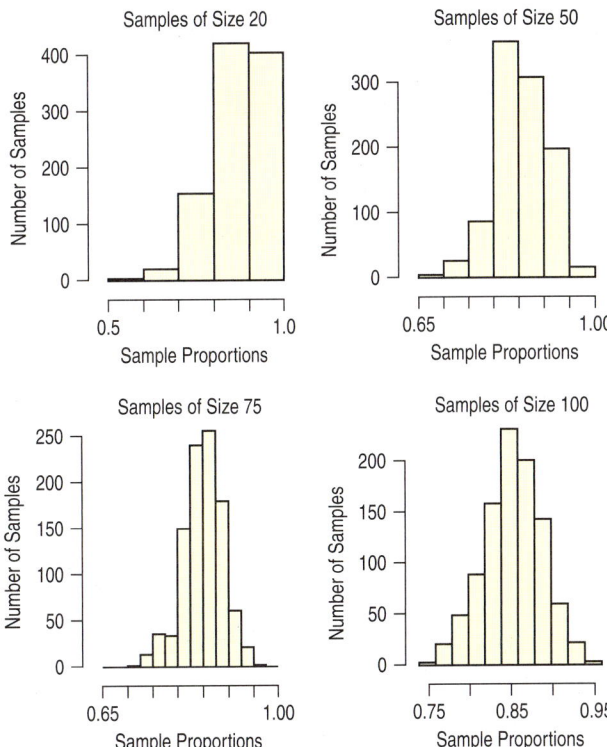

Section 17.2

5. Marriage According to a Pew Research survey, 27% of American adults are pessimistic about the future of marriage and the family. That is based on a random sample of about 1500 people. Is it reasonable for Pew Research to use a Normal model for the sampling distribution of the sample proportion? Why or why not?

6. Campus sample For her final project, Stacy plans on surveying a random sample of 50 students on whether they plan to go to Florida for Spring Break. From past years, she guesses that about 10% of the class goes. Is it reasonable for her to use a Normal model for the sampling distribution of the sample proportion? Why or why not?

7. Send more money The philanthropic organization in Exercise 3 expects about a 5% success rate when they send fundraising letters to the people on their mailing list. In Exercise 3, you looked at the histograms showing distribution of sample proportions from 1000 simulated mailings for samples of size 20, 50, 100, and 200. Here are both the theoretical and observed statistics for these simulated samples:

n	Observed Mean	Theoretical Mean	Observed st. dev	Theoretical st. dev
20	0.0497	0.05	0.0479	0.0487
50	0.0516	0.05	0.0309	0.0308
100	0.0497	0.05	0.0215	0.0218
200	0.0501	0.05	0.0152	0.0154

a) Looking at the histograms of Exercise 3, at what sample size would you be comfortable using the Normal model as an approximation for the sampling distribution?
b) What does the Success/Failure Condition say about the choice you made in part a?

8. Character recognition, again The automatic character recognition device discussed in Exercise 4 successfully reads about 85% of handwritten credit card applications. In Exercise 4, you looked at the histograms showing distributions of sample proportions from 1000 simulated samples of size 20, 50, 75, and 100. The sample statistics and theoretical values for each simulation are as follows:

n	Observed Mean	Theoretical Mean	Observed st.dev.	Theoretical st.dev.
20	0.8481	0.85	0.0803	0.0799
50	0.8507	0.85	0.0509	0.0505
100	0.8481	0.85	0.0406	0.0412
200	0.8488	0.85	0.0354	0.0357

a) Looking at the histograms in Exercise 4, at what sample size would you be comfortable using the Normal model as an approximation for the sampling distribution?
b) What does the Success/Failure Condition say about the choice you made in part a?

Section 17.3

9. Sample maximum The distribution of scores on a Statistics test for a particular class is skewed to the left. The professor wants to predict the maximum score and so wants to understand the distribution of the sample maximum. She simulates the distribution of the maximum of the test for 30 different tests (with $n = 5$). The histogram below shows a simulated sampling distribution of the sample maximum from these tests.

a) Would a Normal model be a useful model for this sampling distribution? Explain.
b) The mean of this distribution is 46.3 and the SD is 3.5. Would you expect about 95% of the samples to have their maximums within 7 of 46.3? Why or why not?

10. Soup A machine is supposed to fill cans with 16 oz of soup. Of course there, will be some variation in the amount actually dispensed, and measurement errors are often approximately normally distributed. The manager would like to understand the variability of the variances of the samples, so he collects information from the last 250 batches of size 10 and plots a histogram of the variances:

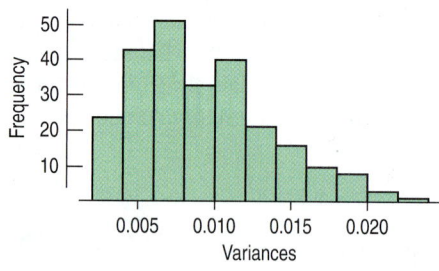

a) Would a Normal model be a useful model for this sampling distribution? Explain.
b) The mean of this distribution is 0.009 and the SD is 0.004. Would you expect about 95% of the samples to have their variances within 0.008 of 0.009? Why or why not?

Section 17.4

11. Market research A marketing researcher for a phone company surveys 100 people and finds that the

proportion of customers who are likely to switch providers when their contract expires is 15%.
   a) What is the standard deviation of the sampling distribution of the proportion?
   b) If she wants to reduce the standard deviation by half, how large a sample would she need?

12. **Market research II** A market researcher for a provider of iPod accessories wants to know the proportion of customers who own cars to assess the market for a new iPod car charger. A survey of 500 customers indicates that 76% own cars.
    a) What is the standard deviation of the sampling distribution of the proportion?
    b) How large would the standard deviation have been if he had surveyed only 125 customers (assuming the proportion is about the same)?

13. **Tips** A waiter believes the distribution of his tips has a model that is slightly skewed to the right, with a mean of $9.60 and a standard deviation of $5.40.
    a) Explain why you cannot determine the probability that a given party will tip him at least $20.
    b) Can you estimate the probability that the next 4 parties will tip an average of at least $15? Explain.
    c) Is it likely that his 10 parties today will tip an average of at least $15? Explain.

14. **Groceries** A grocery store's receipts show that Sunday customer purchases have a skewed distribution with a mean of $32 and a standard deviation of $20.
    a) Explain why you cannot determine the probability that the next Sunday customer will spend at least $40.
    b) Can you estimate the probability that the next 10 Sunday customers will spend an average of at least $40? Explain.
    c) Is it likely that the next 50 Sunday customers will spend an average of at least $40? Explain.

15. **More tips** The waiter in Exercise 13 usually waits on about 40 parties over a weekend of work.
    a) Estimate the probability that he will earn at least $500 in tips.
    b) How much does he earn on the best 10% of such weekends?

16. **More groceries** Suppose the store in Exercise 14 had 312 customers this Sunday.
    a) Estimate the probability that the store's revenues were at least $10,000.
    b) If, on a typical Sunday, the store serves 312 customers, how much does the store take in on the worst 10% of such days?

17. **Coin tosses** In a large class of introductory Statistics students, the professor has each student toss a coin 16 times and calculate the proportion of his or her tosses that were heads. The students then report their results, and the professor plots a histogram of these several proportions.
    a) What shape would you expect this histogram to be? Why?
    b) Where do you expect the histogram to be centered?
    c) How much variability would you expect among these proportions?
    d) Explain why a Normal model should not be used here.

18. **M&M's** The candy company claims that 10% of the M&M's it produces are green. Suppose that the candies are packaged at random in small bags containing about 50 M&M's. A class of elementary school students learning about percents opens several bags, counts the various colors of the candies, and calculates the proportion that are green.
    a) If we plot a histogram showing the proportions of green candies in the various bags, what shape would you expect it to have?
    b) Can that histogram be approximated by a Normal model? Explain.
    c) Where should the center of the histogram be?
    d) What should the standard deviation of the proportion be?

19. **More coins** Suppose the class in Exercise 17 repeats the coin-tossing experiment.
    a) The students toss the coins 25 times each. Use the 68–95–99.7 Rule to describe the sampling distribution model.
    b) Confirm that you can use a Normal model here.
    c) They increase the number of tosses to 64 each. Draw and label the appropriate sampling distribution model. Check the appropriate conditions to justify your model.
    d) Explain how the sampling distribution model changes as the number of tosses increases.

20. **Bigger bag** Suppose the class in Exercise 18 buys bigger bags of candy, with 200 M&M's each. Again the students calculate the proportion of green candies they find.
    a) Explain why it's appropriate to use a Normal model to describe the distribution of the proportion of green M&M's they might expect.
    b) Use the 68–95–99.7 Rule to describe how this proportion might vary from bag to bag.
    c) How would this model change if the bags contained even more candies?

21. **Just (un)lucky?** One of the students in the introductory Statistics class in Exercise 19 claims to have tossed her coin 200 times and found only 42% heads. What do you think of this claim? Explain.

Chapter Exercises

22. **Too many green ones?** In a really large bag of M&M's, the students in Exercise 20 found 500 candies, and 12% of them were green. Is this an unusually large proportion of green M&M's? Explain.

23. **Speeding** State police believe that 70% of the drivers traveling on a major interstate highway exceed the speed limit. They plan to set up a radar trap and check the speeds of 80 cars.
    a) Using the 68–95–99.7 Rule, draw and label the distribution of the proportion of these cars the police will observe speeding.
    b) Do you think the appropriate conditions necessary for your analysis are met? Explain.

24. **Smoking 2014** The Centers for Disease Control and Prevention estimated in 2014 that 18.1% of American adults smoked cigarettes. Using the 68–95–99.7 Rule, describe the sampling distribution model for the proportion of smokers among a randomly selected group of 60 adults. Be sure to discuss your assumptions and conditions.

25. **Vision** It is generally believed that nearsightedness affects about 12% of all children. A school district has registered 170 incoming kindergarten children.
    a) Can you apply the Central Limit Theorem to describe the sampling distribution model for the sample proportion of children who are nearsighted? Check the conditions and discuss any assumptions you need to make.
    b) Sketch and clearly label the sampling model, based on the 68–95–99.7 Rule.
    c) How many of the incoming students might the school expect to be nearsighted? Explain.

26. **Mortgages 2013** In December, 2013, Lender Processing Services* reported that although mortgage defaults had fallen, in New York state, 12.4% of mortgages were still "noncurrent," meaning that payments had been missed. Suppose a New York bank holds 1731 mortgages.
    a) Can you apply the Central Limit Theorem to describe the sampling distribution model for the sample proportion of noncurrent mortgages? Check the conditions and discuss any assumptions you need to make.
    b) Sketch and clearly label the sampling model, based on the 68–95–99.7 Rule.
    c) How many of these homeowners might the bank expect will default on their mortgages? Explain.

27. **Loans** Based on past experience, a bank believes that 7% of the people who receive loans will not make payments on time. The bank has recently approved 200 loans.
    a) What are the mean and standard deviation of the proportion of clients in this group who may not make timely payments?
    b) What assumptions underlie your model? Are the conditions met? Explain.
    c) What's the probability that over 10% of these clients will not make timely payments?

28. **Contacts** Assume that 30% of students at a university wear contact lenses.
    a) We randomly pick 100 students. Let $\hat{p}$ represent the proportion of students in this sample who wear contacts. What's the appropriate model for the distribution of $\hat{p}$? Specify the name of the distribution, the mean, and the standard deviation. Be sure to verify that the conditions are met.
    b) What's the approximate probability that more than one third of this sample wear contacts?

29. **Back to school? 2013** Best known for its testing program, ACT, Inc., also compiles data on a variety of issues in education. In 2013, the company reported that the national college freshman-to-sophomore retention rate that had held steady at 74% over the previous several years had recently fallen from 74% to 72.2%. (http://www.act.org/research/policymakers/pdf/retain_2013.pdf). Consider random samples of 400 freshmen who took the ACT. Use the 68–95–99.7 Rule to describe the sampling distribution model for the percentage of those students we expect to return to that school for their sophomore years. Do you think the appropriate conditions are met?

30. **Binge drinking** As we learned in Chapter 14, a national study found that 44% of college students engage in binge drinking (5 drinks at a sitting for men, 4 for women). Use the 68–95–99.7 Rule to describe the sampling distribution model for the proportion of students in a randomly selected group of 200 college students who engage in binge drinking. Do you think the appropriate conditions are met?

31. **Back to school, again** Based on the 72.2% national retention rate described in Exercise 29, does a college where 522 of the 603 freshmen returned the next year as sophomores have a right to brag that it has an unusually high retention rate? Explain.

32. **Binge sample** After hearing of the national result that 44% of students engage in binge drinking (5 drinks at a sitting for men, 4 for women), a professor surveyed a random sample of 244 students at his college and found that 96 of them admitted to binge drinking in the past week. Should he be surprised at this result? Explain.

33. **Polling** Just before a referendum on a school budget, a local newspaper polls 400 voters in an attempt to predict whether the budget will pass. Suppose that the budget actually has the support of 52% of the voters. What's the probability the newspaper's sample will lead them to predict defeat? Be sure to verify that the assumptions and conditions necessary for your analysis are met.

34. **Seeds** Information on a packet of seeds claims that the germination rate is 92%. What's the probability that more than 95% of the 160 seeds in the packet will germinate? Be sure to discuss your assumptions and check the conditions that support your model.

35. **Gaydar** Exercise 13 in Chapter 1 describes a study that showed that heterosexual women, during ovulation, were significantly better at correctly identifying the sexual orientation of a man from a photograph of his face than

---

*http://www.lpsvcs.com/LPSCorporateInformation/CommunicationCenter/DataReports/MortgageMonitor/201312MortgageMonitor/MortgageMonitorDecember2013.pdf

women who were not ovulating. In other words, ovulation improves a woman's "gaydar." Near ovulation, on average women correctly identified the orientation of about 65% of the 100 men shown to them. If this is the probability of correctly identifying the orientation of a man in any given photograph, what is the probability a woman would correctly classify 80 or more of the men (as two women in the study did)?

36. **Genetic defect** It's believed that 4% of children have a gene that may be linked to juvenile diabetes. Researchers hoping to track 20 of these children for several years test 732 newborns for the presence of this gene. What's the probability that they find enough subjects for their study?

37. **"No Children" section** Some restaurant owners, at the request of some of their less tolerant customers, have stopped allowing children into their restaurant. This, naturally, outrages other customers. One restaurateur hopes to please both sets of customers by having a "no children" section. She estimates that in her 120-seat restaurant, about 30% of her seats, on average, are taken by families with children. How many seats should be in the "children allowed" area in order to be very sure of having enough seating there? Comment on the assumptions and conditions that support your model, and explain what "very sure" means to you.

38. **Meals** A restaurateur anticipates serving about 180 people on a Friday evening, and believes that about 20% of the patrons will order the chef's steak special. How many of those meals should he plan on serving in order to be pretty sure of having enough steaks on hand to meet customer demand? Justify your answer, including an explanation of what "pretty sure" means to you.

39. **Sampling** A sample is chosen randomly from a population that can be described by a Normal model.
    a) What's the sampling distribution model for the sample mean? Describe shape, center, and spread.
    b) If we choose a larger sample, what's the effect on this sampling distribution model?

40. **Sampling, part II** A sample is chosen randomly from a population that was strongly skewed to the left.
    a) Describe the sampling distribution model for the sample mean if the sample size is small.
    b) If we make the sample larger, what happens to the sampling distribution model's shape, center, and spread?
    c) As we make the sample larger, what happens to the expected distribution of the data in the sample?

41. **Waist size** A study measured the *Waist Size* of 250 men, finding a mean of 36.33 inches and a standard deviation of 4.02 inches. Here is a histogram of these measurements.

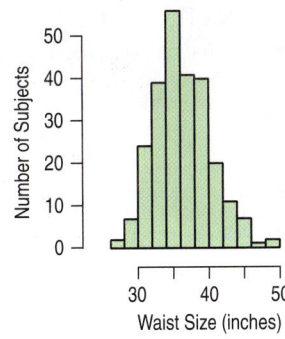

a) Describe the histogram of *Waist Size*.
b) To explore how the mean might vary from sample to sample, they simulated by drawing many samples of size 2, 5, 10, and 20, with replacement, from the 250 measurements. Here are histograms of the sample means for each simulation. Explain how these histograms demonstrate what the Central Limit Theorem says about the sampling distribution model for sample means.

42. **CEO compensation** The total compensation of the chief executive officers (CEOs) of the 800 largest U.S. companies (the Fortune 800) averaged (in thousands of dollars) 10,307.31 with a standard deviation (also in

(continued)

$1000) of 17,964.62. Here is a histogram of their annual compensations (in $1000):

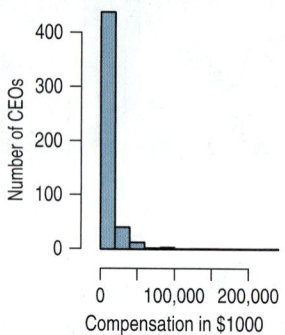

a) Describe the histogram of *Total Compensation*. A research organization simulated sample means by drawing samples of 30, 50, 100, and 200, with replacement, from the 800 CEOs. The histograms show the distributions of means for many samples of each size.

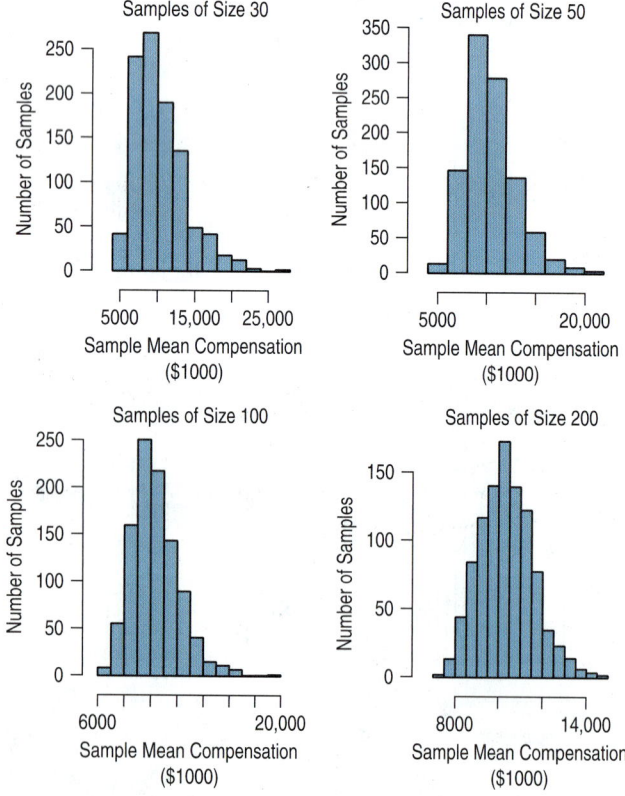

b) Explain how these histograms demonstrate what the Central Limit Theorem says about the sampling distribution model for sample means. Be sure to talk about shape, center, and spread.

c) Comment on the "rule of thumb" that "With a sample size of at least 30, the sampling distribution of the mean is Normal."

43. **Waist size, revisited** Researchers measured the *Waist Sizes* of 250 men in a study on body fat. The true mean and standard deviation of the *Waist Sizes* for the 250 men are 36.33 in and 4.019 inches, respectively. In Exercise 41, you looked at the histograms of simulations that drew samples of sizes 2, 5, 10, and 20 (with replacement). The summary statistics for these simulations were as follows:

n	Mean	st. dev.
2	36.314	2.855
5	36.314	1.805
10	36.341	1.276
20	36.339	0.895

a) According to the Central Limit Theorem, what should the theoretical mean and standard deviation be for each of these sample sizes?
b) How close are the theoretical values to what was observed in the simulation?
c) Looking at the histograms in Exercise 41, at what sample size would you be comfortable using the Normal model as an approximation for the sampling distribution?
d) What about the shape of the distribution of *Waist Size* explains your choice of sample size in part c?

44. **CEOs, revisited** In Exercise 42, you looked at the annual compensation for 800 CEOs, for which the true mean and standard deviation were (in thousands of dollars) 10,307.31 and 17,964.62, respectively. A simulation drew samples of sizes 30, 50, 100, and 200 (with replacement) from the total annual compensations of the Fortune 800 CEOs. The summary statistics for these simulations were as follows:

n	Mean	st. dev.
30	10,251.73	3359.64
50	10,343.93	2483.84
100	10,329.94	1779.18
200	10,340.37	1230.79

a) According to the Central Limit Theorem, what should the theoretical mean and standard deviation be for each of these sample sizes?
b) How close are the theoretical values to what was observed from the simulation?

45. **GPAs** A college's data about the incoming freshmen indicates that the mean of their high school GPAs was 3.4, with a standard deviation of 0.35; the distribution was roughly mound-shaped and only slightly skewed. The students are randomly assigned to freshman writing seminars in groups of 25. What might the mean GPA of one of these seminar groups be? Describe the appropriate

sampling distribution model—shape, center, and spread—with attention to assumptions and conditions. Make a sketch using the 68–95–99.7 Rule.

46. **Home values** Assessment records indicate that the value of homes in a small city is skewed right, with a mean of $140,000 and standard deviation of $60,000. To check the accuracy of the assessment data, officials plan to conduct a detailed appraisal of 100 homes selected at random. Using the 68–95–99.7 Rule, draw and label an appropriate sampling model for the mean value of the homes selected.

47. **Lucky spot?** A reporter working on a story about the New York lottery contacted one of the authors of this book, wanting help analyzing data to see if some ticket sales outlets were more likely to produce winners. His data for each of the 966 New York lottery outlets are graphed below; the scatterplot shows the ratio *Total-Paid/TotalSales* vs. *TotalSales* for the state's "instant winner" games for all of 2007.

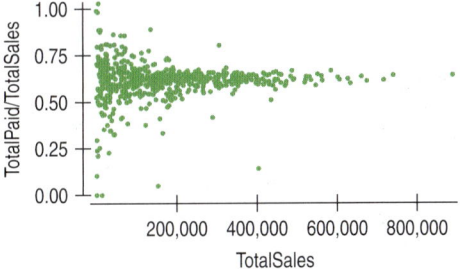

The reporter thinks that by identifying the outlets with the highest fraction of bets paid out, players might be able to increase their chances of winning. Typically—but not always—instant winners are paid immediately (instantly) at the store at which they are purchased. However, the fact that tickets may be scratched off and then cashed in at any outlet may account for some outlets paying out more than they take in. The few with very low payouts may be on interstate highways where players may purchase cards but then leave.
a) Explain why the plot has this funnel shape.
b) Explain why the reporter's idea wouldn't have worked anyway.

48. **Safe cities** Allstate Insurance Company identified the 10 safest and 10 least-safe U.S. cities from among the 200 largest cities in the United States, based on the mean number of years drivers went between automobile accidents. The cities on both lists were all smaller than the 10 largest cities. Using facts about the sampling distribution model of the mean, explain why this is not surprising.

49. **Pregnancy** Assume that the duration of human pregnancies can be described by a Normal model with mean 266 days and standard deviation 16 days.

a) What percentage of pregnancies should last between 270 and 280 days?
b) At least how many days should the longest 25% of all pregnancies last?
c) Suppose a certain obstetrician is currently providing prenatal care to 60 pregnant women. Let $\bar{y}$ represent the mean length of their pregnancies. According to the Central Limit Theorem, what's the distribution of this sample mean, $\bar{y}$? Specify the model, mean, and standard deviation.
d) What's the probability that the mean duration of these patients' pregnancies will be less than 260 days?

50. **Rainfall** Statistics from Cornell's Northeast Regional Climate Center indicate that Ithaca, New York, gets an average of 35.4″ of rain each year, with a standard deviation of 4.2″. Assume that a Normal model applies.
a) During what percentage of years does Ithaca get more than 40″ of rain?
b) Less than how much rain falls in the driest 20% of all years?
c) A Cornell University student is in Ithaca for 4 years. Let $\bar{y}$ represent the mean amount of rain for those 4 years. Describe the sampling distribution model of this sample mean, $\bar{y}$.
d) What's the probability that those 4 years average less than 30″ of rain?

51. **Pregnant again** The duration of human pregnancies may not actually follow the Normal model described in Exercise 49.
a) Explain why it may be somewhat skewed to the left.
b) If the correct model is in fact skewed, does that change your answers to parts a, b, and c of Exercise 49? Explain why or why not for each.

52. **At work** Some business analysts estimate that the length of time people work at a job has a mean of 6.2 years and a standard deviation of 4.5 years.
a) Explain why you suspect this distribution may be skewed to the right.
b) Explain why you could estimate the probability that 100 people selected at random had worked for their employers an average of 10 years or more, but you could not estimate the probability that an individual had done so.

53. **Dice and dollars** You roll a die, winning nothing if the number of spots is odd, $1 for a 2 or a 4, and $10 for a 6.
a) Find the expected value and standard deviation of your prospective winnings.
b) You play twice. Find the mean and standard deviation of your total winnings.
c) You play 40 times. What's the probability that you win at least $100?

54. New game You pay $10 and roll a die. If you get a 6, you win $50. If not, you get to roll again. If you get a 6 this time, you get your $10 back.
a) Create a probability model for this game.
b) Find the expected value and standard deviation of your prospective winnings.
c) You play this game five times. Find the expected value and standard deviation of your average winnings.
d) 100 people play this game. What's the probability the person running the game makes a profit?

55. AP Stats 2013 The College Board reported the score distribution shown in the table for all students who took the 2013 AP Statistics exam.

Score	Percent of Students
5	12.6
4	20.2
3	25.0
2	18.8
1	23.4

a) Find the mean and standard deviation of the scores.
b) If we select a random sample of 40 AP Statistics students, would you expect their scores to follow a Normal model? Explain.
c) Consider the mean scores of random samples of 40 AP Statistics students. Describe the sampling model for these means (shape, center, and spread).

56. Museum membership A museum offers several levels of membership, as shown in the table.

Member Category	Amount of Donation ($)	Percent of Members
Individual	50	41
Family	100	37
Sponsor	250	14
Patron	500	7
Benefactor	1000	1

a) Find the mean and standard deviation of the donations.
b) During their annual membership drive, they hope to sign up 50 new members each day. Would you expect the distribution of the donations for a day to follow a Normal model? Explain.
c) Consider the mean donation of the 50 new members each day. Describe the sampling model for these means (shape, center, and spread).

57. AP Stats 2013, again An AP Statistics teacher had 63 students preparing to take the AP exam discussed in Exercise 55. Though they were obviously not a random sample, he considered his students to be "typical" of all the national students. What's the probability that his students will achieve an average score of at least 3?

58. Joining the museum One of the phone volunteers for the museum in exercise 56 sets a personal goal of getting an average donation of at least $100 from the new members she enrolls during the membership drive. If she gets 80 new members and they can be considered a random sample of all the museum's members, what is the probability that she can achieve her goal?

59. Fuel economy The combined gas mileage of midsize cars varies with mean 24 miles per gallon (mpg) and a standard deviation of about 5.5 mpg. A particular rental car agency typically has 150 midsize cars in its lot. Let $\bar{y}$ represent the mean combined gas mileage for all cars in the lot.
a) What's the approximate model for the distribution of $\bar{y}$? Explain.
b) Estimate the probability that $\bar{y}$ is between 25 and 27 mpg.
c) There is only a 5% chance that the mean combined gas mileage is greater than what value?

60. Potato chips The weight of potato chips in a medium-size bag is stated to be 10 ounces. The amount that the packaging machine puts in these bags is believed to have a Normal model with mean 10.2 ounces and standard deviation 0.12 ounces.
a) What fraction of all bags sold are underweight?
b) Some of the chips are sold in "bargain packs" of 3 bags. What's the probability that none of the 3 is underweight?
c) What's the probability that the mean weight of the 3 bags is below the stated amount?
d) What's the probability that the mean weight of a 24-bag case of potato chips is below 10 ounces?

61. IQs Suppose that IQs of East State University's students can be described by a Normal model with mean 130 and standard deviation 8 points. Also suppose that IQs of students from West State University can be described by a Normal model with mean 120 and standard deviation 10.
a) We select a student at random from East State. Find the probability that this student's IQ is at least 125 points.
b) We select a student at random from each school. Find the probability that the East State student's IQ is at least 5 points higher than the West State student's IQ.
c) We select 3 West State students at random. Find the probability that this group's average IQ is at least 125 points.
d) We also select 3 East State students at random. What's the probability that their average IQ is at least 5 points higher than the average for the 3 West Staters?

62. Milk Although most of us buy milk by the quart or gallon, farmers measure daily production in pounds. Ayrshire cows average 55 pounds of milk a day, with a standard deviation of 6 pounds. For Jersey cows, the mean daily production is 52 pounds, with a standard deviation of 5 pounds. Assume that Normal models describe milk production for these breeds.

a) We select an Ayrshire at random. What's the probability that she averages more than 58 pounds of milk a day?

b) What's the probability that a randomly selected Ayrshire gives more milk than a randomly selected Jersey?

c) A farmer has 20 Jerseys. What's the probability that the average production for this small herd exceeds 54 pounds of milk a day, assuming they can be treated as a random sample of all Jersey cows?

d) A neighboring farmer has 10 Ayrshires. What's the probability that his herd average is at least 5 pounds higher than the average for part c's Jersey herd?

Just Checking ANSWERS

1. A Normal model (approximately).
2. At the actual proportion of all students who are in favor.
3. $SD(\hat{p}) = \sqrt{\dfrac{(0.5)(0.5)}{100}} = 0.05$
4. No, this is a histogram of individuals. It may or may not be Normal, but we can't tell from the information provided.
5. A Normal model (approximately).
6. 266 days
7. $\dfrac{10}{\sqrt{100}} = 1.0$ day

chapter 18 Confidence Intervals for Proportions

- 18.1 A Confidence Interval
- 18.2 Interpreting Confidence Intervals: What Does 95% Confidence Really Mean?
- 18.3 Margin of Error: Certainty vs. Precision
- 18.4 Assumptions and Conditions

Where are we going?
Recently, the Pew Research Center surveyed 1286 cell phone users.[1] They found that, overall, 22% admitted that "When I'm on my cell phone I'm not always truthful about exactly where I am." Pew reports that there is a "3% margin of error" for this result. Among younger respondents, those 18 to 29 years old, the percentage increased to 39%, and Pew warns that the margin of error for this group is larger, too. What does that mean? Are 18- to 29-year-olds less reliable (as well as, apparently, less honest)? What exactly is a margin of error, how do we find one, and how should we interpret them? We'll see what they are all about in this chapter.

Who	Facebook users age 18–22
What	Behavior on Facebook
When	2010
Where	United States
Why	Better understand the impact of social media

Social networking sites such as Facebook and Twitter have become fixtures in the social lives of many people around the world. As a result, there's been a growing interest in the impact that they have on our daily lives and relationships. Late in 2010, Pew Research surveyed U.S. residents to ask about their use of social networking sites.[2] Of the 156 respondents aged 18 to 22 who use Facebook, 48 said that they update their status at least daily.

The authors of the report discuss this finding as if these 156 respondents tell us about how the millions of Facebook users in this age bracket behave. Of course, that would be more interesting than knowing about 156 particular Facebook users, however carefully they may have been sampled. What can a study like this tell us about 18- to 22-year-old Facebook users in general?

We have a sample proportion, which we write as $\hat{p}$, of 48/156 or 30.8%. Our first guess might be that this observed proportion is close to the population proportion, p. But we also know that because of natural sampling variability, if the researchers had drawn a second sample of 156 young Facebook users at roughly the same time, the proportion who update daily from that sample probably wouldn't have been exactly 30.8%.

What *can* we say about the population proportion, p? To start to answer this question, think about how different the sample proportion might have been if we'd taken another random sample from the same population. But wait. Remember—we aren't actually going to take more samples. We just want to *imagine* how the sample proportions might vary from sample to sample. In other words, we want to know about the *sampling distribution* of the sample proportion who update daily.

---

[1] www.pewinternet.org/Reports/2006/Americans-and-their-cell-phones.aspx
[2] "Social networking sites and our lives," Hampton, Goulet, Rainie, and Purcell, June 16, 2011, www.pewinternet.org/Reports/2011/Technology-and-social-networks/Summary.aspx.

18.1 A Confidence Interval

*Activity: **Confidence Intervals and Sampling Distributions.** Simulate the sampling distribution, and see how it gives a confidence interval.*

Let's look at our model for the sampling distribution. We know it's approximately Normal (under certain assumptions, which we should be careful to check) and that its mean is the proportion of all young Facebook users who update their status daily. Is that proportion 30.8%? No, that's not the proportion from the entire population, that's just $\hat{p}$, our estimate. We don't know the proportion, p, of all young Facebook daily updaters; that's what we're trying to find out. We do know, though, that the sampling distribution model of $\hat{p}$ is centered at p, and we know that the standard deviation of the sampling distribution is $\sqrt{\dfrac{pq}{n}}$.

Now we have a problem. Since we don't know p, we can't find the true standard deviation of the sampling distribution model. We do know the observed proportion, $\hat{p}$, so, for want of something better, we just use what we know, and we estimate. That may not seem like a big deal, but it gets a special name. When we estimate the standard deviation of a sampling distribution, we call it a **standard error**.[3] For a sample proportion, $\hat{p}$, the standard error is

$$SE(\hat{p}) = \sqrt{\dfrac{\hat{p}\hat{q}}{n}}.$$

For the Facebook users, then:

$$SE(\hat{p}) = \sqrt{\dfrac{\hat{p}\hat{q}}{n}} = \sqrt{\dfrac{(0.308)(0.692)}{156}} = 0.037 = 3.7\%.$$

Now we know that the sampling model for $\hat{p}$ should look like this:

■ **NOTATION ALERT**

Remember that $\hat{p}$ is our sample-based estimate of the true proportion p. Recall also that q is just shorthand for $1 - p$, and $\hat{q} = 1 - \hat{p}$.

When we use $\hat{p}$ to estimate the standard deviation of the sampling distribution model, we call that the standard error and write

$$SE(\hat{p}) = \sqrt{\dfrac{\hat{p}\hat{q}}{n}}.$$

Figure 18.1
The sampling distribution model for $\hat{p}$ is Normal with a mean of p and a standard deviation we estimate to be 0.037.

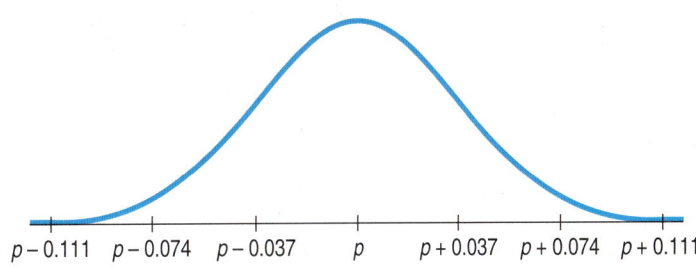

$p - 0.111 \quad p - 0.074 \quad p - 0.037 \quad p \quad p + 0.037 \quad p + 0.074 \quad p + 0.111$

What does that tell us? Well, because it's Normal, it says that about 68% of all samples of 156 young Facebook users will have $\hat{p}$'s within 1 SE, 0.037, of p. And about 95% of all these samples will be within $p \pm 2$ SEs. But where is *our* sample proportion in this picture? And what value does p have? We still don't know!

We do know that for 95% of random samples, the sample proportion $\hat{p}$ will be no more than 2 SEs away from the true population proportion p. So let's look at this from $\hat{p}$'s point of view. If I'm $\hat{p}$, there's a 95% chance that p is no more than 2 SEs away from me. If I reach out 2 SEs, or 2×0.037, away from me on both sides, I'm 95% sure that p will be within my grasp. Now I've got him! Probably. Of course, even if my interval does catch p, I still don't know its true value. The best I can do is an interval, and even then I can't be positive it contains p.

---

[3] Some statisticians (and books) refer to the standard deviation of the sampling distribution as the (theoretical) standard error, and refer to our estimate as the (estimated) standard error. But your statistics software (whatever software you are using) just calls the estimated version the "standard error." We think that's easier, so we will too.

Figure 18.2
Reaching out 2 SEs on either side of $\hat{p}$ makes us 95% confident that we'll trap the true proportion, p.

So what can we really say about p? Here's a list of things we'd like to be able to say, in order of strongest to weakest and the reasons we can't say most of them:

1. "30.8% of *all* Facebook users between the ages of 18 and 22 update their status daily." No. It would be nice to be able to make absolute statements about population values with certainty, but we just don't have enough information to do that. There's no way to be sure that the population proportion is the same as the sample proportion; in fact, it almost certainly isn't. Observations vary. Another sample would yield a different sample proportion.

2. "It is *probably* truly that 30.8% of all Facebook users between the ages of 18 and 22 update their status daily." No again. In fact, we can be pretty sure that whatever the true proportion is, it's not exactly 30.8%. So the statement is not true.

3. "We don't know exactly what proportion of Facebook users between the ages of 18 and 22 update their status daily, but we *know* that it's within the interval 30.8% ± 2 × 3.7%. That is, it's between 23.4% and 38.2%. This is getting closer, but it's still wrong. We just can't be certain. We can't know *for sure* that the true proportion is in this interval—or in any particular interval.

4. "We don't know exactly what proportion of young Facebook users update their status daily, but the interval from 23.4% to 38.2% *probably* contains the true proportion." Now we're getting closer. We've admitted that we're unsure about two things. First, we need an interval, not just a value, to try to capture the true proportion. Second, we aren't even certain that the true proportion is in that interval, but we're "pretty sure" that it is.

That last statement may be true, but we can tighten it up by quantifying what we mean by "probably." We saw that 95% of the time when we reach out 2 SEs from $\hat{p}$, we capture p, *so we can be 95% confident that this is one of those times*. After putting a number on the probability that this interval covers the true proportion, we've given our best guess of where the parameter is and how certain we are that it's within some range.

5. "We are 95% confident that between 23.4% and 38.2% of Facebook users between the ages of 18 and 22 update their status at least daily." Statements like these are called **confidence intervals**. They're the best we can do.

Each confidence interval discussed in the book has a name. You'll see many different kinds of confidence intervals in the following chapters. Some will be about more than *one* sample, some will be about statistics other than *proportions*, and some will use models other than the Normal. The interval calculated and interpreted here is sometimes called a **one-proportion z-interval**.[4]

A | S *Activity:* Can We Estimate a Parameter? Consider these four interpretations of a confidence interval by simulating to see whether they could be right.

> "Far better an approximate answer to the right question, . . . than an exact answer to the wrong question."
>
> —John W. Tukey

---

[4] In fact, this confidence interval is so standard for a single proportion that you may see it simply called a "confidence interval for the proportion."

Just Checking

The Pew Research study regarding cell phones asked other questions about cell phone experience. One growing concern is unsolicited advertising in the form of text messages. Pew asked cell phone owners, "Have you ever received unsolicited text messages on your cell phone from advertisers?" and 17% reported that they had. Pew estimates a 95% confidence interval to be 0.17 ± 0.04, or between 13% and 21%.

Are the following statements about people who have cell phones correct interpretations of that confidence interval? Explain.

1. In Pew's sample, somewhere between 13% and 21% of respondents reported that they had received unsolicited advertising text messages.

2. We can be 95% confident that 17% of U.S. cell phone owners have received unsolicited advertising text messages.

3. We are 95% confident that between 13% and 21% of all U.S. cell phone owners have received unsolicited advertising text messages.

4. We know that between 13% and 21% of all U.S. cell phone owners have received unsolicited advertising text messages.

5. Ninety-five percent of all U.S. cell phone owners have received unsolicited advertising text messages.

18.2 Interpreting Confidence Intervals: What Does 95% Confidence Really Mean?

A S *Activity:* **Confidence Intervals for Proportions.** This new interactive tool makes it easy to construct and experiment with confidence intervals. We'll use this tool for the rest of the course—sure beats calculating by hand!

What do we mean when we say we have 95% confidence that our interval contains the true proportion? Formally, what we mean is that "95% of samples of this size will produce confidence intervals that capture the true proportion." This is correct, but a little long winded, so we sometimes say, "we are 95% confident that the true proportion lies in our interval." Our uncertainty is about whether the particular sample we have at hand is one of the successful ones or one of the 5% that fail to produce an interval that captures the true value.

We know that proportions vary from sample to sample. If other researchers select their own samples of Facebook users who update status daily each person's sample proportion will almost certainly differ. When they each try to estimate the true rate of status updating in the entire population, they'll center *their* confidence intervals at the proportions they observed in their own samples. Each of us will end up with a different interval.

Our interval guessed the true proportion of daily updaters to be between about 23.4% and 38.2%. Another survey might find 52 of 156 daily updaters and get a confidence interval from 25.8% to 40.9%. Still another might find only 43 of 156 who update daily for a confidence interval from 20.4% to 34.7%. And so on. We don't expect all of these intervals to contain the true proportion of young Facebook daily updaters. Some will be duds, missing the population proportion entirely. But we expect, on average, that 95% of them will be winners and capture the true proportion.

On the next page, you'll see confidence intervals produced by simulating 20 different random samples. The red dots are the proportions of daily updaters in each sample, and the blue segments show the confidence intervals found for each. The green line represents the true rate of daily updating in the population, so you can see that most of the intervals caught it—but a few missed. (And notice again that it is the *intervals* that vary from sample to sample; the green line doesn't move.)

The horizontal green line shows the true percentage of all young Facebook users who update status daily. Most of the 20 simulated samples produced confidence intervals that captured the true value, but a few missed.

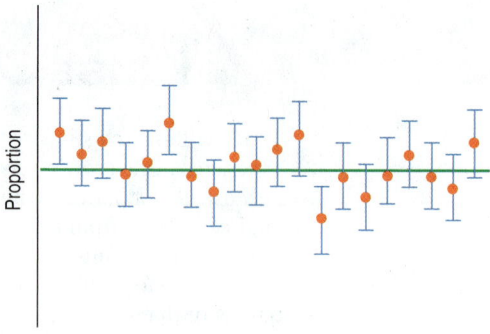

Of course, there's a huge number of possible samples that *could* be drawn, each with its own sample proportion. These are just some of them. Each sample proportion can be used to make a confidence interval. That's a large pile of possible confidence intervals, and ours is just one of those in the pile. Did *our* confidence interval "work"? We can never be sure, because we'll never know the true proportion.[5] However, the Central Limit Theorem assures us that (in the long run) 95% of the intervals are winners, covering the true value, and only 5% are duds. *That's* why we're 95% confident that our interval is a winner!

> **For Example** POLLS AND MARGIN OF ERROR
>
> In April and May 2011, the Yale Project on Climate Change Communication and the George Mason University Center for Climate Change Communication interviewed 1010 U.S. adults about American's global warming beliefs and attitudes.[6]
>
> **QUESTION:** It is standard among pollsters to use a 95% confidence level unless otherwise stated. Given that, what do these researchers mean by their confidence interval in this context?
>
> **ANSWER:** If this polling were done repeatedly, 95% of all random samples would yield confidence intervals that contain the true proportion of all U.S. adults who believe that there's a lot of disagreement among scientists about global warming.

So, What *Can* I Say?

Confidence intervals are based on random samples, so the interval is random, too. The Central Limit Theorem tells us that 95% of the random samples will yield intervals that capture the true value. That's what we mean by being 95% confident.

Technically, we should say, "I am 95% confident that the interval from 23.4% and 38.2% captures the true proportion of U.S. Facebook users between the ages of 18 and 22 who update their profiles daily." That formal phrasing emphasizes that *our confidence (and our uncertainty) is about the interval, not the true proportion*. But you may choose a more casual phrasing like "I am 95% confident that between 23.4% and 38.2% of U.S. Facebook users between the ages of 18 and 22 update their profiles daily." Because you've

---

[5] Actually, Facebook could probably find the true value for a particular short time interval, but even that value would fluctuate from one time to another. And they aren't likely to tell us, so we'll have to estimate it like this.

[6] Among their questions, they asked what respondents thought were the views of scientists. Among respondents, 40% agreed with the alternative "There is a lot of disagreement among scientists about whether or not global warming is happening." The investigators provide a confidence interval from 37% to 43%.

made it clear that the uncertainty is yours and you didn't suggest that the randomness is in the true proportion, this is OK. Keep in mind that it's the interval that's random and is the focus of both our confidence and doubt.

18.3 Margin of Error: Certainty vs. Precision

We've just claimed that with a certain confidence we've captured the true proportion of all young Facebook users. Our confidence interval had the form

$$\hat{p} \pm 2\, SE(\hat{p}).$$

The extent of the interval on either side of $\hat{p}$ is called the **margin of error** (*ME*). We'll want to use the same approach for many other situations besides estimating proportions. In fact, almost any population parameter—a proportion, a mean, or a regression slope, for example—can be estimated with some margin of error. The margin of error is a way to describe our uncertainty in estimating the population value. We'll see how to find a margin of error for each of these values and for others.

For all of those statistics, regardless of how we calculate the margin of error, we'll be able to construct a confidence interval that looks like this:

$$Estimate \pm ME.$$

The margin of error for our 95% confidence interval was 2 *SE*. What if we wanted to be more confident? To be more confident, we'd need to capture *p* more often, and to do that we'll need to make the interval wider. For example, if we want to be 99.7% confident, the margin of error will have to be 3 *SE*.

Figure 18.3

Reaching out 3 *SE*s on either side of $\hat{p}$ makes us 99.7% confident we'll trap the true proportion *p*. Compare with Figure 18.2.

The more confident we want to be, the larger the margin of error must be. We can be 100% confident that the proportion of Facebook users who update daily is between 0% and 100%, but this isn't very useful. On the other hand, we could give a confidence interval from 30.7% to 30.9%, but we can't be very confident about a precise statement like this. Every confidence interval is a balance between certainty and precision.

GARFIELD © 1999 Paws, Inc. Distributed by Universal Uclick. Reprinted with permission. All rights reserved.

Activity: Balancing Precision and Certainty. What percent of parents expect their kids to pay for college with a student loan? Investigate the balance between the precision and the certainty of a confidence interval.

The tension between certainty and precision is always there. Fortunately, in most cases we can be both sufficiently certain and sufficiently precise to make useful statements. There is no simple answer to the conflict. You must choose a confidence level yourself. The data can't do it for you. The choice of confidence level is somewhat arbitrary. The most commonly chosen confidence levels are 90%, 95%, and 99%, but, in theory, any percentage can be used. (In practice, though, using something like 92.9% or 97.2% is likely to make people think you're up to something.)

For Example FINDING THE MARGIN OF ERROR (TAKE 1)

RECAP: An April 2011 Yale/George Mason poll of 1010 U.S. adults asking questions about current topics reported a margin of error of 3% It is a convention among pollsters to use a 95% confidence level and to report the "worst case" margin of error, based on $p = 0.5$.

QUESTION: How did the researchers calculate their margin of error?

ANSWER: Assuming $p = 0.5$, for random samples of $n = 1010$,

$$SD(\hat{p}) = \sqrt{\frac{pq}{n}} = \sqrt{\frac{(0.5)(0.5)}{1010}} = 0.0157.$$

For a 95% confidence level, $ME = 2(0.0157) = 0.031$, so their margin of error is just a bit over 3%.

Critical Values

■ **NOTATION ALERT**

We'll put an asterisk on a letter to indicate a critical value, so z^* is always a critical value from a Normal model.

In our Facebook example, we used $2SE$ to give us a 95% confidence interval. To change the confidence level, we'd need to change the *number* of *SE*s so that the size of the margin of error corresponds to the new level. This number of *SE*s is called the **critical value**. Here it's based on the Normal model, so we denote it z^*. For any confidence level, we can find the corresponding critical value from a computer, a calculator, or a Normal probability table, such as Table Z.

For a 95% confidence interval, you'll find the precise critical value is $z^* = 1.96$. That is, 95% of a Normal model is found within ± 1.96 standard deviations of the mean. We've been using $z^* = 2$ from the 68–95–99.7 Rule because it's easy to remember.

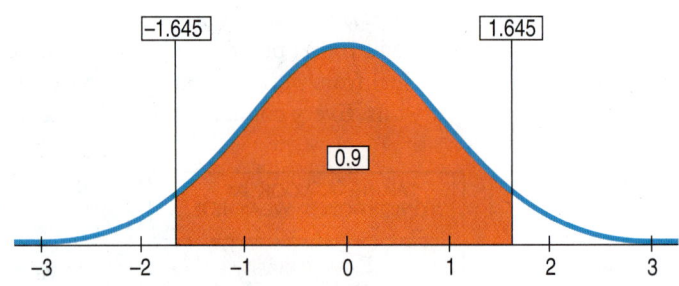

Figure 18.4
For a 90% confidence interval, the critical value is 1.645, because, for a Normal model, 90% of the values are within 1.645 standard deviations from the mean.

For Example FINDING THE MARGIN OF ERROR (TAKE 2)

RECAP: In April 2011, a Yale/George Mason poll of 1010 U.S. adults found that 40% of the respondents believed that scientists disagreed about whether global warming exists. They reported a 95% confidence interval with a margin of error of 3%.

> **QUESTIONS:** Using the critical value of *z* and the standard error based on the observed proportion, what would be the margin of error for a 90% confidence interval? What's good and bad about this change?
>
> With $n = 1010$ and $\hat{p} = 0.40$, $SE(\hat{p}) = \sqrt{\dfrac{\hat{p}\hat{q}}{n}} = \sqrt{\dfrac{(0.40)(0.60)}{1010}} = 0.0154$
>
> For a 90% confidence level, $z^* = 1.645$, so $ME = 1.645(0.0154) = 0.025$.
>
> **ANSWER:** Now the margin of error is only about 2%, producing a narrower interval. What's good about the change is that we now have a smaller interval, but what's bad is that we are less certain that the interval actually contains the true proportion of adults who think that scientists disagree about global warming.

✓ Just Checking

Think some more about the 95% confidence interval originally created for the proportion of U.S. adults who believe that scientists disagree about whether global warming exists.

6. If the researchers wanted to be 98% confident, would their confidence interval need to be wider or narrower?

7. The study's margin of error was about 3%. If the researchers wanted to reduce it to 2%, would their level of confidence be higher or lower?

8. If the researchers had polled more people, would the interval's margin of error have been larger or smaller?

18.4 Assumptions and Conditions

We've just made some pretty sweeping statements about Facebook users. Those statements were possible because we used a Normal model for the sampling distribution. But is that model appropriate?

We've also said that the same basic ideas will work for other statistics. One difference for those statistics is that some will have sampling distribution models that are different from the Normal. But the background theory (which we won't bother with) is so similar for all of those models that they share the same basic assumptions and conditions about independence and sample size. Then for each statistic, we usually tack on a special-case assumption or two. We'll deal with those as we get to them in later chapters.

We saw the assumptions and conditions for using the Normal to model the sampling distribution for a proportion in the last chapter. Because they are so crucial to making sure our confidence interval is useful, we'll repeat them here.

Independence Assumption

Activity: **Assumptions and Conditions.** Here's an animated review of the assumptions and conditions.

Independence Assumption: If the data were sampled at random or generated from a properly randomized experiment, they satisfy the **Randomization Condition**. Otherwise, we need to think about whether the independence assumption is plausible. We often look for reasons to suspect that it fails. We wonder whether there is any reason to believe that the data values somehow affect each other. (For example, might your Facebook friends, seeing your frequent status updates, be encouraged to update their own status more often?) Whether you decide that the **Independence Assumption** is plausible depends on your knowledge of the situation. It's not one you can check by looking at the data.

However, now that we have data, there are two conditions that we can check:

Randomization Condition: Were the data sampled at random or generated from a properly randomized experiment? Proper randomization can help ensure independence.

10% Condition: If you sample more than 10% of a population, the formula for the standard error won't be quite right. There is a special formula (found in advanced books) that corrects for this, but it isn't a common problem unless your population is small.

Sample Size Assumption

The model we use for inference for proportions is based on the Central Limit Theorem. We need to know whether the sample is large enough to make the sampling model for the sample proportions approximately Normal. It turns out that we need more data as the proportion gets closer and closer to either extreme (0 or 1). That's why we check the:

Success/Failure Condition: We must expect at least 10 "successes" and at least 10 "failures." Recall that by tradition we arbitrarily label one alternative (usually the outcome being counted) as a "success" even if it's something bad (like getting a disease). The other alternative is then, of course, a "failure."

A S Activity: A Confidence Interval for p. View the video story of pollution in Chesapeake Bay, and make a confidence interval for the analysis with the interactive tool.

> **One-proportion z-interval** When the conditions are met, we are ready to find a level C confidence interval for the population proportion, p. The confidence interval is $\hat{p} \pm z^* \times SE(\hat{p})$ where the standard deviation of the proportion is estimated by
> $$SE(\hat{p}) = \sqrt{\frac{\hat{p}\hat{q}}{n}}$$
> and the critical value, z^*, specifies the number of SEs needed for C% of random samples to yield confidence intervals that capture the true parameter value.

Step-by-Step Example A CONFIDENCE INTERVAL FOR A PROPORTION

In October 2010, the Gallup Poll[7] asked 510 randomly sampled adults the question "Generally speaking, do you believe the death penalty is applied fairly or unfairly in this country today?" Of these, 58% answered "Fairly," 36% said "Unfairly," and 7% said they didn't know (percentages add up to 101% due to rounding).

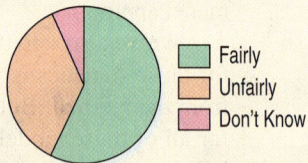

Who	Adults in the United States
What	Response to a question about the death penalty
When	October 2010
Where	United States
How	510 adults were randomly sampled and asked by the Gallup Poll
Why	Public opinion research

Question: From this survey, what can we conclude about the opinions of *all* adults?

To answer this question, we'll build a confidence interval for the proportion of all U.S. adults who believe the death penalty is applied fairly. There are four steps to building a confidence interval for proportions: Plan, Model, Mechanics, and Conclusion.

[7] www.gallup.com/poll/1606/death-penalty.aspx

THINK ➡ **Plan** State the problem and the W's. Identify the *parameter* you wish to estimate. Identify the *population* about which you wish to make statements. Choose and state a confidence level. **Model** Think about the assumptions and check the conditions. State the sampling distribution model for the statistic. Choose your method.	I want to find an interval that is likely, with 95% confidence, to contain the true proportion, p, of U.S. adults who think the death penalty is applied fairly. I have a random sample of 510 U.S. adults. ✓ **Randomization Condition:** Gallup drew a random sample from all U.S. adults. I can be confident that the respondents are independent. ✓ **10% Condition:** The sample is certainly less than 10% of the population. ✓ **Success/Failure Condition:** $n\hat{p} = 510(58\%) = 296 \geq 10$ and $n\hat{q} = 510(42\%) = 214 \geq 10$, so the sample appears to be large enough to use the Normal model. The conditions are satisfied, so I can use a Normal model to find a **one-proportion z-interval**.
SHOW ➡ **Mechanics** Construct the confidence interval. First find the standard error. (Remember: It's called the "standard error" because we don't know p and have to use $\hat{p}$ instead.) Next find the margin of error. We could informally use 2 for our critical value, but 1.96 (found from a table or technology) is more accurate. Write the confidence interval (CI). **REALITY CHECK** The CI is centered at the sample proportion and about as wide as we might expect for a sample of 500.	$n = 510, \hat{p} = 0.58$, so $SE(\hat{p}) = \sqrt{\dfrac{\hat{p}\hat{q}}{n}} = \sqrt{\dfrac{(0.58)(0.42)}{510}} = 0.022.$ Because the sampling model is Normal, for a 95% confidence interval, the critical value $z^* = 1.96$. The margin of error is $ME = z^* \times SE(\hat{p}) = 1.96(0.022) = 0.043.$ So the 95% confidence interval is 0.58 ± 0.043 or $(0.537, 0.623).$
TELL ➡ **Conclusion** Interpret the confidence interval in the proper context. We're 95% confident that our interval captured the true proportion.	I am 95% confident that between 53.7% and 62.3% of all U.S. adults think that the death penalty is applied fairly.

Choosing Your Sample Size

The question of how large a sample to take is an important step in planning any study. We weren't ready to make that calculation when we first looked at study design in Chapter 11, but now we can—and we always should.

Suppose a candidate is planning a poll and wants to estimate voter support within 3% with 95% confidence. How large a sample does she need?

Let's look at the margin of error:

$$ME = z^*\sqrt{\frac{\hat{p}\hat{q}}{n}}$$

$$0.03 = 1.96\sqrt{\frac{\hat{p}\hat{q}}{n}}.$$

We want to find n, the sample size. To find n we need a value for $\hat{p}$. We don't know $\hat{p}$ because we don't have a sample yet, but we can probably guess a value. To be safe, we can take the value that makes $\hat{p}\hat{q}$ (and therefore n) largest. That value is $\hat{p} = \hat{q} = 0.50$, so if we use that value for $\hat{p}$, we'll certainly be safe. Our candidate probably expects to be near 50% anyway.

Our equation, then, is

$$0.03 = 1.96\sqrt{\frac{(0.5)(0.5)}{n}}.$$

To solve for n, we first multiply both sides of the equation by $\sqrt{n}$ and then divide by 0.03:

$$0.03\sqrt{n} = 1.96\sqrt{(0.5)(0.5)}$$

$$\sqrt{n} = \frac{1.96\sqrt{(0.5)(0.5)}}{0.03} \approx 32.67$$

Notice that evaluating this expression tells us the *square root* of the sample size. We need to square that result to find n:

$$n \approx (32.67)^2 \approx 1067.1$$

When computing sample sizes, we always round up (to ensure the confidence level we wanted) and conclude that we need at least 1068 respondents to keep the margin of error as small as 3% with a confidence level of 95%.

> **WHAT DO I USE INSTEAD OF $\hat{p}$?**
>
> Often we have an estimate of the population proportion based on experience or perhaps a previous study. If so, use that value as $\hat{p}$ in calculating what size sample you need. If not, the cautious approach is to use $\hat{p} = 0.5$ in the sample size calculation; that will determine the largest sample necessary regardless of the true proportion.

For Example CHOOSING A SAMPLE SIZE

RECAP: The Yale/George Mason poll that estimated that 40% of all voters believed that scientists disagree about whether global warming exists had a margin of error of ±3%. Suppose an environmental group planning a follow-up survey of voters' opinions on global warming wants to determine a 95% confidence interval with a margin of error of no more than ±2%.

QUESTION: How large a sample do they need? (You could take $p = 0.5$, but we have data that indicates $p = 0.40$, so we can use that.)

ANSWER:
$$ME = z^*\sqrt{\frac{\hat{p}\hat{q}}{n}}$$

$$0.02 = 1.96\sqrt{\frac{(0.40)(0.60)}{n}}$$

$$\sqrt{n} = \frac{1.96\sqrt{(0.40)(0.60)}}{0.02} \approx 48.01$$

$$n = 48.01^2 = 2304.96$$

The environmental group's survey will need at least 2305 respondents.

Unfortunately, bigger samples cost more money and more effort. Because the standard error declines only with the *square root* of the sample size, to cut the standard error (and thus the *ME*) in half, we must *quadruple* the sample size.

> **How big should a margin of error be?**
> Public opinion polls often sample 1000 people, which gives an ME of 3% when $\hat{p} = 0.5$. But businesses and nonprofit organizations typically use much larger samples to estimate the proportion who will accept a direct mail offer. Why? Because that proportion is very low—often far below 5%. An ME of 3% wouldn't be precise enough. An ME like 0.1% would be more useful, and that requires a very large sample size.

Generally, a margin of error of 5% or less is acceptable, but different circumstances call for different standards. For a pilot study, a margin of error of 10% may be fine, so a sample of 100 will do quite well. In a close election, a polling organization might want to get the margin of error down to 2%. Drawing a large sample to get a smaller ME, however, can run into trouble. It takes time to survey 2400 people, and a survey that extends over a week or more may be trying to hit a target that moves during the time of the survey. An important event can change public opinion in the middle of the survey process. Professional polling organizations often report both an instantaneous estimate and a rolling average of several recent polls in order to increase the sample size.

The sample size for a survey is the number of respondents, not the number of people to whom questionnaires were sent or whose phone numbers were dialed. And keep in mind that a low response rate turns any study essentially into a voluntary response study, which is of little value for inferring population values. It's almost always better to spend resources on increasing the response rate than on surveying a larger group. A full or nearly full response by a modest-size sample can yield more useful results.

Surveys are not the only place where proportions pop up. Banks sample huge mailing lists to estimate what proportion of people will accept a credit card offer. Even pilot studies may mail offers to over 50,000 customers. Most don't respond; but, in this case, that doesn't make the sample smaller—they simply said "No thanks." Those who do respond want the card. To the bank, the response rate[8] is $\hat{p}$. With a typical success rate around 0.5%, the bank needs a very small margin of error—often as low as 0.1%—to make a sound business decision. That calls for a large sample, and the bank must take care in estimating the size needed. For our election poll calculation, we used $\hat{p} = 0.5$, both because it's safe and because we honestly believed the true p to be near 0.5. If the bank used that value, they'd get an absurd answer. Instead, they base their calculation on a proportion closer to the one they expect to find.

For Example SAMPLE SIZE REVISITED

A credit card company is about to send out a mailing to test the market for a new credit card. From that sample, they want to estimate the true proportion of people who will sign up for the card nationwide. A pilot study suggests that about 0.5% of the people receiving the offer will accept it.

QUESTION: To be within a tenth of a percentage point (0.001) of the true rate with 95% confidence, how big does the test mailing have to be?

ANSWER: Using the estimate $\hat{p} = 0.5\%$:

$$ME = 0.001 = z^* \sqrt{\frac{\hat{p}\hat{q}}{n}} = 1.96 \sqrt{\frac{(0.005)(0.995)}{n}}$$

$$(0.001)^2 = 1.96^2 \frac{(0.005)(0.995)}{n} \Rightarrow n = \frac{1.96^2(0.005)(0.995)}{(0.001)^2}$$

$$= 19{,}111.96 \text{ or } 19{,}112$$

That's a lot, but it's actually a reasonable size for a trial mailing such as this. Note, however, that if they had assumed 0.50 for the value of p, they would have found

$$ME = 0.001 = z^* \sqrt{\frac{pq}{n}} = 1.96 \sqrt{\frac{(0.5)(0.5)}{n}}$$

$$(0.001)^2 = 1.96^2 \frac{(0.5)(0.5)}{n} \Rightarrow n = \frac{1.96^2(0.5)(0.5)}{(0.001)^2} = 960{,}400.$$

Quite a different (and unreasonable) result.

---

[8] In marketing studies, every mailing yields a response—"yes" or "no"—and "response rate" means the proportion of customers who accept an offer. That's not the way we use the term for survey response.

WHAT CAN GO WRONG?

Confidence intervals are powerful tools. Not only do they tell what we know about the parameter value, but—more important—they also tell what we *don't* know. To use confidence intervals effectively, you must be clear about what you say about them.

Don't Misstate What the Interval Means

- **Don't claim that other samples will agree with yours.** Keep in mind that the confidence interval makes a statement about the true population proportion. An interpretation such as "In 95% of samples of U.S. adults, the proportion who think marijuana should be decriminalized will be between 42.7% and 51.3%" is just wrong. The interval isn't about sample proportions but about the population proportion.

- **Don't be certain about the parameter.** Saying "Between 23.4% and 38.2% of Facebook users between the ages of 18 and 22 update their profiles daily" asserts that the population proportion cannot be outside that interval. Of course, we can't be absolutely certain of that. (Just reasonably sure.)

- **Don't forget: It's about the parameter.** Don't say, "I'm 95% confident that $\hat{p}$ is between 23.4% and 38.2%." Of course you are—in fact, we calculated that $\hat{p} = 30.8\%$ so we know the sample proportion is in the interval! The confidence interval is about the (unknown) population parameter, p.

- **Don't claim to know too much.** Don't say, "I'm 95% confident that between 23.4% and 38.2% of all Facebook users in the world update their profiles daily." This survey was about U.S. users between the ages of 18 and 22.

- **Do take responsibility.** Confidence intervals are about *un*certainty. *You* are the one who is uncertain, not the parameter. You have to accept the responsibility and consequences of the fact that not all the intervals you compute will capture the true value. In fact, about 5% of the 95% confidence intervals you find will fail to capture the true value of the parameter. You *can* say, "I am 95% confident that between 23.4% and 38.2% of U.S. Facebook users between 18 and 22 update their profiles daily."[9]

- **Don't suggest that the parameter varies.** A statement like "There is a 95% chance that the true proportion is between 42.7% and 51.3%" sounds as though you think the population proportion wanders around and sometimes happens to fall between 42.7% and 51.3%. When you interpret a confidence interval, make it clear that *you* know that the population parameter is fixed and that it is the interval that varies from sample to sample.

- **Do treat the whole interval equally.** Although a confidence interval is a set of plausible values for the parameter, don't think that the values in the middle of a confidence interval are somehow "more plausible" than the values near the edges. Your interval provides no information about where in your current interval (if at all) the parameter value is most likely to be hiding.

- **Beware of margins of error that are too large to be useful.** A confidence interval that says that the percentage of daily updaters is between 10% and 90% wouldn't be of much use. Most likely, you have some sense of how large a margin of error you can tolerate. What can you do?

  One way to make the margin of error smaller is to reduce your level of confidence. But that may not be a useful solution. It's a rare study that reports confidence levels lower than 80%. Levels of 95% or 99% are more common.

---

[9] When we are being very careful we say, "95% of samples of this size will produce confidence intervals that capture the true proportion of U.S. Facebook users between the ages of 18 and 22 who update their profiles daily."

The time to think about whether your margin of error is small enough to be useful is when you design your study. Don't wait until you compute your confidence interval. To get a narrower interval without giving up confidence, you need to have less variability in your sample proportion. How can you do that? Choose a larger sample.

Confidence intervals and margins of error are often reported along with poll results and other analyses. But it's easy to misuse them and wise to be aware of the ways things can go wrong.

- **Watch out for biased sampling.** Don't forget about the potential sources of bias in surveys that we discussed in Chapter 11. Just because we have more statistical machinery now doesn't mean we can forget what we've already learned. A questionnaire that finds that most people enjoy filling out surveys suffers from nonresponse bias if only 10% of the surveys are returned. Even though now we're able to put confidence intervals around this (biased) estimate, the nonresponse has made it useless.

- **Think about independence.** The assumption that the values in our sample are mutually independent is one that we usually cannot check. It always pays to think about it, though. For example, if we sampled Facebook users and all of the friends in their network, we might get a biased estimate because friends in the same network may behave similarly. That would violate the Independence Assumption and could severely affect our sample proportion. It could be that the proportion of daily updaters is actually quite small, but the researchers just happened to find a network of friends who spend all their time on Facebook. To avoid this, the researchers should be careful in their sample design to ensure randomization.

CONNECTIONS

Now we can see a practical application of sampling distributions. To find a confidence interval, we lay out an interval measured in standard deviations. We're using the standard deviation as a ruler again. But now the standard deviation we need is the standard deviation of the sampling distribution. That's the one that tells how much the proportion varies. (And when we estimate it from the data, we call it a standard error.)

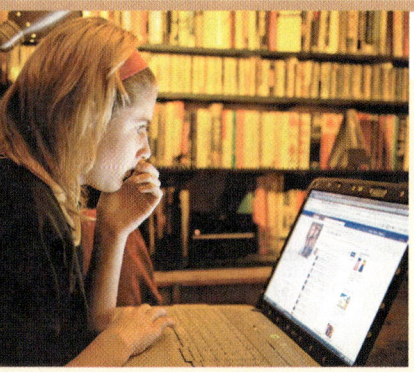

What Have We Learned?

Learning Objectives

Construct a confidence interval for a proportion, p, as the statistic, $\hat{p}$, plus and minus a **margin of error**.

- The margin of error consists of a **critical value** based on the sampling model times a **standard error** based on the sample.
- The critical value is found from the Normal model.
- The standard error of a sample proportion is calculated as $\sqrt{\dfrac{\hat{p}\hat{q}}{n}}$.

Interpret a confidence interval correctly.

- You can claim to have the specified level of confidence that the interval you have computed actually covers the true value.

Understand the relationship of the sample size, n, to both the certainty (confidence level) and precision (margin of error).

- For the same sample size and true population proportion, more certainty means less precision (wider interval) and more precision (narrower interval) implies less certainty.

Know and check the assumptions and conditions for finding and interpreting confidence intervals.
- Independence Assumption or Randomization Condition
- 10% Condition
- Success/Failure Condition

Be able to invert the calculation of the margin of error to find the sample size required, given a proportion, a confidence level, and a desired margin of error.

Review of Terms

Standard error When we estimate the standard deviation of a sampling distribution using statistics found from the data, the estimate is called a standard error (p. 473).
$$SE(\hat{p}) = \sqrt{\frac{\hat{p}\hat{q}}{n}}$$

Confidence interval A level C confidence interval for a model parameter is an interval of values usually of the form
$$\text{estimate} \pm \text{margin of error}$$
found from data in such a way that C% of all random samples will yield intervals that capture the true parameter value (p. 474).

One-proportion z-interval A confidence interval for the true value of a proportion. The confidence interval is
$$\hat{p} \pm z^* SE(\hat{p}),$$
where z^* is a critical value from the Standard Normal model corresponding to the specified confidence level (p. 474).

Margin of error In a confidence interval, the extent of the interval on either side of the observed statistic value is called the margin of error. A margin of error is typically the product of a critical value from the sampling distribution and a standard error from the data. A small margin of error corresponds to a confidence interval that pins down the parameter precisely. A large margin of error corresponds to a confidence interval that gives relatively little information about the estimated parameter. For a proportion (p. 477):
$$ME = z^* \sqrt{\frac{\hat{p}\hat{q}}{n}}$$

Critical value The number of standard errors to move away from the mean of the sampling distribution to correspond to the specified level of confidence. The critical value, for a normal sampling distribution, denoted z^*, is usually found from a table or with technology (p. 478).

On the Computer CONFIDENCE INTERVALS FOR PROPORTIONS

Confidence intervals for proportions are so easy and natural that many statistics packages don't offer special commands for them. Most statistics programs want the "raw data" for computations. For proportions, the raw data are the "success" and "failure" status for each case. Usually, these are given as 1 or 0, but they might be category names like "yes" and "no." Often, we just know the proportion of successes, $\hat{p}$, and the total count, n. Computer packages don't usually deal with summary data like this easily, but the statistics routines found on many graphing calculators allow you to create confidence intervals from summaries of the data—usually all you need to enter are the number of successes and the sample size.

In some programs, you can reconstruct variables of 0's and 1's with the given proportions. But even when you have (or can reconstruct) the raw data values, you may not get *exactly* the same margin of error from a computer package as you would find working by hand. The reason is that some packages make approximations or use other methods. The result is very close but not exactly the same. Fortunately, Statistics means never having to say you're certain, so the approximate result is good enough.

DATA DESK

Data Desk does not offer built-in methods for inference with proportions.

COMMENTS

For summarized data, open a Scratchpad to compute the standard deviation and margin of error by typing the calculation. Then use *z*-interval for individual μs.

EXCEL

Inference methods for proportions are not part of the standard Excel tool set.

COMMENTS

For summarized data, type the calculation into any cell and evaluate it.

JMP

For a **categorical** variable that holds category labels, the **Distribution** platform includes tests and intervals for proportions. For summarized data, put the category names in one variable and the frequencies in an adjacent variable. Designate the frequency column to have the **role** of **frequency**. Then use the **Distribution** platform.

COMMENTS

JMP uses slightly different methods for proportion inferences than those discussed in this text. Your answers are likely to be slightly different, especially for small samples.

MINITAB

- Choose **Basic Statistics** from the **Stat** menu.
- Choose **1Proportion** from the Basic Statistics submenu.
- If the data are category names in a variable, assign the variable from the variable list box to the **Samples in columns** box. If you have summarized data, click the **Summarized Data** button and fill in the number of trials and the number of successes.
- Click the **Options** button and specify the remaining details.
- If you have a large sample, check **Use test and interval based on normal distribution**.
  Click the **OK** button.

COMMENTS

When working from a variable that names categories, MINITAB treats the last category as the "success" category. You can specify how the categories should be ordered.

R

The standard libraries in R do not contain a function for the confidence of a proportion, but a simple function can be written to do so. For example:

```
pconfint=function(phat,n,conf=.95)
  {
    se=sqrt(phat*(1-phat)/n)
    al2=1-(1-conf)/2
    zstar=qnorm(al2)
    ul=phat+zstar*se
    ll=phat-zstar*se
    return(c(ll,ul))
  }
```

For example, pconfint(0.3,500) will give a 95% confidence interval for *p* based on 150 successes out of 500.

SPSS

SPSS does not find confidence intervals for proportions.

STATCRUNCH

To create a confidence interval for a proportion using summaries:
- Click on **Stat**.
- Choose **Proportions » One sample » with summary**.
- Enter the **Number of successes** (x) and **Number of observations** (n).
- Click on **Next**.
- Indicate **Confidence Interval** (Standard-Wald), and then enter the **Level** of confidence.
- Click on **Calculate**.

To create a confidence interval for a proportion using data:
- Click on **Stat**.
- Choose **Proportions » One sample » with data**.
- Choose the variable **Column** listing the **Outcomes**.
- Enter the outcome to be considered a **Success**.
- Click on **Next**.
- Indicate **Confidence Interval**, and then enter the **Level** of confidence.
- Click on **Calculate**.

TI-83/84 PLUS

To calculate a confidence interval for a population proportion:
- Go to the **STATS TESTS** menu and select **A:1-PropZInt**.
- Enter the number of successes observed and the sample size.
- Specify a confidence level.
- Calculate the interval.

COMMENTS

Beware: When you enter the value of *x*, you need the count, not the percentage. The count must be a whole number. If the number of successes are given as a percentage, you must first multiply *np* and round the result.

Exercises

Section 18.1

1. **Lying about age** Pew Research, in November 2011, polled a random sample of 799 U.S. teens about Internet use. 49% of those teens reported that they had misrepresented their age online to gain access to websites and online services.
   a) Explain the meaning of $\hat{p} = 0.49$ in the context of this situation.
   b) Calculate the standard error of $\hat{p}$.
   c) Explain what this standard error means in the context of this situation.

2. **How's life?** Gallup regularly conducts a poll asking people to imagine a ladder with 10 rungs. Rung 0 represents the worst possible life and rung 10 represents the best possible life. Respondents are asked what rung they would say they are on. Responses are classified as "Thriving" (standing on rung 7 or higher, and expecting to be on rung 8 or higher five years from now), "Suffering" (standing on rung 4 or lower and expecting to be on rung 4 or lower 5 years from now), or "Struggling" (not thriving or suffering). The poll for November 22–25, 2011, found 4% of respondents to be suffering. The polls all use a sample size of approximately 1500 people. (www.gallup.com/poll/152033/Fewer-Americans-Thriving-2011-2010.aspx)
   a) Explain the meaning of "$\hat{p} = 0.04$" in the context of this situation.
   b) Calculate the standard error of $\hat{p}$.
   c) Explain what this standard error means in the context of this situation.

Section 18.2

3. **Lying about age again** The 95% confidence interval for the number of teens in Exercise 1 who reported that

they had misrepresented their age online is from 45.6% to 52.5%.

a) Interpret the interval in this context.
b) Explain the meaning of "95% confident" in this context.

4. **Single moms** In a 2010 Pew Research study on trends in marriage and family, 4% of randomly selected respondents said that the trend of more single women having children is "a good thing." The 95% confidence interval is from 2.9% to 5.1% ($n = 1229$).

a) Interpret this interval in this context.
b) Explain the meaning of "95% confident" in this context.

Section 18.3

5. **Wrong direction** St. Norbert's College in Green Bay, Wisconsin, and Wisconsin Public Radio conduct an annual poll of Wisconsinites about political opinions. The Fall 2011 survey asked a random sample of 402 adult Wisconsin residents whether they think things in the country are going in the right direction or in the wrong direction. 66% said that things were going in the wrong direction.

a) Calculate the margin of error for the proportion of all adult Wisconsin residents who think things are going in the wrong direction for 90% confidence.
b) Would the margin of error be larger or smaller for 95% confidence? Explain.

6. **I asked, "How's life?"** In the Gallup poll described in Exercise 2, 49% of those polled considered themselves to be "thriving."

a) Calculate the margin of error for the proportion of all American adults who were rated as "thriving" for 99% confidence.
b) Would the margin of error be larger or smaller for 95% confidence? Explain.

Section 18.4

7. **Wrong direction again** Consider the St. Norbert's College poll of Exercise 5.

a) Are the assumptions and conditions met?
b) How many people would need to be surveyed for a 90% confidence interval to ensure the margin of error would be less than 2%?

8. **More single moms** In Exercise 4, we saw that 4% of people think the trend of more single women having children is a "good thing."

a) Are the conditions for constructing a confidence interval met?
b) How many people would need to be surveyed for a 95% confidence interval to ensure the margin of error would be less than 3%? (Hint: Don't use $\hat{p} = 0.5$.)

Chapter Exercises

9. **Margin of error** A TV newscaster reports the results of a poll of voters, and then says, "The margin of error is plus or minus 4%." Explain carefully what that means.

10. **Another margin of error** A medical researcher estimates the percentage of children exposed to lead-based paint, adding that he believes his estimate has a margin of error of about 3%. Explain what the margin of error means.

11. **Conditions** For each situation described below, identify the population and the sample, explain what p and $\hat{p}$ represent, and tell whether the methods of this chapter can be used to create a confidence interval.

a) Police set up an auto checkpoint at which drivers are stopped and their cars inspected for safety problems. They find that 14 of the 134 cars stopped have at least one safety violation. They want to estimate the percentage of all cars that may be unsafe.
b) A TV talk show asks viewers to register their opinions on prayer in schools by logging on to a website. Of the 602 people who voted, 488 favored prayer in schools. We want to estimate the level of support among the general public.
c) A school is considering requiring students to wear uniforms. The PTA surveys parent opinion by sending a questionnaire home with all 1245 students; 380 surveys are returned, with 228 families in favor of the change.
d) A college admits 1632 freshmen one year, and four years later, 1388 of them graduate on time. The college wants to estimate the percentage of all their freshman enrollees who graduate on time.

12. **More conditions** Consider each situation described. Identify the population and the sample, explain what p and $\hat{p}$ represent, and tell whether the methods of this chapter can be used to create a confidence interval.

a) A consumer group hoping to assess customer experiences with auto dealers surveys 167 people who recently bought new cars; 3% of them expressed dissatisfaction with the salesperson.
b) What percent of college students have cell phones? 2883 students were asked as they entered a football stadium, and 2430 said they had phones with them.
c) Two hundred forty potato plants in a field in Maine are randomly checked, and only 7 show signs of blight. How severe is the blight problem for the U.S. potato industry?
d) Twelve of the 309 employees of a small company suffered an injury on the job last year. What can the company expect in future years?

13. **Conclusions** A catalog sales company promises to deliver orders placed on the Internet within 3 days. Follow-up calls to a few randomly selected customers show that a 95% confidence interval for the proportion

of all orders that arrive on time is 88% ± 6%. What does this mean? Are these conclusions correct? Explain.

a) Between 82% and 94% of all orders arrive on time.
b) Ninety-five percent of all random samples of customers will show that 88% of orders arrive on time.
c) Ninety-five percent of all random samples of customers will show that 82% to 94% of orders arrive on time.
d) We are 95% sure that between 82% and 94% of the orders placed by the sampled customers arrived on time.
e) On 95% of the days, between 82% and 94% of the orders will arrive on time.

14. **More conclusions** In January 2002, two students made worldwide headlines by spinning a Belgian euro 250 times and getting 140 heads—that's 56%. That makes the 90% confidence interval (51%, 61%). What does this mean? Are these conclusions correct? Explain.

a) Between 51% and 61% of all euros are unfair.
b) We are 90% sure that in this experiment this euro landed heads on between 51% and 61% of the spins.
c) We are 90% sure that spun euros will land heads between 51% and 61% of the time.
d) If you spin a euro many times, you can be 90% sure of getting between 51% and 61% heads.
e) Ninety percent of all spun euros will land heads between 51% and 61% of the time.

15. **Confidence intervals** Several factors are involved in the creation of a confidence interval. Among them are the sample size, the level of confidence, and the margin of error. Which statements are true?

a) For a given sample size, higher confidence means a smaller margin of error.
b) For a specified confidence level, larger samples provide smaller margins of error.
c) For a fixed margin of error, larger samples provide greater confidence.
d) For a given confidence level, halving the margin of error requires a sample twice as large.

16. **Confidence intervals, again** Several factors are involved in the creation of a confidence interval. Among them are the sample size, the level of confidence, and the margin of error. Which statements are true?

a) For a given sample size, reducing the margin of error will mean lower confidence.
b) For a certain confidence level, you can get a smaller margin of error by selecting a bigger sample.
c) For a fixed margin of error, smaller samples will mean lower confidence.
d) For a given confidence level, a sample 9 times as large will make a margin of error one third as big.

17. **Cars** What fraction of cars is made in Japan? The computer output below summarizes the results of a random sample of 50 autos. Explain carefully what it tells you.

z-Interval for proportion
With 90.00% confidence,
$0.29938661 < P(japan) < 0.46984416$

18. **Parole** A study of 902 decisions (to grant parole or not) made by the Nebraska Board of Parole produced the following computer output. Assuming these cases are representative of all cases that may come before the Board, what can you conclude?

z-Interval for proportion
With 95.00% confidence,
$0.56100658 < P(parole) < 0.62524619$

19. **Mislabeled seafood** In December 2011, *Consumer Reports* published their study of labeling of seafood sold in New York, New Jersey, and Connecticut. They purchased 190 pieces of seafood from various kinds of food stores and restaurants in the three states and genetically compared the pieces to standard gene fragments that can identify the species. Laboratory results indicated that 22% of these packages of seafood were mislabeled, incompletely labeled, or misidentified by store or restaurant employees.

a) Construct a 95% confidence interval for the proportion of all seafood packages in those three states that are mislabeled or misidentified.
b) Explain what your confidence interval says about seafood sold in these three states.
c) A 2009 report by the Government Accountability Board says that the Food and Drug Administration has spent very little time recently looking for seafood fraud. Suppose an official said, "That's only 190 packages out of the billions of pieces of seafood sold in a year. With the small number tested, I don't know that one would want to change one's buying habits." (An official was quoted similarly in a different but similar context). Is this argument valid? Explain.

20. **Mislabeled seafood, second course** The December 2011 *Consumer Reports* study described in Exercise 19 also found that 12 of the 22 "red snapper" packages tested were a different kind of fish.

a) Are the conditions for creating a confidence interval satisfied? Explain.
b) Construct a 95% confidence interval.
c) Explain what your confidence interval says about "red snapper" sold in these three states.

21. **Baseball fans** In a poll taken in December 2012, Gallup asked 1006 national adults whether they were baseball fans; 48% said they were. Three years earlier, in February 2008, only 35% of a similar-size sample had reported being baseball fans.

a) Find the margin of error for the 2012 poll if we want 90% confidence in our estimate of the percent of national adults who are baseball fans.
b) Explain what that margin of error means.
c) If we wanted to be 99% confident, would the margin of error be larger or smaller? Explain.

d) Find that margin of error.
e) In general, if all other aspects of the situation remain the same, will smaller margins of error produce greater or less confidence in the interval?

22. **Still lying about age** The Pew Research poll described in Exercise 1 found that 49% of a sample of 799 teens admitted to misrepresenting their age online to access websites and online services. (Treat this as a simple random sample.)
    a) Find the margin of error for this poll if we want 95% confidence in our estimate of the percent of American teens who have misrepresented their age online.
    b) Explain what that margin of error means.
    c) If we only need to be 90% confident, will the margin of error be larger or smaller? Explain.
    d) Find that margin of error.
    e) In general, if all other aspects of the situation remain the same, would smaller samples produce smaller or larger margins of error?

23. **Contributions, please** The Paralyzed Veterans of America is a philanthropic organization that relies on contributions. They send free mailing labels and greeting cards to potential donors on their list and ask for a voluntary contribution. To test a new campaign, they recently sent letters to a random sample of 100,000 potential donors and received 4781 donations.
    a) Give a 95% confidence interval for the true proportion of their entire mailing list who may donate.
    b) A staff member thinks that the true rate is 5%. Given the confidence interval you found, do you find that percentage plausible?

24. **Take the offer** First USA, a major credit card company, is planning a new offer for their current cardholders. The offer will give double airline miles on purchases for the next 6 months if the cardholder goes online and registers for the offer. To test the effectiveness of the campaign, First USA recently sent out offers to a random sample of 50,000 cardholders. Of those, 1184 registered.
    a) Give a 95% confidence interval for the true proportion of those cardholders who will register for the offer.
    b) If the acceptance rate is only 2% or less, the campaign won't be worth the expense. Given the confidence interval you found, what would you say?

25. **Teenage drivers** An insurance company checks police records on 582 accidents selected at random and notes that teenagers were at the wheel in 91 of them.
    a) Create a 95% confidence interval for the percentage of all auto accidents that involve teenage drivers.
    b) Explain what your interval means.
    c) Explain what "95% confidence" means.
    d) A politician urging tighter restrictions on drivers' licenses issued to teens says, "In one of every five auto accidents, a teenager is behind the wheel." Does your confidence interval support or contradict this statement? Explain.

26. **Junk mail** Direct mail advertisers send solicitations (a.k.a. "junk mail") to thousands of potential customers in the hope that some will buy the company's product. The acceptance rate is usually quite low. Suppose a company wants to test the response to a new flyer, and sends it to 1000 people randomly selected from their mailing list of over 200,000 people. They get orders from 123 of the recipients.
    a) Create a 90% confidence interval for the percentage of people the company contacts who may buy something.
    b) Explain what this interval means.
    c) Explain what "90% confidence" means.
    d) The company must decide whether to now do a mass mailing. The mailing won't be cost-effective unless it produces at least a 5% return. What does your confidence interval suggest? Explain.

27. **Safe food** Some food retailers propose subjecting food to a low level of radiation in order to improve safety, but sale of such "irradiated" food is opposed by many people. Suppose a grocer wants to find out what his customers think. He has cashiers distribute surveys at checkout and ask customers to fill them out and drop them in a box near the front door. He gets responses from 122 customers, of whom 78 oppose the radiation treatments. What can the grocer conclude about the opinions of all his customers?

28. **Local news** The mayor of a small city has suggested that the state locate a new prison there, arguing that the construction project and resulting jobs will be good for the local economy. A total of 183 residents show up for a public hearing on the proposal, and a show of hands finds only 31 in favor of the prison project. What can the city council conclude about public support for the mayor's initiative?

29. **Death penalty, again** In the survey on the death penalty you read about in the chapter, the Gallup Poll actually split the sample at random, asking 510 respondents the question quoted earlier, "Generally speaking, do you believe the death penalty is applied fairly or unfairly in this country today?" The other 510 were asked, "Generally speaking, do you believe the death penalty is applied unfairly or fairly in this country today?" Seems like the same question, but sometimes the order of the choices matters. Suppose that for the second way of phrasing it, 64% said they thought the death penalty was fairly applied. (Recall that 58% of the original 510 thought the same thing.)
    a) What kind of bias may be present here?
    b) If we combine them, considering the overall group to be one larger random sample of 1020 respondents, what is a 95% confidence interval for the proportion

of the general public that thinks the death penalty is being fairly applied?
c) How does the margin of error based on this pooled sample compare with the margins of error from the separate groups? Why?

30. **Gambling** A city ballot includes a local initiative that would legalize gambling. The issue is hotly contested, and two groups decide to conduct polls to predict the outcome. The local newspaper finds that 53% of 1200 randomly selected voters plan to vote "yes," while a college Statistics class finds 54% of 450 randomly selected voters in support. Both groups will create 95% confidence intervals.
   a) Without finding the confidence intervals, explain which one will have the larger margin of error.
   b) Find both confidence intervals.
   c) Which group concludes that the outcome is too close to call? Why?

31. **Rickets** Vitamin D, whether ingested as a dietary supplement or produced naturally when sunlight falls on the skin, is essential for strong, healthy bones. The bone disease rickets was largely eliminated in England during the 1950s, but now there is concern that a generation of children more likely to watch TV or play computer games than spend time outdoors is at increased risk. A recent study of 2700 children randomly selected from all parts of England found 20% of them deficient in vitamin D.
   a) Find a 98% confidence interval.
   b) Explain carefully what your interval means.
   c) Explain what "98% confidence" means.

32. **Teachers** A 2011 Gallup poll found that 76% of Americans believe that high achieving high school students should be recruited to become teachers. This poll was based on a random sample of 1002 Americans.
   a) Find a 90% confidence interval for the proportion of Americans who would agree with this.
   b) Interpret your interval in this context.
   c) Explain what "90% confidence" means.
   d) Do these data refute a pundit's claim that 2/3 of Americans believe this statement? Explain.

33. **Privacy or Security?** In January 2014 AP-GfK polled 1060 U.S. adults to find if people were more concerned with privacy or security. Privacy concerns outweighed concerns about being safe from terrorists for 646 out the 1060 polled. Of the 1060 adults about 180 are 65 and older whose concerns may be different from the rest of the population.
   a) Do you expect the 95% confidence interval for the true proportion of those 65 and over who are more concerned with privacy to be wider or narrower than the 95% confidence interval for the true proportion of all U.S. consumers? Explain.
   b) If 13% of the 1060 polled were 18–24 years old, would you expect the margin of error for this group to be larger or smaller than for the 65 and older group? Explain.

34. **Back to campus, again** ACT, Inc. reported that 74% of 1644 randomly selected college freshmen returned to college the next year. The study was stratified by type of college—public or private. The retention rates were 71.9% among 505 students enrolled in public colleges and 74.9% among 1139 students enrolled in private colleges.
   a) Will the 95% confidence interval for the true national retention rate in private colleges be wider or narrower than the 95% confidence interval for the retention rate in public colleges? Explain.
   b) Do you expect the margin of error for the overall retention rate to be larger or smaller? Explain.

35. **Deer ticks** Wildlife biologists inspect 153 deer taken by hunters and find 32 of them carrying ticks that test positive for Lyme disease.
   a) Create a 90% confidence interval for the percentage of deer that may carry such ticks.
   b) If the scientists want to cut the margin of error in half, how many deer must they inspect?
   c) What concerns do you have about this sample?

36. **Back to campus III** Suppose ACT, Inc. wants to update their information from Exercise 34 on the percentage of freshmen that return for a second year of college.
   a) They want to cut the stated margin of error in half. How many college freshmen must be surveyed?
   b) Do you have any concerns about this sample? Explain.

37. **Graduation** It's believed that as many as 25% of adults over 50 never graduated from high school. We wish to see if this percentage is the same among the 25 to 30 age group.
   a) How many of this younger age group must we survey in order to estimate the proportion of non-grads to within 6% with 90% confidence?
   b) Suppose we want to cut the margin of error to 4%. What's the necessary sample size?
   c) What sample size would produce a margin of error of 3%?

38. **Hiring** In preparing a report on the economy, we need to estimate the percentage of businesses that plan to hire additional employees in the next 60 days.
   a) How many randomly selected employers must we contact in order to create an estimate in which we are 98% confident with a margin of error of 5%?
   b) Suppose we want to reduce the margin of error to 3%. What sample size will suffice?
   c) Why might it not be worth the effort to try to get an interval with a margin of error of only 1%?

39. Graduation, again As in Exercise 37, we hope to estimate the percentage of adults aged 25 to 30 who never graduated from high school. What sample size would allow us to increase our confidence level to 95% while reducing the margin of error to only 2%?

40. Better hiring info Editors of the business report in Exercise 38 are willing to accept a margin of error of 4% but want 99% confidence. How many randomly selected employers will they need to contact?

41. Pilot study A state's environmental agency worries that many cars may be violating clean air emissions standards. The agency hopes to check a sample of vehicles in order to estimate that percentage with a margin of error of 3% and 90% confidence. To gauge the size of the problem, the agency first picks 60 cars and finds 9 with faulty emissions systems. How many should be sampled for a full investigation?

42. Another pilot study During routine screening, a doctor notices that 22% of her adult patients show higher than normal levels of glucose in their blood—a possible warning signal for diabetes. Hearing this, some medical researchers decide to conduct a large-scale study, hoping to estimate the proportion to within 4% with 98% confidence. How many randomly selected adults must they test?

43. Approval rating A newspaper reports that the governor's approval rating stands at 65%. The article adds that the poll is based on a random sample of 972 adults and has a margin of error of 2.5%. What level of confidence did the pollsters use?

44. Amendment A TV news reporter says that a proposed constitutional amendment is likely to win approval in the upcoming election because a poll of 1505 likely voters indicated that 52% would vote in favor. The reporter goes on to say that the margin of error for this poll was 3%.
a) Explain why the poll is actually inconclusive.
b) What confidence level did the pollsters use?

Just Checking ANSWERS

1. No. We know that in the sample 17% said "yes"; there's no need for a margin of error.
2. No, we are 95% confident that the percentage falls in some interval, not exactly on a particular value.
3. Yes. That's what the confidence interval means.
4. No. We don't know for sure that's true; we are only 95% confident.
5. No. That's our level of confidence, not the proportion of people receiving unsolicited text messages. The sample suggests the proportion is much lower.
6. Wider.
7. Lower.
8. Smaller.

chapter 19
Testing Hypotheses About Proportions

- 19.1 Hypotheses
- 19.2 P-Values
- 19.3 The Reasoning of Hypothesis Testing
- 19.4 Alternative Alternatives
- 19.5 P-Values and Decisions: What to Tell About a Hypothesis Test

> *Where are we going?*
> Do people ages 18–24 really prefer Pepsi to Coke? Does this new allergy medication really reduce symptoms more than a placebo? There are times when we want to make a *decision*. To do that, we'll propose a model for the situation at hand and test a hypothesis about that model. The result will help us answer the real-world question.

A S **Activity: Testing a Claim.** Can we really draw a reasonable conclusion from a random sample? Run this simulation before you read the chapter, and you'll gain a solid sense of what we're doing here.

"Half the money I spend on advertising is wasted; the trouble is I don't know which half."

—John Wanamaker (attributed)

Ingots are huge pieces of metal, sometimes weighing more than 20,000 pounds, made in a giant mold. They must be cast in one large piece for use in fabricating large structural parts for cars and planes. As the liquid metal cools, cracks can develop on the surface of the ingot, which can propagate into the zone required for the part, compromising its integrity. Airplane manufacturers insist that metal for their planes be defect-free, so the ingot must be made over if any cracking is detected.

Even though the metal from the cracked ingot is recycled, the cost runs into the tens of thousands of dollars to recast an ingot, not to mention the energy waste. About 2/3 of all aluminum ingots produced in the United States use a process called the "direct chill" method designed to reduce recasting. Metal manufacturers would like to avoid cracking if at all possible. But the casting process is complicated and not everything is completely under control. It's estimated that about 5% of aluminum ingots need to be recast because of cracking. That rate depends on the size of the ingot cast. In one plant that specializes in very large (over 30,000 lb) ingots designed for the airplane industry, about 20% of the ingots have had some kind of surface crack. In an attempt to reduce the cracking proportion, the plant engineers and chemists recently tried out some changes in the casting process. Since then, 400 ingots have been cast and only 17% of them have cracked. Has the new method worked? Has the cracking rate really decreased, or was 17% just due to luck?

We can treat the 400 ingots cast with the new method as a random sample. We know that each random sample will have a somewhat different proportion of cracked ingots. Is the 17% we observe merely a result of natural sampling variability, or is this lower cracking rate strong enough evidence to assure management that the true cracking rate now is really below 20%?

People want answers to questions like these all the time. Has the president's approval rating changed since last month? Has teenage smoking decreased in the past five years? Is the global temperature increasing? Did the Super Bowl ad we bought actually increase sales? To answer such questions, we test *hypotheses* about models.

19.1 Hypotheses

> **Hypothesis n.;** pl. {Hypotheses}. A supposition; a proposition or principle which is supposed or taken for granted, in order to draw a conclusion or inference for proof of the point in question; something not proved, but assumed for the purpose of argument.
> — *Webster's Unabridged Dictionary, 1913*

■ **NOTATION ALERT**

Capital H is the standard letter for hypotheses. H_0 always labels the null hypothesis, and H_A labels the alternative hypothesis.

© 2013 Randall Munroe. Reprinted with permission. All rights reserved.

We want to know if the changes made by the engineers have lowered the cracking rate from 20%. Humans are natural skeptics, so to test whether the changes have worked, we'll assume that they haven't. We'll make the hypothesis that the cracking rate is still 20% and see if the data can convince us otherwise. Hypotheses are models that we adopt temporarily—until we can test them once we have data. This starting hypothesis to be tested is called the **null hypothesis**—null because it assumes that nothing has changed. We denote it H_0. It specifies a parameter value—here that the cracking rate is 20%—which we usually write in the form H_0: *parameter = hypothesized value*. So, for the ingots we would write: $H_0: p = 0.20$.

The **alternative hypothesis**, which we denote H_A, contains the values of the parameter that we consider plausible if we reject the null hypothesis. Our null hypothesis is that $p = 0.20$. What's the alternative? Management is interested in *reducing* the cracking rate, so their alternative is $H_A: p < 0.20$.

What would convince you that the cracking rate had actually gone down? If only 4 out of the next 400 ingots crack (for a rate of 1%), most folks would conclude that the changes helped. But if the sample cracking rate is 19.8% instead of 20%, you should be skeptical. After all, observed proportions do vary, so we wouldn't be surprised to see some difference. How much smaller must the cracking rate be before we *are* convinced that it has changed? That's the crucial question in a hypothesis test. As usual in statistics, when we think about how big the change has to be, we think of using the standard deviation as the ruler to measure that change. So let's start by finding the standard deviation of the sample cracking rate.

Since the company changed the process, 400 new ingots have been cast of which 68 have visible surface cracks, for a sample proportion of 17%. The sample size of 400 is big enough to satisfy the **Success/Failure Condition**. (We expect $0.20 \times 400 = 80$ ingots to crack.) Although not a random sample, the engineers think that whether an ingot cracks should be independent from one ingot to the next, so the Normal sampling distribution model should work well. The standard deviation of the sampling model is

$$SD(\hat{p}) = \sqrt{\frac{pq}{n}} = \sqrt{\frac{(0.20)(0.80)}{400}} = 0.02.$$

> **WHY IS THIS A STANDARD DEVIATION AND NOT A STANDARD ERROR?**
> Remember that we reserve the term standard error for the *estimate* of the standard deviation of the sampling distribution. But we're not estimating here—we have a value of p from our null hypothesis model. To remind us that the parameter value comes from the null hypothesis, it is sometimes written as p_0 and the standard deviation as $SD(\hat{p}) = \sqrt{p_0 q_0/n}$. That's different than when we found a confidence interval for p. In that case we couldn't assume that we knew its value, so we estimated the standard deviation from the sample value $\hat{p}$.

Now we know both parameters of the Normal sampling distribution model: $p = 0.20$ and $SD(\hat{p}) = 0.02$, so we can find out how likely it would be to see the observed value of $\hat{p} = 0.17$. Since we are using a Normal model, we find the z-score:

$$z = \frac{0.17 - 0.20}{0.02} = -1.5$$

Figure 19.1
How likely is a z-score of -1.5 (or lower)? This is what it looks like. The red area is 0.067 of the total area under the curve.

Then we ask, "How likely is it to observe a value at least 1.5 standard deviations below the mean of a Normal model?" The answer (from a calculator, computer program, or the Normal table) is about 0.067. This is the probability of observing a cracking rate of 17% or less in a sample of 400 if the null hypothesis is true and the true cracking rate is still 20%. Management now must decide whether an event that would happen 6.7% of the time by chance is strong enough evidence to conclude that the true cracking proportion has decreased.

A Trial as a Hypothesis Test

Does the reasoning of hypothesis tests seem backward? That could be because we usually prefer to think about getting things right rather than getting them wrong. You have seen this reasoning before because it's the logic of jury trials.

Let's suppose a defendant has been accused of robbery. In British common law and those systems derived from it (including U.S. law), the null hypothesis is that the defendant is innocent. Instructions to juries are quite explicit about this.

How is the null hypothesis tested? The prosecution first collects evidence. ("If the defendant were innocent, wouldn't it be remarkable that the police found him at the scene of the crime with a bag full of money in his hand, a mask on his face, and a getaway car parked outside?") For us, the data is the evidence.

The next step is to judge the evidence. Evaluating the evidence is the responsibility of the jury in a trial, but it falls on your shoulders in hypothesis testing. The jury considers the evidence in light of the *presumption* of innocence and judges whether the evidence against the defendant would be plausible *if the defendant were in fact innocent*.

Like the jury, you ask, "Could these data plausibly have happened by chance if the null hypothesis were true?" If they are very unlikely to have occurred, then the evidence raises a reasonable doubt about the null hypothesis.

Ultimately, you must make a decision. The standard of "beyond a reasonable doubt" is wonderfully ambiguous because it leaves the jury to decide the degree to which the evidence contradicts the hypothesis of innocence. Juries don't explicitly use probability to help them decide whether to reject that hypothesis. But when you ask the same question of your null hypothesis, you have the advantage of being able to quantify exactly how surprising the evidence would be if the null hypothesis were true.

How unlikely is unlikely? Some people set rigid standards, like 1 time out of 20 (0.05) or 1 time out of 100 (0.01). But if *you* have to make the decision, you must judge for yourself in each situation whether the probability of observing your data is small enough to constitute "reasonable doubt."

> **Activity: The Reasoning of Hypothesis Testing.** Our reasoning is based on a rule of logic that dates back to ancient scholars. Here's a modern discussion of it.

19.2 P-Values

The fundamental step in our reasoning is the question "Are the data surprising, given the null hypothesis?" The key calculation is to determine exactly how likely the data we observed would be given that the null hypothesis were a true model of the world. Specifically, we want to find the *probability* of seeing data like these (or something even less likely) *given* that the null hypothesis is true. This probability tells us how surprised we'd be to see the data we collected if the null hypothesis is true. It's so important that it gets a special name: it's called the **P-value**.[1]

When the P-value *is* small enough, it says that we are very surprised. It means that it's very unlikely we'd observe data like these if our null hypothesis were true. The model we started with (the null hypothesis) and the data we collected are at odds with each other, so we have to make a choice. Either the null hypothesis is correct and we've just seen something remarkable, or the null hypothesis is wrong, and we were wrong to use it as the

> **Beyond a reasonable doubt** We ask whether the data were unlikely beyond a reasonable doubt. We've just calculated that probability. The probability that the observed statistic value (or an even more extreme value) could occur if the null model were true—in this case, 0.067—is the P-value.

---

[1] You'd think if it were that special it would have a better name, but "P-value" is about as creative as statisticians get.

■ **NOTATION ALERT**

We have many P's to keep straight. We use an uppercase P for probabilities, as in $P(\mathbf{A})$, and for the special probability we care about in hypothesis testing, the P-value.

We use lowercase p to denote our model's underlying proportion parameter and $\hat{p}$ to denote our observed proportion statistic.

"If the People fail to satisfy their burden of proof, you must find the defendant not guilty."

—*NY state jury instructions*

DON'T "ACCEPT" THE NULL HYPOTHESIS

Think about the null hypothesis that H_0: All swans are white. Does collecting a sample of 100 white swans prove the null hypothesis? The data are *consistent* with this hypothesis and seem to lend support to it, but they don't *prove* it. In fact, all we can do is disprove the null hypothesis—for example, by finding just one non-white swan.

basis for computing our P-value. On the other hand, if you believe in data more than in assumptions, then, given that choice, you should reject the null hypothesis.

When the P-value is high, we haven't seen anything unlikely or surprising at all. Events that have a high probability of happening happen often. The data are consistent with the model from the null hypothesis, and we have no reason to reject the null hypothesis. But many other similar hypotheses could also account for the data we've seen, so *we haven't proven that the null hypothesis is true*. The most we can say is that it doesn't appear to be false. Formally, we "fail to reject" the null hypothesis. That's a pretty weak conclusion, but it's all we can do with a high P-value.

What to Do with an "Innocent" Defendant

If the evidence is not strong enough to reject the defendant's presumption of innocence, what verdict does the jury return? They say "not guilty." Notice that they do not say that the defendant is innocent. All they say is that they have not seen sufficient evidence to convict, to reject innocence. The defendant may, in fact, be innocent, but the jury has no way to be sure.

Said statistically, the jury's null hypothesis is H_0: innocent defendant. If the evidence is too unlikely given this assumption—if the P-value is too small—the jury rejects the null hypothesis and finds the defendant guilty. But—and this is an important distinction—if there is *insufficient evidence* to convict the defendant, the jury does not decide that H_0 is true and declare the defendant innocent. Juries can only *fail to reject* the null hypothesis and declare the defendant "not guilty."

In the same way, if the data are not particularly unlikely under the assumption that the null hypothesis is true, then the most we can do is to "fail to reject" our null hypothesis. We never declare the null hypothesis to be true (or "accept" the null), because we simply do not know whether it's true or not. (After all, more evidence may come along later.)

In the trial, the burden of proof is on the prosecution. In a hypothesis test, the burden of proof is on the unusual claim. The null hypothesis is the ordinary state of affairs, so it's the alternative to the null hypothesis that we consider unusual and for which we must marshal evidence.

Imagine a clinical trial testing the effectiveness of a new headache remedy. In Chapter 12, we saw the value of comparing such treatments to a placebo. The null hypothesis, then, is that the new treatment is no more effective than the placebo. This is important because some patients will improve even when administered the placebo treatment. If we use only six people to test the drug, the results are likely *not to be clear* and we'll be unable to reject the hypothesis. Does this mean the drug doesn't work? Of course not. It simply means that we don't have enough evidence to reject our assumption. That's why we don't start by assuming that the drug *is more effective*. If we were to do that, then we could test just a few people, find that the results aren't clear, and claim that since we've been unable to reject our original assumption the drug must be effective. The FDA is unlikely to be impressed by that argument.

 Just Checking

1. A research team wants to know if aspirin helps to thin blood. The null hypothesis says that it doesn't. They test 12 patients, observe the proportion with thinner blood, and get a P-value of 0.32. They proclaim that aspirin doesn't work. What would you say?

2. An allergy drug has been tested and found to give relief to 75% of the patients in a large clinical trial. Now the scientists want to see if the new, improved version works even better. What would the null hypothesis be?

3. The new drug is tested and the P-value is 0.0001. What would you conclude about the new drug?

19.3 The Reasoning of Hypothesis Testing

> "The null hypothesis is never proved or established, but is possibly disproved, in the course of experimentation. Every experiment may be said to exist only in order to give the facts a chance of disproving the null hypothesis."
>
> —*Sir Ronald Fisher,*
> The Design of Experiments

Hypothesis tests follow a carefully structured path. To avoid getting lost as we navigate down it, we divide that path into four distinct sections.

1. Hypotheses

First, we state the null hypothesis. That's usually the skeptical claim that nothing's different. Are we considering a (New! Improved!) possibly better method? The null hypothesis says, "Oh yeah? Convince me!" To convert a skeptic, we must pile up enough evidence against the null hypothesis that we can reasonably reject it.

In statistical hypothesis testing, hypotheses are almost always about model parameters. To assess how unlikely our data may be, we need a null model. The null hypothesis specifies a particular parameter value to use in our model. In the usual shorthand, we write H_0: *parameter = hypothesized value*. The alternative hypothesis, H_A, contains the values of the parameter we consider plausible when we reject the null.

> **HOW TO SAY IT**
> You might think that the 0 in H_0 should be pronounced as "zero" or "O," but it's actually pronounced "naught" as in "all is for naught."

For Example WRITING HYPOTHESES

A large city's Department of Motor Vehicles claimed that 80% of candidates pass driving tests, but a newspaper reporter's survey of 90 randomly selected local teens who had taken the test found only 61 who passed.

QUESTION: Does this finding suggest that the passing rate for teenagers is lower than the DMV reported? Write appropriate hypotheses.

ANSWER: I'll assume that the passing rate for teenagers is the same as the DMV's overall rate of 80%, unless there's strong evidence that it's lower.

$$H_0: p = 0.80$$
$$H_A: p < 0.80$$

2. Model

To plan a statistical hypothesis test, specify the *model* you will use to test the null hypothesis and the parameter of interest. Of course, all models require assumptions, so you will need to state them and check any corresponding conditions.

Your Model step should end with a statement such as

Because the conditions are satisfied, I can model the sampling distribution of the proportion with a Normal model.

Watch out, though. Your Model step could end with

Because the conditions are not satisfied, I can't proceed with the test.

If that's the case, stop and reconsider.

> **WHEN THE CONDITIONS FAIL . . .**
> You might proceed with caution, explicitly stating your concerns. Or you may need to do the analysis with and without an outlier, or on different subgroups, or after re-expressing the response variable. Or you may not be able to proceed at all.

Each test in the book has a name that you should include in your report. We'll see many tests in the chapters that follow. Some will be about more than one sample, some will involve statistics other than proportions, and some will use models other than the Normal (and so will not use *z*-scores). The test about proportions is called a **one-proportion z-test**.[2]

---
[2] It's also called the "one-sample test for a proportion."

A S *Activity:* **Was the Observed Outcome Unlikely?** Complete the test you started in the first activity for this chapter. The narration explains the steps of the hypothesis test.

> **ONE-PROPORTION z-TEST**
> The conditions for the one-proportion z-test are the same as for the one-proportion z-interval. We test the hypothesis $H_0: p = p_0$ using the statistic $z = \dfrac{(\hat{p} - p_0)}{SD(\hat{p})}$. We use the hypothesized proportion to find the standard deviation, $SD(\hat{p}) = \sqrt{\dfrac{p_0 q_0}{n}}$.
>
> When the conditions are met and the null hypothesis is true, this statistic follows the standard Normal model, so we can use that model to obtain a P-value.

For Example CHECKING THE CONDITIONS

RECAP: A large city's DMV claimed that 80% of candidates pass driving tests. A reporter has results from a survey of 90 randomly selected local teens who had taken the test.

QUESTION: Are the conditions for inference satisfied?

- ✓ **Randomization Condition:** The 90 teens surveyed were a random sample of local teenage driving candidates.
- ✓ **10% Condition:** 90 is fewer than 10% of the teenagers who take driving tests in a large city.
- ✓ **Success/Failure Condition:** We expect $np_0 = 90(0.80) = 72$ successes and $nq_0 = 90(0.20) = 18$ failures. Both are at least 10.

ANSWER: The conditions are satisfied, so it's okay to use a Normal model and perform a one-proportion z-test.

3. Mechanics

Under "Mechanics," we place the actual calculation of our test statistic from the data. Different tests we encounter will have different formulas and different test statistics. Usually, the mechanics are handled by a statistics program or calculator, but it's good to have the formulas recorded for reference and to know what's being computed. The ultimate goal of the calculation is to find out how surprising our data would be if the null hypothesis were true. We measure this by the P-value—the probability that the observed statistic value (or an even more extreme value) occurs if the null model is correct. If the P-value is small enough, we'll reject the null hypothesis.

> **CONDITIONAL PROBABILITY**
> Did you notice that a P-value is a conditional probability? It's the probability that the observed results could have happened *if (or given that) the null hypothesis were true.*

For Example FINDING A P-VALUE

RECAP: A large city's DMV claimed that 80% of candidates pass driving tests, but a survey of 90 randomly selected local teens who had taken the test found only 61 who passed.

QUESTION: What's the P-value for the one-proportion z-test?

ANSWER: I have $n = 90$, $x = 61$, and a hypothesized $p = 0.80$.

$$\hat{p} = \frac{61}{90} \approx 0.678$$

$$SD(\hat{p}) = \sqrt{\frac{p_0 q_0}{n}} = \sqrt{\frac{(0.8)(0.2)}{90}} \approx 0.042$$

$$z = \frac{\hat{p} - p_0}{SD(\hat{p})} = \frac{0.678 - 0.800}{0.042} \approx -2.90$$

$$\text{P-value} = P(z < -2.90) = 0.002$$

(continued)

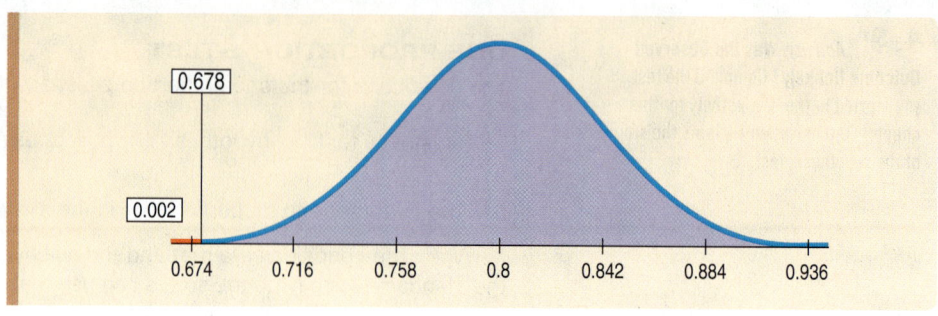

4. Conclusion

The conclusion in a hypothesis test is always a statement about the null hypothesis. The conclusion must state either that we reject or that we fail to reject the null hypothesis. And, as always, the conclusion should be stated in context.

Your conclusion about the null hypothesis should never be the end of a testing procedure. Often, there are actions to take or policies to change. In our ingot example, management must decide whether to continue the changes proposed by the engineers. The decision always includes the practical consideration of whether the new method is worth the cost. Suppose management decides to reject the null hypothesis of 20% cracking in favor of the alternative that the percentage has been reduced. They must still evaluate how much the cracking rate has been reduced and how much it cost to accomplish the reduction. The *size of the effect* is always a concern when we test hypotheses. A good way to look at the **effect size** is to examine a confidence interval.

"... They make things
admirably plain,
But one hard question will
remain:
If one hypothesis you lose,
Another in its place you
choose ..."

—*James Russell Lowell,*
Credidimus Jovem Regnare

> ### HOW MUCH DOES IT COST?
> Formal tests of a null hypothesis base the decision of whether to reject the null hypothesis solely on the size of the P-value. But in real life, we want to evaluate the costs of our decisions as well. How much would you be willing to pay for a faster computer? Shouldn't your decision depend on how much faster? And on how much more it costs? Costs are not just monetary either. Would you use the same standard of proof for testing the safety of an airplane as for the speed of your new computer?

For Example STATING THE CONCLUSION

RECAP: A large city's DMV claimed that 80% of candidates pass driving tests. Data from a reporter's survey of randomly selected local teens who had taken the test produced a P-value of 0.002.

QUESTION: What can the reporter conclude? And how might the reporter explain what the P-value means for the newspaper story?

ANSWER: Because the P-value of 0.002 is very small, I reject the null hypothesis. These survey data provide strong evidence that the passing rate for teenagers taking the driving test is lower than 80%.

If the passing rate for teenage driving candidates were actually 80%, we'd expect to see success rates this low in only about 1 in 500 (0.2%) samples of this size. This seems quite unlikely, casting doubt that the DMV's stated success rate applies to teens.

19.4 Alternative Alternatives

Activity: The Alternative Hypotheses. This interactive tool provides easy ways to visualize how one- and two-tailed alternative hypotheses work.

Tests on the ingot data can be viewed in two different ways. We know the old cracking rate is 20%, so the null hypothesis is

$$H_0: p = 0.20.$$

But we have a choice of alternative hypotheses. A metallurgist working for the company might be interested in *any* change in the cracking rate due to the new process. Even if the rate got worse, she might learn something useful from it. In that case, she's interested in possible changes on both sides of the null hypothesis. So she would write her alternative hypothesis as

$$H_A: p \neq 0.20.$$

An alternative hypothesis such as this is known as a **two-sided alternative**[3] because we are equally interested in deviations on either side of the null hypothesis value. For two-sided alternatives, the P-value is the probability of deviating in *either* direction from the null hypothesis value.

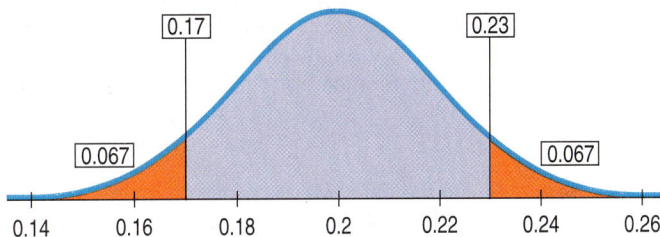

But management is really interested only in *lowering* the cracking rate below 20%. The scientific value of knowing how to *increase* the cracking rate may not appeal to them. The only alternative of interest to them is that the cracking rate *decreases*. So, they would write their alternative hypothesis as

$$H_A: p < 0.20.$$

An alternative hypothesis that focuses on deviations from the null hypothesis value in only one direction is called a **one-sided alternative**.

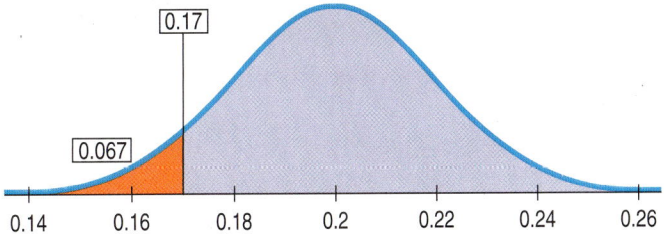

For a hypothesis test with a one-sided alternative, the P-value is the probability of deviating *only in the direction of the alternative* away from the null hypothesis value. For the same data, the one-sided P-value is half the two-sided P-value. So, a one-sided test will reject the null hypothesis more often. If you aren't sure which to use, a two-sided test is always more conservative. Be sure you can justify the choice of a one-sided test from the *Why* of the situation.

---

[3] It is also called a **two-tailed alternative**, because the probabilities we care about are found in both tails of the sampling distribution.

Step-by-Step Example: TESTING A HYPOTHESIS

Advances in medical care such as prenatal ultrasound examination now make it possible to determine a child's sex early in a pregnancy. There is a fear that in some cultures some parents may use this technology to select the sex of their children. A study from Punjab, India[4] reports that, in one hospital, 56.9% of the 550 live births that year were boys. It's a medical fact that male babies are slightly more common than female babies. The study's authors report a baseline for this region of 51.7% male live births.

Question: Is there evidence that the proportion of male births is different for this hospital?

THINK

Plan State what we want to know.

Define the variables and discuss the W's.

Hypotheses The null hypothesis makes the claim of no difference from the baseline.

Before seeing the data, we were interested in any change in male births, so the alternative hypothesis is two-sided.

Model Think about the assumptions and check the appropriate conditions.

For testing proportions, the conditions are the same ones we had for making confidence intervals, except that we check the **Success/Failure Condition** with the *hypothesized* proportions rather than with the *observed* proportions.

Specify the sampling distribution model.

Tell what test you plan to use.

I want to know whether the proportion of male births in this hospital is different from the established baseline of 51.7%. The data are the recorded sexes of the 550 live births from a hospital in Punjab, India, in 1993, collected for a study on fetal sex determination. The parameter of interest, p, is the proportion of male births:

$$H_O: p = 0.517$$
$$H_A: p \neq 0.517$$

✓ **Independence Assumption:** There is no reason to think that the sex of one baby can affect the sex of other babies, so births can reasonably be assumed to be independent with regard to the sex of the child.

✓ **Randomization Condition:** The 550 live births are not a random sample, so I must be cautious about any general conclusions. I hope that this is a representative year, and I think that the births at this hospital may be typical of this area of India.

✓ **10% Condition:** I would like to be able to make statements about births at similar hospitals in India. These 550 births are fewer than 10% of all of those births.

✓ **Success/Failure Condition:** Both $np_O = 550(0.517) = 284.35$ and $nq_O = 550(0.483) = 265.65$ are greater than 10; I expect the births of at least 10 boys and at least 10 girls, so the sample is large enough.

The conditions are satisfied, so I can use a Normal model and perform a **one-proportion z-test**.

---

[4] E. E. Booth, M. Verma, and R. S. Beri, "Fetal Sex Determination in Infants in Punjab, India: Correlations and Implications," *BMJ* 309: 1259–1261.

SHOW ▶ Mechanics

The null model gives us the mean, and (because we are working with proportions) the mean gives us the standard deviation.

We find the z-score for the observed proportion to find out how many standard deviations it is from the hypothesized proportion.

Make a picture. Sketch a Normal model centered at $p_0 = 0.517$. Shade the region to the right of the observed proportion, and because this is a two-tail test, also shade the corresponding region in the other tail.

From the z-score, we can find the P-value, which tells us the probability of observing a value that extreme (or more). Use technology or a table (see below).

Because this is a two-tail test, the P-value is the probability of observing an outcome more than 2.44 standard deviations from the mean of a Normal model *in either direction*. We must therefore *double* the probability we find in the upper tail.

The null model is a Normal distribution with a mean of 0.517 and a standard deviation of

$$SD(\hat{p}) = \sqrt{\frac{p_0 q_0}{n}} = \sqrt{\frac{(0.517)(1 - 0.517)}{550}}$$
$$= 0.0213.$$

The observed proportion, $\hat{p}$, is 0.569, so

$$z = \frac{\hat{p} - p_0}{SD(\hat{p})} = \frac{0.569 - 0.517}{0.0213} = 2.44.$$

The sample proportion lies 2.44 standard deviations above the mean.

$P = 2P(z > 2.44) = 2(0.0073) = 0.0146$

TELL ▶ Conclusion

State your conclusion in context.

This P-value is roughly 1 time in 70. That's clearly significant, but don't jump to other conclusions. We can't be sure how this deviation came about. For instance, we don't know whether this hospital is typical, or whether the time period studied was selected at random.

The P-value of 0.0146 says that if the true proportion of male babies were still at 51.7%, then an observed proportion as different as 56.9% male babies would occur at random only about 15 times in 1000. With a P-value this small, I reject H_0. This is strong evidence that the proportion of boys is not equal to the baseline for the region. It appears that the proportion of boys may be larger.

Here's a portion of a Normal table that gives the probability we needed for the hypothesis test. At $z = 2.44$, the table gives the percentile as 0.9927. The upper-tail probability (shaded red) is, therefore, $1 - 0.9927 = 0.0073$; so, for our two-sided test, the P-value is $2(0.0073) = 0.0146$.

Figure 19.2
A portion of a Normal table with the look-up for $z = 2.44$ indicated. Since $1 - 0.9927 = 0.0073$, the P-value we want is $2(0.0073) = 0.0146$.

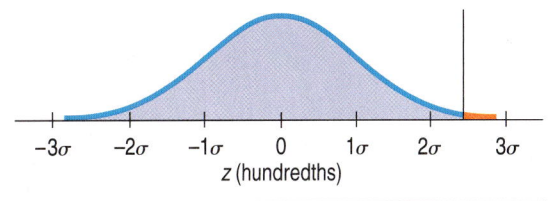

z	0.00	0.01	0.02	0.03	0.04	0.05
1.9	0.9713	0.9719	0.9726	0.9732	0.9738	0.9744
2.0	0.9772	0.9778	0.9783	0.9788	0.9793	0.9798
2.1	0.9821	0.9826	0.9830	0.9834	0.9838	0.9842
2.2	0.9861	0.9864	0.9868	0.9871	0.9875	0.9878
2.3	0.9893	0.9896	0.9898	0.9901	0.9904	0.9906
2.4	0.9918	0.9920	0.9922	0.9925	0.9927	0.9929
2.5	0.9938	0.9940	0.9941	0.9943	0.9945	0.9946
2.6	0.9953	0.9955	0.9956	0.9957	0.9959	0.9960

19.5 P-Values and Decisions: What to Tell About a Hypothesis Test

Hypothesis tests are particularly useful when we must make a decision. Is the defendant guilty or not? Should we choose print advertising or television? The absolute nature of the hypothesis test decision, however, makes some people (including the authors) uneasy. Whenever possible, it's a good idea to report a confidence interval for the parameter of interest as well.

How small should the P-value be to reject the null hypothesis? A jury needs enough evidence to show the defendant guilty "beyond a reasonable doubt." How does that translate to P-values? The answer is that there is no good, universal answer. How small the P-value has to be to reject the null hypothesis is highly context-dependent. When we're screening for a disease and want to be sure we treat all those who are sick, we may be willing to reject the null hypothesis of no disease with a P-value as large as 0.10. That would mean that 10% of the healthy people would be treated as sick and subjected to further testing. We might rather treat (or recommend further testing for) the occasional healthy person than fail to treat someone who was really sick. But a long-standing hypothesis, believed by many to be true, needs stronger evidence (and a correspondingly small P-value) to reject it.

See if you require the same P-value to reject each of the following null hypotheses:

> **Don't we want to reject the null?** Often the folks who collect the data or perform the experiment hope to reject the null. (They hope the new drug is better than the placebo, or the new ad campaign is better than the old one.) But when we practice Statistics, we can't allow that hope to affect our decision. The essential attitude for a hypothesis tester is skepticism. Until we become convinced otherwise, we cling to the null's assertion that there's nothing unusual, no effect, no difference, etc. As in a jury trial, the burden of proof rests with the alternative hypothesis—innocent until proven guilty. When you test a hypothesis, you must act as judge and jury, but you are not the prosecutor.

- A renowned musicologist claims that she can distinguish between the works of Mozart and Haydn simply by hearing a randomly selected 20 seconds of music from any work by either composer. What's the null hypothesis? If she's just guessing, she'll get 50% of the pieces correct, on average. So our null hypothesis is that p equals 50%. If she's for real, she'll get more than 50% correct. Now, we present her with 10 pieces of Mozart or Haydn chosen at random. She gets 9 out of 10 correct. It turns out that the P-value associated with that result is 0.011. (In other words, if you tried to just guess, you'd get at least 9 out of 10 correct only about 1% of the time.) What would *you* conclude? Most people would probably reject the null hypothesis and be convinced that she has some ability to do as she claims. Why? Because the P-value is small and we don't have any particular reason to doubt the alternative.
- On the other hand, imagine a student who bets that he can make a flipped coin land the way he wants just by thinking hard. To test him, we flip a fair coin 10 times. Suppose he gets 9 out of 10 right. This also has a P-value of 0.011. Are you willing now to reject this null hypothesis? Are you convinced that he's not just lucky? What amount of evidence *would* convince you? We require more evidence if rejecting the null hypothesis would contradict long-standing beliefs or other scientific results. Of course, with sufficient evidence we would revise our opinions (and scientific theories). That's how science makes progress.

> "An extraordinary claim requires extraordinary proof."
> —Marcello Truzzi
>
> This saying is often quoted by scientists without attributing it to Truzzi. But he appears to have published it first (in On the Extraordinary: An Attempt at Clarification, *Zetetic Scholar*, Vol. 1, No. 1, p. 11, 1978)

Another factor in choosing a P-value is the importance of the issue being tested. Consider the following two tests:

- A researcher claims that the proportion of college students who hold part-time jobs now is higher than the proportion known to hold such jobs a decade ago. You might be willing to believe the claim (and reject the null hypothesis of no change) with a P-value of 0.05.
- An engineer claims that even though there were several problems with the rivets holding the wing on an airplane in their fleet, they've retested the proportion of faulty rivets and now the P-value is small enough to reject the null hypothesis that the proportion is the same. What P-value would be small enough to get you to fly on that plane?

> **A S** **Activity:** *Hypothesis Tests for Proportions.* You've probably noticed that the tools for confidence intervals and for hypothesis tests are similar. See how tests and intervals for proportions are related—and an important way in which they differ.

Your conclusion about any null hypothesis should always be accompanied by the P-value of the test. Don't just declare the null hypothesis rejected or not rejected. Report the P-value to show the strength of the evidence against the hypothesis and the effect size. This will let each reader decide whether or not to reject the null hypothesis and whether or not to consider the result important if it is statistically significant.

To complete your analysis, follow your test with a confidence interval for the parameter of interest, to report the size of the effect.

Just Checking

4. A bank is testing a new method for getting delinquent customers to pay their past-due credit card bills. The standard way was to send a letter (costing about $0.40) asking the customer to pay. That worked 30% of the time. They want to test a new method that involves sending a DVD to customers encouraging them to contact the bank and set up a payment plan. Developing and sending the video costs about $10.00 per customer. What is the parameter of interest? What are the null and alternative hypotheses?

5. The bank sets up an experiment to test the effectiveness of the DVD. They mail it out to several randomly selected delinquent customers and keep track of how many actually do contact the bank to arrange payments. The bank's statistician calculates a P-value of 0.003. What does this P-value suggest about the DVD?

6. The statistician tells the bank's management that the results are clear and that they should switch to the DVD method. Do you agree? What else might you want to know?

A S *Activity:* **Practice with Testing Hypotheses About Proportions.** Here's an interactive tool that makes it easy to see what's going on in a hypothesis test.

Step-by-Step Example TESTS AND INTERVALS

Anyone who plays or watches sports has heard of the "home field advantage." Tournaments in many sports are designed to try to neutralize the advantage of the home team or player. Most people believe that teams tend to win more often when they play at home. But do they?

If there were no home field advantage, the home teams would win about half of all games played. To test this, we'll use the games in the Major League Baseball 2013 season. That year, there were 2431 regular-season games. It turns out that the home team won 1308 of the 2431 games, or 53.81% of the time.

Question: Could this deviation from 50% be explained just from natural sampling variability, or is it evidence to suggest that there really is a home field advantage, at least in professional baseball?

THINK ▶ Plan State what we want to know.

Define the variables and discuss the W's.

I want to know whether the home team in professional baseball is more likely to win. The data are all 2431 games from the 2013 Major League Baseball season. The variable is whether or not the home team won. The parameter of interest is the proportion of home team wins. If there's no advantage, I'd expect that proportion to be 0.50.

Hypotheses The null hypothesis makes the claim of no difference from the baseline. Here, that means no home field advantage.

We are interested only in a home field *advantage*, so the alternative hypothesis is one-sided.

$H_0: p = 0.50$

$H_A: p > 0.50$

Model Think about the assumptions and check the appropriate conditions. This is not a random sample. If we wanted to talk only about this season there would be no inference. So, we view the 2431 games here not as a random sample, but as a representative collection of games. Our inference is about all years of Major League Baseball.

Specify the sampling distribution model.

State what test you plan to use.

✓ **Independence Assumption:** Generally, the outcome of one game has no effect on the outcome of another game. But this may not be strictly true. For example, if a key player is injured, the probability that the team will win in the next couple of games may decrease slightly, but independence is still roughly true. The data come from one entire season, but I expect other seasons to be similar.

I'm not just interested in 2013, and those games, while not randomly selected, should be a reasonable representative sample of all Major League Baseball games in the recent past and near future.

✓ **10% Condition:** We are interested in home field advantage for Major League Baseball for all seasons. While not a random sample, these 2431 games are fewer than 10% of all games played over the years.

✓ **Success/Failure Condition:** Both $np_0 = 2431(0.50) = 1215.5$ and $nq_0 = 2431(0.50) = 1215.5$ are at least 10.

Because the conditions are satisfied, I'll use a Normal model for the sampling distribution of the proportion and do a **one-proportion z-test**.

SHOW ➡ Mechanics The null model gives us the mean, and (because we are working with proportions) the mean gives us the standard deviation.

Next, we find the z-score for the observed proportion, to find out how many standard deviations it is from the hypothesized proportion.

From the z-score, we can find the P-value, which tells us the probability of observing a value that extreme (or more).

The probability of observing a value 3.75 or more standard deviations above the mean of a Normal model can be found by computer, calculator, or table to be less than 0.0001.

The null model is a Normal distribution with a mean of 0.50 and a standard deviation of

$$SD(\hat{p}) = \sqrt{\frac{p_0 q_0}{n}} = \sqrt{\frac{(0.5)(1-0.5)}{2431}}$$

$$= 0.010141.$$

The observed proportion, $\hat{p}$, is 0.53805.

So the z-value is

$$z = \frac{0.53805 - 0.5}{0.010141} = 3.75.$$

The sample proportion lies 3.75 standard deviations above the mean.

The corresponding P-value is less than 0.0001.

TELL → Conclusion State your conclusion about the parameter—in context, of course!

The P-value of 0.0001 says that if the true proportion of home team wins were 0.50, then an observed value of 0.53805 (or larger) would occur less than one time in 1000. With a P-value so small, I reject H_0. I have reasonable evidence that the true proportion of home team wins is greater than 50%. It appears there is a home field advantage in Major League Baseball.

Question: OK, but how big a difference are we talking about? Just knowing that there is an effect is only part of the answer. Let's find a confidence interval for the home field advantage.

THINK → Model Think about the assumptions and check the conditions.

The conditions are identical to those for the hypothesis test, with one difference: Now we are not given a hypothesized proportion, p_0, so we must instead work with the observed results.

Specify the sampling distribution model.

Tell what method you plan to use.

✓ **Success/Failure Condition:** There were 1308 home team wins and 1123 losses, both at least 10.

The conditions are satisfied, so I can model the sampling distribution of the proportion with a Normal model and find a **one-proportion z-interval**.

SHOW → Mechanics We can't find the sampling model standard deviation from the null model proportion. (In fact, we've just rejected it.) Instead, we find the standard error of $\hat{p}$ from the *observed* proportions. Other than that substitution, the calculation looks the same as for the hypothesis test.

With this large a sample size, the difference is negligible, but in smaller samples, it could make a bigger difference.

$$SE(\hat{p}) = \sqrt{\frac{\hat{p}\hat{q}}{n}} = \sqrt{\frac{(0.53805)(1 - 0.53805)}{2431}}$$
$$= 0.01011$$

The sampling model is Normal, so for a 95% confidence interval, the critical value $z^* = 1.96$.

The margin of error is

$ME = z^* \times SE(\hat{p}) = 1.96 \times 0.01011 = 0.0198$.

So the 95% confidence interval is

0.53805 ± 0.0198 or $(0.5182, 0.5579)$.

TELL → Conclusion Confidence intervals help us think about the size of the effect. Here we can see that the home field advantage may affect enough games to make a real difference.

I am 95% confident that, in professional baseball, home teams win between 51.82% and 55.59% of the games.

In a season of 162 games, the low end of this interval, 51.82% of the 81 home games, would mean nearly one and a half extra home victories, on average. The upper end, 55.79%, would mean more than 4 extra wins.

WHAT CAN GO WRONG?

Hypothesis tests are so widely used—and so widely misused—that we've devoted all of Chapter 21 to discussing the pitfalls involved, but there are a few issues that we can talk about already.

- **Don't base your null hypothesis on what you see in the data.** You are not allowed to look at the data first and then adjust your null hypothesis so that it will be rejected. When your sample value turns out to be $\hat{p} = 51.8\%$, with a standard deviation of 1%, don't form a null hypothesis like $H_0: p = 49.8\%$, knowing that you can reject it. You should always *Think* about the situation you are investigating and make your null hypothesis describe the "nothing interesting" or "nothing has changed" scenario. No peeking at the data!

- **Don't base your alternative hypothesis on the data, either.** Again, you need to *Think* about the situation. Are you interested only in knowing whether something has *increased*? Then write a one-sided (upper-tail) alternative. Or would you be equally interested in a change in either direction? Then you want a two-sided alternative. You should decide whether to do a one- or two-sided test based on what results would be of interest to you, not what you see in the data.

- **Don't make your null hypothesis what you want to show to be true.** Remember, the null hypothesis is the status quo, the nothing-is-strange-here position a skeptic would take. You wonder whether the data cast doubt on that. You can reject the null hypothesis, but you can never "accept" or "prove" the null.

- **Don't forget to check the conditions.** The reasoning of inference depends on randomization. No amount of care in calculating a test result can recover from biased sampling. The probabilities we compute depend on the independence assumption. And the sample must be large enough to justify the use of a Normal model.

- **Don't accept the null hypothesis.** You may not have found enough evidence to reject it, but you surely have *not* proven it's true!

- **If you fail to reject the null hypothesis, don't think that a bigger sample would be more likely to lead to rejection.** If the results you looked at were "almost" significant, it's enticing to think that because you would have rejected the null had these same observations come from a larger sample, then a larger sample would surely lead to rejection. Don't be misled. Remember, each sample is different, and a larger sample won't necessarily duplicate your current observations. Indeed, the Central Limit Theorem tells us that statistics will vary *less* in larger samples. We should therefore expect such results to be less extreme. Maybe they'd be statistically significant but maybe (perhaps even probably) not. Even if you fail to reject the null hypothesis, it's a good idea to examine a confidence interval. If none of the plausible parameter values in the interval would matter to you (for example, because none would be *practically* significant), then even a larger study with a correspondingly smaller standard error is unlikely to be worthwhile.

CONNECTIONS

Hypothesis tests and confidence intervals share many of the same concepts. Both rely on sampling distribution models, and because the models are the same and require the same assumptions, both check the same conditions. They also calculate many of the same statistics. Like confidence intervals, hypothesis tests use the standard deviation of the sampling distribution as a ruler, as we first saw in Chapter 5.

For testing, we find ourselves looking once again at *z*-scores, and we compute the P-value by finding the distance of our test statistic from the center of the null model. P-values are conditional probabilities. They give the probability of observing the result we have seen (or one even more extreme) *given* that the null hypothesis is true.

The Standard Normal model is here again as our connection between *z*-score values and probabilities.

What Have We Learned?

Learning Objectives

Know how to formulate a null and an alternative hypothesis for a question of interest.
- The null hypothesis specifies a parameter and a (null) value for that parameter.
- The alternative hypothesis specifies a range of plausible values should we reject the null.

Be able to perform a hypothesis test for a proportion.
- The null hypothesis has the form $H_0: p = p_0$.
- We find the standard deviation of the sampling distribution of the sample proportion by assuming that the null hypothesis is true:

$$SD(\hat{p}) = \sqrt{\frac{p_0 q_0}{n}}$$

- We refer the statistic $z = \dfrac{\hat{p} - p_0}{SD(\hat{p})}$ to the standard Normal model.

Understand P-values.
- A P-value is the estimated probability of observing a statistic value at least as far from the (null) hypothesized value as the one we have actually observed.
- A small P-value indicates that the statistic we have observed would be unlikely were the null hypothesis true. That leads us to doubt the null.
- A large P-value just tells us that we have insufficient evidence to doubt the null hypothesis. In particular, it does not prove the null to be true.

Know the reasoning of hypothesis testing.
- State the **hypotheses**.
- Determine (and check assumptions for) the sampling distribution **model**.
- Calculate the test statistic—the **mechanics**.
- State your **conclusions and decisions**.

Be able to decide on a two-sided or one-sided alternative hypothesis, and justify your decision.

Know that confidence intervals and hypothesis tests go hand in hand in helping us think about models.
- A hypothesis test makes a yes/no decision about the plausibility of the value of a parameter value.
- A confidence interval shows us the range of plausible values for the parameter.

Review of Terms

Hypothesis   A model or proposition that we adopt in order to test (p. 495).

Null hypothesis   The claim being assessed in a hypothesis test that states "no change from the traditional value," "no effect," "no difference," or "no relationship." For a claim to be a testable null hypothesis, it must specify a value for some population parameter that can form the basis for assuming a sampling distribution for a test statistic (p. 495).

Alternative hypothesis   The alternative hypothesis proposes what we should conclude if we reject the null hypothesis (p. 495).

P-value   The probability of observing a value for a test statistic at least as far from the hypothesized value as the statistic value actually observed if the null hypothesis is true. A small P-value indicates either that the observation is improbable or that the probability calculation was based on incorrect assumptions. The assumed truth of the null hypothesis is the assumption under suspicion (p. 496).

One-proportion z-test   A test of the null hypothesis that the proportion of a single sample equals a specified value ($H_0: p = p_0$) by referring the statistic $z = \dfrac{\hat{p} - p_0}{SD(\hat{p})}$ to a Standard Normal model (p. 498).

Effect size	The difference between the null hypothesis value and the true value of a model parameter (p. 500).
Two-sided alternative (Two-tailed alternative)	An alternative hypothesis is two-sided ($H_A: p \neq p_0$) when we are interested in deviations in *either* direction away from the hypothesized parameter value (p. 501).
One-sided alternative (One-tailed alternative)	An alternative hypothesis is one-sided (e.g., $H_A: p > p_0$ or $H_A: p < p_0$) when we are interested in deviations in *only one* direction away from the hypothesized parameter value (p. 501).

On the Computer HYPOTHESIS TESTS FOR PROPORTIONS

Hypothesis tests for proportions are so easy and natural that many statistics packages don't offer special commands for them. Most statistics programs want to know the "success" and "failure" status for each case. Usually these are given as 1 or 0, but they might be category names like "yes" and "no." Often you just know the proportion of successes, $\hat{p}$, and the total count, n. Computer packages don't usually deal naturally with summary data like this, but the statistics routines found on many graphing calculators do. These calculators allow you to test hypotheses from summaries of the data—usually, all you need to enter are the number of successes and the sample size.

In some programs you can reconstruct the original values. But even when you have reconstructed (or can reconstruct) the raw data values, often you won't get exactly the same test statistic from a computer package as you would find working by hand. The reason is that when the packages treat the proportion as a mean, they make some approximations. The result is very close, but not exactly the same.

DATA DESK

Data Desk does not offer built-in methods for inference with proportions. The **Replicate Y by X** command in the **Manip** menu will "reconstruct" summarized count data so that you can display it.

COMMENTS

For summarized data, open a Scratchpad to compute the standard deviation and margin of error by typing the calculation. Then perform the test with the **z-test for individual μs** found in the Test command.

EXCEL

Inference methods for proportions are not part of the standard Excel tool set.

COMMENTS

For summarized data, type the calculation into any cell and evaluate it.

JMP

For a **categorical** variable that holds category labels, the **Distribution** platform includes tests and intervals of proportions. For summarized data:

- Put the category names in one variable and the frequencies in an adjacent variable.
- Designate the frequency column to have the **role** of **frequency**. Then use the **Distribution** platform.

COMMENTS

JMP uses slightly different methods for proportion inferences than those discussed in this text. Your answers are likely to be slightly different.

MINITAB

Choose **Basic Statistics** from the **Stat** menu:

- Choose **1Proportion** from the Basic Statistics submenu.
- If the data are category names in a variable, assign the variable from the variable list box to the **Samples in columns** box.
- If you have summarized data, click the **Summarized Data** button and fill in the number of trials and the number of successes.
- Click the **Options** button and specify the remaining details.
- If you have a large sample, check **Use test and interval based on Normal distribution.**
- Click the **OK** button.

COMMENTS

When working from a variable that names categories, Minitab treats the last category as the "success" category. You can specify how the categories should be ordered.

R

In library(stats):

- prop.test(X, n, p = NULL, alternative = c("two.sided," "less," "greater"), conf.level = 0.95, correct=FALSE)

will test the hypothesis that $p = p_0$ against various alternatives. For example with 260 successes out of 500, to test that $p = 0.5$ vs. $p \neq 0.5$, use:

prop.test(260,500,0.5,"two.sided," correct=FALSE)

SPSS

SPSS does not offer hypothesis tests for proportions.

STATCRUNCH

To test a hypothesis for a proportion using summaries:
- Click on **Stat**.
- Choose **Proportions » One sample » with summary**.
- Enter the **Number of successes** (x) and **Number of observations** (n).
- Click on **Next**.
- Indicate **Hypothesis Test**, then enter the hypothesized **Null** proportion, and choose the **Alternative** hypothesis.
- Click on **Calculate**.

To test a hypothesis for a proportion using data:
- Click on **Stat**.
- Choose **Proportions » One sample » with data**.
- Choose the variable **Column** listing the **Outcomes**.
- Enter the outcome to be considered a **Success**.
- Click on **Next**.
- Indicate **Hypothesis Test**, then enter the hypothesized **Null** proportion, and choose the **Alternative** hypothesis.
- Click on **Calculate**.

TI-83/84 PLUS

To do the mechanics of a hypothesis test for a proportion:
- Select **5:1-PropZTest** from the **STAT TESTS** menu.
- Specify the hypothesized proportion.
- Enter the observed value of *x*.
- Specify the sample size.
- Indicate what kind of test you want: one-tail lower tail, two-tail, or one-tail upper tail.
- Calculate the result.

COMMENTS

Beware: When you enter the value of *x*, you need the *count*, not the percentage. The count must be a whole number. If the number of successes is given as a percent, you must first multiply *np* and round the result to obtain *x*.

Exercises

Section 19.1

1. **Better than aspirin?** A very large study showed that aspirin reduced the rate of first heart attacks by 44%. A pharmaceutical company thinks they have a drug that will be more effective than aspirin, and plans to do a randomized clinical trial to test the new drug.
   a) What is the null hypothesis the company will use?
   b) What is their alternative hypothesis?

2. **Psychic** A friend of yours claims to be psychic. You are skeptical. To test this you take a stack of 100 playing cards and have your friend try to identify the suit (hearts, diamonds, clubs, or spades), without looking, of course!
   a) State the null hypothesis for your experiment.
   b) State the alternative hypothesis.

Section 19.2

3. **Better than aspirin 2?** A clinical trial compares the new drug described in Exercise 1 to aspirin. The group using the new drug had somewhat fewer heart attacks than those in the aspirin group.
   a) The P-value from the hypothesis test was 0.28. What do you conclude?
   b) What would you have concluded if the P-value had been 0.004?

4. **Psychic after all?** The "psychic" friend from Exercise 2 correctly identified more than 25% of the cards.
   a) A hypothesis test gave a P-value of 0.014. What do you conclude?
   b) What would you conclude if the P-value had been 0.245?

Section 19.3

5. **Hispanic origin** According to the 2010 Census, 16% of the people in the United States are of Hispanic or Latino origin. One county supervisor believes her county has a different proportion of Hispanic people than the nation as a whole. She looks at their most recent survey data, which was a random sample of 437 county residents, and found that 44 of those surveyed are of Hispanic origin.
   a) State the hypotheses.
   b) Name the model and check appropriate conditions for a hypothesis test.
   c) Draw and label a sketch, and then calculate the test statistic and P-value.
   d) State your conclusion.

6. **Empty houses** According to the 2010 Census, 11.4% of all housing units in the United States were vacant. A county supervisor wonders if her county is different from this. She randomly selects 850 housing units in her county and finds that 129 of the housing units are vacant.
   a) State the hypotheses.
   b) Name the model and check appropriate conditions for a hypothesis test.
   c) Draw and label a sketch, and then calculate the test statistic and P-value.
   d) State your conclusion.

Section 19.4

7. **Psychic again (you should have seen this coming)** If you were to do a hypothesis test on your experiment with your "psychic" friend from Exercise 2, would your alternative hypothesis be one-sided or two-sided? Explain why.

8. **Hispanic origin II** The county supervisor in Exercise 5 is going to perform a hypothesis test to see if she has evidence that the proportion of people in her county that are of Hispanic or Latino origin is different from that for the nation. Should her alternative hypothesis be one-sided or two-sided? Explain.

Section 19.5

9. **Bad medicine** Occasionally, a report comes out that a drug that cures some disease turns out to have a nasty side effect. For example, some antidepressant drugs may cause suicidal thoughts in younger patients. A researcher wants to study such a drug and look for evidence that such side effects exist.
   a) If the test yields a low P-value and the researcher rejects the null hypothesis, but there is actually no ill side effect of the drug, what are the consequences of such an error?
   b) If the test yields a high P-value and the researcher fails to reject the null hypothesis, but there *is* a bad side effect of the drug, what are the consequences of such an error?

10. **Expensive medicine** Developing a new drug can be an expensive process, resulting in high costs to patients. A pharmaceutical company has developed a new drug to reduce cholesterol, and it will conduct a clinical trial to compare the effectiveness to the most widely used current treatment. The results will be analyzed using a hypothesis test.
    a) If the test yields a low P-value and the researcher rejects the null hypothesis that the new drug is not more effective, but it actually is not better, what are the consequences of such an error?
    b) If the test yields a high P-value and the researcher fails to reject the null hypothesis, but the new drug *is* more effective, what are the consequences of such an error?

Chapter Exercises

11. **Hypotheses** Write the null and alternative hypotheses you would use to test each of the following situations:
    a) A governor is concerned about his "negatives"—the percentage of state residents who express disapproval of his job performance. His political committee pays for a series of TV ads, hoping that they can keep the negatives below 30%. They will use follow-up polling to assess the ads' effectiveness.
    b) Is a coin fair?
    c) Only about 20% of people who try to quit smoking succeed. Sellers of a motivational tape claim that listening to the recorded messages can help people quit.

12. **More hypotheses** Write the null and alternative hypotheses you would use to test each situation.
    a) In the 1950s, only about 40% of high school graduates went on to college. Has the percentage changed?
    b) Twenty percent of cars of a certain model have needed costly transmission work after being driven between 50,000 and 100,000 miles. The manufacturer hopes that a redesign of a transmission component has solved this problem.
    c) We field-test a new-flavor soft drink, planning to market it only if we are sure that over 60% of the people like the flavor.

13. **Negatives** After the political ad campaign described in Exercise 11, part a, pollsters check the governor's negatives. They test the hypothesis that the ads produced no change against the alternative that the negatives are now below 30% and find a P-value of 0.22. Which conclusion is appropriate? Explain.
    a) There's a 22% chance that the ads worked.
    b) There's a 78% chance that the ads worked.
    c) There's a 22% chance that their poll is correct.
    d) There's a 22% chance that natural sampling variation could produce poll results like these if there's really no change in public opinion.

14. **Dice** The seller of a loaded die claims that it will favor the outcome 6. We don't believe that claim, and roll the die 200 times to test an appropriate hypothesis. Our P-value turns out to be 0.03. Which conclusion is appropriate? Explain.
    a) There's a 3% chance that the die is fair.
    b) There's a 97% chance that the die is fair.
    c) There's a 3% chance that a loaded die could randomly produce the results we observed, so it's reasonable to conclude that the die is fair.
    d) There's a 3% chance that a fair die could randomly produce the results we observed, so it's reasonable to conclude that the die is loaded.

15. **Relief** A company's old antacid formula provided relief for 70% of the people who used it. The company tests a new formula to see if it is better and gets a P-value of 0.27. Is it reasonable to conclude that the new formula and the old one are equally effective? Explain.

16. **Cars** A survey investigating whether the proportion of today's high school seniors who own their own cars is higher than it was a decade ago finds a P-value of 0.017. Is it reasonable to conclude that more high schoolers have cars? Explain.

17. **He cheats?** A friend of yours claims that when he tosses a coin he can control the outcome. You are skeptical and want him to prove it. He tosses the coin, and you call heads; it's tails. You try again and lose again.
    a) Do two losses in a row convince you that he really can control the toss? Explain.
    b) You try a third time, and again you lose. What's the probability of losing three tosses in a row if the process is fair?
    c) Would three losses in a row convince you that your friend controls the outcome? Explain.
    d) How many times in a row would you have to lose to be pretty sure that this friend really can control the toss? Justify your answer by calculating a probability and explaining what it means.

18. **Candy** Someone hands you a box of a dozen chocolate-covered candies, telling you that half are vanilla creams and the other half peanut butter. You pick candies at random and discover the first three you eat are all vanilla.
    a) If there really were 6 vanilla and 6 peanut butter candies in the box, what is the probability that you would have picked three vanillas in a row?
    b) Do you think there really might have been 6 of each? Explain.
    c) Would you continue to believe that half are vanilla if the fourth one you try is also vanilla? Explain.

19. **Smartphones** Many people have trouble setting up all the features of their smartphones, so a company has developed what it hopes will be easier instructions. The goal is to have at least 96% of customers succeed. The company tests the new system on 200 people, of whom 188 were successful. Is this strong evidence that the new system fails to meet the company's goal? A student's test of this hypothesis is shown. How many mistakes can you find?

    $H_0: \hat{p} = 0.96$

    $H_A: \hat{p} \neq 0.96$

    SRS, $0.96(200) > 10$

    $\dfrac{188}{200} = 0.94; \quad SD(\hat{p}) = \sqrt{\dfrac{(0.94)(0.06)}{200}} = 0.017$

    $z = \dfrac{0.96 - 0.94}{0.017} = 1.18$

    $P = P(z > 1.18) = 0.12$

    There is strong evidence the new instructions don't work.

20. **Obesity 2008** In 2008, the Centers for Disease Control and Prevention reported that 34% of adults in the United States are obese. A county health service planning a new awareness campaign polls a random sample of 750 adults living there. In this sample, 228 people were found to be obese based on their answers to a health questionnaire. Do these responses provide strong evidence that the 34% figure is not accurate for this region? Correct the mistakes you find in a student's attempt to test an appropriate hypothesis.

    $H_0: \hat{p} = 0.34$

    $H_A: \hat{p} < 0.34$

    SRS, $750 \geq 10$

    $\dfrac{228}{750} = 0.304; \quad SD(\hat{p}) = \sqrt{\dfrac{(0.304)(0.696)}{750}} = 0.017$

    $z = \dfrac{0.304 - 0.34}{0.017} = -2$

    $P = P(z > -2) = 0.977$

    There is more than a 97% chance that the stated percentage is correct for this region.

21. **Dowsing** In a rural area, only about 30% of the wells that are drilled find adequate water at a depth of 100 feet or less. A local man claims to be able to find water by "dowsing"—using a forked stick to indicate where the well should be drilled. You check with 80 of his customers and find that 27 have wells less than 100 feet deep. What do you conclude about his claim?
    a) Write appropriate hypotheses.
    b) Check the necessary assumptions and conditions.
    c) Perform the mechanics of the test. What is the P-value?
    d) Explain carefully what the P-value means in context.
    e) What's your conclusion?

22. **Abnormalities** In the 1980s, it was generally believed that congenital abnormalities affected about 5% of the nation's children. Some people believe that the increase in the number of chemicals in the environment has led to an increase in the incidence of abnormalities. A recent study examined 384 children and found that 46 of them showed signs of an abnormality. Is this strong evidence that the risk has increased?
    a) Write appropriate hypotheses.
    b) Check the necessary assumptions and conditions.
    c) Perform the mechanics of the test. What is the P-value?
    d) Explain carefully what the P-value means in context.
    e) What's your conclusion?
    f) Do environmental chemicals cause congenital abnormalities?

23. **Absentees** The National Center for Education Statistics monitors many aspects of elementary and secondary education nationwide. Their 1996 numbers are often used as a baseline to assess changes. In 1996, 34% of students had not been absent from school even once during the

previous month. In the 2000 survey, responses from 8302 students showed that this figure had slipped to 33%. Officials would, of course, be concerned if student attendance were declining. Do these figures give evidence of a change in student attendance?

a) Write appropriate hypotheses.
b) Check the assumptions and conditions.
c) Perform the test and find the P-value.
d) State your conclusion.
e) Do you think this difference is meaningful? Explain.

24. **Educated mothers** The National Center for Education Statistics monitors many aspects of elementary and secondary education nationwide. Their 1996 numbers are often used as a baseline to assess changes. In 1996, 31% of students reported that their mothers had graduated from college. In 2000, responses from 8368 students found that this figure had grown to 32%. Is this evidence of a change in education level among mothers?

a) Write appropriate hypotheses.
b) Check the assumptions and conditions.
c) Perform the test and find the P-value.
d) State your conclusion.
e) Do you think this difference is meaningful? Explain.

25. **Contributions, please, part II** In Exercise 23 of Chapter 8, you learned that the Paralyzed Veterans of America is a philanthropic organization that relies on contributions. They send free mailing labels and greeting cards to potential donors on their list and ask for a voluntary contribution. To test a new campaign, the organization recently sent letters to a random sample of 100,000 potential donors and received 4781 donations. They've had a contribution rate of 5% in past campaigns, but a staff member worries that the rate will be lower if they run this campaign as currently designed.

a) What are the hypotheses?
b) Are the assumptions and conditions for inference met?
c) Do you think the rate would drop? Explain.

26. **Take the offer, part II** In Exercise 24 of Chapter 8, you learned that First USA, a major credit card company, is planning a new offer for their current cardholders. First USA will give double airline miles on purchases for the next 6 months if the cardholder goes online and registers for this offer. To test the effectiveness of this campaign, the company recently sent out offers to a random sample of 50,000 cardholders. Of those, 1184 registered. A staff member suspects that the success rate for the full campaign will be comparable to the standard 2% rate that they are used to seeing in similar campaigns. What do you predict?

a) What are the hypotheses?
b) Are the assumptions and conditions for inference met?
c) Do you think the rate would change if they use this fundraising campaign? Explain.

27. **Law school 2011** According to the Law School Admission Council, in the fall of 2011, 71% of law school applicants were accepted to some law school.[5] The training program *LSATisfaction* claims that 174 of the 240 students trained in 2011 were admitted to law school. You can safely consider these trainees to be representative of the population of law school applicants. Has *LSATisfaction* demonstrated a real improvement over the national average?

a) What are the hypotheses?
b) Check the conditions and find the P-value.
c) Would you recommend this program based on what you see here? Explain.

28. **Med school 2013** According to the Association of American Medical Colleges, only 42% of medical school applicants were admitted to a medical school in the fall of 2013.[6] Upon hearing this, the trustees of Striving College expressed concern that only 34 of the 81 students in their class of 2013 who applied to medical school were admitted. The college president assured the trustees that this was just the kind of year-to-year fluctuation in fortunes that is to be expected and that, in fact, the school's success rate was consistent with the national average. Who is right?

a) What are the hypotheses?
b) Check the conditions and find the P-value.
c) Are the trustees right to be concerned, or is the president correct? Explain.

29. **Pollution** A company with a fleet of 150 cars found that the emissions systems of 7 out of the 22 they tested failed to meet pollution control guidelines. Is this strong evidence that more than 20% of the fleet might be out of compliance? Test an appropriate hypothesis and state your conclusion. Be sure the appropriate assumptions and conditions are satisfied before you proceed.

30. **Scratch and dent** An appliance manufacturer stockpiles washers and dryers in a large warehouse for shipment to retail stores. Sometimes in handling them the appliances get damaged. Even though the damage may be minor, the company must sell those machines at drastically reduced prices. The company goal is to keep the level

---

[5] As reported by the Cornell office of career services in their *Class of 2011 Postgraduate Report*.

[6] www.aamc.org/data/facts/applicantmatriculant/

of damaged machines below 2%. One day an inspector randomly checks 60 washers and finds that 5 of them have scratches or dents. Is this strong evidence that the warehouse is failing to meet the company goal? Test an appropriate hypothesis and state your conclusion. Be sure the appropriate assumptions and conditions are satisfied before you proceed.

31. **Twins** A national vital statistics report indicated that about 3% of all births produced twins. Is the rate of twin births the same among very young mothers? Data from a large city hospital found that only 7 sets of twins were born to 469 teenage girls. Test an appropriate hypothesis and state your conclusion. Be sure the appropriate assumptions and conditions are satisfied before you proceed.

32. **Football 2013** During the 2013 season, the home team won 153 of the 245 regular-season National Football League games. Is this strong evidence of a home field advantage in professional football? Test an appropriate hypothesis and state your conclusion. Be sure the appropriate assumptions and conditions are satisfied before you proceed.

33. **WebZine** A magazine is considering the launch of an online edition. The magazine plans to go ahead only if it's convinced that more than 25% of current readers would subscribe. The magazine contacted a simple random sample of 500 current subscribers, and 137 of those surveyed expressed interest. What should the company do? Test an appropriate hypothesis and state your conclusion. Be sure the appropriate assumptions and conditions are satisfied before you proceed.

34. **Seeds** A garden center wants to store leftover packets of vegetable seeds for sale the following spring, but the center is concerned that the seeds may not germinate at the same rate a year later. The manager finds a packet of last year's green bean seeds and plants them as a test. Although the packet claims a germination rate of 92%, only 171 of 200 test seeds sprout. Is this evidence that the seeds have lost viability during a year in storage? Test an appropriate hypothesis and state your conclusion. Be sure the appropriate assumptions and conditions are satisfied before you proceed.

35. **Women executives** A company is criticized because only 13 of 43 people in executive-level positions are women. The company explains that although this proportion is lower than it might wish, it's not a surprising value given that only 40% of all its employees are women. What do you think? Test an appropriate hypothesis and state your conclusion. Be sure the appropriate assumptions and conditions are satisfied before you proceed.

36. **Jury** Census data for a certain county show that 19% of the adult residents are Hispanic. Suppose 72 people are called for jury duty and only 9 of them are Hispanic. Does this apparent underrepresentation of Hispanics call into question the fairness of the jury selection system? Explain.

37. **Dropouts** Some people are concerned that new tougher standards and high-stakes tests adopted in many states have driven up the high school dropout rate. The National Center for Education Statistics reported that the high school dropout rate for the year 2004 was 10.3%. One school district whose dropout rate has always been very close to the national average reports that 210 of their 1782 high school students dropped out last year. Is this evidence that their dropout rate may be increasing? Explain.

38. **Acid rain** A study of the effects of acid rain on trees in the Hopkins Forest shows that 25 of 100 trees sampled exhibited some sort of damage from acid rain. This rate seemed to be higher than the 15% quoted in a recent *Environmetrics* article on the average proportion of damaged trees in the Northeast. Does the sample suggest that trees in the Hopkins Forest are more susceptible than trees from the rest of the region? Comment, and write up your own conclusions based on an appropriate confidence interval as well as a hypothesis test. Include any assumptions you made about the data.

39. **Lost luggage** An airline's public relations department says that the airline rarely loses passengers' luggage. It further claims that on those occasions when luggage is lost, 90% is recovered and delivered to its owner within 24 hours. A consumer group that surveyed a large number of air travelers found that only 103 of 122 people who lost luggage on that airline were reunited with the missing items by the next day. Does this cast doubt on the airline's claim? Explain.

40. **TV ads** A start-up company is about to market a new computer printer. It decides to gamble by running commercials during the Super Bowl. The company hopes that name recognition will be worth the high cost of the ads. The goal of the company is that over 40% of the public recognize its brand name and associate it with computer equipment. The day after the game, a pollster contacts 420 randomly chosen adults and finds that 181 of them know that this company manufactures printers. Would you recommend that the company continue to advertise during Super Bowls? Explain.

41. **John Wayne** Like a lot of other Americans, John Wayne died of cancer. But is there more to this story? In 1955, Wayne was in Utah shooting the film *The Conqueror*. Across the state line, in Nevada, the United States

military was testing atomic bombs. Radioactive fallout from those tests drifted across the filming location. A total of 46 of the 220 people working on the film eventually died of cancer. Cancer experts estimate that one would expect only about 30 cancer deaths in a group this size.

a) Is the death rate among the movie crew unusually high?
b) Does this prove that exposure to radiation increases the risk of cancer?

42. AP Stats The College Board reported that 57.9% of all students who took the 2013 AP Statistics exam earned scores of 3 or higher. One teacher wondered if the performance of her school was better. She believed that year's students to be typical of those who will take AP Stats at that school and was pleased when 34 of her 54 students achieved scores of 3 or better. Can she claim that her school is better? Explain.

Just Checking ANSWERS

1. You can't conclude that the null hypothesis is true. You can conclude only that the experiment was unable to reject the null hypothesis. They were unable, on the basis of 12 patients, to show that aspirin was effective.

2. The null hypothesis is $H_0: p = 0.75$.

3. With a P-value of 0.0001, this is very strong evidence against the null hypothesis. We can reject H_0 and conclude that the improved version of the drug gives relief to a higher proportion of patients.

4. The parameter of interest is the proportion, p, of all delinquent customers who will pay their bills. $H_0: p = 0.30$ and $H_A: p > 0.30$.

5. The very low P-value leads us to reject the null hypothesis. There is strong evidence that the DVD is more effective in getting people to start paying their debts than just sending a letter had been.

6. All we know is that there is strong evidence to suggest that $p > 0.30$. We don't know how much higher than 30% the new proportion is. We'd like to see a confidence interval to see if the new method is worth the cost.

chapter 20
Inferences About Means

20.1 Getting Started: The Central Limit Theorem (Again)

20.2 Gosset's *t*

20.3 Interpreting Confidence Intervals

20.4 A Hypothesis Test for the Mean

20.5 Choosing the Sample Size

> **Where are we going?**
> We've learned how to generalize from the data at hand to the world at large for proportions. But not all data are as simple as Yes or No. In this chapter, we'll learn how to make confidence intervals and test hypotheses for the mean of a quantitative variable.

Psychologists Jim Maas and Rebecca Robbins, in their book *Sleep for Success!*, say that

> In general, high school and college students are the most pathologically sleep-deprived segment of the population. Their alertness during the day is on a par with that of untreated narcoleptics and those with untreated sleep apnea. Not surprisingly, teens are also 71 percent more likely to drive drowsy and/or fall asleep at the wheel compared to other age groups. (Males under the age of twenty-six are particularly at risk.)

They report that adults require between 7 and 9 hours of sleep each night and claim that college students require 9.25 hours of sleep to be fully alert. They note that "There is a 19 percent memory deficit in sleep-deprived individuals" (p. 35).

A student surveyed students at a small school in the northeast U.S. and asked, among other things, how much they had slept the previous night. Here's a histogram and the data for 25 of the students selected at random from the survey (Data in **Sleep**).

We're interested both in estimating the mean amount slept by college students and in testing whether it is less than the minimum recommended amount of 7 hours. These data were collected in a suitably randomized survey so we can treat them as representative of students at that college, and possibly as representative of college students in general.

These data differ from data on proportions in one important way. Proportions are summaries of individual responses, which had two possible values such as "yes" and "no," "male" and "female," or "1" and "0." Quantitative data, though, usually report a quantitative value for each individual. When you have a quantitative value for each individual, you should remember the three rules of data analysis and plot the data, as we have done here.

20.1 Getting Started: The Central Limit Theorem (Again)

You've learned how to create confidence intervals and test hypotheses about proportions. We always center confidence intervals at our best guess of the unknown parameter. Then we add and subtract a margin of error. For proportions, we write $\hat{p} \pm ME$.

We found the margin of error as the product of the standard error, $SE(\hat{p})$, and a critical value, z^*, from the Normal table. So we had $\hat{p} \pm z^* SE(\hat{p})$.

We knew we could use z because the Central Limit Theorem told us (back in Chapter 17) that the sampling distribution model for proportions is Normal.

Now we want to do exactly the same thing for means, and fortunately, the Central Limit Theorem (still in Chapter 17) told us that the same Normal model works as the sampling distribution for means.

> **THE CENTRAL LIMIT THEOREM**
>
> When a random sample is drawn from any population with mean μ and standard deviation σ, its sample mean, $\bar{y}$, has a sampling distribution with the same *mean* μ but whose *standard deviation* is $\frac{\sigma}{\sqrt{n}}$ (and we write $\sigma(\bar{y}) = SD(\bar{y}) = \frac{\sigma}{\sqrt{n}}$).
>
> No matter what population the random sample comes from, the *shape* of the sampling distribution is approximately Normal as long as the sample size is large enough. The larger the sample used, the more closely the Normal approximates the sampling distribution for the mean.

For Example USING THE CENTRAL LIMIT THEOREM (AS IF WE KNEW σ)

Based on weighing thousands of animals, the American Angus Association reports that mature Angus cows have a mean weight of 1309 pounds with a standard deviation of 157 pounds. This result was based on a very large sample of animals from many herds over a period of 15 years, so let's assume that these summaries are the population parameters and that the distribution of the weights was unimodal and reasonably symmetric.

QUESTION: What does the Central Limit Theorem (CLT) predict about the mean weight seen in random samples of 100 mature Angus cows?

ANSWER: It's given that weights of all mature Angus cows have $\mu = 1309$ and $\sigma = 157$ pounds. Because $n = 100$ animals is a fairly large sample, I can apply the Central Limit Theorem. I expect the resulting sample means $\bar{y}$ will average 1309 pounds and have a standard deviation of $SD(\bar{y}) = \frac{\sigma}{\sqrt{n}} = \frac{157}{\sqrt{100}} = 15.7$ pounds.

(continued)

> The CLT also says that the distribution of sample means follows a Normal model, so the 68–95–99.7 Rule applies. I'd expect that
> - ✓ in 68% of random samples of 100 mature Angus cows, the mean weight will be between 1309 − 15.7 = 1293.3 and 1309 + 15.7 = 1324.7 pounds;
> - ✓ in 95% of such samples, $1277.6 \le \bar{y} \le 1340.4$ pounds;
> - ✓ in 99.7% of such samples, $1261.9 \le \bar{y} \le 1356.1$ pounds.

STANDARD ERROR
Because we estimate the standard deviation of the sampling distribution model from the data, it's a *standard error*. So we use the $SE(\bar{y})$ notation. Remember, though, that it's just the estimated standard deviation of the sampling distribution model for means.

Activity: Estimating the Standard Error. What's the average age at which people have heart attacks? A confidence interval gives a good answer, but we must estimate the standard deviation from the data to construct the interval.

The CLT says that all we need to model the sampling distribution of $\bar{y}$ is a random sample of quantitative data.

And the true population standard deviation, σ.

Uh oh. That could be a problem. How are we supposed to know σ? With proportions, we had a link between the proportion value and the standard deviation of the sample proportion: $SD(\hat{p}) = \sqrt{\frac{pq}{n}}$. And there was an obvious way to estimate the standard deviation from the data: $SE(\hat{p}) = \sqrt{\frac{\hat{p}\hat{q}}{n}}$. But for means, $SD(\bar{y}) = \frac{\sigma}{\sqrt{n}}$, so knowing $\bar{y}$ doesn't tell us anything about $SD(\bar{y})$. We know n, the sample size, but the population standard deviation, σ, could be *anything*. So what should we do? We do what any sensible person would do: We estimate the population parameter σ with s, the sample standard deviation based on the data. The resulting standard error is $SE(\bar{y}) = \frac{s}{\sqrt{n}}$.

A century ago, people used this standard error with the Normal model, assuming it would work. And for large sample sizes it *did* work pretty well. But they began to notice problems with smaller samples. The sample standard deviation, s, like any other statistic, varies from sample to sample. And this extra variation in the standard error was messing up the P-values and margins of error.

William S. Gosset is the man who investigated this fact. He realized that not only do we need to allow for the extra variation with larger margins of error and P-values, but we even need a new sampling distribution model. In fact, we need a whole *family* of models, depending on the sample size, *n*. These models are unimodal, symmetric, bell-shaped models, but the smaller our sample, the more we must stretch out the tails. Gosset's work transformed Statistics, but most people who use his work don't even know his name.

20.2 Gosset's *t*

Gosset had a job that made him the envy of many. He was the chief Experimental Brewer for the Guinness Brewery in Dublin, Ireland. The brewery was a pioneer in scientific brewing and Gosset's job was to meet the demands of the brewery's many discerning customers by developing the best stout (a thick, dark beer) possible.

Gosset's experiments often required as much as a day to make the necessary chemical measurements or a full year to grow a new crop of hops. For these reasons, not to mention his health, his samples sizes were small—often as small as 3 or 4.

When he calculated means of these small samples, Gosset wanted to compare them to a target mean to judge the quality of the batch. To do so, he followed common statistical practice of the day, which was to calculate *z*-scores and compare them to the Normal model. But Gosset noticed that with samples of this size, his tests weren't quite right. He knew this because when the batches that he rejected were sent back to the laboratory for more extensive testing, too often they turned out to be OK. In fact, about 3 times more often than he expected. Gosset knew something was wrong, and it bugged him.

Guinness granted Gosset time off to earn a graduate degree in the emerging field of Statistics, and naturally he chose this problem to work on. He figured out that when he

To find the sampling distribution of $\frac{\bar{y} - \mu}{s/\sqrt{n}}$, Gosset simulated it *by hand*. He drew 750 samples of size 4 by shuffling 3000 cards on which he'd written the heights of some prisoners and computed the means and standard deviations with a mechanically cranked calculator. (He knew μ because he was simulating and knew the population from which his samples were drawn.) Today you could repeat in seconds on a computer the experiment that took him over a year. Gosset's work was so meticulous that not only did he get the shape of the new histogram approximately right, but he even figured out the exact *formula* for it from his sample. The formula was not confirmed mathematically until years later by Sir R. A. Fisher.

used the standard error, $\frac{s}{\sqrt{n}}$, as an estimate of the standard deviation of the mean, the shape of the sampling model changed. He even figured out what the new model should be.

The Guinness Company may have been ahead of its time in using statistical methods to manage quality, but they also had a policy that forbade their employees to publish. Gosset pleaded that his results were of no specific value to brewers and was allowed to publish under the pseudonym "Student," chosen for him by Guinness's managing director. Accounts differ about whether the use of a pseudonym was to avoid ill feelings within the company or to hide from competing brewers the fact that Guinness was using statistical methods. In fact, Guinness was alone in applying Gosset's results in their quality assurance operations.

It was a number of years before the true value of "Student's" results was recognized. By then, statisticians knew Gosset well, as he continued to contribute to the young field of Statistics. But this important result is still widely known as **Student's t**.

Gosset's sampling distribution model is always bell-shaped, but the details change with different sample sizes. When the sample size is very large, the model is nearly Normal, but when it's small the tails of the distribution are much heavier than the Normal. That means that values far from the mean are more common and that can be important for small samples (see Figure 20.1). So the Student's *t*-models form a whole *family* of related distributions that depend on a parameter known as **degrees of freedom**. The degrees of freedom of a distribution represent the number of independent quantities that are left after we've estimated the parameters. Here it's simply the number of data values, n, minus the number of estimated parameters. For means, that's just $n - 1$. We often denote degrees of freedom as *df* and the model as t_{df}, with the degrees of freedom as a subscript.

Figure 20.1
The *t*-model (solid curve) on 2 degrees of freedom has fatter tails than the Normal model (dashed curve). So the 68–95–99.7 Rule doesn't work for *t*-models with only a few degrees of freedom. It may not look like a big difference, but a *t* with 2 df is more than 4 times as likely to have a value greater than 2 than a standard Normal.

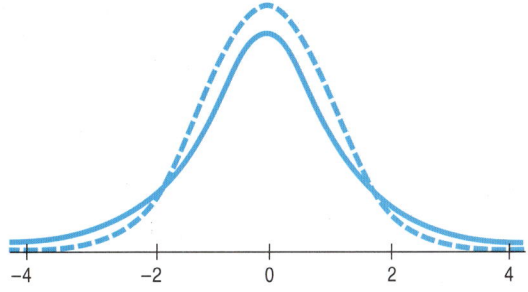

A S *Activity: Student's Distributions.* Interact with Gosset's family of *t*-models. Watch the shape of the model change as you slide the degrees of freedom up and down.

Degrees of Freedom

When we introduced the formula for the standard deviation, we promised to explain why we divide by $n - 1$ rather than by n. The reason is closely tied to the concept of degrees of freedom.

If only we knew the true population mean, μ, we would use it in our formula for the sample standard deviation:

$$s = \sqrt{\frac{\sum(y - \mu)^2}{n}} \quad \text{(Equation 20.1)}[1]$$

> **Don't divide by *n*** Some calculators offer an alternative button for standard deviation that divides by *n* instead of by $n - 1$. Why don't you stick a wad of gum over the "*n*" button so you won't be tempted to use it? Use $n - 1$.

But we don't know μ, so we naturally use $\bar{y}$ in its place. And that causes a small problem. For any sample, the data values will generally be closer to their own sample mean, $\bar{y}$, than to the true population mean, μ. Why is that? Imagine that we take a random sample of 10 high school seniors. The mean SAT verbal score is 500 in the United States. But the sample mean, $\bar{y}$, for *these* 10 seniors won't be exactly 500. Are the 10 seniors' scores closer to 500 or $\bar{y}$? They'll always be closer to their own average $\bar{y}$. So, when we calculate s using $\sum(y - \bar{y})^2$ instead of $\sum(y - \mu)^2$ in Equation 20.1, our standard deviation estimate is too small. The amazing mathematical fact is that we can fix it by dividing by $n - 1$ instead of by n. This difference is much more important when n is small than when

---

[1] Statistics textbooks usually have equation numbers so they can talk about equations by name. We haven't needed equation numbers yet, but we admit it's useful here, so this is our first.

the sample size is large. The *t*-distribution inherits this same number and we call $n - 1$ the degrees of freedom.

What Did Gosset See?

We can reproduce the simulation experiment that Gosset performed to get an idea of what he saw and how he reached some of his conclusions. Gosset drew 750 samples of size 4 from data on heights of 3000 convicts. That population looks like this:[2]

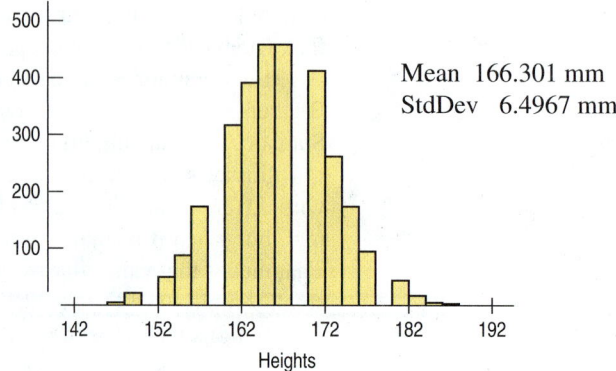

Following Gosset's example, we drew 750 independent random samples of size 4 and found their means and standard deviations.[3] As we (and Gosset) expected, the distribution of the means was even more Normal.[4]

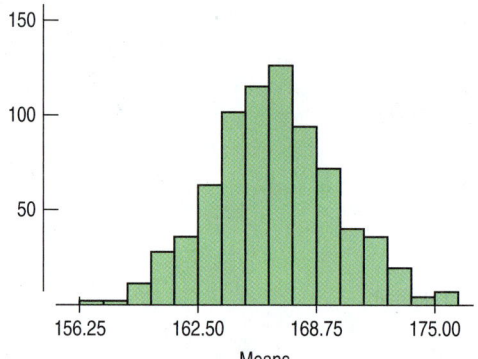

Gosset's concern was for the distribution of $\dfrac{\bar{y} - \mu}{s/\sqrt{n}}$. We know $\mu = 166.301$ mm because we know the population, and $n = 4$. The values of $\bar{y}$ and s we find for each sample. Here's what the distribution looks like:

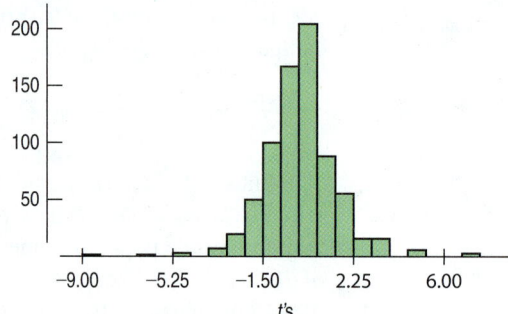

---

[2] You might have noticed some gaps in the histogram. Gosset's height data were rounded to the nearest 1/8 inch, which made for some gaps. Gosset noted that flaw in his paper.

[3] In fact, Gosset shuffled 3000 cards with the numbers on them and then dealt them into 750 piles of 4 each. That's not quite the same thing as drawing independent samples, but it was quite good enough for his purpose. We've drawn these samples in the same way.

[4] Of course, we don't know the means that Gosset actually got because we randomized using a computer and he shuffled 3000 cards, but this is one of the distributions he might have gotten, and we're pretty sure most of the others look like this as well.

It's easy to see that this distribution is much thinner in the middle and longer in the tails than a Normal model we saw for the means themselves. This was Gosset's principal result.

A Confidence Interval for Means

■ **NOTATION ALERT**
Ever since Gosset, *t* has been reserved in Statistics for his distribution.

To make confidence intervals or test hypotheses for means, we need to use Gosset's model. Which one? Well, for means, it turns out the right value for degrees of freedom is $df = n - 1$.

> **A PRACTICAL SAMPLING DISTRIBUTION MODEL FOR MEANS**
> When certain assumptions and conditions[5] are met, the standardized sample mean,
> $$t = \frac{\bar{y} - \mu}{SE(\bar{y})},$$
> follows a Student's *t*-model with $n - 1$ degrees of freedom. We estimate the standard deviation with
> $$SE(\bar{y}) = \frac{s}{\sqrt{n}}.$$

When Gosset corrected the model to take account of the extra uncertainty, the margin of error got bigger, as you might guess. When you use Gosset's model instead of the Normal model, your confidence intervals will be just a bit wider and your P-values just a bit larger. That's the correction you need. By using the *t*-model, you've compensated for the extra variability in precisely the right way.[6]

■ **NOTATION ALERT**
When we found critical values from a Normal model, we called them z^*. When we use a Student's *t*-model, we'll denote the critical values t^*.

A | S Activity: Student's *t* in Practice. Use a statistics package to find a *t*-based confidence interval; that's how it's almost always done.

> **ONE-SAMPLE *t*-INTERVAL FOR THE MEAN**
> When the assumptions and conditions[7] are met, we are ready to find the confidence interval for the population mean, μ. The confidence interval is
> $$\bar{y} \pm t^*_{n-1} \times SE(\bar{y}),$$
> where the standard error of the mean is $SE(\bar{y}) = \frac{s}{\sqrt{n}}.$
> The critical value t^*_{n-1} depends on the particular confidence level, C, that you specify and on the number of degrees of freedom, $n - 1$, which we get from the sample size.

For Example A ONE-SAMPLE *t*-INTERVAL FOR THE MEAN

In 2004, a team of researchers published a study of contaminants in farmed salmon.[8] Fish from many sources were analyzed for 14 organic contaminants. The study expressed concerns about the level of contaminants found. One of those was the insecticide mirex, which has been shown to be carcinogenic and is suspected to be toxic to the liver, kidneys, and endocrine system. One farm in particular produced salmon

(continued)

---

[5] You can probably guess what they are. We'll see them in the next section.

[6] Gosset, as the first to recognize the consequence of using *s* rather than σ, was also the first to give the sample standard deviation, *s*, a different letter than the population standard deviation, σ.

[7] Yes, the same ones, and they're still coming in the next section.

[8] Ronald A. Hites, Jeffery A. Foran, David O. Carpenter, M. Coreen Hamilton, Barbara A. Knuth, and Steven J. Schwager, "Global Assessment of Organic Contaminants in Farmed Salmon," *Science* 9 January 2004: Vol. 303, no. 5655, pp. 226–229.

> with very high levels of mirex. After those outliers are removed, summaries for the mirex concentrations (in parts per million) in the rest of the farmed salmon are (Data in **Farmed Salmon**):
>
> $$n = 150 \quad \bar{y} = 0.0913 \text{ ppm} \quad s = 0.0495 \text{ ppm}.$$
>
> **QUESTION:** What does a 95% confidence interval say about mirex?
>
> **ANSWER:** $df = 150 - 1 = 149 \qquad t^*_{149} \approx 1.977$ (from Table T, using 140 df)
> (actually, $t^*_{149} \approx 1.976$ from technology)
>
> $$SE(\bar{y}) = \frac{s}{\sqrt{n}} = \frac{0.0495}{\sqrt{150}} = 0.0040$$
>
> So the confidence interval for μ is $\bar{y} \pm t^*_{149} \times SE(\bar{y}) = 0.0913 \pm 1.977(0.0040)$
> $$= 0.0913 \pm 0.0079$$
> $$= (0.0834, 0.0992)$$
>
> I'm 95% confident that the mean level of mirex concentration in farm-raised salmon is between 0.0834 and 0.0992 parts per million.

Student's *t*-models are unimodal, symmetric, and bell-shaped, just like the Normal. But *t*-models with only a few degrees of freedom have longer tails and a larger standard deviation than the Normal. (That's what makes the margin of error bigger.) As the degrees of freedom increase, the *t*-models look more and more like the standard Normal. In fact, the *t*-model with infinite degrees of freedom is exactly Normal.[9] This is great news if you happen to have an infinite number of data values, but that's not likely. Fortunately, above a few hundred degrees of freedom it's very hard to tell the difference. Of course, in the rare situation that we *know* σ, it would be foolish not to use that information. And if we don't have to estimate σ, we can use the Normal model.

> **z OR t?**
> If you know σ, use z. (That's rare!) Whenever you use s to estimate σ, use t.

> **When σ is known** Administrators of a hospital were concerned about the prenatal care given to mothers in their part of the city. To study this, they examined the gestation times of babies born there. They drew a sample of 25 babies born in their hospital in the previous 6 months. Human gestation times for healthy pregnancies are thought to be well-modeled by a Normal with a mean of 280 days and a standard deviation of 10 days. The hospital administrators wanted to test the mean gestation time of their sample of babies against the known standard. For this test, they should use the established value for the standard deviation, 10 days, rather than estimating the standard deviation from their sample. Because they use the model parameter value for σ, they should base their test on the Normal model rather than Student's *t*.

Assumptions and Conditions

Gosset initially found the *t*-model by simulation and then made calculations to derive its form. He made some assumptions so his math would work out. Years later, when Sir Ronald A. Fisher showed mathematically that Gosset was right, he confirmed that Gosset's assumptions were necessary for the math to work. These are the assumptions we need to use the Student's *t*-models.

Independence Assumption
The data values should be mutually independent. There's really no way to check independence of the data by looking at the sample, but you should think about whether the assumption is reasonable.

Randomization Condition: This condition is satisfied if the data arise from a random sample or suitably randomized experiment. Randomly sampled data—and especially data from a Simple Random Sample—are almost surely independent. If the data don't satisfy the **Randomization Condition** then you should think about whether the values are likely to be independent for the variables you are concerned with and whether the sample you have is likely to be representative of the population you wish to learn about.

Sir Ronald Fisher (1890–1962) was one of the founders of modern Statistics.

---

[9] Formally, in the limit as *n* goes to infinity.

In the rare case that you have a sample that is a more than 10% of the population, you may want to consider using special formulas that adjust for that. But that's not a common concern for means.

Normal Population Assumption
Student's t-models won't work for small samples that are badly skewed. How skewed is too skewed? Well, formally, we assume that the data are from a population that follows a Normal model. Practically speaking, we can't be sure this is true.

And it's almost certainly *not* true. Models are idealized; real data are, well, real—*never* exactly Normal. The good news, however, is that even for small samples, it's sufficient to check the . . .

Nearly Normal Condition: The data come from a distribution that is unimodal and symmetric.

Check this condition by making a histogram or Normal probability plot. Normality is less important for larger sample sizes. Just our luck: It matters most when it's hardest to check.[10]

For very small samples ($n < 15$ or so), the data should follow a Normal model pretty closely. Of course, with so little data, it can be hard to tell. But if you do find outliers or strong skewness, don't use these methods.

For moderate sample sizes (n between 15 and 40 or so), the t methods will work well as long as the data are unimodal and reasonably symmetric. Make a histogram.

When the sample size is larger than 40 or 50, the t methods are safe to use unless the data are extremely skewed. Be sure to make a histogram. If you find outliers in the data, it's always a good idea to perform the analysis twice, once with and once without the outliers, even for large samples. Outliers may well hold additional information about the data, but you may decide to give them individual attention and then summarize the rest of the data. If you find multiple modes, you may well have different groups that should be analyzed and understood separately.

> **WHEN THE ASSUMPTIONS FAIL**
>
> When you check conditions you usually hope to make a meaningful analysis of your data. The conditions serve as *disqualifiers*—keep going unless there's a serious problem. If you find minor issues, note them and express caution about your results. If the sample is not a simple random sample (SRS), but you believe it's representative of some population, limit your conclusions accordingly. If there are outliers, perform the analysis both with and without them. If the sample looks bimodal, try to analyze subgroups separately. Only when there's major trouble—like a strongly skewed small sample or an obviously nonrepresentative sample—are you unable to proceed at all.

For Example CHECKING ASSUMPTIONS AND CONDITIONS FOR STUDENT'S t

RECAP: Researchers purchased whole farmed salmon from 51 farms in eight regions in six countries. The histogram shows the concentrations of the insecticide mirex in 150 farmed salmon.

QUESTION: Are the assumptions and conditions for inference satisfied?

ANSWER:

✓ **Independence Assumption:** The fish were not a random sample because no simple population existed to sample from. But they were raised in many different places, and samples were purchased independently from several sources, so they were likely to be independent and to represent the population of farmed salmon worldwide.

✓ **Nearly Normal Condition:** The histogram of the data is unimodal. Although it may be somewhat skewed to the right, this is not a concern with a sample size of 150.

It's okay to use these data for inference about farm-raised salmon.

---

[10] There are formal tests of Normality, but they don't really help. When we have a small sample—just when we really care about checking Normality—these tests are not very effective. So it doesn't make much sense to use them in deciding whether to perform a t-test. We don't recommend that you use them.

Just Checking

Every 10 years, the United States takes a census. The census tries to count every resident. There have been two forms, known as the "short form," answered by most people, and the "long form," slogged through by about one in six or seven households chosen at random. (For the 2010 Census, the long form was replaced by the American Community Survey.) According to the Census Bureau (www.census.gov), ". . . each estimate based on the long form responses has an associated confidence interval."

1. Why does the Census Bureau need a confidence interval for long-form information but not for the questions that appear on both the long and short forms?

2. Why must the Census Bureau base these confidence intervals on t-models?

The Census Bureau goes on to say, "These confidence intervals are wider . . . for geographic areas with smaller populations and for characteristics that occur less frequently in the area being examined (such as the proportion of people in poverty in a middle-income neighborhood)."

3. Why is this so? For example, why should a confidence interval for the mean amount families spend monthly on housing be wider for a sparsely populated area of farms in the Midwest than for a densely populated area of an urban center? How does the formula show this will happen?

To deal with this problem, the Census Bureau reports long-form data only for ". . . geographic areas from which about two hundred or more long forms were completed—which are large enough to produce good quality estimates. If smaller weighting areas had been used, the confidence intervals around the estimates would have been significantly wider, rendering many estimates less useful. . . ."

4. Suppose the Census Bureau decided to report on areas from which only 50 long forms were completed. What effect would that have on a 95% confidence interval for, say, the mean cost of housing? Specifically, which values used in the formula for the margin of error would change? Which would change a lot and which would change only slightly?

5. Approximately how much wider would that confidence interval based on 50 forms be than the one based on 200 forms?

Step-by-Step Example A ONE-SAMPLE t-INTERVAL FOR THE MEAN

Let's build a 90% confidence interval for the mean amount that college students sleep in a night.

Question: What can we say about the mean amount of sleep that college students get?

THINK ➡ Plan State what we want to know. Identify the parameter of interest.

Identify the variables and review the W's.

I want to find a 90% confidence interval for the mean, μ, of hours slept by college students. I have data on the number of hours that 25 students slept.

Make a picture. Check the distribution shape and look for skewness, multiple modes, and outliers.

REALITY CHECK The histogram centers around 7 hours, and the data lie between 4 and 9 hours. We'd expect a confidence interval to place the population mean within an hour or so of 7.

Here's a histogram of the 25 observed amounts that students slept.

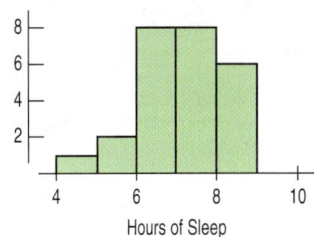

✓ **Randomization Condition:** These are data from a randomized survey, so respondents are likely to be independent.

✓ **Nearly Normal Condition:** The histogram of the *Hours of Sleep* is unimodal and slightly skewed, but not enough to be a concern.

Model Think about the assumptions and check the conditions.

Because this was a randomized survey, we check the randomization condition.

State the sampling distribution model for the statistic.

Choose your method.

The conditions are satisfied, so I will use a Student's t-model with

$$n - 1 = 24 \text{ degrees of freedom}$$

and find a **one-sample t-interval for the mean.**

SHOW ▶ Mechanics Construct the confidence interval.

Be sure to include the units along with the statistics.

Calculating from the data (see page 518):

$n = 25$ students

$\bar{y} = 6.64$ hours

$s = 1.075$ hours

The standard error of $\bar{y}$ is

$$SE(\bar{y}) = \frac{s}{\sqrt{n}} = \frac{1.075}{\sqrt{25}} = 0.215 \text{ hours.}$$

The critical value we need to make a 90% interval comes from a Student's t table such as Table T at the back of the book, a computer program, or a calculator. We have $25 - 1 = 24$ degrees of freedom. The selected confidence level says that we want 90% of the probability to be caught in the middle, so we exclude 5% in *each* tail, for a total of 10%. The degrees of freedom and 5% tail probability are all we need to know to find the critical value.

The 90% critical value is $t^*_{24} = 1.711$, so the margin of error is

$$ME = t^*_{24} \times SE(\bar{y})$$
$$= 1.711(0.215)$$
$$= 0.368 \text{ hours.}$$

The 90% confidence interval for the mean number of sleep hours is 6.64 ± 0.368 hours $= (6.272, 7.008)$ hours.

REALITY CHECK The result looks plausible and in line with what we thought.

TELL ▶ Conclusion Interpret the confidence interval in the proper context.

When we construct confidence intervals in this way, we expect 90% of them to cover the true mean and 10% to miss the true value. That's what "90% confident" means.

I am 90% confident that the interval from 6.272 to 7.008 hours contains the true mean number of hours that college students sleep.

Part of Table T.

Using Table T to Find *t*-Values

The Student's *t*-model is different for each value of degrees of freedom. Usually we find critical values and margins of error for Student's *t*-based intervals and tests with technology. Calculators or statistics programs can give critical values for a *t*-model for any number of degrees of freedom and for any confidence level you please.

But you can also use tables such as Table T at the back of this book. The tables run down the page for as many degrees of freedom as can fit. As the degrees of freedom increase, the *t*-model gets closer and closer to the Normal, so the tables give a final row with the critical values from the Normal model and label it "∞ df."

These tables are only a portion of the full tables, such as the one we used for the Normal model. We could have printed a book with a table like Table Z for every df, but that's a lot of pages, and it's not likely to be a best seller. To shorten the book, we limit ourselves to only a few confidence values. Although it might be nice to be able to get a critical value for a 93.4% confidence interval with 179 df, in practice we usually limit ourselves to 90%, 95%, 99%, and 99.9% and selected degrees of freedom. So, Table T fits on a single page with columns for selected confidence levels and rows for selected df's.

For confidence intervals, the values in the table are usually enough to cover most cases of interest. If you can't find a row for the df you need, just use the next smaller df in the table.[11]

Here's the part of the Student's *t* table that gives the critical value we needed for the Step-by-Step confidence interval. (See Table T in the back of the book.) To find a critical value, locate the row of the table corresponding to the degrees of freedom and the column corresponding to the probability you want. Our 90% confidence interval leaves 5% of the values on either side, so look for 0.05 at the top of the column or 90% at the bottom. The value in the table at that intersection is the critical value we need: 1.711.

> **HIGHER DEGREES OF FREEDOM AND THE *t*-TABLE**
>
> As degrees of freedom increase, the shape of Student's *t*-models changes more gradually. Table T at the back of the book includes degrees of freedom between 100 and 1000 selected so that you can pin down the P-value for just about any df. If your df's aren't listed, take the cautious approach by using the next lower value, or use technology.

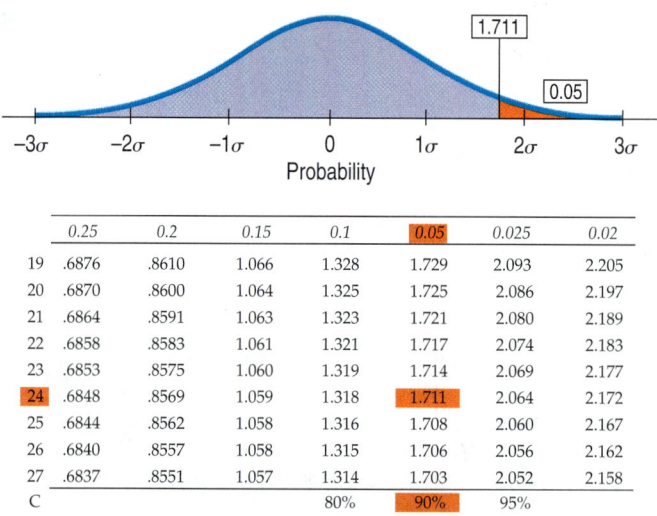

Of course, you can also create the confidence interval with computer software or a calculator.

Make a Picture, Make a Picture, Make a Picture

The only reasonable way to check the Nearly Normal Condition is with graphs of the data. Make a histogram of the data and verify that its distribution is unimodal and symmetric and that it has no outliers. It's also a good idea to make a Normal probability plot to see that it's reasonably straight. You'll be able to spot deviations from the Normal model more easily with a Normal probability plot, but it's easier to understand the particular nature of the deviations from a histogram.

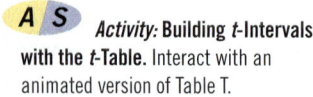

Activity: Building *t*-Intervals with the *t*-Table. Interact with an animated version of Table T.

Figure 20.2
A Normal probability plot of sleep hours is reasonably straight. The data are integers. That is why there are flat regions of the plot. Nevertheless, the plot looks reasonably straight.

---
[11] You can also find tables on the Internet. Search for terms like "statistical tables *z t*."

If you use a computer or graphing calculator to do the work, there's no excuse not to look at *both* displays as part of checking the Nearly Normal Condition.

Just Checking

In discussing estimates based on the long-form samples, the Census Bureau notes, "The disadvantage . . . is that . . . estimates of characteristics that are also reported on the short form will not match the [long-form estimates]."

The short-form estimates are values from a complete census, so they are the "true" values—something we don't usually have when we do inference.

6. Suppose we use long-form data to make 95% confidence intervals for the mean age of residents for each of 100 of the Census-defined areas. How many of these 100 intervals should we expect will fail to include the true mean age (as determined from the complete short-form Census data)?

20.3 Interpreting Confidence Intervals

Activity: Intuition for *t*-based Intervals. A narrated review of Student's *t*.

Confidence intervals for means offer new, tempting, wrong interpretations. Here are some things you *shouldn't* say:

- *Don't say*, "90% of all *students* sleep between 6.272 and 7.008 hours per night." The confidence interval is about the *mean* sleep, not about the sleep of *individual* students.
- *Don't say*, "We are 90% confident that *a randomly selected student* will sleep between 6.272 and 7.008 hours per night." This false interpretation is also about individual students rather than about the *mean*. We are 90% confident that the *mean* amount of sleep is between 6.272 and 7.008 hours per night.
- *Don't say*, "The mean amount students sleep is 6.64 hours 90% of the time." That's about means, but still wrong. It implies that the true mean varies, when in fact it is the confidence interval that would have been different had we gotten a different sample.
- Finally, *don't say*, "90% *of all samples* will have mean sleep between 6.272 and 7.008 hours per night." That statement suggests that *this* interval somehow sets a standard for every other interval. In fact, this interval is no more (or less) likely to be correct than any other. You could say that 90% of all possible samples will produce intervals that actually do contain the true mean sleep. (The problem is that, because we'll never know where the true mean sleep really is, we can't know if our sample was one of those 90%.)
- *Do say*, "90% of intervals that could be found in this way would cover the true value." Or make it more personal and say, "I am 90% confident that the true mean amount that students sleep is between 6.272 and 7.008 hours per night."

SO WHAT SHOULD WE SAY?

Since 90% of random samples yield an interval that captures the true mean, we *should* say, "I am 90% confident that the interval from 6.272 and 7.008 hours per night contains the mean amount that students sleep." It's also okay to say something less formal: "I am 90% confident that the average amount that students sleep is between 6.272 and 7.008 hours per night." Remember: *Our uncertainty is about the interval, not the true mean.* The interval varies randomly. The true mean sleep is neither variable nor random—just unknown.

20.4 A Hypothesis Test for the Mean

We are told that adults need between 7 and 9 hours of sleep. Do college students get what they need? Can we say that the mean amount college students sleep is at least 7 hours? A question like this calls for a hypothesis test called the **one-sample *t*-test for the mean**.

You already know enough to construct this test. The test statistic looks just like the others we've seen. It compares the difference between the observed statistic and a hypothesized value to the standard error of the observed statistic. We've seen that, for means, the appropriate probability model to use for P-values is Student's t with $n - 1$ degrees of freedom.

Activity: A t-Test for Wind Speed. Watch the video in the preceding activity, and then use the interactive tool to test whether there's enough wind for electricity generation at a site under investigation.

> **ONE-SAMPLE t-TEST FOR THE MEAN**
> The assumptions and conditions for the one-sample t-test for the mean are the same as for the one-sample t-interval. We test the hypothesis $H_0: \mu = \mu_0$ using the statistic
> $$t_{n-1} = \frac{\bar{y} - \mu_0}{SE(\bar{y})}.$$
> The standard error of $\bar{y}$ is $SE(\bar{y}) = \dfrac{s}{\sqrt{n}}$.
>
> When the conditions are met and the null hypothesis is true, this statistic follows a Student's t-model with $n - 1$ degrees of freedom. We use that model to obtain a P-value.

For Example A ONE-SAMPLE t-TEST FOR THE MEAN

RECAP: Researchers tested 150 farm-raised salmon for organic contaminants. They found the mean concentration of the carcinogenic insecticide mirex to be 0.0913 parts per million, with standard deviation 0.0495 ppm. As a safety recommendation to recreational fishers, the Environmental Protection Agency's (EPA) recommended "screening value" for mirex is 0.08 ppm.

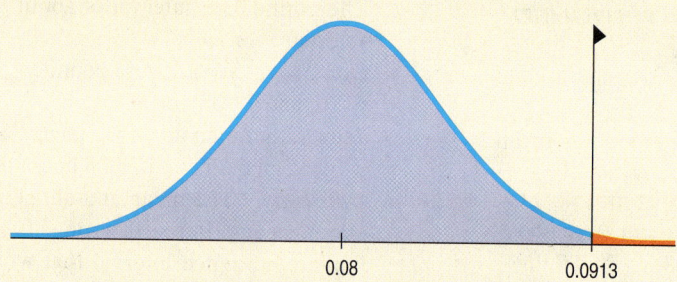

QUESTION: Are farmed salmon contaminated beyond the level permitted by the EPA?

ANSWER: (We've already checked the conditions; see page 524.)

$$H_0: \mu = 0.08$$
$$H_A: \mu > 0.08$$

These data satisfy the conditions for inference; I'll do a one-sample t-test for the mean:

$$n = 150, df = 149$$
$$\bar{y} = 0.0913, s = 0.0495$$
$$SE(\bar{y}) = \frac{0.0495}{\sqrt{150}} = 0.0040$$
$$t_{149} = \frac{0.0913 - 0.08}{0.0040} = 2.825$$
$$P(t_{149} > 2.825) = 0.0027 \text{ (from technology)}$$

With a P-value that low, I reject the null hypothesis and conclude that, in farm-raised salmon, the mirex contamination level does exceed the EPA screening value.

Step-by-Step Example: A ONE-SAMPLE t-TEST FOR THE MEAN

Let's apply the one-sample t-test to the student sleep survey. It is clear at a glance that students don't get as much sleep as Maas and Robbins recommend. Do they even make it to the minimum for adults? The Sleep Foundation (www.sleepfoundation.org) says that adults should get at least 7 hours of sleep each night.

Question: Is the mean amount that college students sleep at least as much as the 7-hour minimum recommended for adults?

THINK

Plan State what we want to test. Make clear what the population and parameter are.

Identify the variables and review the W's.

Hypotheses The null hypothesis is that the true mean sleep is equal to the minimum recommended. Because we're interested in whether students get enough sleep, the alternative is one-sided.

Make a picture. Check the distribution for skewness, multiple modes, and outliers.

REALITY CHECK The histogram of the observed hours of sleep is clustered around a value less than 7. But is this enough evidence to suggest that the mean for all college students is less than 7?

Model Think about the assumptions and check the conditions.

State the sampling distribution model. (Be sure to include the degrees of freedom.)

Choose your method.

I want to know whether the mean amount that college students sleep meets or exceeds the recommended minimum of 7 hours per night. I have a sample of 25 student reports of their sleep amounts.

H_0: Mean sleep, $\mu = 7$ hours

H_A: Mean sleep, $\mu < 7$ hours

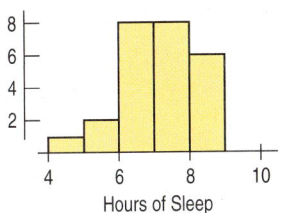

✓ **Randomization Condition:** The students were sampled in a randomized survey, so the amounts they sleep are likely to be mutually independent.

✓ **Nearly Normal Condition:** The histogram of the speeds is unimodal and reasonably symmetric.

The conditions are satisfied, so I'll use a Student's t-model with $(n - 1) = 24$ degrees of freedom to do a **one-sample t-test for the mean**.

SHOW

Mechanics Be sure to include the units when you write down what you know from the data.

We use the null model to find the P-value. Make a picture of the t-model centered at $\mu = 7$. Since this is a lower-tail test, shade the region to the left of the observed mean speed.

The t-statistic calculation is just a standardized value, like z. We subtract the hypothesized mean and divide by the standard error.

From the data,

$$n = 25 \text{ students}$$
$$\bar{y} = 6.64 \text{ hours}$$
$$s = 1.075 \text{ hours}$$
$$SE(\bar{y}) = \frac{s}{\sqrt{n}} = \frac{1.075}{\sqrt{25}} = 0.215 \text{ hour}$$

The P-value is the probability of observing a sample mean as small as 6.64 (or smaller) *if* the true mean were 7, as the null hypothesis states. We can find this P-value from a table, calculator, or computer program.

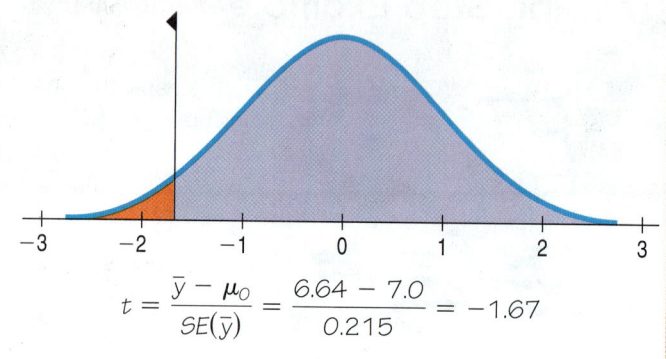

$$t = \frac{\bar{y} - \mu_0}{SE(\bar{y})} = \frac{6.64 - 7.0}{0.215} = -1.67$$

The observed mean is 1.67 standard errors below the hypothesized value. Using technology, I find

$$P\text{-value} = P(t_{24} < -1.67) = 0.054.$$

REALITY CHECK The *t*-statistic is negative because the observed mean is below the hypothesized value. That makes sense because we suspect students get too little sleep, not too much.

TELL ➡ Conclusion Link the P-value to your decision about H_0, and state your conclusion in context.

The P-value of 0.054 says that if the true mean student nightly sleep amount were 7 hours, samples of 25 students can be expected to have an observed mean of 6.64 hours or less by random chance about 54 times in 1000. If we use 0.05 as the cutoff value, then this P-value is not quite small enough to provide reasonable evidence to reject the hypothesis that the true mean is at least 7 hours. The 90% confidence interval (6.272, 7.008) shows the plausible values for the mean number of hours that college students get, and it does include 7. Although we are unable to reject the hypothesis that the mean is 7 hours, the P-value is very close to 0.05 and the confidence interval shows that most of the plausible values lie below 7. If some of those values would result in negative health impacts to students, it might be worth collecting more data to reduce the margin of error and obtain evidence on which to base some action.

For hypothesis tests, the computed *t*-statistic can take on any value, so the value you get is not likely to be one found in the table. The best we can do is to trap a calculated *t*-value between two columns. Just look across the row with the appropriate degrees of freedom to find where the *t*-statistic falls. The P-value will lie between the two values at the heads of the columns. Report that the P-value falls between these two values. Usually that's good enough.

Intervals and Tests

The 90% confidence interval for the mean number of sleep hours was 6.64 hours ± 0.368, or (6.272 to 7.008 hours.) If someone hypothesized that the mean sleep amount was really 8 hours, how would you feel about it? How about 6 hours?

Because the confidence interval includes 7 hours, it certainly looks like 7 hours might be a plausible value for the true mean sleep time. A hypothesized mean of 7 hours lies *within the confidence interval*. It's one of the plausible values for the mean.

Confidence intervals and significance tests are built from the same calculations. In fact, they are complementary ways of looking at the same question. Here's the connection: The confidence interval contains all the null hypothesis values we can't reject with these data.

Confidence intervals and hypothesis tests look at the same problem from two different perspectives. A hypothesis test starts with a *proposed parameter value* and asks if the *data* are consistent with that value. If the observed statistic is too far from the proposed parameter value, that makes it less plausible that the proposed value is the truth. So we reject the null hypothesis. By contrast, a confidence interval starts with the *data* and finds an interval of plausible values for where the parameter may lie.

To understand the connection between the two, focus on the boundary of the confidence interval and on the P-value. Our 90% confidence interval based on the sample mean of 6.64 hours was 6.272 to 7.008 hours. This 90% interval leaves 5% of the plausible values beyond the interval boundaries on each side. All the null hypothesis values inside the interval have proposed values that—by our definition—are plausible. We wouldn't reject a hypothesis that was plausible. But null hypothesis values outside the interval are too far away. We should reject such hypotheses.

How is the confidence level related to the P-value? To be precise, a level C confidence interval contains *all* of the plausible null hypothesis values that would *not* be rejected if you use a (two-sided) P-value of $(1 - C)$ as the cutoff for deciding to reject H_0.

Confidence intervals are naturally two-sided, so they correspond to two-sided P-values. When, as in our example, the hypothesis is one-sided, the interval contains values that would not be rejected using a cutoff P-value of $(1 - C)/2$.

In our case, because our test was one-sided, any null hypothesis value smaller than 7.008 would *not* be rejected according to the $(1 - 0.9)/2 = 0.05$ P-value cutoff. That's why the statistic resulted in a P-value of 0.054 for the hypothesized value of 7 hours. Because 7 is just below 7.008, the P-value is just above 0.05.

When you've performed a hypothesis test, the corresponding confidence interval can provide additional information. By providing the plausible values for the parameter, it can help you judge the importance of your result. If your hypothesized value was far from the observed statistic (and you therefore rejected your null hypothesis), the null value has not information about your data, so a confidence interval provides the *only* information about plausible values.

> **Fail to reject** Our 90% confidence interval was 6.272 to 7.008 hours. If any of these values had been the null hypothesis for the mean, then the corresponding hypothesis test using a P-value cutoff of 0.05 (because $\frac{1 - 0.90}{2} = 0.05$) would not have been able to reject the null. So, we would not reject any hypothesized value between 6.272 and 7.008 hours.

The Special Case of Proportions

The relationship between confidence intervals and hypothesis tests works for almost every inference procedure we'll see in this book. We use the standard error of the statistic as a ruler for both. But proportions are special. When we test a hypothesis about a proportion, we use the hypothesized null value rather than the observed proportion to find the P-value. Because we don't estimate the value, we call the standard deviation of the proportion $SD(\hat{p})$.

But when we construct a confidence interval, there's no null hypothesis value. So, we use the observed $\hat{p}$ value, and calculate a standard error, $SE(\hat{p})$. When the hypothesized null value is close to the observed value, the SE and SD are similar, so the usual relationship between the test and the interval works reasonably well. But if the hypothesized value is quite far from the observed value the relationship between test and interval breaks down. In that case, the confidence interval is a more appropriate way to describe the population proportion.

20.5 Choosing the Sample Size

Activity: The Real Effect of Small Sample Size. We know that smaller sample sizes lead to wider confidence intervals, but is that just because they have fewer degrees of freedom?

SOLVING FOR n

$$ME = t^*_{n-1} \frac{s}{\sqrt{n}}$$

$$\sqrt{n} = t^*_{n-1} \frac{s}{ME}$$

$$n = \left(t^*_{n-1} \frac{s}{ME} \right)^2$$

How large a sample do we need? The simple answer is "more." But more data cost money, effort, and time, so how much is enough? Suppose your computer just took half an hour to download a movie you wanted to watch. You're not happy. You hear about a program that claims to download movies in less than 15 minutes. You're interested enough to spend $29.95 for it, but only if it really delivers. So you get the free evaluation copy and test it by downloading that movie 5 different times. Of course, the mean download time is not exactly 15 minutes as claimed. Observations vary. If the margin of error were 8 minutes, you might find it hard to decide whether the software is worth the money. Doubling the sample size would require another 5 hours of testing and would reduce your margin of error to just less than 6 minutes. You'll need to decide whether that's worth the effort.

Armed with the *ME* that you decide is necessary to draw a conclusion and confidence level, you can find the sample size you need. Approximately.

For a mean, $ME = t^*_{n-1} \times SE(\bar{y})$ and $SE(\bar{y}) = \frac{s}{\sqrt{n}}$, so to find the sample size, solve this equation for *n* (see the margin note).

$$ME = t^*_{n-1} \frac{s}{\sqrt{n}}$$

But the equation needs *s* and before we collect the data, we don't know *s*. A guess is often good enough, but if you have no idea what the standard deviation might be, or if the sample size really matters (for example, because each additional individual is very expensive to sample or experiment on), a small *pilot study* can provide some information about the standard deviation.

That's not all. Without knowing *n*, you don't know the degrees of freedom so we can't find the critical value, t^*_{n-1}. One common approach is to use the corresponding z^* value from the Normal model. If you've chosen a 95% confidence level, then just use 2, following the 68–95–99.7 Rule. If your estimated sample size is, say, 60 or more, it's probably okay—z^* was a good guess. If it's smaller than that, you may want to add a step, using z^* at first, finding *n*, and then replacing z^* with the corresponding t^*_{n-1} and calculating the sample size once more.

Sample size calculations are *never* exact. But it's always a good idea to know whether your sample size is large enough to give you a good chance of being able to tell you what you want to know before you collect your data.

For Example FINDING SAMPLE SIZE

A company claims its program will allow your computer to download movies quickly. We'll test the free evaluation copy by downloading a movie several times, hoping to estimate the mean download time with a margin of error of only 8 minutes. We think the standard deviation of download times is about 10 minutes.

QUESTION: How many trial downloads must we run if we want 95% confidence in our estimate with a margin of error of only 8 minutes?

ANSWER: Using $z^* = 2$ (from the 68–95–99.7 Rule), solve

$$8 = 2 \frac{10}{\sqrt{n}}$$

$$\sqrt{n} = \frac{20}{8} = 2.5$$

$$n = 2.5^2 = 6.25.$$

That's a small sample size, so I'll use $6 - 1 = 5$ degrees of freedom[12] to substitute an appropriate t^* value. At 95%, $t_5^* = 2.571$. Solving the equation one more time:

$$8 = 2.571 \frac{10}{\sqrt{n}}$$

$$\sqrt{n} = \frac{2.571 \times 10}{8} \approx 3.214$$

$$n = (3.214)^2 \approx 10.33$$

To make sure the *ME* is no larger, we'll round *up*, which gives $n = 11$ runs. So, to get an *ME* of 8 minutes, we'll find the downloading times for 11 movies.

WHAT CAN GO WRONG?

The most fundamental issue you face is knowing when to use Student's *t* methods.

- **Don't confuse proportions and means.** When you treat your data as categorical, counting successes and summarizing with a sample proportion, make inferences using the Normal model methods you learned about in Chapters 18 and 19. When you treat your data as quantitative, summarizing with a sample mean, make your inferences using Student's *t* methods.

Student's *t* methods work only when the Normality Assumption is true. Naturally, many of the ways things can go wrong turn out to be different ways that the Normality Assumption can fail. It's always a good idea to look for the most common kinds of failure. It turns out that you can even fix some of them.

- **Beware of multimodality.** The Nearly Normal Condition clearly fails if a histogram of the data has two or more modes. When you see this, look for the possibility that your data come from two groups. If so, your best bet is to try to separate the data into different groups. (Use the variables to help distinguish the modes, if possible. For example, if the modes seem to be composed mostly of men in one and women in the other, split the data according to sex.) Then you could analyze each group separately.

- **Beware of skewed data.** Make a Normal probability plot and a histogram of the data. If the data are very skewed, you might try re-expressing the variable. Re-expressing may yield a distribution that is unimodal and symmetric, more appropriate for Student's *t* inference methods for means. Re-expression cannot help if the sample distribution is not unimodal. Some people may object to re-expressing the data, but unless your sample is very large, you just can't use the methods of this chapter on skewed data.

- **Set outliers aside.** Student's *t* methods are built on the mean and standard deviation, so we should beware of outliers when using them. When you make a histogram to check the Nearly Normal Condition, be sure to check for outliers as well. If you find some, consider doing the analysis twice, both with the outliers excluded and with them included in the data, to get a sense of how much they affect the results.

  The suggestion that you can perform an analysis with outliers removed may be controversial in some disciplines. Setting aside outliers is seen by some as "cheating." But an analysis of data with outliers left in place is *always* wrong. The outliers violate the Nearly Normal Condition and also the implicit assumption of a homogeneous population, so they invalidate inference procedures. An analysis of the nonoutlying points, along with a separate discussion of the outliers, is often much more informative and can reveal important aspects of the data.

> **DON'T IGNORE OUTLIERS**
>
> As tempting as it is to get rid of annoying values, you can't just throw away outliers and not discuss them. It isn't appropriate to lop off the highest or lowest values just to improve your results.

---

[12]Ordinarily, we'd round the sample size *up*. But at this stage of the calculation, rounding *down* is the safer choice. Can you see why?

How can you tell whether there are outliers in your data? The "outlier nomination rule" of boxplots can offer some guidance, but it's just a rule of thumb and not an absolute definition. The best practical definition is that a value is an outlier if removing it substantially changes your conclusions about the data. You won't want a single value to determine your understanding of the world unless you are very, very sure that it is correct and similar in nature to the other cases in your data. Of course, when the outliers affect your conclusion, this can lead to the uncomfortable state of not really knowing what to conclude. Such situations call for you to use your knowledge of the real world and your understanding of the data you are working with.[13]

Of course, Normality issues aren't the only risks you face when doing inferences about means. Remember to *Think* about the usual suspects.

- **Watch out for bias.** Measurements of all kinds can be biased. If your observations differ from the true mean in a systematic way, your confidence interval may not capture the true mean. And there is no sample size that will save you. A bathroom scale that's 5 pounds off will be 5 pounds off even if you weigh yourself 100 times and take the average. We've seen several sources of bias in surveys, and measurements can be biased, too. Be sure to think about possible sources of bias in your measurements.

- **Make sure cases are independent.** Student's *t* methods also require the sampled values to be mutually independent. We check for random sampling. You should also think hard about whether there are likely violations of independence in the data collection method. If there are, be very cautious about using these methods.

- **Make sure that data are from an appropriately randomized sample.** Ideally, all data that we analyze are drawn from a simple random sample or generated by a randomized experiment. When they're not, be careful about making inferences from them. You may still compute a confidence interval correctly, or get the mechanics of the P-value right, but this might not save you from making a serious mistake in inference.

- **Interpret your confidence interval correctly.** Many statements that sound tempting are, in fact, misinterpretations of a confidence interval for a mean. You might want to have another look at some of the common mistakes, explained on page 528. Keep in mind that a confidence interval is about the mean of the population, not about the means of samples, individuals in samples, or individuals in the population.

CONNECTIONS

The steps for finding a confidence interval or hypothesis test for means are just like the corresponding steps for proportions. Even the form of the calculations is similar. As the *z*-statistic did for proportions, the *t*-statistic tells us how many standard errors our sample mean is from the hypothesized mean. For means, though, we have to estimate the standard error separately. This added uncertainty changes the model for the sampling distribution from *z* to *t*.

As with all of our inference methods, the randomization applied in drawing a random sample or in randomizing a comparative experiment is what generates the sampling distribution. Randomization is what makes inference in this way possible at all.

The new concept of degrees of freedom connects back to the denominator of the sample standard deviation calculation, as shown earlier.

There's just no escaping histograms and Normal probability plots. The Nearly Normal Condition required to use Student's *t* can be checked best by making appropriate displays of the data. Back when we first used histograms, we looked at their shape and, in particular, checked whether they were unimodal and symmetric, and whether they showed any outliers. Those are just the features we check for here. The Normal probability plot zeros in on the Normal model a little more precisely.

---

[13] An important reason for *you* to know Statistics rather than let someone else analyze your data.

What Have We Learned?

Learning Objectives

Know the sampling distribution of the mean.
- To apply the Central Limit Theorem for the mean in practical applications, we must estimate the standard deviation. This *standard error* is
$$SE(\bar{y}) = \frac{s}{\sqrt{n}}.$$
- When we use the SE, the sampling distribution that allows for the additional uncertainty is Student's *t*.

Construct confidence intervals for the true mean, μ.
- A confidence interval for the mean has the form $\bar{y} \pm ME$.
- The Margin of Error is $ME = t^*_{df} SE(\bar{y})$.

Perform hypothesis tests for the mean using the standard error of $\bar{y}$ and a critical value from Student's *t* model on $n - 1$ degrees of freedom.
- Find critical values by technology or from tables.
- When constructing confidence intervals for means, the correct degrees of freedom is $n - 1$.
- Check the Assumptions and Conditions before using any sampling distribution for inference.

Write clear summaries to interpret a confidence interval or state a hypothesis test's conclusion.

Find the sample size needed to produce a given margin of error.

Review of Terms

Student's t A family of distributions indexed by its degrees of freedom. The *t*-models are unimodal, symmetric, and bell shaped, but have fatter tails and a narrower center than the Normal model. As the degrees of freedom increase, *t*-distributions approach the Normal (p. 521).

Degrees of freedom for Student's *t*-distribution For the *t*-distribution, the degrees of freedom are equal to $n - 1$, where *n* is the sample size (p. 521).

One-sample *t*-interval for the mean A one-sample *t*-interval for the population mean is (p. 523)
$$\bar{y} \pm t_{n-1} \times SE(\bar{y}), \text{ where } SE(\bar{y}) = \frac{s}{\sqrt{n}}.$$
The critical value t^*_{n-1} depends on the particular confidence level, C, that you specify and on the number of degrees of freedom, $n - 1$.

One-sample *t*-test for the mean The one-sample *t*-test for the mean tests the hypothesis $H_0: \mu = \mu_0$ using the statistic (p. 529)
$$t_{n-1} = \frac{\bar{y} - \mu_0}{SE(\bar{y})}.$$
The standard error of $\bar{y}$ is
$$SE(\bar{y}) = \frac{s}{\sqrt{n}}.$$

On the Computer INFERENCE FOR MEANS

Statistics packages offer convenient ways to make histograms of the data. Even better for assessing near-Normality is a Normal probability plot. When you work on a computer, there is simply no excuse for skipping the step of plotting the data to check that it is nearly Normal. *Beware:* Statistics packages don't agree on whether to place the Normal scores on the *x*-axis (as we have done) or the *y*-axis. Read the axis labels.

Any standard statistics package can compute a hypothesis test. Here's what the package output might look like in general (although no package we know gives the results in exactly this form):[14]

Activity: Student's *t* in Practice. We almost always use technology to do inference with Student's *t*. Here's a chance to do that as you investigate several questions.

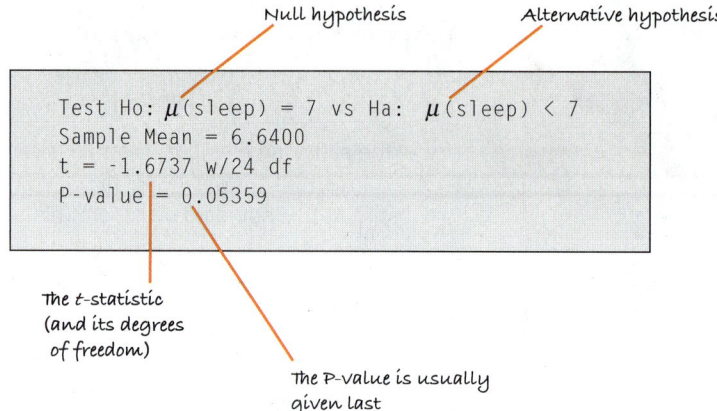

The package computes the sample mean and sample standard deviation of the variable and finds the P-value from the *t*-distribution based on the appropriate number of degrees of freedom. All modern statistics packages report P-values. The package may also provide additional information such as the sample mean, sample standard deviation, *t*-statistic value, and degrees of freedom. These are useful for interpreting the resulting P-value and telling the difference between a meaningful result and one that is merely statistically significant. Statistics packages that report the estimated standard deviation of the sampling distribution usually label it "standard error" or "*SE*."

Inference results are also sometimes reported in a table. You may have to read carefully to find the values you need. Often, test results and the corresponding confidence interval bounds are given together. And often you must read carefully to find the alternative hypotheses. Here's an example of that kind of output:

---

[14] Many statistics packages keep as many as 16 digits for all intermediate calculations. If we had kept as many, our results in the Step-by-Step section would have been closer to these.

The commands to do inference for means on common statistics programs and calculators are not always obvious. (By contrast, the resulting output is usually clearly labeled and easy to read.) The guides for each program can help you start navigating.

DATA DESK

- Select variables.
- From the **Calc** menu, choose **Estimate** for confidence intervals or **Test** for hypothesis tests.
- Select the interval or test from the drop-down menu and make other choices in the dialog.

EXCEL

Specify formulas. Find t^* with the TINV(alpha, df) function.

COMMENTS

Not really automatic. There's no easy way to find P-values in Excel. For the examples in this chapter, substitute 0.05 for "alpha" in the TINV command.

JMP

- From the **Analyze** menu, select **Distribution**.
- For a confidence interval, scroll down to the "Moments" section to find the interval limits.
- For a hypothesis test, click the red triangle next to the variable's name and choose **Test Mean** from the menu.
- Then fill in the resulting dialog.

COMMENTS

"Moment" is a fancy statistical term for means, standard deviations, and other related statistics.

MINITAB

- From the **Stat** menu, choose the **Basic Statistics** submenu.
- From that menu, choose **1-sample t....**
- Then fill in the dialog.

COMMENTS

The dialog offers a clear choice between confidence interval and test.

R

To test the hypothesis that $\mu =$ mu (default is mu = 0) against an alternative (default is two-sided) and to produce a confidence interval (default is 95%), create a vector of data in x and then:

- t.test(x, alternative = c("two.sided", "less", "greater"), mu = 0, conf.level = 0.95)

provides the t-statistic, P-value, degrees of freedom, and the confidence interval for a specified alternative.

SPSS

- From the **Analyze** menu, choose the **Compare Means** submenu.
- From that, choose the **One-Sample t-test** command.

COMMENTS

The commands suggest neither a single mean nor an interval. But the results provide both a test and an interval.

STATCRUNCH

To do inference for a mean using summaries:
- Click on **Stat**.
- Choose **T Statistics » One sample » with summary**.
- Enter the **Sample mean**, **Sample std dev**, and **Sample size**.
- Click on **Next**.
- Indicate **Hypothesis Test**, then enter the hypothesized **Null mean**, and choose the **Alternative** hypothesis.

   OR

   Indicate **Confidence Interval**, and then enter the **Level** of confidence.
- Click on **Calculate**.

To do inference for a mean using data:
- Click on **Stat**.
- Choose **T Statistics » One sample » with data**.
- Choose the variable **Column**.
- Click on **Next**.
- Indicate **Hypothesis Test**, then enter the hypothesized **Null mean**, and choose the **Alternative** hypothesis.

   OR

   Indicate **Confidence Interval**, and then enter the **Level** of confidence.
- Click on **Calculate**.

TI-83/84 PLUS

Finding a confidence interval:
- In the **STAT TESTS** menu, choose **8:TInterval**.
- Specify whether you are using data stored in a list or whether you will enter the mean, standard deviation, and sample size.
- You must also specify the desired level of confidence.

Testing a hypothesis:
- In the **STAT TESTS** menu, choose **2:T-Test**. You may specify that you are using data stored in a list, or you may enter the mean, standard deviation, and size of your sample. You must also specify the hypothesized model mean and whether the test is to be two-tail, lower-tail, or upper-tail.

Exercises

Section 20.1

1. **Salmon** A specialty food company sells whole King Salmon to various customers. The mean weight of these salmon is 35 pounds with a standard deviation of 2 pounds. The company ships them to restaurants in boxes of 4 salmon, to grocery stores in cartons of 16 salmon, and to discount outlet stores in pallets of 100 salmon. To forecast costs, the shipping department needs to estimate the standard deviation of the mean weight of the salmon in each type of shipment
   a) Find the standard deviations of the mean weight of the salmon in each type of shipment.
   b) The distribution of the salmon weights turns out to be skewed to the high end. Would the distribution of shipping weights be better characterized by a Normal model for the boxes or pallets? Explain.

2. **LSAT** The LSAT (a test taken for law school admission) has a mean score of 151 with a standard deviation of 9 and a unimodal, symmetric distribution of scores. A test preparation organization teaches small classes of 9 students at a time. A larger organization teaches classes of 25 students at a time. Both organizations publish the mean scores of all their classes.
   a) What would you expect the distribution of mean class scores to be for each organization?
   b) If either organization has a graduating class with a mean score of 160, they'll take out a full-page ad in the local school paper to advertise. Which organization is more likely to have that success? Explain.
   c) Both organizations advertise that if any class has an average score below 145, they'll pay for everyone to retake the LSAT. Which organization is at greater risk to have to pay?

Section 20.2

3. ***t*-models, part I** Using the *t* tables, software, or a calculator, estimate
   a) the critical value of *t* for a 90% confidence interval with df = 17.

b) the critical value of t for a 98% confidence interval with df = 88.

4. **t-models, part II** Using the t tables, software, or a calculator, estimate
   a) the critical value of t for a 95% confidence interval with df = 7.
   b) the critical value of t for a 99% confidence interval with df = 102.

5. **t-models, part III** Describe how the shape, center, and spread of t-models change as the number of degrees of freedom increases.

6. **t-models, part IV** Describe how the critical value of t for a 95% confidence interval changes as the number of degrees of freedom increases.

Section 20.3

7. **Home sales** The housing market has recovered slowly from the economic crisis of 2008. Recently, in one large community, realtors randomly sampled 36 bids from potential buyers to estimate the average loss in home value. The sample showed the average loss from the peak in 2008 was $9,560 with a standard deviation of $1500.
   a) What assumptions and conditions must be checked before finding a confidence interval? How would you check them?
   b) Find a 95% confidence interval for the mean loss in value per home.
   c) Interpret this interval and explain what 95% confidence means in this context.

8. **Home sales again** In the previous exercise, you found a 95% confidence interval to estimate the average loss in home value.
   a) Suppose the standard deviation of the losses had been $3000 instead of $1500. What would the larger standard deviation do to the width of the confidence interval (assuming the same level of confidence)?
   b) Your classmate suggests that the margin of error in the interval could be reduced if the confidence level were changed to 90% instead of 95%. Do you agree with this statement? Why or why not?
   c) Instead of changing the level of confidence, would it be more statistically appropriate to draw a bigger sample?

Section 20.4

9. **t-models, again** Using the t tables, software, or a calculator, estimate
   a) the P-value for $t \geq 2.09$ with 4 degrees of freedom.
   b) the P-value for $|t| > 1.78$ with 22 degrees of freedom.

10. **t-models, last time** Using the t tables, software, or a calculator, estimate
    a) the P-value for $t \leq 2.19$ with 41 degrees of freedom.
    b) the P-value for $|t| > 2.33$ with 12 degrees of freedom.

11. **Home prices** The last census of the entire community from which the sample in Exercise 7 was taken was 2010 when the mean loss was $10,200. From the sample in Exercise 7, use a t-test to test whether the mean loss has decreased.

12. **Home prices II** Suppose the standard deviation of home price losses had been $3000, as in Exercise 8? What would your conclusion be then?

Section 20.5

13. **Jelly** A consumer advocate wants to collect a sample of jelly jars and measure the actual weight of the product in the container. He needs to collect enough data to construct a confidence interval with a margin of error of no more than 2 grams with 99% confidence. The standard deviation of these jars is usually 4 grams. What do you recommend for his sample size?

14. **A good book** An English professor is attempting to estimate the mean number of novels that the student body reads during their time in college. He is conducting an exit survey with seniors. He hopes to have a margin of error of 3 books with 95% confidence. From reading previous studies, he expects a large standard deviation and is going assume it is 10. How many students should he survey?

Chapter Exercises

15. **Cattle** Livestock are given a special feed supplement to see if it will promote weight gain. Researchers report that the 77 cows studied gained an average of 56 pounds, and that a 95% confidence interval for the mean weight gain this supplement produces has a margin of error of ± 11 pounds. Some students wrote the following conclusions. Did anyone interpret the interval correctly? Explain any misinterpretations.
    a) 95% of the cows studied gained between 45 and 67 pounds.
    b) We're 95% sure that a cow fed this supplement will gain between 45 and 67 pounds.
    c) We're 95% sure that the average weight gain among the cows in this study was between 45 and 67 pounds.
    d) The average weight gain of cows fed this supplement will be between 45 and 67 pounds 95% of the time.
    e) If this supplement is tested on another sample of cows, there is a 95% chance that their average weight gain will be between 45 and 67 pounds.

16. **Teachers** Software analysis of the salaries of a random sample of 288 Nevada teachers produced the confidence interval shown below. Which conclusion is correct? What's wrong with the others?

    t-Interval for μ: with 90.00% Confidence,
    $43454 < \mu(\text{TchPay}) < 45398$

(continued)

a) If we took many random samples of 288 Nevada teachers, about 9 out of 10 of them would produce this confidence interval.
b) If we took many random samples of Nevada teachers, about 9 out of 10 of them would produce a confidence interval that contained the mean salary of all Nevada teachers.
c) About 9 out of 10 Nevada teachers earn between $43,454 and $45,398.
d) About 9 out of 10 of the teachers surveyed earn between $43,454 and $45,398.
e) We are 90% confident that the average teacher salary in the United States is between $43,454 and $45,398.

17. **Meal plan** After surveying students at Dartmouth College, a campus organization calculated that a 95% confidence interval for the mean cost of food for one term (of three in the Dartmouth trimester calendar) is ($1372, $1562). Now the organization is trying to write its report and is considering the following interpretations. Comment on each.

a) 95% of all students pay between $1372 and $1562 for food.
b) 95% of the sampled students paid between $1372 and $1562.
c) We're 95% sure that students in this sample averaged between $1372 and $1562 for food.
d) 95% of all samples of students will have average food costs between $1372 and $1562.
e) We're 95% sure that the average amount all students pay is between $1372 and $1562.

18. **Snow** Based on meteorological data for the past century, a local TV weather forecaster estimates that the region's average winter snowfall is 23", with a margin of error of ± 2 inches. Assuming he used a 95% confidence interval, how should viewers interpret this news? Comment on each of these statements:

a) During 95 of the past 100 winters, the region got between 21" and 25" of snow.
b) There's a 95% chance the region will get between 21" and 25" of snow this winter.
c) There will be between 21" and 25" of snow on the ground for 95% of the winter days.
d) Residents can be 95% sure that the area's average snowfall is between 21" and 25".
e) Residents can be 95% confident that the average snowfall during the past century was between 21" and 25" per winter.

19. **Pulse rates** A medical researcher measured the pulse rates (beats per minute) of a sample of randomly selected adults and found the following Student's t-based confidence interval:

With 95.00% Confidence,
$70.887604 < \mu(Pulse) < 74.497011$

a) Explain carefully what the software output means.
b) What's the margin of error for this interval?
c) If the researcher had calculated a 99% confidence interval, would the margin of error be larger or smaller? Explain.

20. **Crawling** Data collected by child development scientists produced this confidence interval for the average age (in weeks) at which babies begin to crawl:

t-Intervval for μ
(95.00% Confidence): $29.202 < \mu(age) < 31.844$

a) Explain carefully what the software output means.
b) What is the margin of error for this interval?
c) If the researcher had calculated a 90% confidence interval, would the margin of error be larger or smaller? Explain.

21. **CEO compensation** A sample of 20 CEOs from the Forbes 500 shows total annual compensations ranging from a minimum of $0.1 to $62.24 million. The average for these 20 CEOs is $7.946 million. Here's a histogram:

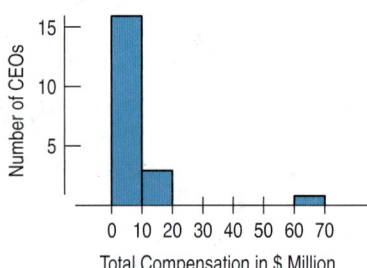

Based on these data, a computer program found that a 95% confidence interval for the mean annual compensation of all Forbes 500 CEOs is (1.69, 14.20) $ million. Why should you be hesitant to trust this confidence interval?

22. **Credit card charges** A credit card company takes a random sample of 100 cardholders to see how much they charged on their card last month. Here's a histogram.

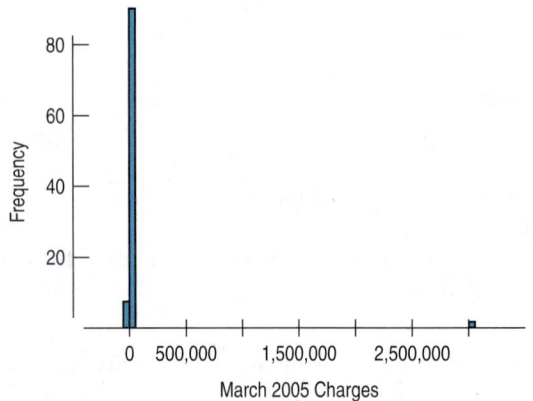

A computer program found that the resulting 95% confidence interval for the mean amount spent in

March 2011 is (−$28,366.84, $90,691.49). Explain why the analysts didn't find the confidence interval useful, and explain what went wrong.

23. **Normal temperature** The researcher described in Exercise 19 also measured the body temperatures of that randomly selected group of adults. Here are summaries of the data he collected. We wish to estimate the average (or "normal") temperature among the adult population.

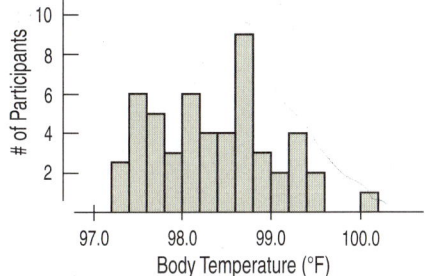

a) Check the conditions for creating a *t*-interval.
b) Find a 98% confidence interval for mean body temperature.
c) Explain the meaning of that interval.
d) Explain what "98% confidence" means in this context.
e) 98.6°F is commonly assumed to be "normal." Do these data suggest otherwise? Explain.

24. **Parking** Hoping to lure more shoppers downtown, a city builds a new public parking garage in the central business district. The city plans to pay for the structure through parking fees. During a two-month period (44 weekdays), daily fees collected averaged $126, with a standard deviation of $15.
a) What assumptions must you make in order to use these statistics for inference?
b) Write a 90% confidence interval for the mean daily income this parking garage will generate.
c) Interpret this confidence interval in context.
d) Explain what "90% confidence" means in this context.
e) The consultant who advised the city on this project predicted that parking revenues would average $130 per day. Based on your confidence interval, do you think the consultant was correct? Why?

25. **Normal temperatures, part II** Consider again the statistics about human body temperature in Exercise 23.

a) Would a 90% confidence interval be wider or narrower than the 98% confidence interval you calculated before? Explain. (Don't compute the new interval.)
b) What are the advantages and disadvantages of the 98% confidence interval?
c) If we conduct further research, this time using a sample of 500 adults, how would you expect the 98% confidence interval to change? Explain.
d) How large a sample might allow you to estimate the mean body temperature to within 0.1 degrees with 98% confidence?

26. **Parking II** Suppose that, for budget planning purposes, the city in Exercise 24 needs a better estimate of the mean daily income from parking fees.
a) Someone suggests that the city use its data to create a 95% confidence interval instead of the 90% interval first created. How would this interval be better for the city? (You need not actually create the new interval.)
b) How would the 95% interval be worse for the planners?
c) How could they achieve an interval estimate that would better serve their planning needs?
d) How many days' worth of data should they collect to have 95% confidence of estimating the true mean to within $3?

27. **Speed of light** In 1882, Michelson measured the speed of light (usually denoted c as in Einstein's famous equation $E = mc^2$). His values are in km/sec and have 299,000 subtracted from them. He reported the results of 23 trials with a mean of 756.22 and a standard deviation of 107.12.
a) Find a 95% confidence interval for the true speed of light from these statistics.
b) State in words what this interval means. Keep in mind that the speed of light is a physical constant that, as far as we know, has a value that is true throughout the universe.
c) What assumptions must you make in order to use your method?

28. **Better light** After his first attempt to determine the speed of light (described in Exercise 27), Michelson conducted an "improved" experiment. In 1897, he reported results of 100 trials with a mean of 852.4 and a standard deviation of 79.0.
a) What is the standard error of the mean for these data?
b) Without computing it, how would you expect a 95% confidence interval for the second experiment to differ from the confidence interval for the first? Note at least three specific reasons why they might differ, and indicate the ways in which these differences would change the interval.
c) According to Stigler (who reports these values), the true speed of light is 299,710.5 km/sec, corresponding to a value of 710.5 for Michelson's 1897 measurements. What does this indicate about Michelson's two experiments? Find a new confidence interval and explain using your confidence interval.

29. Flight on time 2013 What are the chances your flight will leave on time? The U.S. Bureau of Transportation Statistics of the Department of Transportation publishes information about airline performance. We saw the data in a Just Checking exercise in Chapter 4. Here are a histogram and summary statistics for the percentage of flights departing on time each month from 1994 thru October 2013. (www.transtats.bts.gov/HomeDrillChart.asp)

There is no evidence of a trend over time.
a) Check the assumptions and conditions for inference.
b) Find a 90% confidence interval for the true percentage of flights that depart on time.
c) Interpret this interval for a traveler planning to fly.

30. Flight on time 2013 revisited Will your flight get you to your destination on time? The U.S. Bureau of Transportation Statistics reported the percentage of flights that were delayed each month from 1994 through October of 2013. Here's a histogram, along with some summary statistics:

We can consider these data to be a representative sample of all months. There is no evidence of a time trend ($r = 0.07$).
a) Check the assumptions and conditions for inference about the mean.
b) Find a 99% confidence interval for the true percentage of flights that arrive late.
c) Interpret this interval for a traveler planning to fly.

31. Farmed salmon, second look This chapter's For Examples looked at mirex contamination in farmed salmon. We first found a 95% confidence interval for the mean concentration to be 0.0834 to 0.0992 parts per million. Later we rejected the null hypothesis that the mean did not exceed the EPA's recommended safe level of 0.08 ppm based on a P-value of 0.0027. Explain how these two results are consistent. Your explanation should discuss the confidence level, the P-value, and the decision.

32. Hot dogs A nutrition lab tested 40 hot dogs to see if their mean sodium content was less than the 325-mg upper limit set by regulations for "reduced sodium" franks. The lab failed to reject the hypothesis that the hot dogs did not meet this requirement, with a P-value of 0.142. A 90% confidence interval estimated the mean sodium content for this kind of hot dog at 317.2 to 326.8 mg. Explain how these two results are consistent. Your explanation should discuss the confidence level, the P-value, and the decision.

33. Pizza A researcher tests whether the mean cholesterol level among those who eat frozen pizza exceeds the value considered to indicate a health risk. She gets a P-value of 0.07. Explain in this context what the "7%" represents.

34. Golf balls The United States Golf Association (USGA) sets performance standards for golf balls. For example, the initial velocity of the ball may not exceed 250 feet per second when measured by an apparatus approved by the USGA. Suppose a manufacturer introduces a new kind of ball and provides a sample for testing. Based on the mean speed in the test, the USGA comes up with a P-value of 0.34. Explain in this context what the "34%" represents.

35. Fuel economy 2014 revisited In Chapter 6, we examined the average fuel economy of several 2014 model vehicles.
a) Find and interpret a 95% confidence interval for the gas mileage of 2014 vehicles.
b) Do you think that this confidence interval captures the mean gas mileage for all 2014 vehicles?

36. Computer lab fees The technology committee has stated that the average time spent by students per lab visit has increased, and the increase supports the need for increased lab fees. To substantiate this claim, the committee randomly samples 12 student lab visits and notes the amount of time spent using the computer. The times in minutes are as follows:

Time	
52	74
57	53
54	136
76	73
62	8
52	62

a) Plot the data. Are any of the observations outliers? Explain.

b) The previous mean amount of time spent using the lab computer was 55 minutes. Find a 95% confidence interval for the true mean. What do you conclude about the claim? If there are outliers, find intervals with and without the outliers present.

37. **Marriage** In 1960, census results indicated that the age at which American men first married had a mean of 23.3 years. It is widely suspected that young people today are waiting longer to get married. We want to find out if the mean age of first marriage has increased during the past 40 years.
   a) Write appropriate hypotheses.
   b) We plan to test our hypothesis by selecting a random sample of 40 men who married for the first time last year. Do you think the necessary assumptions for inference are satisfied? Explain.
   c) Describe the approximate sampling distribution model for the mean age in such samples.
   d) The men in our sample married at an average age of 24.2 years, with a standard deviation of 5.3 years. What's the P-value for this result?
   e) Explain (in context) what this P-value means.
   f) What's your conclusion?

38. **Saving gas** Congress regulates corporate fuel economy and sets an annual gas mileage for cars. A company with a large fleet of cars hopes to meet the 2011 goal of 30.2 mpg or better for their fleet of cars. To see if the goal is being met, they check the gasoline usage for 50 company trips chosen at random, finding a mean of 32.12 mpg and a standard deviation of 4.83 mpg. Is this strong evidence that they have attained their fuel economy goal?
   a) Write appropriate hypotheses.
   b) Are the necessary assumptions to make inferences satisfied?
   c) Describe the sampling distribution model of mean fuel economy for samples like this.
   d) Find the P-value.
   e) Explain what the P-value means in this context.
   f) State an appropriate conclusion.

39. **Ruffles** Students investigating the packaging of potato chips purchased 6 bags of Lay's Ruffles marked with a net weight of 28.3 grams. They carefully weighed the contents of each bag, recording the following weights (in grams): 29.3, 28.2, 29.1, 28.7, 28.9, 28.5.
   a) Do these data satisfy the assumptions for inference? Explain.
   b) Find the mean and standard deviation of the weights.
   c) Create a 95% confidence interval for the mean weight of such bags of chips.
   d) Explain in context what your interval means.
   e) Comment on the company's stated net weight of 28.3 grams.

40. **Doritos** Some students checked 6 bags of Doritos marked with a net weight of 28.3 grams. They carefully weighed the contents of each bag, recording the following weights (in grams): 29.2, 28.5, 28.7, 28.9, 29.1, 29.5.
   a) Do these data satisfy the assumptions for inference? Explain.
   b) Find the mean and standard deviation of the weights.
   c) Create a 95% confidence interval for the mean weight of such bags of chips.
   d) Explain in context what your interval means.
   e) Comment on the company's stated net weight of 28.3 grams.

41. **Popcorn** Yvon Hopps ran an experiment to test optimum power and time settings for microwave popcorn. His goal was to find a combination of power and time that would deliver high-quality popcorn with less than 10% of the kernels left unpopped, on average. After experimenting with several bags, he determined that power 9 at 4 minutes was the best combination. To be sure that the method was successful, he popped 8 more bags of popcorn (selected at random) at this setting. All were of high quality, with the following percentages of uncooked popcorn: 7, 13.2, 10, 6, 7.8, 2.8, 2.2, 5.2. Does this provide evidence that he met his goal of an average of no more than 10% uncooked kernels? Explain.

42. **Ski wax** Bjork Larsen was trying to decide whether to use a new racing wax for cross-country skis. He decided that the wax would be worth the price if he could average less than 55 seconds on a course he knew well, so he planned to test the wax by racing on the course 8 times. His 8 race times were 56.3, 65.9, 50.5, 52.4, 46.5, 57.8, 52.2, and 43.2 seconds. Should he buy the wax? Explain.

43. **Chips Ahoy!** In 1998, as an advertising campaign, the Nabisco Company announced a "1000 Chips Challenge," claiming that every 18-ounce bag of their Chips Ahoy! cookies contained at least 1000 chocolate chips. Dedicated Statistics students at the Air Force Academy (no kidding) purchased some randomly selected bags of cookies and counted the chocolate chips. Some of their data are given below. (Source: *Chance*, 12, no. 1[1999])

| 1219 | 1214 | 1087 | 1200 | 1419 | 1121 | 1325 | 1345 |
| 1244 | 1258 | 1356 | 1132 | 1191 | 1270 | 1295 | 1135 |

   a) Check the assumptions and conditions for inference. Comment on any concerns you have.
   b) Create a 95% confidence interval for the average number of chips in bags of Chips Ahoy! cookies.
   c) What does this evidence say about Nabisco's claim? Use your confidence interval to test an appropriate hypothesis and state your conclusion.

44. **Yogurt** *Consumer Reports* tested 11 brands of vanilla yogurt and found these numbers of calories per serving:

130 160 150 120 120 110 170 160 110 130 90

a) Check the assumptions and conditions for inference.
b) Create a 95% confidence interval for the average calorie content of vanilla yogurt.
c) A diet guide claims that you will get an average of 120 calories from a serving of vanilla yogurt. What does this evidence indicate? Use your confidence interval to test an appropriate hypothesis and state your conclusion.

T 45. Maze Psychology experiments sometimes involve testing the ability of rats to navigate mazes. The mazes are classified according to difficulty, as measured by the mean length of time it takes rats to find the food at the end. One researcher needs a maze that will take rats an average of about one minute to solve. He tests one maze on several rats, collecting the data shown.

Time (sec)	
38.4	57.6
46.2	55.5
62.5	49.5
38.0	40.9
62.8	44.3
33.9	93.8
50.4	47.9
35.0	69.2
52.8	46.2
60.1	56.3
55.1	

a) Plot the data. Do you think the conditions for inference are satisfied? Explain.
b) Test the hypothesis that the mean completion time for this maze is 60 seconds. What is your conclusion?
c) Eliminate the outlier, and test the hypothesis again. What is your conclusion?
d) Do you think this maze meets the "one-minute average" requirement? Explain.

46. Braking A tire manufacturer is considering a newly designed tread pattern for its all-weather tires. Tests have indicated that these tires will provide better gas mileage and longer tread life. The last remaining test is for braking effectiveness. The company hopes the tire will allow a car traveling at 60 mph to come to a complete stop within an average of 125 feet after the brakes are applied. They will adopt the new tread pattern unless there is strong evidence that the tires do not meet this objective. The distances (in feet) for 10 stops on a test track were 129, 128, 130, 132, 135, 123, 102, 125, 128, and 130. Should the company adopt the new tread pattern? Test an appropriate hypothesis and state your conclusion. Explain how you dealt with the outlier and why you made the recommendation you did.

T 47. Golf drives III 2013 In Chapter 3 we looked at distance that 155 professional golfers drove the ball (in yds) during a week in 2013. Here is a histogram of those drives (espn.go.com/golf/statistics/_/year/2013/type/expanded). Assuming this is a representative sample of all professional golfers,

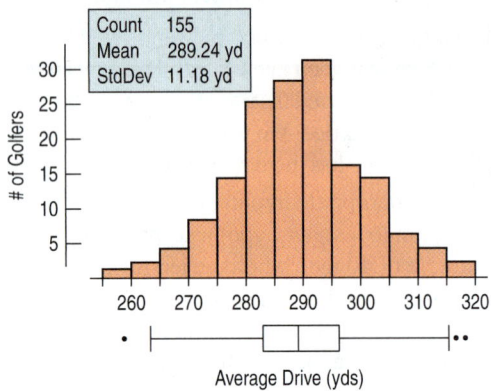

a) Find a 95% confidence interval for the mean drive distance.
b) Interpreting this interval raises some problems. Discuss.
c) Suppose the data were actually the mean driving distance for each golfer. Would that be a concern in interpreting the interval? (*Hint:* Review the What Can Go Wrong warnings of Chapter 8. Chapter 8?! Yes, Chapter 8.)

T 48. Wind power Should you generate electricity with your own personal wind turbine? That depends on whether you have enough wind on your site. To produce enough energy, your site should have an annual average wind speed above 8 miles per hour, according to the Wind Energy Association. One candidate site was monitored for a year, with wind speeds recorded every 6 hours. A total of 1114 readings of wind speed averaged 8.019 mph with a standard deviation of 3.813 mph. You've been asked to make a statistical report to help the landowner decide whether to place a wind turbine at this site.

a) Discuss the assumptions and conditions for using Student's *t* inference methods with these data. Here are some plots that may help you decide whether the methods can be used:

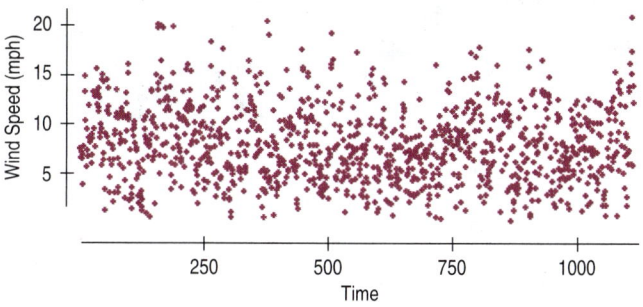

b) What would you tell the landowner about whether this site is suitable for a small wind turbine? Explain.

Just Checking ANSWERS

1. Questions on the short form are answered by everyone in the population. This is a census, so means or proportions *are* the true population values. The long forms are given just to a sample of the population. When we estimate parameters from a sample, we use a confidence interval to take sample-to-sample variability into account.

2. They don't know the population standard deviation, so they must use the sample standard deviation as an estimate. The additional uncertainty is taken into account by *t*-models.

3. The margin of error for a confidence interval for a mean depends, in part, on the standard error,

$$SE(\bar{y}) = \frac{s}{\sqrt{n}}.$$

Since n is in the denominator, smaller sample sizes lead to larger *SE*s and correspondingly wider intervals. Long forms returned by one in every six or seven households in a less populous area will be a smaller sample.

4. The critical values for *t* with fewer degrees of freedom would be slightly larger. The $\sqrt{n}$ part of the standard error changes a lot, making the *SE* much larger. Both would increase the margin of error.

5. The smaller sample is one fourth as large, so the confidence interval would be roughly twice as wide.

6. We expect 95% of such intervals to cover the true value, so we would expect about 5 of the 100 intervals to miss.

Chapter 21: More About Tests and Intervals

21.1 Choosing Hypotheses
21.2 How to Think About P-Values
21.3 Alpha Levels
21.4 Critical Values for Hypothesis Tests
21.5 Errors

Where are we going?

A news headline reports a "statistically significant" increase in global temperatures. Another says that studies have found no "statistically significant" benefits of taking vitamin C to prevent colds. What does significance really mean? Can failing to reject the null hypothesis sometimes be as important as rejecting it? Knowing what the hypotheses were and how the researchers arrived at their conclusion can help you decide whether you agree with the findings of their studies. We'll look at hypothesis testing in more depth in this chapter.

Who	Florida motorcycle riders aged 20 and younger involved in motorcycle accidents
What	% wearing helmets
When	2001–2003
Where	Florida
Why	Assessment of injury rates commissioned by the National Highway Traffic Safety Administration (NHTSA)

In 2000, Florida changed its motorcycle helmet law. No longer are riders 21 and older required to wear helmets. Under the new law, those under 21 still must wear helmets, but a report by the Preusser Group (www.preussergroup.com) suggests that helmet use may have declined in this group, too.

It isn't practical to survey young motorcycle riders. (For example, how can you construct a sampling frame? If you contacted licensed riders, would they admit to riding illegally without a helmet?) The researchers adopted a different strategy. Police reports of motorcycle accidents record whether the rider wore a helmet and give the rider's age. Before the change in the helmet law, 60% of youths involved in a motorcycle accident had been wearing their helmets. The Preusser study looked at accident reports during 2001–2003, the three years following the law change, considering these riders to be a representative sample of the larger population. They observed 781 young riders who were involved in accidents. Of these, 396 (or 50.7%) were wearing helmets. Is this evidence of a decline in helmet-wearing, or just the natural fluctuation of such statistics?

21.1 Choosing Hypotheses

How do we choose the null hypothesis? The appropriate null arises directly from the context of the problem. It is dictated, not by the data, but by the situation. One good way to identify both the null and alternative hypotheses is to think about why the study is being done and what we hope to learn from the test. We are interested in seeing whether the proportion using a helmet has decreased, but our null hypothesis is that the proportion is unchanged at 60%. The alternative is that the proportion has decreased.

The same reasoning works for means. If we wanted to test athletes' strength training with new equipment, we would be interested in whether the new equipment *increases* their strength. In that case, the null hypothesis would be that their mean strength hasn't changed. The alternative would be that the mean has increased.

To write a null hypothesis, identify a parameter and choose a null value that relates to the question at hand. Even though the null usually means no difference or no change, you can't automatically interpret "null" to mean zero. A claim that "nobody" wears a motorcycle helmet would be absurd, so we don't write $H_0: p = 0$. The null hypothesis for the Florida study is that the true proportion of helmet use remained unchanged at $p = 0.60$ among young riders after the law changed. You need to find the value for the parameter in the null hypothesis from the context of the problem.

There is a temptation to state your *claim* as the null hypothesis. As we have seen, however, you cannot prove a null hypothesis true any more than you can prove a defendant innocent. So, it makes more sense to use what you want to show as the *alternative*. This way, if the data support your claim, the P-value will be small. When you then reject the null, you will be left with what you want to show.

For Example WRITING HYPOTHESES

The diabetes drug Avandia® was approved to treat Type 2 diabetes in 1999. But an article in the *New England Journal of Medicine* (*NEJM*)[1] raised concerns that the drug might carry an increased risk of heart attack. This study combined results from a number of other separate studies to obtain an overall sample of 4485 diabetes patients taking Avandia. People with Type 2 diabetes are known to have about a 20.2% chance of suffering a heart attack within a seven-year period. According to the article's author, Dr. Steven E. Nissen,[2] the risk found in the *NEJM* study was equivalent to a 28.9% chance of heart attack over seven years. The FDA is the government agency responsible for relabeling Avandia to warn of the risk if it is judged to be unsafe. Although the statistical methods they use are more sophisticated, we can get an idea of their reasoning with the tools we have learned.

QUESTION: What null hypothesis and alternative hypothesis about seven-year heart attack risk would you test? Explain.

ANSWER:

$$H_0: p = 0.202$$
$$H_A: p > 0.202$$

The parameter of interest is the proportion of diabetes patients suffering a heart attack in seven years. The FDA is concerned only with whether Avandia *increases* the seven-year risk of heart attacks above the baseline value of 20.2%, so a one-sided upper-tail test is appropriate.

> **One-sided or two?** In the 1930s, a series of experiments was performed at Duke University in an attempt to see whether humans were capable of extrasensory perception, or ESP. Psychologist Karl Zener designed a set of cards with 5 symbols later made infamous in the movie *Ghostbusters*:
>
>
>
> In the experiment, the "sender" selects one of the 5 cards at random from a deck and then concentrates on it. The "receiver" tries to determine which card it is. If we let p be the proportion of correct responses, what's the null hypothesis? The null hypothesis is that ESP makes no difference. Without ESP, the receiver would just be guessing, and since there are 5 possible responses, there would be a 20% chance of guessing each card correctly. So, H_0 is $p = 0.20$. What's the alternative? It seems that it should be $p > 0.20$, a one-sided alternative. But some ESP researchers have expressed the claim that if the proportion guessed were much *lower* than expected, that would show an "interference" and should be considered evidence for ESP as well. So they argue for a two-sided alternative.

---

[1] Steven E. Nissen, M.D., and Kathy Wolski, M.P.H., "Effect of Rosiglitazone on the Risk of Myocardial Infarction and Death from Cardiovascular Causes," *NEJM* 2007; 356.

[2] Interview reported in the *New York Times* [May 26, 2007].

21.2 How to Think About P-Values

WHICH CONDITIONAL?

Suppose that as a political science major you are offered the chance to be a White House intern. There would be a very high probability that next summer you'd be in Washington, D.C. That is, $P(\text{Washington} \mid \text{Intern})$ would be high. But if we find a student in Washington, D.C., is it likely that he's a White House intern? Almost surely not; $P(\text{Intern} \mid \text{Washington})$ is low. You can't switch around conditional probabilities. The P-value is $P(\text{data} \mid H_0)$. We might wish we could report $P(H_0 \mid \text{data})$, but these two quantities are NOT the same.

A P-value is a conditional probability. It tells us the probability of getting results at least as unusual as the observed statistic, *given* that the null hypothesis is true. We can write P-value $= P(\text{observed statistic value [or even more extreme]} \mid H_0)$.

Writing the P-value this way helps to make clear that the P-value is *not* the probability that the null hypothesis is true. It is a probability about the data. Let's say that again:

The P-value is not the probability that the null hypothesis is true.

The P-value is not even the conditional probability that the null hypothesis is true given the data. We would write that probability as $P(H_0 \mid \text{observed statistic value})$. This is a conditional probability but in reverse. It would be nice to know this probability, but it's impossible to calculate without making additional assumptions. As we saw in Chapter 14, reversing the order in a conditional probability is difficult, and the results can be counterintuitive.

We can find the P-value, $P(\text{observed statistic value} \mid H_0)$, because H_0 gives the parameter values that we need to calculate the required probability. But there's no direct way to find $P(H_0 \mid \text{observed statistic value})$.[3] As tempting as it may be to say that a P-value of 0.03 means there's a 3% chance that the null hypothesis is true, that just isn't right. All we can say is that, given the null hypothesis, there's a 3% chance of observing the statistic value that we have actually observed (or one more unlike the null value).

What to Do with a Small P-Value

We know that a small P-value means that the result we just observed is unlikely to occur if the null hypothesis is true. So we have evidence against the null hypothesis. An even smaller P-value implies stronger evidence against the null hypothesis, but it doesn't mean that the null hypothesis is "less true" (see "How Guilty Is the Suspect" on page 551).

How small the P-value has to be for you to reject the null hypothesis depends on a lot of things, not all of which can be precisely quantified. Your belief in the null hypothesis will influence your decision. Your trust in the data, in the experimental method if the data come from a planned experiment, in the survey protocol if the data come from a designed survey, all influence your decision. The P-value should serve as a measure of the strength of the evidence against the null hypothesis, but should never serve as a hard and fast rule for decisions. You have to take that responsibility on yourself.

As a review, let's look at the helmet law example from the chapter opener. Did helmet wearing among young riders decrease after the law allowed older riders to ride without helmets? What is the evidence?

Step-by-Step Example ANOTHER ONE-PROPORTION z-TEST

Question: Has helmet use in Florida declined among riders under the age of 21 subsequent to the change in the helmet laws?

---

[3] The approach to statistical inference known as Bayesian Statistics addresses the question in just this way, but it requires more advanced mathematics and more assumptions. See p. 379 for more about the founding father of this approach.

THINK **Plan** State the problem and discuss the variables and the W's. **Hypotheses** The null hypothesis is established by the rate set before the change in the law. The study was concerned with safety, so they'll want to know of any decline in helmet use, making this a lower-tail test.		I want to know whether the rate of helmet wearing among Florida's motorcycle riders under the age of 21 decreased after the law changed to allow older riders to go without helmets. The proportion before the law was passed was 60% so I'll use that as my null hypothesis value. The alternative is one-sided because I'm interested only in seeing if the rate decreased. I have data from accident records showing 396 of 781 young riders were wearing helmets. $H_0: p = 0.60$ $H_A: p < 0.60$
SHOW **Model** Check the conditions. Specify the sampling distribution model and name the test.		✓ **Independence Assumption:** The data are for riders involved in accidents during a three-year period. Individuals are independent of one another. ✓ **Randomization Condition:** The Risky Behavior Surveillance survey is in fact a complex, multi-stage sample, but it is randomized and great effort is taken to make it representative. It is safe to treat it as though it were a random sample. ✓ **Success/Failure Condition:** We'd expect $np = 781(0.6) = 468.6$ helmeted riders and $nq = 781(0.4) = 312.4$ non-helmeted. Both are at least 10. The conditions are satisfied, so I can use a Normal model and perform a **one-proportion z-test**.
SHOW **Mechanics** Find the standard deviation of the sampling model using the hypothesized proportion. Find the z-score for the observed proportion. Make a picture. Sketch a Normal model centered at the hypothesized helmet rate of 60%. This is a lower-tail test, so shade the region to the left of the observed rate. Given this z-score, the P-value is obviously very low.		There were 396 helmet wearers among the 781 accident victims. $\hat{p} = \dfrac{396}{781} = 0.507$ $SD(\hat{p}) = \sqrt{\dfrac{p_0 q_0}{n}} = \sqrt{\dfrac{(0.60)(0.40)}{781}} = 0.0175$ $z = \dfrac{\hat{p} - p_0}{SD(\hat{p})} = \dfrac{0.507 - 0.60}{0.0175} = -5.31$ 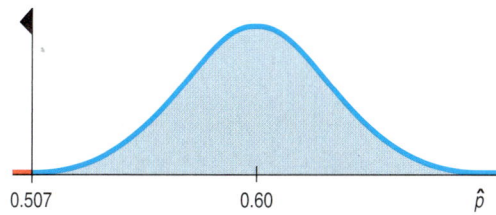 The observed helmet rate is 5.31 standard deviations below the former rate. The corresponding P-value is less than 0.001.

> **TELL ➡ Conclusion** Link the P-value to your decision about the null hypothesis, and then state your conclusion in context.
>
> The very small P-value says that if the true rate of helmet-wearing among riders under 21 were still 60%, the probability of observing a rate no higher than 50.7% in a sample like this is less than 1 chance in 1000, so I reject the null hypothesis. There is strong evidence that there has been a decline in helmet use among riders under 21.

The P-value in the helmet example is quite small—less than 0.001. That's strong evidence to suggest that the rate has decreased since the law was changed. But it doesn't say that it was "a lot lower." To answer that question, you'd need to construct a confidence interval:

$$\hat{p} \pm z^* \sqrt{\frac{\hat{p}\hat{q}}{n}} = 0.507 \pm 1.96(0.0175) = (0.473, 0.541)$$

(using 95% confidence).

There is strong evidence that the rate is no longer 60%, but the small P-value by itself says nothing about how much lower the rate might be. The confidence interval provides that information; the rate seems to be closer to 50% now. Whether a change from 60% to 50% makes an important difference in safety is a judgment that depends on the situation, but not on the P-value. Not coincidentally, on July 1, 2008, Florida required a motorcycle "endorsement" for all motorcycle riders. For riders under 21, that requires a motorcycle safety course. Although only about 70% of motorcycle riders are endorsed, the percentage of unendorsed riders involved in crashes dropped considerably after 2008.[4]

> **HOW GUILTY IS THE SUSPECT?**
> We might like to know $P(H_0 | \text{data})$, but when you think about it, we can't talk about the probability that the null hypothesis is true. The null is not a random event, so either it is true or it isn't. The data, however, are random in the sense that if we were to repeat a randomized experiment or draw another random sample, we'd get different data and expect to find a different statistic value. So we can talk about the probability of the data given the null hypothesis, and that's the P-value.
>
> But it does make sense that the smaller the P-value, the more confident we can be in declaring that we doubt the null hypothesis. Think again about the jury trial. Our null hypothesis is that the defendant is innocent. Then the evidence starts rolling in. A car the same color as his was parked in front of the bank. Well, there are lots of cars that color. The probability of that happening (given his innocence) is pretty high, so we're not persuaded that he's guilty. The bank's security camera showed the robber was male and about the defendant's height and weight. Hmmm. Could that be a coincidence? If he's innocent, then it's a little less likely that the car and description would *both* match, so our P-value goes down. We're starting to question his innocence a little. Witnesses said the robber wore a blue jacket just like the one the police found in a garbage can behind the defendant's house. Well, if he's innocent, then that doesn't seem very likely, does it? If he's really innocent, the probability that all of these could have happened is getting pretty low. Now our P-value may be small enough to be called "beyond a reasonable doubt" and lead to a conviction. Each new piece of evidence strains our skepticism a bit more. The more compelling the evidence—the more *unlikely* it would be were he innocent—the more convinced we become that he's guilty.
>
> But even though it may make *us* more confident in declaring him guilty, additional evidence does not make *him* any guiltier. Either he robbed the bank or he didn't. Additional evidence (like the teller picking him out of a police lineup) just makes us more confident that we did the right thing when we convicted him. The lower the P-value, the more comfortable we feel about our decision to reject the null hypothesis, but the null hypothesis doesn't get any more false.

"The wise man proportions his belief to the evidence."
—David Hume,
"Enquiry Concerning Human Understanding," 1748

"You're so guilty now."
—Rearview Mirror

---

[4] www.ridesmartflorida.com

For Example THINKING ABOUT THE P-VALUE

RECAP: A *New England Journal of Medicine* paper reported that the seven-year risk of heart attack in diabetes patients taking the drug Avandia was increased from the baseline of 20.2% to an estimated risk of 28.9% and said the P-value was 0.03.

QUESTION: How should the P-value be interpreted in this context?

ANSWER: The P-value = $P(\hat{p} \geq 28.9\% \mid p = 20.2\%)$. That is, it's the probability of seeing such a high heart attack rate among the people studied if, in fact, taking Avandia really didn't increase the risk at all.

What to Do with a High P-Value

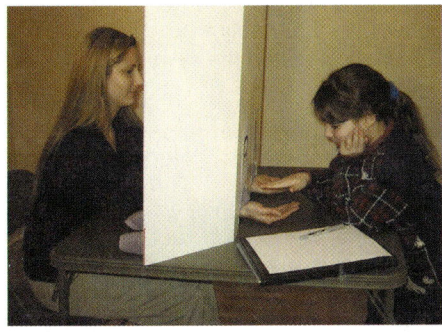

Video: Is There Evidence for Therapeutic Touch? This video shows the experiment and tells the story.

Activity: Testing Therapeutic Touch. Perform the one-proportion z-test using *ActivStats* technology. The test in *ActivStats* is two-sided. Do you think this is the appropriate choice?

Therapeutic touch (TT), taught in many schools of nursing, is a therapy in which the practitioner moves her hands near, but does not touch, a patient in an attempt to manipulate a "human energy field." Therapeutic touch practitioners believe that by adjusting this field they can promote healing. However, no instrument has ever detected a human energy field, and no experiment has ever shown that TT practitioners can detect such a field.

In 1998, the *Journal of the American Medical Association* published a paper reporting work by a then nine-year-old girl.[5] She had performed a simple experiment in which she challenged 15 TT practitioners to detect whether her unseen hand was hovering over their left or right hand (selected by the flip of a coin).

The practitioners "warmed up" with a period during which they could see the experimenter's hand, and each said that they could detect the girl's human energy field. Then a screen was placed so that the practitioners could not see the girl's hand, and they attempted 10 trials each. Overall, of 150 trials, the TT practitioners were successful only 70 times—a success proportion of 46.7%.

The null hypothesis here is that the TT practitioners were just guessing. If that were the case, since the hand was chosen using a coin flip, the practitioners would guess correctly 50% of the time. So the null hypothesis is that $p = 0.5$ and the alternative that they could actually detect a human energy field is (one-sided) $p > 0.5$.

What would constitute evidence that they weren't guessing? Certainly, a very high proportion of correct guesses out of 150 would convince most people. Exactly how high the proportion of correct guesses has to be for you to reject the null hypothesis depends on how small a P-value you need to be convinced (Which, in turn, depends on how often you're willing to make mistakes—a topic we'll discuss later in the chapter).

But let's look again at the TT practitioners' proportion. Does it provide any evidence that they weren't guessing? The proportion of correct guesses is 46.7%—that's *less* than the hypothesized value, not greater! When we find $SD(\hat{p}) = 0.041$ (or 4.1%) we can see that 46.7% is almost 1 SD *below* the hypothesized proportion:

$$SD(\hat{p}) = \sqrt{\frac{p_0 q_0}{n}} = \sqrt{\frac{(0.5)(0.5)}{150}} \approx 0.041$$

The observed proportion, $\hat{p}$, is 0.467.

$$z = \frac{\hat{p} - p_0}{SD(\hat{p})} = \frac{0.467 - 0.5}{0.041} = -0.805$$

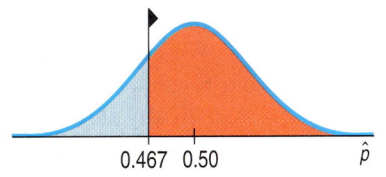

---

[5] L. Rosa, E. Rosa, L. Sarner, and S. Barrett, "A Close Look at Therapeutic Touch," *JAMA* 279(13) [1 April 1998]: 1005–1010.

The observed success rate is 0.805 standard deviations below the hypothesized mean.

$$P = P(z > -0.805) = 0.790$$

If the practitioners had been highly successful, we would have seen a low P-value. In that case, we would then have concluded that they could actually detect a human energy field.

But that's not what happened. What we observed was a $\hat{p} = 0.467$ success rate. The P-value for this proportion is greater than 0.5 because the observed value is on the "wrong" side of the null hypothesis value. Wrong, that is, if you wanted to find evidence against the null hypothesis. To convince us, the practitioners should be doing better than guessing, not worse!

Obviously, we won't be rejecting the null hypothesis; for us to reject it, the P-value would have to be quite small. But a P-value of 0.790 seems so big it is almost strange. With a success rate even lower than chance, we could have concluded right away that we have no evidence for rejecting H_0. In fact, for a one-sided test, any time our test statistic is on the "wrong" side of the hypothesis, we know immediately that the P-value is at least 0.50 and we know that we have no evidence to reject the null hypothesis.

Big P-values just mean that what we've observed isn't surprising. That is, the results are in line with our assumption that the null hypothesis models the world, so we have no reason to reject it. A big P-value doesn't prove that the null hypothesis is true, but it certainly offers no evidence that it's *not* true. When we see a large P-value, all we can say is that we "don't reject the null hypothesis."

For Example   MORE ABOUT P-VALUES

RECAP: The question of whether the diabetes drug Avandia increased the risk of heart attack was raised by a study in the *New England Journal of Medicine*. This study estimated the seven-year risk of heart attack to be 28.9% and reported a P-value of 0.03 for a test of whether this risk was higher than the baseline seven-year risk of 20.2%. An earlier study (the ADOPT study) had estimated the seven-year risk to be 26.9% and reported a P-value of 0.27.

QUESTION: Why did the researchers in the ADOPT study not express alarm about the increased risk they had seen?

ANSWER: A P-value of 0.27 means that a heart attack rate at least as high as the one they observed could be expected in 27% of similar experiments even if, in fact, there were no increased risk from taking Avandia. That's not remarkable enough to reject the null hypothesis. In other words, the ADOPT study wasn't convincing.

21.3 Alpha Levels

Activity: Rejecting the Null Hypothesis. See alpha levels at work in the animated hypothesis-testing tool.

■ **NOTATION ALERT**
The first Greek letter, α, is used in Statistics for the threshold value of a hypothesis test. You'll hear it referred to as the alpha level. Common values are 0.10, 0.05, 0.01, and 0.001.

Sometimes we need to make a firm decision about whether or not to reject the null hypothesis. A jury must *decide* whether the evidence reaches the level of "beyond a reasonable doubt." A business must *select* a Web design. You need to decide which section of Statistics to enroll in.

When the P-value is small, it tells us that our data are rare, *given the null hypothesis*. As humans, we are suspicious of rare events. If the data are "rare enough," we just don't think that could have happened due to chance. Since the data *did* happen, something must be wrong. All we can do now is reject the null hypothesis.

But how rare is "rare"?

We can define "rare event" arbitrarily by setting a threshold for our P-value. If our P-value falls below that point, we'll reject the null hypothesis. We call such results **statistically significant**. The threshold is called an **alpha level**. Not surprisingly, it's labeled with the Greek letter α. Common α-levels are 0.10, 0.05, 0.01, and 0.001. You have the option—almost the *obligation*—to consider your alpha level carefully and choose an appropriate one for the situation. If you're assessing the safety of air bags, you'll want a low alpha level; even 0.01 might not be low enough. If you're just wondering whether folks prefer their pizza with or without pepperoni, you might be happy with $\alpha = 0.10$. It can be hard to justify your choice of α, though, so often we arbitrarily choose 0.05. Note, however: You must select the alpha level *before* you look at the data. Otherwise you can be accused of cheating by tuning your alpha level to suit the data.

> **Where did the value 0.05 come from?** In 1931, in a famous book called *The Design of Experiments*, Sir Ronald Fisher (see p. 523) discussed the amount of evidence needed to reject a null hypothesis. He said that it was *situation dependent*, but remarked, somewhat casually, that for many scientific applications, 1 out of 20 *might be* a reasonable value. Since then, some people—indeed some entire disciplines—have treated the number 0.05 as sacrosanct.

The alpha level is also called the **significance level**. When we reject the null hypothesis, we say that the test is "significant at that level." For example, we might say that we reject the null hypothesis "at the 5% level of significance."

What can you say if the P-value does not fall below α?

When you have not found sufficient evidence to reject the null according to the standard you have established, you should say that "The data have failed to provide sufficient evidence to reject the null hypothesis." Don't say that you "accept the null hypothesis." You certainly haven't proven or established it; it was merely assumed to begin with. Say that you've failed to reject it.

Think again about the therapeutic touch example. The P-value was 0.788. This is so much larger than any reasonable alpha level that we can't reject H_0. For this test, we'd conclude, "We fail to reject the null hypothesis. There is insufficient evidence to conclude that the practitioners are performing better than they would if they were just guessing."

The automatic nature of the reject/fail-to-reject decision when we use an alpha level may make you uncomfortable. If your P-value falls just slightly above your alpha level, you're not allowed to reject the null. Yet a P-value just barely below the alpha level leads to rejection. If this bothers you, you're in good company. Many statisticians think it better to report the P-value than to base a decision on an arbitrary alpha level.

> **IT'S IN THE STARS**
> Some disciplines carry the idea further and code P-values by their size. In this scheme, a P-value between 0.05 and 0.01 gets highlighted by *. A P-value between 0.01 and 0.001 gets **, and a P-value less than 0.001 gets ***. This can be a convenient summary of the weight of evidence against the null hypothesis if it's not taken too literally. But we warn you against taking the distinctions too seriously and against making a black-and-white decision near the boundaries. The boundaries are a matter of tradition, not science; there is nothing special about 0.05. A P-value of 0.051 should be looked at very seriously and not casually thrown away just because it's larger than 0.05, and one that's 0.009 is not very different from one that's 0.011.

> **It could happen to you!** Of course, if the null hypothesis *is* true, no matter what alpha level you choose, you still have a probability α of rejecting the null hypothesis by mistake. This is the rare event we want to protect ourselves against. When we do reject the null hypothesis, no one ever thinks that *this* is one of those rare times. As statistician Stu Hunter notes, "*The statistician says 'rare events do happen—but not to me!'*"

When you decide to declare a verdict, it's always a good idea to report the P-value as an indication of the strength of the evidence. Sometimes it's best to report that the conclusion is not yet clear and to suggest that more data be gathered. (In a trial, a jury may "hang" and be unable to return a verdict.) In these cases, the P-value is the best summary we have of what the data say or fail to say about the null hypothesis.

Practical vs. Statistical Significance

What do we mean when we say that a test is statistically significant? All we mean is that the test statistic had a P-value lower than our alpha level. Don't be lulled into thinking that statistical significance carries with it any sense of practical importance or impact.

For large samples, even small, unimportant ("insignificant") deviations from the null hypothesis can be statistically significant. On the other hand, if the sample is not large enough, even large financially or scientifically "significant" differences may not be statistically significant.

It's good practice to report a confidence interval for the parameter along with the P-value on which we base statistical significance to indicate the range of plausible values for the parameter.

> **Statistically significant? Yes, but is it important?** A large insurance company mined its data and found a statistically significant ($P = 0.04$) difference between the mean value of policies sold in 2013 and 2014. The difference in the mean values was $9.83. Even though it was statistically significant, management did not see this as an important difference when a typical policy sold for more than $1000. On the other hand, even a clinically important improvement of 10% in cure rate with a new treatment is not likely to be statistically significant in a study of fewer than 225 patients. A small clinical trial would probably not be conclusive.

For Example ARE THEY SPEEDING?

Motor vehicle crashes are a leading cause of death for people between 4 and 33 years old. In the past 10 years motor vehicle accidents have claimed the lives of over 40,000 people a year in the United States. This means that, on average, motor vehicle crashes resulted in about 110 deaths each day, or 1 death every 13 minutes. Speeding is a contributing factor in 31% of all fatal accidents, according to the National Highway Traffic Safety Administration.

College Terrace is a neighborhood of Palo Alto, California on the edge of Stanford University. The town has implemented a series of traffic-calming measures over the past decade including building speed tables, center median islands, and traffic circles and reducing the number of through streets. The posted speed limit in College Terrace is 25 mph. The town reports that speeds and traffic volumes in the neighborhood have been "noticeably reduced" so they are seeking approval to make the changes permanent.

A Palo Alto resident, convinced that cars were still speeding by his house, measured speeds of every car going by his house by using two infrared photoelectric gates connected to a microcontroller. We can examine 250 of the speeds randomly selected from more than 2000 cars that he recorded.

QUESTION: Is the mean speed of the cars greater than 25 mph?

ANSWER:

$$H_0: \mu = 25 \text{ mph}$$
$$H_A: \mu > 25 \text{ mph}$$

Independence Assumption: The data were a census of cars over several days. We have drawn a random sample of about 10% of the data. That choice makes it reasonable to assume that the car speeds in our sample are independent, although cars driving near each other may not have been. That random selection makes independence a safe assumption.

Nearly Normal Condition: The histogram of the data is unimodal and fairly symmetric. With a sample size of 250, the sampling model for the mean will not be affected much by the slight skew to the high side. There are no outliers on either side.

$$n = 250, \quad df = 249$$
$$\bar{y} = 25.55; \quad s = 3.618$$

$$SE(\bar{y}) = \frac{3.618}{\sqrt{250}} = 0.2288$$

$$t_{249} = \frac{25.55 - 25}{0.2288} = 2.404$$

$$P(t_{249} > 2.404) = 0.0085$$

A 95% confidence interval for the true mean speed is (25.099, 26.001).

With a P-value of 0.0085, I reject the null hypothesis and conclude that the cars are, on average, going faster than 25 mph. A confidence interval shows that a plausible set of values for the true mean speed is between 25.1 and 26.0 mph.

In this example, the hypothesis test leads us to conclude that the true mean speed is greater than 25 mph. But, would the town of Palo Alto want to act on that conclusion and try further traffic calming measures? As the confidence interval shows, cars are not on average going very *much* faster than 25 mph. In fact, the true mean speed is probably no more than 26 mph. The difference between the speed limit of 25 mph and the sample mean of 25.55 mph is "statistically significant," but it doesn't seem meaningful. For example, would it be a large enough difference to convince the town of Palo Alto to spend more money on traffic control? Probably not. It's always a good idea when you test a hypothesis to check the confidence interval and think about the likely values for the mean.

21.4 Critical Values for Hypothesis Tests

When we make a confidence interval, we find a **critical value**, z^* or t^*, to correspond to our selected confidence level. Critical values can also be used as a shortcut for hypothesis tests. Before computers and calculators were common, P-values were hard to find. It was easier to select a few common alpha levels (0.05, 0.01, 0.001, for example) and find the corresponding critical values for the Normal model (that is, the critical values corresponding to confidence levels 0.95, 0.99, and 0.999, respectively).

For *t*-values, tables (like Table T) were used with several confidence levels for each of the many selected degrees of freedom. Any *z*- or *t*-score larger in magnitude (that is, more extreme) than a particular critical value has to be less likely, so it will have a P-value smaller than the corresponding alpha. If we are willing to settle for a flat reject/fail-to-reject decision, comparing an observed score with the critical value for a specified alpha level gives a shortcut path to that decision. For the motorcycle helmet example, if we choose $\alpha = 0.05$, then, in order to reject H_0, our z-score has to be less

> **1.96 or 2** If you want to make a decision on the fly without technology, remember "2." That's our old friend from the 68–95–99.7 Rule. It's roughly the critical value for testing a hypothesis against a two-sided alternative at $\alpha = 0.05$.
>
> A more exact critical value for z is 1.96. For t, the value is 2.000 at 60 *df*. As the number of degrees of freedom increases, this critical value approaches 1.96. If the number of degrees of freedom is less than 60, the critical value is larger. If you are performing a hypothesis test for a mean with only three values, the 0.05 critical value for t is 4.303. But by 28 *df*, it rounds to 2.0, and even for samples bigger than 15, using "2" for the 0.05 critical value is good enough for most decisions.

Figure 21.1
When the alternative is one-sided, the critical value puts all of α on one side.

Figure 21.2
When the alternative is two-sided, the critical value splits α equally into two tails.

than the one-sided critical value of -1.645. Our calculated z-score is -5.31, so clearly we can reject the null hypothesis. For the car speeds, we could have looked up the one-sided 0.05, 0.01, and 0.001 critical values for t with 249 df which are (from Table T): 1.65, 2.34, and 3.12, respectively. Because our value was 2.40, we could have decided to reject the null hypothesis at the 0.01 (since it's bigger than 2.34) but not at the 0.001 (since it's smaller than 3.12) level. This way of thinking is perfectly correct and does give us a yes/no decision, but it gives us less information about the hypothesis because we don't have the P-value to think about. Technology makes it easy to find P-values. And since they give more information about the strength of the evidence, you should always report them.

Here are the traditional critical values from the Normal model. The corresponding t critical values are found in Table T.[6]

α	1-Sided	2-Sided
0.05	1.645	1.96
0.01	2.33	2.576
0.001	3.09	3.29

*A Confidence Interval for Small Samples

We've mentioned that there are special "exact" methods that don't require the 10% assumption so they can be used when your sample is a large fraction of the population. How about at the other extreme when you have very few successes or failures? When the **Success/Failure Condition** fails, all is not lost. A simple adjustment to the calculation lets us make a 95% confidence interval anyway.

All we do is add four *phony* observations—two to the successes, two to the failures. So instead of the proportion $\hat{p} = \dfrac{y}{n}$, we use the adjusted proportion $\widetilde{p} = \dfrac{y+2}{n+4}$ and, for convenience, we write $\widetilde{n} = n + 4$. We modify the interval by using these adjusted values for both the center of the interval *and* the margin of error. Now the adjusted interval is

$$\widetilde{p} \pm z^* \sqrt{\dfrac{\widetilde{p}(1-\widetilde{p})}{\widetilde{n}}}.$$

This adjusted form gives better performance overall[7] and works much better for proportions near 0 or 1. It has the additional advantage that we no longer need to check the **Success/Failure Condition** that $n\hat{p}$ and $n\hat{q}$ are greater than 10. This interval is called the Agresti-Coull interval or the "plus-four" interval. The plus-four method is not the most common way to find a confidence interval for a proportion, but you can use it safely for any sample size—as long as you explain clearly what you have done.

---

[6]In a sense, these are the flip side of the 68–95–99.7 Rule. When we studied that rule, we chose simple statistical distances from the mean and recalled the areas of the tails. Now we select convenient tail areas (0.05, 0.01, and 0.001, either on one side or adding both together) and record the corresponding statistical distances.

[7]By "better performance," we mean that a 95% confidence interval has more nearly a 95% chance of covering the true population proportion. Simulation studies have shown that our original, simpler confidence interval in fact is less likely than 95% to cover the true population proportion when the sample size is small or the proportion very close to 0 or 1. The original idea for this method can be attributed to E. B. Wilson. The simpler approach discussed here was proposed by Agresti and Coull (Source: A. Agresti and B. A. Coull, "Approximate Is Better Than 'Exact' for Interval Estimation of Binomial Proportions," *The American Statistician*, 52[1998]: 119–129).

For Example AN AGRESTI-COULL "PLUS-FOUR" INTERVAL

Surgeons examined their results to compare two methods for a surgical procedure used to alleviate pain on the outside of the wrist. A new method was compared with the traditional "freehand" method for the procedure. Of 45 operations using the "freehand" method, three were unsuccessful, for a failure rate of 6.7%. With only 3 failures, the data don't satisfy the **Success/Failure Condition**, so we can't use a standard confidence interval.

QUESTION: What's the confidence interval using the "plus-four" method?

ANSWER: There were 42 successes and 3 failures. Adding 2 "pseudo-successes" and 2 "pseudo-failures," we find

$$\tilde{p} = \frac{3+2}{45+4} = 0.102.$$

A 95% confidence interval (for example) is then

$$0.102 \pm 1.96\sqrt{\frac{0.102(1-0.102)}{49}} = 0.102 \pm 0.085 \text{ or } (0.017, 0.187).$$

Notice that although the observed failure rate of 0.067 is contained in the interval, it is not at the center of the interval—something we haven't seen with any of the other confidence intervals we've considered.

Confidence Intervals and Hypothesis Tests

For the motorcycle helmet example, a 95% confidence interval would give $0.507 \pm 1.96 \times 0.0175 = (0.473, 0.541)$, or 47.3% to 54.1%. If the previous rate of helmet compliance had been, say, 50%, we would not have been able to reject the null hypothesis because 50% is in the interval, so it's a plausible value. Indeed, *any* hypothesized value for the true proportion of helmet wearers in this interval is consistent with the data. Any value outside the confidence interval would make a null hypothesis that we would reject, but we'd feel more strongly about values far outside the interval.

Confidence intervals and hypothesis tests are built from the same calculations.[8] They have the same assumptions and conditions. As we have just seen, you can approximate a hypothesis test by examining the confidence interval. Just ask whether the null hypothesis value is consistent with a confidence interval for the parameter at the corresponding confidence level. Because confidence intervals are naturally two-sided, they correspond to two-sided tests. For example, a 95% confidence interval corresponds to a two-sided hypothesis test at $\alpha = 5\%$. In general, a confidence interval with a confidence level of $C\%$ corresponds to a two-sided hypothesis test with an α level of $100 - C\%$.

The relationship between confidence intervals and one-sided hypothesis tests is a little more complicated. For a one-sided test with $\alpha = 5\%$, the corresponding confidence interval has a confidence level of 90%—that's 5% in each tail. In general, a confidence interval with a confidence level of $C\%$ corresponds to a one-sided hypothesis test with an α level of $\frac{1}{2}(100 - C)\%$.

---

[8] As we saw in Chapter 19, this is not exactly true for proportions. For a confidence interval, we estimate the standard deviation of $\hat{p}$ from $\hat{p}$ itself. Because we estimate it from the data, we have a *standard error*. For the corresponding hypothesis test, we use the model's standard deviation for $\hat{p}$, based on the null hypothesis value p_0. When $\hat{p}$ and p_0 are close, these calculations give similar results. When they differ, you're likely to reject H_0 (because the observed proportion is far from your hypothesized value). In that case, you're better off building your confidence interval with a standard error estimated from the data.

For Example MAKING A DECISION BASED ON A CONFIDENCE INTERVAL

RECAP: The baseline seven-year risk of heart attacks for diabetics is 20.2%. A *NEJM* study reported a 95% confidence interval equivalent to 20.8% to 40.0% for the risk among patients taking the diabetes drug Avandia.

QUESTION: What did this confidence interval suggest to the FDA about the safety of the drug?

ANSWER: The FDA could be 95% confident that the interval from 20.8% to 40.0% included the true risk of heart attack for diabetes patients taking Avandia. Because the lower limit of this interval was higher than the baseline risk of 20.2%, there was evidence of an increased risk..

Just Checking

1. An experiment to test the fairness of a roulette wheel gives a *z*-score of 0.62. What would you conclude?

2. In Chapter 19, we encountered a bank that wondered if it could get more customers to make payments on delinquent balances by sending them a DVD urging them to set up a payment plan. Well, the bank just got back the results on their test of this strategy. A 90% confidence interval for the success rate is (0.29, 0.45). Their old send-a-letter method had worked 30% of the time. Can you reject the null hypothesis that the proportion is still 30% at $\alpha = 0.05$? Explain.

3. Given the confidence interval the bank found in their trial of DVDs, what would you recommend that they do? Should they scrap the DVD strategy?

21.5 Errors

 Activity: Type I and Type II Errors. View an animated exploration of Type I and Type II errors—a good backup for the reading in this section.

False-positive consequences Some false-positive results mean no more than an unnecessary chest X-ray. But for a drug test or a disease like AIDS, a false-positive result that is not kept confidential could have serious consequences.

Nobody's perfect. Even with lots of evidence, we can still make the wrong decision. In fact, when we perform a hypothesis test, we can make mistakes in *two* ways:

I. The null hypothesis is true, but we mistakenly reject it.
II. The null hypothesis is false, but we fail to reject it.

These two types of errors are known as **Type I** and **Type II errors**. One way to keep the names straight is to remember that we start by assuming the null hypothesis is true, so a Type I error is the first kind of error we could make.

In medical disease testing, the null hypothesis is usually the assumption that a person is healthy. The alternative is that he or she has the disease we're testing for. So a Type I error is a *false positive*: A healthy person is diagnosed with the disease. A Type II error, in which an infected person is diagnosed as disease free, is a *false negative*. These errors have other names, depending on the particular discipline and context.

Which type of error is more serious depends on the situation. In the jury trial, a Type I error occurs if the jury convicts an innocent person. A Type II error occurs if the jury fails to convict a guilty person. Which seems more serious? In medical diagnosis, a false negative could mean that a sick patient goes untreated. A false positive might mean that the person must undergo further tests. In a Statistics final exam (with H_0: the student has learned only 60% of the material), a Type I error would be passing a student who in fact learned less than 60% of the material, while a Type II error would be failing a student who knew enough to pass. Which of these errors seems more serious? It depends on the situation, the cost, and your point of view.

Here's an illustration of the situations:

		The Truth	
		H_0 True	H_0 False
My Decision	Reject H_0	Type I Error	OK
	Fail to Reject H_0	OK	Type II Error

> ■ **NOTATION ALERT**
>
> In Statistics, α is almost always saved for the alpha level. But β has already been used for the parameters of a linear model. Fortunately, it's usually clear whether we're talking about a Type II error probability or the slope or intercept of a regression model.

How often will a Type I error occur? It happens when the null hypothesis is true but we've had the bad luck to draw an unusual sample. To reject H_0, the P-value must fall below α. When H_0 is true, that happens *exactly* with probability α. So when you choose level α, you're setting the probability of a Type I error to α.

What if H_0 is not true? Then we can't possibly make a Type I error. You can't get a false positive from a sick person. A Type I error can happen only when H_0 is true.

When H_0 is false but we *fail* to reject it, we have made a Type II error. We assign the letter β to the probability of this mistake. What's the value of β? That's harder to assess than α because we don't know the true value of the parameter. When H_0 is true, it specifies a single parameter value. But when H_0 is false, we don't know the parameter value and there are many possible values. We can compute the probability β for any parameter value in H_A. But the one we should choose depends on the situation.

> **FINDING β**
>
> The null hypothesis specifies a single value for the parameter. So it's easy to calculate the probability of a Type I error. But the alternative gives a whole range of possible values, and we may want to find a β for several of them.

We could reduce β for *all* alternative parameter values by increasing α. By making it easier to reject the null, we'd be more likely to reject it whether it's true or not. So we'd reduce β, the chance that we fail to reject a false null—but we'd make more Type I errors. This tension between Type I and Type II errors is inevitable. In the political arena, think of the ongoing debate between those who favor provisions to reduce Type I errors in the courts (supporting defendants' rights, requiring warrants for wiretaps, providing legal representation for those who can't afford it) and those who advocate changes to reduce Type II errors (admitting into evidence confessions made when no lawyer is present, eavesdropping on conferences with lawyers, restricting paths of appeal, etc.).

The only way to reduce *both* types of error is to collect more evidence or, in statistical terms, to collect more data. Too often, studies fail because their sample sizes are too small to detect the change they are looking for.

Of course, what we really want to do is to detect a false null hypothesis. When H_0 is false and we reject it, we have done the right thing. A test's ability to detect a false null hypothesis is called the power of the test. In a jury trial, power is the ability of the criminal justice system to convict people who are guilty—a good thing!

> **Finding the sample size** We have seen ways to find a sample size by specifying the margin of error. Choosing the sample size to achieve a specified β (for a particular alternative value) is sometimes more appropriate, but the calculation is more complex and lies beyond the scope of this book.

> **For Example** THINKING ABOUT ERRORS
>
> **RECAP:** A published study found the risk of heart attack to be increased in patients taking the diabetes drug Avandia. The issue of the *New England Journal of Medicine* (*NEJM*) in which that study appeared also included an editorial that said, in part, "A few events either way might have changed the findings for myocardial infarction [medical jargon for "heart attack"] or for death from cardiovascular causes. In this setting, the possibility that the findings were due to chance cannot be excluded."
>
> **QUESTION:** What kind of error would the researchers have made if, in fact, their findings were due to chance? What could be the consequences of this error?
>
> **ANSWER:** The null hypothesis said the risk didn't change, but the researchers rejected that model and claimed evidence of a higher risk. If these findings were just due to chance, they rejected a true null hypothesis—a Type I error.
>
> If, in fact, Avandia carried no extra risk, then patients might be deprived of its benefits for no good reason.

> **Fisher and $\alpha = 0.05$** Why did Sir Ronald Fisher suggest 0.05 as a criterion for testing hypotheses? It turns out that he had in mind small initial studies. Fisher was concerned that they might make too many Type II errors—failing to discover an important effect—if too strict a criterion were used.
>
> The increased risk of Type I errors arising from a generous criterion didn't concern him as much because he expected a significant result to be followed with a replication or a larger study. The probability that two independent studies tested at $\alpha = 0.05$ would both make Type I errors is $0.05 \times 0.05 = 0.0025$, so Fisher didn't think that Type I errors in initial studies were a major concern.
>
> The widespread use of the relatively generous 0.05 criterion even in large studies is most likely not what Fisher had in mind.

Power

When we failed to reject the null hypothesis about TT practitioners, did we prove that they were just guessing? No, it could be that they actually *can* discern a human energy field but we just couldn't tell. For example, suppose they really have the ability to get 53% of the trials right but just happened to get only 47% in our experiment. Our confidence interval shows that with these data we wouldn't have rejected the null. And if we retained the null even though the true proportion was actually greater than 50%, we would have made a Type II error because we failed to detect their ability.

Remember, we can never prove a null hypothesis true. We can only fail to reject it. But when we fail to reject a null hypothesis, it's natural to wonder whether we looked hard enough. Might the null hypothesis actually be false and our test too weak to tell?

When the null hypothesis actually *is* false, we hope our test is strong enough to reject it. We'd like to know how likely we are to succeed. The power of the test gives us a way to think about that. The **power** of a test is the probability that it correctly rejects a false null hypothesis. When the power is high, we can be confident that we've looked hard enough. We know that β is the probability that a test *fails* to reject a false null hypothesis, so the power of the test is the probability that it *succeeds* in rejecting the false null hypothesis: $1 - \beta$.

Whenever a study fails to reject its null hypothesis, the test's power comes into question. Was the sample size big enough to detect an effect had there been one? Might we have missed an effect large enough to be interesting just because we failed to gather sufficient data or because there was too much variability in the data we could gather? The therapeutic touch experiment failed to reject the null hypothesis that the TT practitioners were just guessing. Might the problem be that the experiment simply lacked adequate power to detect their ability?

Activity: **Hypothesis Tests Are Random.** Simulate hypothesis tests and watch Type I errors occur. When you conduct real hypothesis tests you'll never know, but simulation can tell you when you've made an error.

> **For Example** ERRORS AND POWER
>
> **RECAP:** The study of Avandia published in the *NEJM* combined results from 47 different trials—a method called *meta-analysis*. The drug's manufacturer, GlaxoSmithKline (GSK), issued a statement that pointed out, "Each study is designed differently and looks at unique questions: For example, individual studies vary in size and length, in the type of patients who participated, and in the outcomes they investigate." Nevertheless, by combining data from many studies, meta-analyses can achieve a much larger sample size.
>
> **QUESTION:** How could this larger sample size help?
>
> **ANSWER:** If Avandia really did increase the seven-year heart attack rate, doctors needed to know. To overlook that would have been a Type II error (failing to detect a false null hypothesis), resulting in patients being put at greater risk. Increasing the sample size could increase the power of the analysis, making it more likely that researchers will detect the danger if there is one.

Effect Size

Activity: **The Power of a Test.** Power is a concept that's much easier to understand when you can visualize what's happening.

When we think about power, we imagine that the null hypothesis is false. The value of the power depends on how far the truth lies from the value we hypothesize. We call the distance between the null hypothesis value (for example), p_0, and the truth, p, the **effect size**. Not knowing the true value, we estimate the effect size as the difference between the null and the observed value.

The effect size is central to how we think about the power of a hypothesis test. A larger effect is easier to see and results in larger power. Small effects are naturally more difficult to detect. They'll result in more Type II errors and therefore lower power. Knowing the effect size and the sample size helps us determine the power. But when we design a study we won't know the effect size, so we can only imagine possible effect sizes and look at their consequences. How can we decide what effect sizes to look at?

One way to think about the effect size is to ask "*How big a difference would matter?*" The answer to this question depends on who is asking it and why. For example, if therapeutic touch practitioners could detect a human energy field, but only 53% of the time, that might not be a sufficient improvement over chance for health insurers to conclude it was worth covering. But *any* real ability to detect a previously unknown human energy field would be of great interest to scientists.[9]

The power of a test depends on the size of the effect and the standard deviation. For proportions, our test statistic is $\hat{p}$, and its standard deviation is $\sqrt{\frac{pq}{n}}$. When researchers design a study they often know what effect size would matter to their conclusions and have some idea of the proportion they are estimating. Once they know both of those, a bit of algebra gives them an estimate of n, the sample size they'll need. This is a common calculation to perform when designing a study so you can be sure to gather a large enough sample (and, speaking practically, to gather enough funds to pay for that large a sample). In Chapter 18, we made a similar calculation of a sample size based on how large a margin of error we could accept. This is essentially the same idea.

Effect size and power are also an important part of your conclusions whenever you fail to reject the null hypothesis. The natural question to wonder about in that event is whether you tried hard enough—whether you had sufficient power to discern a difference. After all, if you based your test on only three observations, you might easily have missed even a large effect.

---
[9] And would probably lead to a Nobel prize.

Whenever a hypothesis test fails to reject the null, the question of power can arise. For example, in the therapeutic touch experiment, with a sample size of 150, if the researchers took an increase to 75% as a reasonably interesting effect size (keeping in mind that 50% is the level of guessing) and used an α-level of 0.05, they could determine that the TT experiment would have been able to detect such an ability with a power of 99.99%. So there is only a very small chance that their study would have failed to detect a practitioner's ability at that level, had it existed. If, on the other hand, they thought that even an increase to 51% was interesting, then they would need a sample size of over 20,000 trials to have a 90% chance of detecting it!

Effect size is a concern whenever we use statistical inference. The details of the calculations depend on the situation and the appropriate statistic, but as we will see in the following chapters, the concept is the same.

Just Checking

4. Remember the bank that's sending out DVDs to try to get customers to make payments on delinquent loans? It is looking for evidence that the costlier DVD strategy produces a higher success rate than the letters it has been sending. Explain what a Type I error is in this context and what the consequences would be to the bank.

5. What's a Type II error in the bank experiment context, and what would the consequences be?

6. For the bank, which situation has higher power: a strategy that works really well, actually getting 60% of people to pay off their balances, or a strategy that barely increases the payoff rate to 32%? Explain briefly.

A Picture Worth $\frac{1}{P(z > 3.09)}$ Words

It makes intuitive sense that the larger the effect size, the easier it should be to see it. Obtaining a larger sample size decreases the probability of a Type II error, so it increases the power. It also makes sense that the more we're willing to accept a Type I error, the less likely we will be to make a Type II error.

Figure 21.3 shows a good way to visualize the relationships among these concepts. It uses a simple hypothesis test about a proportion for illustration, but a similar figure could be drawn for testing means. Suppose we are testing $H_0: p = p_0$ against the alternative $H_A: p > p_0$. We'll reject the null if the observed proportion, $\hat{p}$, is big enough. By big enough, we mean $\hat{p} > p^*$ for some critical value, p^* (shown as the red region in the right tail of the upper curve). For example, we might be willing to believe the ability of therapeutic touch practitioners if they were successful in 65% of our trials. This is what the upper model shows. It's a picture of the sampling distribution model for the proportion if the null hypothesis were true. We'd make a Type I error whenever the sample gave us $\hat{p} > p^*$, because we would reject the (true) null hypothesis. And unusual samples like that would happen only with probability α.

In reality, though, the null hypothesis is rarely *exactly* true. The lower probability model supposes that H_0 is not true. In particular, it supposes that the true value is p, not p_0. (Perhaps the TT practitioner really can detect the human energy field 72% of the time.)

■ **NOTATION ALERT**
We've attached symbols to many of the p's. Let's keep them straight. p is a true proportion parameter. p_0 is a hypothesized value of p. $\hat{p}$ is an observed proportion. p^* is a critical value of a proportion corresponding to a specified α.

Figure 21.3

The power of a test is the probability that it rejects a false null hypothesis. The upper figure shows the null hypothesis model for a hypothesis test about a proportion. We'd reject the null in a one-sided test if we observed a value of $\hat{p}$ in the red region to the right of the critical value, p^*. The lower figure shows the true model. If the true value of p is greater than p_0, then we're more likely to observe a value that exceeds the critical value and make the correct decision to reject the null hypothesis. The power of the test is the blue region on the right of the lower figure. Of course, even drawing samples whose observed proportions are distributed around p, we'll sometimes get a value in the red region on the left and make a Type II error of failing to reject the null.

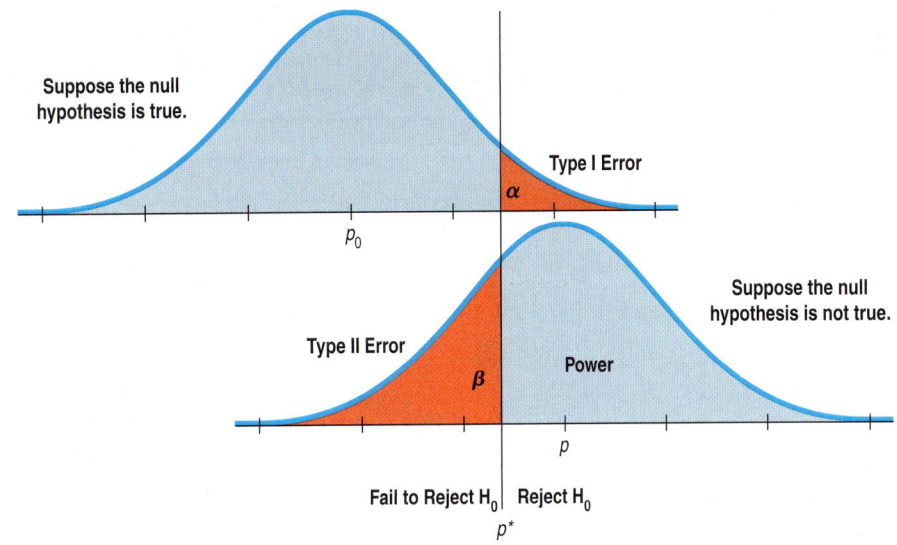

It shows a distribution of possible observed $\hat{p}$ values around this true value. Because of sampling variability, sometimes $\hat{p} < p^*$ and we fail to reject the (false) null hypothesis. Suppose a TT practitioner with a true ability level of 72% is actually successful on fewer than 65% of our tests. Then we'd make a Type II error. The area under the curve to the left of p^* in the bottom model represents how often this happens. The probability is β. In this picture, β is less than half, so most of the time we *do* make the right decision. The *power* of the test—the probability that we make the right decision—is shown as the region to the right of p^*. It's $1 - \beta$.

We calculate p^* based on the upper model because p^* depends only on the null model and the alpha level. No matter what the true proportion, no matter whether the practitioners can detect a human energy field 90%, 53%, or 2% of the time, p^* doesn't change. After all, we don't *know* the truth, so we can't use it to determine the critical value. But we always reject H_0 when $\hat{p} > p^*$.

How often we correctly reject H_0 when it's *false* depends on the effect size. We can see from the picture that if the effect size were larger (the true proportion were farther above the hypothesized value), the bottom curve would shift to the right, making the power greater.

We can see several important relationships from this figure:

- Power = $1 - \beta$.
- Reducing α to lower the chance of committing a Type I error will move the critical value, p^*, to the right (in this example). This will have the effect of increasing β, the probability of a Type II error, and correspondingly reducing the power.
- The larger the real difference between the hypothesized value, p_0, and the true population value, p, the smaller the chance of making a Type II error and the greater the power of the test. If the two proportions are very far apart, the two models will barely overlap, and we will not be likely to make any Type II errors at all—but then, we are unlikely to really need a formal hypothesis-testing procedure to see such an obvious difference. If the TT practitioners were successful almost all the time, we'd be able to see that with even a small experiment.

> **Sensitivity and specificity** The terms *sensitivity* and *specificity* often appear in medical studies. In these studies, the null hypothesis is that the person is healthy and the alternative is that the person is sick. The *specificity* of a test measures its ability to correctly identify only the healthy (the ones who should test negative). It's defined as
>
> number of true negatives/(number of true negatives + number of false positives)
>
> So *specificity* = $1 - \alpha$.
>
> Specificity gives you the probability that the test will correctly identify you as healthy when you are healthy.
>
> The *sensitivity* is the ability to detect the disease. It's defined as:
>
> number of true positives/(number of true positives + number of false negatives)
>
> So *sensitivity* = $1 - \beta$ = power of the test. Sensitivity gives you the probability that the test will correctly identify you as sick when you are sick.
>
> In Figure 21.3, the sensitivity and specificity are the blue areas of the corresponding Normal curves.

Reducing Both Type I and Type II Errors

Figure 21.3 seems to show that if we reduce Type I error, we automatically must increase Type II error. But there is a way to reduce both. Can you think of it?

If we can make both curves narrower, as shown in Figure 21.4, then both the probability of Type I errors and the probability of Type II errors will decrease, and the power of the test will increase.

A S **Activity: Power and Sample Size.** Investigate how the power of a test changes with the sample size. The interactive tool is really the only way you can see this easily.

Figure 21.4
Making the standard deviations smaller increases the power without changing the corresponding critical value. The means are just as far apart as in Figure 21.3, but the error rates are reduced.

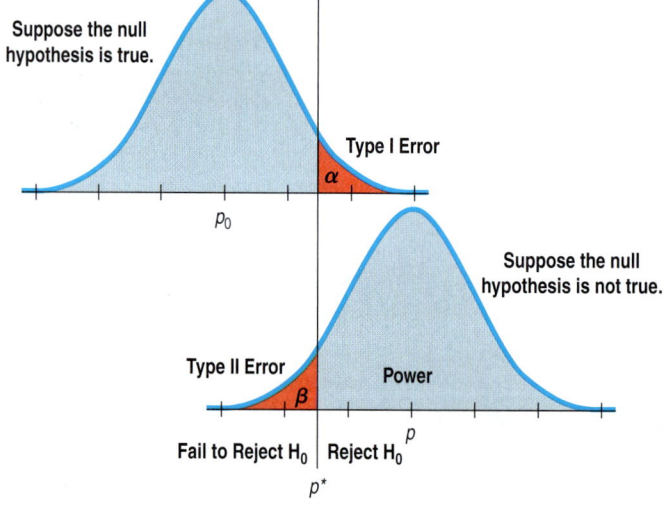

How can we accomplish that? The only way is to reduce the standard deviations which we can do by increasing the sample size. Compare the distributions in Figure 21.4 to those in Figure 21.3. You'll see that both sampling distributions are much narrower. The means haven't moved, but if we keep α the same size, the critical value, p^*, moves closer to p_0 and farther from p. That means that larger sample size has increased the power of the test as well, as you can see by the smaller β.

> **For Example** SAMPLE SIZE, ERRORS, AND POWER
>
> **RECAP:** The meta-analysis of the risks of heart attacks in patients taking the diabetes drug Avandia combined results from 47 smaller studies. As GlaxoSmithKline (GSK), the drug's manufacturer, pointed out in their rebuttal, "Data from the ADOPT clinical trial did show a small increase in reports of myocardial infarction among the *Avandia*-treated group . . . however, the number of events is too small to reach a reliable conclusion about the role any of the medicines may have played in this finding."

> **QUESTION:** Why would this smaller study have been less likely to detect the difference in risk? What are the appropriate statistical concepts for comparing the smaller studies? What are the possible drawbacks of a very large sample?
>
> **ANSWER:** Smaller studies are subject to greater sampling variability; that is, the sampling distributions they estimate have a larger standard deviation for the sample proportion. That gives small studies less power: They'd be less able to discern whether an apparently higher risk was merely the result of chance variation or evidence of real danger. The FDA doesn't want to restrict the use of a drug that's safe and effective (Type I error), nor do they want patients to continue taking a medication that puts them at risk (Type II error). Larger sample sizes can reduce the risk of both kinds of error. Greater power (the probability of rejecting a false null hypothesis) means a better chance of spotting a genuinely higher risk of heart attacks. Of course, larger studies are more expensive. Their larger power also makes them able to detect differences that may not be clinically or financially significant. This tension among the power, cost, and sample size lies at the heart of most clinical trial designs.
>
> On September 23, 2010, "The U.S. Food and Drug Administration announced that it will significantly restrict the use of the diabetes drug Avandia (rosiglitazone) to patients with Type 2 diabetes who cannot control their diabetes on other medications. These new restrictions are in response to data that suggest an elevated risk of cardiovascular events, such as heart attack and stroke, in patients treated with Avandia." (www.fda.gov/drugs/DrugSafety/PostmarketDrugSafetyInformationforPatientsandProviders/ucm226956.htm)

WHAT CAN GO WRONG?

- **Don't interpret the P-value as the probability that H_0 is true.** The P-value is about the data, not the hypothesis. It's the probability of observing data this unusual, *given* that H_0 is true, not the other way around.

- **Don't believe too strongly in arbitrary alpha levels.** There's not really much difference between a P-value of 0.051 and a P-value of 0.049, but sometimes it's regarded as the difference between night (having to refrain from rejecting H_0) and day (being able to shout to the world that your results are "statistically significant"). It may just be better to report the P-value and a confidence interval and let the world decide along with you.

- **Don't confuse practical and statistical significance.** A large sample size can make it easy to discern even a trivial change from the null hypothesis value. On the other hand, an important difference can be missed if your test lacks sufficient power.

- **Don't forget that in spite of all your care, you might make a wrong decision.** We can never reduce the probability of a Type I error (α) or of a Type II error (β) to zero (but increasing the sample size helps).

CONNECTIONS

All of the hypothesis tests we'll see boil down to the same question: "Is the difference between two quantities large enough to draw conclusions that go beyond this particular sample?" We always measure "how large" by finding a ratio of this difference to the standard deviation of the sampling distribution of the statistic. Using the standard deviation as our ruler for inference is one of the core ideas of statistical thinking.

We've discussed the close relationship between hypothesis tests and confidence intervals. They are two sides of the same coin.

This chapter also has natural links to the discussion of probability, to the Normal model, and to the three previous chapters on inference.

What Have We Learned?

Learning Objectives

Review: Know the reasoning of hypothesis testing.
- State the **hypotheses**.
- Determine (and check assumptions for) the sampling distribution **model**.
- Calculate the test statistic—the **mechanics**.
- State your **conclusions and decisions**.

Know how to formulate a null and alternative hypothesis for a question of interest.
- The null hypothesis specifies a parameter and a (null) value for that parameter.
- The alternative hypothesis specifies a range of plausible values should we fail to reject the null.
- Be able to decide on a two-sided or one-sided alternative hypothesis, and justify your decision.

Understand P-values.
- A P-value is the estimated probability of observing a statistic value at least as far from the (null) hypothesized value as the one we have actually observed.
- A small P-value indicates that the statistic we have observed would be unlikely were the null hypothesis true. That leads us to doubt the null.
- A large P-value just tells us that we have insufficient evidence to doubt the null hypothesis. In particular, it does not prove the null to be true.

Know that α is the probability of making a Type I error—rejecting the null hypothesis when it is, in fact, true.

Know that statistical significance simply implies strong evidence that the null hypothesis is false, not that the difference is important.

Know that sometimes P-values are compared to a predetermined α-level, called a critical value, to decide whether to reject the null hypothesis.

Know the value of estimating and reporting the effect size.
- A test may be statistically significant, but practically meaningless if the estimated effect is of trivial importance.

Be aware of the risks of making errors when testing hypotheses.
- A Type I error can occur when rejecting a null hypothesis if that hypothesis is, in fact, true. The probability of this is α.
- A Type II error can occur when failing to reject a null hypothesis if that hypothesis is, in fact, false. The probability of this is β.

Understand the concept of the power of a test.
- We are particularly concerned with power when we fail to reject a null hypothesis.
- The power of a test reports, for a specified effect size, the probability that the test would reject a false null hypothesis.
- Remember that increasing the sample size will generally improve the power of any test.

Review of Terms

Statistically significant When the P-value falls below the alpha level, we say that the test is "statistically significant" at that alpha level (p. 554).

Alpha level The threshold P-value that determines when we reject a null hypothesis. If we observe a statistic whose P-value based on the null hypothesis is less than α, we reject that null hypothesis (p. 554).

Significance level The alpha level is also called the significance level, most often in a phrase such as a conclusion that a particular test is "significant at the 5% significance level" (p. 555).

Critical value The value in the sampling distribution model of the statistic whose P-value is equal to the alpha level. The critical value is often denoted with an asterisk, as z^*, and t^*, for example (p. 557).

Type I error	The error of rejecting a null hypothesis when in fact it is true (also called a "false positive"). The probability of a Type I error is α (p. 560).
Type II error	The error of failing to reject a null hypothesis when in fact it is false (also called a "false negative") (p. 560).
β	The probability of a Type II error is commonly denoted β and depends on the effect size (p. 561).
Power	The probability that a hypothesis test will correctly reject a false null hypothesis is the power of the test. To find power, we must specify a particular alternative parameter value as the "true" value. For any specific value in the alternative, the power is $1 - \beta$ (p. 562).
Effect size	The difference between the null hypothesis value and true value of a model parameter is called the effect size (p. 563).

On the Computer HYPOTHESIS TESTS

Reports about hypothesis tests generated by technologies don't follow a standard form. Most will name the test and provide the test statistic value, its standard deviation, and the P-value. But these elements may not be labeled clearly. For example, the expression "Prob $> |z|$" means the probability (the "Prob") of observing a test statistic whose magnitude (the absolute value tells us this) is larger than that of the one (the "z") found in the data (which, because it is written as "z," we know follows a Normal model). That is a fancy (and not very clear) way of saying P-value. In some packages, you can specify that the test be one-sided. Others might report three P-values, covering the ground for both one-sided tests and the two-sided test.

Sometimes a confidence interval and hypothesis test are automatically given together. The CI ought to be for the corresponding confidence level: $1 - \alpha$ for 2-tailed tests, $1 - 2\alpha$ for 1-tailed tests.

Often, the standard deviation of the statistic is called the "standard error," and usually that's appropriate because we've had to estimate its value from the data. That's not the case for proportions, however: We get the standard deviation for a proportion from the null hypothesis value. Nevertheless, you may see the standard deviation called a "standard error" even for tests with proportions.

It's common for statistics packages and calculators to report more digits of "precision" than could possibly have been found from the data. You can safely ignore them. Round values such as the standard deviation to one digit more than the number of digits reported in your data.

Here are the kind of results you might see. This is not from any program or calculator we know of, but it shows some of the things you might see in typical computer output.

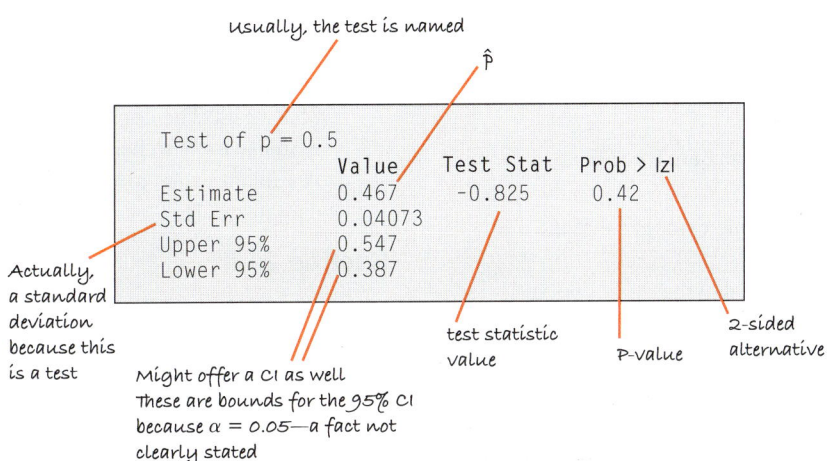

For information on hypothesis testing with particular statistics packages, see the Hypothesis Tests for Proportions, On the Computer section in Chapter 19 on pages 509–511.

Exercises

Section 21.1

1. **Parameters and hypotheses** For each of the following situations, define the parameter (proportion or mean) and write the null and alternative hypotheses in terms of parameter values. Example: We want to know if the proportion of up days in the stock market is 50%. Answer: Let $p =$ the proportion of up days. $H_0: p = 0.5$ vs. $H_A: p \neq 0.5$.
   a) A casino wants to know if their slot machine really delivers the 1 in 100 win rate that it claims.
   b) Last year, customers spent an average of $35.32 per visit to the company's website. Based on a random sample of purchases this year, the company wants to know if the mean this year has changed.
   c) A pharmaceutical company wonders if their new drug has a cure rate different from the 30% reported by the placebo.
   d) A bank wants to know if the percentage of customers using their website has changed from the 40% that used it before their system crashed last week.

2. **Hypotheses and parameters** As in Exercise 1, for each of the following situations, define the parameter and write the null and alternative hypotheses in terms of parameter values.
   a) Seat-belt compliance in Massachusetts was 65% in 2008. The state wants to know if it has changed.
   b) Last year, a survey found that 45% of the employees were willing to pay for on-site day care. The company wants to know if that has changed.
   c) Regular card customers have a default rate of 6.7%. A credit card bank wants to know if that rate is different for their Gold card customers.
   d) Regular card customers have been with the company for an average of 17.3 months. The credit card bank wants to know if their Gold card customers have been with the company on average the same amount of time.

Section 21.2

3. **P-values** Which of the following are true? If false, explain briefly.
   a) A very high P-value is strong evidence that the null hypothesis is false.
   b) A very low P-value proves that the null hypothesis is false.
   c) A high P-value shows that the null hypothesis is true.
   d) A P-value below 0.05 is always considered sufficient evidence to reject a null hypothesis.

4. **More P-values** Which of the following are true? If false, explain briefly.
   a) A very low P-value provides evidence against the null hypothesis.
   b) A high P-value is strong evidence in favor of the null hypothesis.
   c) A P-value above 0.10 shows that the null hypothesis is true.
   d) If the null hypothesis is true, you can't get a P-value below 0.01.

5. **Hypotheses** For each of the following, write out the null and alternative hypotheses, being sure to state whether the alternative is one-sided or two-sided.
   a) A company reports that last year, 40% of their reports in accounting were on time. From a random sample this year, they want to know if that proportion has changed.
   b) A company wants to know if average click-through rates (in minutes) are shorter than the 5.4 minutes that customers spent on their old website before making a purchase.
   c) Last year, 42% of the employees enrolled in at least one wellness class at the company's site. Using a survey, they want to know if a greater percentage is planning to take a wellness class this year.
   d) A political candidate wants to know from recent polls if she's going to garner a majority of votes in next week's election.

6. **More hypotheses** For each of the following, write out the alternative hypothesis, being sure to indicate whether it is one-sided or two-sided.
   a) *Consumer Reports* discovered that 20% of a certain computer model had warranty problems over the first three months. From a random sample, the manufacturer wants to know if a new model has improved that rate.
   b) The last time a philanthropic agency requested donations, 4.75% of people responded. From a recent pilot mailing, they wonder if that rate has increased.
   c) The average age of a customer of a clothing store is 35.2 years. The company wants to know if customers who use their website are younger on average.
   d) A student wants to know if other students on her campus prefer Coke or Pepsi.

Section 21.3

7. **Alpha true and false** Which of the following statements are true? If false, explain briefly.
   a) Using an alpha level of 0.05, a P-value of 0.04 results in rejecting the null hypothesis.
   b) The alpha level depends on the sample size.
   c) With an alpha level of 0.01, a P-value of 0.10 results in rejecting the null hypothesis.
   d) Using an alpha level of 0.05, a P-value of 0.06 means the null hypothesis is true.

8. **Alpha false and true** Which of the following statements are true? If false, explain briefly.
   a) It is better to use an alpha level of 0.05 than an alpha level of 0.01.
   b) If we use an alpha level of 0.01, then a P-value of 0.001 is statistically significant.
   c) If we use an alpha level of 0.01, then we reject the null hypothesis if the P-value is 0.001.
   d) If the P-value is 0.01, we reject the null hypothesis for any alpha level greater than 0.01.

Section 21.4

9. **Critical values** For each of the following situations, find the critical value(s) for z or t.
   a) $H_0: p = 0.5$ vs. $H_A: p \neq 0.5$ at $\alpha = 0.05$.
   b) $H_0: p = 0.4$ vs. $H_A: p > 0.4$ at $\alpha = 0.05$.
   c) $H_0: \mu = 10$ vs. $H_A: \mu \neq 10$ at $\alpha = 0.05; n = 36$.
   d) $H_0: p = 0.5$ vs. $H_A: p > 0.5$ at $\alpha = 0.01; n = 345$.
   e) $H_0: \mu = 20$ vs. $H_A: \mu < 20$ at $\alpha = 0.01; n = 1000$.

10. **More critical values** For each of the following situations, find the critical value for z or t.
    a) $H_0: \mu = 105$ vs. $H_A: \mu \neq 105$ at $\alpha = 0.05; n = 61$.
    b) $H_0: p = 0.05$ vs. $H_A: p > 0.05$ at $\alpha = 0.05$.
    c) $H_0: p = 0.6$ vs. $H_A: p \neq 0.6$ at $\alpha = 0.01$.
    d) $H_0: p = 0.5$ vs. $H_A: p < 0.5$ at $\alpha = 0.01; n = 500$.
    e) $H_0: p = 0.2$ vs. $H_A: p < 0.2$ at $\alpha = 0.01$.

Section 21.5

11. **Errors** For each of the following situations, state whether a Type I, a Type II, or neither error has been made. Explain briefly.
    a) A bank wants to know if the enrollment on their website is above 30% based on a small sample of customers. They test $H_0: p = 0.3$ vs. $H_A: p > 0.3$ and reject the null hypothesis. Later they find out that actually 28% of all customers enrolled.
    b) A student tests 100 students to determine whether other students on her campus prefer Coke or Pepsi and finds no evidence that preference for Coke is not 0.5. Later, a marketing company tests all students on campus and finds no difference.
    c) A human resource analyst wants to know if the applicants this year score, on average, higher on their placement exam than the 52.5 points the candidates averaged last year. She samples 50 recent tests and finds the average to be 54.1 points. She fails to reject the null hypothesis that the mean is 52.5 points. At the end of the year, they find that the candidates this year had a mean of 55.3 points.
    d) A pharmaceutical company tests whether a drug lifts the headache relief rate from the 25% achieved by the placebo. They fail to reject the null hypothesis because the P-value is 0.465. Further testing shows that the drug actually relieves headaches in 38% of people.

12. **More errors** For each of the following situations, state whether a Type I, a Type II, or neither error has been made.
    a) A test of $H_0: \mu = 25$ vs. $H_A: \mu > 25$ rejects the null hypothesis. Later it is discovered that $\mu = 24.9$.
    b) A test of $H_0: p = 0.8$ vs. $H_A: p < 0.8$ fails to reject the null hypothesis. Later it is discovered that $p = 0.9$.
    c) A test of $H_0: p = 0.5$ vs. $H_A: p \neq 0.5$ rejects the null hypothesis. Later is it discovered that $p = 0.65$.
    d) A test of $H_0: p = 0.7$ vs. $H_A: p < 0.7$ fails to reject the null hypothesis. Later is it discovered that $p = 0.6$.

Chapter Exercises

13. **One sided or two?** In each of the following situations, is the alternative hypothesis one-sided or two-sided? What are the hypotheses?
    a) A business student conducts a taste test to see whether students prefer Diet Coke or Diet Pepsi.
    b) PepsiCo recently reformulated Diet Pepsi in an attempt to appeal to teenagers. They run a taste test to see if the new formula appeals to more teenagers than the standard formula.
    c) A budget override in a small town requires a two-thirds majority to pass. A local newspaper conducts a poll to see if there's evidence it will pass.
    d) One financial theory states that the stock market will go up or down with equal probability. A student collects data over several years to test the theory.

14. **Which alternative?** In each of the following situations, is the alternative hypothesis one-sided or two-sided? What are the hypotheses?
    a) A college dining service conducts a survey to see if students prefer plastic or metal cutlery.
    b) In recent years, 10% of college juniors have applied for study abroad. The dean's office conducts a survey to see if that's changed this year.
    c) A pharmaceutical company conducts a clinical trial to see if more patients who take a new drug experience headache relief than the 22% who claimed relief after taking the placebo.
    d) At a small computer peripherals company, only 60% of the hard drives produced passed all their performance tests the first time. Management recently invested a lot of resources into the production system and now conducts a test to see if it helped.

15. **P-value** A medical researcher tested a new treatment for poison ivy against the traditional ointment. He concluded that the new treatment is more effective. Explain what the P-value of 0.047 means in this context.

16. **Another P-value** Have harsher penalties and ad campaigns increased seat-belt use among drivers and passengers? Observations of commuter traffic failed to find evidence of a significant change compared with three years ago. Explain what the study's P-value of 0.17 means in this context.

17. Alpha A researcher developing scanners to search for hidden weapons at airports has concluded that a new device is significantly better than the current scanner. He made this decision based on a test using $\alpha = 0.05$. Would he have made the same decision at $\alpha = 0.10$? How about $\alpha = 0.01$? Explain.

18. Alpha, again Environmentalists concerned about the impact of high-frequency radio transmissions on birds found that there was no evidence of a higher mortality rate among hatchlings in nests near cell towers. They based this conclusion on a test using $\alpha = 0.05$. Would they have made the same decision at $\alpha = 0.10$? How about $\alpha = 0.01$? Explain.

19. Significant? Public health officials believe that 98% of children have been vaccinated against measles. A random survey of medical records at many schools across the country found that, among more than 13,000 children, only 97.4% had been vaccinated. A statistician would reject the 98% hypothesis with a P-value of $P < 0.0001$.
   a) Explain what the P-value means in this context.
   b) The result is statistically significant, but is it important? Comment.

20. Significant again? A new reading program may reduce the number of elementary school students who read below grade level. The company that developed this program supplied materials and teacher training for a large-scale test involving nearly 8500 children in several different school districts. Statistical analysis of the results showed that the percentage of students who did not meet the grade-level goal was reduced from 15.9% to 15.1%. The hypothesis that the new reading program produced no improvement was rejected with a P-value of 0.023.
   a) Explain what the P-value means in this context.
   b) Even though this reading method has been shown to be significantly better, why might you not recommend that your local school adopt it?

21. Groceries In January 2011, Yahoo surveyed 2400 U.S. men. 1224 of the men identified themselves as the primary grocery shopper in their household.
   a) Estimate the percentage of all American males who identify themselves as the primary grocery shopper. Use a 98% confidence interval. Check the conditions first.
   b) A grocery store owner believed that only 45% of men are the primary grocery shopper for their family, and targets his advertising accordingly. He wishes to conduct a hypothesis test to see if the fraction is in fact higher than 45%. What does your confidence interval indicate? Explain.
   c) What is the level of significance of this test? Explain.

22. Is the Euro fair? Soon after the Euro was introduced as currency in Europe, it was widely reported that someone had spun a Euro coin 250 times and gotten heads 140 times. We wish to test a hypothesis about the fairness of spinning the coin.
   a) Estimate the true proportion of heads. Use a 95% confidence interval. Don't forget to check the conditions.
   b) Does your confidence interval provide evidence that the coin is unfair when spun? Explain.
   c) What is the significance level of this test? Explain.

23. Approval 2014 In March 2014, Barack Obama's approval rating stood at 40% in Gallup's daily tracking poll of 1500 randomly surveyed U.S. adults. (http://www.gallup.com/poll/113980/gallup-daily-obama-job-approval.aspx)
   a) Make a 95% confidence interval for his approval rating by all U.S. adults.
   b) Based on the confidence interval, test the null hypothesis that Obama's approval rating was no worse than his November 2011 approval rating of 45%.

24. Hard times In June 2010, a random poll of 800 working men found that 9% had taken on a second job to help pay the bills. (www.careerbuilder.com)
   a) Estimate the true percentage of men that are taking on second jobs by constructing a 95% confidence interval.
   b) A pundit on a TV news show claimed that only 6% of working men had a second job. Use your confidence interval to test whether his claim is plausible given the poll data.

25. Dogs Canine hip dysplasia is a degenerative disease that causes pain in many dogs. Sometimes advanced warning signs appear in puppies as young as 6 months. A veterinarian checked 42 puppies whose owners brought them to a vaccination clinic, and she found 5 with early hip dysplasia. She considers this group to be a random sample of all puppies.
   a) Explain why we cannot use this information to construct a confidence interval for the rate of occurrence of early hip dysplasia among all 6-month-old puppies.
   *b) Construct a "plus-four" confidence interval and interpret it in this context.

26. Fans A survey of 81 randomly selected people standing in line to enter a football game found that 73 of them were home team fans.
   a) Explain why we cannot use this information to construct a confidence interval for the proportion of all people at the game who are fans of the home team.
   *b) Construct a "plus-four" confidence interval and interpret it in this context.

27. Loans Before lending someone money, banks must decide whether they believe the applicant will repay the loan. One strategy used is a point system. Loan officers assess information about the applicant, totaling points they award for the person's income level, credit history, current debt burden, and so on. The higher the point total, the more convinced the bank is that it's safe to make the loan. Any applicant with a lower point total than a certain cutoff score is denied a loan.

We can think of this decision as a hypothesis test. Since the bank makes its profit from the interest collected on repaid loans, their null hypothesis is that the applicant will repay the loan and therefore should get the money. Only if the person's score falls below the minimum cutoff will the bank reject the null and deny the loan. This system is reasonably reliable, but, of course, sometimes there are mistakes.

   a) When a person defaults on a loan, which type of error did the bank make?
   b) Which kind of error is it when the bank misses an opportunity to make a loan to someone who would have repaid it?
   c) Suppose the bank decides to lower the cutoff score from 250 points to 200. Is that analogous to choosing a higher or lower value of α for a hypothesis test? Explain.
   d) What impact does this change in the cutoff value have on the chance of each type of error?

28. **Spam** Spam filters try to sort your e-mails, deciding which are real messages and which are unwanted. One method used is a point system. The filter reads each incoming e-mail and assigns points to the sender, the subject, key words in the message, and so on. The higher the point total, the more likely it is that the message is unwanted. The filter has a cutoff value for the point total; any message rated lower than that cutoff passes through to your inbox, and the rest, suspected to be spam, are diverted to the junk mailbox.
    We can think of the filter's decision as a hypothesis test. The null hypothesis is that the e-mail is a real message and should go to your inbox. A higher point total provides evidence that the message may be spam; when there's sufficient evidence, the filter rejects the null, classifying the message as junk. This usually works pretty well, but, of course, sometimes the filter makes a mistake.

   a) When the filter allows spam to slip through into your inbox, which kind of error is that?
   b) Which kind of error is it when a real message gets classified as junk?
   c) Some filters allow the user (that's you) to adjust the cutoff. Suppose your filter has a default cutoff of 50 points, but you reset it to 60. Is that analogous to choosing a higher or lower value of α for a hypothesis test? Explain.
   d) What impact does this change in the cutoff value have on the chance of each type of error?

29. **Second loan** Exercise 27 describes the loan score method a bank uses to decide which applicants it will lend money. Only if the total points awarded for various aspects of an applicant's financial condition fail to add up to a minimum cutoff score set by the bank will the loan be denied.

   a) In this context, what is meant by the power of the test?
   b) What could the bank do to increase the power?
   c) What's the disadvantage of doing that?

30. **More spam** Consider again the points-based spam filter described in Exercise 28. When the points assigned to various components of an e-mail exceed the cutoff value you've set, the filter rejects its null hypothesis (that the message is real) and diverts that e-mail to a junk mailbox.

   a) In this context, what is meant by the power of the test?
   b) What could you do to increase the filter's power?
   c) What's the disadvantage of doing that?

31. **Homeowners 2013** In 2013, the U.S. Census Bureau reported that 65% of American families owned their homes—the lowest rate in 18 years. Census data reveal that the ownership rate in one small city is much lower. The city council is debating a plan to offer tax breaks to first-time home buyers in order to encourage people to become homeowners. They decide to adopt the plan on a 2-year trial basis and use the data they collect to make a decision about continuing the tax breaks. Since this plan costs the city tax revenues, they will continue to use it only if there is strong evidence that the rate of home ownership is increasing.

   a) In words, what will their hypotheses be?
   b) What would a Type I error be?
   c) What would a Type II error be?
   d) For each type of error, tell who would be harmed.
   e) What would the power of the test represent in this context?

32. **Alzheimer's** Testing for Alzheimer's disease can be a long and expensive process, consisting of lengthy tests and medical diagnosis. A group of researchers (Solomon et al., 1998) devised a 7-minute test to serve as a quick screen for the disease for use in the general population of senior citizens. A patient who tested positive would then go through the more expensive battery of tests and medical diagnosis. The authors reported a false positive rate of 4% and a false negative rate of 8%.

   a) Put this in the context of a hypothesis test. What are the null and alternative hypotheses?
   b) What would a Type I error mean?
   c) What would a Type II error mean?
   d) Which is worse here, a Type I or Type II error? Explain.
   e) What is the power of this test?

33. **Testing cars** A clean air standard requires that vehicle exhaust emissions not exceed specified limits for various pollutants. Many states require that cars be tested annually to be sure they meet these standards. Suppose state regulators double-check a random sample of cars that a suspect repair shop has certified as okay. They will revoke the shop's license if they find significant evidence that the shop is certifying vehicles that do not meet standards.

   a) In this context, what is a Type I error?
   b) In this context, what is a Type II error?
   c) Which type of error would the shop's owner consider more serious?
   d) Which type of error might environmentalists consider more serious?

34. Quality control Production managers on an assembly line must monitor the output to be sure that the level of defective products remains small. They periodically inspect a random sample of the items produced. If they find a significant increase in the proportion of items that must be rejected, they will halt the assembly process until the problem can be identified and repaired.
   a) In this context, what is a Type I error?
   b) In this context, what is a Type II error?
   c) Which type of error would the factory owner consider more serious?
   d) Which type of error might customers consider more serious?

35. Cars, again As in Exercise 33, state regulators are checking up on repair shops to see if they are certifying vehicles that do not meet pollution standards.
   a) In this context, what is meant by the power of the test the regulators are conducting?
   b) Will the power be greater if they test 20 or 40 cars? Why?
   c) Will the power be greater if they use a 5% or a 10% level of significance? Why?
   d) Will the power be greater if the repair shop's inspectors are only a little out of compliance or a lot? Why?

36. Production Consider again the task of the quality control inspectors in Exercise 34.
   a) In this context, what is meant by the power of the test the inspectors conduct?
   b) They are currently testing 5 items each hour. Someone has proposed that they test 10 instead. What are the advantages and disadvantages of such a change?
   c) Their test currently uses a 5% level of significance. What are the advantages and disadvantages of changing to an alpha level of 1%?
   d) Suppose that, as a day passes, one of the machines on the assembly line produces more and more items that are defective. How will this affect the power of the test?

37. Equal opportunity? A company is sued for job discrimination because only 19% of the newly hired candidates were minorities when 27% of all applicants were minorities. Is this strong evidence that the company's hiring practices are discriminatory?
   a) Is this a one-tailed or a two-tailed test? Why?
   b) In this context, what would a Type I error be?
   c) In this context, what would a Type II error be?
   d) In this context, what is meant by the power of the test?
   e) If the hypothesis is tested at the 5% level of significance instead of 1%, how will this affect the power of the test?
   f) The lawsuit is based on the hiring of 37 employees. Is the power of the test higher than, lower than, or the same as it would be if it were based on 87 hires?

38. Stop signs Highway safety engineers test new road signs, hoping that increased reflectivity will make them more visible to drivers. Volunteers drive through a test course with several of the new- and old-style signs and rate which kind shows up the best.
   a) Is this a one-tailed or a two-tailed test? Why?
   b) In this context, what would a Type I error be?
   c) In this context, what would a Type II error be?
   d) In this context, what is meant by the power of the test?
   e) If the hypothesis is tested at the 1% level of significance instead of 5%, how will this affect the power of the test?
   f) The engineers hoped to base their decision on the reactions of 50 drivers, but time and budget constraints may force them to cut back to 20. How would this affect the power of the test? Explain.

39. Software for learning A Statistics professor has observed that for several years students score an average of 105 points out of 150 on the semester exam. A salesman suggests that he try a statistics software package that gets students more involved with computers, predicting that it will increase students' scores. The software is expensive, and the salesman offers to let the professor use it for a semester to see if the scores on the final exam increase significantly. The professor will have to pay for the software only if he chooses to continue using it.
   a) Is this a one-tailed or two-tailed test? Explain.
   b) Write the null and alternative hypotheses.
   c) In this context, explain what would happen if the professor makes a Type I error.
   d) In this context, explain what would happen if the professor makes a Type II error.
   e) What is meant by the power of this test?

40. Ads A company is willing to renew its advertising contract with a local radio station only if the station can prove that more than 20% of the residents of the city have heard the ad and recognize the company's product. The radio station conducts a random phone survey of 400 people.
   a) What are the hypotheses?
   b) The station plans to conduct this test using a 10% level of significance, but the company wants the significance level lowered to 5%. Why?
   c) What is meant by the power of this test?
   d) For which level of significance will the power of this test be higher? Why?
   e) They finally agree to use $\alpha = 0.05$, but the company proposes that the station call 600 people instead of the 400 initially proposed. Will that make the risk of Type II error higher or lower? Explain.

41. Software, part II 203 students signed up for the Stats course in Exercise 39. They used the software suggested by the salesman, and scored an average of 108 points on the final with a standard deviation of 8.7 points.

a) Should the professor spend the money for this software? Support your recommendation with an appropriate test.
b) Does this improvement seem to be practically significant?

42. Testing the ads The company in Exercise 40 contacts 600 people selected at random, and only 133 remember the ad.
a) Should the company renew the contract? Support your recommendation with an appropriate test.
b) Explain what your P-value means in this context.

43. TV safety The manufacturer of a metal stand for home TV sets must be sure that its product will not fail under the weight of the TV. Since some larger sets weigh nearly 300 pounds, the company's safety inspectors have set a standard of ensuring that the stands can support an average of over 500 pounds. Their inspectors regularly subject a random sample of the stands to increasing weight until they fail. They test the hypothesis $H_0: \mu = 500$ against $H_A: \mu > 500$, using the level of significance $\alpha = 0.01$. If the sample of stands fails to pass this safety test, the inspectors will not certify the product for sale to the general public.
a) Is this an upper-tail or lower-tail test? In the context of the problem, why do you think this is important?
b) Explain what will happen if the inspectors commit a Type I error.
c) Explain what will happen if the inspectors commit a Type II error.

44. Catheters During an angiogram, heart problems can be examined via a small tube (a catheter) threaded into the heart from a vein in the patient's leg. It's important that the company that manufactures the catheter maintain a diameter of 2.00 mm. (The standard deviation is quite small.) Each day, quality control personnel make several measurements to test $H_0: \mu = 2.00$ against $H_A: \mu \neq 2.00$ at a significance level of $\alpha = 0.05$. If they discover a problem, they will stop the manufacturing process until it is corrected.
a) Is this a one-sided or two-sided test? In the context of the problem, why do you think this is important?
b) Explain in this context what happens if the quality control people commit a Type I error.
c) Explain in this context what happens if the quality control people commit a Type II error.

45. TV safety, revisited The manufacturer of the metal TV stands in Exercise 43 is thinking of revising its safety test.
a) If the company's lawyers are worried about being sued for selling an unsafe product, should they increase or decrease the value of α? Explain.
b) In this context, what is meant by the power of the test?
c) If the company wants to increase the power of the test, what options does it have? Explain the advantages and disadvantages of each option.

46. Catheters, again The catheter company in Exercise 44 is reviewing its testing procedure.
a) Suppose the significance level is changed to $\alpha = 0.01$. Will the probability of a Type II error increase, decrease, or remain the same?
b) What is meant by the power of the test the company conducts?
c) Suppose the manufacturing process is slipping out of proper adjustment. As the actual mean diameter of the catheters produced gets farther and farther above the desired 2.00 mm, will the power of the quality control test increase, decrease, or remain the same?
d) What could they do to improve the power of the test?

47. Two coins In a drawer are two coins. They look the same, but one coin produces heads 90% of the time when spun while the other one produces heads only 30% of the time. You select one of the coins. You are allowed to spin it *once* and then must decide whether the coin is the 90%- or the 30%-head coin. Your null hypothesis is that your coin produces 90% heads.
a) What is the alternative hypothesis?
b) Given that the outcome of your spin is tails, what would you decide? What if it were heads?
c) How large is α in this case?
d) How large is the power of this test? (*Hint:* How many possibilities are in the alternative hypothesis?)
e) How could you lower the probability of a Type I error and increase the power of the test at the same time?

48. Faulty or not? You are in charge of shipping computers to customers. You learn that a RAM chip was put into some of the machines. There's a simple test you can perform, but it's not perfect. All but 4% of the time, a good chip passes the test, but unfortunately, 35% of the bad chips pass the test, too. You have to decide on the basis of one test whether the chip is good or bad. Make this a hypothesis test.
a) What are the null and alternative hypotheses?
b) Given that a computer fails the test, what would you decide? What if it passes the test?
c) How large is α for this test?
d) What is the power of this test? (*Hint:* How many possibilities are in the alternative hypothesis?)

49. Hoops A basketball player with a poor foul-shot record practices intensively during the off-season. He tells the coach that he has raised his proficiency from 60% to 80%. Dubious, the coach asks him to take 10 shots, and is surprised when the player hits 9 out of 10. Did the player prove that he has improved?
a) Suppose the player really is no better than before—still a 60% shooter. What's the probability he can hit at least 9 of 10 shots anyway? (*Hint:* Use a Binomial model.)
b) If that is what happened, now the coach thinks the player has improved when he has not. Which type of error is that?

(continued)

c) If the player really can hit 80% now, and it takes at least 9 out of 10 successful shots to convince the coach, what's the power of the test?

d) List two ways the coach and player could increase the power to detect any improvement.

50. **Pottery** An artist experimenting with clay to create pottery with a special texture has been experiencing difficulty with these special pieces. About 40% break in the kiln during firing. Hoping to solve this problem, she buys some more expensive clay from another supplier. She plans to make and fire 10 pieces and will decide to use the new clay if at most one of them breaks.

    a) Suppose the new, expensive clay really is no better than her usual clay. What's the probability that this test convinces her to use it anyway? (*Hint:* Use a Binomial model.)

    b) If she decides to switch to the new clay and it is no better, what kind of error did she commit?

    c) If the new clay really can reduce breakage to only 20%, what's the probability that her test will not detect the improvement?

    d) How can she improve the power of her test? Offer at least two suggestions.

Just Checking ANSWERS

1. With a *z*-score of 0.62, you can't reject the null hypothesis. The experiment shows no evidence that the wheel is not fair.

2. At $\alpha = 0.05$, you can't reject the null hypothesis because 0.30 is contained in the 90% confidence interval—it's plausible that sending the DVDs is no more effective than just sending letters.

3. The confidence interval is from 29% to 45%. The DVD strategy is more expensive and may not be worth it. We can't distinguish the success rate from 30% given the results of this experiment, but 45% would represent a large improvement. The bank should consider another trial, increasing their sample size to get a narrower confidence interval.

4. A Type I error would mean deciding that the DVD success rate is higher than 30% when it really isn't. They would adopt a more expensive method for collecting payments that's no better than the less expensive strategy.

5. A Type II error would mean deciding that there's not enough evidence to say that the DVD strategy works when in fact it does. The bank would fail to discover an effective method for increasing their revenue from delinquent accounts.

6. 60%; the larger the effect size, the greater the power. It's easier to detect an improvement to a 60% success rate than to a 32% rate.

Review of Part V

From the Data at Hand to the World at Large

Quick Review

What do samples really tell us about the populations from which they are drawn? Are the results of an experiment meaningful, or are they just sampling error? Statistical inference based on our understanding of sampling models can help answer these questions. Here's a brief summary of the key concepts and skills:

- Sampling models describe the variability of sample statistics using a remarkable result called the Central Limit Theorem.
  - When the number of trials is sufficiently large, proportions found in different samples vary according to an approximately Normal model.
  - When samples are sufficiently large, the means of different samples vary, with an approximately Normal model.
  - The variability of sample statistics decreases as sample size increases.
  - Statistical inference procedures are based on the Central Limit Theorem.
  - No inference procedure is valid unless the underlying assumptions are true. Always check the conditions before proceeding.
- Statistical inference procedures for means are also based on the Central Limit Theorem, but we don't usually know the population standard deviation. Student's t-models take the additional uncertainty of independently estimating the standard deviation into account.
- A confidence interval uses a sample statistic (such as a proportion or mean) to estimate a range of plausible values for the parameter of a population model.
  - All confidence intervals involve an estimate of the parameter, a margin of error, and a level of confidence.
  - For confidence intervals based on a given sample, the greater the margin of error, the higher the confidence.
  - At a given level of confidence, the larger the sample, the smaller the margin of error.
- A hypothesis test proposes a model for the population, then examines the observed statistics to see if that model is plausible.
  - A null hypothesis suggests a parameter value for the population model. Usually, we assume there is nothing interesting, unusual, or different about the sample results.
  - The alternative hypothesis states what we will believe if the sample results turn out to be inconsistent with our null model.
  - We compare the difference between the statistic and the hypothesized value with the standard deviation of the statistic. It's the sampling distribution of this ratio that gives us a P-value.
  - The P-value of the test is the conditional probability that the null model could produce results at least as extreme as those observed in the sample or the experiment just as a result of sampling error.
  - A low P-value indicates evidence against the null model. If it is sufficiently low, we reject the null model.
  - A high P-value indicates that the sample results are not inconsistent with the null model, so we cannot reject it. However, this does not prove the null model is true.
  - Sometimes we will mistakenly reject the null hypothesis even though it's actually true—that's called a Type I error. If we fail to reject a false null hypothesis, we commit a Type II error.
  - The power of a test measures its ability to detect a false null hypothesis.
  - You can lower the risk of a Type I error by requiring a higher standard of proof (lower P-value) before rejecting the null hypothesis. But this will raise the risk of a Type II error and decrease the power of the test.
  - The only way to increase the power of a test while decreasing the chance of committing either error is to design a study based on a larger sample.

And now for some opportunities to review these concepts and skills . . .

Review Exercises

1. **Crohn's disease** Omega-3 fatty acids have been tested as a means to prevent relapse of Crohn's disease. Two large, randomized, placebo-controlled studies have shown no such benefit from omega-3 fatty acids. Suppose you are asked to design an experiment to further study this claim. Imagine that you have collected data on Crohn's relapses in subjects who have used these omega-3 fatty acids and similar subjects who have not used them and that you can measure incidences of relapse for these subjects. State the null and alternative hypotheses you would use in your study.

2. **Color-blind** Medical literature says that about 8% of males are color-blind. A university's introductory psychology course is taught in a large lecture hall. Among the students, there are 325 males. Each semester when the professor discusses visual perception, he shows the class a test for color blindness. The percentage of males who are color-blind varies from semester to semester.
   a) Is the sampling distribution model for the sample proportion likely to be Normal? Explain.
   b) What are the mean and standard deviation of this sampling distribution model?
   c) Sketch the sampling model, using the 68–95–99.7 Rule.
   d) Write a few sentences explaining what the model says about this professor's class.

3. **Hamsters** How large are hamster litters? Among 47 golden hamster litters recorded, there were an average of 7.72 baby hamsters, with a standard deviation of 2.5 hamsters per litter.
   a) Create and interpret a 90% confidence interval.
   b) Would a 98% confidence interval have a larger or smaller margin of error? Explain.
   c) How many litters must be used to estimate the average litter size to within 1 baby hamster with 95% confidence?

4. **Polling 2004** The 2004 U.S. presidential election was the last presidential election that had a signficant third-party candidate. The official results showed that George W. Bush received 50.7% of the vote and John Kerry received 48.3%. Ralph Nader, running as a third-party candidate, picked up only 0.4%. After the election, there was much discussion about exit polls, which had initially indicated a different result. Suppose you had taken a random sample of 1000 voters in an exit poll and asked them for whom they had voted.
   a) Would you always get 507 votes for Bush and 483 for Kerry?
   b) In 95% of such polls, your sample proportion of voters for Bush should be between what two values?
   c) What might be a problem in finding a 95% confidence interval for the true proportions of Nader voters from this sample?
   d) Would you expect the sample proportion of Nader votes to vary more, less, or about the same as the sample proportion of Bush votes? Why?

5. **Leaky gas tanks** Nationwide, it is estimated that 40% of service stations have gas tanks that leak to some extent. California is trying to lessen the prevalence of these leaks. We want to assess the effectiveness of their efforts by seeing if the percentage of service stations whose tanks leak has decreased. To do this, we randomly sample 27 service stations in California and determine whether there is any evidence of leakage. In our sample, only 7 of the stations exhibit any leakage. Is there evidence that their efforts are effective?
   a) What are the null and alternative hypotheses?
   b) Check the assumptions necessary for inference.
   c) Test the null hypothesis.
   d) What do you conclude (in plain English)?
   e) If California's program actually works, have you made an error? What kind?
   f) What two things could you do to decrease the probability of making this kind of error?
   g) What are the advantages and disadvantages of taking those two courses of action?

6. **Babies** The National Perinatal Statistics Unit of the Sydney Children's Hospital reports that the mean birth weight of all babies born in birth centers in Australia in 2002 was 3564 grams—about 7.86 pounds. A Missouri hospital reports that the average weight of 112 babies born there last year was 7.68 pounds, with a standard deviation of 1.31 pounds. If we believe the Missouri babies fairly represent American newborns, is there any evidence that U.S. babies and Australian babies do not weigh the same amount at birth?

7. **Scrabble** Using a computer to play many simulated games of Scrabble, researcher Charles Robinove found that the letter "A" occurred in 54% of the hands. This study had a margin of error of $\pm 10\%$. (Source: *Chance*, 15, no. 1 [2002])

a) Explain what the margin of error means in this context.
b) Why might the margin of error be so large?
c) Probability theory predicts that the letter "A" should appear in 63% of the hands. Does this make you concerned that the simulation might be faulty? Explain.

8. **Dice** When one die is rolled, the number of spots showing has a mean of 3.5 and a standard deviation of 1.7. Suppose you roll 10 dice. What's the approximate probability that your total is between 30 and 40 (that is, the average for the 10 dice is between 3 and 4)? Specify the model you use and the assumptions and conditions that justify your approach.

9. **Net-Newsers** In June 2008, the Pew Research Foundation sampled 3615 U.S. adults and asked about their choice of news sources. They identified 13% as "Net-Newsers" who regularly get their news from online sources rather than TV or newspapers.
   a) Pew reports a margin of error of ±2% for this result. Explain what the margin of error means.
   b) Pew's survey included 2802 respondents contacted by landline and 813 contacted by cell phone. If the percentage of Net-Newsers is the same in both groups and Pew estimated those percentages separately, which group would have the larger margin of error? Explain.
   c) Pew reports that 82% of the 470 Net-Newsers in their survey get news during the course of the day, far more than other respondents. Find a 95% confidence interval for this statistic.
   d) How does the margin of error for your confidence interval compare with the values in parts a and b? Explain why.

10. **Gay marriage** In February 2014, Pew Research asked a random sample of 1821 U.S. adults, "Do you strongly favor, favor, oppose, or strongly oppose allowing gays and lesbians to marry legally?" (http://www.pewsocialtrends.org/files/2014/03/2014-03-07_generations-appendix-b-Topline-Questionnaire.pdf)

    Of those polled, 54% said they favored marriage equality, the highest percentage to date.
    a) Create a 95% confidence interval for the percentage of all American adults who support marriage equality.
    b) Based on your confidence interval, can you conclude that a majority of U.S. adults support legally recognizing gay marriages? Explain.
    c) If pollsters wanted to follow up on this poll with another survey that could determine the level of support for gay marriage to within 2% with 98% confidence, how many people should they poll?

11. **Bimodal** We are sampling randomly from a distribution known to be bimodal.
    a) As our sample size increases, what's the expected shape of the sample's distribution?
    b) What's the expected value of our sample's mean? Does the size of the sample matter?
    c) How is the variability of sample means related to the standard deviation of the population? Does the size of the sample matter?
    d) How is the shape of the sampling distribution model affected by the sample size?

12. **Vitamin D 2012** In 2012, the *American Journal of Clinical Nutrition* reported that 31% of Australian adults over age 25 have a vitamin D deficiency. The data came from the AusDiab study of 11,218 Australians.
    a) Do these data meet the assumptions necessary for inference? What would you like to know that you don't?
    b) Create a 95% confidence interval.
    c) Interpret the interval in this context.
    d) Explain in this context what "95% confidence" means.

13. **Archery** A champion archer can generally hit the bull's-eye 80% of the time. Suppose she shoots 200 arrows during competition. Let $\hat{p}$ represent the percentage of bull's-eyes she gets (the sample proportion).
    a) What are the mean and standard deviation of the sampling distribution model for $\hat{p}$?
    b) Is a Normal model appropriate here? Explain.
    c) Sketch the sampling model, using the 68–95–99.7 Rule.
    d) What's the probability that she gets at least 85% bull's-eyes?

14. **Eggs** The ISA Babcock Company supplies poultry farmers with hens, advertising that a mature B300 Layer produces eggs with a mean weight of 60.7 grams. Suppose that egg weights follow a Normal model with standard deviation 3.1 grams.
    a) What fraction of the eggs produced by these hens weigh more than 62 grams?
    b) What's the probability that a dozen randomly selected eggs average more than 62 grams?
    c) Using the 68–95–99.7 Rule, sketch a model of the total weights of a dozen eggs.

15. **Polling disclaimer** A newspaper article that reported the results of an election poll included the following explanation:

    The Associated Press poll on the 2012 presidential campaign is based on telephone interviews with 798 randomly selected registered voters from all states except Alaska and Hawaii.

    (*continued*)

The results were weighted to represent the population by demographic factors such as age, sex, region, and education.

No more than 1 time in 20 should chance variations in the sample cause the results to vary by more than 4 percentage points from the answers that would be obtained if all Americans were polled.

The margin of sampling error is larger for responses of subgroups, such as income categories or those in political parties. There are other sources of potential error in polls, including the wording and order of questions.

a) Did they describe the 5 W's well?
b) What kind of sampling design could take into account the several demographic factors listed?
c) What was the margin of error of this poll?
d) What was the confidence level?
e) Why is the margin of error larger for subgroups?
f) Which kinds of potential bias did they caution readers about?

16. Enough eggs? One of the important issues for poultry farmers is the production rate—the percentage of days on which a given hen actually lays an egg. Ideally, that would be 100% (an egg every day), but realistically, hens tend to lay eggs on about 3 of every 4 days. ISA Babcock wants to advertise the production rate for the B300 Layer (see Exercise 14) as a 95% confidence interval with a margin of error of ±2%. How many hens must they collect data on?

17. Teen deaths 2005 Traffic accidents are the leading cause of death among people aged 16 to 20. In May 2005, the National Highway Traffic Safety Administration reported that even though only 6.3% of licensed drivers are between 16 and 20 years old, they were involved in 12.6% of all fatal crashes. Insurance companies have long known that teenage boys were high risks, but what about teenage girls? One insurance company found that the driver was a teenage girl in 44 of the 388 fatal accidents they investigated. Is this strong evidence that the accident rate is lower for girls than for teens in general?

a) Test an appropriate hypothesis and state your conclusion.
b) Explain what your P-value means in this context.

18. Largemouth bass Organizers of a fishing tournament believe that the lake holds a sizable population of largemouth bass. They assume that the weights of these fish have a model that is skewed to the right with a mean of 3.5 pounds and a standard deviation of 2.2 pounds.

a) Explain why a skewed model makes sense here.
b) Explain why you cannot determine the probability that a largemouth bass randomly selected ("caught") from the lake weighs over 3 pounds.
c) Each fisherman in the contest catches 5 fish each day. Can you determine the probability that someone's catch averages over 3 pounds? Explain.
d) The 12 fishermen competing each caught the limit of 5 fish. What's the probability that the total catch of 60 fish averaged more than 3 pounds?

19. Cheating A Rutgers University study found that many high school students cheat on tests. The researchers surveyed a random sample of 4500 high school students nationwide; 74% of them said they had cheated at least once.

a) Create a 90% confidence interval for the level of cheating among high school students. Don't forget to check the appropriate conditions.
b) Interpret your interval.
c) Explain what "90% confidence" means.
d) Would a 95% confidence interval be wider or narrower? Explain without actually calculating the interval.

20. Language Neurological research has shown that in about 80% of people language abilities reside in the brain's left side. Another 10% display right-brain language centers, and the remaining 10% have two-sided language control. (The latter two groups are mainly left-handers.) (Source: *Science News*, 161, no. 24 [2002])

a) We select 60 people at random. Is it reasonable to use a Normal model to describe the possible distribution of the proportion of the group that has left-brain language control? Explain.
b) What's the probability that our group has at least 75% left-brainers?
c) If the group had consisted of 100 people, would that probability be higher, lower, or about the same? Explain why, without actually calculating the probability.
d) How large a group would almost certainly guarantee at least 75% left-brainers? Explain.

21. Cigarettes 2009 In 1999, the Centers for Disease Control and Prevention estimated that about 34.8% of high school students smoked cigarettes. They established a national health goal of reducing that figure to 16% by the year 2010. To that end, they would be on track if they achieved a reduction to 17.7% by 2009. In 2009, they released a research study in which 19.5% of a random sample of 5080 high school students said they were current smokers. Is this evidence that progress toward the goal is off track?

a) Write appropriate hypotheses.
b) Verify that the appropriate assumptions are satisfied.
c) Find the P-value of this test.
d) Explain what the P-value means in this context.
e) State an appropriate conclusion.
f) Of course, your conclusion may be incorrect. If so, which kind of error did you commit?

22. Church going In December 2011, Pew Research found that 921 American adults out of 2303 surveyed are active in a church, religious, or spiritual organization.
a) Create a 95% confidence interval for the proportion of all American adults who are active in a church, religious, or spiritual organization.
b) Interpret your interval in context.
c) Explain carefully what "95% confidence" means in this context.

23. Teen smoking 2009 The Centers for Disease Control and Prevention say that about 19.5% of teenagers smoke tobacco (down from a high of 38% in 1997). A college has 522 students in its freshman class. Is it likely that more than 25% of them are smokers? Explain.

24. Alcohol abuse Growing concern about binge drinking among college students has prompted one large state university to conduct a survey to assess the size of the problem on its campus. The university plans to randomly select students and ask how many have been drunk during the past week. If the school hopes to estimate the true proportion among all its students with 90% confidence and a margin of error of $\pm 4\%$, how many students must be surveyed?

25. Errors An auto parts company advertises that its special oil additive will make the engine "run smoother, cleaner, longer, with fewer repairs." An independent laboratory decides to test part of this claim. It arranges to use a taxicab company's fleet of cars. The cars are randomly divided into two groups. The company's mechanics will use the additive in one group of cars but not in the other. At the end of a year the laboratory will compare the percentage of cars in each group that required engine repairs.
a) What kind of a study is this?
b) Will they do a one-tailed or a two-tailed test?
c) Explain in this context what a Type I error would be.
d) Explain in this context what a Type II error would be.
e) Which type of error would the additive manufacturer consider more serious?
f) If the cabs with the additive do indeed run significantly better, can the company conclude it is an effect of the additive? Can they generalize this result and recommend the additive for all cars? Explain.

26. Safety Observers in Texas watched children at play in eight communities. Of the 814 children seen biking, roller skating, or skateboarding, only 14% wore a helmet.

a) Create and interpret a 95% confidence interval.
b) What concerns do you have about this study that might make your confidence interval unreliable?
c) Suppose we want to do this study again, picking various communities and locations at random, and hope to end up with a 98% confidence interval having a margin of error of $\pm 4\%$. How many children must we observe?

27. Fried PCs A computer company recently experienced a disastrous fire that ruined some of its inventory. Unfortunately, during the panic of the fire, some of the damaged computers were sent to another warehouse, where they were mixed with undamaged computers. The engineer responsible for quality control would like to check out each computer in order to decide whether it's undamaged or damaged. Each computer undergoes a series of 100 tests. The number of tests it fails will be used to make the decision. If it fails more than a certain number, it will be classified as damaged and then scrapped. From past history, the distribution of the number of tests failed is known for both undamaged and damaged computers. The probabilities associated with each outcome are listed in the table below:

Number of Tests Failed	0	1	2	3	4	5	> 5
Undamaged (%)	80	13	2	4	1	0	0
Damaged (%)	0	10	70	5	4	1	10

The table indicates, for example, that 80% of the undamaged computers have no failures, while 70% of the damaged computers have 2 failures.
a) To the engineers, this is a hypothesis-testing situation. State the null and alternative hypotheses.
b) Someone suggests classifying a computer as damaged if it fails any of the tests. Discuss the advantages and disadvantages of this test plan.
c) What number of tests would a computer have to fail in order to be classified as damaged if the engineers want to have the probability of a Type I error equal to 5%?
d) What's the power of the test plan in part c?
e) A colleague points out that by increasing α just 2%, the power can be increased substantially. Explain.

28. Power We are replicating an experiment. How will each of the following changes affect the power of our test? Indicate whether it will increase, decrease, or remain the same, assuming that all other aspects of the situation remain unchanged.
a) We increase the number of subjects from 40 to 100.
b) We require a higher standard of proof, changing from $\alpha = 0.05$ to $\alpha = 0.01$.

29. **Approval 2008** Of all the post–World War II presidents, Richard Nixon had the highest *disapproval* rating near the end of his presidency. His disapproval rating peaked at 66% in July 1974, just before he resigned. This percentage has been considered by some pundits as a high water mark for presidential disapproval. However, in April 2008, George W. Bush's disapproval rating peaked at 69%, according to a Gallup poll of 1016 voters. Pundits started discussing whether his rating was discernibly worse than the previous high water mark of 66%? What do you think?

30. **Grade inflation** In 1996, 20% of the students at a major university had an overall grade point average of 3.5 or higher (on a scale of 4.0). In 2012, a random sample of 1100 student records found that 25% had a GPA of 3.5 or higher. Is this evidence of grade inflation?

31. **Name recognition** An advertising agency won't sign an athlete to do product endorsements unless it is sure the person is known to more than 25% of its target audience. The agency always conducts a poll of 500 people to investigate the athlete's name recognition before offering a contract. Then it tests $H_0: p = 0.25$ against $H_A: p > 0.25$ at a 5% level of significance.
    a) Why does the company use upper tail tests in this situation?
    b) Explain what Type I and Type II errors would represent in this context, and describe the risk that each error poses to the company.
    c) The company is thinking of changing its test to use a 10% level of significance. How would this change the company's exposure to each type of risk?

32. **Name recognition, part II** The advertising company described in Exercise 31 is thinking about signing a WNBA star to an endorsement deal. In its poll, 27% of the respondents could identify her.
    a) Fans who never took Statistics can't understand why the company did not offer this WNBA player an endorsement contract even though the 27% recognition rate in the poll is above the 25% threshold. Explain it to them.
    b) Suppose that further polling reveals that this WNBA star really is known to about 30% of the target audience. Did the company initially commit a Type I or Type II error in not signing her?
    c) Would the power of the company's test have been higher or lower if the player were more famous? Explain.

33. **Dropouts** One study comparing various treatments for the eating disorder anorexia nervosa initially enlisted 198 subjects, but found overall that 105 failed to complete their assigned treatment programs. Construct and interpret an appropriate confidence interval. Discuss any reservations you have about this inference.

34. **Women** The U.S. Census Bureau reports that 26% of all U.S. businesses are owned by women. A Colorado consulting firm surveys a random sample of 410 businesses in the Denver area and finds that 115 of them have women owners. Should the firm conclude that its area is unusual? Test an appropriate hypothesis and state your conclusion.

35. **Speeding** A newspaper report in August 2002 raised the issue of racial bias in the issuance of speeding tickets. The following facts were noted:
    - Sixteen percent of drivers registered in New Jersey are black.
    - Of the 324 speeding tickets issued in one month on a 65 mph section of the New Jersey Turnpike, 25% went to black drivers.
    a) Is the percentage of speeding tickets issued to blacks unusually high compared with the state's registration information?
    b) Does this prove that racial profiling was used?
    c) What other statistics would you like to know about this situation?

36. **Petitions** To get a voter initiative on a state ballot, petitions that contain at least 250,000 valid voter signatures must be filed with the Elections Commission. The board then has 60 days to certify the petitions. A group wanting to create a statewide system of universal health insurance has just filed petitions with a total of 304,266 signatures. As a first step in the process, the Board selects an SRS of 2000 signatures and checks them against local voter lists. Only 1772 of them turn out to be valid.
    a) What percent of the sample signatures were valid?
    b) What percent of the petition signatures submitted must be valid in order to have the initiative certified by the Elections Commission?
    c) What will happen if the Elections Commission commits a Type I error?
    d) What will happen if the Elections Commission commits a Type II error?
    e) Does the sample provide evidence in support of certification? Explain.
    f) What could the Elections Commission do to increase the power of the test?

37. **Meals** A college student is on a "meal program." His budget allows him to spend an average of $10 per day for the semester. He keeps track of his daily food expenses for 2 weeks; the data are given in the table at the top of the following page. Is there strong evidence that he will overspend his food allowance? Explain.

Date	Cost ($)
7/29	15.20
7/30	23.20
7/31	3.20
8/1	9.80
8/2	19.53
8/3	6.25
8/4	0
8/5	8.55
8/6	20.05
8/7	14.95
8/8	23.45
8/9	6.75
8/10	0
8/11	9.01

38. Occupy Wall Street In 2011, the Occupy Wall Street movement protested the concentration of wealth and power in the United States. A 2012 University of Delaware survey asked a random sample of 901 American adults whether they agreed or disagreed with the following statement:

> The Occupy Wall Street protesters offered new insights on social issues.

Of those asked, 59.9% said they strongly or somewhat agreed with this statement. We know that if we could ask the entire population of American adults, we would not find that exactly 59.9% think that Wall Street workers would be willing to break the law to make money. Construct a 95% confidence interval for the true percentage of American adults who agree with the statement.

39. Streams Researchers in the Adirondack Mountains collect data on a random sample of streams each year. One of the variables recorded is the substrate of the stream—the type of soil and rock over which they flow. The researchers found that 69 of the 172 sampled streams had a substrate of shale. Construct a 95% confidence interval for the proportion of Adirondack streams with a shale substrate. Clearly interpret your interval in this context.

40. Skin cancer In February 2012, MedPage Today reported that researchers used vemurafenib to treat metastatic melanoma (skin cancer). Out of 152 patients, 53% had a partial or complete response to vemurafenib.
a) Write a 95% confidence interval for the proportion helped by the treatment, and interpret it in this context.
b) If researchers subsequently hope to produce an estimate (with 95% confidence) of treatment effectiveness for metastatic melanoma that has a margin of error of only 6%, how many patients should they study?

41. Bread Clarksburg Bakery is trying to predict how many loaves to bake. In the past 100 days, the bakery has sold between 95 and 140 loaves per day. Here are a histogram and the summary statistics for the number of loaves sold for the past 100 days.

Summary of Sales	
Mean	103
Median	100
StdDev	9.000
Min	95
Max	140
Lower 25th %tile	97
Upper 25th %tile	105.5

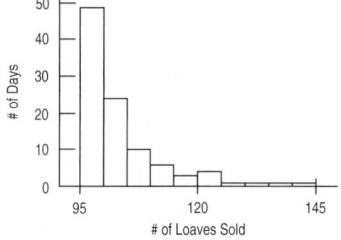

a) Can you use these data to estimate the number of loaves sold on the busiest 10% of all days? Explain.
b) Explain why you can use these data to construct a 95% confidence interval for the mean number of loaves sold per day.
c) Calculate a 95% confidence interval and carefully interpret what that confidence interval means.
d) If the bakery would have been satisfied with a confidence interval whose margin of error was twice as wide, how many days' data could they have used?
e) When the bakery opened, the owners estimated that they would sell an average of 100 loaves per day. Does your confidence interval provide strong evidence that this estimate was incorrect? Explain.

42. Fritos® As a project for an Introductory Statistics course, students checked 6 bags of Fritos marked with a net weight of 35.4 grams. They carefully weighed the contents of each bag, recording the following weights (in grams): 35.5, 35.3, 35.1, 36.4, 35.4, 35.5. Is there evidence that the mean weight of bags of Fritos is less than advertised?
a) Write appropriate hypotheses.
b) Do these data satisfy the assumptions for inference? Explain.
c) Test your hypothesis using all 6 weights.
d) Retest your hypothesis with the one unusually high weight removed.
e) What would you conclude about the stated net weight?

43. **And it means?** Every statement about a confidence interval contains two parts—the level of confidence and the interval. Suppose that an insurance agent estimating the mean loss claimed by clients after home burglaries created the 95% confidence interval ($1644, $2391).
   a) What's the margin of error for this estimate?
   b) Carefully explain what the interval means.
   c) Carefully explain what the 95% confidence level means.

44. **Batteries** We work for the "Watchdog for the Consumer" consumer advocacy group. We've been asked to look at a battery company that claims its batteries last an average of 100 hours under normal use. There have been several complaints that the batteries don't last that long, so we decide to test them. To do this we select 16 batteries and run them until they die. They lasted a mean of 97 hours, with a standard deviation of 12 hours.
   a) One of the editors of our newsletter (who does not know statistics) says that 97 hours is a lot less than the advertised 100 hours, so we should reject the company's claim. Explain to him the problem with doing that.
   b) What are the null and alternative hypotheses?
   c) What assumptions must we make in order to proceed with inference?
   d) At a 5% level of significance, what do you conclude?
   e) Suppose that, in fact, the average life of the company's batteries is only 98 hours. Has an error been made in part d? If so, what kind?

Chapter 22

Comparing Groups

22.1 The Standard Deviation of a Difference

22.2 Assumptions and Conditions for Comparing Proportions

22.3 A Confidence Interval for the Difference Between Two Proportions

22.4 The Two Sample z-Test: Testing for the Difference Between Proportions

22.5 A Confidence Interval for the Difference Between Two Means

22.6 The Two-Sample t-Test: Testing for the Difference Between Two Means

22.7 The Pooled t-Test: Everyone into the Pool?

> ### Where are we going?
> Do people really feel better about the economy this month compared to last month, or was that increase just sampling variation? Do students who use technology learn Statistics better than those who do not? In practice, it's much more common to compare two groups than to test whether a single proportion or mean is equal to a given number. Comparing two groups is very much like testing one. The standard errors are different, but the concepts are really the same.

Who	6971 male drivers
What	Seat-belt use
Why	Highway safety
When	2007
Where	Massachusetts

Do men take more risks than women? Psychologists have documented that in many situations, men choose riskier behavior than women do. But what is the effect of having a woman in the passenger's seat? A recent seat-belt observation study in Massachusetts found that, not surprisingly, male drivers wear seat belts less often than women do. The study also noted that men's belt-wearing jumped more than 16 percentage points when they had a female passenger. Seat-belt use was recorded at 161 locations in Massachusetts, using random-sampling methods developed by the National Highway Traffic Safety Administration (NHTSA). Female drivers wore belts more than 70% of the time, regardless of the sex of their passengers. Of 4208 male drivers with female passengers, 2777 (66.0%) were belted. But among 2763 male drivers with male passengers only, 1363 (49.3%) wore seat belts. This was a random sample, but it suggests there may be a shift in men's risk-taking behavior when women are present. What would we estimate the true size of that difference to be?

Sometimes, instead of a proportion, we have a quantitative variable measured on two groups and we want to compare their means. If you bought a used camera in good condition from a friend, would you pay the same as you would if you bought the same

586 PART VI Accessing Associations Between Variables

Who University students
What Prices offered for a used camera
Units $
Why Study of the effects of friendship on transactions
When 1990s
Where Cornell University

camera from a stranger? A researcher at Cornell University[1] wanted to know how friendship might affect simple sales such as this. She randomly divided subjects into two groups and gave each group descriptions of items they might want to buy. One group was told to imagine buying from a friend whom they expected to see again. The other group was told to imagine buying from a stranger.

Here are the prices they offered for a used camera in good condition:

Price Offered for a Used Camera ($)	
Buying from a Friend	Buying from a Stranger
275	260
300	250
260	175
300	130
255	200
275	225
290	240
300	

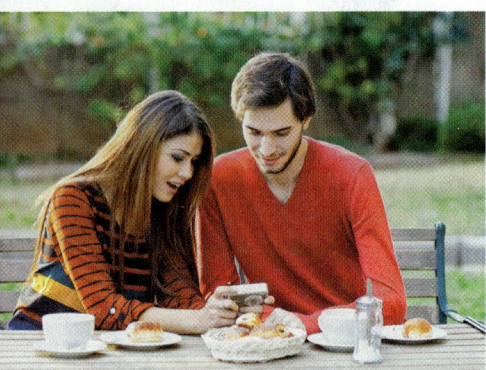

Those buying from a friend offered an average of $281.88 and those buying from a stranger offered an average of $211.43. But that's just from these 15 subjects. How big a difference should we expect in general?

22.1 The Standard Deviation of a Difference

The difference we *observe* between two groups is specific to the data at hand. But what's the *true* difference for the general population? We know that our estimate probably isn't exactly right. To say more, we need a way to judge how large the difference is. That means we'll need to estimate its standard deviation.

How do we find this estimate? We know how to estimate the standard deviation of the proportion or mean in each group, but how about their difference? The answer comes from realizing that we have to start with the variances and the Pythagorean Theorem of Statistics from Chapter 15

The variance of the sum or difference of two independent random variables is the sum of their variances.

This is such an important (and powerful) idea in Statistics that it's worth pausing a moment to review the reasoning. Here's some intuition about why variation increases even when we subtract two random quantities.

Let's say everyone in your classroom has a full box of cereal that's labeled 16 ounces. We know that's not exact: there's some variation from box to box. Now everyone pours a 2-ounce bowl of cereal. There'll be some variation in how much each person pours, too. How much variation will there be in the amount of cereal that is left in each box? Even though the amount left in the box is the *difference* between the original amount and the amount that was poured, there will be even *more* variation than there was in either the original amount or the amount poured. This is because the amount left depends on *both* the variation in how much was originally in the box and the variation in the amount poured.

According to the rule, the variance of the amount of cereal left in the box would now be the *sum* of the two *variances*.

---

[1] J. J. Halpern, "The Transaction Index: A Method for Standardizing Comparisons of Transaction Characteristics Across Different Contexts," *Group Decision and Negotiation*, 6: 557–572.

We want a standard deviation, not a variance, but that's just a square root away. We can write symbolically what we've just said:

$$Var(X - Y) = Var(X) + Var(Y), \text{ so}$$
$$SD(X - Y) = \sqrt{SD^2(X) + SD^2(Y)} = \sqrt{Var(X) + Var(Y)}.$$

Be careful, though—this simple formula applies only when X and Y are independent. Just as the Pythagorean Theorem[2] works only for right triangles, the formula works only for independent random variables. Always check for independence before using it.

> For independent random variables, **variances add**.

The Standard Deviation of the Difference Between Two Proportions

We'll apply what we've learned about the variance of a difference to proportions first. Fortunately, proportions observed in independent random samples *are* independent, so we can put the two proportions in for X and Y and add their variances. We just need to use careful notation to keep things straight.

When we have two samples, each can have a different size and proportion value, so we keep them straight with subscripts. Often we choose subscripts that remind us of the groups. For our example, we might use "$_M$" and "$_F$", but generically we'll just use "$_1$" and "$_2$." We will represent the two sample proportions as $\hat{p}_1$ and $\hat{p}_2$, and the two sample sizes as n_1 and n_2.

> Combining independent random quantities always *increases* the overall variation, so even for *differences* of independent random variables, **variances add**.

The standard deviations of the sample proportions are $SD(\hat{p}_1) = \sqrt{\dfrac{p_1 q_1}{n_1}}$ and $SD(\hat{p}_2) = \sqrt{\dfrac{p_2 q_2}{n_2}}$, so the variance of the difference in the proportions is

$$Var(\hat{p}_1 - \hat{p}_2) = \left(\sqrt{\dfrac{p_1 q_1}{n_1}}\right)^2 + \left(\sqrt{\dfrac{p_2 q_2}{n_2}}\right)^2 = \dfrac{p_1 q_1}{n_1} + \dfrac{p_2 q_2}{n_2}.$$

The standard deviation is the square root of that variance:

$$SD(\hat{p}_1 - \hat{p}_2) = \sqrt{\dfrac{p_1 q_1}{n_1} + \dfrac{p_2 q_2}{n_2}}.$$

We usually don't know the true values of p_1 and p_2. When we have the sample proportions in hand from the data, we use them to estimate the variances. So the standard error is

$$SE(\hat{p}_1 - \hat{p}_2) = \sqrt{\dfrac{\hat{p}_1 \hat{q}_1}{n_1} + \dfrac{\hat{p}_2 \hat{q}_2}{n_2}}.$$

For Example FINDING THE STANDARD ERROR OF A DIFFERENCE IN PROPORTIONS

A recent survey of 886 randomly selected teenagers (aged 12–17) found that more than half of them had online profiles.[3] Some researchers and privacy advocates are concerned about the possible access to personal information about teens in public places on the Internet. There appear to be differences between boys and girls in their online behavior. Among teens aged 15–17, 57% of the 248 boys had posted profiles, compared to 70% of the 256 girls. Let's start the process of estimating how large the true gender gap might be.

(continued)

---

[2] It is not coincidental that Pythagorus needs a right angle and we need independence. The mathematics behind this fact shows that they are the same condition.

[3] Princeton Survey Research Associates International for the Pew Internet & American Life Project.

> **QUESTION:** What's the standard error of the difference in sample proportions?
>
> **ANSWER:** Because the boys and girls were selected at random, it's reasonable to assume their behaviors are independent, so it's ok to use the Pythagorean Theorem of Statistics and add the variances:
>
> $$SE(\hat{p}_{boys}) = \sqrt{\frac{0.57 \times 0.43}{248}} = 0.0314 \quad SE(\hat{p}_{girls}) = \sqrt{\frac{0.70 \times 0.30}{256}} = 0.0286$$
>
> $$SE(\hat{p}_{girls} - \hat{p}_{boys}) = \sqrt{0.0314^2 + 0.0286^2} = 0.0425$$

22.2 Assumptions and Conditions for Comparing Proportions

Before we complete our example, we need to check assumptions and conditions.

Independence

Independence Assumption: Within each group, the data should be based on results for independent individuals. We can't check that for certain, but we *can* check the following:

Randomization Condition: Ideally, the data in each group should be drawn independently and at random from a population or generated by a randomized comparative experiment.

The 10% Condition: If the data are sampled without replacement, the sample should not exceed 10% of the population without an adjustment to the standard deviation.

Because we are comparing two groups in this way, we need an additional Independence Assumption. In fact, this is the most important of these assumptions. If it is violated, these methods just won't work.

Independent Groups Assumption: The two groups we're comparing must also be independent *of each other*. Usually, the independence of the groups from each other is evident from the way the data were collected.

Why is the Independent Groups Assumption so important? If we compare husbands with their wives, or a group of subjects before and after some treatment, we can't just add the variances. Subjects' performance before a treatment might very well be related to their performance after the treatment. So the proportions are not independent and the Pythagorean-style variance formula does not hold. We'll see a way to compare a common kind of nonindependent sample in a later chapter.

Sample Size

Each of the groups must be big enough. As with individual proportions, we need larger groups to estimate proportions that are near 0% or 100%. We usually check the Success/Failure Condition for each group.

Success/Failure Condition: Both groups are big enough that at least 10 successes and at least 10 failures have been observed in each.

> ### For Example CHECKING ASSUMPTIONS AND CONDITIONS
>
> **RECAP:** Among randomly sampled teens aged 15–17, 57% of the 248 boys had posted online profiles, compared to 70% of the 256 girls.
>
> **QUESTION:** Can we use these results to make inferences about all 15–17-year-olds?
>
> **ANSWER:**
>
> ✓ **Randomization Condition:** The sample of boys and the sample of girls were both chosen randomly.

> ✓ **10% Condition:** 248 boys and 256 girls are each less than 10% of all teenage boys and girls.
> ✓ **Independent Groups Assumption:** Because the samples were selected at random, it's reasonable to believe the boys' online behaviors are independent of the girls' online behaviors.
> ✓ **Success/Failure Condition:** Among the boys, $248(0.57) = 141$ had online profiles and the other $248(0.43) = 107$ did not. For the girls, $256(0.70) = 179$ successes and $256(0.30) = 77$ failures. All counts are at least 10.
>
> Because all the assumptions and conditions are satisfied, it's ok to proceed with inference for the difference in proportions.
>
> (Note that when we find the *observed* counts of successes and failures, we round off to whole numbers. We're using the reported percentages to recover the actual counts.)

22.3 A Confidence Interval for the Difference Between Two Proportions

WHY NORMAL?
One of the special properties of the Normal model is that sums and differences of independent Normal random variables also follow a Normal model. That's the reason we use a Normal model for the difference of two independent proportions.

We're almost there. We just need one more fact about proportions. We already know that for large enough samples, each of our proportions has an approximately Normal sampling distribution. The same is true of their difference.

> **THE SAMPLING DISTRIBUTION MODEL FOR A DIFFERENCE BETWEEN TWO INDEPENDENT PROPORTIONS**
> Provided that the sampled values are independent, the samples are independent, and the sample sizes are large enough, the sampling distribution of $\hat{p}_1 - \hat{p}_2$ is modeled by a Normal model with mean $\mu = p_1 - p_2$ and standard deviation
> $$SD(\hat{p}_1 - \hat{p}_2) = \sqrt{\frac{p_1 q_1}{n_1} + \frac{p_2 q_2}{n_2}}.$$

The sampling distribution model and the standard deviation give us all we need to find a margin of error for the difference in proportions—or at least they would if we knew the true proportions, p_1 and p_2. However, we don't know the true values, so we'll work with the observed proportions, $\hat{p}_1$ and $\hat{p}_2$, and use $SE(\hat{p}_1 - \hat{p}_2)$ to estimate the standard deviation. The rest is just like a one-proportion z-interval.

A S *Activity:* **Compare Two Proportions.** Does a preschool program help disadvantaged children later in life?

> **A TWO-PROPORTION z-INTERVAL**
> When the conditions are met, we are ready to find the confidence interval for the difference of two proportions, $p_1 - p_2$. The confidence interval is
> $$(\hat{p}_1 - \hat{p}_2) \pm z^* \times SE(\hat{p}_1 - \hat{p}_2)$$
> where we find the standard error of the difference,
> $$SE(\hat{p}_1 - \hat{p}_2) = \sqrt{\frac{\hat{p}_1 \hat{q}_1}{n_1} + \frac{\hat{p}_2 \hat{q}_2}{n_2}},$$
> from the observed proportions.
> The critical value z^* depends on the particular confidence level, C, that we specify.

For Example FINDING A TWO-PROPORTION z-INTERVAL

RECAP: Among randomly sampled teens aged 15–17, 57% of the 248 boys had posted online profiles, compared to 70% of the 256 girls. We calculated the standard error for the difference in sample proportions to be $SE(\hat{p}_{girls} - \hat{p}_{boys}) = 0.0425$ and found that the assumptions and conditions required for inference checked out ok.

QUESTION: What does a confidence interval say about the difference in online behavior?

ANSWER:

A 95% confidence interval for $p_{girls} - p_{boys}$ is $(\hat{p}_{girls} - \hat{p}_{boys}) \pm z^* SE(\hat{p}_{girls} - \hat{p}_{boys})$.

$$(0.70 - 0.57) \pm 1.96(0.0425)$$
$$0.13 \pm 0.083$$
$$(4.7\%, 21.3\%)$$

We can be 95% confident that among teens aged 15–17, the proportion of girls who post online profiles is between 4.7 and 21.3 percentage points higher than the proportion of boys who do. It seems clear that teen girls are more likely to post profiles than are boys the same age.

Step-by-Step Example A TWO-PROPORTION z-INTERVAL

Now we are ready to be more precise about the passenger-based gap in male drivers' seat-belt use. We'll estimate the difference with a confidence interval using a method called the **two-proportion z-interval** and follow the four confidence interval steps.

Question: How much difference is there in the proportion of male drivers who wear seat belts when sitting next to a male passenger and the proportion who wear seat belts when sitting next to a female passenger?

THINK ➡ **Plan** State what you want to know. Discuss the variables and the W's.

Identify the parameter you wish to estimate. (It doesn't usually matter mathematically in which direction we subtract, but it is much easier for people to understand the meaning of the interval when we choose the direction with a positive difference)

Choose and state a confidence level.

Model Think about the assumptions and check the conditions.

I want to estimate the true difference in the population proportion, p_M, of male drivers who wear seat belts when sitting next to a man and p_F, the proportion who wear seat belts when sitting next to a woman. The data are from a random sample of drivers in Massachusetts in 2007, observed according to procedures developed by the NHTSA. The parameter of interest is the difference $p_F - p_M$.

I will find a 95% confidence interval for this parameter.

✓ **Randomization Condition:** The NHTSA methods are more complex than an SRS, but they result in a suitable random sample so it's reasonable to assume that the driver behavior is independent from car to car.

✓ **10% Condition:** The samples include far fewer than 10% of all male drivers accompanied by male or by female passengers.

The Success/Failure Condition must hold for each group.	✓ **Independent Groups Assumption:** It's reasonable to believe that seat-belt use among drivers with male passengers and those with female passengers are independent.
	✓ **Success/Failure Condition:** Among male drivers with female passengers, 2777 wore seat belts and 1431 did not; of those driving with male passengers, 1363 wore seat belts and 1400 did not. Each group contained far more than 10 successes and 10 failures.
State the sampling distribution model for the statistic. Choose your method.	Under these conditions, the sampling distribution of the difference between the sample proportions is approximately Normal, so I'll find a **two-proportion z-interval**.

SHOW ➤ Mechanics Construct the confidence interval.

I know
$$n_F = 4208, n_M = 2763.$$
The observed sample proportions are
$$\hat{p}_F = \frac{2777}{4208} = 0.660, \hat{p}_M = \frac{1363}{2763} = 0.493.$$

As often happens, the key step in finding the confidence interval is estimating the standard deviation of the sampling distribution model of the statistic. Here the statistic is the difference in the proportions of men who wore seat belts when they had a female passenger and those who had a male passenger. Substitute the data values into the formula.

I'll estimate the SD of the difference with
$$SE(\hat{p}_F - \hat{p}_M) = \sqrt{\frac{\hat{p}_F \hat{q}_F}{n_F} + \frac{\hat{p}_M \hat{q}_M}{n_M}}$$
$$= \sqrt{\frac{(0.660)(0.340)}{4208} + \frac{(0.493)(0.507)}{2763}}$$
$$= 0.012.$$

The sampling distribution is Normal, so the critical value for a 95% confidence interval, z^*, is 1.96. The margin of error is the critical value times the SE.

$$ME = z^* \times SE(\hat{p}_F - \hat{p}_M)$$
$$= 1.96(0.012) = 0.024$$

The confidence interval is the statistic $\pm ME$.

The observed difference in proportions is $\hat{p}_F - \hat{p}_M = 0.660 - 0.493 = 0.167$, so the 95% confidence interval is

$$0.167 \pm 0.024$$

or 14.3% to 19.1%.

TELL ➤ Conclusion Interpret your confidence interval in the proper context. (Remember: We're 95% confident that our interval captured the true difference.)

I am 95% confident that the proportion of male drivers who wear seat belts when driving next to a female passenger is between 14.3 and 19.1 percentage points higher than the proportion who wear seat belts when driving next to a male passenger.

(continued)

This is an interesting result—but be careful not to try to say too much! In Massachusetts, overall seat-belt use is lower than the national average, so we can't be certain that these results generalize to other states. And these were two different groups of men, so we can't say that, individually, men are more likely to buckle up when they have a woman passenger. You can probably think of several alternative explanations; we'll suggest just a couple. Perhaps age is a lurking variable: Maybe older men are more likely to wear seat belts and also more likely to be driving with their wives. Or maybe men who don't wear seat belts have trouble attracting women!

Just Checking

A public broadcasting station plans to launch a special appeal for additional contributions from current members. Unsure of the most effective way to contact people, they run an experiment. They randomly select two groups of current members. They send the same request for donations to everyone, but it goes to one group by e-mail and to the other group by regular mail. The station was successful in getting contributions from 26% of the members they e-mailed but only from 15% of those who received the request by regular mail. A 90% confidence interval estimated the difference in donation rates to be 11% ± 7%.

1. Interpret the confidence interval in this context.

2. Based on this confidence interval, what conclusion would we reach if we tested the hypothesis that there's no difference in the response rates to the two methods of fund-raising? Explain.

22.4 The Two Sample z-Test: Testing for the Difference Between Proportions

The 2011 Sleep in America Poll surveyed 1508 randomly sampled U.S. residents aged 13 to 65 asking about their sleep habits and, in particular, their use of technology around the time they try to go to sleep. The researchers report that:

> This research shows that people who regularly use their computers or laptops in the hour before trying to go to sleep are *less* likely to report getting a good night's sleep . . . , more likely to be categorized as "sleepy" . . . , and more likely to drive drowsy . . . than their counterparts.[4]

The poll found that 205 of 293 or 70% of those 19–29 years old ("Gen-Y") reported using the Internet in the hour before trying to go to sleep at least a few nights a week. By contrast, 235 of 469 or 50% of those aged 30–45 ("Gen-X") reported Internet use before sleep. Is this difference of 20% real or is it likely to be due only to natural fluctuations in the sample?

The question calls for a hypothesis test. Now the parameter of interest is the true *difference* between the (reported) pre-sleep surfing rates of the two age groups.

What's the appropriate null hypothesis? That's easy here. We hypothesize that there is no difference in the proportions. This is such a natural null hypothesis that we rarely consider any other. But as we did for means, instead of writing $H_0: p_1 = p_2$, we usually express like this so it relates directly to the *difference* between the proportions

$$H_0: p_1 - p_2 = 0.$$

Who	Randomly selected U.S. adults aged 13–65
What	Proportion who use their laptops or computers in the hour before sleep, categorized by age
When	2011
Where	United States
Why	To study sleep behaviors

---
[4] Excerpt from "The 2011 Sleep in America Poll" from National Sleep Foundation. Copyright © 2011, published by National Sleep Foundation.

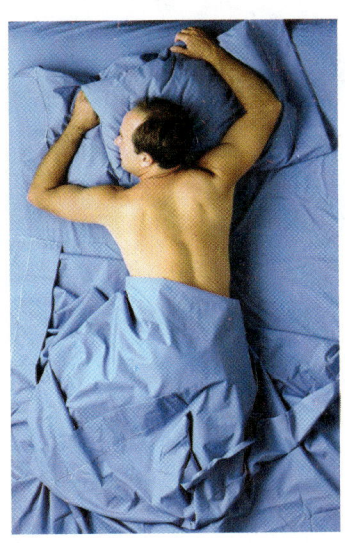

We know that the standard error of the difference in proportions is

$$SE(\hat{p}_1 - \hat{p}_2) = \sqrt{\frac{\hat{p}_1\hat{q}_1}{n_1} + \frac{\hat{p}_2\hat{q}_2}{n_2}},$$

and we could just plug in the numbers, but we can do slightly better. There are two proportions in the standard error formula—but look at the null hypothesis. It says that these proportions are equal. To do a hypothesis test, we *assume* that the null hypothesis is true. So there should be just a single value of $\hat{p}$ in the SE formula (and, of course, $\hat{q}$ is just $1 - \hat{p}$).

So, we should somehow combine the samples into one and come up with a single proportion for both groups to use in the standard deviation formula.[5] How would we do this for the sleep example? If the null hypothesis is true, then the two groups have the same proportion. Overall, we saw $205 + 235 = 440$ surfers out of a total of $293 + 469 = 762$ respondents. The overall proportion of surfers was $440/762 = 0.5774$.

Combining the counts like this to get an overall proportion is called **pooling**. Whenever we have data from different sources or different groups but we believe that they really came from the same underlying population, we pool them to get better estimates.

In general, we find the pooled proportion as

$$\hat{p}_{pooled} = \frac{Success_1 + Success_2}{n_1 + n_2},$$

where $Success_1$ is the number of successes in group 1 and $Success_2$ is the number of successes in group 2. That's the overall proportion of success.

When you have only proportions and not the counts, you'll have to reconstruct the number of successes by multiplying the sample sizes by the proportions:

$$Success_1 = n_1\hat{p}_1 \quad \text{and} \quad Success_2 = n_2\hat{p}_2.$$

If these calculations don't come out to whole numbers, round them first. There must have been a whole number of successes, after all. (This is the *only* time you should round values in the middle of a calculation.)

We then put this combined value into the formula for the standard error, substituting it for *both* sample proportions:

$$SE_{pooled}(\hat{p}_1 - \hat{p}_2) = \sqrt{\frac{\hat{p}_{pooled}\hat{q}_{pooled}}{n_1} + \frac{\hat{p}_{pooled}\hat{q}_{pooled}}{n_2}}$$

$$= \sqrt{\frac{0.5774 \times (1 - 0.5774)}{293} + \frac{0.5774 \times (1 - 0.5774)}{469}}.$$

This comes out to 0.0368.

> **ROUNDING**
>
> When finding the number of successes, round the values to integers. For example, we were told that 70% of the 293 Gen-Yers reported using the Internet before sleep. But $293 \times 0.7 = 205.1$. So, we round back to the nearest whole number to find the count that could have yielded the rounded percent we were given.

A S *Activity:* Test for a Difference Between Two Proportions. Is premium-brand chicken less likely to be contaminated than store-brand chicken?

> **TWO-PROPORTION z-TEST**
>
> The conditions for the two-proportion z-test are the same as for the two-proportion z-interval. We are testing the hypothesis
>
> $$H_0: p_1 - p_2 = 0.$$
>
> Because we hypothesize that the proportions are equal, we pool the groups to find
>
> $$\hat{p}_{pooled} = \frac{Success_1 + Success_2}{n_1 + n_2}.$$

---

[5] We didn't do this for the confidence interval, because we weren't assuming the two proportions were equal. Combining them doesn't usually make a big practical difference, but we should take advantage of it for the hypothesis test.

and use that pooled value to estimate the standard error:

$$SE_{pooled}(\hat{p}_1 - \hat{p}_2) = \sqrt{\frac{\hat{p}_{pooled}\hat{q}_{pooled}}{n_1} + \frac{\hat{p}_{pooled}\hat{q}_{pooled}}{n_2}}.$$

Now we find the test statistic,

$$z = \frac{(\hat{p}_1 - \hat{p}_2) - 0}{SE_{pooled}(\hat{p}_1 - \hat{p}_2)}.$$

When the conditions are met and the null hypothesis is true, this statistic follows the standard Normal model, so we can use that model to obtain a P-value.

Step-by-Step Example A TWO-PROPORTION z-TEST

Question: Are the pre-sleep surfing rates of the two age groups really different?

THINK ▶ Plan State what you want to know. Discuss the variables and the W's.

I want to know whether pre-sleep Internet surfing rates differ for "Generation Y's" (19- to 29-year-olds) relative to "Generation X's" (30-45). The data are from a random sample of 1508 U.S. adults surveyed in the 2011 Sleep in America Poll. Of these, 293 Gen-Y's and 469 Gen-X's responded to the question about surfing, indicating whether or not they had surfed the Internet at least a few nights a week before trying to go to sleep.

Hypotheses The study simply broke down the responses by age, so there is no sense that either alternative was preferred. A two-sided alternative hypothesis is appropriate.

H_0: There is no difference in pre-sleep surfing rates in the two age groups:

$$p_{GenY} - p_{GenX} = 0.$$

H_A: The rates are different: $p_{GenY} - p_{GenX} \neq 0$.

Model Think about the assumptions and check the conditions.

✓ **Randomization Condition:** The respondents were randomly selected by telephone number and stratified by sex and region, so they should be independent.

✓ **10% Condition:** The number of adults surveyed in each age group is certainly far less than 10% of that population.

✓ **Independent Groups Assumption:** The two groups are independent of each other because the sample was selected at random.

✓ **Success/Failure Condition:** The observed numbers of both successes and failures are much more than 10 for both groups.

State the null model.

Choose your method.

Because the conditions are satisfied, I'll use a Normal model and perform a **two-proportion z-test**.

SHOW → Mechanics

The hypothesis is that the proportions are equal, so pool the sample data.

Use the pooled SE to estimate $SD(p_{GenY} - p_{GenX})$.

$n_{GenY} = 293$, $y_{GenY} = 205$, $\hat{p}_{GenY} = 0.700$

$n_{GenX} = 469$, $y_{GenX} = 235$, $\hat{p}_{GenX} = 0.501$

$$\hat{p}_{pooled} = \frac{y_{GenY} + y_{GenX}}{n_{GenY} + n_{GenX}} = \frac{205 + 235}{293 + 469} = 0.5774$$

$$SE_{pooled}(\hat{p}_{GenY} - \hat{p}_{GenX})$$

$$= \sqrt{\frac{\hat{p}_{pooled}\hat{q}_{pooled}}{n_{GenY}} + \frac{\hat{p}_{pooled}\hat{q}_{pooled}}{n_{GenX}}}$$

$$= \sqrt{\frac{0.5774 \times 0.4226}{293} + \frac{0.5774 \times 0.4226}{469}}$$

$$= 0.0368$$

The observed difference in sample proportions is

$\hat{p}_{GenY} - \hat{p}_{GenX} = 0.700 - 0.501 = 0.199.$

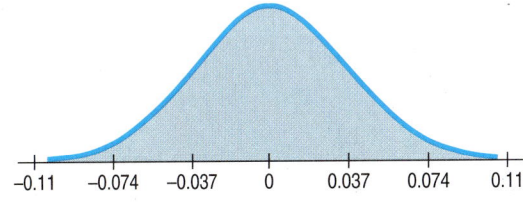

Find the z-score for the observed difference in proportions, 0.199.

Find the P-value using Table Z or technology. Because this is a two-tailed test, we must *double* the probability we find in the upper tail.

$$z = \frac{(\hat{p}_{GenY} - \hat{p}_{GenX}) - 0}{SE_{pooled}(\hat{p}_{GenY} - \hat{p}_{GenX})} = \frac{0.199}{0.0368} = 5.41$$

$P = 2P(z > 5.41) \leq 0.0001$

TELL → Conclusion

Link the P-value to your decision about the null hypothesis, and state your conclusion in context.

The P-value says that if there really were no difference in surfing rates between the two groups, then the difference observed in this study would be very rare indeed. We can conclude that there is in fact a difference in the rate of surfing between GenY and GenX adults.

✓ Just Checking

3. A February 2014 public opinion poll asked 1000 randomly selected adults whether the United States should decrease the amount of immigration allowed; 36% of those responding said "yes." In June 1995, a random sample of 1000 had found that 65% of adults thought immigration should be curtailed. To see if that percentage has decreased, why can't we just use a one-proportion z-test of $H_0: p = 0.65$ and see what the P-value for $\hat{p} = 0.49$ is?

4. For opinion polls like this, which has more variability: the percentage of respondents answering "yes" in either year or the difference in the percentages between the two years?

For Example TWO-PROPORTION z-TEST

Another concern of the study on teens' online profiles was safety and privacy. In the random sample, girls were less likely than boys to say that they are easy to find online from their profiles. Only 19% (62 girls) of 325 teen girls with profiles say that they are easy to find, while 28% (75 boys) of the 268 boys with profiles say the same.

QUESTION: Are these results evidence of a real difference between boys and girls? Perform a two-proportion z-test and discuss what you find.

ANSWER:

$$H_0: p_{boys} - p_{girls} = 0$$
$$H_A: p_{boys} - p_{girls} \neq 0$$

- ✓ **Randomization Condition:** The sample of boys and the sample of girls were both chosen randomly, so they should be independent.
- ✓ **10% Condition:** 268 boys and 325 girls are each less than 10% of all teenage boys and girls with online profiles.
- ✓ **Independent Groups Assumption:** Because the samples were selected at random, it's reasonable to believe the boys' perceptions are independent of the girls'.
- ✓ **Success/Failure Condition:** Among the girls, there were 62 "successes" and 263 failures, and among boys, 75 successes and 193 failures. These counts are at least 10 for each group.

Because all the assumptions and conditions are satisfied, it's ok to do a **two-proportion z-test**:

$$\hat{p}_{pooled} = \frac{75 + 62}{268 + 325} = 0.231$$

$$SE_{pooled}(\hat{p}_{boys} - \hat{p}_{girls}) = \sqrt{\frac{0.231 \times 0.769}{268} + \frac{0.231 \times 0.769}{325}} = 0.0348$$

$$z = \frac{(0.28 - 0.19) - 0}{0.0348} = 2.59$$

$$P(z > 2.59) = 0.0048$$

This is a two-tailed test, so the P-value = 2(0.0048) = 0.0096. Because this P-value is very small, I reject the null hypothesis. This study provides strong evidence that there really is a difference in the proportions of teen girls and boys who say they are easy to find online.

22.5 A Confidence Interval for the Difference Between Two Means

Comparing two means is just like comparing two proportions. In the camera purchase experiment, for example, the parameter of interest is the difference between the *mean* amount offered to friends and the mean amount offered to strangers, $\mu_1 - \mu_2$.

The rest is essentially the same as before. The statistic of interest is the difference in the two observed means, $\bar{y}_1 - \bar{y}_2$. We'll start with this statistic to build our confidence interval, but we'll need to know its standard deviation and its sampling model. Then we can build confidence intervals and find P-values for hypothesis tests.

Just as we did for proportions, we'll use the Pythagorean Theorem of Statistics: For independent random variables, the variance of their *difference* is the *sum* of their individual variances, $Var(Y - X) = Var(Y) + Var(X)$. To find the standard deviation of the difference between the two independent sample means, we add their variances and then take a square root:

$$SD(\bar{y}_1 - \bar{y}_2) = \sqrt{Var(\bar{y}_1) + Var(\bar{y}_2)}$$
$$= \sqrt{\frac{\sigma_1^2}{n_1} + \frac{\sigma_2^2}{n_2}}.$$

Of course, we still don't know the true standard deviations of the two groups, σ_1 and σ_2, so as usual, we'll use the estimates, s_1 and s_2. Using the estimates gives us the *standard error*:

$$SE(\bar{y}_1 - \bar{y}_2) = \sqrt{\frac{s_1^2}{n_1} + \frac{s_2^2}{n_2}}.$$

We'll use the standard error to judge how big the difference really is. Because we are working with means and estimating the standard error of their difference using the data, we shouldn't be surprised that the sampling model is a Student's *t*.

For Example FINDING THE STANDARD ERROR OF THE DIFFERENCE IN INDEPENDENT SAMPLE MEANS

Can you tell how much you are eating from how full you are? Or do you need visual cues? Researchers[6] constructed a table with two ordinary 18 oz soup bowls and two identical-looking bowls that had been modified to slowly, imperceptibly, refill as they were emptied. They assigned experiment participants to the bowls randomly and served them tomato soup. Those eating from the ordinary bowls had their bowls refilled by ladle whenever they were one-quarter full. If people judge their portions by internal cues, they should eat about the same amount. How big a difference was there in the amount of soup consumed? The table summarizes their results.

	Ordinary Bowl	Refilling Bowl
n	27	27
$\bar{y}$	8.5 oz	14.7 oz
s	6.1 oz	8.4 oz

QUESTION: How much variability do we expect in the difference between the two means? Find the standard error.

ANSWER: Participants were randomly assigned to bowls, so the two groups should be independent. It's ok to add variances.

$$SE(\bar{y}_{refill} - \bar{y}_{ordinary}) = \sqrt{\frac{s_r^2}{n_r} + \frac{s_o^2}{n_o}} = \sqrt{\frac{8.4^2}{27} + \frac{6.1^2}{27}} = 2.0 \ oz.$$

The confidence interval we build is called a **two-sample *t*-interval** (for the difference in means). The corresponding hypothesis test is called a **two-sample *t*-test**. The interval looks just like all the others we've seen—the statistic plus or minus an estimated margin of error:

$$(\bar{y}_1 - \bar{y}_2) \pm ME$$
where $ME = t^* \times SE(\bar{y}_1 - \bar{y}_2)$.

This formula is almost the same as the confidence interval for the difference of two proportions. It's just that here we use a Student's *t*-model instead of a Normal model to find the appropriate critical value corresponding to our chosen confidence level. (So we write it as t^*.)

> **z OR t?**
> If you know σ, use z. (That's rare!) Whenever you use s to estimate σ, use t.

---
[6]Brian Wansink, James E. Painter, and Jill North, "Bottomless Bowls: Why Visual Cues of Portion Size May Influence Intake," *Obesity Research*, Vol. 13, No. 1, January 2005.

What are we missing? Only the degrees of freedom for the Student's t-model. Unfortunately, *that* formula is strange.

The deep, dark secret is that the sampling model isn't *really* Student's t, but only something close. The trick is that by using a special, adjusted degrees-of-freedom value, we can make it so close to a Student's t-model that nobody can tell the difference. The adjustment formula is straightforward but doesn't help our understanding much, so we leave it to the computer or calculator. (If you are curious and really want to see the formula, look in the footnote.[7])

> **A SAMPLING DISTRIBUTION FOR THE DIFFERENCE BETWEEN TWO MEANS**
>
> When the conditions are met, the sampling distribution of the standardized sample difference between the means of two independent groups,
>
> $$t = \frac{(\bar{y}_1 - \bar{y}_2) - (\mu_1 - \mu_2)}{SE(\bar{y}_1 - \bar{y}_2)},$$
>
> can be modeled by a Student's t-model with a number of degrees of freedom found with a special formula. We estimate the standard error with
>
> $$SE(\bar{y}_1 - \bar{y}_2) = \sqrt{\frac{s_1^2}{n_1} + \frac{s_2^2}{n_2}}.$$

Assumptions and Conditions

Independence The independence assumptions are the same as for comparing two proportions. We need independent values in each group and independence between the groups. We see randomization as evidence of independence.

Normal Population One way working with means is different from working with proportions is that, as we did before with Student's t-models, we must assume that the underlying populations are *each* Normally distributed. We check the . . .

Nearly Normal Condition: Check this for *both* groups; a violation by either one violates the condition. As we saw for single sample means, the Normality Assumption matters most when sample sizes are small. For samples of $n < 15$ in either group, you should not use these methods if the histogram or Normal probability plot shows severe skewness. For n's closer to 40, a mildly skewed histogram is OK, but you should remark on any outliers you find and not work with severely skewed data. When both groups are bigger than 40, the Central Limit Theorem starts to kick in no matter how the data are distributed, so the Nearly Normal Condition for the data matters less. Even in large samples, however, you should still be on the lookout for outliers, extreme skewness, and multiple modes.

A Note About Independent Groups

This test is sometimes called the two *independent samples* t-test because it is only appropriate when the two groups are independent of each other. No statistical test can verify this assumption. You have to think about how the data were collected. Paired or matched data of the kind we saw in the previous chapter cannot be analyzed with these methods because paired groups are (kind of obviously) not independent.

A S *Activity:* **Does Restricting Diet Prolong Life?** This activity lets you construct a confidence interval to compare life spans of rats fed two different diets.

---

[7] $df = \dfrac{\left(\dfrac{s_1^2}{n_1} + \dfrac{s_2^2}{n_2}\right)^2}{\dfrac{1}{n_1 - 1}\left(\dfrac{s_1^2}{n_1}\right)^2 + \dfrac{1}{n_2 - 1}\left(\dfrac{s_2^2}{n_2}\right)^2}$

Are you sorry you looked? This formula usually doesn't even give a whole number. If you are using a table, you'll need a whole number, so round down to be safe. If you are using technology, it's even easier. The approximation formulas that computers and calculators use for the Student's t-distribution deal with fractional degrees of freedom automatically.

CHAPTER 22 Comparing Groups

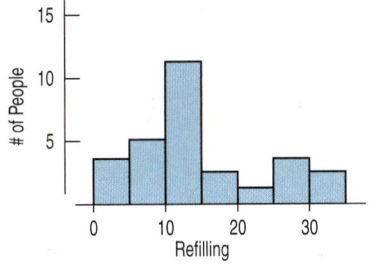

For Example CHECKING ASSUMPTIONS AND CONDITIONS

RECAP: Researchers randomly assigned people to eat soup from one of two bowls: 27 got ordinary bowls that were refilled by ladle, and 27 others bowls that secretly refilled slowly as the people ate.

QUESTION: Can the researchers use their data to make inferences about the role of visual cues in determining how much people eat?

ANSWER:
- ✓ **Randomization Condition:** The fact that subjects were randomly assigned to treatments makes independence between subjects plausible.
- ✓ **Nearly Normal Condition:** The histograms for both groups look unimodal (see the histograms in the margin) but somewhat skewed to the right. I believe both groups are large enough (27) to allow use of *t* methods.
- ✓ **Independent Groups Assumption:** Randomization to treatment groups guarantees this.

It's ok to construct a two-sample *t*-interval for the difference in means.

Note: When you check the Nearly Normal Condition it's important that you include the graphs you looked at (histograms or Normal probability plots).

AN EASIER RULE?

The formula for the degrees of freedom of the sampling distribution of the difference between two means is long, but the number of degrees of freedom is always at *least* the smaller of the two *n*'s, minus 1. Wouldn't it be easier to just use that value? You could, but *that* approximation can be a poor choice because it can give fewer than *half* the degrees of freedom you're entitled to from the correct formula.

TWO-SAMPLE *t*-INTERVAL FOR THE DIFFERENCE BETWEEN MEANS

When the conditions are met, we are ready to find the confidence interval for the difference between means of two independent groups, $\mu_1 - \mu_2$. The confidence interval is

$$(\bar{y}_1 - \bar{y}_2) \pm t^*_{df} \times SE(\bar{y}_1 - \bar{y}_2),$$

where the standard error of the difference of the means

$$SE(\bar{y}_1 - \bar{y}_2) = \sqrt{\frac{s_1^2}{n_1} + \frac{s_2^2}{n_2}}.$$

The critical value t^*_{df} depends on the particular confidence level, *C*, that you specify and on the number of degrees of freedom, which we get from the sample sizes and a special formula.

For Example FINDING A CONFIDENCE INTERVAL FOR THE DIFFERENCE IN SAMPLE MEANS

RECAP: Researchers studying the role of internal and visual cues in determining how much people eat conducted an experiment in which some people ate soup from bowls that secretly refilled. The results are summarized in the table.

	Ordinary Bowl	Refilling Bowl
n	27	27
$\bar{y}$	8.5 oz	14.7 oz
s	6.1 oz	8.4 oz

We've already checked the assumptions and conditions, and have found the standard error for the difference in means to be $SE(\bar{y}_{refill} - \bar{y}_{ordinary}) = 2.0$ oz.

(*continued*)

> **QUESTION:** What does a 95% confidence interval say about the difference in mean amounts eaten?
>
> **ANSWER:** The observed difference in means is $\bar{y}_{refill} - \bar{y}_{ordinary} = 14.7 - 8.5 = 6.2$ oz.
>
> $$df = 47.46 \quad t^*_{47.46} = 2.011 \text{ (Table T gives } t^*_{45} = 2.014.)$$
> $$ME = t^* \times SE(\bar{y}_{refill} - \bar{y}_{ordinary}) = 2.011(2.0) = 4.02 \text{ oz}$$
>
> The 95% confidence interval for $\mu_{refill} - \mu_{ordinary}$ is 6.2 ± 4.02, or (2.18 oz, 10.22 oz.)
> I am 95% confident that people eating from a subtly refilling bowl will eat an average of between 2.18 and 10.22 more ounces of soup than those eating from an ordinary bowl.

Just Checking

Carpal tunnel syndrome (CTS) causes pain and tingling in the hand. It can be bad enough to keep sufferers awake at night and restrict their daily activities. Researchers studied the effectiveness of two alternative surgical treatments for CTS (Mackenzie, Hainer, and Wheatley, *Annals of Plastic Surgery*, 2000). Patients were randomly assigned to have endoscopic or open-incision surgery. Four weeks later, the endoscopic surgery patients demonstrated a mean pinch strength of 9.1 kg compared to 7.6 kg for the open-incision patients.

5. Why is the randomization of the patients into the two treatments important?

6. A 95% confidence interval for the difference in mean strength is about (0.04 kg, 2.96 kg). Explain what this interval means.

7. Why might we want to examine such a confidence interval in deciding between these two surgical procedures?

8. Why might you want to see the data before trusting the confidence interval?

22.6 The Two-Sample *t*-Test: Testing for the Difference Between Two Means

 Activity: The Two-Sample *t*-Test. How different are beef hot dogs and chicken dogs? Test whether measured differences are statistically significant.

■ **NOTATION ALERT**

Δ_0—delta naught—isn't so standard that you can assume everyone will understand it. We use it because Δ is the Greek letter (good for a parameter) "D" for "difference." You should say "delta naught" rather than "delta zero"—that's standard for parameters associated with null hypotheses.

We constructed a confidence interval for the difference in price that someone would offer to a friend versus a stranger. That's fine, but sometimes we want to test whether the population means from two groups could be the same more directly, so we'll need a hypothesis test. Of course, you already know enough to construct this test. The test is called the two-sample *t*-test for the difference between two means.[8] The test statistic looks just like the others we've seen and uses the same standard error as we used for the confidence interval. It finds the difference between the observed group means and compares this with a hypothesized value for that difference.

When we write the null hypothesis, we could write: $H_0: \mu_1 = \mu_2$. But sometimes we'll want to see if the difference is something other than 0, so we usually write it slightly differently. We'll focus on the hypothesized *difference* between the two means and we'll call that hypothesized difference Δ_0 ("delta naught"). It's so common for that hypothesized difference to be zero that we often just assume $\Delta_0 = 0$. In that case, we'll write: $H_0: \mu_1 - \mu_2 = 0$, but in general we'll write: $H_0: \mu_1 - \mu_2 = \Delta_0$.

---

[8] Because it is performed so often, this test is usually just called a "two-sample *t*-test."

TWO-SAMPLE *t*-TEST FOR THE DIFFERENCE BETWEEN MEANS

The conditions for the two-sample *t*-test for the difference between the means of two independent groups are the same as for the two-sample *t*-interval. We test the hypothesis

$$H_0: \mu_1 - \mu_2 = \Delta_0,$$

where the hypothesized difference is almost always 0, using the statistic

$$t = \frac{(\bar{y}_1 - \bar{y}_2) - \Delta_0}{SE(\bar{y}_1 - \bar{y}_2)}.$$

The standard error of $\bar{y}_1 - \bar{y}_2$ is

$$SE(\bar{y}_1 - \bar{y}_2) = \sqrt{\frac{s_1^2}{n_1} + \frac{s_2^2}{n_2}}.$$

When the conditions are met and the null hypothesis is true, this statistic can be closely modeled by a Student's *t*-model with a number of degrees of freedom given by a special formula. We use that model to obtain a P-value.

Step-by-Step Example A TWO-SAMPLE *t*-TEST FOR THE DIFFERENCE BETWEEN TWO MEANS

Recall: We want to know if people will pay the same amount, on average, when buying from a friend or a stranger. The usual null hypothesis is that there's no difference in means. That's just the right null hypothesis for the camera purchase prices.

Question: Is there a difference in the average price people would offer a friend rather than a stranger?

THINK ➡ Plan State what we want to know.

Identify the *parameter* you wish to estimate. Here our parameter is the difference in the means, not the individual group means.

Identify the variables and check the W's.

I want to know whether people are likely to offer a different amount for a used camera when buying from a friend than when buying from a stranger. I plan to test whether the difference between mean amounts is zero. I have bid prices from 8 subjects buying from a friend and 7 buying from a stranger, found in a randomized experiment.

Hypotheses State the null and alternative hypotheses.

The research claim is that friendship changes what people are willing to pay.[9] The natural null hypothesis is that friendship makes no difference.

H_0: The difference in mean price offered to friends and the mean price offered to strangers is zero:

$$\mu_F - \mu_S = 0.$$

---

[9] This claim is a good example of what is called a "research hypothesis" in many social sciences. The only way to formally test a research hypothesis is to deny that it's true and see whether we can reject that null hypothesis.

We didn't start with any knowledge of whether friendship might increase or decrease the price, so we choose a two-sided alternative.

Make a picture. Boxplots are the display of choice for comparing groups. We'll also want to check the distribution of each group. Histograms or Normal probability plots do a better job there.

REALITY CHECK Looks like the prices are higher if you buy from a friend, but we can't be sure. You can't tell from looking at the boxplots whether the difference is statistically significant—the plot shows spreads on the same scale as the data, and we know those don't add. You'll need to add the *variances* to get a suitable ruler for comparing the difference.

Model
Think about the assumptions and check the conditions. (Note that, because this is a randomized experiment, we haven't sampled at all, so the 10% Condition does not apply.)

H_A: The difference in mean prices is not zero:
$$\mu_F - \mu_S \neq 0.$$

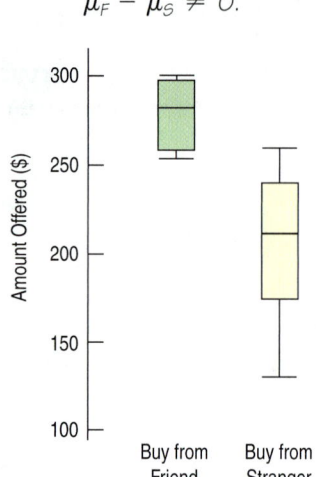

✓ **Randomization Condition:** The experiment was randomized. Subjects were assigned to treatment groups at random and so the prices paid should be independent.

✓ **Independent Groups Assumption:** Randomizing the experiment gives independent groups.

✓ **Nearly Normal Condition:** Histograms of the two sets of prices are reasonably unimodal and symmetric given the small sample sizes:

State the sampling distribution model.

Specify your method.

The assumptions are reasonable and the conditions are ok, so I'll use a Student's t-model to perform a **two-sample t-test**.

SHOW ➡ Mechanics
List the summary statistics. Be sure to use proper notation.

From the data:

$n_F = 8$ $n_S = 7$

$\bar{y}_F = \$281.88$ $\bar{y}_S = \$211.43$

$s_F = \$18.31$ $s_S = \$46.43$

Use the null model to find the P-value. First determine the standard error of the difference between sample means.	For independent groups, $$SE(\bar{y}_F - \bar{y}_S) = \sqrt{SE^2(\bar{y}_F) + SE^2(\bar{y}_S)}$$ $$= \sqrt{\frac{s_F^2}{n_F} + \frac{s_S^2}{n_S}}$$ $$= \sqrt{\frac{18.31^2}{8} + \frac{46.43^2}{7}}$$ $$= 18.70.$$ The observed difference is $$(\bar{y}_F - \bar{y}_S) = 281.88 - 211.43 = \$70.45.$$
Make a picture. Sketch the *t*-model centered at the hypothesized difference of zero. Because this is a two-tailed test, shade the region to the right of the observed difference and the corresponding region in the other tail.	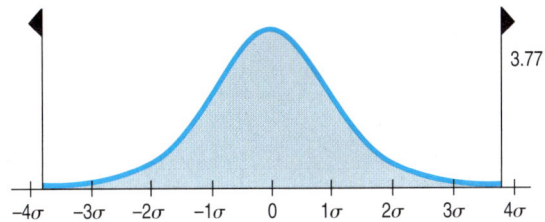
Find the *t*-value. A statistics program or graphing calculator can find the P-value using the fractional degrees of freedom from the approximation formula. If you are doing a test like this without technology, you could use the smaller sample size to determine degrees of freedom. In this case, $n_2 - 1 = 6$.	$$t = \frac{(\bar{y}_F - \bar{y}_S) - (0)}{SE(\bar{y}_F - \bar{y}_S)} = \frac{70.45}{18.70} = 3.77$$ The approximation formula gives 7.62 degrees of freedom.[10] $$\text{P-value} = 2P(t_{7.62} > 3.77) = 0.006$$
TELL ➡ **Conclusion** Link the P-value to your decision about the null hypothesis, and state the conclusion in context. Be cautious about generalizing to items whose prices are outside the range of those in this study.	If there were no difference in the mean prices, a difference this large would occur only 6 times in 1000. That's too rare to believe, so I reject the null hypothesis and conclude that people are likely to offer a friend more than they'd offer a stranger for a used camera (and possibly for other, similar items).

 Just Checking

Recall the experiment comparing patients 4 weeks after surgery for carpal tunnel syndrome. The patients who had endoscopic surgery demonstrated a mean pinch strength of 9.1 kg compared to 7.6 kg for the open-incision patients.

9. What hypotheses would you test?
10. The P-value of the test was less than 0.05. State a brief conclusion.
11. The study reports work on 36 "hands," but there were only 26 patients. In fact, 7 of the endoscopic surgery patients had both hands operated on, as did 3 of the open-incision group. Does this alter your thinking about any of the assumptions? Explain.

---

[10] If you were daring enough to calculate that messy degrees of freedom formula by hand with the values given here, you'd get about 7.74. Computers work with more precision for the standard deviations than we gave in our example. Many computer programs will round the final result down to 7 degrees of freedom. All give about the same result for the P-value, so it doesn't really matter—the conclusion would be the same.

For Example   A TWO-SAMPLE t-TEST

Many office "coffee stations" collect voluntary payments for the food consumed. Researchers at the University of Newcastle upon Tyne performed an experiment to see whether the image of eyes watching would change employee behavior.[11] They alternated pictures of eyes looking at the viewer with pictures of flowers each week on the cupboard behind the "honesty box." They measured the consumption of milk to approximate the amount of food consumed and recorded the contributions (in £) each week per liter of milk. The table summarizes their results.

	Eyes	Flowers
n (# weeks)	5	5
$\bar{y}$	0.417£/l	0.151£/l
s	0.1811£/l	0.067£/l

QUESTION: Do these results provide evidence that there really is a difference in honesty even when it's only photographs of eyes that are "watching"?

ANSWER:
$$H_0: \mu_{eyes} - \mu_{flowers} = 0$$
$$H_A: \mu_{eyes} - \mu_{flowers} \neq 0$$

✓ **Independence Assumption:** The amount paid by one person should be independent of the amount paid by others. However, this is an observational study, not a randomized experiment.

✓ **Nearly Normal Condition:** I don't have the data to check, but it seems unlikely there would be outliers in either group. I could be more certain if I could see histograms for both groups.

✓ **Independent Groups Assumption:** The same workers were recorded each week, but week-to-week independence is plausible.

It's ok to do a two sample t-test for the difference in means:

$$SE(\bar{y}_{eyes} - \bar{y}_{flowers}) = \sqrt{\frac{s^2_{eyes}}{n_{eyes}} + \frac{s^2_{flowers}}{n_{flowers}}} = \sqrt{\frac{0.1811^2}{5} + \frac{0.067^2}{5}} = 0.0864$$

$df = 5.07$

$$t_5 = \frac{(\bar{y}_{eyes} - \bar{y}_{flowers}) - 0}{SE(\bar{y}_{eyes} - \bar{y}_{flowers})} = \frac{0.417 - 0.151}{0.0864} = 3.08$$

$P\text{-value} = 2 \times P(t_5 > 3.08) = 2 \times 0.0137 = 0.027$

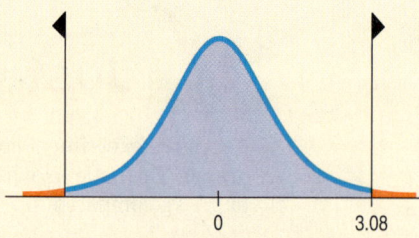

Assuming the data were free of outliers, the very low P-value leads me to reject the null hypothesis. This study provides evidence that people will leave higher average voluntary payments for food if pictures of eyes are "watching."

(Note: In Table T, we can see that at 5 df, $t = 3.08$ lies between the critical values for $P = 0.02$ and $P = 0.05$, so we could report $P < 0.05$.)

---

[11] Melissa Bateson, Daniel Nettle, and Gilbert Roberts, "Cues of Being Watched Enhance Cooperation in a Real-World Setting," Biol. Lett. doi: 10.1098/rsbl.2006.0509.

*Tukey's Quick Test

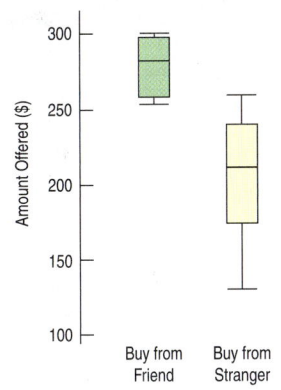

Buying from a Friend	Buying from a Stranger
$275	$260
300	250
260	175
300	130
255	200
275	225
290	240
300	

The famous statistician John Tukey[12] was once challenged to come up with a simpler alternative to the two-sample *t*-test that, like the 68-95-99.7 Rule, had critical values that could be remembered easily. The test he came up with asks you only to count and to remember three numbers: 7, 10, and 13.

When you first looked at the boxplots of the price data, you might have noticed that they didn't overlap very much. That's the basis for Tukey's test.

To use Tukey's test, one group must have the highest value and the other, the lowest. We just count how many values in the high group are higher than *all* the values of the lower group. Add to this the number of values in the low group that are lower than *all* the values of the higher group. (Count ties as 1/2.) Now if this total is 7 or more, we can reject the null hypothesis of equal means at $\alpha = 0.05$. The "critical values" of 10 and 13 give us α's of 0.01 and 0.001.

Let's try it. The "Friend" group has the highest value ($300) and the "Stranger" group has the lowest ($130). Six of the values in the Friend group (shown in red) are higher than the highest value of the Stranger group ($260) and one is a tie. Six of the Stranger values (shown in green) are lower than the lowest value for Friends. That's a total of 12 1/2 values that leak over. That's more than 10, but less than 13. So the P-value is between 0.01 and 0.001—just what we found with the two-sample *t*.

This is a remarkably good test. The only assumption it requires is that the two samples be independent. It's so simple to do that there's no reason not to do one to check your two-sample *t* results. If they disagree, check the assumptions. Tukey's quick test, however, is not as widely known or accepted as the two-sample *t*-test, so you still need to know and use the two-sample *t*.

*A Rank Sum Test

Another distribution-free test of whether two means are equal uses the *ranks* of the data in the two samples. Tukey's test ranks all the data and then counts the excesses on either side. Like Tukey's test, the **Wilcoxon rank sum (or Mann-Whitney) test** first ranks the combined sample from the two groups together from smallest to largest, assigning each observation a rank from 1 to N, where N is $n_1 + n_2$, the total number of observations in both groups. (If there are ties, we use the average rank.) The test statistic, *W*, is the sum of the ranks of the first group. If *W* is too large or too small, it is evidence that the groups do not have equal means (actually, the test is more general and tests whether the two distributions are the same).

The hypothesis test is based on a couple of facts that can be derived mathematically. When the null hypothesis is true, the test statistic, *W*, has mean $\mu_W = \dfrac{n_1(N + 1)}{2}$ and variance $Var(W) = \dfrac{n_1 n_2 (N + 1)}{12}$. Now, for any but very small sample sizes (both larger than about 7 should do), we can use a z-test with $z = \dfrac{W - \mu_W}{SD(W)}$ to test the hypothesis.

Let's try it on the buying from friends data. Here are the data ranked from 1 to 15. There are two ties at ranks 8 and 9, two ties at ranks 10 and 11, and three ties at ranks 13, 14, and 15:

Data	130	175	200	225	240	250	255	260	260	275	275	290	300	300	300
Rank	1	2	3	4	5	6	7	8.5	8.5	10.5	10.5	12	14	14	14
Group	S	S	S	S	S	S	F	S	F	F	F	F	F	F	F

---

[12]You know some of his other work. Tukey originated the stem-and-leaf display and the boxplot. He also developed the Fast Fourier Transform, which is important in engineering, and coined the term "bit" for binary digit.

Now we'll add up all the ranks of the Friends group:

$$W = 7 + 8.5 + 10.5 + 10.5 + 12 + 14 + 14 + 14 = 90.5$$

The mean of W when the null hypothesis is true is $\mu_W = \dfrac{8(15+1)}{2} = 64$ and the SD is

$$SD(W) = \sqrt{Var(W)} = \sqrt{\dfrac{8 \times 7(15+1)}{12}} = 8.64, \text{ so } z = \dfrac{90.5 - 64}{8.64} = 3.07$$

with a two-sided P-value of 0.0021. As with both other tests we looked at, this P-value is between 0.01 and 0.001. We reach the same conclusion as we did with Tukey's test and with the two-sample t-test. Of course, like Tukey's test, the rank sum test has the advantage that it doesn't depend on the Nearly Normal Condition. And, like Tukey's test, it will be less powerful than the two-sample t-test when that condition *is* satisfied because it doesn't use all the information in the data.

22.7 The Pooled *t*-Test: Everyone into the Pool?

You may encounter another way to test whether the means of two groups are different called the **pooled *t*-test**. We present it both for historical reasons and because it's important in the analysis of designed experiments (Chapter 26 and 27). Another, simpler test was used. Although it was simpler, this test had a big assumption—it assumes that the variances of the two groups are the same.

That might not seem like a big deal, but it's a very restrictive assumption. Just because the means of two groups are assumed equal doesn't mean that their variances are as well. Generally, knowing the mean doesn't tell you *anything* about the variance. But there are cases where the assumption makes some sense—especially in the analysis of designed experiments. If you're trying to decide whether the new computer code you designed actually makes downloading files faster, the null hypothesis is that the mean of your method is the same as the standard code. But, if it has no effect on the mean, it might be safe to assume that it has no effect on the variance either. In such a case, the pooled *t*-test may be useful. However, in general, it is better to use the two-sample independent *t*-test.

Pooled *t*-tests have a couple of advantages. They often have a few more degrees of freedom than the corresponding two-sample test and they have a much simpler degrees of freedom formula. But these advantages come at a price: You have to think about another assumption. The assumption of equal variances is a strong one, is often not true, and is difficult to check. For these reasons, we recommend that you use a two-sample *t*-test instead.

The pooled *t*-test is the theoretically correct method only when we have a good reason to believe that the variances are equal. And (as we will see shortly) there are times when this makes sense. Keep in mind, however, that it's never wrong *not* to pool.

Details of the Pooled *t*-Test

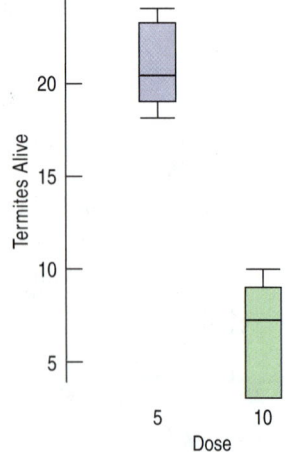

Termites cause billions of dollars of damage each year to homes and other buildings, but some tropical trees seem to be able to resist termite attack. A researcher extracted a compound from the sap of one such tree and tested it by feeding it at two different concentrations to randomly assigned groups of 25 termites.[13] After 5 days, 8 groups fed the lower dose had an average of 20.875 termites alive, with a standard deviation of 2.23. But 6 groups fed the higher dose had an average of only 6.667 termites alive, with a standard deviation of 3.14. Is this a large enough difference to declare the sap compound effective in killing termites? To use the pooled *t*-test, we must make the **Equal Variance Assumption** that the variances of the two populations from which the samples have been drawn are equal. That is, $\sigma_1^2 = \sigma_2^2$. (Of course, we could think about the standard deviations

---

[13] Adam Messer, Kevin McCormick, Sunjaya, H. H. Hagedorm, Ferny Tumbel, and J. Meinwald, "Defensive role of tropical tree resins: Antitermitic sesquiterpenes from Southeast Asian Dipterocarpaceae," *J Chem Ecology*, 16:122, pp. 3333–3352.

being equal instead.) **The corresponding Similar Spreads Condition really just consists of looking at the boxplots to check that the spreads are not wildly different.** We were going to make boxplots anyway, so there's really nothing new here.

Once we decide to pool, we estimate the common variance by combining numbers we already have

$$s^2_{pooled} = \frac{(n_1 - 1)s_1^2 + (n_2 - 1)s_2^2}{(n_1 - 1) + (n_2 - 1)}.$$

$$s^2_{pooled} = \frac{(8-1)2.23^2 + (6-1)3.14^2}{(8-1) + (6-1)} = 7.01$$

(If the two sample sizes are equal, this is just the average of the two variances.)

Now we just substitute this pooled variance in place of each of the variances in the standard error formula.

$$SE_{pooled}(\bar{y}_1 - \bar{y}_2) = \sqrt{\frac{s^2_{pooled}}{n_1} + \frac{s^2_{pooled}}{n_2}} = s_{pooled}\sqrt{\frac{1}{n_1} + \frac{1}{n_2}}$$

$$SE_{pooled}(\bar{y}_1 - \bar{y}_2) = \sqrt{\frac{7.01}{8} + \frac{7.01}{6}} = 1.43$$

The formula for degrees of freedom for the Student's t-model is simpler, too. It was so complicated for the two-sample t that we stuck it in a footnote. Now it's just df $= n_1 + n_2 - 2$.

Substitute the pooled-t estimate of the standard error and its degrees of freedom into the steps of the confidence interval or hypothesis test, and you'll be using the pooled-t method. For the termites, $\bar{y}_1 - \bar{y}_2 = 14.208$, giving a t-value $= 9.935$ with 12 df and a P-value < 0.0001.

$$t = \frac{20.875 - 6.667}{1.43} = 9.935$$

Of course, if you decide to use a pooled-t method, you must defend your assumption that the variances of the two groups are equal.

> **POOLED t-TEST AND CONFIDENCE INTERVAL FOR MEANS**
>
> The conditions for the pooled t-test for the difference between the means of two independent groups (commonly called a "pooled t-test") are the same as for the two-sample t-test with the additional assumption that the variances of the two groups are the same. We test the hypothesis
>
> $$H_0: \mu_1 - \mu_2 = \Delta_0,$$
>
> where the hypothesized difference, Δ_0, is almost always 0, using the statistic
>
> $$t = \frac{(\bar{y}_1 - \bar{y}_2) - \Delta_0}{SE_{pooled}(\bar{y}_1 - \bar{y}_2)}.$$
>
> The standard error of $\bar{y}_1 - \bar{y}_2$ is
>
> $$SE_{pooled}(\bar{y}_1 - \bar{y}_2) = \sqrt{\frac{s^2_{pooled}}{n_1} + \frac{s^2_{pooled}}{n_2}} = s_{pooled}\sqrt{\frac{1}{n_1} + \frac{1}{n_2}},$$
>
> where the pooled variance is
>
> $$s^2_{pooled} = \frac{(n_1 - 1)s_1^2 + (n_2 - 1)s_2^2}{(n_1 - 1) + (n_2 - 1)}.$$
>
> When the conditions are met and the null hypothesis is true, we can model this statistic's sampling distribution with a Student's t-model with $(n_1 - 1) + (n_2 - 1)$ degrees of freedom. We use that model to obtain a P-value for a test or a margin of error for a confidence interval.
>
> The corresponding confidence interval is
>
> $$(\bar{y}_1 - \bar{y}_2) \pm t^*_{df} \times SE_{pooled}(\bar{y}_1 - \bar{y}_2),$$
>
> where the critical value t^* depends on the confidence level and is found with $(n_1 - 1) + (n_2 - 1)$ degrees of freedom.

A S **Activity: The Pooled t-Test.** It's those hot dogs again. The same interactive tool can handle a pooled t-test, too. Take it for a spin here.

Is the Pool All Wet?

We're testing whether the means are equal, so we admit that we don't *know* whether they are equal. Doesn't it seem a bit much to just *assume* that the variances are equal? Well, yes—but there are some special cases to consider. So when *should* you use pooled-*t* methods rather than two-sample *t* methods?

Never.

What, never?

Well, hardly ever.

You see, when the variances of the two groups are in fact equal, the two methods give pretty much the same result. (For the termites, the two-sample *t* statistic is barely different—9.436 with 8 df—and the P-value is still < 0.0001.) Pooled methods have a small advantage (slightly narrower confidence intervals, slightly more powerful tests) mostly because they usually have a few more degrees of freedom, but the advantage is slight.

When the variances are *not* equal, the pooled methods are just not valid and can give poor results. You have to use the two-sample methods instead.

As the sample sizes get bigger, the advantages that come from a few more degrees of freedom make less and less difference. So the advantage (such as it is) of the pooled method is greatest when the samples are small—just when it's hardest to check the conditions. And the difference in the degrees of freedom is greatest when the variances are not equal—just when you can't use the pooled method anyway. Our advice is to use the two-sample *t* methods to compare means.

Pooling

Pooling is natural when we test whether two proportions are equal because the null hypothesis says they are. But why should we bother with it for comparing means?

Pooled methods show up in several places in Statistics. The ones we've seen here are the simplest examples. As you've seen, the most sensitive part of most inference procedures is how we estimate the standard deviation of the test statistic so we can use it as a ruler to judge significance or construct a confidence interval. Anything we can do to improve our standard error estimate pays us back with improved power, shorter confidence intervals, and sometimes, simply with a valid method. So if you think the groups you are comparing have the same standard deviation, you should look for a pooled method that takes advantage of that.

Here's a common situation. In a randomized experiment, we start by assigning our experimental units to treatments at random, as the experimenter did with his termites. Because of this random assignment, each treatment group is a random sample from the same population, so before they were treated, each group had the same variance. The pooled *t*-test assumes that the treatment doesn't change the variances. Our null hypothesis says that the treatment didn't change the means, so it's not much of a stretch to think that the treatments made no difference *at all*. For example, we might suspect that the treatment is no different from the placebo. So it might make sense to assume that the variances have remained equal. It's still an assumption, and there are conditions that need to be checked (make the boxplots, make the boxplots, make the boxplots), but at least it's a plausible assumption.

This line of reasoning is important. The methods used to analyze comparative experiments *do* pool variances in exactly this way and defend the pooling with a version of this argument. Chapter 26 on the DVD introduces these methods.

> **WHEN SHOULD I USE THE POOLED-*t* TEST?**
> Because the advantages of pooling are small, and you are allowed to pool only rarely (when the Equal Variances Assumption is met), *don't*.
> It's never wrong *not* to pool.

> **WHY NOT TEST WHETHER THE VARIANCES ARE EQUAL?**
> There are tests of whether two variances are equal. But they are very sensitive to their assumptions and work poorly for small sample sizes. So they don't help in choosing between two-sample *t* and pooled-*t*.

WHAT CAN GO WRONG?

- **Don't use two-sample methods when the samples aren't independent.** These methods give wrong answers when this assumption of independence is violated. Good random sampling is usually the best assurance of independent groups. Make sure there is no relationship between the two groups. Matched-pairs designs arise often and are important, but can't be analyzed with the methods of this chapter.

- **Look at the plots.** The usual cautions about checking for outliers and non-Normal distributions apply, of course. The simple defense is to make and examine boxplots. You may be surprised how often this simple step saves you from the wrong or even absurd conclusions that can be generated by a single undetected outlier. You don't want to conclude that two methods have very different means just because one observation is atypical.

- **Be cautious if you apply inference methods where there was no randomization.** If the data do not come from representative random samples or from a properly randomized experiment, then the inference about the differences between the groups may be wrong.

- **Don't interpret a significant difference in proportions or means causally.** Studies find that people with higher incomes are more likely to snore. Would surgery to increase your snoring be a wise investment? Probably not. It turns out that older people are more likely to snore, and they are also likely to earn more. In a prospective or retrospective study, there is always the danger that other lurking variables not accounted for are the real reason for an observed difference. Be careful not to jump to conclusions about causality.

> **Do what we say, not what we do . . .** Precision machines used in industry often have a bewildering number of parameters that have to be set, so experiments are performed in an attempt to try to find the best settings. Such was the case for a hole-punching machine used by a well-known computer manufacturer to make printed circuit boards. The data were analyzed by one of the authors, but because he was in a hurry, he didn't look at the boxplots first and just performed *t*-tests on the experimental factors. When he found extremely small P-values even for factors that made no sense, he plotted the data. Sure enough, there was one observation 1,000,000 times bigger than the others. It turns out that it had been recorded in microns (millionths of an inch), while all the rest were in inches.

CONNECTIONS

We looked at contingency tables for two categorical variables. Differences in proportions are just 2×2 contingency tables. You'll often see data presented in this way. For example, the snoring data could be shown as

	Gen-Y	Gen-X	Total
Surf	205	235	440
Don't Surf	88	234	322
Total	293	469	762

We tested whether the column percentages of Web surfers were the same for the two age groups.

It should be clear by now that the inference methods we are learning work in a consistent pattern. Although the methods have different standard errors, the step-by-step procedures are almost identical. In particular, we divide a statistic by its standard error. You should feel right at home.

There we made general statements about the shape, center, and spread of each group. When we compared groups, we asked whether their centers looked different compared to how spread out the distributions were. Here, we've made that kind of thinking precise, with confidence intervals for the difference and tests of whether the means are the same.

We use Student's *t* as we did for single sample means, and for the same reasons: We are using standard errors from the data to estimate the standard deviation of the sample statistic. As before, to work with Student's *t*-models, we need to check the Nearly Normal Condition. Histograms and Normal probability plots are the best methods for such checks.

What Have We Learned?

Learning Objectives

Know how to construct and interpret confidence intervals for the difference between proportions or means of two independent groups.

- The two-proportion z-interval is appropriate for proportions.
- The two-sample t-interval is appropriate for means.
- The Assumptions and Conditions are the same for both, except that for means, we must also assume that the values in each group follow a Normal model.

Be able to perform and interpret a two-sample t-test of the difference between the means of two independent groups.

- Understand the relationship between testing and providing a confidence interval.
- The most common null hypothesis is that the means are equal.

Be able to perform and interpret a two-sample z-test of whether proportions in two independent groups are equal.

- Because we are hypothesizing that the proportions are equal, we pool the counts to find a *pooled standard error* estimate of the standard deviation.

Recognize that in special cases in which it is reasonable to assume equal variances between the groups, we can also use a pooled standard error calculation for testing the difference between means.

- This may make sense particularly for randomized experiments in which the randomization has produced groups with equal variance to start with and the null hypothesis is that a treatment under study has had no effect.

For all comparisons of two groups, keep in mind that we require both independence of observations within each group and independence between the groups themselves.

- These methods are not appropriate for paired or matched data.

Review of Terms

Sampling distribution of the difference between two proportions
The sampling distribution of $\hat{p}_1 - \hat{p}_2$ is, under appropriate assumptions, modeled by a Normal model with mean $\mu = p_1 - p_2$ and standard deviation $SD(\hat{p}_1 - \hat{p}_2) = \sqrt{\dfrac{p_1 q_1}{n_1} + \dfrac{p_2 q_2}{n_2}}$. (p. 589)

Two-proportion z-interval
A two-proportion z-interval gives a confidence interval for the true difference in proportions, $p_1 - p_2$, in two independent groups.

The confidence interval is $(\hat{p}_1 - \hat{p}_2) \pm z^* \times SE(\hat{p}_1 - \hat{p}_2)$, where z^* is a critical value from the standard Normal model corresponding to the specified confidence level. (p. 589)

Pooling
Data from two or more populations may sometimes be combined, or *pooled*, to estimate a statistic (typically a pooled variance) when the estimated value is assumed to be the same in both populations. The resulting larger sample size may lead to an estimate with lower sample variance. However, pooled estimates are appropriate only when the required assumptions are true. (p. 593)

Two-proportion z-test
Test the null hypothesis $H_0: p_1 - p_2 = 0$ by comparing the statistic

$$z = \dfrac{\hat{p}_1 - \hat{p}_2}{SE_{pooled}(\hat{p}_1 - \hat{p}_2)}$$

to a standard Normal model. (p. 593)

Two-sample t-interval for the difference between means
A confidence interval for the difference between the means of two independent groups founds as

$$(\bar{y}_1 - \bar{y}_2) \pm t^*_{df} \times SE(\bar{y}_1 - \bar{y}_2),$$

where
$$SE(\bar{y}_1 - \bar{y}_2) = \sqrt{\frac{s_1^2}{n_1} + \frac{s_2^2}{n_2}}$$
and the number of degrees of freedom is given by a special formula. (p. 597)

Two-sample *t*-test for the difference between means

A hypothesis test for the difference between the means of two independent groups. It tests the null hypothesis
$$H_0: \mu_1 - \mu_2 = \Delta_0,$$
where the hypothesized difference, Δ_0, is almost always 0, using the statistic
$$t_{df} = \frac{(\bar{y}_1 - \bar{y}_2) - \Delta_0}{SE(\bar{y}_1 - \bar{y}_2)},$$
with the number of degrees of freedom given by the special formula. (p. 601)

*Tukey's quick test
*Rank-Sum test

Non-parametric tests of whether the means of two groups are equal. (p. 605)

Pooled *t*-test

A hypothesis test for the difference in the means of two independent groups when we are willing and able to assume that the variances of the groups are equal. It tests the null hypothesis
$$H_0: \mu_1 - \mu_2 = \Delta_0,$$
where the hypothesized difference, Δ_0, is almost always 0, using the statistic
$$t_{df} = \frac{(\bar{y}_1 - \bar{y}_2) - \Delta_0}{SE_{pooled}(\bar{y}_1 - \bar{y}_2)},$$
and the degrees of freedom is $(n_1 - 1) + (n_2 - 1)$. (p. 607)

Pooled *t*-interval

The corresponding confidence interval is
$$(\bar{y}_1 - \bar{y}_2) \pm t_{df}^* \times SE_{pooled}(\bar{y}_1 - \bar{y}_2),$$
where the critical value t^* depends on the confidence level and is found with $(n_1 - 1) + (n_2 - 1)$ degrees of freedom. (p. 607)

On the Computer TWO SAMPLE METHODS FOR PROPORTIONS

It is so common to test against the null hypothesis of no difference between the two true proportions that most statistics programs simply assume this null hypothesis. And most will automatically use the pooled standard deviation. If you wish to test a different null (say, that the true difference is 0.3), you may have to search for a way to do it.

Many statistics packages don't offer special commands for inference for differences between proportions. As with inference for single proportions, most statistics programs want the "success" and "failure" status for each case. Usually these are given as 1 or 0, but they might be category names like "yes" and "no." Often you just know the proportions of successes, $\hat{p}_1$ and $\hat{p}_2$, and the counts, n_1 and n_2. Computer packages don't usually deal with summary data like these easily. Calculators typically do a better job.

In some programs, you can reconstruct the original values. But even when you have (or can reconstruct) the raw data values, often you won't get *exactly* the same test statistic from a computer package as you would find working by hand. The reason is that when the packages treat the proportion as a mean, they make some approximations. The result is very close, but not exactly the same.

DATA DESK

Data Desk does not offer built-in methods for inference with proportions. Use **Replicate Y by X** to construct data corresponding to given proportions and totals.

COMMENTS

For summarized data, open a Scratchpad to compute the standard deviations and margin of error by typing the calculation.

EXCEL

Inference methods for proportions are not part of the standard Excel tool set.

COMMENTS

For summarized data, type the calculation into any cell and evaluate it.

JMP

For a **categorical** variable that holds category labels, the **Distribution** platform includes tests and intervals of proportions:

- For summarized data, put the category names in one variable and the frequencies in an adjacent variable.
- Designate the frequency column to have the **role** of **frequency**. Then use the **Distribution** platform.

COMMENTS

JMP uses slightly different methods for proportion inferences than those discussed in this text. Your answers are likely to be slightly different.

MINITAB

To find a hypothesis test for a proportion:

- Choose **Basic Statistics** from the **Stat** menu.
- Choose **2Proportions...** from the Basic Statistics submenu. If the data are organized as category names in one column and case IDs in another, assign the variables from the variable list box to the **Samples in one column** box.
- If the data are organized as two separate columns of responses, click on **Samples in different columns:** and assign the variables from the variable list box. If you have summarized data, click the **Summarized Data** button and fill in the number of trials and the number of successes for each group.
- Click the **Options** button and specify the remaining details. Remember to click the **Use pooled estimate of p for test** box when testing the null hypothesis of no difference between proportions.
- Click the **OK** button.

COMMENTS

When working from a variable that names categories, MINITAB treats the last category as the "success" category. You can specify how the categories should be ordered.

R

- Create a **vector** of **successes** = c(X1,X2) where X1 is the number of successes in group 1 and X2 is the number of successes in group 2.
- Create a vector of **sample.sizes** = c(n1,n2) where n1 and n2 are the sample sizes for the two groups, respectively.
- To test the equality of the two proportions:
  - prop.test(successes,sample.sizes,alternative = "two.sided",conf.level = 0.95,correct = FALSE)

will test the null hypothesis of equal proportions against the two-sided alternative and give a 95% confidence level.

COMMENTS

R uses the χ^2 statistic instead of the *z*-statistic for testing. The corresponding *z*-statistic can be found by taking the square root of the X-squared (χ^2) statistic in the output. The P-values are the same. Other possibilities for alternative are "greater" and "less".

SPSS

SPSS does not perform hypothesis tests for proportions.

STATCRUNCH

To do inference for the difference between two proportions using summaries:

- Click on **Stat**.
- Choose **Proportions » Two sample » with summary**.
- Enter the **Number of successes** (x) and **Number of observations** (n) in each group.
- Click on **Next**.
- Indicate **Hypothesis Test**, then enter the hypothesized **Null** proportion difference (usually 0), and choose the **Alternative** hypothesis.

   OR

   Indicate **Confidence Interval**, and then enter the **Level** of confidence.
- Click on **Calculate**.

To do inference for the difference between two proportions using data:

- Click on **Stat**.
- Choose **Proportions » Two sample » with data**.
- For each group, choose the variable **Column** listing the **Outcomes**, and enter the outcome to be considered a **Success**.
- Click on **Next**.
- Indicate **Hypothesis Test**, then enter the hypothesized **Null** proportion difference (usually 0), and choose the **Alternative** hypothesis.

   OR

   Indicate **Confidence Interval**, and then enter the **Level** of confidence.
- Click on **Calculate**.

TI-83/84 PLUS

To calculate a confidence interval for the difference between two population proportions:

- Select **B:2-PropZInt** from the **STAT TESTS** menu.
- Enter the observed counts and the sample sizes for both samples.
- Specify a confidence level.
- Calculate the interval.

To do the mechanics of a hypothesis test for equality of population proportions:

- Select **6:2-PropZTest** from the **STAT TESTS** menu.
- Enter the observed counts and sample sizes.

- Indicate what kind of test you want: one-tail upper tail, lower tail, or two-tail.
- Calculate the result.

COMMENTS

Beware: When you enter the value of x, you need the *count*, not the percentage. The count must be a whole number.

TWO-SAMPLE METHODS FOR MEANS

Here's some typical computer package output with comments:

```
2-Sample t-Test of μ1-μ2 = 0 vs ≠ 0

Difference Between Means = 0.99145299  t-Statistic = 1.540
w/196 df
Fail to reject Ho at Alpha = 0.05
P = 0.1251
```

- May just say "difference of means"
- Test Statistic
- Some programs will draw a conclusion about the test. Others just give the P-value and let you decide for yourself.
- df found from approximation formula and rounded down. The unrounded value may be given, or may be used to find the P-value.
- Many programs give far too many digits. Ignore the excess digits.

Most statistics packages compute the test statistic for you and report a P-value corresponding to that statistic. And, of course, statistics packages make it easy to examine the boxplots of the two groups, so you have no excuse for skipping this important check.

Some statistics software automatically tries to test whether the variances of the two groups are equal. Some automatically offer both the two-sample-t and pooled-t results. Ignore the test for the variances; it has little power in any situation in which its results could matter. If the pooled and two-sample methods differ in any important way, you should stick with the two-sample method. Most likely, the Equal Variance Assumption needed for the pooled method has failed.

The degrees of freedom approximation usually gives a fractional value. Most packages seem to round the approximate value down to the next smallest integer (although they may actually compute the P-value with the fractional value, gaining a tiny amount of power).

There are two ways to organize data when we want to compare two independent groups. The data can be in two columns, as in the table for camera prices. Each column can be thought of as a variable. In this method, the variables for the camera example would be *Prices to Friend* and *Prices to Stranger*. Graphing calculators usually prefer this form, and some computer programs can use it as well.

There's another way to think about the data. What is the response variable for the camera experiment? It's the *Price* offered. But the values of this variable are in both columns, and actually there's an experiment factor here, too—namely, whether the *Seller* was a *Friend* or a *Stranger*. So, we could put the data into two different columns, one with the *Prices* in it and one with the *Seller*. Then the data would look like this:

Seller	Price	Seller	Price
Friend	275	Stranger	260
Friend	300	Stranger	250
Friend	260	Stranger	175
Friend	300	Stranger	130
Friend	255	Stranger	200
Friend	275	Stranger	225
Friend	290	Stranger	240
Friend	300		

This way of organizing the data makes sense as well. Now the factor and the response variables are clearly visible. You'll have to see which method your program requires. Some packages even allow you to structure the data either way.

The commands to do inference for two independent groups on common statistics technology are not always found in obvious places. Here are some starting guidelines.

DATA DESK

- Select variables.
- From the **Calc** menu, choose **Estimate** for confidence intervals or **Test** for hypothesis tests.
- Select the interval or test from the drop-down menu and make other choices in the dialog.

COMMENTS

Data Desk expects the two groups to be in separate variables.

EXCEL

- From the Data Tab, Analysis Group, choose **Data Analysis**.
- Alternatively (if the Data Analysis Tool Pack is not installed), in the Formulas Tab, choose More functions > Statistical > TTEST, and specify Type=3 in the resulting dialog.
- Fill in the cell ranges for the two groups, the hypothesized difference, and the alpha level.

COMMENTS

Excel expects the two groups to be in separate cell ranges. Notice that, contrary to Excel's wording, we do not need to assume that the variances are *not* equal; we simply choose not to assume that they *are* equal.

JMP

- From the **Analyze** menu, select **Fit y by x**.
- Select variables: a **Y, Response** variable that holds the data and an **X, Factor** variable that holds the group names. JMP will make a dotplot.
- Click the **red triangle** in the dotplot title, and choose **Unequal variances**. The *t*-test is at the bottom of the resulting table.
- Find the P-value from the Prob > F section of the table (they are the same).

COMMENTS

JMP expects data in one variable and category names in the other. Don't be misled: There is no need for the variances to be unequal to use two-sample *t* methods.

MINITAB

- From the **Stat** menu, choose the **Basic Statistics** submenu.
- From that menu, choose **2-sample t . . .** . Then fill in the dialog.

COMMENTS

The dialog offers a choice of data in two variables, or data in one variable and category names in the other.

R

To test the hypothesis that $\mu_1 = \mu_2$ against an alternative (default is two-sided):

- Create a vector for the data of each group, say *X* and *Y*, and produce the confidence interval (default is 95%):
- t.test(X,Y, alternative = c("two.sided", "less", "greater"), conf.level = 0.95)

will produce the *t*-statistic, degrees of freedom, P-value, and confidence interval for a specified alternative.

COMMENTS

This is the same function as for the one-sample *t*-test, but with two vectors. In R, the default *t*-test does not assume equal variances in the two groups (default is **var.equal = FALSE**). To get the "pooled" *t*-test, add **var.equal = TRUE** to the function call.

SPSS

- From the **Analyze** menu, choose the **Compare Means** submenu.
- From that, choose the **Independent-Samples t-test** command. Specify the data variable and "group variable."
- Then type in the labels used in the group variable. SPSS offers both the two-sample and pooled-*t* results in the same table.

COMMENTS

SPSS expects the data in one variable and group names in the other. If there are more than two group names in the group variable, only the two that are named in the dialog box will be compared.

STATCRUNCH

To do inference for the difference between two means using summaries:

- Click on **Stat**.
- Choose **T Statistics » Two sample » with summary**.
- Enter the **Sample mean, Standard deviation**, and sample **Size** for each group.
- De-select **Pool variances**.
- Click on **Next**.
- Indicate **Hypothesis Test**, then enter the hypothesized **Null mean** difference (usually 0), and choose the **Alternative** hypothesis.

OR

Indicate **Confidence Interval**, and then enter the **Level** of confidence.

- Click on **Calculate**.

To do inference for the difference between two means using data:
- Click on **Stat**.
- Choose **T Statistics » Two sample » with data**.
- Choose the variable Column for each group.
- De-select **Pool variances**.
- Click on **Next**.

- Indicate **Hypothesis Test**, then enter the hypothesized **Null mean** difference (usually 0), and choose the **Alternative** hypothesis.

OR

Indicate **Confidence Interval**, and then enter the **Level** of confidence.

- Click on **Calculate**.

TI-83/84 PLUS

For a confidence interval:
- In the **STAT TESTS** menu, choose **0:2-SampTInt**.
- You may specify that you are using data stored in two lists, or you may enter the means, standard deviations, and sizes of both samples.
- You must also indicate whether to pool the variances (when in doubt, say no) and specify the desired level of confidence.

To test a hypothesis:
- In the **STAT TESTS** menu, choose **4:2-SampTTest**.
- You may specify if you are using data stored in two lists, or you may enter the means, standard deviations, and sizes of both samples.
- You must also indicate whether to pool the variances (when in doubt, say no) and specify whether the test is to be two-tail, lower-tail, or upper-tail.

Exercises

Section 22.1

1. **Canada** Suppose an advocacy organization surveys 960 Canadians and 192 of them reported being born in another country (www.unitednorthamerica.org/simdiff.htm). Similarly, 170 out of 1250 Americans reported being foreign-born. Find the standard error of the difference in sample proportions.

2. **Non-profits** Do people who work for non-profit organizations differ from those who work at for-profit companies when it comes to personal job satisfaction? Separate random samples were collected by a polling agency to investigate the difference. Data collected from 422 employees at non-profit organizations revealed that 377 of them were "highly satisfied." From the for-profit companies, 431 out 518 employees reported the same level of satisfaction. Find the standard error of the difference in sample proportions.

Section 22.2

3. **Canada, deux** If the information in Exercise 1 is to be used to make inferences about the proportion all Canadians and all Americans born in other countries, what conditions must be met before proceeding? Are they met? Explain.

4. **Non-profits, part 2** If the information in Exercise 2 is to be used to make inferences about all people who work at non-profits and for-profit companies, what conditions must be met before proceeding? List them and explain if they are met.

Section 22.3

5. **Canada, trois** The information in Exercise 1 was used to create a 95% two-proportion confidence interval for the difference between Canadians and Americans who were born in foreign countries. Interpret this interval with a sentence in context.

$$95\% \text{ confidence interval for}$$
$$p_{\text{Canadians}} - p_{\text{Americans}} \text{ is } (3.24\%, 9.56\%)$$

6. **Non-profits, part 3** The researchers from Exercise 2 created a 95% two-proportion confidence interval for the difference in those who are "highly satisfied" when comparing people who work at non-profits to people who work at for-profit companies. Interpret the interval with a sentence in context.

$$95\% \text{ confidence interval for}$$
$$p_{\text{non-profits}} - p_{\text{for-profits}} = (1.77\%, 10.50\%)$$

7. **Canada, encore** For the interval given in Exercise 5, explain what "95% confidence" means.

8. **Non-profits, again** For the interval given in Exercise 6, explain what "95% confidence" means.

Section 22.4

9. Canada, la fin The researchers from Exercise 1 want to test if the proportions of foreign born are the same in the United States as in Canada.
   a) What is the difference in the proportions of foreign born residents from both countries?
   b) What is the pooled proportion of foreign born in both countries combined?
   c) What is the standard error of the difference in part a?
   d) What is the value of the z-statistic?
   e) What do you conclude at $\alpha = 0.05$?

10. Non-profits, last one The researchers from Exercise 2 want to test if the proportions of satisfied employees are the same at for-profit companies as at not-for-profit companies.
   a) What is the difference in the proportions of the two types of companies?
   b) What is the pooled proportion of satisfied employees in both types of companies combined?
   c) What is the standard error of the difference in part a?
   d) What is the value of the z-statistic?
   e) What do you conclude at $\alpha = 0.05$?

Section 22.5

11. Cost of shopping Do consumers spend more on a trip to Walmart or Target? Suppose researchers interested in this question collected a systematic sample from 85 Walmart customers and 80 Target customers by asking customers for their purchase amount as they left the stores. The data collected are summarized in the table below.

	Walmart	Target
n	85	80
$\bar{y}$	$45	$53
s	$21	$19

To perform inference on these two samples, what conditions must be met? Are they? Explain.

12. Athlete ages A sports reporter suggests that professional baseball players must be, on average, older than professional football players, since football is a contact sport and players are more susceptible to concussions and serious injuries (www.sports.yahoo.com). One player was selected at random from each team in both professional baseball (MLB) and professional football (NFL). The data are summarized below.

	MLB	NFL
n	30	32
$\bar{y}$	27.5	26.16
s	3.94	2.78

To perform inference on these two samples, what conditions must be met? Are they? Explain.

13. Cost of shopping, again Using the summary statistics provided in Exercise 11, researchers calculated a 95% confidence interval for the mean difference between Walmart and Target purchase amounts. The interval was $(-\$14.15, -\$1.85)$. Explain in context what this interval means.

14. Athlete ages, again Using the summary statistics provided in Exercise 12, the sports reporter calculated the following 95% confidence interval for the mean difference between major league baseball players and professional football players. The 95% interval for $\mu_{MLB} - \mu_{NFL}$ was $(-0.41, 3.09)$. Summarize in context what the interval means.

Section 22.6

15. Cost of shopping, three The researchers in Exercise 11 decide to test the hypothesis that the means are equal. The degrees of freedom formula gives 162.75 df. Test the null hypothesis at $\alpha = 0.05$.

16. Athlete ages, three The researchers in Exercise 12 decide to test the hypothesis. The degrees of freedom formula gives 51.83 df. Test the null hypothesis at $\alpha = 0.05$. Is the alternative one or two-sided?

Section 22.7

17. Cost of shopping, once more Repeat the test you did in Exercise 15, but assume that the variances of purchase amounts is the same at Target and Walmart. Did your conclusion change? Why do you think that is?

18. Athlete ages, one more time Repeat the test you did in Exercise 16, but assume that the variances of ages is the same in the two leagues. Did your conclusion change? Why do you think that is?

19. Cost of shopping, last look Looking back at Exercise 11, instead of comparing two very similar stores, suppose the researchers had compared purchases at two car dealerships: one that specializes in new Italian sports cars and another that carries used domestic vehicles. Would the pooled t-test be a good choice here? Why or why not? Which test or confidence interval would be appropriate for testing whether the mean purchase amounts at the two dealerships is the same?

20. Athletes, final inning Looking back at Exercise 12, instead of comparing the ages of players in Major League Baseball to players in the National Football League, what if they had compared the ages in Major League Baseball to a Little League composed of third graders? Would the pooled t-test be a good choice here? Why or why not? Which test or confidence interval would be appropriate for testing the difference in mean ages in that case?

21. **Online social networking** In September 2013, the Pew Internet and American Life Project surveyed American adults on their Facebook use. It found that 64% visited the site on a daily basis, up from 51% in 2010. What does it mean to say that the difference in proportions is "significant"?

22. **Science news** At the end of 2013, the Pew Project for Excellence in Journalism investigated where people are getting their news. In the study 22% of people 18–29 years old said they still read newspapers as one of their sources of news, while only 18% of people 30–49 said the same. What does it mean to say that the difference is not significant?

Chapter Exercises

23. **Revealing information** Eight hundred eight-six randomly sampled teens were asked which of several personal items of information they thought it ok to share with someone they had just met. Forty-four percent said it was ok to share their e-mail addresses, but only 29% said they would give out their cell phone numbers. A researcher claims that a two-proportion z-test could tell whether there was a real difference among all teens. Explain why that test would not be appropriate for these data.

24. **Regulating access** When a random sample of 935 parents were asked about rules in their homes, 77% said they had rules about the kinds of TV shows their children could watch. Among the 790 of those parents whose teenage children had Internet access, 85% had rules about the kinds of Internet sites their teens could visit. That looks like a difference, but can we tell? Explain why a two-sample z-test would not be appropriate here.

25. **Gender gap** A presidential candidate fears he has a problem with women voters. His campaign staff plans to run a poll to assess the situation. They'll randomly sample 300 men and 300 women, asking if they have a favorable impression of the candidate. Obviously, the staff can't know this, but suppose the candidate has a positive image with 59% of males but with only 53% of females.
    a) What kind of sampling design is his staff planning to use?
    b) What difference would you expect the poll to show?
    c) Of course, sampling error means the poll won't reflect the difference perfectly. What's the standard deviation for the difference in the proportions?
    d) Sketch a sampling model for the size difference in proportions of men and women with favorable impressions of this candidate that might appear in a poll like this.
    e) Could the campaign be misled by the poll, concluding that there really is no gender gap? Explain.

26. **Buy it again?** A consumer magazine plans to poll car owners to see if they are happy enough with their vehicles that they would purchase the same model again. They'll randomly select 450 owners of American-made cars and 450 owners of Japanese models. Obviously, the actual opinions of the entire population couldn't be known, but suppose 76% of owners of American cars and 78% of owners of Japanese cars would purchase another.
    a) What kind of sampling design is the magazine planning to use?
    b) What difference would you expect their poll to show?
    c) Of course, sampling error means the poll won't reflect the difference perfectly. What's the standard deviation for the difference in the proportions?
    d) Sketch a sampling model for the difference in proportions that might appear in a poll like this.
    e) Could the magazine be misled by the poll, concluding that owners of American cars are much happier with their vehicles than owners of Japanese cars? Explain.

27. **Arthritis** The Centers for Disease Control and Prevention reported a survey of randomly selected Americans age 65 and older, which found that 411 of 1012 men and 535 of 1062 women suffered from some form of arthritis.
    a) Are the assumptions and conditions necessary for inference satisfied? Explain.
    b) Create a 95% confidence interval for the difference in the proportions of senior men and women who have this disease.
    c) Interpret your interval in this context.
    d) Does this confidence interval suggest that arthritis is more likely to afflict women than men? Explain.

28. **Graduation** In October 2000, the U.S. Department of Commerce reported the results of a large-scale survey on high school graduation. Researchers contacted more than 25,000 Americans aged 24 years to see if they had finished high school; 84.9% of the 12,460 males and 88.1% of the 12,678 females indicated that they had high school diplomas.
    a) Are the assumptions and conditions necessary for inference satisfied? Explain.
    b) Create a 95% confidence interval for the difference in graduation rates between males and females.
    c) Interpret your confidence interval.
    d) Does this provide strong evidence that girls are more likely than boys to complete high school? Explain.

29. **Pets** Researchers at the National Cancer Institute released the results of a study that investigated the effect of weed-killing herbicides on house pets. They examined 827 dogs from homes where an herbicide was used on a regular basis, diagnosing malignant lymphoma in 473 of them. Of the 130 dogs from homes where no herbicides were used, only 19 were found to have lymphoma.
    a) What's the standard error of the difference in the two proportions?
    b) Construct a 95% confidence interval for this difference.
    c) State an appropriate conclusion.

30. **Carpal tunnel** The painful wrist condition called carpal tunnel syndrome can be treated with surgery or less invasive wrist splints. Recently, *Time* magazine reported on a study of 176 patients. Among the half that had surgery, 80% showed improvement after three months, but only 48% of those who used the wrist splints improved.
    a) What's the standard error of the difference in the two proportions?
    b) Construct a 95% confidence interval for this difference.
    c) State an appropriate conclusion.

31. **Prostate cancer** There has been debate among doctors over whether surgery can prolong life among men suffering from prostate cancer, a type of cancer that typically develops and spreads very slowly. Recently, *The New England Journal of Medicine* published results of some Scandinavian research. Men diagnosed with prostate cancer were randomly assigned to either undergo surgery or not. Among the 347 men who had surgery, 16 eventually died of prostate cancer, compared with 31 of the 348 men who did not have surgery.
    a) Was this an experiment or an observational study? Explain.
    b) Create a 95% confidence interval for the difference in rates of death for the two groups of men.
    c) Based on your confidence interval, is there evidence that surgery may be effective in preventing death from prostate cancer? Explain.

32. **Race and smoking 2013** Data collected in 2013 by the Behavioral Risk Factor Surveillance System revealed that in the state of New Jersey, 14.7% of whites and 18.8% of blacks were cigarette smokers. Suppose these proportions were based on samples of 11,112 whites and 1916 blacks.
    a) Create a 90% confidence interval for the difference in the percentage of smokers between black and white adults in New Jersey.
    b) Does this survey indicate a race-based difference in smoking among American adults? Explain, using your confidence interval to test an appropriate hypothesis.
    c) What alpha level did your test use?

33. **Ear infections** A new vaccine was recently tested to see if it could prevent the painful and recurrent ear infections that many infants suffer from. *The Lancet*, a medical journal, reported a study in which babies about a year old were randomly divided into two groups. One group received vaccinations; the other did not. During the following year, only 333 of 2455 vaccinated children had ear infections, compared to 499 of 2452 unvaccinated children in the control group.
    a) Are the conditions for inference satisfied?
    b) Find a 95% confidence interval for the difference in rates of ear infection.
    c) Use your confidence interval to explain whether you think the vaccine is effective.

34. **Anorexia** The *Journal of the American Medical Association* reported on an experiment intended to see if the drug Prozac® could be used as a treatment for the eating disorder anorexia nervosa. The subjects, women being treated for anorexia, were randomly divided into two groups. Of the 49 who received Prozac, 35 were deemed healthy a year later, compared to 32 of the 44 who got the placebo.
    a) Are the conditions for inference satisfied?
    b) Find a 95% confidence interval for the difference in outcomes.
    c) Use your confidence interval to explain whether you think Prozac is effective.

35. **Another ear infection** In Exercise 33, you used a confidence interval to examine the effectiveness of a vaccine against ear infections in babies. Suppose that instead you had conducted a hypothesis test. (Answer these questions *without* actually doing the test.)
    a) What hypotheses would you test?
    b) State a conclusion based on your confidence interval.
    c) If that conclusion is wrong, which type of error did you make?
    d) What would be the consequences of such an error?

36. **Anorexia again** In Exercise 34, you used a confidence interval to examine the effectiveness of Prozac in treating anorexia nervosa. Suppose that instead you had conducted a hypothesis test. (Answer these questions *without* actually doing the test.)
    a) What hypotheses would you test?
    b) State a conclusion based on your confidence interval.
    c) If that conclusion is wrong, which type of error did you make?
    d) What would be the consequences of such an error?

37. **Teen smoking** A Vermont study published by the American Academy of Pediatrics examined parental influence on teenagers' decisions to smoke. A group of students who had never smoked were questioned about their parents' attitudes toward smoking. These students were questioned again two years later to see if they had started smoking. The researchers found that, among the 284 students who indicated that their parents disapproved of kids smoking, 54 had become established smokers. Among the 41 students who initially said their parents were lenient about smoking, 11 became smokers. Do these data provide strong evidence that parental attitude influences teenagers' decisions about smoking?
    a) What kind of design did the researchers use?
    b) Write appropriate hypotheses.
    c) Are the assumptions and conditions necessary for inference satisfied?
    d) Test the hypothesis and state your conclusion.
    e) Explain in this context what your P-value means.
    f) If it is later found that parental attitudes actually do influence teens decisions to smoke, which type of error did you commit?

g) Create a 95% CI.
h) Interpret your interval.
i) Carefully explain what "95% confidence" means.

38. **Depression** A study published in the *Archives of General Psychiatry* examined the impact of depression on a patient's ability to survive cardiac disease. Researchers identified 450 people with cardiac disease, evaluated them for depression, and followed the group for 4 years. Of the 361 patients with no depression, 67 died. Of the 89 patients with minor or major depression, 26 died. Among people who suffer from cardiac disease, are depressed patients more likely to die than non-depressed ones?
   a) What kind of design was used to collect these data?
   b) Write appropriate hypotheses.
   c) Are the assumptions and conditions necessary for inference satisfied?
   d) Test the hypothesis and state your conclusion.
   e) Explain in this context what your P-value means.
   f) If your conclusion is actually incorrect, which type of error did you commit?
   g) Create a 95% CI.
   h) Interpret your interval.
   i) Carefully explain what "95% confidence" means.

39. **Birthweight** The *Journal of the American Medical Association* reported a study examining the possible impact of air pollution caused by the 9/11 attack on New York's World Trade Center on the weight of babies. Researchers found that 8% of 182 babies born to mothers who were exposed to heavy doses of soot and ash on September 11 were classified as having low birth weight. Only 4% of 2300 babies born in another New York City hospital whose mothers had not been near the site of the disaster were similarly classified. Does this indicate a possibility that air pollution might be linked to a significantly higher proportion of low-weight babies?
   a) Test an appropriate hypothesis at $\alpha = 0.01$ and state your conclusion.
   b) If you concluded there is a difference, estimate that difference with a confidence interval and interpret that interval in context.

40. **Politics and sex** One month before the election, a poll of 630 randomly selected voters showed 54% planning to vote for a certain candidate. A week later, it became known that he had had an extramarital affair, and a new poll showed only 51% of 1010 voters supporting him. Do these results indicate a decrease in voter support for his candidacy?
   a) Test an appropriate hypothesis and state your conclusion.
   b) If you concluded there was a difference, estimate that difference with a confidence interval and interpret your interval in context.

41. **Mammograms** It's widely believed that regular mammogram screening may detect breast cancer early, resulting in fewer deaths from that disease. One study that investigated this issue over a period of 18 years was published during the 1970s. Among 30,565 women who had never had mammograms, 196 died of breast cancer, while only 153 of 30,131 who had undergone screening died of breast cancer.
   a) Do these results suggest that mammograms may be an effective screening tool to reduce breast cancer deaths?
   b) If your conclusion is incorrect, what type of error have you committed?

42. **Mammograms redux** In 2001, the conclusion of the study outlined in Exercise 41 was questioned. A new 9-year study was conducted in Sweden, comparing 21,088 women who had mammograms with 21,195 who did not. Of the women who underwent screening, 63 died of breast cancer, compared to 66 deaths among the control group. (Source: *The New York Times*, Dec 9, 2001)
   a) Do these results support the effectiveness of regular mammograms in preventing deaths from breast cancer?
   b) If your conclusion is incorrect, what kind of error have you committed?

43. **Pain** Researchers comparing the effectiveness of two pain medications randomly selected a group of patients who had been complaining of a certain kind of joint pain. They randomly divided these people into two groups, then administered the pain killers. Of the 112 people in the group who received medication A, 84 said this pain reliever was effective. Of the 108 people in the other group, 66 reported that pain reliever B was effective.
   a) Write a 95% confidence interval for the percent of people who may get relief from this kind of joint pain by using medication A. Interpret your interval.
   b) Write a 95% confidence interval for the percent of people who may get relief by using medication B. Interpret your interval.
   c) Do the intervals for A and B overlap? What do you think this means about the comparative effectiveness of these medications?
   d) Find a 95% confidence interval for the difference in the proportions of people who may find these medications effective. Interpret your interval.
   e) Does this interval contain zero? What does that mean?
   f) Why do the results in parts c and e seem contradictory? If we want to compare the effectiveness of these two pain relievers, which is the correct approach? Why?

44. **Gender gap** Candidates for political office realize that different levels of support among men and women may be a crucial factor in determining the outcome of an election. One candidate finds that 52% of 473 men

polled say they will vote for him, but only 45% of the 522 women in the poll express support.

a) Write a 95% confidence interval for the percent of male voters who may vote for this candidate. Interpret your interval.

b) Write a 95% confidence interval for the percent of female voters who may vote for him. Interpret your interval.

c) Do the intervals for males and females overlap? What do you think this means about the gender gap?

d) Find a 95% confidence interval for the difference in the proportions of males and females who will vote for this candidate. Interpret your interval.

e) Does this interval contain zero? What does that mean?

f) Why do the results in parts c and e seem contradictory? If we want to see if there is a gender gap among voters with respect to this candidate, which is the correct approach? Why?

45. Convention bounce Political pundits talk about the "bounce" that a presidential candidate gets after his party's convention. In the past 40 years, it has averaged about 6 percentage points. Just before the 2004 Democratic convention, Rasmussen Reports polled 1500 likely voters at random and found that 47% favored John Kerry. Just afterward, they took another random sample of 1500 likely voters and found that 49% favored Kerry. That's a two percentage point increase, but the pollsters claimed that there was no bounce. Explain.

46. Stay-at-home dads A *Time* magazine article about a survey of men's attitudes reported that 11 of 161 black respondents and 20 of 358 Latino respondents responded "Yes" to the question "Are you a stay-at-home dad?" How big is the difference in proportions in the two populations?

a) Construct and interpret an appropriate confidence interval.

b) Overall, the survey contacted 1302 men and claims a margin of error of ±2.9%. Why is the margin of error different for your confidence interval?

47. Sensitive men In the same article from Exercise 46, *Time* magazine, reporting on a survey of men's attitudes, noted that "Young men are more comfortable than older men talking about their problems." The survey reported that 80 of 129 surveyed 18- to 24-year-old men and 98 of 184 25- to 34-year-old men said they were comfortable. What do you think? Is *Time*'s interpretation justified by these numbers?

48. Carbs Recently, the Gallup Poll asked 1005 U.S. adults if they actively try to avoid carbohydrates in their diet. That number increased to 27% from 20% in a similar 2002 poll. Is this a statistically significant increase? Explain.

49. Food preference GfK Roper Consulting gathers information on consumer preferences around the world to help companies monitor attitudes about health, food, and healthcare products. They asked people in many different cultures how they felt about the following statement:

I have a strong preference for regional or traditional products and dishes from where I come from.

In a random sample of 800 respondents, 417 of 646 people who live in urban environments agreed (either completely or somewhat) with that statement, compared to 78 out of 154 people who live in rural areas.

Based on this sample, is there evidence that the percentage of people agreeing with the statement about regional preferences differs between all urban and rural dwellers?

50. Fast food The global survey we learned about in Exercise 49 also asked respondents how they felt about the statement "I try to avoid eating fast foods." The random sample of 800 included 411 people 35 years old or younger, and of those, 197 agreed (completely or somewhat) with the statement. Of the 389 people over 35 years old, 246 people agreed with the statement. Is there evidence that the percentage of people avoiding fast food is different in the two age groups?

51. Hot dogs In the July 2007 issue, *Consumer Reports* examined the calorie content of two kinds of hot dogs: meat (usually a mixture of pork, turkey, and chicken) and all beef. The researchers purchased samples of several different brands. The meat hot dogs averaged 111.7 calories, compared to 135.4 for the beef hot dogs. A test of the null hypothesis that there's no difference in mean calorie content yields a P-value of 0.124. Would a 95% confidence interval for $\mu_{Meat} - \mu_{Beef}$ include 0? Explain.

52. Washers In the June 2007 issue, *Consumer Reports* also examined the relative merits of top-loading and front-loading washing machines, testing samples of several different brands of each type. Suppose the study tested the null hypothesis that top- and front-loading machines don't differ in their mean costs, and the test had a P-value of 0.32. Would a 95% confidence interval for contain 0? Explain.

53. Hot dogs, second helping The *Consumer Reports* article described in Exercise 51 also listed the fat content (in grams) for samples of beef and meat hot dogs. The resulting 90% confidence interval for $\mu_{Meat} - \mu_{Beef}$ is $(-6.5, -1.4)$.

a) The endpoints of this confidence interval are negative numbers. What does that indicate?

b) What does the fact that the confidence interval does not contain 0 indicate?
c) If we use this confidence interval to test the hypothesis that $\mu_{Meat} - \mu_{Beef} = 0$, what's the corresponding alpha level?

54. **Second load of wash** The *Consumer Reports* article described in Exercise 52 continued their investigation of washing machines. One of the variables the article reported was "cycle time," the number of minutes it took each machine to wash a load of clothes. Among the machines rated good to excellent, the 98% confidence interval for the difference in mean cycle time $(\mu_{Top} - \mu_{Front})$ is $(-40, -22)$.
   a) The endpoints of this confidence interval are negative numbers. What does that indicate?
   b) What does the fact that the confidence interval does not contain 0 indicate?
   c) If we use this confidence interval to test the hypothesis that $\mu_{Top} - \mu_{Front} = 0$, what's the corresponding alpha level?

55. **Hot dogs, last one** In Exercise 53, we saw a 90% confidence interval of $(-6.5, -1.4)$ grams for $\mu_{Meat} - \mu_{Beef}$, the difference in mean fat content for meat vs. all-beef hot dogs. Explain why you think each of the following statements is true or false:
   a) If I eat a meat hot dog instead of a beef dog, there's a 90% chance I'll consume less fat.
   b) 90% of meat hot dogs have between 1.4 and 6.5 grams less fat than a beef hot dog.
   c) I'm 90% confident that meat hot dogs average between 1.4 and 6.5 grams less fat than the beef hot dogs.
   d) If I were to get more samples of both kinds of hot dogs, 90% of the time the meat hot dogs would average between 1.4 and 6.5 grams less fat than the beef hot dogs.
   e) If I tested more samples, I'd expect about 90% of the resulting confidence intervals to include the true difference in mean fat content between the two kinds of hot dogs.

56. **Third load of wash** In Exercise 54, we saw a 98% confidence interval of $(-40, -22)$ minutes for $\mu_{Top} - \mu_{Front}$, the difference in time it takes top-loading and front-loading washers to do a load of clothes. Explain why you think each of the following statements is true or false:
   a) 98% of top loaders are 22 to 40 minutes faster than front loaders.
   b) If I choose the laundromat's top loader, there's a 98% chance that my clothes will be done faster than if I had chosen the front loader.
   c) If I tried more samples of both kinds of washing machines, in about 98% of these samples I'd expect the top loaders to be an average of 22 to 40 minutes faster.
   d) If I tried more samples, I'd expect about 98% of the resulting confidence intervals to include the true difference in mean cycle time for the two types of washing machines.
   e) I'm 98% confident that top loaders wash clothes an average of 22 to 40 minutes faster than front-loading machines.

57. **Learning math** The Core Plus Mathematics Project (CPMP) is an innovative approach to teaching Mathematics that engages students in group investigations and mathematical modeling. After field tests in 36 high schools over a three-year period, researchers compared the performances of CPMP students with those taught using a traditional curriculum. In one test, students had to solve applied algebra problems using calculators. Scores for 320 CPMP students were compared to those of a control group of 273 students in a traditional math program. Computer software was used to create a confidence interval for the difference in mean scores. (Source: *Journal for Research in Mathematics Education,* 31, no. 3)

   Conf level: 95%   Variable: Mu(CPMP)−Mu(Ctrl)
   Interval: (5.573, 11.427)

   a) What's the margin of error for this confidence interval?
   b) If we had created a 98% CI, would the margin of error be larger or smaller?
   c) Explain what the calculated interval means in this context.
   d) Does this result suggest that students who learn mathematics with CPMP will have significantly higher mean scores in algebra than those in traditional programs? Explain.

58. **Stereograms test** In Exercise 50, Chapter 4, we looked at data from an experiment to determine whether visual information about an image helped "see" the image in 3D.

   2-Sample *t*-Interval for $\mu 1 - \mu 2$
   Conf level = 90%   df = 70
   $\mu(NV) - \mu(VV)$ interval: (0.55, 5.47)

   a) Interpret your interval in context.
   b) Does it appear that viewing a picture of the image helps people "see" the 3D image in a stereogram?
   c) What's the margin of error for this interval?
   d) Explain carefully what the 90% confidence level means.
   e) Would you expect a 99% confidence interval to be wider or narrower? Explain.
   f) Might that change your conclusion in part b? Explain.

59. CPMP, again During the study described in Exercise 57, students in both CPMP and traditional classes took another algebra test that did not allow them to use calculators. The table below shows the results. Are the mean scores of the two groups significantly different?

Math Program	n	Mean	SD
CPMP	312	29.0	18.8
Traditional	265	38.4	16.2

Performance on Algebraic Symbolic Manipulation Without Use of Calculators

a) Write appropriate hypotheses.
b) Do you think the assumptions for inference are satisfied? Explain.
c) Here is computer output for this hypothesis test. Explain what the P-value means in this context.

2-Sample t-Test of $\mu_1 - \mu_2 \neq 0$
t-Statistic = -6.451 w/574.8761 df
$P < 0.0001$

d) State a conclusion about the CPMP program.

60. CPMP and word problems The study of the new CPMP Mathematics methodology described in Exercise 57 also tested students' abilities to solve word problems. This table shows how the CPMP and traditional groups performed. (The df are 590.049.) What do you conclude?

Math Program	n	Mean	SD
CPMP	320	57.4	32.1
Traditional	273	53.9	28.5

61. Commuting A man who moves to a new city sees that there are two routes he could take to work. A neighbor who has lived there a long time tells him Route A will average 5 minutes faster than Route B. The man decides to experiment. Each day, he flips a coin to determine which way to go, driving each route 20 days. He finds that Route A takes an average of 40 minutes, with standard deviation 3 minutes, and Route B takes an average of 43 minutes, with standard deviation 2 minutes. Histograms of travel times for the routes are roughly symmetric and show no outliers.

a) Find a 95% confidence interval for the difference in average commuting time for the two routes. (From technology, df = 33.1.)
b) Should the man believe the old-timer's claim that he can save an average of 5 minutes a day by always driving Route A? Explain.

62. Pulse rates A researcher wanted to see whether there is a significant difference in resting pulse rates for men and women. The data she collected are displayed in the boxplots and summarized in the next column.

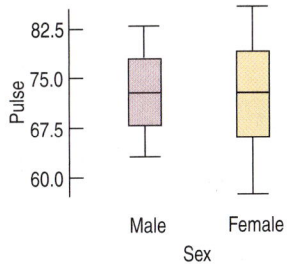

a) What do the boxplots suggest about differences between male and female pulse rates?
b) Is it appropriate to analyze these data using the methods of inference discussed in this chapter? Explain.
c) Create a 90% confidence interval for the difference in mean pulse rates. (From technology, df = 40.2.)
d) Does the confidence interval confirm your answer to part a? Explain.

63. View of the water How much extra is having a waterfront property worth? A student took a random sample of 170 recently sold properties in Upstate New York to examine the question. Here are her summaries and boxplots of the two groups of prices:

	Non-Waterfront Properties		Waterfront Properties
n	100	n	70
$\bar{y}$	$219,896.60	$\bar{y}$	$319,906.40
s	$94,627.15	s	$153,303.80

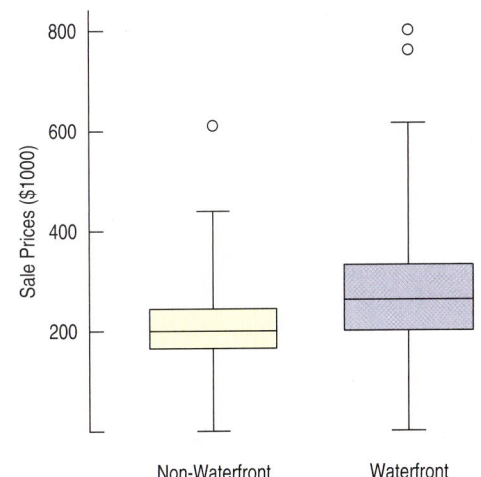

Construct and interpret a 95% confidence interval for the mean additional amount that waterfront property is worth. (From technology, df = 105.48.)

64. **New construction** The sample of 100 house sales we looked at in Exercise 63 also listed whether the home was new construction or not. Find and interpret a 95% confidence interval for how much more an agent can expect to sell a new home for. (From technology, df = 197.8.) Here are the summaries and boxplots of the *Sale Prices*:

	Old Construction		New Construction
n	100	n	100
$\bar{y}$	$201,707.50	$\bar{y}$	$267,878.10
s	$96,116.88	s	$93,302.18

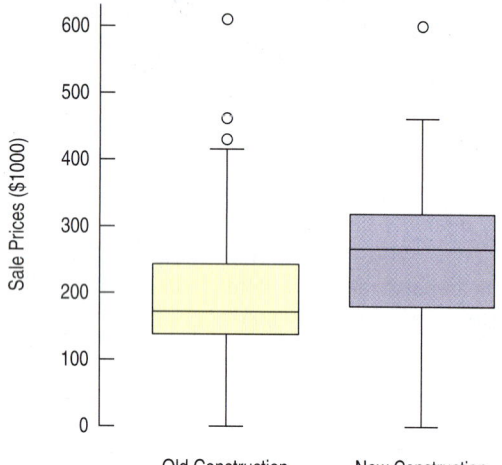

65. **Cereal** The data below show the sugar content (as a percentage of weight) of several national brands of children's and adults' cereals. Create and interpret a 95% confidence interval for the difference in mean sugar content. Be sure to check the necessary assumptions and conditions.

Children's cereals: 40.3, 55, 45.7, 43.3, 50.3, 45.9, 53.5, 43, 44.2, 44, 47.4, 44, 33.6, 55.1, 48.8, 50.4, 37.8, 60.3, 46.6

Adults' cereals: 20, 30.2, 2.2, 7.5, 4.4, 22.2, 16.6, 14.5, 21.4, 3.3, 6.6, 7.8, 10.6, 16.2, 14.5, 4.1, 15.8, 4.1, 2.4, 3.5, 8.5, 10, 1, 4.4, 1.3, 8.1, 4.7, 18.4

66. **Egyptians** Some archaeologists theorize that ancient Egyptians interbred with several different immigrant populations over thousands of years. To see if there is any indication of changes in body structure that might have resulted, they measured 30 skulls of male Egyptians dated from 4000 B.C.E. and 30 others dated from 200 B.C.E. (Source: A. Thomson and R. Randall-Maciver, *Ancient Races of the Thebaid,* Oxford: Oxford University Press, 1905)

a) Are these data appropriate for inference? Explain.

b) Create a 95% confidence interval for the difference in mean skull breadth between these two eras.

c) Do these data provide evidence that the mean breadth of males' skulls changed over this period? Explain.

*d) Perform Tukey's test for the difference. Do your conclusions of part c change?

*e) Perform a rank sum test for the difference. Do your conclusions of part c change?

Maximum Skull Breadth (mm)			
4000 B.C.E.	4000 B.C.E.	200 B.C.E.	200 B.C.E.
131	131	141	131
125	135	141	129
131	132	135	136
119	139	133	131
136	132	131	139
138	126	140	144
139	135	139	141
125	134	140	130
131	128	138	133
134	130	132	138
129	138	134	131
134	128	135	136
126	127	133	132
132	131	136	135
141	124	134	141

67. **Reading** An educator believes that new reading activities for elementary school children will improve reading comprehension scores. She randomly assigns third graders to an eight-week program in which some will use these activities and others will experience traditional teaching methods. At the end of the experiment, both groups take a reading comprehension exam. Their scores are shown in the back-to-back stem-and-leaf display. Do these results suggest that the new activities are better? Test an appropriate hypothesis and state your conclusion. (*Would Tukey's test be appropriate here? Explain. *Use a rank sum test to test an appropriate hypothesis as well. Compare your conclusions with your previous one.)

New Activities		Control
	1	07
4	2	068
3	3	377
96333	4	12222238
9876432	5	355
721	6	02
1	7	
	8	5

68. Streams Researchers collected samples of water from streams in the Adirondack Mountains to investigate the effects of acid rain. They measured the pH (acidity) of the water and classified the streams with respect to the kind of substrate (type of rock over which they flow). A lower pH means the water is more acidic. Here is a plot of the pH of the streams by substrate (limestone, mixed, or shale):

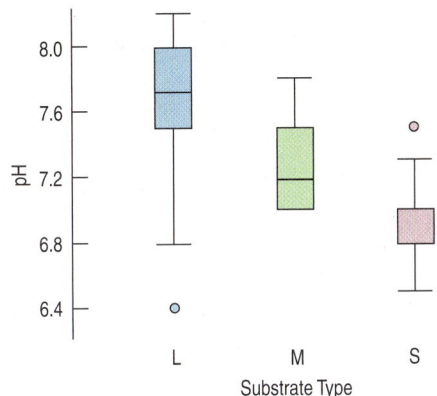

Here are selected parts of a software analysis comparing the pH of streams with limestone and shale substrates:

2-Sample t-Test of $\mu_1 - \mu_2$
Difference Between Means = 0.735
t-Statistic = 16.30 w/133 df
p ≤ 0.0001

a) State the null and alternative hypotheses for this test.
b) From the information you have, do the assumptions and conditions appear to be met?
c) What conclusion would you draw?

69. Hurricanes to 2013 The data below show the number of hurricanes (category 3 or higher) recorded annually before and after 1970. Create an appropriate visual display, and determine whether these data are appropriate for testing whether there has been a change in the frequency of hurricanes.

1944–1969	1970–2013
3, 2, 1, 2, 4, 3, 7, 2, 3, 3, 2, 5, 2, 2, 4, 2, 2, 6, 0, 2, 5, 1, 3, 1, 0, 3	2, 1, 0, 1, 2, 3, 2, 1, 2, 2, 2, 3, 1, 1, 1, 3, 0, 1, 3, 2, 1, 2, 1, 1, 0, 5, 6, 1, 3, 5, 3, 3, 2, 3, 6, 7, 2, 2, 5, 2, 5, 4, 2, 0

70. Memory Does ginkgo biloba enhance memory? In an experiment to find out, subjects were assigned randomly to take ginkgo biloba supplements or a placebo. Their memory was tested to see whether it improved. Here are boxplots comparing the two groups and in the next column is some computer output from a two-sample t-test computed for the data.

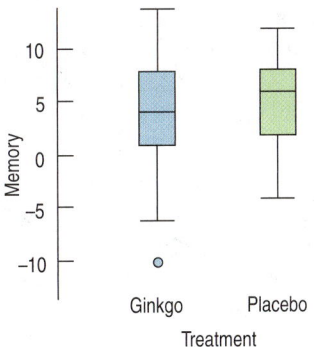

2-Sample t-Test of $\mu_G - \mu_P > 0$
Difference Between Means = −0.9914
t-Statistic = −1.540 w/196 df
P = 0.9374

a) Explain in this context what the P-value means.
b) State your conclusion about the effectiveness of ginkgo biloba.
c) Proponents of ginkgo biloba continue to insist that it works. What type of error do they claim your conclusion makes? Explain.

71. Home runs 2013 American League baseball teams play their games with the designated hitter rule, meaning that pitchers do not bat. The league believes that replacing the pitcher, traditionally a weak hitter, with another player in the batting order produces more runs and generates more interest among fans. Below are the average numbers of home runs hit per game in American League and National League stadiums for the 2013 season (www.hit-trackeronline.com/stadiums.php).

American		National	
2.86	1.94	2.30	1.80
2.64	1.94	2.28	1.67
2.31	1.93	2.17	1.60
2.21	1.89	2.16	1.35
2.10	1.89	1.96	1.33
2.06	1.75	1.84	1.31
2.04	1.54	1.84	1.04
1.94		1.83	

a) Create an appropriate display of these data. What do you see?
b) With a 95% confidence interval, estimate the mean number of home runs hit in American League games.
c) Coors Field, in Denver, stands a mile above sea level, an altitude far greater than that of any other major league ball park. Some believe that the thinner air makes it harder for pitchers to throw curve balls and easier for batters to hit the ball a long way. Do you

think the 1.96 home runs hit per game at Coors is unusual? (Denver is a National League team.) Explain.

d) Explain why you should not use two separate confidence intervals to decide whether the two leagues differ in average number of runs scored.

e) Using a 95% confidence interval, estimate the difference between the mean number of home runs hit in American and National League games.

f) Interpret your interval.

g) Does this interval suggest that the two leagues may differ in average number of home runs hit per game?

T 72. Hard water Derby In an investigation of environmental causes of disease, data were collected on the annual mortality rate (deaths per 100,000) for males in 61 large towns in England and Wales. In addition, the water hardness was recorded as the calcium concentration (parts per million, ppm) in the drinking water. The data set also notes, for each town, whether it was south or north of Derby. Is there a significant difference in mortality rates in the two regions? Here are the summary statistics.

Summary of:	mortality			
For categories in:	Derby			
Group	Count	Mean	Median	StdDev
North	34	1631.59	1631	138.470
South	27	1388.85	1369	151.114

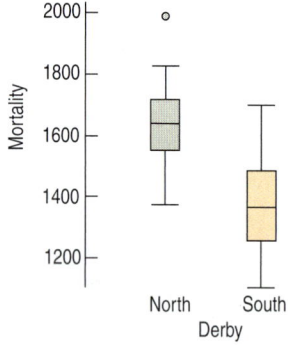

a) Test appropriate hypotheses and state your conclusion.

b) The boxplots of the two distributions show an outlier among the data north of Derby. What effect might that have had on your test?

T 73. Job satisfaction A company institutes an exercise break for its workers to see if this will improve job satisfaction, as measured by a questionnaire that assesses workers' satisfaction. Scores for 10 randomly selected workers before and after implementation of the exercise program are shown. The company wants to assess the effectiveness of the exercise program. Explain why you can't use the methods discussed in this chapter to do that.

Worker Number	Job Satisfaction Index	
	Before	After
1	34	33
2	28	36
3	29	50
4	45	41
5	26	37
6	27	41
7	24	39
8	15	21
9	15	20
10	27	37

74. Summer school Having done poorly on their math final exams in June, six students repeat the course in summer school, then take another exam in August. If we consider these students representative of all students who might attend this summer school in other years, explain why you can't use the methods discussed in this chapter to test whether the class was beneficial.

June	54	49	68	66	62	62
Aug.	50	65	74	64	68	72

75. Sex and violence The *Journal of Applied Psychology* reported on a study that examined whether the content of TV shows influenced the ability of viewers to recall brand names of items featured in the commercials. The researchers randomly assigned volunteers to watch one of three programs, each containing the same nine commercials. One of the programs had violent content, another sexual content, and the third neutral content. After the shows ended, the subjects were asked to recall the brands of products that were advertised. Results are summarized below.

	Program Type		
	Violent	Sexual	Neutral
No. of Subjects	108	108	108
Brands Recalled			
Mean	2.08	1.71	3.17
SD	1.87	1.76	1.77

a) Do these results indicate that viewer memory for ads may differ depending on program content? A test of the hypothesis that there is no difference in ad memory between programs with sexual content and those with violent content has a P-value of 0.136. State your conclusion.

b) Is there evidence that viewer memory for ads may differ between programs with sexual content and those with neutral content? Test an appropriate hypothesis and state your conclusion.

76. Ad campaign You are a consultant to the marketing department of a business preparing to launch an ad campaign for a new product. The company can afford to run ads during one TV show, and has decided not to sponsor a show with sexual content. You read the study described in Exercise 75, then use a computer to create a confidence interval for the difference in mean number of brand names remembered between the groups watching violent shows and those watching neutral shows.

   TWO-SAMPLE T
   95% CI FOR MUviol − MUneut: (−1.578, −0.602)

   a) At the meeting of the marketing staff, you have to explain what this output means. What will you say?
   b) What advice would you give the company about the upcoming ad campaign?

77. Hungry? Researchers investigated how the size of a bowl affects how much ice cream people tend to scoop when serving themselves.[14] At an "ice cream social," people were randomly given either a 17 oz or a 34 oz bowl (both large enough that they would not be filled to capacity). They were then invited to scoop as much ice cream as they liked. Did the bowl size change the selected portion size? Here are the summaries:

Small Bowl		Large Bowl	
n	26	n	22
$\bar{y}$	5.07 oz	$\bar{y}$	6.58 oz
s	1.84 oz	s	2.91 oz

Test an appropriate hypothesis and state your conclusions. For assumptions and conditions that you cannot test, you may assume that they are sufficiently satisfied to proceed.

78. Thirsty? Researchers randomly assigned participants either a tall, thin "highball" glass or a short, wide "tumbler," each of which held 355 ml. Participants were asked to pour a shot (1.5 oz $= 44.3$ ml) into their glass. Did the shape glass make a difference in how much liquid they poured?[15] Here are the summaries:

Highball		Tumbler	
n	99	n	99
$\bar{y}$	42.2 ml	$\bar{y}$	60.9 ml
s	16.2 ml	s	17.9 ml

Test an appropriate hypothesis and state your conclusions. For assumptions and conditions that you cannot test, you may assume that they are sufficiently satisfied to proceed.

---
[14] Brian Wansink, Koert van Ittersum, and James E. Painter, "Ice Cream Illusions: Bowls, Spoons, and Self-Served Portion Sizes," *Am J Prev Med* 2006.

[15] Brian Wansink and Koert van Ittersum, "Shape of Glass and Amount of Alcohol Poured: Comparative Study of Effect of Practice and Concentration," *BMJ* 2005;331;1512–1514.

79. Swim the Lake 2013 revisited As we saw in chapter 8, Exercise 40, between 1954 and 2013, swimmers have crossed Lake Ontario 58 times. Both women and men have made the crossing. Here are some plots (we've omitted a crossing by Vikki Keith, who swam a round trip—North to South to North—in 3390 minutes):

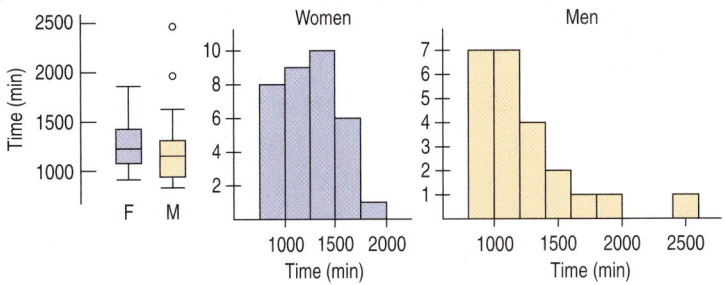

Summary statistics are as follows:

	Summary of Time (min)		
Group	Count	Mean	StdDev
F	34	1257.09	261.10
M	23	1226.04	382.85

Is there a difference between the mean amount of time (in minutes) it takes female and male swimmers to swim the lake?

a) Construct and interpret a 95% confidence interval for the difference between female and male crossing times (technology gives 35.7 df).
b) Comment on the assumptions and conditions and your conclusions.

80. Still swimming Here's some additional information about the Ontario crossing times presented in Exercise 79. It is generally thought to be harder to swim across the lake from north to south. Indeed, this has been done only 5 times. Every one of those crossings was by a woman. If we omit those 5 crossings, the boxplots look like this:

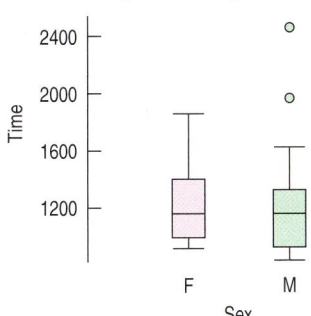

Although not designated as an outlier, the slowest female time belongs to Vikki Keith, who crossed the lake swimming only butterfly stroke. Omitting that extraordinary swim gives the following summary statistics:

	Summary of Time (min)		
SEX	Count	Mean	StdDev
F	29	1195.34	216.326
M	23	1226.04	382.85

a) Do women or men appear to be faster at swimming across the lake? Support your answer by interpreting a confidence interval (Use 32 df).
b) Some might argue that the men are being unfairly slowed down by two extraordinarily slow swims. Here are the summaries without those two swims. Repeat a) based on these data. Does your conclusion change? (Use 42 df).

	Summary of Time (min)		
SEX	Count	Mean	StdDev
F	29	1195.34	216.326
M	21	1132.14	220.36

c) Vikki Keith was responsible for two of the more remarkable crossings, but she also swam Lake Ontario two other times. In fact, of the 50 crossings in this analysis, 7 were repeat crossings by a swimmer who'd crossed the lake before. How does this fact affect your thoughts about the confidence interval?

81. Running heats London In Olympic running events, preliminary heats are determined by random draw, so we should expect that the abilities of runners in the various heats to be about the same, on average. Here are the times (in seconds) for the 400-m women's run in the 2012 Olympics in London for preliminary heats 2 and 5. Is there any evidence that the mean time to finish is different for randomized heats? Explain. Be sure to include a discussion of assumptions and conditions for your analysis. (Note: one runner in heat 2 did not finish and one runner in heat 5 did not start.)
(*Use Tukey's test to compare the mean times in each heat. Do your conclusions change? *Use a rank sum test to compare the mean mean times in each heat. Do your conclusions change?)

Country	Name	Time	Heat
BOT	Amantle Montsho	50.4	2
JAM	Christine Day	51.05	2
GBR	Shana Cox	52.01	2
GUY	Aliann Pompey	52.1	2
ANT	Afia Charles	54.25	2
NCA	Ingrid Narvaez	59.55	2
BAH	Shaunae Miller	DNF	2
RUS	Antonina Krivoshapka	50.75	5
UKR	Alina Lohvynenko	52.08	5
GBR	Lee McConnell	52.23	5
SWE	Moa Hjelmer	52.86	5
MAW	Ambwene Simukonda	54.2	5
FIJ	Danielle Alakija	56.77	5
GRN	Kanika Beckles	DNS	5

82. Swimming heats London In Exercise 81, we looked at the times in two different heats for the 400-m women's run from the 2012 Olympics. Unlike track events, swimming heats are *not* determined at random. Instead, swimmers are seeded so that better swimmers are placed in later heats. Here are the times (in seconds) for the women's 400-m freestyle from heats 2 and 5. Do these results suggest that the mean times of seeded heats are not equal? Explain. Include a discussion of assumptions and conditions for your analysis.
(*Use Tukey's test to compare the mean times in each heat. Do your conclusions change? *Use a rank sum test to compare the mean mean times in each heat. Do your conclusions change?)

Country	Name	Time	Heat
BUL	Nina Rangelova	251.7	2
CHI	Kristel Köbrich	252.0	2
JPN	Aya Takano	252.3	2
LIE	Julia Hassler	253.0	2
MEX	Susana Escobar	254.8	2
THA	Nathanan Junkrajang	256.5	2
SIN	Lynette Lim	258.6	2
KOR	Kim Ga-Eul	283.5	2
FRA	Camille Muffat	243.3	5
USA	Allison Schmitt	243.3	5
NZL	Lauren Boyle	243.6	5
DEN	Lotte Friis	244.2	5
ESP	Melanie Costa	246.8	5
AUS	Bronte Barratt	248.0	5
HUN	Boglarka Kapas	250.0	5
RUS	Elena Sokolova	252.2	5

83. Tees Does it matter what kind of tee a golfer places the ball on? The company that manufactures "Stinger" tees claims that the thinner shaft and smaller head will lessen resistance and drag, reducing spin and allowing the ball to travel farther. In August 2003, Golf Laboratories, Inc., compared the distance traveled by golf balls hit off regular wooden tees to those hit off Stinger tees. All the balls were struck by the same golf club using a robotic device set to swing the club head at approximately 95 miles per hour. Summary statistics from the test are shown in the table below. Assume that 6 balls were hit off each tee and that the data were suitable for inference.

		Total Distance (yards)	Ball Velocity (mph)	Club Velocity (mph)
Regular Tee	Avg.	227.17	127.00	96.17
	SD	2.14	0.89	0.41
Stinger Tee	Avg.	241.00	128.83	96.17
	SD	2.76	0.41	0.52

Is there evidence that balls hit off the Stinger tees would have a higher initial velocity?

84. **Golf again** Given the test results on golf tees described in Exercise 83, is there evidence that balls hit off Stinger tees would travel farther? Again, assume that 6 balls were hit off each tee and that the data were suitable for inference.

85. **Music and memory** Is it a good idea to listen to music when studying for a big test? In a study conducted by some Statistics students, 62 people were randomly assigned to listen to rap music, music by Mozart, or no music while attempting to memorize objects pictured on a page. They were then asked to list all the objects they could remember. Here are summary statistics for each group:

	Rap	Mozart	No Music
Count	29	20	13
Mean	10.72	10.00	12.77
SD	3.99	3.19	4.73

a) Does it appear that it is better to study while listening to Mozart than to rap music? Test an appropriate hypothesis and state your conclusion.

b) Create a 90% confidence interval for the mean difference in memory score between students who study to Mozart and those who listen to no music at all. Interpret your interval.

86. **Rap** Using the results of the experiment described in Exercise 85, does it matter whether one listens to rap music while studying, or is it better to study without music at all?

a) Test an appropriate hypothesis and state your conclusion.

b) If you concluded there is a difference, estimate the size of that difference with a confidence interval and explain what your interval means.

Just Checking ANSWERS

1. We're 90% confident that if members are contacted by e-mail, the donation rate will be between 4 and 18 percentage points higher than if they received regular mail.

2. Since a difference of 0 is not in the confidence interval, we'd reject the null hypothesis. There is evidence that more members will donate if contacted by e-mail.

3. The proportion from the sample in 1995 has variability, too. If we do a one-proportion z-test, we won't take that variability into account and our P-value will be incorrect.

4. The difference in the proportions between the two years has more variability than either individual proportion. The variance of the difference is the sum of the two variances.

5. Randomization should balance unknown sources of variability in the two groups of patients and helps us believe the two groups are independent.

6. We can be 95% confident that after 4 weeks endoscopic surgery patients will have a mean pinch strength between 0.04 kg and 2.96 kg higher than open-incision patients.

7. The lower bound of this interval is close to 0, so the difference may not be great enough that patients could actually notice the difference. We may want to consider other issues such as cost or risk in making a recommendation about the two surgical procedures.

8. Without data, we can't check the Nearly Normal Condition.

9. H_0: Mean pinch strength is the same after both surgeries. $(\mu_E - \mu_0 = 0)$

   H_A: Mean pinch strength is different after the two surgeries. $(\mu_E - \mu_0 \neq 0)$

10. With a P-value this low, we reject the null hypothesis. We can conclude that mean pinch strength differs after 4 weeks in patients who undergo endoscopic surgery vs. patients who have open-incision surgery. Results suggest that the endoscopic surgery patients may be stronger, on average.

11. If some patients contributed two hands to the study, then the groups may not be internally independent. It is reasonable to assume that two hands from the same patient might respond in similar ways to similar treatments.

CHAPTER 23
Paired Samples and Blocks

23.1 Paired Data
23.2 Assumptions and Conditions
23.3 Confidence Intervals for Matched Pairs
23.4 Blocking

Where are we going?
How much will an LSAT prep course raise scores? Are boys better at computer games than their sisters? Questions like these look at paired variables. When pairs of observations go together naturally, they can't be independent. In this chapter, you'll see what to do with paired data.

Who	Olympic speed-skaters
What	Time for women's 1500 m
Units	Seconds
When	2006
Where	Torino, Italy
Why	To see whether one lane is faster than the other

Speed-skating races are run in pairs. Two skaters start at the same time, one on the inner lane and one on the outer lane. Halfway through the race, they cross over, switching lanes so that each will skate the same distance in each lane. Even though this seems fair, at the 2006 Olympics some fans thought there might have been an advantage to starting on the outside. After all, the winner, Cindy Klassen, started on the outside and skated a remarkable 1.47 seconds faster than the silver medalist.

Here are the data for the women's 1500-m race:

Inner Lane		Outer Lane	
Name	Time	Name	Time
OLTEAN Daniela	129.24	(no competitor)	
ZHANG Xiaolei	125.75	NEMOTO Nami	122.34
ABRAMOVA Yekaterina	121.63	LAMB Maria	122.12
REMPEL Shannon	122.24	NOH Seon Yeong	123.35
LEE Ju-Youn	120.85	TIMMER Marianne	120.45
ROKITA Anna Natalia	122.19	MARRA Adelia	123.07
YAKSHINA Valentina	122.15	OPITZ Lucille	122.75
BJELKEVIK Hedvig	122.16	HAUGLI Maren	121.22
ISHINO Eriko	121.85	WOJCICKA Katarzyna	119.96
RANEY Catherine	121.17	BJELKEVIK Annette	121.03
OTSU Hiromi	124.77	LOBYSHEVA Yekaterina	118.87
SIMIONATO Chiara	118.76	JI Jia	121.85
ANSCHUETZ THOMS Daniela	119.74	WANG Fei	120.13
BARYSHEVA Varvara	121.60	van DEUTEKOM Paulien	120.15
GROENEWOLD Renate	119.33	GROVES Kristina	116.74
RODRIGUEZ Jennifer	119.30	NESBITT Christine	119.15
FRIESINGER Anni	117.31	KLASSEN Cindy	115.27
WUST Ireen	116.90	TABATA Maki	120.77

Figure 23.1
Using boxplots to compare times in the inner and outer lanes shows little because it ignores the fact that the skaters raced in pairs.

We can view this skating event as an experiment testing whether the lanes were equally fast. Skaters were assigned to lanes randomly. The boxplots of times recorded in the inner and outer lanes don't show much difference. But that's not the right way to compare these times. Conditions can change during the day. The question raised is about the difference between starting lanes, so it is those differences we should examine.

23.1 Paired Data

Data such as these are called **paired**. We have the times for skaters in each lane for each race. We want to compare the inner and outer lanes across all the races, so what we're interested in is the *differences* in times for each racing pair.

Paired data arise in a number of ways. Perhaps the most common way is to compare subjects with themselves before and after a treatment. When pairs arise from an experiment, the pairing is a type of *blocking*. When they arise from an observational study, it is a form of *matching*.

For Example   IDENTIFYING PAIRED DATA

Do flexible schedules reduce the demand for resources? The Lake County, Illinois, Health Department experimented with a flexible four-day workweek. For a year, the department recorded the mileage driven by 11 field workers on an ordinary five-day workweek. Then it changed to a flexible four-day workweek and recorded mileage for another year.[1] The data are shown.

Name	5-Day Mileage	4-Day Mileage
Jeff	2798	2914
Betty	7724	6112
Roger	7505	6177
Tom	838	1102
Aimee	4592	3281
Greg	8107	4997
Larry G.	1228	1695
Tad	8718	6606
Larry M.	1097	1063
Leslie	8089	6392
Lee	3807	3362

QUESTION: Why are these data paired?

ANSWER: The mileage data are paired because each driver's mileage is measured before and after the change in schedule. I'd expect drivers who drove more than others before the schedule change to continue to drive more afterwards, so the two sets of mileages can't be considered independent.

Pairing isn't a problem; it's an opportunity. If you know the data are paired, you can take advantage of that fact—in fact, you *must* take advantage of it. Remember: Those methods rely on the Pythagorean Theorem of Statistics, and that requires the two

---

[1] Charles S. Catlin, "Four-day Work Week Improves Environment," *Journal of Environmental Health*, Denver, 59:7.

samples be independent. Paired data aren't. There is no test to determine whether the data are paired. You must determine that from understanding how they were collected and what they mean (check the W's).

Once we recognize that the speed-skating data are matched pairs, it makes sense to consider the difference in times for each two-skater race. So we look at the *pairwise differences*:

A S *Activity:* **Differences in Means of Paired Groups.** Are married couples typically the same age, or do wives tend to be younger than their husbands, on average?

Skating Pair	Inner Time	Outer Time	Inner − Outer
1	129.24		.
2	125.75	122.34	3.41
3	121.63	122.12	−0.49
4	122.24	123.35	−1.11
5	120.85	120.45	0.40
6	122.19	123.07	−0.88
7	122.15	122.75	−0.60
8	122.16	121.22	0.94
9	121.85	119.96	1.89
10	121.17	121.03	0.14
11	124.77	118.87	5.90
12	118.76	121.85	−3.09
13	119.74	120.13	−0.39
14	121.60	120.15	1.45
15	119.33	116.74	2.59
16	119.30	119.15	0.15
17	117.31	115.27	2.04
18	116.90	120.77	−3.87

The first skater raced alone, so we'll omit that race. Because it is the *differences* we care about, we'll treat them as if *they* were the data, ignoring the original two columns. Now that we have only one column of values to consider, we can use a one-sample *t*-test. Mechanically, a **paired *t*-test** is just a one-sample *t*-test for the means of these pairwise differences. The sample size is the number of pairs.

So you've already seen the *Show*.

23.2 Assumptions and Conditions

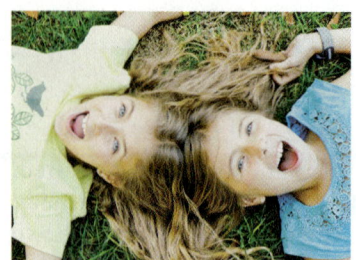

Paired Data Condition

Paired Data Condition: The data must be paired. You can't just decide to pair data when in fact the samples are independent. When you have two groups with the same number of observations, it may be tempting to match them up. Don't, unless you are prepared to justify your claim that the data are paired.

On the other hand, be sure to recognize paired data when you have them. Remember, two-sample *t* methods aren't valid without independent groups, and paired groups aren't independent. Although this is a strictly required assumption, it is one that can be easy to check if you understand how the data were collected.

Independence Assumption

Independence Assumption: If the data are paired, the *groups* are not independent. For these methods, it's the *differences* that must be independent of each other. There's no reason to believe that the difference in speeds of one pair of races could affect the difference in speeds for another pair.

An excellent way to be assured of independence is to generate or sample the data with suitable randomization. With paired data, randomness can arise in many ways. The pairs may be a random sample. In an experiment, the order of the two treatments may be randomly assigned, or the treatments may be randomly assigned to one member of each pair. In a before-and-after study, we may believe that the observed differences are a representative sample from a population of interest. If we have any doubts, we'll need to include a control group to be able to draw conclusions. What we want to know usually focuses our attention on where the randomness should be. In our example, skaters were assigned to the lanes at random.

Normal Population Assumption

We need to assume that the population of *differences* follows a Normal model. We don't need to check the individual groups.

Nearly Normal Condition: This condition can be checked with a histogram or Normal probability plot of the *differences*—but not of the individual groups. As with the one-sample *t*-methods, this assumption matters less the more pairs we have to consider. You may be pleasantly surprised when you check this condition. Even if your original measurements are skewed or bimodal, the *differences* may be nearly Normal. After all, the individual who was way out in the tail on an initial measurement is likely to still be out there on the second one, giving a perfectly ordinary difference.

For Example CHECKING ASSUMPTIONS AND CONDITIONS

RECAP: Field workers for a health department compared driving mileage on a five-day work schedule with mileage on a new four-day schedule. To see if the new schedule changed the amount of driving they did, we'll look at paired differences in mileages before and after.

Name	5-Day Mileage	4-Day Mileage	Difference
Jeff	2798	2914	−116
Betty	7724	6112	1612
Roger	7505	6177	1328
Tom	838	1102	−264
Aimee	4592	3281	1311
Greg	8107	4997	3110
Larry G.	1228	1695	−467
Tad	8718	6606	2112
Larry M.	1097	1063	34
Leslie	8089	6392	1697
Lee	3807	3362	445

(continued)

> **QUESTION:** Is it okay to use these data to test whether the new schedule changed the amount of driving?
>
> **ANSWER:**
> ✓ **Paired Data Condition:** The data are paired because each value is the mileage driven by the same person before and after a change in work schedule.
> ✓ **Independence Assumption:** The driving behavior of any individual worker is independent of the others, so the differences are mutually independent.
> ✓ **Randomization Condition:** The mileages are the sums of many individual trips, each of which experienced random events that arose while driving. Repeating the experiment in two new years would give randomly different values.
> ✓ **Nearly Normal Condition:** The histogram of the mileage differences is unimodal and symmetric:
>
>
>
> Since the assumptions and conditions are satisfied, it's okay to use paired t methods for these data.

The steps in testing a hypothesis for paired differences are very much like the steps for a one-sample t-test for a mean.

> ### THE PAIRED t-TEST
> When the conditions are met, we are ready to test whether the mean of paired differences is significantly different from zero. We test the hypothesis
>
> $$H_0: \mu_d = \Delta_0,$$
>
> where the d's are the pairwise differences and Δ_0 is almost always 0.
> We use the statistic
>
> $$t_{n-1} = \frac{\bar{d} - \Delta_0}{SE(\bar{d})},$$
>
> where $\bar{d}$ is the mean of the pairwise differences, n is the number of *pairs*, and
>
> $$SE(\bar{d}) = \frac{s_d}{\sqrt{n}}.$$
>
> $SE(\bar{d})$ is the ordinary standard error for the mean, applied to the differences.
> When the conditions are met and the null hypothesis is true, we can model the sampling distribution of this statistic with a Student's t-model with $n - 1$ degrees of freedom, and use that model to obtain a P-value.

Step-by-Step Example A PAIRED t-TEST

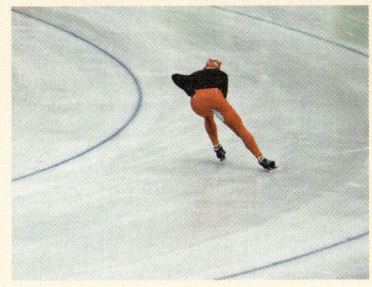

Question: Was there a difference in speeds between the inner and outer speed-skating lanes at the 2006 Winter Olympics?

THINK ➡	
Plan State what we want to know. Identify the *parameter* we wish to estimate. Here our parameter is the mean difference in race times. Identify the variables and check the W's.	I want to know whether there really was a difference in the speeds of the two lanes for speed skating at the 2006 Olympics. I have data for 17 pairs of racers at the women's 1500-m race.
Hypotheses State the null and alternative hypotheses. Although fans suspected one lane was faster, we can't use the data we have to specify the direction of a test. We (and Olympic officials) would be interested in a difference in either direction, so we'd better test a two-sided alternative.	H_0: Neither lane offered an advantage: $$\mu_d = 0.$$ H_A: The mean difference is different from zero: $$\mu_d \neq 0.$$
REALITY CHECK The individual differences are all in seconds. We should expect the mean difference to be comparable in magnitude.	

Model Think about the assumptions and check the conditions.

State why you think the data are paired. Simply having the same number of individuals in each group and displaying them in side-by-side columns doesn't make them paired.

Think about what we hope to learn and where the randomization comes from. Here, the randomization comes from the racer pairings and lane assignments.

Make a picture—just one. Don't plot separate distributions of the two groups—that entirely misses the pairing. For paired data, it's the Normality of the *differences* that we care about. Treat those paired differences as you would a single variable, and check the Nearly Normal Condition with a histogram or a Normal probability plot.

Specify the sampling distribution model.

Choose the method.

✓ **Paired Data Condition:** The data are paired because racers compete in pairs.

✓ **Independence Assumption:** Each race is independent of the others, so the differences are mutually independent. Skaters are assigned to lanes at random.

✓ **Nearly Normal Condition:** The histogram of the differences is unimodal and symmetric:

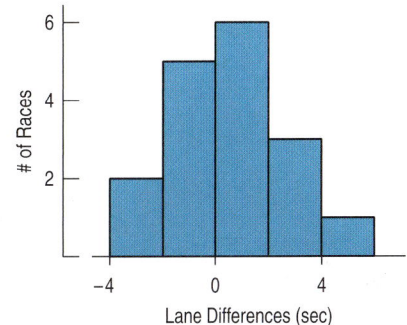

The conditions are met, so I'll use a Student's t-model with $(n - 1) = 16$ degrees of freedom, and perform a **paired t-test**.

SHOW ▶ Mechanics

n is the number of *pairs*—in this case, the number of races.

$\bar{d}$ is the mean difference.

s_d is the standard deviation of the differences.

Find the standard error and the *t*-score of the observed mean difference. There is nothing new in the mechanics of the paired *t* methods. These are the mechanics of the *t*-test for a mean applied to the differences.

The data give

$$n = 17 \text{ pairs}$$
$$\bar{d} = 0.499 \text{ seconds}$$
$$s_d = 2.333 \text{ seconds}.$$

I estimate the standard deviation of $\bar{d}$ using

$$SE(\bar{d}) = \frac{s_d}{\sqrt{n}} = \frac{2.333}{\sqrt{17}} = 0.5658$$

So $t_{16} = \dfrac{\bar{d} - 0}{SE(\bar{d})} = \dfrac{0.499}{0.5658} = 0.882$

Make a picture. Sketch a *t*-model centered at the hypothesized mean of 0. Because this is a two-tail test, shade both the region to the right of the observed mean difference of 0.499 seconds and the corresponding region in the lower tail.

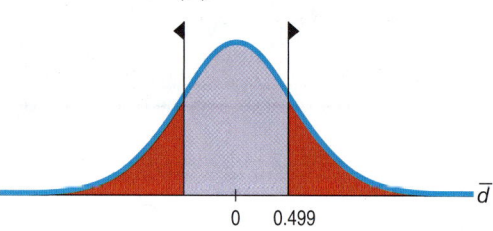

Find the P-value, using technology.

P-value = $2P(t_{16} > 0.882) = 0.39$

REALITY CHECK The mean difference is 0.499 seconds. That may not seem like much, but a smaller difference determined the Silver and Bronze medals. The standard error is about this big, so a *t*-value less than 1.0 isn't surprising. Nor is a large P-value.

TELL ▶ Conclusion
Link the P-value to your decision about H_0, and state your conclusion in context.

The P-value is large. Events that happen more than a third of the time are not remarkable. So, even though there is an observed difference between the lanes, I can't conclude that it isn't due simply to random chance. It appears the fans may have interpreted a random fluctuation in the data as favoring one lane. There's insufficient evidence to declare any lack of fairness.

For Example DOING A PAIRED *t*-TEST

RECAP: We want to test whether a change from a five-day workweek to a four-day workweek could change the amount driven by field workers of a health department. We've already confirmed that the assumptions and conditions for a paired *t*-test are met.

QUESTION: Is there evidence that a four-day workweek would change how many miles workers drive?

ANSWER: H_0: The change in the health department workers' schedules didn't change the mean mileage driven; the mean difference is zero:

$$\mu_d = 0.$$

H_A: The mean difference is different from zero:

$$\mu_d \neq 0.$$

The conditions are met, so I'll use a Student's *t*-model with $(n-1) = 10$ degrees of freedom and perform a **paired *t*-test**.

The data give

$$n = 11 \text{ pairs}$$
$$\bar{d} = 982 \text{ miles}$$
$$s_d = 1139.6 \text{ miles}.$$

$$SE(\bar{d}) = \frac{s_d}{\sqrt{n}} = \frac{1139.6}{\sqrt{11}} = 343.6$$

So $t_{10} = \dfrac{\bar{d} - 0}{SE(\bar{d})} = \dfrac{982.0}{343.6} = 2.86$

P-value $= 2P(t_{10} > 2.86) = 0.017$

The P-value is small, so we reject the null hypothesis and conclude that the change in workweek did lead to a change in average driving mileage. It appears that changing the work schedule may reduce the mileage driven by workers.

Note: We should propose a course of action, but it's hard to tell from the hypothesis test whether the reduction matters. Is the difference in mileage important in the sense of reducing air pollution or costs, or is it merely statistically significant? To help make that decision, we should look at a confidence interval. If the difference in mileage proves to be large in a practical sense, then we might recommend a change in schedule for the rest of the department.

23.3 Confidence Intervals for Matched Pairs

In developed countries, the average age of women is generally higher than that of men. After all, women tend to live longer. But if we look at *married couples*, husbands tend to be slightly older than wives. How much older, on average, are husbands? We have data from a random sample of 200 British couples, the first 7 of which are shown below. Only 170 couples provided ages for both husband and wife, so we can work only with that many pairs. Let's form a confidence interval for the mean difference of husband's and wife's ages for these 170 couples. Here are the first 7 pairs:

Who   170 randomly sampled couples
What Ages
Units Years
When Recently
Where Britain

Wife's Age	Husband's Age	Difference (husband − wife)
43	49	6
28	25	−3
30	40	10
57	52	−5
52	58	6
27	32	5
52	43	−9
⋮	⋮	⋮

638 PART VI Accessing Associations Between Variables

Clearly, these data are paired. The survey selected *couples* at random, not individuals. We're interested in the mean age difference within couples. How would we construct a confidence interval for the true mean difference in ages?

> **PAIRED t-INTERVAL**
>
> When the conditions are met, we are ready to find the confidence interval for the mean of the paired differences. The confidence interval is
>
> $$\bar{d} \pm t^*_{n-1} \times SE(\bar{d}),$$
>
> where the standard error of the mean difference is $SE(\bar{d}) = \dfrac{s_d}{\sqrt{n}}$.
>
> The critical value t^* from the Student's t-model depends on the particular confidence level, C, that you specify and on the degrees of freedom, $n - 1$, which is based on the number of pairs, n.

Making confidence intervals for matched pairs follows exactly the steps for a one-sample t-interval.

Step-by-Step Example A PAIRED t-INTERVAL

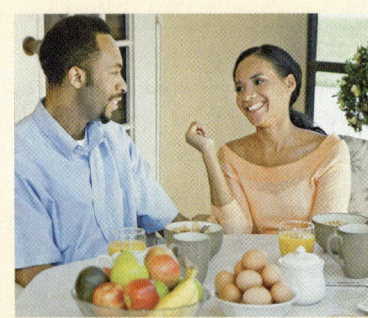

Question: How big a difference is there, on average, between the ages of husbands and wives?

THINK ➡ Plan State what we want to know.

Identify the variables and check the W's.

Identify the parameter you wish to estimate. For a paired analysis, the parameter of interest is the mean of the differences. The population of interest is the population of differences.

I want to estimate the mean difference in age between husbands and wives. I have a random sample of 200 British couples, 170 of whom provided both ages.

Model Think about the assumptions and check the conditions.

✓ **Paired Data Condition:** The data are paired because they are on members of married couples.

✓ **Independence Assumption:** The data are from a randomized survey, so couples should be independent of each other.

Make a picture. We focus on the differences, so a histogram or Normal probability plot is best here. **REALITY CHECK** The histogram shows husbands are often older than wives (because most of the differences are greater than 0). The mean difference seen here of about 2 years is reasonable.	✓ **Nearly Normal Condition:** The histogram of the husband − wife differences is unimodal and symmetric: 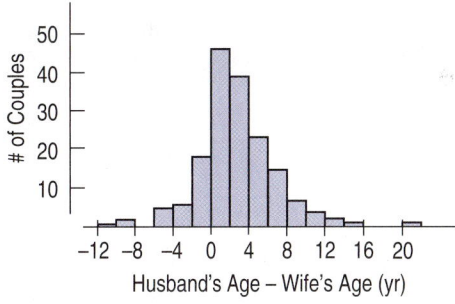
State the sampling distribution model. Choose your method.	The conditions are met, so I can use a Student's t-model with $(n - 1) = 169$ degrees of freedom and find a **paired t-interval**.

SHOW ➡ Mechanics

n is the number of *pairs*, here, the number of couples. $\bar{d}$ is the mean difference. s_d is the standard deviation of the differences. Be sure to include the units along with the statistics. The critical value we need to make a 95% interval comes from a Student's t table, a computer program, or a calculator. **REALITY CHECK** This result makes sense. Our everyday experience confirms that an average age difference of about 2 years is reasonable.	$n = 170$ couples $\bar{d} = 2.2$ years $s_d = 4.1$ years I estimate the standard error of $\bar{d}$ as $$SE(\bar{d}) = \frac{s_d}{\sqrt{n}} = \frac{4.1}{\sqrt{170}} = 0.31 \text{ years.}$$ The df for the t-model is $n - 1 = 169$. The 95% critical value for t_{169} (from the table) is 1.97. The margin of error is $$ME = t^*_{169} \times SE(\bar{d}) = 1.97(0.31) = 0.61.$$ So the 95% confidence interval is $$2.2 \pm 0.6 \text{ years,}$$ or an interval of $(1.6, 2.8)$ years.

TELL ➡ Conclusion

Interpret the confidence interval in context.	I am 95% confident that British husbands are, on average, 1.6 to 2.8 years older than their wives.

Effect Size

When we examined the speed-skating times, we failed to reject the null hypothesis, so we couldn't be certain whether there really was a difference between the lanes. Maybe there wasn't any difference, or maybe whatever difference there might have been was just too small to matter at all. Were the fans right to be concerned?

We can't tell from the hypothesis test, but using the same summary statistics, we can find that the corresponding 95% confidence interval for the mean difference is $(-0.70 < \mu_d < 1.70)$ seconds.

A confidence interval is a good way to get a sense for the size of the effect we're trying to understand. That gives us a plausible range of values for the true mean difference in lane times. If differences of 1.7 seconds were too small to matter in 1500-m Olympic speed skating, we'd be pretty sure there was no need for concern.

But in fact, except for the Gold − Silver gap, the successive gaps between each skater and the next-faster one were *all* less than the high end of this interval, and most were right around the middle of the interval.

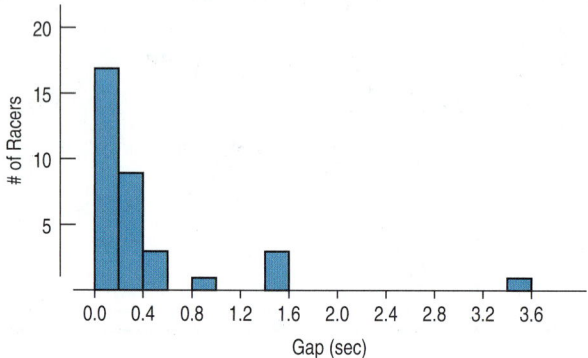

So even though we were unable to discern a real difference, the confidence interval shows that the effects we're considering may be big enough to be important. We may want to continue this investigation by checking out other races on this ice and being alert for possible differences at other venues.

For Example LOOKING AT EFFECT SIZE WITH A PAIRED t CONFIDENCE INTERVAL

RECAP: We know that, on average, the switch from a five-day workweek to a four-day workweek reduced the amount driven by field workers in that Illinois health department. However, finding that there is a significant difference doesn't necessarily mean that difference is meaningful or worthwhile. To assess the size of the effect, we need a confidence interval. We already know the assumptions and conditions are met.

QUESTION: By how much, on average, might a change in workweek schedule reduce the amount driven by workers?

ANSWER:
$$\bar{d} = 982 \text{ mi} \quad SE(\bar{d}) = 343.6 \quad t^*_{10} = 2.228 \ (for\ 95\%)$$
$$ME = t^*_{10} \times SE(\bar{d}) = 2.228(343.6) = 765.54$$

So the 95% confidence interval for μ_d is 982 ± 765.54 or (216.46, 1747.54) fewer miles.

With 95% confidence, I estimate that by switching to a four-day workweek employees would drive an average of between 216 and 1748 fewer miles per year. With high gas prices, this could save a lot of money.

23.4 Blocking

Because the sample of British husbands and wives includes both older and younger couples, there's a lot of variation in the ages of the men and in the ages of the women. In fact, that variation is so great that a boxplot of the two groups would show little difference. But that would be the wrong plot. It's the *difference* we care about. Pairing isolates the extra variation and allows us to focus on the individual differences. In Chapter 12, we saw how to design an experiment with blocking to isolate the variability between identifiable groups of subjects. Blocking makes it easier to see the variability among treatment groups that is attributable to their responses to the treatment. A paired design is an example of blocking.

When we pair, we have roughly half the degrees of freedom of a two-sample test. You may see discussions that suggest that in "choosing" a paired analysis we "give up" these

Figure 23.2
This display is worthless. It does no good to compare all the wives as a group with all the husbands. We care about the paired differences.

degrees of freedom. This isn't really true, though. If the data are paired, then there never were additional degrees of freedom, and we have no "choice." The fact of the pairing determines how many degrees of freedom are available.

Matching pairs generally removes so much extra variation that it more than compensates for having only half the degrees of freedom, so it is usually a good choice when you design a study. Of course, inappropriate matching when the groups are in fact independent (say, by matching on the first letter of the last name of subjects) would cost degrees of freedom without the benefit of reducing the variance. When you design a study or experiment, you should consider using a paired design if possible.

What's Independent?

The methods of this chapter require the pairs to be independent of each other. They make no assumptions about the individuals—only about the pairs. By contrast, many other inference methods require independence among all the individuals.

You will sometimes see the paired t methods of this chapter referred to as methods for "dependent" samples to differentiate them from the independence we required in the previous chapter. But in fact, the only "dependence" in paired data is the pairing itself. That's not a failure of statistical independence, it's a feature of the design for the data collection.

 Just Checking

Think about each of the situations described below.

- Would you use a two-sample t or paired t method (or neither)? Why? or Why not?
- Would you perform a hypothesis test or find a confidence interval?

1. Random samples of 50 men and 50 women are asked to imagine buying a birthday present for their best friend. We want to estimate the difference in how much they are willing to spend.

2. Mothers of twins were surveyed and asked how often in the past month strangers had asked whether the twins were identical.

3. Are parents equally strict with boys and girls? In a random sample of families, researchers asked a brother and sister from each family to rate how strict their parents were.

4. Forty-eight overweight subjects are randomly assigned to either aerobic or stretching exercise programs. They are weighed at the beginning and at the end of the experiment to see how much weight they lost.
   a) We want to estimate the mean amount of weight lost by those doing aerobic exercise.
   b) We want to know which program is more effective at reducing weight.

5. Couples at a dance club were separated and each person was asked to rate the band. Do men or women like this band more?

*The Sign Test Again

Because we have paired data, we've been using a simple t-test for the paired differences. This suggests that if we want a distribution-free method, it would be natural to compute a sign test on the paired differences and test whether the *median* of the differences is 0. That's exactly what we do. The test is very simple. We record a 0 for every paired difference that's negative and a 1 for each positive difference, ignoring pairs for which the difference is exactly 0. We test the associated proportion against $H_0: p = 0.5$ using a z-test if the number of pairs is at least 20 (so that we expect at least 10 successes and failures), or compute the exact Binomial probabilities if $n < 20$ (as discussed in Chapter 16).

Let's try it on the married couples data. Of the 170 couples, there are 119 with the husband older, 32 with the wife older, and 19 that have the same age. So we test $p = 0.5$, with a sample proportion of $119/151 = 0.788$. The Success/Failure Condition is easily satisfied. Applying the one-proportion z-test to the differences gives a z-score of 7.08, with a P-value < 0.00001. We can be pretty confident that the median is not 0. (Just for comparison, the t-statistic for testing $\mu_d = 0$ is 7.152 with 169 df—almost the identical result.)

As with other distribution-free tests, the advantage of the **sign test for matched pairs** is that we don't require the Nearly Normal Condition for the paired differences. Because it looks only at the direction of the difference, the sign test isn't affected by outliers—extraordinarily large differences—which can be an advantage. On the other hand, when the assumptions of the *paired t-test* are met, the paired *t*-test is more powerful than the sign test.

WHAT CAN GO WRONG?

- **Don't use a two-sample *t*-test when you have paired data.** See the What Can Go Wrong? discussion in Chapter 22.
- **Don't use a paired *t* method when the samples aren't paired.** Just because two groups have the same number of observations doesn't mean they can be paired, even if they are shown side by side in a table. We might have 25 men and 25 women in our study, but they might be completely independent of one another. If they were siblings or spouses, we might consider them paired. Remember that you cannot *choose* which method to use based on your preferences. If the data are from two independent samples, use two-sample *t* methods. If the data are from an experiment in which observations were paired, you must use a paired method. If the data are from an observational study, you must be able to defend your decision to use matched pairs or independent groups.
- **Don't forget outliers.** The outliers we care about now are in the differences. A subject who is extraordinary both before and after a treatment may still have a perfectly typical difference. But one outlying difference can completely distort your conclusions. Be sure to plot the differences (even if you also plot the data).
- **Don't look for the difference between the means of paired groups with side-by-side boxplots.** The point of the paired analysis is to remove extra variation. The boxplots of each group still contain that variation. Comparing them is likely to be misleading.

CONNECTIONS

The most important connection is to the concept of blocking that we first discussed when we considered designed experiments in Chapter 12. Pairing is a basic and very effective form of blocking.

Of course, the details of the mechanics for paired *t*-tests and intervals are identical to those for the one-sample *t* methods. Everything we know about those methods applies here.

The connection to the two-sample methods of Chapter 22 is that when the data are naturally paired, those methods are not appropriate because paired data fail the required condition of independence.

What Have We Learned?

Learning Objectives

Recognize when data are paired or matched.

Know how to construct a confidence interval for the mean difference in paired data.

Know how to perform a hypothesis test about the mean difference (usually with a null of zero representing no difference between the groups.)

Review of Terms

Paired data — Data are paired when the observations are collected in pairs or the observations in one group are naturally related to observations in the other. The simplest form of pairing is to measure each subject twice—often before and after a treatment is applied. More sophisticated forms of pairing in experiments are a form of blocking and arise in other contexts. Pairing in observational and survey data is a form of matching (p. 631).

Paired *t*-test — A hypothesis test for the mean of the pairwise differences of two groups. It tests the null hypothesis

$$H_0: \mu_d = \Delta_0,$$

where the hypothesized difference is almost always 0, using the statistic

$$t = \frac{\bar{d} - \Delta_0}{SE(\bar{d})}$$

with $n - 1$ degrees of freedom, where $SE(\bar{d}) = \frac{s_d}{\sqrt{n}}$, and n is the number of pairs (p. 632).

Paired *t* confidence interval — A confidence interval for the mean of the pairwise differences between paired groups found as

$$\bar{d} \pm t^*_{n-1} \times SE(\bar{d}), \text{ where } SE(\bar{d}) = \frac{s_d}{\sqrt{n}}$$

and n is the number of pairs (p. 638).

On the Computer PAIRED *t*

Most statistics programs can compute paired *t* analyses. Some may want you to find the differences yourself and use the one-sample *t* methods. Those that perform the entire procedure will need to know the two variables to compare. The computer, of course, cannot verify that the variables are naturally paired. Most programs will check whether the two variables have the same number of observations, but some stop there, and that can cause trouble. Most programs will automatically omit any pair that is missing a value for either variable (as we did with the British couples). You must look carefully to see whether that has happened.

As we've seen with other inference results, some packages pack a lot of information into a simple table, but you must locate what you want for yourself. Here's a generic example with comments:

Other packages try to be more descriptive. It may be easier to find the results, but you may get less information from the output table.

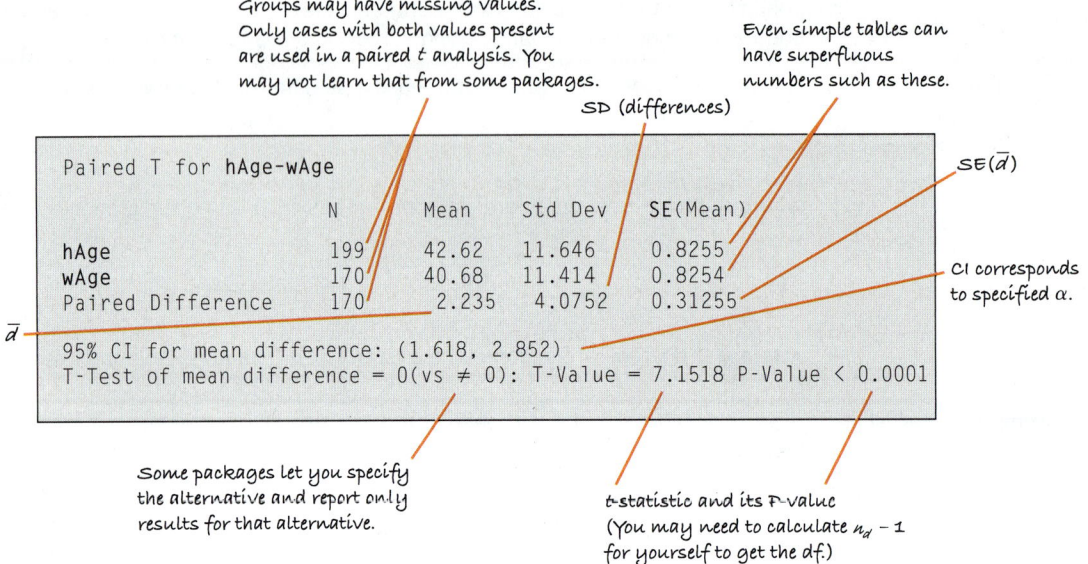

Computers make it easy to examine the boxplots of the two groups and the histogram of the differences—both important steps. Some programs offer a scatterplot of the two variables. That can be helpful. In terms of the scatterplot, a paired *t*-test is about whether the points tend to be above or below the line $y = x$. (Note, though, that pairing says nothing about whether the scatterplot should be straight. That doesn't matter for our *t* methods.)

DATA DESK

- Select variables.
- From the **Calc** menu, choose **Estimate** for confidence intervals or **Test** for hypothesis tests.
- Select the interval or test from the drop-down menu, and make other choices in the dialog.

COMMENTS

Data Desk expects the two groups to be in separate variables and in the same "Relation"—that is, about the same cases.

EXCEL

- In Excel 2003 and earlier, select **Data Analysis** from the **Tools** menu.
- In Excel 2007, select **Data Analysis** from the **Analysis** Group on the **Data** Tab.
- From the **Data Analysis** menu, choose **t-test: paired two-sample for Means**. Fill in the cell ranges for the two groups, the hypothesized difference, and the alpha level.

COMMENTS

Excel expects the two groups to be in separate cell ranges.

Warning: Do not compute this test in Excel without checking for missing values. If there are any missing values (empty cells), Excel will usually give a wrong answer. Excel compacts each list, pushing values up to cover the missing cells, and then checks only that it has the same number of values in each list. The result is mismatched pairs and an entirely wrong analysis.

JMP

- From the **Analyze** menu, select **Matched Pairs**.
- Specify the columns holding the two groups in the **Y Paired Response** dialog.
- Click **OK**.

MINITAB

- From the **Stat** menu, choose the **Basic Statistics** submenu.
- From that menu, choose **Paired t . . .**
- Then fill in the dialog.

COMMENTS

Minitab takes "First sample" minus "Second sample."

R

To test the hypothesis that $\mu_1 = \mu_2$ for paired data against an alternative (default is two-sided), create a vector for the data of each group, such as x and y and produce the confidence interval (default is 95%):

- t.test(x,y, alternative = c("two.sided", "less", "greater"), paired = TRUE, conf.level = 0.95)

will produce the *t*-statistic, degrees of freedom, P-value, and confidence interval for a specified alternative.

COMMENTS

This is the same function as for both the one- and two-sample *t*-test, but with paired=True.

SPSS

- From the **Analyze** menu, choose the **Compare Means** submenu.
- From that, choose the **Paired-Samples *t*-test** command.
- Select pairs of variables to compare, and click the arrow to add them to the selection box.

COMMENTS

You can compare several pairs of variables at once. Options include the choice to exclude cases missing in any pair from all tests.

STATCRUNCH

To do inference for the mean of paired differences:
- Click on **Stat**.
- Choose **T Statistics » Paired**.
- Choose the **Column** for each variable.
- Check **Save differences** so you can look at a histogram to be sure the Nearly Normal condition is satisfied.
- Click on **Next**.

- Indicate **Hypothesis Test**, then enter the hypothesized **Null mean** difference (usually 0), and choose the **Alternative** hypothesis.

OR

Indicate **Confidence Interval**, and then enter the **Level** of confidence.
- Click on **Calculate**.

TI-83/84 PLUS

If the data are stored in two lists, say, L1 and L2, create a list of the differences:

- L1 − L2 → L3. (The arrow is the STO button.)
- Since inference for paired differences uses one-sample *t*-procedures, select **2:T-Test** or **8:TInterval** from the **STAT TESTS** menu.

- Specify as the data the list of differences you just created in L3, and apply the procedure.

Exercises

Section 23.1

1. **Which method?** Which of the following scenarios should be analyzed as paired data?
   a) Students take a MCAT prep course. Their before and after scores are compared.
   b) 20 male and 20 females students in class take a midterm. We compare their scores.
   c) A group of college freshmen are asked about the quality of the university cafeteria. A year later, the same students are asked about the cafeteria again. Do student's opinions change during their time at school?

2. **Which method II?** Which of the following scenarios should be analyzed as paired data?
   a) Spouses are asked about the number of hours of sleep they get each night. We want to see if husbands get more sleep than wives.
   b) 50 insomnia patients are given a placebo and 50 are given a mild sedative. Which subjects sleep more hours?
   c) A group of college freshmen and a group of sophomores are asked about the quality of the university cafeteria. Do students' opinions change during their time at school?

Section 23.2

3. **Cars and trucks** We have data on the city and highway fuel efficiency of 316 cars and 316 trucks.

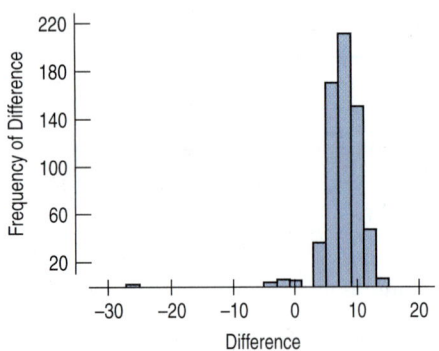

   a) Would it be appropriate to use paired *t* methods to compare the cars and the trucks?
   b) Would it be appropriate to use paired *t* methods to compare the city and highway fuel efficiencies of these vehicles?
   c) A histogram of the differences (highway − city) is shown. Are the conditions for inference satisfied?

4. **Weighing trucks** One kind of scale for weighing trucks can measure their weight as they drive across a plate. Is this method consistent with the traditional method of static weighing? Are the conditions for matched pairs inference satisfied? Weights are in 1000s of pounds.

Weight-in-Motion	Static Weight	Diff (Static − Motion)
26.0	27.9	−1.9
29.9	29.1	0.8
39.5	38.0	1.5
25.1	27.0	−1.9
31.6	30.3	1.3
36.2	34.5	1.7
25.1	27.8	−2.7
31.0	29.6	1.4
35.6	33.1	2.5
40.2	35.5	4.7

Section 23.3

5. **Cars and trucks again** In Exercise 3, after deleting an outlying value of −27, the mean difference in fuel efficiencies for the 632 vehicles was 7.37 mpg with a standard deviation of 2.52 mpg. Find a 95% confidence interval for this difference and interpret it in context.

6. **Weighing trucks II** Find a 98% confidence interval of the weight differences in Exercise 4. Interpret this interval in context.

Section 23.4

7. **Blocking cars and trucks** Thinking about the data on fuel efficiency in Exercise 3, why is the blocking accomplished by a matched pairs analysis particularly important for a sample that has both cars and trucks?

8. **Weighing trucks III** Consider the weights from Exercise 4. The side-by-side boxplots below show little difference between the two groups. Should this be sufficient to draw a conclusion about the accuracy of the weigh-in-motion scale?

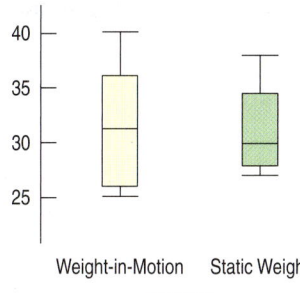

Chapter Exercises

9. **More eggs?** Can a food additive increase egg production? Agricultural researchers want to design an experiment to find out. They have 100 hens available. They have two kinds of feed: the regular feed and the new feed with the additive. They plan to run their experiment for a month, recording the number of eggs each hen produces.
   a) Design an experiment that will require a two-sample t procedure to analyze the results.
   b) Design an experiment that will require a matched-pairs t procedure to analyze the results.
   c) Which experiment would you consider the stronger design? Why?

10. **Music** Some students do homework with music playing in their headphones. (Anyone come to mind?) Some researchers want to see if people can work as effectively with as without distraction. The researchers will time some volunteers to see how long it takes them to complete some relatively easy crossword puzzles. During some of the trials, the room will be quiet; during other trials in the same room, subjects will wear headphones and listen to a Pandora channel.
    a) Design an experiment that will require a two-sample t procedure to analyze the results.
    b) Design an experiment that will require a matched-pairs t procedure to analyze the results.
    c) Which experiment would you consider the stronger design? Why?

11. **Sex sells?** Ads for many products use sexual images to try to attract attention to the product. But do these ads bring people's attention to the item that was being advertised? We want to design an experiment to see if the presence of sexual images in an advertisement affects people's ability to remember the product.
    a) Describe an experiment design requiring a matched-pairs t procedure to analyze the results.
    b) Describe an experiment design requiring an independent sample procedure to analyze the results.

12. **Freshman 15?** Many people believe that students gain weight as freshmen. Suppose we plan to conduct a study to see if this is true.
    a) Describe a study design that would require a matched-pairs t procedure to analyze the results.
    b) Describe a study design that would require a two-sample t procedure to analyze the results.

13. **Women** Values for the labor force participation rate of women (LFPR) are published by the U.S. Bureau of Labor Statistics. We are interested in whether there was a difference between female participation in 1968 and 1972, a time of rapid change for women. We check LFPR values for 19 randomly selected cities for 1968 and 1972. Shown below is software output for two possible tests:

    Paired t-Test of $\mu(1 - 2)$
    Test Ho: $\mu(1972-1968) = 0$ vs Ha: $\mu(1972-1968) \neq 0$
    Mean of Paired Differences = 0.0337
    t-Statistic = 2.458 with 18 df
    $p = 0.0244$

    2-Sample t-Test of $\mu 1 - \mu 2$
    Ho: $\mu 1 - \mu 2 = 0$ Ha: $\mu 1 - \mu 2 \neq 0$
    Test Ho: $\mu(1972) - \mu(1968) = 0$ vs
    Ha: $\mu(1972) - \mu(1968) \neq 0$
    Difference Between Means = 0.0337
    t-Statistic = 1.496 with 35 df
    $p = 0.1434$

    a) Which of these tests is appropriate for these data? Explain.
    b) Using the test you selected, state your conclusion.

14. **Cloud seeding** Simpson, Alsen, and Eden (*Technometrics* 1975) report the results of trials in which clouds were seeded and the amount of rainfall recorded. The authors report on 26 seeded and 26 unseeded clouds in order of the amount of rainfall, largest amount first. Here are two possible tests to study the question of whether cloud seeding works.

    Paired t-Test of $\mu(1 - 2)$
    Mean of Paired Differences = -277.39615
    t-Statistic = -3.641 with 25 df
    $p = 0.0012$

    2-Sample t-Test of $\mu 1 - \mu 2$
    Difference Between Means = -277.4
    t-Statistic = -1.998 with 33 df
    $p = 0.0538$

    a) Which of these tests is appropriate for these data? Explain.
    b) Using the test you selected, state your conclusion.

15. Friday the 13th, traffic The *British Medical Journal* published an article titled, "Is Friday the 13th Bad for Your Health?" Researchers in Britain examined how Friday the 13th affects human behavior. One question was whether people tend to stay at home more on Friday the 13th. The data below are the number of cars passing Junctions 9 and 10 on the M25 motorway for consecutive Fridays (the 6th and 13th) for five different periods.

Year	Month	6th	13th
1990	July	134,012	132,908
1991	September	133,732	131,843
1991	December	121,139	118,723
1992	March	124,631	120,249
1992	November	117,584	117,263

Here are summaries of two possible analyses:

Paired t-Test of $\mu(1 - 2) = 0$ vs. $\mu(1 - 2) > 0$
Mean of Paired Differences: 2022.4
t-Statistic = 2.9377 with 4 df
P = 0.0212

2-Sample t-Test of $\mu 1 = \mu 2$ vs. $\mu 1 > \mu 2$
Difference Between Means: 2022.4
t-Statistic = 0.4273 with 7.998 df
P = 0.3402

a) Which of the tests is appropriate for these data? Explain.
b) Using the test you selected, state your conclusion.
c) Are the assumptions and conditions for inference met?

16. Friday the 13th, accidents The researchers in Exercise 15 also examined the number of people admitted to emergency rooms for vehicular accidents on 12 Friday evenings (6 each on the 6th and 13th).

Year	Month	6th	13th
1989	October	9	13
1990	July	6	12
1991	September	11	14
1991	December	11	10
1992	March	3	4
1992	November	5	12

Based on these data, is there evidence that more people are admitted, on average, on Friday the 13th? Here are two possible analyses of the data:

Paired t-Test of $\mu(1 - 2) = 0$ vs. $\mu(1 - 2) < 0$
Mean of Paired Differences = −3.333
t-Statistic = −2.7116 with 5 df
P = 0.0211

2-Sample t-Test of $\mu 1 = \mu 2$ vs. $\mu 1 < \mu 2$
Difference Between Means = −3.333
t-Statistic = −1.6644 with 9.940 df
P = 0.0636

a) Which of these tests is appropriate for these data? Explain.
b) Using the test you selected, state your conclusion.
c) Are the assumptions and conditions for inference met?

17. Online insurance I After seeing countless commercials claiming one can get cheaper car insurance from an online company, a local insurance agent was concerned that he might lose some customers. To investigate, he randomly selected profiles (type of car, coverage, driving record, etc.) for 10 of his clients and checked online price quotes for their policies. The comparisons are shown in the table below. His statistical software produced the following summaries (where *PriceDiff* = *Local* − *Online*):

Variable	Count	Mean	StdDev
Local	10	799.200	229.281
Online	10	753.300	256.267
PriceDiff	10	45.9000	175.663

Local	Online	PriceDiff
568	391	177
872	602	270
451	488	−37
1229	903	326
605	677	−72
1021	1270	−249
783	703	80
844	789	55
907	1008	−101
712	702	10

At first, the insurance agent wondered whether there was some kind of mistake in this output. He thought the Pythagorean Theorem of Statistics should work for finding the standard deviation of the price differences—in other words, that

$$SD(Local - Online) = \sqrt{SD^2(Local) + SD^2(Online)}.$$

But when he checked, he found that
$\sqrt{(229.281)^2 + (256.267)^2} = 343.864$, not 175.663 as given by the software. Tell him where his mistake is.

18. Wind speed, part I To select the site for an electricity-generating wind turbine, wind speeds were recorded at several potential sites every 6 hours for a year. Two sites not far from each other looked good. Each had a mean wind speed high enough to qualify, but we should choose the site with a higher average daily wind speed. Because the sites are near each other and the wind speeds were recorded at the same times, we should view the speeds as paired. Here are the summaries of the speeds (in miles per hour):

Variable	Count	Mean	StdDev
site2	1114	7.452	3.586
site4	1114	7.248	3.421
site2 − site4	1114	0.204	2.551

Is there a mistake in this output? Why doesn't the Pythagorean Theorem of Statistics work here? In other words, shouldn't

$$SD(site2 - site4) = \sqrt{SD^2(site2) + SD^2(site4)}?$$

But $\sqrt{(3.586)^2 + (3.421)^2} = 4.956$, not 2.551 as given by the software. Explain why this happened.

T 19. Online insurance II In Exercise 17, we saw summary statistics for 10 drivers' car insurance premiums quoted by a local agent and an online company. Here are displays for each company's quotes and for the difference (*Local − Online*):

a) Which of the summaries would help you decide whether the online company offers cheaper insurance? Why?
b) The standard deviation of *PriceDiff* is quite a bit smaller than the standard deviation of prices quoted by either the local or online companies. Discuss why.
c) Using the information you have, discuss the assumptions and conditions for inference with these data.

T 20. Wind speed, part II In Exercise 18, we saw summary statistics for wind speeds at two sites near each other, both being considered as locations for an electricity-generating wind turbine. The data, recorded every 6 hours for a year, showed each of the sites had a mean wind speed high enough to qualify, but how can we tell which site is best? Here are some displays:

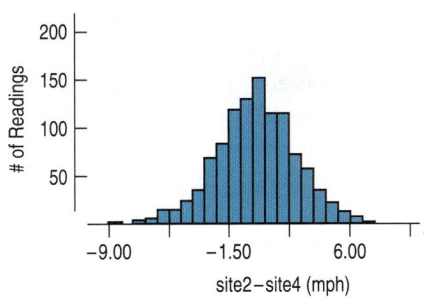

a) The boxplots show outliers for each site, yet the histogram shows none. Discuss why.
b) Which of the summaries would you use to select between these sites? Why?
c) Using the information you have, discuss the assumptions and conditions for paired *t* inference for these data. (*Hint:* Think hard about the independence assumption in particular.)

T 21. Online insurance III Exercises 17 and 19 give summaries and displays for car insurance premiums quoted by a local agent and an online company. Test an appropriate hypothesis to see if there is evidence that drivers might save money by switching to the online company.

T 22. Wind speed, part III Exercises 18 and 20 give summaries and displays for two potential sites for a wind turbine. Test an appropriate hypothesis to see if there is evidence that either of these sites has a higher average wind speed.

T 23. City Temperatures The following table gives the average daily high temperatures in January and July for several cities. Find a 95% confidence interval for the mean temperature difference between summer and winter.

a) Check the assumptions and conditions. If you find a city that doesn't belong with the others, set it aside.
b) Compute your interval for the data (adjusted, if necessary), and explain what it means.

Location	Jan.	July	Location	Jan.	July
Amsterdam (Netherlands)	41	69	Madrid (Spain)	50	89
Athens (Greece)	54	90	Montreal (Canada)	22	79
Auckland (New Zealand)	73	56	Moscow (Russia)	21	76
Beijing (China)	35	87	Nassau (Bahamas)	77	88
Belgrade (Yugoslavia)	38	81	Oslo (Norway)	30	73
Berlin (Germany)	35	74	Paris (France)	42	76
Cairo (Egypt)	65	96	Prague (Czech Republic)	34	74
Calcutta (India)	80	89	Quebec (Canada)	19	77
Copenhagen (Denmark)	36	72	Rome (Italy)	54	88
Dublin (Ireland)	47	67	Seoul (Korea)	33	84
Glasgow (Scotland)	43	66	Stockholm (Sweden)	31	70
Hamilton (Bermuda)	68	85	Taipei (Taiwan)	66	92
Helsinki (Finland)	27	71	Tokyo (Japan)	48	84
Hong Kong (China)	67	90	Toronto (Canada)	30	79
Istanbul (Turkey)	48	78	Vancouver (Canada)	42	71
Jerusalem (Israel)	55	87	Vienna (Austria)	34	75
Lisbon (Portugal)	56	79	Zurich (Switzerland)	36	77
London (United Kingdom)	44	73			

24. NY Marathon 2013 The table below shows the winning times (in minutes) for men and women in the New York City Marathon between 1978 and 2013. (The race was not run in 2012 because of Superstorm Sandy.). Assuming that performances in the Big Apple resemble performances elsewhere, we can think of these data as a sample of performance in marathon competitions. Create a 90% confidence interval for the mean difference in winning times for male and female marathon competitors. (www.nycmarathon.org)

Year	Men	Women	Year	Men	Women
1978	132.2	152.5	1996	129.9	148.3
1979	131.7	147.6	1997	128.2	148.7
1980	129.7	145.7	1998	128.8	145.3
1981	128.2	145.5	1999	129.2	145.1
1982	129.5	147.2	2000	130.2	145.8
1983	129.0	147.0	2001	127.7	144.4
1984	134.9	149.5	2002	128.1	145.9
1985	131.6	148.6	2003	130.5	142.5
1986	131.1	148.1	2004	129.5	143.2
1987	131.0	150.3	2005	129.5	144.7
1988	128.3	148.1	2006	130.0	145.1
1989	128.0	145.5	2007	129.1	143.2
1990	132.7	150.8	2008	128.7	143.9
1991	129.5	147.5	2009	129.3	148.9
1992	129.5	144.7	2010	128.3	148.3
1993	130.1	146.4	2011	125.1	143.3
1994	131.4	147.6	2012	<cancelled>	
1995	131.1	148.1	2013	128.4	140.1

25. Push-ups Every year, the students at Gossett High School take a physical fitness test during their gym classes. One component of the test asks them to do as many push-ups as they can. Results for one class are shown below, separately for boys and girls. Assuming that students at Gossett are assigned to gym classes at random, create a 90% confidence interval for how many more push-ups boys can do than girls, on average, at that high school.

Boys	17	27	31	17	25	32	28	23	25	16	11	34
Girls	24	7	14	16	2	15	19	25	10	27	31	8

26. Brain waves An experiment was performed to see whether sensory deprivation over an extended period of time has any effect on the alpha-wave patterns produced by the brain. To determine this, 20 subjects, inmates in a Canadian prison, were randomly split into two groups. Members of one group were placed in solitary confinement. Those in the other group were allowed to remain in their own cells. Seven days later, alpha-wave frequencies were measured for all subjects, as shown in the following table. (Source: P. Gendreau et al., "Changes in EEG Alpha Frequency and Evoked Response Latency During Solitary Confinement," *Journal of Abnormal Psychology* 79 [1972]: 54–59)

Nonconfined	Confined
10.7	9.6
10.7	10.4
10.4	9.7
10.9	10.3
10.5	9.2
10.3	9.3
9.6	9.9
11.1	9.5
11.2	9.0
10.4	10.9

a) What are the null and alternative hypotheses? Be sure to define all the terms and symbols you use.
b) Are the assumptions necessary for inference met?
c) Perform the appropriate test, indicating the formula you used, the calculated value of the test statistic, the df, and the P-value.
d) State your conclusion.

27. Job satisfaction (When you first read about this exercise break plan in Chapter 22, you did not have an inference method that would work. Try again now.) A company institutes an exercise break for its workers to see if it will improve job satisfaction, as measured by a questionnaire that assesses workers' satisfaction. Scores for 10 randomly selected workers before and after the implementation of the exercise program are shown in the table below.

a) Identify the procedure you would use to assess the effectiveness of the exercise program, and check to see if the conditions allow the use of that procedure.
b) Test an appropriate hypothesis and state your conclusion.
c) If your conclusion turns out to be incorrect, what kind of error did you commit?
*d) Use a matched-pairs sign test to test the appropriate hypothesis. Do your conclusions change from those in part b?

Worker Number	Job Satisfaction Index	
	Before	After
1	34	33
2	28	36
3	29	50
4	45	41
5	26	37
6	27	41
7	24	39
8	15	21
9	15	20
10	27	37

28. Summer school (When you first read about the summer school issue in Chapter 22, you did not have an inference method that would work. Try again now.) Having done poorly on their Math final exams in June, six students repeat the course in summer school and take another exam in August.

June	54	49	68	66	62	62
Aug.	50	65	74	64	68	72

a) If we consider these students to be representative of all students who might attend this summer school in other years, do these results provide evidence that the program is worthwhile?
b) This conclusion, of course, may be incorrect. If so, which type of error was made?

29. Yogurt Is there a significant difference in calories between servings of strawberry and vanilla yogurt? Based on the data shown in the table, test an appropriate hypothesis and state your conclusion. Don't forget to check assumptions and conditions!

	Calories per Serving	
Brand	Strawberry	Vanilla
America's Choice	210	200
Breyer's Lowfat	220	220
Columbo	220	180
Dannon Light'n Fit	120	120
Dannon Lowfat	210	230
Dannon la Crème	140	140
Great Value	180	80
La Yogurt	170	160
Mountain High	200	170
Stonyfield Farm	100	120
Yoplait Custard	190	190
Yoplait Light	100	100

30. Gasoline Many drivers of cars that can run on regular gas actually buy premium in the belief that they will get better gas mileage. To test that belief, we use 10 cars from a company fleet in which all the cars run on regular gas. Each car is filled first with either regular or premium gasoline, decided by a coin toss, and the mileage for that tankful is recorded. Then the mileage is recorded again for the same cars for a tankful of the other kind of gasoline. We don't let the drivers know about this experiment.

Here are the results (miles per gallon):

Car #	1	2	3	4	5	6	7	8	9	10
Regular	16	20	21	22	23	22	27	25	27	28
Premium	19	22	24	24	25	25	26	26	28	32

a) Is there evidence that cars get significantly better fuel economy with premium gasoline?
b) How big might that difference be? Check a 90% confidence interval.
c) Even if the difference is significant, why might the company choose to stick with regular gasoline?
d) Suppose you had done a "bad thing." (We're sure you didn't.) Suppose you had mistakenly treated these data as two independent samples instead of matched pairs. What would the significance test have found? Carefully explain why the results are so different.
*e) Use a matched-pairs sign test to test the appropriate hypothesis. Do your conclusions change from those in part a?

31. Braking test A tire manufacturer tested the braking performance of one of its tire models on a test track. The company tried the tires on 10 different cars, recording the stopping distance for each car on both wet and dry pavement. Results are shown in the table.

	Stopping Distance (ft)	
Car #	Dry Pavement	Wet Pavement
1	150	201
2	147	220
3	136	192
4	134	146
5	130	182
6	134	173
7	134	202
8	128	180
9	136	192
10	158	206

a) Write a 95% confidence interval for the mean dry pavement stopping distance. Be sure to check the appropriate assumptions and conditions, and explain what your interval means.
b) Write a 95% confidence interval for the mean increase in stopping distance on wet pavement. Be sure to check the appropriate assumptions and conditions, and explain what your interval means.

32. Braking test 2 For another test of the tires in Exercise 31, a car made repeated stops from 60 miles per hour. The test was run on both dry and wet pavement, with results as shown in the table. (Note that actual *braking distance*, which takes into account the driver's reaction time, is much longer, typically nearly 300 feet at 60 mph!)

a) Write a 95% confidence interval for the mean dry pavement stopping distance. Be sure to check the appropriate assumptions and conditions, and explain what your interval means.

b) Write a 95% confidence interval for the mean increase in stopping distance on wet pavement. Be sure to check the appropriate assumptions and conditions, and explain what your interval means.

Stopping Distance (ft)	
Dry Pavement	Wet Pavement
145	211
152	191
141	220
143	207
131	198
148	208
126	206
140	177
135	183
133	223

33. **Tuition 2013** How much more do public colleges and universities charge out-of-state students for tuition per year? A random sample of 19 public colleges and universities listed at www.collegeboard.com yielded the following data for students entering as Freshmen in Fall 2013.

Institution	Resident	Nonresident
Univ of Akron (OH)	10,130	18,494
Athens State (AL)	5,760	10,770
Ball State (IN)	9,610	24,124
Bloomsburg U (PA)	8,582	18,516
UC Irvine (CA)	13,149	36,027
Central State (OH)	6,058	13,510
Clarion U (PA)	9,404	12,716
Dakota State	8,286	10,286
Fairmont State (WV)	5,824	12,288
Johnson State (VT)	10,286	21,950
Lock Haven U (PA)	8,898	16,832
New College of Florida	6,783	29,812
Oakland U (MI)	10,613	23,873
U Pittsburgh	17,100	27,106
Savannah State (GA)	6,340	18,542
Louisiana State	5,241	10,973
W Liberty University (WV)	6,226	13,540
Central Texas College	2,130	6,270
Worcester State (MA)	8,157	14,230

a) Create a 90% confidence interval for the mean difference in cost. Be sure to justify your procedure.
b) Interpret your interval in context.
c) A national magazine claims that public institutions charge state residents an average of $7000 less than out-of-staters for tuition each year. What does your confidence interval indicate about this assertion?

34. **Sex sells, part II** In Exercise 11, you considered the question of whether sexual images in ads affected people's abilities to remember the item being advertised. To investigate, a group of Statistics students cut ads out of magazines. They were careful to find two ads for each of 10 similar items, one with a sexual image and one without. They arranged the ads in random order and had 39 subjects look at them for one minute. Then they asked the subjects to list as many of the products as they could remember. Their data are shown in the table. Is there evidence that the sexual images mattered?

Subject Number	Ads Remembered Sexual Image	No Sex	Subject Number	Ads Remembered Sexual Image	No Sex
1	2	2	21	2	3
2	6	7	22	4	2
3	3	1	23	3	3
4	6	5	24	5	3
5	1	0	25	4	5
6	3	3	26	2	4
7	3	5	27	2	2
8	7	4	28	2	4
9	3	7	29	7	6
10	5	4	30	6	7
11	1	3	31	4	3
12	3	2	32	4	5
13	6	3	33	3	0
14	7	4	34	4	3
15	3	2	35	2	3
16	7	4	36	3	3
17	4	4	37	5	5
18	1	3	38	3	4
19	5	5	39	4	3
20	2	2			

35. **Strikes** Advertisements for an instructional video claim that the techniques will improve the ability of Little League pitchers to throw strikes and that, after undergoing the training, players will be able to throw strikes on at least 60% of their pitches. To test this claim, we have 20 Little Leaguers throw 50 pitches each, and we record the number of strikes. After the players participate in the training program, we repeat the test. The table (on the next page) shows the number of strikes each player threw before and after the training.
a) Is there evidence that after training players can throw strikes more than 60% of the time?
b) Is there evidence that the training is effective in improving a player's ability to throw strikes?
*c) Use a matched-pairs sign test to test the appropriate hypothesis. Do your conclusions change from those in part a?

Number of Strikes (out of 50)		Number of Strikes (out of 50)	
Before	After	Before	After
28	35	33	33
29	36	33	35
30	32	34	32
32	28	34	30
32	30	34	33
32	31	35	34
32	32	36	37
32	34	36	33
32	35	37	35
33	36	37	32

36. Freshman 15, revisited In Exercise 12, you thought about how to design a study to see if it's true that students tend to gain weight during their first year in college. Well, Cornell Professor of Nutrition David Levitsky did just that. He recruited students from two large sections of an introductory health course. Although they were volunteers, they appeared to match the rest of the freshman class in terms of demographic variables such as sex and ethnicity. The students were weighed during the first week of the semester, then again 12 weeks later. Based on Professor Levitsky's data, estimate the mean weight gain in first-semester freshmen and comment on the "freshman 15." (Weights are in pounds.)

Subject Number	Initial Weight	Terminal Weight	Subject Number	Initial Weight	Terminal Weight
1	171	168	21	157	158
2	110	111	22	119	126
3	134	136	23	113	114
4	115	119	24	120	128
5	150	155	25	135	139
6	104	106	26	148	150
7	142	148	27	110	112
8	120	124	28	160	163
9	144	148	29	220	224
10	156	154	30	132	133
11	114	114	31	145	147
12	121	123	32	141	141
13	122	126	33	158	160
14	120	115	34	135	134
15	115	118	35	148	150
16	110	113	36	164	165
17	142	146	37	137	138
18	127	127	38	198	201
19	102	105	39	122	124
20	125	125	40	146	146

Subject Number	Initial Weight	Terminal Weight	Subject Number	Initial Weight	Terminal Weight
41	150	151	55	106	108
42	187	192	56	185	188
43	94	96	57	125	128
44	105	105	58	125	126
45	127	130	59	155	158
46	142	144	60	118	120
47	140	143	61	149	150
48	107	107	62	149	149
49	104	105	63	122	121
50	111	112	64	155	158
51	160	162	65	160	161
52	134	134	66	115	119
53	151	151	67	167	170
54	127	130	68	131	131

37. Wheelchair marathon 2013 The Boston Marathon has had a wheelchair division since 1977. Who do you think is typically faster, the men's marathon winner on foot or the women's wheelchair marathon winner? Because the conditions differ from year to year, and speeds have improved over the years, it seems best to treat these as paired measurements. Here are summary statistics for the pairwise differences in finishing time (in minutes): (www.boston.com/sports/marathon/history/champions/)

Summary of wheelchrF − runM
N = 37
Mean = −4.88
SD = 35.227

a) Comment on the assumptions and conditions.
b) Assuming that these times are representative of such races and the differences appeared acceptable for inference, construct and interpret a 95% confidence interval for the mean difference in finishing times.

c) Would a hypothesis test at $\alpha = 0.05$ reject the null hypothesis of no difference? What conclusion would you draw?

T 38. Marathon start-up years 2013 When we considered the Boston Marathon in Exercise 37, we were unable to check the Nearly Normal Condition. Here's a histogram of the differences:

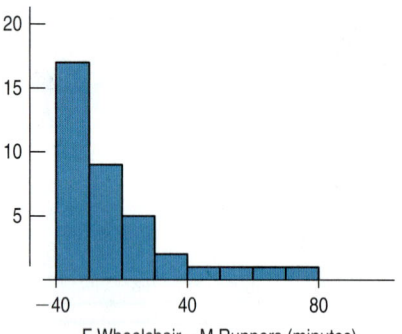

The three largest differences are the first three years of wheelchair competition: 1977, 1978, and 1979. Often the start-up years of new events are different; later on, more athletes train and compete. If we omit those three years, the summary statistics change as follows:

Summary of wheelchrF − runM
n = 34
Mean = −13.40
SD = 20.57

a) Comment on the assumptions and conditions.
b) Assuming that these times are representative of such races, construct and interpret a 95% confidence interval for the mean difference in finishing time.
c) Would a hypothesis test at $\alpha = 0.05$ reject the null hypothesis of no difference? What conclusion would you draw?

39. BST Many dairy cows now receive injections of BST, a hormone intended to spur greater milk production. After the first injection, a test herd of 60 Ayrshire cows increased their mean daily production from 47 pounds to 61 pounds of milk. The standard deviation of the increases was 5.2 pounds. We want to estimate the mean increase a farmer could expect in his own cows.

a) Check the assumptions and conditions for inference.
b) Write a 95% confidence interval.
c) Explain what your interval means in this context.
d) Given the cost of BST, a farmer believes he cannot afford to use it unless he is sure of attaining at least a 25% increase in milk production. Based on your confidence interval, what advice would you give him?

40. BST II In the experiment about hormone injections in cows described in Exercise 39, a group of 52 Jersey cows increased average milk production from 43 pounds to 52 pounds per day, with a standard deviation of 4.8 pounds. Is this evidence that the hormone may be more effective in one breed than the other? Test an appropriate hypothesis and state your conclusion. Be sure to discuss any assumptions you make.

Just Checking ANSWERS

1. These are independent groups sampled at random, so use a two-sample t confidence interval to estimate the size of the difference.

2. There is only one sample. Use a one-sample t-interval.

3. A brother and sister from the same family represent a matched pair. The question calls for a paired t-test.

4. a) A before-and-after study calls for paired t methods. To estimate the loss, find a confidence interval for the before–after differences.

   b) The two treatment groups were assigned randomly, so they are independent. Use a two-sample t-test to assess whether the mean weight losses differ.

5. Sometimes it just isn't clear. Most likely, couples would discuss the band or even decide to go to the club because they both like a particular band. If we think that's likely, then these data are paired. But maybe not. If we asked them their opinions of, say, the decor or furnishings at the club, the fact that they were couples might not affect the independence of their answers.

Chapter 24
Comparing Counts

24.1 Goodness-of-Fit Tests

24.2 Chi-Square Test of Homogeneity

24.3 Examining the Residuals

24.4 Chi-Square Test of Independence

Where are we going?
Is your favorite color related to how much education you've had? A survey found a higher percentage of those naming blue and a lower percentage saying red among adults with only a high school education compared with adults with more education. Could this be just random fluctuation, or is the distribution of color preference different for different education levels? We saw tables of counts and percentages in Chapter 2. In this chapter, we'll see how to test the strength of the patterns we saw in those tables.

Who	Executives of Fortune 400 companies
What	Zodiac birth sign
Why	Maybe the researcher was a Gemini and naturally curious?

A S *Activity:* **Children at Risk.** See how a contingency table helps us understand the different risks to which an incident exposed children.

Does your zodiac sign predict how successful you will be later in life? *Fortune* magazine collected the zodiac signs of 256 heads of the largest 400 companies. The table shows the number of births for each sign.

Births	Sign	Births	Sign
23	Aries	18	Libra
20	Taurus	21	Scorpio
18	Gemini	19	Sagittarius
23	Cancer	22	Capricorn
20	Leo	24	Aquarius
19	Virgo	29	Pisces

Birth totals by sign for 256 Fortune 400 executives.

We can see some variation in the number of births per sign, and there *are* more Pisces, but is that enough to claim that successful people are more likely to be born under some signs than others?

24.1 Goodness-of-Fit Tests

If these 256 births were distributed uniformly across the year, we would expect about 1/12 of them to occur under each sign of the zodiac. That suggests 256/12, or about 21.3 births per sign. How closely do the observed numbers of births per sign fit this simple "null" model?

655

A hypothesis test to address this question is called a test of **"goodness-of-fit."** The name suggests a certain badness-of-grammar, but it is quite standard. After all, we are asking whether the model that births are uniformly distributed over the signs fits the data good, ... er, well. Goodness-of-fit involves testing a hypothesis. We have specified a model for the distribution and want to know whether it fits. There is no single parameter to estimate, so a confidence interval wouldn't make much sense.

If the question were about only one astrological sign (for example, "Are executives more likely to be Pisces?"[1]), we could use a one-proportion z-test and ask if the true proportion of executives with that sign is equal to 1/12. However, here we have 12 hypothesized proportions, one for each sign. We need a test that considers all of them together and gives an overall idea of whether the observed distribution differs from the hypothesized one.

For Example FINDING EXPECTED COUNTS

Birth month may not be related to success as a CEO, but what about on the ball field? It has been proposed by some researchers that children who are the older ones in their class at school naturally perform better in sports and that these children then get more coaching and encouragement. Could that make a difference in who makes it to the professional level in sports?

Month	Ballplayer Count	National Birth %	Month	Ballplayer Count	National Birth %
1	137	8%	7	102	9%
2	121	7%	8	165	9%
3	116	8%	9	134	9%
4	121	8%	10	115	9%
5	126	8%	11	105	8%
6	114	8%	12	122	9%
			Total	1478	100%

Baseball is a remarkable sport, in part because so much data are available, including the birth date of every player who ever played in a major league game. Since the effect we're suspecting may be due to relatively recent policies (and to keep the sample size moderate), we'll consider the birth months of 1478 major league players born since 1975. We can also look up the national demographic statistics to find what percentage of people were born in each month. Let's test whether the observed distribution of ballplayers' birth months shows just random fluctuations or whether it represents a real deviation from the national pattern.

QUESTION: How can we find the expected counts?

ANSWER: There are 1478 players in this set of data. I found the national percentage of births in each month. Based on the national birth percentages, I'd expect 8% of players to have been born in January, and 1478(0.08) = 118.24. I won't round off, because expected "counts" needn't be integers. Multiplying 1478 by each of the birth percentages gives the expected counts shown in the table in the margin.

Month	Expected	Month	Expected
1	118.24	7	133.02
2	103.46	8	133.02
3	118.24	9	133.02
4	118.24	10	133.02
5	118.24	11	118.24
6	118.24	12	133.02

---

[1] A question someone actually asked us. We suspect he's a Pisces.

Assumptions and Conditions

These data are organized in tables as we saw in Chapter 2, and the assumptions and conditions reflect that. Rather than having an observation for each individual, we typically work with summary counts in categories. In our example, we don't see the birth signs of each of the 256 executives, only the totals for each sign.

Counted Data Condition The data must be *counts* for the categories of a categorical variable. This might seem a simplistic, even silly condition. But many kinds of values can be assigned to categories, and it is unfortunately common to find the methods of this chapter applied incorrectly to proportions, percentages, or measurements just because values happen to be organized in a table. So check to be sure the values in each **cell** really are counts.

Independence Assumption The counts in the cells should be independent of each other. The easiest case is when the individuals who are counted in the cells are sampled independently from some population. That's what we'd like to have if we want to draw conclusions about that population. Randomness can arise in other ways, though. For example, these Fortune 400 executives are not a random sample of company executives, but we have no reason to think that their birth dates should not be randomly distributed throughout the year. If we want to generalize to a larger population, we should check the Randomization Condition.

We can use chi-square tests to assess patterns in a table provided the individuals counted are independent. But if we want to generalize our conclusions to a larger population, the individuals who have been counted should be representative of that population. The easiest way to be sure of this is if they are a random sample from the population of interest.

Sample Size Assumption We must have enough data for the methods to work, so we usually check the **Expected Cell Frequency Condition**. We should expect to see at least 5 individuals in each cell.

The Expected Cell Frequency Condition sounds like—and is, in fact, quite similar to—the condition that np and nq be at least 10 when we tested proportions. In our astrology example, assuming equal births in each zodiac sign leads us to expect 21.3 births per sign, so the condition is easily met here.

For Example CHECKING ASSUMPTIONS AND CONDITIONS

RECAP: Are professional baseball players more likely to be born in some months than in others? We have observed and expected counts for 1478 players born since 1975.

QUESTION: Are the assumptions and conditions met for performing a goodness-of-fit test?

ANSWER:
- ✓ **Counted Data Condition:** I have month-by-month counts of ballplayer births.
- ✓ **Independence Assumption:** These births were independent.
  Even though these players are not from a random sample, we can think of them as representative of players past and future. I'll proceed with caution.
- ✓ **Expected Cell Frequency Condition:** The expected counts range from 103.46 to 133.02, all much greater than 5.

It's okay to use these data for a goodness-of-fit test.

Calculations

We have observed a count in each category from the data, and have an expected count for each category from the hypothesized proportions. Are the differences just natural sampling variability, or are they so large that they indicate something important? It's natural to look

■ **NOTATION ALERT**

We compare the counts *observed* in each cell with the counts we *expect* to find. The usual notation uses O's and E's or abbreviations such as those we've used here. The method for finding the expected counts depends on the model.

■ **NOTATION ALERT**

The only use of the Greek letter χ (chi) in Statistics is to represent this statistic and the associated sampling distribution. This is another violation of our "rule" that Greek letters represent population parameters. Here we are using a Greek letter simply to name a family of distribution models and a statistic.

at the *differences* between these observed and expected counts, denoted $(Obs - Exp)$. We'd like to think about the total of the differences, but just adding them won't work because some differences are positive; others negative. We've been in this predicament before—once when we looked at deviations from the mean and again when we dealt with residuals. In fact, these *are* residuals. They're just the differences between the observed data and the counts given by the (null) model. We handle these residuals in essentially the same way we did in regression: We square them. That gives us positive values and focuses attention on any cells with large differences from what we expected. Because the differences between observed and expected counts generally get larger the more data we have, we also need to get an idea of the *relative* sizes of the differences. To do that, we divide each squared difference by the expected count for that cell.

The test statistic, called the **chi-square** (or chi-squared) **statistic**, is found by adding up the squares of the deviations between the observed and expected counts divided by the expected counts:

$$\chi^2 = \sum_{all\ cells} \frac{(Obs - Exp)^2}{Exp}.$$

The chi-square statistic is denoted χ^2, where χ is the Greek letter chi (pronounced "ky" as in "sky"). It refers to a family of sampling distribution models we have not seen before called (remarkably enough) the **chi-square models**.

This family of models, like the Student's t-models, differ only in the number of degrees of freedom. The number of degrees of freedom for a goodness-of-fit test is $n - 1$. Here, however, n is *not* the sample size, but instead is the number of categories. For the zodiac example, we have 12 signs, so our χ^2 statistic has 11 degrees of freedom.

Chi-Square P-Values

The chi-square statistic for tables of counts is used only for testing hypotheses, not for constructing confidence intervals. If the observed counts don't match the expected, the statistic will be large. It can't be "too small." That would just mean that our model *really* fit the data well. So the chi-square test is always one-sided. If the calculated statistic value is large enough, we'll reject the null hypothesis. What could be simpler?

If you don't have technology handy, it's easy to read the χ^2 table (Table X in Appendix D).

A portion of Table X.

Right-Tail Probability		0.10	0.05	0.025	0.01	0.005
	df					
	1	2.706	3.841	5.024	6.635	7.879
Values of χ^2_α	2	4.605	5.991	7.378	9.210	10.597
	3	6.251	7.815	9.348	11.345	12.838
	4	7.779	9.488	11.143	13.277	14.860
	5	9.236	11.070	12.833	15.086	16.750
	6	10.645	12.592	14.449	16.812	18.548
	7	12.017	14.067	16.013	18.475	20.278
	8	13.362	15.507	17.535	20.090	21.955
	9	14.684	16.919	19.023	21.666	23.589
	10	15.987	18.307	20.483	23.209	25.188
	11	17.275	19.675	21.920	24.725	26.757
	12	18.549	21.026	23.337	26.217	28.300
	13	19.812	23.362	24.736	27.688	29.819
	14	21.064	23.685	26.119	29.141	31.319

The usual selected P-values are at the top of the columns. As with the *t*-tables, we have only selected probabilities, so the best we can do is to trap a P-value between two of the values in the table. Just find the row for the correct number of degrees of freedom and read across to find where your calculated χ^2 value falls. Of course, technology can find an exact P-value, and that's usually what we'll see.

Even though its mechanics work like a one-sided test, the interpretation of a chi-square test is in some sense *many*-sided. With more than two proportions, there are many ways the null hypothesis can be wrong. By squaring the differences, we made all the deviations positive, whether our observed counts were higher or lower than expected. There's no direction to the rejection of the null model. All we know is that it doesn't fit.

For Example DOING A GOODNESS-OF-FIT TEST

RECAP: The birth months of major league baseball players are appropriate for performing a χ^2 test.

QUESTIONS: What are the hypotheses, and what does the test show?

ANSWER: H_0: The distribution of birth months for major league ballplayers is the same as that for the general population.

H_A: The distribution of birth months for major league ballplayers differs from that of the rest of the population.

$$df = 12 - 1 = 11$$

$$\chi^2 = \sum \frac{(Obs - Exp)^2}{Exp}$$

$$= \frac{(137 - 118.24)^2}{118.24} + \frac{(121 - 103.46)^2}{103.46} + \cdots$$

$$= 26.48 \text{ (by technology)}$$

$$\text{P-value} = P(\chi^2_{11} \geq 26.48) = 0.0055 \text{ (by technology)}$$

Because of the small P-value, I reject H_0; there's evidence that birth months of major league ballplayers have a different distribution from the rest of us.

Step-by-Step Example A CHI-SQUARE TEST FOR GOODNESS-OF-FIT

We have counts of 256 executives in 12 zodiac sign categories. The natural null hypothesis is that birth dates of executives are divided equally among all the zodiac signs. The test statistic looks at how closely the observed data match this idealized situation.

Question: Are zodiac signs of CEOs distributed uniformly?

THINK → Plan
State what you want to know.

Identify the variables and check the W's.

I want to know whether births of successful people are uniformly distributed across the signs of the zodiac. I have counts of 256 Fortune 400 executives, categorized by their birth sign.

Hypotheses State the null and alternative hypotheses. For χ^2 tests, it's usually easier to do that in words than in symbols.

H_0: Births are uniformly distributed over zodiac signs.[2]

H_A: Births are not uniformly distributed over zodiac signs.

Model Make a picture. The null hypothesis is that the frequencies are equal, so a bar chart (with a line at the hypothesized "equal" value) is a good display.

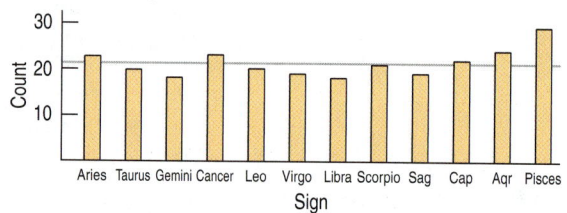

The bar chart shows some variation from sign to sign, and Pisces is the most frequent. But it is hard to tell whether the variation is more than I'd expect from random variation.

Think about the assumptions and check the conditions.

We don't need randomization to assert that birth dates are independent. But we do want our sample to be representative of executives so we can draw inferences about the population of executives. That's a reasonable assumption about this sample.

✓ **Counted Data Condition:** I have counts of the number of executives in 12 categories.

✓ **Independence Assumption:** The birth dates of executives should be independent of each other. This is a convenience sample of executives. It is random, but not a SRS. However, there's no reason to suspect bias.

✓ **Expected Cell Frequency Condition:** The null hypothesis expects that 1/12 of the 256 births, or 21.333, should occur in each sign. These expected values are all at least 5, so the condition is satisfied.

Specify the sampling distribution model.

Name the test you will use.

The conditions are satisfied, so I'll use a χ^2 model with $12 - 1 = 11$ degrees of freedom and do a **chi-square goodness-of-fit test**.

SHOW → Mechanics
Each cell contributes an $\frac{(Obs - Exp)^2}{Exp}$ value to the chi-square sum.

We add up these components for each zodiac sign. If you do it by hand, it can be helpful to arrange the calculation in a table. We show that after this Step-by-Step.

The expected value for each zodiac sign is 21.333.

$$\chi^2 = \sum \frac{(Obs - Exp)^2}{Exp} = \frac{(23 - 21.333)^2}{21.333}$$

$$+ \frac{(20 - 21.333)^2}{21.333} + \cdots$$

$$= 5.094 \text{ for all 12 signs.}$$

---

[2] It may seem that we have broken our rule of thumb that null hypotheses should specify parameter values. If you want to get formal about it, the null hypothesis is that

$$p_{Aries} = p_{Taurus} = \cdots = p_{Pisces}.$$

That is, we hypothesize that the true proportions of births of CEOs under each sign are equal. The role of the null hypothesis is to specify the model so that we can compute the test statistic. That's what this one does.

The P-value is the area in the upper tail of the χ^2 model above the computed χ^2 value. The χ^2 models are skewed to the high end, and change shape depending on the degrees of freedom. The P-value considers only the right tail. Large χ^2 statistic values correspond to small P-values, which lead us to reject the null hypothesis.	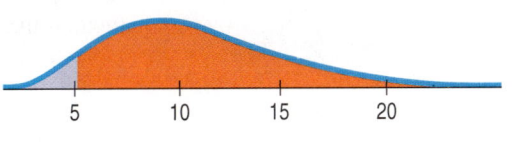 P-value = $P(\chi^2 > 5.094) = 0.926$
TELL ➔ Conclusion Link the P-value to your decision. Remember to state your conclusion in terms of what the data mean, rather than just making a statement about the distribution of counts.	The P-value of 0.926 says that if the zodiac signs of executives were in fact distributed uniformly, an observed chi-square value of 5.09 or higher would occur about 93% of the time. This certainly isn't unusual, so I fail to reject the null hypothesis, and conclude that these data show virtually no evidence of nonuniform distribution of zodiac signs among executives.

A S *Activity:* **Calculating Standardized Residuals.** The incident of the earlier ActivStats activity in which children were placed at risk, also put women at risk. Standardized residuals help us understand the relative risks.

A S *Activity:* **The Chi-Square Test.** This animation completes the calculation of the chi-square statistic and the hypothesis test based on it.

The Chi-Square Calculation

Let's make the chi-square procedure very clear. Here are the steps:

1. **Find the expected values.** These come from the null hypothesis model. Every model gives a hypothesized proportion for each cell. The expected value is the product of the total number of observations times this proportion.
   For our example, the null model hypothesizes *equal* proportions. With 12 signs, 1/12 of the 256 executives should be in each category. The expected number for each sign is 21.333.
2. **Compute the residuals.** Once you have expected values for each cell, find the residuals, *Observed* − *Expected*.
3. **Square the residuals.**
4. **Compute the components.** Now find the **component**, $\frac{(Observed - Expected)^2}{Expected}$, for each cell.
5. **Find the sum of the components.** That's the chi-square statistic.
6. **Find the degrees of freedom.** It's equal to the number of cells minus one. For the zodiac signs, that's $12 - 1 = 11$ degrees of freedom.
7. **Test the hypothesis.** Large chi-square values mean lots of deviation from the hypothesized model, so they give small P-values. Look up the critical value from a table of chi-square values, or use technology to find the P-value directly.

The steps of the chi-square calculations are often laid out in tables. If you don't have a statistics program handy, these calculations are particularly well-suited to being performed on a spreadsheet. Use one row for each category, and columns for observed counts, expected counts, residuals, squared residuals, and the contributions to the chi-square total like this:

Sign	Observed	Expected	Residual = (Obs − Exp)	(Obs − Exp)²	Component = $\frac{(Obs - Exp)^2}{Exp}$
Aries	23	21.333	1.667	2.778889	0.130262
Taurus	20	21.333	−1.333	1.776889	0.083293
Gemini	18	21.333	−3.333	11.108889	0.520737
Cancer	23	21.333	1.667	2.778889	0.130262
Leo	20	21.333	−1.333	1.776889	0.083293
Virgo	19	21.333	−2.333	5.442889	0.255139
Libra	18	21.333	−3.333	11.108889	0.520737
Scorpio	21	21.333	−0.333	0.110889	0.005198
Sagittarius	19	21.333	−2.333	5.442889	0.255139
Capricorn	22	21.333	0.667	0.444889	0.020854
Aquarius	24	21.333	2.667	7.112889	0.333422
Pisces	29	21.333	7.667	58.782889	2.755491
					$\Sigma = 5.094$

How big is big? When we calculated χ^2 for the zodiac sign example, we got 5.094. That value would have been big for z or t, leading us to reject the null hypothesis. Not here. Were you surprised that $\chi^2 = 5.094$ had a huge P-value of 0.926? What *is* big for a χ^2 statistic, anyway?

Think about how χ^2 is calculated. In every cell, any deviation from the expected count contributes to the sum. Large deviations generally contribute more, but if there are a lot of cells, even small deviations can add up, making the χ^2 value larger. So the more cells there are, the higher the value of χ^2 has to get before it becomes noteworthy. For χ^2, then, the decision about how big is big depends on the number of degrees of freedom.

Unlike the Normal and t families, χ^2 models are skewed. Curves in the χ^2 family change both shape and center as the number of degrees of freedom grows. Here, for example, are the χ^2 curves for 5 and 9 degrees of freedom.

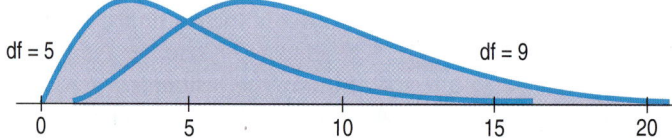

Notice that the value $\chi^2 = 10$ might seem somewhat extreme when there are 5 degrees of freedom, but appears to be rather ordinary for 9 degrees of freedom. Here are two simple facts to help you think about χ^2 models:

- The mode is at $\chi^2 = df - 2$. (Look back at the curves; their peaks are at 3 and 7, see?)
- The expected value (mean) of a χ^2 model is its number of degrees of freedom. That's a bit to the right of the mode—as we would expect for a skewed distribution.

Our test for zodiac birthdays had 11 df, so the relevant χ^2 curve peaks at 9 and has a mean of 11. Knowing that, we might have easily guessed that the calculated χ^2 value of 5.094 wasn't going to be significant.

A S *Lesson:* **The Chi-Square Family of Curves.** (Not an activity like the others, but there's no better way to see how χ^2 changes with more df.) Click on the Lesson Book's Resources tab and open the chi-square table. Watch the curve at the top as you click on a row and scroll down the degrees-of freedom column.

The Trouble with Goodness-of-Fit Tests: What's the Alternative?

Goodness-of-fit tests are likely to be performed by people who have a theory of what the proportions *should* be in each category and who believe their theory to be true. Unfortunately, the only *null* hypothesis available for a goodness-of-fit test is that the theory is true. And as we know, the hypothesis-testing procedure allows us only to *reject* the null or *fail to reject* it. We can never confirm that a theory is in fact true, which is often what people want to do.

Unfortunately, they're stuck. At best, we can point out that the data are consistent with the proposed theory. But this doesn't *prove* the theory. The data *could* be consistent with the model even if the theory were wrong. In that case, we fail to reject the null hypothesis but can't conclude anything for sure about whether the theory is true.

And we can't fix the problem by turning things around. Suppose we try to make our favored hypothesis the alternative. Then it is impossible to pick a single null. For example, suppose, as a doubter of astrology, you want to prove that the distribution of executive births is uniform. If you choose uniform as the null hypothesis, you can only *fail* to reject it. So you'd like uniformity to be your alternative hypothesis. Which particular violation of equally distributed births would you choose as your null? The problem is that the model can be wrong in many, many ways. There's no way to frame a null hypothesis the other way around. There's just no way to prove that a favored model is true.

> **WHY CAN'T WE PROVE THE NULL?**
> A biologist wanted to show that her inheritance theory about fruit flies is valid. It says that 10% of the flies should be type 1, 70% type 2, and 20% type 3. After her students collected data on 100 flies, she did a goodness-of-fit test and found a P-value of 0.07. She started celebrating, since her null hypothesis wasn't rejected—that is, until her students collected data on 100 more flies. With 200 flies, the P-value dropped to 0.02. Although she knew the answer was probably no, she asked the statistician somewhat hopefully if she could just ignore half the data and stick with the original 100. By this reasoning we could always "prove the null" just by not collecting much data. With only a little data, the chances are good that they'll be consistent with almost anything. But they also have little chance of disproving anything either. In this case, the test has no power. Don't be lured into this scientist's reasoning. With data, more is always better. But you can't ever prove that your null hypothesis is true.

24.2 Chi-Square Test of Homogeneity

Many universities survey graduating classes to determine the plans of the graduates. We might wonder whether the plans of students are the same at different colleges within a university. Here's a **two-way table** for Class of 2011 graduates from several colleges at one university. Each cell of the table shows how many students from a particular college made a certain choice.

Table 24.1
Postgraduation activities of the class of 2011 for several colleges of a large university. (Note: ILR is the School of Industrial and Labor Relations.)

	Agriculture	Arts & Sciences	Engineering	ILR	Total
Employed	209	198	177	101	685
Grad School	104	171	158	33	466
Other	135	115	39	16	305
Total	448	484	374	150	1456

Because class sizes are so different, we see differences better by examining the proportions for each class rather than the counts:

Table 24.2
Activities of graduates as a percentage of respondents from each college.

	Agriculture	Arts & Sciences	Engineering	ILR	Total
Employed	46.7%	40.9%	47.3%	67.3%	47.0%
Grad School	23.2%	35.3%	42.2%	22.0%	32.0%
Other	30.1%	23.8%	10.4%	10.7%	20.9%
Total	100.0%	100.0%	100.0%	100.0%	100.0%

Who	Graduates from 4 colleges at an upstate New York university
What	Postgraduation activities
When	2011
Why	Survey for general information

Video: The Incident. You may have guessed which famous incident put women and children at risk. Here you can view the story complete with rare film footage.

We already know how to test whether *two* proportions are the same. For example, we could use a two-proportion z-test to see whether the proportion of students choosing graduate school is the same for Agriculture students as for Engineering students. But now we have more than two groups. We want to test whether the students' choices are the same across all four colleges. The z-test for two proportions generalizes to a **chi-square test of homogeneity**.

Chi-square again? It turns out that the mechanics of this test are *identical* to the chi-square test for goodness-of-fit that we just saw. (How similar can you get?) Why a different name, then? The tests are really quite different. The goodness-of-fit test compared counts with a theoretical model. But here we're asking whether the distribution of choices is the same among different groups, so we find the expected counts for each category directly from the data. As a result, we count the degrees of freedom slightly differently as well.

The term "homogeneity" means that things are the same. Here, we ask whether the post-graduation choices made by students are the *same* for these four colleges. The homogeneity test comes with a built-in null hypothesis: We hypothesize that the distribution does not change from group to group. The test looks for differences large enough to step beyond what we might expect from random sample-to-sample variation. It can reveal a large deviation in a single category or small, but persistent, differences over all the categories—or anything in between.

Assumptions and Conditions

The assumptions and conditions are the same as for the chi-square test for goodness-of-fit. The **Counted Data Condition** says that these data must be counts. You can't do a test of homogeneity on proportions, so you have to work with the counts of graduates given in the first table. Also, you can't do a chi-square test on measurements. For example, if we had recorded GPAs for these same groups, we wouldn't be able to determine whether the mean GPAs were different using this test.[3]

Ideally, when we compare the proportions across several groups, we would like the cases within each group to be selected randomly. We need to know whether the **Independence Assumption** both within and across groups is reasonable. As usual, randomization is a good condition to check to make the assumption plausible.

We still must be sure we have enough data for this method to work. The **Expected Cell Frequency Condition** says that the expected count in each cell must be at least 5. We'll confirm that as we do the calculations.

Calculations

The null hypothesis says that the distribution of the proportions of graduates choosing each alternative is the same for all four colleges, so we can estimate those overall proportions by pooling our data from the four colleges together. Within each college, the expected proportion for each choice is just the overall proportion of all students making that choice. The expected counts are those proportions applied to the number of students in each graduating class.

For example, overall, 685, or about 47.0%, of the 1456 students who responded to the survey were employed. If the distributions are homogeneous (as the null hypothesis asserts), then 47% of the 448 Agriculture school graduates (or about 210.76 students) should be employed. Similarly, 47% of the 374 Engineering grads (or about 175.95) should be employed.

---

[3] To do that, you'd use a method called Analysis of Variance (Chapter 26).

Working in this way, we (or, more likely, the computer) can fill in expected values for each cell. Because these are theoretical values, they don't have to be integers. The expected values look like this:

Table 24.3
Expected values for the '11 graduates.

	Agriculture	Arts & Sciences	Engineering	ILR	Total
Employed	210.769	227.706	175.955	70.570	685
Grad School	143.385	154.907	119.701	48.008	466
Other	93.846	101.387	78.345	31.422	305
Total	448	484	374	150	1456

Now check the **Expected Cell Frequency Condition**. Indeed, there are at least 5 individuals expected in each cell.

Following the pattern of the goodness-of-fit test, we compute the component for each cell of the table. For the highlighted cell, employed students graduating from the Ag school, that's

$$\frac{(Obs - Exp)^2}{Exp} = \frac{(209 - 210.769)^2}{210.769} = 0.0148.$$

Summing these components across all cells gives

$$\chi^2 = \sum_{all\ cells} \frac{(Obs - Exp)^2}{Exp} = 93.66.$$

How about the degrees of freedom? We don't really need to calculate all the expected values in the table. We know there is a total of 685 employed students, so once we find the expected values for three of the colleges, we can determine the expected number for the fourth by just subtracting. Similarly, we know how many students graduated from each college, so after filling in two rows, we can find the expected values for the remaining row by subtracting. To fill out the table, we need to know the counts in only $R - 1$ rows and $C - 1$ columns. So the table has $(R - 1)(C - 1)$ degrees of freedom.

In our example, we need to calculate only 2 choices in each column and counts for 3 of the 4 colleges, for a total of $2 \times 3 = 6$ degrees of freedom. We'll need the degrees of freedom to find a P-value for the chi-square statistic.

■ **NOTATION ALERT**
For a contingency table, R represents the number of rows and C the number of columns.

Step-by-Step Example A CHI-SQUARE TEST FOR HOMOGENEITY

We have reports from four colleges on the post-graduation activities of their 2011 graduating classes.

Question: Are the distributions of students' choices of post-graduation activities the same across all the colleges?

THINK ➡ Plan State what you want to know.
Identify the variables and check the W's.

I want to test whether post-graduation choices are the same for students from each of four colleges. I have a table of counts classifying each college's Class of 2011 respondents according to their activities.

Hypotheses State the null and alternative hypotheses.

H_0: Students' post-graduation activities are distributed in the same way for all four colleges.

H_A: Students' plans *do not* have the same distribution.

Model Make a picture: A side-by-side bar chart shows the four distributions of post-graduation activities. Plot column percents to remove the effect of class size differences. A split bar chart would also be an appropriate choice.

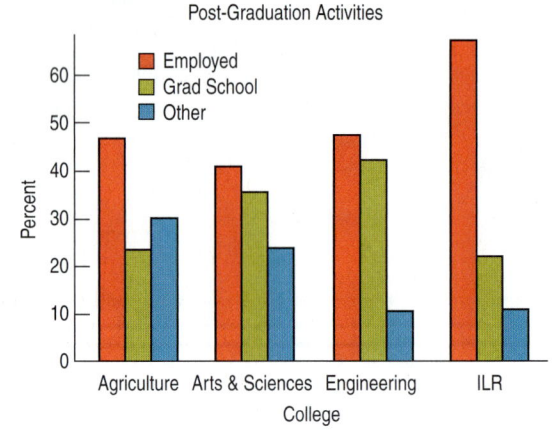

A side-by-side bar chart shows how the distributions of choices differ across the four colleges.

Think about the assumptions and check the conditions.

✓ **Counted Data Condition:** I have counts of the number of students in categories.

✓ **Independence Assumption:** Even though this isn't a random sample, student plans should be largely independent of each other. The occasional friends who decide to join Teach for America together or couples who make grad school decisions together are too rare to affect this analysis.

✓ **Expected Cell Frequency Condition:** The expected values (shown below) are all at least 5.

State the sampling distribution model and name the test you will use.

The conditions seem to be met, so I can use a χ^2 model with $(3 - 1) \times (4 - 1) = 6$ degrees of freedom and do a **chi-square test of homogeneity**.

SHOW ➡ Mechanics

Show the expected counts for each cell of the data table. You could make separate tables for the observed and expected counts, or put both counts in each cell as shown here. While observed counts must be whole numbers, expected counts rarely are—don't be tempted to round those off.

	Ag	A&S	Eng	ILR
Empl.	209 210.769	198 227.706	177 175.955	101 70.570
Grad School	104 143.385	171 154.907	158 119.701	33 48.008
Other	135 93.846	115 101.387	39 78.345	16 31.422

Calculate χ^2.	$\chi^2 = \sum_{\text{all cells}} \dfrac{(Obs - Exp)^2}{Exp}$ $= \dfrac{(209 - 210.769)^2}{210.769} + \ldots$ $= 93.66$
The shape of a χ^2 model depends on the degrees of freedom. A χ^2 model with 6 df is skewed to the high end. The P-value considers only the right tail. Here, the calculated value of the χ^2 statistic is off the scale, so the P-value is quite small.	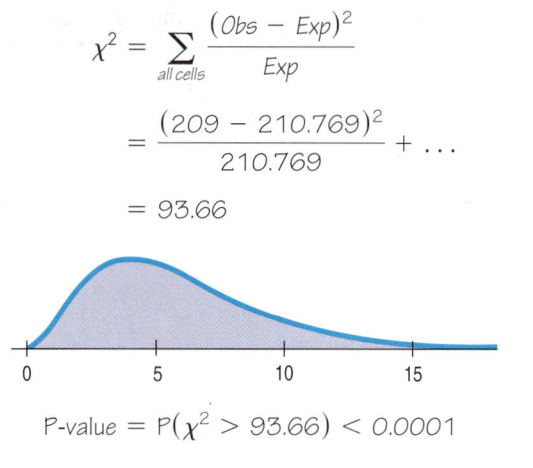 P-value = $P(\chi^2 > 93.66) < 0.0001$
TELL ▶ Conclusion State your conclusion in the context of the data. You should specifically talk about whether the distributions for the groups appear to be different.	The P-value is very small, so I reject the null hypothesis and conclude that there's evidence that the post-graduation activities of students from these four colleges don't have the same distribution.

If you find that simply rejecting the hypothesis of homogeneity is a bit unsatisfying, you're in good company. OK, so the post-graduation plans are different. What we'd really like to know is what the differences are, where they're the greatest, and where they're smallest. The test for homogeneity doesn't answer these interesting questions, but it does provide some evidence that can help us.

24.3 Examining the Residuals

Whenever we reject the null hypothesis, it's a good idea to examine residuals. (We don't need to do that when we fail to reject because when the χ^2 value is small, all of its components must have been small.) For chi-square tests, we want to compare residuals for cells that may have very different counts. So we're better off standardizing the residuals. We know the mean residual is zero,[4] but we need to know each residual's standard deviation. When we tested proportions, we saw a link between the expected proportion and its standard deviation. For counts, there's a similar link. To standardize a cell's residual, we just divide by the square root of its expected value:

$$c = \dfrac{(Obs - Exp)}{\sqrt{Exp}}.$$

Notice that these **standardized residuals** are just the square roots of the **components** we calculated for each cell, and their sign indicates whether we observed more cases than we expected, or fewer.

The standardized residuals give us a chance to think about the underlying patterns and to consider the ways in which the distribution of post-graduation plans may differ from college to college. Now that we've subtracted the mean (the residual, $Obs - Exp$, has mean 0,

---

[4]Residual = Observed − Expected. Because the total of the expected values is set to be the same as the observed total, the residuals must sum to zero.

so it's already subtracted) and divided by their standard deviations, these are z-scores. If the null hypothesis were true, we could even appeal to the Central Limit Theorem, think of the Normal model, and use the 68–95–99.7 Rule to judge how extraordinary the large ones are.

Here are the standardized residuals for the Class of '11 data:

	Ag	A&S	Eng	ILR
Employed	−0.121866	−1.96860	0.078805	3.62235
Grad School	−3.28909	1.29304	3.50062	−2.16607
Other	4.24817	1.35192	−4.44511	−2.75117

Table 24.4
Standardized residuals can help show how the table differs from the null hypothesis pattern.

The column for Engineering students immediately attracts our attention. It holds both the largest positive and the largest negative standardized residuals. It looks like Engineering college graduates are more likely to go on to graduate work and very unlikely to take time off for "volunteering and travel, among other activities" (as the "Other" category is explained). By contrast, Ag school graduates seem to be readily employed and less likely to pursue graduate work immediately after college.

For Example LOOKING AT χ^2 RESIDUALS

RECAP: Some people suggest that schoolchildren who are the older ones in their class naturally perform better in sports and therefore get more coaching and encouragement. To see if there's any evidence for this, we looked at major league baseball players born since 1975. A goodness-of-fit test found their birth months to have a distribution that's significantly different from the rest of us. The table shows the standardized residuals.

Month	Residual	Month	Residual
1	1.73	7	−2.69
2	1.72	8	2.77
3	−0.21	9	0.08
4	0.25	10	−1.56
5	0.71	11	−1.22
6	−0.39	12	−0.96

QUESTION: What's different about the distribution of birth months among major league ballplayers?

ANSWER: It appears that, compared to the general population, fewer ballplayers than expected were born in July and more than expected in August. Either month would make them the younger kids in their grades in school, so these data don't offer support for the conjecture that being older is an advantage in terms of a career as a professional baseball player.

Just Checking

Tiny black potato flea beetles can damage potato plants in a vegetable garden. These pests chew holes in the leaves, causing the plants to wither or die. They can be killed with an insecticide, but a canola oil spray has been suggested as a non-chemical "natural" method of controlling the beetles. To conduct an experiment to test the effectiveness of the natural spray, we gather 500 beetles and place them in three Plexiglas® containers. Two hundred beetles go in the first container, where we spray them with the canola oil mixture. Another 200 beetles go in the second container; we spray them with the insecticide. The remaining 100 beetles in the last container serve as a control group; we simply spray them with water. Then we wait 6 hours and count the number of surviving beetles in each container.

1. Why do we need the control group?
2. What would our null hypothesis be?
3. After the experiment is over, we could summarize the results in a table as shown. How many degrees of freedom does our χ^2 test have?

	Natural Spray	Insecticide	Water	Total
Survived				
Died				
Total	200	200	100	500

4. Suppose that, all together, 125 beetles survived. (That's the first-row total.) What's the expected count in the first cell—survivors among those sprayed with the natural spray?
5. If it turns out that only 40 of the beetles in the first container survived, what's the calculated component of χ^2 for that cell?
6. If the total calculated value of χ^2 for this table turns out to be around 10, would you expect the P-value of our test to be large or small? Explain.

24.4 Chi-Square Test of Independence

A study from the University of Texas Southwestern Medical Center examined 626 people being treated for non–blood-related diseases to see whether the risk of hepatitis C was related to whether people had tattoos and to where they got their tattoos. Hepatitis C causes about 10,000 deaths each year in the United States but often goes undetected for years after infection.

The data from this study can be summarized in a two-way table, as follows:

Table 24.5
Counts of patients classified by their hepatitis C test status according to whether they had a tattoo from a tattoo parlor or from another source, or had no tattoo.

	Hepatitis C	No Hepatitis C	Total
Tattoo, Parlor	17	35	52
Tattoo, Elsewhere	8	53	61
None	22	491	513
Total	47	579	626

Who Patients being treated for non–blood-related disorders
What Tattoo status and hepatitis C status
When 1991, 1992
Where Texas

These data differ from the kinds of data we've considered before in this chapter because they categorize subjects from a single group on two categorical variables rather than on only one. The categorical variables here are *Hepatitis C Status* ("Hepatitis C" or "No Hepatitis C") and *Tattoo Status* ("Parlor," "Elsewhere," "None"). We've seen counts classified by two categorical variables displayed like this in Chapter 2, so we know such tables are called contingency tables. **Contingency tables** categorize counts on two (or more) variables so that we can see whether the distribution of counts on one variable is contingent on the other.

> **Activity: Independence and Chi-Square.** This unusual simulation shows how independence arises (and fails) in contingency tables.

The natural question to ask of these data is whether the chance of having hepatitis C is *independent* of tattoo status. Recall that for events **A** and **B** to be independent $P(\mathbf{A})$ must equal $P(\mathbf{A}|\mathbf{B})$. Here, this means the probability that a randomly selected patient has hepatitis C should be the same regardless of the patient's tattoo status. If *Hepatitis Status* is independent of tattoos, we'd expect the proportion of people testing positive for hepatitis to be the same for the three levels of *Tattoo Status*. Of course, for real data we won't expect them to be exactly the same, so we look to see how close they are. This sounds a lot like the test of homogeneity. In fact, the mechanics of the calculation are identical. The difference is that now we have two categorical variables measured on a single population. For the homogeneity test, we had a single categorical variable measured independently on two or more populations. But now we ask a different question: "Are the variables independent?" rather than "Are the groups homogeneous?" These are subtle differences, but they are important when we state hypotheses and draw conclusions. ==When we ask whether two variables measured on the same population are independent we're performing a **chi-square test of independence**.==

For Example   WHICH χ^2 TEST?

Many states and localities now collect data on traffic stops regarding the race of the driver. The initial concern was that Black drivers were being stopped more often (the "crime" ironically called "Driving While Black"). With more data in hand, attention has turned to other issues. For example, data from 2533 traffic stops in Cincinnati[5] report the race of the driver (Black, White, or Other) and whether the traffic stop resulted in a search of the vehicle.

		Race			
		Black	White	Other	Total
Search	No	787	594	27	1408
	Yes	813	293	19	1125
	Total	1600	887	46	2533

QUESTION: Which test would be appropriate to examine whether race is a factor in vehicle searches? What are the hypotheses?

ANSWER: These data represent one group of traffic stops in Cincinnati, categorized on two variables, Race and Search. I'll do a chi-square test of independence.

H_0: Whether or not police search a vehicle is independent of the race of the driver.

H_A: Decisions to search vehicles are not independent of the driver's race.

Assumptions and Conditions

> **Activity: Chi-Square Tables.** Work with *ActivStats*' interactive chi-square table to perform a hypothesis test.

Of course, we still need counts and enough data so that the expected values are at least 5 in each cell.

If we're interested in the independence of variables, we usually want to generalize from the data to some population. In that case, we'll need to check that the data are a representative random sample from, and fewer than 10% of, that population.

---

[5] John E. Eck, Lin Liu, and Lisa Growette Bostaph, Police Vehicle Stops in Cincinnati, Oct. 1, 2003, available at www.cincinnati-oh.gov. Data for other localities can be found by searching from www.racialprofilinganalysis.neu.edu.

Step-by-Step Example   A CHI-SQUARE TEST FOR INDEPENDENCE

We have counts of 626 individuals categorized according to their "tattoo status" and their "hepatitis status."

Question: Are tattoo status and hepatitis status independent?

THINK ➤ Plan State what you want to know.

Identify the variables and check the W's.

Hypotheses State the null and alternative hypotheses.

We perform a test of independence when we suspect the variables may not be independent. We are on the familiar ground of making a claim (in this case, that knowing *Tattoo Status* will change probabilities for *Hepatitis C Status*) and testing the null hypothesis that it is *not* true.

Model Make a picture. Because these are only two categories—Hepatitis C and No Hepatitis C—a simple bar chart of the distribution of tattoo sources for Hep C patients shows all the information.

Think about the assumptions and check the conditions.

I want to test whether the categorical variables Tattoo Status and Hepatitis Status are statistically independent. I have a contingency table of 626 Texas patients under treatment for a non–blood-related disease.

H_O: Tattoo Status and Hepatitis Status are independent.[6]

H_A: Tattoo Status and Hepatitis Status are not independent.

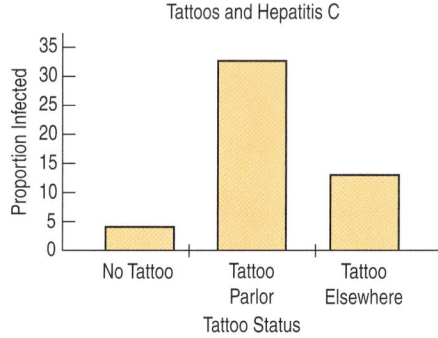

The bar chart suggests strong differences in Hepatitis C risk based on tattoo status.

✓ **Counted Data Condition:** I have counts of individuals categorized on two variables.

✓ **Independence Assumption:** These data are from a retrospective study of patients being treated for something unrelated to hepatitis. Although they are not an SRS, they are likely to be independent of each other.

---

[6] Once again, parameters are hard to express. The hypothesis of independence itself tells us how to find expected values for each cell of the contingency table. That's all we need.

This table shows both the observed and expected counts for each cell. The expected counts are calculated exactly as they were for a test of homogeneity; in the first cell, for example, we expect $\frac{52}{626}$ (that's 8.3%) of 47.

Warning: Be wary of proceeding when there are small expected counts. If we see expected counts that fall far short of 5, or if many cells violate the condition, we should not use χ^2. (We will soon discuss ways you can fix the problem.) If you do continue, always check the residuals to be sure those cells did not have a major influence on your result.

Specify the model.

Name the test you will use.

✗ **Expected Cell Frequency Condition:** The expected values do not meet the condition that all are at least 5.

	Hepatitis C	No Hepatitis C	Total
Tattoo, Parlor	17 3.904	35 48.096	52
Tattoo, Elsewhere	8 4.580	53 56.420	61
None	22 38.516	491 474.484	513
Total	47	579	626

Although the Expected Cell Frequency Condition is not satisfied, the values are close to 5. I'll go ahead, but I'll check the residuals carefully. I'll use a χ^2 model with $(3 - 1) \times (2 - 1) = 2$ df and do a **chi-square test of independence**.

SHOW → Mechanics Calculate χ^2.

The shape of a chi-square model depends on its degrees of freedom. With 2 df, the model looks quite different, as you can see here. We still care only about the right tail.

$$\chi^2 = \sum_{all\ cells} \frac{(Obs - Exp)^2}{Exp}$$

$$= \frac{(17 - 3.094)^2}{3.094} + \cdots = 57.91$$

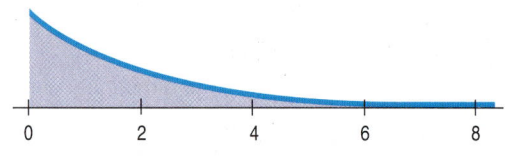

P-Value = $P(\chi^2 > 57.91) < 0.0001$

TELL → Conclusion Link the P-value to your decision. State your conclusion about the independence of the two variables.

(We should be wary of this conclusion because of the small expected counts. A complete solution must include the additional analysis, recalculation, and final conclusion discussed in the following section.)

The P-value is very small, so I reject the null hypothesis and conclude that *Hepatitis Status* is not independent of *Tattoo Status*. Because the Expected Cell Frequency Condition was violated, I need to check that the two cells with small expected counts did not influence this result too greatly.

For Example CHI-SQUARE MECHANICS

RECAP: We have data that allow us to investigate whether police searches of vehicles they stop are independent of the driver's race.

QUESTIONS: What are the degrees of freedom for this test? What is the expected frequency of searches for the Black drivers who were stopped? What's that cell's component in the χ^2 computation? And how is the standardized residual for that cell computed?

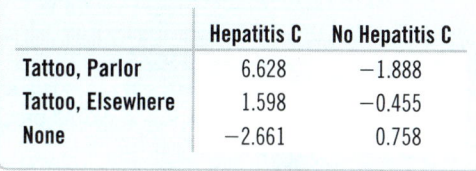

ANSWER: This is a 2 × 3 contingency table, so df = (2 − 1)(3 − 1) = 2. Overall, 1125 of 2533 vehicles were searched. If searches are conducted independent of race, then I'd expect $\frac{1125}{2533}$ of the 1600 Black drivers to have been searched: $\frac{1125}{2533} \times 1600 \approx 710.62$.

That cell's term in the χ^2 calculation is $\frac{(Obs - Exp)^2}{Exp} = \frac{(813 - 710.62)^2}{710.62} = 14.75$.

The standardized residual for that cell is $\frac{Obs - Exp}{\sqrt{Exp}} = \frac{813 - 710.62}{\sqrt{710.62}} = 3.84$.

Examine the Residuals

Each cell of the contingency table contributes a term to the chi-square sum. As we did earlier, we should examine the residuals because we have rejected the null hypothesis. In this instance, we have an additional concern that the cells with small expected frequencies not be the ones that make the chi-square statistic large.

Our interest in the data arises from the potential for improving public health. If patients with tattoos are more likely to test positive for hepatitis C, perhaps physicians should be advised to suggest blood tests for such patients.

The standardized residuals look like this:

	Hepatitis C	No Hepatitis C
Tattoo, Parlor	6.628	−1.888
Tattoo, Elsewhere	1.598	−0.455
None	−2.661	0.758

Table 24.6
Standardized residuals for the hepatitis and tattoos data. Are any of them particularly large in magnitude?

The chi-square value of 57.91 is the sum of the squares of these six values. The cell for people with tattoos obtained in a tattoo parlor who have hepatitis C is large and positive, indicating there are more people in that cell than the null hypothesis of independence would predict. Maybe tattoo parlors are a source of infection or maybe those who go to tattoo parlors also engage in risky behavior.

The second-largest component is a negative value for those with no tattoos who test positive for hepatitis C. A negative value says that there are fewer people in this cell than independence would expect. That is, those who have no tattoos are less likely to be infected with hepatitis C than we might expect if the two variables were independent.

What about the cells with small expected counts? The formula for the chi-square standardized residuals divides each residual by the square root of the expected frequency. Too small an expected frequency can arbitrarily inflate the residual and lead to an inflated chi-square statistic. Any expected count close to the arbitrary minimum of 5 calls for checking that cell's standardized residual to be sure it is not particularly large. In this case, the standardized residual for the "Hepatitis C and Tattoo, Elsewhere" cell is not particularly large, but the standardized residual for the "Hepatitis C and Tattoo, Parlor" cell is large.

We might choose not to report the results because of concern with the small expected frequency. Alternatively, we could include a warning along with our report of the results. Yet another approach is to combine categories to get a larger category total and correspondingly larger expected frequencies, if there are some categories that can be appropriately combined. Here, we might naturally combine the two rows for tattoos, obtaining a 2 × 2 table:

Table 24.7
Combining the two tattoo categories gives a table with all expected counts greater than 5.

	Hepatitis C	No Hepatitis C	Total
Tattoo	25	88	113
None	22	491	513
Total	47	579	626

This table has expected values of at least 5 in every cell, and a chi-square value of 42.42 on 1 degree of freedom. The corresponding P-value is <0.0001.

We conclude that *Tattoo Status* and *Hepatitis C Status* are not independent. The data *suggest* that tattoo parlors may be a particular problem, but we haven't enough data to draw that conclusion.

For Example WRITING CONCLUSIONS FOR χ^2 TESTS

RECAP: We're looking at Cincinnati traffic stop data to see if police decisions about searching cars show evidence of racial bias. With 2 df, technology calculates $\chi^2 = 73.25$, a P-value less than 0.0001, and these standardized residuals:

		Race		
		Black	White	Other
Search	No	−3.43	4.55	0.28
	Yes	3.84	−5.09	−0.31

QUESTION: What's your conclusion?

ANSWER: The very low P-value leads me to reject the null hypothesis. There's strong evidence that police decisions to search cars at traffic stops are associated with the driver's race.

The largest residuals are for White drivers, who are searched less often than independence would predict. It appears that Black drivers' cars are searched more often.

Chi-Square and Causation

Chi-square tests are common. Tests for independence are especially widespread. Unfortunately, many people interpret a small P-value as proof of causation. We know better. Just as correlation between quantitative variables does not demonstrate causation, a failure of independence between two categorical variables does not show a cause-and-effect relationship between them, nor should we say that one variable *depends* on the other.

The chi-square test for independence treats the two variables symmetrically. There is no way to differentiate the direction of any possible causation from one variable to the other. In our example, it is unlikely that having hepatitis causes one to crave a tattoo, but other examples are not so clear.

In this case, it's easy to imagine that lurking variables are responsible for the observed lack of independence. Perhaps the lifestyles of some people include both tattoos and behaviors that put them at increased risk of hepatitis C, such as body piercings or even drug use. Even a small subpopulation of people with such a lifestyle among those with tattoos might be enough to create the observed result. After all, we observed only 25 patients with both tattoos and hepatitis.

In some sense, a failure of independence between two categorical variables is less impressive than a strong, consistent, linear association between quantitative variables. Two categorical variables can fail the test of independence in many ways, including ways that show no consistent pattern of failure. Examination of the chi-square standardized residuals can help you think about the underlying patterns.

Just Checking

Which of the three chi-square tests—goodness-of-fit, homogeneity, or independence—would you use in each of the following situations?

7. A restaurant manager wonders whether customers who dine on Friday nights have the same preferences among the four "chef's special" entrées as those who dine on Saturday nights. One weekend he has the wait staff record which entrées were ordered each night. Assuming these customers to be typical of all weekend diners, he'll compare the distributions of meals chosen Friday and Saturday.

8. Company policy calls for parking spaces to be assigned to everyone at random, but you suspect that may not be so. There are three lots of equal size: lot A, next to the building; lot B, a bit farther away; and lot C, on the other side of the highway. You gather data about employees at middle management level and above to see how many were assigned parking in each lot.

9. Is a student's social life affected by where the student lives? A campus survey asked a random sample of students whether they lived in a dormitory, in off-campus housing, or at home, and whether they had been out on a date 0, 1–2, 3–4, or 5 or more times in the past two weeks.

WHAT CAN GO WRONG?

- **Don't use chi-square methods unless you have counts.** All three of the chi-square tests apply only to counts. Other kinds of data can be arrayed in two-way tables. Just because numbers are in a two-way table doesn't make them suitable for chi-square analysis. Data reported as proportions or percentages can be suitable for chi-square procedures, *but only after they are converted to counts*. If you try to do the calculations without first finding the counts, your results will be wrong.

A S *Simulation:* **Sample Size and Chi-Square.** Chi-square statistics have a peculiar problem. They don't respond to increasing the sample size in quite the same way you might expect.

- **Beware large samples.** Beware *large* samples?! That's not the advice you're used to hearing. The chi-square tests, however, are unusual. Be wary of chi-square tests performed on very large samples. No hypothesized distribution fits perfectly, no two groups are exactly homogeneous, and two variables are rarely perfectly independent. The degrees of freedom for chi-square tests don't grow with the sample size. With a sufficiently large sample size, a chi-square test can always reject the null hypothesis. But we have no measure of how far the data are from the null model. There are no confidence intervals to help us judge the effect size.

- **Don't say that one variable "depends" on the other just because they're not independent.** Dependence suggests a pattern and implies causation, but variables can fail to be independent in many different ways. When variables fail the test for independence, you might just say they are "associated."

CONNECTIONS

Chi-square methods relate naturally to inference methods for proportions. We can think of a test of homogeneity as stepping from a comparison of two proportions to a question of whether three or more proportions are equal. The standard deviations of the residuals in each cell are linked to the expected counts much like the standard deviations we found for proportions.

Independence is, of course, a fundamental concept in Statistics. But chi-square tests do not offer a general way to check on independence for all those times when we have had to assume it.

Stacked bar charts or side-by-side pie charts can help us think about patterns in two-way tables. A histogram or boxplot of the standardized residuals can help locate extraordinary values.

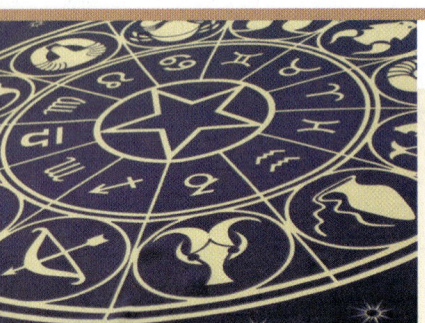

What Have We Learned?

Learning Objectives

Recognize when a chi-square test of goodness of fit, homogeneity, or independence is appropriate.

For each test, find the expected cell frequencies.

For each test, check the assumptions and corresponding conditions and know how to complete the test.

- Counted data condition.
- Independence assumption; randomization makes independence more plausible.
- Sample size assumption with the expected cell frequency condition; expect at least 5 observations in each cell.

Interpret a chi-square test.

- Even though we might believe the model, we cannot prove that the data fit the model with a chi-square test because that would mean confirming the null hypothesis.

Examine the standardized residuals to understand what cells were responsible for rejecting a null hypothesis.

State the null hypothesis for a test of independence and understand how that is different from the null hypothesis for a test of homogeneity.

- Both are computed the same way. You may not find both offered by your technology. You can use either one as long as you interpret your result correctly.

Review of Terms

Chi-square goodness-of-fit test A test of whether the distribution of counts in one categorical variable matches the distribution predicted by a model is called a test of goodness-of-fit. In a chi-square goodness-of-fit test, the expected counts come from the predicting model. The test finds a P-value from a chi-square model with $n - 1$ degrees of freedom, where n is the number of categories in the categorical variable (p. 656).

Cell A cell is one element of a table corresponding to a specific row and a specific column. Table cells can hold counts, percentages, or measurements on other variables. Or they can hold several values (p. 657).

Chi-square statistic The chi-square statistic can be used to test whether the observed counts in a frequency distribution or contingency table match the counts we would expect according to some model. It is calculated as

$$\chi^2 = \sum_{\text{all cells}} \frac{(Obs - Exp)^2}{Exp}.$$

Chi-square statistics differ in how expected counts are found, depending on the question asked (p. 658).

CHAPTER 24 Comparing Counts

Chi-square model Chi-square models are skewed to the right. They are parameterized by their degrees of freedom and become less skewed with increasing degrees of freedom (p. 658).

Chi-square component The components of a chi-square calculation are

$$\frac{(Observed - Expected)^2}{Expected},$$

found for each cell of the table (p. 661, 667).

Two-way table Each *cell* of a two-way table shows counts of individuals. One way classifies a sample according to a categorical variable. The other way can classify different groups of individuals according to the same variable or classify the same individuals according to a different categorical variable (p. 663).

Chi-square test of homogeneity A test comparing the distribution of counts for *two or more groups* on the same categorical variable. A chi-square test of homogeneity finds expected counts based on the overall frequencies, adjusted for the totals in each group under the (null hypothesis) assumption that the distributions are the same for each group. We find a P-value from a chi-square distribution with (#*Rows* − 1) × (#*Cols* − 1) degrees of freedom, where #*Rows* gives the number of categories and #*Cols* gives the number of independent groups (p. 664).

Standardized residual In each cell of a two-way table, a standardized residual is the square root of the chi-square component for that cell with the sign of the *Observed* − *Expected* difference:

$$\frac{(Obs - Exp)}{\sqrt{Exp}}.$$

When we reject a chi-square test, an examination of the standardized residuals can sometimes reveal more about how the data deviate from the null model (p. 667).

Contingency table A two-way table that classifies individuals according to two categorical variables (p. 669).

Chi-square test of independence A test of whether two categorical variables are independent examines the distribution of counts for *one group of individuals* classified according to both variables. A chi-square test of *independence* finds expected counts by assuming that knowing the marginal totals tells us the cell frequencies, assuming that there is no association between the variables. This turns out to be the same calculation as a test of homogeneity. We find a P-value from a chi-square distribution with (#*Rows* − 1) × (#*Cols* − 1) degrees of freedom, where #*Rows* gives the number of categories in one variable and #*Cols* gives the number of categories in the other (p. 670).

On the Computer CHI-SQUARE

Most statistics packages associate chi-square tests with contingency tables. Often chi-square is available as an option only when you make a contingency table. This organization can make it hard to locate the chi-square test and may confuse the three different roles that the chi-square test can take. In particular, chi-square tests for goodness-of-fit may be hard to find or missing entirely. Chi-square tests for homogeneity are computationally the same as chi-square tests for independence, so you may have to perform the mechanics as if they were tests of independence and interpret them afterwards as tests of homogeneity.

Most statistics packages work with data on individuals rather than with the summary counts. If the only information you have is the table of counts, you may find it more difficult to get a statistics package to compute chi-square. Some packages offer a way to reconstruct the data from the summary counts so that they can then be passed back through the chi-square calculation, finding the cell counts again. Many packages offer chi-square standardized residuals (although they may be called something else).

DATA DESK

- Select variables.
- From the **Calc** menu, choose **Contingency Table**.
- From the table's HyperView menu choose **Table Options**. (Or Choose **Calc > Calculation Options > Table Options**.)
- In the dialog, check the boxes for **Chi Square** and for **Standardized Residuals**. Data Desk will display the chi-square and its P-value below the table, and the standardized residuals within the table.

COMMENTS

Data Desk automatically treats variables selected for this command as categorical variables even if their elements are numerals. The **Compute Counts** command in the table's HyperView menu will make variables that hold the table contents (as selected in the Table Options dialog), including the standardized residuals.

EXCEL

Excel offers the function
CHITEST(actual_range, expected_range), which computes a chi-square value for homogeneity. Both ranges are of the form UpperLeftcell:LowerRightCell, specifying two rectangular tables that must hold counts (although Excel will not check for integer values). The two tables must be of the same size and shape.

COMMENTS

Excel's documentation claims this is a test for independence and labels the input ranges accordingly, but Excel offers no way to find expected counts, so the function is not particularly useful for testing independence. You can use this function only if you already know both tables of counts or are willing to program additional calculations.

JMP

- From the **Analyze** menu, select **Fit Y by X**.
- Choose one variable as the Y, response variable, and the other as the X, factor variable. Both selected variables must be Nominal or Ordinal.
- **JMP** will make a plot and a contingency table. Below the contingency table, **JMP** offers a **Tests** panel. In that panel, the Chi Square for independence is called a **Pearson ChiSquare**. The table also offers the P-value.
- Click on the **Contingency Table** title bar to drop down a menu that offers to include a **Deviation** and **Cell Chi square** in each cell of the table.

COMMENTS

JMP will choose a chi-square analysis for a **Fit Y by X** if both variables are nominal or ordinal (marked with an N or O), but not otherwise. Be sure the variables have the right type.

Deviations are the observed−expected differences in counts. Cell chi-squares are the squares of the standardized residuals. Refer to the deviations for the sign of the difference.

Look under **Distributions** in the **Analyze** menu to find a chi-square test for goodness-of-fit.

MINITAB

- From the **Stat** menu, choose the **Tables** submenu.
- From that menu, choose **Chi Square Test . . . .**
- In the dialog, identify the columns that make up the table. Minitab will display the table and print the chi-square value and its P-value.

COMMENTS

Alternatively, select the **Cross Tabulation . . .** command to see more options for the table, including expected counts and standardized residuals.

R

Goodness-of-fit test:
To test the probabilities in a vector prob with observed values in a vector x,
- **chisq.test**(x,p=prob)

Test of Independence or homogeneity:
With counts in a contingency table (a matrix called say, con.table), the test of independence (or homogeneity) is found by
- **chisq.test**(con.table)

COMMENTS

Using the function xtabs you can create a contingency table from two variables x and y in a data frame called mydata by
- con.table = xtabs(~x+y,data=mydata)
  then
- **chisq.test**(con.table)

SPSS

- From the **Analyze** menu, choose the **Descriptive Statistics** submenu.
- From that submenu, choose **Crosstabs**. . . .
- In the **Crosstabs** dialog, assign the row and column variables from the variable list. Both variables must be categorical.
- Click the **Cells** button to specify that standardized residuals should be displayed.
- Click the **Statistics** button to specify a chi-square test.

COMMENTS

SPSS offers only variables that it knows to be categorical in the variable list for the Crosstabs dialog. If the variables you want are missing, check that they have the right type.

STATCRUNCH

To perform a Goodness-of-Fit test:
- Enter the observed counts in one column of a data table, and the expected counts in another.
- Click on **Stat**.
- Choose **Goodness-of-fit » Chi-Square test**.
- Choose the **Observed Column** and the **Expected Column**.
- Click on **Calculate**.

COMMENTS

These Chi-square tests may also be performed using the actual data table instead of summary counts. See the StatCrunch **Help page** for details.

To perform a Test of Homogeneity or Independence:
- Create a table (without totals):
- Name the first column as one variable, enter the categories underneath.
- Name the adjacent columns as the categories of the other variable, entering the observed counts underneath.
- Click on **Stat**.
- Choose **Tables » Contingency » with summary**.
- Choose the **Columns** holding counts.
- Choose the **Row labels column**.
- Enter the **Column variable** name.
- Click on **Next** to ask for **expected counts**.
- Click on **Calculate**.

TI-83/84 PLUS

To test goodness of fit:
- Enter the observed counts in one list and the expected counts in another list. Expected counts can be entered as n*p and the calculator will compute them for you.
- From the **STAT TESTS** menu, select χ^2 **GOF-Test**. Specify the lists where you stored the observed and expected counts.
- Enter the number of degrees of freedom, and then hit **Calculate**.

COMMENTS

The cell-by-cell components will be calculated automatically and stored in a list named **CNTRB**.

Exercises

Section 24.1

1. **Human births** If there is no seasonal effect on human births, we would expect equal numbers of children to be born in each season (winter, spring, summer, and fall). A student takes a census of her statistics class and finds that of the 120 students in the class, 25 were born in winter, 35 in spring, 32 in summer, and 28 in fall. She wonders if the excess in the spring is an indication that births are not uniform throughout the year.
   a) What is the expected number of births in each season if there is no "seasonal effect" on births?
   b) Compute the χ^2 statistic.
   c) How many degrees of freedom does the χ^2 statistic have?

2. **Bank cards** At a major credit card bank, the percentages of people who historically apply for the Silver, Gold, and Platinum cards are 60%, 30%, and 10%, respectively. In a recent sample of customers responding to a promotion, of 200 customers, 110 applied for Silver, 55 for Gold, and 35 for Platinum. Is there evidence to suggest that the percentages for this promotion may be different from the historical proportions?
   a) What is the expected number of customers applying for each type of card in this sample if the historical proportions are still true?
   b) Compute the χ^2 statistic.
   c) How many degrees of freedom does the χ^2 statistic have?

3. **Human births, again** For the births in Exercise 1,
   a) If there is no seasonal effect, about how big, on average, would you expect the χ^2 statistic to be (what is the mean of the χ^2 distribution)?
   b) Does the statistic you computed in Exercise 1 seem large in comparison to this mean? Explain briefly.
   c) What does that say about the null hypothesis?
   d) Find the $\alpha = 0.05$ critical value for the χ^2 distribution with the appropriate number of df.
   e) Using the critical value, what do you conclude about the null hypothesis at $\alpha = 0.05$?

4. **Bank cards, again** For the customers in Exercise 2,
   a) If the customers apply for the three cards according to the historical proportions, about how big, on average, would you expect the χ^2 statistic to be (what is the mean of the χ^2 distribution)?
   b) Does the statistic you computed in Exercise 2 seem large in comparison to this mean? Explain briefly.
   c) What does that say about the null hypothesis?
   d) Find the $\alpha = 0.05$ critical value for the χ^2 distribution with the appropriate number of df.
   e) Using the critical value, what do you conclude about the null hypothesis at $\alpha = 0.05$?

Section 24.2

5. **Customer ages** An analyst at a local bank wonders if the age distribution of customers coming for service at his branch in town is the same as at the branch located near the mall. He selects 100 transactions at random from each branch and researches the age information for the associated customer. Here are the data:

	Age			
	Less Than 30	30–55	56 or Older	Total
In-Town Branch	20	40	40	100
Mall Branch	30	50	20	100
Total	50	90	60	200

   a) What is the null hypothesis?
   b) What type of test is this?

c) What are the expected numbers for each cell if the null hypothesis is true?
d) Find the χ^2 statistic.
e) How many degrees of freedom does it have?
f) Find the P-value.
g) What do you conclude?

6. **Bank cards, once more** A market researcher working for the bank in Exercise 2 wants to know if the distribution of applications by card is the same for the past three mailings. She takes a random sample of 200 from each mailing and counts the number applying for Silver, Gold, and Platinum. The data follow:

	Type of Card			
	Silver	Gold	Platinum	Total
Mailing 1	120	50	30	200
Mailing 2	115	50	35	200
Mailing 3	105	55	40	200
Total	340	155	105	600

a) What is the null hypothesis?
b) What type of test is this?
c) What are the expected numbers for each cell if the null hypothesis is true?
d) Find the χ^2 statistic.
e) How many degrees of freedom does it have?
f) Find the P-value.
g) What do you conclude?

Section 24.3

7. **Human births, last time** For the data in Exercise 1,
   a) Compute the standardized residual for each season.
   b) Are any of these particularly large? (Compared to what?)
   c) Why should you have anticipated the answer to part b?

8. **Bank cards, last time** For the data in Exercise 2,
   a) Compute the standardized residual for each type of card.
   b) Are any of these particularly large? (Compared to what?)
   c) What does the answer to part b say about this new group of customers?

Section 24.4

9. **Internet use poll** A Pew Research poll in April 2009 from a random sample of U.S. adults asked the questions "Did you use the Internet yesterday?" and "Are you White, Black, or Hispanic/Other?" Is the response to the question about the Internet independent of race? The data follow:

	Did You Use the Internet Yesterday?	
	Yes	No
White	2546	856
Black	314	146
Hispanic/Other	431	174

a) Under the null hypothesis, what are the expected values?
b) Compute the χ^2 statistic.
c) How many degrees of freedom does it have?
d) Find the P-value.
e) What do you conclude?

10. **Internet use poll, II** The same poll as in Exercise 9 also asked the questions "Did you use the Internet yesterday?" and "What is your educational level?" Is the response to the question about the internet independent of educational level? The data follow:

	Did You Use the Internet Yesterday?	
	Yes	No
Less Than High School	209	131
High School	932	550
Some College	958	346
College Grad	1447	247

a) Under the null hypothesis, what are the expected values?
b) Compute the χ^2 statistic.
c) How many degrees of freedom does it have?
d) Find the P-value.
e) What do you conclude?

Chapter Exercises

11. **Which test?** For each of the following situations, state whether you'd use a chi-square goodness-of-fit test, a chi-square test of homogeneity, a chi-square test of independence, or some other statistical test:
    a) A brokerage firm wants to see whether the type of account a customer has (Silver, Gold, or Platinum) affects the type of trades that customer makes (in person, by phone, or on the Internet). It collects a random sample of trades made for its customers over the past year and performs a test.
    b) That brokerage firm also wants to know if the type of account affects the size of the account (in dollars). It performs a test to see if the mean size of the account is the same for the three account types.
    c) The academic research office at a large community college wants to see whether the distribution of courses chosen (Humanities, Social Science, or Science) is different for its residential and nonresidential students. It assembles last semester's data and performs a test.

12. **Which test, again?** For each of the following situations, state whether you'd use a chi-square goodness-of-fit test, a chi-square test of homogeneity, a chi-square test of independence, or some other statistical test:
    a) Is the quality of a car affected by what day it was built? A car manufacturer examines a random sample of the warranty claims filed over the past two years to test whether defects are randomly distributed across days of the workweek.

(continued)

b) A medical researcher wants to know if blood cholesterol level is related to heart disease. She examines a database of 10,000 patients, testing whether the cholesterol level (in milligrams) is related to whether or not a person has heart disease.

c) A student wants to find out whether political leaning (liberal, moderate, or conservative) is related to choice of major. He surveys 500 randomly chosen students and performs a test.

13. **Dice** After getting trounced by your little brother in a children's game, you suspect the die he gave you to roll may be unfair. To check, you roll it 60 times, recording the number of times each face appears. Do these results cast doubt on the die's fairness?

Face	Count
1	11
2	7
3	9
4	15
5	12
6	6

a) If the die is fair, how many times would you expect each face to show?
b) To see if these results are unusual, will you test goodness-of-fit, homogeneity, or independence?
c) State your hypotheses.
d) Check the conditions.
e) How many degrees of freedom are there?
f) Find χ^2 and the P-value.
g) State your conclusion.

14. **M&M's** As noted in an earlier chapter, Mars Inc. says that until very recently yellow candies made up 20% of its milk chocolate M&M's, red another 20%, and orange, blue, and green 10% each. The rest are brown. On his way home from work the day he was writing these exercises, one of the authors bought a bag of plain M&M's. He got 29 yellow ones, 23 red, 12 orange, 14 blue, 8 green, and 20 brown. Is this sample consistent with the company's stated proportions? Test an appropriate hypothesis and state your conclusion.

a) If the M&M's are packaged in the stated proportions, how many of each color should the author have expected to get in his bag?
b) To see if his bag was unusual, should he test goodness-of-fit, homogeneity, or independence?
c) State the hypotheses.
d) Check the conditions.
e) How many degrees of freedom are there?
f) Find χ^2 and the P-value.
g) State a conclusion.

15. **Nuts** A company says its premium mixture of nuts contains 10% Brazil nuts, 20% cashews, 20% almonds, and 10% hazelnuts, and the rest are peanuts. You buy a large can and separate the various kinds of nuts. On weighing them, you find there are 112 grams of Brazil nuts, 183 grams of cashews, 207 grams of almonds, 71 grams of hazelnuts, and 446 grams of peanuts. You wonder whether your mix is significantly different from what the company advertises.

a) Explain why the chi-square goodness-of-fit test is not an appropriate way to find out.
b) What might you do instead of weighing the nuts in order to use a χ^2 test?

16. **Mileage** A salesman who is on the road visiting clients thinks that, on average, he drives the same distance each day of the week. He keeps track of his mileage for several weeks and discovers that he averages 122 miles on Mondays, 203 miles on Tuesdays, 176 miles on Wednesdays, 181 miles on Thursdays, and 108 miles on Fridays. He wonders if this evidence contradicts his belief in a uniform distribution of miles across the days of the week. Explain why it is not appropriate to test his hypothesis using the chi-square goodness-of-fit test.

17. **NYPD and race** Census data for New York City indicate that 29.2% of the under-18 population is white, 28.2% black, 31.5% Latino, 9.1% Asian, and 2% other ethnicities. The New York Civil Liberties Union points out that, of 26,181 police officers, 64.8% are white, 14.5% black, 19.1% Latino, and 1.4% Asian. Do the police officers reflect the ethnic composition of the city's youth? Test an appropriate hypothesis and state your conclusion.

18. **Violence against women 2009** In its study *When Men Murder Women: An Analysis of 2009 Homicide Data*, 2011, the Violence Policy Center (www.vpc.org) reported that 1818 women were murdered by men in 2009. Of these victims, a weapon could be identified for 1654 of them. Of those for whom a weapon could be identified, 861 were killed by guns, 364 by knives or other cutting instruments, 214 by other weapons, and 215 by personal attack (battery, strangulation, etc.). The FBI's Uniform Crime Report says that, among all murders nationwide, the weapon use rates were as follows: guns 63.4%, knives 13.1%, other weapons 16.8%, personal attack 6.7%. Is there evidence that violence against women involves different weapons than other violent attacks in the United States?

19. **Fruit flies** Offspring of certain fruit flies may have yellow or ebony bodies and normal wings or short wings. Genetic theory predicts that these traits will appear in the ratio 9:3:3:1 (9 yellow, normal: 3 yellow, short: 3 ebony, normal: 1 ebony, short). A researcher checks 100 such flies and finds the distribution of the traits to be 59, 20, 11, and 10, respectively.

a) Are the results this researcher observed consistent with the theoretical distribution predicted by the genetic model?

b) If the researcher had examined 200 flies and counted exactly twice as many in each category—118, 40, 22, 20—what conclusion would he have reached?

c) Why is there a discrepancy between the two conclusions?

20. **Pi** Many people know the mathematical constant π is approximately 3.14. But that's not exact. To be more precise, here are 20 decimal places: 3.14159265358979323846. Still not exact, though. In fact, the actual value is irrational, a decimal that goes on forever without any repeating pattern. But notice that there are no 0's and only one 7 in the 20 decimal places above. Does that pattern persist, or do all the digits show up with equal frequency? The table shows the number of times each digit appears in the first million digits. Test the hypothesis that the digits 0 through 9 are uniformly distributed in the decimal representation of π.

The First Million Digits of π

Digit	Count
0	99,959
1	99,758
2	100,026
3	100,229
4	100,230
5	100,359
6	99,548
7	99,800
8	99,985
9	100,106

21. **Hurricane frequencies** The National Hurricane Center provides data that list the numbers of large (category 3, 4, or 5) hurricanes that have struck the United States, by decade since 1851 (www.nhc.noaa.gov/dcmi.shtml). The data are given below.

Decade	Count	Decade	Count
1851–1860	6	1931–1940	8
1861–1870	1	1941–1950	10
1871–1880	7	1951–1960	9
1881–1890	5	1961–1970	6
1891–1900	8	1971–1980	4
1901–1910	4	1981–1990	4
1911–1920	7	1991–2000	5
1921–1930	5	2001–2010	7

Recently, there's been some concern that perhaps the number of large hurricanes has been increasing. The natural null hypothesis would be that the frequency of such hurricanes has remained constant.

a) With 96 large hurricanes observed over the 16 periods, what are the expected value(s) for each cell?

b) What kind of chi-square test would be appropriate?

c) State the null and alternative hypotheses.

d) How many degrees of freedom are there?

e) The value of χ^2 is 12.67. What's the P-value?

f) State your conclusion.

22. **Lottery numbers** The fairness of the South African lottery was recently challenged by one of the country's political parties. The lottery publishes historical statistics at its website (www.nationallottery.co.za). Here is a table of the number of times each number appeared in the lottery and as the "Powerball" number as of June 2007:

Number	Lotto	Powerball	Number	Lotto	Powerball
1	81	14	26	78	12
2	91	16	27	83	16
3	78	14	28	76	7
4	77	12	29	76	12
5	67	16	30	99	16
6	87	12	31	78	10
7	88	15	32	73	15
8	90	16	33	81	14
9	80	9	34	81	13
10	77	19	35	77	15
11	84	12	36	73	8
12	68	14	37	64	17
13	79	9	38	70	11
14	90	12	39	67	14
15	82	9	40	75	13
16	103	15	41	84	11
17	78	14	42	79	8
18	85	14	43	74	14
19	67	18	44	87	14
20	90	13	45	82	19
21	77	13	46	91	10
22	78	17	47	86	16
23	90	14	48	88	21
24	80	8	49	76	13
25	65	11			

We wonder if all the numbers are equally likely to be the "Powerball."

a) What kind of test should we perform?

b) There are 655 Powerball observations. What are the appropriate expected value(s) for the test?

c) State the null and alternative hypotheses.

(continued)

d) How many degrees of freedom are there?
e) The value of χ^2 is 34.5. What's the P-value?
f) State your conclusion.

23. **Childbirth, part 1** There is some concern that if a woman has an epidural to reduce pain during childbirth, the drug can get into the baby's bloodstream, making the baby sleepier and less willing to breastfeed. In December 2006, the *International Breastfeeding Journal* published results of a study conducted at Sydney University. Researchers followed up on 1178 births, noting whether the mother had an epidural and whether the baby was still nursing after 6 months. Below are their results.
    a) What kind of test would be appropriate?
    b) State the null and alternative hypotheses.

		Epidural?		
		Yes	No	Total
Breastfeeding at 6 Months?	Yes	206	498	704
	No	190	284	474
	Total	396	782	1178

24. **Does your doctor know?** A survey[7] of articles from the *New England Journal of Medicine (NEJM)* classified them according to the principal statistics methods used. The articles recorded were all noneditorial articles appearing during the indicated years. Let's just look at whether these articles used statistics at all.

	Publication Year			
	1978–79	1989	2004–05	Total
No Stats	90	14	40	144
Stats	242	101	271	614
Total	332	115	311	758

Has there been a change in the use of statistics?
a) What kind of test would be appropriate?
b) State the null and alternative hypotheses.

25. **Childbirth, part 2** In Exercise 23, the table shows results of a study investigating whether aftereffects of epidurals administered during childbirth might interfere with successful breastfeeding. We're planning to do a chi-square test.
    a) How many degrees of freedom are there?
    b) The smallest expected count will be in the epidural/no breastfeeding cell. What is it?
    c) Check the assumptions and conditions for inference.

26. **Does your doctor know? (part 2)** The table in Exercise 24 shows whether *NEJM* medical articles during various time periods included statistics or not. We're planning to do a chi-square test.
    a) How many degrees of freedom are there?
    b) The smallest expected count will be in the 1989/No cell. What is it?
    c) Check the assumptions and conditions for inference.

27. **Childbirth, part 3** In Exercises 23 and 25, we've begun to examine the possible impact of epidurals on successful breastfeeding.
    a) Calculate the component of chi-square for the epidural/no breastfeeding cell.
    b) For this test, $\chi^2 = 14.87$. What's the P-value?
    c) State your conclusion.

28. **Does your doctor know? (part 3)** In Exercises 24 and 26, we've begun to examine whether the use of statistics in *NEJM* medical articles has changed over time.
    a) Calculate the component of chi-square for the 1989/No cell.
    b) For this test, $\chi^2 = 25.28$. What's the P-value?
    c) State your conclusion.

29. **Childbirth, part 4** In Exercises 23, 25, and 27, we've tested a hypothesis about the impact of epidurals on successful breastfeeding. The following table shows the test's residuals.

		Epidural?	
		Yes	No
Breastfeeding at 6 Months?	Yes	−1.99	1.42
	No	2.43	−1.73

a) Show how the residual for the epidural/no breastfeeding cell was calculated.
b) What can you conclude from the standardized residuals?

30. **Does your doctor know? (part 4)** In Exercises 24, 26, and 28, we've tested a hypothesis about whether the use of statistics in *NEJM* medical articles has changed over time. The table shows the test's residuals.

	1978–79	1989	2004–05
No Stats	3.39	−1.68	−2.48
Stats	−1.64	0.81	1.20

a) Show how the residual for the 1989/No cell was calculated.
b) What can you conclude from the patterns in the standardized residuals?

31. **Childbirth, part 5** In Exercises 23, 25, 27, and 29, we've looked at a study examining epidurals as one factor that might inhibit successful breastfeeding of newborn babies. Suppose a broader study included several

---
[7]Suzanne S. Switzer and Nicholas J. Horton, "What Your Doctor Should Know about Statistics (but Perhaps Doesn't)," *Chance*, 20:1, 2007.

additional issues, including whether the mother drank alcohol, whether this was a first child, and whether the parents occasionally supplemented breastfeeding with bottled formula. Why would it not be appropriate to use chi-square methods on the 2 × 8 table with yes/no columns for each potential factor?

32. Does your doctor know? (part 5) In Exercises 24, 26, 28, and 30, we considered data on articles in the *NEJM*. The original study listed 23 different Statistics methods. (The list read: *t*-tests, contingency tables, linear regression, ...) Why would it not be appropriate to use a chi-square test on the 23 × 3 table with a row for each method?

33. Titanic Here is a table we first saw in Chapter 2 showing who survived the sinking of the *Titanic* based on whether they were crew members, or passengers booked in first-, second-, or third-class staterooms:

	Crew	First	Second	Third	Total
Alive	212	202	118	178	710
Dead	673	123	167	528	1491
Total	885	325	285	706	2201

a) If we draw an individual at random, what's the probability that we will draw a member of the crew?
b) What's the probability of randomly selecting a third-class passenger who survived?
c) What's the probability of a randomly selected passenger surviving, given that the passenger was a first-class passenger?
d) If someone's chances of surviving were the same regardless of their status on the ship, how many members of the crew would you expect to have lived?
e) Are *Survival* and *Ticket Class* related? State the null and alternative hypotheses.
f) Give the degrees of freedom for the test.
g) The chi-square value for the table is 187.8, and the corresponding P-value is barely greater than 0. State your conclusions about the hypotheses.

34. NYPD and sex discrimination The table below shows the rank attained by male and female officers in the New York City Police Department (NYPD). Do these data indicate that men and women are equitably represented at all levels of the department?

		Male	Female
Rank	Officer	21,900	4,281
	Detective	4,058	806
	Sergeant	3,898	415
	Lieutenant	1,333	89
	Captain	359	12
	Higher Ranks	218	10

a) What's the probability that a person selected at random from the NYPD is a female?
b) What's the probability that a person selected at random from the NYPD is a detective?
c) Assuming no bias in promotions, how many female detectives would you expect the NYPD to have?
d) To see if there is evidence of differences in ranks attained by males and females, will you test goodness-of-fit, homogeneity, or independence?
e) State the hypotheses.
f) Check the conditions.
g) How many degrees of freedom are there?
h) The chi-square value for the table is 290.1 and the P-value is less than 0.0001. State your conclusion about the hypotheses.

35. Titanic, again Examine and comment on this table of the standardized residuals for the chi-square test you looked at in Exercise 33.

	Crew	First	Second	Third
Alive	−4.35	9.49	2.72	−3.30
Dead	3.00	−6.55	−1.88	2.27

36. NYPD again Examine and comment on this table of the standardized residuals for the chi-square test you looked at in Exercise 34.

	Male	Female
Officer	−2.34	5.57
Detective	−1.18	2.80
Sergeant	3.84	−9.14
Lieutenant	3.58	−8.52
Captain	2.46	−5.86
Higher Ranks	1.74	−4.14

37. Cranberry juice It's common folk wisdom that drinking cranberry juice can help prevent urinary tract infections in women. In 2001, the *British Medical Journal* reported the results of a Finnish study in which three groups of 50 women were monitored for these infections over 6 months. One group drank cranberry juice daily, another group drank a lactobacillus drink, and the third drank neither of those beverages, serving as a control group. In the control group, 18 women developed at least one infection, compared to 20 of those who consumed the lactobacillus drink and only 8 of those who drank cranberry juice. Does this study provide supporting evidence for the value of cranberry juice in warding off urinary tract infections?

(continued)

a) Is this a survey, a retrospective study, a prospective study, or an experiment? Explain.
b) Will you test goodness-of-fit, homogeneity, or independence?
c) State the hypotheses.
d) Check the conditions.
e) How many degrees of freedom are there?
f) Find χ^2 and the P-value.
g) State your conclusion.
h) If you concluded that the groups are not the same, analyze the differences using the standardized residuals of your calculations.

T 38. Car origins A random survey of autos parked in the student lot and the staff lot at a large university classified the brands by country of origin, as seen in the table. Are there differences in the national origins of cars driven by students and staff?

		Driver	
		Student	Staff
Origin	American	107	105
	European	33	12
	Asian	55	47

a) Is this a test of independence or homogeneity?
b) Write appropriate hypotheses.
c) Check the necessary assumptions and conditions.
d) Find the P-value of your test.
e) State your conclusion and analysis.

T 39. Montana A poll conducted by the University of Montana classified respondents by whether they were male or female and political party, as shown in the table. We wonder if there is evidence of an association between being male or female and party affiliation.

	Democrat	Republican	Independent
Male	36	45	24
Female	48	33	16

a) Is this a test of homogeneity or independence?
b) Write an appropriate hypothesis.
c) Are the conditions for inference satisfied?
d) Find the P-value for your test.
e) State a complete conclusion.

T 40. Fish diet Medical researchers followed 6272 Swedish men for 30 years to see if there was any association between the amount of fish in their diet and prostate cancer. (Source: "Fatty Fish Consumption and Risk of Prostate Cancer," *Lancet*, June 2001)

Fish Consumption	Total Subjects	Prostate Cancers
Never/Seldom	124	14
Small Part of Diet	2621	201
Moderate Part	2978	209
Large Part	549	42

a) Is this a survey, a retrospective study, a prospective study, or an experiment? Explain.
b) Is this a test of homogeneity or independence?
c) Do you see evidence of an association between the amount of fish in a man's diet and his risk of developing prostate cancer?
d) Does this study prove that eating fish does not prevent prostate cancer? Explain.

T 41. Montana, revisited The poll described in Exercise 39 also investigated the respondents' party affiliations based on what area of the state they lived in. Test an appropriate hypothesis about this table and state your conclusions.

	Democrat	Republican	Independent
West	39	17	12
Northeast	15	30	12
Southeast	30	31	16

T 42. Working parents In April 2009, Gallup published results from data collected from a large sample of adults in the 27 European Union member states. One of the questions asked was, "Which is the most practicable and realistic option for child care, taking into account the need to earn a living?" The counts below are representative of the entire collection of responses.

	Male	Female
Both Parents Work Full Time	161	140
One Works Full Time, Other Part Time	259	308
One Works Full Time, Other Stays Home for Kids	189	161
Both Parents Work Part Time	49	63
No Opinion	42	28

Source: www.gallup.com/poll/117358/Work-Life-Balance-Tilts-Against-Women-Single-Parents.aspx

a) Is this a survey, a retrospective study, a prospective study, or an experiment?
b) Will you test goodness-of-fit, homogeneity, or independence?
c) Based on these results, do you think men and women have differing opinions when it comes to raising children?

43. Grades Two different professors teach an introductory Statistics course. The table shows the distribution of final grades they reported. We wonder whether one of these professors is an "easier" grader.

	Prof. Alpha	Prof. Beta
A	3	9
B	11	12
C	14	8
D	9	2
F	3	1

a) Will you test goodness-of-fit, homogeneity, or independence?
b) Write appropriate hypotheses.
c) Find the expected counts for each cell, and explain why the chi-square procedures are not appropriate.

T 44. Full moon Some people believe that a full moon elicits unusual behavior in people. The table shows the number of arrests made in a small town during weeks of six full moons and six other randomly selected weeks in the same year. We wonder if there is evidence of a difference in the types of illegal activity that take place.

	Full Moon	Not Full
Violent (murder, assault, rape, etc.)	2	3
Property (burglary, vandalism, etc.)	17	21
Drugs/Alcohol	27	19
Domestic Abuse	11	14
Other Offenses	9	6

a) Will you test goodness-of-fit, homogeneity, or independence?
b) Write appropriate hypotheses.
c) Find the expected counts for each cell, and explain why the chi-square procedures are not appropriate.

45. Grades, again In some situations where the expected cell counts are too small, as in the case of the grades given by Professors Alpha and Beta in Exercise 43, we can complete an analysis anyway. We can often proceed after combining cells in some way that makes sense and also produces a table in which the conditions are satisfied. Here, we create a new table displaying the same data, but calling D's and F's "Below C":

	Prof. Alpha	Prof. Beta
A	3	9
B	11	12
C	14	8
Below C	12	3

a) Find the expected counts for each cell in this new table, and explain why a chi-square procedure is now appropriate.
b) With this change in the table, what has happened to the number of degrees of freedom?
c) Test your hypothesis about the two professors, and state an appropriate conclusion.

T 46. Full moon, next phase In Exercise 44, you found that the expected cell counts failed to satisfy the conditions for inference.

a) Find a sensible way to combine some cells that will make the expected counts acceptable.
b) Test a hypothesis about the full moon and state your conclusion.

T 47. Racial steering A subtle form of racial discrimination in housing is "racial steering." Racial steering occurs when real estate agents show prospective buyers only homes in neighborhoods already dominated by that family's race. This violates the Fair Housing Act of 1968. According to an article in *Chance* magazine (Vol. 14, no. 2 [2001]), tenants at a large apartment complex recently filed a lawsuit alleging racial steering. The complex is divided into two parts: Section A and Section B. The plaintiffs claimed that white potential renters were steered to Section A, while African-Americans were steered to Section B. The table displays the data that were presented in court to show the locations of recently rented apartments. Do you think there is evidence of racial steering?

	New Renters		
	White	Black	Total
Section A	87	8	95
Section B	83	34	117
Total	170	42	212

T 48. Survival on the *Titanic* Newspaper headlines at the time, and traditional wisdom in the succeeding decades, have held that women and children escaped the *Titanic* in greater proportions than men. Here's a table with the relevant data. Do you think that survival was independent of whether the person was male or female? Explain.

	Female	Male	Total
Alive	343	367	710
Dead	127	1364	1491
Total	470	1731	2201

(*continued*)

49. Steering, revisited You could have checked the data in Exercise 47 for evidence of racial steering using two-proportion z procedures.
   a) Find the z-value for this approach, and show that when you square your z-value, you get the value of χ^2 you calculated in Exercise 47.
   b) Show that the resulting P-values are the same.

50. Survival on the *Titanic*, one more time In Exercise 48, you could have checked for a difference in the chances of survival for men and women using two-proportion z procedures.
   a) Find the z-value for this approach.
   b) Show that the square of your calculated value of z is the value of χ^2 you calculated in Exercise 48.
   c) Show that the resulting P-values are the same.

51. Pregnancies Most pregnancies are full term, but some are preterm (less than 37 weeks). Of those that are preterm, the Centers for Disease Control and Prevention classifies them as early (less than 34 weeks) and late (34 to 36 weeks). A December 2010 National Vital Statistics Report examined those outcomes in the United States during 2008, broken down by age of the mother. The table shows counts consistent with that report. Is there evidence that the outcomes are not independent of age group?

	Early Preterm	Late Preterm
Under 20	129	270
20 to 29	243	612
30 to 39	165	424
40 or Over	18	39

Source: www.cdc.gov/nchs/data/nvsr/nvsr59/nvsr59_01.pdf

52. Education by age Use the survey results in the table to investigate differences in education level attained among different age groups in the United States.

Education Level	Age Group				
	25–34	35–44	45–54	55–64	≥65
Not HS Grad	27	50	52	71	101
HS	82	19	88	83	59
1–3 Years College	43	56	26	20	20
≥4 Years College	48	75	34	26	20

Just Checking ANSWERS

1. We need to know how well beetles can survive 6 hours in a Plexiglas® box so that we have a baseline to compare the treatments.

2. There's no difference in survival rate in the three groups.

3. $(2 - 1)(3 - 1) = 2$ df

4. 50

5. 2

6. The mean value for a χ^2 with 2 df is 2, so 10 seems pretty large. The P-value is probably small.

7. This is a test of homogeneity. The clue is that the question asks whether the distributions are alike.

8. This is a test of goodness-of-fit. We want to test the model of equal assignment to all lots against what actually happened.

9. This is a test of independence. We have responses on two variables for the same individuals.

CHAPTER 25
Inferences for Regression

- 25.1 The Population and the Sample
- 25.2 Assumptions and Conditions
- 25.3 Intuition About Regression Inference
- 25.4 Regression Inference
- 25.5 Standard Errors for Predicted Values
- 25.6 Confidence Intervals for Predicted Values
- *25.7 Logistic Regression

Where are we going?
A scatterplot of IQ vs. brain size shows a mildly positive association. Could this just be due to chance? A hypothesis test is clearly what we need. We can estimate the slope, but how reliable is our estimate? In this chapter we'll apply what we know about tests and confidence intervals for means to regression.

Who	250 male subjects
What	Body fat and waist size
Units	% Body fat and inches
When	1990s
Where	United States
Why	Scientific research

Three percent of a man's body is essential fat. (For a woman, the percentage is closer to 12.5%.) As the name implies, essential fat is necessary for a normal, healthy body. Fat is stored in small amounts throughout your body. Too much body fat, however, can be dangerous to your health. For men between 18 and 39 years old, a healthy percent body fat ranges from 8% to 19%. (For women of the same age, it's 21% to 32%.)

Measuring body fat can be tedious and expensive. The "standard reference" measurement is by dual-energy X-ray absorptiometry (DEXA), which involves two low-dose X-ray generators and takes from 10 to 20 minutes.

How close can we get to a useable prediction of body fat from easily measurable variables such as *Height*, *Weight*, or *Waist* size? Here's a scatterplot of *%Body Fat* plotted against *Waist* size for a sample of 250 males of various ages. (Data in **Body fat complete**).

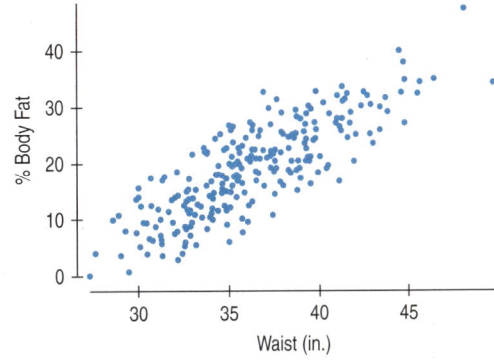

Figure 25.1

Percent Body Fat vs. *Waist* size for 250 men of various ages. The scatterplot shows a strong, positive, linear relationship.

Back in Chapter 7, we modeled relationships like this by fitting a least squares line. The plot is clearly straight, so we can find that line. The equation of the least squares line for these data is

$$\widehat{\%Body\ Fat} = -42.7 + 1.7\ Waist.$$

The slope says that, on average, *%Body Fat* is greater by 1.7 percent in men who have an additional inch around the waist.

How useful is this model? When we fit linear models before, we used them to describe the relationship between the variables and we interpreted the slope and intercept as descriptions of the data. Now we'd like to understand what the regression model can tell us beyond the 250 men in this study. To do that, we'll want to make confidence intervals and test hypotheses about the slope and intercept of the regression line.

25.1 The Population and the Sample

When we found a confidence interval for a mean, we could imagine a single, true underlying value for the mean. When we tested whether two means or two proportions were equal, we imagined a true underlying difference. But what does it mean to do inference for regression? We know better than to think that the data would line up perfectly on a straight line even if we knew every population value. After all, even in our sample, not all men who have 38-inch waists have the same *%Body Fat*. In fact, there's a whole distribution of *%Body Fat* for these men:

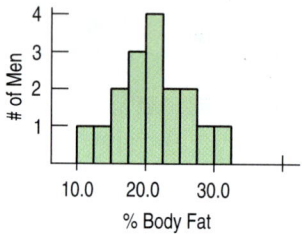

Figure 25.2
The distribution *of %Body Fat* for men with a *Waist* size of 38 inches is unimodal and symmetric.

This is true at each *Waist* size. In fact, we could depict the distribution of *%Body Fat* at different *Waist* sizes like this:

Figure 25.3
There's a distribution of *%Body Fat* for each value of *Waist* size. We'd like the means of these distributions to line up.

But we want to *model* the relationship between *%Body Fat* and *Waist* size for all men. To do that, we imagine an idealized regression line. The model assumes that the *means of the distributions of %Body Fat* for each *Waist* size fall along the line, even though the individuals are scattered around it. We know that this model is not a perfect description of how the variables are associated, but it may be useful for predicting *%Body Fat* and for understanding how it's related to *Waist* size.

If only we had all the values in the population, we could find the slope and intercept of this *idealized regression line* explicitly by using least squares. Following our usual conventions, we write the idealized line with Greek letters and consider the coefficients

■ **NOTATION ALERT**
This time we used up only one Greek letter for two things. Lower-case Greek β (beta) is the natural choice to correspond to the *b*'s in the regression equation. We used β before for the probability of a Type II error, but there's little chance of confusion here.

(the slope and intercept) to be *parameters*: β_0 is the intercept and β_1 is the slope. Corresponding to our fitted line of $\hat{y} = b_0 + b_1 x$, we write

$$\mu_y = \beta_0 + \beta_1 x.$$

Why μ_y instead of $\hat{y}$? Because this is a model. There is a distribution of *%Body Fat* for each *Waist* size. The model places the *means* of the distributions of *%Body Fat* for each *Waist* size on the same straight line.

Of course, not all the individual *y*'s are at these means. (In fact, the line will miss most—and quite possibly all—of the plotted points.) Some individuals lie above and some below the line, so, like all models, this one makes **errors**. Lots of them. In fact, one at each point. These errors are random and, of course, can be positive or negative. They are model errors, so we use a Greek letter and denote them by ε.

When we put the errors into the equation, we can account for each individual *y*:

$$y = \beta_0 + \beta_1 x + \varepsilon.$$

This equation is now true for each data point (since there is an ε to soak up the deviation), so the model gives a value of *y* for any value of *x*.

For the body fat data, an idealized model such as this provides a summary of the relationship between *%Body Fat* and *Waist* size. Like all models, it simplifies the real situation. We know there is more to predicting body fat than waist size alone. But the advantage of a model is that the simplification might help us to think about the situation and assess how well *%Body Fat* can be predicted from simpler measurements.

We estimate the β's by finding a regression line, $\hat{y} = b_0 + b_1 x$, as we did in Chapter 7. The residuals, $e = y - \hat{y}$, are the sample-based versions of the errors, ε. We'll use them to help us assess the regression model.

We know that least squares regression will give reasonable estimates of the parameters of this model from a random sample of data. Our challenge is to account for our uncertainty in how well they do. For that, we need to make some assumptions about the model and the errors.

25.2 Assumptions and Conditions

A | S *Activity:* **Conditions for Regression Inference.** View an illustrated discussion of the conditions for regression inference.

Back in Chapter 7 when we fit lines to data, we needed to check only the Straight Enough Condition. Now, when we want to make inferences about the coefficients of the line, we'll have to make more assumptions. Fortunately, we can check conditions to help us judge whether these assumptions are reasonable for our data. And as we've done before, we'll make some checks *after* we find the regression equation.

Also, we need to be careful about the order in which we check conditions. If our initial assumptions are not true, it makes no sense to check the later ones. So now we number the assumptions to keep them in order.

CHECK THE SCATTERPLOT

The shape must be linear or we can't use linear regression at all.

1. Linearity Assumption

If the true relationship is far from linear and we use a straight line to fit the data, our entire analysis will be useless, so we always check this first.

The **Straight Enough Condition** is satisfied if a scatterplot looks straight. It's generally not a good idea to draw a line through the scatterplot when checking. That can fool your eyes into seeing the plot as more straight. Sometimes it's easier to see violations of the Straight Enough Condition by looking at a scatterplot of the residuals against *x* or against the predicted values, $\hat{y}$. That plot will have a horizontal direction and should have no pattern if the condition is satisfied.

If the scatterplot is straight enough, we can go on to some assumptions about the errors. If not, stop here, or consider re-expressing the data (see Chapter 6) to make the scatterplot more nearly linear. For the *%Body Fat* data, the scatterplot is beautifully linear. Of course, the data must be quantitative for this to make sense. Check the **Quantitative Data Condition**.

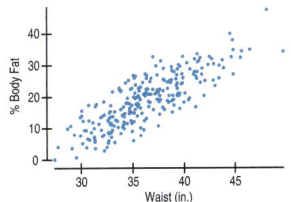

Figure 25.4

The relationship between *%Body Fat* and *Waist* size is straight enough.

> **CHECK THE RESIDUALS PLOT (1)**
> The residuals should appear to be randomly scattered.

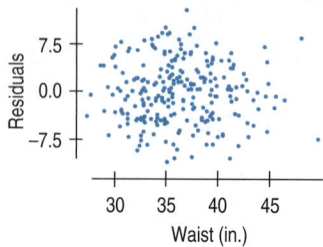

Figure 25.5
The residuals show only random scatter when plotted against *Waist* size.

> **CHECK THE RESIDUALS PLOT (2)**
> The vertical spread of the residuals should be roughly the same everywhere.

Figure 25.6
A scatterplot of residuals against predicted values can reveal plot thickening. In this plot of the residuals from a regression of diamond prices on carat weight, we see that larger diamonds have more price variation than smaller diamonds. When the Equal Spread Assumption is violated, we can't summarize how the residuals vary with a single number.

> **CHECK A HISTOGRAM OF THE RESIDUALS**
> The distribution of the residuals should be unimodal and symmetric.

2. Independence Assumption

Independence Assumption: The errors in the true underlying regression model (the ε's) must be independent of each other. As usual, there's no way to be sure that the Independence Assumption is true.

Usually when we care about inference for the regression parameters, it's because we think our regression model might apply to a larger population. In such cases, we can check that the individuals are a representative (ideally, random) sample from that population.

We can also check displays of the regression residuals for evidence of patterns, trends, or clumping, any of which would suggest a failure of independence. In the special case when the *x*-variable is related to time, a common violation of the Independence Assumption is for the errors to be correlated. (The error our model makes today may be similar to the one it made for yesterday.) This violation can be checked by plotting the residuals against the *x*-variable and looking for patterns. A further check can be made by plotting the residuals against those same residuals offset by one time position.[1] Neither plot should show patterns.

The *%Body Fat* data were collected on a sample of men taken to be representative. The subjects were not related in any way, so we can be pretty sure that their measurements are independent. The residuals plot shows no pattern.

3. Equal Variance Assumption

The variability of *y* should be about the same for all values of *x*. In Chapter 7, we looked at the standard deviation of the residuals (s_e) to measure the size of the scatter. Now we'll need this standard deviation to build confidence intervals and test hypotheses. The standard deviation of the residuals is the building block for the standard errors of all the regression parameters. But it makes sense only if the scatter of the residuals is the same everywhere. In effect, the standard deviation of the residuals "pools" information across all of the individual distributions at each *x*-value, and pooled estimates are appropriate only when they combine information for groups with the same variance.

Practically, what we can check is the **Does the Plot Thicken? Condition**. A scatterplot of *y* against *x* offers a visual check. Fortunately, we've already made one. Make sure the spread around the line is nearly constant. Be alert for a "fan" shape or other tendency for the variation to grow or shrink in one part of the scatterplot (as in the plot on the left). Often it is better to look at the residuals plotted against the predicted values, $\hat{y}$. With the slope of the line removed, it's easier to see patterns left behind. For the body fat data, the spread of *%Body Fat* around the line is remarkably constant across *Waist* sizes from 30 inches to about 45 inches.

If the plot is straight enough, the data are independent, and the plot doesn't thicken, you can now move on to the final assumption.

4. Normal Population Assumption

We assume the errors around the idealized regression line at each value of *x* follow a Normal model. We need this assumption so that we can use a Student's *t*-model for inference.

As we have at other times when we've used Student's *t*, we'll settle for the residuals satisfying the **Nearly Normal Condition** and the **Outlier Condition**. Look at a histogram or Normal probability plot of the residuals.[2]

The histogram of residuals in the *%Body Fat* regression certainly looks nearly Normal. As we have noted before, the Normality Assumption becomes less important as the sample size grows, because the model is about means and the Central Limit Theorem takes over.

---

[1] Such an offset version is said to be "lagged."

[2] *This* is why we have to check the conditions in order. We have to check that the residuals are independent and that the variation is the same for all *x*'s so that we can lump all the residuals together for a single check of the Nearly Normal Condition.

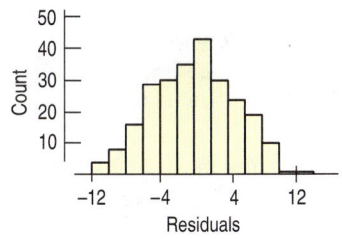

Figure 25.7
A histogram of the residuals is one way to check whether they are Nearly Normal. Alternatively, we can look at a Normal probability plot.

If all four assumptions were true, the idealized regression model would look like this:

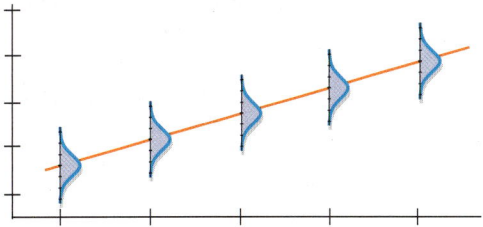

Figure 25.8
The regression model has a distribution of *y*-values for each *x*-value. These distributions follow a Normal model with means lined up along the line and with the same standard deviations.

At each value of *x*, there is a distribution of *y*-values that follows a Normal model, and each of these Normal models is centered on the line and has the same standard deviation. Of course, we don't expect the assumptions to be exactly true, and we know that all models are wrong, but the linear model is often close enough to be very useful.

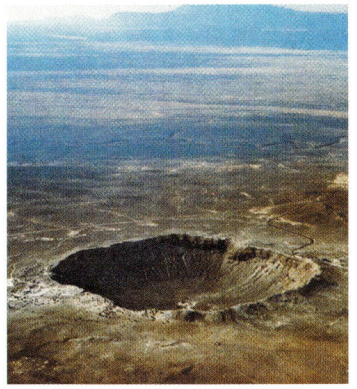

Who	39 impact craters
What	Diameter and age
Units	km and millions of years ago
When	Past 35 million years
Where	Worldwide
Why	Scientific research

For Example CHECKING ASSUMPTIONS AND CONDITIONS

Look at the moon with binoculars or a telescope, and you'll see craters formed by thousands of impacts. The earth, being larger, has been hit even more often. Meteor Crater in Arizona was the first recognized impact crater and was identified as such only in the 1920s. With the help of satellite images, more and more craters have been identified; now more than 180 are known. These, of course, are only a small sample of all the impacts the earth has experienced: Only 29% of earth's surface is land, and many craters have been covered or eroded away. Astronomers have recognized a roughly 35 million-year cycle in the frequency of cratering, although the cause of this cycle is not fully understood. Here's a scatterplot of the known impact craters from the most recent 35 million years.[3] We've taken logs of both age (in millions of years ago) and diameter (km) to make the relationship simpler. (See Chapter 6.)

QUESTION: Are the assumptions and conditions satisfied for fitting a linear regression model to these data?

ANSWER:

✓ **Linearity Assumption:** The scatterplot satisfies the Straight Enough Condition.

✓ **Independence Assumption:** Sizes of impact craters are likely to be generally independent. However, these are the only known craters, and may differ from

---

[3] Data, pictures, and much more information at the Earth Impact Database found at www.passc.net/EarthImpactDatabase/index.html

others that have disappeared or not yet been found. we'll need to be careful not to generalize my conclusions too broadly.

- ✓ **Does the Plot Thicken? Condition:** After fitting a linear model, we find the residuals shown.

  Two points seem to give the impression that the residuals may be more variable for higher predicted values than for lower ones, but this doesn't seem to be a serious violation of the Equal Variance Assumption.

- ✓ **Nearly Normal Condition:** A Normal probability plot suggests a bit of skewness in the distribution of residuals, and the histogram confirms that.

  There are no violations severe enough to stop my regression analysis, but we'll be cautious about my conclusions.

Which Come First: The Conditions or the Residuals?

In regression, there's a little catch. The best way to check many of the conditions is with the residuals, but we get the residuals only *after* we compute the regression. Before we compute the regression, however, we should check at least one of the conditions.

So we work in this order:

> "Truth will emerge more readily from error than from confusion."
>
> —*Francis Bacon (1561–1626)*

1. Make a scatterplot of the data to check the Straight Enough Condition. (If the relationship is curved, try re-expressing the data. Or stop.)
2. If the data are straight enough, fit a regression and find the predicted values, $\hat{y}$, and the residuals, e.
3. Make a scatterplot of the residuals against x or against the predicted values. This plot should have no pattern. Check in particular for any bend (which would suggest that the data weren't all that straight after all), for any thickening (or thinning), and, of course, for any outliers. (If there are outliers, and you can correct them or justify removing them, do so and go back to step 1, or consider performing two regressions—one with and one without the outliers.)
4. If the data are measured over time, plot the residuals against time to check for evidence of patterns that might suggest they are not independent.
5. If the scatterplots look OK, then make a histogram and Normal probability plot of the residuals to check the Nearly Normal Condition.
6. If all the conditions seem to be reasonably satisfied, go ahead with inference.

CHAPTER 25 Inferences for Regression

Step-by-Step Example REGRESSION INFERENCE

If our data can jump through all these hoops, we're ready to do regression inference. Let's see how much more we can learn about body fat and waist size from a regression model.

Question: What is the relationship between *%Body Fat* and *Waist* size in men? What model best predicts body fat from waist size, and how well does it do the job?

THINK ▶ Plan Specify the question of interest.

Name the variables and report the W's.

Identify the parameters you want to estimate.

Model Think about the assumptions and check the conditions.

Make pictures. For regression inference, you'll need a scatterplot, a residuals plot, and either a histogram or a Normal probability plot of the residuals.

(We've seen plots of the residuals already. See Figures 25.6 and 25.7.)

I have quantitative body measurements on 250 adult males from the BYU Human Performance Research Center. I want to understand the relationship between *%Body Fat* and *Waist* size.

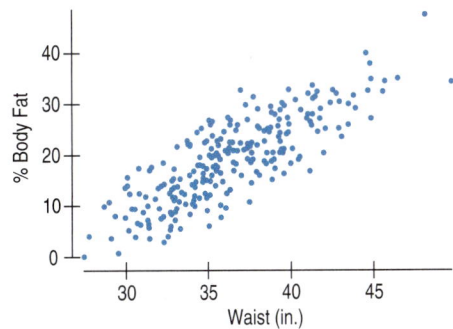

✓ **Straight Enough Condition:** There's no obvious bend in the original scatterplot of the data or in the plot of residuals against predicted values.

✓ **Independence Assumption:** These data are not collected over time, and there's no reason to think that the *%Body Fat* of one man influences the *%Body Fat* of another.

✓ **Does the Plot Thicken? Condition:** Neither the original scatterplot nor the residual scatterplot shows any changes in the spread about the line.

✓ **Nearly Normal Condition, Outlier Condition:** A histogram of the residuals is unimodal and symmetric. The Normal probability plot of the residuals is quite straight, indicating that the Normal model is reasonable for the errors.

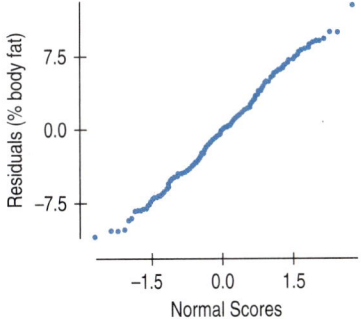

Choose your method.	Under these conditions a **regression model** is appropriate.
SHOW ▶ Mechanics Let's just "push the button" and see what the regression looks like. The formula for the regression equation can be found in Chapter 7, and the standard error formulas will be shown a bit later, but regressions are almost always computed with a computer program or calculator. Write the regression equation.	Here's the computer output for this regression: Dependent variable is %BF R-squared = 67.8% s = 4.713 with 250 − 2 = 248 degrees of freedom | Variable	Coeff	SE(Coeff)	t-Ratio	P-Value
---	---	---	---	---
Intercept	−42.734	2.717	−15.7	<0.0001
Waist	1.70	0.0743	22.9	<0.0001	 The estimated regression equation is $$\widehat{\%Body\ Fat} = -42.73 + 1.70\ Waist.$$
TELL ▶ Conclusion Interpret your results in context. **More Interpretation** We haven't worked it out in detail yet, but the output gives us numbers labeled as *t*-statistics and corresponding P-values, and we have a general idea of what those mean. (Now it's time to learn more about regression inference so we can figure out what the rest of the output means.)	The R^2 for the regression is 67.8%. *Waist* size seems to account for about 2/3 of the *%Body Fat* variation in men. The slope of the regression says that *%Body Fat* increases by about 1.7 percentage points per inch of *Waist* size, on average. The standard error of 0.07 for the slope is much smaller than the slope itself, so it looks like the estimate is reasonably precise. And there are a couple of *t*-ratios and P-values given. Because the P-values are small, it appears that some null hypotheses can be rejected.

25.3 Intuition About Regression Inference

A|S *Simulation:* **Simulate the Sampling Distribution of a Regression Slope.** Draw samples repeatedly to see for yourself how slope can vary from sample to sample. This simulation experiment lets you build up a histogram to see the sampling distribution.

Wait a minute! We've just pulled a fast one. We've pushed the "regression button" on our computer or calculator but haven't discussed where the standard errors for the slope or intercept come from. We know that if we had collected similar data on a different random sample of men, the slope and intercept would be different. Each sample would have produced its own regression line, with slightly different b_0's and b_1's. This sample-to-sample variation is what generates the sampling distributions for the coefficients.

There's only one regression model; each sample regression is trying to estimate the same parameters, β_0 and β_1. We expect any sample to produce a b_1 whose expected value is the true slope, β_1. What about its standard deviation? What aspects of the data affect how much the slope (and intercept) vary from sample to sample?

- **Spread around the line.** Here are two situations in which we might do regression. Which situation would be more likely to yield a consistent slope? That is, if we were to sample over and over from the two underlying populations that these samples come from and compute all the slopes, which group of slopes would vary less?

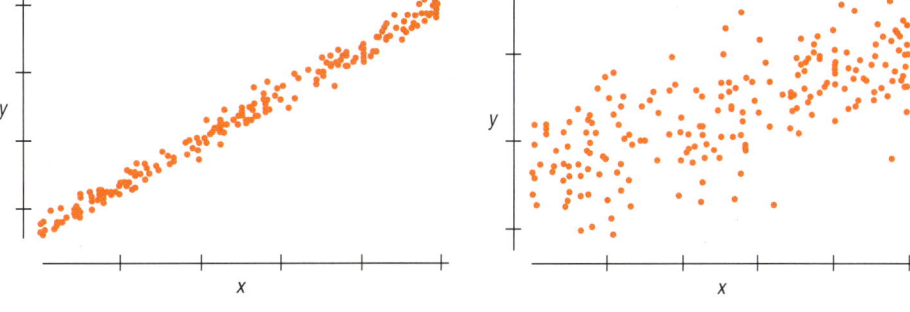

Figure 25.9
Which of these scatterplots shows a situation that would give the more consistent regression slope estimate if we were to sample repeatedly from its underlying population?

Clearly, data like those in the left plot will give more consistent slopes estimates.

Less scatter around the line means the slope will be more consistent from sample to sample. The spread around the line is measured with the **residual standard deviation**, s_e. You can always find s_e in the regression output, often just labeled s. Sometimes it is called the "standard error" although, as we know, that's not quite right—probably a misinterpretation of the s_e notation. You're not likely to calculate the residual standard deviation by hand. When we first saw this formula in Chapter 7, we said that it looks a lot like the standard deviation of y, only subtracting the predicted values rather than the mean and dividing by $n - 2$ instead of $n - 1$:

$$s_e = \sqrt{\frac{\sum (y - \hat{y})^2}{n - 2}}.$$

> **n − 2?**
> For standard deviation (in Chapter 3), we divided by $n - 1$ because we didn't know the true mean and had to estimate it. Now it's later in the course and there's even more we don't know. Here we don't know *two* things: the slope and the intercept. If we knew them both, we'd divide by n and have n degrees of freedom. When we *estimate* both, however, we adjust by subtracting 2, so we divide by $n - 2$ and (as we will see soon) have 2 fewer degrees of freedom.

The less scatter around the line, the smaller the residual standard deviation and the stronger the relationship between x and y.

Some people prefer to assess the strength of a regression by looking at s_e rather than R^2. After all, s_e has the same units as y, and because it's the standard deviation of the errors around the line, it tells you how close the data are to our model. By contrast, R^2 is the proportion of the variation of y accounted for by x. We say, why not look at both?

- **Spread of the x's.** Here are two more situations. Which of these would yield more consistent slope estimates?

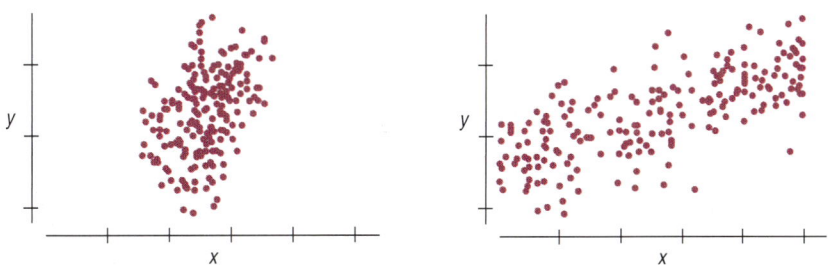

Figure 25.10
Which of these scatterplots shows a situation that would give the more consistent regression slope estimates if we were to sample repeatedly from the underlying population?

A plot like the one on the right has a broader range of x-values, so it gives a more stable base for the slope. We'd expect the slopes of samples from situations like that to vary less from sample to sample. If s_x, the standard deviation of x is large, it provides a more stable regression.

Figure 25.11
Which of these scatterplots shows a situation that would give the more consistent regression slope estimates if we were to sample repeatedly from the underlying population?

- **Sample size.** What about these two?

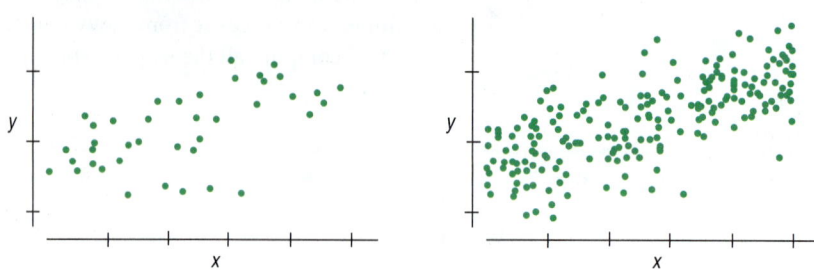

It shouldn't be a surprise that having a larger sample size, n, gives more consistent estimates from sample to sample.

Standard Error for the Slope

Three aspects of the scatterplot, then, affect the standard error of the regression slope:

- Spread around the line: s_e
- Spread of x values: s_x
- Sample size: n

These are in fact the *only* things that affect the standard error of the slope. Although you'll probably never have to calculate it by hand, the formula for the standard error is

$$SE(b_1) = \frac{s_e}{\sqrt{n-1}\, s_x}.$$

A|S *Activity:* **Regression Slope Standard Error.** See how $SE(b_1)$ is constructed and where the values used in the formula are found in the regression output table.

The error standard deviation, s_e, is in the *numerator*, since spread around the line *increases* the slope's standard error. The denominator has both a sample size term, $\sqrt{n-1}$, and s_x, because increasing either of these *decreases* the slope's standard error.

We know the b_1's vary from sample to sample. As you'd expect, their sampling distribution model is centered at β_1, the slope of the idealized regression line. Now we can estimate its standard deviation with $SE(b_1)$. What about its shape? Here the Central Limit Theorem and "Wild Bill" Gosset come to the rescue again. When we standardize the slopes by subtracting the model mean and dividing by their standard error, we get a Student's t-model, this time with $n-2$ degrees of freedom:

$$\frac{b_1 - \beta_1}{SE(b_1)} \sim t_{n-2}$$

A|S *Simulation:* **x-Variance and Slope Variance.** You don't have to just imagine how the variability of the slope depends on the spread of the x's.

■ NOTATION ALERT

Don't confuse the standard deviation of the residuals, s_e, with the standard error of the slope, $SE(b_1)$. The first measures the scatter around the line, and the second tells us how reliably we can estimate the slope.

> **A SAMPLING DISTRIBUTION FOR REGRESSION SLOPES**
>
> When the conditions are met, the standardized estimated regression slope,
>
> $$t = \frac{b_1 - \beta_1}{SE(b_1)},$$
>
> follows a Student's t-model with $n-2$ degrees of freedom. We estimate the standard error with
>
> $$SE(b_1) = \frac{s_e}{\sqrt{n-1}\, s_x},$$
>
> where $s_e = \sqrt{\dfrac{\sum(y-\hat{y})^2}{n-2}},$
>
> n is the number of data values, and s_x is the ordinary standard deviation of the x-values.

For Example FINDING STANDARD ERRORS

RECAP: Recent terrestrial impact craters seem to show a relationship between age and size that is linear when re-expressed using logarithms (see Chapter 6).

Here are summary statistics and regression output:

Variable	Count	Mean	StdDev
LogAge	39	−0.656310	1.57682
LogDiam	39	0.012600	1.04104

Dependent variable is LogDiam
R-squared = 63.6%
s = 0.6362 with 39 − 2 = 37 degrees of freedom

Variable	Coefficient	Se(coeff)	t-Ratio	P-Value
Intercept	0.358262	0.1106	3.24	0.0025
LogAge	0.526674	0.0655	8.04	≤0.0001

QUESTIONS: How are the standard error of the slope and the *t*-ratio for the slope calculated? (And aren't you glad the software does this for you?)

ANSWER:

$$SE(b_1) = \frac{s_e}{\sqrt{n-1} \times s_x} = \frac{0.6362}{\sqrt{39-1} \times 1.57682} = 0.0655$$

To test the hypothesis of no linear association ($\beta_1 = 0$),

$$t_{37} = \frac{b_1 - \beta_1}{SE(b_1)} = \frac{0.526674 - 0}{0.0655} = 8.04.$$

What About the Intercept?

The same reasoning applies for the intercept. We could write

$$\frac{b_0 - \beta_0}{SE(b_0)} \sim t_{n-2}$$

and use it to construct confidence intervals and test hypotheses, but often the value of the intercept isn't something we care about—or have a natural null hypothesis for. The intercept usually isn't interesting. Most hypothesis tests and confidence intervals for regression are about the slope.

25.4 Regression Inference

Now that we have the standard error of the slope and its sampling distribution, we can test a hypothesis about it and make confidence intervals. The usual null hypothesis about the slope is that it's equal to 0. Why? Well, a slope of zero would say that *y* doesn't tend to change linearly when *x* changes—in other words, that there is no linear association between the two variables. If the slope were zero, there wouldn't be much left of our regression equation.

So a null hypothesis of a zero slope questions the entire claim of a linear relationship between the two variables—and often that's just what we want to know. In fact, every

> **WHAT IF THE SLOPE WERE 0?**
>
> If $b_1 = 0$, our prediction is $\hat{y} = b_0 + 0x$. The equation collapses to just $\hat{y} = b_0$. Now x is nowhere in sight, so y doesn't depend on x at all.
>
> And b_0 would turn out to be $\bar{y}$. Why? We know that $b_0 = \bar{y} - b_1\bar{x}$, but when $b_1 = 0$, that becomes simply $b_0 = \bar{y}$. It turns out, then, that when the slope is 0, the equation is just $\hat{y} = \bar{y}$; at every value of x, we always predict the mean value for y.

software package or calculator that does regression simply assumes that you want to test the null hypothesis that the slope is really zero.

To test $H_0: \beta_1 = 0$, we find

$$t_{n-2} = \frac{b_1 - 0}{SE(b_1)}.$$

This is just like every t-test we've seen: a difference between the statistic and its hypothesized value, divided by its standard error.

For our body fat data, the computer found the slope (1.7), its standard error (0.0743), and the ratio of the two: $\frac{1.7 - 0}{0.0743} = 22.9$ (see p. 696). Nearly 23 standard errors from the hypothesized value certainly seems big. The P-value (<0.0001) confirms that a t-ratio this large would be very unlikely to occur if the true slope were zero.

Maybe the standard null hypothesis isn't all that interesting here. Did you have any doubts that *%Body Fat* is related to *Waist* size? A more sensible use of these same values might be to make a confidence interval for the slope instead.

We can build a confidence interval in the usual way, as an estimate plus or minus a margin of error. As always, the margin of error is just the product of the standard error and a critical value. Here the critical value comes from the t-distribution with $n - 2$ degrees of freedom, so a 95% **confidence interval for β** is

$$b_1 \pm t^*_{n-2} \times SE(b_1).$$

For the body fat data, $t^*_{248} = 1.970$, so that comes to $1.7 \pm 1.97 \times 0.074$, or an interval from 1.55 to 1.85 *%Body Fat* per inch of *Waist* size.

For Example   INTERPRETING A REGRESSION MODEL

RECAP: On a log scale, there seems to be a linear relationship between the diameter and the age of recent terrestrial impact craters. We have regression output from statistics software:

Dependent variable is LogDiam
R-squared = 63.6%
s = 0.6362 with 39 − 2 = 37 degrees of freedom

Variable	Coefficient	Se(coeff)	t-Ratio	P-Value
Intercept	0.358262	0.1106	3.24	0.0025
LogAge	0.526674	0.0655	8.05	≤0.0001

QUESTION: What's the regression model, and what can it tell us?

ANSWER: For terrestrial impact craters younger than 35 million years, the logarithm of *Diameter* grows linearly with the logarithm of *Age*:

$$\widehat{logDiam} = 0.358 + 0.527\ logAge.$$

The P-value for each coefficient's t-statistic is very small, so we can be quite confident that neither coefficient is zero. Based on the regression model, we can conclude that, on average, the older a crater is, the larger it tends to be. This model accounts for 63.6% of the variation in *logDiam*.

Although it is possible that impacts (and their craters) are getting smaller, it is more likely that we are seeing the effects of age on craters. Small craters are probably more likely to erode or become buried or otherwise be difficult to find as they age. Larger craters may survive the huge expanses of geologic time more successfully.

Just Checking

Researchers in Food Science studied how big people's mouths tend to be. They measured mouth volume by pouring water into the mouths of subjects who lay on their backs. Unless this is your idea of a good time, it would be helpful to have a model to estimate mouth volume more simply. Fortunately, mouth volume is related to height. (Mouth volume is measured in cubic centimeters and height in meters.)

The data were checked and deemed suitable for regression. Take a look at the computer output.

1. What does the t-ratio of 3.27 for the slope tell about this relationship? How does the P-value help your understanding?

2. Would you say that measuring a person's height could reliably be used as a substitute for the wetter method of determining how big a person's mouth is? What numbers in the output helped you reach that conclusion?

3. What does the value of s_e add to this discussion?

Summary of	Mouth Volume
Mean	60.2704
StdDev	16.8777

Dependent variable is Mouth Volume
R-squared = 15.3%
s = 15.66 with 61 − 2 = 59 degrees of freedom

Variable	Coefficient	SE(coeff)	t-Ratio	P-Value
Intercept	−44.7113	32.16	−1.39	0.1697
Height	61.3787	18.77	3.27	0.0018

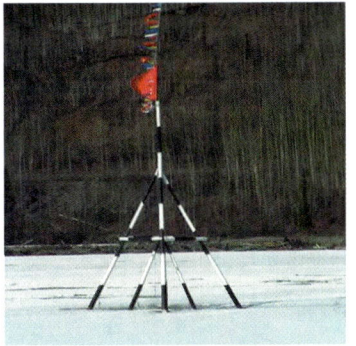

Activity: A Hypothesis Test for the Regression Slope. View an animated discussion of testing the standard null hypothesis for slope.

Another Example: Breaking Up Is Hard to Predict

Every spring, Nenana, Alaska, hosts a contest in which participants try to guess the exact minute that a wooden tripod placed on the frozen Tanana River will fall through the breaking ice. The contest started in 1917 as a diversion for railroad engineers, with a jackpot of $800 for the closest guess. It has grown into an event in which hundreds of thousands of entrants enter their guesses on the Internet[4] and vie for a prize that in recent years has exceeded $300,000.

Because so much money and interest depends on the time of breakup, it has been recorded to the nearest minute with great accuracy ever since 1917. And because a standard measure of breakup has been used throughout this time, the data are consistent. An article in *Science*[5] used the data to investigate global warming—whether greenhouse gasses and other human actions have been making the planet warmer. Others might just want to make a good prediction of next year's breakup time.

Of course, we can't use regression to tell the *causes* of any change. But we can estimate the *rate* of change (if any) and use it to make better predictions.

---

[4] www.nenanaakiceclassic.com
[5] "Climate Change in Nontraditional Data Sets." *Science* 294 [26 October 2001]: 811.

Here are some of the data:

Who	Years
What	Year, day, and hour of ice breakup
Units	x is in years since 1900. y is in days after midnight Dec. 31.
When	1917–present
Where	Nenana, Alaska
Why	Wagering, but proposed to look at global warming (Data in **Nenana 2014**)

Year (since 1900)	Breakup Date (days after Jan. 1)	Year (since 1900)	Breakup Date (days after Jan. 1)
17	119.4792	30	127.7938
18	130.3979	31	129.3910
19	122.6063	32	121.4271
20	131.4479	33	127.8125
21	130.2792	34	119.5882
22	131.5556	35	134.5639
23	128.0833	36	120.5403
24	131.6319	37	131.8361
25	126.7722	38	125.8431
26	115.6688	39	118.5597
27	131.2375	40	110.6437
28	126.6840	41	122.0764
29	124.6535	⋮	⋮

Step-by-Step Example A REGRESSION SLOPE *t*-TEST

The slope of the regression gives the change in Nenana ice breakup date per year.
Questions: Is there sufficient evidence to claim that ice breakup times are changing? If so, how rapid is the change?

THINK → Plan State what you want to know.

Identify the *parameter* you wish to estimate. Here our parameter is the slope.

Identify the variables and review the W's.

Hypotheses Write your null and alternative hypotheses.

Model Think about the assumptions and check the conditions.

Make pictures. Because the scatterplot seems straight enough, we can find and plot the residuals.

I wonder whether the date of ice breakup in Nenana has changed over time. The slope of that change might indicate climate change. I have the date of ice breakup annually for 98 years starting in 1917, recorded as the number of days and fractions of a day until the ice breakup.

H_0: There is no change in the date of ice breakup: $\beta_1 = 0$
H_A: Yes, there is: $\beta_1 \neq 0$

✓ **Straight Enough Condition:** I have quantitative data with no obvious bend in the scatterplot.

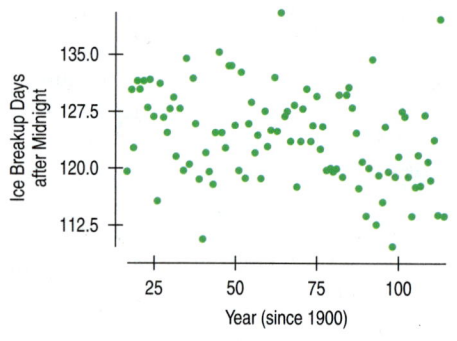

Usually, we check for suggestions that the Independence Assumption fails by plotting the residuals against the predicted values. Patterns and clusters in that plot raise our suspicions. But when the data are measured over time, it is always a good idea to plot residuals against time to look for trends and oscillations.

✓ **Independence Assumption:** These data are a time series, which raises my suspicions that they may not be independent. To check, here's a plot of the residuals against time, the x-variable of the regression:

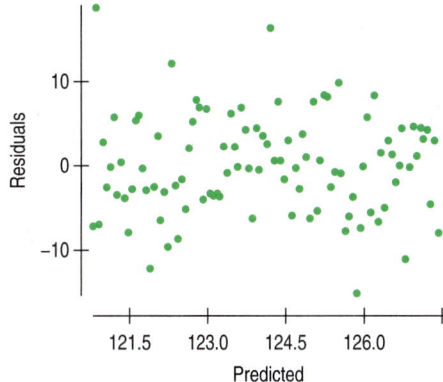

I see a hint that the data oscillate up and down, which suggests some failure of independence, but not so strongly that I can't proceed with the analysis. These data are not a random sample, so I'm reluctant to extend my conclusions beyond this river and these years.

✓ **Does the Plot Thicken? Condition:** The residuals plot shows no obvious trends in the spread.

✓ **Nearly Normal Condition, Outlier Condition:** A histogram of the residuals is unimodal and symmetric.

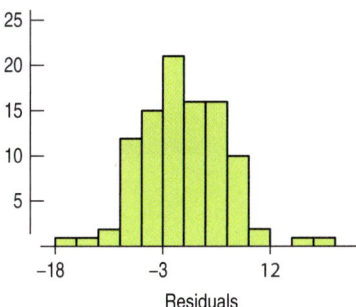

State the sampling distribution model.

Under these conditions, the sampling distribution of the regression slope can be modeled by a Student's t-model with $(n - 2) = 96$ degrees of freedom.

Choose your method.

I'll do a **regression slope t-test**.

SHOW → Mechanics

The regression equation can be found from the formulas in Chapter 7, but regressions are almost always found from a computer program or calculator.

The P-values given in the regression output table are from the Student's t-distribution on $(n - 2) = 93$ degrees of freedom. They are appropriate for two-sided alternatives.

Here's the computer output for this regression:

Response variable is   Ice Breakup Days After Midnight
R-squared = 9.8%
s = 5.925 with 98 − 2 = 96 degrees of freedom

Variable	Coefficient	SE(Coeff)	t-Ratio	P-Value
Intercept	128.581	1.510	85.2	<0.0001
Years Since 1900	−0.068340	0.0212	−3.23	0.0017

The estimated regression equation is

$$\widehat{Date} = 128.581 - 0.068\, YearSince1900.$$

TELL → Conclusion

Link the P-value to your decision and state your conclusion in the proper context.

The P-value of 0.0017 means that the association we see in the data is unlikely to have occurred by chance even though the R^2 is not particularly strong. I reject the null hypothesis, and conclude that there is strong evidence that, on average, the ice breakup is occurring earlier each year. But the oscillation pattern in the residuals raises concerns.

SHOW MORE → Create a confidence interval for the true slope

A 95% confidence interval for β_1 is

$$b_1 \pm t^*_{96} \times SE(b_1)$$
$$-0.068 \pm (1.985)(0.0212)$$
$$\text{or } (-0.11, -0.03) \text{ days per year.}$$

TELL MORE → Interpret the interval

Simply rejecting the standard null hypothesis doesn't guarantee that the size of the effect is large enough to be important. Whether we want to know the breakup time to the nearest minute or are interested in global warming, a change measured in hours each year is big enough to be interesting.

I am 95% confident that the ice has been breaking up, on average, between 0.03 days (about 40 minutes) and 0.11 days (about 3 hours) earlier each year since 1900.

But is it global warming? So the ice is breaking up earlier. Temperatures are higher. Must be global warming, right?

Maybe.

An article challenging the original analysis of the Nenana data proposed a possible confounding variable. It noted that the city of Fairbanks is upstream from Nenana and suggested that the growth of Fairbanks could have warmed the river. So maybe it's not global warming.

Or maybe global warming is a lurking variable, leading more people to move to a now balmier Fairbanks and also leading to generally earlier ice breakup in Nenana.

Or maybe there's some other variable or combination of variables at work. We can't set up an experiment, so we may never really know.

Only one thing is for sure. When you try to explain an association by claiming cause and effect, you're bound to be on thin ice.[6]

---

[6] How *do* scientists sort out such messy situations? Even though they can't conduct an experiment, they *can* look for replications elsewhere. A number of studies of ice on other bodies of water have also shown earlier ice breakup times in recent years. That suggests they need an explanation that's more comprehensive than just Fairbanks and Nenana.

25.5 Standard Errors for Predicted Values

Once we have a useful regression, how can we indulge our natural desire to predict, without being irresponsible? We know how to compute predicted values of y for any value of x. We first did that in Chapter 7. This predicted value would be our best estimate, but it's still just an informed guess.

Now, however, we have standard errors. We can use those to construct a confidence interval for the predictions and to report our uncertainty honestly.

From our model of %Body Fat and Waist size, we might want to use Waist size to get a reasonable estimate of %Body Fat. A confidence interval can tell us how precise that prediction will be. The precision depends on the question we ask, however, and there are two questions: Do we want to know the mean %Body Fat for *all* men with a Waist size of, say, 38 inches? Or do we want to estimate the %Body Fat for a particular man with a 38-inch Waist without making him climb onto the X-ray table?

What's the difference between the two questions? The predicted %Body Fat is the same, but one question leads to an answer much more precise than the other. We can predict the *mean* %Body Fat for *all* men whose Waist size is 38 inches with a lot more precision than we can predict the %Body Fat of a *particular individual* whose Waist size happens to be 38 inches. Both are interesting questions.

We start with the same prediction in both cases. We are predicting the value for a new individual, one that was not part of the original data set. To emphasize this, we'll call his x-value "x sub new" and write it x_ν.[7] Here, x_ν is 38 inches. The regression equation predicts %Body Fat as $\hat{y}_\nu = b_0 + b_1 x_\nu$.

Now that we have the predicted value, we construct both intervals around this same number. Both intervals take the form

$$\hat{y}_\nu \pm t^*_{n-2} \times SE.$$

Even the t^* value is the same for both. It's the critical value (from Table T or technology) for $n - 2$ degrees of freedom and the specified confidence level. The intervals differ because they have different standard errors. Our choice of ruler depends on which interval we want.

Each of these factors contributes uncertainty—that is, variability—to the estimate. When there's more spread around the line, the standard error will be larger. If we're less certain of the slope, we'll be less certain of our predictions. If we have more data (n larger), our predictions will be more precise. Finally, predictions farther from the mean of x will have more variability than predictions closer to it. This last factor is new but makes intuitive sense: It's a lot easier to predict a data point near the middle of the data set than far from the center. Because the factors are independent of each other, we can add their variances to find the total variability. The resulting formula for the standard error of the predicted *mean* value explicitly takes into account each of the factors:

$$SE(\hat{\mu}_\nu) = \sqrt{SE^2(b_1) \times (x_\nu - \bar{x})^2 + \frac{s_e^2}{n}}.$$

Individual values vary more than means, so the standard error for a single predicted value has to be larger than the standard error for the mean. In fact, the standard error of a single predicted value has an *extra* source of variability: the variation of individuals

> **MEAN VS. INDIVIDUAL PREDICTIONS**
>
> For the Nenana Ice Classic, someone who planned to place a bet would want to predict this year's breakup time. By contrast, scientists studying global warming are likely to be more interested in the mean breakup time. If you want to gamble, be sure to take into account that the variability is greater when predicting for a single year.

---

[7] Yes, this is a bilingual pun. The Greek letter ν is called "nu." Don't blame me; my co-author suggested this.

around the predicted mean. That appears as the extra variance term, s_e^2, at the end under the square root:

$$SE(\hat{y}_\nu) = \sqrt{SE^2(b_1) \times (x_\nu - \bar{x})^2 + \frac{s_e^2}{n} + s_e^2}.$$

Remember to keep this distinction between the two kinds of standard errors when looking at computer output. The smaller one is for the predicted *mean* value, and the larger one is for a predicted *individual* value.[8]

25.6 Confidence Intervals for Predicted Values

Now that we have standard errors, we can ask how well our analysis can predict the mean *%Body Fat* for men with 38-inch *Waists*. The regression output table (still on page 696) provides most of the numbers we need:

$$s_e = 4.713$$
$$n = 250$$
$$SE(b_1) = 0.074, \text{ and from the data we need to know that}$$
$$\bar{x} = 36.3$$

The regression model gives a predicted value at $x_\nu = 38$ of

$$\hat{y}_\nu = -42.7 + 1.7(38) = 21.9\%.$$

Let's find the 95% confidence interval for the mean *%Body Fat* for all men with 38-inch *Waists*. We find the standard error from the formula:

$$SE(\hat{\mu}_\nu) = \sqrt{0.074^2 \times (38 - 36.3)^2 + \frac{4.713^2}{250}} = 0.32\%.$$

The t^* value that excludes 2.5% in either tail with $250 - 2 = 248$ df is (according to the tables) 1.97.

Putting it all together, we find the margin of error as

$$ME = 1.97(0.32) = 0.63\%.$$

So, we are 95% confident that the mean *%Body Fat* for men with 38-inch *Waists* is

$$21.9\% \pm 0.63\%.$$

Suppose, instead, we want to predict the *%Body Fat* for an individual man with a 38-inch *Waist*. We need the larger standard error:

$$SE(\hat{y}_\nu) = \sqrt{0.074^2 \times (38 - 36.3)^2 + \frac{4.713^2}{250} + 4.713^2} = 4.72\%.$$

The corresponding margin of error is

$$ME = 1.97(4.72) = 9.30\%,$$

so the **prediction interval** is

$$21.9\% \pm 9.30\%.$$

---

[8] You may see the standard error expressions written in other, equivalent ways. The most common alternatives are

$$SE(\hat{\mu}_\nu) = s_e\sqrt{\frac{1}{n} + \frac{(x_\nu - \bar{x})^2}{\sum(x - \bar{x})^2}} \quad \text{and} \quad SE(\hat{y}_\nu) = s_e\sqrt{1 + \frac{1}{n} + \frac{(x_\nu - \bar{x})^2}{\sum(x - \bar{x})^2}}.$$

Figure 25.12
A scatterplot of *%Body Fat* vs. *Waist* size with a least squares regression line. The solid green lines near the regression line show the extent of the 95% confidence intervals for mean *%Body Fat* at each *Waist* size. Notice that there is more uncertainty the farther we are from the mean of the *x* values. The dashed red lines show the prediction intervals. Most of the points are contained within the prediction intervals, but not within the narrower confidence intervals.

We can think of this interval as having a 95% chance of capturing the true *%Body Fat* of a randomly selected man whose *Waist* is 38 inches.[9] Notice how much wider this interval is than the previous one. As we've known since Chapter 17, the mean is much less variable than a randomly selected individual value.

Keep in mind this distinction between the two kinds of confidence intervals: The narrower interval is a **confidence interval for the predicted mean value** at x_ν, and the wider interval is a **prediction interval for an individual** with that *x*-value.

For Example INTERVALS FOR PREDICTIONS

RECAP: We've found a linear model for the relationship between the *logDiam* and *logAge* for terrestrial impact craters:

$$\widehat{logDiam} = 0.358 + 0.527 \, logAge.$$

Let's look at some confidence intervals, keeping in mind that the model is in terms of the logs of these variables, so we must work with logs and then transform back to original units when we are done. (Many of the values substituted into the formulas are from the regression tables shown in the For Examples on pages 699 and 700.)

QUESTION #1: If we wish to hunt for new craters that are 5 million years old, what's a 95% confidence interval for the mean size we'd expect them to be?

ANSWER #1:

$$\widehat{logDiam} = 0.358 + 0.527 \, logAge$$
$$= 0.358 + 0.527 \, log(5)$$
$$= 0.726$$

$$SE(\hat{\mu}) = \sqrt{SE^2(b_1) \times (x - \bar{x})^2 + \frac{s_e^2}{n}}$$

$$= \sqrt{0.0655^2 \times (log(5) - (-0.65631))^2 + \frac{0.6362^2}{39}}$$

$$= 0.1351$$

From Table T, $t^*_{37} = 2.030$, so a 95% CI is $0.726 \pm 2.030(0.1351)$, or $(0.452, 1.00)$. Since these represent logs of the diameters, I'll find the original units:

$$10^{0.452} = 2.8 \text{ km and } 10^{1.00} = 10 \text{ km}.$$

I'm 95% confident that terrestrial craters 5 million years old average between 2.8 and 10 kilometers in diameter.

---

[9]Technically, it's a little more complicated, but it's very close to this.

> **QUESTION #2:** There's just been a news report announcing the discovery of a new crater that is 10 million years old. How large do you expect it to be? What's a 95% prediction interval for your estimate?
>
> **ANSWER #2:** For a crater 10 million years old, $\widehat{logDiam} = 0.358 + 0.527\, log(10) = 0.885$ and $10^{0.885} = 7.674$. I expect the new crater to be about 7.7 km in diameter.
>
> $$SE(\hat{y}) = \sqrt{SE^2(b_1) \times (x - \bar{x})^2 + \frac{s_e^2}{n} + s_e^2}$$
> $$= \sqrt{0.0655^2 \times (log(10) - (-0.65631))^2 + \frac{0.6362^2}{39} + 0.6362^2}$$
> $$= 0.653$$
>
> A 95% CI is $0.885 \pm 2.030(0.653) = (-0.441, 2.211)$; in original units, that's $(0.36, 162.55)$.
>
> I'm 95% confident that the newly discovered 10 million-year-old crater is between 0.36 and 162.55 kilometers in diameter. (Without a lot of data and a very strong association, prediction intervals often aren't very precise!)

*25.7 Logistic Regression

The Pima Indians of southern Arizona are a unique community. Their ancestors were among the first people to cross over into the Americas some 30,000 years ago. For at least two millennia, they have lived in the Sonoran Desert near the Gila River. Known throughout history as a generous people, they have given of themselves for the past 30 years helping researchers at the National Institutes of Health study certain diseases like diabetes and obesity. Young Pima Indians often marry other Pimas, making them an ideal group for genetic researchers to study. Pimas also have an extremely high incidence of diabetes.

Researchers investigating factors for increased risk of diabetes examined data on 768 adult women of Pima Indian heritage.[10] One possible predictor is the body mass index, *BMI*, calculated as weight/height^2, where weight is measured in kilograms and height in meters. We are interested in the relationship between *BMI* and the incidence of diabetes. We might start by looking at boxplots of *BMI* for each group:

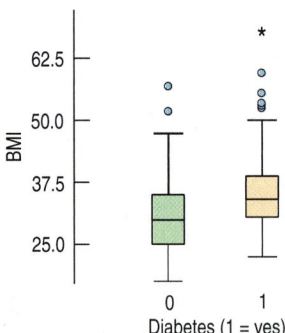

Figure 25.13

Side-by-side boxplots for the two *Diabetes* groups (1 = has diabetes; 0 = doesn't have) show significantly elevated body mass index (BMI) for the women who have diabetes.

From the boxplots, we see that the group with diabetes has a higher mean *BMI*. (A *t*-test would show the difference to be more than 9 SEs from 0 with a P-value < 0.0001.) There is clearly a relationship. Here we've displayed *BMI* as the response and *Diabetes* as the predictor. But the researchers are interested in predicting the increased risk of *Diabetes* due to increased *BMI*, not the other way around.

---

[10] Data can be found at the UCI Machine Learning Repository http://archive.ics.uci.edu/ml/datasets/Pima+Indians+Diabetes.

Diabetes is usually treated as a categorical variable. It might seem unwise at first, but coding having *Diabetes* as 1 and *No Diabetes* as 0 gives rise to a useful model as we'll soon see. Using this coding, we can reverse roles, treat *Diabetes* as the response, and make this scatterplot:

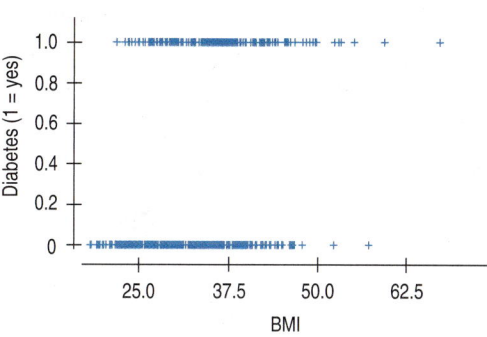

Figure 25.14
A scatterplot of *Diabetes* by *BMI*, using the 0–1 coding for *Diabetes*, shows a shift in *BMI* for the two groups.

We now have the problem of how to model this relationship. Fitting a linear regression to these data has some obvious problems:

Figure 25.15
A regression of *Diabetes* on *BMI* using the 0–1 coding for *Diabetes*. The linear regression is not appropriate because even if the predicted values are interpreted as the probability of having the disease, this line will predict probabilities greater than 1 and less than 0.

The equation says: Predicted *Diabetes* = −0.351 + 0.022 *BMI*. Does this make *any* sense? Suppose someone had a *BMI* of 44. The equation predicts 0.62. What would you guess is the chance that she will have diabetes? If you said about 62%, then you're using the line to model the *probability* of having *Diabetes*.

There are some obvious problems with this model, though. What's the probability that someone with a *BMI* of 10 has diabetes? It's low, but the equation predicts −0.13, obviously an impossible probability. And if we imagined someone with a *BMI* of 70, we might suspect that the probability of her having diabetes is pretty high, but not the predicted value of 1.19.

One simple fix to this problem might be to set all negative probabilities to 0 and all probabilities greater than 1 to 1. That would give a model that looked like this:

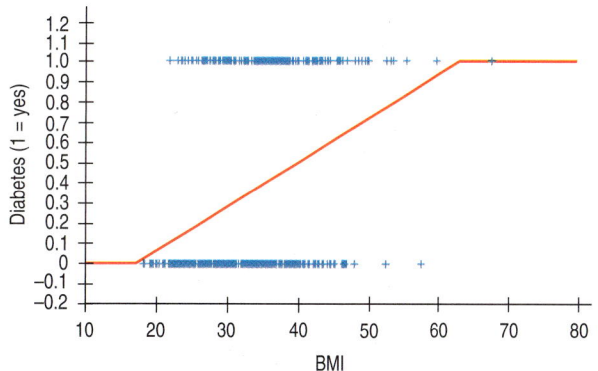

Figure 25.16
This model eliminates the problem of probability values that are negative or that exceed 1, but the corners are neither aesthetically pleasing nor do they make scientific sense.

Figure 25.17
The smooth curve models the probability of having *Diabetes* as a function of *BMI* in a sensible way. The logistic curve shown here is just one of a number of choices for the form of the curve, but all are fairly similar in shape and in the resulting predicted probabilities.

This avoids one problem, but it can't really be correct. The occurrence of *Diabetes* can't be either certain ($p = 1$) or impossible ($p = 0$) based only on *BMI*. And it's not aesthetically pleasing. Real-world changes are likely to be smooth, so we prefer models with smooth transitions rather than corners. That makes good predictive sense, too. There's no reason to expect sharp changes at certain *BMI* values. So instead, we can use a smooth curve to model the probability of having *Diabetes*, like this:

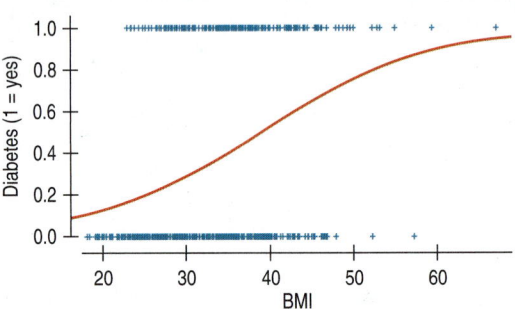

This smoother version is a sensible way to model the probability of having *Diabetes* as a function of *BMI*. There are many curves in mathematics with shapes like this that we might use for our model. One of the most common is the *logistic curve*. The regression based on this curve is called **logistic regression**.

The equation for logistic regression can be written like an ordinary regression on a transformed *y*-variable:

$$\ln\left(\frac{\hat{p}}{1-\hat{p}}\right) = b_0 + b_1 x.$$

The expression on the left-hand side gives the *logistic* curve. The logarithm used is the natural, or base *e*, log, although that doesn't really matter to the shape of the model.

It turns out that the logistic curve has a particularly useful interpretation. Racetrack enthusiasts know that when p is a probability, $\frac{p}{1-p}$ is the *odds* in favor of a success. For example, when the probability of success, p, = 1/3, we'd get the ratio $\frac{1/3}{2/3} = \frac{1}{2}$. We'd say that the odds in favor of success are 1:2 (or we'd probably say the odds *against* it are 2:1). Logistic regression models the *logarithm* of the odds as a linear function of *x*. In fact, nobody really thinks in terms of the log of the odds ratio. But it's the combination of that ratio and the logarithm that gets us the nice S-curved shape. What is important is that we can work backward from a log odds ratio to get the probability—which is often easier to think about.

Because we're not fitting a straight line to the data directly, we can't use ordinary least squares to estimate the parameters b_0 and b_1.[11] Instead, we use special *nonlinear* methods that require a good deal more computation—but computers take care of that for us.

For the Pima Indians data set, the equation is

$$\ln\left(\frac{\hat{p}}{1-\hat{p}}\right) = -4.00 + 0.1025\, BMI.$$

A computer output will typically provide a table like the following:

Term	Estimate	Std Error	ChiSquare	P-Value
Intercept	−3.9967	0.4288	86.8576	<0.00001
BMI	0.1025	0.0126	66.0872	<0.00001

[11]This is a tricky point. We've fit regressions with transformed *y*'s before without special methods. If our data consisted of observed *proportions* of people, we could transform the data using log $(p/(1-p))$ and use linear regression to fit the equation. But all we have are individual 0's and 1's. To be able to fit this equation with our raw data, we need special nonlinear methods.

We usually don't interpret the slope itself (unless you happen to think naturally in log odds), but we can perform a test on whether the slope is 0, similar to the *t*-test we did for linear regression. Unlike linear regression, the ratio of the estimate to its standard error does not have a *t* distribution, but the square of that ratio has a χ^2 distribution.[12] The P-values for both the slope and the intercept clearly indicate that neither is 0.

Once we have decided that the slope is not 0, we can use the model to predict the probabilities. Solving the equation $\ln\left(\dfrac{\hat{p}}{1-\hat{p}}\right) = b_0 + b_1 x$ for $\hat{p}$ gives

$$\hat{p} = \frac{1}{1 + e^{-(b_0 + b_1 x)}}$$

and the logistic equation guarantees that the estimate of $\hat{p}$ will be between 0 and 1. Fortunately, technology can provide these probability estimates, producing the curve shown previously and giving an estimate for the probability at any *BMI* value.

Response variables that are dichotomous (having only two possible values) like the variable *Diabetes* are common, so there's a widespread need to model data like these. Phone companies want to predict who will switch carriers, credit card companies want to know which transactions are likely to be fraudulent, and loan companies want to know who is most likely to declare bankruptcy. All of these are potential application areas for logistic regression. It's not surprising that logistic regression has become an important modeling tool in the toolbox of analysts in science and industry in the past decade. By understanding the basics of what logistic regression can do, you can expand your ability to apply regression to many other important applications.

WHAT CAN GO WRONG?

In this chapter, we've added inference to the regression explorations that we did in Chapters 7 and 8. Everything covered in those chapters that could go wrong with regression can still go wrong. It's probably a good time to review Chapter 8. Take your time; we'll wait.

With inference, we've put numbers on our estimates and predictions, but these numbers are only as good as the model. Here are the main things to watch out for:

- **Don't fit a linear regression to data that aren't straight.** The linearity assumption is the most fundamental assumption. If the relationship between *x* and *y* isn't approximately linear, there's no sense in fitting a straight line to it.

- **Watch out for the plot thickening.** The common part of confidence and prediction intervals is the estimate of the error standard deviation, the spread around the line. If it changes with *x*, the estimate won't make sense. Imagine making a prediction interval for these data.

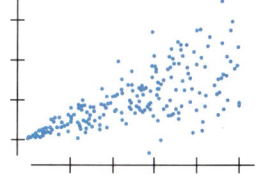

When *x* is small, we can predict *y* precisely, but as *x* gets larger, it's much harder to pin *y* down. Unfortunately, if the spread changes, the single value of s_e won't pick that up. The prediction interval will use the average spread around the line, with the result that we'll be too pessimistic about our precision for low *x*-values and too optimistic for high *x*-values. A re-expression of *y* is often a good fix for changing spread.

- **Make sure the errors are Normal.** When we make a prediction interval for an individual, the Central Limit Theorem can't come to our rescue. For us to believe the prediction interval, the errors must be from the Normal model. Check the histogram and Normal probability plot of the residuals to see if this assumption looks reasonable.

---

[12] Here's that nonlinear fitting popping up again. Don't let it bother you. You've seen χ^2 models before, and all you really need to look at for a test is the P-value.

- **Watch out for extrapolation.** It's tempting to think that because we have prediction *intervals*, they'll take care of all our uncertainty so we don't have to worry about extrapolating. Wrong. The interval is only as good as the model. The uncertainty our intervals predict is correct only if our model is true. There's no way to adjust for wrong models. That's why it's always dangerous to predict for *x*-values that lie far from the center of the data.

- **Watch out for influential points and outliers.** We always have to be on the lookout for a few points that have undue influence on our estimated model—and regression is certainly no exception.

- **Watch out for one-tailed tests.** Because tests of hypotheses about regression coefficients are usually two-tailed, software packages report two-tailed P-values. If you are using software to conduct a one-tailed test about slope, you'll need to divide the reported P-value in half.

CONNECTIONS

Regression inference is connected to almost everything we've done so far. Scatterplots are essential for checking linearity and whether the plot thickens. Histograms and normal probability plots come into play to check the Nearly Normal condition. And we're still thinking about the same attributes of the data in these plots as we were back in the first part of the book.

Regression inference is also connected to just about every inference method we have seen for measured data. The assumption that the spread of data about the line is constant is essentially the same as the assumption of equal variances required for the pooled-*t* methods. Our use of all the residuals together to estimate their standard deviation is a form of pooling.

Inference for regression is closely related to inference for means, so your understanding of means transfers directly to your understanding of regression. Here's a table that displays the similarities:

	Means	Regresssion Slope
Parameter	μ	β_1
Statistic	$\bar{y}$	b_1
Population Spread Estimate	$s_y = \sqrt{\dfrac{\sum(y-\bar{y})^2}{n-1}}$	$s_e = \sqrt{\dfrac{\sum(y-\hat{y})^2}{n-2}}$
Standard Error of the Statistic	$SE(\bar{y}) = \dfrac{s_y}{\sqrt{n}}$	$SE(b_1) = \dfrac{s_e}{s_x\sqrt{n-1}}$
Test Statistic	$\dfrac{\bar{y}-\mu_0}{SE(\bar{y})} \sim t_{n-1}$	$\dfrac{b_1-\beta_1}{SE(b_1)} \sim t_{n-2}$
Margin of Error	$ME = t^*_{n-1} \times SE(\bar{y})$	$ME = t^*_{n-2} \times SE(b_1)$

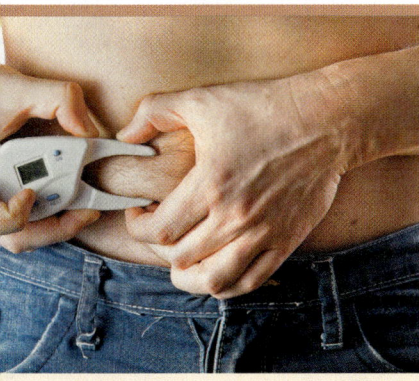

What Have We Learned?

Learning Objectives

Apply your understanding of inference for means using Student's t to inference about regression coefficients.

Know the Assumptions and Conditions for inference about regression coefficients and how to check them, in this order:

- **Linearity Assumption**, checked with the Straight Enough Condition by examining a scatterplot of y vs. x or of the residuals plotted against the predicted values.
- **Independence Assumption**, which can't be checked, but is more plausible if the data were collected with appropriate randomization—the Randomization Condition.
- **Equal Variance Assumption**, which requires that the spread around the regression model be the same everywhere. We check it with the Does the Plot Thicken? Condition, assessed with a scatterplot of the residuals versus the predicted values.
- **Normal Population Assumption**, which is required to use Student's t-models unless the sample size is large. Check it with the Nearly Normal Condition by making a histogram or normal probability plot of the residuals.

Know the components of the standard error of the slope coefficient:

- The standard deviation of the residuals, $s_e = \sqrt{\dfrac{\sum(y - \hat{y})^2}{n - 2}}$
- The standard deviation of x, $s_x = \sqrt{\dfrac{(x - \bar{x})^2}{n - 1}}$
- The sample size, n

Be able to find and interpret the standard error of the slope.

- $SE(b_1) = \dfrac{s_e}{s_x \sqrt{n - 1}}$
- The standard error of the slope is the estimated standard deviation of the sampling distribution of the slope.

State and test the standard null hypothesis on the slope.

- $H_0: \beta_1 = 0$. This would mean that x and y are not linearly related.
- We test this null hypothesis using the t-statistic $t = \dfrac{b_1 - 0}{SE(b_1)}$

Know how to use a t-test to test whether the true correlation is zero.

- $t = r\sqrt{\dfrac{n - 2}{1 - r^2}}$

Construct and interpret a confidence interval for the predicted mean value corresponding to a specified value, x_ν.

- $\hat{y}_\nu \pm t^*_{n-2} \times SE(\hat{\mu}_\nu)$, where $SE(\hat{\mu}_\nu) = \sqrt{SE^2(b_1) \times (x_\nu - \bar{x})^2 + \dfrac{s_e^2}{n}}$

Construct and interpret a confidence interval for an individual predicted value corresponding to a specified value, x_ν.

- $\hat{y}_\nu \pm t^*_{n-2} \times SE(\hat{y}_\nu)$, where $SE(\hat{y}_\nu) = \sqrt{SE^2(b_1) \times (x_\nu - \bar{x})^2 + \dfrac{s_e^2}{n} + s_e^2}$

Review of Terms

Residual standard deviation The spread of the data around the regression line is measured with the residual standard deviation, s_e:

$$s_e = \sqrt{\frac{\sum(y - \hat{y})^2}{n - 2}} = \sqrt{\frac{\sum e^2}{n - 2}} \quad \text{(p. 697)}.$$

t-test for the regression slope When the assumptions are satisfied, we can perform a test for the slope coefficient. We usually test the null hypothesis that the true value of the slope is zero against the alternative that it is not. A zero slope would indicate a complete absence of linear relationship between y and x.

To test $H_0: \beta_1 = 0$, we find

$$t = \frac{b_1 - 0}{SE(b_1)}$$

where

$$SE(b_1) = \frac{s_e}{\sqrt{n - 1}\, s_x}, \quad s_e = \sqrt{\frac{\sum(y - \hat{y})^2}{n - 2}},$$

n is the number of cases, and s_x is the standard deviation of the x-values. We find the P-value from the Student's t-model with $n - 2$ degrees of freedom (p. 700).

Confidence interval for the regression slope When the assumptions are satisfied, we can find a confidence interval for the slope parameter from $b_1 \pm t^*_{n-2} \times SE(b_1)$. The critical value, t^*_{n-2}, depends on the confidence level specified and on Student's t-model with $n - 2$ degrees of freedom (p. 700).

Confidence interval for a predicted mean value Different samples will give different estimates of the regression model and, so, different predicted values for the same value of x. We find a confidence interval for the mean of these predicted values at a specified x-value, x_ν, as $\hat{y}_\nu \pm t^*_{n-2} \times SE(\hat{\mu}_\nu)$, where

$$SE(\hat{\mu}_\nu) = \sqrt{SE^2(b_1) \times (x_\nu - \bar{x})^2 + \frac{s_e^2}{n}}.$$

The critical value, t^*_{n-2}, depends on the specified confidence level and the Student's t-model with $n - 2$ degrees of freedom (p. 705, 707).

Prediction interval for an individual Different samples will give different estimates of the regression model and, so, different predicted values for the same value of x. We can make a confidence interval to capture a certain percentage of the entire distribution of predicted values. This makes it much wider than the corresponding confidence interval for the mean. The confidence interval takes the form $\hat{y}_\nu \pm t^*_{n-2} \times SE(\hat{y}_\nu)$, where

$$SE(\hat{y}_\nu) = \sqrt{SE^2(b_1) \times (x_\nu - \bar{x})^2 + \frac{s_e^2}{n} + s_e^2}.$$

The critical value, t^*_{n-2}, depends on the specified confidence level and the Student's t-model with $n - 2$ degrees of freedom (p. 706, 707).

*Logistic regression A regression model suitable for a categorical response variable with two categories (p. 710).

On the Computer REGRESSION ANALYSIS

All statistics packages make a table of results for a regression. These tables differ slightly from one package to another, but all are essentially the same. We've seen two examples of such tables already.

All packages offer analyses of the residuals. With some, you must request plots of the residuals as you request the regression. Others let you find the regression first and then analyze the residuals afterward. Either way, your analysis is not complete if you don't check the residuals with a histogram or Normal probability plot and a scatterplot of the residuals against x or the predicted values.

You should, of course, always look at the scatterplot of your two variables before computing a regression.

Regressions are almost always found with a computer or calculator. The calculations are too long to do conveniently by hand for data sets of any reasonable size. No matter how the regression is computed, the results are usually presented in a table that has a standard form. Here's a portion of a typical regression results table, along with annotations showing where the numbers come from:

A S *Activity:* **Regression on the Computer.** How fast is the universe expanding? And how old is it? A prominent astronomer used regression to astound the scientific community. Read the story, analyze the data, and interactively learn about each of the numbers in a typical computer regression output table.

The regression table gives the coefficients (once you find them in the middle of all this other information), so we can see that the regression equation is

$$\widehat{\%BF} = -42.73 + 1.7\, Waist$$

and that the R^2 for the regression is 67.8%. (Is accounting for 68% of the variation in %*Body Fat* good enough to be useful? Is a prediction ME of more than 9% good enough? Health professionals might not be satisfied.)

The column of *t*-ratios gives the test statistics for the respective null hypotheses that the true values of the coefficients are zero. The corresponding P-values are also usually reported.

DATA DESK

- Select *Y*- and *X*-variables.
- From the **Calc** menu, choose **Regression**.
- Data Desk displays the regression table.
- Select plots of residuals from the Regression table's HyperView menu.

COMMENTS

You can change the regression by dragging the icon of another variable over either the *Y*- or *X*-variable name in the table and dropping it there. The regression will recompute automatically.

EXCEL

- In Excel 2003 and earlier, select Data Analysis from the **Tools** menu. In Excel 2007, select Data Analysis from the **Analysis Group** on the Data Tab.
- Select Regression from the **Analysis Tools** list.
- Click the **OK** button.
- Enter the data range holding the Y-variable in the box labeled "Y-range."
- Enter the range of cells holding the X-variable in the box labeled "X-range."
- Select the **New Worksheet Ply** option.
- Select **Residuals** options. Click the **OK** button.

COMMENTS

The Y and X ranges do not need to be in the same rows of the spreadsheet, although they must cover the same number of cells. But it is a good idea to arrange your data in parallel columns as in a data table.

Although the dialog offers a Normal probability plot of the residuals, the data analysis add-in does not make a correct probability plot, so don't use this option.

Excel calls the standard deviation of the residuals the "Standard Error." This is a common error. Don't be confused; it is not $SE(y)$, but rather s_e.

JMP

- From the **Analyze** menu, select **Fit Y by X**.
- Select variables: a Y, Response variable, and an X, Factor variable. Both must be continuous (quantitative).
- JMP makes a scatterplot.
- Click on the red triangle beside the heading labeled **Bivariate Fit . . .** and choose **Fit Line**. JMP draws the least squares regression line on the scatterplot and displays the results of the regression in tables below the plot.
- The portion of the table labeled "Parameter Estimates" gives the coefficients and their standard errors, t-ratios, and P-values.

COMMENTS

JMP chooses a regression analysis when both variables are "Continuous." If you get a different analysis, check the variable types.

The Parameter table does not include the residual standard deviation s_e. You can find that as Root Mean Square Error in the Summary of Fit panel of the output.

MINITAB

- Choose **Regression** from the **Stat** menu.
- Choose **Regression . . .** from the **Regression** submenu.
- In the Regression dialog, assign the Y-variable to the Response box and assign the X-variable to the Predictors box.
- Click the **Graphs** button.
- In the Regression-Graphs dialog, select **Standardized residuals**, and check **Normal plot of residuals** and **Residuals versus fits**.
- Click the **OK** button to return to the Regression dialog.
- Click the **OK** button to compute the regression.

COMMENTS

You can also start by choosing a Fitted Line plot from the **Regression** submenu to see the scatterplot first—usually good practice.

R

From a data frame called mydata with variables x and y, to find the linear model:

- mylm=lm($y \sim x$,mydata)
- summary(mylm) # shows a summary of the model including estimates, SEs, and the F-statistic

To get confidence or prediction intervals use:

- predict(mylm,interval="confidence")

or

- predict(mylm, interval="prediction")

COMMENTS

Predictions on points not found in the original data frame can be found from predict as well. In the new data frame (called, say, mynewdata), there must be a predictor variable with the same name as the original "x" variable. Then predict(mylm,newdata=mynewdata) will produce predictions at all the "x" values of mynewdata.

SPSS

- Choose **Regression** from the **Analyze** menu.
- Choose **Linear** from the **Regression** submenu.
- In the Linear Regression dialog that appears, select the Y-variable and move it to the dependent target. Then move the X-variable to the independent target.
- Click the **Plots** button.
- In the Linear Regression Plots dialog, choose to plot the *SRESIDs against the *ZPRED values.
- Click the **Continue** button to return to the Linear Regression dialog.
- Click the **OK** button to compute the regression.

STATCRUNCH

- Click on **Stat**.
- Choose **Regression » Simple Linear**.
- Choose X and Y variable names from the list of columns.
- Click on **Next**, and then click on **Next** again.
- Indicate that you want to see a residuals plot and a histogram of the residuals.
- Click on **Calculate**.
- Click on **Next** to see any plots you chose.

COMMENTS

Be sure to check the conditions for regression inference by looking at both the residuals plot and a histogram of the residuals.

Although StatCrunch will not create a confidence interval for β, the regression output reports the information necessary: the observed b, $SE(b)$, and the number of degrees of freedom.

TI-83/84 PLUS

Under **STAT TESTS** choose **LinRegTTest**. Specify the two lists where the data are stored and (usually) choose the two-tail option. In addition to reporting the calculated value of t and the P-value, the calculator will tell you the coefficients of the regression equation (a and b), the values of r^2 and r, and the value of s you need for confidence or prediction intervals.

Exercises

Section 25.2

1. **Graduation rates 2013** A prestigious college is interested in factors that might be associated with better graduation rates. The administrators wonder whether there is a relationship between acceptance rates and graduation rates. Before proceeding with their regression inference, what conditions and assumptions must be satisfied? Are they? (Source: *US News and World Report* National University Rankings October 2013. http://colleges.usnews.rankingsandreviews.com/best-colleges/rankings/national-universities/data/spp%2B50/page+1)

Response variable is Grad%
R-squared = 75.2%
s = 0.0254 with 50 − 2 = 48 degrees of freedom

Variable	Coefficient	SE(Coeff)	t-Ratio	P-Value
Intercept	0.975175	0.0074	131	≤0.0001
Acceptance %	−0.284895	0.0236	−12.1	≤0.0001

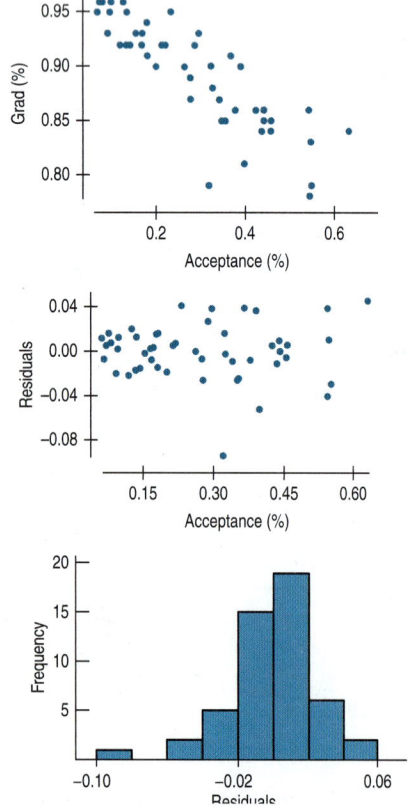

2. **Shoot to score** A college hockey coach collected data from the 2010–2011 National Hockey League season. He hopes to convince his players that the number of shots taken has an affect on the number of goals scored. The coach performed a preliminary analysis, using the scoring statistics from 293 offensive players who play professional hockey. He predicts *Goals* from number of *Shots* (taken for the season). Discuss each of the conditions and assumptions required for him to proceed with the regression analysis. (www.nhl.com)

Dependent variable is Goals
R-squared = 63.4%
s = 5.13 with 293 − 2 = 291 degrees of freedom

Variable	Coefficient	SE(Coeff)	t-Ratio	P-Value
Intercept	−1.77095	0.9087	−1.9488	0.052
Shots	0.11938	0.0053	22.460	<0.0001

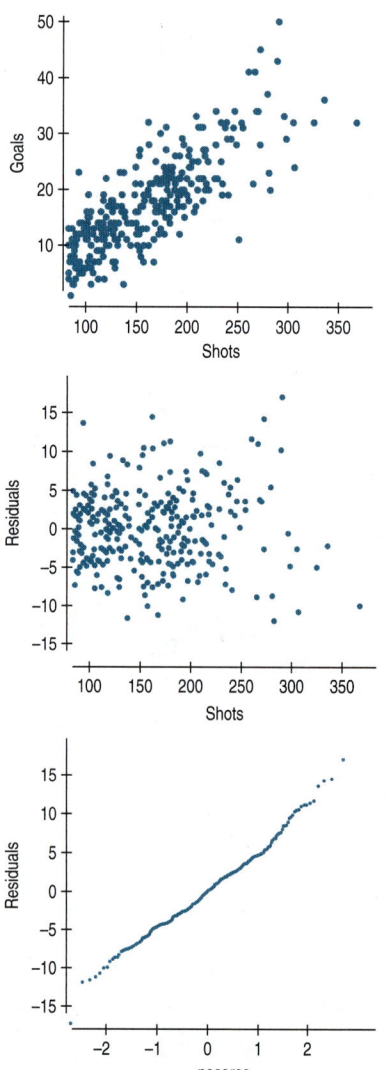

Section 25.3

3. **Graduation rates 2013, part II** Using the regression output in Exercise 1, identify the error standard deviation and explain what it means in the context of the problem.

4. **Shoot to score another one** Using the regression output from Exercise 2, identify the error standard deviation and explain its meaning with a sentence in context.

5. **Graduation rates 2013, part III** Continuing with the regression of Exercise 1, write a sentence that explains the meaning of the standard error of the slope of the regression line, $SE(b_1) = 0.0240$.

6. **Shoot to score, hat trick** Returning to the results of Exercise 2, write a sentence to explain the meaning of the standard error of the slope of the regression line, $SE(b_1) = 0.0053$.

Section 25.4

7. **Graduation 2013, part IV** The college administrators in Exercise 1 tested the hypotheses $H_0: \beta_1 = 0$ vs. $H_A: \beta_1 \neq 0$ and rejected the null hypothesis because the P-value was less than 0.0001. What can they conclude about the relationship between admission rates and graduation rates?

8. **Shoot to score, number four** What can the hockey coach in Exercise 2 conclude about shooting and scoring goals from the fact that the P-value < 0.0001 for the slope of the regression line? Write a sentence in context.

9. **Graduation 2013, part V** The college administrators in Exercise 1 constructed a 95% confidence interval for the slope of their regression line. Interpret the meaning of their interval $(-0.332, -0.238)$ within the context of the problem.

10. **Shoot to score, overtime** The coach in Exercise 2 found a 95% confidence interval for the slope of his regression line. Recall that he is trying to predict total goals scored based on shots taken. Interpret with a sentence the meaning of the interval 0.12 ± 0.01.

Section 25.5

11. **Graduation 2013, part VI** The college described in Exercise 1 admits about 33% of applicants each year. Which will be larger, the standard error of the predicted *6-Year Graduation Rate* of this college, or the standard error of the mean *6-Year Graduation Rate* of all colleges admitting 33% of their applicants? Explain briefly.

12. **Shoot to score, double overtime** One of the coach's best players has been in a slump, and the coach is trying to convince him to take more shots. Last year, the player took 104 shots. Would the standard error of the predicted number of goals he shot be larger or smaller than the standard error of the mean predicted number of all players who took 104 shots? Explain briefly.

Section 25.6

13. **Graduation 2013, part VII** As in Exercise 11, the college described in Exercise 1 admits about 33% of applicants each year. Using their regression model, the administrators calculated a 95% confidence interval for the mean *6-Year Graduation Rate* for all top colleges who admit 33% of students. The interval was (87.35%, 88.88%). Interpret the meaning of this interval with a sentence.

14. **Shoot to score, triple overtime** As in Exercise 12, one of the coach's best players has been in a slump, and the coach is trying to convince him to take more shots. Last year, the player took 104 shots. The coach calculated a 95% confidence interval for the mean number of goals a player would score if 104 shots were taken. The interval turned out to be 9.8 to 11.5 goals. Summarize the interval with a sentence in context.

15. **Graduation 2013, part VIII** The administrators from Exercise 1 also determined a 95% prediction interval for the graduation rate of a school admitting 33% of applicants to be (82.96%, 93.28%). Carefully explain what the 95% indicates for this interval.

16. **Shoot to score, again** The coach, still working with his superstar from Exercise 12, went on to suggest the player shoot more. Hopeful for 150 shots for the season, he calculated a 95% prediction interval for the number of goals this player might score. If the prediction interval for goals scored was (6.03, 24.25), interpret the meaning of this interval with a sentence in context.

Chapter Exercises

17. **Graduation 2013 party** Would it be wise for the administrators at the college described in Exercise 1 to conclude that lowering their admission rate (i.e., making it harder to get in) would help to increase their graduation rates? Why or why not?

18. **Shoot to score, shootout** The coach from Exercise 2 called a team meeting to summarize the results from his study. Would it be a good strategy to tell the players that all they need to do is to shoot more and the goals will follow?

19. **Tracking hurricanes 2012** In Chapter 6, we looked at data from the National Oceanic and Atmospheric Administration about their success in predicting hurricane tracks. On the following page is a scatterplot of the error (in nautical miles) for predicting hurricane locations 24 hours in the future vs. the year in which the prediction (and the hurricane) occurred.

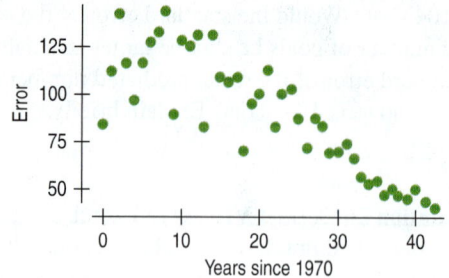

In Chapter 6, we could describe this relationship only in general terms. Now we can learn more. Here is the regression analysis:

Response variable is   24 Error
R-squared = 72.5% R-squared (adjusted) = 71.8%
s = 16.13 with 43 − 2 = 41 degrees of freedom

Variable	Coefficient	SE(Coeff)	t-Ratio	P-Value
Intercept	133.024	4.835	27.5	≤0.0001
Years Since 1970	−2.06016	0.1982	−10.4	≤0.0001

a) Explain in words and numbers what the regression says.
b) State the hypothesis about the slope (both numerically and in words) that describes how hurricane prediction quality has changed.
c) Assuming that the assumptions for inference are satisfied, perform the hypothesis test and state your conclusion. Be sure to state it in terms of prediction errors and years.
d) Explain what the R-squared means in terms of this regression.

20. Drug use 2013 The *2013 World Drug Report* investigated the prevalence of drug use as a percentage of the population aged 15 to 64. Data from 32 European countries are shown in the following scatterplot and regression analysis. (Source: *World Drug Report*, 2013. www.unodc.org/unodc/en/data-and-analysis/WDR-2013.html)

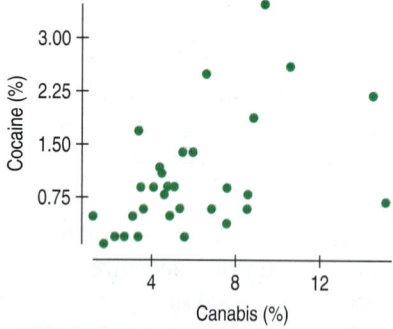

Response variable is Cocaine
R-squared = 25.8%
s = 0.7076 with 32 − 2 = 30 degrees of freedom

Variable	Coefficient	SE(Coeff)	t-Ratio	P-Value
Intercept	0.296899	0.2627	1.13	0.2673
Canabis	0.123638	0.0382	3.23	0.0030

a) Explain in context what the regression says.
b) State the hypothesis about the slope (both numerically and in words) that describes how use of marijuana is associated with other drugs.
c) Assuming that the assumptions for inference are satisfied, perform the hypothesis test and state your conclusion in context.
d) Explain what R-squared means in context.
e) Do these results indicate that marijuana use leads to the use of harder drugs? Explain.

21. Movie budgets[13] How does the cost of a movie depend on its length? Data on the cost (millions of dollars) and the running time (minutes) for major release films of 2005 are summarized in these plots and computer output:

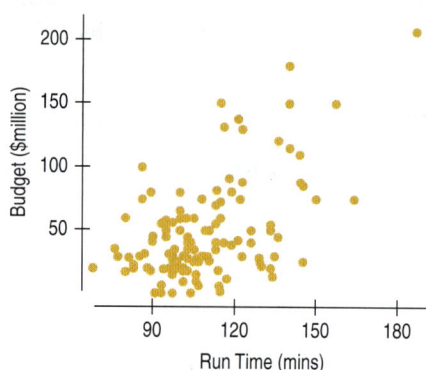

Dependent variable is Budget($million)
R-squared = 27.3%
s = 32.95 with 120 − 2 = 118 degrees of freedom

Variable	Coefficient	SE(Coeff)	t-Ratio	P-Value
Intercept	−63.9981	17.12	−3.74	0.0003
Run Time	1.02648	0.1540	6.66	≤0.0001

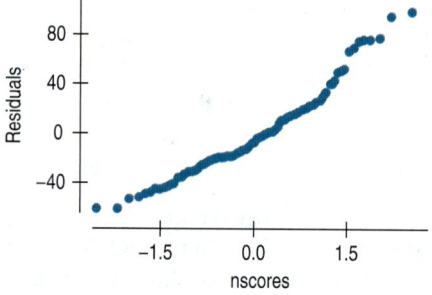

(continued)

[13]Data have been corrected since previous editions.

a) Explain in context what the regression says.
b) The intercept is negative. Discuss its value.
c) The output reports $s = 32.95$. Explain what that means in this context.
d) What's the value of the standard error of the slope of the regression line?
e) Explain what that means in this context.

22. **Saratoga house prices** How does the price of a house depend on its size? Data from Saratoga, New York, on 1064 randomly selected houses that had been sold include data on price ($1000s) and size ($1000s ft^2), producing the following graphs and computer output:

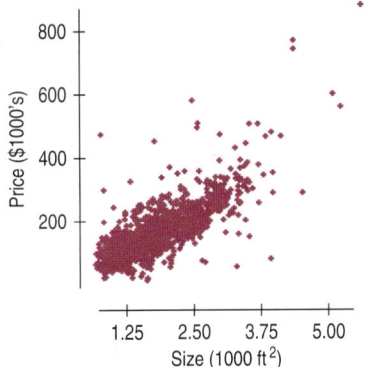

Dependent variable is Price
R-squared = 59.5%
$s = 53.79$ with $1064 - 2 = 1062$ degrees of freedom

Variable	Coefficient	SE(Coeff)	t-Ratio	P-Value
Intercept	−3.11686	4.688	−0.665	0.5063
Size	94.4539	2.393	39.5	≤0.0001

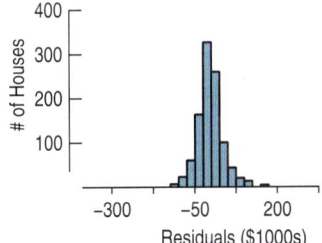

a) Explain in context what the regression says.
b) The intercept is negative. Discuss its value, taking note of its P-value.
c) The output reports $s = 53.79$. Explain what that means in this context.

d) What's the value of the standard error of the slope of the regression line?
e) Explain what that means in this context.

23. **Movie budgets, the sequel** Exercise 21 shows computer output examining the association between the length of a movie and its cost.
a) Check the assumptions and conditions for inference.
b) Find a 95% confidence interval for the slope and interpret it in context.

24. **Second home** Exercise 22 shows computer output examining the association between the sizes of houses and their sale prices.
a) Check the assumptions and conditions for inference.
b) Find a 95% confidence interval for the slope and interpret it in context.

25. **Hot dogs** Healthy eating probably doesn't include hot dogs, but if you are going to have one, you'd probably hope it's low in both calories and sodium. Recently, *Consumer Reports* listed the number of calories and sodium content (in milligrams) for 13 brands of all-beef hot dogs it tested. Examine the association, assuming that the data satisfy the conditions for inference.

Dependent variable is Sodium
R-squared = 60.5%
$s = 59.66$ with $13 - 2 = 11$ degrees of freedom

Variable	Coefficient	SE(Coeff)	t-Ratio	P-Value
Constant	90.9783	77.69	1.17	0.2663
Calories	2.29959	0.5607	4.10	0.0018

a) State the appropriate hypotheses about the slope.
b) Test your hypotheses and state your conclusion in the proper context.

26. **Cholesterol** Does a person's cholesterol level tend to change with age? Data collected from 1406 adults aged 45 to 62 produced the regression analysis shown. Assuming that the data satisfy the conditions for inference, examine the association between age and cholesterol level.

Dependent variable is Chol
$s = 46.16$

Variable	Coefficient	SE(Coeff)	t-Ratio	P-Value
Intercept	194.232	13.55	14.3	≤0.0001
Age	0.771639	0.2574	3.00	0.0056

a) State the appropriate hypothesis for the slope.
b) Test your hypothesis and state your conclusion in the proper context.

27. **Second frank** Look again at Exercise 25's regression output for the calorie and sodium content of hot dogs.
a) The output reports $s = 59.66$. Explain what that means in this context.
b) What's the value of the standard error of the slope of the regression line?
c) Explain what that means in this context.

28. More cholesterol Look again at Exercise 26's regression output for age and cholesterol level.
   a) The output reports $s = 46.16$. Explain what that means in this context.
   b) What's the value of the standard error of the slope of the regression line?
   c) Explain what that means in this context.

29. Last dog Based on the regression output seen in Exercise 25, create a 95% confidence interval for the slope of the regression line and interpret your interval in context.

30. Cholesterol, finis Based on the regression output seen in Exercise 26, create a 95% confidence interval for the slope of the regression line and interpret it in context.

31. Marriage age 2011 The scatterplot suggests a decrease in the difference in ages at first marriage for men and women since 1975. We want to examine the regression to see if this decrease is significant.

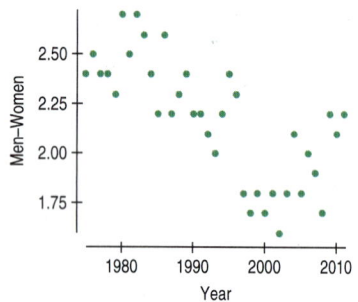

Response variable is Men-Women
R-squared = 51.9%
s = 0.2143 with 37 − 2 = 35 degrees of freedom

Variable	Coefficient	SE(Coeff)	t-Ratio	P-Value
Intercept	42.5662	6.578	6.47	≤0.0001
Year	−0.020270	0.0033	−6.14	≤0.0001

a) Write appropriate hypotheses.
b) Here is the residuals plot and a histogram of the residuals. Do you think the conditions for inference are satisfied? Explain.

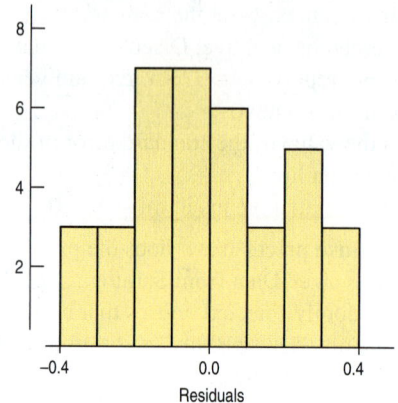

c) Test the hypothesis and state your conclusion about the trend in age at first marriage.

32. Used cars 2010 Vehix.com offered several used Toyota Corollas for sale. Listed below are the ages of the cars and the advertised prices.

Age (yr)	Price ($)	Age (yr)	Price ($)
1	15988	6	9995
1	13988	6	11988
2	14488	7	8990
3	10995	8	9488
3	13998	8	8995
4	13622	9	5990
4	12810	10	4100
5	9988	12	2995

a) Make a scatterplot for these data.
b) Do you think a linear model is appropriate? Explain.
c) Find the equation of the regression line.
d) Check the residuals to see if the conditions for inference are met.

33. Marriage age 2011, again Based on the analysis of marriage ages since 1975 given in Exercise 31, find a 95% confidence interval for the rate at which the age gap is closing. Explain what your confidence interval means.

34. Used cars 2010, again Based on the analysis of used car prices you did for Exercise 32, create a 95% confidence interval for the slope of the regression line and explain what your interval means in context.

35. Fuel economy
A consumer organization has reported test data for 50 car models. We will examine the association between the weight of the car (in thousands of pounds) and the fuel efficiency (in miles per gallon). Here are the scatterplot, summary statistics, and regression analysis:

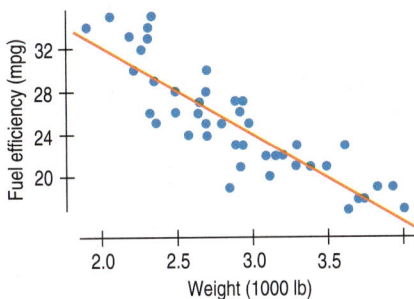

Variable	Count	Mean	StdDev
MPG	50	25.0200	4.83394
Wt(1000's)	50	2.88780	0.511656

Dependent variable is MPG
R-squared = 75.6%
s = 2.413 with 50 − 2 = 48 df

Variable	Coefficient	SE(Coeff)	t-Ratio	P-Value
Intercept	48.7393	1.976	24.7	≤0.0001
Weight	−8.2136	0.674	−12.2	≤0.0001

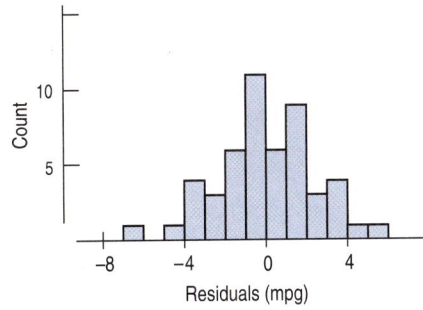

a) Is there strong evidence of an association between the weight of a car and its gas mileage? Write an appropriate hypothesis.
b) Are the assumptions for regression satisfied?
c) Test your hypothesis and state your conclusion.

36. SAT scores
How strong was the association between student scores on the Math and Verbal sections of the old SAT? Scores on each ranged from 200 to 800 and were widely used by college admissions offices. Here are summaries and plots of the scores for a graduating class at Ithaca High School:

Variable	Count	Mean	Median	StdDev	Range	IntQRange
Verbal	162	596.296	610	99.5199	490	140
Math	162	612.099	630	98.1343	440	150

Dependent variable is Math
R-squared = 46.9%
s = 71.75 with 162 − 2 = 160 df

Variable	Coefficient	SE(Coeff)	t-Ratio	P-Value
Intercept	209.554	34.35	6.10	≤0.0001
Verbal	0.675	0.0568	11.9	≤0.0001

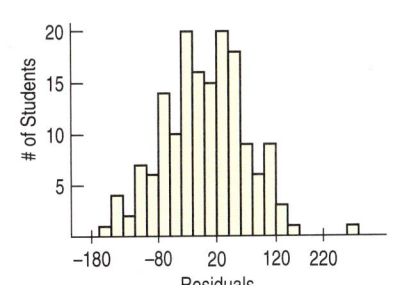

a) Is there evidence of an association between Math and Verbal scores? Write an appropriate hypothesis.
b) Discuss the assumptions for inference.
c) Test your hypothesis and state an appropriate conclusion.

37. Fuel economy, part II Consider again the data in Exercise 35 about the gas mileage and weights of cars.
   a) Create a 95% confidence interval for the slope of the regression line.
   b) Explain in this context what your confidence interval means.

38. SATs, part II Consider the high school SAT scores data from Exercise 36.
   a) Find a 90% confidence interval for the slope of the true line describing the association between Math and Verbal scores.
   b) Explain in this context what your confidence interval means.

39. Fuel economy, part III Consider again the data in Exercise 35 about the gas mileage and weights of cars.
   a) Create a 95% confidence interval for the average fuel efficiency among cars weighing 2500 pounds, and explain what your interval means.
   b) Create a 95% prediction interval for the gas mileage you might get driving your new 3450-pound SUV, and explain what that interval means.

40. SATs, again Consider the high school SAT scores data from Exercise 36 once more.
   a) Find a 90% confidence interval for the mean SAT-Math score for all students with an SAT-Verbal score of 500.
   b) Find a 90% prediction interval for the Math score of the senior class president if you know she scored 710 on the Verbal section.

41. Cereals A healthy cereal should be low in both calories and sodium. Data for 77 cereals were examined and judged acceptable for inference. The 77 cereals had between 50 and 160 calories per serving and between 0 and 320 mg of sodium per serving. Here's the regression analysis:

Dependent variable is Sodium
R-squared = 9.0%
s = 80.49 with 77 − 2 = 75 degrees of freedom

Variable	Coefficient	SE(Coeff)	t-Ratio	P-Value
Intercept	21.4143	51.47	0.416	0.6786
Calories	1.29357	0.4738	2.73	0.0079

   a) Is there an association between the number of calories and the sodium content of cereals? Explain.
   b) Do you think this association is strong enough to be useful? Explain.

42. Brain size Does your IQ depend on the size of your brain? A group of female college students took a test that measured their verbal IQs and also underwent an MRI scan to measure the size of their brains (in 1000s of pixels). The scatterplot and regression analysis are shown, and the assumptions for inference were satisfied.

Dependent variable is IQ_Verbal
R-squared = 6.5%

Variable	Coefficient	SE(Coeff)
Intercept	24.1835	76.38
Size	0.0988	0.088

   a) Test an appropriate hypothesis about the association between brain size and IQ.
   b) State your conclusion about the strength of this association.

43. Cereals, part 2 Further analysis of the data for the breakfast cereals in Exercise 41 looked for an association between *Fiber* content and *Calories* by attempting to construct a linear model. Here are several graphs. Which of the assumptions for inference are violated? Explain.

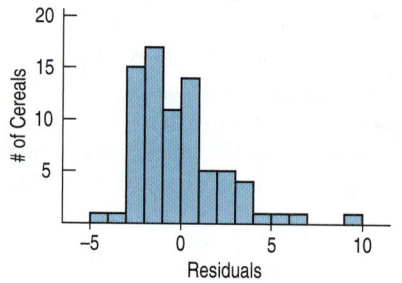

44. Winter The output shows an attempt to model the association between average *January Temperature* (in degrees Fahrenheit) and *Latitude* (in degrees north of the equator) for 59 U.S. cities. Which of the assumptions for inference do you think are violated? Explain.

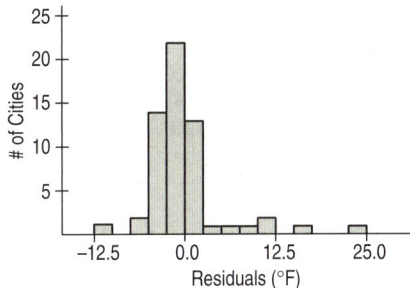

45. Streams Biologists studying the effects of acid rain on wildlife collected data from 172 streams in the Adirondack Mountains. They recorded the *pH* (acidity) of the water and the *BCI*, a measure of biological diversity. Here's a scatterplot of *BCI* against *pH* for the 163 streams for which we have these data:

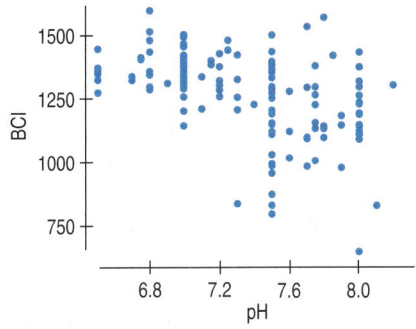

And here is part of the regression analysis:

Dependent variable is BCI
R-squared = 27.1%
s = 140.4 with 163 − 2 = 161 degrees of freedom

Variable	Coefficient	SE(Coeff)
Intercept	2733.37	187.9
pH	−197.694	25.57

a) State the null and alternative hypotheses under investigation.
b) Assuming that the assumptions for regression inference are reasonable, find the *t*- and P-values.
c) State your conclusion.

46. Climate change 2013 Data collected from around the globe show that the earth is getting warmer. The most common theory relates climate change to an increase in atmospheric levels of carbon dioxide (CO_2), a greenhouse gas. The mean annual CO_2 concentration in the atmosphere (parts per million) is measured at the top of Mauna Loa in Hawaii (away from any local contaminants) and available at ftp://aftp.cmdl.noaa.gov/products/trends/co2/co2_annmean_mlo.txt

The mean surface air temperature is recorded as the change in °C relative to a base period of 1951 to 1980. It is available at data.giss.nasa.gov/gistemp/graphs_v3/

Here are a scatterplot and regression for the years from 1970 to 2013:

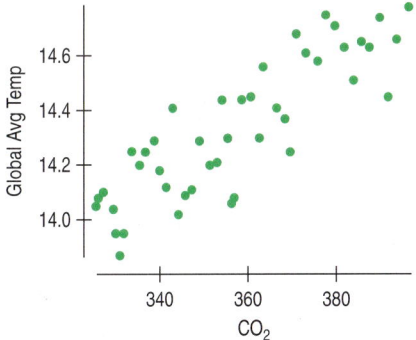

Response variable is Global Avg Temp
R-squared = 73.6%   R-squared (adjusted) = 72.9%
s = 0.1316 with 44 − 2 = 42 degrees of freedom

Variable	Coefficient	SE(Coeff)	t-Ratio	P-Value
Intercept	10.6442	0.3418	31.1	≤0.0001
CO_2	0.010325	0.0010	10.8	≤0.0001

a) Write the equation of the regression line.
b) Is there evidence of an association between CO_2 level and global temperature?
c) Do you think predictions made by this regression will be very accurate? Explain.
d) Does this regression prove that increasing CO_2 levels are causing global warming? Discuss.

47. Ozone and population The Environmental Protection Agency is examining the relationship between the ozone level (in parts per million) and the population (in millions) of U.S. cities. Part of the regression analysis is shown.

Dependent variable is Ozone
R-squared $= 84.4\%$
$s = 5.454$ with $16 - 2 = 14$ df

Variable	Coefficient	SE(Coeff)
Intercept	18.892	2.395
Pop	6.650	1.910

a) We suspect that the greater the population of a city, the higher its ozone level. Is the relationship statistically significant? Assuming the conditions for inference are satisfied, test an appropriate hypothesis and state your conclusion in context.
b) Do you think that the population of a city is a useful predictor of ozone level? Use the values of both R^2 and s in your explanation.

48. Sales and profits A business analyst was interested in the relationship between a company's sales and its profits. She collected data (in millions of dollars) from a random sample of Fortune 500 companies and created the regression analysis and summary statistics shown. The assumptions for regression inference appeared to be satisfied.

	Profits	Sales
Count	79	79
Mean	209.839	4178.29
Variance	635,172	49,163,000
Std Dev	796.977	7011.63

Dependent variable is Profits
R-squared $= 66.2\%$ $s = 466.2$

Variable	Coefficient	SE(Coeff)
Intercept	−176.644	61.16
Sales	0.092498	0.0075

a) Is there a statistically significant association between sales and profits? Test an appropriate hypothesis and state your conclusion in context.
b) Do you think that a company's sales serve as a useful predictor of its profits? Use the values of both R^2 and s in your explanation.

49. Ozone, again Consider again the relationship between the population and ozone level of U.S. cities that you analyzed in Exercise 47.

a) Give a 90% confidence interval for the approximate increase in ozone level associated with each additional million city inhabitants.
b) For the cities studied, the mean population was 1.7 million people. The population of Boston is approximately 0.6 million people. Predict the mean ozone level for cities of that size with an interval in which you have 90% confidence.

50. More sales and profits Consider again the relationship between the sales and profits of Fortune 500 companies that you analyzed in Exercise 48.

a) Find a 95% confidence interval for the slope of the regression line. Interpret your interval in context.

b) Last year, the drug manufacturer Eli Lilly, Inc., reported gross sales of $23 billion (that's $23,000 million). Create a 95% prediction interval for the company's profits, and interpret your interval in context.

51. Tablet computers 2014 cnet.com tests tablet computers and continuously updates its list. As of January 2014, the list included the battery life (in hours) and luminous intensity (i.e., screen brightness, in cd/m^2). We want to know if brighter screens are related to the maximum *Screen Brightness*. (http://reviews.cnet.com/8301-19736_7-20080768-251/cnet-updates-tablet-test-results/?tag=contentBody;contentHighlights)

Response variable is Battery life (hrs)
R-squared $= 11.3\%$
$s = 2.128$ with $34 - 2 = 32$ degrees of freedom

Variable	Coefficient	SE(Coeff)	t-Ratio	P-Value
Intercept	5.38719	1.727	3.12	0.0038
Max Brightness	0.00904	0.0045	2.02	0.0522

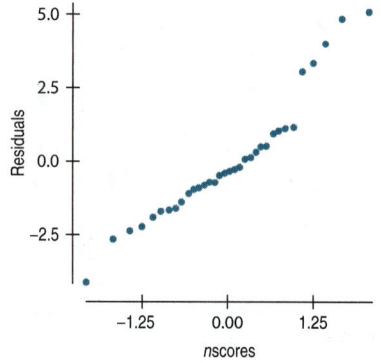

(continued)

a) How many tablet computers were tested?
b) Are the conditions for inference satisfied? Explain.
c) Is there evidence of an association between maximum brightness of the screen and battery life? Test an appropriate hypothesis and state your conclusion.
d) Is the association strong? Explain.
e) What is the equation of the regression line?
f) Create a 90% confidence interval for the slope of the true line.
g) Interpret your interval in this context.

52. Crawling Researchers at the University of Denver Infant Study Center wondered whether temperature might influence the age at which babies learn to crawl. Perhaps the extra clothing that babies wear in cold weather would restrict movement and delay the age at which they started crawling. Data were collected on 208 boys and 206 girls. Parents reported the month of the baby's birth and the age (in weeks) at which their child first crawled. The table gives the average *Temperature* (°F) when the babies were 6 months old and average *Crawling Age* (in weeks) for each month of the year. Make the plots and compute the analyses necessary to answer the following questions.

Birth Month	6-Month Temperature	Average Crawling Age
Jan.	66	29.84
Feb.	73	30.52
Mar.	72	29.70
April	63	31.84
May	52	28.58
June	39	31.44
July	33	33.64
Aug.	30	32.82
Sept.	33	33.83
Oct.	37	33.35
Nov.	48	33.38
Dec.	57	32.32

a) Would this association appear to be weaker, stronger, or the same if data had been plotted for individual babies instead of using monthly averages? Explain.
b) Is there evidence of an association between *Temperature* and *Crawling Age*? Test an appropriate hypothesis and state your conclusion. Don't forget to check the assumptions.
c) Create and interpret a 95% confidence interval for the slope of the true relationship.

53. Body fat Do the data shown in the table below indicate an association between *Waist* size and *%Body Fat*?
a) Test an appropriate hypothesis and state your conclusion.
b) Give a 95% confidence interval for the mean *%Body Fat* found in people with a 40-inch *Waist*.

Waist (in.)	Weight (lb)	Body Fat (%)	Waist (in.)	Weight (lb)	Body Fat (%)
32	175	6	33	188	10
36	181	21	40	240	20
38	200	15	36	175	22
33	159	6	32	168	9
39	196	22	44	246	38
40	192	31	33	160	10
41	205	32	41	215	27
35	173	21	34	159	12
38	187	25	34	146	10
38	188	30	44	219	28

54. Body fat, again Use the data from Exercise 53 to examine the association between *Weight* and *%Body Fat*.
a) Find a 90% confidence interval for the slope of the regression line of *%Body Fat* on *Weight*.
b) Interpret your interval in context.
c) Give a 95% prediction interval for the *%Body Fat* of an individual who weighs 165 pounds.

55. Grades The data set below shows midterm scores from an Introductory Statistics course.

First Name	Midterm 1	Midterm 2	Homework
Timothy	82	30	61
Karen	96	68	72
Verena	57	82	69
Jonathan	89	92	84
Elizabeth	88	86	84
Patrick	93	81	71
Julia	90	83	79
Thomas	83	21	51
Marshall	59	62	58
Justin	89	57	79
Alexandra	83	86	78
Christopher	95	75	77
Justin	81	66	66
Miguel	86	63	74
Brian	81	86	76
Gregory	81	87	75
Kristina	98	96	84
Timothy	50	27	20

First Name	Midterm 1	Midterm 2	Homework
Jason	91	83	71
Whitney	87	89	85
Alexis	90	91	68
Nicholas	95	82	68
Amandeep	91	37	54
Irena	93	81	82
Yvon	88	66	82
Sara	99	90	77
Annie	89	92	68
Benjamin	87	62	72
David	92	66	78
Josef	62	43	56
Rebecca	93	87	80
Joshua	95	93	87
Ian	93	65	66
Katharine	92	98	77
Emily	91	95	83
Brian	92	80	82
Shad	61	58	65
Michael	55	65	51
Israel	76	88	67
Iris	63	62	67
Mark	89	66	72
Peter	91	42	66
Catherine	90	85	78
Christina	75	62	72
Enrique	75	46	72
Sarah	91	65	77
Thomas	84	70	70
Sonya	94	92	81
Michael	93	78	72
Wesley	91	58	66
Mark	91	61	79
Adam	89	86	62
Jared	98	92	83
Michael	96	51	83
Kathryn	95	95	87
Nicole	98	89	77
Wayne	89	79	44
Elizabeth	93	89	73
John	74	64	72
Valentin	97	96	80
David	94	90	88
Marc	81	89	62
Samuel	94	85	76
Brooke	92	90	86

a) Fit a model predicting the second midterm score from the first.
b) Comment on the model you found, including a discussion of the assumptions and conditions for regression. Is the coefficient for the slope statistically significant?
c) A student comments that because the P-value for the slope is very small, Midterm 2 is very well predicted from Midterm 1. So, he reasons, next term the professor can give just one midterm. What do you think?

T 56. Grades? The professor teaching the Introductory Statistics class discussed in Exercise 55 wonders whether performance on homework can accurately predict midterm scores.
a) To investigate it, she fits a regression of the sum of the two midterms scores on homework scores. Fit the regression model.
b) Comment on the model including a discussion of the assumptions and conditions for regression. Is the coefficient for the slope "statistically significant"?
c) Do you think she can accurately judge a student's performance without giving the midterms? Explain.

T 57. Strike two Remember the Little League instructional video discussed in Chapter 23? Ads claimed it would improve the performances of Little League pitchers. To test this claim, 20 Little Leaguers threw 50 pitches each, and we recorded the number of strikes. After the players participated in the training program, we repeated the test. The table shows the number of strikes each player threw before and after the training. A test of paired differences failed to show that this training improves ability to throw strikes. Is there any evidence that the effectiveness of the video (*After – Before*) depends on the player's initial ability to throw strikes (*Before*)? Test an appropriate hypothesis and state your conclusion. Propose an explanation for what you find.

Number of Strikes (out of 50)			
Before	After	Before	After
28	35	33	33
29	36	33	35
30	32	34	32
32	28	34	30
32	30	34	33
32	31	35	34
32	32	36	37
32	34	36	33
32	35	37	35
33	36	37	32

T 58. All the efficiency money can buy 2011 A sample of 84 model-2011 cars from an online information service was examined to see how fuel efficiency (as highway mpg) relates to the cost (Manufacturer's

(*continued*)

Suggested Retail Price in dollars) of cars. Here are displays and computer output:

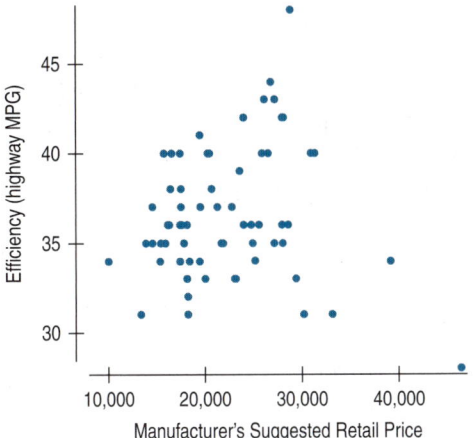

Dependent variable is MPG
R-squared = 0.0216%
s = 3.54

Variable	Coefficient	SE(Coeff)	t-Ratio	P-Value
Intercept	36.514	1.496	24.406	<0.0001
Slope	−8.089E−6	6.439E−5	−0.1256	0.900

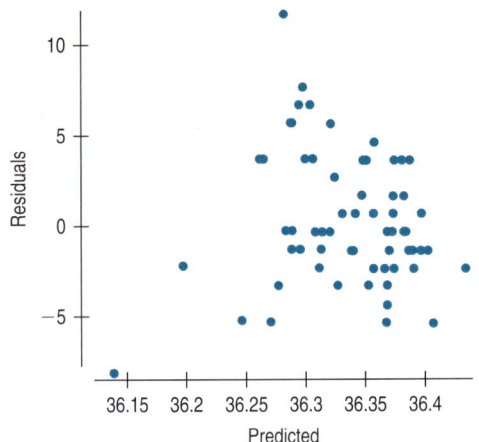

a) State what you want to know, identify the variables, and give the appropriate hypotheses.
b) Check the assumptions and conditions.
c) If the conditions are met, complete the analysis.

T 59. Education and mortality The following software output is based on the mortality rate (deaths per 100,000 people) and the education level (average number of years in school) for 58 U.S. cities.

Variable	Count	Mean	StdDev
Mortality	58	942.501	61.8490
Education	58	11.0328	0.793480

Dependent variable is Mortality
R-squared = 41.0%
s = 47.92 with 58 − 2 = 56 degrees of freedom

Variable	Coefficient	SE(Coeff)
Intercept	1493.26	88.48
Education	−49.9202	8.000

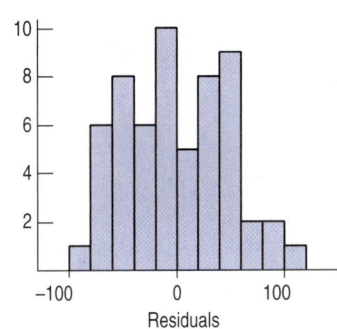

a) Comment on the assumptions for inference.
b) Is there evidence of a strong association between the level of *Education* in a city and the *Mortality* rate? Test an appropriate hypothesis and state your conclusion.
c) Can we conclude that getting more education is likely (on average) to prolong your life? Why or why not?
d) Find a 95% confidence interval for the slope of the true relationship.
e) Explain what your interval means.
f) Find a 95% confidence interval for the average *Mortality* rate in cities where the adult population completed an average of 12 years of school.

T 60. Property assessments The following software outputs provide information about the *Size* (in square feet) of 18 homes in Ithaca, New York, and the city's assessed *Value* of those homes.

Variable	Count	Mean	StdDev	Range
Size	18	2003.39	264.727	890
Value	18	60946.7	5527.62	19710

Dependent variable is Value
R-squared = 32.5%
s = 4682 with 18 − 2 = 16 degrees of freedom

Variable	Coefficient	SE(Coeff)
Intercept	37108.8	8664
Size	11.8987	4.290

a) Explain why inference for linear regression is appropriate with these data.
b) Is there a significant association between the *Size* of a home and its assessed *Value*? Test an appropriate hypothesis and state your conclusion.
c) What percentage of the variability in assessed *Value* is explained by this regression?
d) Give a 90% confidence interval for the slope of the true regression line, and explain its meaning in the proper context.
e) From this analysis, can we conclude that adding a room to your house will increase its assessed *Value*? Why or why not?
f) The owner of a home measuring 2100 square feet files an appeal, claiming that the $70,200 assessed *Value* is too high. Do you agree? Explain your reasoning.

*61. **Right-to-work laws** Are so-called "right-to-work" laws related to the percent of public sector employees in unions and the percent of private sector employees in unions? This data set looks at these percentages for the states in the United States in 1982. The dependent variable is whether the state had a right-to-work law or not. The computer output for the logistic regression is given here. (Source: N. M. Meltz, "Interstate and Interprovincial Differences in Union Density," *Industrial Relations*, 28:2 [Spring 1989], 142–158 by way of DASL.)

Logistic Regression Table

Predictor	Coeff	SE(Coeff)	Chisq	P
Intercept	6.1995	1.78724	12.04	0.001
publ	−0.1062	0.0475	5.02	0.025
pvt	−0.2230	0.0811	7.56	0.006

a) Write out the estimated regression equation.
b) The following are scatterplots of the response variable against each of the explanatory variables. Does logistic regression seem appropriate here? Explain.

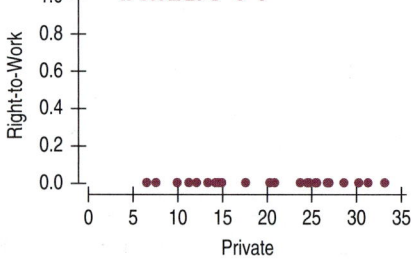

*62. **Cost of higher education** Are there fundamental differences between liberal arts colleges and universities? In this case, we have information on the top 25 liberal arts colleges and the top 25 universities in the Unites States. We will consider the type of school as our response variable and will use the percent of students who were in the top 10% of their high school class and the amount of money spent per student by the college or university as our explanatory variables. The output from this logistic regression is given on the following page.

(*continued*)

Logistic Regression Table

Predictor	Coeff	SE(Coeff)	Chisq	P
Intercept	−13.1461	3.98629	10.89	0.001
Top 10%	0.0845	0.0396345	4.54	0.033
$/Student	0.0003	0.0000860	9.12	0.003

a) Write out the estimated regression equation.
b) Is percent of students in the top 10% of their high school class statistically significant in predicting whether or not the school is a university? Explain.
c) Is the amount of money spent per student statistically significant in predicting whether or not the school is a university? Explain.

Just Checking ANSWERS

1. A high t-ratio of 3.27 indicates that the slope is different from zero—that is, that there is a linear relationship between height and mouth size. The small P-value says that a slope this large would be very unlikely to occur by chance if, in fact, there was no linear relationship between the variables.

2. Not really. The R^2 for this regression is only 15.3%, so height doesn't account for very much of the variability in mouth size.

3. The value of s tells the standard deviation of the residuals. Mouth sizes have a mean of 60.3 cubic centimeters. A standard deviation of 15.7 in the residuals indicates that the errors made by this regression model can be quite large relative to what we are estimating. Errors of 15 to 30 cubic centimeters would be common.

Review of Part VI

Assessing Associations Between Variables

Quick Review

We now turn our focus to relationships between variables. This is where the effectiveness of Statistics becomes clear. Statistical inference about how variables are related is essential to science, social science, health, and business. In this part we have considered comparing groups with means and proportions, comparing groups that have a natural pairing, comparing proportions in several categories and between two categorical variables, and inferential decisions about linear regression. That may seem like a wide variety of ideas, but they are all held together by the central concept that by understanding the ways in which variables and groups relate to each other, we can understand how the world works. Here's a brief summary of the key concepts and skills:

- A confidence interval uses a sample statistic to estimate a range of possible values for the parameter of a population model.

- A hypothesis test proposes a model for the population, then examines the observed statistics to see if the model is plausible.

- Statistical inference procedures for comparing proportions are based on the Central Limit Theorem. We can make inferences about the difference of two proportions using Normal models.

- Statistical inference procedures for comparing means and for estimating regression coefficients are also based on the Central Limit Theorem, but we don't usually know the population standard deviation. Student's t-models take the additional uncertainty of independently estimating the standard deviation into account.

  - We can make inferences about the difference of two independent means, the mean of paired differences, or the intercept or slope of a linear regression using t-models.
  - No inference procedure is valid unless the underlying assumptions are true. Always think about the assumptions and check the conditions before proceeding. For regression, take care to check them in the right order.
  - Because t-models assume that samples are drawn from Normal populations, data in the sample (or, for regression, the residuals) should appear to be nearly Normal. Skewness and outliers are particularly problematic.
  - To identify the appropriate statistic for comparing the means of groups, you must think carefully about how the data were collected. You may use two-sample t procedures only if the groups are independent.
  - Unless there is some obvious reason to suspect that two independent populations have the same standard deviation, you should not pool the variances. It is never wrong to use unpooled t procedures.
  - If two groups are paired, the data are *not* from independent groups. You must use matched-pairs t procedures and test the mean difference rather than the difference in the means.
  - Linear regression inference is only valid if the relationship between the two variables is straight. Examine a scatterplot.

- Not all sampling distributions are unimodal, symmetric, or bell-shaped. Inferences about distributions of counts use chi-square models, which are unimodal but skewed to the high end. Nevertheless, the sampling distribution plays the same role in inference, helping us to translate between probabilities and values based on data.

  - To see if an observed distribution is consistent with a proposed model, use a chi-square goodness-of-fit test.
  - To see if two or more observed distributions could have arisen from populations with the same model, use a test of homogeneity.
  - To see if two categorical variables are independent, perform a chi-square test of independence.

- You can now use statistical inference to answer questions about means, proportions, distributions, associations, and linear regression models.

  - No inference procedure is valid unless the underlying assumptions are true. Always check the conditions before proceeding.
  - You can make inferences about the difference between two proportions using Normal models.
  - You can make inferences about the difference between two independent means, or about the mean of paired differences using t-models.
  - You can make inferences about distributions using chi-square models.
  - You can make inferences about association between categorical variables using chi-square models.
  - You can make inferences about the coefficients in a linear regression model using t-models.

Now for some opportunities to review these concepts. Be careful. You have a lot of thinking to do. These review exercises mix questions about proportions, means, chi square, and regression. You have to determine which of our inference procedures is appropriate in each situation. Then you have to check the proper assumptions and conditions. Keeping track of those can be difficult, so first we summarize the many procedures with their corresponding assumptions and conditions on the next page. Look them over carefully . . . then, on to the Exercises!

Assumptions for Inference	And the Conditions that Support or Override them
Proportions (z)	
■ **One sample**	
1. Individuals are independent.	1. SRS and $n < 10\%$ of the population.
2. Sample is sufficiently large.	2. Successes and failures ≥ 10.
■ **Two sample**	
1. Samples are independent.	1. (Think about how the data were collected.)
2. Data in each sample are independent.	2. Both are SRSs and $n < 10\%$ of populations OR random allocation.
3. Both samples are sufficiently large.	3. Successes and failures ≥ 10 for both.
Means (t)	
■ **One sample** (df $= n - 1$)	
1. Individuals are independent.	1. SRS and $n < 10\%$ of the population.
2. Population has a Normal model.	2. Histogram is unimodal and symmetric.*
■ **Two independent Samples** (df from technology)	
1. Samples are independent.	1. (Think about the design.)
2. Data in each sample are independent.	2. SRSs and $n < 10\%$ OR random allocation.
3. Both populations are Normal.	3. Both histograms are unimodal and symmetric.*
■ **Matched pairs** (df $= n - 1$)	
1. Data are matched; n pairs.	1. (Think about the design.)
2. Individuals are independent.	2. SRSs and $n < 10\%$ OR random allocation.
3. Population of differences is Normal.	3. Histogram of differences is unimodal and symmetric.
Distributions/Association (χ^2)	
■ **Goodness of fit** [df $=$ # of cells -1; one variable, one sample compared with population model]	
1. Data are counts.	1. (Are they?)
2. Data in sample are independent.	2. SRS and $n < 10\%$ of the population.
3. Sample is sufficiently large.	3. All expected counts ≥ 5.
■ **Homogeneity** [df $= (r-1)(c-1)$; samples from many populations compared on one variable]	
1. Data are counts.	1. (Are they?)
2. Data in samples are independent.	2. SRSs and $n < 10\%$ OR random allocation.
3. Groups are sufficiently large.	3. All expected counts ≥ 5.
■ **Independence** [df $= (r-1)(c-1)$; sample from one population classified on two variables]	
1. Data are counts.	1. (Are they?)
2. Data are independent.	2. SRSs and $n < 10\%$ of the population.
3. Group is sufficiently large.	3. All expected counts ≥ 5.
Regression with One Quantitative Predictor (t, df $= n - 2$)	
1. Form of relationship is linear.	1. Scatterplot of y against x is straight enough. Scatterplot of residuals against predicted values shows no special structure.
2. Errors are independent.	2. No apparent pattern in plot of residuals against predicted values.
3. Variability of errors is constant.	3. Plot of residuals against predicted values has constant spread, doesn't "thicken."
4. Errors follow a Normal model.	4. Histogram of residuals is approximately unimodal and symmetric, or Normal probability plot is resonably straight.*

*Less critical as n increases

Quick Guide to Inference

	Think			Show			Tell?
Inference about?	One sample or two?	Procedure	Model	Parameter	Estimate	SE	Chapter
Proportions	One sample	1-Proportion z-Interval	z	p	$\hat{p}$	$\sqrt{\dfrac{\hat{p}\hat{q}}{n}}$	18
		1-Proportion z-Test				$\sqrt{\dfrac{p_0 q_0}{n}}$	19
	Two independent groups	2-Proportion z-Interval	z	$p_1 - p_2$	$\hat{p}_1 - \hat{p}_2$	$\sqrt{\dfrac{\hat{p}_1 \hat{q}_1}{n_1} + \dfrac{\hat{p}_2 \hat{q}_2}{n_2}}$	22
		2-Proportion z-Test				$\sqrt{\dfrac{\hat{p}\hat{q}}{n_1} + \dfrac{\hat{p}\hat{q}}{n_2}}, \hat{p} = \dfrac{y_1 + y_2}{n_1 + n_2}$	22
Means	One sample	t-Interval t-Test	t df $= n-1$	μ	$\bar{y}$	$\dfrac{s}{\sqrt{n}}$	20
	Two independent groups	2-Sample t-Test 2-Sample t-Interval	t df from technology	$\mu_1 - \mu_2$	$\bar{y}_1 - \bar{y}_2$	$\sqrt{\dfrac{s_1^2}{n_1} + \dfrac{s_2^2}{n_2}}$	22
	n Matched pairs	Paired t-Test Paired t-Interval	t df $= n-1$	μ_d	$\bar{d}$	$\dfrac{s_d}{\sqrt{n}}$	23
Distributions (one categorical variable)	One sample	Goodness of fit	χ^2 df $=$ cells $- 1$			$\displaystyle\sum \dfrac{(\text{obs} - \text{exp})^2}{\text{exp}}$	24
	Many independent samples	Homogeneity χ^2 Test					
Independence (two categorical variables)	One sample	Independence χ^2 Test	χ^2 df $= (r-1)(c-1)$				
Association (two quantitative variables)	One sample	Linear Regression t-Test or Confidence Interval for β		β_1	b_1	$\dfrac{s_e}{s_x \sqrt{n-1}}$ (compute with technology)	25
		Confidence Interval for μ_v	t df $= n-2$	μ_v	$\hat{y}_v$	$\sqrt{SE^2(b_1) \cdot (x_v - \bar{x})^2 + \dfrac{s_e^2}{n}}$	
		Prediction Interval for y_v		y_v	$\hat{y}_v$	$\sqrt{SE^2(b_1) \cdot (x_n - \bar{x})^2 + \dfrac{s_e^2}{n} + s_e^2}$	
Inference about?	One sample or two?	Procedure	Model	Parameter	Estimate	SE	Chapter

Review Exercises

1. **Herbal cancer** A report in the *New England Journal of Medicine* notes growing evidence that the herb *Aristolochia fangchi* can cause urinary tract cancer in those who take it. Suppose you are asked to design an experiment to study this claim. Imagine that you have data on urinary tract cancers in subjects who have used this herb and similar subjects who have not used it and that you can measure incidences of cancer and precancerous lesions in these subjects. State the null and alternative hypotheses you would use in your study.

2. **Birth days** During a 2-month period, 72 babies were born at the Tompkins Community Hospital in upstate New York. The table shows how many babies were born on each day of the week.

Day	Births
Mon.	7
Tues.	17
Wed.	8
Thurs.	12
Fri.	9
Sat.	10
Sun.	9

   a) If births are uniformly distributed across all days of the week, how many would you expect on each day?
   b) Test the hypothesis that babies are equally likely to be born on any of the days of the week.
   c) Given the results of part b, do you think that the 7 births on Monday and 17 births on Tuesday indicate that women might be less likely to give birth on Monday, or more likely to give birth on Tuesday?
   d) Can you think of any reasons why births may not occur completely at random?

3. **Surgery and germs** Joseph Lister (for whom Listerine is named!) was a British physician who was interested in the role of bacteria in human infections. He suspected that germs were involved in transmitting infection, so he tried using carbolic acid as an operating room disinfectant. In 75 amputations, he used carbolic acid 40 times. Of the 40 amputations using carbolic acid, 34 of the patients lived. Of the 35 amputations without carbolic acid, 19 patients lived. The question of interest is whether carbolic acid is effective in increasing the chances of surviving an amputation.
   a) What kind of a study is this?
   b) What do you conclude? Support your conclusion by testing an appropriate hypothesis.
   c) What reservations do you have about the design of the study?

4. **Free throws 2011** During the 2010–2011 NBA season, Stephen Curry led the league by making 212 of 227 free throws, for a success rate of 93.39%. But Chauncey Billups was close behind, with 384 of 419 (91.65%).
   a) Find a 95% confidence interval for the difference in their free throw percentages.
   b) Based on your confidence interval, is it certain that Curry is better than Billups at making free throws?

5. **Twins** There is some indication in medical literature that doctors may have become more aggressive in inducing labor or doing preterm cesarean sections when a woman is carrying twins. Records at a large hospital show that, of the 43 sets of twins born in 2000, 20 were delivered before the 37th week of pregnancy. In 2010, 26 of 48 sets of twins were born preterm. Does this indicate an increase in the incidence of early births of twins? Test an appropriate hypothesis and state your conclusion.

6. **Eclampsia** It's estimated that 50,000 pregnant women worldwide die each year of eclampsia, a condition involving elevated blood pressure and seizures. A research team from 175 hospitals in 33 countries investigated the effectiveness of magnesium sulfate in preventing the occurrence of eclampsia in at-risk patients. Results are summarized below. (Source: *Lancet*, June 1, 2002)

Treatment	Total Subjects	Reported Side Effects	Developed Eclampsia	Deaths
Magnesium Sulfate	4999	1201	40	11
Placebo	4993	228	96	20

   a) Write a 95% confidence interval for the increase in the proportion of women who may develop side effects from this treatment. Interpret your interval.
   b) Is there evidence that the treatment may be effective in preventing the development of eclampsia? Test an appropriate hypothesis and state your conclusion.

7. **Eclampsia deaths** Refer again to the research summarized in Exercise 6. Is there any evidence that when eclampsia does occur, the magnesium sulfide treatment may help prevent the woman's death?
   a) Write an appropriate hypothesis.
   b) Check the assumptions and conditions.
   c) Find the P-value of the test.
   d) What do you conclude about the magnesium sulfide treatment?
   e) If your conclusion is wrong, which type of error have you made?

f) Name two things you could do to increase the power of this test.
g) What are the advantages and disadvantages of those two options?

8. **Perfect pitch** A recent study of perfect pitch tested 2700 students in American music conservatories. It found that 7% of non-Asian and 32% of Asian students have perfect pitch. A test of the difference in proportions resulted in a P-value of < 0.0001.
a) What are the researchers' null and alternative hypotheses?
b) State your conclusion.
c) Explain in this context what the P-value means.
d) The researchers claimed that the data prove that genetic differences between the two populations cause a difference in the frequency of occurrence of perfect pitch. Do you agree? Why or why not?

9. **More errors** A corporation with a fleet of vehicles wanted to test the cost-effectiveness of using Motor Silk oil additive. For the study, 6100 delivery and passenger vehicles were tested for the same 3-month period in 2007 and then again in 2008. In 2007, the fleet was driven without Motor Silk. Then in 2008, Motor Silk was used according to the standard instructions. The average fuel economy increased from 18.97 mpg to 21.72 mpg.
a) What kind of a study is this?
b) Will they do a one-tailed or a two-tailed test?
c) Explain in this context what a Type I error would be.
d) Explain in this context what a Type II error would be.
e) Which type of error would the additive manufacturer consider more serious?
f) If the vehicles with the additive are indeed statistically significantly better, can the company conclude it is an effect of the additive? Can they generalize this result and recommend the additive for all cars? Explain.

10. **Preemies** Among 242 Cleveland-area children born prematurely at low birth weights between 1977 and 1979, only 74% graduated from high school. Among a comparison group of 233 children of normal birth weight, 83% were high school graduates. (Source: "Outcomes in Young Adulthood for Very-Low-Birth-Weight Infants," *New England Journal of Medicine*, 346, no. 3)
a) Create a 95% confidence interval for the difference in graduation rates between children of normal and children of very low birth weights. Be sure to check the appropriate assumptions and conditions.
b) Does this provide evidence that premature birth may be a risk factor for not finishing high school? Use your confidence interval to test an appropriate hypothesis.
c) Suppose your conclusion is incorrect. Which type of error did you make?

11. **Crawling** A study found that babies born at different times of the year may develop the ability to crawl at different ages. The author of the study suggested that these differences may be related to the temperature at the time the infant is 6 months old. (Source: Benson and Janette, *Infant Behavior and Development*)
a) The study found that 32 babies born in January crawled at an average age of 29.84 weeks, with a standard deviation of 7.08 weeks. Among 21 July babies, crawling ages averaged 33.64 weeks, with a standard deviation of 6.91 weeks. Is this difference significant?
b) For 26 babies born in April, the mean and standard deviation were 31.84 and 6.21 weeks, while for 44 October babies the mean and standard deviation of crawling ages were 33.35 and 7.29 weeks. Is this difference significant?
c) Are these results consistent with the researcher's claim?

12. **Mazes and smells** Can pleasant smells improve learning? Researchers timed 21 subjects as they tried to complete paper-and-pencil mazes. Each subject attempted a maze both with and without the presence of a floral aroma. Subjects were randomized with respect to whether they did the scented trial first or second. Some of the data collected are shown in the table. Is there any evidence that the floral scent improved the subjects' ability to complete the mazes? (Source: A. R. Hirsch and L. H. Johnston, "Odors and Learning." Chicago: Smell and Taste Treatment and Research Foundation)

Time to Complete the Maze (sec)			
Unscented	Scented	Unscented	Scented
25.7	30.2	61.5	48.4
41.9	56.7	44.6	32.0
51.9	42.4	35.3	48.1
32.2	34.4	37.2	33.7
64.7	44.8	39.4	42.6
31.4	42.9	77.4	54.9
40.1	42.7	52.8	64.5
43.2	24.8	63.6	43.1
33.9	25.1	56.6	52.8
40.4	59.2	58.9	44.3
58.0	42.2		

13. **Pottery** Archaeologists can use the chemical composition of clay found in pottery artifacts to determine whether different sites were populated by the same ancient people. They collected five samples of Romano–British pottery from each of two sites in Great Britain—the Ashley Rails site and the New Forest site—and measured the percentage of aluminum oxide in each. Based on these data, do you think the same people used these two kiln sites? Base your conclusion on a 95% confidence interval for the difference in aluminum oxide content of pottery made at the

sites. (Source: A. Tubb, A. J. Parker, and G. Nickless, "The Analysis of Romano–British Pottery by Atomic Absorption Spectrophotometry." *Archaeometry*, 22[1980]:153–171)

Ashley Rails	19.1	14.8	16.7	18.3	17.7
New Forest	20.8	18.0	18.0	15.8	18.3

14. Grant writing Does race matter when applying for National Institutes of Health grants? A new study found that of 58,148 applications submitted by white researchers, 15,700 were accepted and funded by the NIH. Additionally, 198 of the 1164 applications submitted by black researchers were funded. (Source: *Science,* August 19, 2011)
   a) Is there evidence that the chance for funding is higher for white researchers?
   b) What kind of study is this? How does that affect the inference you made in part a?

15. Feeding fish In the midwestern United States, a large aquaculture industry raises largemouth bass. Researchers wanted to know whether the fish would grow better if fed a natural diet of fathead minnows or an artificial diet of food pellets. They stocked six ponds with bass fingerlings weighing about 8 grams. For one year, the fish in three of the ponds were fed minnows, and the others were fed the commercially prepared pellets. The fish were then harvested, weighed, and measured. The bass fed a natural food source had a higher average length (19.6 cm) and weight (95.9 g) than those fed the commercial fish food (17.3 cm and 72.0 g, respectively). The researchers reported P-values for both measurements to be less than 0.001.
   a) Explain to someone who has not studied Statistics what the P-values mean here.
   b) What advice should the researchers give the people who raise largemouth bass?
   c) If that advice turns out to be incorrect, what type of error occurred?

16. Driving fatalities 2011 The U.S. Census compiled data from each state in 2011 to record the number of deaths per 100 million vehicle miles traveled. Wondering if the death rate is different in New England when compared to the Mountain states, we hope to create a 95% confidence interval for the difference in death rates. Are these data appropriate for inference? If so, create the interval and make a concluding remark.

New England	0.71	0.95	0.62	0.71	0.84	0.77		
Mountain	1.38	0.96	1.05	1.79	1.02	1.38	0.92	1.46

17. Age In a study of how depression may affect one's ability to survive a heart attack, the researchers reported the ages of the two groups they examined. The mean age of 2397 patients without cardiac disease was 69.8 years (SD = 8.7 years), while for the 450 patients with cardiac disease, the mean and standard deviation of the ages were 74.0 and 7.9, respectively.
   a) Create a 95% confidence interval for the difference in mean ages of the two groups.
   b) How might an age difference confound these research findings about the relationship between depression and ability to survive a heart attack?

18. Smoking In the depression and heart attack research described in Exercise 17, 32% of the diseased group were smokers, compared with only 23.7% of those free of heart disease.
   a) Create a 95% confidence interval for the difference in the proportions of smokers in the two groups.
   b) Is this evidence that the two groups in the study were different? Explain.
   c) Could this be a problem in analyzing the results of the study? Explain.

19. Eating disorders A study conducted in the multicultural Spanish city of Ceuta investigated the relationship between religion and the prevalence of eating disorders. Students aged 12–20 were selected from three public schools. In the study, suppose there were 150 Muslim students and 46 were diagnosed with eating disorders. Of the 200 Christian students, 17% were diagnosed the same way.
   a) Create a 95% confidence interval for the difference in eating disorder rates when comparing Muslims to Christians.
   b) Based on your interval, are you convinced that there is a true difference when comparing rates between the religions? Explain.

20. Cesareans Some people fear that differences in insurance coverage can affect healthcare decisions. A survey of several randomly selected hospitals found that 16.6% of 223 recent births in Vermont involved cesarean deliveries, compared to 18.8% of 186 births in New Hampshire. Is this evidence that the rate of cesarean births in the two states is different?

21. Teach for America Several programs attempt to address the shortage of qualified teachers by placing uncertified instructors in schools with acute needs—often in inner cities. A study compared students taught by certified teachers to others taught by uncertified teachers in the same schools. Reading scores of the students of certified teachers averaged 35.62 points with standard deviation 9.31. The scores of students instructed by uncertified teachers had mean 32.48 points with standard deviation 9.43 points on the same test.

There were 44 students in each group. The appropriate *t* procedure has 86 degrees of freedom. Is there evidence of lower scores with uncertified teachers? Discuss. (Source: *The Effectiveness of "Teach for America" and Other Under-certified Teachers on Student Academic Achievement: A Case of Harmful Public Policy.* Education Policy Analysis Archives)

22. Legionnaires' disease In 1974, the Bellevue-Stratford Hotel in Philadelphia was the scene of an outbreak of what later became known as legionnaires' disease. The cause of the disease was finally discovered to be bacteria that thrived in the air-conditioning units of the hotel. Owners of the Rip Van Winkle Motel, hearing about the Bellevue-Stratford, replace their air-conditioning system. The following data are the bacteria counts in the air of eight rooms, before and after a new air-conditioning system was installed (measured in colonies per cubic foot of air). The objective is to find out whether the new system has succeeded in lowering the bacterial count. You are the statistician assigned to report to the hotel whether the strategy has worked. Base your analysis on a confidence interval. Be sure to list all your assumptions, methods, and conclusions.

Room Number	Before	After
121	11.8	10.1
163	8.2	7.2
125	7.1	3.8
264	14.0	12.0
233	10.8	8.3
218	10.1	10.5
324	14.6	12.1
325	14.0	13.7

23. Teach for America, part II The study described in Exercise 21 also looked at scores in mathematics and language. Here are software outputs for the appropriate tests. Explain what they show.

Mathematics
T-TEST OF Mu(1) − Mu(2) = 0
Mu(Cert) − Mu(NoCert) = 4.53   t(86) = 2.95   p = 0.002

Language
T-TEST OF Mu(1) − Mu(2) = 0
Mu(Cert) − Mu(NoCert) = 2.13   t(84) = 1.71   p = 0.045

24. Irises Can measurements of the petal length of flowers be of value when you need to determine the species of a certain flower? Here are the summary statistics from measurements of the petals of two species of irises. (Source: R. A. Fisher, "The Use of Multiple Measurements in Axonomic Problems." *Annals of Eugenics* 7 [1936]:179–188)

	Species	
	Versicolor	Virginica
Count	50	50
Mean	55.52	43.22
Median	55.50	44.00
SD	5.519	5.362
Min	45	30
Max	69	56
Lower Quartile	51	40
Upper Quartile	59	47

a) Make parallel boxplots of petal lengths for the two species.
b) Describe the differences seen in the boxplots.
c) Write a 95% confidence interval for this difference.
d) Explain what your interval means.
e) Based on your confidence interval, is there evidence of a difference in petal length? Explain.

25. Insulin and diet A study published in the *Journal of the American Medical Association* examined people to see if they showed any signs of IRS (insulin resistance syndrome) involving major risk factors for Type 2 diabetes and heart disease. Among 102 subjects who consumed dairy products more than 35 times per week, 24 were identified with IRS. In comparison, IRS was identified in 85 of 190 individuals with the lowest dairy consumption, fewer than 10 times per week.

a) Is this strong evidence that IRS risk is different in people who frequently consume dairy products than in those who do not?
b) Does this prove that diary consumption influences the development of IRS? Explain.

26. Rainmakers? In an experiment to determine whether seeding clouds with silver iodide increases rainfall, 52 clouds were randomly assigned to be seeded or not. The amount of rain they generated was then measured (in acre-feet). Create a 95% confidence interval for the average amount of additional rain created by seeding clouds. Explain what your interval means.

	Unseeded Clouds	Seeded Clouds
Count	26	26
Mean	164.588	441.985
Median	44.200	221.600
SD	278.426	650.787
IntQRange	138.600	337.600
25 %ile	24.400	92.400
75 %ile	163	430

27. Genetics Two human traits controlled by a single gene are the ability to roll one's tongue and whether one's ear lobes are free or attached to the neck. Genetic theory says that people will have neither, one, or both of these traits in the ratio 1:3:3:9 (1 attached, noncurling; 3 attached, curling; 3 free, noncurling; 9 free, curling). An Introductory Biology class of 122 students collected the data shown. Are they consistent with the genetic theory? Test an appropriate hypothesis and state your conclusion.

	Trait			
	Attached, Noncurling	Attached, Curling	Free, Noncurling	Free, Curling
Count	10	22	31	59

28. Tableware Nambe Mills manufactures plates, bowls, and other tableware made from an alloy of several metals. Each item must go through several steps, including polishing. To better understand the production process and its impact on pricing, the company checked the polishing time (in minutes) and the retail price (in US$) of these items. The regression analysis is shown below. The scatterplot showed a linear pattern, and residuals were deemed suitable for inference.

Dependent variable is Price
R-squared = 84.5%
s = 20.50 with 59 − 2 = 57 degrees of freedom

Variable	Coefficient	SE(Coeff)
Intercept	−2.89054	5.730
Time	2.49244	0.1416

a) How many different products were included in this analysis?
b) What fraction of the variation in retail price is explained by the polishing time?
c) Create a 95% confidence interval for the slope of this relationship.
d) Interpret your interval in this context.

29. Hard water In an investigation of environmental causes of disease, data were collected on the annual mortality rate (deaths per 100,000) for males in 61 large towns in England and Wales. In addition, the water hardness was recorded as the calcium concentration (parts per million, or ppm) in the drinking water. Here are the scatterplot and regression analysis of the relationship between mortality and calcium concentration.

Dependent variable is Mortality
R-squared = 43%
s = 143.0 with 61 − 2 = 59 degrees of freedom

Variable	Coefficient	SE(Coeff)
Intercept	1676	29.30
Calcium	−3.23	0.48

a) Is there an association between the hardness of the water and the mortality rate? Write the appropriate hypothesis.
b) Assuming the assumptions for regression inference are met, what do you conclude?
c) Create a 95% confidence interval for the slope of the true line relating calcium concentration and mortality.
d) Interpret your interval in context.

30. Wealth distribution The following table is based on a June 2011 Gallup Poll in which Americans were classified as high income (over $75,000), middle income ($30k–$75k), or low income (less than $30k). Those polled were asked for their views on redistributing U.S. wealth by heavily taxing the rich. The data are summarized in the table below.

	Should Redistribute Wealth	Should Not	No Opinion
High Income	170	371	9
Middle Income	306	282	12
Low Income	362	184	29

Is there any evidence that income level is associated with feelings toward the wealth distribution in the United States? Test an appropriate hypothesis about this table, and state your conclusions.

31. Wild horses Large herds of wild horses can become a problem on some federal lands in the West. Researchers hoping to improve the management of these herds collected data to see if they could predict the number of foals that would be born based on the size of the current herd. Their attempt to model this herd growth is summarized in the output shown.

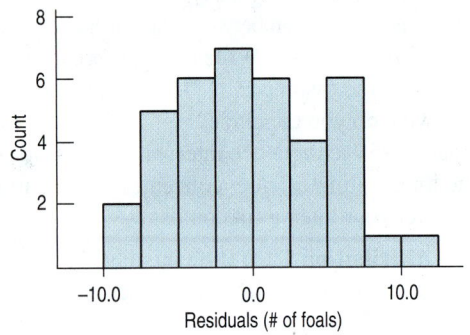

Variable	Count	Mean	StdDev
Adults	38	110.237	71.1809
Foals	38	15.3947	11.9945

Dependent variable is Foals
R-squared = 83.5%
s = 4.941 with 38 − 2 = 36 degrees of freedom

Variable	Coefficient	SE(Coeff)	t-Ratio	P-Value
Intercept	−1.57835	1.492	−1.06	0.2970
Adults	0.153969	0.0114	13.5	≤0.0001

a) How many herds of wild horses were studied?
b) Are the conditions necessary for inference satisfied? Explain.
c) Create a 95% confidence interval for the slope of this relationship.
d) Explain in this context what that slope means.
e) Suppose that a new herd with 80 adult horses is located. Estimate, with a 90% prediction interval, the number of foals that may be born.

32. **Lefties and music** In an experiment to see if left- and right-handed people have different abilities in music, subjects heard a tone and were then asked to identify which of several other tones matched the first. Of 76 right-handed subjects, 38 were successful in completing this test, compared with 33 of 53 lefties. Is this strong evidence of a difference in musical abilities based on handedness?

33. **AP Statistics scores 2010** In 2010, almost 130,000 Statistics students nationwide took the Advanced Placement Examination in Statistics. The national distribution of scores and the results at Ithaca High School are shown in the table.

		Ithaca High School	
Score	National Distribution	Number of Boys	Number of Girls
5	12.8%	13	13
4	22.4%	21	15
3	23.5%	6	13
2	18.2%	7	3
1	23.1%	4	2

a) Is the distribution of scores at this high school significantly different from the national results?
b) Was there a significant different between the performances of boys and girls at this school?

34. **Twins** In 2000, the *Journal of the American Medical Association* published a study that examined a sample of pregnancies that resulted in the birth of twins. Births were classified as preterm with intervention (induced labor or cesarean), preterm without such procedures, or term or postterm. Researchers also classified the pregnancies by the level of prenatal medical care the mother received (inadequate, adequate, or intensive). The data, from the years 1995–1997, are summarized in the table below. Figures are in thousands of births. (Source: JAMA 284 [2000]: 335–341)

		Twin Births, 1995–1997 (in thousands)			
		Preterm (induced or cesarean)	Preterm (without procedures)	Term or Postterm	Total
Level of Prenatal Care	Intensive	18	15	28	61
	Adequate	46	43	65	154
	Inadequate	12	13	38	63
	Total	76	71	131	278

Is there evidence of an association between the duration of the pregnancy and the level of care received by the mother?

35. **Twins, again** After reading of the JAMA study in Exercise 34, a large city hospital examined their records of twin births for several years and found the data summarized in the table below. Is there evidence that the way the hospital deals with pregnancies involving twins may have changed?

	2000	2005	2010
Outcome of Pregnancy Preterm (induced or cesarean)	11	13	19
Preterm (without procedures)	13	14	18
Term or Postterm	27	26	32

36. Retirement planning According to the 2011 Retirement Confidence Survey, run by the Employee Benefit Research Institute, 37% of men reported being "a lot behind schedule" while 43% of women answered in the same way. Is this evidence that in general, more women are feeling this way when it comes to retirement planning? Assuming the study had 722 men and 701 women, run the appropriate hypothesis test.

T 37. Age and party 2011 The Pew Research Center conducted a representative telephone survey during 2011. Among the reported results was the following table concerning the preferred political party affiliation of respondents and their ages for white voters. Is there evidence of age-based differences in party affiliation in the United States for white voters?

	Leaning Republican	Leaning Democrat	Other	Total
18–29	274	216	36	526
30–49	888	581	146	1615
50–64	1173	962	211	2346
65+	1062	812	209	2083
Total	3397	2571	602	6570

a) Will you conduct a test of homogeneity or independence? Why?
b) Test an appropriate hypothesis.
c) State your conclusion, including an analysis of differences you find (if any).

T 38. Eye and hair color A survey of 1021 school-age children was conducted by randomly selecting children from several large urban elementary schools. Two of the questions concerned eye and hair color. In the survey, the following codes were used:

Hair Color	Eye Color
1 = Blond	1 = Blue
2 = Brown	2 = Green
3 = Black	3 = Brown
4 = Red	4 = Grey
5 = Other	5 = Other

The Statistics students analyzing the data were asked to study the relationship between eye and hair color.

a) One group of students produced the output shown below. What kind of analysis is this? What are the null and alternative hypotheses? Is the analysis appropriate? If so, summarize the findings, being sure to include any assumptions you've made and/or limitations to the analysis. If it's not an appropriate analysis, state explicitly why not.

Dependent variable is Eyes
R-squared = 3.7%
s = 1.112 with 1021 − 2 = 1019 degrees of freedom

Variable	Coefficient	SE(Coeff)	t-Ratio	P-Value
Intercept	1.99541	0.08346	23.9	≤0.0001
Hair	0.211809	0.03372	0.28	≤0.0001

b) A second group of students used the same data to produce the output shown below. The table displays counts and standardized residuals in each cell. What kind of analysis is this? What are the null and alternative hypotheses? Is the analysis appropriate? If so, summarize the findings, being sure to include any assumptions you've made and/or limitations to the analysis. If it's not an appropriate analysis, state explicitly why not.

		Eye Color				
		1	2	3	4	5
Hair Color	1	143 7.67540	30 0.41799	58 −5.88169	15 −0.63925	12 −0.31451
	2	90 −2.57141	45 0.29019	215 1.72235	30 0.49189	20 −0.08246
	3	28 −5.39425	15 −2.34780	190 6.28154	10 −1.76376	10 −0.80382
	4	30 2.06116	15 2.71589	10 −4.05540	10 2.37402	5 0.75993
	5	10 −0.52195	5 0.33262	15 −0.94192	5 1.36326	5 2.07578

$$\sum \frac{(Observed - Expected)^2}{Expected} = 223.6 \quad P\text{-value} < 0.00001$$

39. Depression and the Internet The September 1998 issue of the *American Psychologist* published an article reporting on an experiment examining "the social and psychological impact of the Internet on 169 people in 73 households during their first 1 to 2 years online." In the experiment, a sample of households was offered free Internet access for one or two years in return for allowing their time and activity online to be tracked. The members of the households who participated in the study were also given a battery of tests at the beginning and again at the end of the study. One of the tests measured the subjects' levels of depression on a 4-point scale, with higher numbers meaning the person was more depressed. Internet usage was measured in average number of hours per week. The regression analysis examines the association between the subjects' depression levels and the amounts of Internet use. The conditions for inference were satisfied.

Dependent variable is Depression After
R-squared = 4.6%
s = 0.4563 with 162 − 2 = 160 degrees of freedom

Variable	Coefficient	SE(coeff)	t-Ratio	Prob
Constant	0.565485	0.0399	14.2	≤0.0001
Intr_use	0.019948	0.0072	2.76	0.0064

a) Do these data indicate that there is an association between Internet use and depression? Test an appropriate hypothesis and state your conclusion clearly.
b) One conclusion of the study was that those who spent more time online tended to be more depressed at the end of the experiment. News headlines said that too much time on the Internet can lead to depression. Does the study support this conclusion? Explain.
c) As noted, the subjects' depression levels were tested at both the beginning and end of this study; higher scores indicated the person was more depressed. Results are summarized in the table. Is there evidence that the depression level of the subjects changed during this study?

Depression Level
162 subjects

Variable	Mean	StdDev
DeprBfore	0.730370	0.487817
DeprAfter	0.611914	0.461932
Difference	−0.118457	0.552417

40. Pregnancy A San Diego reproductive clinic reported 42 live births to 157 women under the age of 38, but only 7 successes for 89 clients aged 38 and older. Is this evidence of a difference in the effectiveness of the clinic's methods for older women?
a) Test the appropriate hypotheses, using the two-proportion z-procedure.
b) Repeat the analysis, using an appropriate chi-square procedure.
c) Explain how the two results are equivalent.

41. Family planning A 1954 study of 1438 pregnant women examined the association between the woman's education level and the occurrence of unplanned pregnancies, producing these data:

	Education Level		
	<3 Yr HS	3+ Yr HS	Some College
Number of Pregnancies	591	608	239
% Unplanned	66.2%	55.4%	42.7%

Do these data provide evidence of an association between family planning and education level? (Source: *Fertility Planning and Fertility Rates by Socio-Economic Status, Social and Psychological Factors Affecting Fertility*, 1954)

42. Old Faithful Old Faithful eruptions do not occur at uniform intervals and may vary greatly. Can we improve our chances of predicting the time of the next eruption if we know how long the previous eruption lasted?
a) Describe what you see in this scatterplot.

b) Write an appropriate hypothesis.
c) Here are a histogram of the residuals and the residuals plot. Do you think the assumptions for inference are met? Explain.

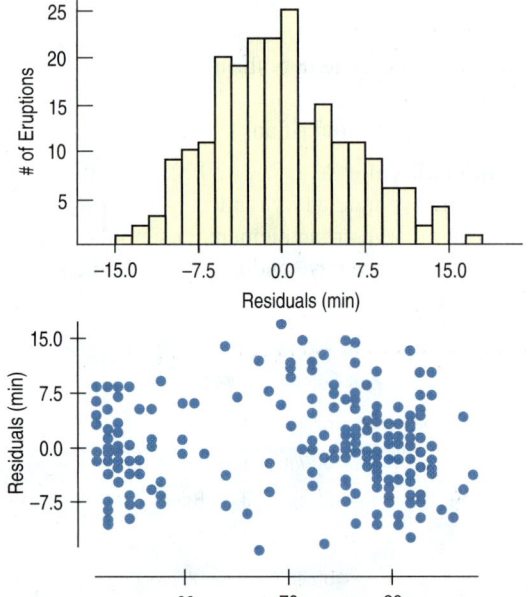

d) State a conclusion based on this regression analysis:

Dependent variable is Interval
R-squared = 77.0%
s = 6.159 with 222 − 2 = 220 degrees of freedom

Variable	Coefficient	SE(Coeff)	t-Ratio	P-Value
Intercept	33.9668	1.428	23.8	≤0.0001
Duration	10.3582	0.3822	27.1	≤0.0001

Variable	Mean	StdDev
Duration	3.57613	1.08395
Interval	71.0090	12.7992

e) The second table shows the summary statistics for the two variables. Create a 95% confidence interval for the mean length of time that will elapse following a 2-minute eruption.

f) You arrive at Old Faithful just as an eruption ends. Witnesses say it lasted 4 minutes. Create a 95% prediction interval for the length of time you will wait to see the next eruption.

43. **Togetherness** Are good grades in high school associated with family togetherness? A simple random sample of 142 high-school students was asked how many meals per week their families ate together. Their responses produced a mean of 3.78 meals per week, with a standard deviation of 2.2. Researchers then matched these responses against the students' grade point averages. The scatterplot appeared to be reasonably linear, so they went ahead with the regression analysis, seen below. No apparent pattern emerged in the residuals plot.

Dependent variable is GPA
R-squared = 11.0%
s = 0.6682 with 142 − 2 = 140 df

Variable	Coefficient	SE(Coeff)
Intercept	2.7288	0.1148
Meals/wk	0.1093	0.0263

a) Is there evidence of an association? Test an appropriate hypothesis and state your conclusion.
b) Do you think this association would be useful in predicting a student's grade point average? Explain.
c) Are your answers to parts a and b contradictory? Explain.

44. **Learning math** Developers of a new math curriculum called "Accelerated Math" compared performances of students taught by their system with control groups of students in the same schools who were taught using traditional instructional methods and materials. Statistics about pretest and posttest scores are shown in the table. (Source: J. Ysseldyke and S. Tardrew, *Differentiating Math Instruction*, Renaissance Learning)

a) Did the groups differ in average math score at the start of this study?
b) Did the group taught using the Accelerated Math program show a significant improvement in test scores?
c) Did the control group show a significant improvement in test scores?
d) Were gains significantly higher for the Accelerated Math group than for the control group?

		Instructional Method	
		Acc. Math	Control
Number of Students		231	245
Pretest	Mean	560.01	549.65
	St. Dev	84.29	74.68
Posttest	Mean	637.55	588.76
	St. Dev	82.9	83.24
Individual Gain	Mean	77.53	39.11
	St. Dev.	78.01	66.25

45. **Juvenile offenders** According to a 2011 article in the *Journal of Consulting and Clinical Psychology*, Charles Borduin pioneered a treatment called Multisystemic Therapy (MST) as a way to prevent serious mental health problems in adolescents. The therapy involves a total support network including family and community, rather than the more common individual therapy (e.g., visits to a therapist). After a 22-year-long study, one notable fact was that while 15.5% of juveniles who received individual therapy were arrested for a violent felony, only 4.3% of the juveniles treated with MST had done so.

a) Suppose the results are based on sample sizes of 125 juveniles in each group. Create a 99% confidence interval for the reduction in violent felony rate when comparing MST to the traditional individual therapy.
b) Using your interval, is there evidence of a true reduction for the whole population? Which population is the study investigating?

T 46. **Dairy sales** Peninsula Creameries sells both cottage cheese and ice cream. The CEO recently noticed that in months when the company sells more cottage cheese, it seems to sell more ice cream as well. Two of his aides were assigned to test whether this is true or not. The first aide's plot and analysis of sales data for the past 12 months (in millions of pounds for cottage cheese and for ice cream) appear below.

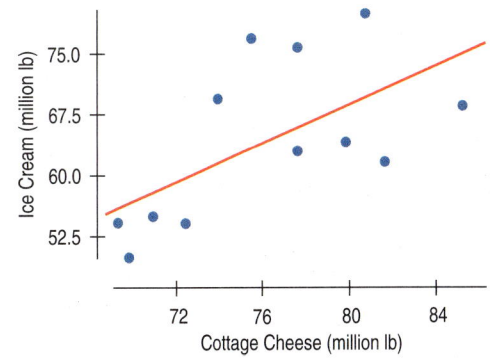

Dependent variable is Ice cream
R-squared = 36.9%
s = 8.320 with 12 − 2 = 10 degrees of freedom

Variable	Coefficient	SE(Coeff)	t-Ratio	P-Value
Intercept	−26.5306	37.68	−0.704	0.4975
Cottage C ...	1.19334	0.4936	2.42	0.0362

The other aide looked at the differences in sales of ice cream and cottage cheese for each month and created the following output:

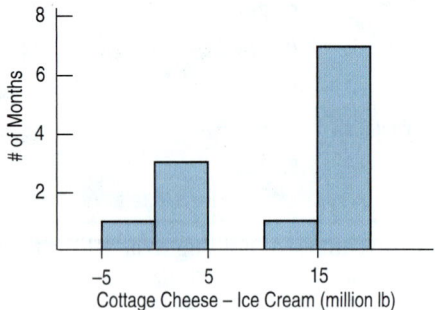

Cottage Cheese − Ice Cream

Count	12
Mean	11.8000
Median	15.3500
StdDev	7.99386
IntQRange	14.3000
25th %tile	3.20000
75th %tile	17.5000

Test H_0: $\mu(CC - IC) = 0$ vs H_a: $\mu(CC - IC) \neq 0$
Sample Mean = 11.800000 t-Statistic = 5.113 w/11 df
Prob = 0.0003
Lower 95% bound = 6.7209429
Upper 95% bound = 16.879057

a) Which analysis would you use to answer the CEO's question? Why?
b) What would you tell the CEO?
c) Which analysis would you use to test whether the company sells more cottage cheese or ice cream in a typical year? Why?
d) What would you tell the CEO about this other result?
e) What assumptions are you making in the analysis you chose in part a? What assumptions are you making in the analysis in part c?
f) Next month's cottage cheese sales are 82 million pounds. Ice cream sales are not yet available. How much ice cream do you predict Peninsula Creameries will sell?
g) Give a 95% confidence interval for the true slope of the regression equation of ice cream sales by cottage cheese sales.
h) Explain what your interval means.

47. **Infliximab** In an article appearing in the journal *The Lancet*, medical researchers reported on the experimental use of the arthritis drug infliximab in treating Crohn's disease. In a trial, 573 patients were given initial 5-mg injections of the drug. Two weeks later, 335 had responded positively. These patients were then randomly assigned to three groups. Group I received continued injections of a placebo, Group II continued with 5 mg of infliximab, and Group III received 10 mg of the drug. After 30 weeks, 23 of 110 Group I patients were in remission, compared with 44 of 113 Group II and 50 of 112 Group III patients. Do these data indicate that continued treatment with infliximab is of value for Crohn's disease patients who exhibit a positive initial response to the drug?

48. **Education vs. income 2009** The information below examines the median income and education level (percent of population with at least a Bachelor's degree) for several large U.S. cities in 2009.

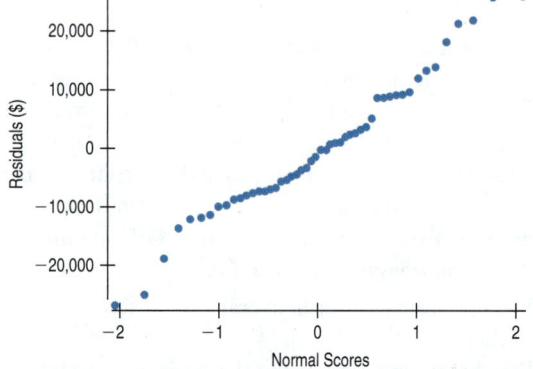

Summary Statistics:

Column	n	Mean	Std. Dev.
Median Household Income	50	46521.22	15455.175
Percentage Bachelor's Degrees	50	28.402	11.756

Dependent variable is Median Household Income
R-squared = 42.1%
s = 11,881.87 with 50 − 2 = 48 degrees of freedom

Variable	Coefficient	SE(Coeff)	t-Ratio	P-Value
Intercept	22294.24	4431.67	5.03	<0.0001
Percentage Bachelor's Degrees	853.00	144.38	5.91	<0.0001

a) Do you think the assumptions for inference are met? Explain.
b) Does there appear to be an association between education and income levels in these cities?
c) Would this association appear to be weaker, stronger, or the same if data were plotted for individual people rather than for cities in aggregate? Explain.
d) Create and interpret a 95% confidence interval for the slope of the true line that describes the association between income and education.
e) Predict the median household income for cities where an average of 25% of residents have at least a Bachelor's degree. Describe your estimate with a 90% confidence interval, and interpret that result.

T 49. Diet Thirteen overweight women volunteered for a study to determine whether eating specially prepared crackers before a meal could help them lose weight. The subjects were randomly assigned to eat crackers with different types of fiber (bran fiber, gum fiber, both, and a control cracker). Unfortunately, some of the women developed uncomfortable bloating and upset stomachs. Researchers suspected that some of the crackers might be at fault. The contingency table of "Cracker" versus "Bloat" shows the relationship between the four different types of crackers and the reported bloating. The study was paid for by the manufacturers of the gum fiber. What would you recommend to them about the prospects for marketing their new diet cracker?

		Bloat	
		Little/None	Moderate/Severe
Cracker	Bran	11	2
	Gum	4	9
	Combo	7	6
	Control	8	4

T 50. Cramming Students in two basic Spanish classes were required to learn 50 new vocabulary words. One group of 45 students received the list on Monday and studied the words all week. Statistics summarizing this group's scores on Friday's quiz are given. The other group of 25 students did not get the vocabulary list until Thursday. They also took the quiz on Friday, after "cramming" Thursday night. Then, when they returned to class the following Monday, they were retested—without advance warning. Both sets of test scores for these students are shown.

Group 1
Fri.
Number of students = 45
Mean = 43.2 (of 50)
StDev = 3.4
Students passing (score ≥ 40) = 33%

Group 2

Fri.	Mon.	Fri.	Mon.
42	36	50	47
44	44	34	34
45	46	38	31
48	38	43	40
44	40	39	41
43	38	46	32
41	37	37	36
35	31	40	31
43	32	41	32
48	37	48	39
43	41	37	31
45	32	36	41
47	44		

a) Did the week-long study group have a mean score significantly higher than that of the overnight crammers?
b) Was there a significant difference in the percentages of students who passed the quiz on Friday?
c) Is there any evidence that when students cram for a test, their "learning" does not last for 3 days?
d) Use a 95% confidence interval to estimate the mean number of words that might be forgotten by crammers.
e) Is there any evidence that how much students forget depends on how much they "learned" to begin with?

T 51. Hearing Fitting someone for a hearing aid requires assessing the patient's hearing ability. In one method of assessment, the patient listens to a tape of 50 English words. The tape is played at low volume, and the patient is asked to repeat the words. The patient's hearing ability score is the number of words perceived correctly. Four tapes of equivalent difficulty are available so that each ear can be tested with more than one hearing aid. These lists were created to be equally difficult to perceive in silence, but hearing aids must work in the presence of background noise. Researchers had 24 subjects with normal hearing compare two of the tapes when a background noise was present, with the order of the tapes randomized. Is it reasonable to assume that the two lists are still equivalent for purposes of the hearing test when there is background noise? Base your decision on a confidence interval for the mean difference in the number of words people might misunderstand. (Source: Faith Loven, *A Study of the Interlist Equivalency of the CID W-22 Word List Presented in Quiet and in Noise*. University of Iowa [1981])

Subject	List A	List B
1	24	26
2	32	24
3	20	22
4	14	18
5	32	24
6	22	30
7	20	22
8	26	28
9	26	30
10	38	16
11	30	18
12	16	34
13	36	32
14	32	34
15	38	32
16	14	18
17	26	20
18	14	20
19	38	40
20	20	26
21	14	14
22	18	14
23	22	30
24	34	42

52. **Newspapers** Who reads the newspaper more, men or women? Eurostat, an agency of the European Union (EU), conducts surveys on several aspects of daily life in EU countries. Recently, the agency asked samples of 1000 respondents in each of 14 European countries whether they read the newspaper on a daily basis. Below are the data by country and gender.

Country	% Reading a Newspaper Daily	
	Men	Women
Belgium	56.3	45.5
Denmark	76.8	70.3
Germany	79.9	76.8
Greece	22.5	17.2
Spain	46.2	24.8
Ireland	58.0	54.0
Italy	50.2	29.8
Luxembourg	71.0	67.0
Netherlands	71.3	63.0
Austria	78.2	74.1
Portugal	58.3	24.1
Finland	93.0	90.0
Sweden	89.0	88.0
U.K.	32.6	30.4

a) Examine the differences in the percentages for each country. Which of these countries seem to be outliers? What do they have in common?
b) After eliminating the outliers, is there evidence that in Europe men are more likely than women to read the newspaper?

Chapter 26

Analysis of Variance

26.1 Testing Whether the Means of Several Groups Are Equal

26.2 The ANOVA Table

26.3 Assumptions and Conditions

26.4 Comparing Means

26.5 ANOVA on Observational Data

> **Where are we going?**
> In Chapter 4, we looked at the performance of four brands of cups to see which was the most effective at keeping coffee hot. Are all these brands equally good? How can we compare them all? We could run a *t*-test for each of the 6 head-to-head comparisons, but we'll learn a better way to compare more than two groups in this chapter.

Who	Hand washings by four different methods, assigned randomly and replicated 8 times each
What	Number of bacteria colonies
How	Sterile media plates incubated at 36°C for 2 days

Figure 26.1
Boxplots of the bacteria colony counts for the four different washing methods suggest some differences between treatments. (Data in **Bacterial Soap**)

Did you wash your hands with soap before eating? You've undoubtedly been asked that question a few times in your life. Mom knows that washing with soap eliminates most of the germs you've managed to collect on your hands. Or does it? A student decided to investigate just how effective washing with soap is in eliminating bacteria. To do this she tested four different methods—washing with water only, washing with regular soap, washing with antibacterial soap (ABS), and spraying hands with antibacterial spray (AS) (containing 65% ethanol as an active ingredient). Her experiment consisted of one experimental factor, the washing *Method*, at four levels.

She suspected that the number of bacteria on her hands before washing might vary considerably from day to day. To help even out the effects of those changes, she generated random numbers to determine the order of the four treatments. Each morning, she washed her hands according to the treatment randomly chosen. Then she placed her right hand on a sterile media plate designed to encourage bacteria growth. She incubated each plate for 2 days at 36°C, after which she counted the bacteria colonies. She replicated this procedure 8 times for each of the four treatments.

A side-by-side boxplot of the numbers of colonies seems to show some differences among the treatments:

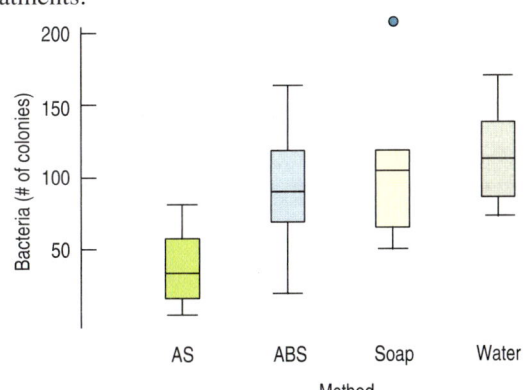

When we first looked at a quantitative variable measured for each of several groups in Chapter 4, we displayed the data this way with side-by-side boxplots. And when we compared the boxes, we asked whether the centers seemed to differ, using the spreads of the boxes to judge the size of the differences. Now we want to make this more formal by testing a hypothesis. We'll make the same kind of comparison, comparing the variability among the means with the spreads of the boxes. It looks like the alcohol spray has lower bacteria counts, but as always, we're skeptical. Could it be that the four methods really have the same mean counts and we just *happened* to get a difference like this because of natural sampling variability?

What is the null hypothesis here? It seems natural to start with the hypothesis that *all the group means are equal.* That would say it doesn't matter what method you use to wash your hands because the mean bacteria count will be the same. We know that even if there were no differences at all in the *means* (for example, if someone replaced all the solutions with water) there would still be sample-to-sample differences. We want to see, statistically, whether differences as large as those observed in the experiment could naturally occur by chance in groups that have equal means. If we find that the differences in washing *Methods* are so large that they would occur only very infrequently in groups that actually have the same mean, then, as we've done with other hypothesis tests, we'll reject the null hypothesis and conclude that the washing *Methods* really have different means.[1]

For Example

Contrast baths are a treatment commonly used in hand clinics to reduce swelling and stiffness after surgery. Patients' hands are immersed alternately in warm and cool water. (That's the *contrast* in the name.) Sometimes, the treatment is combined with mild exercise. Although the treatment is widely used, it had never been verified that it would accomplish the stated outcome goal of reducing swelling.

Researchers[2] randomly assigned 59 patients who had received carpal tunnel release surgery to one of three treatments: contrast bath, contrast bath with exercise, and (as a control) exercise alone. Hand therapists who did not know how the subjects had been treated measured hand volumes before and after treatments in milliliters by measuring how much water the hand displaced when submerged. The change in hand volume (after treatment minus before) was reported as the outcome.

QUESTION: Specify the details of the experiment's design. Identify the subjects, the sample size, the experiment factor, the treatment levels, and the response. What is the null hypothesis? Was randomization employed? Was the experiment blinded? Was it double-blinded?

ANSWER: Subjects were patients who received carpal tunnel release surgery. Sample size is 59 patients. The factor was contrast bath treatment with three levels: contrast baths alone, contrast baths with exercise, and exercise alone. The response variable is the change in hand volume. The null hypothesis is that the mean changes in hand volume will be the same for the three treatment levels. Patients were randomly assigned to treatments. The study was single-blind because the evaluators were blind to the treatments. It was not (and could not be) double-blind because the patients had to be aware of their treatments.

---

[1] The alternative hypothesis is that "the means are not *all* equal." Be careful not to confuse that with "all the means are different." With 11 groups we could have 10 means equal to each other and 1 different. The null hypothesis would still be false.

[2] Janssen, Robert G., Schwartz, Deborah A., and Velleman, Paul F., "A Randomized Controlled Study of Contrast Baths on Patients with Carpal Tunnel Syndrome," *Journal of Hand Therapy,* 22:3, pp. 200–207. The data reported here differ slightly from those in the original paper because they include some additional subjects and exclude some outliers.

26.1 Testing Whether the Means of Several Groups Are Equal

We saw in Chapter 22 how to use a *t*-test to see whether two groups have equal means. We compared the difference in the means to a standard error estimated from all the data. And when we were willing to assume that the underlying group variances were equal, we pooled the data from the two groups to find the standard error.

Now we have more groups, so we can't just look at differences in the means.[3] But all is not lost. Even if the null hypothesis were true, and the means of the populations underlying the groups were equal, we'd still expect the sample means to vary a bit. We could measure that variation by finding the variance of the means. How much should they vary? Well, if we look at how much the data themselves vary, we can get a good idea of how much the means should vary. And if the underlying means are actually different, we'd expect that variation to be larger.

It turns out that we can build a hypothesis test to check whether the variation in the means is bigger than we'd expect it to be just from random fluctuations. We'll need a new sampling distribution model, called the *F*-model, but that's just a different table to look at (Table F can be found at the end of this chapter).

To get an idea of how it works, let's start by looking at the following two sets of boxplots:

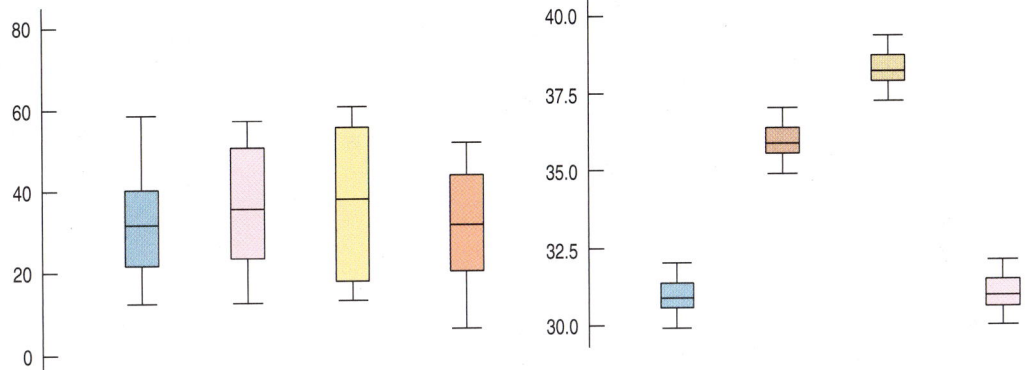

Figure 26.2

It's hard to see the difference in the means in these boxplots because the spreads are large relative to the differences in the means.

Figure 26.3

In contrast with Figure 26.2, the smaller variation makes it much easier to see the differences among the group means. (Notice also that the scale of the *y*-axis is considerably different from the plot on the left.)

We're trying to decide if the means are different enough for us to reject the null hypothesis. If they're close, we'll attribute the differences to natural sampling variability. What do you think? It's easy to see that the means in the second set differ. It's hard to imagine that the means could be that far apart just from natural sampling variability alone. How about the first set? It looks like these observations *could* have occurred from treatments with the same means.[4] This much variation among groups does seem consistent with equal group means.

Believe it or not, the two sets of treatment means in both figures are the same. (They are 31, 36, 38, and 31, respectively.) Then why do the figures look so different? In the second figure, the variation *within* each group is so small that the differences *between* the means stand out. This is what we looked for when we compared boxplots by eye back in Chapter 4. And it's the central idea of the *F*-test. We compare the differences *between* the means of the groups with the variation *within* the groups. When the differences between means are large compared with the variation within the groups, we reject the null hypothesis and conclude

---

[3] You might think of testing all pairs, but that method generates too many Type I errors. We'll see more about this later in the chapter.

[4] Of course, with a large enough sample, we can detect any differences that we like. For experiments with the same sample size, it's easier to detect the differences when the variation *within* each box is smaller.

that the means are not equal. In the first figure, the differences among the means look as though they could have arisen just from natural sampling variability from groups with equal means, so there's not enough evidence to reject H_0.

How can we make this comparison more precise statistically? All the tests we've seen have compared differences of some kind with a ruler based on an estimate of variation. And we've always done that by looking at the ratio of the statistic to that variation estimate. Here, the differences among the means will show up in the numerator, and the ruler we compare them with will be based on the underlying standard deviation—that is, on the variability *within* the treatment groups.

For Example

RECAP: Fifty-nine postsurgery patients were randomly assigned to one of three treatment levels. Changes in hand volume were measured. Here are the boxplots. The recorded values are volume after treatment − volume before treatment, so positive values indicate swelling. Some swelling is to be expected. (Data in **Contrast baths**)

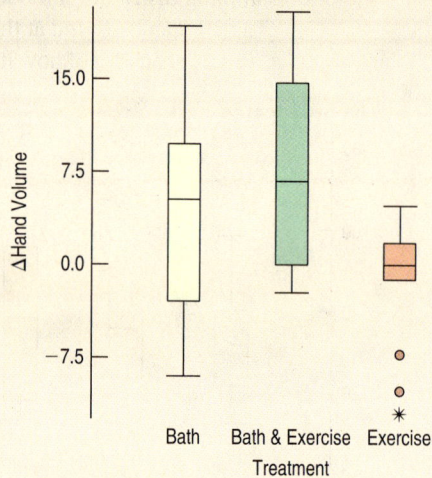

QUESTION: What do the boxplots say about the results?

ANSWER: There doesn't seem to be much difference between the two contrast bath treatments. The exercise only treatment may result in less swelling.

How Different Are They?

The challenge here is that we can't take a simple difference as we did when comparing two groups. In the hand-washing experiment, we have differences in mean bacteria counts across *four* treatments. How should we measure how different the four group means are? With only two groups, we naturally took the difference between their means as the numerator for the *t*-test. It's hard to imagine what else we could have done. How can we generalize that to more than two groups? When we've wanted to know how different many observations were, we measured how much they vary, and that's what we do here.

How much natural variation should we expect among the means if the null hypothesis were true? If the null hypothesis were *true*, then each of the treatment means would estimate the *same* underlying mean. If the washing methods are all the same, it's as if we're just estimating the mean bacteria count on hands that have been washed with plain water. And we have several (in our experiment, four) different, independent estimates of this mean. Here comes the clever part. We can treat these estimated means as if they were observations and simply calculate their (sample) variance. This variance is the measure we'll use to assess how different the group means are from each other. It's the generalization of the difference between means for only two groups.

WHY VARIANCES?

We've usually measured variability with standard deviations. Standard deviations have the advantage that they're in the same units as the data. Variances have the advantage that for independent variables, the variances add. Because we're talking about sums of variables, we'll stay with variances before we get back to standard deviations.

Level	n	Mean
Alcohol Spray	8	37.5
Antibacterial Soap	8	92.5
Soap	8	106.0
Water	8	117.0

The more the group means resemble each other, the smaller this variance will be. The more they differ (perhaps because the treatments actually have an effect), the larger this variance will be.

For the bacteria counts, the four means are listed in the table to the left. If you took those four values, treated them as observations, and found their sample variance, you'd get 1245.08. That's fine, but how can we tell whether it is a big value? Now we need a model, and the model is based on our null hypothesis that all the group means are equal. Here, the null hypothesis is that it doesn't matter what washing method you use; the mean bacteria count will be about the same:

$$H_0: \mu_1 = \mu_2 = \mu_3 = \mu_4 = \mu.$$

As always when testing a null hypothesis, we'll start by assuming that it is true. And if the group means are equal, then there's an overall mean, μ—the bacteria count you'd expect all the time after washing your hands in the morning. And each of the observed group means is just a sample-based estimate of that underlying mean.

We know how sample means vary. The variance of a sample mean is σ^2/n. With eight observations in a group, that would be $\sigma^2/8$. The estimate that we've just calculated, 1245.08, should estimate this quantity. If we want to get back to the variance of the *observations*, σ^2, we need to multiply it by 8. So $8 \times 1245.08 = 9960.64$ should estimate σ^2.

Is 9960.64 large for this variance? How can we tell? We'll need a hypothesis test. You won't be surprised to learn that there is just such a test. The details of the test, due to Sir Ronald Fisher in the early 20th century, are truly ingenious, and may be the most amazing statistical result of that century.

The Ruler Within

We need a suitable ruler for comparison—one based on the underlying variability in our measurements. That variability is due to the day-to-day differences in the bacteria count even when the same soap is used. Why would those counts be different? Maybe the experimenter's hands were not equally dirty, or she washed less well some days, or the plate incubation conditions varied. We randomized just so we could see past such things.

We need an independent estimate of σ^2, one that doesn't depend on the null hypothesis being true, one that won't change if the groups have different means. As in many quests, the secret is to look "within." We could look in *any* of the treatment groups and find its variance. But which one should we use? The answer is, *all* of them!

At the start of the experiment (when we randomly assigned experimental units to treatment groups), the units were drawn randomly from the same pool, so each treatment group had a sample variance that estimated the same σ^2. If the null hypothesis is true, then not much has happened to the experimental units—or at least, their means have not moved apart. It's not much of a stretch to believe that their variances haven't moved apart much either. (If the washing methods are equivalent, then the choice of method would not affect the mean *or* the variability.) So each group variance still estimates a common σ^2.

We're assuming that the null hypothesis is true. If the group variances are equal, then the common variance they all estimate is just what we've been looking for. Since all the group variances estimate the same σ^2, we can pool them to get an overall estimate of σ^2. Recall that we pooled to estimate variances when we tested the null hypothesis that two proportions were equal—and for the same reason. It's also exactly what we did in a pooled *t*-test. The variance estimate we get by pooling we'll denote, as before, by s_p^2.

For the bacteria counts, the standard deviations and variances are listed below.

Level	n	Mean	Std Dev	Variance
Alcohol Spray	8	37.5	26.56	705.43
Antibacterial Soap	8	92.5	41.96	1760.64
Soap	8	106.0	46.96	2205.24
Water	8	117.0	31.13	969.08

If we pool the four variances (here we can just average them because all the sample sizes are equal), we'd get $s_p^2 = 1410.10$. In the pooled variance, each variance is taken around its *own* treatment mean, so the pooled estimate doesn't depend on the treatment means being equal. But the estimate in which we took the four means as observations and took their variance does. That estimate gave 9960.64. That seems a lot bigger than 1410.10. Might this be evidence that the four means are not equal?

Let's see what we have. We have an estimate of σ^2 from the variation *within* groups of 1410.10. That's just the variance of the *residuals* pooled across all groups. Because it's a pooled variance, we could write it as s_p^2. Traditionally this quantity is also called the **Error Mean Square**, or sometimes the **Within Mean Square** and denoted by MS_E. These names date back to the early 20th century when the methods were developed. If you think about it, the names do make sense—variances are means of squared differences.[5]

But we also have a *separate* estimate of σ^2 from the variation *between* the groups because we know how much means ought to vary. For the hand-washing data, when we took the variance of the four means and multiplied it by n we got 9960.64. We expect this to estimate σ^2 too, *as long as we assume the null hypothesis is true*. We call this quantity the **Treatment Mean Square** (or sometimes the **Between Mean Square**[6]) and denote by MS_T.

The *F*-Statistic

Now we have two different estimates of the underlying variance. The first one, the MS_T, is based on the differences *between* the group means. If the group means are equal, as the null hypothesis asserts, it will estimate σ^2. But, if they are not, it will give some bigger value. The other estimate, the MS_E, is based only on the variation *within* the groups around each of their own means, and doesn't depend at all on the null hypothesis being true.

So, how do we test the null hypothesis? When the null hypothesis is true, the treatment means are equal, and both MS_E and MS_T estimate σ^2. Their ratio, then, should be close to 1.0. But, when the null hypothesis is false, the MS_T will be *larger* because the treatment means are not equal. The MS_E is a pooled estimate in which the variation within each group is found around its own group mean, so differing means won't inflate it. That makes the ratio MS_T/MS_E perfect for testing the null hypothesis. When the null hypothesis is true, the ratio should be near 1. If the treatment means really are different, the numerator will tend to be larger than the denominator, and the ratio will tend to be bigger than 1.

Of course, even when the null hypothesis *is* true, the ratio will vary around 1 just due to natural sampling variability. How can we tell when it's big enough to reject the null hypothesis? To be able to tell, we need a sampling distribution model for the ratio. Sir Ronald Fisher found the sampling distribution model of the ratio in the early 20th century. In his honor, we call the distribution of MS_T/MS_E the ***F*-distribution**. And we call the ratio MS_T/MS_E the ***F*-statistic**. By comparing this statistic with the appropriate *F*-distribution we (or the computer) can get a P-value.

The resulting ***F*-test** is one-tailed because any differences in the means make the *F*-statistic larger. Larger differences in the treatments' effects lead to the means being more variable, making the MS_T bigger. That makes the *F*-ratio grow. So the test is significant if the *F*-ratio is big enough. In practice, we find a P-value, and big *F*-statistic values go with small P-values.

The entire analysis is called the Analysis of Variance, commonly abbreviated **ANOVA** (and pronounced uh-NŌ-va). You might think that it should be called the analysis of means, since it's the equality of the means we're testing. But we use the *variances* within and between the groups for the test.

■ **NOTATION ALERT**
Capital *F* is used only for this distribution model and statistic. Fortunately, Fisher's name didn't start with a Z, a T, or an R.

---

[5] Well, actually, they're sums of squared differences divided by their degrees of freedom—$(n-1)$ for the first variance we saw back in Chapter 3, and other degrees of freedom for each of the others we've seen. But even back in Chapter 3, we said this was a "kind of" mean, and indeed, it still is.

[6] Grammarians would probably insist on calling it the *Among Mean Square*, since the variation is among *all* the group means. Traditionally, though, it's called the *Between Mean Square* and we have to talk about the variation between all the groups (as bad as that sounds).

Like Student's t-models, the F-models are a family. F-models depend on not one, but two, degrees of freedom parameters. The degrees of freedom come from the two variance estimates and are sometimes called the *numerator df* and the *denominator df*. The *Treatment Mean Square*, MS_T, is the sample variance of the observed treatment means. If we think of them as observations, then since there are k groups, this variance has $k - 1$ degrees of freedom. The *Error Mean Square*, MS_E, is the pooled estimate of the variance within the groups. If there are n observations in each group, then we get $n - 1$ degrees of freedom from each, for a total of $k(n - 1)$ degrees of freedom.

A simpler way of tracking the degrees of freedom is to start with all the cases. We'll call that N. Each group has its own mean, costing us a degree of freedom—k in all. So we have $N - k$ degrees of freedom for the error. When the groups all have equal sample size, that's the same as $k(n - 1)$, but this way works even if the group sizes differ.

We say that the F-statistic, MS_T/MS_E, has $k - 1$ and $N - k$ degrees of freedom.

> **NOTATION ALERT**
> What, first little n and now big N? In an experiment, it's standard to use N for *all* the cases and n for the number in each treatment group.

Back to Bacteria

For the hand-washing experiment, the $MS_T = 9960.64$. The $MS_E = 1410.14$. If the treatment means were equal, the *Treatment Mean Square* should be about the same size as the *Error Mean Square*, about 1410. But it's 9960.64, which is 7.06 times bigger. In other words, $F = 7.06$. This F-statistic has $4 - 1 = 3$ and $32 - 4 = 28$ degrees of freedom.

An F-value of 7.06 is bigger than 1, but we can't tell for sure whether it's big enough to reject the null hypothesis until we check the $F_{3,28}$ model to find its P-value. (Usually, that's most easily done with technology, but we can use printed tables.) It turns out the P-value is 0.0011. In other words, if the treatment means were actually equal, we would expect the ratio MS_T/MS_E to be 7.06 or larger about 11 times out of 10,000, just from natural sampling variability. That's not very likely, so we reject the null hypothesis and conclude that the means are different. We have strong evidence that the four different methods of hand washing are not equally effective at eliminating germs.

26.2 The ANOVA Table

You'll often see the mean squares and other information put into a table called the **ANOVA table**. Here's the table for the washing experiment:

Analysis of Variance Table

Source	DF	Sum of Squares	Mean Square	F-Ratio	P-Value
Method	3	29882	9960.64	7.0636	0.0011
Error	28	39484	1410.14		
Total	31	69366			

> **Calculating the ANOVA table** This table has a long tradition stretching back to when ANOVA calculations were done by hand. Major research labs had rooms full of mechanical calculators operated by women. (Yes, always women; women were thought—by the men in charge, at least—to be more careful at such an exacting task.) Three women would perform each calculation, and if any two of them agreed on the answer, it was taken as the correct value.

The ANOVA table was originally designed to organize the calculations. With technology, we have much less use for that. We'll show how to calculate the sums of squares later in the chapter, but the most important quantities in the table are the F-statistic and its associated P-value. When the F-statistic is large, the Treatment (here *Method*) Mean Square is large compared to the Error Mean Square (MS_E), and provides evidence that in fact the means of the groups are not all equal.

You'll almost always see ANOVA results presented in a table like this. After nearly a century of writing the table this way, statisticians (and their technology) aren't going to change. Even though the table was designed to facilitate hand calculation, computer programs that compute ANOVAs still present the results in this form. Usually the P-value is found next to the F-ratio. The P-value column may be labeled with a title such as "Prob > F," "sig," or "Prob." Don't let that confuse you; it's just the P-value.

You'll sometimes see the two mean squares referred to as the *Mean Square Between* and the *Mean Square Within*—especially when we test data from observational studies rather than experiments. ANOVA is often used for such observational data, and as long as certain conditions are satisfied, there's no problem with using it in that context.

For Example

RECAP: An experiment to determine the effect of contrast bath treatments on swelling in postsurgical patients recorded hand volume changes for patients who had been randomly assigned to one of three treatments.

Here is the Analysis of Variance for these data:

Analysis of Variance for Hand Volume Change

Source	df	Sum of Squares	Mean Square	F-Ratio	P-Value
Treatment	2	716.159	358.080	7.4148	0.0014
Error	56	2704.38	48.2926		
Total	58	3420.54			

QUESTION: What does the ANOVA say about the results of the experiment? Specifically, what does it say about the null hypothesis?

ANSWER: The F-ratio of 7.4148 has a P-value that is quite small. We can reject the null hypothesis that the mean change in hand volume is the same for all three treatments.

The *F*-Table

Usually, you'll get the P-value for the *F*-statistic from technology. Any software program performing an ANOVA will automatically "look up" the appropriate one-sided P-value for the *F*-statistic. If you want to do it yourself, you'll need an *F*-table. *F*-tables are usually printed only for a few values of α, often 0.05, 0.01, and 0.001. They give the critical value of the *F*-statistic with the appropriate number of degrees of freedom determined by your data, for the α level that you select. If your *F*-statistic is greater than that value, you know that its P-value is less than that α level. So, you'll be able to tell whether the P-value is greater or less than 0.05, 0.01, or 0.001, but to be more precise, you'll need technology (or an interactive table like the one in the *ActivStats* program online at pearsonhighered.com/deveaux).

Here's an excerpt from an *F*-table for $\alpha = 0.05$:

Figure 26.4

Part of an *F*-table showing critical values for $\alpha = 0.05$ and highlighting the critical value, 2.947, for 3 and 28 degrees of freedom. We can see that only 5% of the values will be greater than 2.947 with this combination of degrees of freedom.

Notice that the critical value for 3 and 28 degrees of freedom at $\alpha = 0.05$ is 2.947. Since our *F*-statistic of 7.06 is larger than this critical value, we know that the P-value is less than 0.05. We could also look up the critical value for $\alpha = 0.01$ and find that it's 4.568 and the critical value for $\alpha = 0.001$ is 7.193. So our *F*-statistic sits between the two critical values 0.01 and 0.001, and our P-value is slightly *greater* than 0.001. Technology can find the value precisely. It turns out to be 0.0011.

Just Checking

A student conducted an experiment to see which, if any, of four different paper airplane designs results in the longest flights (measured in inches). The boxplots look like this (with the overall mean shown in red): (Data in **Paper planes**)

The ANOVA table shows:

Analysis of Variance

Source	DF	Sum of Squares	Mean Square	F-Ratio	Prob > F
Design	3	51991.778	17330.6	37.4255	<0.0001
Error	32	14818.222	463.1		
Total	35	66810.000			

1. What is the null hypothesis?
2. From the boxplots, do you think that there is evidence that the mean flight distances of the four designs differ?
3. Does the *F*-test in the ANOVA table support your preliminary conclusion in (2)?
4. The researcher concluded that "there is substantial evidence that all four of the designs result in different mean flight distances." Do you agree?

The ANOVA Model

To understand the ANOVA table, let's start by writing a model for what we observe. We start with the simplest interesting model: one that says that the only differences of interest among the groups are the differences in their means. This **one-way ANOVA model** characterizes each observation in terms of its mean and assumes that any variation around that mean is just random error:

$$y_{ij} = \mu_j + \varepsilon_{ij}.$$

That is, each observation is the sum of the mean for the treatment it received plus a random error. Our null hypothesis is that the treatments made no difference—that is, that the means are all equal:

$$H_0: \mu_1 = \mu_2 = \cdots = \mu_k.$$

It will help our discussion if we think of the overall mean of the experiment and consider the treatments as adding or subtracting an effect to this overall mean. Thinking in this way, we could write μ for the overall mean and τ_j for the deviation from this mean to get to the *j*th treatment mean—the *effect* of the treatment (if any) in moving that group away from the overall mean:

$$y_{ij} = \mu + \tau_j + \varepsilon_{ij}.$$

Thinking in terms of the effects, we could also write the null hypothesis in terms of these treatment *effects* instead of the means:

$$H_0: \tau_1 = \tau_2 = \cdots = \tau_k = 0.$$

We now have three different kinds of parameters: the overall mean, the treatment effects, and the errors. We'll want to estimate them from the data. Fortunately, we can do that in a straightforward way.

To estimate the overall mean, μ, we use the mean of all the observations: $\bar{\bar{y}}$ (called the "grand mean"[7]). To estimate each treatment effect, we find the difference between the mean of that particular treatment and the grand mean:

$$\hat{\tau}_j = \bar{y}_j - \bar{\bar{y}}.$$

There's an error, ε_{ij}, for each observation. We estimate those with the residuals from the treatment means: $e_{ij} = y_{ij} - \bar{y}_j$.

We can write each observation as the sum of three quantities that correspond to our model:

$$y_{ij} = \bar{\bar{y}} + (\bar{y}_j - \bar{\bar{y}}) + (y_{ij} - \bar{y}_j).$$

What this says is simply that we can write each observation as the sum of

- the grand mean,
- the effect of the treatment it received, and
- the residual

Or:

Observations = Grand mean + Treatment effect + Residual.

If we look at the equivalent equation

$$y_{ij} = \bar{\bar{y}} + (\bar{y}_j - \bar{\bar{y}}) + (y_{ij} - \bar{y}_j)$$

closely, it doesn't really seem like we've done anything. In fact, collecting terms on the right-hand side will give back just the observation, y_{ij} again. But this decomposition is actually the secret of the Analysis of Variance. We've split each observation into "sources"—the grand mean, the treatment effect, and error.

> **Where does the residual term come from?** Think of the annual report from any Fortune 500 company. The company spends billions of dollars each year and at the end of the year, the accountants show where each penny goes. How do they do it? After accounting for salaries, bonuses, supplies, taxes, etc., etc., etc., what's the last line? It's always labeled "other" or miscellaneous. Using "other" as the difference between all the sources they know and the total they start with, they can always make it add up perfectly. The residual is just the statisticians' "other." It takes care of all the other sources we didn't think of or don't want to consider, and makes the decomposition work by adding (or subtracting) back in just what we need.

---

[7] The father of your father is your grandfather. The mean of the group means should probably be the grandmean, but we usually spell it as two words.

Let's see what this looks like for our hand-washing data. Here are the data again, displayed a little differently:

	Alcohol	AB Soap	Soap	Water
	51	70	84	74
	5	164	51	135
	19	88	110	102
	18	111	67	124
	58	73	119	105
	50	119	108	139
	82	20	207	170
	17	95	102	87
Treatment Means	37.5	92.5	106	117

The grand mean of all observations is 88.25. Let's put that into a similar table:

Alcohol	AB Soap	Soap	Water
88.25	88.25	88.25	88.25
88.25	88.25	88.25	88.25
88.25	88.25	88.25	88.25
88.25	88.25	88.25	88.25
88.25	88.25	88.25	88.25
88.25	88.25	88.25	88.25
88.25	88.25	88.25	88.25
88.25	88.25	88.25	88.25

The treatment *means* are 37.5, 92.5, 106, and 117, respectively, so the treatment *effects* are those minus the grand mean (88.25). Let's put the treatment effects into their table:

Alcohol	AB Soap	Soap	Water
−50.75	4.25	17.75	28.75
−50.75	4.25	17.75	28.75
−50.75	4.25	17.75	28.75
−50.75	4.25	17.75	28.75
−50.75	4.25	17.75	28.75
−50.75	4.25	17.75	28.75
−50.75	4.25	17.75	28.75
−50.75	4.25	17.75	28.75

Finally, we compute the residuals as the differences between each observation and its treatment mean:

Alcohol	AB Soap	Soap	Water
13.5	−22.5	−22	−43
−32.5	71.5	−55	18
−18.5	−4.5	4	−15
−19.5	18.5	−39	7
20.5	−19.5	13	−12
12.5	26.5	2	22
44.5	−72.5	101	53
−20.5	2.5	−4	−30

Now we have four tables for which

$$\text{Observations} = \text{Grand Mean} + \text{Treatment Effect} + \text{Residual}.$$

(You can verify, for example, that the first observation, $51 = 88.25 + (-50.75) + 13.5$).

Why do we want to think in this way? Think back to the boxplots in Figures 26.2 and 26.3. To *test* the hypothesis that the treatment effects are zero, we want to see whether the treatment effects are large *compared* to the errors. Our eye looks at the variation *between* the treatment means and compares it to the variation *within* each group.

The ANOVA separates those two quantities into the Treatment Effects and the Residuals. Sir Ronald Fisher's insight was how to turn those quantities into a statistical test. We want to see if the Treatment Effects are large compared with the Residuals. To do that, we first compute the Sums of Squares of each table. Fisher's insight was that dividing these sums of squares by their respective degrees of freedom lets us test their ratio by a distribution that he found (which was later named the F in his honor). When we divide a sum of squares by its degrees of freedom we get the associated *mean square*.

When the Treatment Mean Square is *large* compared to the Error Mean Square, this provides evidence that the treatment means are different. And we can use the F-distribution to see how large "large" is.

The sums of squares for each table are easy to calculate. Just take every value in the table, square it, and add them all up. For the *Methods*, the Treatment Sum of Squares, $SS_T = (-50.75)^2 + (-50.75)^2 + \cdots + (28.75)^2 = 29882$. There are four treatments, and so there are 3 degrees of freedom. So,

$$MS_T = SS_T/3 = 29882/3 = 9960.64.$$

In general, we could write the Treatment Sum of Squares as

$$SS_T = \sum\sum (\bar{y}_j - \bar{\bar{y}})^2.$$

Be careful to note that the summation is over the *whole* table, rows and columns. That's why there are two summation signs.

And,

$$MS_T = SS_T/(k-1).$$

The table of residuals shows the variation that remains after we remove the overall mean and the treatment effects. These are what's left over after we account for what we're interested in—in this case the treatments. Their variance is the variance *within* each group that we see in the boxplots of the four groups. To find its value, we first compute the Error Sum of Squares, SS_E, by summing up the squares of every element in the residuals table. To get the Mean Square (the variance) we have to divide it by $N - k$ rather than by $N - 1$ because we found them by subtracting each of the k treatment means.

So,

$$SS_E = (13.5)^2 + (-32.5)^2 + \cdots + (-30)^2 = 39484$$

and

$$MS_E = SS_E/(32 - 4) = 1410.14.$$

As equations:

$$SS_E = \sum\sum (y_{ij} - \bar{y}_j)^2,$$

and

$$MS_E = SS_E/(N - k).$$

Now where are we? To test the null hypothesis that the treatment means are all equal we find the F-statistic:

$$F_{k-1, N-k} = MS_T/MS_E$$

and compare that value to the F-distribution with $k - 1$ and $N - k$ degrees of freedom. When the F-statistic is large enough (and its associated P-value small) we reject the null hypothesis and conclude that at least one mean is different.

There's another amazing result hiding in these tables. If we take each of these tables, square every observation, and add them up, the sums add as well!

$$SS_{Observations} = SS_{Grand\ Mean} + SS_T + SS_E$$

The $SS_{Observations}$ is usually very large compared to SS_T and SS_E, so when ANOVA was originally done by hand, or even by calculator, it was hard to check the calculations using this fact. The first sum of squares was just too big. So, usually the ANOVA table uses the "Corrected Total" sum of squares. If we write

$$Observations = Grand\ Mean + Treatment\ Effect + Residual,$$

we can naturally write

$$Observations - Grand\ Mean = Treatment\ Effect + Residual.$$

Mathematically, this is the same statement, but numerically this is more stable. What's amazing is that the sums of the squares still add up. That is, if you make the first table of observations with the grand mean subtracted from each, square those, and add them up, you'll have the SS_{Total} and

$$SS_{Total} = SS_T + SS_E.$$

That's what the ANOVA table shows. If you find this surprising, you must be following along. The tables add up, so sums of their elements must add up. But it is not at all obvious that the sums of the *squares* of their elements should add up, and this is another great insight of the Analysis of Variance.

Back to Standard Deviations

We've been using the variances because they're easier to work with. But when it's time to think about the data, we'd really rather have a standard deviation because it's in the units of the response variable. The natural standard deviation to think about is the standard deviation of the residuals.

The variance of the residuals is staring us in the face. It's the MS_E. All we have to do to get the **residual standard deviation** is take the square root of MS_E:

$$s_p = \sqrt{MS_E} = \sqrt{\frac{\sum e^2}{(N-k)}}.$$

The p subscript is to remind us that this is a *pooled* standard deviation, combining residuals across all k groups. The denominator in the fraction shows that finding a mean for each of the k groups costs us one degree of freedom for each.

This standard deviation should "feel" right. That is, it should reflect the kind of variation you expect to find in any of the experimental groups. For the hand-washing data, $s_p = \sqrt{1410.14} = 37.6$ bacteria colonies. Looking back at the boxplots of the groups, we see that 37.6 seems to be a reasonable compromise standard deviation for all four groups.

26.3 Assumptions and Conditions

Just as you would never find a linear regression without looking at the scatterplot of y vs. x, you should never embark on an ANOVA without first examining side-by-side boxplots of the data comparing the responses for all of the groups. You already know what to look for—we talked about that back in Chapter 4. Check for outliers within any of the groups and correct them if there are errors in the data. Get an idea of whether the groups have similar spreads (as we'll need) and whether the centers seem to be alike (as the null hypothesis claims) or different. If the spreads of the groups are very different—and especially if they seem to grow consistently as the means grow—consider re-expressing the response variable to make the spreads more nearly equal. Doing so is likely to make the analysis more powerful and more correct. Likewise, if the boxplots are skewed in the same direction, you may be able to make the distributions more symmetric with a re-expression.

Don't ever carry out an Analysis of Variance without looking at the side-by-side boxplots first. The chance of missing an important pattern or violation is just too great.

When we checked assumptions and conditions for regression we had to take care to perform our checks in order. Here we have a similar concern. For regression we found that displays of the residuals were often a good way to check the corresponding conditions. That's true for ANOVA as well.

Independence Assumptions
The groups must be independent of each other. No test can verify this assumption. You have to think about how the data were collected. The assumption would be violated, for example, if we measured subjects' performance before some treatment, again in the middle of the treatment period, and then again at the end.[8]

The data *within* each treatment group must be independent as well. The data must be drawn independently and at random from a homogeneous population, or generated by a randomized comparative experiment.

We check the **Randomization Condition:** Were the data collected with suitable randomization? For surveys, are the data drawn from each group a representative random sample of that group? For experiments, were the treatments assigned to the experimental units at random?

We were told that the hand-washing experiment was randomized.

Equal Variance Assumption
The ANOVA requires that the variances of the treatment groups be equal. After all, we need to find a pooled variance for the MS_E. To check this assumption, we can check that the groups have similar variances:

Similar Spread Condition: There are some ways to see whether the variation in the treatment groups seems roughly equal:

- Look at side-by-side boxplots of the groups to see whether they have roughly the same spread. It can be easier to compare spreads across groups when they have the same center, so consider making side-by-side boxplots of the residuals. If the groups have differing spreads, it can make the pooled variance—the MS_E—larger, reducing the *F*-statistic value and making it less likely that we can reject the null hypothesis. So the ANOVA will usually fail on the "safe side," rejecting H_0 less often than it should. Because of this, we usually require the spreads to be quite different from each other before we become concerned about the condition failing. If you've rejected the null hypothesis, this is especially true.
- Look at the original boxplots of the response values again. In general, do the spreads seem to change *systematically* with the centers? One common pattern is for the boxes with bigger centers to have bigger spreads. This kind of systematic trend in the variances is more of a problem than random differences in spread among the groups and should not be ignored. Fortunately, such systematic violations are often helped by re-expressing the data. (If, in addition to spreads that grow with the centers, the boxplots are skewed with the longer tail stretching off to the high end, then the data are pleading for a re-expression. Try taking logs of the dependent variable for a start. You'll likely end up with a much cleaner analysis.)
- Look at the residuals plotted against the predicted values. Often, larger predicted values lead to larger magnitude residuals. This is another sign that the condition is violated. (This may remind you of the **Does the Plot Thicken? Condition** of

---

[8] There is a modification of ANOVA, called *repeated measures* ANOVA, that deals with such data. (If the design reminds you of a paired-*t* situation, you're on the right track, and the lack of independence is the same kind of issue we discussed in Chapter 23.)

regression. And it should.) When the plot thickens (to one side or the other), it's usually a good idea to consider re-expressing the response variable. Such a systematic change in the spread is a more serious violation of the equal variance assumption than slight variations of the spreads across groups.

Let's check the conditions for the hand-washing data. Here's a boxplot of residuals by group and a scatterplot of residuals by predicted value:

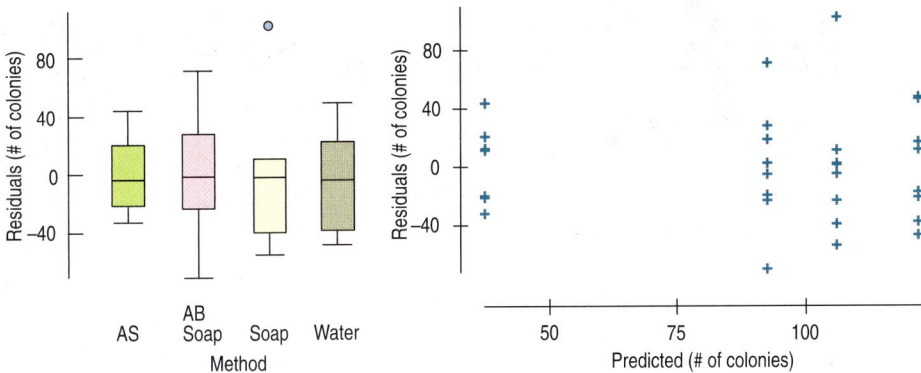

Figure 26.5
Boxplots of residuals for the four washing methods and a plot of residuals vs. predicted values. There's no evidence of a systematic change in variance from one group to the other or by predicted value.

Neither plot shows a violation of the condition. The IQRs (the box heights) are quite similar and the plot of residuals vs. predicted values does not show a pronounced widening to one end. The pooled estimate of 37.6 colonies for the error standard deviation seems reasonable for all four groups.

Normal Population Assumption Like Student's t-tests, the F-test requires the underlying errors to follow a Normal model. As before when we've faced this assumption, we'll check a corresponding **Nearly Normal Condition**.

Technically, we need to assume that the Normal model is reasonable for the populations underlying *each* treatment group. We can (and should) look at the side-by-side boxplots for indications of skewness. Certainly, if they are all (or mostly) skewed in the same direction, the Nearly Normal Condition fails (and re-expression is likely to help).

In experiments, we often work with fairly small groups for each treatment, and it's nearly impossible to assess whether the distribution of only six or eight numbers is Normal (though sometimes it's so skewed or has such an extreme outlier that we can see that it's not). Here we are saved by the Equal Variance Assumption (which we've already checked). The residuals have their group means subtracted, so the mean residual for each group is 0. If their variances are equal, we can group all the residuals together for the purpose of checking the Nearly Normal Condition.

Check Normality with a histogram or a Normal probability plot of all the residuals together. The hand-washing residuals look nearly Normal in the Normal probability plot, although, as the boxplots showed, there's a possible outlier in the Soap group.

Because we really care about the Normal model within each group, the Normal Population Assumption is violated if there are outliers in any of the groups. Check for outliers in the boxplots of the values for each treatment group. The Soap group of the hand-washing data shows an outlier, so we might want to compute the analysis again without that observation. (For these data, it turns out to make little difference.)

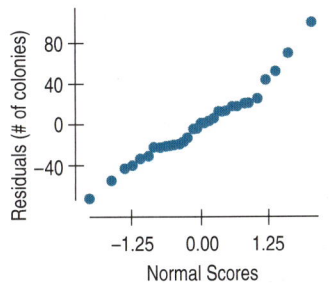

Figure 26.6
The hand-washing residuals look nearly Normal in this Normal probability plot.

> **One-way ANOVA F-test** We test the null hypothesis $H_0: \mu_1 = \mu_2 = \cdots = \mu_k$ against the alternative that the group means are not all equal. We test the hypothesis with the F-statistic, $F = \dfrac{MS_T}{MS_E}$, where MS_T is the Treatment Mean Square, found from the variance of the means of the treatment groups, and MS_E is the Error Mean Square, found by pooling the variances within each of the treatment groups. If the F-statistic is large enough, we reject the null hypothesis.

Step-by-Step Example ANALYSIS OF VARIANCE

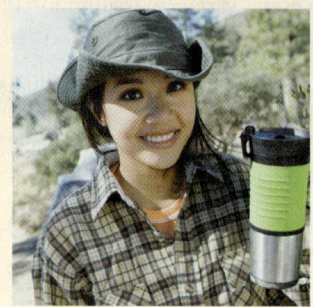

In Chapter 4, we looked at side-by-side boxplots of four different containers for holding hot beverages. The experimenter wanted to know which type of container would keep his hot beverages hot longest. To test it, he heated water to a temperature of 180°F, placed it in the container, and then measured the temperature of the water again 30 minutes later. He randomized the order of the trials and tested each container 8 times. His response variable was the difference in temperature (in °F) between the initial water temperature and the temperature after 30 minutes. (Data in **Cups**)

Question: Do the four containers maintain temperature equally well?

THINK

Plot Plot the side-by-side boxplots of the data.

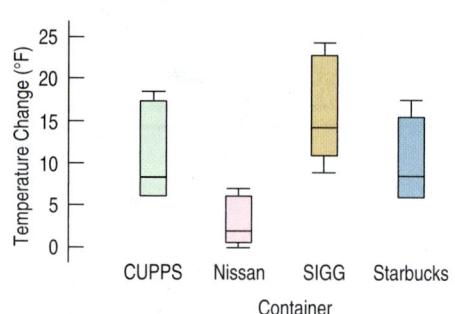

Plan State what you want to know and the null hypothesis you wish to test. For ANOVA, the null hypothesis is that all the treatment groups have the same mean. The alternative is that at least one mean is different.

I want to test whether there is any difference among the four containers in their ability to maintain the temperature of a hot liquid for 30 minutes. I'll write μ_k for the mean temperature difference for container k, so the null hypothesis is that these means are all the same:

$$H_0: \mu_1 = \mu_2 = \mu_3 = \mu_4.$$

The alternative is that the group means are not all equal.

Think about the assumptions and check the conditions.

✓ **Randomization Condition:** The experimenter performed the trials in random order, so it's reasonable to assume that the performance of one tested cup is independent of other cups.

✓ **Similar Spread Condition:** The Nissan cup variation seems to be a bit smaller than the others. I'll look later at the plot of residuals vs. predicted values to see if the plot thickens.

SHOW

Mechanics Fit the ANOVA model.

Analysis of Variance

Source	DF	Sum of Squares	Mean Square	F-Ratio	P-Value
Container	3	714.1875	238.063	10.713	<0.0001
Error	28	622.1875	22.221		
Total	31	1336.3750			

✓ **Nearly Normal Condition, Outlier Condition:** The Normal probability plot is not very straight, but there are no outliers.

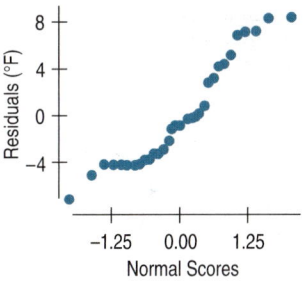

The histogram shows that the distribution of the residuals is skewed to the right:

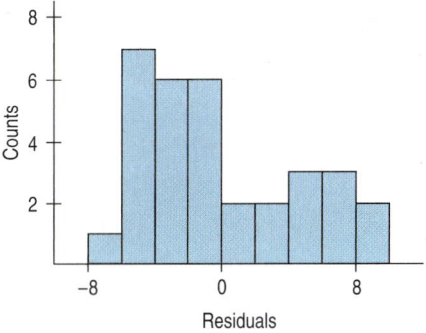

The table of means and SDs (below) shows that the standard deviations grow along with the means. Possibly a re-expression of the data would improve matters.

Under these circumstances, I cautiously find the P-value for the F-statistic from the F-model with 3 and 28 degrees of freedom.

The ratio of the mean squares gives an F-ratio of 10.7134 with a P-value of <0.0001.

SHOW ➡ Show the table of means.

From the ANOVA table, the Error Mean Square, MS_E, is 22.22, which means that the standard deviation of all the errors is estimated to be $\sqrt{22.22} = 4.71$ degrees F.

This seems like a reasonable value for the error standard deviation in the four treatments (with the possible exception of the Nissan cup).

Level	n	Mean	Std Dev
CUPPS	8	10.1875	5.20259
Nissan	8	2.7500	2.50713
SIGG	8	16.0625	5.90059
Starbucks	8	10.2500	4.55129

TELL → Interpretation Tell what the F-test means.	An F-ratio this large would be very unlikely if the containers all had the same mean temperature difference.
THINK → State your conclusions. (You should be more worried about the changing variance if you fail to reject the null hypothesis.) More specific conclusions might require a re-expression of the data.	**Conclusions:** Even though some of the conditions are mildly violated, I still conclude that the means are not all equal and that the four cups do not maintain temperature equally well.

The Balancing Act

The two examples we've looked at so far share a special feature. Each treatment group has the same number of experimental units. For the hand-washing experiment, each washing method was tested 8 times. For the cups, there were also 8 trials for each cup. This feature (the equal numbers of cases in each group, not the number 8) is called **balance**, and experiments that have equal numbers of experimental units in each treatment are said to be balanced or to have balanced designs.

Balanced designs are a bit easier to analyze because the calculations are simpler, so we usually try for balance. But in the real world, we often encounter unbalanced data. Participants drop out or become unsuitable, plants die, or maybe we just can't find enough experimental units to fit a particular criterion.

Everything we've done so far works just fine for unbalanced designs except that the calculations get a bit more complicated. Where once we could write n for the number of experimental units in a treatment, now we have to write n_k and sum more carefully. Where once we could pool variances with a simple average, now we have to adjust for the different n's. Technology clears these hurdles easily, so you're safe thinking about the analysis in terms of the simpler balanced formulas and trusting that the technology will make the necessary adjustments.

For Example

RECAP: An ANOVA for the contrast baths experiment had a statistically significant F-value.

Here are summary statistics for the three treatment groups:

Group	Count	Mean	StdDev
Bath	22	4.54545	7.76271
Bath & Exercise	23	8	7.03885
Exercise	14	−1.07143	5.18080

QUESTION: What can you conclude about these results?

ANSWER: We can be confident that there is a difference. However, it is the exercise treatment that appears to reduce swelling and not the contrast bath treatments. We might conclude (as the researchers did) that contrast bath treatments are of limited value.

26.4 Comparing Means

When we reject H_0, it's natural to ask which means are different. No one would be happy with an experiment to test 10 cancer treatments that concluded only with "We can reject H_0—the treatments are different!" We'd like to know more, but the F-statistic doesn't offer that information.

What can we do? If we can't reject the null, we've got to stop. There's no point in further testing. If we've rejected the simple null hypothesis, however, we *can* do more. In particular, we can test whether any pairs or combinations of group means differ. For example, we might want to compare treatments against a control or a placebo, or against the current standard treatment.

In the hand-washing experiment, we could consider plain water to be a control. Nobody would be impressed with (or want to pay for) a soap that did no better than water alone. A test of whether the antibacterial soap (for example) was different from plain water would be a simple test of the difference between two group means. To be able to perform an ANOVA, we first check the Similar Variance Condition. If things look OK, we assume that the variances are equal. If the variances *are* equal then a pooled t-test is appropriate. Even better (this is the special part), we already have a pooled estimate of the standard deviation based on *all* of the tested washing methods. That's s_p, which, for the hand-washing experiment, was equal to 37.55 bacteria colonies.

The null hypothesis is that there is no difference between water and the antibacterial soap. As we did in Chapter 20, we'll write that as a hypothesis about the difference in the means:

$$H_0: \mu_W - \mu_{ABS} = 0. \text{ The alternative is}$$
$$H_0: \mu_W - \mu_{ABS} \neq 0.$$

The natural test statistic is $\bar{y}_W - \bar{y}_{ABS}$, and the (pooled) standard error is

$$SE(\mu_W - \mu_{ABS}) = s_p\sqrt{\frac{1}{n_W} + \frac{1}{n_{ABS}}}.$$

Level	n	Mean	Std Dev
Alcohol Spray	8	37.5	26.56
Antibacterial Soap	8	92.5	41.96
Soap	8	106.0	46.96
Water	8	117.0	31.13

The difference in the observed means is $117.0 - 92.5 = 24.5$ colonies. The standard error comes out to 18.775. The t-statistic, then, is $t = \frac{24.5}{18.775} = 1.31$. To find the P-value we consult the Student's t-distribution on $N - k = 32 - 4 = 28$ degrees of freedom. The P-value is about 0.2—not small enough to impress us. So we can't discern a significant difference between washing with the antibacterial soap and just using water.

Our t-test asks about a simple difference. We could also ask a more complicated question about groups of differences. Does the average of the two soaps differ from the average of three sprays, for example? Complex combinations like these are called *contrasts*. Finding the standard errors for contrasts is straightforward but beyond the scope of this book. We'll restrict our attention to the common question of comparing pairs of treatments after H_0 has been rejected.

Bonferroni Multiple Comparisons

Our hand-washing experimenter *was* pretty sure that alcohol would kill the germs even before she started the experiment. But alcohol dries the skin and leaves an unpleasant smell. She was hoping that one of the antibacterial soaps would work as well as alcohol so she could use that instead. That means she really wanted to compare *each* of the other treatments against the alcohol spray. We know how to compare two of the means with a t-test. But now we want to do several tests, and each test poses the risk of a Type I error. As we do more and more tests, the risk that we might make a Type I error grows bigger than the α level of each individual test. With each additional test, the risk of making an error grows. If we do enough tests, we're almost sure to reject one of the null hypotheses by mistake—and we'll never know which one.

There is a defense against this problem. In fact, there are several defenses. As a class, they are called **methods for multiple comparisons**. All multiple comparisons methods

Carlo Bonferroni (1892–1960) was a mathematician who taught in Florence. He wrote two papers in 1935 and 1936 setting forth the mathematics behind the method that bears his name.

require that we first be able to reject the overall null hypothesis with the ANOVA's *F*-test. Once we've rejected the overall null, then we can think about comparing several—or even all—pairs of group means.

Let's look again at our test of the water treatment against the antibacterial soap treatment. This time we'll look at a confidence interval instead of the pooled *t*-test. We did a test at significance level $\alpha = 0.05$. The corresponding confidence level is $1 - \alpha = 95\%$. For *any* pair of means, a confidence interval for their difference is $(\bar{y}_1 - \bar{y}_2) \pm ME$, where the margin of error is

$$ME = t^* \times s_p \sqrt{\frac{1}{n_1} + \frac{1}{n_2}}.$$

As we did in the previous section, we get s_p as the pooled standard deviation found from *all* the groups in our analysis. Because s_p uses the information about the standard deviation from *all* the groups it's a better estimate than we would get by combining the standard deviation of just two of the groups. This uses the **Equal Variance Assumption** and "borrows strength" in estimating the common standard deviation of all the groups. We find the critical value t^* from the Student's *t*-model corresponding to the specified confidence level found with $N - k$ degrees of freedom, and the n_k's are the number of experimental units in each of the treatments.

To reject the null hypothesis that the two group means are equal, the difference between them must be larger than the ME. That way 0 won't be in the confidence interval for the difference. When we use it in this way, we call the margin of error the **least significant difference** (**LSD** for short). If two group means differ by more than this amount, then they are significantly different at level α *for each individual test*.

For our hand-washing experiment, each group has $n = 8$, $s_p = 37.55$, and df $= 32 - 4 = 28$. From technology or Table T, we can find that t^* with 28 df (for a 95% confidence interval) is 2.048. So

$$LSD = 2.048 \times 37.55 \times \sqrt{\frac{1}{8} + \frac{1}{8}} = 38.45 \text{ colonies},$$

and we could use this margin of error to make a 95% confidence interval for any difference between group means. Any two washing methods whose means differ by more than 38.45 colonies could be said to differ at $\alpha = 0.05$ by this method.

Of course, we're still just examining individual pairs. If we want to examine *many* pairs simultaneously, there are several methods that adjust the critical t^*-value so that the resulting confidence intervals provide appropriate tests for all the pairs. And, in spite of making *many* such intervals, the overall Type I error rate stays at (or below) α.

One such method is called the **Bonferroni method**. This method adjusts the LSD to allow for making many comparisons. The result is a wider margin of error called the **minimum significant difference**, or **MSD**. The MSD is found by replacing t^* with a slightly larger number. That makes the confidence intervals wider for each contrast and the corresponding Type I error rates lower for *each* test. And it keeps the *overall* Type I error rate at or below α.

The Bonferroni method distributes the error rate equally among the confidence intervals. It divides the error rate among J confidence intervals, finding each interval at confidence level $1 - \frac{\alpha}{J}$ instead of the original $1 - \alpha$. To signal this adjustment, we label the critical value t^{**} rather than t^*. For example, to make the three confidence intervals comparing the alcohol spray with the other three washing methods, and preserve our overall α risk at 5%, we'd construct each with a confidence level of

$$1 - \frac{0.05}{3} = 1 - 0.01667 = 0.98333.$$

The only problem with this is that *t*-tables don't have a column for 98.33% confidence (or, correspondingly, for $\alpha = 0.01667$). Fortunately, technology has no such constraints. For the hand-washing data, if we want to examine the three confidence intervals comparing each of the other methods with the alcohol spray, the t^{**}-value (on 28 degrees of freedom) turns out to be 2.546. That's somewhat larger than the individual t^*-value of 2.048 that we would have used for a single confidence interval. And the corresponding ME is 47.69 colonies (rather than 38.45 for a single comparison). The larger critical value along with correspondingly wider intervals is the price we pay for making multiple comparisons.

Many statistics packages assume that you'd like to compare all pairs of means. Some will display the result of these comparisons in a table like this:

Level	n	Mean	Groups
Alcohol Spray	8	37.5	A
Antibacterial Soap	8	92.5	B
Soap	8	106.0	B
Water	8	117.0	B

This table shows that the alcohol spray is in a class by itself and that the other three hand-washing methods are indistinguishable from one another.

26.5 ANOVA on Observational Data

So far we've applied ANOVA only to data from designed experiments. That's natural for several reasons. The primary one is that, as we saw in Chapter 12, randomized comparative experiments are specifically designed to compare the results for different treatments. The overall null hypothesis, and the subsequent tests on pairs of treatments in ANOVA, address such comparisons directly. In addition, as we discussed earlier, the **Equal Variance Assumption** (which we need for all of the ANOVA analyses) is often plausible in a randomized experiment because the treatment groups start out with sample variances that all estimate the same underlying variance of the collection of experimental units.

Sometimes, though, we just can't perform an experiment. When ANOVA is used to test equality of group means from observational data, there's no *a priori* reason to think the group variances might be equal at all. Even if the null hypothesis of equal means were true, the groups might easily have different variances. But if the side-by-side boxplots of responses for each group show roughly equal spreads and symmetric, outlier-free distributions, you can use ANOVA on observational data.

Observational data tend to be messier than experimental data. They are much more likely to be unbalanced. If you aren't assigning subjects to treatment groups, it's harder to guarantee the same number of subjects in each group. And because you are not controlling conditions as you would in an experiment, things tend to be, well, less controlled. The only way we know to avoid the effects of possible lurking variables is with control and randomized assignment to treatment groups, and for observational data, we have neither.

ANOVA is often applied to observational data when an experiment would be impossible or unethical. (We can't randomly break some subjects' legs, but we *can* compare pain perception among those with broken legs, those with sprained ankles, and those with stubbed toes by collecting data on subjects who have already suffered those injuries.) In such data, subjects are already in groups, but not by random assignment.

Be careful; if you have not assigned subjects to treatments randomly, you can't draw *causal* conclusions even when the F-test is significant. You have no way to control for lurking variables or confounding, so you can't be sure whether any differences you see among groups are due to the grouping variable or to some other unobserved variable that may be related to the grouping variable.

Because observational studies often are intended to estimate parameters, there is a temptation to use pooled confidence intervals for the group means for this purpose. Although these confidence intervals are statistically correct, be sure to think carefully about the population that the inference is about. The relatively few subjects that happen to be in a group may not be a simple random sample of any interesting population, so their "true" mean may have only limited meaning.

Step-by-Step Example   ONE MORE EXAMPLE

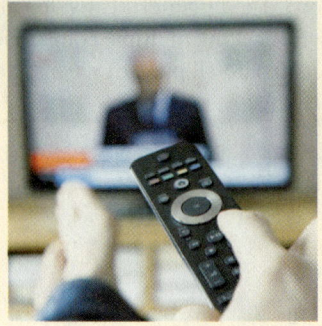

Here's an example that exhibits many of the features we've been discussing. It gives a fair idea of the kinds of challenges often raised by real data.

A study at a liberal arts college attempted to find out who watches more TV at college. Men or women? Varsity athletes or non-athletes? Student researchers asked 200 randomly selected students questions about their backgrounds and about their television-viewing habits and received 197 legitimate responses. The researchers found that men watch, on average, about 2.5 hours per week more TV than women, and that varsity athletes watch about 3.5 hours per week more than those who are not varsity athletes. But is this the whole story? To investigate further, they divided the students into four groups: male athletes (MA), male non-athletes (MNA), female athletes (FA), and female non-athletes (FNA). (Data in **TV watching**)

Question: Do these four groups of students spend about the same amount of time watching TV?

THINK ➡ Variables
Name the variables, report the W's, and specify the questions of interest.

I have the number of hours spent watching TV in a week for 197 randomly selected students. We know their sex and whether they are varsity athletes or not. I wonder whether TV watching differs according to sex and athletic status.

Plot
Always start an ANOVA with side-by-side boxplots of the responses in each of the groups. Always.

These data offer a good example why.

Here are the side-by-side boxplots of the data:

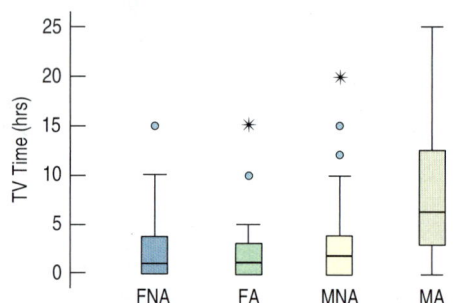

This plot suggests problems with the data. Each box shows a distribution skewed to the high end, and outliers pepper the display, including some extreme outliers. The box with the highest center (MA) also has the largest spread. These data just don't pass our first screening for suitability. This sort of pattern calls for a re-expression.

The responses are counts—numbers of TV hours. You may recall from Chapter 6 that a good re-expression to try first for counts is the square root.

Here are the boxplots for the square root of TV hours.

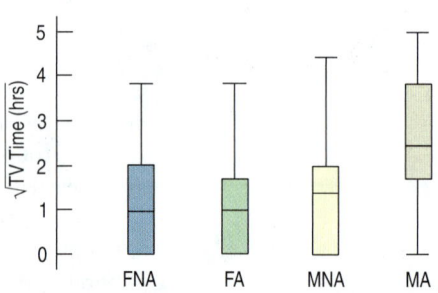

Think about the assumptions and check the conditions.	The spreads in the four groups are now more similar and the individual distributions more symmetric. And now there are no outliers. ✓ **Randomization Condition:** The data come from a random sample of students. The responses should be independent. It might be a good idea to see if the number of athletes and men are representative of the campus population. ✓ **Similar Spread Condition:** The boxplots show similar spreads. I may want to check the residuals later.
Fit the ANOVA model.	The ANOVA table looks like this: | Source | DF | Sum of Squares | Mean Square | F-Ratio | P-Value | |---|---|---|---|---|---| | Group | 3 | 47.24733 | 15.7491 | 12.8111 | <0.0001 | | Error | 193 | 237.26114 | 1.2293 | | | | Total | 196 | 284.50847 | | | | **Nearly Normal Condition, Outlier Condition:** A histogram of the residuals looks reasonably Normal: 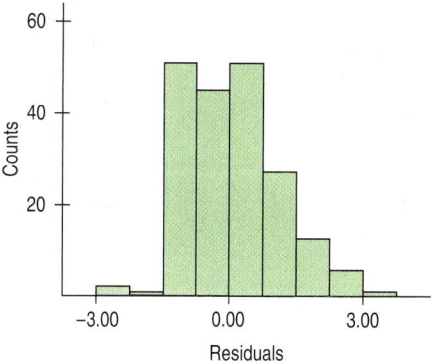 Interestingly, the few cases that seem to stick out on the low end are male athletes who watched no TV, making them different from all the other male athletes. Under these conditions, it's appropriate to use Analysis of Variance.
TELL ➡ Interpretation	The F-statistic is large and the corresponding P-value small. I conclude that the TV-watching behavior is not the same among these groups.

So Do Male Athletes Watch More TV?

Here's a Bonferroni comparison of all pairs of groups:

	Difference	Std. Err.	P-Value
FA–FNA	0.049	0.270	0.9999
MNA–FNA	0.205	0.182	0.8383
MNA–FA	0.156	0.268	0.9929
MA–FNA	1.497	0.250	<0.0001
MA–FA	1.449	0.318	<0.0001
MA–MNA	1.292	0.248	<0.0001

> **DIFFERING STANDARD ERRORS?**
> In case you were wondering ... The standard errors are different because this isn't a balanced design. Differing numbers of experimental units in the groups generate differing standard errors.

Three of the differences are very significant. It seems that among women there's little difference in TV watching between varsity athletes and others. Among men, though, the corresponding difference is large. And among varsity athletes, men watch significantly more TV than women.

But wait. How far can we extend the inference that male athletes watch more TV than other groups? The data came from a random sample of students made during the week of March 21. If the students carried out the survey correctly using a simple random sample, we should be able to make the inference that the generalization is true for the entire student body during that week.

Is it true for other colleges? Is it true throughout the year? The students conducting the survey followed up the survey by collecting anecdotal information about TV watching of male athletes. It turned out that during the week of the survey, the NCAA men's basketball tournament was televised. This could explain the increase in TV watching for the male athletes. It could be that the increase extends to other students at other times, but we don't know that. Always be cautious in drawing conclusions too broadly. Don't generalize from one population to another.

WHAT CAN GO WRONG?

- **Watch out for outliers.** One outlier in a group can change both the mean and the spread of that group. It will also inflate the Error Mean Square, which can influence the F-test. The good news is that ANOVA fails on the safe side by losing power when there are outliers. That is, you are less likely to reject the overall null hypothesis if you have (and leave) outliers in your data. But they are not likely to cause you to make a Type I error.

- **Watch out for changing variances.** The conclusions of the ANOVA depend crucially on the assumptions of independence and constant variance, and (somewhat less seriously as n increases) on Normality. If the conditions on the residuals are violated, it may be necessary to re-express the response variable to approximate these conditions more closely. ANOVA benefits so greatly from a judiciously chosen re-expression that the choice of a re-expression might be considered a standard part of the analysis.

- **Be wary of drawing conclusions about causality from observational studies.** ANOVA is often applied to data from randomized experiments for which causal conclusions are appropriate. If the data are not from a designed experiment, however, the Analysis of Variance provides no more evidence for causality than any other method we have studied. Don't get into the habit of assuming that ANOVA results have causal interpretations.

- **Be wary of generalizing** to situations other than the one at hand. Think hard about how the data were generated to understand the breadth of conclusions you are entitled to draw.

- **Watch for multiple comparisons.** When rejecting the null hypothesis, you can conclude that the means are not *all* equal. But you can't start comparing every pair of treatments in your study with a t-test. You'll run the risk of inflating your Type I error rate. Use a multiple comparisons method when you want to test many pairs.

CONNECTIONS

We first learned about side-by-side boxplots in Chapter 4. There we made general statements about the shape, center, and spread of each group. When we compared groups, we asked whether their centers looked different compared with how spread out the distributions were. Now we've made that kind of thinking precise. We've added confidence intervals for the difference and tests of whether the means are the same.

We pooled data to find a standard deviation when we tested the hypothesis of equal proportions. For that test, the assumption of equal variances was a consequence of the null hypothesis that the proportions were equal, so it didn't require an extra assumption. Means don't have a linkage with their corresponding variances, so to use pooled methods we must make the additional assumption of equal variances. In a randomized experiment, that's a plausible assumption.

Chapter 12 offered a variety of designs for randomized comparative experiments. Each of those designs can be analyzed with a variant or extension of the ANOVA methods discussed in this chapter. Entire books and courses deal with these extensions, but all follow the same fundamental ideas presented here.

ANOVA is closely related to the regression analyses we saw in Chapter 25. (In fact, most statistics packages offer an ANOVA table as part of their regression output.) The assumptions are similar—and for good reason. The analyses are, in fact, related at a deep conceptual (and computational) level, but those details are beyond the scope of this book.

The pooled two-sample t-test for means is a special case of the ANOVA F-test. If you perform an ANOVA comparing only two groups, you'll find that the P-value of the F-statistic is exactly the same as the P-value of the corresponding pooled t-statistic. That's because in this special case the F-statistic is just the square of the t-statistic. The F-test is more general. It can test the hypothesis that several group means are equal.

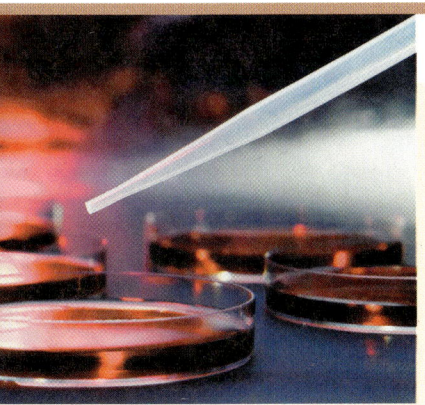

What Have We Learned?

Learning Objectives

Recognize when to use an Analysis of Variance (ANOVA) to compare the means of several groups.

Know how to read an ANOVA table to locate the degrees of freedom, the Mean Squares, and the resulting F-statistic.

Know how to check the three conditions required for an ANOVA:
- Independence of the groups from each other and of the individual cases within each group.
- Equal variance of the groups.
- Normal error distribution.

Know how to create and interpret confidence intervals for the differences between each pair of group means, recognizing the need to adjust the confidence interval for the number of comparisons made.

Review of Terms

Error (or Within) Mean Square (MS_E)   The Error Mean Square (MS_E) is the estimate of the error variance obtained by *pooling* the variances of each treatment group. The square root of the (MS_E) is the estimate of the error standard deviation, s_p (p. 752).

Treatment (or Between) Mean Square (MS_T)   The Treatment Mean Square (MS_T) is the estimate of the error variance under the assumption that the treatment means are all equal. If the (null) assumption is not true, the MS_T will be larger than the error variance (p. 752).

F-distribution	The *F*-distribution is the sampling distribution of the *F*-statistic when the null hypothesis that the treatment means are equal is true. It has two degrees of freedom parameters, one for the numerator, $(k-1)$, and one for the denominator, $(N-k)$, where *N* is the total number of observations and *k* is the number of groups (p. 752).
F-statistic	The *F*-statistic is the ratio MS_T/MS_E. When the *F*-statistic is sufficiently large, we reject the null hypothesis that the group means are equal (p. 752).
F-test	The *F*-test tests the null hypothesis that all the group means are equal against the one-sided alternative that they are not all equal. We reject the hypothesis of equal means if the *F*-statistic exceeds the critical value from the *F*-distribution corresponding to the specified significance level and degrees of freedom (p. 752).
ANOVA	An analysis method for testing equality of means across treatment groups (p. 752).
ANOVA table	The ANOVA table is convenient for showing the degrees of freedom, the Treatment Mean Square, the Error Mean Square, their ratio, the *F*-statistic, and its P-value. There are usually other quantities of lesser interest included as well (p. 753).
One-way ANOVA model	The model for a one-way (one response, one factor) ANOVA is $$y_{ij} = \mu_j + \varepsilon_{ij}.$$ Estimating with $y_{ij} = \bar{y}_j + e_{ij}$ gives predicted values $\hat{y}_{ij} = \bar{y}_j$ and residuals $e_{ij} = y_{ij} - \bar{y}_j$ (p. 755).
Residual standard deviation	The residual standard deviation, $$s_p = \sqrt{MS_E} = \sqrt{\frac{\sum e^2}{N-k}},$$ gives an idea of the underlying variability of the response values (p. 759).
Balance	An experiment's design is balanced if each treatment level has the same number of experimental units. Balanced designs make calculations simpler and are generally more powerful (p. 764).
Methods for multiple comparisons	If we reject the null hypothesis of equal means, we often then want to investigate further and compare pairs of treatment group means to see if they differ. If we want to test several such pairs, we must adjust for performing several tests to keep the overall risk of a Type I error from growing too large. Such adjustments are called methods for multiple comparisons (p. 765).
Least significant difference (LSD)	The standard margin of error in the confidence interval for the difference of two means is called the least significant difference. It has the correct Type I error rate for a single test, but not when performing more than one comparison (p. 766).
Bonferroni method	One of many methods for adjusting the length of the margin of error when testing the differences between several group means (p. 766).
Minimum significant difference (MSD)	The Bonferroni method's margin of error for the confidence interval for the difference of two group means is called the minimum significant difference. This can be used to test differences of several pairs of group means. If their difference exceeds the MSD, they are different at the overall rate (p. 766).

On the Computer ANOVA

Most analyses of variance are found with computers. And all statistics packages present the results in an ANOVA table much like the one we discussed. Technology also makes it easy to examine the side-by-side boxplots and check the residuals for violations of the assumptions and conditions.

Statistics packages offer different choices among possible multiple comparisons methods (although Bonferroni is quite common). This is a specialized area. Get advice or read further if you need to choose a multiple comparisons method.

As we saw in Chapter 4, there are two ways to organize data recorded for several groups. We can put all the response values in a single variable and use a second, "factor," variable to hold the group identities. This is sometimes called *stacked format*. The alternative is to place the data for each group in its own column or variable. Then the variable identities become the group identifiers.

Most statistics packages expect the data to be in stacked format because this form also works for more complicated experimental designs. Some packages can work with either form, and some use one form for some things and the other for others. (Be careful, for example, when you make side-by-side boxplots; be sure to give the appropriate version of the command to correspond to the structure of your data.)

Most packages offer to save residuals and predicted values and make them available for further tests of conditions. In some packages you may have to request them specifically.

DATA DESK

- Select the response variable as Y and the factor variable as X.
- From the **Calc** menu, choose **ANOVA**.
- Data Desk displays the ANOVA table.
- Select plots of residuals from the ANOVA table's HyperView menu.

COMMENTS

Data Desk expects data in "stacked" format. You can change the ANOVA by dragging the icon of another variable over either the Y or X variable name in the table and dropping it there. The analysis will recompute automatically.

EXCEL

- In Excel 2003 and earlier, select **Data Analysis** from the Tools menu.
- In Excel 2007, select **Data Analysis** from the Analysis Group on the Data Tab.
- Select **ANOVA Single Factor** from the list of analysis tools.
- Click the **OK** button.
- Enter the data range in the box provided.
- Check the **Labels in First Row** box, if applicable.
- Enter an alpha level for the F-test in the box provided.
- Click the **OK** button.

COMMENTS

The data range should include two or more columns of data to compare. Unlike all other statistics packages, Excel expects each column of the data to represent a different level of the factor. However, it offers no way to label these levels. The columns need not have the same number of data values, but the selected cells must make up a rectangle large enough to hold the column with the most data values.

JMP

- From the **Analyze** menu select **Fit Y by X**.
- Select variables: a quantitative Y, Response variable, and a categorical X, Factor variable.
- JMP opens the **Oneway** window.
- Click on the red triangle beside the heading, select **Display Options**, and choose **Boxplots**.
- From the same menu choose the **Means/ANOVA.t-test** command.
- JMP opens the oneway ANOVA output.

COMMENTS

JMP expects data in "stacked" format with one response and one factor variable.

MINITAB

- Choose **ANOVA** from the **Stat** menu.
- Choose **One-way** . . . from the **ANOVA** submenu.
- In the One-way ANOVA dialog, assign a quantitative Y variable to the Response box and assign a categorical X variable to the Factor box.
- Check the **Store Residuals** check box.
- Click the **Graphs** button.
- In the ANOVA-Graphs dialog, select **Standardized residuals**, and check **Normal plot of residuals** and **Residuals versus fits**.
- Click the **OK** button to return to the ANOVA dialog.
- Click the **OK** button to compute the ANOVA.

COMMENTS

If your data are in unstacked format, with separate columns for each treatment level, choose **One-way (unstacked)** from the **ANOVA** submenu.

R

To perform an analysis of variance of a variable y on a categorical variable (factor) x:

- myaov = aov(y~x)
- summary(myaov) # gives the ANOVA table

To get confidence or prediction intervals use:

- predict(myaov,interval = "confidence")

or

- predict(myaov, interval = "prediction")

SPSS

- Choose **Compare Means** from the **Analyze** menu.
- Choose **One-way ANOVA** from the **Compare Means** submenu.
- In the One-Way ANOVA dialog, select the Y-variable and move it to the dependent target. Then move the X-variable to the independent target.
- Click the **OK** button.

COMMENTS

SPSS expects stacked format. The **Contrasts** and **Post Hoc** buttons offer ways to test contrasts and perform multiple comparisons. See your SPSS manual for details.

STATCRUNCH

To compute an ANOVA:

- Click on **Stat**.
- Choose **ANOVA > One Way**.
- Choose the **Columns** for all groups. (After the first one, you may hold down the ctrl or command key to choose more.)

OR

Choose the **single column** containing the data (Responses) and the column containing the **Factors**.

- Click on **Calculate**.

Exercises

Section 26.1

1. **Popcorn** A student runs an experiment to test four different popcorn brands, recording the number of kernels left unpopped. She pops measured batches of each brand 4 times, using the same popcorn popper and randomizing the order of the brands. After collecting her data and analyzing the results, she reports that the F-ratio is 13.56.

   a) What are the null and alternative hypotheses?
   b) How many degrees of freedom does the treatment sum of squares have? How about the error sum of squares?

c) Assuming that the conditions required for ANOVA are satisfied, what is the P-value? What would you conclude?
d) What else about the data would you like to see in order to check the assumptions and conditions?

2. **Skating** A figure skater tried various approaches to her Salchow jump in a designed experiment using 5 different places for her focus (arms, free leg, midsection, takeoff leg, and free). She tried each jump 6 times in random order, using two of her skating partners to judge the jumps on a scale from 0 to 6. After collecting the data and analyzing the results, she reports that the F-ratio is 7.43.
a) What are the null and alternative hypotheses?
b) How many degrees of freedom does the treatment sum of squares have? How about the error sum of squares?
c) Assuming that the conditions are satisfied, what is the P-value? What would you conclude?
d) What else about the data would you like to see in order to check the assumptions and conditions?

3. **Gas mileage** A student runs an experiment to study the effect of three different mufflers on gas mileage. He devises a system so that his Jeep Wagoneer uses gasoline from a one-liter container. He tests each muffler 8 times, carefully recording the number of miles he can go in his Jeep Wagoneer on one liter of gas. After analyzing his data, he reports that the F-ratio is 2.35 with a P-value of 0.1199.
a) What are the null and alternative hypotheses?
b) How many degrees of freedom does the treatment sum of squares have? How about the error sum of squares?
c) What would you conclude?
d) What else about the data would you like to see in order to check the assumptions and conditions?
e) If your conclusion in part c is wrong, what type of error have you made?

4. **Darts** A student interested in improving her dart-throwing technique designs an experiment to test 4 different stances to see whether they affect her accuracy. After warming up for several minutes, she randomizes the order of the 4 stances, throws a dart at a target using each stance, and measures the distance of the dart in centimeters from the center of the bull's-eye. She replicates this procedure 10 times. After analyzing the data she reports that the F-ratio is 1.41.
a) What are the null and alternative hypotheses?
b) How many degrees of freedom does the treatment sum of squares have? How about the error sum of squares?
c) What would you conclude?
d) What else about the data would you like to see in order to check the assumptions and conditions?
e) If your conclusion in part c is wrong, what type of error have you made?

Section 26.2

5. **Activating baking yeast** To shorten the time it takes him to make his favorite pizza, a student designed an experiment to test the effect of sugar and milk on the activation times for baking yeast. Specifically, he tested four different recipes and measured how many seconds it took for the same amount of dough to rise to the top of a bowl. He randomized the order of the recipes and replicated each treatment 4 times.

Here are the boxplots of activation times from the four recipes:

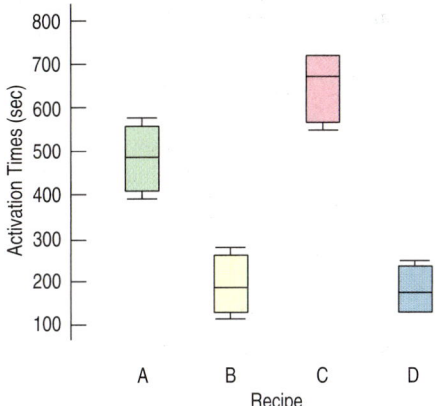

The ANOVA table follows:

Source	DF	Sum of Squares	Mean Square	F-Ratio	P-Value
Recipe	3	638967.69	212989	44.7392	<0.0001
Error	12	57128.25	4761		
Total	15	696095.94			

a) State the hypotheses about the recipes (both numerically and in words).
b) Assuming that the assumptions for inference are satisfied, perform the hypothesis test and state your conclusion. Be sure to state it in terms of activation times and recipes.
c) Would it be appropriate to follow up this study with multiple comparisons to see which recipes differ in their mean activation times? Explain.

6. **Frisbee throws** A student performed an experiment with three different grips to see what effect it might have on the distance of a backhanded Frisbee throw. She tried it with her normal grip, with one finger out, and with the Frisbee inverted. She measured in paces how far her throw went. The boxplots and the ANOVA table for the three grips are shown below:

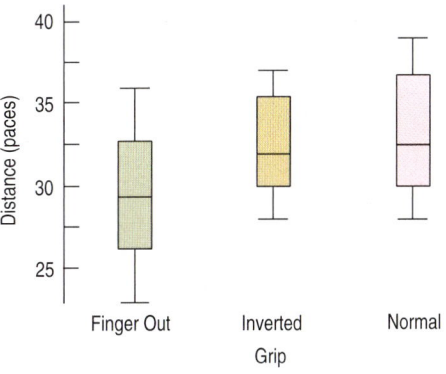

(continued)

Source	DF	Sum of Squares	Mean Square	F-Ratio	P-Value
Grip	2	58.58333	29.2917	2.0453	0.1543
Error	21	300.75000	14.3214		
Total	23	359.33333			

a) State the hypotheses about the grips.
b) Assuming that the assumptions for inference are satisfied, perform the hypothesis test and state your conclusion. Be sure to state it in terms of Frisbee grips and distance thrown.
c) Would it be appropriate to follow up this study with multiple comparisons to see which grips differ in their mean distance thrown? Explain.

Section 26.3

7. **Fuel economy by cylinders, revisited** Here are boxplots, first seen in Chapter 4 Exercise 28, that show the relationship between the number of cylinders a car's engine has and the car's fuel economy.

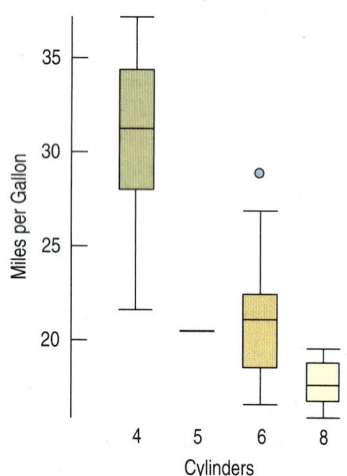

a) State the null and alternative hypotheses that you might consider for these data.
b) Do the conditions for an ANOVA seem to be met here? Why or why not?

8. **Finger Lakes wines 2014** Here are bottle prices (in dollars) of wines produced by wineries along three of the Finger Lakes.

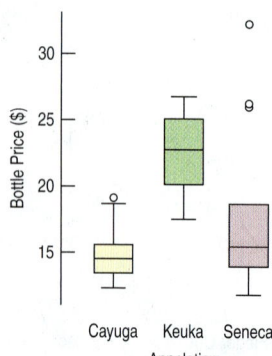

a) What null and alternative hypotheses would you test for these data? Talk about prices and location, not symbols.

b) Do the conditions for an ANOVA seem to be met here? Why or why not?

Section 26.4

9. **Tellers** A bank is studying the time that it takes 6 of its tellers to serve an average customer. Customers line up in the queue and then go to the next available teller. Here is a boxplot of the last 200 customers and the times it took each teller:

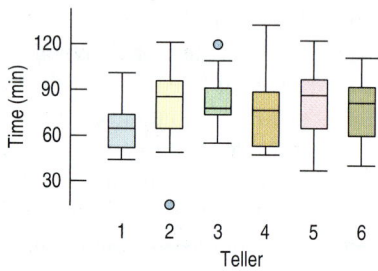

ANOVA Table

Source	DF	Sum of Squares	Mean Square	F-Ratio	P-Value
Teller	5	3315.32	663.064	1.508	0.1914
Error	134	58919.1	439.695		
Total	139	62234.4			

a) What are the null and alternative hypotheses?
b) What do you conclude?
c) Would it be appropriate to run a multiple comparisons test (for example, a Bonferroni test) to see which tellers differ from each other? Explain.

10. **Hearing** A researcher investigated four different word lists for use in hearing assessment. She wanted to know whether the lists were equally difficult to understand in the presence of a noisy background. To find out, she tested 96 subjects with normal hearing randomly assigning 24 to each of the four word lists and measured the number of words perceived correctly in the presence of background noise. Here are the boxplots of the four lists:

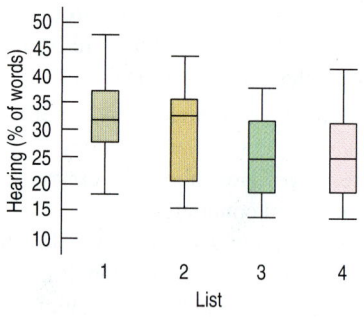

ANOVA Table

Source	DF	Sum of Squares	Mean Square	F-Ratio	P-Value
List	3	920.4583	306.819	4.9192	0.0033
Error	92	5738.1667	62.371		
Total	95	6658.6250			

a) What are the null and alternative hypotheses?
b) What do you conclude?
c) Would it be appropriate to run a multiple comparisons test (for example, a Bonferroni test) to see which lists differ from each other in terms of mean percent correct? Explain.

Section 26.5

11. Eye and hair color In Chapter 4, Exercise 32, we saw a survey of 1021 school-age children conducted by randomly selecting children from several large urban elementary schools. Two of the questions concerned eye and hair color. In the survey, the following codes were used:

Hair Color	Eye Color
1 = Blond	1 = Blue
2 = Brown	2 = Green
3 = Black	3 = Brown
4 = Red	4 = Grey
5 = Other	5 = Other

The students analyzing the data were asked to study the relationship between eye and hair color. They produced this plot:

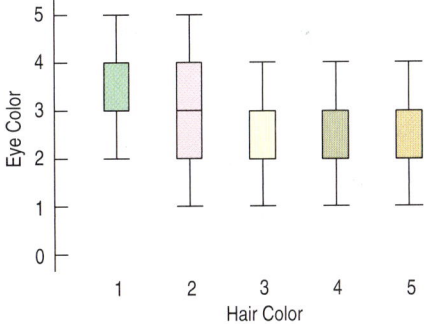

They then ran an Analysis of Variance with *Eye Color* as the response and *Hair Color* as the factor. The ANOVA table they produced follows:

Source	DF	Sum of Squares	Mean Square	F-Ratio	P-Value
Hair Color	4	1.46946	0.367365	0.4024	0.8070
Error	1016	927.45317	0.912848		
Total	1020	928.92263			

What suggestions do you have for the Statistics students? What alternative analysis might you suggest?

12. ZIP codes, revisited The intern from the marketing department at the Holes R Us online piercing salon (Chapter 3, Exercise 55) has recently finished a study of the company's 500 customers. He wanted to know whether people's ZIP codes vary by the last product they bought. They have 16 different products, and the ANOVA table of ZIP code by product showed the following:

Source	DF	Sum of Squares	Mean Square	F-Ratio	P-Value
Product	15	3.836e10	2.55734e9	4.9422	<0.0001
Error	475	2.45787e11	517445573		
Total	490	2.84147e11			

(Nine customers were not included because of missing ZIP code or product information.)

What criticisms of the analysis might you make? What alternative analysis might you suggest?

Chapter Exercises

13. Yogurt An experiment to determine the effect of several methods of preparing cultures for use in commercial yogurt was conducted by a food science research group. Three batches of yogurt were prepared using each of three methods: traditional, ultrafiltration, and reverse osmosis. A trained expert then tasted each of the 9 samples, presented in random order, and judged them on a scale from 1 to 10. A partially completed Analysis of Variance table of the data follows.

Source	Sum of Squares	Degrees of Freedom	Mean Square	F-Ratio
Treatment	17.300			
Residual	0.460			
Total	17.769			

a) Calculate the mean square of the treatments and the mean square of the error.
b) Form the *F*-statistic by dividing the two mean squares.
c) The P-value of this *F*-statistic turns out to be 0.000017. What does this say about the null hypothesis of equal means?
d) What assumptions have you made in order to answer part c?
e) What would you like to see in order to justify the conclusions of the *F*-test?
f) What is the average size of the error standard deviation in the judge's assessment?

14. Smokestack scrubbers Particulate matter is a serious form of air pollution often arising from industrial production. One way to reduce the pollution is to put a filter, or scrubber, at the end of the smokestack to trap the particulates. An experiment to determine which smokestack scrubber design is best was run by placing four scrubbers of different designs on an industrial stack in random order. Each scrubber was tested 5 times. For each run, the same material was produced, and the particulate emissions coming out of the scrubber were measured (in parts per billion). A partially completed Analysis of Variance table of the data follows on the next page.

(continued)

Source	Sum of Squares	Degrees of Freedom	Mean Square	F-Ratio
Treatment	81.2			
Residual	30.8			
Total	112.0			

a) Calculate the mean square of the treatments and the mean square of the error.
b) Form the F-statistic by dividing the two mean squares.
c) The P-value of this F-statistic turns out to be 0.0000949. What does this say about the null hypothesis of equal means?
d) What assumptions have you made in order to answer part c?
e) What would you like to see in order to justify the conclusions of the F-test?
f) What is the average size of the error standard deviation in particulate emissions?

15. Eggs A student wants to investigate the effects of real vs. substitute eggs on his favorite brownie recipe. He enlists the help of 10 friends and asks them to rank each of 8 batches on a scale from 1 to 10. Four of the batches were made with real eggs, four with substitute eggs. The judges tasted the brownies in random order. Here is a boxplot of the data:

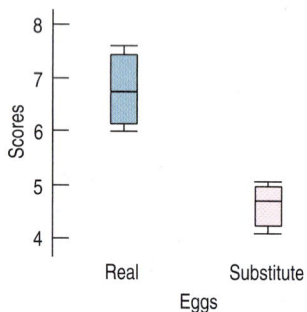

ANOVA Table

Source	DF	Sum of Squares	Mean Square	F-Ratio	P-Value
Eggs	1	9.010013	9.01001	31.0712	0.0014
Error	6	1.739875	0.28998		
Total	7	10.749883			

The mean score for the real eggs was 6.78 with a standard deviation of 0.651. The mean score for the substitute eggs was 4.66 with a standard deviation of 0.395.

a) What are the null and alternative hypotheses?
b) What do you conclude from the ANOVA table?
c) Do the assumptions for the test seem to be reasonable?
d) Perform a two-sample pooled t-test of the difference. What P-value do you get? Show that the square of the t-statistic is the same (to rounding error) as the F-ratio.

16. Auto noise filters In a statement to a Senate Public Works Committee, a senior executive of Texaco, Inc., cited a study on the effectiveness of auto filters on reducing noise. Because of concerns about performance, two types of filters were studied, a standard silencer and a new device developed by the Associated Octel Company. Here are the boxplots from the data on noise reduction (in decibels) of the two filters. Type 1 = standard; Type 2 = Octel.

ANOVA Table

Source	DF	Sum of Squares	Mean Square	F-Ratio	P-Value
Type	1	6.31	6.31	0.7673	0.3874
Error	33	271.47	8.23		
Total	34	722.78			

Means and Std Deviations

Level	n	Mean	StdDev
Standard	18	81.5556	3.2166
Octel	17	80.7059	2.43708

a) What are the null and alternative hypotheses?
b) What do you conclude from the ANOVA table?
c) Do the assumptions for the test seem to be reasonable?
d) Perform a two-sample pooled t-test of the difference. What P-value do you get? Show that the square of the t-statistic is the same (to rounding error) as the F-ratio.

17. School system A school district superintendent wants to test a new method of teaching arithmetic in the fourth grade at his 15 schools. He plans to select 8 students from each school to take part in the experiment, but to make sure they are roughly of the same ability, he first gives a test to all 120 students. Here are the scores of the test by school:

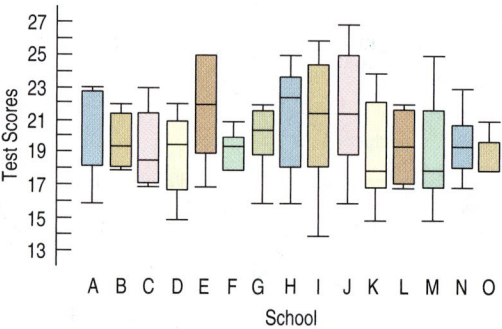

The ANOVA table shows:

Source	DF	Sum of Squares	Mean Square	F-Ratio	P-Value
School	14	108.800	7.7714	1.0735	0.3899
Error	105	760.125	7.2392		
Total	119	868.925			

a) What are the null and alternative hypotheses?
b) What does the ANOVA table say about the null hypothesis? (Be sure to report this in terms of scores and schools.)
c) An intern reports that he has done t-tests of every school against every other school and finds that several of the schools seem to differ in mean score. Does this match your finding in part b? Give an explanation for the difference, if any, of the two results.

T 18. Fertilizers A biology student is studying the effect of 10 different fertilizers on the growth of mung bean sprouts. She sprouts 12 beans in each of 10 different petri dishes, and adds the same amount of fertilizer to each dish. After one week she measures the heights of the 120 sprouts in millimeters. Here are boxplots and an ANOVA table of the data:

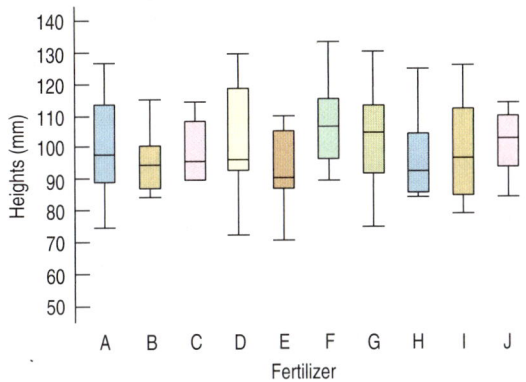

Source	DF	Sum of Squares	Mean Square	F-Ratio	P-Value
Fertilizer	9	2073.708	230.412	1.1882	0.3097
Error	110	21331.083	193.919		
Total	119	23404.791			

a) What are the null and alternative hypotheses?
b) What does the ANOVA table say about the null hypothesis? (Be sure to report this in terms of heights and fertilizers).
c) Her lab partner looks at the same data and says that he did t-tests of every fertilizer against every other fertilizer and finds that several of the fertilizers seem to have significantly higher mean heights. Does this match your finding in part b? Give an explanation for the difference, if any, between the two results.

T 19. Cereals once more We first saw data on breakfast cereals in Chapter 7. Supermarkets often place similar types of cereal on the same supermarket shelf. We have data on the shelf as well as the sugar, sodium, and calorie content of 77 cereals. Does sugar content vary by shelf? At the top of the next column is a boxplot and an ANOVA table for the 77 cereals.

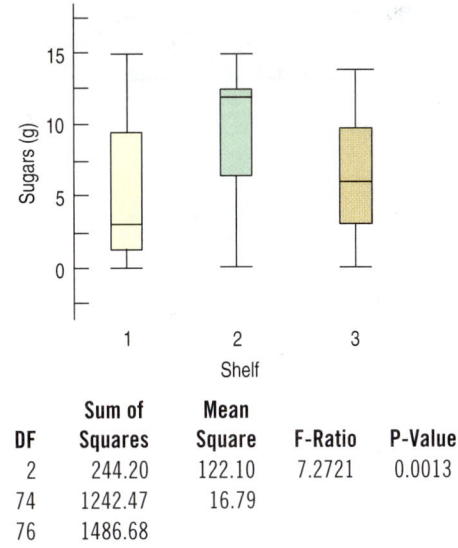

Source	DF	Sum of Squares	Mean Square	F-Ratio	P-Value
Shelf	2	244.20	122.10	7.2721	0.0013
Error	74	1242.47	16.79		
Total	76	1486.68			

Means and Std Deviations

Level	n	Mean	StdDev
1	20	4.8500	4.51051
2	21	9.61905	4.12888
3	36	6.52778	3.83582

a) What are the null and alternative hypotheses?
b) What does the ANOVA table say about the null hypothesis? (Be sure to report this in terms of *Sugars* and *Shelves*.)
c) Can we conclude that cereals on shelf 2 have a higher mean sugar content than cereals on shelf 3? Can we conclude that cereals on shelf 2 have a higher mean sugar content than cereals on shelf 1? What *can* we conclude?
d) To check for significant differences between the shelf means, we can use a Bonferroni test, whose results are shown below. For each pair of shelves, the difference is shown along with its standard error and significance level. What does it say about the questions in part c?

Dependent Variable: SUGARS

	(I) SHELF	(J) SHELF	Mean Difference (I-J)	Std. Error	P-Value	95% Confidence Interval Lower Bound	95% Confidence Interval Upper Bound
Bonferroni	1	2	−4.769(*)	1.3524	0.0033	−8.155	−1.383
		3	−1.678	1.1941	0.5070	−4.683	1.328
	2	1	4.769(*)	1.3524	0.0033	1.383	8.155
		3	3.091(*)	1.1048	0.0237	0.329	5.853
	3	1	1.678	1.1941	0.5070	−1.328	4.483
		2	−3.091(*)	1.1048	0.0237	−5.853	−0.329

*The mean difference is significant at the 0.05 level.

20. **Cereals, redux** We also have data on the protein content of the cereals in Exercise 19 by their shelf number. Here are the boxplot and ANOVA table:

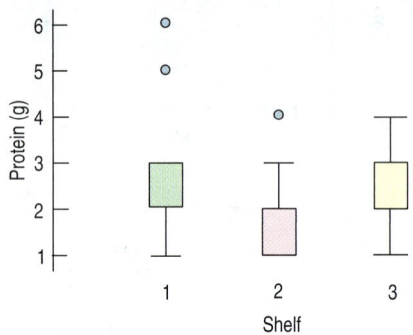

Source	DF	Sum of Squares	Mean Square	F-Ratio	P-Value
Shelf	2	12.4258	6.2129	5.8445	0.0044
Error	74	78.6650	1.0630		
Total	76	91.0909			

Means and Std Deviations

Level	n	Mean	StdDev
1	20	2.65000	1.46089
2	21	1.90476	0.99523
3	36	2.86111	0.72320

a) What are the null and alternative hypotheses?
b) What does the ANOVA table say about the null hypothesis? (Be sure to report this in terms of protein content and shelves.)
c) Can we conclude that cereals on shelf 2 have a lower mean protein content than cereals on shelf 3? Can we conclude that cereals on shelf 2 have a lower mean protein content than cereals on shelf 1? What *can* we conclude?
d) To check for significant differences between the shelf means we can use a Bonferroni test, whose results are shown below. For each pair of shelves, the difference is shown along with its standard error and significance level. What does it say about the questions in part c?

Dependent Variable: PROTEIN

	(I) SHELF	(J) SHELF	Mean Difference (I-J)	Std. Error	P-Value	95% Confidence Interval	
						Lower Bound	Upper Bound
Bonferroni	1	2	0.75	0.322	0.070	−0.04	1.53
		3	−0.21	0.288	1.000	−0.92	0.49
	2	1	−0.75	0.322	0.070	−1.53	0.04
		3	−0.96(*)	0.283	0.004	−1.65	−0.26
	3	1	0.21	0.288	1.000	−0.49	0.92
		2	0.96(*)	0.283	0.004	0.26	1.65

*The mean difference is significant at the 0.05 level.

21. **Downloading** To see how much of a difference time of day made on the speed at which he could download files, a college sophomore performed an experiment. He placed a file on a remote server and then proceeded to download it at three different time periods of the day. He downloaded the file 48 times in all, 16 times at each *Time of Day*, and recorded the *Time* in seconds that the download took.

a) State the null and alternative hypotheses, being careful to talk about download *Time* and *Time of Day* as well as parameters.

Time of Day	Time (sec)	Time of Day	Time (sec)	Time of Day	Time (sec)
Early (7 a.m.)	68	Evening (5 p.m.)	299	Late Night (12 a.m.)	216
Early (7 a.m.)	138	Evening (5 p.m.)	367	Late Night (12 a.m.)	175
Early (7 a.m.)	75	Evening (5 p.m.)	331	Late Night (12 a.m.)	274
Early (7 a.m.)	186	Evening (5 p.m.)	257	Late Night (12 a.m.)	171
Early (7 a.m.)	68	Evening (5 p.m.)	260	Late Night (12 a.m.)	187
Early (7 a.m.)	217	Evening (5 p.m.)	269	Late Night (12 a.m.)	213
Early (7 a.m.)	93	Evening (5 p.m.)	252	Late Night (12 a.m.)	221
Early (7 a.m.)	90	Evening (5 p.m.)	200	Late Night (12 a.m.)	139
Early (7 a.m.)	71	Evening (5 p.m.)	296	Late Night (12 a.m.)	226
Early (7 a.m.)	154	Evening (5 p.m.)	204	Late Night (12 a.m.)	128
Early (7 a.m.)	166	Evening (5 p.m.)	190	Late Night (12 a.m.)	236
Early (7 a.m.)	130	Evening (5 p.m.)	240	Late Night (12 a.m.)	128
Early (7 a.m.)	72	Evening (5 p.m.)	350	Late Night (12 a.m.)	217
Early (7 a.m.)	81	Evening (5 p.m.)	256	Late Night (12 a.m.)	196
Early (7 a.m.)	76	Evening (5 p.m.)	282	Late Night (12 a.m.)	201
Early (7 a.m.)	129	Evening (5 p.m.)	320	Late Night (12 a.m.)	161

b) Perform an ANOVA on these data. What can you conclude?
c) Check the assumptions and conditions for an ANOVA. Do you have any concerns about the experimental design or the analysis?
d) (Optional) Perform a multiple comparisons test to determine which times of day differ in terms of mean download time.

T 22. Analgesics A pharmaceutical company tested three formulations of a pain relief medicine for migraine headache sufferers. For the experiment, 27 volunteers were selected and 9 were randomly assigned to one of three drug formulations. The subjects were instructed to take the drug during their next migraine headache episode and to report their pain on a scale of $1 =$ no pain to $10 =$ extreme pain 30 minutes after taking the drug.

Drug	Pain	Drug	Pain	Drug	Pain
A	4	B	6	C	6
A	5	B	8	C	7
A	4	B	4	C	6
A	3	B	5	C	6
A	2	B	4	C	7
A	4	B	6	C	5
A	3	B	5	C	6
A	4	B	8	C	5
A	4	B	6	C	5

a) State the null and alternative hypotheses, being careful to talk about *Drug* and *Pain* levels as well as parameters.
b) Perform an ANOVA on these data. What can you conclude?
c) Check the assumptions and conditions for an ANOVA. Do you have any concerns about the experimental design or the analysis?
d) (Optional) Perform a multiple comparisons test to determine which drugs differ in terms of mean pain level reported.

Just Checking ANSWERS

1. The null hypothesis is that the mean flight distance for all four designs is the same.
2. Yes, it looks as if the variation between the means is greater than the variation within each boxplot.
3. Yes, the *F*-test rejects the null hypothesis with a P-value <0.0001.
4. No. The alternative hypothesis is that *at least* one mean is different from the other three. Rejecting the null hypothesis does not imply that all four means are different.

Chapter 27
Multifactor Analysis of Variance

27.1 A Two-Factor ANOVA Model
27.2 Assumptions and Conditions
27.3 Interactions

Where are we going?
When production problems forced a computer manufacturer to halt operations, teams of engineers, scientists, production workers, and managers were formed to solve the problem. They found themselves overwhelmed with many potential causes. But experiments that test one factor at a time would have been far too inefficient. And they might have missed important combinations of factors. In this chapter, we'll see how to look at two factors at a time, and we'll discover why these two-factor designs are so much more interesting than one-factor experiments.

Who	Attempts at hitting a bull's-eye in a dartboard
What	Distance from the center of the target
Units	Inches
How	Two-factor experiment, randomized order of runs

How accurately can you throw a dart? It probably depends on which hand you use and on how far from the target you're standing. A student designed an experiment to test the effects of both factors. He used three levels of the factor *Distance: Near, Middle,* and *Far*—and the two natural levels *Left* and *Right* for *Hand.* (He was right-handed.) In random order, he threw six darts at a dartboard under each of the six treatment conditions, and measured the distance each dart landed from the bull's-eye in inches.[1]

It makes sense that dart throws would be more accurate from nearer distances, especially when using his regular throwing hand. Side-by-side boxplots of the distance from the bull's-eye by each factor seem to support our intuition:

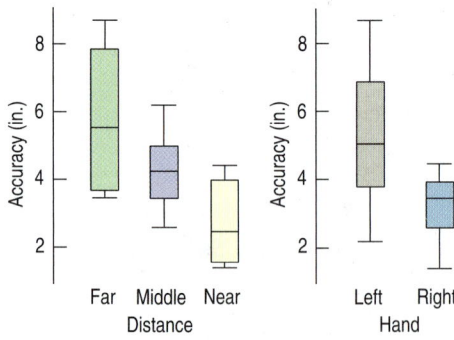

Figure 27.1
Boxplots of *Accuracy* (distance in inches from the bull's-eye) plotted for each experimental factor, *Distance* and *Hand*.

---

[1] He called the response variable *Accuracy,* although it measures inches *away* from the target. That means that bigger values correspond to *less* accuracy, not more.

27.1 A Two-Factor ANOVA Model

It appears that both the distance and the hand he used affected the accuracy of his throw, and by about the same amount. How can we measure the effect of each variable? And how can we test whether the observed differences are larger than we'd expect just from sampling variation?

Two Factors at Once?!

Unlike the experiments we've seen so far, this experiment has not one but two factors. You might expect this to confuse things, but the surprising fact is that it actually improves the experiment and its analysis. It also lets us see something we'd never be able to see if we ran two different one-factor experiments.

True, with two factors, we have two hypothesis tests. Each of those hypotheses asks whether the mean of the response variable is the same for each of the treatment levels. In comparing means, we know that any excess variation just gets in the way. Back in Chapter 12, we considered ways to remove or avoid extra variation in designing experiments.

A two-factor experiment does just that. If both factors do in fact have an effect on the response (as we'd expect hand choice and distance to do for dart accuracy), then removing the effects of one factor should make it easier to see and assess the effects of the other. We'll have removed variability due to an identifiable cause.

For Example

Students set up an experiment to investigate factors that determine how far a paper airplane will fly. Among the factors are whether the plane is constructed of copier paper or heavier construction paper, whether it is a "dart" or a glider, and whether the ends of the wings are folded up into "flaps." We'll combine the paper and shape factor into a single *Design* factor with four levels; *Dart/Copier, Dart/Construction, Glider/Copier,* and *Glider/Construction*. The second factor, *Flaps*, has two levels: *Flaps* and *None*.

Four planes were constructed of each of the four designs. For each design, two planes were given flaps and two were not. Each plane was flown 4 times and the distance of the flight carefully measured.

QUESTION: What are the details of the experiment design?

We have two factors: *Design* and *Flaps*. *Design* has 4 levels, Dart/Copier, Dart/Construction, Glider/Copier and Glider/Construction. *Flaps* has 2: Flaps and None. There are thus 4 × 2 = 8 treatment groups. Each treatment group has 2 planes and each plane was flown 4 times for a total of 8 observations. The response variable is the distance flown.

An ANOVA Model

The model for one-way ANOVA represented each observation as the sum of simple components. It broke each observation into the sum of three effects: the grand mean, the treatment effect, and an error:

$$y_{ij} = \mu + \tau_j + \varepsilon_{ij}.$$

Now we have two factors, *Hand* and *Distance*. So each treatment is a combination of a *Hand* assignment and a *Distance*. For example, throwing with the left hand from a near distance would be one treatment. Our model should reflect the effects of both factors. We've already used τ for the treatment effects for the first factor, so for the second factor, we'll denote the effects by γ. Now, for each observation, we write:

$$y_{ijk} = \mu + \tau_j + \gamma_k + \varepsilon_{ijk}.$$

Here we see the i-th observation at level j of factor A and level k of factor B. Why the i subscript? We're told that the student recorded six trials at each treatment, so the subscript i denotes the different observations at each combination of the two factors. The error has three subscripts because we associate a possibly different error with each observation. (If there were only one observation at each treatment, we wouldn't need the extra subscript.)

When you write out a model, you'll usually find it clearer to name the factors rather than using Greek letters. So the model will look like this:

$$y_{ijk} = \mu + \text{Hand effect}_j + \text{Distance effect}_k + \text{Error}_{ijk}.$$

To estimate the effect of each treatment level for each factor, we take the *difference* between the mean response of that treatment and the grand mean, $\bar{\bar{y}}$. So, to find the effect of throwing with the left hand, we'd find the mean accuracy of *all* runs using the left hand and subtract the grand average from that.

In general, we'd like to know whether the mean dart accuracy changes with changes in the levels of either factor. The model helps us to think about this question. Our null hypothesis on each factor is that the effects of that treatment are all zero. So we can write two null hypotheses we'd like to test as

$$H_0: \tau_1 = \tau_2 \quad \text{and} \quad H_0: \gamma_1 = \gamma_2 = \gamma_3,$$

where the τ's (taus) denote the effects due to the two levels of *Hand* and the γ's (gammas) denote the effects due to the three levels of *Distance*.

The alternative hypotheses are that the treatment effects are not all equal. The details of the analysis are pretty much the same as for a one-factor ANOVA. We want to compare the differences across the treatment effects with the underlying variability within the treatment. But now with two factors, that underlying variability has the effect of both factors removed from it. So the error term holds the variability that's left after removing the effects of both factors. Here's the ANOVA table for the dart experiment:

Analysis of Variance for Accuracy

Source	DF	Sum of Squares	Mean Square	F-Ratio	Prob
Distance	2	51.0439	25.5219	28.561	<0.0001
Hand	1	39.6900	39.6900	44.416	<0.0001
Error	32	28.5950	0.893594		
Total	35	119.329			

This analysis splits up the total variation into three sources: the variation due to changes in *Distance*, the variation due to changes in *Hand*, and the error variation. The square root of the Error Mean Square, $\sqrt{0.894} = 0.946$ inches, estimates the standard deviation we'd expect if we made repeated observations at any treatment condition. Both F-ratios are large. Looking back at the boxplots, we shouldn't be surprised. It was clear that there were differences in accuracy among *near*, *middle*, and *far* even without accounting for the fact that each level contained both hands. The F-ratio for *Distance* is 28.56. The probability of an F-ratio that big occurring by chance is very small (<0.0001). So we can reject the hypothesis that the mean effects of *near*, *middle*, and *far* distances are equal.

The means of *left* and *right* don't look the same either. The F-ratio of 44.42 with a P-value of <0.0001 leads us to reject the null hypothesis about *Hand* effects as well.

Of course, there are assumptions and conditions to check. The good news is that most are the same as for one-way ANOVA.

27.2 Assumptions and Conditions

Even before checking any assumptions, you should have already plotted the data. Boxplots give a first look that draws our attention to the aspects tested by the ANOVA.

Plot the Data . . .

You could start by examining side-by-side boxplots of the data across levels of *each factor* as we did earlier, looking for outliers within any of the groups and correcting them if they are errors, or consider omitting them if we can't correct them.[2] (There don't seem to be any.) We'd like to get an idea of whether the groups have similar spreads (as we'll want) and whether the centers seem to be alike (as the null hypotheses claim) or different.

The problem of just looking at the boxplots is that the responses at each level of one factor contain *all levels of the other factor*. For example, the responses for the *left* hand were measured at all three distances. As a result, the boxplots show more variability than they need to. Our model will deal with both factors, so it will take care of this extra variability but the simple side-by-side boxplots can't do that.

On the other hand, if you do see a difference here, that's a difference *in spite of this extra variation*, so it's very likely that the F-test will reject the null hypothesis for that factor. Just be careful not to rule out the possibility of differences that aren't apparent.

A better alternative might be to make boxplots for each factor level *after removing the effects of the other factor*. How would you do that? You could compute a one-way ANOVA on one factor and find the residuals. Then make boxplots of those residuals for each level of the other factor. We might call such a display a **partial boxplot**.[3] By removing the variability due to the other factor, you should see differences in the responses among the levels of the factor you are graphing more readily. For example, here are the boxplots of dart toss *Accuracy* plotted for the two levels of *Hand* both in the original data and after removing the effects of *Distance*:

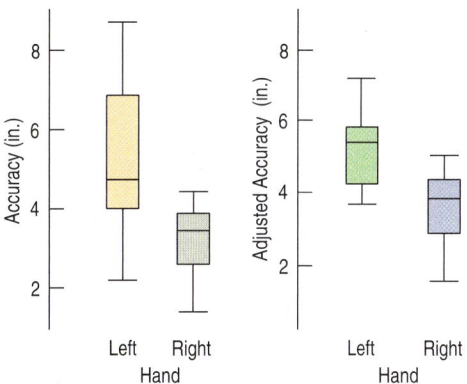

Figure 27.2
Boxplots of *Accuracy* by *Hand* (as in Figure 27.1) show the effect of changing *Hand* less clearly than the corresponding partial boxplots on the right, which show the effects of *Hand* after the effect of *Distance* has been removed. The effect of changing hands is much easier to see without the unwanted variation caused by changing the other factor, *Distance*.

Removing the variation due to *Distance* makes the differences between the two *Hands* in the partial boxplots much clearer. We'll want to do this for each factor, removing the effects of the other factor by computing the one-way ANOVA and finding the residuals.

---

[2] Unlike regression coefficients, which can be changed drastically by outliers, ANOVA tests tend to fail on the "safe side." Outliers generally inflate the Error Mean Square, which reduces the F-statistics and increases the P-values. That makes it harder to reject the null hypotheses, so we are more likely to make Type II errors. That's generally thought to be safer than making Type I errors. For example, we are not likely to be mistakenly led to approve a drug that isn't really effective. But it also means that we might fail to approve a drug that really is effective. Generally, it is best to remove outliers and report them separately.

[3] This is not a standard term. You are not likely to find it in a statistics package. But it is a natural term because we've removed *part* of the variability.

On the following page are the side-by-side and the partial boxplots for *Distance* removing the variability due to *Hand:*

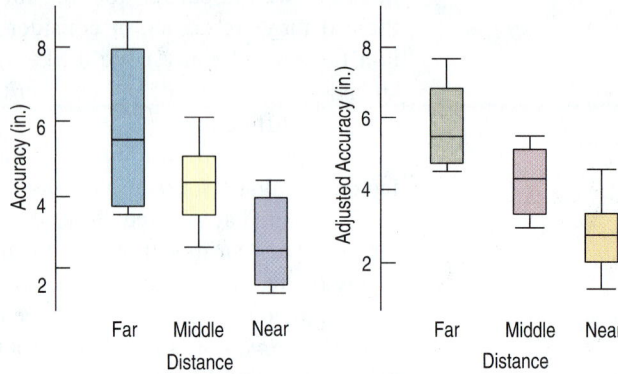

Figure 27.3
Partial boxplots after removing the effect of *Hand* (on the right) show the differences in *Accuracy* among the three distances more clearly than the original boxplots of *Accuracy* by *Distance* (on the left).

We'll use the partial boxplots for diagnostic purposes as we used the side-by-side boxplots in a one-way analysis. If the partial boxplots were all skewed in the same direction, or if their spreads changed systematically from level to level across a factor, we would have considered re-expressing the response variable to make them more symmetric. Re-expression would be likely to make the analysis more powerful and more correct.

Additivity

Each observation is assigned a treatment that combines two factor levels, one from each factor. Our model assumes that we can just *add* the effects of these two factor levels together:

$$y_{ijk} = \mu + \text{Hand effect}_j + \text{Distance effect}_k + \text{Error}_{ijk}.$$

That's a nice simple model, but it may not be a good one for our data. Just like linearity, **additivity** is an assumption. Like all assumptions, it helps us by making the model simpler, but it may be wrong for our data. Of course, we can't know for sure, but we can check the **Additive Enough Condition.** You won't be surprised that we check it with displays.

For the effects of *Hand* and *Distance* to be additive, changing hands must make the *same* difference in accuracy no matter what distance you throw from (or *vice versa*). Is this reasonable for this experiment? Alternatively, we could ask whether moving away from the target makes a greater difference to accuracy when throwing with the weaker hand. That's a *conditional* question, so we look at how *Accuracy* changes by *Distance* for each *Hand* separately.

Figure 27.4
Boxplots of *Accuracy* by *Distance* conditional on each *Hand*. Changing the *Distance* seems to have a greater effect on left *Hand* accuracies.

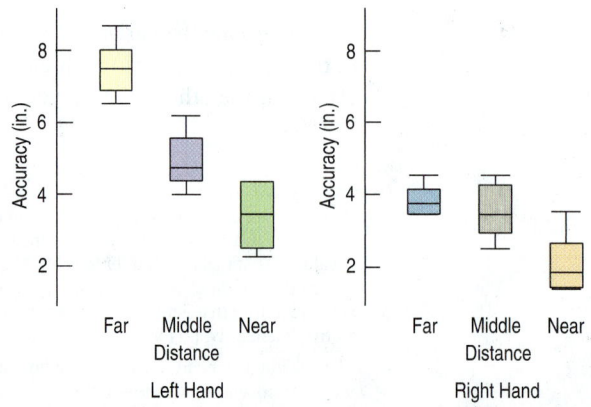

The changes in accuracy due to *Distance* do look larger for the *left* hand, especially at the *far* distance.

When the effects of one factor *change* for different levels of another factor, we say that there's an **interaction**. To show the interaction, we often plot the treatment means in a single display, called an **interaction plot**. This plot shows the average of the observations at *each* level of one factor broken up by the levels of the other factor.

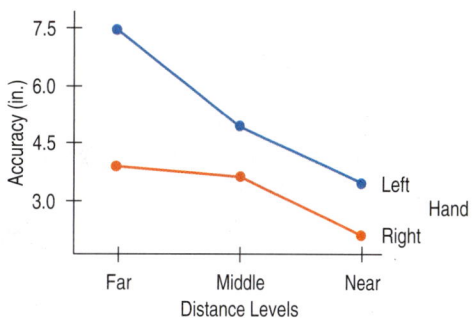

Figure 27.5
An interaction plot showing the mean accuracy at the six combinations of *Distance* and *Hand*. Connecting lines for right and left hands shows the greater effect of the far *Distance* on left-hand *Accuracy* than on right-hand *Accuracy*.

Here it's easier to see that throwing from the Far distance has more of an effect for the Left hand. If the effect of *Distance* were constant, the lines in this plot would be parallel. If the lines aren't parallel, that's evidence that the effects are not additive. But how parallel is enough? Are these lines sufficiently far from parallel for us to doubt whether our data satisfy the **Additive Enough Condition?** A graphical check like this is good enough to proceed. Later in the chapter, we'll see a way to include a term for non-additivity in our model and test for it.

Independence Assumptions

The **Independence Assumption** for the two-factor model is the same as in one-way ANOVA. That is, the observations *within each treatment group* must be independent of each other. As usual, no test can verify this assumption. Ideally, the data should be generated by a randomized comparative experiment or at least be drawn independently and at random from a homogeneous population. Without randomization of some sort, our P-values are meaningless. We have no sampling distribution models and no valid inference.

> "Without randomization, your inferences haven't got a chance."
>
> —George Cobb, famous statistician

Check the **Randomization Condition.** Were the data collected with suitable randomization? For surveys, are the data for each group a representative random sample of that group? For experiments, was the experiment randomized?

We were told that the order of treatment conditions in the dart experiment was randomized. This is especially crucial for an experiment like this because of the possibility of either learning or tiring. Your accuracy might increase after a few throws as you warm up, but after an hour or two, your concentration—and your aim—might start to wander.

If we are satisfied with the Additivity and Independence Assumptions, then we can look at the ANOVA table. For the dart data, we might have some doubts about the additivity, but we'll go ahead for now.

Analysis of Variance for Accuracy

Source	DF	Sum of Squares	Mean Square	F-Ratio	Prob
Distance	2	51.0439	25.5219	28.561	<0.0001
Hand	1	39.600	39.6900	44.416	<0.0001
Error	32	28.5950	0.893594		
Total	35	119.329			

These may look like small P-values, but inference requires additional assumptions. Before we believe the F-tests and use the P-values to make decisions, we have other assumptions to think about.

Equal Variance Assumption

Like the one-way ANOVA, the two-factor ANOVA requires that the variances of all treatment groups be equal. It's the residuals after fitting *both* effects that we'll pool for the Error Mean Square, so it's their variance whose stability we care about most. We check this assumption by checking the **Similar Variance Condition.**

In one-way ANOVA, we looked at the side-by-side boxplots of the groups to see whether they had roughly the same spread. But here, we need to check for equal spread across all treatment groups. The easiest way to check this is to plot the residuals from the two-way ANOVA model.

- Look at the residuals plotted against the predicted values. If the variance is constant, the plot should be patternless. If the plot thickens (to one side or the other), it's a sign that the variance is changing *systematically*. Then it's usually a good idea to consider re-expressing the response variable.

Figure 27.6
A scatterplot of residuals vs. predicted values from the two-way ANOVA model for *Accuracy* shows a U-shaped pattern. This suggests that a condition is violated.

This plot doesn't seem to show changing variance, but it's certainly not patternless. Residuals at both ends of the predicted values are higher than those in the middle. This could mean that something is missing from our model. Could that nagging suspicion we had about the additivity be resurfacing? Stay tuned.

- We can also plot the residuals grouped by each factor. Here's a boxplot of residuals by treatment level for each factor in the dart experiment:

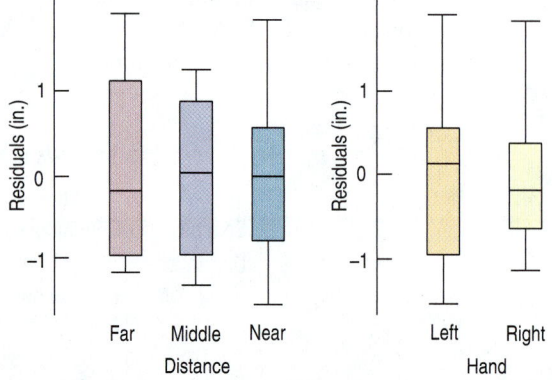

Figure 27.7
Boxplots of the residuals from the two-way ANOVA of the dart accuracy experiment show roughly equal variability when plotted for each factor.

Remember, we can't use the original boxplots to check changing spread. From those it seemed that the spread of the *near* level was smaller than that of the *far* level. But much of that difference was due to changes in the way the factor *Hand* behaves at different distances rather than a change in the underlying variance. By contrast, these boxplots of residuals look fine.

Normal Error Assumption

As with one-way ANOVA, the *F*-tests require that the underlying errors follow a Normal model. We'll check a corresponding **Nearly Normal Condition** with a Normal probability plot or histogram of the residuals.

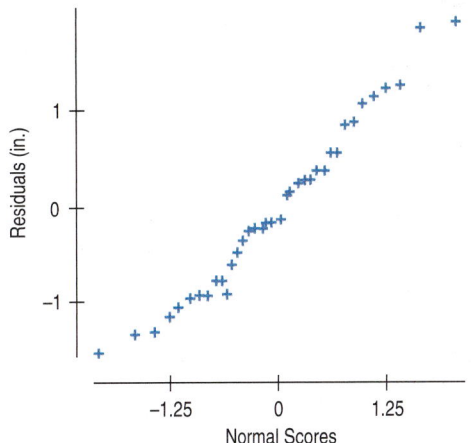

Figure 27.8
A Normal probability plot of the residuals from the two-way ANOVA model for dart accuracy seems reasonably straight.

This condition appears to be satisfied.

What Have We Learned so Far?

The ANOVA table confirmed what we saw in the boxplots. Both *Distance* and *Hand* affect accuracy of throwing a dart, at least at the levels chosen in this experiment. The small P-values give us confidence that the differences we see are due to the factors and not to chance. We did worry that the Additive Enough Condition might not be satisfied, and there seems to be something amiss in the plot of residuals versus predicted values. Before we examine that more, let's look at a different example with two factors, Step-By-Step.

Step-By-Step Example TWO-FACTOR ANALYSIS OF VARIANCE

Another student, who prefers the great outdoors to damp pub basements, wonders whether leaving her tennis balls in the trunk of her car for several days after the can is opened affects their performance, especially in the winter when it can get quite cold. She also wonders if the more expensive brand might retain its bounce better. To investigate, she performed a two-factor experiment on *Brand* and *Temperature*, using two *Brands* and three levels of *Temperature*. She bounced three balls under each of the six treatment conditions by first randomly selecting a *Brand* and for that ball, randomly selecting whether to leave it at room temperature or to put it in the refrigerator or the freezer. After 8 hours she dropped the balls from a height of 1 meter to a concrete floor, recording the Height of the bounce in centimeters.

This is a completely randomized replicated design in two factors. The factor *Brand* has two levels: *Premium* and *Standard*. The factor *Temperature* has three levels: *Room, Fridge,* and *Freezer*. The null hypotheses are that *Brand* has no effect and that *Temperature* has no effect.

Who   Tennis balls of two *Brands* allocated to three *Temperatures*.

What *Height* of the ball after it bounces from an initial height of 100 centimeters

Units Centimeters

How   Two-factor experiment, randomized order of runs

Brand	Temperature	Bounce Height
Standard	Freezer	37
Standard	Fridge	59
Standard	Room	59
Premium	Freezer	45
Premium	Fridge	60
Premium	Room	63
Standard	Freezer	37
Standard	Fridge	58
Standard	Room	60
Premium	Freezer	39
Premium	Fridge	64
Premium	Room	62
Standard	Freezer	41
Standard	Fridge	60
Standard	Room	61
Premium	Freezer	37
Premium	Fridge	63
Premium	Room	61

THINK ➡ State what we want to know and the null hypotheses we wish to test. For two-factor ANOVA, the null hypotheses are that all the treatment groups have the same mean for each factor. The alternatives are that the means are not all equal.

I want to know whether the storage *Temperature* affects the mean *Bounce Height* of tennis balls. I also want to know whether the two *Brands* have different mean *Bounce Heights*. Writing τ_j for the effect of Brand j, the null hypothesis is that the two brand effects are the same:

$$H_0: \tau_P = \tau_S.$$

Writing γ_k for the effect of Temperature level k, then the other null hypothesis is

$$H_0: \gamma_{room} = \gamma_{fridge} = \gamma_{freezer}.$$

The alternative for the first hypothesis is that the effects of the two *Brands* on the *Bounce Height* are different. The alternative for the second is that the effects of the *Temperature* levels are not all equal.

Plot Examine the side-by-side partial boxplots of the data.

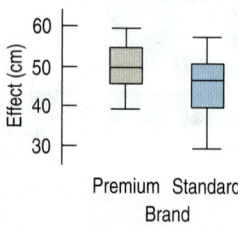

Plan Think about the assumptions and check the appropriate conditions.

It's pretty clear that *Temperature* has an effect, but we can't be sure about the effect of *Brand* just from the plots.

✓ **Independence Assumption:** Because the order was randomized, I can assume the observations are independent. However, I might want to check to see if the balls of each brand came from the same box. It might be better to have selected balls of the same brand from different boxes.

✓ **Randomization Condition:** The experimenter performed the trials in random order.

✓ **Additive Enough Condition:** Here's the plot of the mean *Bounce Heights* at each treatment condition:

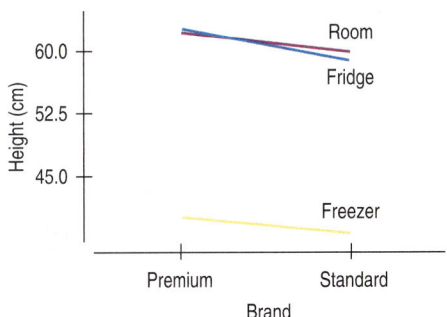

The lines look reasonably parallel. Now I can fit the additive model in two factors:

$$y_{ijk} = \mu + \text{Brand effect}_j + \text{Temp effect}_k + \text{Error}_{ijk}.$$

Under these conditions it's appropriate to fit the two-way Analysis of Variance model.

Show the ANOVA table.

Analysis of Variance For Height

Source	DF	Sum of Squares	Mean Square	F-ratio	P-value
Temp	2	1849.33	924.665	209.55	<0.0001
Brand	1	26.8889	26.8889	6.0935	0.0271
Error	14	61.7778	4.41270		
Total	17	1938.00			

The Error Mean Square is 4.41, which means that the common error standard deviation is estimated to be $\sqrt{4.41}$, a bit more than 2 centimeters. This seems like a reasonable accuracy for measuring *Bounce Heights* of tennis balls dropped from a height of 1 meter.

Before testing the hypotheses, I need to check some more conditions:

✓ **Similar Variance Condition:** A plot of residuals vs. predicted values shows some increased spread on the left, but with only 5 points it's hard to say.

✓ **Nearly Normal Condition, Outlier Condition:** A Normal probability plot of the residuals looks OK too:

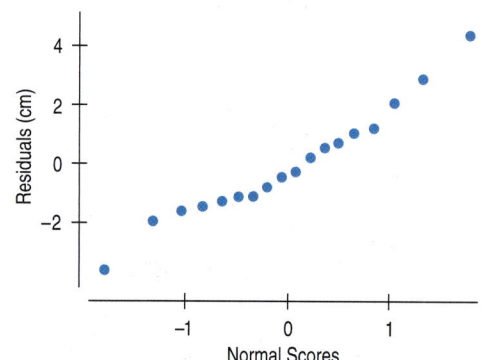

Under these conditions, it is appropriate to perform and interpret the F-test.

Show → Mechanics Discuss the tests.

Display the effects for each level of the significant factors. In the next section, we'll show how they're calculated. Remember, significance does not guarantee that the differences are meaningful.

The F-ratio for *Brand* is 6.09, with a P-value of 0.0271. The F-ratio for *Temperature* is 209.55, with a P-value less than 0.0001.

The overall mean *Bounce Height* is 53.67 cm.

The *Brand* effects are
Premium   1.22
Standard   −1.22

The *Temperature* effects are
Room   7.33
Fridge   7.00
Freezer   −14.33

TELL ➡ Interpretation Tell what the F-tests mean.

F-ratios this large would be very unlikely if the factors had no effect. I conclude that the means of the two *Brand* levels are not equal. I also conclude that the means of the three *Temperature* levels are not all equal.

Overall, the premium brand bounces about 2.44 cm higher than the standard brand, averaging across the temperature ranges. Generally, there seems to be little difference between the average *Bounce Height* for room and fridge temperatures, but tennis balls from the freezer bounce almost 22 cm less on average.

Conclusions: It appears that the premium brand outperforms the standard brand overall. It looks from the plots like our student needn't be too worried about playing tennis on a cold day, but if the temperature drops below freezing, she should try to warm the tennis balls first.[4]

For Example

RECAP: In an experiment to investigate the effects of paper airplane design on flight distance two factors were tested.
Here are boxplots of distance by each of the factors:

QUESTION: What can we learn from the boxplots?

It appears that the Darts travel farther than the Gliders. There does not appear to be an effect for flaps. The spreads of all the groups are comparable. There are two observations plotted as outliers for the Glider/Copier design, but they don't seem to be very extreme and are probably not a problem.

---

[4] It's quite obvious from the plots and the tables of effects that there's a difference between *freezer* and the other temperatures, but little difference between *room* and *fridge*. A Bonferroni test would confirm this.

*How ANOVA Works

You'll probably never carry out a two-way ANOVA by hand. But to understand the model it can help to see exactly how it decomposes each observation into components. Our tennis ball model says

$$y_{ijk} = \mu + \text{Brand effect}_j + \text{Temp effect}_k + \text{Error}_{ijk}.$$

To apply this model to data, we need to *estimate* each of these quantities:

$$y_{ijk} = \bar{\bar{y}} + \widehat{\text{Brand effect}_j} + \widehat{\text{Temp effect}_k} + \text{Residual}_{ijk}.$$

Here are the data again, displayed a little differently:

		Temperature		
		Room	Fridge	Freezer
Brand	Premium	63	60	45
		62	64	39
		61	63	37
	Standard	59	59	37
		60	58	37
		61	60	41

The first term in the model is μ, the overall mean. It's reasonable to estimate that with the grand mean, $\bar{\bar{y}}$, which is just the mean of all the observations. For the tennis balls, the average bounce was 53.67 cm. The grand mean has the same value for every observation, which we can lay out as a table like this:

		Temperature		
		Room	Fridge	Freezer
Brand	Premium	53.67	53.67	53.67
		53.67	53.67	53.67
		53.67	53.67	53.67
	Standard	53.67	53.67	53.67
		53.67	53.67	53.67
		53.67	53.67	53.67

Now we turn to the *Brand* effects. To estimate the effect of, say, the *premium Brand*, we find the mean of all *premium* balls and subtract the grand mean. That's 54.89 cm − 53.67 cm = 1.22 cm. In other words, the *premium Brand* balls bounced an average of 1.22 cm higher than the overall average. The *standard Brand* balls bounced an average of 52.444 cm, so their effect is 52.444 cm − 53.667 cm = −1.22 cm. Of course, with only two levels, the fact that these two effects are equal and opposite is no coincidence. Effects *always sum to* 0.

We can put the *Brand* effects into a table like this:

		Temperature		
		Room	Fridge	Freezer
Brand	Premium	1.22	1.22	1.22
		1.22	1.22	1.22
		1.22	1.22	1.22
	Standard	−1.22	−1.22	−1.22
		−1.22	−1.22	−1.22
		−1.22	−1.22	−1.22

We can find a *Temperature* effect table in a similar way. We take the treatment average at each level of *Temperature* and subtract the grand mean. Here's the table:

		Temperature		
		Room	Fridge	Freezer
Brand	Premium	7.33	7.00	−14.33
		7.33	7.00	−14.33
		7.33	7.00	−14.33
	Standard	7.33	7.00	−14.33
		7.33	7.00	−14.33
		7.33	7.00	−14.33

We're missing only one table, the *Residuals*. Finding the residuals is easy, too. The additive model represents each observation as the sum of these three components. As always, the residuals are the differences between the observed values and those given by the model. Look at the first ball in the table. It was a *premium Brand* ball at *room Temperature*. From the effects, we'd predict it to bounce 53.67 (grand mean) +1.22 (from being premium) +7.33 (from being at room temperature) = 62.22 cm. Because it actually bounced 63 cm, its residual is 63 − 62.22 = 0.78 cm. If we compute all the residuals this way (or have the computer do it), we'd find the residual table to be

		Temperature		
		Room	Fridge	Freezer
Brand	Premium	0.78	−1.89	4.44
		−0.22	2.11	−1.56
		−1.22	1.11	−3.56
	Standard	−0.78	−0.44	−1.11
		0.22	−1.44	−1.11
		1.22	0.56	2.89

What we have done with all these tables is to decompose each original bounce height into the three estimated components plus a residual:

Bounce Height = mean *Bounce Height* + *Brand effect* + *Temperature effect* + *Residual*.

These tables actually display the workings of the ANOVA. Each term in the ANOVA model corresponds to one of these tables. Because the overall mean doesn't vary, we usually leave it out of the ANOVA table. Each Sum of Squares term in the ANOVA is simply the sum of the squares of all the values in its table.

The degrees of freedom are easy to find too. Look at the grand mean table. Because it has the same value for all observations, it has 1 degree of freedom. The *Brand* effect table seems to have 2 degrees of freedom, but we know that the two effects always sum to 0. So it really has only 1. In general, the degrees of freedom for a factor is one less than its number of levels. So *Temperature* has 2 degrees of freedom.

The original table has 18 observations and so 18 degrees of freedom. But what about the residuals? Remarkably enough, the degrees of freedom from the tables add. That means that we can find the degrees of freedom for the residuals by subtraction. They must have $18 - (1 + 1 + 2) = 14$ degrees of freedom.

The mean squares are just the sum of squares for each effect divided by the degrees of freedom. The *F*-ratios are the ratios of each Treatment Mean Square to the Error Mean Square. You can check the degrees of freedom and the mean squares in the step-by-step ANOVA table. The original table has 18 observations and so 18 degrees of freedom, but most programs choose not to display the overall mean and subtract out the degree of freedom associated with it to give what's sometimes called a "corrected" total or just "total." That's why our ANOVA table has only 17 degrees of freedom in "total."

Blocks

Back in Chapter 12, we thought about a *randomized block design*.[5] We imagined obtaining tomato plants from two stores and randomly assigning plants from each store to one of three fertilizer treatments. We can view the *Store* as a factor in this design. We now have a two-factor model, with *Store* and *Fertilizer* as factors. The ANOVA looks just the same, but now we're not likely to be interested in testing the *Store* effect. We *assumed* that the plants might be different from the different stores—that's why we used a blocked design.

The two-way ANOVA takes advantage of the blocking to remove the unwanted variation due to *Store*. As in a completely randomized two-factor design, removing the unwanted variation makes the effects of the other factor easier to see.

We wouldn't call a blocked design *completely randomized,* because we can't assign the plants at random to which store they come from. We must randomize *within* the blocks when we assign plants to levels of the *Fertilizer* factor.

27.3 Interactions

Up to now, we have assumed that whatever effects the two factors have on the response can be modeled by simply adding the separate effects together. What if that's not good enough? In the dart-throwing experiment, we worried that the experimenter's accuracy when throwing from farther away may deteriorate *even more* when he uses his nondominant hand. Looking at the interaction plot added evidence to that suspicion.

---

[5]You may have noticed in the tennis ball example that balls couldn't be randomly assigned to a *Brand. Brand* is actually a blocking factor, and the experiment is a randomized block design. But it doesn't change the analysis at all.

Figure 27.9
Interaction plot of *Accuracy* by *Hand* and *Distance*. The effect of *Distance* appears to be greater for the left hand.

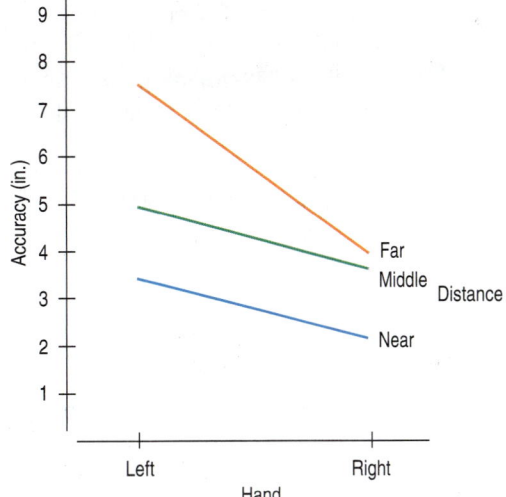

It doesn't look like the effect of *Distance* is constant for the two hands. In particular, moving from *middle* to *far* has a much larger effect when using the *left Hand*. It's not unusual for the effect of a factor's level to depend on the level of the other factor. Unfortunately, our model doesn't take this into account. So, we need a new model.

When the effect of one factor *changes* depending on the levels of the other factor, the factors are said to *interact*. We can model this interaction by adding a term to our model and testing whether the adjustment is worthwhile with an *F*-test.

We include a new term to our model, adding an interaction effect at every *combination* of the levels of the other two factors. The model can now be written:

$$y_{ijk} = \mu + \tau_j + \gamma_k + \omega_{jk} + \varepsilon_{ijk}.$$

The new term, ω_{jk}, represents the *interaction* effect at level *j* of factor 1 and level *k* of factor 2.

We add a corresponding line to the ANOVA table. Of course, we'll have to spend some degrees of freedom on this new term. We'll have to take them from the error term where we parked all the leftover degrees of freedom after fitting the two factors. We'll need the *product* of each term's degrees of freedom. For the darts, this is only 2 degrees of freedom (1 from *Hand* and 2 from *Distance*), but for factors with more levels, this could get expensive.

Here's the ANOVA table for the dart experiment:

ANOVA Table for Accuracy

Source	DF	Sum of Squares	Mean Square	F-ratio	P-value
Distance	2	51.044	25.522	41.977	<0.0001
Hand	1	39.690	39.690	65.28	<0.0001
Distance × Hand	2	10.355	5.178	8.516	0.0012
Error	30	18.240	0.608		
Total	35	119.329			

What has changed? Nearly everything. (Of course, the Total Sum of Squares remains the same because it was calculated from the observations minus the overall mean, and the overall mean is still the same.) The interaction term is significant. Because we've now removed a significant source of variability, our Error Mean Square is smaller in spite of the loss of the 2 degrees of freedom devoted to the interaction term.

For Example

RECAP: An experiment on paper airplanes considered two factors; *Design* and *Flaps*. Here is the two-factor ANOVA:

Source	DF	Sum of Squares	Mean Square	F-ratio	P-value
Design	3	1549.50	516.499	56.610	<0.0001
Flap	1	27.0140	27.0140	2.9608	0.0905
Error	59	538.303	9.12378		
Total	63	2114.81			

State and test the null hypotheses. What can you conclude?

The first null hypothesis is that the mean distance flown will be the same for each design, after allowing for any effects of flaps. That hypothesis is rejected. As we saw in the boxplots, the darts fly farther.

The second null hypothesis is that the mean distance will be the same whether or not flaps are present, after allowing for the differences due to design. The P-value for that factor is too large to allow us to reject the null hypothesis.

Inference When Variables Are Related

The *F*-test for the interaction leads us to reject the null hypothesis of no interaction effect and indicates that the effect of *Distance* really does change depending on which *Hand* is used. (Or, equivalently the effect of *Hand* changes depending on the *Distance*.)

What about the main effects? When a significant interaction term is present, it's difficult to *Tell* about the main effect of each factor. How much does switching from *right* hand to *left* hand matter? It depends on the *Distance*. It no longer makes sense to talk about the main effects of each factor separately. The main effects are just averages of the effects over all the levels of the other factor. As we saw with Simpson's paradox (back in Chapter 2), it can be misleading to average over very different situations.

When a significant interaction is present, the best way to display the results is with an interaction plot.

Now that we have the interaction term in our model, we recalculate the residuals. They're the values that are left over after accounting for the overall mean, the effect of each factor, and the interaction effect. Here's a plot of residuals versus predicted values for the two-factor model with interaction:

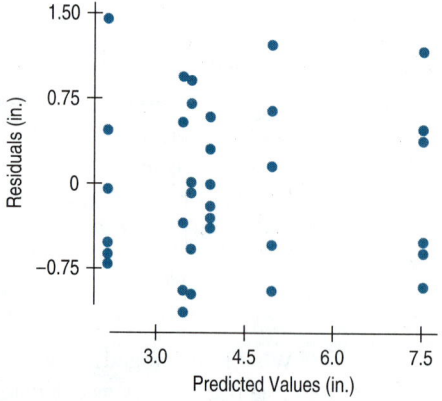

Figure 27.10
Residuals for the two-way ANOVA of dart accuracy with an interaction term included. Now, unlike Figure 27.6, there is no U-shaped pattern.

How has the interaction term helped? These residuals have less structure. Fitting the interaction term succeeded in removing structure from the error. This new model seems to satisfy the assumptions more successfully, and so our inferences are likely to be closer to the truth.

Why Not Always Start by Including an Interaction Term?

With two factors in the model, there's always the possibility that they interact. We checked the Additive Enough Condition with an interaction plot. Now we know we can include the interaction term and use the F-test for added information. If the interaction plot shows parallel lines and the F-test for the interaction term is not significant, you can comfortably proceed with a two-way ANOVA model without interaction. This will put back into the error term the degrees of freedom you used for the interaction.

On the other hand, if the interaction is significant, the model *must* contain more than just the main effects. In this case, the interpretation of the analysis depends crucially on the interaction term, and the interaction plot shows most of what to *Tell* about the data. When there's a significant interaction term, the effect of each factor depends on the level of the other factor, so it may not make sense to talk about the effects of a factor by itself. However, we may still be able to talk about the main effects. For example, we can see that the *right* hand is better *on average* than the *left* at any distance, and the closer you are to the target, the better the accuracy, no matter which hand you use. So both main effects are significant, and we can discuss them. A significant interaction effect tells us more. It says that *how much* better the performance is at near distances *depends* on which hand you use.

But if the lines in the interaction plot cross, you need to be careful. Here's an interaction plot from an experiment on fuels and alternative engine designs:

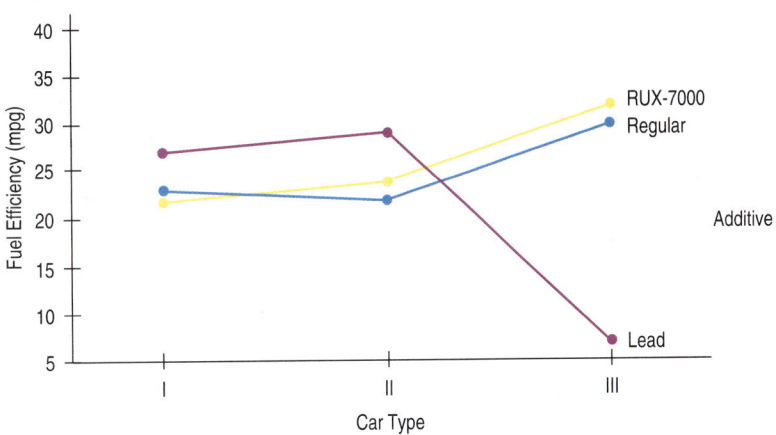

Figure 27.11
An interaction plot of gas mileage by *Car Type* and *Gas Additive*. If the lines cross, as they do here, be careful of interpreting the main effects.

It looks like the fuel with lead is great for two of the engine designs, but a disaster for the third. Suppose someone asked you to summarize the effect of adding lead. Nothing simpler than saying it *depends* on the engine type makes sense. *On average*, it lowers gas mileage, but is that relevant? The real message is that sometimes it improves mileage and sometimes it hurts. You won't want to *Tell anything* about the main effect. Its F-test is irrelevant.

Whether the lines cross or not, including an interaction term in our model is always a good way to start. But sometimes we just can't. Suppose we had run the dart experiment with only one dart at each treatment. A design like that is called an **unreplicated two-factor design**. Imagine that the throw at the combination of left hand and far distance missed the bull's-eye by a mile. Would this be evidence of an interaction between *Distance* and *Hand*, or was it just a bad throw? How could we tell? The obvious thing to do would be to *repeat* the treatment. If accuracy was consistently poor at this combination of factors, we'd want to model the interaction. If the other throws were no worse than we'd expect from the two main effects, then we might set the outlier aside. We can't distinguish an interaction effect from error unless we replicate the experiment. In fact, the design tells us that. Without replication, if we try to fit an interaction term, there are exactly 0 degrees of freedom left for the error. That makes any inference or testing impossible. If we are willing to *assume* that the additive model is adequate, then we can reduce the number of runs required and use an unreplicated design. In that case, although the residuals are indistinguishable from the interaction effects, the residual plot may help reveal whether the assumption made sense.

When you do replicate, it's best to replicate *all* treatment conditions equally. (Unbalanced and other more complicated designs can be analyzed, but they are beyond the scope of this book.)

Just Checking

The student conducting the paper airplane experiment decides to do a follow-up experiment to investigate two other factors in more detail. Using his favorite design, he uses one of three different size weights and places the weight either on the nose or on the tail of the plane. At each combination, he performs four test flights (randomizing the order of the flights). An interaction plot looks like this:

The ANOVA table shows:

Analysis of Variance

Source	DF	Sum of Squares	Mean Square	F-Ratio	P-value
Position	1	13371.9314	13371.9314	53.3868719	<0.0001
Weight	2	12625.5486	6312.7743	25.2034851	<0.0001
Position × Weight	2	10.6208	5.3104	0.02120155	0.9790
Error	18	4508.501	250.47		
Total	23	30516.602			

1. What are the null hypotheses?
2. From the interaction plot, do the effects appear to be additive?
3. Does the *F*-test in the ANOVA table support your conclusion in (2)?
4. Does the position of the weight seem to matter?
5. Does the amount of weight appear to affect flight distance?
6. What would you recommend to the student to increase flight distance?

Step-By-Step Example: TWO-FACTOR ANOVA WITH INTERACTION

In Chapter 26 we looked at how much TV four groups of students watched on average. Let's look at their grade point averages (GPAs). Back in that chapter, we treated the four groups (male athletes, female athletes, male non-athletes, and female non-athletes) as four levels of the factor *Group*. Now we recognize that there are really two factors: the factor *Sex* with levels *male* and *female* and the factor *Varsity* with levels *yes* and *no*. Let's analyze the GPA data with a two-factor ANOVA.

As with many Social Science studies, this isn't an experiment. (Some fans might object if we assigned students at random to varsity teams.[6]) Instead, it is an application of ANOVA to an observational study.

THINK State what we want to know and the null hypotheses we wish to test. For two-factor ANOVA with interaction, the null hypotheses are that all the treatment groups have the same mean for each factor and that the interaction effect is 0. The alternatives are that at least one effect is not 0.

I want to know whether the mean GPA is the same for men and women, whether the mean GPA is the same for athletes and non-athletes, and whether the factors interact. Writing τ_j for the effect of *Sex* level j, the null hypothesis is that the effects are the same:

$$H_0: \tau_M = \tau_F$$

Writing γ_k for the effect of *Varsity* level k, the other null hypothesis is

$$H_0: \gamma_Y = \gamma_N$$

Finally, the null hypothesis for interaction is that the interaction terms are all 0.

The alternative for the first hypothesis is that the mean GPA is different for men and women. The alternative for the second is that the mean GPA is different for *Varsity* athletes than for other students. The alternative for the last is that there is an interaction effect.

Plot Examine the side-by-side partial boxplots of the data.

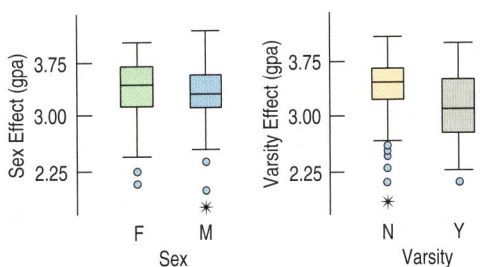

Plan Think about the assumptions and check the conditions.

It appears that there is no difference in mean GPA between men and women (after accounting for varsity status), but there may be a difference between varsity athletes and non-athletes (after accounting for sex).

---

[6] And randomly assigning the *other* factor would be problematic.

The study is a random sample, so students' GPAs should be independent of each other.

✓ **Independence Assumption, Randomization Condition:** The study was based on a random sample of students.

✓ **Additive Enough Condition:** Here's the interaction plot of the means at each treatment condition:

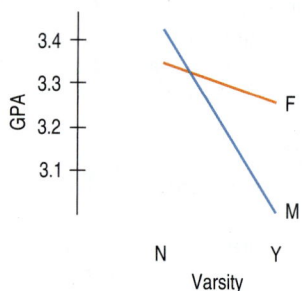

The lines cross, so I'll add an interaction term to the model and fit the two-way ANOVA with interaction:

$$y_{ijk} = \mu + \text{Sex effect}_j + \text{Varsity effect}_k + \text{Interaction}_{jk} + \text{Error}_{ijk}.$$

Show the ANOVA table.

Analysis of Variance for gpa

Source	DF	Sum of Squares	Mean Square	F-ratio	P-value
Sex	1	0.3040	0.3040	1.7681	0.1852
Varsity	1	2.4345	2.4345	14.1871	0.0002
Sex × Varsity	1	1.0678	1.0678	6.2226	0.0134
Error	196	33.6397	0.1716		
Total	199	37.4898			

✓ **Similar Variance Condition:** A plot of residuals vs. predicted shows no thickening or other patterns:

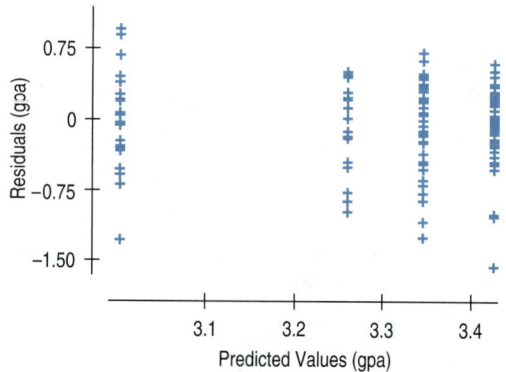

✓ **Nearly Normal Condition, Outlier Condition:** The Normal probability plot of the residuals is reasonable as well.

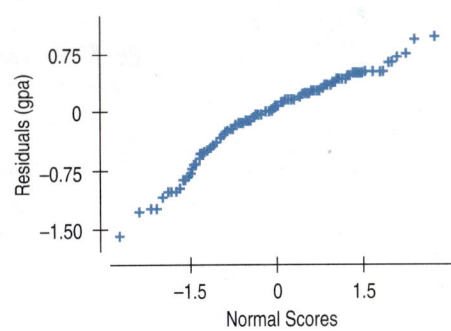

Under these conditions it is appropriate to interpret the F-tests and their P-values.

SHOW ➡ Mechanics

The error mean square is 0.172 points, which means that the common error standard deviation is estimated to be $\sqrt{0.172} = 0.415$ points (nearly half a letter grade).

The F-ratio for the interaction term is 6.223, with a P-value of 0.0134. Based on this, I reject the hypothesis of no interaction. Because there is interaction, I am cautious in testing the main effects, but will examine the means further to understand the factor effects.

Overall mean GPA = 3.32.

Level of Varsity	Effect
N	0.1281
Y	−0.1281

The main effect of Varsity is significant. Varsity athletes have lower GPA by about a quarter of a grade point averaged over all, but this effect is much greater for men than for women. Here are the four group mean GPAs:

Show the table of means.

Group	Mean GPA
Female, non-athlete	3.35
Female, varsity	3.26
Male, non-athlete	3.43
Male, varsity	3.00

TELL ➡ Interpretation   Tell what the F-test means.

The interaction term is significant, and it appears that most of the effect we see is due to a difference of 0.43 in mean GPAs between male athletes and non-athletes. It doesn't look like any of the other differences is large enough to be important. Probably none of them is statistically significant either.

For Example

RECAP: The ANOVA for paper airplane distance on two factors, *Design* and *Flaps*, showed a significant effect for *Design*, but not for *Flaps*. But maybe there is an interaction. Here is the ANOVA with the interaction term and an interaction plot.

Source	DF	Sum of Squares	Mean Square	F-ratio	P-value
Design	3	1549.50	516.499	81.065	<0.0001
Flaps	1	27.0140	27.0140	4.2399	0.0441
Design × Flaps	3	181.503	60.5009	9.4956	<0.0001
Error	56	356.800	6.37144		
Total	63	2114.81			

QUESTION: Interpret the interactions. What do they say about paper airplane design?

Including the interaction term reveals that the *Flaps* factor is significant. But it matters in different ways for different designs. Flaps help improve the distance of both kinds of gliders. But flaps hinder the performance of darts made of lighter copier paper, although they help darts made of construction paper.

What Next?

This chapter introduces some of the issues that arise in advanced experimental design and analysis. We have scratched the surface of this large and important area by showing the analysis of a balanced two-way design with interaction.

Of course, there's no reason to stop with two factors. And if we have more than two, there are many ways in which they may interact. As you can imagine, real-life experiments and observational studies may pose many interesting challenges. Anyone who is serious about analyzing or designing multifactor studies should consult a professional statistician. You now know the vocabulary and principles, so you are well equipped to do that.

WHAT CAN GO WRONG?

- **Beware of unreplicated designs unless you are sure there is no interaction.** Without replicating the experiment for each treatment combination, there is no way to distinguish the interaction terms from the residuals. If you are designing a two-factor experiment, you must be willing to assume that there is no interaction if you choose not to replicate. In such a case, you can fit an additive model only in the two factors. If there is an interaction, it will show up in the error term. You should examine a residual plot to help reveal a possible interaction effect. You must be prepared to defend the assumption of no interaction and the decision not to replicate.

- **Don't attempt to fit an interaction term to an unreplicated two-factor design.** If you have an unreplicated two-factor experiment or observational study, you'll find that if you try to fit an interaction term, you'll get a strange ANOVA table. The design exhausts the degrees of freedom for error, so fitting the interactions leaves no degrees of freedom for residuals. That wipes out the mean square errors, *F*-ratios, and P-values, which may appear in the computer output as dots, dashes, or some other indication of things gone wrong. Remove the interaction term from the model and try again.

- **Be sure to fit an interaction term when it exists.** When the design is replicated, it is always a good idea to fit an interaction term. If it turns out not to be statistically significant, you can then fit a simpler two-factor main effects model instead.

- **When the interaction effect is significant, don't interpret the main effects.** Main effects can be very misleading in the presence of interaction terms. Look at this interaction plot.

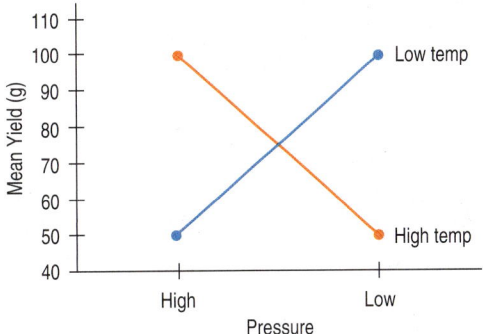

Figure 27.12
An interaction plot of *Yield* by *Temperature* and *Pressure*. The main effects are misleading. There is no (main) effect of *Pressure* because the average *Yield* at the two pressures is the same. That doesn't mean that *Pressure* has no effect on the *Yield*. In the presence of an interaction effect, be careful when interpreting the main effects.

The experiment was run at two temperatures and two pressure levels. High amounts of material were produced at high pressure with high temperature and at low pressure with low temperature. What's the effect of *Temperature*? Of *Pressure*? Both main effects are 0, but it would be silly (and wrong) to say that neither *Temperature* nor *Pressure* was important. The real story is in the interaction.

- **Always check for outliers.** As in any analysis, outliers can distort your conclusions. An outlier can inflate the Error Mean Square so much that it may be hard to discern any effect, whether it exists or not. Use the partial boxplots to search for outliers. Consider setting outliers aside and re-analyzing the results. An outlier can make an interaction term appear significant. For example, a single male varsity athlete with a very low GPA could account for the results we saw.

- **Check for skewness.** If the underlying data distributions are skewed, you should consider a transformation to make them more symmetric.

- **Beware of unbalanced designs and designs with empty cells.** We've been assuming that the data are *balanced* over the design—that we have equal numbers of observations in each cell. There are methods that will easily compensate for small amounts of imbalance, but empty cells and other more serious violations of balance require different methods and additional assumptions.

CONNECTIONS

We first discussed designing experiments in Chapter 12. Now we know how to analyze a two-factor design. In Chapter 26, we saw that the one-way ANOVA generalized the pooled *t*-test to more than two levels of one factor. Now we've added a second factor.

Of course, we're still relying on boxplots, Normal probability plots, and scatterplots of residuals to help us check conditions and understand relationships. What we first said we'd look for in these displays back in Chapters 3, 4, 5, and 6 is still what we're concerned with here.

What Have We Learned?

In Chapter 26, we learned that the Analysis of Variance is a natural way to extend the *t*-test for testing the means of two groups to compare several groups. Often, those groups are different levels of a single factor. Now we've learned that we can extend the Analysis of Variance to designs with more than one factor. We've learned that partial boxplots are a good way to examine the effect of each factor on the response. We've seen that separate tests of hypotheses can be carried out for each of the factors, testing whether the factor has any discernible effect on the response across any of the levels tested.

We've also learned that sometimes factors can interact with each other. When we have at least two observations at each combination of factor levels, we can add an interaction term to our model to account for the possible interaction.

Of course, we still need to check the appropriate assumptions and conditions as we did for simple ANOVA. All the methods we learned for displaying data in the early chapters are still important. And we've begun to see the deep connection between the design concepts we learned in Chapter 13, the exploratory methods we learned in the early chapters, and the inference methods of the latter part of the book.

Review of Terms

Two-way ANOVA model (main effects model)
The model for a two-way (one response, two factors) ANOVA is

$$y_{ijk} = \mu + \tau_j + \gamma_k + \varepsilon_{ijk},$$

where τ_j represents the effect of (level *j*) of factor one and γ_k represents the effect of (level *k*) of factor two. The subscript *i* designates the *i*th replication of the treatment combination (p. 784).

Partial boxplot
Partial boxplots are found by plotting the response at each of the levels of one factor with the effects of the second factor removed. A simple way to generate partial boxplots of factor A is to first save the residuals from a one-way ANOVA of the response on factor B. Add the overall mean to the residuals and make boxplots of the resulting data against factor A (p. 785).

Additivity, additive model
A model in two factors is additive if it simply adds the effects of the two factors and doesn't include an interaction term (p. 786).

Interaction
When the effects of the levels of one factor change depending on the level of the other factor, the two factors are said to interact. When interaction terms are present, it is misleading to talk about the main effect of one factor because how large it is *depends* on the level of the other factor (p. 787).

Interaction plot A plot that shows the means at each treatment combination, highlighting the factor effects and their behavior at all the combinations (p. 787).

Two-way ANOVA model with interaction (main effects model) For a replicated two-way design, an interaction term can also be fit. The resulting model is

$$y_{ijk} = \mu + \tau_j + \gamma_k + \omega_{jk} + \varepsilon_{ijk},$$

where τ_j represents the effect of (level j) of factor 1 and γ_k represents the effect of (level k) of factor 2. Now ω_{jk} represents the effect of the interaction at levels j of factor one and k of factor two. The subscript i designates the ith replication of the treatment combination (p. 797).

On the Computer TWO-WAY ANOVA

Some statistics packages distinguish between models with one factor and those with two or more factors. You must be alert to these differences when analyzing a two-factor ANOVA. It's not unusual to find ANOVA models in several different places in the same package.

Usually, you must specify the interaction term yourself. That's because these features of statistics packages typically are designed for models with three or more factors where the number of interactions can explode unless the data analyst carefully selects the interactions to be included in the model.

DATA DESK

- Select the response variable as Y and the factor variable as X.
- From the **Calc** menu, choose **ANOVA > ANOVA** or **ANOVA > ANOVA with Interactions**.
- Data Desk displays the ANOVA table.
- Select plots of residuals from the ANOVA table's HyperView menu.

COMMENTS

Data Desk expects data in "stacked" format. You can change the ANOVA by dragging the icon of another variable over either the Y or the X variable name in the table and dropping it there. The analysis will recompute automatically.

EXCEL

- The Excel Data Analysis Add-in offers a two-way ANOVA "with and without replication."

COMMENTS

Excel requires that the data be in a special format and cannot deal with unbalanced data, estimate interactions, or make boxplots.

JMP

- From the **Analyze** menu select **Fit Model**.
- Select variables: and **Add** them to the **Construct Model Effects** box.
- To specify an interaction, select both factors and press the **Cross** button.
- Click **Run Model**.
- JMP opens a **Fit Least Squares** window.
- Click on the red triangle beside each effect to see the means plots for that factor. For the interaction term, this is the interaction plot.
- Consult the JMP documentation for information about other features.

COMMENTS

JMP expects data in "stacked" format with one continuous response and two nominal factor variables.

MINITAB

- Choose **ANOVA** from the **Stat** menu.
- Choose **Two-way...** from the **ANOVA** submenu.
- In the Two-way ANOVA dialog, assign a quantitative Y variable to the Response box and assign the categorical X factors to the Factor box.
- Specify interactions.
- Check the **Store Residuals** check box.
- Click the **Graphs** button.
- In the ANOVA-Graphs dialog, select **Standardized residuals**, and check **Normal plot of residuals** and **Residuals versus fits**.

COMMENTS

Minitab expects "stacked" format data for two-way designs.

R

To perform an analysis of variance of a variable y on factors x_1 and x_2:

- myaov = aov($y \sim x_1 + x_2$)
- anova(myaov) # gives the ANOVA table

To include the interaction of x_1 and x_2:

- myaov = aov($y \sim x_1 + x_2 + x_1{:}x_2$) or equivalently
- myaov = aov($y \sim x_1{*}x_2$)

- summary(myaov) # shows the individual effects
- anova(myaov) # gives the ANOVA table

COMMENTS

To get confidence or prediction intervals use:

- predict(myaov, interval = "confidence")

or

- predict(myaov, interval = "prediction")

SPSS

- Choose **Analyze > General Linear Model > Univariate**.
- Assign the response variable to the **Dependent Variable** box.
- Assign the two factors to the **Fixed Factor(s)** box. This will fit the model with interactions by default.
- To omit interactions, click on **Model**. Select **Custom**. Highlight the factors. Select **Main Effects** under the **Build Terms** arrow and click the arrow.
- Click **Continue** and **OK** to compute the model.

STATCRUNCH

To find an ANOVA of response variable y on two factors, x_1 and x_2, from the Stat menu,

- choose ANOVA > Two Way
- In the dialog, specify the response variable.
- Specify the factors. It really doesn't matter which you specify as the "row factor" and which as the "column factor".

StatCrunch automatically includes an interaction term in your model. Choose the "Fit Additive Model" option to omit the interaction term.

A table of cell means is available.

CHAPTER 27 Multifactor Analysis of Variance

TI-83/84 PLUS

COMMENTS

You need a special program to compute a two-way ANOVA on the TI-83.

Exercises

Section 27.1

1. **Student TV time** In the previous chapter we considered TV watching by male and female student athletes. In that example, we categorized the students into four groups, but now we have seen that these data could be analyzed with two factors, *Sex* and *Athlete*. Write the ANOVA model for the design in this form.

2. **Cookies** A student performed an experiment to compare chocolate chip cookie recipes. He baked batches of cookies with different amounts of *Sugar*: 0.5, 0.375, and 0.25 cups, and with different kinds of *Chips*: milk, semisweet, and dark chocolate. Cookie quality was scored on a 0 to 10 scale by other students in the class, with the mean score being the result for each recipe. Write the ANOVA model for the design of this experiment.

Section 27.2

3. **Student TV assumptions** The TV watching study of Exercise 1 was collected as a survey of students at a small college. Do the assumptions of ANOVA appear to be met? Here are some displays to help in your decision:

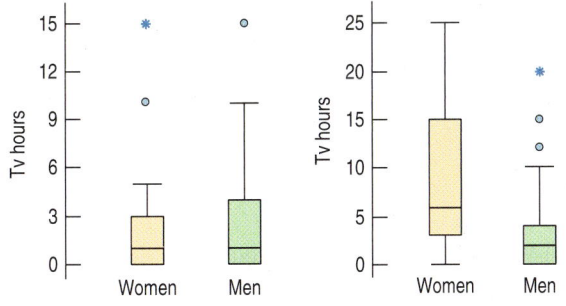

4. **Cookie assumptions** The student performing the chocolate chip cookie experiment of Exercise 2 planned to analyze his results with an Analysis of Variance on two factors. Here are some displays. Do you think the Assumptions for ANOVA are satisfied?

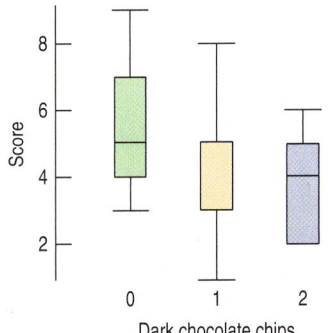

Section 27.3

5. **Student TV interactions** Here is an ANOVA table with an interaction term and the corresponding interaction plot for the TV watching data of Exercise 1. What does the interaction term mean here?

Analysis of Variance For TVhours
No Selector

Source	DF	Sums of Squares	Mean Square	F-ratio	P-value
Intercept	1	2635.13	2635.13	134.51	<0.0001
Sex	1	464.839	464.839	23.728	<0.0001
Ath	1	374.634	374.634	19.123	<0.0001
Sex x Ath	1	312.877	312.877	15.971	<0.0001
Error	193	3780.97	19.5905		
Total	196	4765.12			

6. **Have ANOVA cookie** Here is the ANOVA table for the cookie experiment of Exercise 2 along with an interaction plot. What does the interaction term say about the cookie recipes?

Analysis of Variance For Score

Source	DF	Sums of Squares	Mean Square	F-ratio	P-value
Intercept	1	2515.16	2515.16	677.74	<0.0001
Ch	2	27.8990	13.9495	3.7588	0.0270
Sgr	2	27.6566	13.8283	3.7262	0.0279
Ch x Sgr	4	38.2828	9.57071	2.5789	0.0426
Error	90	334.000	3.71111		
Total	98	427.838			

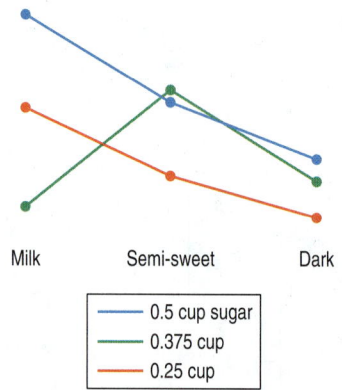

Chapter Exercises

7. **Popcorn revisited** In the previous chapter we saw a two-factor experiment to test how microwave power and temperature affect popping. She chooses 3 levels of *Power* (*low*, *medium*, and *high*) and 3 *Times* (*3 minutes*, *4 minutes*, and *5 minutes*), running one bag at each condition. She counts the number of uncooked kernels as the response variable.
   a) What are the null and alternative hypotheses for the main effects?
   b) How many degrees of freedom does each factor sum of squares have? How about the error sum of squares?
   c) Should she consider fitting an interaction term to the model? Why or why not?

8. **Gas mileage revisited** In Chapter 26 we saw a student's experiment to study the effect of *Tire Pressure* and *Acceleration* on gas mileage. He devises a system so that his Jeep Wagoneer uses gasoline from a one-liter container. He uses 3 levels of *Tire Pressure* (*low*, *medium*, and *full*) and 2 levels of *Acceleration*, either holding the pedal *steady* or *pumping* it every few seconds. He randomizes the trials, performing 4 runs under each treatment condition, carefully recording the number of miles he can go in his Jeep Wagoneer on one liter of gas.
   a) What are the null and alternative hypotheses for the main effects?
   b) How many degrees of freedom does each treatment sum of squares have? How about the error sum of squares?
   c) Should he consider fitting an interaction term to the model? Why might it be a good idea?
   d) If he fits an interaction term, how many degrees of freedom would it have?

9. **Popcorn again** Refer to the experiment in Exercise 7. After collecting her data and analyzing the results, the student reports that the F-ratio for *Power* is 13.56 and the F-ratio for *Time* is 9.36.
   a) What are the P-values?
   b) What would you conclude?
   c) What else about the data would you like to see in order to check the assumptions and conditions?

10. **Gas mileage again** Refer to the experiment in Exercise 8. After analyzing his data the student reports that the F-ratio for *Tire Pressure* is 4.29 with a P-value of 0.030, the F-ratio for *Acceleration* is 2.35 with a P-value of 0.143, and the F-ratio for the *Interaction effect* is 1.54 with a P-value of 0.241.
    a) What would you conclude?
    b) What else about the data would you like to see in order to check the assumptions and conditions?
    c) If your conclusion about the *Acceleration* factor in part a is wrong, what type of error have you made?

11. **Crash analysis** The National Highway Transportation Safety Administration runs crash tests in which stock automobiles are crashed into a wall at 35 mph with dummies in both the passenger and the driver's seats. The THOR Alpha crash dummy is capable of recording 134 channels of data on the impact of the crash at various

sites on the dummy. In this test 335 cars are crashed. The response variable is a measure of head injury. Researchers want to know if which seat the dummy is in affects head injury severity, as well as whether the type of car affects severity.

Here are partial boxplots for the 2 different *Seat* (*driver, passenger*) and the 6 different *Size* classifications (*compact, light, medium, mini, pickup, van*):

An interaction plot shows:

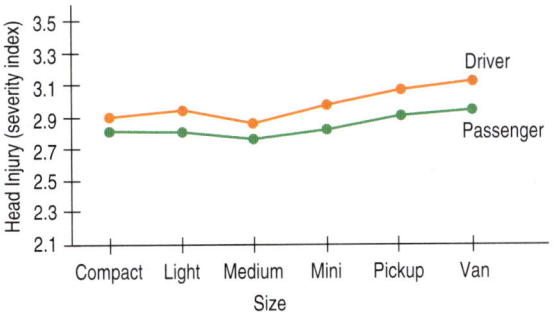

A scatterplot of residuals vs. predicted values shows:

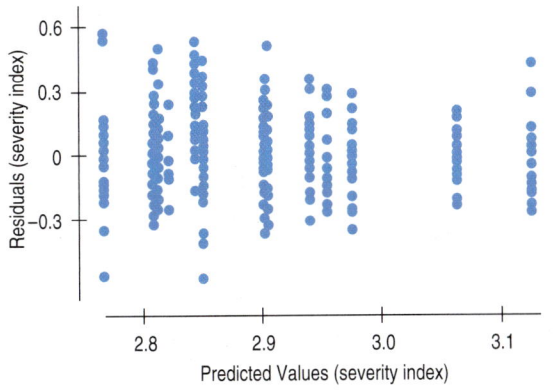

The ANOVA table follows:

Analysis of Variance for Head Severity: 293 cases

Source	DF	Sum of Squares	Mean Square	F-ratio	P-value
Seat	1	0.88713	0.88713	25.501	<0.0001
Size	5	1.49253	0.29851	8.581	<0.0001
Seat × Size	5	0.07224	0.01445	0.415	0.838
Error	282	9.8101	0.03479		
Total	293	12.3853			

a) State the hypotheses about the main effects (both numerically and in words).
b) Are the conditions for two-way ANOVA met?
c) If so, perform the hypothesis tests and state your conclusion. Be sure to state it in terms of head injury severity, seats, and vehicle types.

12. **Sprouts** An experiment on mung beans was performed to investigate the environmental effects of salinity and water temperature on sprouting. Forty beans were randomly allocated to each of 36 petri dishes that were subject to one of four levels of *Salinity* (0, 4, 8, and 12 *ppm*) and one of three *Temperatures* (32°, 34°, or 36° C). After 48 hours, the biomass of the sprouts was measured.

Here are partial boxplots of *Biomass* on *Salinity* and *Temperature*:

The interaction plot shows:

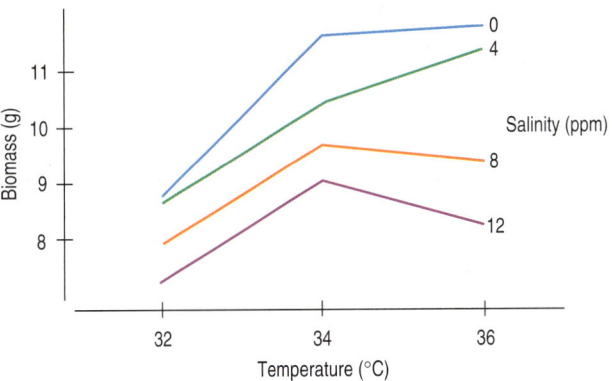

A two-way ANOVA model is fit, and the following ANOVA table results:

Analysis of Variance for Biomass (g)

Source	DF	Sum of Squares	Mean Square	F-ratio	P-value
Salinity	3	36.4701	12.1567	16.981	<0.0001
Temp	2	34.7168	17.3584	24.247	<0.0001
Salinity × Temp	6	5.2972	0.8829	1.233	0.3244
Error	24	17.1816	0.7159		
Total	35	93.6656			

The plot of residuals vs. predicted values shows:

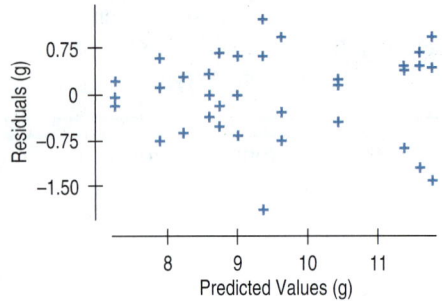

a) State the hypotheses about the factors (both numerically and in words).
b) Are the conditions for two-way ANOVA met?
c) Perform the hypothesis tests and state your conclusions. Be sure to state your conclusions in terms of biomass, salinity, and water temperature.

13. **Baldness and heart disease** A retrospective study examined the link between baldness and the incidence of heart disease. In the study, 1435 middle-aged men were selected at random and examined to see whether they showed signs of *Heart Disease* (or not) and what amount of *Baldness* they exhibited (*none, little, some,* or *much*).

Baldness	Heart Disease	Number of Men
None	Yes	251
None	No	331
Little	Yes	165
Little	No	221
Some	Yes	195
Some	No	185
Much	Yes	52
Much	No	35

A student runs a two-factor ANOVA on these data and finds the following ANOVA table:

Source	DF	Sum of Squares	Mean Square	F-Ratio	P-value
Baldness	3	62441.375	20813.792	17.956	0.020
Heart disease	1	1485.125	1485.125	1.281	0.340
Error	3	3477.375	1159.125		

a) Comment on her analysis. What problems, if any, do you find with the analysis?
b) What sort of analysis might you do instead?

14. **Fish and prostate** The Chapter 3 Step-By-Step looked at a Swedish study that asked 6272 men how much fish they ate and whether or not they had prostate cancer. Here are the data:

		Prostate Cancer?	
		No	Yes
Eat Fish?	Never/seldom	110	14
	Small part of diet	2420	201
	Moderate part	2769	209
	Large part	507	42

Armed with the methods of this chapter, a student performs a two-way ANOVA on the data. Here is her ANOVA table:

Source	DF	Sum of Squares	Mean Square	F-ratio	P-value
Fish	3	3110203.0	1036734.3	1.3599	0.4033
Prostate cancer	1	3564450.0	3564450.0	4.6756	0.1193
Error	3	2287051.0	762350.0		

a) Comment on her analysis. What problems, if any, do you find with the analysis?
b) What sort of analysis might you do instead?

15. **Baldness and heart disease again** Refer back to Exercise 13. Perform your own analysis of the data to see if baldness and heart disease are related. Do your conclusions support the claim that baldness is a cause of heart disease? Explain.

16. **Fish and prostate** Refer back to Exercise 14. Perform your own analysis of the data to see if eating fish and contracting prostate cancer are related.

T 17. **Basketball shots** A student performed an experiment to see if her favorite sneakers and the time of day might affect her free throw percentage. She tried shooting with and without her favorite sneakers and in the early morning and at night. For each treatment combination, she shot 50 baskets on 4 different occasions, recording the number of shots made each time. She randomized the treatment conditions by drawing numbers out of a hat. Here are her data:

Time of Day	Shoes	Shots Made
Morning	Others	25
Morning	Others	26
Night	Others	27
Night	Others	27
Morning	Favorite	32
Morning	Favorite	22
Night	Favorite	30
Night	Favorite	34
Morning	Others	35
Morning	Others	34
Night	Others	33
Night	Others	30
Morning	Favorite	33
Morning	Favorite	37
Night	Favorite	36
Night	Favorite	38

a) What are the null and alternative hypotheses?
b) Write a short report on your findings, being sure to include diagnostic analysis as well as practical conclusions.

18. **Washing** For his final project, Jonathan examined the effects of two factors on how well stains are removed when washing clothes. On each of 16 new white handkerchiefs, he spread a teaspoon of dirty motor oil (obtained from a local garage). He chose 4 *Temperature* settings (each of which is a combination of wash and rinse: *cold-cold, cold-warm, warm-hot,* and *hot-hot*) and 4 *Cycle* lengths (*short, med short, med long,* and *long*). After its washing, each handkerchief was dried in a dryer for 20 minutes and hung up. He rounded up 10 family members to judge the handkerchiefs for cleanliness on a scale of 1 to 10 and used the average score as his response. Here are the data:

Temp	Cycle	Score
Cold-cold	Med long	3.7
Warm-hot	Med long	6.5
Cold-warm	Med long	4.9
Hot-hot	Med long	6.5
Cold-cold	Long	4.6
Warm-hot	Long	8.3
Cold-warm	Long	4.7
Hot-hot	Long	9.1
Cold-cold	Short	3.4
Warm-hot	Short	5.6
Cold-warm	Short	3.8
Hot-hot	Short	7.1
Cold-cold	Med short	3.1
Warm-hot	Med short	6.3
Cold-warm	Med short	5
Hot-hot	Med short	6.1

You may assume, as Jonathan did, that interactions between *Temperature* and *Cycle* are negligible. Write a report showing what you found about washing factors and stain removal.

19. **Sprouts again** The students running the sprouts experiment (Exercise 12) also kept track of the number of beans sprouted (out of 40) for each of the 36 dishes. Here are the partial boxplots of *Sprouts* plotted against *Salinity* and *Temperature:*

An ANOVA table shows:

Source	DF	Sum of Squares	Mean Square	F-ratio	P-value
Salinity	3	2014.750	671.583	23.657	<0.0001
Temp	2	57.556	28.778	1.014	0.3779
Salinity × Temp	6	96.000	16.000	0.564	0.7549
Error	24	681.333	28.389		
Total	35	2849.639			

An interaction plot shows:

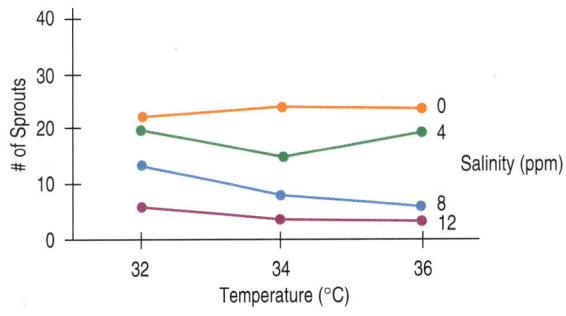

A plot of residuals vs. predicted values shows:

a) State the hypotheses about the factors (both numerically and in words).
b) Perform the hypothesis tests and state your conclusions. Be sure to check conditions.

T 20. Containers revisited Building on the cup experiment of the Chapter 4 Step-By-Step, a student selects one type of container and designs an experiment to see whether the type of *Liquid* stored and the outside *Environment* affect the ability of a cup to maintain temperature. He randomly chooses an experimental condition and runs each twice:

Liquid	Environment	Change in Temperature
Water	Room	13
Water	Room	14
Water	Outside	31
Water	Outside	31
Coffee	Room	11
Coffee	Room	11
Coffee	Outside	27
Coffee	Outside	29

After fitting a two-way ANOVA model, he obtains the following interaction plot, ANOVA table, and effects table:

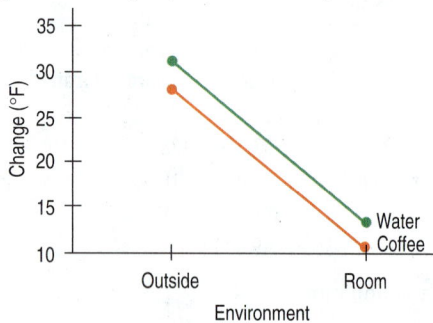

Source	DF	Sum of Squares	Mean Square	F-ratio	P-value
Liquid	1	15.125	15.125	24.2	0.0079
Environ	1	595.125	595.125	952.2	<0.0001
Interaction	1	0.125	0.125	0.2	0.6779
Error	4	2.500	0.625		
Total	7	612.875			

Term	Estimate
Overall mean	20.875
Liquid II [Coffee]	−1.375
Liquid II [Water]	1.375
Environment II [Outside]	8.625
Environment II [Room]	−8.625

a) State the null and alternative hypotheses.
b) Test the hypotheses at $\alpha = 0.05$.
c) Perform a residual analysis.
d) Summarize your findings.

21. Gas additives An experiment to test a new gasoline additive, *Gasplus*, was performed on three different cars: a sports car, a minivan, and a hybrid. Each car was tested with both *Gasplus* and regular gas on 10 different occasions and their gas mileage was recorded. Here are the partial boxplots:

A two-way ANOVA with interaction model was run, and the following ANOVA table resulted:

Source	DF	Sum of Squares	Mean Square	F-ratio	P-value
Type	2	23175.4	11587.7	2712.2	<0.0001
Additive	1	92.1568	92.1568	21.57	<0.0001
Type × Additive	2	51.8976	25.9488	6.0736	0.0042
Error	54	230.711	4.27242		
Total	59	23550.2			

A plot of the residuals vs. predicted values showed:

What conclusions about the additive and car types do you draw? Do you see any potential problems with the analysis?

22. Chromatography A gas chromatograph is an instrument that measures the amounts of various compounds contained in a sample by separating the various constituents. Because different components are flushed through the system at different rates, chromatographers are able to both measure and distinguish the various constituents of the sample. A counter is placed somewhere along the instrument that records how much material is passing at various times. By looking at the counts at various times, the chemist is able to reconstruct the amounts of various compounds present. The total number of counts is proportional to the amount of the compound present.

An experiment was performed to see whether slowing down the flow rate would increase total counts. A mixture was produced with three different *Concentration* levels: *low*, *medium*, and *high*. The two *Flow Rates* used were *slow* and *fast*. Each mixture was run 5 times and the total counts recorded each time. Partial boxplots for *Concentration* and *Flow Rate* show:

A two-way ANOVA with interaction model was run, and the following ANOVA table resulted:

Source	DF	Sum of Squares	Mean Square	F-ratio	P-value
Conc	2	483655E5	241828E5	1969.44	<0.0001
Flow rate	1	364008E3	364008E3	29.65	<0.0001
Interaction	2	203032E3	101516E3	8.27	0.0019
Error	24	294698E3	122791E3		
Total	29	492272E5			

A plot of residuals vs. predicted values showed:

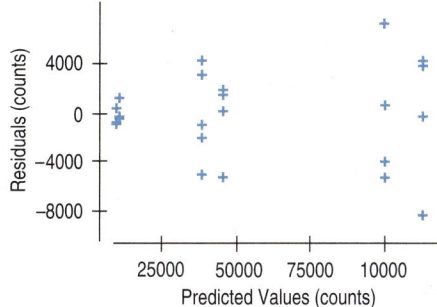

What conclusions about the effect of flow rate do you draw? Do you see any potential problems with the analysis?

23. Gas additives again Refer back to the experiment in Exercise 21. Instead of *mpg* redo the analysis using log(*mpg*) as the response. Do your conclusions change? How? Are the assumptions of the model better satisfied?

24. Chromatography again Refer back to the experiment in Exercise 22. Instead of *Total counts*, redo the analysis using log(*Total counts*) as the response. Do your conclusions change? How? Are the assumptions of the model better satisfied?

25. Batteries A student experiment was run to test the performance of 4 brands of batteries under 2 different *Environments* (*room temperature* and *cold*). For each of the 8 treatments, 2 batteries of a particular brand were put into a flashlight. The flashlight was then turned on and allowed to run until the light went out. The number of minutes the flashlight stayed on was recorded. Each treatment condition was run twice.

Partial boxplots showed:

An interaction plot showed:

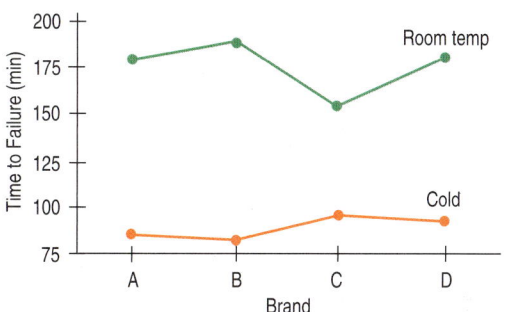

An ANOVA table showed:

Source	DF	Sum of Squares	Mean Square	F-ratio	P-value
Envir	1	30363.1	30363.1	789.93	<0.0001
Brand	3	338.187	112.729	2.9328	0.0994
Interaction	3	1278.19	426.063	11.085	0.0032
Error	8	307.5	38.4375		
Total	15	32286.9			

a) What are the main effect null and alternative hypotheses?
b) From the partial boxplots, do you think that the *Brand* has an effect on the time the batteries last? How about the condition?
c) Do the conclusions of the ANOVA table match your intuition?

d) What does the interaction plot say about the performance of the brands?
e) Why might you be uncomfortable with a recommendation to go with the cheapest battery (brand C)?

26. **Peas** In an experiment on growing sweet peas, a team of students selected 2 factors at 4 levels each and recorded *Weight, Stem Length,* and *Root Length* after $6\frac{1}{2}$ days of growth. They grew plants using various amounts of *Water* and *Quickgrow* solution, a fertilizer designed to help plants grow faster. Each factor was run at 4 levels: little, some, moderate, and full. They grew 2 plants under each of the 16 conditions.

An interaction plot of *Weight* in mg (*x*-axis is *Quickgrow*—levels are water) shows:

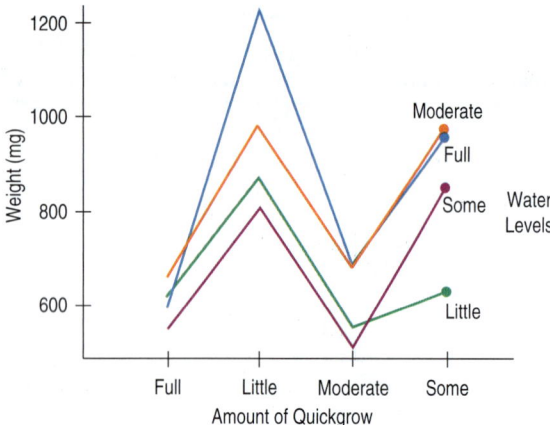

Because of this, a two-way ANOVA with interaction model was fit to the data, resulting in the following ANOVA table:

Analysis of Variance for Weight

Source	DF	Sum of Squares	Mean Square	F-ratio	P-value
Water	3	246255	82084.8	2.5195	0.0948
QS	3	827552	275851	8.4669	0.0013
Water × QS	8	176441	22055.1	0.6770	0.7130
Error	16	521275	32579.7		
Total	31	1771523			

Residuals plots showed no violations of the variance or Normality conditions. A table of effects for *Quickgrow* shows:

Level of Quickgrow	Effect
Little	213.8
Some	97.1
Moderate	−155.5
Full	−155.5

A table of effects for *Water* shows:

Level of Water	Effect
Little	−91.1
Some	−81.9
Moderate	66.4
Full	106.0

Write a report summarizing what the students have learned about the effects of *Water* and *Quickgrow* solution on the early stages of sweet pea growth as measured by *Weight*.

27. **Batteries once more** Another student analyzed the battery data from Exercise 25, using a one-way ANOVA. He considered the experimental factor to be an 8-level factor consisting of the 8 possible combinations of *Brand* and *Environment*. Here are the boxplots for the 8 treatments and a one-way ANOVA:

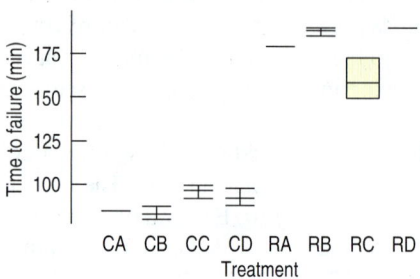

Source	DF	Sum of Squares	Mean Square	F-ratio	P-value
Treatment	7	31895.8	4556.54	93.189	<0.0001
Error	8	391.167	48.8958		
Total	15	32286.9			

Compare this analysis with the one performed in Exercise 25. Which one provides a better understanding of the data? Explain.

28. **Containers one more time** Another student performs a one-way ANOVA on the container data of Exercise 20, using the 4 treatments *water room, water outside, coffee room,* and *coffee outside.* Perform this analysis and comment on the differences between this analysis and the one in Exercise 20.

Just Checking ANSWERS

1. The null hypotheses are that the mean flight distance for all three *Weights* is the same, that the mean flight distance for both *Positions* is the same, and that the interaction effects are all 0.

2. Yes, the effects appear to be additive. The lines are nearly parallel.

3. Yes, the *F*-test provides no evidence to reject the null hypothesis with a P-value of >0.9.

4. Yes, we reject the null hypothesis with a P-value <0.0001.

5. Yes, we reject the null hypothesis with a P-value <0.0001.

6. Because both factors are significant, it appears that using the light *Weight* in the rear *Position* may result in the longest mean flight distance.

chapter 28 Multiple Regression

28.1 What Is Multiple Regression?
28.2 Interpreting Multiple Regression Coefficients
28.3 The Multiple Regression Model—Assumptions and Conditions
28.4 Multiple Regression Inference
28.5 Comparing Multiple Regression Models

Where are we going?
We've seen that the top wind speed in a hurricane depends on the central barometric pressure. But what about the sea surface temperature? Can we include other variables in our model? Linear models are often useful, but the world is usually not so simple that a two-variable model does the trick. For a more realistic understanding, we need models with several variables.

In Chapter 25 we tried to predict the percent body fat of male subjects from their waist size, and we did pretty well. The R^2 of 67.8% says that we accounted for almost 68% of the variability in *%Body Fat* by knowing only the *Waist* size. We completed the analysis by performing hypothesis tests on the coefficients and looking at the residuals.

But that remaining 32% of the variance has been bugging us. Couldn't we do a better job of accounting for *%Body Fat* if we weren't limited to a single predictor? In the full data set there were 15 other measurements on the 250 men. We might be able to use other predictor variables to help us account for the leftover variation that wasn't accounted for by waist size.

What about *Height*? Does *Height* help to predict *%Body Fat*? Men with the same *Waist* size can vary from short and corpulent to tall and emaciated. Knowing a man has a 50-inch waist suggests that he's likely to carry a lot of body fat. If we found out that he was 7 feet tall, that might change our impression of his body type. Knowing his *Height* as well as his *Waist* size might help us to make a more accurate prediction.

28.1 What Is Multiple Regression?

Who	250 Male subjects
What	Body fat and waist size
Units	%Body fat and inches
When	1990s
Where	United States
Why	Scientific research

Does a regression with *two* predictors even make sense? It does—and that's fortunate because the world is too complex a place for linear regression to model it with a single predictor. A regression with two or more predictor variables is called a **multiple regression**. (When we need to note the difference, a regression on a single predictor is called a *simple regression*.) We'd never try to find a regression by hand, and even calculators aren't really up to the task. This is a job for a statistics program on a computer. If you know how to find the regression of *%Body Fat* on *Waist* size with a statistics package, you can usually just add *Height* to the list of predictors without having to think hard about how to do it. (Data in **Body fat complete**)

817

A NOTE ON TERMINOLOGY
When we have two or more predictors and fit a linear model by least squares, we are formally said to fit a least squares linear multiple regression. Most folks just call it "multiple regression." You may also see the abbreviation OLS used with this kind of analysis. It stands for "Ordinary Least Squares."

For simple regression, we found the **Least Squares** solution, the one whose coefficients made the sum of the squared residuals as small as possible. For multiple regression, we'll do the same thing but this time with more coefficients. Remarkably enough, we can still solve this problem. Even better, a statistics package can find the coefficients of the least squares model easily.

Here's a typical example of a multiple regression table:

Dependent variable is: %Body Fat
R-squared = 71.3%   R-squared (adjusted) = 71.1%
s = 4.460 with 250 − 3 = 247 degrees of freedom

Variable	Coefficient	SE(Coeff)	t-ratio	P-value
Intercept	−3.10088	7.686	−0.403	0.6870
Waist	1.77309	0.0716	24.8	≤0.0001
Height	−0.60154	0.1099	−5.47	≤0.0001

You should recognize most of the numbers in this table. Most of them mean what you expect them to.

R^2 gives the fraction of the variability of *%Body Fat* accounted for by the *multiple* regression model. (With *Waist* alone predicting *%Body Fat*, the R^2 was 67.8%.) The multiple regression model accounts for 71.3% of the variability in *%Body Fat*. We shouldn't be surprised that R^2 has gone up. It was the hope of accounting for some of the leftover variability that led us to try a second predictor.

The standard deviation of the residuals is still denoted s (or sometimes s_e to distinguish it from the standard deviation of y).

The degrees of freedom calculation follows our rule of thumb: The degrees of freedom is the number of observations (250) minus 1 for each coefficient estimated—for this model, 3.

For each predictor we have a coefficient, its standard error, a *t*-ratio, and the corresponding P-value. As with simple regression, the *t*-ratio measures how many standard errors the coefficient is away from 0. So, we can find a P-value from a Student's *t*-model to test the null hypothesis that the true value of the coefficient is 0.

Using the coefficients from this table, we can write the regression model:

$$\widehat{\%Body\ Fat} = -3.10 + 1.77\ Waist - 0.60\ Height.$$

As before, we define the residuals as

$$Residuals = \%Body\ Fat - \widehat{\%Body\ Fat}.$$

We've fit this model with the same least squares principle: The sum of the squared residuals is as small as possible for any choice of coefficients.

So what's different? With so much of the multiple regression looking just like simple regression, why devote an entire chapter (or two) to the subject?

There are several answers to this question. First—and most important—the *meaning* of the coefficients in the regression model has changed in a subtle but important way. Because that change is not obvious, multiple regression coefficients are often misinterpreted. This chapter will show some examples to help make the meaning clear.

Second, multiple regression is an extraordinarily versatile calculation, underlying many widely used Statistics methods. A sound understanding of the multiple regression model will help you to understand these other applications.

Third, multiple regression offers our first glimpse into statistical models that use more than two quantitative variables. The real world is complex. Simple models of the kind we've seen so far are a great start, but often they're just not detailed enough to be useful for understanding, predicting, and decision making. Models that use several variables can be a big step toward realistic and useful modeling of complex phenomena and relationships.

For Example REAL ESTATE

As a class project, students in a large Statistics class collected publicly available information on recent home sales in their hometowns. There are 894 properties. These are not a random sample, but they may be representative of home sales during a short period of time, nationwide. (Data in **Real estate project**)

Variables available include the price paid, the size of the living area (sq ft), the number of bedrooms, the number of bathrooms, the year of construction, the lot size (acres), and a coding of the location as urban, suburban, or rural made by the student who collected the data.

Here's a regression to model the sale price from the living area (sq ft) and the number of bedrooms.

Dependent variable is: Price
R squared = 14.6%   R squared (adjusted) = 14.4%
s = 266899 with 894 − 3 = 891 degrees of freedom

Variable	Coefficient	SE(Coeff)	t-ratio	P-value
Intercept	308100	41148	7.49	<0.0001
Living area	135.089	11.48	11.8	<0.0001
bedrooms	−43346.8	12844	−3.37	0.0008

QUESTION: How should we interpret the regression output?

The model is

$$\widehat{Price} = 308{,}100 + 135\, Living\ Area - 43{,}346\, Bedrooms$$

The R-squared says that this model accounts for 14.6% of the variation in *Price*. But the value of s leads us to doubt that this model would provide very good predictions because the standard deviation of the residuals is more than $266,000. Nevertheless, we may be able to learn about home prices because the P-values of the coefficients are all very small, so we can be quite confident that none of them is really zero.

28.2 Interpreting Multiple Regression Coefficients

We said that height might be important in predicting body fat in men. What's the relationship between *%Body Fat* and *Height* in men? We know how to approach this question; we follow the three rules. Here's the scatterplot:

Figure 28.1

The scatterplot of *%Body Fat* against *Height* seems to say that there is little relationship between these variables.

It doesn't look like *Height* tells us much about *%Body Fat*. You just can't tell much about a man's *%Body Fat* from his *Height*. Or can you? Remember, in the multiple regression model, the coefficient of *Height* was −0.60, had a *t*-ratio of −5.47, and had a very small P-value. So it *did* contribute to the *multiple* regression model. How could that be?

The answer is that the multiple regression coefficient of *Height* takes account of the other predictor, *Waist* size, in the regression model.

To understand the difference, let's think about all men whose waist size is about 37 inches—right in the middle of our sample. If we think only about *these* men, what do we expect the relationship between *Height* and *%Body Fat* to be? Now a negative association makes sense because taller men probably have less body fat than shorter men *who have the same waist size*. Let's look at the plot:

Figure 28.2

When we restrict our attention to men with waist sizes between 36 and 38 inches (points in blue), we can see a relationship between *%Body Fat* and *Height*.

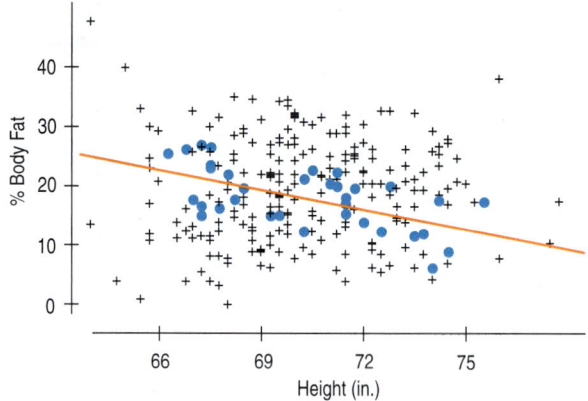

Here we've highlighted the men with waist sizes between 36 and 38 inches. Overall, there's little relationship between *%Body Fat* and *Height,* as we can see from the full set of points. But when we focus on *particular* waist sizes, there *is* a relationship between body fat and height. This relationship is *conditional* because we've restricted our set to only those men within a certain range of waist size. For men with that waist size, an extra inch of height is associated with a decrease of about 0.60% in body fat. If that relationship is consistent for each *Waist* size, then the multiple regression coefficient will estimate it. The simple regression coefficient simply couldn't see it.

> As their name reminds us, residuals are what's left over after we fit a model. That lets us remove the effects of some variables. The residuals are what's left.

We've picked one particular *Waist* size to highlight. How could we look at the relationship between *%Body Fat* and height conditioned on *all waist* sizes *at the same time*? Once again, residuals come to the rescue.

We plot the residuals of *%Body Fat* after a regression on *Waist* size against the residuals of *Height* after regressing *it* on *Waist* size. This display is called a *partial regression plot*. It shows us just what we asked for: the relationship of *%Body Fat* to *Height* after removing the linear effects of *Waist* size.

Figure 28.3

A partial regression plot for the coefficient of *Height* in the regression model has a slope equal to the coefficient value in the multiple regression model.

A **partial regression plot** for a particular predictor has a slope that is the same as the *multiple* regression coefficient for that predictor. Here, it's −0.60. It also has the same residuals as the full multiple regression, so you can spot any outliers or influential points and tell whether they've affected the estimation of this particular coefficient.

Many modern statistics packages offer partial regression plots as an option for any coefficient of a multiple regression. For the same reasons that we always look at a scatterplot before interpreting a simple regression coefficient, it's a good idea to make a partial regression plot for any multiple regression coefficient that you hope to understand or interpret.

28.3 The Multiple Regression Model—Assumptions and Conditions

We can write a multiple regression model like this, numbering the predictors arbitrarily (we don't care which one is x_1), writing β's for the model coefficients (which we will estimate from the data), and including the errors in the model:

$$y = \beta_0 + \beta_1 x_1 + \beta_2 x_2 + \varepsilon.$$

Of course, the multiple regression model is not limited to two predictor variables, and regression model equations are often written to indicate summing any number (a typical letter to use is k) of predictors. That doesn't really change anything, so we'll often stick with the two-predictor version just for simplicity. But don't forget that we can have many predictors.

The assumptions and conditions for the multiple regression model sound nearly the same as for simple regression, but with more variables in the model, we'll have to make a few changes.

Linearity Assumption

We are fitting a linear model.[1] For that to be the right kind of model, we need an underlying linear relationship. But now we're thinking about several predictors. To see whether the assumption is reasonable, we'll check the Straight Enough Condition for *each* of the predictors.

Straight Enough Condition: Scatterplots of y against each of the predictors are reasonably straight. As we have seen with *Height* in the body fat example, the scatterplots need not show a strong (or any!) slope; we just check that there isn't a bend or other nonlinearity. For the body fat data, the scatterplot is beautifully linear in *Waist* as we saw in Chapter 25. For *Height*, we saw no relationship at all, but at least there was no bend.

As we did in simple regression, it's a good idea to check the residuals for linearity after we fit the model. It's good practice to plot the residuals against the predicted values and check for patterns, especially bends or other nonlinearities. (We'll watch for other things in this plot as well.)

If we're willing to assume that the multiple regression model is reasonable, we can fit the regression model by least squares. But we must check the other assumptions and conditions before we can interpret the model or test any hypotheses.

> **CHECK THE RESIDUAL PLOT (PART 1)**
> The residuals should appear to have no pattern with respect to the predicted values.

Independence Assumption

As with simple regression, the errors in the true underlying regression model must be independent of each other. As usual, there's no way to be sure that the Independence Assumption is true. Fortunately, even though there can be many predictor variables, there

---

[1] By *linear* we mean that each x appears simply multiplied by its coefficient and added to the model. No x appears in an exponent or some other more complicated function. That means that as we move along any x-variable, our prediction for y will change at a constant rate (given by the coefficient) if nothing else changes.

is only one response variable and only one set of errors. The Independence Assumption concerns the errors, so you should check the corresponding conditions on the residuals.

Randomization Condition: The data should arise from a random sample or randomized experiment. Randomization assures us that the data are representative of some identifiable population. If you can't identify the population, you can interpret the regression model only as a description of the data you have, and you can't interpret the hypothesis tests at all because they are about a regression model for that population. Regression methods are often applied to data that were not collected with randomization. Regression models fit to such data may still do a good job of modeling the data at hand, but without some reason to believe that the data are representative of a particular population, you should be reluctant to believe that the model generalizes to other situations.

> **CHECK THE RESIDUAL PLOT (PART 2)**
> The residuals should appear to be randomly scattered and show no patterns or clumps when plotted against the predicted values.

You should also check displays of the regression residuals for evidence of patterns, trends, or clumping, any of which would suggest a failure of independence. In the special case when one of the *x*-variables is related to time, be sure that the residuals do not have a pattern when plotted against that variable or against *Time*.

The body fat data were collected on a sample of men. The men were not related in any way, so we can be pretty sure that their measurements are independent.

Equal Variance Assumption

The variability of the errors should be about the same for all values of *each* predictor. To see if this is reasonable, we look at scatterplots.

Does the Plot Thicken? Condition: Scatterplots of the regression residuals against each *x* or against the predicted values, $\hat{y}$, offer a visual check. The spread around the line should be nearly constant. Be alert for a "fan" shape or other tendency for the variability to grow or shrink in one part of the scatterplot.

> **CHECK THE RESIDUAL PLOT (PART 3)**
> The spread of the residuals should be uniform when plotted against any of the *x*'s or against the predicted values.

Here are the residuals plotted against *Waist* and *Height*. Neither plot shows patterns that might indicate a problem.

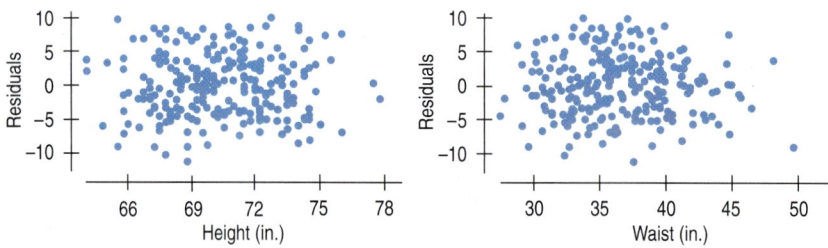

Figure 28.4
Residuals plotted against each predictor show no pattern. That's a good indication that the Straight Enough Condition and the Does the Plot Thicken? Condition are satisfied.

If residual plots show no pattern, if the data are plausibly independent, and if the plots don't thicken, we can feel good about interpreting the regression model. Before we test hypotheses, however, we must check one final assumption.

Normality Assumption

We assume that the errors around the idealized regression model at any specified values of the *x*-variables follow a Normal model. We need this assumption so that we can use a Student's *t*-model for inference. As with other times when we've used Student's *t*, we'll settle for the residuals satisfying the Nearly Normal Condition.

Nearly Normal Condition: Because we have only one set of residuals, this is the same set of conditions we had for simple regression. Look at a histogram or Normal probability plot of the residuals. The histogram of residuals in the body fat regression certainly looks Nearly Normal, and the Normal probability plot is fairly straight. And, as we have said before, the Normality Assumption becomes less important as the sample size grows.

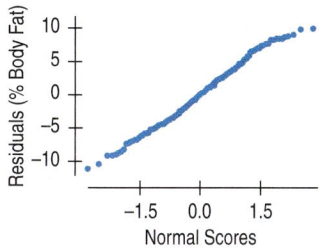

Figure 28.5
Check a histogram of the residuals. The distribution of the residuals should be unimodal and symmetric. Or check a Normal probability plot to see whether it is straight.

Let's summarize all the checks of conditions that we've made and the order that we've made them:

1. Check the Straight Enough Condition with scatterplots of the *y*-variable against each *x*-variable.
2. If the scatterplots are straight enough (that is, if it looks like the regression model is plausible), fit a multiple regression model to the data. (Otherwise, either stop or consider re-expressing an *x*- or the *y*-variable.)
3. Find the residuals and predicted values.
4. Make a scatterplot of the residuals against the predicted values.[2] This plot should look patternless. Check in particular for any bend (which would suggest that the data weren't all that straight after all) and for any thickening. If there's a bend and especially if the plot thickens, consider re-expressing the *y*-variable and starting over.
5. Think about how the data were collected. Was suitable randomization used? Are the data representative of some identifiable population? If the data are measured over time, check for evidence of patterns that might suggest they're not independent by plotting the residuals against time to look for patterns.
6. If the conditions check out this far, feel free to interpret the regression model and use it for prediction. If you want to investigate a particular coefficient, make a partial regression plot for that coefficient.
7. If you wish to test hypotheses about the coefficients or about the overall regression, then make a histogram and Normal probability plot of the residuals to check the Nearly Normal Condition.

Step-By-Step Example MULTIPLE REGRESSION

Question: How should we model %*Body Fat* in terms of *Height* and *Waist* size?

THINK ▶ Variables Name the variables, report the W's, and specify the questions of interest.

I have quantitative body measurements on 250 adult males from the BYU Human Performance Research Center. I want to understand the relationship between %*Body Fat*, *Height*, and *Waist* size.

Plan Think about the assumptions and check the conditions.

✓ **Straight Enough Condition:** There is no obvious bend in the scatterplots of %*Body Fat* against either *x*-variable. The scatterplot of residuals against predicted values below shows no patterns that would suggest nonlinearity.

---

[2] In Chapter 25 we noted that a scatterplot of residuals against the predicted values looked just like the plot of residuals against *x*. But for a multiple regression, there are several *x*'s. Now the predicted values, $\hat{y}$, are a combination of the *x*'s—in fact, they're the combination given by the regression equation we have computed. So they combine the effects of all the *x*'s in a way that makes sense for our particular regression model. That makes them a good choice to plot against.

✓ **Independence Assumption:** These data are not collected over time, and there's no reason to think that the *%Body Fat* of one man influences that of another. I don't know whether the men measured were sampled randomly, but the data are presented as being representative of the male population of the United States.

Now we can find the regression and examine the residuals.

✓ **Does the Plot Thicken? Condition:** The scatterplot of residuals against predicted values shows no obvious changes in the spread about the line.

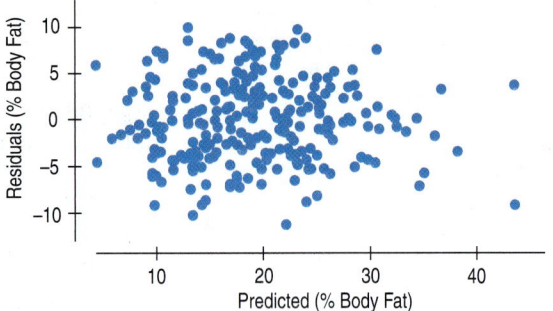

Actually, we need the Nearly Normal Condition only if we want to do inference.

✓ **Nearly Normal Condition, Outlier Condition:** A histogram of the residuals is unimodal and symmetric.

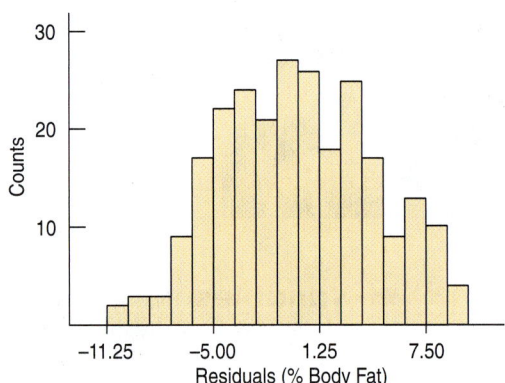

The Normal probability plot of the residuals is reasonably straight:

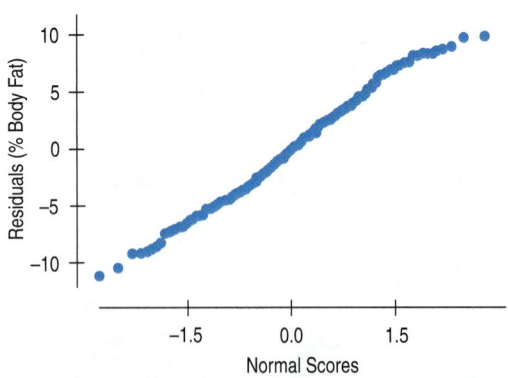

Choose your method.	Under these conditions a full multiple regression analysis is appropriate.
SHOW → Mechanics	Here is the computer output for the regression: Dependent variable is: %Body Fat R-squared = 71.3% R-squared (adjusted) = 71.1% s = 4.460 with 250 − 3 = 247 degrees of freedom 	Source	Sum of Squares	DF	Mean Square	F-ratio	P-value
---	---	---	---	---	---
Regression	12216.6	2	6108.28	307	<0.0001
Residual	4912.26	247	19.8877		| | Variable	Coefficient	SE(Coeff)	t-ratio	P-value
---	---	---	---	---
Intercept	−3.10088	7.686	−0.403	0.6870
Waist	1.77309	0.0716	24.8	<0.0001
Height	−0.60154	0.1099	−5.47	<0.0001	 The estimated regression equation is $$\widehat{\%Body\,Fat} = -3.10 + 1.77\,Waist - 0.60\,Height.$$
TELL → Interpretation	The R^2 for the regression is 71.3%. *Waist* size and *Height* together account for about 71% of the variation in *%Body Fat* among men. The regression equation indicates that each inch in *Waist* size is associated with about a 1.77 increase in *%Body Fat* among men who are of a particular *Height*. Each inch of *Height* is associated with a decrease in *%Body Fat* of about 0.60 among men with a particular *Waist* size.
More Interpretation	The standard errors for the slopes of 0.07 (*Waist*) and 0.11 (*Height*) are both small compared with the slopes themselves, so it looks like the coefficient estimates are fairly precise. The residuals have a standard deviation of 4.46%, which gives an indication of how precisely we can predict *%Body Fat* with this model.

28.4 Multiple Regression Inference

There are several hypothesis tests in the multiple regression output, but all of them talk about the same thing. Each is concerned with whether the underlying model parameters are actually zero.

The first of these hypotheses is one we skipped over for simple regression (for reasons that will be clear in a minute). Now that we've looked at ANOVA (in Chapter 26),[3] we can recognize the ANOVA table sitting in the middle of the regression output. Where'd that come from?

---

[3] If you skipped over Chapter 26, you can just take our word for this and read on.

The answer is that now that we have more than one predictor, there's an overall test we should consider before we do more inference on the coefficients. We ask the global question "Is this multiple regression model any good at all?" That is, would we do as well using just $\bar{y}$ to model y? What would that mean in terms of the regression? Well, if all the coefficients (except the intercept) were zero, we'd have

$$\hat{y} = b_0 + 0x_1 + \cdots + 0x_k$$

and we'd just set $b_0 = \bar{y}$.

To address the overall question, we'll test

$$H_0: \beta_1 = \beta_2 = \cdots = \beta_k = 0.$$

(That null hypothesis looks very much like the null hypothesis we tested with an F-test in the Analysis of Variance in Chapter 26.)

We can test this hypothesis with a statistic that is labeled with the letter F (in honor of Sir Ronald Fisher, the developer of Analysis of Variance). In our example, the F-value is 307 on 2 and 247 degrees of freedom. The alternative hypothesis is just that the slope coefficients aren't all equal to zero, and the test is one-sided—bigger F-values mean smaller P-values. If the null hypothesis were true, the F-statistic would be near 1. The F-statistic here is quite large, so we can easily reject the null hypothesis and conclude that the multiple regression model is better than just using the mean.[4]

Why didn't we do this for simple regression? Because the null hypothesis would have just been that the lone model slope coefficient was zero, and we were already testing that with the t-statistic for the slope. In fact, the *square* of that t-statistic is equal to the F-statistic for the simple regression, so it really was the identical test.

Once we check the F-test and reject the null hypothesis—and, if we are being careful, *only* if we reject that hypothesis—we can move on to checking the test statistics for the individual coefficients. Those tests look like what we did for the slope of a simple regression in Chapter 25. For each coefficient, we test

$$H_0: \beta_j = 0$$

against the (two-sided) alternative that it isn't zero. The regression table gives a standard error for each coefficient and the ratio of the estimated coefficient to its standard error. If the assumptions and conditions are met (and now we need the Nearly Normal Condition), these ratios follow a Student's t-distribution.

$$t_{n-k-1} = \frac{b_j - 0}{SE(b_j)}$$

How many degrees of freedom? We have a rule of thumb and it works here. The degrees of freedom is the number of data values minus the number of predictors (counting the intercept term). For our regression on two predictors, that's $n - 3$. You shouldn't have to look up the t-values. Almost every regression report includes the corresponding P-values.

We can build a confidence interval in the usual way, as an estimate ± a margin of error. As always, the margin of error is just the product of the standard error and a critical value. Here the critical value comes from the t-distribution on $n - k - 1$ degrees of freedom. So a confidence interval for β_j is

$$b_j \pm t^*_{n-k-1} SE(b_j).$$

The tricky parts of these tests are that the standard errors of the coefficients now require harder calculations (so we leave it to the technology) and the meaning of a coefficient, as we have seen, depends on all the *other* predictors in the multiple regression model.

---

[4]There are F-tables in Table F at the end of the book, and they work pretty much as you'd expect. Most regression tables include a P-value for the F-statistic, but there's almost never a need to perform this particular test in a multiple regression. Usually we just glance at the F-statistic to see that it's reasonably far from 1.0, the value it would have if the true coefficients were really all zero.

That last bit is important. If we fail to reject the null hypothesis for a multiple regression coefficient, it does **not** mean that the corresponding predictor variable has no linear relationship to *y*. It means that the corresponding predictor contributes nothing to modeling *y after allowing for all the other predictors.*

Be Careful Interpreting Coefficients

This last point bears repeating. The multiple regression model looks so simple and straightforward:

$$y = \beta_0 + \beta_1 x_1 + \cdots + \beta_k x_k + \varepsilon.$$

It *looks* like each β_j tells us the effect of its associated predictor, x_j, on the response variable, *y*. But that is not so. This is, without a doubt, the most common error that people make with multiple regression:

- It is possible for there to be no simple relationship between *y* and x_j, and yet β_j in a *multiple* regression can be significantly different from 0. We saw this happen for the coefficient of *Height* in our example.
- It is also possible for there to be a strong two-variable relationship between *y* and x_j, and yet β_j in a multiple regression can be almost 0 with a large P-value so that we cannot reject the null hypothesis that the true coefficient is zero. If we're trying to model the horsepower of a car, using both its weight and its engine size, it may turn out that the coefficient for *Engine Size* is nearly 0. That *doesn't* mean that engine size isn't important for understanding horsepower. It simply means that after allowing for the weight of the car, the engine size doesn't give much *additional* information.
- It is even possible for there to be a significant linear relationship between *y* and x_j in one direction, and yet β_j can be of the *opposite* sign and strongly significant in a multiple regression. More expensive cars tend to be bigger, and since bigger cars have worse fuel efficiency, the price of a car has a slightly negative association with fuel efficiency. But in a multiple regression of *Fuel Efficiency* on *Weight* and *Price,* the coefficient of *Price* may be positive. If so, it means that *among cars of the same weight,* more expensive cars have better fuel efficiency. The simple regression on *Price,* though, has the opposite direction because, *overall,* more expensive cars are bigger. This switch in sign may seem a little strange at first, but it's not really a contradiction at all. It's due to the change in the *meaning* of the coefficient of *Price* when it is in a multiple regression rather than a simple regression.

So we'll say it once more: The coefficient of x_j in a multiple regression depends as much on the *other* predictors as it does on x_j. Remember that when you interpret a multiple regression model.

For Example INTERPRETING COEFFICIENTS

We looked at a multiple regression to predict the price of a house from its living area and the number of bedrooms. We found the model

$$\widehat{Price} = 308,100 + 135\ Living\ Area - 43,347\ Bedrooms.$$

However, common sense says that houses with more bedrooms are usually worth more. And, in fact, the simple regression of *Price* on *Bedrooms* finds the model

$$\widehat{Price} = 338,975 + 40,234\ Bedrooms$$

and the P-value for the slope coefficient is 0.0005.

QUESTION: How should we understand the coefficient of *Bedrooms* in the multiple regression?

The coefficient of *Bedrooms* in the multiple regression does not mean that houses with more bedrooms are generally worth less. It must be interpreted taking account of the other predictor (*Living area*) in the regression. If we consider houses with a given amount of living area, those that devote more of that area to bedrooms either must have smaller bedrooms or less living area for other parts of the house. Those differences could result in reducing the home's value.

Just Checking

Recall the regression example in Chapter 7 to predict hurricane maximum wind speed from central barometric pressure. Another researcher, interested in the possibility that global warming was causing hurricanes to become stronger, added the variable Year as a predictor and obtained the following regression:

Dependent variable is: Max. Winds (kn)
R squared = 80.33%   R squared (adjusted) = 80.14%
s = 8.13 with 210 − 3 = 207 degrees of freedom

Source	Sum of Squares	df	Mean Square	F-ratio
Regression	55867.0	2	27933.5	422.65
Residual	13681.1	207	66.1	

Variable	Coefficient	SE(Coeff)	t-ratio	P-value
Intercept	1035.34	39.74	26.05	<0.0001
Central Pressure	−0.97050	0.003372	−28.78	<0.0001
Year	−0.00438	0.009375	−0.47	0.6406

1. Interpret the R^2 of this regression.
2. Interpret the coefficient of *Central Pressure*.
3. The researcher concluded that "There has been no change over time in the strength of Atlantic hurricanes." Is this conclusion a sound interpretation of the regression model?

Another Example: Modeling Infant Mortality

Infant Mortality is often used as a general measure of the quality of health care for children and mothers. It is reported as the rate of deaths of newborns per 1000 live births. Data recorded for each of the 50 states of the United States may allow us to build regression models to help understand or predict infant mortality. The variables available for our model are *Child Deaths* (deaths per 100,000 children aged 1–14), percent of teens (ages 16–19) who drop out of high school (*HS Drop%*), percent of low-birth-weight babies (*Low BW%*), *Teen Births* (births per 100,000 females aged 15–17), and *Teen Deaths* by accident, homicide, and suicide (deaths per 100,000 teens ages 15–19).[5] (Data in **Infant mortality**)

All of these variables were displayed and found to have no outliers and Nearly Normal distributions.[6] One useful way to check many of our conditions is with a **scatterplot matrix**. Figure 28.6 shows an array of scatterplots set up so that the plots in each row have the same variable on their y-axis and those in each column have the same variable on their

Who U.S. states
What Various measures relating to children and teens
When 1999
Why Research and policy

---

[5]The data are available from the Kids Count section of the Annie E. Casey Foundation (http://datacenter.kidscount.org/), and are all for 1999.
[6]In the interest of complete honesty, we should point out that the original data include the District of Columbia, but it proved to be an outlier on several of the variables, so we've restricted attention to the 50 states here.

Figure 28.6
A scatterplot matrix shows a scatterplot of each pair of variables arrayed so that the vertical and horizontal axes are consistent across rows and down columns. You can tell which variable is plotted on the x-axis of any plot by reading down to the diagonal and looking to the left. The diagonal cells may hold Normal probability plots (as they do here), histograms, or just the names of the variables. These are a great way to check the Straight Enough Condition and to check for simple outliers.

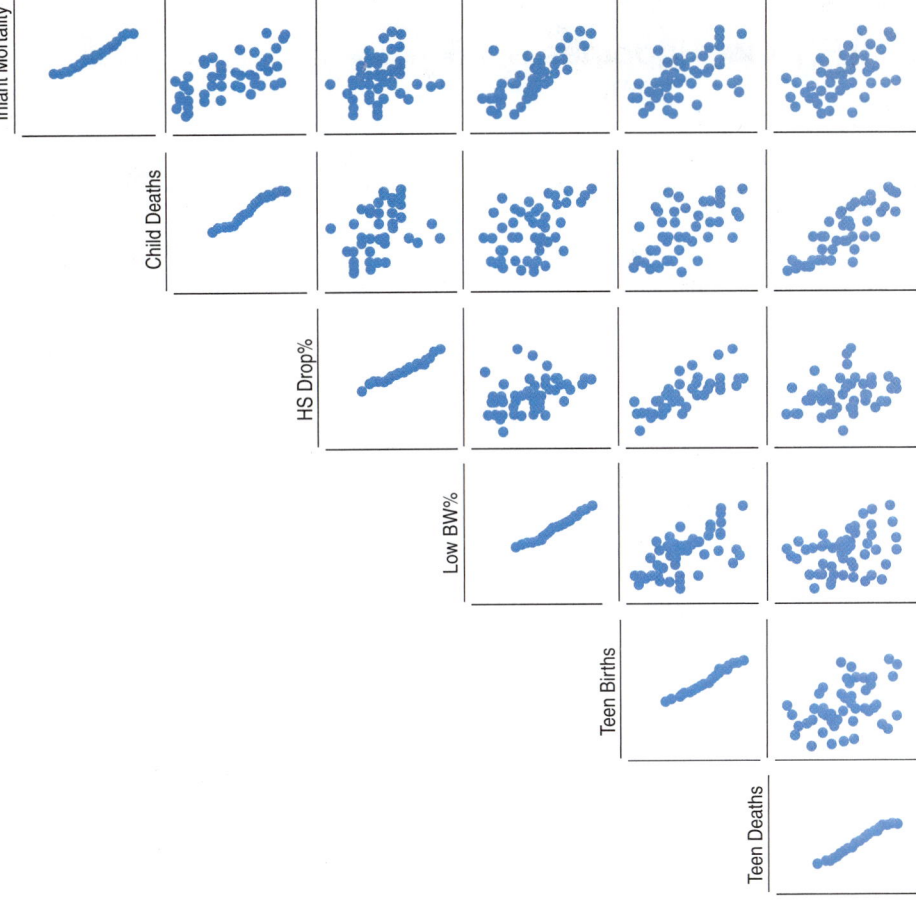

x-axis. This way every pair of variables is graphed. On the diagonal, rather than plotting a variable against itself, you'll usually find either a Normal probability plot or a histogram of the variable to help us assess the **Nearly Normal Condition**.

The individual scatterplots show at a glance that each of the relationships is straight enough for regression. There are no obvious bends, clumping, or outliers. And the plots don't thicken. So it looks like we can examine some multiple regression models with inference.

Step-By-Step Example   INFERENCE FOR MULTIPLE REGRESSION

Question: How should we model *Infant Mortality* using the available predictors?

THINK **Hypotheses** State what we want to know. (Hypotheses on the intercept are not particularly interesting for these data.) **Plan** State the null model. Think about the assumptions and check the conditions.	I wonder whether all or some of these predictors contribute to a useful model for *Infant Mortality*. First, I'll check the overall null hypothesis that asks whether the entire model is better than just modeling y with its mean: H_0: The model itself contributes nothing useful, and all the slope coefficients are zero: $$\beta_1 = \beta_2 = \cdots = \beta_k = 0.$$ H_A: At least one of the β_j is not 0. If I reject this hypothesis, then I'll test a null hypothesis for each of the coefficients of the form: H_0: The j-th variable contributes nothing useful, after allowing for the other predictors in the model: $\beta_j = 0$. H_A: The j-th variable makes a useful contribution to the model: $\beta_j \neq 0$. ✓ **Straight Enough Condition, Outlier Condition:** The scatterplot matrix shows no bends, clumping, or outliers. ✓ **Independence Assumption:** These data are based on random samples and can be considered independent. These conditions are enough to compute the regression model and find residuals. ✓ **Does the Plot Thicken? Condition:** The residual plot shows no obvious trends in the spread: 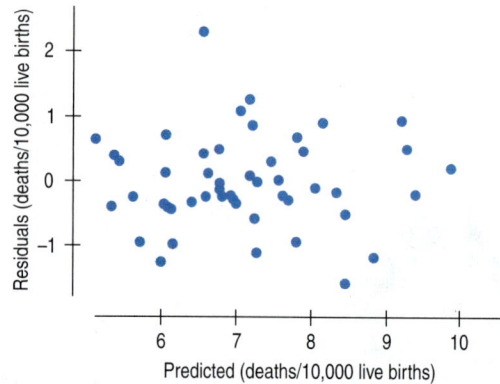

✓ **Nearly Normal Condition:** A histogram of the residuals is unimodal and symmetric.

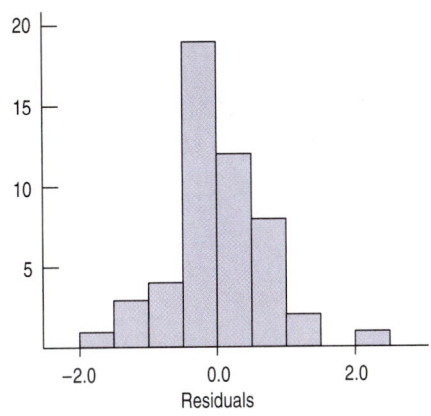

The one possible outlier is South Dakota. I may want to repeat the analysis after removing South Dakota to see whether it changes substantially.

Under these conditions I can continue with the multiple regression analysis.

Choose your method.

SHOW ➤ Mechanics
Multiple regressions are always found from a computer program.

Computer output for this regression looks like this:

Dependent variable is: Infant mort
R-squared = 71.3%   R-squared (adjusted) 68.0%
s = 0.7520 with 50 − 6 = 44 degrees of freedom

Source	Sum of Squares	DF	Mean Square	F-ratio
Regression	61.7319	5	12.3464	21.8
Residual	24.8843	44	0.565553	

Variable	Coefficient	SE(Coeff)	t-ratio	P-value
Intercept	1.63168	0.9124	1.79	0.0806
Child Deaths	0.03123	0.0139	2.25	0.0292
HS Drop%	−0.09971	0.0610	−1.63	0.1096
Low BW%	0.66103	0.1189	5.56	<0.0001
Teen Births	0.01357	0.0238	0.57	0.5713
Teen Deaths	0.00556	0.0113	0.49	0.6245

The P-values given in the regression output table are from the Student's t-distribution on $(n - 6) = 44$ degrees of freedom. They are appropriate for two-sided alternatives.

Consider the hypothesis tests.

The F-ratio of 21.8 on 5 and 44 degrees of freedom is certainly large enough to reject the default null hypothesis that the regression model is no better than using the mean infant mortality rate. So I will examine the individual coefficients.

Under the assumptions we're willing to accept, and considering the conditions we've checked, the individual coefficients follow Student's t-distributions on 44 degrees of freedom.

Most of these coefficients have relatively small t-ratios, so I can't be sure that their underlying values are not zero. Two of the coefficients, *Child Deaths* and *Low BW%*, have P-values less than 5%. So I can be confident that in this model both of these variables are unlikely to really have zero coefficients.

TELL → Interpretation

Overall the R^2 indicates that more than 71% of the variability in *Infant Mortality* can be accounted for with this regression model.

After allowing for the linear effects of the other variables in the model, an increase in *Child Deaths* of 1 death per 100,000 is associated with an increase of 0.03 deaths per 1000 live births in the *Infant Mortality* rate. And an increase of 1% in the percentage of live births that are low birth weight is associated with an increase of 0.66 deaths per 1000 live births.

28.5 Comparing Multiple Regression Models

There may be even more variables available to model *Infant Mortality*. Moreover, several of those we tried don't seem to contribute to the model. How do we know that some other choice of predictors might not provide a better model? What exactly *would* make an alternative model better?

These are not easy questions. There is no simple measure of the success of a multiple regression model. Many people look at the R^2 value, and certainly we are not likely to be happy with a model that accounts for only a small fraction of the variability of y. But that's not enough. You can always drive the R^2 up by piling on more and more predictors, but models with many predictors are hard to understand. Keep in mind that the meaning of a regression coefficient depends on all the *other* predictors in the model, so it is best to keep the number of predictors as small as possible.

Regression models should make sense. Predictors that are easy to understand are usually better choices than obscure variables. Similarly, if there is a known mechanism by which a predictor has an effect on the response variable, that predictor is usually a good choice for the regression model.

How can we know whether we have the best possible model? The simple answer is that we can't. There's always the chance that some other predictors might bring an improvement (in higher R^2 or fewer predictors or simpler interpretation).

Adjusted R^2

You may have noticed that the full regression tables shown in this chapter include another statistic we haven't discussed. It is called adjusted R^2 and sometimes appears in computer output as R^2 (adjusted). The **adjusted R^2** statistic is a rough attempt to adjust for the simple fact that when we add another predictor to a multiple regression, the R^2 can't go down and will most likely go up. Only if we were to add a predictor whose coefficient turned out to be exactly zero would the R^2 remain the same. This fact complicates the comparison of alternative regression models that have different numbers of predictors.

We can write a formula for R^2 using the sums of squares in the ANOVA table portion of the regression output table:

$$R^2 = \frac{SS_{Regression}}{SS_{Regression} + SS_{Residual}} = 1 - \frac{SS_{Residual}}{SS_{Total}}.$$

Adjusted R^2 simply substitutes the corresponding *Mean Squares* for the SS's:[7]

$$R^2_{adj} = 1 - \frac{MS_{Residual}}{MS_{Total}}.$$

Because the Mean Squares are Sums of Squares divided by degrees of freedom, they are adjusted for the number of predictors in the model. As a result, the adjusted R^2 value won't necessarily increase when a new predictor is added to the multiple regression model. That's fine. But adjusted R^2 no longer tells the fraction of variability accounted for by the model, and it isn't even bounded by 0 and 100%, so it can be awkward to interpret.

Comparing alternative regression models is a challenge, especially when they have different numbers of predictors. The search for a summary statistic to help us choose among models is the subject of much contemporary research in Statistics. Adjusted R^2 is one common—but not necessarily the best—choice often found in computer regression output tables. Don't use it as the sole decision criterion when you compare different regression models.

WHAT CAN GO WRONG?

Interpreting Coefficients

- **Don't claim to "hold everything else constant" for a single individual.** It's often meaningless to say that a regression coefficient says what we expect to happen if all variables but one were held constant for an individual and the predictor in question changed. While it's mathematically correct, it often just doesn't make any sense. We can't gain a year of experience or have another child without getting a year older. Instead, we *can* think about all those who fit given criteria on some predictors and ask about the conditional relationship between y and one x for those individuals. The coefficient -0.60 of *Height* for predicting *%Body Fat* says that among men of the same *Waist* size, those who are one inch taller in *Height* tend to be, on average, 0.60% lower in *%Body Fat*. The multiple regression coefficient measures that average conditional relationship.

- **Don't interpret regression causally.** Regressions are usually applied to observational data. Without deliberately assigned treatments, randomization, and control, we can't draw conclusions about causes and effects. We can never be certain that there are no variables lurking in the background, causing everything we've seen. Don't interpret b_1, the coefficient of x_1 in the multiple regression, by saying, "If we were to change an individual's x_1 by 1 unit (holding the other x's constant) it would change his y by b_1 units." We have no way of knowing what applying a change to an individual would do.

- **Be cautious about interpreting a regression model as predictive.** Yes, we do call the x's predictors, and you can certainly plug in values for each of the x's and find a corresponding *predicted value*, $\hat{y}$. But the term "prediction" suggests extrapolation into the future or beyond the data, and we know that we can get into trouble when we use models to estimate $\hat{y}$ values for x's not in the range of the data. Be careful not to extrapolate very far from the span of your data. In simple regression it was easy to tell when you extrapolated. With many predictor variables, it's often harder to know when you are outside the bounds of your original data.[8] We usually think of fitting models to the data more as modeling than as prediction, so that's often a more appropriate term.

---

[7] We learned about Mean Squares in Chapter 26. A Mean Square is just a Sum of Squares divided by its appropriate degrees of freedom. Mean Squares are variances.

[8] With several predictors it is easy to wander beyond the data because of the *combination* of values even when individual values are not extraordinary. For example, both 28-inch waists and 76-inch heights can be found in men in the body fat study, but a single individual with both these measurements would not be at all typical. The model we fit is probably not appropriate for predicting the *%Body Fat* for such a tall and skinny individual.

- **Don't think that the sign of a coefficient is special.** Sometimes our primary interest in a predictor is whether it has a positive or negative association with y. As we have seen, though, the sign of the coefficient also depends on the other predictors in the model. Don't look at the sign in isolation and conclude that "the direction of the relationship is positive (or negative)." Just like the value of the coefficient, the sign is about the relationship after allowing for the linear effects of the other predictors. The sign of a variable can change depending on which other predictors are in or out of the model. For example, in the regression model for infant mortality, the coefficient of *HS Drop%* was negative and its P-value was fairly small, but the simple association between *Dropout Rate* and *Infant Mortality* is positive. (Check the plot matrix.)

- **If a coefficient's *t*-statistic is not significant, don't interpret it at all.** You can't be sure that the value of the corresponding parameter in the underlying regression model isn't really zero.

WHAT ELSE CAN GO WRONG?

- **Don't fit a linear regression to data that aren't straight.** Linearity is the most fundamental regression assumption. If the relationship between y and the x's isn't approximately linear, there's no sense in fitting a linear model to it. What we mean by "linear" is a model of the form we have been writing for the regression. When we have two predictors, this is the equation of a plane, which is linear in the sense of being flat in all directions. With more predictors, the geometry is harder to visualize, but the simple structure of the model is consistent; the predicted values change consistently with equal size changes in any predictor.

    Usually we're satisfied when plots of y against each of the x's are straight enough. We'll also check a scatterplot of the residuals against the predicted values for evidence of nonlinearity.

- **Watch out for the plot thickening.** The estimate of the error standard deviation shows up in all the inference formulas. But that estimate assumes that the error standard deviation is the same throughout the range of the x's so that we can combine (pool, actually) all the residuals when we estimate it. If s_e changes with any x, these estimates won't make sense. The most common check is a plot of the residuals against the predicted values. If plots of residuals against several of the predictors all show a thickening, and especially if they also show a bend, then consider re-expressing y. If the scatterplot against only one predictor shows thickening, consider re-expressing that predictor.

- **Make sure the errors are nearly Normal.** All of our inferences require that the true errors be modeled well by a Normal model. Check the histogram and Normal probability plot of the residuals to see whether this assumption looks reasonable.

- **Watch out for high-influence points and outliers.** We always have to be on the lookout for a few points that have undue influence on our model, and regression is certainly no exception. Partial regression plots are a good place to look for influential points and to understand how they affect each of the coefficients.

CONNECTIONS

We would never consider a regression analysis without first making scatterplots. The aspects of scatterplots that we always look for—their direction, form, and strength—relate directly to regression, and we assess the **Nearly Normal Condition** by examining the shape of a residual histogram or with a Normal probability plot.

Regression inference is connected to just about every inference method we have seen for measured data. The assumption that the spread of data about the line is constant is essentially the same as the assumption of equal variances required for the pooled-t methods. Our use of all the residuals together to estimate their standard deviation is a form of pooling.

Of course, the ANOVA table in the regression output connects to our consideration of ANOVA in Chapter 26. This, too, is not coincidental. Multiple Regression, ANOVA, pooled t-tests, and inference for means are all part of a more general statistical model known as the General Linear Model (often just called the GLM).

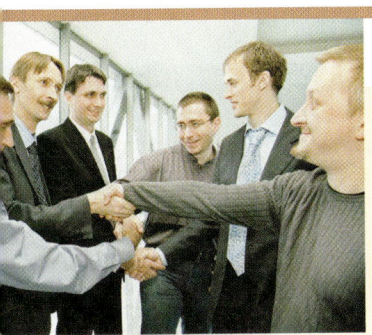

What Have We Learned?

In Chapter 25 we learned to apply our inference methods to linear regression models. Now we've seen that much of what we know about those models is also true for multiple regression:

- The assumptions and conditions are the same: linearity (checked now with scatterplots of y against each x), independence (think about it), constant variance (checked with the scatterplot of residuals against predicted values), and Nearly Normal residuals (checked with a histogram or probability plot).
- R^2 is still the fraction of the variation in y accounted for by the regression model.
- s_e is still the standard deviation of the residuals—a good indication of the precision of the model.
- The degrees of freedom (in the denominator of s_e and for each of the t-tests) follows the same rule: n minus the *number of parameters estimated*.
- The regression table produced by any statistics package shows a row for each coefficient, giving its estimate, a standard error, a t-statistic, and a P-value.
- If all the conditions are met, we can test each coefficient against the null hypothesis that its parameter value is zero with a Student's t-test.

And we've learned some new things that are useful now that we have multiple predictors:

- We can perform an overall test of whether the multiple regression model provides a better summary for y than its mean by using the F-distribution we saw in the previous chapters.
- We learned that R^2 may not be appropriate for comparing multiple regression models with different numbers of predictors. Adjusted R^2 is one approach to this problem.

Finally, we've learned that multiple regression models extend our ability to model the world to many more situations, but that we must take great care when we interpret its coefficients. To interpret a coefficient of a multiple regression model, remember that it estimates the linear relationship between y and that predictor *after allowing for the linear effects of all the other predictors on both y and that x.*

Review of Terms

Multiple regression   A linear regression with two or more predictors whose coefficients are found to minimize the sum of the squared residuals is a least squares linear multiple regression. But it is usually just called a multiple regression. When the distinction is needed, a least squares linear regression with a single predictor is called a simple regression. The multiple regression model is (p. 817)

$$y = \beta_0 + \beta_1 x_1 + \cdots + \beta_k x_k + \varepsilon.$$

Least Squares   We still fit multiple regression models by choosing the coefficients that make the sum of the squared residuals as small as possible, the method of least squares (p. 818).

Partial regression plot   The partial regression plot for a specified coefficient is a display that helps in understanding the meaning of that coefficient in a multiple regression. It has a slope equal to the coefficient value and shows the influences of each case on that value. Partial regression plots display the residuals when y is regressed on the *other* predictors against the residuals when the specified x is regressed on the other predictors (p. 821).

ANOVA   The Analysis of Variance table that is ordinarily part of the multiple regression results offers an F-test to test the null hypothesis that the overall regression is no improvement over just modeling y with its mean:

$$H_0: \beta_1 = \beta_2 = \cdots = \beta_k = 0.$$

If this null hypothesis is not rejected, then do not proceed to test the individual coefficients (p. 825).

***t*-ratios for the coefficients**   The t-ratios for the coefficients can be used to test the null hypotheses that the true value of each coefficient is zero against the alternative that it is not (p. 826).

Scatterplot matrix   A scatterplot matrix displays scatterplots for all pairs of a collection of variables, arranged so that all the plots in a row have the same variable displayed on their y-axis and all plots in a column have the same variable on their x-axis. Usually, the diagonal holds a display of a single variable such as a histogram or Normal probability plot, and identifies the variable in its row and column (p. 828).

Adjusted R^2   An adjustment to the R^2 statistic that attempts to allow for the number of predictors in the model. It is sometimes used when comparing regression models with different numbers of predictors (p. 832).

$$R^2_{adj} = 1 - \frac{MS_{Residual}}{MS_{Total}}$$

On the Computer REGRESSION ANALYSIS

All statistics packages make a table of results for a regression. If you can read a package's regression output table for simple regression, then you can read its table for a multiple regression. You'll want to look at the ANOVA table, and you'll see information for each of the coefficients, not just for a single slope.

Most packages offer to plot residuals against predicted values. Some will also plot residuals against the x's. With some packages you must request plots of the residuals when you request the regression. Others let you find the regression first and then analyze the residuals afterward. Either way, your analysis is not complete if you don't check the residuals with a histogram or Normal probability plot and a scatterplot of the residuals against the x's or the predicted values.

One good way to check assumptions before embarking on a multiple regression analysis is with a scatterplot matrix. This is sometimes abbreviated SPLOM in commands.

Multiple regressions are always found with a computer or programmable calculator. Before computers were available, a full multiple regression analysis could take months or even years of work.

DATA DESK

- Select Y- and X-variable icons.
- From the **Calc** menu, choose **Regression.**
- Data Desk displays the regression table.
- Select plots of residuals from the Regression table's HyperView menu.

COMMENTS

You can change the regression by dragging the icon of another variable over either the Y- or an X-variable name in the table and dropping it there. You can add a predictor by dragging its icon into that part of the table. The regression, predicted values, residuals, and diagnostic statistics will recompute automatically.

EXCEL

- In Excel 2003 and earlier, select **Data Analysis** from the **Tools** menu.
- In Excel 2007, select **Data Analysis** from the **Analysis Group** on the Data Tab.
- Select **Regression** from the **Analysis Tools** list.
- Click the **OK** button.
- Enter the data range holding the Y-variable in the box labeled "Y-range."
- Enter the range of cells holding the X-variables in the box labeled "X-range."
- Select the **New Worksheet Ply** option.
- Select **Residuals** options. Click the **OK** button.

COMMENTS

The Y and X ranges do not need to be in the same rows of the spreadsheet, although they must cover the same number of cells. But it is a good idea to arrange your data in parallel columns as in a data table. The X-variables must be in adjacent columns. No cells in the data range may hold non-numeric values.

Although the dialog offers a Normal probability plot of the residuals, the data analysis add-in does not make a correct probability plot, so don't use this option.

JMP

- From the **Analyze** menu select **Fit Model.**
- Specify the response, Y. Assign the predictors, X, in the **Construct Model Effects** dialog box.
- Click on **Run Model.**

COMMENTS

JMP chooses a regression analysis when the response variable is "Continuous." The predictors can be any combination of quantitative or categorical. If you get a different analysis, check the variable types.

MINITAB

- Choose **Regression** from the **Stat** menu.
- Choose **Regression...** from the **Regression** submenu.
- In the Regression dialog, assign the Y-variable to the Response box and assign the X-variables to the Predictors box.
- Click the **Graphs** button.
- In the Regression-Graphs dialog, select **Standardized residuals,** and check **Normal plot of residuals** and **Residuals versus fits.**
- Click the **OK** button to return to the Regression dialog.
- Click the **OK** button to compute the regression.

R

Suppose the response variable y and predictor variables $x_1, \ldots, x_k$ are in a data frame called mydata. To fit a multiple regression of y on x_1 and x_2:

- mylm = lm(y ~ x_1 + x_2, data = mydata)
- summary(mylm) # gives the details of the fit, including the ANOVA table
- plot(mylm) # gives a variety of plots

To fit the model with *all* the predictors in the data frame,

- mylm = lm(y ~., data = mydata) # The period means use all other variables

COMMENTS

To get confidence or prediction intervals use:

- predict(mylm, interval = "confidence")

or

- predict(mylm, interval = "prediction")

SPSS

- Choose **Regression** from the **Analyze** menu.
- Choose **Linear** from the **Regression** submenu.
- When the Linear Regression dialog appears, select the Y-variable and move it to the dependent target. Then move the X-variables to the independent target.
- Click the **Plots** button.
- In the Linear Regression Plots dialog, choose to plot the *SRESIDs against the *ZPRED values.
- Click the **Continue** button to return to the Linear Regression dialog.
- Click the **OK** button to compute the regression.

STATCRUNCH

To find a multiple regression of a response variable y on two factors, x_1 and x_2, from the Stat menu, choose Regression > Multiple Linear.

In the dialog, specify the response (Y) variable. Specify the predictors (X variables) and Interactions (optional).

Click Compute! to perform the regression.

COMMENTS

You can save residuals and predicted values among other quantities with the save options.

TI-83/84 PLUS

COMMENTS

You need a special program to compute a multiple regression on the TI-83.

Exercises

Section 28.1

1. House prices The following regression model was found for houses in an area of upstate New York:

$$\widehat{Price} = 20{,}986.09 - 7483.10\,Bedrooms + 93.84\,Living\,Area.$$

a) Find the predicted price of a 2 bedroom, 1000-sq-ft house from this model.

b) The house just sold for $135,000. Find the residual corresponding to this house.

c) What does that residual say about this transaction?

2. Candy sales A candy maker surveyed chocolate bars available in a local supermarket and found the following least squares regression model:

$$\widehat{Calories} = 28.4 + 11.37\,Fat(g) + 2.91\,Sugar(g).$$

a) The hand-crafted chocolate she makes has 15g of fat and 20g of sugar. How many calories does the model predict for a serving?
b) In fact, a laboratory test shows that her candy has 227 calories per serving. Find the residual corresponding to this candy. (Be sure to include the units.)
c) What does that residual say about her candy?

Section 28.2

3. **Movie profits** What can predict how much a motion picture will make? We have data on a number of recent releases that includes the *USGross* (in $M), the *Budget* ($M), the *Run Time* (minutes), and the average number of *Stars* awarded by reviewers. The first several entries in the data table look like this:

Movie	USGross ($M)	Budget ($M)	Run Time (minutes)	Stars
White Noise	56.094360	30	101	2
Coach Carter	67.264877	45	136	3
Elektra	24.409722	65	100	2
Racing Stripes	49.772522	30	110	3
Assault on Precinct 13	20.040895	30	109	3
Are We There Yet?	82.674398	20	94	2
Alone in the Dark	5.178569	20	96	1.5
Indigo	51.100486	25	105	3.5

We want a regression model to predict *USGross*. Parts of the regression output computed in Excel look like this:

Dependent variable is: USGross($)
R squared = 47.4%   R squared (adjusted) = 46.0%
s = 46.41 with 120 − 4 = 116 degrees of freedom

Variable	Coefficient	SE(Coeff)	t-ratio	P-value
Intercept	−22.9898	25.70	−0.895	0.3729
Budget($)	1.13442	0.1297	8.75	<0.0001
Stars	24.9724	5.884	4.24	<0.0001
Run Time	−0.403296	0.2513	−1.60	0.1113

a) Write the multiple regression equation.
b) What is the interpretation of the coefficient of *Budget* in this regression model?

4. **Movie profits again** A middle manager at an entertainment company, upon seeing the analysis of Exercise 3, concludes that longer movies make less money. He argues that his company's films should all be cut by 30 minutes to improve their gross. Explain the flaw in his interpretation of this model.

Section 28.3

5. **More movies** For the movies examined in Exercise 4, here is a scatterplot of *USGross* vs. *Budget*:

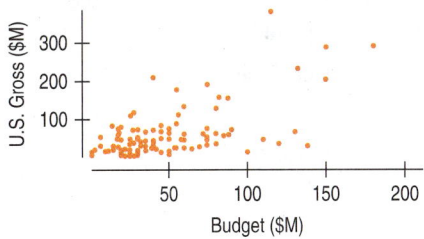

What (if anything) does this scatterplot tell us about the following Assumptions and Conditions for the regression?
a) Linearity condition
b) Equal Spread condition
c) Normality assumption

6. **Movie residuals** For the movies regression, here is a histogram of the residuals. What does it tell us about these Assumptions and Conditions?

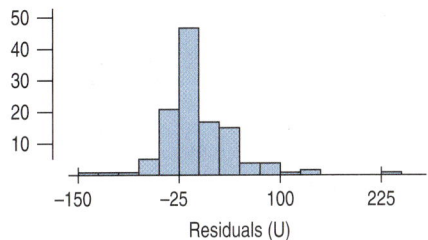

a) Linearity condition
b) Nearly Normal condition
c) Equal Spread condition

Section 28.4

7. **Movie tests** In the regression output for the movies of Exercise 3,
a) What is the null hypothesis tested for the coefficient of *Stars* in this table?
b) What is the *t*-statistic corresponding to this test?
c) What is the P-value corresponding to this *t*-statistic?
d) Complete the hypothesis test. Do you reject the null hypothesis?

8. **More movie tests**
a) What is the null hypothesis tested for the coefficient of *Run Time* in the regression of Exercise 3?
b) What is the *t*-statistic corresponding to this test?
c) Why is this *t*-statistic negative?
d) What is the P-value corresponding to this *t*-statistic?
e) Complete the hypothesis test. Do you reject the null hypothesis?

Section 28.5

9. Movie R-squared In the regression model of Exercise 3,
a) What is the R^2 for this regression? What does it mean?
b) Why is the "Adjusted R Square" in the table different from the "R Square"?

10. Movie output Here is another part of the regression output for the movies in Exercise 3:

Source	Sum of Squares	df	Mean Square	F-ratio
Regression	224995	3	74998.4	34.8
Residual	249799	116	2153.44	

a) Using the values from the table, show how the value of R^2 could be computed. Don't try to do the calculation, just show what is computed.
b) What is the F-statistic value for this regression?
c) What null hypothesis can you test with it?
d) Would you reject that null hypothesis?

Chapter Exercises

11. Interpretations A regression performed to predict the selling price of houses found the equation
$\widehat{Price} = 169{,}328 + 35.3\,Area + 0.718\,Lotsize - 6543\,Age$
where *Price* is in dollars, *Area* is in square feet, *Lotsize* is in square feet, and *Age* is in years. The R^2 is 92%. One of the interpretations below is correct. Which is it? Explain what's wrong with the others.

a) Each year a house *Age*s it is worth $6543 less.
b) Every extra square foot of *Area* is associated with an additional $35.30 in average price, for houses with a given *Lotsize* and *Age*.
c) Every dollar in price means *Lotsize* increases 0.718 square feet.
d) This model fits 92% of the data points exactly.

12. More interpretations A household appliance manufacturer wants to analyze the relationship between total sales and the company's three primary means of advertising (television, magazines, and radio). All values were in millions of dollars. They found the regression equation
$\widehat{Sales} = 250 + 6.75\,TV + 3.5\,Radio + 2.3\,Magazines.$

One of the interpretations below is correct. Which is it? Explain what's wrong with the others.

a) If they did no advertising, their income would be $250 million.
b) Every million dollars spent on radio makes sales increase $3.5 million, all other things being equal.
c) Every million dollars spent on magazines increases TV spending $2.3 million.
d) Sales increase on average about $6.75 million for each million spent on TV, after allowing for the effects of the other kinds of advertising.

13. Predicting final exams How well do exams given during the semester predict performance on the final? One class had three tests during the semester. Computer output of the regression gives

Dependent variable is Final
s = 13.46   R-Sq = 77.7%   R-Sq(adj) = 74.1%

Predictor	Coeff	SE(Coeff)	t	P-value
Intercept	−6.72	14.00	−0.48	0.636
Test1	0.2560	0.2274	1.13	0.274
Test2	0.3912	0.2198	1.78	0.091
Test3	0.9015	0.2086	4.32	<0.0001

Analysis of Variance

Source	DF	SS	MS	F	P-value
Regression	3	11961.8	3987.3	22.02	<0.0001
Error	19	3440.8	181.1		
Total	22	15402.6			

a) Write the equation of the regression model.
b) How much of the variation in final exam scores is accounted for by the regression model?
c) Explain in context what the coefficient of *Test3* scores means.
d) A student argues that clearly the first exam doesn't help to predict final performance. She suggests that this exam not be given at all. Does Test1 have no effect on the final exam score? Can you tell from this model? (*Hint*: Do you think test scores are related to each other?)

T 14. Scottish hill races 2008 Hill running—races up and down hills—has a written history in Scotland dating back to the year 1040. Races are held throughout the year at different locations around Scotland. A recent compilation of information for 91 races (for which full information was available and omitting two unusual races) includes the *Distance* (km), the *Climb* (m), and the *Record Time* (minutes). A regression to predict the men's records as of 2008 looks like this:

Dependent variable is: Time (mins)
R-squared = 98.0%   R-squared (adjusted) = 98.0%
s = 6.623 with 90 − 3 = 87 degrees of freedom

Source	Sum of Squares	df	Mean Square	F-ratio
Regression	189204	2	94602.1	2157
Residual	3815.92	87	43.8612	

Variable	Coefficient	SE(Coeff)	t-ratio	P-value
Intercept	−10.3723	1.245	−8.33	<0.0001
Climb (m)	0.034227	0.0022	15.7	<0.0001
Distance (km)	4.04204	0.1448	27.9	<0.0001

a) Write the regression equation. Give a brief report on what it says about men's record times in hill races.
b) Interpret the value of R^2 in this regression.
c) What does the coefficient of *Climb* mean in this regression?

15. Home prices Many variables have an impact on determining the price of a house. A few of these are *Size* of the house (square feet), *Lotsize,* and number of *Bathrooms.* Information for a random sample of homes for sale in the Statesboro, GA, area was obtained from the Internet. Regression output modeling the *Asking Price* with *Square Footage* and number of *Bathrooms* gave the following result:

Dependent Variable is: Asking Price
s = 67013   R-Sq = 71.1%   R-Sq (adj) = 64.6%

Predictor	Coeff	SE(Coeff)	t-ratio	P-value
Intercept	−152037	85619	−1.78	0.110
Baths	9530	40826	0.23	0.821
Sq ft	139.87	46.67	3.00	0.015

Analysis of Variance

Source	DF	SS	MS	F	P-value
Regression	2	99303550067	49651775033	11.06	0.004
Residual	9	40416679100	4490742122		
Total	11	1.39720E+11			

a) Write the regression equation.
b) How much of the variation in home asking prices is accounted for by the model?
c) Explain in context what the coefficient of *Square Footage* means.
d) The owner of a construction firm, upon seeing this model, objects because the model says that the number of bathrooms has no effect on the price of the home. He says that when *he* adds another bathroom, it increases the value. Is it true that the number of bathrooms is unrelated to house price? (*Hint:* Do you think bigger houses have more bathrooms?)

T 16. More hill races 2008 Here is the regression for the women's records for the same Scottish hill races we considered in Exercise 14:

Dependent variable is: Women's Time (mins)
R-squared = 96.7%   R-squared (adjusted) = 96.7%
s = 10.06 with 90 − 3 = 87 degrees of freedom

Source	Sum of Squares	df	Mean Square	F-ratio
Regression	261029	2	130515	1288
Residual	8813.02	87	101.299	

Variable	Coefficient	SE(Coeff)	t-ratio	P-value
Intercept	−11.6545	1.891	−6.16	<0.0001
Climb (m)	0.045195	0.0033	13.7	<0.0001
Distance	4.43427	0.2200	20.2	<0.0001

a) Compare the regression model for the women's records with that found for the men's records in Exercise 14.

Here's a scatterplot of the residuals for this regression:

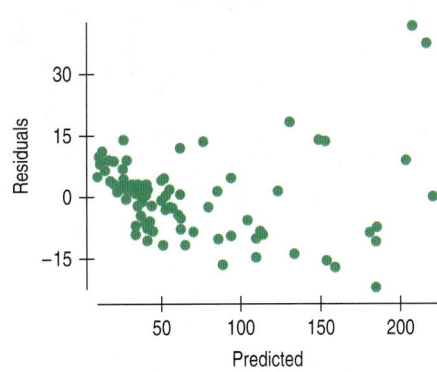

b) Discuss the residuals and what they say about the assumptions and conditions for this regression.

17. Predicting finals II Here are some diagnostic plots for the final exam data from Exercise 13. These were generated by a computer package and may look different from the plots generated by the packages you use. (In particular, note that the axes of the Normal probability plot are swapped relative to the plots we've made in the text. We only care about the pattern of this plot, so it shouldn't affect your interpretation.) Examine these plots and discuss whether the assumptions and conditions for the multiple regression seem reasonable.

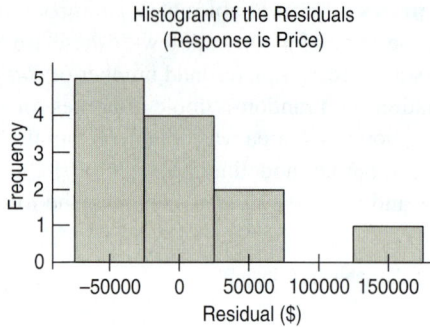

18. Home prices II Here are some diagnostic plots for the home prices data from Exercise 15. These were generated by a computer package and may look different from the plots generated by the packages you use. (In particular, note that the axes of the Normal probability plot are swapped relative to the plots we've made in the text. We only care about the pattern of this plot, so it shouldn't affect your interpretation.) Examine these plots and discuss whether the assumptions and conditions for the multiple regression seem reasonable.

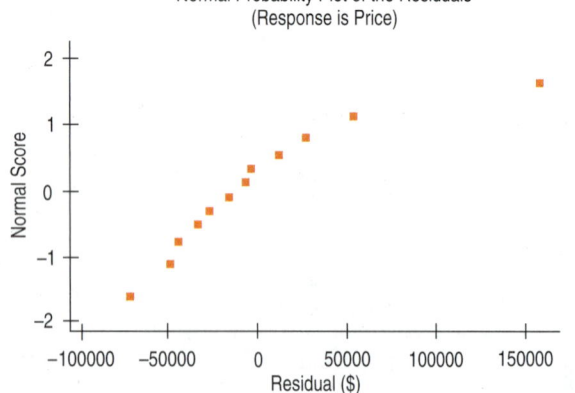

19. Admin performance The AFL-CIO has undertaken a study of the yearly salaries (in thousands of dollars) of 30 administrative assistants. The organization wants to predict salaries from several other variables.

The variables considered to be potential predictors of salary are:

$X1$ = months of service
$X2$ = years of education
$X3$ = score on standardized test
$X4$ = words per minute (wpm) typing speed
$X5$ = ability to take dictation in words per minute

A multiple regression model with all five variables was run on a computer package, resulting in the following output:

Variable	Coefficient	Std. Error	t-value
Intercept	9.788	0.377	25.960
X1	0.110	0.019	5.178
X2	0.053	0.038	1.369
X3	0.071	0.064	1.119
X4	0.004	0.307	0.013
X5	0.065	0.038	1.734

$s = 0.430 \quad R^2 = 0.863$

Assume that the residual plots show no violations of the conditions for using a linear regression model.

a) What is the regression equation?
b) From this model, what is the predicted *Salary* (in thousands of dollars) of a secretary with 10 years (120 months) of experience, 9th grade education (9 years of education), a 50 on the standardized test, 60 wpm typing speed, and the ability to take 30 wpm dictation?
c) Test whether the coefficient for words per minute of typing speed ($X4$) is significantly different from zero at $\alpha = 0.05$.
d) How might this model be improved?
e) A correlation of *Age* with *Salary* finds $r = 0.682$, and the scatterplot shows a moderately strong positive linear association. However, if $X6 = Age$ is added to the multiple regression, the estimated coefficient of *Age* turns out to be $b_6 = -0.154$. Explain some possible causes for this apparent change of direction in the relationship between age and salary.

20. GPA and SATs A large section of Stat 101 was asked to fill out a survey on grade point average and SAT scores. A regression was run to find out how well Math and Verbal SAT scores could predict academic performance as measured by GPA. The regression was run on a computer package with the following output:

Response: GPA

Variable	Coefficient	SE(Coeff)	t-ratio	P-value
Intercept	0.574968	0.253874	2.26	0.0249
SAT Verbal	0.001394	0.000519	2.69	0.0080
SAT Math	0.001978	0.000526	3.76	0.0002

a) What is the regression equation?
b) From this model, what is the predicted GPA of a student with an SAT Verbal score of 500 and an SAT Math score of 550?
c) What else would you want to know about this regression before writing a report about the relationship between SAT scores and grade point averages? Why would these be important to know?

21. Body fat revisited The data set on body fat contains 15 body measurements on 250 men from 22 to 81 years old. Is average *%Body Fat* related to *Weight?* Here's a scatterplot:

And here's the simple regression:

Dependent variable is: Pct BF
R-squared = 38.1% R-squared (adjusted) = 37.9%
s = 6.538 with 250 − 2 = 248 degrees of freedom

Variable	Coefficient	SE(Coeff)	t-ratio	P-value
Intercept	−14.6931	2.760	−5.32	<0.0001
Weight	0.18937	0.0153	12.4	<0.0001

a) Is the coefficient of *%Body Fat* on *Weight* statistically distinguishable from 0? (Perform a hypothesis test.)
b) What does the slope coefficient mean in this regression?

We saw before that the slopes of both *Waist* size and *Height* are statistically significant when entered into a multiple regression equation. What happens if we add *Weight* to that regression? Recall that we've already checked the assumptions and conditions for regression on *Waist* size and *Height* in the chapter. Here is the output from a regression on all three variables:

Dependent variable is: Pct BF
R-squared = 72.5% R-squared (adjusted) = 72.2%
s = 4.376 with 250 − 4 = 246 degrees of freedom

Source	Sum of Squares	df	Mean Square	F-ratio
Regression	12418.7	3	4139.57	216
Residual	4710.11	246	19.1468	

Variable	Coefficient	SE(Coeff)	t-ratio	P-value
Intercept	−31.4830	11.54	−2.73	0.0068
Waist	2.31848	0.1820	12.7	<0.0001
Height	−0.224932	0.1583	−1.42	0.1567
Weight	−0.100572	0.0310	−3.25	0.0013

c) Interpret the slope for *Weight*. How can the coefficient for *Weight* in this model be negative when its coefficient was positive in the simple regression model?
d) What does the P-value for *Height* mean in this regression? (Perform the hypothesis test.)

22. Breakfast cereals We saw in Chapter 7 that the calorie content of a breakfast cereal is linearly associated with its sugar content. Is that the whole story? Here's the output of a regression model that regresses *Calories* for each serving on its *Protein(g)*, *Fat(g)*, *Fiber(g)*, *Carbohydrate(g)*, and *Sugars(g)* content.

Dependent variable is: Calories
R-squared = 93.6% R-squared (adjusted) = 93.1%
s = 5.113 with 77 − 6 = 71 degrees of freedom

Source	Sum of Squares	df	Mean Square	F-ratio
Regression	26995.9	5	5399.18	207
Residual	1856.03	71	26.1412	

Variable	Coefficient	SE(Coeff)	t-ratio	P-value
Intercept	−0.879994	4.383	−0.201	0.8414
Protein	3.60495	0.6977	5.17	<0.0001
Fat	8.56877	0.6625	12.9	<0.0001
Fiber	0.309180	0.3337	0.927	3.572
Carbo	4.13996	0.2049	20.2	<0.0001
Sugars	4.00677	0.1719	23.3	<0.0001

Assuming that the conditions for multiple regression are met,

a) What is the regression equation?
b) Do you think this model would do a reasonably good job at predicting calories? Explain.
c) To check the conditions, what plots of the data might you want to examine?
d) What does the coefficient of *Fat* mean in this model?

23. Body fat one more time Chest size might be a good predictor of body fat. Here's a scatterplot of *%Body Fat* vs. *Chest Size*.

A regression of *%Body Fat* on *Chest Size* gives the following equation:

Dependent variable is: Pct BF
R-squared = 49.1%   R-squared (adjusted) = 48.9%
S = 5.930 with 250 − 2 = 248 degrees of freedom

Variable	Coefficient	SE(Coeff)	t-ratio	P-value
Intercept	−52.7122	4.654	−11.3	<0.0001
Chest Size	0.712720	0.0461	15.5	<0.0001

a) Is the slope of *%Body Fat* on *Chest Size* statistically distinguishable from 0? (Perform a hypothesis test.)
b) What does the answer in part a mean about the relationship between *%Body Fat* and *Chest Size*?

We saw before that the slopes of both *Waist* size and *Height* are statistically significant when entered into a multiple regression equation. What happens if we add *Chest Size* to that regression? Here is the output from a regression on all three variables:

Dependent variable is: Pct BF
R-squared = 72.2%   R-squared (adjusted) = 71.9%
s = 4.399 with 250 − 4 = 246 degrees of freedom

Source	Sum of Squares	df	Mean Square	F-ratio	P
Regression	12368.9	3	4122.98	213	<0.0001
Residual	4759.87	246	19.3491		

Variable	Coefficient	SE(Coeff)	t-ratio	P-value
Intercept	2.07220	7.802	0.266	0.7908
Waist	2.19939	0.1675	13.1	<0.0001
Height	−0.561058	0.1094	−5.13	<0.0001
Chest Size	−0.233531	0.0832	−2.81	0.0054

c) Interpret the coefficient for *Chest Size*.
d) Would you consider removing any of the variables from this regression model? Why or why not?

24. Grades The following table shows the five scores from an Introductory Statistics course. Find a model for final exam score by trying all possible models with two predictor variables. Which model would you choose? Be sure to check the conditions for multiple regression.

Name	Final	Midterm 1	Midterm 2	Project	Homework
Timothy F.	117	82	30	10.5	61
Karen E.	183	96	68	11.3	72
Verena Z.	124	57	82	11.3	69
Jonathan A.	177	89	92	10.5	84
Elizabeth L.	169	88	86	10.6	84
Patrick M.	164	93	81	10	71
Julia E.	134	90	83	11.3	79
Thomas A.	98	83	21	11.2	51
Marshall K.	136	59	62	9.1	58
Justin E.	183	89	57	10.7	79
Alexandra E.	171	83	86	11.5	78
Christopher B.	173	95	75	8	77
Justin C.	164	81	66	10.7	66
Miguel A.	150	86	63	8	74
Brian J.	153	81	86	9.2	76
Gregory J.	149	81	87	9.2	75
Kristina G.	178	98	96	9.3	84
Timothy B.	75	50	27	10	20
Jason C.	159	91	83	10.6	71
Whitney E.	157	87	89	10.5	85
Alexis P.	158	90	91	11.3	68
Nicholas T.	171	95	82	10.5	68
Amandeep S.	173	91	37	10.6	54
Irena R.	165	93	81	9.3	82
Yvon T.	168	88	66	10.5	82
Sara M.	186	99	90	7.5	77
Annie P.	157	89	92	10.3	68
Benjamin S.	177	87	62	10	72
David W.	170	92	66	11.5	78
Josef H.	78	62	43	9.1	56
Rebecca S.	191	93	87	11.2	80
Joshua D.	169	95	93	9.1	87
Ian M.	170	93	65	9.5	66
Katharine A.	172	92	98	10	77
Emily R.	168	91	95	10.7	83
Brian M.	179	92	80	11.5	82
Shad M.	148	61	58	10.5	65
Michael R.	103	55	65	10.3	51
Israel M.	144	76	88	9.2	67
Iris J.	155	63	62	7.5	67
Mark G.	141	89	66	8	72
Peter H.	138	91	42	11.5	66
Catherine R.M.	180	90	85	11.2	78
Christina M.	120	75	62	9.1	72
Enrique J.	86	75	46	10.3	72
Sarah K.	151	91	65	9.3	77
Thomas J.	149	84	70	8	70
Sonya P.	163	94	92	10.5	81

(Continued)

Name	Final	Midterm 1	Midterm 2	Project	Home-work
Michael B.	153	93	78	10.3	72
Wesley M.	172	91	58	10.5	66
Mark R.	165	91	61	10.5	79
Adam J.	155	89	86	9.1	62
Jared A.	181	98	92	11.2	83
Michael T.	172	96	51	9.1	83
Kathryn D.	177	95	95	10	87
Nicole M.	189	98	89	7.5	77
Wayne E.	161	89	79	9.5	44
Elizabeth S.	146	93	89	10.7	73
John R.	147	74	64	9.1	72
Valentin A.	160	97	96	9.1	80
David T.O.	159	94	90	10.6	88
Marc I.	101	81	89	9.5	62
Samuel E.	154	94	85	10.5	76
Brooke S.	183	92	90	9.5	86

25. Fifty states 2009 Here is a data set on various measures of the 50 United States. The *Murder* rate is per 100,000, *HS Graduation* rate is in %, *Income* is per capita income in dollars, *Illiteracy* rate is per 1000, and *Life Expectancy* is in years. Find a regression model for *Life Expectancy* with three predictor variables by trying all four of the possible models.

a) Which model appears to do the best?
b) Would you leave all three predictors in this model?
c) Does this model mean that by changing the levels of the predictors in this equation, we could affect life expectancy in that state? Explain.
d) Be sure to check the conditions for multiple regression. What do you conclude?

State	Life expectancy	Murder rate08	Graduation Rate	Income/cap07	Illiteracy
Alabama	74.4	7.6	62%	27557	0.15
Alaska	77.1	4.1	67	34316	0.09
Arizona	77.5	6.3	59	28088	0.13
Arkansas	75.2	5.7	72	25563	0.14
California	78.2	5.8	68	35352	0.23
Colorado	78.2	3.2	68	34902	0.10
Connecticut	78.7	3.5	75	46021	0.09
Delaware	76.8	6.5	73	34533	0.11
Florida	77.5	6.4	59	32693	0.20
Georgia	75.3	6.6	54	28452	0.17
Hawaii	80.0	1.9	69	33369	0.16
Idaho	77.9	1.5	78	26530	0.11
Illinois	76.4	6.1	78	34290	0.13
Indiana	76.1	5.1	74	28587	0.08
Iowa	78.3	2.5	93	29784	0.07
Kansas	77.3	4	76	31268	0.08
Kentucky	75.2	4.6	71	26457	0.12
Louisiana	74.2	11.9	69	29557	0.16
Maine	77.6	2.4	78	28677	0.07
Maryland	76.3	8.8	75	39136	0.11
Massachusetts	78.4	2.6	75	41740	0.10
Michigan	76.3	5.4	75	29837	0.08
Minnesota	78.8	2.1	82	34896	0.06
Mississippi	73.6	8.1	62	24530	0.16
Missouri	75.9	7.7	75	29245	0.07
Montana	77.2	2.4	83	27602	0.09
Nebraska	77.8	3.8	85	31015	0.07
Nevada	75.8	6.3	58	34424	0.16
New Hampshire	78.3	1	71	35302	0.06
New Jersey	77.5	4.3	75	41835	0.17
New Mexico	77.0	7.2	65	26766	0.16
New York	77.7	4.3	70	40296	0.22
North Carolina	75.8	6.5	63	28604	0.14
North Dakota	78.3	0.5	88	29633	0.06
Ohio	76.2	4.7	77	29657	0.09
Oklahoma	75.2	5.8	74	29044	0.12
Oregon	77.8	2.2	67	29580	0.10
Pennsylvania	76.7	5.6	82	32986	0.13
Rhode Island	78.3	2.8	72	33560	0.08
South Carolina	74.8	6.8	62	26374	0.15
South Dakota	77.7	3.2	80	28833	0.07
Tennessee	75.1	6.6	60	28301	0.13
Texas	76.7	5.6	67	31624	0.19
Utah	78.7	1.4	81	26523	0.09
Vermont	78.2	2.7	84	31184	0.07
Virginia	76.8	4.7	74	35162	0.12
Washington	78.2	2.9	70	34368	0.10
West Virginia	75.1	3.3	82	25118	0.13
Wisconsin	77.9	2.6	85	30655	0.07
Wyoming	76.7	1.9	81	36760	0.09

26. Breakfast cereals again We saw in Chapter 7 that the calorie count of a breakfast cereal is linearly associated with its sugar content. Can we predict the calories of a serving from its vitamin and mineral content? Here's a multiple regression model of *Calories* per serving on its *Sodium* (mg), *Potassium* (mg), and *Sugars* (g):

Dependent variable is: Calories
R-squared = 38.4%   R-squared (adjusted) = 35.9%
s = 15.60 with 77 − 4 = 73 degrees of freedom

Source	Sum of Squares	dF	Mean Square	F-ratio	P-value
Regression	11091.8	3	3697.28	15.2	<0.0001
Residual	17760.1	73	243.289		

Variable	Coefficient	SE(Coeff)	t-ratio	P-value
Intercept	83.0469	5.198	16.0	<0.0001
Sodium	0.05721	0.0215	2.67	0.0094
Potass	−0.01933	0.0251	−0.769	0.4441
Sugars	2.38757	0.4066	5.87	<0.0001

Assuming that the conditions for multiple regression are met,

a) What is the regression equation?
b) Do you think this model would do a reasonably good job at predicting calories? Explain.
c) Would you consider removing any of these predictor variables from the model? Why or why not?
d) To check the conditions, what plots of the data might you want to examine?

27. Burger King Recall the Burger King Items menu data from Chapter 7. BK's nutrition sheet lists many variables. Here's a multiple regression to predict calories for Burger King foods from *Protein* content *(g)*, *Total Fat (g)*, *Carbohydrate (g)*, and *Sodium (mg)* per serving:

Dependent variable is: Calories
R-squared = 99.88%   R-squared (adjusted) = 99.88%
s = 8.178 with 122 − 5 = 117 degrees of freedom

Source	Sum of Squares	dF	Mean Square	F-ratio
Regression	6652972.7	4	1663243.2	24861.63
Residual	7824.2	117	66.9	

Variable	Coefficient	SE(Coeff)	t-ratio	P-value
Intercept	0.27812	1.79532	0.15492	0.87715
Protein	3.71339	0.09533	38.95356	<0.0001
Fat	9.13441	0.08212	111.23529	<0.0001
Carbs	3.82954	0.04400	87.03665	<0.0001
Sodium	0.00879	0.00201	4.37341	<0.0001

a) Do you think this model would do a good job of predicting calories for a new BK menu item? Why or why not?
b) The mean of *Calories* is 453.98 with a standard deviation of 234.63. Discuss what the value of *s* in the regression means about how well the model fits the data.
c) Does the R^2 value of mean that the residuals are all almost equal to zero?

Just Checking ANSWERS

1. 80.33% of the variation in *Maximum Wind Speed* can be accounted for by multiple regression on *Central Pressure* and *Year*.

2. In any given year, hurricanes with a *Central Pressure* that is 1 mb lower can be expected to have, on average, winds that are 0.971 kn faster.

3. First, the researcher is trying to prove his null hypothesis for this coefficient and, as we know, statistical inference won't permit that. Beyond that problem, we can't even be sure we understand the relationship of *Wind Speed* to *Year* from this analysis. For example, both *Central Pressure* and *Wind Speed* might be changing over time, but their relationship might well stay the same during any given year.

Review of Part VII

Inference When Variables Are Related

Quick Review

This part introduces inference for the most widely used class of statistical models: linear models. You may have read one, two, or all of the chapters in this part. They have a consistent central theme, and it is one you'll likely see if you study more Statistics.

- Linear models predict a single quantitative variable from one or more other variables. The predictor variables can be quantitative or categorical. Linear models require only that the effects of the predictor variables be simply added together (rather than, for example, being multiplied or exponential).

- When the predictors are categorical, we are finding an Analysis of Variance (ANOVA).

- When the predictors are quantitative, we are finding a regression model. With two or more predictors, we have a *multiple* regression model.

- The overall fit of a linear model is tested with an F-test. The null hypothesis for this test is that we'd do better just modeling y with its mean.

- If the overall F-test is significant, we can move on to consider the individual coefficients. For regression, this means considering the t-statistics. For ANOVA, it means first considering the F-statistics for individual factors. We may then move on to consider *contrasts* among different levels of significant factors.

- Models with two or more predictors can fail conditions in even more ways than those with only one, and some of these can be quite subtle. These models require care in checking assumptions and conditions.

Remember (have we said this often enough yet?): Never use any inference procedure without first checking the assumptions and conditions. We summarized those in the last unit; have another look.

Here come more opportunities to practice using these concepts and skills, but we've also sprinkled in some exercises that call for methods learned in earlier chapters, so keep alert.

PART VII Inference When Variables Are Related

Quick Guide to Inference

Inference about?	One sample or two?	Procedure	Model	Parameter	Estimate	SE	Chapter
Proportions	One sample	1-Proportion z-Interval	z	p	$\hat{p}$	$\sqrt{\dfrac{\hat{p}\hat{q}}{n}}$	19
		1-Proportion z-Test				$\sqrt{\dfrac{p_0 q_0}{n}}$	20, 21
	Two independent groups	2-Proportion z-Interval	z	$p_1 - p_2$	$\hat{p}_1 - \hat{p}_2$	$\sqrt{\dfrac{\hat{p}_1\hat{q}_1}{n_1} + \dfrac{\hat{p}_2\hat{q}_2}{n_2}}$	22
		2-Proportion z-Test				$\sqrt{\dfrac{\hat{p}\hat{q}}{n_1} + \dfrac{\hat{p}\hat{q}}{n_2}},\ \hat{p} = \dfrac{y_1 + y_2}{n_1 + n_2}$	22
Means	One sample	t-Interval t-Test	t df $= n - 1$	μ	$\bar{y}$	$\dfrac{s}{\sqrt{n}}$	23
	Two independent groups	2-Sample t-Test 2-Sample t-Interval	t df from technology	$\mu_1 - \mu_2$	$\bar{y}_1 - \bar{y}_2$	$\sqrt{\dfrac{s_1^2}{n_1} + \dfrac{s_2^2}{n_2}}$	24
	n Matched pairs	Paired t-Test Paired t-Interval	t df $= n - 1$	μ_d	$\bar{d}$	$\dfrac{s_d}{\sqrt{n}}$	25
Distributions (one categorical variable)	One sample	Goodness of fit	χ^2 df $=$ cells $- 1$			$\sum \dfrac{(\text{obs} - \text{exp})^2}{\text{exp}}$	26
	Many independent samples	Homogeneity χ^2 Test					
Independence (two categorical variables)	One sample	Independence χ^2 Test	χ^2 df $= (r - 1)(c - 1)$				
Association (two quantitative variables)	One sample	Linear Regression t-Test or Confidence Interval for β		β_1	b_1	$\dfrac{s_e}{s_x\sqrt{n - 1}}$ (compute with technology)	27
		Confidence Interval for μ_v	t df $= n - 2$	μ_v	$\hat{y}_v$	$\sqrt{SE^2(b_1) \cdot (x_v - \bar{x})^2 + \dfrac{s_e^2}{n}}$	
		Prediction Interval for y_v		y_v	$\hat{y}_v$	$\sqrt{SE^2(b_1) \cdot (x_n - \bar{x})^2 + \dfrac{s_e^2}{n} + s_e^2}$	
Inference about?	One sample or two?	Procedure	Model	Parameter	Estimate	SE	Chapter

Review Exercises

Note: Exercises marked with a ‡ refer to material found in chapter 29, which is located on the DVD and at pearsonhighered.com/dvb

1. **Mutual funds 2013** We have a sample of mutual funds of various types that report their return (%) in the previous year and in the previous 3 years.
   a) Create a 95% confidence interval for the difference in rate of return for the 1- and 3-year periods covered by these data. Clearly explain what your interval means.
   b) It's common for advertisements to carry the disclaimer that "past returns may not be indicative of future performance," but do these data indicate that there was an association between 1-year and 3-year rates of return?

Symbol	Fund	1 Yr	3 Yr
ALVIX	American Century Large Co Val Inv	19.09	12.85
SINAX	ClearBridge Large Cap Value A	19.27	12.91
INDZX	Columbia Diversified Equity Income A	16.76	11.71
CFVLX	Commerce Value	18.23	15.76
EHSTX	Eaton Vance Large-Cap Value A	17.45	9.66
HFCVX	Hennessy Cornerstone Value Investor	17.15	13.61
HDCAX	Huntington Dividend Capture A	11.95	12.32
JIVAX	JPMorgan Intrepid Value A	18.77	13.25
NPRTX	Neuberger Berman Large Cap Value Inv	19.05	9.51
NQCAX	Nuveen NWQ Large-Cap Value A	7.96	8.61
PQIAX	Principal Equity Income A	14.77	14.28
PEYAX	Putnam Equity Income A	20.75	14.39
STVTX	RidgeWorth Large Cap Value Equity I	18.66	14.03

2. **Polling** How accurate are pollsters in predicting the outcomes of Congressional elections? The table shows the actual number of Democrat seats in the House of Representatives and the number predicted by the Gallup organization for nonpresidential election years since World War II.
   a) Is there a significant difference between the number of seats predicted for the Democrats and the number they actually held? Test an appropriate hypothesis and state your conclusions.
   b) Is there a strong association between the pollsters' predictions and the outcomes of the elections? Test an appropriate hypothesis and state your conclusions.

Democratic Congressmen		
Year	Predicted	Actual
1946	190	188
1950	235	234
1954	232	232
1958	272	283
1962	259	258
1966	247	248
1970	260	255
1974	292	291
1978	277	277
1982	275	269
1986	264	258
1990	260	267
1994	201	204
1998	211	211

3. **Football** A student runs an experiment to test four different grips on his football throwing distance, recording the distance in yards that he can throw the football using each grip. He randomizes the grip used each time by drawing numbers out of a hat until each grip has been used 5 times. After collecting his data and analyzing the results, he reports that the P-value of his test is 0.0032.
   a) What kind of test should he have performed?
   b) What are the null and alternative hypotheses?
   c) Assuming that the conditions required for the test are satisfied, what would you conclude?
   d) What else about the data would you like to see in order to check the assumptions and conditions?
   e) What might he want to test next?

4. **Golf** A student runs an experiment to test four different clubs on her putting accuracy, recording the distance in centimeters from a small target that she places on the green. She randomizes the club used each time by drawing numbers out of a hat until each club has been used 6 times. After collecting her data and analyzing the results, she reports that the P-value of her test is 0.0245.
   a) What kind of test should she have performed?
   b) What are the null and alternative hypotheses?
   c) Assuming that the conditions required for the test are satisfied, what would you conclude?
   d) What else about the data would you like to see in order to check the assumptions and conditions?
   e) What might she want to test next?

‡5. **Horses again, a bit less wild** In an attempt to control the growth of the herds of wild horses discussed in Review VI Exercise 31, managers sterilized some of the stallions. The variable *Sterilized* is coded 0 for herds and

years in which no stallions were sterilized and 1 for herds and years in which some stallions were sterilized. The resulting regression looks like this:

Dependent variable is: Foals
R-squared = 84.8%   R-squared (adjusted) = 83.9%
s = 4.814 with 38 − 3 = 35 degrees of freedom

Source	Sum of squares	DF	Mean square	F-ratio
Regression	4511.85	2	2255.92	97.3
Residual	811.229	35	23.1780	

Variable	Coefficient	SE(Coeff)	t-ratio	P-value
Intercept	6.23667	4.801	1.30	0.2024
Adults	0.112271	0.0268	4.18	0.0002
Sterilized	−6.43657	3.769	−1.71	0.0965

a) What is the name for the kind of variable used here to represent sterilizing?
b) What is the interpretation of the coefficient of *Sterilized*?
c) Does sterilizing appear to work? That is, do these data show a statistically significant effect of sterilizing stallions on the number of foals born?

‡6. **North and hard** The data on water hardness and mortality considered in Part VI Review Exercise 29 also included information on whether the town was north or south of Derby. When that information is coded in a variable as 1 for "north of Derby" and 0 for "south of Derby," and that variable is included in the regression, the resulting table looks like this:

Dependent variable is: Mortality
R-squared = 56.2%   R-squared (adjusted) = 54.7%
s = 126.4 with 61 − 3 = 58 degrees of freedom

Variable	Coefficient	SE(Coeff)	t-ratio	P-value
Intercept	1537.50	42.02	36.6	<0.0001
Calcium	−2.16011	0.4979	−4.34	<0.0001
North of Derby	158.892	37.87	4.20	<0.0001

a) What is the name for the kind of variable that the 0/1 variable is for being north of Derby?
b) What does the coefficient of that variable mean in this regression?
c) Would you prefer this regression to the one of Review VI. Explain why or why not.

7. **Lost baggage** The Bureau of Transportation Statistics of the U.S. Department of Transportation reports statistics about airline performance. For 2005 they report the following number of bags lost per 1000 passengers.

Airline	Lost Bags	Airline	Lost Bags
Continental	4.12	Northwest	4.84
Southwest	4.25	American	5.92
United	4.26	Delta	7.09
America West	4.33	USAir	9.62

Are the airlines roughly equal in their baggage performance? Perform a chi-square goodness-of-fit test, or explain why that would not be appropriate.

8. **Bowling** A student runs an experiment to test how different factors affect her bowling performance. She uses 3 levels for the weight of the ball (low, medium, and high) and 2 approaches (standing and walking), throwing 4 balls at each condition and choosing the conditions in random order. She counts the number of pins knocked down as the response variable.
   a) What are the null and alternative hypotheses for the main effects?
   b) How many degrees of freedom does each factor sum of squares have? How about the error sum of squares?
   c) Should she consider fitting an interaction term to the model? Why or why not? How many degrees of freedom would it take?

9. **Video racing** A student runs an experiment to test how different factors affect his reaction time while playing video games. He uses a specified race course with random hazards and times how long it takes him (in seconds) to finish 4 laps under a variety of experimental conditions. As factors, he uses three different types of *Mouse* (cordless ergonomic, cordless regular, or corded) and keeps the *Lights* on or off. He measures his time once at each condition. To avoid a learning effect, he runs the race 5 times before selecting the conditions in random order.
   a) What are the null and alternative hypotheses for the main effects?
   b) How many degrees of freedom does each factor sum of squares have? How about the error sum of squares?
   c) Should he consider fitting an interaction term to the model? Why or why not? How many degrees of freedom would it take?

10. **Resume fraud** In 2002 the Veritas Software company found out that its chief financial officer did not actually have the MBA he had listed on his resume. They fired him, and the value of the company's stock dropped 19%. Kroll, Inc., a firm that specializes in investigating such matters, said that they believe as many as 25 percent of background checks might reveal false information. How many such random checks would they have to do to estimate the true percentage of people who misrepresent their backgrounds to within ±5% with 98% confidence?

11. Paper planes In preparation for a regional paper airplane competition, a student tried out her latest design. The distances her plane traveled (in feet) in 11 trial flights are given here. (The world record is an astounding 193.01 feet!) The data were 62, 52, 68, 23, 34, 45, 27, 42, 83, 56, and 40 feet.

Here are some summaries:

Count	11
Mean	48.3636
Median	45
StdDev	18.0846
StdErr	5.45273
IntQRange	25
25th %tile	35.5000
75th %tile	60.5000

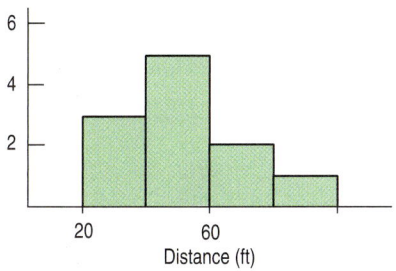

a) Construct a 95% confidence interval for the mean distance.
b) Based on your confidence interval, is it plausible that the mean distance is 40 ft? Explain.
c) How would a 99% confidence interval for the mean distance differ from your answer in part a? Explain briefly, without actually calculating a new interval.
d) How large a sample size would the student need to get a confidence interval half as wide as the one you got in part a, at the same confidence level?

12. Nuclear plants Here are data on 32 light water nuclear power plants. The variables are:

Cost: In $100,000, adjusted to 1976 base.

Date: Date that construction permit was issued in years after 1900. Thus, 68.58 is roughly halfway through 1968.

Mwatts: Power plant net capacity in megawatts.

We are interested in the *Cost* of the plants as a function of *Date* and *Mwatts*.

Cost	Mwatts	Date	Cost	Mwatts	Date
345.39	514	67.92	712.27	845	69.50
460.05	687	68.58	289.66	530	68.42
452.99	1065	67.33	881.24	1090	69.17
443.22	1065	67.33	490.88	1050	68.92
652.32	1065	68.00	567.79	913	68.75
642.23	1065	68.00	665.99	828	70.92
272.37	822	68.17	621.45	786	69.67
317.21	457	68.42	608.80	821	70.08
457.12	822	68.42	473.64	538	70.42
690.19	792	68.33	697.14	1130	71.08
350.63	560	68.58	207.51	745	67.25
402.59	790	68.75	288.48	821	67.17
412.18	530	68.42	284.88	886	67.83
495.58	1050	68.92	280.36	886	67.83
394.36	850	68.92	217.38	745	67.25
423.32	778	68.42	270.71	886	67.83

a) Examine the relationships between *Cost* and *Mwatts* and between *Cost* and *Date*. Make appropriate displays and interpret them with a sentence or two.
b) Find the regression of *Cost* on *Mwatts*. Write a sentence that explains the relationship as described by the regression.
c) Make a scatterplot of residuals vs. predicted values and discuss what it shows. Make a Normal probability plot or histogram of the residuals. Discuss the four assumptions needed for regression analysis and indicate whether you think they are satisfied here. Give your reasons.
d) State the standard null hypothesis for the slope coefficient and complete the *t*-test at the 5% level. State your conclusion.
e) Estimate the cost of a 1000-mwatt plant. Show your work.
f) Compute the residuals for this regression. Discuss the meaning of the *R*-squared in this regression. Plot the residuals against *Date*. Does it appear that *Date* can account for some of the remaining variability?
g) Compute the multiple regression of *Cost* on both *Mwatts* and *Date*. Compare the coefficient in this regression with those you have found for each of these predictors.
‡h) Would you expect *Mwatts* and *Date* to be correlated? Why or why not? Examine the relationship between *Mwatts* and *Date*. Make a scatterplot and find the correlation coefficient, for example. It's only because of the extraordinary nature of this relationship that the relationships you saw at earlier steps were this simple.

13. **Barbershop music** At a barbershop music singing competition, choruses are judged on three scales: *Music* (quality of the arrangement, etc.), *Performance*, and *Singing*. The scales are supposed to be independent of each other, and each is scored by a different judge, but a friend claims that he can predict a chorus's singing score from the other two scores. He offers the following regression based on the scores of all 34 choruses in a recent competition:

Dependent variable is: Singing
R-squared = 90.9% R-squared (adjusted) = 90.3%
s = 6.483 with 34 − 3 = 31 degrees of freedom

Variable	Coefficient	SE(Coeff)	t-ratio	P-value
Intercept	2.08926	7.973	0.262	0.7950
Performance	0.793407	0.0976	8.13	≤0.0001
Music	0.219100	0.1196	1.83	0.0766

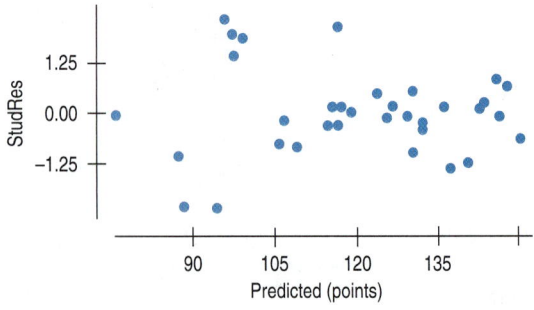

a) What do you think of your friend's claim? Can he predict singing scores? Explain.
b) State the standard null hypothesis for the coefficient of *Performance* and complete the *t*-test at the 5% level. State your conclusion.
c) Complete the analysis. Check assumptions and conditions to the extent you can with the information provided.

14. **Sleep** Using a simple random sample, a student group asked 450 students about their sleep and study habits and received about 200 responses. The group wanted to know if the average amount of sleep varied by sex (F or M) or by year (Freshman, Sophomore, Junior, or Senior) of the respondent. Partial boxplots of the amount of sleep last night by the two factors are shown below with the interaction plot:

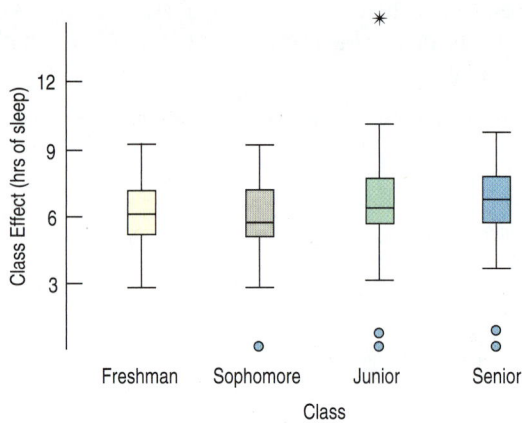

The ANOVA table shows:

Source	DF	Sum of squares	F-ratio	P-value
Sex	1	8.0658	2.2821	0.1325
Year	3	8.8990	0.8393	0.4739
Sex × Year	3	19.4075	1.8303	0.1431
Total	196	707.0180		

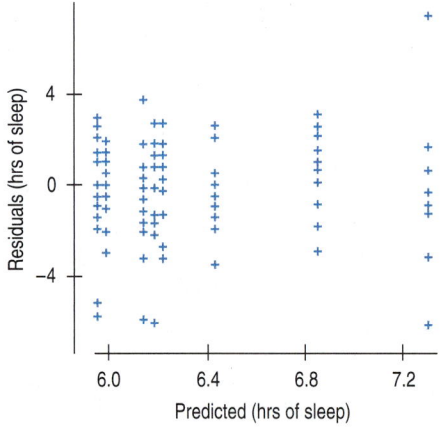

a) What are the null and alternative hypotheses for the main effects?
b) Should the group consider fitting an interaction term to the model? Why or why not?
c) What effects appear to be significant?
d) What, if any, reservations do you have about the conclusions?

15. **Study habits** The survey in Exercise 14 also asked students about the number of hours they studied. Those doing the survey wanted to know if the average amount of studying varied by sex or by class of the respondent. Partial boxplots of the *Hours Studied* last night by the two factors are shown below:

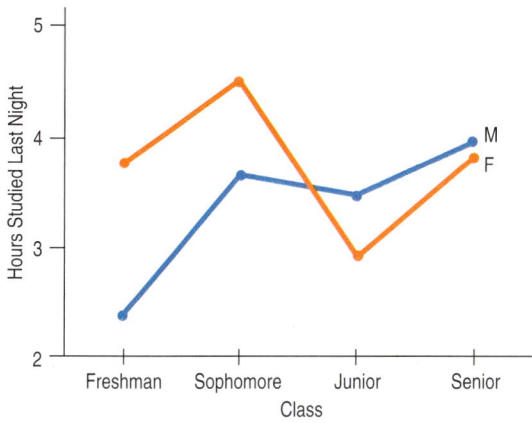

Source	DF	Sum of squares	F-ratio	P-value
Sex	1	7.1861	1.0744	0.3013
Year	3	34.9922	1.7440	0.1595
Sex × Year	3	27.8188	1.3865	0.2483
Total	196	1329.8020		

A plot of residuals vs. predicted shows:

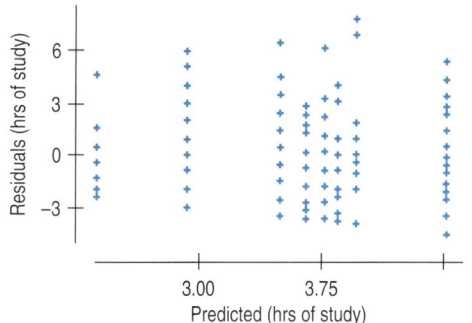

a) What are the null and alternative hypotheses for the main effects?
b) Should those doing the survey consider fitting an interaction term to the model? Why or why not?
c) What effects appear to be significant?
d) What, if any, reservations do you have about the conclusions?

16. **Is Old Faithful getting older?** The data on Old Faithful eruptions that we saw in Review VI Exercise 42 includes another variable recording the *Day* on which the eruption occurred (where 1 is the first day and each successive day just counts one more). The correlation of *Interval* (minutes until the next eruption) with *Day* is -0.004. But when we include *Day* in a multiple regression along with the *Duration* (in minutes) of the previous eruption, we get the following model:

Dependent variable is: Interval
R-squared = 77.6% R-squared (adjusted) = 77.3%
s = 6.092 with $222 - 3 = 219$ degrees of freedom

Variable	Coefficient	SE(Coeff)	t-ratio	P-value
Intercept	35.2463	1.509	23.4	≤0.0001
Duration	10.4348	0.3794	27.5	≤0.0001
Day	−0.126316	0.0523	−2.42	0.0166

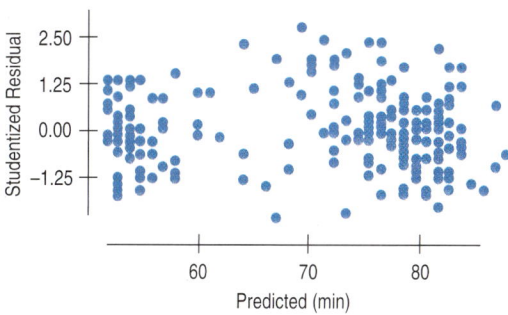

a) What is the model fit by this regression?
b) Is the *Interval* changing over time? Perform a formal test of the relevant hypothesis.
c) Doesn't the small P-value for *Day* contradict the correlation between *Interval* and *Day* being virtually zero? Explain.
d) Is the amount of change in *Interval* due to *Day* meaningful?

17. **Teen traffic deaths 2007** The Insurance Institute for Highway Safety publishes data on a variety of traffic-related risks. One report gives the numbers of male and female teenagers killed in highway accidents during each

year from 1975 to 2007. Here is a regression predicting *Female Deaths* by *Year*:

Dependent variable is: Female deaths
R-squared = 57.4%   R-squared (adjusted) = 56.1%
s = 182.6 with 33 − 2 = 31 degrees of freedom

Source	Sum of squares	df	Mean square	F-ratio
Regression	1396025	1	1396025	41.9
Residual	1033964	31	33353.7	

Variable	Coefficient	SE(Coeff)	t-ratio	P-value
Intercept	45074.0	6648	6.78	<0.0001
Year	−21.6006	3.339	−6.47	<0.0001

a) Here's a scatterplot of residuals vs. predicted values for this regression. Discuss the assumptions and conditions of the regression.

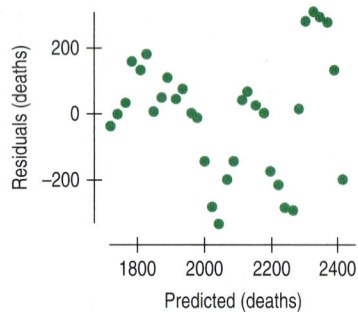

b) What is the meaning of the R^2 value of 57.4% for this regression?
c) What is the regression equation?
d) Give an interpretation of the coefficient of *Year* in this regression.

18. **Births** The National Vital Statistics Report provides information on live *Births* (per 1000 women), according to the age of the woman (in 5-year brackets—*Age* used here is the midpoint of the bracket) and the *Year* from 1990 to 1999. The report isolates births to women younger than 20 as a separate category. Looking only at women 20 years old and older, we find the following regression:

Dependent variable is: Births
R-squared = 98.1%   R-squared (adjusted) = 98.0%
s = 15.55 with 50 − 3 = 47 degrees of freedom

Variable	Coefficient	SE(Coeff)	t-ratio	P-value
Intercept	4422.98	1527	2.90	0.0057
Age	−15.1898	0.3110	−48.8	<0.0001
Year	−1.89830	0.7656	−2.48	0.0168

a) Write out the regression model.
b) How would you interpret the coefficient of *year* in this regression? What happened to pregnancy rates during the decade of the 1990s?
Here's a scatterplot of the Studentized residuals against the predicted values.

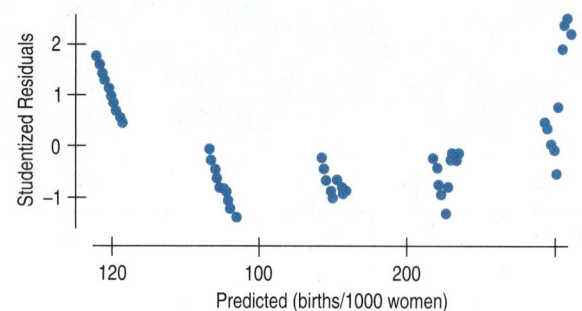

c) What might you do to improve the regression model?

19. **Typing** For a class project, Nick M. designed and carried out an experiment to see if the room *Temperature* and the wearing of *Gloves* affected his typing speed. He ran each combination of hot and cold temperature and gloves on and off 8 times, recording the net number of words typed (words typed minus mistakes). Partial boxplots and interaction plots show:

The ANOVA table shows:

Source	DF	Sum of squares	F-ratio	P-value
Gloves	1	120.1250	57.2511	<0.0001
Temperature	1	18.0000	8.5787	0.0067
Gloves*Temp	1	15.1250	7.2085	0.0121
Error	28	58.75		

A scatterplot of residuals vs. predicted values shows:

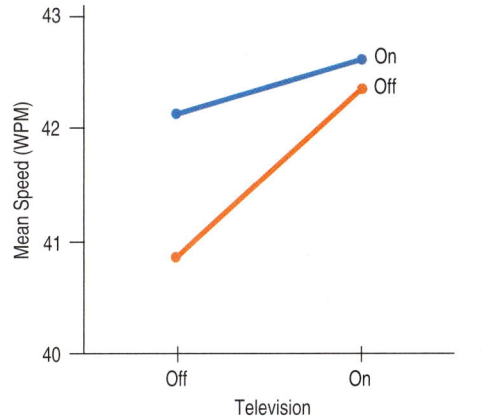

The ANOVA table shows:

Source	DF	Sum of squares	F-ratio	P-value
Music	1	4.5000	0.6380	0.4312
Television	1	8.0000	1.1342	0.2960
Interaction	1	2.0000	0.2835	0.5986
Error	28	197.5000		

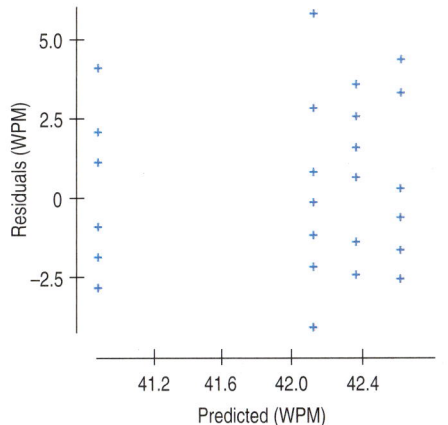

a) What are the null and alternative hypotheses for the main effects?
b) Given the partial boxplots, what do you suspect the F-test will say about the two main effects?
c) Should he consider fitting an interaction term to the model? Why or why not?
d) If he does fit an interaction term, do you suspect it will be significant on the basis of the interaction plot? Explain.
e) Which effects appear to be significant?
f) Describe the effects of the factors.
g) What is the size of the estimated standard deviation of the errors? Does this seem reasonable given the partial boxplots? Explain.
h) If Nick wants to increase his typing speed, what recommendations would you give him?
i) What reservations, if any, do you have about the conclusions?

20. **Typing again** Nick (see Exercise 19) designed a follow-up experiment to see if having *Music* or *Television* on would affect his typing speed. In particular, he'd like to know if he can type just as effectively with the music and/or the TV on while he types. He ran each combination of *Music* and *Television* on or off 8 times, recording the net number of words typed (words typed minus mistakes). Partial boxplots and interaction plots show:

a) What are the null and alternative hypotheses for the main effects?
b) Given the partial boxplots, what do you suspect the *F*-test will say about the two main effects?
c) Should he consider fitting an interaction term to the model? Why or why not?
d) If he does fit an interaction term, do you suspect it will be significant on the basis of the interaction plot? Explain.
e) What effects appear to be significant?
f) Describe the effects of the factors.
g) What is the size of the estimated standard deviation of the errors? Does this seem reasonable given the partial boxplots? Explain.
h) If Nick wants to increase his typing speed, what recommendations would you give him?
i) What reservations, if any, do you have about the conclusions?

T 21. NY Marathon The *New York Times* reported the results of the 2003 NY Marathon by listing time brackets and the number of racers who finished within that bracket. Because the brackets are of different sizes, we look at the number of racers finishing *per minute* against the time at the middle of the time bracket. The resulting regression looks like this:

Dependent variable is: #/minute
R-squared = 10.9%   R-squared (adjusted) = 9.5%
s = 78.45 with 63 − 2 = 61 degrees of freedom

Variable	Coefficient	SE(Coeff)	t-ratio	P-value
Intercept	45.1407	51.74	0.873	0.3863
Mid time ...	0.519037	0.1899	2.73	0.0082

a) How would you interpret the coefficient of *Mid Time*?

Here's a scatterplot of the Studentized residuals against the predicted values for this regression:

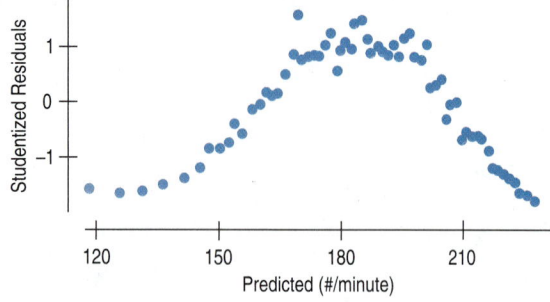

b) Comment on the regression in the light of this plot.

22. Learning math Developers of a new math curriculum called "Accelerated Math" compared performances of students taught by their system with control groups of students in the same schools who were taught using traditional instructional methods and materials. Statistics about pretest and posttest scores are shown in the table. (J. Ysseldyke and S. Tardrew, *Differentiating Math Instruction*, Renaissance Learning, 2002)
a) Did the groups differ in average math score at the start of this study?
b) Did the group taught using the Accelerated Math program show a significant improvement in test scores?
c) Did the control group show significant improvement in test scores?
d) Were gains significantly higher for the Accelerated Math group than for the control group?

		Instructional Method	
		Acc. math	Control
Number of students		231	245
Pretest	Mean	560.01	549.65
	St. Dev	84.29	74.68
Posttest	Mean	637.55	588.76
	St. Dev	82.9	83.24
Individual gain	Mean	77.53	39.11
	St. Dev.	78.01	66.25

23. Pesticides A study published in the journal *Environmental Health Perspectives* examined the sex ratios of children born to workers exposed to dioxin in Russian pesticide factories. The data covered the years 1961 to 1988 in the city of Ufa, Bashkortostan, Russia. Of 227 children born to workers exposed to dioxin, only 40% were male. Overall in the city of Ufa the proportion of males was 51.2%. Is this evidence that human exposure to dioxin may results in the birth of more girls? (An interesting note: It appeared that paternal exposure was most critical; 51% of babies born to mothers exposed to the chemical were boys.)

24. Video pinball A student runs an experiment to test how different factors affect his score while playing video pinball. Here are the results of 16 runs of an experiment performed in random order. Factor *Eyes* has two levels: *both* open and *right eye* closed. Factor *Tilt* has two levels: tilt *on* and tilt *off*. The response is the score of one ball at a combination of factors. Each combination was repeated 4 times in random order.
a) What are the null and alternative hypotheses for the main effects?
b) Analyze the data and write a short report on your findings. Include appropriate graphics and diagnostic plots.

Eyes	Tilt	Score
Both	On	67059
Right eye	Off	21036
Both	On	59520
Right eye	Off	3100
Both	Off	61272
Right eye	On	55957
Both	Off	72472
Right eye	On	18460
Right eye	On	16556
Both	Off	89553
Right eye	On	37950
Both	Off	74336
Right eye	Off	700
Both	On	79037
Right eye	Off	36591
Both	On	74610

25. Javelin Brianna, a member of the track team, runs an experiment to test how different factors affect her javelin throw. She wants to know if the more expensive (*premium*) javelin is worth the extra price and is curious to know how much warming up helps her distance. She tries all four combinations of the two *Javelins* (*standard* and *premium*) and *Preparation* (no warm-up—*cold*—and *warm-up*), repeating each combination twice in random order. She measures the distance of her throw to the nearest meter. Here are the results of the 8 runs.

Javelin	Preparation	Distance (meters)
Premium	Warm-up	46
Standard	Warm-up	39
Premium	Cold	37
Standard	Cold	30
Premium	Warm-up	45
Standard	Warm-up	40
Premium	Cold	35
Standard	Cold	32

a) What are the null and alternative hypotheses for the main effects?
b) Analyze the data and write a short report on your findings. Include appropriate graphics and diagnostic plots and a recommendation for Brianna to optimize her javelin distance.

26. LA rainfall 2012 The Los Angeles Almanac website reports recent annual rainfall (in inches), as shown in the table.
a) Create a 90% confidence interval for the mean annual rainfall in LA.
b) If you wanted to estimate the mean annual rainfall with a margin of error of only 2 inches, how many years' data would you need?

c) Do these data suggest any change in annual rainfall as time passes? Check for an association between rainfall and year.

Year	Rainfall (in.)	Year	Rainfall (in.)
1990	7.35	2002	4.42
1991	11.99	2003	16.42
1992	21.00	2004	9.25
1993	27.36	2005	37.96
1994	8.11	2006	13.19
1995	24.35	2007	3.21
1996	12.44	2008	13.53
1997	12.40	2009	9.08
1998	31.01	2010	16.36
1999	9.09	2011	20.20
2000	11.57	2012	8.69
2001	17.94		

27. TV and athletics Using a simple random sample, a student asked 200 students questions about their study and workout habits, and received 124 responses. One of the questions asked, "Do you participate in intramural (IM) athletics, varsity athletics, or no athletics?" while the second asked, "How many hours of television did you watch last week?" The student wants to see if participation in athletics is associated with amount of TV watched. Here are the boxplots of *TV Watching* by *Athletic Participation*:

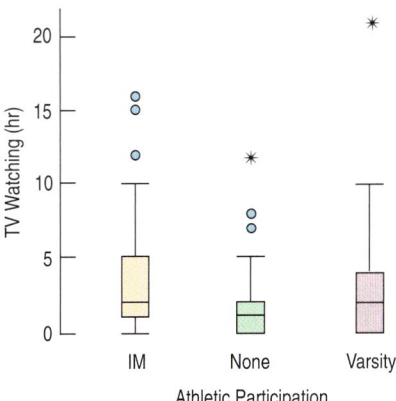

Analysis of Variance

Source	DF	Sum of squares	Mean square	F-ratio	P-value
Athl. Part.	2	128.271	64.1353	4.2343	0.0167
Error	121	1832.72	15.1465		
Total	123	1960.99			

a) State the hypothesis about the students (both numerically and in words).
b) Do the conditions for ANOVA appear to be satisfied? What concerns, if any, do you have?
c) Assuming that the assumptions for inference are satisfied, perform the hypothesis test and state your

conclusion. Be sure to state it in terms of television watching and athletic participation.

d) An analysis of the data with all the points highlighted as outliers removed was performed. The *F*-test showed a P-value of 0.0049. How does this affect your answer to part b?

28. **Weight and athletics** Using the same survey as in Exercise 27, the student examined the relationship between *Athletic Participation* and *Weight*. Here are the boxplots of *Weight* by *Athletic Participation*:

Analysis of Variance

Source	DF	Sum of squares	Mean square	F-ratio	P-value
Athl. Part.	2	9025.55	4512.78	5.7368	0.0042
Error	121	95183.2	786.638		
Total	123	104209			

a) State the null hypothesis about the students (both numerically and in words).
b) Do the conditions for ANOVA appear to be satisfied? What concerns, if any, do you have?
c) Assuming that the assumptions for inference are satisfied, perform the hypothesis test and state your conclusion. Be sure to state it in terms of *Weight* and *Athletic Participation*. What might explain the apparent relationship?
d) An analysis of the data with the points highlighted as an outlier removed was performed. The *F*-test showed a P-value of 0.0030. How does this affect your answer to part b?

29. **Weight loss** A weight loss clinic advertises that its program of diet and exercise will allow clients to lose 10 pounds in one month. A local reporter investigating weight reduction gets permission to interview a randomly selected sample of clients who report the given weight losses during their first month in this program. Create a confidence interval to test the clinic's claim that typical weight loss is 10 pounds.

Pounds Lost	
9.5	9.5
13	9
9	8
10	7.5
11	10
9	7
5	8
9	10.5
12.5	10.5
6	9

30. **Cramming** Students in two basic Spanish classes were required to learn 50 new vocabulary words. One group of 45 students received the list on Monday and studied the words all week. Statistics summarizing this group's scores on Friday's quiz are given. The other group of 25 students did not get the vocabulary list until Thursday. They also took the quiz on Friday after "cramming" Thursday night. Then, when they returned to class the following Monday they were retested—without advance warning. Both sets of test scores for these students are shown.

Group 1
Fri.
Number of students = 45
Mean = 43.2 (of 50)
StDev = 3.4
Students passing (score $\geq$ 40) = 33%

a) Did the week-long study group have a mean score significantly higher than that of the overnight crammers?
b) Was there a significant difference in the percentages of students who passed the quiz on Friday?
c) Is there any evidence that when students cram for a test their "learning" does not last for 3 days?
d) Use a 95% confidence interval to estimate the mean number of words that might be forgotten by crammers.
e) Is there any evidence that how much students forget depends on how much they "learned" to begin with?

Group 2			
Fri.	Mon.	Fri.	Mon.
42	36	50	47
44	44	34	34
45	46	38	31
48	38	43	40
44	40	39	41
43	38	46	32
41	37	37	36
35	31	40	31
43	32	41	32
48	37	48	39
43	41	37	31
45	32	36	41
47	44		

Appendix A: Answers

Here are the "answers" to the exercises for the chapters and the unit reviews. The answers are outlines of the complete solution. Your solution should follow the model of the Step-by-Step examples, where appropriate. You should explain the context, show your reasoning and calculations, and draw conclusions. For some problems, what you decide to include in an argument may differ somewhat from the answers here. But, of course, the numerical part of your answer should match the numbers in the answers shown.

Chapter 1

1. Retailers, and suppliers to whom they sell the information, will use the information about what products consumers buy to target their advertisements to customers more likely to buy their products.

3. The individual games.

5. a) Sample—A principal wants to know how many from each grade will be attending a performance of a school play.
   b) Sample—A principal wants to see the trend in the amount of mathematics learned as students leave each grade.

7. Categorical.

9. Quantitative.

11. Answers will vary.

13. *Who*—40 undergraduate women; *What*—Ability to differentiate gay men from straight men; *Population*—All women.

15. *Who*—2500 cars; *What*—Distance from car to bicycle; *Population*—All cars passing bicyclists.

17. *Who*—Coffee drinkers at a Newcastle University coffee station; *What*—Amount of money contributed; *Population*—All people in honor system payment situations.

19. *Who*—474 participants in the San Antonio Longitudinal Study of Aging; *What*—Diet soda consumption and waist size change; *Population*—All diet soda drinkers.

21. *Who*—54 bears; *Cases*—Each bear is a case; *What*—Weight, neck size, length, and sex; *When*—Not specified; *Where*—Not specified; *Why*—To estimate weight from easier-to-measure variables; *How*—Researchers collected data on 54 bears they were able to catch.
    Variable—Weight; *Type*—Quantitative; *Units*—Not specified; *Variable*—Neck size; *Type*—Quantitative; *Units*—Not specified; *Variable*—Length; *Type*—Quantitative; *Units*—Not specified; *Variable*—Sex; *Type*—Categorical.

23. *Who*—Arby's sandwiches; *Cases*—Each sandwich is a case; *What*—Type of meat, number of calories, and serving size; *When*—Not specified; *Where*—Arby's restaurants; *Why*—To assess nutritional value of sandwiches; *How*—Report by Arby's restaurants; *Variable*—Type of meat; *Type*—Categorical; *Variable*—Number of calories; *Type*—Quantitative; *Units*—Calories; *Variable*—Serving size; *Type*—Quantitative; *Units*—Ounces.

25. *Who*—882 births; *Cases*—Each of the 882 births is a case; *What*—Mother's age, length of pregnancy, type of birth, level of prenatal care, birth weight of baby, sex of baby, and baby's health problems; *When*—1998–2000; *Where*—Large city hospital; *Why*—Researchers were investigating the impact of prenatal care on newborn health; *How*—Not specified exactly, but probably from hospital records; *Variable*—Mother's age; *Type*—Quantitative; *Units*—Not specified, probably years; *Variable*—Length of pregnancy; *Type*—Quantitative; *Units*—Weeks; *Variable*—Birth weight of baby; *Type*—Quantitative; *Units*—Not specified, probably pounds and ounces; *Variable*—Type of birth; *Type*—Categorical; *Variable*—Level of prenatal care; *Type*—Categorical; *Variable*—Sex; *Type*—Categorical; *Variable*—Baby's health problems; *Type*—Categorical.

27. *Who*—Experiment subjects; *Cases*—Each subject is a case; *What*—Treatment (herbal cold remedy or sugar solution) and cold severity; *When*—Not specified; *Where*—Not specified; *Why*—To test efficacy of herbal remedy on common cold; *How*—The scientists set up an experiment; *Variable*—Treatment; *Type*—Categorical; *Variable*—Cold severity rating; *Type*—Quantitative (perhaps ordinal categorical); *Units*—Scale from 0 to 5; *Concerns*—The severity of a cold seems subjective and difficult to quantify. Scientists may feel pressure to report negative findings of herbal product.

29. *Who*—Streams; *Cases*—Each stream is a case; *What*—Name of stream, substrate of the stream, acidity of the water, temperature, BCI; *When*—Not specified; *Where*—Upstate New York; *Why*—To study ecology of streams; *How*—Not specified; *Variable*—Stream name; *Type*—Identifier; *Variable*—Substrate; *Type*—Categorical; *Variable*—Acidity of water; *Type*—Quantitative; *Units*—pH; *Variable*—Temperature; *Type*—Quantitative; *Units*—Degrees Celsius; *Variable*—BCI; *Type*—Quantitative; *Units*—Not specified

31. *Who*—353 refrigerator models; *Cases*—Each of the 353 refrigerator models is a case; *What*—Brand, cost, size, type, estimated annual energy cost, overall rating, and repair history; *When*—2013; *Where*—United States; *Why*—To provide information to the readers of *Consumer Reports*; *How*—Not specified; *Variable*—Brand; *Type*—Categorical; *Variable*—Cost; *Type*—Quantitative; *Units*—Not specified (dollars); *Variable*—Size; *Type*—Quantitative; *Units*—Cubic feet; *Variable*—Type; *Type*—Categorical; *Variable*—Estimated annual energy cost; *Type*—Quantitative; *Units*—Not specified (dollars); *Variable*—Overall rating; *Type*—Categorical (ordinal); *Variable*—Percent requiring repair in the past 5 years; *Type*—Quantitative; *Units*—Percent.

A-1

33. *Who*—Kentucky Derby races; *What*—Date, winner, jockey, trainer, owner, and time; *When*—1875 to 2013; *Where*—Churchill Downs, Louisville, Kentucky; *Why*—Not specified (To see trends in horse racing?); *How*—Official statistics collected at race; *Variable*—Year; *Type*—Identifier and Quantitative; *Units*—Year; *Variable*—Winner; *Type*—Identifier; *Variable*—Jockey; *Type*—Categorical; *Variable*—Trainer; *Type*—Categorical; *Variable*—Owner; *Type*—Categorical; *Variable*—Time; *Type*—Quantitative; *Units*—Minutes and seconds.

Chapter 2

1. Subcompact and mini 0.1128
Compact 0.3163
Intermediate 0.3380
Full 0.2193
Unknown 0.0137

3. 1996: B, 2006: C, 2008: A, 2012: D

5. a) 170/348 = 48.9%
b) 41/348 = 11.8%
c) 41/170 = 24.1%
d) 41/90 = 45.6%

7. Answers will vary.

9. Answers will vary.

11. a) Yes; each is categorized in a single genre.
b) Other.

13. a) SciFi/Fantasy
b) It is easy to tell from either chart; sometimes differences are easier to see with the bar chart because slices of the pie chart are too similar in size.

15. 1755 students applied for admission to the magnet schools program. 53% were accepted, 17% were wait-listed, and the other 30% were turned away.

17. a) Yes. We can add because these categories do not overlap. (Each person is assigned only one cause of death.)
b) 37.7%
c) Either a bar chart or pie chart with "other" added would be appropriate. A bar chart is shown.

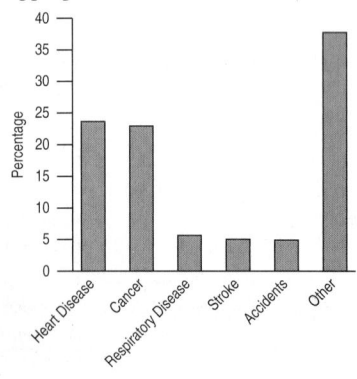

19. a) The bar chart shows that grounding and collision are the most frequent causes of oil spills with spills due to hull or equipment failures or due to fire & explosions much less common. Relatively few have unknown causes.
b) A pie chart seems appropriate as well.
c) Much easier to see from the bar chart.

21. There's no title, the percentages total only 93%, and the three-dimensional display distorts the sizes of the regions.

23. In both the South and West, about 58% of the eighth-grade smokers preferred Marlboro. Newport was the next most popular brand, but was far more popular in the South than in the West, where Camel was cited nearly 3 times as often as in the South. Nearly twice as many smokers in the West as in the South indicated that they had no usual brand (12.9% to 6.7%).

25. a) The column totals are 100%.
b) 25.86%   c) 28.57%
d) i. 27.36%;   ii. can't tell;   iii. 0%;   iv. can't tell

27. a) 82.5%   b) 12.9%   c) 11.1%
d) 13.4%   e) 85.7%

29. a) 73.9% 4-yr college, 13.4% 2-year college, 1.5% military, 5.2% employment, 6.0% other.
b) 77.2% 4-yr college, 10.5% 2-year college, 1.8% military, 5.3% employment, 5.3% other.
c) Many charts are possible. Here is a side-by-side bar chart.

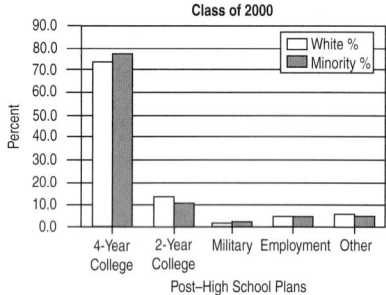

d) The white and minority students' plans are very similar. The small differences should be interpreted with caution because the total number of minority students is small. There is little evidence of an association between race and plans.

31. a) 16.6%   b) 11.8%
c) 37.7%   d) 53.0%

33. 1755 students applied for admission to the magnet schools program: 53% were accepted, 17% were wait-listed, and the other 30% were turned away. While the overall acceptance rate was 53%, 93.8% of blacks and Hispanics were accepted, compared to only 37.7% of Asians and 35.5% of whites. Overall, 29.5% of applicants were black or Hispanic, but only 6% of those turned away were. Asians accounted for 16.6% of all applicants, but 25.4% of those were turned away. Whites were 54% of the applicants and 68.5% of those who were turned away. It appears that the admissions decisions were not independent of the applicant's ethnicity.

35. a) 9.3%   b) 24.7%   c) 80.8%
d) No, there appears to be no association between weather and ability to forecast weather. On days it rained, his forecast was correct 79.4% of the time. When there was no rain, his forecast was correct 81.0% of the time. Although these are not exactly the same, there are pretty close.

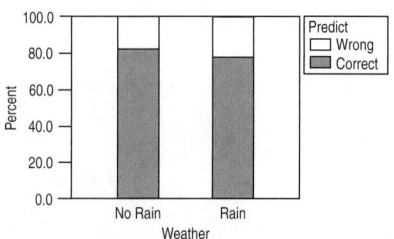

37. a) Low 20.0%, Normal 48.9%, High 31.0%.
b)

	Under 30	30–49	Over 50
Low	27.6%	20.7%	15.7%
Normal	49.0%	50.8%	47.2%
High	23.5%	28.5%	37.1%

(Blood Pressure)

c) Blood Pressure by Age bar chart (High, Normal, Low across Under 30, 30–49, over 50).

d) As age increases, the percent of adults with high blood pressure increases. By contrast, the percent of adults with low blood pressure decreases.
e) No, but it gives an indication that it might. There might be additional reasons that explain the differences in blood pressures.

39. No, there's no evidence that Prozac is effective. The relapse rates were nearly identical: 28.6% among the people treated with Prozac, compared to 27.3% among those who took the placebo.

41. a) 4.8%
b) 49.7%
c) There are about 50% of each sex in each age group. Below age 25, there are more male drivers in each age group. Above age 45, there are more female drivers in each group.
d) There is a slight association. As the age increases, there is a small increase in the percentage of female drivers.

43. a) 160 of 1300, or 12.3%
b) Yes. Major surgery: 15.3% vs. minor surgery: 6.7%.
c) Large hospital: 13%; small hospital: 10%.
d) Large hospital: Major 15% vs. minor 5%. Small hospital: Major 20% vs. minor 8%.
e) No. Smaller hospitals have a higher rate for both kinds of surgery, even though it's lower "overall."
f) The small hospital has a larger percentage of minor surgeries (83.3%) than the large hospital (20%). Minor surgeries have a lower delay rate, so the small hospital looks better "overall."

45. a) 42.6%
b) A higher percentage of males than females were admitted: Males: 47.2% to females: 30.9%.
c) Program 1: Males 61.9%, females 82.4%.
Program 2: Males 62.9%, females 68.0%.
Program 3: Males 33.7%, females 35.2%.
Program 4: Males 5.9%, females 7.0%.
d) The comparisons in part c show that males have a lower admittance rate in every program, even though the overall rate shows males with a higher rate of admittance. This is an example of Simpson's paradox.

Chapter 3

1. The boxplot does not show clusters and gaps, so bimodality (or multimodality) will not generally be apparent in a boxplot.

3. $IQR = 116 - 98 = 18$. $Q3 + 1.5 \times IQR = 116 + 1.5 \times 18 = 143$. $Q1 - 1.5 \times IQR = 71$. So any point above 143 or below 71 is an outlier. Because the Min and Max are both outside these fences, there is at least one outlier on both the low end and the high end.

5. The mean will be higher. The distribution is unimodal and skewed to the right, so the mean will be pulled by the tail toward the higher values.

7. The IQR will not be affected at all. The standard deviation will increase.

9. Answers will vary.

11. Answers will vary.

13. a) Unimodal (near 0) and skewed to the right. Many seniors will have 0 or 1 speeding tickets. Some may have several, and a few may have more than that.
b) Probably unimodal and slightly skewed to the right. It is easier to score 15 strokes over the mean than 15 strokes under the mean.
c) Probably unimodal and symmetric. Weights may be equally likely to be over or under the average.
d) Probably bimodal. Men's and women's distributions may have different modes. It may also be skewed to the right, since it is possible to have very long hair, but hair length can't be negative.

15. a) Bimodal. Looks like two groups. Modes are near 6% and 46%. No real outliers.
b) Looks like two groups of cereals, a low-sugar and a high-sugar group.

17. a) 78%
b) Skewed to the right with at least one high outlier. Most of the vineyards are less than 90 acres with a few high ones. The mode is between 0 and 30 acres.

19. a) Because the distribution is skewed to the right, we expect the mean to be larger.
b) Bimodal and skewed to the right. Center mode near 8 days. Another mode at 1 day (may represent patients who didn't survive). Most of the patients stay between 1 and 15 days. There are some extremely high values above 25 days.
c) The median and IQR, because the distribution is strongly skewed.

21. a) 45 points
b) 35 points and 54.5 (or 55) points
c) In the Super Bowl, teams typically score a total of about 45 points, with half the games totaling between 35 and 55 points. In only one-fourth of the games have the teams scored fewer than 35 points, and they once totaled 75.

23. a) 144.4   b) 130, 150, 150   c) 60, 20

25. a) Median will probably be unaffected. The mean will be larger.
b) The range and standard deviation will increase; IQR will be unaffected.

27. a) The standard deviation will be larger for set 2, since the values are more spread out. SD(set 1) = 2.2, SD(set 2) = 3.2.
b) The standard deviation will be larger for set 2, since 11 and 19 are farther from 15 than are 14 and 16. Other numbers are the same. SD(set 1) = 3.6, SD(set 2) = 4.5.
c) The standard deviation will be the same for both sets, since the values in the second data set are just the values in the first data set +80. The spread has not changed. SD(set 1) = 4.2, SD(set 2) = 4.2.

29. The mean and standard deviation because the distribution is unimodal and symmetric.

31. a) The mean is closest to $2.60 because that's the balancing point of the histogram.
b) The standard deviation is closest to $0.15 since that's a typical distance from the mean. There are no prices as far as $0.50 or $1.00 from the mean.

33. a) About 105 minutes
b) Yes, only 3 of these 150 movies, or 2% of them, run that long.
c) Overall, the distribution is skewed a bit to the higher values, but the outlier on the lower end would also affect the mean. They are probably about equal.

35. a) i. The middle 50% of movies ran between 98 and 116 minutes.
ii. Typically, movie lengths varied from the mean run time by 16.6 minutes.
b) We should be cautious in using the standard deviation because the distribution of run times is skewed to the right, and the outlier might inflate the value.

37. The publication is using the median; the watchdog group is using the mean, pulled higher by the several very successful movies in the long right tail.

39. a) Mean $525, median $450
b) Two employees earn more than the mean.
c) The median because of the outlier.
d) The IQR will be least sensitive to the outlier of $1200, so it would be the best to report.

41. a)
```
Stem | Leaf
31   | 1
31   | 5
32   | 1233
32   | 6678
33   |
33   | 9
34   | 23
34   | 556
```
32|1 = $3.21/gallon

b) The distribution of gas prices is bimodal and skewed to the left (lower values) with peaks around $3.25 and $3.45. The lowest and highest prices were $3.11 and $3.46.
c) There is a rather large gap between the $3.28 and $3.39 prices.

43. a) Since these data are strongly skewed, the median and IQR are the best statistics to report.
b) The mean will be larger than the median because the data are skewed to the high end.
c) The median is 4.5 million. The IQR is 5 million (Q3 = 7 million, Q1 = 2 million).
d) The distribution of populations of the states is unimodal and skewed to the right. The median population is 4.5 million. One state is an outlier, with a population of 37 million. Another, at 25 million, could be considered an outlier.

45. Reasonably symmetric except for second mode at the low end (around 10 hr).

47. a)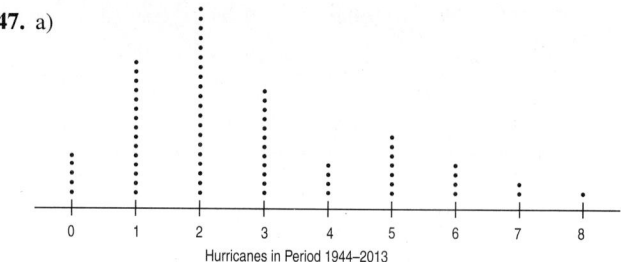

b) Slightly skewed to the right. Unimodal, mode near 2. Possibly a second mode near 5. No outliers.

49. a) This is not a histogram. The horizontal axis should split the number of home runs hit in each year into bins. The vertical axis should show the number of years in each bin.
b)

51. Skewed to the right, possibly bimodal with one fairly symmetric group near 4.4, another at 5.6. Two outliers in middle seem not to belong to either group.

```
Stem | Leaf
57   | 8
56   | 27
55   | 1
54   |
53   |
52   | 9
51   |
50   | 8
49   |
48   | 2
47   | 3
46   | 034
45   | 267
44   | 015
43   | 0199
42   | 669
41   | 22
```
41|2 = 4.12 pH

53. Histogram bins are too wide to be useful.

55. Neither appropriate nor useful. ZIP codes are categorical data, not quantitative. But they do contain *some* information. The leading digit gives a rough East-to-West placement in the United States. So, we see that they have almost no customers in the Northeast, but a bar chart by leading digit would be more appropriate.

57. a) Median 284, IQR 10, Mean 282.98, SD 7.60
b) Because it's skewed to the left, probably better to report Median and IQR.
c) Skewed to the left. The center is around 284. The middle 50% of states scored between 278 and 288. Mississippi's was much lower than other states' scores.

59. The histogram shows that the distribution of *Percent Change* is unimodal and skewed to the right. The states vary from a minimum of −0.6% (Michigan) to 35.1% (Nevada) growth in the decade. The median was 7.8% and half of the states had growth between 4.3% and 14.1%. Not surprisingly, the top three states in terms of growth were all from the West: Nevada (35.1%), Arizona (24.6%), and Utah (23.8%).

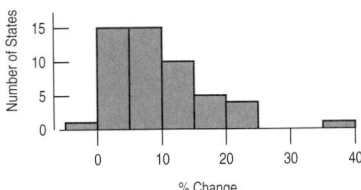

Chapter 4

1. Both distributions are unimodal and skewed to the left. The lowest international value may be an outlier. Because the distributions are skewed, we choose to compare medians and IQRs. The medians are very similar. The IQRs show that the domestic load factors vary a bit more.

3. Load factors are generally highest and least variable in the summer months (June–August). They are lower and more variable in the winter and spring. There are several months with low outliers and October has a far low outlier that is the minimum of all the data.

5. Air travel immediately after the events of 9/11 was not typical of air travel in general. If we want to analyze monthly patterns, it might be best to set these months aside.

7. a) After a period of little change in 2000–2001, load factors have been increasing steadily.
   b) No. These are percentages, so they can't exceed 100%. Moreover, we have no reason to believe this trend will continue.

9. Data with a distribution this skewed are difficult to summarize. The extremely large values will dominate any summary or description.

11. Answers will vary.

13. Answers will vary.

15. a) Prices appear to be both higher on average and more variable in Baltimore than in the other three cities. Prices in Chicago may be slightly higher than in Dallas and Denver, but the difference is very small.
    b) There are outliers on the low end in Baltimore and Chicago and one high outlier in Dallas, but these do not affect the overall conclusions reached in part a.

17. a) Essentially symmetric, very slightly skewed to the right with two high outliers at 36 and 48. Most victims are between the ages of 16 and 24.
    b) The slight increase between ages 22 and 24 is apparent in the histogram but not in the boxplot. It may be a second mode.
    c) The median would be the most appropriate measure of center because of the slight skew and the outliers.
    d) The IQR would be the most appropriate measure of spread because of the slight skew and the outliers.

19. a) About 59%
    b) Bimodal.
    c) Some cereals are very sugary; others are healthier low-sugar brands.
    d) Yes.
    e) Although the ranges appear to be comparable for both groups (about 28%), the IQR is larger for the adult cereals, indicating that there's more variability in the sugar content of the middle 50% of adult cereals.

21. a)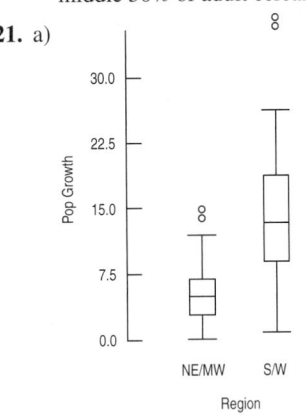
    b) Growth rates in NE/MW states are tightly clustered near 5%, but with two high outliers. S/W states are more variable, with a median near 13% and two outliers as well. The distributions are fairly symmetric.

23. a) They should be put on the same scale, from 0 to 20 days.
    b) Lengths of men's stays appear to vary more than for women. Men have a mode at 1 day and then taper off from there. Women have a mode near 5 days, with a sharp drop afterward.
    c) A possible reason is childbirth.

25. a) Both girls have a median score of about 17 points per game, but Scyrine is much more consistent. Her IQR is about 2 points, while Alexandra's is over 10.
    b) If the coach wants a consistent performer, she should take Scyrine. She'll almost certainly deliver somewhere between 15 and 20 points. But if she wants to take a chance and needs a "big game," she should take Alexandra. Alex scores over 24 points about a quarter of the time. (On the other hand, she scores under 11 points as often.)

27. Women appear to marry about 3 years younger than men, but the two distributions are very similar in shape and spread.

29. In general, fuel economy is higher in cars than in either SUVs or pickup trucks. The top 50% of cars get higher fuel economy than 75% of SUVs and all pickups. The distribution for pickups shows less spread.

31. The class A is 1, class B is 2, and class C is 3.

33. a) Probably slightly left skewed. The mean is slightly below the median, and the 25th percentile is farther from the median than the 75th percentile.
    b) No, all data are within the fences.
    c)
    d) The 48 universities graduate, on average, about 68% of freshmen "on time," with percents ranging from 43% to 87%. The middle 50% of these universities graduate between 59% and 75% of their freshmen in 4 years.

35. a) *Who*—Student volunteers; *What*—Memory test; *Where, when*—Not specified; *How*—Students took memory test 2 hours after drinking caffeine-free, half-dose caffeine, or high-caffeine soda; *Why*—To see if caffeine makes you more alert and aids memory retention.
b) Drink: categorical; Test score: quantitative.
c)
d) The participants scored about the same with no caffeine and low caffeine. The medians for both were 21 points, with slightly more variation for the low-caffeine group. The high-caffeine group generally scored lower than the other two groups on all measures of the 5-number summary: min, lower quartile, median, upper quartile, and max.

37. a) About 36 mph
b) Q_1 about 35 mph and Q_3 about 37 mph.
c) The range appears to be about 7 mph, from about 31 to 38 mph. The IQR is about 2 mph.
d) We can't know exactly, but the boxplot may look something like this:

e) The median winning speed has been about 36 mph, with a max of about 38 and a min of about 31 mph. Half have run between about 35 and 37 mph, for an IQR of 2 mph.

39. a) Boys.
b) Boys.
c) Girls.
d) Girls. Their median and upper quartiles are larger. The lower quartile is slightly lower, but close.
e) $[14(4.2) + 11(4.6)]/25 = 4.38$

41.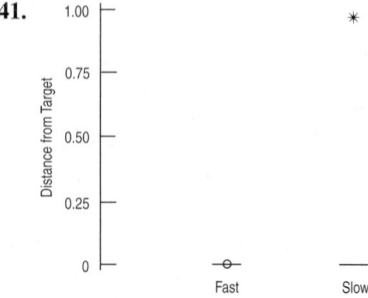

There appears to be an outlier! This point should be investigated. We'll proceed by redoing the plots with the outlier omitted:

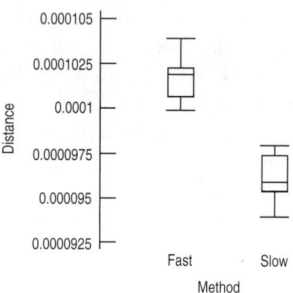

It appears that slow speed provides much greater accuracy. But the outlier should be investigated. It is possible that slow speed can induce an infrequent very large distance.

43. a)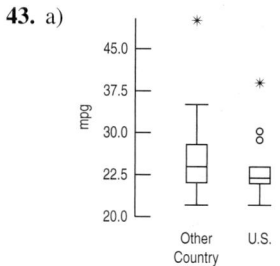

b) Mileage for U.S. models is typically lower, and less variable than for cars made elsewhere. The median for U.S. models is around 22 mpg, compared to 24 for the others. Both groups have some high outliers—most likley hybrid vehicles. (Other answers possible.)

45. a) Day 16 (but any estimate near 20 is okay).
b) Day 65 (but anything around 60 is okay).
c) Around day 50

47. a) Most of the data are found in the far left of this histogram. The distribution is very skewed to the right.
b) Re-expressing the data by, for example, logs or square roots might help make the distribution more nearly symmetric.

49. a) The logarithm makes the histogram more symmetric. It is easy to see that the center is around 3.5 in log assets.
b) That has a value of around 2500 million dollars.
c) That has a value of around 1000 million dollars.

51. a) Fusion time and group.
b) Fusion time is quantitative (units = seconds). Group is categorical.
c) Both distributions are skewed to the right with high outliers. The boxplot indicates that visual information may reduce fusion time. The median for the Verbal/Visual group seems to be about the same as the lower quartile of the No/Verbal group.

Chapter 5

1. 65

3. In January, a high of 55° is not quite 2 standard deviations above the mean, whereas in July a high of 55° is more than 2 standard deviations lower than the mean. So it's less likely to happen in July.

5. a) 72 oz., 40 oz.   b) 4.5 lb, 2.5 lb

7. a)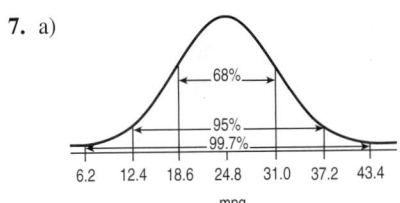
b) 18.6 to 31.0 mpg   c) 16%
d) 13.5%   e) Less than 12.4 mpg

9. a) 12.2%   b) 71.6%   c) 23.3%

11. a) No. The plot is not straight.   b) It is skewed to the right.

13. a) Skewed to the right; mean is higher than median.
b) $350 and $950
c) Minimum $350. Mean $750. Median $550. Range $1200. IQR $600. Q1 $400. SD $400.
d) Minimum $330. Mean $770. Median $550. Range $1320. IQR $660. Q1 $385. SD $440.

15. Lowest score = 910. Mean = 1230. SD = 120. Q3 = 1350. Median = 1270. IQR = 240.

17. Your score was 2.2 standard deviations higher than the mean score in the class.

19. 1.202

21. The z-scores, which account for the difference in the distributions of the two tests, are 1.5 and 0 for Derrick and 0.5 and 2 for Julie. Derrick's total is 1.5, which is less than Julie's 2.5.

23. a) Megan.   b) Anna.

25. a) About 1.81 standard deviations below the mean.
b) 1000 ($z = -1.81$) is more unusual than 1250 ($z = 1.17$).

27. a) Mean = 1152 − 1000 = 152 pounds; SD is unchanged at 84 pounds.
b) Mean = 0.40(1152) = $460.80; SD = 0.40(84) = $33.60.

29. Min = 0.40(980) − 20 = $372;
median = 0.40(1140) − 20 = $436;
SD = 0.40(84) = $33.60; IQR = 0.40(102) = $40.80.

31. College professors can have between 0 and maybe 40 (or possibly 50) years' experience. A standard deviation of 1/2 year is impossible, because many professors would be 10 or 20 SDs away from the mean, whatever it is. An SD of 16 years would mean that 2 SDs on either side of the mean is plus or minus 32, for a range of 64 years. That's too high. So, the SD must be 6 years.

33. Any weight more than 2 standard deviations below the mean, or less than 1152 − 2(84) = 984 pounds, is unusually low. We expect to see a steer below 1152 − 3(84) = 900 pounds only rarely.

35. a)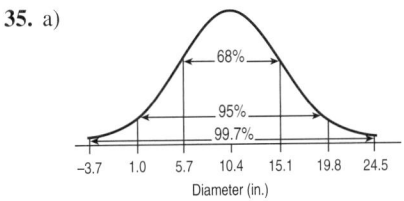
b) Between 1.0 and 19.8 inches   c) 2.5%
d) 34%   e) 16%

37. Since the histogram is not unimodal and symmetric, it is not wise to have faith in numbers from the Normal model.

39. a) 16%   b) 13.56%
c) Because the Normal model doesn't fit perfectly.
d) Distribution is skewed to the right.

41. a) 2.5%
b) 2.5% of the 181 receivers, or 4.5 of them should gain more than 426.98 + 2 × 408.34 = 1244 yards. (In fact, 12 receivers exceeded this value.)
c) Data are strongly skewed to the right, not symmetric.

43. a) 1259.7 lb   b) 1081.3 lb   c) 1108 lb to 1196 lb

45. a) 1130.7 lb   b) 1347.4 lb   c) 113.3 lb

47. a)
b) 30.85%   c) 17.00%   d) 32 points
e) 212.9 points

49. a) 11.1%   b) (35.9, 40.5) inches
c) 40.5 inches

51. a) 5.3 grams   b) 6.4 grams
c) Younger because SD is smaller.

Part I Review

1. a)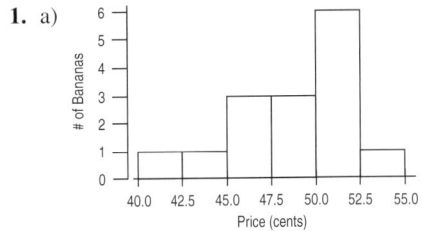
b) Median 49 cents, IQR 6 cents.
c) The distribution is unimodal and left skewed. The center is near 50 cents; values range from 42 cents to 53 cents.

3. a) If enough sopranos have a height of 65 inches, this can happen.
b) The distribution of heights for each voice part is roughly symmetric. The basses are slightly taller than the tenors. The sopranos and altos have about the same median height. Heights of basses and sopranos are more consistent than those of altos and tenors.

5. a) It means their heights are also more variable.
b) The z-score for women to qualify is 2.40, compared with 1.75 for men, so it is harder for women to qualify.

7. a) *Who*—People who live near State University; *What*—Age, attended college? Favorable opinion of State?; *When*—Not stated; *Where*—Region around State U; *Why*—To report to the university's directors; *How*—Sampled and phoned 850 local residents.
b) Age—Quantitative (years); attended college?—categorical; favorable opinion?—categorical.
c) The fact that the respondents know they are being interviewed by the university's staff may influence answers.

9. a) These are categorical data, so mean and standard deviation are meaningless.
b) Not appropriate. Even if it fits well, the Normal model is meaningless for categorical data.

11. a)

b) The scores on Friday were higher by about 5 points on average. This is a drop of more than 10% off the average score and shows that students fared worse on Monday after preparing for the test on Friday. The spreads are about the same, but the scores on Monday are a bit skewed to the right.

c)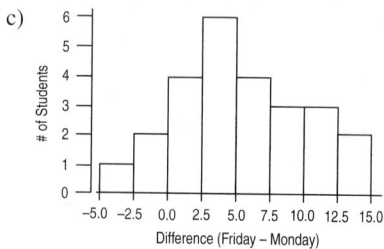

d) The changes (Friday–Monday) are unimodal and centered near 4 points, with a spread of about 5 (SD). They are fairly symmetric, but slightly skewed to the right. Only 3 students did better on Monday (had a negative difference).

13. a) Categorical.
b) Go fish. All you need to do is match the denomination. The denominations are not ordered. (Answers will vary.)
c) Gin rummy. All cards are worth their value in points (face cards are 10 points). (Answers will vary.)

15. a) Annual mortality rate for males (quantitative) in deaths per 100,000 and water hardness (quantitative) in parts per million.
b) Calcium is skewed right, possibly bimodal. There looks to be a mode down near 12 ppm that is the center of a fairly tight symmetric distribution and another mode near 62.5 ppm that is the center of a much more spread out, symmetric (almost uniform) distribution. Mortality, however, appears unimodal and symmetric with the mode near 1500 deaths per 100,000.

17. a) They are on different scales.
b) January's values are lower and more spread out.
c) Roughly symmetric but slightly skewed to the left. There are more low outliers than high ones. Center is around 40 degrees with an IQR of around 7.5 degrees.

19. a) Bimodal with modes near 2 and 4.5 minutes. Fairly symmetric around each mode.
b) Because there are two modes, which probably correspond to two different groups of eruptions, an average might not make sense.
c) The intervals between eruptions are longer for long eruptions. There is very little overlap. More than 75% of the short eruptions had intervals less than about an hour (62.5 minutes), while more than 75% of the long eruptions had intervals longer than about 75 minutes. Perhaps the interval could even be used to predict whether the next eruption will be long or short.

21. a)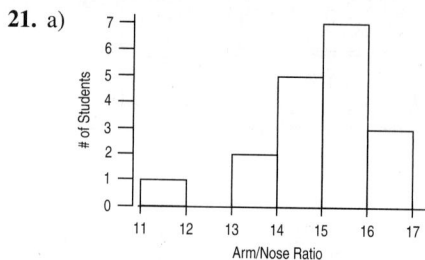

The distribution is left skewed with a center of about 15. It has an outlier between 11 and 12.
b) Even though the distribution is somewhat skewed, the mean and median are close. The mean is 15.0 and the SD is 1.25.
c) Yes. 11.8 is already an outlier. 9.3 is more than 4.5 SDs below the mean. It is a very low outlier.

23. If we look only at the overall statistics, it appears that the follow-up group is insured at a much lower rate than those not traced (11.1% of the time compared with 16.6%). But most of the follow-up group were black, who have a lower rate of being insured. When broken down by race, the follow-up group actually has a higher rate of being insured for both blacks and whites. So the overall statistic is misleading and is attributable to the difference in race makeup of the two groups.

25. a)

b) According to the model, reaction times are symmetric with center at 1.5 seconds. About 95% of all reaction times are between 1.14 and 1.86 seconds.
c) 8.2%
d) 24.1%
e) Quartiles are 1.38 and 1.62 seconds, so the IQR is 0.24 second.
f) The slowest 1/3 of all drivers have reaction times of 1.58 seconds or more.

27. a)

b) Mean 100.25, SD 25.54 pieces of mail.
c) The distribution is somewhat symmetric and unimodal, but the center is rather flat, almost uniform.
d) 64%. The Normal model seems to work reasonably well, since it predicts 68%.

29. a) *Who*—100 health food store customers; *What*—Have you taken a cold remedy?, and Effectiveness (scale 1 to 10); *When*—Not stated; *Where*—Not stated; *Why*—Promotion of herbal medicine; *How*—In-person interviews.

b) Have you taken a cold remedy?—categorical. Effectiveness—categorical or ordinal.
c) No. Customers are not necessarily representative, and the Council had an interest in promoting the herbal remedy.

31. a) 38 cars
b) Possibly because the distribution is skewed to the right.
c) Center—median is 148.5 cubic inches. Spread—IQR is 126 cubic inches.
d) No. It's bigger than average, but smaller than more than 25% of cars. The upper quartile is at 231 inches.
e) No. 1.5 IQR is 189, and 105 − 189 is negative, so there can't be any low outliers. 231 + 189 = 420. There aren't any cars with engines bigger than this, since the maximum has to be at most 105 (the lower quartile) + 275 (the range) = 380.
f) Because the distribution is skewed to the right, this is probably not a good approximation.
g) Mean, median, range, quartiles, IQR, and SD all get multiplied by 16.4.

33. a) 44.0%
b) If this were a random sample of all voters, yes.
c) 38.4%
d) 0.87%
e) 9.96%
f) 8.96%

35. a) Republican—3705, Democrat—3976, Neither—733; or Republican—44.0%, Democrat—47.3%, Neither—8.7%.
b)
c) It appears that the older the voter, the less likely they are to lean Democratic and the more likely to lean Republican.
d) No. There is an association between age and affiliation. Younger voters tend to be more Democratic and less Republican.

37. a) 0.43 hour
b) 1.4 hours
c) 0.89 hour (or 53.4 minutes)
d) Survey results vary, and the mean and the SD may have changed.

Chapter 6

1. a) Weight in ounces: explanatory; Weight in grams: response. (Could be other way around.) To predict the weight in grams based on ounces. Scatterplot: positive, straight, strong (perfectly linear relationship).
b) Circumference: explanatory. Weight: response. To predict the weight based on the circumference. Scatterplot: positive, linear, moderately strong.
c) Shoe size: explanatory; GPA: response. To try to predict GPA from shoe size. Scatterplot: no direction, no form, very weak.
d) Miles driven: explanatory; Gallons remaining: response. To predict the gallons remaining in the tank based on the miles driven since filling up. Scatterplot: negative, straight, moderate.

3. a)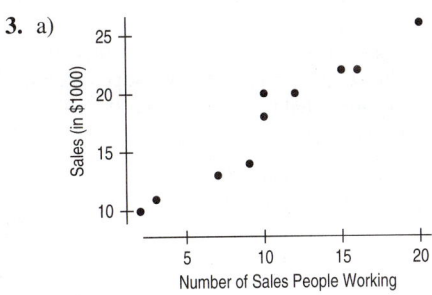
b) Positive.
c) Linear.
d) Strong.
e) No.

5. a) True.
b) False. It will not change the correlation.
c) False. Correlation has no units.

7. Correlation does not demonstrate causation. The analyst's argument is that sales staff cause sales. However, the data may reflect the store hiring more people as sales increase, so any causation would run the other way.

9. 4, 5, and 6. Yes, it will be straight.

11. a) Altitude: explanatory; Temperature: response. (Other way around possible as well.) To predict the temperature based on the altitude. Scatterplot: negative, possibly straight, weak to moderate.
b) Ice cream cone sales: explanatory. Air conditioner sales: response—although the other direction would work as well. To predict one from the other. Scatterplot: positive, straight, moderate.
c) Age: explanatory; Grip strength: response. To predict the grip strength based on age. Scatterplot: curved down, moderate. Very young and elderly would have grip strength less than that of adults.
d) Reaction time: explanatory; Blood alcohol level: response. To predict blood alcohol level from reaction time test. (Other way around is possible.) Scatterplot: positive, nonlinear, moderately strong.

13. a) None   b) 3 and 4   c) 2, 3, and 4
d) 2   e) 3 and 1

15. There seems to be a very weak—or possibly no—relationship between brain size and performance IQ.

17. a)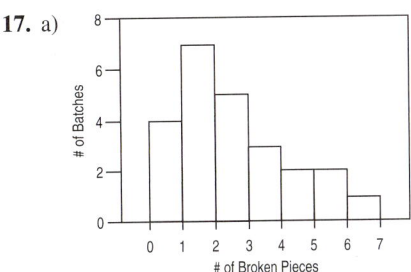
b) Unimodal, skewed to the right. The skew.
c) The positive, somewhat linear relationship between batch number and broken pieces.

19. a) 0.006   b) 0.777   c) −0.923   d) −0.487

21. There may be an association, but not a correlation unless the variables are quantitative. There could be a correlation between average number of hours of TV watched per week per person and number of crimes committed per year. Even if there is a relationship, it doesn't mean one causes the other.

23. a) Yes. It shows a linear form and no outliers.
    b) There is a strong, positive, linear association between drop and speed; the greater the coaster's initial drop, the higher the top speed.

25. The scatterplot is not linear; correlation is not appropriate.

27. The correlation may be near 0. We expect nighttime temperatures to be low in January, increase through spring and into the summer months, then decrease again in the fall and winter. The relationship is not linear.

29. The correlation coefficient won't change, because it's based on z-scores. The z-scores of the prediction errors are the same whether they are expressed in nautical miles or miles.

31. a) Assuming the relation is linear, a correlation of -0.772 shows a strong relation in a negative direction.
    b) Continent is a categorical variable. Correlation does not apply.

33. a) Actually, yes, taller children will tend to have higher reading scores, but this doesn't imply causation.
    b) Older children are generally both taller and are better readers. Age is the lurking variable.

35. a) No. We don't know this from the correlation alone. There may be a nonlinear relationship or outliers.
    b) No. We can't tell from the correlation what the form of the relationship is.
    c) No. We don't know from the correlation coefficient.
    d) Yes, the correlation doesn't depend on the units used to measure the variables.

37. This is categorical data even though it is represented by numbers. The correlation is meaningless.

39. a) The association is positive, moderately strong, and roughly straight, with several states whose HCI seems high for their median income and one state whose HCI appears low given its median income.
    b) The correlation would still be 0.65.
    c) The correlation wouldn't change.
    d) DC would be a moderate outlier whose HCI is high for its median income. It would lower the correlation slightly.
    e) No. We can only say that higher median incomes are associated with higher housing costs, but we don't know why. There may be other economic variables at work.
    *f) No. We can say that there is a consistent monotone pattern, but correlation—even nonparametric correlation—does not demonstrate causation.

41. a)
    b) Negative, linear, strong, with one high outlier.
    c) -848

d) There is a fairly strong linear relation in a negative direction between engine size and gas mileage. Lower fuel efficiency is generally associated with larger engines.

43. a)

    (Plot could have explanatory and predictor variables swapped.) Correlation is 0.199 (with the outlier included). There does not appear to be a relationship between sodium and fat content in burgers, especially without the low-fat, low-sodium item. Without the outlier, the correlation is -0.325, but the relationship is still weak.
    b) Spearman's rho is slightly negative. Using ranks doesn't allow the outlier to have as strong an influence and the remaining points have little or no association.

45. a) Yes, although the association is weak, the form does seem to be somewhat linear.
    b) As the number of runs increases, the attendance also increases.
    c) There is a slight positive association, but even if it were stronger it does not *prove* that more fans would come if the number of runs increased. Association does not indicate causality.

47. There is a fairly strong, positive, linear relationship between the length of the track and the duration of the ride. In general, a longer track corresponds with a longer ride, with a correlation of 0.698.

49. a) We would expect that as one variable (say length of ride) increases, the rank will improve, which means it will decrease.
    b) Drop has the strongest correlation ($r = -0.193$), but even that correlation is weak. And the scatterplot is not convincing. There appear to be other factors that influence the rank of coaster more than any of the ones measured in this data set.

51. a)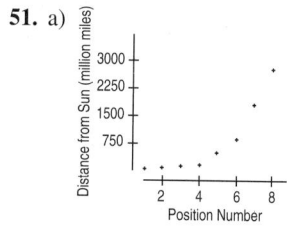

    The relationship between position and distance is nonlinear, with a positive direction. There is very little scatter from the trend.
    b) The relationship is not linear.
    c)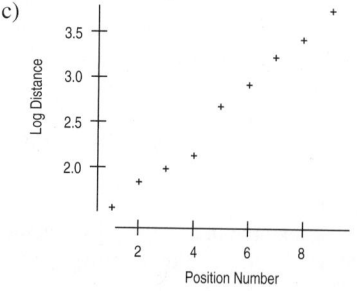

The relationship between position number and log of distance appears to be roughly linear.
d) Kendall's tau is 1.0 because the points are perfectly monotonically increasing.

Chapter 7

The calculations in this chapter are particularly sensitive to the rounding. If your answers differ from those here even in the second place, they may still be correct in the sense that you calculated them by a correct method, buy you may have used values that were rounded differently than those used to find these answers. In general, we use the original data and do no intermediate rounding. Don't be concerned about minor differences—what is important is your interpretations of the results.

1. a) False. The line usually touches none of the points. We minimize the sum of the squared errors.
   b) True.
   c) False. It is the sum of the squares of all the residuals that is minimized.

3. The weight of a newborn boy can be predicted as -5.94 kg plus 0.1875 kg per cm of length. This is a model fit to data. No particular baby should be expected to fit this model exactly.

5. a) $b_1 = 0.914$ if found by hand. $b_1 = 0.913$ if found by technology. (Difference is due to rounding error.)
   b) It means that an additional 0.914 ($1000) or $914 of sales is associated with each additional sales person working.
   c) $b_0 = 8.10$
   d) It would mean that, on average, we expect sales of 8.10 ($1000) or $8100 with 0 sales people working. Doesn't really make sense in this context.
   e) $\widehat{Sales} = 8.10 + 0.914$ *Number of Sales People Working*
   f) $24.55 ($1000) or $24,550. (24,540 if using the technology solution.)
   g) 0.45 ($1000) or $450. ($460 with technology.)
   h) Underestimated.

7. The winners may be suffering from regression to the mean. Perhaps they weren't really better than other rookie executives, but just happened to have a lucky year.

9. a) Thousands of dollars
   b) 2.77 (the largest residual in magnitude)
   c) 0.07 (the smallest residual in magnitude)

11. $R^2 = 93.12\%$ About 93% of the variance in *Sales* can be accounted for by the regression of *Sales* on *Number of Sales Workers*.

13. a) Linearity assumption.
    b) Outlier condition.
    c) Equal spread condition.

15. 281 milligrams

17. The potassium content is actually lower than the model predicts for a cereal with that much fiber.

19. The model predicts that cereals will have approximately 27 more milligrams of potassium, on average, for every additional gram of fiber.

21. 81.5%

23. The true potassium contents of cereals vary from the predicted amounts with a standard deviation of 30.77 milligrams.

25.

	$\bar{x}$	s_x	$\bar{y}$	s_y	r	$\hat{y} = b_0 + b_1 x$
a)	10	2	20	3	0.5	$\hat{y} = 12.5 + 0.75x$
b)	2	0.06	7.2	1.2	-0.4	$\hat{y} = 23.2 - 8x$
c)	12	6	152	30	-0.8	$\hat{y} = 200 - 4x$
d)	2.5	1.2	25	100	0.6	$\hat{y} = -100 + 50x$

27. a) Model is appropriate.
    b) Model is not appropriate. Relationship is nonlinear.
    c) Model may not be appropriate. Spread is changing.

29. a) *Price* (in thousands of dollars) is y and *Size* (in square feet) is x.
    b) Slope is thousands of $ per square foot.
    c) Positive. Larger homes should cost more.

31. 300 pounds/foot. If a "typical" car is 15 feet long, all of 3, 30, and 3000 would give ridiculous weights.

33. A linear model on *Size* accounts for 71.4% of the variation in home *Price*.

35. a) R^2 does not tell whether the model is appropriate. High R^2 could also be due to an outlier.
    b) Predictions based on a regression line are for average values of y for a given x. The actual wingspan will vary around the prediction.

37. a) 0.845
    b) Price should be 0.845 SDs above the mean in price.
    c) Price should be 1.690 SDs below the mean in price.

39. a) Probably not. Your score is better than about 97.5% of people, assuming scores follow the Normal model. Your next score is likely to be closer to the mean.
    b) The friend probably should retake the test. His score is better than only about 16% of people. His next score is likely to be closer to the mean.

41. a) *Price* increases by about $0.061 per 1000 square feet or $61.00 per square foot.
    b) 230.82 thousand, or $230,820.
    c) $115,020; $6000 is the residual.

43. a) Probably. The residuals show some initially low points, but there is no clear curvature.
    b) The linear model on *Tar* content accounts for 92.4% of the variability in *Nicotine*.

45. a) $r = 0.961$
    b) Nicotine should be 1.922 SDs below average.
    c) Tar should be 0.961 SDs above average.

47. a) $\widehat{Nicotine} = 0.15403 + 0.065052$ *Tar*
    b) 0.414 mg
    c) Nicotine content increases by 0.065 mg of nicotine per milligram of tar.
    d) We'd expect a cigarette with no tar to have 0.154 mg of nicotine.
    e) 0.1094 mg

49. a) Yes. The relationship is straight enough, with a few outliers. The spread increases a bit for states with large median incomes, but we can still fit a regression line.
    b) From summary statistics: $\widehat{HCI} = -156.50 + 0.0107$ *MFI*; from original data: $\widehat{HCI} = -157.64 + 0.0107$ *MFI*

c) From summary statistics: predicted HCI = 324.93; from original data: 323.79.
d) 223.09
e) $\widehat{z_{HCI}} = 0.65 z_{MFI}$
f) $\widehat{z_{MFI}} = 0.65 z_{HCI}$

51. a) $\widehat{Total} = 539.803 + 1.103\, Age$
b) Yes. Both variables are quantitative; the plot is straight (although flat); there are no apparent outliers; the plot does not appear to change spread throughout the range of *Age*.
c) $559.65; $594.94
d) 0.14%
e) No. The plot is nearly flat. The model explains almost none of the variation in *Total Yearly Purchases*.

53. a) Moderately strong, fairly straight, and positive. Possibly some outliers (higher-than-expected math scores).
b) The student with 500 verbal and 800 math.
c) Positive, fairly strong linear relationship. 46.9% of variation in math scores is explained by verbal scores.
d) $\widehat{Math} = 217.7 + 0.662 \times Verbal$
e) Every point of verbal score adds 0.662 points to the predicted average math score.
f) 548.5 points
g) 53.0 points

55. a) 0.685
b) $\widehat{Verbal} = 162.1 + 0.71\, Math$
c) The observed verbal score is higher than predicted from the math score.
d) 516.7 points
e) 559.6 points
f) Regression to the mean. Someone whose math score is below average is predicted to have a verbal score below average, but not as far (in SDs). So if we use *that* verbal score to predict math, they will be even closer to the mean in predicted math score than their observed math score. If we kept cycling back and forth, eventually we would predict the mean of each and stay there.

57. a) The relationship is straight enough, but very weak. In fact, there may be no relationship at all between these variables.
b) The number of wildfires has been increasing by about 78 per year.
c) Yes, the intercept estimates the number of wildfires in 1985 as about 75,609.
d) The residuals are distributed around zero with a standard deviation of 11,609 fires. Compared to the observed values, most of which are between 60,000 and 90,000 fires, this residual standard deviation in our model's predictions is quite large.
e) Less than 1% of the variation in the number of wildfires can be accounted for by the linear model on *Year*. This confirms the impression from the scatterplot that there is very little association between these variables—that is, that there has been little change in the number of wildfires during this period. The average number might provide as good a prediction as the model.

59. a)

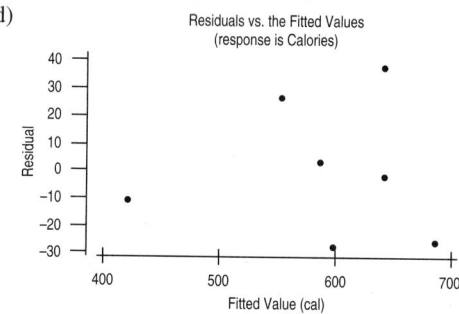

b) Negative, linear, strong.
c) Yes.
d) −0.867
e) *Age* accounts for 75.2% of the variation in *Advertised Price*.
f) Other factors contribute—options, condition, mileage, etc.

61. a) $\widehat{Price} = 17{,}674 - 844.5\, Years$
b) Every extra year of age decreases average value by $844.50.
c) The average new Corolla costs $17,674.
d) $11,762.50
e) Negative residual. Its price is below the predicted value for its age.
f) −$729
g) No. As of age 21, the model predicts negative prices. The relationship is no longer linear.

63. a)

b) 92.3% of the variation in calories can be accounted for by the fat content.
c) $\widehat{Calories} = 211.0 + 11.06\, Fat$
d)

Residuals show no clear pattern, so the model seems appropriate.
e) Could say a fat-free burger still has 211.0 calories, but this is extrapolation (no data close to 0).
f) Every gram of fat adds 11.06 calories, on average.
g) 553.5 calories

65. a) The regression was for predicting calories from fat, not the other way around.
b) $\widehat{Fat} = -15.0 + 0.083\, Calories$. Predict 34.8 grams of fat.

67. a) This model predicts that every bridge will be deficient. The intercept is less than 5 and the slope is negative, so all predicted conditions are less than 5.
b) This model says bridges in New York City are decreasing in condition at only 0.0048 per year—less rapidly than bridges in Tompkins County.
c) The R^2 for this model is only 2.6%. I don't think it has much predictive value.

69. a) 0.858
b) CO_2 levels account for 73.6% of the variation in mean temperature.
c) $\widehat{Mean\ Temperature} = 10.64 + 0.0103\, CO_2$

d) The predicted mean temperature has been increasing at an average rate of 0.0103 degrees (C)/ppm of CO_2.
e) One *could* say that with no CO_2 in the atmosphere, there would be a temperature of 10.64 degrees Celsius, but this is extrapolation to a nonsensical point.
f) No.
g) Predicted 14.76 degrees C
h) No. The model indicates only that CO_2 and Temperature are associated. It doesn't imply any causal relationship.

71. a) $\widehat{\% \text{ Body Fat}} = -27.4 + 0.25 \text{ Weight}$
b) Residuals look randomly scattered around 0, so conditions are satisfied.
c) % Body Fat increases, on average, by 0.25 percent per pound of Weight.
d) Reliable is relative. R^2 is 48.5%, but residuals have a standard deviation of 7%, so variation around the line is large.
e) 0.9 percent

73. a) $\widehat{\text{Highjump}} = 2.681 - 0.00671 \times 800m \text{ Time}$. High-jump height is lower, on average, by 0.00671 meters per second of 800-m race time.
b) 16.4%
c) Yes, the slope is negative. Faster runners tend to jump higher as well.
d) There is a slight tendency for less variation in high-jump height among the slower runners than among the faster runners.
e) Not especially. The residual standard deviation is 0.060 meters, which is not much smaller than the SD of all high jumps (0.062 meters). The model doesn't appear to do a very good job of predicting.

75. a) As calcium levels increase, mortality rate decreases. Relationship is fairly strong, negative, and linear.
b) $\widehat{\text{Mortality}} = 1676.0 - 3.23 \text{ Calcium}$
c) Mortality decreases 3.23 deaths per 100,000, on average, for each part per million of calcium. The intercept indicates a baseline mortality of 1676 deaths per 100,000 with no calcium, but this is extrapolation.
d) Exeter has 348.6 fewer deaths per 100,000 than the model predicts.
e) 1353 deaths per 100,000
f) Calcium concentration accounts for 43.0% of the variation in death rate per 100,000 people.

77. The line $\hat{y} = 7.0 + 1.1x$ minimizes the sum of squared vertical distances from the points to the line.

Chapter 8

1. The different segments are not scattered at random throughout the scatterplot. Each segment may have a different relationship.

3. Yes, it is clear that the relationship between January and December spending is not the same for all five segments. Using one overall model to predict January spending would be very misleading.

5. Your friend is extrapolating. It is impossible to know if a trend like this will continue so far into the future.

7. a) $354,472
b) An extrapolation this far from the data is unreliable.

9. This observation was influential. After it was removed, the R^2 value and the slope of the regression line both changed by a large amount.

11. No. In warm weather, more children will go outside and play.

13. Individual student scores will vary greatly. The class averages will have much less variability and may disguise important trends.

15. a) The trend appears to be somewhat linear up to about 1940, but from 1940 to about 1970 the trend appears to be nonlinear with a large drop from 1940 to 1950. From 1975 or so to the present, the trend appears to be linear.
b) Relatively strong for certain periods.
c) No, as a whole the graph is clearly nonlinear. Within certain periods (ex: 1975 to the present) the correlation is high.
d) Overall, no. You could fit a linear model to the period from 1975 to 2003, but why? You don't need to interpolate, since every year is reported, and extrapolation seems dangerous.

17. a) The relationship is not straight.
b) It will be curved downward.

19. a) No. We need to see the scatterplot first to see if the conditions are satisfied, and models are always wrong.
b) No, the linear model might not fit the data everywhere.

21. a) Millions of dollars per minute of run time.
b) Costs for movies increase at approximately the same rate per minute.
c) On average dramas cost about $20 million less for the same runtime.

23. a) There are several features to comment on in this plot. There is a strong monthly pattern around the general trend. From 1997 to 2008, passengers increased fairly steadily with a notable exception of Sept. 2001, probably due to the attack on the twin towers. Then sometime in late 2008, departures dropped dramatically, possibly due to the economic crisis. Recently, they have been recovering, but not at the same rate as their previous increase.
b) The trend was fairly linear until late 2008 then passengers dropped suddenly.
c) The trend since 2009 has been linear (overall, ignoring monthly oscillations) If the increase continues to be linear, the predictions should be reasonable for the short term.

25. a) 1) High leverage, small residual.
2) No, not influential for the slope.
3) Correlation would decrease because outlier has large z_x and z_y, increasing correlation.
4) Slope wouldn't change much because outlier is in line with other points.
b) 1) High leverage, probably small residual.
2) Yes, influential.
3) Correlation would weaken, increasing toward zero.
4) Slope would increase toward 0, since outlier makes it negative.
c) 1) Some leverage, large residual.
2) Yes, somewhat influential.
3) Correlation would increase, since scatter would decrease.
4) Slope would increase slightly.

d) 1) Little leverage, large residual.
2) No, not influential.
3) Correlation would become stronger and become more negative because scatter would decrease.
4) Slope would change very little.

27. 1) e   2) d   3) c   4) b   5) a

29. Perhaps high blood pressure causes high body fat, high body fat causes high blood pressure, or both could be caused by a lurking variable such as a genetic or lifestyle issue.

31. a) The graph shows that, on average, students progress at about one reading level per year. This graph shows averages for each grade. The linear trend has been enhanced by using averages.
b) Very close to 1.
c) The individual data points would show much more scatter, and the correlation would be lower.
d) A slope of 1 would indicate that for each 1-year grade level increase, the average reading level is increasing by 1 year.

33. a) *Cost* decreases by $2.13 per degree of average daily *Temp*. So warmer temperatures indicate lower costs.
b) For an avg. monthly temperature of 0°F, the cost is predicted to be $133.
c) Too high; the residuals (observed − predicted) around 32°F are negative, showing that the model overestimates the costs.
d) $111.70
e) About $105
f) No, the residuals show a definite curved pattern. The data are probably not linear.
g) No, there would be no difference. The relationship does not depend on the units.

35. a) 0.88
b) Interest rates during this period grew at about 0.25% per year, starting from an interest rate of about 0.61%.
c) Substituting 50 in the model yields a predicted rate of about 13%.
d) Not really. Extrapolating 20 years beyond the end of these data would be dangerous and unlikely to be accurate.

37. a) Interest rates peaked about 1980 and decreased afterward. This regression has a negative slope and high intercept.
b) This model predicts the interest rate in 2000 to be 3.71%, much lower than the other model predicts.
c) Separating the data into parts will make the linear model predict better on each part, but the model is still not appropriate. There is no way of knowing if the linear trend will continue.
d) The first predicts 17.96%, the second −2.35%
e) Both predictions are absurd. Neither model can extrapolate that far. The first would be historically high and the second is negative.

39. a) Stronger. Both slope and correlation would increase.
b) Restricting the study to nonhuman animals would justify it.
c) Moderately strong.
d) For every year increase in life expectancy, the gestation period increases by about 15.5 days, on average.
e) About 270.5 days

41. a) Removing hippos would make the association stronger, since hippos are more of a departure from the pattern.
b) Increase.
c) No, there must be a good reason for removing data points.
d) Yes, removing it lowered the slope from 15.5 to 11.6 days per year.

43. a) Answers may vary. Using the data for 1955–2010 results in a scatterplot that is relatively linear with some curvature. You might use the data after 1955 only to predict 2015, but that would still call for extrapolation and would not be safe. The prediction is 27.19.
b) Not much, since the data are not truly linear and 2020 is 10 years from the last data point (extrapolating is risky).
c) No, that extrapolation of more than 50 years would be absurd. There's no reason to believe the trend from 1955 to 2005 will continue.

45. a)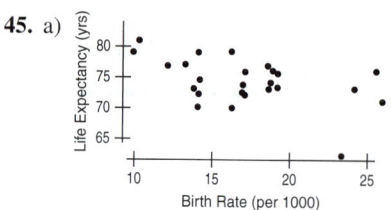
There is a moderate, negative, linear relationship between birth rate and life expectancy.
b) Haiti is an outlier with a much lower than expected life expectancy.
c) $\widehat{lifeexp} = 79.504 - 0.264\, Birth\ Rate$
d) The residual plot is reasonably scattered.
e) Yes.
f) 14.1% of the variation in life expectancy is accounted for by the regression on birth rate.
g) While there is an association, there is no reason to expect causality. Lurking variables may be involved.

47. a) The scatterplot is clearly nonlinear; however, the last few years—say, from 1970 on—do appear to be linear.
b) Using the data from 1970 to 2011 gives $r = 0.999$ and $\widehat{CPI} = -8967 + 4.57\, Year$. Predicted CPI in 2020 = 263 (an extrapolation of doubtful accuracy).

Chapter 9

1. a) No re-expression needed.
b) Re-express to straighten the relationship.
c) Re-express to equalize spread.

3. Improve homoscedasticity (more nearly equal variances between groups.)

5. Yes. The square root doesn't make the spreads as nearly equal. The reciprocal clearly goes too far on the Ladder of Powers.

7. a) There's an annual pattern in when people fly, so the residuals cycle up and down. There was also a sudden decrease in passenger traffic after 2008.
b) No, this kind of pattern can't be helped by re-expression.

9. a) 16.44   b) 7.84   c) 0.36   d) 1.75   e) 27.59

11. a) Fairly linear, negative, strong.
b) Gas mileage decreases an average 7.652 mpg for each thousand pounds of weight.
c) No. Residuals show a curved pattern.

13. a) Residuals are more randomly spread around 0, with some low outliers.
b) $\widehat{Fuel\ Consumption} = 0.625 + 1.178 \times Weight$.
c) For each additional 1000 pounds of *Weight*, an additional 1.178 gallons will be needed to drive 100 miles.
d) 21.06 miles per gallon.

15. a) Although nearly 97% of the variation in GDP can be accounted for by this model, we should examine a scatterplot of the residuals to see if it's appropriate.
b) No. The residuals show clear curvature.

17. No. The residual pattern shows that this model is still not appropriate.

19. a) The plot looks straight.

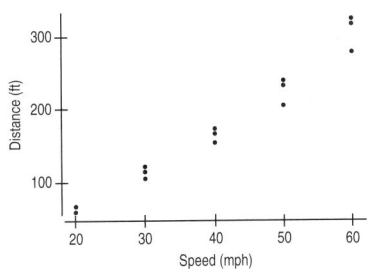

But if we fit a linear model, the residuals show both a bend and increasing spread

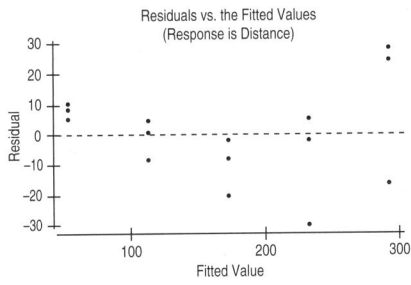

b) Square root seems best among the ladder of powers choices.
c) $\widehat{\sqrt{Distance}} = 3.30 + 0.235\ Speed$

d) $16.225^2 = 263.25$ ft
e) $19.75^2 = 390.06$ ft
f) The prediction for 55mph is in the middle of the data and should be good. R^2 is large and s_e is small. The prediction for 70mph is an extrapolation beyond the data and is likely to be less reliable.

21. a) No. Salaries were flat for 100 years and then increased rapidly.

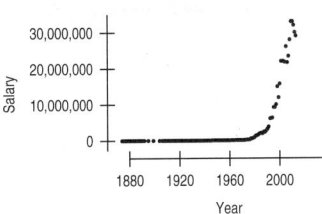

b) No. The plot is still very bent.

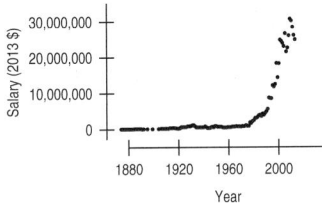

c) log(*Adjusted Salary*) works well.

d) $\widehat{\log(Adjusted\ Salary)} = -30.022 + 0.0185\ Year$
e) $R^2 = 87.7\%$
f)

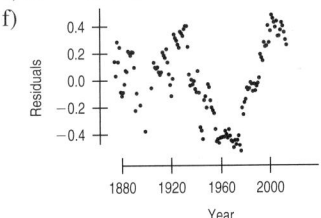

g) The residuals show a cyclic pattern. Salaries were lower than the model predicts at the beginning of the 20th century and again in the 1960's. They may have recently hit a high point and started turning lower, but it is difficult to tell.

23. a) $\widehat{\log(Distance)}$ against position works pretty well.
$\widehat{\log(Distance)} = 1.245 + 0.271 \times Position\ number$.

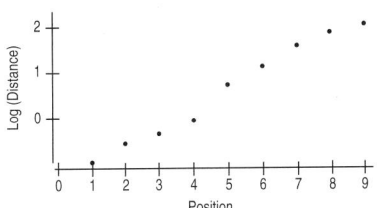

b) Pluto's residual is not especially large in the log scale. However, a model without Pluto predicts the 9th planet should be 5741 million miles. Pluto, at "only" 3707 million miles, doesn't fit very well, giving support to the argument that Pluto doesn't behave like a planet.

25. The predicted log (*Distance*) of Eris is 3.956, corresponding to a distance of 9036 million miles. That's more than the actual average distance of 6300 million miles.

27. a)

$\widehat{\sqrt{Bdft}} = -4 + diam$
The model is exact.
b) 36 board feet.
c) 1024 board feet.

29. $\widehat{\log_{10} Life} = 1.646 + 0.214 \log_{10} Decade$.

31. The relationship cannot be made straight by the methods of this chapter.

33. a) $\widehat{\sqrt{Left}} = 8.79 - 0.0733(Age)$.
b) 55.81 years
c) No; the residuals plot still shows a pattern.

Part II Review

1. % over 50, 0.69.
% under 20, −0.71.
% Graduating on time, −0.51.
% Full-time Faculty, 0.09

3. a) There does not appear to be a linear relationship.
b) Nothing, there is no reason to believe that the results for the Finger Lakes region are representative of the vineyards of the world.
c) $\widehat{CasePrice} = 92.77 + 0.567 \times Years$.
d) Only 2.7% of the variation in case price is accounted for by the ages of vineyards. Most of that is due to two outliers. We are better off using the mean price rather than this model.

5. a) $\widehat{TwinBirthsRate} = 17.77 + 0.551 \, Years \, Since \, 1980$.
b) Each year, the birth rate of twins increases, on average, by approximately 0.55.
c) 36.50 births per 1000 births. The scatterplot appears to be somewhat linear, but there is some curvature in the pattern. There is no reason to believe that the increase will continue to be linear 5 years beyond the data.
d) The residuals plot shows a definite curved pattern, so the relation is not linear. There is no reason to think the rate will continue to climb. The past 4 years have been constant.

7. a) −0.520
b) Negative, not strong, somewhat linear, but with more variation as pH increases.
c) The BCI would also be average.
d) The predicted *BCI* will be 1.56 SDs of *BCI* below the mean *BCI*.

9. a) $\widehat{Manatee \, Deaths} = -51.87 + 0.1429 \, Powerboat \, Registrations$ (in 1000s).
b) According to the model, for each increase of 10,000 powerboat registrations, the number of manatees killed increases by approximately 1.43 on average.

c) If there were 0 powerboat registrations, the number of manatee deaths would be −51.87. This is obviously a silly extrapolation.
d) The predicted number is 83.72 deaths. The actual number of deaths was 97. The residual is 97 − 83.72 = 13.28. The model underestimated the number of deaths by 13.28 manatees.
e) Negative residuals would suggest that the actual number of deaths was lower than the predicted number.
f) As the number of powerboat registrations has increased the number of manatee kills has increased. We don't know whether powerboat registrations will continue to increase. And even if they do, we don't know whether the trend will continue. Even though these manatees were killed by powerboats, we have no way of knowing if the relationship will continue as it is.

11. a) −0.984
b) 96.9%
c) 32.95 mph
d) 1.66 mph
e) Slope will increase.
f) Correlation will weaken (become less negative).
g) Correlation is the same, regardless of units.

13. a) Weight (but unable to verify linearity).
b) As weight increases, mileage decreases.
c) *Weight* accounts for 81.5% of the variation in *Fuel Efficiency*.

15. a) $\widehat{Horsepower} = 3.50 + 34.314 \times Weight$.
b) Thousands. For the equation to have predicted values between 60 and 160, the *X* values would have to be in thousands of pounds.
c) Yes. The residual plot does not show any pattern.
d) 115.0 horsepower.

17. a) The scatterplot shows a fairly strong linear relation in a positive direction. There seem to be two distinct clusters of data.
b) $\widehat{Interval} = 33.967 + 10.358 \times Duration$.
c) The time between eruptions increases by about 10.4 minutes per minute of *Duration* on average.
d) Since 77% of the variation in *Interval* is accounted for by *Duration* and the error standard deviation is 6.16 minutes, the prediction will be relatively accurate.
e) 75.4 minutes.
f) A residual is the observed value minus the predicted value. So the residual = 79 − 75.4 = 3.6 minutes, indicating that the model underestimated the interval in this case.

19. a) *r* = 0.888. Although *r* is high, you must look at the scatterplot and verify that the relation is linear in form.
b)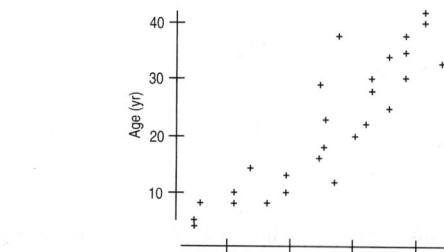

The association between diameter and age appears to be strong, somewhat linear, and positive.

c) $\widehat{Age} = -0.97 + 2.21 \times Diameter$.

d)

The residuals show a curved pattern (and two outliers).

e) The residuals for five of the seven largest trees (15 in. or larger) are positive, indicating that the predicted values underestimate the age.

21. TVs are 19 to 70 inches. A TV with a screen that was 10 inches larger would be predicted to cost 10(0.03) = $30, 10(0.3) = $300, 10(3) = $3000, or 10(30) = $30,000. Only $300 is reasonable, so the slope must be 0.3.

23. a) The model predicts % smoking from year, not the other way around.
    b) $\widehat{Year} = 2027.17 - 1.9823 \times \%\ Smoking$.
    c) The model predicts 2027, but an extrapolation to $x = 0$ is probably too far from the given data. The prediction is not very reliable in spite of the strong correlation.

25. The relation shows a negative direction, with a somewhat linear form, but perhaps with some slight curvature. There are several model outliers.

27. a) 71.9%
    b) As latitude increases, the average January temperature decreases.
    c) $\widehat{January\ Temperature} = 108.80 - 2.111 \times Latitude$.
    d) As the latitude increases by 1 degree, the average January temperature drops by about 2.11 degrees, on average.
    e) The y-intercept would indicate that the average January temperature is 108.8 when the latitude is 0. However, this is extrapolation and may not be meaningful.
    f) 24.4 degrees.
    g) The equation underestimates the average January temperature.

29. a) The scatterplot shows a strong, linear, positive association.
    b) There is an association, but it is likely that training and technique have increased over time and affected both jump performances.
    c) Neither; the change in units does not affect the correlation.
    d) I would predict the winning long jump to be 0.912 SDs above the mean long jump.

31. a) No relation; the correlation would probably be close to 0.
    b) The relation would have a positive direction and the correlation would be strong, assuming that students were studying French in each grade level. Otherwise, no correlation.
    c) No relation; correlation close to 0.
    d) The relation would have a positive direction and the correlation would be strong, since vocabulary would increase with each grade level.

33. $\widehat{Calories} = 560.7 - 3.08 \times Time$.
    Each minute extra at the table results in 3.08 fewer calories being consumed, on average. Perhaps the hungry children eat fast and eat more.

35. There seems to be a strong, positive, linear relationship with one high-leverage point (Northern Ireland) that makes the overall R^2 quite low. Without that point, the R^2 increases to 61.5%. Of course, these data are averaged across thousands of households, so the correlation appears to be higher than it would be for individuals. Any conclusions about individuals would be suspect.

37. a) 3.842
    b) 501.187
    c) 4.0

39. a) 30,818 pounds.
    b) 1302 pounds.
    c) 31,187.6 pounds.
    d) I would be concerned about using this relation if we needed accuracy closer than 1000 pounds or so, as the residuals are more than ± 1000 pounds.
    e) Negative residuals will be more of a problem, as the predicted weight would overestimate the weight of the truck; trucking companies might be inclined to take the ticket to court.

41. The original data are nonlinear, with a significant curvature. Using reciprocal square root of diameter gave a scatterplot that is nearly linear:
    $\widehat{1/\sqrt{Drain\ Time}} = 0.0024 + 0.219\ Diameter$.

Chapter 10

1. a) Yes, who takes out the trash cannot be predicted before the flip of a coin.
   b) No, it is not random, since you will probably name a favorite team.
   c) Yes, your new roommate cannot be predicted before names are drawn.

3. Rolling the pair of dice is the component.

5. To simulate, you could roll dice and note whether or not "doubles" came up. A trial would be completed once "doubles" came up. You would count up the number of rolls until "doubles" for the response variable. Alternatively, you could use the digits 1, 2, 3, 4, 5, 6 on a random digits table and disregard digits 7, 8, 9, and 0. Using the table, note the first legal digit for the first die, and then note the next legal digit for the second die. A double would indicate rolling doubles.

7. A machine pops up numbered balls. If it were truly random, the outcome could not be predicted and the outcomes would be equally likely. It is random only if the balls generate numbers in equal frequencies.

9. Use two-digit numbers 00–99; let 00–02 = defect, 03–99 = no defect.

11. a) 45, 10   b) 17, 22

13. If the lottery is random, it doesn't matter which numbers you play; all are equally likely to win.

15. a) The outcomes are not equally likely; for example, tossing 5 heads does not have the same probability as tossing 0 or 9 heads, but the simulation assumes they are equally likely.
    b) The even-odd assignment assumes that the player is equally likely to score or miss the shot. In reality, the likelihood of making the shot depends on the player's skill.

c) The likelihood for the first ace in the hand is not the same as for the second or third or fourth. But with this simulation, the likelihood is the same for each. (And it allows you to get 5 aces, which could get you in trouble in a real poker game!)

17. The conclusion should indicate that the simulation *suggests* that the average length of the line would be about 3.2 people. Future results might not match the simulated results exactly.

19. a) The component is one voter voting. An outcome is a vote for our candidate or not. Use two random digits, giving 00–54 a vote for your candidate and 55–99 for the underdog.
    b) A trial is 100 votes. Examine 100 two-digit random numbers, and count how many people voted for each candidate. Whoever gets the majority of votes wins that trial.
    c) The response variable is whether the underdog wins or not.

21. Answers will vary, but average answer will be about 51%.

23. Answers will vary, but average answer will be about 26%.

25. a) Answers will vary, but you should win about 10% of the time.
    b) You should win at the same rate with any number.

27. Answers will vary, but you should win about 10% of the time.

29. Answers will vary, but average answer will be about 1.9 tests.

31. Answers will vary, but average answer will be about 1.24 points.

33. Do the simulation in two steps. First, simulate the payoffs. Then, count until $500 is reached. Answers will vary, but average should be near 10.2 customers.

35. Answers will vary, but average answer will be about 3 children.

37. Answers will vary, but average answer will be about 7.5 rolls.

39. No, it will happen about 40% of the time.

41. Answers will vary, but average answer will be about 37.5%.

43. Three women will be selected about 7.8% of the time.

Chapter 11

1. The A&M administrators should take a survey. They should sample a part of the student body, selecting respondents with a randomization method. They should be sure to draw a sufficiently large sample.

3. The proportion in the sample is a statistic. The proportion of all students is the parameter of interest. The statistic estimates that parameter, but is not likely to be exactly the same.

5. This is not an SRS. Although each student may have an equal chance to be in the survey, groups of friends who choose to sit together will either all be in or out of the sample, so the selection is not independent.

7. a) Cluster sample.
   b) Stratified sample.
   c) Systematic sample.

9. Several terms are poorly defined. The survey needs to specify the meaning of "family" for this purpose and the meaning of "higher education." The term "seek" may also be poorly defined (for example, would applying to college but not being admitted qualify for seeking more education?).

11. a) This would suffer from voluntary response bias.
    b) This would be a convenience sample.

13. a) No. It would be nearly impossible to get exactly 500 males and 500 females from every country by random chance.
    b) A stratified sample, stratified by whether the respondent is male or female.

15. a) A cluster sample, with teams being the clusters.
    b) It is a reasonable solution to the problem of randomly sampling players because you can sample an entire team at once relatively easily but couldn't efficiently draw a random sample of all players on the same day.

17. a) Population—U.S. adults?
    b) Parameter—Proportion who have used and benefited from alternative medicine.
    c) Sampling Frame—All Consumers Union subscribers.
    d) Sample—Those who responded.
    e) Method—Questionnaire to all (nonrandom).
    f) Left Out—Those who are not Consumers Union subscribers.
    g) Bias—Nonresponse. Those who respond may have strong feelings one way or another.

19. a) Population—City voters.
    b) Parameter—Not clear; percentages of voters favoring issues?
    c) Sampling Frame—All city residents.
    d) Sample—Stratified sample; one block from each district.
    e) Method—Convenience sample within each stratum.
    f) Left Out—People not home during survey.
    g) Bias—Parameter(s) of interest not clear. Sampling within clusters is not random and may bias results.

21. a) Population—Cars.
    b) Parameter—Proportion with up-to-date registration, insurance, and safety inspections.
    c) Sampling Frame—All cars on that road.
    d) Sample—Those actually stopped by roadblock.
    e) Method—Cluster sample of location; census within cluster.
    f) Left Out—Local drivers that do not take that road.
    g) Bias—Time of day and location may not be representative of all cars.

23. a) Population—Dairy farms.
    b) Parameter—Proportion passing inspection.
    c) Sampling Frame—All dairy farms?
    d) Sample—Not clear. Perhaps a random sample of farms and then milk within each farm?
    e) Method—Multistage sampling.
    f) Left Out—Nothing.
    g) Bias—Should be unbiased if farms and milk at each farm are randomly selected.

25. Sampling error. The description of the sampling method suggests that samples should be representative of the voting population. Nonetheless, random chance in selecting the individuals who were polled means that sample statistics will vary from the population parameter, perhaps by quite a bit.

27. a) Voluntary response. Only those who see the show *and* feel strongly will respond.
    b) Possibly more representative than part a, but only strongly motivated parents go to PTA meetings (voluntary response).
    c) Multistage sampling, with cluster sample within each school. Probably a good design if most of the parents respond.
    d) Systematic sampling. Probably a reasonable design.

29. They will get responses only from people who come to the park to use the playground. Parents who are dissatisfied with the playground may not come.

31. The first sentence points out problems the respondent may not have noticed and might lead them to feel they should agree. However, the last phrase mentions higher fees, which could make people reject improvements to the playground.

33. a) This is a voluntary response survey. Even a large sample size can't make it representative.
    b) The wording seems fair enough. It states the facts and gives voice to both sides of the issue.
    c) The sampling frame is, at best, those who visit this particular site and even then depends on their volunteering to respond to the question.
    d) This is a true statement.

35. a) Seems neutral.
    b) Biased toward yes because of "great tradition." Better to ask, "Do you favor continued funding for the space program?"

37. Cell phones may be more likely among younger citizens, which could introduce a bias. As cell phones grow in use, this problem will be lessened.

39. a) Mean gas mileage for the last six fill-ups.
    b) Mean gas mileage for the vehicle.
    c) Recent driving conditions may not be typical.
    d) Mean gas mileage for all cars of this make and model.

41. a) Most likely that all laborers are selected, no managers, and few foremen. Bias may be introduced because the company itself is conducting the survey.
    b) Assign a number from 001 to 439 to each employee. Use a random-number table or software to select the sample.
    c) Still heavily favors the laborers.
    d) Stratify by job type (proportionately to the members in each).
    e) Answers will vary. Assign numbers 01 to 14 to each person; use a random-number table or software to do the selection.

43. What conclusions they may be able to make will depend on whether fish with discolored scales are equally likely to be caught as those without. It also depends on the level of compliance by fishermen. If fish are not equally likely to be caught, or fishermen are more disposed to bringing discolored fish, the results will be biased.

45. a) Petition may bias people to say they support the playground. Also, many may not be home on Saturday afternoon.
    b) If the food at the largest cafeteria is representative, this should be OK. However, those who really don't like the food won't be eating there.

Chapter 12

1. This is an observational study because the sports writer is not randomly assigning players to take steroids or not take steroids; the writer is merely observing home run totals between two eras. It would be unwise to conclude steroids caused any increases in home runs because we need to consider other factors besides steroids—factors possibly leading to more home runs include better equipment, players training more in the off-season, smaller ballparks, better scouting techniques, etc.

3. Each of the 40 deliveries is an experimental unit. He has randomized the experiment by flipping a coin to decide whether or not to phone.

5. The factor is calling, and the levels are whether or not he calls the customer. The response variable is the tip percentage for each delivery.

7. By calling some customers but not others during the same run, the driver has controlled many variables, such as day of the week, season, and weather. The experiment was randomized because he flipped a coin to determine whether or not to phone and it was replicated because he did this for 40 deliveries.

9. Because customers don't know about the experiment, those that are called don't know that others are not, and vice versa. Thus, the customers are blind. That would make this a single-blind study. It can't be double-blind because the delivery driver must know whether or not he phones.

11. Yes. Driver is now a block. The experiment is randomized within each block. This is a good idea because some drivers might generally get higher tips than others, but the goal of the experiment is to study the effect of phone calls.

13. Answers will vary. The cost or size of a delivery may confound his results. Larger orders may generally tip a higher or lower percentage of the bill.

15. a) No. There are no manipulated factors. Observational study.
    b) There may be lurking variables that are associated with both parental income and performance on the SAT.

17. a) This is a retrospective observational study.
    b) That's appropriate because MS is a relatively rare disease.
    c) The subjects were U.S. military personnel, some of whom had developed MS.
    d) The variables were the vitamin D blood levels and whether or not the subject developed MS.

19. a) This was a randomized, placebo-controlled experiment.
    b) Yes, such an experiment is the right way to determine whether black cohosh has an effect.
    c) 351 women aged 45 to 55 who reported at least two hot flashes a day.
    d) The treatments were black cohosh, a multiherb supplement with black cohosh, a multiherb supplement plus advice, estrogen, and a placebo. The response was the women's symptoms (presumably frequency of hot flashes).

21. a) Experiment.
    b) Bipolar disorder patients.
    c) Omega-3 fats from fish oil, two levels.
    d) 2 treatments
    e) Improvement (fewer symptoms?).
    f) Design not specified.
    g) Blind (due to placebo), unknown if double-blind.
    h) Individuals with bipolar disease improve with high-dose omega-3 fats from fish oil.

23. a) Observational study.
    b) Prospective.
    c) Men and women with moderately high blood pressure and normal blood pressure, unknown selection process.
    d) Memory and reaction time.
    e) As there is no random assignment, there is no way to know that high blood pressure *caused* subjects to do worse on memory and reaction-time tests. A lurking variable may also be the cause.

25. a) Experiment.
b) Postmenopausal women.
c) Alcohol—2 levels; blocking variable—estrogen supplements (2 levels).
d) 1 factor (alcohol) at 2 levels = 2 treatments.
e) Increase in estrogen levels.
f) Blocked.
g) Not blind.
h) Indicates that alcohol consumption *for those taking estrogen supplements* may increase estrogen levels.

27. a) Observational study.
b) Retrospective.
c) Women in Finland, unknown selection process with data from church records.
d) Women's life spans.
e) As there is no random assignment, there is no way to know that having sons or daughters shortens or lengthens the life span of mothers.

29. a) Observational study.
b) Prospective.
c) People with or without depression, unknown selection process.
d) Frequency of crying in response to sad situations.
e) There is no apparent difference in crying response (to sad movies) for depressed and nondepressed groups.

31. a) Experiment.
b) People experiencing migraines.
c) 2 factors (pain reliever and water temperature), 2 levels each.
d) 4 treatments
e) Level of pain relief.
f) Completely randomized over 2 factors.
g) Blind, as subjects did not know if they received the pain medication or the placebo, but not blind, as the subjects will know if they are drinking regular or ice water.
h) It may indicate whether pain reliever alone or in combination with ice water gives pain relief, but patients are not blinded to ice water, so placebo effect may also be the cause of any relief seen caused by ice water.

33. a) Experiment.
b) Athletes with hamstring injuries.
c) 1 factor: type of exercise program (2 levels).
d) 2 treatments
e) Time to return to sports.
f) Completely randomized.
g) No blinding—subjects must know what kind of exercise they do.
h) Can determine which of the two exercise programs is more effective.

35. They need to compare omega-3 results to something. Perhaps bipolarity is seasonal and would have improved during the experiment anyway.

37. a) Subjects' responses might be related to many other factors (diet, exercise, genetics, etc). Randomization should equalize the two groups with respect to unknown factors.
b) More subjects would minimize the impact of individual variability in the responses, but the experiment would become more costly and time consuming.

39. People who engage in regular exercise might differ from others with respect to bipolar disorder, and that additional variability could obscure the effectiveness of this treatment.

41. Answers may vary. Use a random-number generator to randomly select 24 numbers from 01 to 24 without replication. Assign the first 8 numbers to the first group, the second 8 numbers to the second group, and the third 8 numbers to the third group.

43. a) First, they are using athletes who have a vested interest in the success of the shoe by virtue of their sponsorship. They should choose other athletes. Second, they should randomize the order of the runs, not run all the races with their shoes second. They should blind the athletes by disguising the shoes if possible, so they don't know which is which. The timers shouldn't know which athletes are running with which shoes, either. Finally, they should replicate several times, since times will vary under both shoe conditions.
b) Because of the problems in part a, the results they obtain may favor their shoes. In addition, the results obtained for Olympic athletes may not be the same as for the general runner.

45. a) Allowing athletes to self-select treatments could confound the results. Other issues such as severity of injury, diet, age, etc., could also affect time to heal; randomization should equalize the treatment groups with respect to any such variables.
b) A control group could have revealed whether either exercise program was better (or worse) than just letting the injury heal.
c) Doctors who evaluated the athletes to approve their return to sports should not know which treatment the subject had.
d) It's hard to tell. The difference of 15 days seems large, but the standard deviations indicate that there was a great deal of variability in the times.

47. a) The differences among the Mozart and quiet groups were more than would have been expected from sampling variation.
b)

c) The Mozart group seems to have the smallest median difference and thus the *least* improvement, but there does not appear to be a significant difference.
d) No, if anything, there is less improvement, but the difference does not seem significant compared with the usual variation.

49. a) Observational, prospective study.
b) The supposed relation between health and wine consumption might be explained by the confounding variables of income and education.
c) None of these. While the variables have a relation, there is no causality indicated for the relation.

51. a) Arrange the 20 containers in 20 separate locations. Use a random-number generator to identify the 10 containers that should be filled with water.
b) Guessing, the dowser should be correct about 50% of the time. A record of 60% (12 out of 20) does not appear to be significantly different.
c) Answers may vary. You would need to see a high level of success—say, 90% to 100%, that is, 18 to 20 correct.

53. Randomly assign half the reading teachers in the district to use each method. Students should be randomly assigned to teachers as well. Make sure to block both by school and grade (or control grade by using only one grade). Construct an appropriate reading test to be used at the end of the year, and compare scores.

55. a) They mean that the difference is higher than they would expect from normal sampling variability.
b) Observational study.
c) No. Perhaps the differences are attributable to some confounding variable (e.g., people are more likely to engage in riskier behaviors on the weekend) rather than the day of admission.
d) Perhaps people have more serious accidents and traumas on weekends and are thus more likely to die as a result.

57. Answers may vary. This experiment has 1 factor (pesticide), at 3 levels (pesticide A, pesticide B, no pesticide), resulting in 3 treatments. The response variable is the number of beetle larvae found on each plant. Randomly select a third of the plots to be sprayed with pesticide A, a third with pesticide B, and a third with no pesticide (since the researcher also wants to know whether the pesticides even work at all). To control the experiment, the plots of land should be as similar as possible with regard to amount of sunlight, water, proximity to other plants, etc. If not, plots with similar characteristics should be blocked together. If possible, use some inert substance as a placebo pesticide on the control group, and do not tell the counters of the beetle larvae which plants have been treated with pesticides. After a given period of time, count the number of beetle larvae on each plant and compare the results.

59. Answers may vary. Find a group of volunteers. Each volunteer will be required to shut off the machine with his or her left hand and right hand. Randomly assign the left or right hand to be used first. Complete the first attempt for the whole group. Now repeat the experiment with the alternate hand. Check the differences in time for the left and right hands.

61. a) Jumping with or without a parachute.
b) Volunteer skydivers (the dimwitted ones).
c) A parachute that looks real but doesn't work.
d) A good parachute and a placebo parachute.
e) Whether parachutist survives the jump (or extent of injuries).
f) All should jump from the same altitude in similar weather conditions and land on similar surfaces.
g) Randomly assign people the parachutes.
h) The skydivers (and the people involved in distributing the parachute packs) shouldn't know who got a working chute. And the people evaluating the subjects after the jumps should not be told who had a real parachute either!

Part III Review

1. Observational prospective study. Indications of behavior differences can be seen in the two groups. May show a link between premature birth and behavior, but there may be lurking variables involved.

3. Experiment, matched by gender and weight, randomization within blocks of two pups of same gender and weight. Factor: type of diet. Treatments: low-calorie diet and allowing the dog to eat all it wants. Response variable: length of life. Can conclude that, on average, dogs with a lower-calorie diet live longer.

5. Completely randomized experiment, with the treatment being receiving folic acid or not (one factor, two levels). Treatments assigned randomly and the response variable is the number of/ occurrence of additional precancerous growths. Neither blocking nor matching are mentioned, but in a study such as this one, it is likely that researchers and patients are blinded. Since treatments were randomized, it seems reasonable to generalize results to all people with precancerous polyps, though caution is warranted since these results contradict a previous study.

7. Sampling. Probably a simple random sample, although may be stratified by type of firework. Population is all fireworks produced each day. Parameter is proportion of duds. Can determine if the day's production is ready for sale.

9. Observational retrospective study. Researcher can conclude that for anyone's lunch, even when packed with ice, food temperatures are rising to unsafe levels.

11. Experiment, with a control group being the genetically engineered mice who received no antidepressant and the treatment group being the mice who received the drug. The response variable is the amount of plaque in their brains after one dose and after four months. There is no mention of blinding or matching. Conclusions can be drawn to the general population of mice and we should assume treatments were randomized. To conclude the same for humans would be risky, but researchers might propose an experiment on humans based on this study.

13. Experiment. Factor is gene therapy. Hamsters were randomized to treatments. Treatments were gene therapy or not. Response variable is heart muscle condition. Can conclude that gene therapy is beneficial (at least in hamsters).

15. Sampling. Population is all oranges on the truck. Parameter is proportion of unsuitable oranges. Procedure is probably stratified random sampling with regions inside the truck being the strata. Can conclude whether or not to accept the truckload.

17. Observational retrospective study performed as a telephone-based randomized survey. Based on the excerpt, it seems reasonable to conclude that more education is associated with a higher Emotional Health Index score, but the insist on causality would be faulty reasoning.

19. Answers will vary. This is a simulation problem. Using a random digits table or software, call 0–4 a loss and 5–9 a win for the gambler on a game. Use blocks of 5 digits to simulate a week's pick.

21. Answers will vary.

23. a) Experiment. Actively manipulated candy giving, diners were randomly assigned treatments, control group was those with no candy, lots of dining parties.
b) It depends on when the decision was made. If early in the meal, the server may give better treatment to those who will receive candy—biasing the results.
c) A difference in response so large it cannot be attributed to natural sampling variability.

25. There will be voluntary response bias, and results will mimic those only of the visitors to sodahead.com and not the general U.S. population. The question is leading responders to answer "yes" though many might understand that the president's timing for his vacation had nothing to do with the events of the week.

27. a) Simulation results will vary. Average will be around 5.8 points.
    b) Simulation results will vary. Average will also be around 5.8 points.
    c) Answers will vary.

29. a) Yes.
    b) No. Residences without phones are excluded. Residences with more than one phone number had a higher chance.
    c) No. People who respond to the survey may be of age but not registered voters.
    d) No. Households who answered the phone may be more likely to have someone at home when the phone call was generated. These may not be representative of all households.

31. a) Does not prove it. There may be other confounding variables. Only way to prove this would be to do a controlled experiment.
    b) Alzheimer's usually shows up late in life. Perhaps smokers have died of other causes before Alzheimer's is evident.
    c) An experiment would be unethical. One could design a prospective study in which groups of smokers and nonsmokers are followed for many years and the incidence of Alzheimer's is tracked.

33.

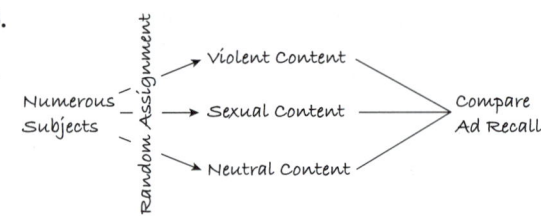

Numerous subjects will be randomly assigned to see shows with violent, sexual, or neutral content. They will see the same commercials. After the show, they will be interviewed for their recall of brand names in the commercials.

35. a) May have been a simple random sample, but given the relative equality in age group, it may have been stratified.
    b) 38.2%
    c) We don't know. If data were collected from voting precincts that are primarily Democratic or primarily Republican, that would bias the results. Because the survey was comissioned by NBC News, we can assume the data collected are probably OK.
    d) Do party affiliations differ for different age groups?

37. The factor in the experiment will be type of bird control. I will have three treatments: scarecrow, netting, and no control. I will randomly assign several different areas in the vineyard to one of the treatments, taking care that there is sufficient separation that the possible effect of the scarecrow will not be confounded. At the end of the season, the response variable will be the proportion of bird-damaged grapes.

39. a) This was the control group. Subjects needed to have some sort of treatment to perceive they were getting help for their lower back pain.
    b) Even though patients were volunteers, their treatments were randomized. To generalize the results, we would need to assume these volunteers have the same characteristics of the general population of those with lower back pain (probably reasonable).
    c) It means that the success rates for the acupuncture groups were higher than could be attributed to chance when compared to the conventional treatment group. In other words, researchers concluded that both proper and "fake" acupuncture reduced back pain.

41. a) Use stratified sampling to select 2 first-class passengers and 12 from coach.
    b) Number passengers alphabetically, 01 = Bergman to 20 = Testut. Read in blocks of two, ignoring any numbers more than 20. This gives 65, 43, 67, 11 (selects Fontana), 27, 04 (selects Castillo).
    c) Number passengers alphabetically from 001 to 120. Use the random-number table to find three-digit numbers in this range until 12 different values have been selected.

43. Simulation results will vary. (Use integers 00 to 99 as a basis. Use integers 00 to 69 to represent a tee shot on the fairway. If on the fairway, use digits 00 to 79 to represent on the green. If off the fairway, use 00 to 39 to represent getting on the green. If not on the green, use digits 00 to 89 to represent landing on the green. For the first putt, use digits 00 to 19 to represent making the shot. For subsequent putts, use digits 00 to 89 to represent making the shot.)

45. a) All members of the U.S. military.
    b) Answers vary. The survey should have been given out randomly and to all branches of the military. If not, there will be undercoverage and the data collected may not represent the opinions of all members of the military.
    c) Answers may vary.

Chapter 13

1. In the long run, a fair coin will generate 50% heads and 50% tails, approximately. But for each flip we cannot predict the outcome.

3. There is no law of averages for the short run. The first five flips do not affect the sixth flip.

5. a) 0.30   b) 0.80

7. a) 0.237   b) 0.763   c) 0.999

9. a) S = {HH, HT, TH, TT}, equally likely
   b) S = {0, 1, 2, 3}, not equally likely
   c) S = {H, TH, TTH, TTT}, not equally likely
   d) S = {1, 2, 3, 4, 5, 6}, not equally likely

11. In this context, "truly random" should mean that every number is equally likely to occur and outcomes are mutually independent.

13. There is no "Law of Averages." She would be wrong to think they are "due" for a harsh winter.

15. There is no "Law of Averages." If his shooting is independent, he could very well shoot just as poorly the next day.

17. a) There is some chance you would have to pay out much more than the $1500.
    b) Many customers pay for insurance. The small risk for any one customer is spread among all.

19. a) Legitimate.   b) Legitimate.
    c) Not legitimate (sum more than 1).   d) Legitimate.
    e) Not legitimate (can't have negatives or values more than 1).

21. A family may own both a computer and an HD TV. The events are not disjoint, so the Addition Rule does not apply.

23. When cars are traveling close together, their speeds are not independent, so the Multiplication Rule does not apply.

25. a) He has multiplied the two probabilities.
b) He assumes that being accepted at the colleges are independent events.
c) No. Colleges use similar criteria for acceptance, so the decisions are not independent.

27. a) 0.72   b) 0.89   c) 0.28

29. a) 0.5184   b) 0.0784   c) 0.4816

31. a) Repair needs for the two cars must be independent.
b) Maybe not. An owner may treat the two cars similarly, taking good (or poor) care of both. This may decrease (or increase) the likelihood that each needs to be repaired.

33. a) $164/529 = 0.310$
b) $37/529 + 16/529 = 0.100$

35. a) 0.205   b) 0.805
c) Responses are independent.
d) People were polled at random.

37. a) 0.5332   b) 0.9132
c) $(1 - 0.62) + 0.62(1 - 0.14)$

39. a) 1) 0.30   2) 0.30   3) 0.90   4) 0.0
b) 1) 0.027   2) 0.128   3) 0.512   4) 0.271

41. a) Disjoint (can't be both red and orange).
b) Independent (unless you're drawing from a small bag).
c) No. Once you know that one of a pair of disjoint events has occurred, the other is impossible.

43. a) 0.0046   b) 0.125   c) 0.296
d) 0.421   e) 0.995

45. a) 0.027   b) 0.063   c) 0.973   d) 0.014

47. a) 0.024   b) 0.250   c) 0.543

49. 0.266

51. a) For any day with a valid three-digit date, the chance is 0.001, or 1 in 1000. For many dates in October through December, the probability is 0. (No three digits will make 10/15, for example.)
b) There are 65 days when the chance to match is 0. (Oct. 10–31, Nov. 10–30, and Dec. 10–31.) The chance for no matches on the remaining 300 days is 0.741.
c) 0.259   d) 0.049

Chapter 14

1. 0.42

3. 0.675

5. 0.135

7. No $P(S) = 0.323$ but $P(S|FC) = 0.625$. These are not the same.

9. We have joint and marginal probabilities, not conditional probabilities, so a table is better.

	United States	Not United States	Total
Log On Every Day	0.20	0.30	0.50
Do Not Log On Every Day	0.10	0.40	0.50
Total	0.30	0.70	1.00

11. A tree is better because we have conditional and marginal probabilities. The joint probabilities are found at the end of the branches.

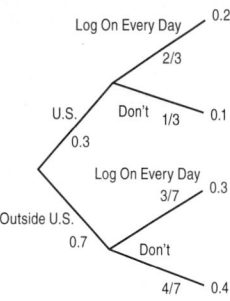

13. $P(\text{US}|\text{Log on every day}) = P(\text{US and Log on every day})/P(\text{Log on every day}) = 0.2/(0.2 + 0.3) = 0.4$. It has increased.

15. a) 0.98   b) 0.02   c) 0.25

17. a) 0.31   b) 0.48   c) 0.31

19. a) 0.2025   b) 0.7023   c) 0.2404   d) 0.0402

21. a) 0.50   b) 1.00   c) 0.077   d) 0.333

23. a) 0.11   b) 0.27   c) 0.407   d) 0.344

25. a) 0.011   b) 0.222   c) 0.054   d) 0.337
e) 0.436

27. 0.21

29. a) 0.145   b) 0.118   c) 0.414   d) 0.217

31. a) 0.318   b) 0.955   c) 0.071   d) 0.009

33. a) 32%   b) 0.135
c) No. 7% of juniors have taken both.
d) No. The probability that a junior has taken a computer course is 0.23. The probability that a junior has taken a computer course *given* he or she has taken a Statistics course is 0.135.

35. a) 0.795
b) No, not quite. 79.5% of homes with phones have cell phones, but 83% of people have cell phones.
c) No. 58% of homes have both.

37. Yes. $P(\text{Ace}) = 1/13$. $P(\text{Ace}|\text{Spade}) = 1/13$. And the same is true for all four suits.

39. a) 0.33
b) No. 9% of the chickens had both contaminants.
c) No. $P(C|S) = 0.64 \neq P(C)$. If a chicken is contaminated with salmonella, it's more likely also to have campylobacter.

41. No. 28.8% of men with OK blood pressure have high cholesterol, but 40.7% of those with high blood pressure do.

43. a) 72.8%
b) Probably. 71.5% of people with cell phones had landlines, and 72.8% of all people did.

45. No. Only 34% of men were Democrats, but over 39% of all voters were.

47. a) No. The probability that the luggage arrives on time depends on whether the flight is on time. The probability is 0.95 if the flight is on time and only 0.65 if not.
b) 0.695

49. 0.975

51. a) No. The probability of missing work for day-shift employees is 0.01. It is 0.02 for night-shift employees. The probability depends on whether they work day or night shift.
    b) 1.4%
53. 57.1%
55. a) 0.20 b) 0.272 c) 0.353 d) 0.033
57. 0.563
59. 0.9995

Chapter 15

1. 33 oranges
3. 4.58 oranges
5. $1700, $141
7. 0.055
9. a) 19 b) 4.2
11. a)

Amount won	$0	$5	$10	$30
P(Amount won)	26/52	13/52	12/52	1/52

    b) $4.13 c) $4 or less (answers may vary)
13. a)

Children	1	2	3
P(Children)	0.5	0.25	0.25

    b) 1.75 children c) 0.875 boys

Boys	0	1	2	3
P(Boys)	0.5	0.25	0.125	0.125

15. $27,000
17. a) 7 b) 1.89
19. $5.44
21. 0.83
23. a) 1.7 b) 0.9
25. $\mu = 0.64, \sigma = 0.93$
27. a) $50 b) $100
29. a) No. The probability of winning the second depends on the outcome of the first.
    b) 0.42 c) 0.08
    d)

Games won	0	1	2
P(Games won)	0.42	0.50	0.08

    e) $\mu = 0.66, \sigma = 0.62$
31. a)

Number good	0	1	2
P(Number good)	0.067	0.467	0.467

    b) 1.40 c) 0.61
33. a) $\mu = 30, \sigma = 6$
    b) $\mu = 26, \sigma = 5$
    c) $\mu = 30, \sigma = 5.39$
    d) $\mu = -10, \sigma = 5.39$
    e) $\mu = 20, \sigma = 2.83$
35. a) $\mu = 240, \sigma = 12.80$
    b) $\mu = 140, \sigma = 24$
    c) $\mu = 720, \sigma = 34.18$
    d) $\mu = 60, \sigma = 39.40$
    e) $\mu = 600, \sigma = 22.63$

37. a) 1.8 b) 0.87
    c) Cartons are independent of each other.
39. $\mu = 13.6, \sigma = 2.55$ (assuming the hours are independent of each other).
41. a) $\mu = 23.4, \sigma = 2.97$
    b) We assume each truck gets tickets independently.
43. a) There will be many gains of $150 with a few large losses.
    b) $\mu = \$300, \sigma = \$8485.28$
    c) $\mu = \$1,500,000, \sigma = \$600,000$
    d) Yes. $0 is 2.5 SDs below the mean for 10,000 policies.
    e) Losses are independent of each other. A major catastrophe with many policies in an area would violate the assumption.
45. a) 1 oz b) 0.5 oz
    c) 0.023 d) $\mu = 4$ oz, $\sigma = 0.5$ oz
    e) 0.159 f) $\mu = 12.3$ oz, $\sigma = 0.54$ oz
47. a) 12.2 oz b) 0.51 oz
    c) 0.058
49. a) $\mu = 200.57$ sec, $\sigma = 0.46$ sec
    b) No, $z = \dfrac{199.48 - 200.57}{0.461} = -2.36$. There is only 0.009 probability of swimming that fast or faster.
51. a) A = price of a pound of apples; P = price of a pound of potatoes; Net Income = $100A + 50P - 2$
    b) $63.00
    c) $20.62
    d) Mean—no; SD—yes (independent sales prices).
53. a) $\mu = 1920, \sigma = 48.99; P(T > 2000) = 0.051$
    b) $\mu = 220, \sigma = 11.09$; No—$300 is more than 7 SDs above the mean.
    c) $P(D - \tfrac{1}{2}C > 0) \approx 0.26$

Chapter 16

1. a) No. More than two outcomes are possible.
   b) Yes, assuming the people are unrelated to each other.
   c) No. The chance of a heart changes as cards are dealt.
   d) No, 500 is more than 10% of 3000.
   e) If packages in a case are independent of each other, yes; otherwise, no.
3. 0.0596
5. 0.011
7. 0.140
9. a) A uniform; all numbers should be equally likely to be selected.
   b) 0.02 c) 0.10
11. a) $\lambda = 1/3$. b) 0.811
13. a) Use single random digits. Let 0, 1 = Hope. Count the number of random numbers until a 0 or 1 occurs.
    c) Results will vary.
    d)

x	1	2	3	4	5	6	7	8	≥9
P(x)	0.2	0.16	0.128	0.102	0.082	0.066	0.052	0.042	0.168

15. a) Use single random digits. Let 0, 1 = Hope. Examine random digits in groups of five, counting the number of 0's and 1's.

c) Results will vary.
d)

x	0	1	2	3	4	5
P(x)	0.33	0.41	0.20	0.05	0.01	0.0

17. Departures from the same airport during a 2-hour interval may not be independent. All could be delayed by weather, for example.
19. a) 0.0819   b) 0.0064   c) 0.992
21. 5   23. 20 calls
25. a) 25   b) 0.185   c) 0.217   d) 0.693
27. a) 50   b) $E(X) = np = 100(0.5) = 50$.
29. a) 0.0745   b) 0.502   c) 0.211
    d) 0.0166   e) 0.0179   f) 0.9987
31. a) 0.65   b) 0.75   c) 7.69 picks
33. a) $\mu = 10.44, \sigma = 1.16$
    b) 0.812   c) 0.475   d) 0.00193   e) 0.998
35. $\mu = 20.28, \sigma = 4.22$
37. a) 0.118   b) 0.324   c) 0.744   d) 0.580
39. a) $\mu = 56, \sigma = 4.10$
    b) Yes, $np = 56 \geq 10, nq = 24 \geq 10$, serves are independent.
    c) In a match with 80 serves, approximately 68% of the time she will have between 51.9 and 60.1 good serves, approximately 95% of the time she will have between 47.8 and 64.2 good serves, and approximately 99.7% of the time she will have between 43.7 and 68.3 good serves.
    d) Normal, approx.: 0.014; Binomial, exact: 0.016
41. a) Assuming apples fall and become blemished independently of each other, Binom(300, 0.06) is appropriate. Since $np \geq 10$ and $nq \geq 10$, $N(18, 4.11)$ is also appropriate.
    b) Normal, approx.: 0.072; Binomial, exact: 0.085
    c) No, 50 is 7.8 SDs above the mean.
43. Normal, approx.: 0.053; Binomial, exact: 0.061
45. The mean number of sales should be 24 with SD 4.60. Ten sales is more than 3.0 SDs below the mean. He was probably misled.
47. a) 0.0863   b) 0.0365
49. a) 4 cases   b) 0.9817
51. a) 5   b) 0.066   c) 0.107
    d) $\mu = 24, \sigma = 2.19$
    e) Normal, approx.: 0.819; Binomial, exact: 0.848
53. $\mu = 20, \sigma = 4$. I'd want *at least* 32 (3 SDs above the mean). (Answers will vary.)
55. Probably not. There's a more than 9% chance that he could hit 4 shots in a row, so he can expect this to happen nearly once in every 10 sets of 4 shots he takes. That does not seem unusual.
57. Yes. We'd expect him to make 22 shots, with a standard deviation of 3.15 shots. 32 shots is more than 3 standard deviations above the expected value, an unusually high rate of success.
59. a) The Poission model
    b) 0.9502   c) 0.0025
61. a) The exponential model
    b) 1/3 minutes   c) 0.0473

Part IV Review

1. a) 0.34   b) 0.27   c) 0.069
   d) No, 2% of cars have both types of defects.
   e) Of all cars with cosmetic defects, 6.9% have functional defects. Overall, 7.0% of cars have functional defects. The probabilities here are estimates, so these are probably close enough to say the defects are independent.
3. a) C = Price to China; F = Price to France; Total = $3C + 5F$
   b) $\mu = \$5500, \sigma = \$672.68$
   c) $\mu = \$500, \sigma = \$180.28$
   d) Means—no. Standard deviations—yes; ticket prices must be independent of each other for different countries, but all tickets to the same country are at the same price.
5. a) $\mu = -\$0.20, \sigma = \$1.89$
   b) $\mu = -\$0.40, \sigma = \$2.67$
7. a) 0.106   b) 0.651   c) 0.442
9. a) 0.590   b) 0.328   c) 0.00856
11. a) $\mu = 15.2, \sigma = 3.70$
    b) Yes, $np \geq 10$ and $nq \geq 10$
    c) Normal, approx.: 0.080; Binomial, exact: 0.097
13. a) 0.0173
    b) 0.591
    c) Left: 960; right: 120; both: 120
    d) $\mu = 120, \sigma = 10.39$
    e) About 68% chance of between 110 and 130; about 95% between 99 and 141; about 99.7% between 89 and 151.
15. a) Men's heights are more variable than women's.
    b) Men
    c) M = Man's height; W = Woman's height; $M - W$ is how much taller the man is.
    d) 5.1″
    e) 3.75″
    f) 0.913
    g) If independent, it should be about 91.3%. We are told 92%. This difference seems small and may be due to natural sampling variability.
17. a) The chance is 1.6×10^{-7}.   b) 0.952
    c) 0.063
19. −$2080.00
21. a) 0.717   b) 0.588
23. a) $\mu = 100, \sigma = 8$ b) $\mu = 1000, \sigma = 60$
    c) $\mu = 100, \sigma = 8.54$ d) $\mu = -50, \sigma = 10$
    e) $\mu = 100, \sigma = 11.31$
25. a) Many do both, so the two categories can total more than 100%.
    b) No. They can't be disjoint. If they were, the total would be 100% or less.
    c) No. Probabilities are different for boys and girls.
    d) 0.0524
27. a) 21 days   b) 1649.73 som
    c) 3300 som extra. About 157-som "cushion" each day.
29. No, you'd expect 541.2 homeowners, with an SD of 13.56. 523 is 1.34 SDs below the mean; not unusual.

31. a) 0.027 b) 0.240 c) 0.30
c) Answer b would become 0.0026.

33. a) 6 b) 15 c) 0.402

35. a) 34% b) 35% c) 31.4%
d) 31.4% of classes that used calculators used computer assignments, while in classes that didn't use calculators, 30.6% used computer assignments. These are close enough to think the choice is probably independent.

37. a) A Poisson model
b) 0.1 failures
c) 0.090
d) 0.095

39. a) 1/11 b) 7/22 c) 5/11
d) 0 e) 19/66

41. a) Expected number of stars with planets.
b) Expected number of planets with intelligent life.
c) Probability of a planet with a suitable environment having intelligent life.
d) f_l: If a planet has a suitable environment, the probability that life develops.
f_i: If a planet develops life, the probability that the life evolves intelligence.
f_c: If a planet has intelligent life, the probability that it develops radio communication.

43. 0.991

Chapter 17

1. a) Unimodal and symmetric (roughly Normal).
b) 0.36
c) 0.034

3. All the histograms are centered near 0.05. As n gets larger, the histograms approach the Normal shape, and the variability in the sample proportions decreases.

5. Check conditions: The data come from a random sample, so the randomization condition is met. We don't know the exact value of p, but if the sample is anywhere near correct, $np \approx 1500 \times 0.27 = 405$, and $nq \approx 1500 \times 0.73 = 1095$. So there are well over 10 successes and 10 failures, meeting the Success/Failure Condition. Since there are more than $10 \times 1500 = 15{,}000$ adults in the United States, the 10% Condition is met. A Normal model is appropriate for the sampling distribution of the sample proportion.

7. a) The histogram is unimodal and symmetric for $n = 200$.
b) The success/failure condition says that np and nq should both be at least 10, which is not satisfied until $n = 200$ for $p = 0.05$. The theory predicted my choice.

9. a) No. The sampling distribution of the maximum is skewed to the left.
b) No. The 95% rule is based on the Normal distribution.

11. a) 0.0357 b) 400

13. a) Can't use a Normal model to estimate probabilities. The distribution is skewed right—not Normal.
b) 4 is probably not a large enough sample to say the average follows the Normal model.
c) No. This is 3.16 SDs above the mean.

15. a) 0.0003. Model is $N(384, 34.15)$.
b) $427.77 or more ($427.71 using tables)

17. a) Symmetric, because probability of heads and tails is equal.
b) 0.5
c) $s = 0.125$
d) $np = 8 < 10$

19. a) About 68% should have proportions between 0.4 and 0.6, about 95% between 0.3 and 0.7, and about 99.7% between 0.2 and 0.8.
b) $np = 12.5, nq = 12.5$; both are ≥ 10.
c)

Proportion

$np = nq = 32$; both are ≥ 10.
d) Becomes narrower (less spread around 0.5).

21. This is a fairly unusual result: about 2.26 SDs below the mean. The probability of that is about 0.012. So, in a class of 100 this is certainly a reasonable possibility.

23. a)

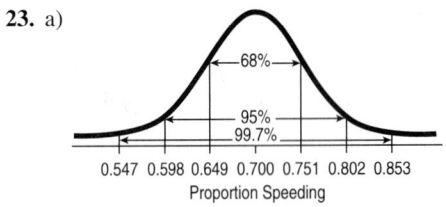
Proportion Speeding

b) Both $np = 56$ and $nq = 24 \geq 10$. Drivers *may* be independent of each other, but if flow of traffic is very fast, they may not be. Or weather conditions may affect all drivers. In these cases, they may get more or fewer speeders than they expect.

25. a) Assume that these children are typical of the population. They represent fewer than 10% of all children. We expect 20.4 nearsighted and 149.6 not; both are at least 10.
b)

Proportion

c) Probably between 12 and 29

27. a) $\mu = 7\%, \sigma = 1.8\%$
b) Assume that clients pay independently of each other, that we have a random sample of all possible clients, and that these represent less than 10% of all possible clients. $np = 14$ and $nq = 186$ are both at least 10.
c) 0.048

29.

Retention Rate

These are random samples of students who took the ACT, who may not be representative of all college freshmen. Based on the parameters given, $np = 288.8$, $nq = 111.2$ are both at least 10, and the number of college freshmen is more than $10 \times 400 = 4000$. But the samples are not random samples from this population.

31. Yes. If their students were typical, a retention rate of $522/603 = 86.6$ would be over 7 standard deviations above the expected rate of 72.2%.

33. 0.212. Reasonable that those polled are independent of each other and represent less than 10% of all potential voters. We assume the sample was selected at random. Success/Failure Condition met: $np = 208$, $nq = 192$. Both ≥ 10.

35. 0.00089 using $N(0.65, 0.048)$ model.

37. Answers will vary. Using $\mu + 3\sigma$ for "very sure," the restaurant should have 51 family-friendly seats. Assumes customers at any time are independent of each other, a random sample, and represent less than 10% of all potential customers. $np = 36$, $nq = 84$, so Normal model is reasonable ($\mu = 0.30$, $\sigma = 0.042$).

39. a) Normal, center at μ, standard deviation $\sigma/\sqrt{n}$.
    b) Standard deviation will be smaller. Center will remain the same.

41. a) The histogram is unimodal and slightly skewed to the right, centered at 36 inches with a standard deviation near 4 inches.
    b) All the histograms are centered near 36 inches. As n gets larger, the histograms approach the Normal shape and the variability in the sample means decreases. The histograms are fairly normal by the time the sample reaches size 5.

43. a)

n	Observed Mean	Theoretical Mean	Observed st. dev.	Theoretical st. dev.
2	36.314	36.33	2.855	2.842
5	36.314	36.33	1.805	1.797
10	36.341	36.33	1.276	1.271
20	36.339	36.33	0.895	0.899

   b) They are all very close to what we would expect.
   c) For samples as small as 5, the sampling distribution of sample means is unimodal and very symmetric.
   d) The distribution of the original data is nearly unimodal and symmetric, so it doesn't take a very large sample size for the distribution of sample means to be approximately Normal.

45.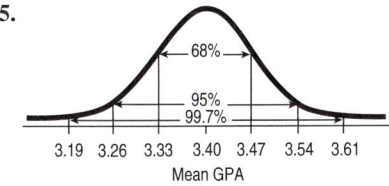

Normal, $\mu = 3.4$, $\sigma = 0.07$. We assume that the students are randomly assigned to the seminars and represent less than 10% of all possible students, and that individuals' GPAs are independent of one another.

47. a) As the CLT predicts, there is more variability in the smaller outlets.
    b) If the lottery is random, all outlets are equally likely to sell winning tickets.

49. a) 21.1% b) 276.8 days or more
    c) $N(266, 2.07)$ d) 0.002

51. a) There are more premature births than very long pregnancies. Modern practice of medicine stops pregnancies at about 2 weeks past normal due date.
    b) Parts (a) and (b)—yes—we can't use Normal model if it's very skewed. Part (c)—no—CLT guarantees a Normal model for this large sample size.

53. a) $\mu = \$2.00$, $\sigma = \$3.61$
    b) $\mu = \$4.00$, $\sigma = \$5.10$
    c) 0.191. Model is $N(80, 22.83)$.

55. a) $\mu = 2.798$, $\sigma = 1.338$
    b) No. The score distribution in the sample should resemble that in the population, somewhat uniform for scores 1–4 and about half as many 5's.
    c) Approximately $N\left(2.798, \dfrac{1.338}{\sqrt{40}}\right)$

57. About 11.5%, based on $N\left(2.798, \dfrac{1.338}{\sqrt{63}}\right)$.

59. a) $N(24, 0.45)$, if the cars in the lot are equivalent to a simple random sample of all midsize cars.
    b) 0.0131
    c) 24.7 mpg

61. a) 0.734
    b) 0.652. Model is $N(10, 12.81)$.
    c) 0.193. Model is $N(120, 5.774)$.
    d) 0.751. Model is $N(10, 7.394)$.

Chapter 18

1. a) This means that 49% of the 799 teens in the sample said they have misrepresented their age online. This is our best estimate of p, the proportion of *all* U.S. teens who would say they have done so.
   b) $SE(\hat{p}) = \sqrt{\dfrac{0.49 \times 0.51}{799}} \approx 0.018$
   c) Because we don't know p, we use $\hat{p}$ to estimate the standard deviation of the sampling distribution. So the standard error is our estimate of the amount of variation in the sample proportion we expect to see from sample to sample when we ask 799 teens whether they've misrepresented their age online.

3. a) We are 95% confident that, if we were to ask *all* U.S. teens whether they have misrepresented their age online, between 45.6% and 52.5% of them would say they have.
   b) If we were to collect many random samples of 799 teens, about 95% of the confidence intervals we construct would contain the proportion of all U.S. teens who admit to misrepresenting their age online.

5. a) $0.039 = 3.9\%$
   b) The *ME* for 95% confidence would be larger. The critical value of 1.96 is greater than the 1.645 needed for 90% confidence.

7. a) The sample is a simple random sample. Both $n\hat{p} = 402 \times 0.66 = 265.32$ and $n\hat{q} = 402 \times 0.34 = 136.68$ are at least ten. The sample size of 402 is less than 10% of the adult population of Wisconsin. Conditions for the confidence interval are met.
   b) $n = 1519$ (using $p = 0.5$, $n = 1692$)

9. She believes the true proportion is within 4% of her estimate, with some (probably 95%) degree of confidence.

11. a) Population—all cars; sample—those actually stopped at the checkpoint; p—proportion of all cars with safety problems; $\hat{p}$—proportion actually seen with safety problems (10.4%); if sample (a cluster sample) is representative, then the methods of this chapter will apply.
b) Population—general public; sample—those who logged onto the website; p—population proportion of those who favor prayer in school; $\hat{p}$—proportion of those who voted in the poll who favored prayer in school (81.1%); can't use methods of this chapter—sample is biased and nonrandom.
c) Population—parents of students at the school; sample—those who returned the questionnaire; p—proportion of all parents who favor uniforms; $\hat{p}$—proportion of respondents who favor uniforms (60%); should not use methods of this chapter, since not SRS. (Possible nonresponse bias.)
d) Population—students at the college; sample—the 1632 students who entered that year; p—proportion of all students who will graduate on time; $\hat{p}$—proportion of that year's students who graduate on time (85.0%); can use methods of this chapter if that year's students (a cluster sample) are viewed as a representative sample of all possible students at the school.

13. a) Not correct. This implies certainty.
b) Not correct. Different samples will give different results. Many fewer than 95% will have 88% on-time orders.
c) Not correct. The interval is about the population proportion, not the sample proportion in different samples.
d) Not correct. In this sample, we *know* 88% arrived on time.
e) Not correct. The interval is about the parameter, not the days.

15. a) False.
b) True.
c) True.
d) False.

17. On the basis of this sample, we are 90% confident that the proportion of Japanese cars is between 29.9% and 47.0%.

19. a) $(0.162, 0.280)$
b) We're 95% confident that between 16.2% and 28% of all seafood packages sold in stores and restaurants in these three states are misidentified.
c) The size of the population is irrelevant. If *Consumer Reports* had a random sample, 95% of intervals generated by studies like this will capture the true fraud level.

21. a) 0.026
b) We're 90% confident that this poll's estimate is within $\pm 2.6\%$ of the true proportion of people who are baseball fans.
c) Larger. To be more certain, we must be less precise.
d) 0.040
e) Less confidence.

23. a) $(0.0465, 0.0491)$. The assumptions and conditions for constructing a confidence interval are satisfied.
b) The confidence interval gives the set of plausible values (with 95% confidence). Since 0.05 is above the interval, that seems to be a bit too optimistic.

25. a) $(12.7\%, 18.6\%)$
b) We are 95% confident, based on this sample, that the proportion of all auto accidents that involve teenage drivers is between 12.7% and 18.6%.
c) About 95% of all random samples of this size will produce confidence intervals that contain the true population proportion.
d) Contradicts. The interval is completely below 20%.

27. Probably nothing. Those who bothered to fill out the survey may be a biased sample.

29. a) Response bias (wording).
b) $(0.58, 0.64)$
c) Smaller. The sample size was larger.

31. a) $(18.2\%, 21.8\%)$
b) We are 98% confident, based on the sample, that between 18.2% and 21.8% of English children are deficient in vitamin D.
c) About 98% of all random samples of this size will produce a confidence interval that contains the true proportion of English children deficient in vitamin D.

33. a) Wider. The sample size is probably about one fourth of the sample size for all adults, so we'd expect the confidence interval to be about twice as wide.
b) There are fewer "young" people than seniors, so the CI for them would be expected to be a bit wider.

35. a) $(15.5\%, 26.3\%)$
b) 612
c) Sample may not be random or representative. Deer that are legally hunted may not represent all sexes and ages.

37. a) 141 (using $\hat{p} = 0.25$)
b) 318
c) 564

39. 1801 (using $\hat{p} = 0.25$)

41. 384 total, using $p = 0.15$

43. 90%

Chapter 19

1. a) The new drug is not more effective than aspirin.
b) The new drug is more effective than aspirin.

3. a) There is not sufficient evidence to conclude that the new drug is better than aspirin.
b) There is evidence that the new drug is more effective than aspirin.

5. a) H_0: The proportion, p, of people in the county that are of Hispanic or Latino origin is 0.16.

$$H_a: p \neq 0.16$$

b) This is a one-proportion z-test. The 437 residents were a random sample from the county of interest. 437 is almost certainly less than 10% of the population of a county. We expect $np_0 = 69.9$ successes and $nq_0 = 367.1$ failures, which are both more than 10. The conditions are satisfied, so it's okay to use a Normal model and perform a one-proportion z-test.

c) $\hat{p} = \dfrac{44}{437} = 0.101$;

$$SD(\hat{p}) = \sqrt{\dfrac{p_0 q_0}{n}} = \sqrt{\dfrac{0.16 \times 0.84}{437}} = 0.0175;$$

$$z = \dfrac{\hat{p} - p_0}{SD(\hat{p})} = \dfrac{0.101 - 0.16}{0.0175} = -3.37;$$

P-value $= 2 \cdot P(z < -3.37) < 0.001$

d) Because the P-value is so low, there is evidence that the Hispanic/Latino population in this county differs from that of the nation as a whole.

7. One-sided because the only evidence that would support the friend's claim is guessing more than $1/4$ of the suits correctly.

9. a) The drug may not be approved for use, and people would miss out on a beneficial product.
   b) The drug will go into production and people will suffer the side effect.

11. a) $H_0: p = 0.30$; $H_A: p < 0.30$
    b) $H_0: p = 0.50$; $H_A: p \neq 0.50$
    c) $H_0: p = 0.20$; $H_A: p > 0.20$

13. Statement d is correct.

15. No. We can say only that there is a 27% chance of seeing the observed effectiveness just from natural sampling variation. There is insufficient evidence that the new formula is more effective, but we can't conclude that they are equally effective.

17. a) No. There's a 25% chance of losing twice in a row. That's not unusual.
    b) 0.125
    c) No. We expect that to happen 1 time in 8.
    d) Maybe 5? The chance of 5 losses in a row is only 1 in 32, which seems unusual.

19. 1) Use p, not $\hat{p}$, in hypotheses.
    2) The question was about failing to meet the goal, so H_A should be $p < 0.96$.
    3) Did not check $0.04(200) = 8$. Since $nq < 10$, the Success/Failure Condition is violated. Didn't check 10% Condition.
    4) $188/200 = 0.94$; $SD(\hat{p}) = \sqrt{\dfrac{(0.96)(0.04)}{200}} = 0.014$
    5) z is incorrect; should be $z = \dfrac{0.94 - 0.96}{0.014} = -1.43$.
    6) $P = P(z < -1.43) = 0.076$
    7) There is only weak evidence that the new instructions do not work.

21. a) $H_0: p = 0.30$; $H_A: p > 0.30$
    b) Possibly an SRS; we don't know if the sample is less than 10% of his customers, but it could be viewed as less than 10% of all possible customers; $(0.3)(80) \geq 10$ and $(0.7)(80) \geq 10$. Wells are independent only if customers don't have farms on the same underground springs.
    c) $z = 0.73$; P-value $= 0.232$
    d) If his dowsing is no different from standard methods, there is more than a 23% chance of seeing results as good as those of the dowser's, or better, by natural sampling variation.
    e) These data provide insufficient evidence that the dowser's chance of finding water is any better than normal drilling.

23. a) $H_0: p_{2000} = 0.34$; $H_A: p_{2000} \neq 0.34$
    b) Students were randomly sampled and should be independent. 34% and 66% of 8302 are greater than 10. 8302 students is less than 10% of the entire student population of the United States.
    c) $P = 0.054$ or 0.055 using tables
    d) The P-value provides weak evidence against the null hypothesis.
    e) No. A difference this small, even if statistically significant, is probably not meaningful. We might look at new data in a few years.

25. a) $H_0: p = 0.05$ vs. $H_A: p < 0.05$
    b) We assume the whole mailing list has over 1,000,000 names. This is a random sample, and we expect 5000 successes and 95,000 failures.
    c) $z = -3.178$; P-value $= 0.00074$, so we reject H_0; there is strong evidence that the donation rate would be below 5%.

27. a) $H_0: p = 0.71$; $H_A: p > 0.71$
    b) The sample is representative. $240 < 10\%$ of all law school applicants. We expect $240(0.71) = 170.4$ to be admitted and $240(0.29) = 69.6$ not to be, both at least 10. $z = 0.51$; P-value $= 0.30$.
    c) Because the P-value is high, there is not sufficient evidence to claim that *LSATisfaction* demonstrates improvement over the national average.

29. $H_0: p = 0.20$; $H_A: p > 0.20$. SRS (not clear from information provided); 22 is more than 10% of the population of 150; $(0.20)(22) < 10$. Do not proceed with a test.

31. $H_0: p = 0.03$; $p \neq 0.03$. $\hat{p} = 0.015$. One mother having twins will not affect another, so observations are independent; not an SRS; sample is less than 10% of all births. However, the mothers at this hospital may not be representative of all teenagers; $(0.03)(469) = 14.07 \geq 10$; $(0.97)(469) \geq 10$. $z = -1.92$; P-value $= 0.055$. These data show some (although weak) evidence that the rate of twins born to teenage girls at this hospital may be less than the national rate of 3%. It is not clear whether this can be generalized to all teenagers.

33. $H_0: p = 0.25$; $H_A: p > 0.25$. SRS; sample is less than 10% of all potential subscribers; $(0.25)(500) \geq 10$; $(0.75)(500) \geq 10$. $z = 1.24$; P-value $= 0.1075$. The P-value is high, so do not reject H_0. These data do not provide sufficient evidence that more than 25% of current readers would subscribe; the company should not go ahead with the WebZine solely on the basis of these data.

35. $H_0: p = 0.40$; $H_A: p < 0.40$. Data are for all executives in this company and may not be able to be generalized to all companies; $(0.40)(43) \geq 10$; $(0.60)(43) \geq 10$. $z = -1.31$; P-value $= 0.096$. Because the P-value is high, we fail to reject H_0. These data do not show that the proportion of women executives is less than the 40% of women in the company in general.

37. $H_0: p = 0.103$; $H_A: p > 0.103$. $\hat{p} = 0.118$; $z = 2.06$; P-value $= 0.02$. Because the P-value is low, we reject H_0. These data provide evidence that the dropout rate has increased.

39. $H_0: p = 0.90$; $H_A: p < 0.90$. $\hat{p} = 0.844$; $z = -2.05$; P-value $= 0.020$. Because the P-value is so low, we reject H_0. There is strong evidence that the actual rate at which passengers with lost luggage are reunited with it within 24 hours is less than the 90% claimed by the airline.

41. a) Yes. Assuming this sample to be a typical group of people, P-value $= 0.0008$. This cancer rate is very unusual.
    b) No, this group of people may be atypical for reasons that have nothing to do with the radiation.

Chapter 20

1. a) SD's are 1 lb., 0.5 lbs, and 0.2 lbs respectively.
   b) The distribution of pallets. The CLT tells us that the Normal model is approached in the limit regardless of the underlying distribution. As samples get larger the approximation gets better.

3. a) 1.74
b) 2.37

5. The shape is unimodal, symmetric, and bell-shaped. It becomes closer to Normal. The center is at 0. The standard deviation becomes smaller.

7. a) Houses are independent; randomly sampled. Should check Nearly Normal by making a histogram. Sample size: 36 should be big enough
b) ($9,052.50, $10,067.50)
c) We are 95% confident that the interval ($9052.50, $10,067.50) contains the true mean loss in home value.

9. a) 0.0524
b) 0.0889

11. $t = -2.56$ with 35 df, $P = 0.0075$. We reject the null hypothesis because the P-value is small, and conclude that the loss of home values in this community does appear to be unusual.

13. 27 using z, 32 using t.

15. a) The confidence interval is for the population mean, not the individual cows in the study.
b) The confidence interval is not for individual cows.
c) We *know* the average gain in this study was 56 pounds!
d) The average weight gain of all cows does not vary. It's what we're trying to estimate.
e) No. There is not a 95% chance for another sample to have an average weight gain between 45 and 67 pounds. There is a 95% chance that another sample will have its average weight gain within two standard errors of the true mean.

17. a) No. A confidence interval is not about individuals in the population.
b) No. It's not about individuals in the sample, either.
c) No. We know the mean cost for students in the sample was $1467.
d) No. A confidence interval is not about other sample means.
e) Yes. A confidence interval estimates a population parameter.

19. a) Based on this sample, we can say, with 95% confidence, that the mean pulse rate of adults is between 70.9 and 74.5 beats per minute.
b) 1.8 beats per minute
c) Larger. More confidence requires a larger margin of error.

21. The assumptions and conditions for a t-interval are not met. The distribution is highly skewed to the right and there is a large outlier.

23. a) Yes. Randomly selected group; the histogram is not unimodal and symmetric, but it is not highly skewed and there are no outliers, so with a sample size of 52, the CLT says $\bar{y}$ is approximately Normal.
b) (98.06, 98.51) degrees F
c) We are 98% confident, based on the data, that the average body temperature for an adult is between 98.06°F and 98.51°F.
d) 98% of all such random samples will produce intervals containing the true mean temperature.
e) These data suggest that the true normal temperature is somewhat less than 98.6°F.

25. a) Narrower. A smaller margin of error, so less confident.
b) Advantage: more chance of including the true value. Disadvantage: wider interval.
c) Narrower; due to the larger sample, the SE will be smaller.
d) About 253 (256 if you rounded up at each step)

27. a) (299709.90, 299802.54)
b) With 95% confidence, based on these data, the speed of light is between 299,709.9 and 299,802.5 km/sec.
c) Normal model for the distribution, independent measurements. These seem reasonable here, but it would be nice to see if the Nearly Normal Condition held for the data.

29. a) Given no time trend, the monthly on-time departure rates should be independent. Though not a random sample, these months should be representative. The histogram looks unimodal, but slightly left-skewed; not a concern with this large sample.
b) $77.48\% < \mu(\text{OT Departure\%}) < 78.56\%$
c) We can be 90% confident that the interval from 77.48% to 78.56% holds the true mean monthly percentage of on-time flight departures.

31. The 95% confidence interval lies entirely above the 0.08 ppm limit, evidence that mirex contamination is too high and consistent with rejecting the null. We used an upper-tail test, so the P-value should therefore be smaller than $\frac{1}{2}(1 - 0.95) = 0.025$, and it was.

33. If in fact the mean cholesterol of pizza eaters does not indicate a health risk, then only 7 of every 100 samples would have mean cholesterol levels as high (or higher) as observed in this sample.

35. a) We are 95% confident that the true mean gas mileage for cars like these is between 21.43 mpg and 25.60 mpg.
b) It appears that the data is a mix of small, mid-size, and large vehicles, but without knowing how the data were selected, we are cautious about generalizing to all 2014 cars.

37. a) $H_0: \mu = 23.3$; $H_A: \mu > 23.3$
b) We have a random sample of the population. Population may not be normally distributed, as it would be easier to have a few much older men at their first marriage than some very young men. However, with a sample size of 40, $\bar{y}$ should be approximately Normal. We should check the histogram for severity of skewness and possible outliers.
c) $(\bar{y} - 23.3)/(s/\sqrt{40}) \sim t_{39}$
d) 0.1447
e) If the average age at first marriage is still 23.3 years, there is a 14.5% chance of getting a sample mean of 24.2 years or older simply from natural sampling variation.
f) We lack evidence that the average age at first marriage has increased from the mean of 23.3 years.

39. a) Probably a representative sample; the Nearly Normal Condition seems reasonable. (Show a Normal probability plot or histogram.) The histogram is nearly uniform, with no outliers or skewness.
b) $\bar{y} = 28.78$, $s = 0.40$
c) (28.36, 29.21) grams
d) Based on this sample, we are 95% confident the average weight of the content of Ruffles bags is between 28.36 and 29.21 grams.
e) The company is erring on the safe side, as it appears that, on average, it is putting in slightly more chips than stated.

41. Yes. These are a random sample of bags and the Nearly Normal Condition is met (Show a Normal probability plot or histogram.); $t = -2.51$ with 7 df for a one-sided P-value of 0.0203.

43. a) Random sample; the Nearly Normal Condition seems reasonable from a Normal probability plot. The histogram is roughly unimodal and symmetric with no outliers.
b) (1187.9, 1288.4) chips
c) Based on this sample, the mean number of chips in an 18-ounce bag is between 1187.9 and 1288.4, with 95% confidence. The *mean* number of chips is likely to be greater than 1000. However, if the claim is about individual bags, then it's not necessarily true. If the mean is 1188 and the SD deviation is near 94, then 2.28% of the bags will have fewer than 1000 chips, using the Normal model. If in fact the mean is 1288, the proportion below 1000 will be about 0.1%, but the claim is still false.

45. a) The Normal probability plot is relatively straight, with one outlier at 93.8 sec. Without the outlier, the conditions seem to be met. The histogram is roughly unimodal and symmetric with no other outliers. (Show your plot.)
b) $t = -2.63$; P-value $= 0.0160$. With the outlier included, we might conclude that the mean completion time for the maze is not 60 seconds; in fact, it is less.
c) $t = -4.46$; P-value $= 0.0003$. Because the P-value is so small, we reject H_0. Without the outlier, we see strong evidence that the average completion time for the maze is less than 60 seconds. The outlier here did not change the conclusion.
d) The maze does not meet the "one-minute average" requirement. Both tests rejected a null hypothesis of a mean of 60 seconds.

47. a) $287.47 < \mu(\text{Drive Distance}) < 291.02$
b) These data are not a random sample of golfers. The top professionals are (unfortunately) not representative and were not selected at random. We might consider these data to represent the population of all professional golfers, past, present, and future.
c) If these data were the means for each golfer rather than individual drives, then they would be less variable than individual drives and the inference would be invalid.

Chapter 21

1. a) Let $p =$ probability of winning on the slot machine.
$H_0: p = 0.01$ vs. $H_A: p \neq 0.01$
b) Let $\mu =$ mean spending per customer this year.
$H_0: \mu = \$35.32$ vs. $H_A: \mu \neq \$35.32$
c) Let $p =$ proportion of patients cured by the new drug.
$H_0: p = 0.3$ vs. $H_A: p \neq 0.3$
d) Let $p =$ proportion of clients now using the website.
$H_0: p = 0.4$ vs. $H_A: p \neq 0.4$

3. a) False. A high P-value shows that the data are consistent with the null hypothesis, but provides no evidence for rejecting the null hypothesis.
b) False. It results in rejecting the null hypothesis, but does not prove that it is false.
c) False. A high P-value shows that the data are consistent with the null hypothesis but does not prove that the null hypothesis is true.
d) False. Whether a P-value provides enough evidence to reject the null hypothesis depends on the risk of a type I error that one is willing to assume (the α level).

5. a) $H_0: p = 0.40$ vs. $H_A: p \neq 0.40$. Two-sided.
b) $H_0: \mu = 5.4$ vs. $H_A: \mu < 5.4$. One-sided.
c) $H_0: p = 0.42$ vs. $H_A: p > 0.42$. One-sided.
d) $H_0: p = 0.50$ vs. $H_A: p > 0.50$. One-sided.

7. a) True.
b) False. The alpha level is set independently and does not depend on the sample size.
c) False. The P-value would have to be less than 0.01 to reject the null hypothesis.
d) False. It simply means we do not have enough evidence at that alpha level to reject the null hypothesis.

9. a) $z = \pm 1.96$
b) $z = 1.645$
c) $t = \pm 2.03$
d) $z = 2.33$; n is not relevant for critical values for z.
e) $z = -2.33$

11. a) Type I error. The actual value is not greater than 0.3 but they rejected the null hypothesis.
b) No error. The actual value is 0.50, which was not rejected.
c) Type II error. The actual value was 55.3 points, which is greater than 52.5.
d) Type II error. The null hypothesis was not rejected, but it was false. The true relief rate was greater than 0.25.

13. a) Two sided. Let p be the percentage of students who prefer Diet Pepsi. $H_0: p = 0.5$ vs. $H_A: p \neq 0.5$
b) One sided. Let p be the percentage of teenagers who prefer the new formulation. $H_0: p = 0.5$ vs. $H_A: p > 0.5$
c) One sided. Let p be the percentage of people who intend to vote for the override. $H_0: p = 2/3$ vs. $H_A: p > 2/3$
d) Two sided. Let p be the percentage of days that the market goes up. $H_0: p = 0.5$ vs. $H_A: p \neq 0.5$

15. If there is no difference in effectiveness, the chance of seeing an observed difference this large or larger is 4.7% by natural sampling variation.

17. $\alpha = 0.10$: Yes. The P-value is less than 0.05, so it's less than 0.10. But to reject H_0 at $\alpha = 0.01$, the P-value must be below 0.01, which isn't necessarily the case.

19. a) There is basically zero chance of seeing a sample proportion as low as 97.4% vaccinated by natural sampling variation if 98% have really been vaccinated.
b) We conclude that p is below 0.98, but a 95% confidence interval would suggest that the true proportion is between (0.971, 0.977). Most likely, a decrease from 98% to 97.6% would not be considered important. On the other hand, with 1,000,000 children a year vaccinated, even 0.1% represents about 1000 kids—so this may very well be important.

21. a) SRS so responses are independent, successes and failures both >10; (0.486, 0.534).
b) Because 45% is not in the interval we have strong evidence that more than 45% of men identify themselves as the primary grocery shopper.
c) $\alpha = 0.01$; it's an upper tail test based on a 98% confidence interval.

23. a) (0.375, 0.425)
b) Since 45% is not in the confidence interval, we can reject the hypothesis that $p = 0.45$.

25. a) The Success/Failure Condition is violated: only 5 pups had dysplasia.
b) $(0.048, 0.256)$. We are 95% confident that the true proportion of puppies with dysplasia is between 4.8% and 25.6%.

27. a) Type II error.
b) Type I error.
c) By making it easier to get the loan, the bank has reduced the alpha level.
d) The risk of a Type I error is decreased and the risk of a Type II error is increased.

29. a) Power is the probability that the bank denies a loan that would not have been repaid.
b) Raise the cutoff score.
c) A larger number of trustworthy people would be denied credit, and the bank would miss the opportunity to collect interest on those loans.

31. a) The null is that the level of home ownership remains the same. The alternative is that it rises.
b) The city concludes that home ownership is on the rise, but in fact the tax breaks don't help.
c) The city abandons the tax breaks, but they were helping.
d) A Type I error causes the city to forego tax revenue, while a Type II error withdraws help from those who might have otherwise been able to buy a home.
e) The power of the test is the city's ability to detect an actual increase in home ownership.

33. a) It is decided that the shop is not meeting standards when it is.
b) The shop is certified as meeting standards when it is not.
c) Type I.
d) Type II.

35. a) The probability of detecting a shop that is not meeting standards.
b) 40 cars. Larger n.
c) 10%. More chance to reject H_0.
d) A lot. Larger differences are easier to detect.

37. a) One-tailed. The company wouldn't be sued if "too many" minorities were hired.
b) Deciding the company is discriminating when it is not.
c) Deciding the company is not discriminating when it is.
d) The probability of correctly detecting actual discrimination.
e) Increases power.
f) Lower, since n is smaller.

39. a) One-tailed. The software is supposed to increase the final exam scores.
b) $H_0: \mu = 105; H_A: \mu > 105$
c) He buys the software when it doesn't help students.
d) He doesn't buy the software when it does help students.
e) The probability of correctly deciding the software is helpful.

41. a) $t = 4.91; p < 0.0001$. The change is statistically significant. A 95% confidence interval is $(106.8, 109.2)$. This is clearly higher than 105. The software does appear to improve exam scores.
b) Students' scores improved by only 1 to 4 points on the exam. This might not be enough to be worth the cost.

43. a) Upper-tail. We want to show it will hold 500 pounds (or more) easily.
b) They will decide the stands are safe when they're not.
c) They will decide the stands are unsafe when they are in fact safe.

45. a) Decrease α. This means a smaller chance of declaring the stands safe if they are not.
b) The probability of correctly detecting that the stands are capable of holding more than 500 pounds.
c) Decrease the standard deviation—probably costly. Increase the sample size—takes more time for testing and is costly. Increase α—more Type I errors. Increase the "design load" to be well above 500 pounds—again, costly.

47. a) H_A: The coin is the 30% heads coin.
b) Reject the null hypothesis if the coin comes up tails—otherwise fail to reject.
c) $P(\text{tails given the null hypothesis}) = 0.1 = \alpha$
d) $P(\text{tails given the alternative hypothesis}) = \text{power} = 0.70$
e) Spin the coin more than once and base the decision on the sample proportion of heads.

49. a) 0.0464
b) Type I.
c) 37.6%
d) Increase the number of shots. Or keep the number of shots at 10, but increase alpha by declaring that 8, 9, or 10 will be deemed as having improved.

Part V Review

1. H_0: There is no difference in relapse rates, $p_1 - p_2 = 0$. H_A: The relapse rate in those who use omega-3 fatty acids is lower, $p_1 - p_2 > 0$.

3. a) Based on these data, we are 90% confident that the average hamster litter will have between 7.11 and 8.33 babies.
b) Larger—to be more confident we need a wider interval.
c) About 27 (based on t^*_{24})

5. a) $H_0: p_1 = 0.40; H_A: p_1 < 0.40$
b) Random sample; less than 10% of all California gas stations, $0.4(27) = 10.8, 0.6(27) = 16.2$. Assumptions and conditions are met.
c) $z = -1.49$, P-value $= 0.0677$
d) With a P-value this high, we fail to reject H_0. These data do not provide sufficient evidence that the proportion of leaking gas tanks is less than 40% (or that the new program is effective in decreasing the proportion).
e) Yes, Type II.
f) Increase α, increase the sample size.
g) Increasing α—increases power, lowers chance of Type II error, but increases chance of Type I error. Increasing sample size—increases power, costs more time and money.

7. a) The researcher believes that the true proportion of "A's" is within 10% of the estimated 54%, namely, between 44% and 64%.
b) Small sample.
c) No, 63% is contained in the interval.

9. a) Pew uses a 95% confidence level. We can be 95% confident that the true proportion is within 2% of 13%—that is, that it is between 11% and 15%.
   b) The cell phone group would have the larger ME because its sample size is smaller.
   c) CI = (78.5%, 85.5%)
   d) The ME is 0.035, which is larger than the 2% ME in part a, largely because of the smaller sample size. It is larger than the ME in part b, mostly because 0.82 is smaller than 0.87 and proportions closer to 0.50 have larger MEs.

11. a) Bimodal.
    b) μ, the population mean. Sample size does not matter.
    c) $\sigma/\sqrt{n}$; sample size does matter.
    d) It becomes closer to a Normal model and less variable as the sample size increases.

13. a) $\mu = 0.80, \sigma = 0.028$
    b) Yes. $0.8(200) = 160, 0.2(200) = 40$. Both ≥ 10.
    c)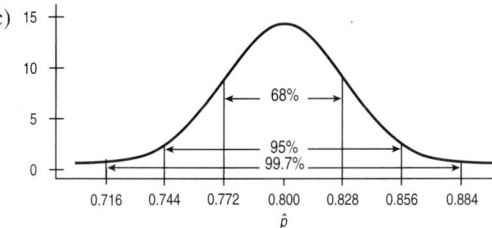
    d) 0.037

15. a) It is not clear what the pollster asked. Otherwise they did fine.
    b) Stratified sampling.
    c) 4%
    d) 95%
    e) Smaller sample size.
    f) Wording and order of questions (response bias).

17. a) H_0: There is no difference, $p = 0.126$. H_A: The fatal accident rate is lower in girls, $p < 0.126$. $z = -0.75$, P-value = 0.227. Because the P-value is high, we fail to reject H_0. These data do not give sufficient evidence to conclude that the fatal accident rate is lower for girls than for teens in general.
    b) If the proportion is really 12.6%, we will see the observed proportion (11.3%) or lower 22.7% of the time by sampling variation.

19. a) Random sample; less than 10% of all high school students; many more than 10 successes and 10 failures. 90% confidence interval for proportion who cheat: 72.9% to 75.1%.
    b) Based on this information, we are 90% confident that the proportion of all high school students who have cheated at least once on a test is between 72.9% and 75.1%.
    c) 90% of all such random samples will produce confidence intervals that contain the true proportion of students who cheat.
    d) Wider. More confidence means a larger margin of error.

21. a) H_0: Progress is on track, $p = 0.177$. H_A: Progress is off track, $p > 0.177$.
    b) Random samples; less than 10% of all high school students; many more than 10 failures and successes.
    c) 0.0004
    d) If the proportion were 17.7%, the probability of seeing a value as high as (or higher than) 19.5% in a random sample this large is about 0.0004.
    e) By 2009, the rate of cigarette smoking in high school students was higher than the target of 17.7%.
    f) Type I.

23. No. The probability of 25% or more is very small: around 7.6×10^{-4} (Round-off could lead to different, but also very small, answers.)

25. a) Experiment.
    b) A one-sided test, since they are interested only in a decrease in percentage needing repairs.
    c) Deciding the additive reduces the number of repairs needed when there really is no difference.
    d) Deciding the additive makes no difference when it really does reduce the number of repairs needed.
    e) Type II.
    f) Given that the two groups received roughly the same use and care, yes. They can't necessarily claim it will work for all cars, only cars similar to their fleet.

27. a) H_0: The computer is undamaged. H_A: The computer is damaged.
    b) 20% of good PCs will be classified as damaged (bad), while all damaged PCs will be detected (good).
    c) 3 or more
    d) 20%
    e) By switching to two or more as the rejection criterion, 7% of the good PCs will be misclassified, but only 10% of the bad ones will, increasing the power from 20% to 90%.

29. The null hypothesis is that Bush's disapproval proportion was 66%—the Nixon benchmark. The one-tailed test has a z-value of 2.02, so the P-value is 0.0217. It looks like there is reasonable evidence to suggest that Bush's April 2008 ratings were worse than the Nixon benchmark low.

31. a) The company is interested only in confirming that the athlete is well known.
    b) Type I: the company concludes that the athlete is well known, but that's not true. It offers an endorsement contract to someone who lacks name recognition. Type II: the company overlooks a well-known athlete, missing the opportunity to sign a potentially effective spokesperson.
    c) Type I would be more likely, Type II less likely.

33. We're 95% confident that between 46% and 60% of anorexia patients will drop out of treatment programs. However, this wasn't a random sample of all patients; they were assigned a treatment rather than choosing one on their own, and they may have had different experiences if they were not part of an experiment.

35. a) $H_0: p = 0.16$; $H_A: p > 0.16$. $z = 4.42$; P-value = 5×10^{-6}. Because the P-value is so low, we reject H_0. These data show the proportion of tickets given to blacks on this section of the New Jersey Turnpike is unusually high.
    b) Doesn't prove it; there may be other factors.
    c) Answers will vary. Possibly what proportion of drivers on this highway are black?

37. $H_0: \mu = \$10$; $H_A: \mu > \$10$. $t = 0.68$; df = 13; P-value = 0.25. With such a high P-value, we do not reject H_0. These data do not provide evidence that he is likely to overspend his budget of \$10 per day, provided that these days are representative.

39. Based on these data we are 95% confident that the proportions of streams in the Adirondacks with shale substrates is between 32.8% and 47.4%.

41. a) From this histogram, about 115 loaves or more. (Not Normal.)
b) Large sample size; CLT says $\bar{y}$ will be approximately Normal.
c) From the data, we are 95% confident that the bakery will sell between 101.2 and 104.8 loaves of bread on an average day.
d) 25
e) Yes, 100 loaves per day is too low—the entire confidence interval is above that.

43. a) $373.50
b) They are 95% confident that the average *Loss* in a home burglary is between $1644 and $2391, based on their sample.
c) 95% of all random samples will produce confidence intervals that contain the true mean *Loss*.

Chapter 22

1. 0.0161

3. We must assume the data were collected randomly and that the Americans selected are independent of the Canadians selected. Both assumptions should be met. Also, for both groups, we have at least 10 national-born and foreign-born citizens and the sample sizes are less than 10% of the population sizes. All conditions for inference are met.

5. We are 95% confident that, based on these data, the proportion of foreign-born Canadians is between 3.24% and 9.56% more than the proportion of foreign-born Americans.

7. If we were to take repeated samples of these sizes of Canadians and Americans, and compute two-proportion confidence intervals, we would expect 95% of the intervals to contain the true difference in the proportion of foreign-born citizens.

9. a) 0.064
b) 0.1638
c) 0.0159
d) 4.030
e) P-value is <0.001. Very strong evidence, so reject the null hypothesis that the proportions are the same in the two countries.

11. We must assume the samples were random or otherwise independent of each other. We also assume that the distributions are roughly Normal, so it would be a good idea to check a histogram to make sure there isn't strong skewness or outliers.

13. We are 95% confident that the mean purchase amount at Walmart is between $1.85 and $14.15 less than the mean purchase amount at Target.

15. The difference is −$8 with an SE of 3.115, so the *t*-stat is −2.569. With 162.75 (or 161) df, the P-value is 0.011 which is less than 0.05. Reject the null hypothesis that the means are equal.

17. The *t*-statistic is still −2.561 using the pooled estimate of the standard deviation. There are 163 df so the P-value is still 0.011. Same conclusion as before. Because the sample standard deviations are nearly the same and the sample sizes are large, the pooled test is essentially the same as the two-sample *t*-test.

19. No. The two-sample test is almost always the safer choice and here the variances are likely to be quite different. The purchase prices of Italian sports cars are much higher and may be more variable than the domestic prices. They should use the two-sample *t*-test.

21. It's very unlikely that samples would show an observed difference this large if in fact there is no real difference in the proportions of American adults using Facebook on a daily basis between the two time periods.

23. The responses are not from two independent groups, but are from the same individuals.

25. a) Stratified.
b) 6% higher among males.
c) 4%
d)

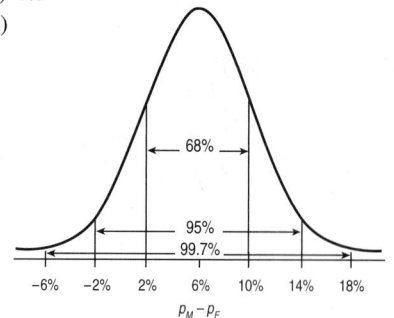

e) Yes. A poll result showing little difference is only 1–2 standard deviations below the expected outcome.

27. a) Yes. Random samples; less than 10% of the population; samples are independent; more than 10 successes and failures in each sample.
b) (0.055, 0.140)
c) We are 95% confident, based on these samples, that the proportion of American women age 65 and older who suffer from arthritis is between 5.5% and 14.0% more than the proportion of American men of the same age who suffer from arthritis.
d) Yes. The entire interval lies above 0.

29. a) 0.035
b) (0.356, 0.495)
c) We are 95% confident, based on these data, that the proportion of pets with a malignant lymphoma in homes where herbicides are used is between 35.6% and 49.5% higher than the proportion of pets with lymphoma in homes where no pesticides are used.

31. a) Experiment. Men were randomly assigned to have surgery or not.
b) (0.006, 0.080)
c) Since the entire interval lies above 0, there is evidence that surgery may be effective in preventing death from prostate cancer.

33. a) Yes. Subjects were randomly divided into independent groups, and more than 10 successes and failures were observed in each group.
b) (4.7%, 8.9%)
c) Yes. We're 95% confident that the rate of infection is 5–9 percentage points lower. That's a meaningful reduction, considering the 20% infection rate among the unvaccinated kids.

35. a) $H_0: p_V - p_{NV} = 0$; $H_A: p_V - p_{NV} < 0$
b) Because 0 is not in the confidence interval, reject the null. There's evidence that the vaccine reduces the rate of ear infections.

c) Type I.
d) Babies would be given ineffective vaccinations.

37. a) Prospective study.
b) $H_0: p_1 - p_2 = 0$; $H_A: p_1 - p_2 \neq 0$ where p_1 is the proportion of teens whose parents disapprove of smoking who become smokers and p_2 is the proportion whose parents are lenient about smoking who become smokers.
c) Yes. We assume the students were randomly selected; they are less than 10% of the population; samples are independent; at least 10 successes and failures in each sample.
d) $z = -1.17$, P-value $= 0.2422$. These samples do not show evidence that parental attitudes influence teens' decisions to smoke.
e) If there is no difference in the proportions, there is about a 24% chance of seeing the observed difference or larger by natural sampling variation.
f) Type II.
g) $(-0.221, 0.065)$
h) We are 95% confident that the proportion of teens whose parents disapprove of smoking who will become established smokers within two years is between 22.1% less and 6.5% more than for teens with parents who are lenient about smoking.
i) 95% of all random samples will produce intervals that contain the true difference.

39. a) $H_0: p_1 - p_2 = 0$; $H_A: p_1 - p_2 > 0$ where p_1 is the proportion of mothers exposed to soot and ash and p_2 is the proportion of mothers not exposed. $z = 2.71$, P-value $= 0.003$ (using $n_1 = 15$ out of 182). With a P-value this low, we reject H_0 at $\alpha = 0.01$. This study shows a significantly higher rate of low-weight babies born to mothers exposed to high levels of air pollution during pregnancy.
b) We are 90% confident that the proportion of low-birth weight babies will be between 0.8% and 7.7% higher for mothers exposed to high levels of air pollution than those who are not.

41. a) $H_0: p_1 - p_2 = 0$; $H_A: p_1 - p_2 > 0$ where p_1 is the proportion of women who never had mammograms who die of breast cancer and p_2 is the proportion of women who undergo screening who die of breast cancer. $z = 2.17$, P-value $= 0.0148$. With a P-value this low, we reject H_0. These data do suggest that mammograms may reduce breast cancer deaths.
b) Type I

43. a) We are 95% confident, based on this study, that between 67.0% and 83.0% of patients with joint pain will find medication A effective.
b) We are 95% confident, based on this study, that between 51.9% and 70.3% of patients with joint pain will find medication B effective.
c) Yes, they overlap. This might indicate no difference in the effectiveness of the medications, although this is not a proper test.
d) We are 95% confident that the proportion of patients with joint pain who will find medication A effective is between 1.7% and 26.1% higher than the proportion who will find medication B effective.
e) No. There is evidence of a difference in the effectiveness of the medications.
f) To estimate the variability in the difference of proportions, we must add variances. The two one-sample intervals do not. The two-sample method is the correct approach.

45. A 95% confidence interval is $(-1.6\%, 5.6\%)$; 0%—or no bounce—is a plausible value. They should have said that there was no evidence of a bounce from their poll, however, since they can't prove there was none at all.

47. The conditions are satisfied to test $H_0: p_{young} - p_{old} = 0$ against $H_A: p_{young} - p_{old} > 0$. $z = 1.54$. The one-sided P-value is 0.0619, so we may not reject the null hypothesis at $\alpha = 0.05$, but would at $\alpha = 0.10$. Although the evidence is not strong, *Time* may be justified in saying that younger men are more comfortable discussing personal problems.

49. $H_0: p_{urban} - p_{rural} = 0$ vs. $H_A: p_{urban} - p_{rural} \neq 0$. Yes. With a low P-value of 0.0014, reject the null hypothesis of no difference. There is strong evidence to suggest that the percentages are different for the two groups: People from urban areas are more likely to agree with the statement than those from rural areas.

51. Yes. The high P-value means that we lack evidence of a difference, so 0 is a plausible value for $\mu_{Meat} - \mu_{Beef}$.

53. a) Plausible values of $\mu_{Meat} - \mu_{Beef}$ are all negative, so the mean fat content is probably higher for beef hot dogs.
b) The difference in sample means is significant.
c) 0.10

55. a) False. The confidence interval is about means, not about individual hot dogs.
b) False. The confidence interval is about means, not about individual hot dogs.
c) True.
d) False. CI's based on other samples will also try to estimate the true difference in population means; there's no reason to expect other samples to conform to this result.
e) True.

57. a) 2.927
b) Larger.
c) Based on this sample, we are 95% confident that students who learn Math using the CPMP method will score, on average, between 5.57 and 11.43 points better on a test solving applied Algebra problems with a calculator than students who learn by traditional methods.
d) Yes. 0 is not in the interval.

59. a) $H_0: \mu_C - \mu_T = 0$ vs. $H_A: \mu_C - \mu_T \neq 0$
b) Yes. Groups are independent, though we don't know if students were randomly assigned to the programs. Sample sizes are large, so CLT applies.
c) If the means for the two programs are really equal, there is less than a 1 in 10,000 chance of seeing a difference as large as or larger than the observed difference just from natural sampling variation.
d) On average, students who learn with the CPMP method do significantly worse on Algebra tests that do not allow them to use calculators than students who learn by traditional methods.

61. a) $(1.36, 4.64)$
b) No. 5 minutes is beyond the high end of the interval.

63. These were random samples, both less than 10% of properties sold. Prices of houses should be independent, and random sampling makes the two groups independent. The boxplots make the price distributions appear to be reasonably symmetric,

and with the large sample sizes the few outliers don't affect the means much. Based on this sample, we're 95% confident that, in New York, having a waterfront is worth, on average, about $59,121 to 140,898 more in sale price.

65.

Random sample—questionable, but probably representative, independent samples, less than 10% of all cereals; boxplot shows no outliers—not exactly symmetric, but these are reasonable sample sizes. Based on these samples, with 95% confidence, children's cereals average between 32.49% and 40.80% more sugar content than adult's cereals.

67. $H_0: \mu_N - \mu_C = 0$ vs. $H_A: \mu_N - \mu_C > 0$; $t = 2.301$; P-value $= 0.0274$; df $= 33.4$. Because of the small P-value, we reject H_0 (at $\alpha = 0.05$). These data do suggest that new activities are better. The mean reading comprehension score for the group with new activities is significantly (at $\alpha = 0.05$) higher than the mean score for the control group.

*Because the Control group has both the highest and lowest values, Tukey's test would not be appropriate. The Rank-sum test gives a P-value of 0.0045, which rejects the null hypothesis.

69.

There are several concerns here. First, we don't have a random sample. We have to assume that the actual number of hurricanes in a given year is a random sample of the hurricanes that might occur under similar weather conditions. Also, the data for 1944–1969 are not symmetric and have four outliers. The outliers will tend to make the average for the period 1944–1969 larger. These data are not appropriate for inference. The boxplots provide little evidence of a change in the mean number of hurricanes in the two periods. Moreover, a t-test of the two periods fails to show evidence of a difference in means with or without the outliers.

71. a)

Both are reasonably symmetric. There are two possible high outliers and one possible low outlier in the American League (depending on how quartiles are calculated).

b) Based on these data, the average number of home runs hit per game in an American League stadium is between 1.89 and 2.25 (or 1.90 to 2.10 with all 3 outliers removed) with 95% confidence.

c) No. The boxplot indicates that there are no National League outliers (and Coors is not even the highest).

d) We want to work directly with the average difference. The two separate confidence intervals do not answer questions about the difference. The difference has a different standard deviation, found by adding variances.

e) $(0.035, 0.573)$ without 3 outliers it is $(0.008, 0.461)$

f) Based on these data, with 95% confidence, American League stadiums average between 0.035 and 0.573 more home runs per game than National League stadiums.

g) Yes, 0 is not in the interval with or without the outliers.

73. These are not two independent samples. These are before and after scores for the same individuals.

75. a) These data do not provide evidence of a difference in ad recall between shows with sexual content and violent content.

b) $H_0: \mu_S - \mu_N = 0$ vs. $H_A: \mu_S - \mu_N \neq 0$. $t = -6.08$; df $= 213.99$; P-value $= 5.5 \times 10^{-9}$. Because the P-value is low, we reject H_0. These data suggest that ad recall between shows with sexual and neutral content is different; those who saw shows with neutral content had higher average recall.

77. $H_0: \mu_{big} - \mu_{small} = 0$ vs. $H_A: \mu_{big} - \mu_{small} \neq 0$; bowl size was assigned randomly; amount scooped by individuals and by the two groups should be independent. With 34 df, $t = 2.104$ and P-value $= 0.0428$. The low P-value leads us to reject the null hypothesis (at $\alpha = 0.05$) and conclude that there is a difference in the average amount of ice cream that people scoop when given a bigger bowl.

79. a) If the assumptions and conditions are met, we can be 95% confident that the interval $(-216.74, 154.64)$ minutes includes the true difference in mean crossing times between men and women. Because the interval includes zero, we cannot be confident that there is any difference at all.

b) Independence Assumption: There is no reason to believe that the swims are not independent or that the two groups are not independent of each other.
Randomization Condition: The swimmers are not a random sample from any identifiable population, they may be representative of swimmers who tackle challenges such as this but we have no way of knowing.
Nearly Normal Condition: the boxplots show two high outliers for the men and some skewness for both. Removing the outliers may make the difference in times larger, but there is

no justification for doing so. The histograms are unimodal; but somewhat skewed to the right. We are reluctant to draw any conclusions about the difference in mean times it takes men or women to swim the lake. The sample is not random, we have no way of knowing if it is representative, and the data are skewed with some outliers.

81. Independent Groups Assumption: The runners are different women, so the groups are independent. The Randomization Condition is satisfied since the runners are selected at random for these heats.

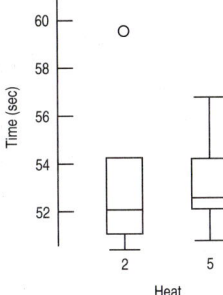

Nearly Normal Condition: The boxplots show an outlier, but we will proceed and then redo the analysis with the outlier deleted. When we include the outlier, $t = 0.0484$ with a two-sided P-value of 0.963. With the outlier deleted, $t = -1.10$ with $P = 0.300$. Either P-value is so large that we fail to reject the null hypothesis of equal means and conclude that there is no evidence of a difference in the mean times for randomly assigned heats.

*There are only two "exceedences," so Tukey's test does not find a significant difference. The Rank-sum test agrees, with a P-value of 0.70

83. H_0: $\mu_{Regular} - \mu_{Stinger} = 0$ vs. H_A: $\mu_{Regular} - \mu_{Stinger} < 0$. With $t = -4.57$ and a very low P-value of 0.0013, we reject the null hypothesis of equal mean velocities. There is strong evidence that golf balls hit off Stinger tees will have a higher mean initial velocity.

85. a) H_0: $\mu_M - \mu_R = 0$ vs. H_A: $\mu_M - \mu_R > 0$. $t = -0.70$; df = 45.88; P-value = 0.7563. Because the P-value is so large, we do not reject H_0 for any reasonable α level. These data provide no evidence that listening to Mozart while studying is better than listening to rap. Now you might have noticed that subjects listening to Mozart recalled fewer items. And so, there is no need for a statistical test. Looking and thinking first can save you some calculation effort. If you did notice, and said there's no evidence of a benefit, without calculation—that's a good answer to this exercise.
b) With 90% confidence, the average difference in score is between 0.189 and 5.351 objects more for those who listen to no music while studying, based on these samples.

Chapter 23

1. a) Paired.
b) Not paired.
c) Paired.

3. a) No. The vehicles have no natural pairing.
b) Possibly. The data are quantitative and paired by vehicle.
c) The sample size is large, but there is at least one extreme outlier that should be investigated before applying these methods.

5. (7.17, 7.57). We are 95% confident that the interval from 7.17 mpg to 7.57 mpg captures the true improvement in highway gas mileage compared to city gas mileage.

7. The difference between fuel efficiency of cars and that of trucks can be large, but isn't relevant to the question asked about highway vs. city driving. Pairing places each vehicle in its own block to remove that variation from consideration.

9. a) Randomly assign 50 hens to each of the two kinds of feed. Compare production at the end of the month.
b) Give all 100 hens the new feed for 2 weeks and the old feed for 2 weeks, randomly selecting which feed the hens get first. Analyze the differences in production for all 100 hens.
c) Matched pairs. Because hens vary in egg production, the matched-pairs design will control for that.

11. a) Show the same people ads with and without sexual images, and record how many products they remember in each group. Randomly decide which ads a person sees first. Examine the differences for each person.
b) Randomly divide volunteers into two groups. Show one group ads with sexual images and the other group ads without. Compare how many products each group remembers.

13. a) Matched pairs—same cities in different periods.
b) There is a significant difference (P-value = 0.0244) in the labor force participation rate for women in these cities; women's participation seems to have increased between 1968 and 1972.

15. a) Use the paired t-test because we have pairs of Fridays in 5 different months. Data from adjacent Fridays within a month may be more similar than data from randomly chosen Fridays.
b) We conclude that there is evidence (P-value = 0.0212) that the mean number of cars found on the M25 motorway on Friday the 13th is less than on the previous Friday.
c) We don't know if these Friday pairs were selected at random. If these are the Fridays with the largest differences, this will affect our conclusion. The Nearly Normal Condition appears to be met by the differences, but the sample size is small.

17. Adding variances requires that the variables be independent. These price quotes are for the same cars, so they are paired. Drivers quoted high insurance premiums by the local company will be likely to get a high rate from the online company, too.

19. a) The histogram—we care about differences in price.
b) Insurance cost is based on risk, so drivers are likely to see similar quotes from each company, making the differences relatively smaller.
c) The price quotes are paired; they were for a random sample of fewer than 10% of the agent's customers; the histogram of differences looks approximately Normal.

21. H_0: $\mu(Local - Online) = 0$ vs. H_A: $\mu(Local - Online) > 0$; with 9 df, $t = 0.83$. With a high P-value of 0.215, we don't

reject the null hypothesis. These data don't provide evidence that online premiums are lower, on average.

23.

a) The histogram shows a low outlier, which turns out to be Auckland, New Zealand. A southern hemisphere city would have seasons reversed, so it doesn't belong in these data.
b) A 95% CI for the remaining 34 cities is (30.94, 39, 53) degrees. We can be 95% certain that this interval contains the true mean difference in temperatures between July and January in cities like these.

Data are paired for each city; cities are independent of each other; boxplot shows the temperature differences are reasonably symmetric, with no outliers. This is probably not a random sample, so we might be wary of inferring that this difference applies to all European cities. Based on these data, we are 90% confident that the average temperature in European cities in July is between 30.9°F and 39.5°F higher than in January.

25. Based on these data, we are 90% confident that boys, on average, can do between 1.6 and 13.0 more push-ups than girls (independent samples—not paired).

27. a) Paired sample test. Data are before/after for the same workers; workers randomly selected; assume fewer than 10% of all this company's workers; boxplot of differences shows them to be symmetric, with no outliers.

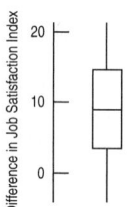

b) $H_0: \mu_D = 0$ vs. $H_A: \mu_D > 0$. $t = 3.60$; P-value $= 0.0029$. Because $P < 0.01$, reject H_0. These data show evidence that average job satisfaction has increased after implementation of the program.
c) Type I.
*d) $P = 0.1094$. This is different from the paired t test.

29. $H_0: \mu_D = 0$ vs. $H_A: \mu_D \neq 0$. Data are paired by brand; brands are independent of each other; fewer than 10% of all yogurts (questionable); boxplot of differences shows an outlier (100) for Great Value:

With the outlier included, the mean difference (Strawberry − Vanilla) is 12.5 calories with a t-stat of 1.332, with 11 df, for a P-value of 0.2098. Deleting the outlier, the difference is even smaller, 4.55 calories with a t-stat of only 0.833 and a P-value of 0.4241. With P-values so large, we do not reject H_0. We conclude that the data do not provide evidence of a difference in mean calories.

31. a) Cars were probably not a simple random sample, but may be representative in terms of stopping distance; boxplot does not show outliers, but does indicate right skewness. A 95% confidence interval for the mean stopping distance on dry pavement is (131.8, 145.6) feet.
b) Data are paired by car; cars were probably not randomly chosen, but representative; boxplot shows an outlier (car 4) with a difference of 12. With deletion of that car, a Normal probability plot of the differences is relatively straight.

Retaining the outlier, we estimate with 95% confidence that the average braking distance is between 38.8 and 62.6 feet more on wet pavement than on dry, based on this sample. (Without the outlier, the confidence interval is 47.2 to 62.8 feet.)

33. a) Paired Data Assumption: Data are paired by college. Randomization Condition: This was a random sample of public colleges and universities. 10% Condition: these are fewer than 10% of all public colleges and universities.

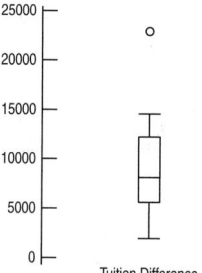

Normal Population Assumption: New College of Florida seems to be an outlier. A search shows that it may be an unusual college. We may want to consider setting it aside.
b) Having deleted the outliers, we are 90% confident, based on the remaining data, that nonresidents pay, on average, between $6821 and $11,082 more than residents. If we retain the outlier, the interval is ($7314, $12,154).
c) Assertion is reasonable; without the outliers, $7000 is in the confidence interval. With the outliers, $7000 is actually lower than the values in the interval.

35. a) 60% is 30 strikes; $H_0: \mu = 30$ vs. $H_A: \mu > 30$. $t = 6.07$; P-value $= 3.92 \times 10^{-6}$. With a very small P-value, we reject H_0. There is very strong evidence that players can

throw more than 60% strikes after training, based on this sample.

b) $H_0: \mu_D = 0$ vs. $H_A: \mu_D > 0$. $t = 0.135$; P-value $= 0.4472$. With such a high P-value, we do not reject H_0. These data provide no evidence that the program has improved pitching in these Little League players.

*c) H_0: Median difference $= 0$; H_A: Median difference > 0; $P = 0.7597$.

37. a) The data are clearly paired. Both races may have improved over time, but the pairwise differences are likely to be independent. We can only check the Nearly Normal Condition by using the computer files.
    b) With 95% confidence we can say the mean time difference is between -16.63 minutes (men are faster) and $+6.86$ minutes (women are faster).
    c) The interval contains 0, so we would not reject the hypothesis of no mean difference at $\alpha = 0.05$. We can't discern a difference between the female wheelchair times and the male running times.

39. a) Same cows before and after injection; the cows should be representative of others of their breed; cows are independent of each other; less than 10% of all cows; don't know about Nearly Normal differences.
    b) $(12.66, 15.34)$
    c) Based on this sample, with 95% confidence, the average increase in milk production for a herd of 60 Ayrshire cows given BST is between 12.66 and 15.34 pounds per day.
    d) $0.25(47) = 11.75$. The average increase is much more than this, so we would recommend he incur the extra expense.

Chapter 24

1. a) $(30, 30, 30, 30)$, 30 for each season.
   b) 1.933
   c) 3

3. a) 3
   b) No. It's smaller than the mean.
   c) It would say that there is not enough evidence to reject the null hypothesis that births are distributed uniformly across the seasons.
   d) 7.815
   e) Do not reject the null hypothesis. As in part c, there is not enough evidence to suggest that births are not distributed uniformly across the seasons.

5. a) The age distributions of customers at the two branches are the same.
   b) Chi-square test of homogeneity.
   c)

	Age			
	Less Than 30	30–55	56 or Older	Total
In-Town Branch	25	45	30	100
Mall Branch	25	45	30	100
Total	50	90	60	200

   d) 9.778
   e) 2
   f) 0.0075
   g) Reject H_0 and conclude that the age distributions at the two branches are not the same.

7. a) $(-0.913, 0.913, 0.365, -0.365)$
   b) No. They are quite small for z-values.
   c) Because we did not reject the null hypothesis, we shouldn't expect any of the standardized residuals to be large.

9. a)

	Did You Use the Internet Yesterday?	
	Yes	No
White	2506.38	895.62
Black	338.90	121.10
Hispanic/Other	445.73	159.27

   b) 11.176
   c) 2
   d) 0.0037
   e) Reject the null hypothesis and conclude that Race and Internet use are not independent.

11. a) Chi-square test of independence. We have one sample and two variables. We want to see if the variable *Account Type* is independent of the variable *Trade Type*.
    b) Other test. *Account Size* is quantitative, not counts.
    c) Chi-square test of homogeneity. We want to see if the distribution of one variable, *Courses*, is the same for two groups (resident and nonresident students).

13. a) 10
    b) Goodness-of-fit.
    c) H_0: The die is fair (all faces have $p = 1/6$). H_A: The die is not fair.
    d) Count data; rolls are random and independent; expected frequencies are all bigger than 5.
    e) 5
    f) $\chi^2 = 5.600$; P-value $= 0.3471$
    g) Because the P-value is high, do not reject H_0. The data show no evidence that the die is unfair.

15. a) Weights are quantitative, not counts.
    b) Count the number of each kind of nut, assuming the company's percentages are based on counts rather than weights.

17. H_0: The police force represents the population (29.2% white, 28.2% black, etc.). H_A: The police force is not representative of the population. $\chi^2 = 16512.7$; df $= 4$; P-value ≤ 0.0001. Because the P-value is so low, we reject H_0. These data show that the police force is not representative of the population. In particular, there are too many white officers in relationship to their membership in the community.

19. a) $\chi^2 = 5.671$; df $= 3$; P-value $= 0.1288$. With a P-value this high, we fail to reject H_0. Yes, these data are consistent with those predicted by genetic theory.
    b) $\chi^2 = 11.342$; df $= 3$; P-value $= 0.0100$. Because of the low P-value, we reject H_0. These data provide evidence that the distribution is not as specified by genetic theory.
    c) With small samples, many more data sets will be consistent with the null hypothesis. With larger samples, small discrepancies will show evidence against the null hypothesis.

21. a) $96/16 = 6$
b) Goodness-of-fit.
c) H_0: The number of large hurricanes remains constant over decades.
H_A: The number of large hurricanes has changed.
d) 15
e) P-value $= 0.63$
f) The very high P-value means these data offer no evidence that the numbers of large hurricanes has changed.

23. a) Independence.
b) H_0: Breastfeeding success is independent of having an epidural. H_A: There's an association between breastfeeding success and having an epidural.

25. a) 1
b) 159.34
c) Breastfeeding behavior should be independent for these babies. They are fewer than 10% of all babies; we assume they are representative. We have counts, and all the expected counts are at least 5.

27. a) 5.90
b) P-value < 0.005
c) The P-value is very low, so reject the null. There's evidence of an association between having an epidural and subsequent success in breastfeeding.

29. a) $\dfrac{(190 - 159.34)}{\sqrt{159.34}} = 2.43$
b) It appears that babies whose mothers had epidurals during childbirth are much less likely to be breastfeeding 6 months later.

31. These factors would not be mutually exclusive. There would be yes or no responses for every baby for each.

33. a) 40.2%
b) 8.1%
c) 62.2%
d) 285.48
e) H_0: Survival was independent of status on the ship. H_A: Survival was not independent of the status.
f) 3
g) We reject the null hypothesis and conclude that survival depended on status.

35. First-class passengers were most likely to survive, while third-class passengers and crew were underrepresented among the survivors.

37. a) Experiment—actively imposed treatments (different drinks).
b) Homogeneity.
c) H_0: The rate of urinary tract infection is the same for all three groups. H_A: The rate of urinary tract infection is different among the groups.
d) Count data; random assignment to treatments; all expected frequencies larger than 5.
e) 2
f) $\chi^2 = 7.776$; P-value $= 0.020$
g) With a P-value this low, we reject H_0 at $\alpha = 0.05$. These data provide reasonably strong evidence that there is a difference in urinary tract infection rates between cranberry juice drinkers, lactobacillus drinkers, and the control group.

h) The standardized residuals are

	Cranberry	Lactobacillus	Control
Infection	−1.87276	1.19176	0.68100
No Infection	1.24550	−0.79259	−0.45291

From the standardized residuals (and the sign of the residuals), it appears those who drank cranberry juice were less likely to develop urinary tract infections; those who drank lactobacillus were more likely to have infections.

39. a) Independence.
b) H_0: *Political Affiliation* is independent of *Sex*. H_A: There is a relationship between *Political Affiliation* and *Sex*.
c) Counted data; probably a random sample, but can't extend results to other states; all expected frequencies greater than 5.
d) $\chi^2 = 4.851$; df $= 2$; P-value $= 0.0884$
e) Because of the high P-value, we do not reject H_0 at $\alpha = 0.05$. These data do not provide evidence of a relationship between *Political Affiliation* and *Sex*.

41. H_0: *Political Affiliation* is independent of *Region*. H_A: There is a relationship between *Political Affiliation* and *Region*. $\chi^2 = 13.849$; df $= 4$; P-value $= 0.0078$. With a P-value this low, we reject H_0 at $\alpha = 0.01$. *Political Affiliation* and *Region* are related. Examination of the residuals shows that those in the West are more likely to be Democrat than Republican; those in the Northeast are more likely to be Republican than Democrat.

43. a) Homogeneity.
b) H_0: The grade distribution is the same for both professors. H_A: The grade distributions are different.
c)

	Prof. Alpha	Prof. Beta
A	6.667	5.333
B	12.778	10.222
C	12.222	9.778
D	6.111	4.889
F	2.222	1.778

Three cells have expected frequencies less than 5.

45. a)

	Prof. Alpha	Prof. Beta
A	6.667	5.333
B	12.778	10.222
C	12.222	9.778
Below C	8.333	6.667

All expected frequencies are now larger than 5.
b) Decreased from 4 to 3.
c) $\chi^2 = 9.306$; P-value $= 0.0255$. Because the P-value is so low, we reject H_0 at $\alpha = 0.05$. The grade distributions for the two professors are different. Dr. Alpha gives fewer A's and more grades below C than Dr. Beta.

47. $\chi^2 = 14.058$; df $= 1$; P-value $= 0.0002$. With a P-value this low, we reject H_0 at $\alpha = 0.001$. There is evidence of racial steering. Blacks are much less likely to rent in Section A than Section B.

49. a) $z = 3.74936$, $z^2 = 14.058$
    b) P-value $(z) = 0.0002$ (same as in Exercise 35)

51. $\chi^2 = 2.699$; df $= 3$; $P = 0.4404$. Because the P-value is > 0.05, these data do not show enough evidence of an association between the mother's age group and the outcome of the pregnancy.

Chapter 25

1. The scatterplot shows a linear relationship with equal spread about a regression line throughout the range of *Acceptance Rates*. The residual plot has no structure, and there are not any striking outliers. The histogram of residuals is symmetric and bell-shaped. All conditions are satisfied to proceed with the regression analysis.

3. $s = 0.0254$ or 2.54 percentage points. Since the range of *Graduation Rates* is only about 17 percentage points, this number is fairly small and helps us to understand the amount of spread about the regression model.

5. The standard error for the slope tells us how much the slope of the regression equation would vary from sample to sample.

7. The administrators can conclude that there is a relationship between *Graduation Rates* and *Admission Rates* for these upper-echelon schools. It seems the lower the *Admission Rate*, the higher the *Graduation Rate*.

9. The administrators can conclude, with 95% confidence, that the slope of the relationship between *Graduation Rates* and *Admission Rates* is captured by the interval from -0.332 to -0.238.

11. The standard error for the mean will be smaller. We can be much more precise about the prediction for the mean graduation rate of all colleges compared to the prediction for just one.

13. The administrators can expect, with 95% confidence, that schools that admit 33% of applicants will average between 87.35% and 88.88% for their *Graduation Rate*.

15. The administrators can be 95% confident that if their *Admission Rate* is 33%, their *Graduation Rate* would be between 82.96% and 93.28%.

17. The sentence in the exercise is implying a cause-and-effect relationship—i.e., if they reduce the admission rate, then the graduation rate must increase. These data don't support that strong a conclusion. We have no evidence of what effects a change in admission rate might have.

19. a) $\widehat{24Error} = 133.0 - 2.06$ *Years since 1970*; according to the model, the error made in predicting a hurricane's path has been declining at a rate of about 2.06 nautical miles per year starting from about 133 nautical miles in 1970.
    b) $H_0: \beta_1 = 0$; there has been no change in prediction accuracy. $H_A: \beta_1 \neq 0$; there has been a change in prediction accuracy.
    c) With a P-value < 0.0001, I reject the null hypothesis and conclude that prediction accuracies have in fact been changing during this period.
    d) 72.5% of the variation in hurricane prediction accuracy is accounted for by this linear model on time.

21. a) $\widehat{Budget} = -63.998 + 1.026$ *RunTime*. The model suggests that movies cost about \$1 million per minute to make.
    b) A negative starting value makes no sense, but it can be a starting value for the model.
    c) Amounts by which movie costs differ from predictions made by this model vary, with a standard deviation of about \$33 million.
    d) 0.154 \$m/min
    e) If we constructed other models based on different samples of movies, we'd expect the slopes of the regression lines to vary, with a standard deviation of about \$154,000 per minute.

23. a) The scatterplot looks straight enough, the residuals look random and nearly normal, and the residuals don't display any clear change in variability.
    b) I'm 95% confident that the cost of making longer movies increases at a rate of between 0.72 and 1.33 million dollars per additional minute.

25. a) $H_0: \beta_1 = 0$; there's no association between calories and sodium content in all-beef hot dogs. $H_A: \beta_1 \neq 0$: there is an association.
    b) Based on the low P-value (0.0018), I reject the null. There is evidence of an association between the number of calories in all-beef hot dogs and their sodium contents.

27. a) Among all-beef hot dogs with the same number of calories, the sodium content varies, with a standard deviation of about 60 mg.
    b) 0.561 mg/cal
    c) If we tested many other samples of all-beef hot dogs, the slopes of the resulting regression lines would be expected to vary, with a standard deviation of about 0.56 mg of sodium per calorie.

29. I'm 95% confident that for every additional calorie, all-beef hot dogs have, on average, between 1.07 and 3.53 mg more sodium.

31. a) H_0: Difference in age at first marriage has not been changing, $\beta_1 = 0$. H_A: Difference in age at first marriage has been changing, $\beta_1 \neq 0$.
    b) Residual plot looks randomly scattered, histogram is unimodal and a bit skewed, but shows no obvious skewness or outliers.
    c) $t = -6.14$, P-value < 0.0001. With such a low P-value, we reject H_0. These data show evidence that difference in age at first marriage is decreasing.

33. Based on these data, we are 95% confident that the average difference in age at first marriage is decreasing at a rate between 0.013 and 0.027 years per year.

35. a) H_0: *Fuel Economy* and *Weight* are not (linearly) related, $\beta_1 = 0$. H_A: *Fuel Economy* changes with *Weight*, $\beta_1 \neq 0$. P-value < 0.0001, indicating strong evidence of an association.
    b) Yes. The conditions seem satisfied. Histogram of residuals is unimodal and symmetric; residual plot looks OK, but some "thickening" of the plot with increasing values.
    c) $t = -12.2$; P-value < 0.0001. These data show evidence that *Fuel Economy* decreases with the *Weight* of the car.

37. a) $(-9.57, -6.86)$ mpg per 1000 pounds

b) Based on these data, we are 95% confident that *Fuel Efficiency* decreases between 6.86 and 9.57 miles per gallon, on average, for each additional 1000 pounds of *Weight*.

39. a) We are 95% confident that 2500-pound cars will average between 27.34 and 29.07 miles per gallon.
b) Based on the regression, a 3450-pound car will get between 15.44 and 25.36 miles per gallon, with 95% confidence.

41. a) Yes. $t = 2.73$; P-value = 0.0079. With a P-value so low, we reject H_0. There is a relationship between *Calories* and *Sodium* content.
b) No. $R^2 = 9\%$ and s appears to be large, although without seeing the data, it is a bit hard to tell.

43. Plot of *Calories* against *Fiber* does not look linear; the residuals plot also shows increasing variance as predicted values get large. The histogram of residuals is right skewed.

45. a) H_0: No (linear) relationship between *BCI* and *pH*, $\beta_1 = 0$. H_A: There seems to be a relationship, $\beta_1 \neq 0$.
b) $t = -7.73$ with 161 df; P-value < 0.0001
c) There seems to be a negative relationship; *BCI* decreases as *pH* increases at an average of 197.7 *BCI* units per increase of 1 *pH*.

47. a) H_0: No linear relationship between *Population* and *Ozone*, $\beta_1 = 0$.
H_A: *Ozone* increases with *Population*, $\beta_1 > 0$. $t = 3.48$; P-value = 0.0018. With a P-value so low, we reject H_0. These data show evidence that *Ozone* increases with *Population*.
b) Yes. *Population* accounts for 84% of the variability in *Ozone* level, and s is just over 5 parts per million.

49. a) Based on this regression, each additional million residents corresponds to an increase in average ozone level of between 3.29 and 10.01 ppm, with 90% confidence.
b) The mean *Ozone* level for cities with 600,000 people is between 18.47 and 27.29 ppm, with 90% confidence.

51. a) 34 tablets.
b) Yes. The scatterplot is roughly linear with lots of scatter; plot of residuals vs. predicted values shows no overt patterns; Normal probability plot of residuals is reasonably straight.
c) H_0: No linear relationship between *Battery Life* and *Screen Brightness*, $\beta_1 = 0$. H_A: *Battery Life* decreases with brighter screens, $\beta_1 < 0$. $t = 2.02$; one-sided P-value = 0.0522. This is not a particularly small P-value. There may be some evidence of a relationship, but it is not strong.
d) Not particularly. $R^2 = 11.3\%$ and $s = 2.128$ hours.
e) *Hours* = 5.387 + 0.009 *Screen Brightness*
f) (0.00142, 0.0166) hours per cd/m² units
g) *Battery life* is longer, on average, between 0.00138 and 0.0166 *hours* per one cd/m² units, with 90% confidence.

53. a) H_0: No linear relationship between *Waist size* and *%Body Fat*, $\beta_1 = 0$. H_A: *%Body Fat* changes with *Waist* size, $\beta_1 \neq 0$. $t = 8.14$; P-value < 0.0001. There's evidence that *%Body Fat* seems to increase with *Waist* size.
b) With 95% confidence, mean *%Body Fat* for people with 40-inch waists is between 23.58 and 29.02, based on this regression.

55. a) The regression model is $\widehat{Midterm2} = 12.005 + 0.721$ *Midterm1*.

	Estimate	Std Error	t-Ratio	P-Value
Intercept	12.00543	15.9553	0.752442	0.454633
Slope	0.72099	0.183716	3.924477	0.000221

RSquare	0.198982
s	16.78107
n	64

b) The scatterplot shows a weak, somewhat linear, positive relationship. There are several outlying points, but removing them only makes the relationship slightly stronger. There is no obvious pattern in the residual plot. The regression model appears appropriate. The small P-value for the slope shows that the slope is statistically distinguishable from 0 even though the R^2 value of 0.199 suggests that the overall relationship is weak.
c) No. The R^2 value is only 0.199 and the value of s of 16.8 points indicates that he would not be able to predict performance on *Midterm2* very accurately.

57. H_0: Slope of *Effectiveness* vs. *Initial Ability* = 0. H_A: Slope $\neq 0$

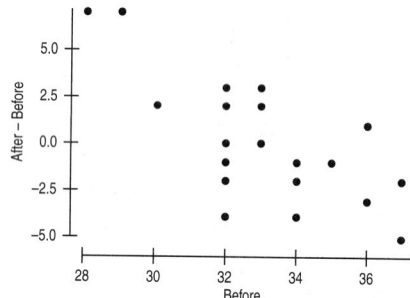

Scatterplot is straight enough. Regression conditions appear to be met. $t = -4.34$; df = 18; P-value = 0.0004. With a P-value this small, we reject the null hypothesis. There is strong evidence that the effectiveness of the video depends on the player's initial ability. The negative slope indicates that the method is more effective for those whose initial performance was poorest and less so for those whose initial performance was better. This looks like a case of regression to the mean. Those who were above average initially tended to be worse after training. Those who were below average initially tended to improve.

59. a) Data plot looks linear; no overt pattern in residuals; histogram of residuals roughly symmetric and unimodal.
b) H_0: No linear relationship between *Education* and *Mortality*, $\beta_1 = 0$. H_A: $\beta_1 \neq 0$. $t = -6.24$; P-value < 0.001. There is evidence that cities in which the mean education level is higher also tend to have a lower mortality rate.
c) No. Data are on cities, not individuals. Also, these are observational data. We cannot predict causal consequences from them.
d) $(-65.95, -33.89)$ deaths per 100,000 people
e) *Mortality rate* decreases, on average, between 33.89 and 65.95 deaths per 100,000 for each extra year of average *Education*.
f) Based on the regression, the average *Mortality rate* for cities with an average of 12 years of *Education* will be between 874.239 and 914.196 deaths per 100,000 people.

61. a) $\widehat{\text{Logit}(\textit{Right-to-work})} = 6.19951 - 0.106155 \, \textit{publ} - 0.222957 \, \textit{pvt}$

b) Yes, logistic regression seems appropriate here. The response variable is dichotomous and there are no outliers in the predictor variables.

Part VI Review

1. H_0: There is no difference in cancer rates, $p_1 - p_2 = 0$. H_A: The cancer rate in those who use the herb is higher, $p_1 - p_2 > 0$.

3. a) Experiment.
b) H_0: There is no difference, $p_1 - p_2 = 0$. H_A: Patients with carbolic acid are more likely to live, $p_1 - p_2 > 0$. $z = 2.91$, P-value $= 0.0018$; with a P-value so low, we reject H_0. These data show that carbolic acid is effective in increasing the chances of surviving an amputation.
c) We are not told whether the patients were randomized to the treatments. We are not told whether the experiment was double-blinded or even blinded at all. This could have biased the results toward a more favorable outcome.

5. H_0: There is no difference, $p_1 - p_2 = 0$. H_A: Early births have increased, $p_1 - p_2 < 0$. $z = -0.729$, P-value $= 0.2329$. Because the P-value is so high, we do not reject H_0. These data do not show an increase in the incidence of early birth of twins.

7. a) H_0: There is no difference, $p_1 - p_2 < 0$. H_A: Treatment prevents deaths from eclampsia, $p_1 - p_2 < 0$.
b) Samples are random and independent; less than 10% of all pregnancies (or eclampsia cases); more than 10 successes and failures in each group.
c) 0.8008
d) There is insufficient evidence to conclude that magnesium sulfate is effective in preventing eclampsia deaths.
e) Type II.
f) Increase the sample size, increase α.
g) Increasing sample size: decreases variation in the sampling distribution, is costly. Increasing α: Increases likelihood of rejecting H_0, increases chance of Type I error.

9. a) Experiment.
b) A one-sided test, since they are interested only in an increase in fuel economy.
c) Deciding the additive increases fuel economy when there really is no difference.
d) Deciding the additive makes no difference when it really does increase fuel economy.
e) Type II.
f) Given that the two groups received roughly the same use and care, yes. They can't necessarily claim it will work for all cars, only cars similar to their fleet.

11. a) H_0: $\mu_{\text{Jan}} - \mu_{\text{Jul}} = 0$; H_A: $\mu_{\text{Jan}} - \mu_{\text{Jul}} \neq 0$. $t = -1.94$, df $= 43.68$, P-value $= 0.0590$. This is mild evidence of a difference, but the P-value is slightly greater than the usual 0.05 cutoff.
b) H_0: $\mu_{\text{Apr}} - \mu_{\text{Oct}} = 0$; H_A: $\mu_{\text{Apr}} - \mu_{\text{Oct}} \neq 0$. $t = -0.92$; df $= 59.40$; P-value $= 0.3610$. Since $P > 0.10$, do not reject the null; these data do not show a significant difference between April and October with regard to the mean age at which crawling begins.
c) There is mild evidence of a possible difference in mean crawling age between January and July babies, but stronger evidence is needed to make a convincing argument. There is no evidence of a difference between mean crawling times for babies born in April and October.

13. Based on these data, we are 95% confident that the mean difference in aluminum oxide content is between -3.37 and 1.65. The means in aluminum oxide content of the pottery made at the two sites could reasonably be the same.

15. a) If there is no difference in the average fish sizes, the chance of seeing an observed difference this large just by natural sampling variation is less than 0.1%.
b) If cost justified, feed them a natural diet.
c) Type I.

17. a) Assuming the conditions are met, from these data we are 95% confident that patients with cardiac disease average between 3.39 and 5.01 years older than those without cardiac disease.
b) Older patients are at greater risk from a variety of other health issues and perhaps more depressed.

19. a) Based on the data, we are 95% confident that the difference in eating disorders between Muslim students and Christian students is between 4.6% and 22.7%.
b) Although caution in generalizing must be used since the study was restricted to the Spanish city of Ceuta, it appears there is a true difference in the prevalence of eating disorders. We can conclude this because the entire interval is above 0.

21. H_0: $\mu_{\text{Cert}} - \mu_{\text{UC}} = 0$; H_A: $\mu_{\text{Cert}} - \mu_{\text{UC}} > 0$. $t = 1.57$; df $= 86$; P-value $= 0.0598$. The P-value is low enough to reject H_0. These data provide evidence that students of certified teachers achieve higher mean reading scores than students of uncertified teachers.

23. There is a significant difference in Math performance; students of certified teachers do better. There is a significant difference in language performance at $\alpha = 0.05$ between students of certified teachers and those of uncertified teachers.

25. a) H_0: $p_{\text{High}} - p_{\text{Low}} = 0$; H_A: $p_{\text{High}} - p_{\text{Low}} \neq 0$. $z = -3.57$; P-value $= 0.0004$. Because the P-value is so low, we reject H_0. These data show the IRS risk is different in the two groups; people who consume dairy products often have a lower risk, on average.
b) Doesn't prove it; this is not an experiment.

27. H_0: The proportions are as specified by the ratio 1:3:3:9; H_A: The proportions are not as stated. $\chi^2 = 5.01$; df $= 3$; P-value $= 0.1711$. Since $P > 0.05$, we fail to reject H_0. These data do not provide evidence to indicate that the proportions are other than 1:3:3:9.

29. a) H_0: *Mortality* and *calcium concentration* in water are not linearly related, $\beta_1 = 0$; H_A: They are linearly related, $\beta_1 \neq 0$.
b) $t = -6.66$; P-value < 0.0001. There is a significant negative relationship between calcium in drinking water and mortality.
c) $(-4.19, -2.26)$ deaths per 100,000 for each ppm calcium
d) Based on the regression, we are 95% confident that mortality (deaths per 100,000) decreases, on average, between 2.26 and 4.19 for each part per million of calcium in drinking water.

31. a) 38
b) Data plot looks linear; residuals plot shows random scatter; histogram of residuals approximately Normal.
c) $(0.131, 0.177)$ foals per adult

d) Based on this regression, we have 95% confidence that for every adult horse, an average of between 0.131 and 0.177 foals will be born.
e) A herd with 80 adult horses will have between 2.27 and 19.21 foals, with 90% confidence.

33. a) Combining boys and girls, $\chi^2 = 40.20$; P-value < 0.0001. Reject H_0. These data show evidence that the distribution of scores at this school is significantly different from national results.
b) Pooling those with scores of 2 or 1, $\chi^2 = 5.59$, P-value $= 0.1336$. There is no significant difference in the distribution of scores for boys and girls.

35. $\chi^2 = 0.69$; P-value $= 0.9526$. Since $P > 0.05$, we do not reject H_0. These data provide no evidence to show that the way the hospital deals with twin pregnancies has changed.

37. a) Independence—individuals classified on two variables.
b) H_0: No relationship between *Age* and *Party*; H_A: There is a relationship. $\chi^2 = 16.54$; P-value $= 0.0111$. With such a small P-value, we reject H_0. These data indicate a relationship between *Age* and *Party*.
c) There is a relationship between age and party: The "largest" standardized residual is for Democrats 30 to 49 (-2.03). The next-largest is for Republicans in the same age group (1.83). These data indicate that people aged 30 to 49 are much more likely to be Republicans than Democrats.

39. a) H_0: There is no (linear) relationship between *Depression* and *Internet usage*. $\beta_1 = 0$. H_A: There is a linear relationship. $\beta_1 \neq 0$. $t = 2.76$; P-value $= 0.0064$. With such a small P-value, we reject H_0. These data provide strong evidence of a relationship between Internet use and depression.
b) The study says nothing about causality. Many other factors may be involved.
c) H_0: $\mu_D = 0$; H_A: $\mu_D \neq 0$. $t = -2.73$; P-value $= 0.007$. With such a small P-value, we reject H_0. These data indicate that mean depression level actually got better (decreased) during the study.

41. $\chi^2 = 40.715$; P-value < 0.0001. These data do indicate an association between education and family planning. More-educated women have fewer unplanned pregnancies.

43. a) H_0: $\beta_1 = 0$; H_A: $\beta_1 \neq 0$. $t = 4.16$, P-value < 0.0001. These data show evidence of a positive relationship between number of meals eaten together and grades.
b) No. R^2 is small and $s = 0.66$ points. So we could predict only to within 1.32 grade points at best.
c) No. The slope is clearly not 0, but that doesn't mean the relationship is strong or the predictions are useful.

45. a) We are 99% confident that the interval 1.78% to 20.6% contains the true difference in violent felony rates when comparing individual therapy to MST.
b) Since the entire interval is above 0, we can conclude that MST is successful in reducing the proportion of juvenile offenders who commit violent felonies. The population of interest is adolescents with mental health problems.

47. $\chi^2 = 14.96$; P-value $= 0.0006$. With such a small P-value, we reject H_0. These data provide evidence that continued treatment with inflixamab is of value.

49. $\chi^2 = 8.23$; P-value $= 0.0414$. There is evidence of an association between *cracker type* and *bloating*. Standardized residuals for the gum cracker are -1.32 and 1.58. Prospects for marketing this cracker are not good.

51. Based on the data, we are 95% confident that the mean difference in words misunderstood is between -3.76 and 3.10. Because 0 is in the confidence interval, we would conclude that the two tapes are not equivalent.

Chapter 26

1. a) H_0: $\mu_1 = \mu_2 = \mu_3 = \mu_4$ vs. the alternative that not all means are equal. Here μ_k refers to the mean number of popcorn kernels left unpopped for brand k.
b) MS_T has 3 df; MS_E has 12.
c) An $F_{3,12}$ of 13.56 has a P-value of 0.00037. The data provide strong evidence that the means are not all equal. The brands do not produce the same mean number of unpopped kernels.
d) I would like to see side-by-side boxplots of the treatment groups, a Normal probability plot of the residuals, a histogram of the residuals, and a plot of the residuals vs. the predicted values.

3. a) The null hypothesis is that the mean mileage driven on one liter of gas using each muffler is the same. The alternative is that not all means are equal.
b) MS_T has 2 df; MS_E has 21 df.
c) We would not reject the null hypothesis of equal means with a P-value of 0.1199. The data from this experiment provide insufficient evidence that the muffler type affects gas mileage.
d) I would like to see side-by-side boxplots of the treatment groups, a normal probability plot of the residuals, a histogram of the residuals, and a plot of the residuals vs. the predicted values.
e) Type II error.

5. a) H_0: $\mu_1 = \mu_2 = \mu_3 = \mu_4$ vs. the alternative that not all means are equal. Here, μ_k refers to the mean activation time using recipe k.
b) The data provide strong evidence with a P-value < 0.0001 to reject the null hypothesis and conclude that the means are not all equal. This experiment provides strong evidence that the mean activation times differ among the recipes.
c) Yes. Because we have rejected the null hypothesis, we can proceed with a multiple comparisons method to compare all the groups.

7. a) H_0: $\mu_4 = \mu_5 = \mu_6 = \mu_8$ vs. the alternative that not all means are equal. Here, μ_k refers to the mean mileage using k cylinders.
b) The spreads of the four groups look very different. The similar variance condition is not met. The response variable might need to be re-expressed before proceeding with an Analysis of Variance.

9. a) H_0: $\mu_1 = \mu_2 = \mu_3 = \mu_4 = \mu_5 = \mu_6$ vs. the alternative that not all means are equal. Here, μ_k refers to the mean *Time* it takes to serve a customer by *Teller k*.
b) The data do not provide evidence that the mean *Time* it takes to serve a customer differs by *Teller*.
c) No. We do not reject the null hypothesis of equal means, so we cannot perform multiple comparisons.

11. An ANOVA is not appropriate because *Eye Color* is a categorical variable. The students should consider analyzing the data with a χ^2 test of independence between eye and hair color.

13. a) $MS_T = 8.65$; $MS_E = 0.0767$
    b) $F = 112.826$
    c) The data provide very strong evidence that the means are not equal.
    d) We have assumed that the experimental runs were performed in random order, that the variances of the treatment groups are similar, and that the residuals are nearly Normal.
    e) A boxplot of the *Scores* by *Method*, a plot of residuals vs. predicted values, a Normal probability plot, and a histogram of the residuals.
    f) $s_p = \sqrt{0.0767} = 0.277$ points

15. a) The null hypothesis is that the mean *Scores* are the same for both *Types* of eggs. The alternative is that they are different.
    b) The data provide strong evidence that the means are different. The real eggs have a higher mean score.
    c) Yes. The conditions look reasonable. It would be good to examine the residuals as well.
    d) The pooled estimate of the variance is equal to MS_E. The *t*-ratio is 5.5727 and $5.5727^2 = 31.05499$, which is approximately the *F*-ratio. The two-sided P-value for the *t* with 6 df is 0.00141, which agrees with the P-value for the *F*-statistic.

17. a) The null hypothesis is that the mean *Test Scores* from all the schools are equal. The alternative is that not all the means are equal.
    b) The data provide insufficient evidence that the mean *Test Scores* differ.
    c) The intern's Type I error rate is higher because he is performing many (over 100) *t*-tests rather than an Analysis of Variance or a multiple comparisons method.

19. a) The null hypothesis is that the mean *Sugar Content* is the same for the cereals on each *Shelf*. The alternative is that not all the means are equal.
    b) The P-value of 0.0013 provides strong evidence that the mean *Sugar Content* is not the same for each *Shelf*.
    c) We cannot conclude that cereals on *Shelf* 2 have a higher mean *Sugar Content* than cereals on *Shelf* 3 or that cereals on *Shelf* 2 have a higher mean *Sugar Content* than cereals on *Shelf* 1. We can conclude only that the means are not all equal.
    d) The Bonferroni rejects the null hypothesis of equal means at $\alpha = 0.05$. Now we can conclude that the mean *Sugar Content* of cereals on *Shelf* 2 is *not* equal to the mean *Sugar Content* on *Shelf* 1 and is *not* equal to the mean *Sugar Content* on *Shelf* 3. In other words, we can conclude from this test what we wanted to conclude in part c.

21. a) The null hypothesis states that the mean *Download Time* (DT) is the same at each *Time of Day*. The alternative is that not all the means are equal.
    b)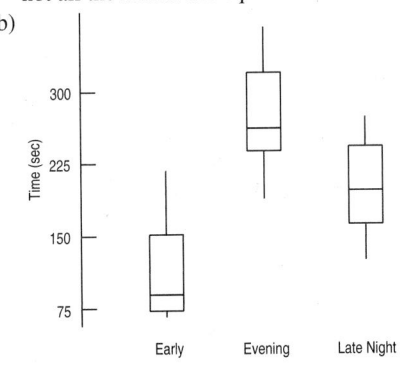

Analysis of Variance for Time (in sec)

Source	DF	Sum of Squares	Mean Square	F-Ratio	P-Value
DT	2	204641	102320	46.035	<0.0001
Error	45	100020	2222.67		
Total	47	304661			

The data provide very strong evidence that mean *Download Time* is not the same at each *Time of Day*.

c) The similar variance condition seems to be met. The residual plots below show no evidence of violating the nearly Normal or similar variance condition. The runs were not randomized because they were in specific time periods. It would be good to check a timeplot of residuals by run order to see if there is dependence from one run to the next.

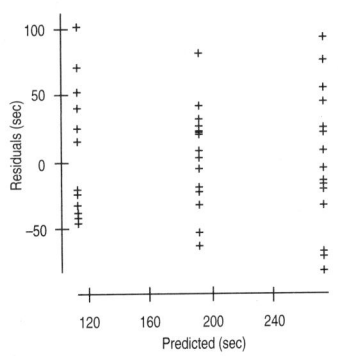

d) A Bonferroni test shows that all three pairs can be distinguished from each other at $\alpha = 0.05$.

Chapter 27

1. $TV_{ijk} = \mu + Sex_j + Athlete_k + \varepsilon_{ijk}$

3. The boxplots show different distributions for Women and Men, so the additivity assumption is probably not satisfied. The plots show that, at least for men, the equal variance assumption seems unlikely. And both plots show outliers, making the normality assumption questionable. If the survey was randomized, the responses may be mutually independent, but we don't have enough information to tell.

5. The interaction term reflects the fact that there is very little difference in TV watching among women between athletes and non-athletes, but there is a difference for men with athletes watching more TV.

7. a) $H_0: \gamma_1 = \gamma_2 = \gamma_3$ where γ represents the effect of the *Power* level vs. H_A: Not all of the *Power* levels have the same effect on the response.
$H_0: \tau_1 = \tau_2 = \tau_3$ where τ represents the effect of *Time* level vs. H_A: Not all of the *Time* levels have the same effect on the response.
b) The *Power* sum of squares has 2 df; the *Time* sum of squares has 2 df; the error sum of squares has 4 df.
c) There are no degrees of freedom left for the interaction term. She must assume that the interaction effects are negligible.

9. a) The P-value for *Power* is 0.0165. The P-value for *Time* is 0.0310.
b) We reject the null hypothesis that *Power* has no effect and conclude that the mean number of uncooked kernels is not equal across all 3 *Power* levels. We also reject the null hypothesis that *Time* has no effect and conclude that the mean number of uncooked kernels is not the same across all 3 *Times*.
c) Partial boxplots, scatterplots of residuals vs. predicted values, and a Normal probability plot of the residuals to check the assumptions.

11. a) $H_0: \gamma_1 = \gamma_2$ where γ represents the effect of the *Seat* level vs. H_A: One of the *Seat* levels has a different effect. The null hypothesis states that *seat* choice (driver vs. passenger) has no effect on mean head injury sustained. The alternative states that it does.
$H_0: \tau_1 = \tau_2 = \tau_3 = \tau_4 = \tau_5 = \tau_6$ where τ represents the effect of the *Size* level vs. H_A: The *Size* level effects are not all equal. The null hypothesis states that *Size* of vehicle (the 6 represented) has no effect on mean *Head Injury* sustained. The alternative states that it does.
b) The conditions appear to be met. The effects are additive enough, the data we assume were collected independently, the boxplots show that the variance is roughly constant, and there are no patterns in the scatterplot of residuals vs. predicted values.
c) There is no significant interaction. The P-values for both *Seat* and *Size* are <0.0001. Thus, we reject the null hypotheses and conclude that both the *Seat* and the *Size* of car affect the severity of head injury. From the partial boxplots we see that the mean *Head Injury* level is higher for the driver's side. The effect of driver's seat seems to be roughly the same for all 6 car sizes.

13. a) A two-factor ANOVA must have a quantitative response variable. Here the response is whether they exhibited baldness or not, which is a categorical variable. A two-factor ANOVA is not appropriate.
b) We could use a chi-square analysis to test whether *Baldness* and *Heart Disease* are independent.

15. a) A chi-square test of independence gives a chi-square statistic of 14.510 with a P-value of 0.0023. We reject the hypothesis that *Baldness* and *Heart Disease* are independent.
b) No, the fact that these are not independent does NOT mean that one causes the other. There could be a lurking variable (such as age) that influences both.

17. a) $H_0: \gamma_1 = \gamma_2$ where γ represents the effect of the *Time of Day* on *Shots Made*. H_A: The means of *Shots Made* differ across the two *Time of Day* levels. The null hypothesis states that *Time of Day* has no effect on *Shots Made*. The alternative states that it does.

$H_0: \tau_1 = \tau_2$ where τ represents the kind of *Shoes* worn vs. H_A: The means of *Shots Made* differ across the two types of *Shoes*. The null hypothesis states that type of *Shoes* worn has no effect on *Shots Made*. The alternative states that it does.
b) Partial boxplots show little effect of either *Time of Day* or *Shoes* on *Shots Made*.

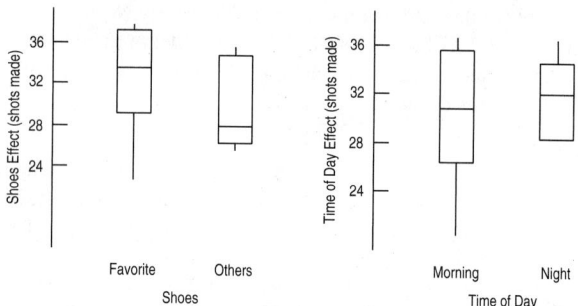

Neither the scatterplot of residuals vs. predicted values nor the Normal probability plots of residuals show any conditions that aren't met. We assume the number of shots made were independent from one treatment condition to the next.

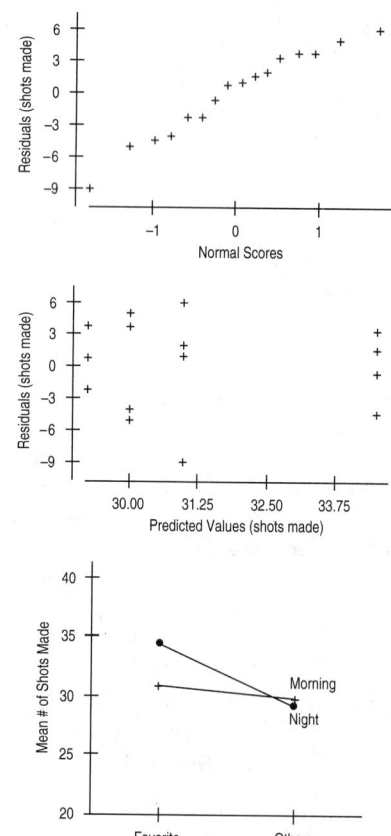

An interaction plot shows a possible interaction effect. It looks as though the favorite shoes may make more of a difference at night. However, the F-test shows that the null hypothesis of no interaction effect is not rejected. None of the effects appears to be significant. It looks as though she cannot conclude that either *Shoes* or *Time of Day* affect her mean *Shots Made*.

19. a) $H_0: \gamma_1 = \gamma_2 = \gamma_3$ where γ represents the effect of the *Temperature* level vs. H_A: Not all the γ are equal. The null

hypothesis states that *Temperature* level (32°, 34°, and 36° C) has no effect on the *Number of Sprouts*. The alternative states that it does.

$H_0: \tau_1 = \tau_2 = \tau_3 = \tau_4$ where τ represents the effect of the *Salinity* level vs. H_A: Not all the τ are equal. The null hypothesis states that *Salinity* level (0, 4, 8, and 12 ppm) has no effect on the *Number of Sprouts*. The alternative states that it does.

b) The partial boxplots show that *Salinity* appears to have an effect on the *Number of Sprouts*, but *Temperature* does not. The ANOVA supports this. The P-value for *Salinity* is $<.0001$, but the P-value for *Temperature* is 0.3779, providing no evidence of a temperature effect over this range of temperatures. There appears to be no interaction as well. There appears to be more spread in *Number of Sprouts* for the lower *Salinity* levels (where the response is higher). This is evident in both the partial boxplot and the residual vs. predicted values. This is cause for some concern, but most likely does not affect the conclusions.

21. The ANOVA table shows that both *Car Type* and *Additive* affect *Gas Mileage*, with P-values <0.0001. There is a significant interaction effect as well that makes interpretation of the main effects problematic. However, the residual plot shows a strong increase in variance, which makes the whole analysis suspect.

23. The ANOVA table now shows only main effects to be significant:

Source	DF	Sum of Squares	Mean Square	F-ratio	P-value
Type	2	10.1254	5.06268	5923.1	<0.0001
Additive	1	0.026092	0.0260915	30.526	<0.0001
Interaction	2	7.57E-05	3.78E-05	0.044265	0.9567
Error	54	0.046156	8.55E-04		
Total	59	10.1977			

The residuals now show *no* violation of the conditions:

We can conclude that both the *Car Type* and *Additive* have an effect on mileage and that the effects are constant (in log10 *Mpg*) over the values of the various levels of the other factor.

25. a) $H_0: \gamma_1 = \gamma_2$ where γ represents the effect of the *Environment* level vs. H_A: The mean *Time* the battery lasted is not the same for the two conditions.
$H_0: \tau_1 = \tau_2 = \tau_3 = \tau_4$ where τ represents the effect of the *Brand* vs. H_A: Not all the τ are equal.

b) From the partial boxplot, it is not clear that *Brand* has an effect. The condition clearly has an effect.

c) Yes, the *Brand* effect has a P-value of 0.099, while the *Environment* effect is clearly significant with a P-value <0.0001.

d) There is also an interaction, however, which makes the statement about *Brands* problematic. Not all brands are affected by the environment in the same way. Brand C, which works best in the cold, performs worst at room temperature.

e) Because it performs the worst of the four at room temperature.

27. In this one-way ANOVA, we can see that the means vary across treatments. (However, boxplots with only 2 observations are not appropriate.) By looking closely, it seems obvious that the four flashlights at room temperature lasted much longer than the ones in the cold. But it is much harder to see whether the means of the four brands are different, or whether they differ by the same amounts across both environmental conditions. The two-way ANOVA with interaction makes these distinctions clear.

Chapter 28
SECTION EXERCISE ANSWERS

1. a) $99,859.89
   b) $35,140.11
   c) The house sold for about $35,100 more than our estimate.

3. a) $\widehat{USGross} = -22.9898 + 1.13442 Budget + 24.9724 Stars - 0.403296 RunTime$
   b) After allowing for the effects of *RunTime* and *Stars*, each million dollars spent making a film yields about 1.13 million dollars in gross revenue.

5. a) Linearity: The plot is reasonably linear with no bends.
   b) Equal spread: The plot fails the Equal Spread condition. It is much more spread out to the right than on the left.
   c) Normality: A scatterplot of two of the variables doesn't tell us anything about the distribution of the residuals.

7. a) $H_0: \beta_{Stars} = 0$
   b) $t = 4.24$
   c) $P < 0.0001$
   d) Reject the null hypothesis and conclude that the coefficient of *Stars* is not zero.

9. a) $R^2 = 0.474$ or 47.4%
   About 47.4% of the variation in *USGross* is accounted for by the least squares regression on *Budget*, *RunTime*, and *Stars*.
   b) Adjusted R^2 accounts for the number of predictors, so it differs from R^2, which does not.

11. a) Doesn't mention the other predictors.
    b) This is correct.
    c) Can't predict from response to predictor.
    d) R^2 is about the fraction of variability accounted for by the regression model, not the fraction of data values.

13. a) $\widehat{Final} = -6.7210 + 0.2560\, Test1 + 0.3912\, Test2 + 0.9015\, Test3$
    b) R^2 is 77.7%, so 77.7% of the variation in final grade is accounted for by the regression model.
    c) After allowing for the linear effects of the other predictors, each point on *Test3* is associated with an average increase of 0.9015 points on the final.
    d) Test scores are probably collinear. If all we are concerned about is predicting final exam score, *Test1* may not add much to the regression. However, we'd expect it to be associated with final exam score.

15. a) $\widehat{Price} = -152{,}037 + 9530\,Baths + 139.87\,Sqft$
b) 71.1% of the variation in asking price is accounted for by this regression model.
c) For homes with the same number of bathrooms, asking price increases, on average, by about $139.87 per square foot.
d) The number of bathrooms is probably associated with the size of the house (even after considering the square footage of the bathroom itself). That association may account for the coefficient of *Baths* not being discernibly different from zero. Moreover, the regression model does not predict what will happen when a house is modified (for example, by converting existing space into a bathroom).

17. The plot of residuals vs. predicted values looks bent, rising in the middle and falling at both ends. This violates the Straight Enough Condition. The Normal probability plot and the histogram of the residuals suggest that the highest five residuals (which we know are in the middle of the predicted value range) are extraordinarily high. These data may benefit from a re-expression.

19. a) $\widehat{Salary} = 9.788 + 0.11\,Service + 0.053\,Educ + 0.071\,Score + 0.004\,Speed + 0.065\,Dictation$
b) 29.2 thousand dollars
c) b_4 has a *t*-statistic of 0.013. That is not significant at any reasonable alpha level.
d) Omitting X4 (typing speed) would simplify the model and might result in one that was almost as good. Other predictors might also be omitted, but we can't tell that from what we know here.
e) *Age* may be collinear with other predictors in the model. In particular, it is likely to be highly associated with X1, months of service.

21. a) $H_0: \beta = 0$; $H_A: \beta \neq 0$; $t = 12.4$; $P < 0.0001$, so we reject the null hypothesis. The coefficient of *Weight* is statistically discernible from zero.
b) Each pound of weight is associated, on average, with an increase of 0.189 in *%Body Fat*.
c) After removing the linear effects of *Waist* and *Height*, each pound of *Weight* is associated on average with a decrease of 0.10 in *%Body Fat*. Alternatively, for men with the same *Waist* and *Height*, each pound of *Weight* is associated, on average, with a decrease of 0.10 in *%Body Fat*. The change in coefficient and sign is a result of including the other predictors. We expect *Weight* to be correlated with both *Waist* and *Height*. It may be collinear with them.
d) The P-value of 0.1567 says that if the coefficient of *Height* in this model is truly zero, we could expect to observe a sample regression coefficient as far from zero as the one we have here about 15.6% of the time.

23. a) P-value is <0.0001, so we can reject the null hypothesis that the true slope is zero.
b) It says that there is a clear relationship between *%Body Fat* and *Chest Size*. In fact, *%Body Fat* grows, on average, by 0.71272% per inch of *Chest Size*.
c) After allowing for the linear effects of *Waist* and *Height*, each inch of *Chest Size* is associated, on average, with a *%Body Fat* that is on average 0.233531 lower. This coefficient is statistically discernible from zero ($P = 0.0054$).
d) Each of the variables appears to contribute to the model. There appears to be no advantage in removing one.

25. a) The only model that seems to do poorly is the one that omits *Murder*. The other three are hard to choose among.
b) Each of the models has at least one coefficient with a large P-value. This one could be omitted to simplify the model without degrading it much.
c) No. A regression model can't be used to predict what effects changes in the predictors might cause.
d) Plots of the residuals highlight some states that may be outliers. You may want to consider setting them aside to see if the model changes.

27. a) With an R^2 of 99.88% it looks like the model would do a good job of prediction but we should check residual plots.
b) The value of *s*, 8.178 calories, is very small compared with the initial standard deviation of *Calories*. This indicates that the model fits the data quite well, leaving very little variation unaccounted for.
c) No, the residuals are not almost zero. Indeed, we know that their standard deviation, *s*, is 8.178. But they are very small compared with the original values.

Part VII Review

1. a) Based on these data, with 95% confidence 3-year yields are between 2.48% and 6.28% higher than 1-year yields on average (paired data).
b) No (at least for this data set). The regression line is 3-year = $7.749 + 0.283 \times$ 1 year. $H_0: \beta_1 = 0$ against $H_A: \beta_1 \neq 0$ has $t = 1.74$; P-value = 0.1095. That's not small enough to reject. (But we don't know that this was an SRS or even a representative sample of large cap funds.)

3. a) An ANOVA and an *F*-test.
b) The null hypothesis is that the mean *Distance* thrown is the same for all 4 grips. The alternative is that it is not.
c) Conclude that the mean *Distance* thrown is not the same for all the grips.
d) Boxplots of *Distance* by *Grip*, residual vs. predicted value scatterplot, and a Normal probability plot of the residuals.
e) A multiple comparison test to see which grip is best.

5. a) Indicator variable.
b) There were, on average, 6.4 fewer foals in herds in which some stallions were sterilized, after allowing for the number of adults in the herd.
c) The P-value for *Sterilized* is 0.096. That's not significant at the 0.05 level.

7. These are not counts, so they fail the Counted Data Condition. We cannot use chi-square methods.

9. a) The null hypotheses are that the mean *Time* is the same for all three *Mice* and for *Lights* on and off. The alternatives are that the means are not all the same for the three *Mice* or for the two lighting conditions.
b) *Mouse* has 2 df; *Light* has 1. The error has 2 df.
c) No, the interaction would use 2 df. There are none left, so the interaction can't be estimated.

11. a) (36.21, 60.51) feet.
b) Yes, 40 is in the interval.
c) Wider; we'd need to have a wider interval to be more confident.

d) Roughly 4 times as big—44 flights.

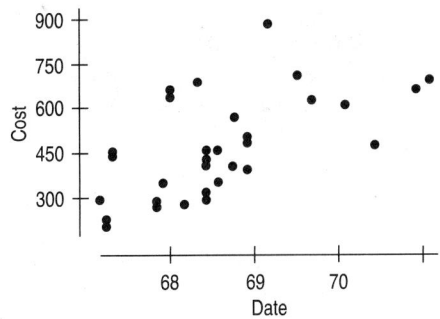

13. a) With an R^2 of 90.9%, your friend is right about being able to predict.
    b) $H_0: \beta_{PRS} = 0$, $t = 8.13$ on 31 df. P is <0.0001, so we reject the null hypothesis.
    c) The plots show evidence that all 4 conditions are met.

15. a) Null hypotheses are that mean hours studied are the same for both sexes and that the mean hours studied are the same for all four classes. Alternatives are that the mean hours studied are not the same for both sexes and that not all four classes have the same mean hours studied.
    b) Yes, an interaction term should be fit because the Additive Enough Condition does not seem to be met.
    c) None appear to be significant.
    d) There are a few outliers. The Constant Variance Condition appears to be met, but we do not have Normal probability plots of residuals. The main concern is that we may not have enough power to detect differences between groups. We would need more data to increase the power.

17. a) Linearity: There's no evidence of nonlinearity in the residual scatterplot.
    Independence: These data are measured over time. We should plot the data and the residuals against time to look for failures of independence.
    Constant variance: The scatterplot of the residuals shows no evidence of thickening.
    Normality: Without a histogram or Normal probability plot of the Studentized residuals, we can't confirm whether the residuals are Nearly Normal.
    b) 57.4% of the variation in female teen traffic deaths is accounted for by the linear trend over time.
    c) Predicted $Deaths = 45074 - 21.6\ Year$.
    d) *Deaths* have been declining at the rate of 21.6 per *Year*. (With a standard deviation of the errors estimated to be 182.6, this doesn't seem like a very meaningful decrease.)

19. a) Null hypotheses are that the mean number of *Words Typed* is the same for both *Temperatures* and for both *Gloves* on and off. The alternatives are that the mean number of *Words Typed* is different for the two *Temperatures* and for *Gloves* on and off.
    b) Based on the partial boxplots, both effects appear to be significant.
    c) Yes, an interaction term should be fit because the Additive Enough Condition does not seem to be met.
    d) The interaction term appears to be real. The temperature effect is only active with gloves off.
    e) All three effects are significant.
    f) *Gloves* decrease *Words Typed* at both *Temperatures*. Hot temperature increases words typed with gloves off, but has little effect with gloves on.
    g) $s = \sqrt{\dfrac{58.75}{28}} = 1.45$ words per minute. Yes, this seems consistent with the size of the variation shown in the partial boxplots.
    h) Tell him to type in a warm room with his gloves off.
    i) Even though we haven't seen a histogram or Normal probability plot of residuals, the other conditions seem to be satisfied. For the levels of the factors that he used, it seems clear that a warm room and not wearing gloves are beneficial for his typing.

21. a) The number of finishers per minute appears to grow at about 0.519 finisher per minute for each additional minute of race time.
    b) Oops! Clearly, the Linearity Assumption is violated. We can't interpret this regression model at all, although there seems to be a strong—but not linear—association.

23. $H_0: p = 0.512$; $H_A: p < 0.512$. $z = -3.35$; P-value $= 0.0004$. With such a small P-value, we reject H_0. These data provide evidence that exposure to dioxin is related to a reduced rate of male births.

25. a) H_0: Both javelin *Brands* give the same average *Distance*.
    H_A: There is a difference in average *Distance* between *Brands*.
    H_0: Both *Preparation* conditions lead to the same average *Distance*.
    H_A: There is a difference in average *Distance* between *Preparation* conditions.
    b) Partial boxplots show that both factors seem to have effects:

No interaction seems to be present. The Additive Enough Condition appears to be satisfied:

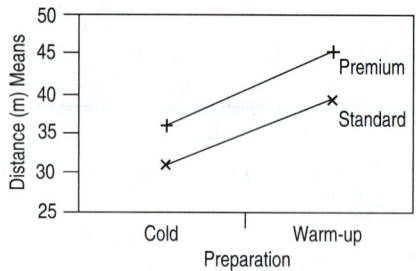

The conditions appear to be met to fit a two-way ANOVA without interaction:

Source	DF	Sum of squares	F-ratio	P-value
Javelin	1	60.58	55	0.0007
Preparation	1	162	147.27	<0.0001
Error	5	5.5		

Both effects are significant. It appears that depending on the cost, the premium javelin may be worth it—it increases the distance about 5.5 meters on average. Warming up increases distance about 9 meters on average. She should always warm up and consider using the premium javelin.

27. a) The null hypothesis is that mean *TV Watching* is the same for all three groups. The alternative is that it is not.
b) The variance for the None group appears to be slightly smaller, and there are outliers in all three groups. We don't have a Normal probability plot of the residuals, but we suspect that the data may not be Normal enough.
c) The *F*-test indicates that the mean amount of *TV Watching* is not the same for the three groups.
d) It seems that the differences are evident even when the outliers are removed. It now seems that the conclusion is valid.

29. Based on these data, the average *Weight Loss* for the clinic is between 8.24 and 10.06 pounds, with 95% confidence. The clinic's claim is plausible.

Chapter 29

1. a) Use an indicator. Code Male = 0, Female = 1 (or the reverse).
b) Treat it as a quantitative predictor.
c) Use an indicator. Code Older than 65 = 1, Younger than 65 = 0.

3. a) Likely high leverage.
b) Likely outlier.

5. No. The predictors are almost certainly collinear, which would affect the coefficient of *Age*.

7. a) They are unimodal and fairly symmetric. Look OK for regression
b) That would not be a correct interpretation. Because CO_2 levels have risen steadily over recent years, CO_2 and Year are likely to be collinear.

9. a) The slope of a partial regression plot is the coefficient of the corresponding predictor—in this case, -1.020.
b) Quaker oatmeal makes the slope more strongly negative. It appears to have substantial influence on this slope.
c) Not surprisingly, omitting Quaker oatmeal changes the coefficient of *Fiber*. It is now positive (although not significantly different from 0). This second regression model has a higher R^2, suggesting that it fits the data better. Without the influential point, the second regression is probably the better model.
d) The coefficient of *Fiber* is not discernibly different from zero. We have no evidence that it contributes significantly to *Calories* after allowing for the other predictors in the model.

11. a) After allowing for the effects of *Sodium* and *Sugars*, each gram of *Potassium* is associated with a 0.019-calorie decrease in *Calories*.
b) Those points pull the slope of the relationship down. Omitting them should increase the value of the coefficient of *Potassium*. It would likely become positive, since the remaining points show a positive slope in the partial regression plot.
c) These appear to be influential points. They have both high leverage and large residual, and the partial regression plot shows their influence.
d) If our goal is to understand the relationships among these variables, then it might be best to omit these cereals because they seem to behave as if they are from a separate subgroup.

13. a) This set of indicators uses *medium* size as its base. The coefficients of *Small*, *Large*, and *Very Large* estimate the average change in amount of *Delay/Person* relative to the amount of *Delay/Person* for *medium* size cities found for each of the other three sizes. If there were an indicator variable for *medium* as well, the four indicators would be collinear, so the coefficients could not be estimated.
b) The costs of congestion per commuter in Large cities are, on average, about $106 greater than in *medium* size cities, after allowing for the effects of *Congested%*.

15. a) An assumption required for indicator variables to be useful is that the regression model's fit for the different groups identified by the indicators be parallel. These lines are not parallel.
b) The coefficient of *SML*C%V* adjusts the slope of the regression model fit for the small cities. We would say that the slope of *Congestion Cost per person* on *Congested%* of motor vehicle miles for *small* cities (after allowing for the linear effects of the other variables in the model) is $27.815 - 4.42 = $23.395.
c) The regression model seems to do a good job. The R^2 shows that 65% of the variability in *Congestion cost per person* is accounted for by the model. Most of the P-values for the coefficients are small.

17. Laredo has high leverage, but it does not have a particularly high Studentized residual. It appears that the point has influence but doesn't exert it. Removing this case from the regression may not result in a large change in the model, so the case is probably not influential.

Appendix B: Photo Acknowledgments

Front Matter
iv David E. Bock; Paul F. Velleman; Richard D. De Veaux

Chapter 1
01 Jannoon028/Shutterstock **02** FRAZZ © 2003 Jef Mallett. Dist. By UNIVERSAL UCLICK. Reprinted with permission. All rights reserved. **05B** Donatas1205/Fotolia **05T** Wavebreakmedia/Shutterstock **07** Thomson Reuters **08B** Jannoon028/ShutterstocK **08T** Toria/Shutterstock

Chapter 2
16 Moviestore collection Ltd/Alamy **19** Randall Monroe/xkcd **21** Nils Jorgensen/Rex Features/Presselect/Alamy **27** Donatas1205/Fotolia **32** Moviestore collection Ltd/Alamy

Chapter 3
45 Thomson Reuters **51** Olivier Juneau/Shutterstock **52** AF archive/Alamy **58** MO_SES Premium/Shutterstock **64** Dmitriy Ystuyjanin/Fotolia **69** Thomson Reuters

Chapter 4
84 Darren Green/Dreamstime **87** Nra/Shutterstock **88** Trosamange/Dreamstime **91** Erin Paul Donovan/Alamy **97** Ithaca Times **98** Darren Green/Dreamstime

Chapter 5
112 Action Plus Sports Images/Alamy **117** Ruslan Kudrin/Fotolia **118** Robert Kneschke/Fotolia **124** David Gallaher/Shutterstock **125** WavebreakMediaMicro/Fotolia **126** Olga Nayashkova/Shutterstock **131** Action Plus Sports Images/Alamy

Chapter 6
151 NASA **155** Randall Monroe/xkcd **158** Randall Monroe/xkcd **159** Monkey Business Images/Dreamstime **167B** NASA **167T** HO/AP Images **180** Kosoff/Shutterstock

Chapter 7
183 NASA **185** Lawrence Sawyer/Getty Images **195** Mara Zemgaliete/Fotolia **199** Kosoff/Shutterstock **203** Paul Velleman **217** Richard D. De Veaux

Chapter 8
221 Randall Monroe/xkcd **224B** Jim Cole/AP Images **224T** Newscom **225** Alamy **229** Moodboard/Fotolia **232** Richard D. De Veaux **245** Joel Saget/Getty Images

Chapter 9
259 Joel Saget/Getty Images **278** Suzanne Tucker/Shutterstock **279** DILBERT © 2001 Scott Adams. Used By permission of UNIVERSAL UCLICK. All rights reserved.

Chapter 10
280B Walter G Arce/Cal Sport Media/Newscom **280c** Imago sportfotodienst/Newscom **280T** Louis Lopez/Cal Sport Media/Newscom **283** Alena Root/Shutterstock **286** Suzanne Tucker/Shutterstock **292** Photodiem/Shutterstock **293** Art Resource

Chapter 11
294 Keystone Pictures USA/Alamy **300** Maxfeld/Shutterstock **301** Jeff Greenberg/Alamy **302** CALVIN AND HOBBES © 1993 Watterson. Reprinted with permission of UNIVERSAL UCLICK. All rights reserved. **308** Photodiem/Shutterstock **316** Pavel Losevsky/Fotolia **317** Shimmo/iStockphoto

Chapter 12
318 Photo Researchers/Alamy **319** Houghton Library **320B** Lurii Davydov/Shutterstock **320T** DILBERT © 2002 Scott Adams. Used By permission of UNIVERSAL UCLICK. All rights reserved. **321** Brian Chase/Shutterstock **326** Chris Wildt/CartoonStock **328** Artem Kursin/Shutterstock **329** iStockphot **331** Pavel Losevsky/Fotolia **346** Jessica Bethke/Shutterstock

Chapter 13
348 Chandler Studios/Pearson Education **349** spxChrome/E+/Getty Images **350** OJO Images Ltd/Alamy **351** John Venn/Pearson Education **355** Carolyn Jenkins/Alamy **358** Jessica Bethke/Shutterstock **365B** US Department of the Treasury **365T** Worradirek/Shutterstock

Chapter 14
368 Aspen Photo/Shutterstock **375** Ventdusud/Shutterstock **378** Alexander Raths/Shutterstock **379** Pearson Education **380** Photos.com/Getty Images **381** Worradirek/Shutterstock **388** Digital Vision/Getty Images

Chapter 15
398 Pearson Education **401** N-Photos/Fotolia **404** Digital Vision/Getty Images **411B** Imago sportfotodienst/Newscom **411T** Visualphotos

Chapter 16
412L Pearson Education **412R** CALVIN AND HOBBES © 1993 Watterson. Reprinted with permission of UNIVERSAL UCLICK. All rights reserved. **418** Keith Brofsky/Pearson Education **423** Lebrecht Music and Arts Photo Library/Alamy **427** Visualphotos **440** Richard D. De Veaux **442** FloridaStock/Shutterstock

Chapter 17
444 Chandler Studios/Pearson Education **448** Pavel L Photo and Video/Shutterstock **451** Pearson education **456** Aispix/Shutterstock **461** FloridaStock/Shutterstock **471** keith morris/Alamy

Chapter 18
476 GARFIELD © 1999 Paws, Inc. Reprinted with permission of UNIVERSAL UCLICK. All rights reserved. **479** Rudall30/Fotolia **484** Keith morris/Alamy **493** Tischenko Irina/ShutterStock

APPENDIX B Photo Acknowledgments

Chapter 19
494 Randall Monroe/xkcd **501** Teamdaddy/ShutterStock **504** Martin Richardson/DK Images **508** Tischenko Irina/Shutterstock **517** Vava Vladimir Jovanovic/ShutterStock

Chapter 20
519 Pearson Education **523** Pearson Education **525** Scott Griessel/Fotolia **530** Moodboard/Fotolia **536B** Vava Vladimir Jovanovic/ShutterStock **536T** Randall Monroe/xkcd **547** Donal Mullins/ShutterStock

Chapter 21
549 AlcelVision/Fotolia **552** Paul Velleman **567** Donal Mullins/ShutterStock **584** Monkey Business Images/Shutterstock

Chapter 22
585 Belushi/Shutterstock **589** Hero/Fancy/Age Fotostock **591** CandyBox Images/Shutterstock **592** Don Tremain/Getty Images **603B** Beth Anderson/Pearson Education **603BC** Beth Anderson/Pearson Education **603C** Beth Anderson/Pearson Education **603T** Beth Anderson/Pearson Education **603TC** Beth Anderson/Pearson Education **609** Monkey Business Images/Shutterstock **629** Yuri Kadobnov/Getty Images

Chapter 23
631 David Bock **634** Vladislav Gajic/Shutterstock **637** Darren Baker/Shutterstock **641** Yuri Kadobnov/Getty Images **652** Andy Crawford/Shutterstock **653** Nomad Soul/Shutterstock

Chapter 24
657 Rob/Fotolia **661** Photodisc/Pearson Education **662** Beth Anderson/Pearson Education **669** Beth Anderson/Pearson Education **674** Nomad Soul/Shutterstock

Chapter 25
689 Ivica Drusany/Shutterstock **693** Getty Images **695** Dean Drobot/Shutterstock **701** Pearson Education **704** Pearson Education **708** Curtis, Edward S./Loc **713** Ivica Drusany/Shutterstock **740** David Bock

Chapter 26
747 Fotolia **762** Bikeriderlondon/Shutterstock **766** Pearson Education **768** Tomas Skopal/Shutterstock **771** Fotolia

Chapter 27
782 dzimin/Fotolia **789** Sebastian Duda/Fotolia **806** dzimin/Fotolia

Chapter 28
817 Pavel L Photo and Video/Shutterstock **823** Jasmin Merdan/Fotolia **829** Amlani/Shutterstock **835** Pavel L Photo and Video/Shutterstock

Chapter 29
859 Bertl123/Shutterstock **861** Digital Vision/Getty Images **868** mahout/Fotolia **871** Andrew Fox/Alamy **874** Oksana Kuzmina/Shutterstock **885B** Bertl123/Shutterstock **885T** Bertl123/Shutterstock

Appendix C: Index

Note: Page numbers in **boldface** indicate chapter-level topics; page numbers in *italics* indicate definitions; pages number with dagger icon † indicate references to Chapter 29 found on the DVD and at pearsonhighered.com/dvb; FE indicates For Example or Step-by-Step references.

Numbers

5-number summary, *56*
   boxplots and, 56–57
   rescaling data, 117
10% Condition, *415*
   comparing proportions, 588
   confidence interval, 480
   independence and, 414–415
   sampling distribution model, 448
68-95-99.7 Rule, *120*
   Central Limit Theorem and, 456
   working with, 120FE–121FE

A

Actuaries, 389
Addition Rule, 355FE
Additive Enough Condition, 786–787
Additivity, *786*, 786–787
Alpha level, *554*
Alternative hypothesis, *495*
   one-sided, 501
   testing, 501
   two-sided, 501
Amazon Standard Identification Number (ASIN)
ANOVA (Analysis of Variance), **741–787**, *752*
   assumptions and conditions, 759–764
   Balancing Act, 764
   Bonferroni multiple comparisons, 765–767
   boxplots, 762FE–764FE
   comparing means, 765–767
   comparing means of groups, 749–751
   on computers, 773–774
   contrast baths experiment, 748FE, 750FE, 754FE, 764FE
   Does the Plot Thicken? Condition, 760
   Equal Variance Assumption, 760–761, 766–767
   Error Mean Square, 752–753
   F-distribution, 752
   F-statistic, 752
   F-table, 754
   F-test, 752
   hand-washing experiment, 753
   Independence Assumptions, 760
   least significant difference, 766
   Between Mean Square, 752
   Within Mean Square, 752
   methods for multiple comparisons, 765
   multifactor. *See* Multifactor ANOVA
   Nearly Normal Condition, 761
   Normal Population Assumption, 761
   on observational data, 767, 770
   one-way ANOVA model, 755
   potential problems, 770
   Randomization Condition, 760
   residual standard deviation, 759
   Similar Spread Condition, 760
   Treatment Mean Square, 752–753
   TV watching example, 768FE–769FE, 770
   two-factor. *See* Two-factor ANOVA
   variability, 751
ANOVA model, 755–759
   one-way, 755
ANOVA table, *753*
Area codes, 5
Area principle, *18*
   bar charts, 18
   categorical data, 17–18, 18–19
   histograms, 53
   stem-and-leaf displays, 49
Association(s), *152*
   between categorical variables, 25FE–26FE
   correlation properties, 161FE
   direction of, 152, 157, 160
   linear, 152, 158, 161
   looking at, 159FE–160FE
   strength of, 151, 153, 155–157
   vs. correlation, 167
Assumptions, *120*
   comparing counts, 657, 657FE
   and conditions, 120
   confidence interval, 479–483
   Equal Variance Assumption. *See* Equal Variance Assumption
   for groups, 588, 588FE–589FE, 598, 599FE
   Independence Assumption. *See* Independence Assumption
   Linearity Assumption, 220
   for linear regression, 196–197
   for mean, 524–525
   for means, 524–525, 525FE
   multifactor ANOVA, 784–796
   multiple regression, 821–825
   Normality Assumption, 404, 428
   Normal Population Assumption. *See* Normal Population Assumption
   paired data, 632–634, 633FE–634FE
   for proportions, 588, 588FE–589FE
   for regression, 196–197
   for regression diagnosis, †872–873
   regression inference, 691–693, 693FE–694FE
   Student's *t* models, 524–525, 525FE

B

Balance, *764*
Bar charts
   area principle, 18
   categorical data condition and, 19
   categorical variable, 19
   relative frequency, 19
   segmented, 26
   Titanic example, 18
Bayes' Rule, 380–381, 381FE
Bernoulli trials, *412*
Between Mean Square, *752*
Bias, *295*
   nonresponse, 309
   response, 309
   voluntary response, 307
Bimodal distribution, *50*
   interquartile range and, 56
   limitations of, 123
Binomial model, *416*
Binomial probability model, *416*, 418FE
   Bernoulli trials, 417
   on computers, 429–431
   Normal Approximation, 419–420, 421FE
   potential problems, 427–428
   spam example, 418FE
   working with, 418FE–419FE
Blinding, 326–327, *327*, 327FE
Blocking, *322*, *329*, 330FE
   independence, 641
   multifactor ANOVA, 796
   and paired data, 640–642
   randomized design, 329
   sign test, 641–642
Bonferroni method, *766*
Bonferroni multiple comparisons, 765–767
Boxplots, *56*
   5-number summary, 56–57
   for ANOVA, 762FE–764FE
   comparing groups with, 86–87, 87FE–89FE
   outliers, 57
   paper airplane example, 793FE
   partial, 785
   rescaling data, 117
Bozo the clown as outlier example, 169

C

Cartesian plane, 154
Cases, *4*
Categorical data, **16–44**
   area principle, 17–18
   bar charts, 18–19
   conditional distributions, 22–25, 23FE
   contingency tables, 20–22
   displaying on computers, 33–35
   frequency table, 18
   pie charts, 19–20
   potential problems, 29–31
   rules of data analysis, 17
   segmented bar charts, 26
Categorical data condition, *19*

Categorical variables, 5
   distribution of, 18
   exploring relationships between, 20–29
   summarizing and displaying, 17–20
Causation
   and Chi-square tests, 674–675
   and correlation, 163–164, 168
   lurking variables and, 229–230
   regression, 229–230
   and regression, 200FE
Census, 297
Center of distributions, 53, 61FE
   median, 53–54
   standardizing z-scores, 113–114, 113FE–114FE
Central Limit Theorem (CLT), 452, 452–456
   assumptions and conditions, 454–455
   Independence Assumption, 454
   for mean, 456FE
   for means, 519, 519FE–520FE
   Normal model and, 455–456
   Randomization Condition, 454
   Sample Size Condition, 455
   using, 519FE–520FE
Certainty vs. precision, 477–478
Chi-square mechanics, 672FE–673FE
Chi-square models, 658
Chi-square statistic, 658
Chi-square tests
   assumptions and conditions, 664, 670
   calculations, 661–662, 664–665
   causation and, 674–675
   on computers, 677–680
   Counted Data Condition, 664
   Expected Cell Frequency Condition, 665
   goodness-of-fit tests, 659FE–661FE
   for homogeneity, 663–665, 665FE–667FE
   Independence, 669–675, 670FE–672FE
   Independence Assumption, 657
   potential problems, 675
   P-values, 658–659
   residuals, 667–668, 668FE, 673–674
   writing conclusions, 674FE
Cluster sampling, 301, 301FE–302FE
Coefficients, multiple regression, 819–821, 827FE–828FE
Collinear, †883
Collinearity, †882–883, †883
Comparative experiments, 319–320
Completely randomized experiment, 323
Compliment Rule, 354FE
Components, 283, 667
   simulation, 283
Computers
   for ANOVA, 773–774
   Chi-square tests, 677–680
   comparing distributions, 99–101
   confidence intervals for proportions, 486–488
   diagnosing regression, 235, †886–887
   displaying categorical data, 33–35
   displaying quantitative data, 71–73
   on experiment, 325
   hypothesis testing, 569
   hypothesis tests for proportions, 510–512
   inferences for means, 538–540
   linear regression, 203–205
   Normal probability plots, 132–134
   paired t-test, 643–645
   probability models, 429–431
   random variables, 406
   re-expression data, 262–263
   regression, 203–205
   regression analysis, 836–838, †886
   regression inference, 715–717
   on sampling, 313
   scatterplots and correlation, 170–172
   simulations on, 288–290
   statistics packages, 10–12
   two-factor ANOVA, 807–809
   two-sample methods for means, 613–616
   two-sample methods for proportions, 611–613
Condition(s), 120
   10% Condition, 415. See 10% Condition
   Additive Enough Condition, 786–787
   categorical data condition, 19–20
   comparing counts, 657, 657FE
   confidence interval, 479–483
   for correlation, 158, 158FE–159FE
   Counted Data Condition. See Counted Data Condition
   Does the Plot Thicken? Condition. See Does the Plot Thicken? Condition
   Expected Cell Frequency Condition. See Expected Cell Frequency Condition
   for groups, 588, 588FE–589FE, 598, 599FE
   for linear regression, 196–197
   Nearly Normal Condition. See Nearly Normal Condition
   No Outliers Condition, 158
   Outlier Condition. See Outlier Condition
   Paired Data Condition, 632
   for proportions, 588, 588FE–589FE
   Quantitative Data Condition. See Quantitative Data Condition
   Quantitative Variables Condition, 158, 196–197
   Randomization Condition. See Randomization Condition
   for regression, 187, 196–197
   for regression diagnosis, †872–873
   regression inference, 691–693, 693FE–694FE
   Sample Size Condition, 455
   Similar Spread Condition. See Similar Spread Condition
   Straight Enough Condition. See Straight Enough Condition
   Student's t models, 524–525, 525FE
   Success/Failure Condition. See Success/Failure Condition
Conditional distributions
   finding, 23FE
   side-by-side bar charts, 24
   Titanic example, 22–25
Conditional probability, 366–388, 371, 371–372
   Bayes' Rule, 380–381, 381FE
   disjoint, 374, 376FE
   finding, 372FE
   General Addition Rule, 366–367, 367FE, 369FE–370FE
   General Multiplication Rule, 372–373
   independence, 372–375, 373FE–374FE
   potential problems, 382
   reversing conditioning, 378–379, 379FE–381FE
   tree diagrams, 377–378
   Venn diagrams, 368FE
Confidence interval(s), **472–493**, 473–474, 474
   10% Condition, 480
   assumptions and conditions, 479–483
   certainty vs. precision, 477–478
   on computers, 486–488
   critical values, 478
   difference between two means, 596–598, 599FE–600FE
   difference between two proportions, 589
   groups, 589
   hypothesis testing, 559
   and hypothesis testing, 505FE–507FE
   Independence Assumption, 479–480, 556FE
   interpreting, 475–477, 529
   interpreting model, 700FE
   margin of error, 477–478
   for matched pairs, 637–640
   for means, 523–524, 529, 607
   one-proportion z-interval, 474, 480
   paired t-interval, 638, 638FE–639FE
   pooled t-test, 607
   potential problems, 484–485
   predicted values, 706–707, 707FE–708FE
   for proportion, 480FE–481FE
   Randomization Condition, 479–480
   regression inference, 700
   Sample Size Assumption, 480
   for small samples, 558
   standard error, 473
   Success/Failure Condition, 480, 558
Confounding, 330, 330–331, 331FE
   lurking variables, 331–332
   two-factor example, 331
Confounding variable, 332
Constant, random variables, 394FE
Context, 4
Contingency tables, 21, 669
   examining, 27FE–28FE
   Titanic example, 20–22
Continuity correction, 421
   using normal distribution, 422FE
Continuous random variables, 390, 400–403, 421
   exponential model, 426–427
   uniform distribution, 425–426
Control, 321, 322FE
Control group, 326
Control treatment, 326
   blinding, 326–327, 327FE
   placebos, 328
Convenience sampling, 308, 308FE
Correlation, 151, 400
   association vs., 167
   categorizing, 160–161
   causation and, 163–164, 168
   changing scales, 161FE
   on computers, 170–172
   computing by hand, 157–158
   conditions, 158, 158FE–159FE
   height/weight example, 155–157
   linear association and, 158, 161
   No Outliers Condition, 158
   outliers and, 158, 164, 168–169

Quantitative Variables Condition, 158
random variables, 399–400
in scatterplots, 155–158, 159FE
straightening scatterplots, 163–165
Straight Enough Condition, 158, 168
Correlation coefficient, *157*
direction of association and, 160
finding by hand, 157–158
linear association and, 170
outliers and, 168–169
properties, 160–161
Correlation tables, 164
Counted Data Condition
Chi-square tests, 664
for comparing counts, 657
Counts, 18, **655–688**
assumptions and conditions, 657, 657FE
calculations, 657–658
Counted Data Condition, 657
Expected Cell Frequency Condition, 657
expected counts, 656FE
goodness-of-fit tests, 655–663
Independence Assumption, 657
potential problems, 675
residuals for, 667–668, 668FE
Sample Size Assumption, 657
Covariance, *399*
random variables, 399–400
Critical values, *557,* 557–559
confidence interval, 478

D

Data, **1–15**
categorical. *See* Categorical data
characteristics about, 3–5, 5FE, 7FE
collection of, 4–5
context, 4
counting, 5
identifier, 6
on Internet, 8
potential problems, 8
rescaling, 116–117, 117FE
shifting, 115–116
technology, 10–12
Database, relational, 6
Data Desk statistical package, 10
ANOVA, 773
Chi-square tests, 678
comparing distributions, 100
confidence intervals for proportions, 487
diagnosing regression, †886
displaying categorical data, 33
displaying quantitative variables, 72
hypothesis tests for proportions, 510
inferences for means, 539
linear regression, 203
Normal probability plots, 133
paired *t* analyses, 644
probability models, 430
re-expression data, 262
regression analysis, 203, 837
regression diagnosis, 235, †886
regression inference, 715
scatterplots and correlation, 170
simulations on, 288
two-factor ANOVA, 807

two-sample methods for means, 614
two-sample methods for proportions, 611
Data table, *3*
Defendant innocent, 497
Degrees of freedom, *521,* 521–522
Delimiter, 10
de Moivre, Abraham, 120n
de Moivre's Rule. *See* 68-95-99.7 Rule
Dependent variables, 155n
Descartes, René, 154
Deviation, *61. See also* Standard deviation
Diagrams, 323
Direction of association, *152,* 157, 160
Discrete random variables, *390,*
393FE–394FE
Disjoint, 374, 376FE
Distributions, *46,* **84–111**
5-number summary, 56
bimodal, 123
categorical variables, 18
center of. *See* Center of distributions
comparing groups with boxplots, 86–87, 87FE–89FE
comparing groups with histograms, 85–86
comparing groups with stem-and-leaf displays, 86FE
comparing on computers, 99–101
displaying quantitative data, 46
flight cancellation example, 58FE–59FE
multimodal, 50
outliers, 51
outliers in, 89–91, 91FE
potential problems, 97–98
of quantitative variable, 46
re-expressing data, 94–97
shapes of. *See* Shapes of distributions
skewed, 51
spread of. *See* Spread
summarizing, 64FE–65FE
symmetric, 59–63, 94–96
tails, 51
tails of, 51
timeplots of, 91–94, 93FE
uniform, 50
unimodal, 50
Does the Plot Thicken? Condition
for ANOVA, 760
for linear regression, 193
multiple regression, 822, †867
for regression, 197
regression inference, 692
scatterplot, 197
Dotplots, 49, *49*
Double-blind experiment, *327*
Dummy variable. *See* Indicator variable

E

Effect size, *500, 563,* 563–564
in hypothesis testing, 563–564
and paired data, 639–640
paired *t*-interval, 640FE
statistical inference, 564
Empirical probability, *350*
Empirical Rule. *See* 68-95-99.7 Rule
Equal Variance Assumption, *193*
for ANOVA, 760–761
multiple regression, 822

pooled *t*-test, 606
regression inference, 692
Error(s), 562FE
in data collection, 90
hypothesis testing, 562FE–563FE, 566FE–567FE
and power, 563FE
Type I. *See* Type I error
Type II. *See* Type II error
Error Mean Square, 752–753
Event, *349*
Excel spreadsheet program, 10
ANOVA, 773
comparing distributions, 100
confidence intervals for proportions, 487
diagnosing regression, †886
displaying categorical data, 33
displaying quantitative variables, 72
hypothesis tests for proportions, 510
linear regression, 203–204
Normal probability plots, 133
paired *t* analyses, 644
probability models, 430
re-expression data, 262
regression analysis, 837
regression diagnosis, 235
regression inference, 716
scatterplots and correlation, 171
simulations on, 289
two-factor ANOVA, 807
two-sample methods for means, 614
two-sample methods for proportions, 612
Expected Cell Frequency Condition
Chi-square tests, 664–665
for comparing counts, 657
Expected counts, 656FE
Expected values, *390,* 391FE
discrete random variables, 393FE–394FE
standard deviation, 393FE–394FE
Experiment(s), **318–347,** *320*
blinding, 326–327, 327FE
blocks, 322, 329, 330FE
comparative, 319–320
completely randomized, 323
on computers, 325
confounding, 330–331, 331FE
control, 321
designing, 323FE–324FE
diagrams, 323
differences in treatment groups, 325
double-blind, 327
lurking variables, 331–332
placebos, 328
potential problems, 332
randomized, 319–320
replicate, 321–322
response variable, 320, 321FE
and samples, 326
single-blind, 327
two-factor example, 331
Experimental design
block, 322
control, 321
principles, 321–322
randomize, 321
replicate, 321–322
Experimental units, *4, 320*
Explanatory variables, *155*

Exponential model, 426–427
    continuous random variables, 426–427
Exponential smoothing, *93*
Exposed to smoke (ETS), 96
Extrapolation
    predicting changes, 225–226
    reaching beyond data, 222–224, 224FE–225FE

F

Factor, *320*
Far outliers, *57*
FDA (Food and Drug Administration), 321
F-distribution, *752*
Fisher, Sir Ronald, 164, 498, 524, 555, 563, 752, 758
Food and Drug Administration (FDA), 321
Formal probability, 353–356
Frequency table, *18*
    relative, 18
F-statistic, *752*
F-table, 754
F-test, *752*

G

Gap, in histogram, *47*
General Addition Rule, 366–367, 367FE, 369FE–370FE
General Multiplication Rule, 372–373, *373*
Generate random numbers, *281*
Geometric probability model, *413*, 414FE
    Bernoulli trials, 412–413
    independence, 414–415
    spam example, 414FE
    working with, 415FE–416FE
Ginkgo biloba, memory-enhancing effect, 328
Goodness-of-fit tests
    chi-square test, 659FE–661FE
    trouble with, 662–663
Gosset, William S., 520–529
Groups, **585–629**
    10% Condition, 588
    assumptions and conditions, 588, 588FE–589FE, 598, 599FE
    comparing, 598
    comparing means of, 749–751
    comparing with boxplots, 86–87, 87FE–89FE
    comparing with histograms, 85–86
    comparing with stem-and-leaf displays, 86FE
    on computers, 611–613
    confidence interval for means, 596–598
    confidence interval for proportions, 589
    equalizing spread across, 96–97
    Independence, 598
    Independence Assumption, 588, 598
    Mann-Whitney test, 605
    Nearly Normal Condition, 598
    Normal Population, 598
    pooled *t*-test, 606–608
    potential problems, 608–609
    Randomization Condition, 588
    sample size, 588
    shifting residuals, 221–222
    standard deviation of difference, 586–587
    Success/Failure Condition, 588
    testing difference between proportions, 592–594
    Tukey's quick test, 605
    two-proportion *z*-interval, 589, 590FE–591FE
    two proportion *z*-test, 593–594
    two-sample *t*-interval difference between means, 599
    two-sample *t*-test for difference between two means, 600–601, 601FE–603FE
    two sample *z*-test, 593–594, 594FE–595FE
    Wilcoxon rank sum test, 605–606

H

Histograms, *46*
    area principle, 53
    bimodal, 50
    calculator tips for, 58FE
    comparing groups with, 85–86
    describing, 52
    designing, 46–47
    displaying outliers, 91
    displaying quantitative variables, 46–47
    gap in, 47
    multimodal, 50
    outliers, 51
    potential problems, 66–68
    relative frequency, 47
    skewed, 51
    symmetric, 51
    uniform, 50
    unimodal, 50
Homogeneity, 663–665, 665FE–667FE
Hurricanes, 185FE
Hypotheses, *495*
    selection, 548–549
    writing, 498FE, 549FE
Hypothesis testing, **494–517**, 502FE–503FE, **548–584**
    alpha level, 554–557
    alternatives, 501
    on computers, 510–512, 569
    conclusion, 500, 500FE
    confidence intervals and, 558–559, 560FE
    critical value and, 557–559
    effect size, 500, *563*, 563–564
    errors, 562FE–563FE, 566FE–567FE
    and intervals, 505FE–507FE
    mechanics, 499–500
    model, 498–499
    potential problems, 508, 567
    power of, 562, 563FE, 566FE–567FE
    practical *vs.* statistical significance, 555–557
    for proportions, 494–517
    P-values and decisions, 504–505
    P-value, 550–554, 550FE–552FE, 553FE
    reasoning of, 498–500
    sample size, 566FE–567FE
    standard deviation, 495
    standard error, 495
    trial as, 496
    Type I error, *560*, 566
    Type II error, *560*, 566

I

Identifier variables, *6*
Independence, *373*, 376FE
    10% Condition, 414–415
    checking for, 373FE–374FE
    Chi-square tests, 669–675, 670FE–672FE
    vs. disjoint, 374
    geometric probability model, 414–415
    for groups, 598
    variables, 405
Independence Assumption, *447*
    for ANOVA, 760
    Central Limit Theorem, 454
    Chi-square tests, 664
    for comparing counts, 657
    comparing proportions, 588
    confidence intervals, 479–480, 556FE
    for means, 524–525
    multifactor ANOVA, 787–788
    multiple regression, 821–822
    paired data, 633
    regression inference, 692
    sampling distribution model, 447–448
Independent events, *350*
Independent variables, *25*
Indicator variables, †*862*, †863FE
Infant mortality, *828*
Inferences
    multifactor ANOVA, 798
    multiple regression, 825–829, 829FE–832FE
    potential problems, 833–834
Influential points, *227*, †868–872, †868FE
    Stepwise Regression and, †874
    Collinearity and, †884
Interaction, *787*
    multifactor ANOVA, 787, 796–800, 801FE–803FE, 804
    two-factor ANOVA, 801FE–803FE
    unreplicated two-factor design, 799
Interaction plot, *787*
Interaction term, †864
Intercept, *184*
    regression inference, 699
Internet, 8
Interquartile range (IQR), *55*
    bimodal distribution and, 56
    boxplots, 57
    rescaling data, 117
    spread of distribution, 54–56

J

JMP statistical package, 11
    ANOVA, 773
    Chi-square tests, 678
    comparing distributions, 100
    confidence intervals for proportions, 487
    diagnosing regression, 235, †887
    displaying categorical data, 34
    displaying quantitative variables, 72
    hypothesis tests for proportions, 510
    inferences for means, 539
    linear regression, 204
    Normal probability plots, 133
    paired *t* analyses, 645
    probability models, 430
    re-expression data, 262
    regression analysis, 837
    regression diagnosis, 235
    regression inference, 716
    scatterplots and correlation, 171
    simulations on, 289
    two-factor ANOVA, 808
    two-sample methods for means, 615
    two-sample methods for proportions, 612

K

Katrina's residual, 192FE–193FE
Keno, casino game, 351

L

Ladder of powers, 165–166, *166*, 167, 253
Law of Averages, 350FE–351FE
Law of Large Numbers (LLN), *350*
Least significant difference, *766*
Least squares, 818
Least squares line, *184*
    finding, 185–187
Levels, *320*
Leverage, 226–228, *227*
    multiple regression, †865–866
Linear association
    correlation and, 158, 161
    correlation coefficient and, 168
    potential problems, 168
    in scatterplots, 152
Linearity Assumption, 220
    multiple regression, 821
    regression inference, 691
Linear model, *183*
    hurricane example, 185FE
    intercept, 184–185
    residual, 183–184
    slope, 184–185
Linear regression, **182–218**. *See also* Multiple Regression
    assumptions and conditions, 196–197
    Burger King example, 182
    causation and, 200FE
    checking reasonableness, 199–200
    on computers, 203–205
    Does the Plot Thicken? Condition, 197
    hurricane example, 185FE
    intercept, 184–185
    least squares line, 185–187
    line of best fit, 184
    Outlier Condition, 197
    potential problems, 200
    Quantitative Variables Condition, 196–197
    R^2, 194–196, 195FE
    residual, 183–184
    slope, 184–185
    Straight Enough Condition, 196
Line of best fit, *184*
LLN (Law of Large Numbers), 350
Logarithms, 96
    re-expression data, 258
Logistic regression, 708–711, *710*
Lower quartiles, *55*
Lurking variables, *163*, *229*, 229–230
    and confounding, 331–332

M

Magnitudes, *46*
Mann-Whitney test, *605*
Marginal distribution, *21*, 22FE
Margin of error, *477*
    Bonferroni multiple comparisons, 766
    confidence interval, 477–478
    finding, 478FE–479FE
    polls and, 476FE
Matched pairs, 637–640
Matching participants, 330

Mean(s), *59*, **518–547**. *See also* Center of distributions
    for ANOVA, 765–767
    assumptions and conditions, 524–525, 525FE
    Central Limit Theorem, 519, 519FE–520FE
    for Central Limit Theorem, 456FE
    on computers, 538–540
    confidence interval, 523–524, 529
    Gosset and, 520–529
    hypothesis testing, 529–533
    Independence Assumption, 524–525
    Nearly Normal Condition, 525
    Normal Population Assumption, 525
    one-sample *t*-interval, 523, 524FE–527FE
    one-sample *t*-test, 529–530, 530FE–532FE
    outliers and, 59–60
    potential problems, 535–536
    for proportions, 533
    Randomization Condition, 524
    regression to, 189–190
    for sample size, 534–535
    for sampling distribution models, 455–456, 457FE–458FE
    and scaling data, 117
    standard error difference, 597FE
    Student's *t* models, 528
    symmetric distributions and, 59–60
Measurement errors, 259
Median, *53*. *See also* Center of distributions
    5-number summary, 54
    finding, 54
    resistant, 60
Mediocrity, 190
Memory-enhancing effect, ginkgo biloba, 328
Methods for multiple comparisons, *765*
Minimum significant difference, *766*
Minitab statistical package, 11
    ANOVA, 774
    Chi-square tests, 678
    comparing distributions, 100
    confidence intervals for proportions, 487
    diagnosing regression, †887
    displaying categorical data, 34
    displaying quantitative variables, 72
    hypothesis tests for proportions, 511
    inferences for means, 539
    linear regression, 204
    Normal probability plots, 133
    paired *t* analyses, 645
    probability models, 430
    re-expression data, 262
    regression analysis, 837
    regression diagnosis, 235
    regression inference, 716
    scatterplots and correlation, 171
    simulations on, 289
    two-factor ANOVA, 808
    two-sample methods for means, 615
    two-sample methods for proportions, 612
Model(s), *183*. *See also* Binomial probability model; Normal models; Probability models; Sampling distribution models; Student's *t*-models
    checking conditions, 499FE
    looking beyond data, 94
    one-proportion *z*-test, 498–499
    parameters in, 119
    for patterns, 84
    usefulness of, 119n

Modeling probability
    personal probability, 352–353
    subjective probability, 352–353
    theoretical probability, 351
    working with, 353
Modes, *50*
Moving Average, *93*
Multicollinearity, †883
Multifactor ANOVA, **782–816**
    Additive Enough Condition, 786–787
    additivity, 786–787
    assumptions and conditions, 784–796
    blocks, 796
    Independence Assumption, 787–788
    inferences, 798
    interaction, 787
    interaction plot, 787
    interactions, 787, 796–800, 801FE–803FE, 804
    Nearly Normal Condition, 789
    Normal Error Assumption, 789
    paper airplane example, 783FE, 793FE, 798FE, 804FE
    partial boxplot, 785
    plotting data, 785–786
    potential problems, 804–805
    Randomization Condition, 787
    Similar Variance Condition, 788
    two-factor model, 783–784
    unreplicated two-factor design, 799
Multimodal distribution, *50*
Multiple regression, *817*, **817–858**, †859–883
    adjusted R^2, 832–833
    assumptions and conditions, 821–825
    Body Fat example, 823FE–825FE
    comparing models, 832–833
    on computers, 836–838, †886–887
    Does the Plot Thicken? Condition, 822
    Equal Variance Assumption, 822
    Independence Assumption, 821–822
    infant mortality, 828–829
    inferences, 825–829, 829FE–832FE
    interpreting coefficients, 819–821, 827FE–828FE, †872
    least squares, 818
    Linearity Assumption, 821
    Nearly Normal Condition, 822, 829
    partial regression plot, 821
    potential problems, 834, †884
    Randomization Condition, 822
    real estate example, 819FE
    scatterplot matrix, 828–829
    Straight Enough Condition, 821
Multiplication Rule, 356FE
Multistage sampling, *302*, 302FE–303FE

N

National Health and Nutrition Examination Survey (NHANES), 115
Nearly Normal Condition, *120*, *121*, 130, 556FE–557FE
    for ANOVA, 761
    comparing groups, 598
    for means, 525
    multiple regression, 822, †867
    regression inference, 692

Negative pattern, *152*
Negative residual, *183*
Nest Egg Index, 86FE
Newton, Isaac, 154
Nominal variables, *5*
Nonexistent Law of Averages, 350
Nonresponse bias, *309*
No Outliers, *187*
No Outliers Condition
   for correlation, 158
Normal Approximation, 419–420, 421FE
Normal distribution, 422FE
Normal Error Assumption
   multifactor ANOVA, 789
Normality Assumption, *119*, 404, 428
Normal model(s), **112–150**, *119*
   68-95-99.7 Rule, 120, 120FE–121FE
   binomial probability, 419–420, 421FE
   and Central Limit Theorem, 455–456
   finding percentiles, 123–125, 124FE–128FE
   Nearly Normal Condition, 120, 121, 130
   Normal probability plots, 129–130
   potential problems, 130–131
   rules for, 121–123
   and sampling distribution model, 445–446
   sketching Normal curves, 121
   spam example, 421FE
   standard, 119
   Success/Failure Condition, 420
   z-scores and, 125
Normal percentiles, *123,* 123–125, 124FE–128FE
Normal Population Assumption
   for ANOVA, 761
   for Student's t models, 525
   paired data, 633
   regression inference, 692
Normal probability plots, *129*
   on computers, 132–134
   how constructed, 129–130
Normal probability tables, 123–125, 124FE–128FE
Normal scores, *130*
Null hypothesis, *495,* 552

O

Observational studies, *318,* 767, 770
   designing, 319FE
   matching participants, 330
   uses, 319
One-proportion z-interval, *474*
One-proportion z-test, *498,* 550FE–552FE, *551*
   hypothesis testing, 499
One-sample t-interval for mean, 523, 524FE–527FE
One-sample t-test for mean, *529,* 529–530, 530FE–532FE
One-sided alternative, *501*
One-way ANOVA F-test, 761
One-way ANOVA model, 755
Ordinal variables, *6*
Outcome, *349*
Outlier Condition
   regression, 197
   regression inference, 692
   scatterplot, 197
Outliers, *51*
   Bozo the clown as, 169
   checking, 91FE

correlation and, 158, 164, 168–169
data analysis of, 89–91
distributions, 89–91
far, 57
No Outliers Condition, 158
potential problems, 97, 130–131
regression, 226–228
reporting, 91
rule of thumb for identifying, 89
in scatterplots, 153

P

Paired data, **630–654,** 631–632
   assumptions and conditions, 632–634, 633FE–634FE
   and blocking, 640–642
   confidence intervals for matched pairs, 637–640
   effect size, 639–640
   identifying, 631FE
   independence, 641
   Independence Assumption, 633
   Normal Population Assumption, 633
   Paired Data Condition, 632
   paired t-interval, 638, 638FE–640FE
   paired t-test, 634, 635FE–637FE
   potential problems, 642
   sign test, 641–642
Paired Data Condition, 632
Paired t-interval, 638, 638FE–639FE
   effect size and, 640FE
Paired t-test, *632*
   on computer, 643–645
   doing, 636FE–637FE
   hypothesis testing, 634
   speed-skating lanes, 635FE–636FE
Parameters, *119*
Partcipants, *320*
Partial boxplot, 785
Partial regression plot, *821*
Participants, *4*
   matching, 330
Percentages, 18, *47*
Percentiles, *55,* 123–125, 124FE–128FE
Personal probability, *352,* 352–353
Pie charts, *19*
   categorical data, 19–20
Pilot study, *307*
Placebo, *328*
Placebo effect, *176,* 328
Poisson model, 423–424
Pooled t-test, *606*
   confidence interval for means, 607
   Equal Variance Assumption, 606
   pooling, 608
   Similar Spreads Condition, 607
Population, *4,* 295
Population parameter, 298
Positive pattern, *152*
Positive residual, *183*
Power
   errors and, 563FE
   in hypothesis testing, 562
Practical significance, 555–557
Predicted values, *183*
   confidence intervals, 706–707, 707FE–708FE
   standard errors, 705–706

Prediction interval, *706*
Prediction interval for individual, *707*
Predictor variable, *155*
Probability(ies), **348–365,** *350,* 357FE–359FE, **366–388**
   Addition Rule, 355FE
   Bayes' Rule, 380–381, 381FE
   Compliment Rule, 354FE
   conditional, 371–372, 372FE
   continuous, 353
   empirical, 350
   formal, 353–356
   General Addition Rule, 366–367, 367FE, 369FE–370FE
   General Multiplication Rule, 372–373
   independence, 372–375, 373FE–374FE
   kind of, 353
   marginal, 375
   modeling, 351–353
   Multiplication Rule, 356FE
   personal, 352–353
   potential problems, 359
   subjective, 352–353
   theoretical, 351
   tree diagram, 377–378
   Venn diagrams, 368FE
   working with, 353
Probability models, *390,* **412–442**
   Bernoulli trials, 412–413
   binomial, 416–417, 418FE–419FE
   on computers, 429–431
   exponential model, 426–427
   geometric, 413–415, 415FE–416FE
   Normal Approximation, 419–420, 421FE
   Poisson model, 423–424
   potential problems, 427–428
   uniform distribution, 425–426
Proportion(s), 18
   10% Condition, 588
   assumptions and conditions, 588, 588FE–589FE
   confidence interval, 480FE–481FE
   fo means, 533
   for hypothesis testing, 494–517
   Independence Assumption, 588
   one-proportion z-test, 498–499
   Randomization Condition, 588
   sample size, 588
   special case for, 533
   standard deviation difference, 587
   standard error difference, 587FE–588FE
   Success/Failure Condition, 588
   two-proportion z-interval, 589, 590FE–591FE
   two proportion z-test, 593–594, 594FE–595FE
Prospective observational study, *319*
Pseudorandom numbers, 281
P-value, *496*
   Chi-square test, 658–659
   finding, 499FE–500FE
   hypothesis testing, 496–497
   in hypothesis testing, 550–554, 550FE–552FE, 553FE
   innocent defendant, 497

Q

Qualitative variables, *5*
see Categorical variables
Quantitative Data Condition, 691
Quantitative variables, *5*, 5–6, **45–83**, *187*
   5-number summary, 56
   boxplots, 56–57
   center of distributions, 53–56
   data analysis considerations, 49–50
   displaying, 45–50
   displaying on computers, 71–73
   dotplots, 49
   histograms, 46–47
   linear association between, 158
   potential problems, 66–68
   Quantitative Data Condition, 50
   scatterplots for, 152
   shapes of distributions, 50–51
   stem-and-leaf displays, 47–48
   summarizing, 63–65
   symmetric distributions, 59–63
   variation in, 62–63
Quantitative Variables Condition, 196–197
Quartiles, *55*
   5-number summary, 56
   boxplots, 56
   finding, 55–56
   lower, 55
   upper, 55

R

R^2
   adjusted, 832–833
   interpreting, 195FE
   linear regression, 194–196, 195FE
Random, *280, 348*
Random assignment, *319*
Randomization Condition, *448*
   for ANOVA, 760
   Central Limit Theorem, 454
   comparing proportions, 588
   confidence interval, 479–480
   for means, 524
   multifactor ANOVA, 787
   multiple regression, 822
Randomized block design, *329*
Randomized comparative experiments, 320
Randomizing, *296*, 321, 322FE
Randomness, **280–293**
   building simulation, 283–284
   card shuffling, 282
   generate random numbers, 281
   meaning of, 280
   potential problems, 287
   pseudorandom numbers, 281
   simulating by hand, 282–284
   simulating dice games, 284FE–285FE
   simulation example, 285FE–287FE
   simulations on computer, 288–290
Random numbers, 303FE
Random phenomenon, *348*, 348–351
Random variables, *389*, **389–411**
   adding constant, 394FE
   combining means and variances, 398FE–399FE
   on computers, 406

   continuous, 390, 400–403, 421
   correlation, 399–400
   covariance, 399–400
   discrete, 390
   packaging stereos example, 402FE–404FE
   potential problems, 404–405
   probability model, 390
   restaurant discount example, 396FE
   shifting and combining, 394–397
   summing a series of outcomes, 397FE–398FE
   working with difference, 397FE
Range, *54*
Re-expressing data, *95, 165,* **247–279,** 691, 760, 788, 823
   comparing, 257FE
   curved pattern, 259
   equalizing spread across groups, 96–97
   finding, 252–259
   goals of, 249–252
   to improve symmetry, 94–96
   ladder of powers, 166, 169, 253
   logarithms, 258
   measurement errors, 259
   multiple benefits, 258–259
   potential problems, 259–260
   recognizing, 251FE
   straightening scatterplot, 248–252, 254FE–256FE
   symmetric distributions, 252
Regression, 197FE–199FE, **219–246**
   assumptions and conditions, 196–197
   causation and, 200FE, 229–230
   on computers, 203–205, 235–236, 715–717, 837–838, †886–887
   conditions for, 187
   Does the Plot Thicken? Condition, 197
   examining residuals, 219–222
   extrapolation, 222–226, 224FE–225FE
   groups in, 221–222
   influential points, 227, †868–872
   leverage, 226–228, †865–866
   Linearity Assumption, 220
   linear model. See Linear regression
   logistic, 708–711, *710*
   lurking variable, 229–230
   to mediocrity, 190
   multiple. See Multiple regression
   No Outliers, 187
   Outlier Condition, 197
   outliers, 226–228
   potential problems, 200, 232–233
   predicting, 196
   quantitative variables, 187
   Quantitative Variables Condition, 196–197
   in reality, 199–200
   re-expressing to straighten, 221
   Straight Enough Condition, 187, 196, 221, 229
   subsets, 222
   summary variables, 230–231
   to the mean, 189–190
   working with multiple methods, 230FE–231FE
Regression diagnosis, †**859–894**
   adjusting for slopes, †863–864
   assumptions and conditions, †872–873
   building models, †872–883, †874FE–880FE
   collinearity, †882–883
   on computers, †886–887

   diagnosing models, †865–872, †866FE–871FE
   Does the Plot Thicken? Condition, †867
   indicators, †861–864
   indicator variables, †862, †863FE
   influential cases, †868
   interaction term, †864
   leverage, †865–866
   Nearly Normal Condition, †867
   potential problems, †884
   residuals, †867
   roles, †881
   seeking automatic models, †873–874
   standardized residuals, †867
   stepwise, †873
   studentized residuals, †867
   three/more level indicators, †881–882
   variables, †860
   wrap-up diagnosis, †871–872
Regression equation
   calculating, 187FE–189FE
   finding, 186FE
Regression inference, **689–746**
   assumptions and conditions, 691–693, 693FE–694FE
   body fat and waist size example, 690–691, 695FE–696FE
   on computers, 715–717
   conditions/residuals, 694
   confidence interval, 700
   Does the Plot Thicken? Condition, 692
   Equal Variance Assumption, 692
   Independence Assumption, 692
   intercept, 699
   intuition, 696–699
   Linearity Assumption, 691
   Nearly Normal Condition, 692
   Normal Population Assumption, 692
   Outlier Condition, 692
   population and sample, 689–691
   potential problems, 711–712
   Quantitative Data Condition, 691
   residual standard deviation, 697
   sample size, 698
   sampling distribution, 698
   slope t-test, 702FE–704FE
   spread around the line, 697
   standard error finding, 699FE
   standard error for slope, 698
   Straight Enough Condition, 691
Regression line, *190*
Regression to the mean, 189–*190*
Relational database, 6
Relative frequency bar charts, *19*
Relative frequency histograms, *47*
Relative frequency table, *18*
Replicate, *321*, 322FE
Replication, *322*
Representative, 299
Rescaling data, *116*, 116–117, 117FE
Residual(s), *183*
   bends in, 220–221
   Chi-square, 667–668, 668FE, 673–674
   examining, 192–194, 219–222, 692, 821–822, †867
   groups in, 221–222
   influential points, 226–228, †868, †868FE–871FE

Residual(s) (*Continued*)
   Katrina's, 192FE–193FE
   linear model, 183–184
   multiple regression, †867
   negative, 183
   positive, 183
   standard deviation, 193–194
   standardized, 667, †867
   Straight Enough Condition, 197
   subsets, 222
Residual standard deviation, 697, *759*
Resistant, median as, *60*
Respondents, *4*
Response bias, *309*
Response variable, *283, 320,* 321FE
Retrospective observational study, *318*
Reversing conditional probability, 378–379, 379FE–381FE
Richter scale, 46, 46n
R statistical package, 11
   ANOVA, 774
   Chi-square tests, 679
   comparing distributions, 101
   confidence intervals for proportions, 487
   diagnosing regression, †887
   displaying categorical data, 34
   displaying quantitative variables, 73
   hypothesis tests for proportions, 511
   inferences for means, 539
   linear regression, 204
   Normal probability plots, 133
   paired *t* analyses, 645
   probability models, 430
   re-expression data, 262
   regression analysis, 838
   regression diagnosis, 235, 887
   regression inference, 717
   scatterplots and correlation, 171
   simulations on, 289
   two-factor ANOVA, 808
   two-sample methods for means, 615
   two-sample methods for proportions, 612

S

Sample(s), *4,* **294–317**, *295,* 303FE–305FE
   bias, 295, 307, 309
   cluster, 301, 301FE–302FE
   common mistakes, 307–310
   on computer, 313
   convenience, 308, 308FE
   determining populations, 298–299
   examining part of the whole, 295–296
   and experiment, 326
   multistage, 302, 302FE–303FE
   potential problems, 310
   randomizing, 296
   representative, 296FE, 299
   sample size, 297
   Simple Random Sample, 299
   stratified, 310–302
   stratified random, 301
   stratified *vs.* cluster, 302
   systematic, 303
   two sample *z*-test, 592–594
   voluntary response, 307, 308FE
Sample size, *297,* 483FE
   comparing proportions, 588
   finding, 561
   for estimating a mean, 534FE–535FE
   regression inference, 698, 825–827
   selection, 481–483, 482FE
   selection for mean, 534
Sample Size Assumption
   for comparing counts, 657
   confidence interval, 480
Sample Size Condition, *455*
Sample space, *349*
Sample statistic, *298*
Sample surveys, **294–317**, *295*
   census considerations, 297–298
   cluster sampling, 301, 301FE–302FE
   determining populations, 298–299
   population parameter, 298
   potential problems, 310
   randomizing, 296
   sample size, 297
   Simple Random Sample, 299
   stratified random sampling, 301
   systematic sampling, 303
   valid, 306–307
Sampling distribution, *444*
Sampling distribution models, **443–471**
   10% Condition, 448
   Central Limit Theorem, 452–456
   difference between two means, 598
   difference between two proportions, 589
   Independence Assumption, 447–448
   for mean, 451–452, 455–456, 457FE–458FE
   Normal model and, 445–446
   other statistics, 451–452
   potential problems, 461
   for proprotion, *443,* 443–444, 444FE–445FE, 448, 448FE–450FE
   Randomization Condition, 448
   for regression slopes, 698
   simulations, 451–452
   Success/Failure Condition, 448
   summarized, 458–460
Sampling error, *446*
Sampling frame, *300, 309*
Sampling variability, *300, 446*
Scatterplot matrix, *828*
Scatterplots, **151–179**, *152*
   association in, 152–153
   axes for, 155
   comparing prices example, 153FE–154FE
   on computers, 170–172
   curved pattern, 259
   direction of, 152
   Does the Plot Thicken? Condition, 197
   emperor penguins example, 220–221
   form of, 152–153
   history, 154
   hurricane example, 152–153
   Outlier Condition, 197
   outliers in, 153
   potential problems, 167–169
   for quantitative variables, 152
   roles for variables in, 154–155
   standardizing, 156
   straightening, 165–167, 248–252, 254FE–256FE
   Straight Enough Condition, 197
   strength of association in, 151, 153, 155–157
   of summary statistics, 230–231
Segmented bar charts, *26*
Sensitivity, 566
Shapes of distributions, *50*
   flight cancellation example, 58FE–59FE
   gaps in, 47
   modes of histograms, 50
   outliers, 51
   symmetric histograms, 51
Shifting data, 115–116, *116*
Significance level, *555*
Similar Spread Condition
   for ANOVA, 760
   pooled *t*-test, 607
Similar Variance Condition, 788
Simple Random Sample, *299*
Simpson's paradox, 31
Simulation(s), **280–293**, *282*
   building, 283–284
   component, 283
   on computer, 288–290
   of dice games, 284FE–285FE
   example, 285FE–287FE
   meaning, 282
   potential problems, 287
   response variable, 283
   sampling distribution models, 451–452
   trials and, 283
Single-blind experiment, *327*
Skewed distribution, *51*
Skewness, 61
Slope, *184*
Smoothing methods
   exponential, 93
   moving average, 93
   for timeplots, 92–93, 93FE
Specificity, 566
Spread, *54*
   comparing, 86
   describing, 63FE
   equalizing across groups, 96–97
   interquartile range, 54–56
   range, 54
   scaling data and, 117–118
   standard deviation, 61–63
   standardizing z-scores, 113–114, 113FE–114FE
SPSS statistical package, 11
   ANOVA, 774
   Chi-square tests, 679
   comparing distributions, 101
   confidence intervals for proportions, 488
   diagnosing regression, †887
   displaying categorical data, 34
   displaying quantitative variables, 73
   hypothesis tests for proportions, 511
   inferences for means, 539
   linear regression, 204
   Normal probability plots, 133
   paired *t* analyses, 645
   probability models, 431
   re-expression data, 263
   regression analysis, 838
   regression diagnosis, 236, †887
   regression inference, 717

scatterplots and correlation, 172
simulations on, 289
two-factor ANOVA, 808
two-sample methods for means, 615
two-sample methods for proportions, 612
SRS (Simple Random Sample), *298, 299*
Standard deviation(s), *62,* **112–150,** *392*
 difference between two proportions, 587
 of a random variable, 392–393, 393FE–394FE
 of difference in groups, 586–587
 discrete random variables, 393FE–394FE
 expected values, 393FE–394FE
 hypothesis testing, 495
 Normal models, rules of, 121–123
 potential problems, 130–131
 rescaling data, 116–117, 117FE
 residual, 193–194, 759
 as ruler, 112–113
 spread, 61–63
 standardized variables, 118FE
 z-scores, 112–113, 113FE–114FE
Standard error(s), *473*
 difference in means, 597FE
 difference in proportions, 587FE–588FE
 finding, 699FE
 predicted values, 705–706
 for slope, 698
Standardized residuals, *667,* †867
Standardized values, *113*
Standardized variables, 118FE
Standardizing
 standard deviation, 112–113, 113FE–114FE
 z-scores, 113–114, 113FE–114FE
Standard Normal distribution, *119*
Standard Normal model, *119*
StatCrunch statistical package, 11
 ANOVA, 774
 Chi-square tests, 679
 comparing distributions, 101
 confidence intervals for proportions, 488
 diagnosing regression, †887
 displaying categorical data, 35
 displaying quantitative variables, 73
 hypothesis tests for proportions, 511
 inferences for means, 540
 linear regression, 205
 Normal probability plots, 134
 paired t analyses, 645
 re-expression data, 263
 regression analysis, 838
 regression diagnosis, 236, †887
 regression inference, 717
 scatterplots and correlation, 172
 simulations on, 289
 two-factor ANOVA, 808
 two-sample methods for means, 615–616
 two-sample methods for proportions, 613
Statistically significant, *325,* 504, *554*–556, 764FE
Statistics, 3, *119*
Stem-and-leaf displays, *48*
 comparing groups, 86FE
 displaying quantitative data, 47–49
Stemplot. *See* Stem-and-leaf display
Stepwise regression, †873

Straight Enough Condition, *158,* 249, 257
 for correlation, 168
 multiple regression, 821
 regression, 187, 196
 for regression, 221, 229
 regression inference, 691
 residual, 197
 scatterplot, 197
Straight lines, 259
Strata, *300*
Stratified random sampling, *301*
Strength of association, 151, *153,* 155–157
Studentized residuals, †867
Student's t models, *521*
 assumptions and conditions, 524–525, 525FE
 Gosset and, 520–529
 Independence Assumption, 524–525
 making pictures, 528–529
 for means, 528
 Nearly Normal Condition, 525
 Normal Population Assumption, 525
 Randomization Condition, 524
Subjective probability, *352,* 352–353
Subjects, *4, 320*
Success/Failure Condition, *420*
 comparing proportions, 588
 confidence interval, 480
 confidence intervals, 558
 hypothesis testing, 495
 sampling distribution model, 448
Surveys. *See* Sample surveys
Symmetric distributions, *51*
 re-expressing data, 94–96, 252
 summarizing, 59–63
Symmetric histogram, *51*
Systematic sampling, *303*

T

Table
 ANOVA, 753
 contingency, 669
 F-, 754
 frequency, 18
 for means, 528
 relative frequency, 18
 two-way, 663
Tails, of distribution, *51*
Theoretical probability, 351
TI-83/84 Plus statistical package, 12
 Chi-square tests, 680
 comparing distributions, 101
 confidence intervals for proportions, 488
 diagnosing regression, †887
 displaying categorical data, 35
 displaying quantitative variables, 73
 hypothesis tests for proportions, 512
 inferences for means, 540
 linear regression, 205
 Normal probability plots, 134
 paired t analyses, 645
 probability models, 431
 random variables, 406
 re-expression data, 263
 regression analysis, 838
 regression diagnosis, 236
 regression inference, 717

scatterplots and correlation, 172
 simulations on, 290
 two-factor ANOVA, 809
 two-sample methods for means, 616
 two-sample methods for proportions, 613
Timeplots, 91–94, *92, 93*FE
t-interval
 one-sample for mean, 523, 524FE–527FE
 paired. *See* Paired t-interval
 two-sample, 597
Treatment, *320*
 determining, 321FE
 differences in, 325
Treatment Mean Square, 752–753
Tree diagram, *377,* 377–378
Trials, *283,* 349, 496. *See also* Bernoulli trials
t-Table, 528
t-test
 one-sample for mean, *529,* 529–530, 530FE–532FE
 paired. *See* Paired t-test
 pooled, 606–608
 two-sample, 597
Tukey, John W., 48, 57, 252, 474
Tukey's quick test, 605
Two-factor ANOVA, 789FE–793FE
 on computers, 807–809
 interactions, 801FE–803FE
Two-proportion z-interval, 589
 confidence interval, 590FE–591FE
 finding, 590FE
Two proportion z-test, 593–594
 pre-sleep surfing rates example, 594FE–595FE
 safety and privacy example, 596FE
Two-sample t-interval, 597
 difference between means, 599
Two-sample t-test, *597,* 604FE
 difference between two means, 600–601, 601FE–603FE
Two-sided alternative, *501*
two-way table, *663*
Type I error, *560,* 566
Type II error, *560,* 566

U

Undercoverage, *309*
Uniform distribution, *50,* 425–426
Unimodal distribution, *50. See also* Nearly Normal Condition
Units, *5*
 correlation and, 160
 experimental, 4
Unreplicated two-factor design, *799*
Upper quartiles, *55*

V

Vague concepts, 160–161
Valid surveys, 306–307
Variables, *5*
 association between, 25–26FE
 categorical, 5
 confounding, 332
 dependent, 155n
 explanatory, 155
 identifier, 6
 independent, 25, 155n, 405
 indicator, †862

Variables (*Continued*)
   inferences, 798
   lurking, 163, 229–230
   nominal, 5
   ordinal, 6
   predictor, 155
   qualitative, 5
   quantitative, 5–6, 45–50, 187
   random. *See* Random variables
   response, 155, 283
   in scatterplots, 154–155
   standardized, 118FE
Variance, *61*
   of random variables, *392*
Variation, 62–63
Venn diagrams, 368FE

Voluntary response bias, *307*
Voluntary response sample, *307,* 308FE

W
Wilcoxon rank sum test, *605,* 605–606
Within Mean Square, *752*
Women's Health Initiative, 320
Writing hypotheses, 498FE

X
x-axis, 155
x-variable, *155*

Y
y-axis, 155
y-variable, *155*

Z
z-interval, two-proportion, 589, 590FE–591FE
ZIP codes, 3
z-scores, *113, 119*
   combining, 114
   Normal percentiles and, 125
   standardizing, 113–114, 113FE–114FE
z-test
   one-proportion, 498–499
   two proportion, 593–594, 594FE–595FE

Index of Applications

BE = Boxed Example; E = Exercises; IE = In-Text Example; JC = Just Checking

Accounting
Audits and Taxes (E), 317, 362, 335; (IE) 307
Budgets (E), 571
Casualty Actuarial Society (BE), 389
Company Assets, Profit, and Revenue (BE), 160; (E), 110–111, 150, 251, 726; (IE), 160
Society of Actuaries (BE), 389

Advertising
Ads (IE), 494
Branding (E), 516, 574-575, 626
Free Products (E), 292
Product Claims (BE), 323–324; (E), 682; (JC), 63
Sexual Images in Advertising (E), 647, 652
Social Networking Sites (IE), 2
Target Audience (E), 582
Text Messaging (JC), 475
Truth in Advertising (E), 652–653

Agriculture
American Angus Association (BE), 519
Beef and Livestock (BE), 519, 541; (E), 135, 137, 139, 497, 654
Crop Damage (E), 341, 489; (JC), 668–669
Farmers' Markets (E), 392
Fertilizers (E), 779, 856
Seeds (E), 390, 466, 516
Vineyards (E), 12, 76, 107, 244, 346

Banking
Bank Tellers (E), 776
Credit Card Charges (BE), 52, 61, 63–64; (E), 542–543
Credit Card Companies (BE), 63–64, 331; (E), 463–464
Credit Card Customers (BE), 52, 61, 63–64; (E), 237, 570, 680–681
Credit Card Debt (E), 492, 570; (JC) 505, 560, 564
Credit Card Fraud (E), 143
Credit Card Offers (BE), 483; (E), 515
Interest Rates (E), 178, 206, 242–244; (IE), 94
Loan Approval (E), 466, 572–573
Mutual funds (E), 849

Business (General)
Awards and Recognition (E), 10, 201
Chief Executives (BE), 659–662; (E), 149–150, 467–468, 542; (IE), 94–95, 130, 455, 461, 655–657
Company Databases (E), 81, 139
Contract Bids (E), 388–389
Employee Benefit Research Institute (E), 741
Forbes 500 Companies (E), 150
Fortune 500 Companies (BE), 756; (IE), 655, 657
Occupy Wall Street (E), 583
Retirement Planning (E), 741
Small Business (IE), 228
Women-Led Businesses (E), 396, 516, 582

Company Names
A.G. Edwards (BE), 86
Allstate Insurance Company (E), 469
Amazon.com (IE), 2–6
American Life Project (E), 618
Arby's (E), 11
AT&T (BE), 5
Buick (E), 176
Burger King (BE), 186; (E), 209, 846; (IE), 182–187, 189–191, 200
Clarksburg Bakery (E), 583
Cleveland Casting Plant (E), 11
Coca-Cola (E), 570–571; (IE), 494
CUPPS (BE), 88
Facebook (E), 383; (IE), 2, 472–479, 484–485
First USA (E), 515
Ford Motor Company (E), 11, 176
GfK Roper (E), 314, 384, 621
GlaxoSmithKline (BE), 563, 566
Guinness Brewery (IE), 520–521
Holes-R-Us (E), 81, 777
Honda (E), 176
Hostess Company (IE), 297
Husqvarna (BE), 232
ISA Babcock Company (E), 579
Lowe's (IE), 228
Monitoring the Future (E), 39
Nabisco Company (E), 546
Nambe Mills (E), 739
The National Strategy for Trusted Identities in Cyberspace (BE), 6
Nissan (BE), 64–65, 88–89; (IE), 129
Peninsula Creameries (E),743–744
PepsiCo (E), 365, 570–571; (IE), 494
Pew Research Center (BE), 297; (E), 36, 39, 149, 338, 346, 363–364, 394, 464, 488–489, 491, 579, 618, 681, 741; (IE), 294, 472; (JC), 357, 379, 475
Pontiac (E), 176
Preusser Group (IE), 548
Progressive Insurance (E), 144
Purina Pet Institute (E), 342
SIGG (BE), 88–89
Starbucks (BE), 88
Toyota (E), 176, 208–209, 722
Twitter (IE), 472

Consumers
Consumer Price Index (CPI) (E), 246
Consumer Spending (BE), 52, 61
Consumers Union (E), 314
Customer Databases (E), 139, 777; (IE), 3–6
Customer Service (E), 135
Nest Egg Index (BE), 86

Demographics
Adoptions (E), 74–75
Age (BE), 296, 638–639; (E), 106, 177, 243–245, 346, 488–489, 491, 545, 680–681, 722, 737, 853; (IE), 186, 366, 637–638, 640–641
American Community Survey (JC), 526
Birth and Death Rates (BE), 829–832; (E), 38,41–42, 105, 141–142, 144–145, 211, 244, 249–250, 340, 388, 516, 575, 620, 625–626, 680–681, 729, 578, 735–737, 739, 845–846, 854; (IE), 828–829, 832
Born in Other Countries (E), 616–617
Census-Taking (JC), 526, 529, 890
Eye Color (E), 107, 741, 777
Eye Color (JC), 26
Family Size/History (E), 292, 361–362, 388, 489, 743; (IE), 296, 306, 366
Income (BE), 61, 296; (E), 177–178, 205–206, 388, 467–468, 739, 744–745, 845–846; (IE), 455, 461; (JC), 63
Left-handedness (BE), 449–450; (E), 363, 394, 740
Licensed Drivers (E), 42–43
Life Expectancy (E), 177, 238, 246, 338, 845–846; (IE), 229–231, 331
Marital Status/Age (BE), 224–225, 296, 638–639; (E), 106, 234 239 243–245, 545, 722; (IE), 637–638, 640–641
Neck Sizes (E), 77–78
Population (E), 79, 82,104–105, 249; (IE), 163–164, 298–299
Race/Ethnicity (BE), 90, 296; (E), 512–513, 582, 737, 619
U.S. Census Bureau (BE), 90, 224, 298; (E), 139, 143, 396, 512, 573, 582, 737; (IE), 298; (JC), 63, 482, 526, 529

Distribution and Operations Management
Delivery Services and Times (E), 43
Mail Delivery (E), 147
Packaging (BE), 84, 381–383; (IE), 380
Product Distribution (BE), 393–394; (E), 391–392
Productivity and Efficiency (BE), 393–394
Shipping (E), 134–135, 388, 515–516, 575; (IE), 5
Waiting Lines (BE), 61; (E), 291, 572; (JC), 371

E-Commerce
Internet Sales/Online Businesses (E), 10, 81, 206, 336, 383, 463; (IE), 2–6
Web Design, Management, and Sales (E), 570–571; (IE), 554

A-63

Economics
Boomtowns (E), 82
Cost of Living (BE), 153–154, 158–159; (E), 103, 209–210
Dow Jones Industrial Average (E), 244–245
Forecasting (IE), 223–224
Gross Domestic Product (E), 177
Human Development Index (E), 239
Inflation Rates (E), 246

Education
Absences from Class (E), 514–515
ACT, Inc. (E), 466, 492
Admissions, College (E), 43, 363, 513, 515; (IE), 31
Association of American Medical Colleges (E), 515
Birth Order (E), 147–148, 384–386
Cheating on Tests (E), 580
College Board (E), 517
College Courses (E), 363; (IE), 554
College Social Life (E), 345, 464, 466; (JC), 675
College Tuition (E), 652
Core Plus Mathematics Project (E), 622–623
Cornell University (E), 80; (IE), 97, 330, 586
Course Evaluations (IE), 6, 330–331
Curriculum (E), 385, 397, 681–682, 743
Dartmouth College (E), 542
Dress Code (E), 489
Dropout Rates (E), 516; (IE), 834
Duke University (BE), 549
Education Levels (E), 688, 744–745; (IE), 354–355, 655
Educational Testing Service (JC), 115
Faculty Ages/Teaching Experience (E), 137; (JC), 53
George Mason University (BE), 476, 478–479, 482
GPA/Grades (E), 10, 81, 238, 241–242, 244–246, 395, 468–469, 574–575, 582, 687, 727–728, 743, 841–845; (IE), 189, 318–320, 330, 350, 374, 561, 664
Graduates and Graduation Rates (E), 107, 387, 489, 492–493, 618, 718–719, 845–846
Harvard School of Public Health (IE), 377
Harvard University (BE), 458
Law School Admission Council (E), 515
Learning Improvement (E), 736, 856, 858
Literacy and Illiteracy Rates (E), 203, 845–846
Magnet Schools (E), 38, 41
McGill University (E), 42
Music and Memory (E), 147, 628–629
National Center for Education Statistics (E), 81, 514–515
Post-Graduation Plans (BE), 665–667; (E), 40–41; (IE), 663–665, 668
Reading and Reading Ability (E), 108, 177, 241–242, 340, 497, 541, 572, 624; (IE), 303
Relationships and Sports (BE), 367–368, 372–374
Retention Rates (E), 466
Rutgers University (E), 580
St. Norbert's College (E), 489
Standardized Tests (BE), 116, 122, 124–125; (E), 110, 134–136, 206–207, 241–242, 336, 341, 466, 470, 517, 540, 646, 723–724, 740, 843; (IE), 123, 155; (JC), 115
Student/Faculty Ratio (E), 243
Student Goals (IE), 371–373
Studying for Exams (E), 143–144, 246, 745, 852–853
Summer School Programs (E), 626, 651
Teacher Shortage (E), 737–738
Teaching Methods (E), 778–779
Test Scores (BE), 458; (E), 12, 75, 81, 107–108, 134–137, 291, 840–845; (IE), 458; (JC), 112, 159
Texas A&M (E), 313–314
Tuition Costs (E), 730–731
Typing (E), 854–856
University of California at Berkeley (IE), 31
University of California at San Diego (IE), 220
University of Delaware (E), 583
University of Denver (E), 727
University of Florida (E), 338
University of Newcastle upon Tyne (BE), 604
University of Texas (IE), 669
University of Utah (IE), 2
U.S. Department of Education (BE), 458
Washington University (E), 344

Energy
Batteries (E), 385, 388–390, 584, 681, 726–727
Energy Information Administration (IE), 224
Fuel Economy (BE), 64–65; (E), 12, 75, 106, 109, 135, 177–178, 242, 246–247, 316, 340, 470, 544–545, 646, 651, 723–724, 728–729, 736, 775–776; (IE), 129, 827; (JC), 63
Gas Prices and Consumption (E), 79, 106, 223, 250
Heat for Homes (E), 242
International Tanker Owners Pollution Federation Limited (E), 38
Leaky Gas Tanks (E), 578
Nuclear Power (IE), 383
Oil (E), 38–39; (IE), 223–224
Wind Speed/Power (BE), 185, 187, 195, 200; (E), 136, 648–649; (IE), 84–87, 89–94, 119, 817; (JC), 828

Environment
Acid Rain (E), 80, 143, 245, 516, 725
Archaeology Sites (E), 736–737
Chemicals in (E), 514
Contaminants and Fish (BE), 523–525, 530; (E), 317, 544
Dowsing (E), 340, 514
Earthquakes (IE), 46–48, 50, 53–57, 59, 375
Emission Standards (E), 250, 449 385, 493, 515, 573–574
Energy Production (E), 363–364
Environmental Protection Agency (E), 12, 135, 315, 726
Exposure to Lead-Based Paint (E), 489
Global Warming (BE), 476, 478–479, 482; (E), 39, 210–211, 725, 888–889; (IE), 476, 494; (JC), 479, 828
Hurricanes (BE), 185, 187, 195, 200; (E), 177, 240–241, 625, 683, 719–720; (IE), 151–153, 155, 223, 375, 817; (JC), 828
Marine Life (BE), 222; (IE), 220–221
National Hurricane Center (E), 240–241, 683; (IE), 151, 155
National Interagency Fire Center (E), 207
National Oceanic and Atmospheric Administration (E), 719
Noise Filters (E), 778
Old Faithful (E), 145, 247, 742–743
Ozone Levels (E), 106–107, 726
Pollution Control (E), 316, 777–778; (IE), 480
Radon Gas (E), 343
Rainfall (E), 857
Scripps Institution of Oceanography (E), 220
Studying Flowers (E), 12, 343, 738
Termite Attacks (IE), 561
Tree Diameters (E), 138, 203, 247–248
Tsunami (IE), 46–48, 53–57, 59
U.S. National Geophysical Data Center (IE), 46
U.S Geological Survey (IE), 47
Volcanoes (E), 397
Weather Conditions (BE), 91, 353; (E), 41, 77, 78–79, 174, 362, 469, 542, 725
Wild Horses (E), 739–740
Wildfires (E), 207–208
Wildlife Studies (BE), 222; (E), 80, 203, 212, 245, 245, 247, 492, 572; (IE), 66–67, 215, 220–221
Yale Project on Climate Change Communication (BE), 476, 478–479, 482

Famous People
Archimedes (BE), 227
Armstrong, Lance (JC), 7
Bacon, Francis (BE), 227; (IE), 694
Barone, Michael (E), 346
Bayes, Thomas (IE), 380
Bernoulli, Daniel (IE), 372
Bernoulli, Jacob (IE), 350, 372
Berra, Yogi (IE), 152, 350, 355,
Billups, Chauncey (E), 735
Bohr, Niels (IE), 223
Bonferroni, Carlo (BE), 766
Borduin, Charles (E), 743
Box, George (IE), 119, 183
Buchanan, Pat (IE), 226–227
Bush, George W. (E), 578, 582; (IE), 226–227, 307
Carroll, Lewis (IE), 377
Castroneves, Hélio (E), 13
Ceci, Stephen (IE), 330–332
Contador, Alberto (JC), 7
Cornet, Henri (JC), 7
Coveyou, Robert R. (IE), 281
Curry, Stephen (E), 735
Dawson, Joe (E), 13
de Moivre, Abraham (IE), 445–447
DePalma, Ralph (E), 13
Descartes, René (BE), 154
Diaconis, Persi (IE), 282
Dixon, Scott (E), 13
Dobrynska, Nataliya (IE), 110–112
Drake, Frank (E), 398
Einstein, Albert (E), 543
Evans, Cadel (JC), 7
Fechner, Gustav (IE), 320
Fisher, Sir Ronald (BE), 562; (IE), 160, 498, 520, 524, 555, 751, 758
Fleet, Frederick (IE), 16
Fountain, Hyleas (IE), 110–112, 131
Franchitti, Dario (E), 13
Gallop, George (IE), 296
Galton, Sir Francis (BE), 189, 191; (IE), 190
Garin, Maurice (JC), 7
Gauss, Carl Friedrich (BE), 184

APPENDIX C Index of Applications A-65

Gill, Colin (E), 346
Gore, Al (IE), 226
Gosset, William S. (IE), 520–523, 698
Goux, Jules (E), 13
Graham, Ronald (IE), 282
Greene, Brian (IE), 283
Gretzky, Wayne (E), 79
Griffin, Blake (IE), 282–283, 291, 371
Halifax, Lord (IE), 319
Harroun, Ray (E), 13
Harvey, William (BE), 226
Hocking, Ian (BE), 280
Howe, Gordie (E), 79
Hunter, Stu (BE), 555
Jastrow, J. (IE), 321
Kantor, W. M. (IE), 282
Keith, Vikki (E), 245, 627
Kerry, John (E), 578, 621
Keynes, John Maynard (BE), 353
Klassen, Cindy (IE), 630
Knox, Pamela Naber (E), 362
Kohavi, Ronny (IE), 2, 4
Landon, Alf (IE), 295–296, 309
Laplace, Pierre-Simon (IE), 452–453
Legendre, Adrien-Marie (BE), 184
Li Lei (BE), 120
Ligety, Ted (BE), 112
Lister, Joseph (E), 735
Lowell, James Russell (IE), 500
Maas, Jim (IE), 518
Maris, Roger (E), 147
McCain, John (BE), 17
McGwire, Mark (E), 147
Meir, Jessica (IE), 220
Mexico, Roger (BE), 379
Michelson, Albert Abraham (E), 543
Miller, Bode (BE), 111–112
Moore, David (IE), 323
Mulford, Ralph (E), 13
Nader, Ralph (E), 578; (IE), 226
Newton, Isaac (BE), 154
Nissen, Steven E. (BE), 549
Nixon, Richard (E), 582
Obama, Barack (BE), 17; (E), 572
Patrick, Dana (IE), 282–283, 291, 371
Peirce, C. S. (IE), 321, 327
Periero, Óscar (JC), 7
Picasso, Pablo (BE), 10
Poisson, Simeon Denis (IE), 378
Ponganis, Paul (IE), 220
Pynchon, Thomas (BE), 379
Robbins, Rebecca (IE), 518
Robinove, Charles (E), 578
Rodriguez, Alex (E), 79–80
Roosevelt, Franklin Delano (IE), 295–296, 309
Ruth, Babe (E), 147
Sagan, Carl (IE), 504
Sastre, Carlos (JC), 7
Saunderson, Nicholas (BE), 381
Solo, Hope (IE), 282–283, 291, 371–373, 376
Stigler, Steven (IE), 381
Thomas, René (E), 13
Trousseller, Louis (JC), 7
Tufte, Edward (IE), 67n
Tukey, John W. (BE), 48, 57
Van Buren, Abigail (IE), 351
Venn, John (IE), 353

Wainer, Howard (BE), 458
Wanamaker, John (IE), 494
Wayne, John (E), 516–517
Weeks, David (E), 345
Wheldon, Dan (E), 13
Wilson, E. B. (IE), 558n
Zener, Karl (BE), 549
Zwerling, Harris (BE), 458

Finance and Investments
401(k) Plans (E), 11
Brokerage Firms (BE), 86; (E), 681
Currency (E), 572; (IE), 366
Day Traders (E), 389
Hedge Funds (BE), 195
Moving Averages (BE), 93
Stock Market and Prices (BE), 163; (E), 395; (IE), 223
Stock Volatility (E), 571; (IE), 379

Food/Drink
Activating Baking Yeast (E), 775
Alcohol (E), 250, 340, 466, 581, 776; (IE), 377–378, 520–521
Bananas (E), 141
Breads (E), 142, 583
Brownies (IE), 168
Candy (BE), 357–359; (E), 210, 344, 364, 465, 514, 682, 839
Carbonated Drinks (E), 343, 365, 513, 570–571; (IE), 327, 450, 494
Cereal (BE), 126–128, 197–199; (E), 76, 104, 202–203, 291, 391, 624, 724, 779–780, 811–814, 816, 844, 846, 889–891; (IE), 199–200, 221–222, 282–284, 371–373, 376, 586
Chips (BE), 297; (E), 470, 545, 583, 745
Coffee (BE), 88–89, 367–368, 604; (E), 11, 108, 175, 387–388, 392
Company Cafeterias/Campus Meal Programs (BE), 604; (E), 396, 542, 582–583
Cookies (E), 546, 809, 810
Dairy Products and Milk (E), 315, 471, 654, 743–744
Dessert (BE), 367–368
Diet Sodas (E), 11
Eggs (E), 140, 390, 580, 647, 778
Fast Food (BE), 186, 197–199; (E), 12, 179, 209, 290, 621, 846; (IE), 182–187, 189–191, 200
Fish/Seafood (BE), 523–525, 530; (E), 317, 540, 544, 580, 686, 737
Grocery Shopping (E), 10
Hot Beverages (BE), 84, 747, 762–764
Hot Dogs (E), 544, 621–622, 721–722
Ice Cream (BE), 51; (E), 627
Irradiated Food (E), 491
Jelly (E), 541
Meats (E), 136
Nutritional Data (BE), 195, 197–199; (E), 209
Nuts (E), 682
Opinions About Food (BE), 301–302, 308; (E), 317
Oranges (E), 343, 387–388
Pie (BE), 51, 302
Pizza (E), 77–78, 103, 336, 544, 889, 892–894; (IE), 555
Popcorn (E), 545, 774–775, 810
Potatoes (JC), 668–669

Predicting Consumption (E), 250
Restaurants (BE), 390–391, 367–371; (E), 465, 467; (JC), 675
Soup (BE), 599; (E), 464; (IE), 295–297
Tomatoes (E), 140, 336, 339
Veggie Burgers (E), 209
Watermelons (E), 396
Yogurt (E), 546, 651, 777

Games
Board Games (E), 290, 292–293, 578–579
Cards (BE), 390–391; (E), 144, 291–292, 365, 384–386, 388, 512; (IE), 280, 282, 351
Carnival Games (E), 388
Casinos (BE), 351; (E), 391, 570
Coin Toss (BE), 351–352; (E), 361–362, 397, 465, 490, 513–514, 575; (IE), 280, 299, 325, 348, 351, 372, 504, 553
Darts (E), 775
Dice (BE), 284–285; (E), 292, 345, 361–362, 365, 388, 397, 469–470, 513, 579, 682; (IE), 280, 383, 451–452, 455
Door Prizes (E), 398
Gambling (E), 344, 492
Keno (IE), 351
Lottery (BE), 285–287; (E), 290–292, 344, 365, 469, 683–684; (IE), 352; (JC), 53, 351
Odds of Winning (E), 291–292
Roulette Wheel (E), 362; (IE), 280; (JC), 560
Slot Machines (E), 365, 570; (IE), 379
Video racing (E), 850

Government, Labor, and Law
AFL-CIO (E), 842–843
Council of Europe (E), 80
Food and Drug Administration (BE), 321, 549, 560, 567; (E), 176; (IE), 318
Internships (BE), 550
Jury Trials (BE), 552; (E), 516; (IE), 496–497, 504, 554–555, 560–561
Labor Force Participation Rate (E), 647
Office of Federal Housing Enterprise Oversight (E), 178, 205–206
Prisons (E), 83, 490–491, 650; (IE), 309
Right-to-Work Laws (E), 730
United Nations (E), 239
U.S. Bureau of Labor Statistics (E), 647
U.S. Department of Commerce (E), 618
U.S. Department of Justice (E), 83

Human Resource Management/Personnel
Absenteeism and Attendance (E), 387
Employee Discrimination (E), 293, 574
Hiring and Recruiting (E), 388, 492, 571, 850
Human Resource Data (E), 384, 393, 470
Job Satisfaction (E), 314, 317, 626, 650
Job Type (E), 393
Part-time/Full-time Employment (IE), 504
Rating Employees (E), 388
Workplace Safety (E), 341, 394

Insurance
Auto Insurance (E), 144, 176, 362, 394, 469, 491, 648–649
Health Insurance (E), 146–147, 317, 386, 389, 737; (JC), 325–326

Homeowner's Insurance (E), 390–391; (IE), 375
Insurance Company Databases (IE), 66, 556
Life Insurance (E), 395; (IE), 389–397
Medicaid (E), 317
National Insurance Crime Bureau (E), 176

Manufacturing
Appliances (E), 388, 515–516, 621–622, 840
Assembly Line Production (IE), 372
Cameras (IE), 161, 167–168, 586
Cars (E), 11, 148–149, 177, 387, 393, 490, 513, 618, 681
Ceramic Pottery (E), 177, 343, 576
Computers and Computer Chips (E), 252; (IE), 564
Cooking and Tableware (E), 739
Electronic Components (E), 387–388
Furniture (E), 140, 149–150
Light Bulbs (E), 365
Metals (IE), 494–496, 500–501
Product Registration (IE), 306
Reporting Production Statistics (JC), 115
Rivets (E), 138, 345; (IE), 504
Shoes (E), 339
Ski Wax (E), 545
Stereos (BE), 381–383; (IE), 380
Tires (E), 139, 546, 651–652

Marketing
Direct Mail (BE), 483; (E), 463–464, 491; (IE), 307

Media and Entertainment
American Journal of Clinical Nutrition (E), 579
American Journal of Health Behavior (IE), 377
American Psychologist (E), 249, 742
Annals of Internal Medicine (E), 388
Archives of General Psychiatry (E), 620
Archives of Internal Medicine (E), 340
Associated Press (E), 579
Barbershop music (E), 852
British Medical Journal (E), 341, 648, 685
Cartoons (IE), 2, 223, 281, 305, 322, 328, 372, 477
Chance (E), 38, 40–41
Circulation (E), 342
CNN (E), 579
Comedians (IE), 169
Concert Venues (E), 237–238
Consumer Reports (BE), 5; (E), 12, 77, 80, 386, 490, 546, 570, 621–622, 721
The Economist (BE), 5
Forbes (E), 110, 150; (IE), 160
Fortune (IE), 655
International Breastfeeding Journal (E), 684
Journal of the American Medical Association (E), 41, 140, 249, 619–620, 738, 740; (IE), 553
Journal of Applied Psychology (E), 626
Journal of Consulting and Clinical Psychology (E), 743
Lancet (E), 619, 744
Literary Digest (IE), 295–296, 309
Magazines (E), 516; (IE), 306
Movies (E), 36, 37, 38, 40, 78, 235, 516, 720–721, ,839–840; (IE), 159, 378
New England Journal of Medicine (BE), 549, 553–554, 560, 562–563; (E), 340, 342 619, 684–685, 735

New York Times (IE), 307
NewGeography.com (E), 82
Newspapers (E), 315, 493, 579, 582, 618, 746; (IE), 294–295
Psychological Science (E), 339
Reader's Digest (E), 345
Rock Concerts (E), 103–104, 137
Science (E), 12, 43, 338, 343; (IE), 701
Science News (E), 338
Seattle Times (BE), 459
Sports Illustrated (E), 204
Television (BE), 303–305, 768–769; (E), 345–346, 362, 489, 493, 575, 579, 618, 626, 647, 809, 810, 857–858; (IE), 229–231, 294, 307, 770, 331, 724
Time (E), 619, 621
Wall Street Journal (IE), 2

Pharmaceuticals, Medicine, and Health
American Academy of Pediatrics (E), 619
American College of Obstetricians and Gynecologists (E), 290
American Medical Association (E), 141–142
Antidepressants/Depression (E), 42, 176, 249, 337–338, 343, 513, 620, 737, 742
Birth Defects (E), 290
Blindness (E), 11
Blood Donors/Blood Type (BE), 375; (E), 292, 364–365; (IE), 376
Blood Pressure (BE), 159–160; (E), 42, 241, 337, 339, 384, 386, 735–736; (IE), 50
Body Fat Percentages (BE), 695–696, 823–825; (E), 211, 241, 727, 843–844; (IE), 155, 404, 448–449, 689–692, 700, 705–707, 709–711, 817–822, 827, 833
Cancer (BE), 320; (E), 338, 343–344, 517, 574, 583, 619, 735, 815; (BE), 27–28; (IE), 160, 378–379
Centers for Disease Control and Prevention (BE), 444–445, 456–457; (E), 38, 42, 396, 580–581, 618; (IE), 29, 113
Child Development (E), 542, 727, 736
Cholesterol (E), 108, 139, 340, 384, 386, 682, 721–722
Color-blindness (E), 247, 290, 578
Contrast Baths (BE), 748, 754, 764
Cranberry Juice and Urinary Tract Infections (E), 685–686
Dangerous Behaviors (BE), 587–589, 596; (E), 241, 342, 626; (IE), 29
Diabetes (BE), 549, 553–554, 562, 566; (E), 493, 738; (IE), 709–711
Diagnostic Testing (E), 620; (IE), 561
Dialysis Data (E), 142
Diseases/Disorders (BE), 566; (E), 291, 336–337, 344–346, 394, 466–467, 492, 572–573, 578, 619, 738, 743–744; (IE), 319–320, 326, 389, 378–379
Drug Tests and Treatments (BE), 549, 553–554, 562–563, 566; (E), 11, 176, 314, 338–341, 512–513, 570–572, 620, 744, 781; (IE), 319, 497; (JC), 325–326, 497
Drug Use/Abuse (E), 80, 178–179, 208, 314, 316, 336, 720; (IE), 484
Eating Disorders/Weight Issues (E), 42, 514, 582, 619, 647, 653, 737, 745; (IE), 709–711

Gestation/Pregnancy/Childbirth (BE), 502–503, 524; (E), 12, 239, 244, 383, 398, 469, 578, 684–685, 688, 735–736, 740–742; (JC), 459–460
Handwashing (IE), 747–748, 750–753, 757, 759, 761, 764–767
Hearing Assessments/Aids (E), 745–746, 776–777
Heart Attacks/Heart Disease (BE), 84, 320, 549, 553–554, 562, 566; (E), 76, 177, 336, 340, 342–344, 512, 575, 737, 682; (IE), 51, 62, 560
Height/Weight Data (BE), 353, 412, 456–458; (E), 136, 142, 336, 395, 467–468, 858; (IE), 76, 79, 113–114, 129, 155–157, 186, 189, 231, 455, 536; (JC), 121
Herbal Compounds (E), 12, 148, 338, 625, 735; (IE), 328
Herbal Medicine Council (E), 148
Hospital Charges and Discharges (E), 105
Injuries/Illnesses (E), 103–104, 238, 338–340, 385, 489, 574, 619–620, 735–736, 812; (IE), 320, 366, 767; (JC), 600, 603
Menopause (E), 337
National Cancer Institute (E), 618
National Center for Chronic Disease Prevention and Health Promotion (IE), 298
National Center for Health Statistics (E), 105, 142, 361; (IE), 113, 129
National Heart Lung and Blood Institute (BE), 448
National Institutes of Health (E),737; (IE), 113, 320
National Sleep Foundation (E), 386; (IE), 592
Number of Doctors (IE), 229–231
Placebo Effect (E), 176, 310 337–338, 345; (IE), 328, 494, 497, 504
Pulse Rates (E), 542, 623; (IE), 48
Sleeping Patterns/Snoring (BE), 444, 526–527, 531–532, 594–595; (E), 338–339, 386, 646; (IE), 306, 326, 383–384, 518, 529, 592–593
Smoking/Nonsmokers (BE), 444–445; (E), 39, 204–205, 239–240, 248, 250, 343, 345, 396, 466, 513, 580–581, 619, 737; (IE), 96–97, 160, 494
Stress Relief/Testing (E), 11, 346
Surgical Procedures (BE), 559, 748, 750, 754; (E), 43, 104, 575, 735, 737; (JC), 600, 603
Sydney Children's Hospital (E), 578
Tattoos (BE), 671–672; (E), 43; (IE), 307, 669–670, 673–675
TB Screening (BE), 379–380
Therapeutic Touch (IE), 553–554, 562–564
Treating Sports-Related Injuries (E), 39
U.S. Surgeon General (IE), 160
Vaccinations (E), 572, 619
Vision (E), 466
Vitamins (E), 336–337, 340, 343, 492, 579; (IE), 319
Women's Health Initiative (BE), 320

Politics and Popular Culture
2000 Elections (IE), 226–227
American Veterinary Association (E), 391
Approval Ratings (E), 513, 572, 582, 620–621; (IE), 494
Belief in Evolution (IE), 443, 446

Candidates (E), 176, 291, 315, 573, 618, 570; (IE), 504, 561
Cooking and Shopping (E), 383
Full Moon Beliefs (E), 687
Hula-hoops (E), 11
Internet Access/Activity (BE), 6, 587–590; (E), 618, 681, 742; (IE), 294
Online Social Networking (BE), 596; (E), 383, 394, 618; (IE), 472–479, 484–485
Pets (BE), 319, 321–322, 327, 330–331; (E), 338, 342–343, 383–384, 386, 390–391, 397, 572, 578, 618
Playgrounds (E), 315–317
Political Parties (E), 149, 346, 365, 385–386, 682, 686, 741; (IE), 296
Predicting the Future (IE), 223, 348, 655
Religion (E), 315, 365, 489, 581
Roller Coasters in Theme Parks (BE), 87, 91, 93; (E), 176, 180, 203–204, 315
Statue of Liberty (E), 146
Text Messaging (BE), 369–370, 376; (IE), 2, 30, 375
Titanic Sinking (E), 383, 685, 687–688; (IE), 16–24, 26, 30
Voting (E), 316, 365, 489, 582
Zodiac Birth Signs (BE), 659–662; (IE), 655–657

Quality Control
Food Inspection and Safety (E), 315, 317, 343, 386, 490, 571
Lifespan of a Product (IE), 389; (JC), 63
Product Defects (E), 363, 365, 386–388, 388–389, 393, 513, 570, 574; (IE), 321–322, 330–331, 372, 494–496, 500–501, 504
Product Inspections and Testing (BE), 319, 321–322, 327, 330–331, 564; (E), 77, 109, 288, 315, 338–339, 341, 343–345, 397, 513, 545, 571, 575, 581, 651–652; (IE), 327, 476, 520–521
Product Ratings and Evaluations (BE), 5; (E), 12, 77
Product Recalls (E), 398
Repair Calls (E), 389–390; (JC), 391

Real Estate
Empty Houses (E), 512
Home Buyers (IE), 308
Home Ownership (BE), 296; (E), 396, 573
Home Sales and Prices (BE), 819, 827–828; (E), 203–204, 362, 469, 541, 623–624, 721, 729–730, 839–842; (IE), 154; (JC), 190, 195
Housing Costs (E), 178, 205–206
Lender Processing Services (E), 466
Mortgages (E), 178, 206, 466
Racial Steering (E), 687–688

Salary and Benefits
Assigned Parking Spaces (JC), 675
Employee Benefits (E), 386, 570
Salaries/Payroll (BE), 153–154, 159; (E), 77, 79, 134–136, 149–150, 174, 423, 467–468, 423–424, 541–542, ,842–843; (IE), 94–95, 129–130, 455, 460
Sick Days (E), 77

Sales and Retail
Bakeries (E), 142, 583
Bicycles (E), 391–392
Book Sales and Stores (E), 173, 201–202; (IE), 2–6
Cars (E), 208–209, 388, 489, 618, 722
Catalog Sales (E), 489–490
Clothing Stores (E), 134, 206
Coffee Shop (E), 177, 337, 392
Cost of Shopping (E), 617
Enjoyment in Shopping (E), 383
Food Store Sales (E), 292, 470, 570
Friendship and Sales (BE), 601–603; (IE), 586, 596, 600–601
Furniture (E), 149–150
Grocery Shopping (E), 10, 465
Hardware (IE), 228
Motorcycles (BE), 231–232
Secret Sales (E), 362
Shopping Malls (IE), 308

Science
Arm Length Measurement (E), 316
Belief in Evolution (IE), 443, 446
Bomb Drop Modeling (IE), 379
Cloud Seeding (E), 108–109, 111, 647, 738
Dexterity Testing (E), 174
Diet Soda and Waist Size (E), 11
ESP Testing (BE), 549; (E), 204
Genetics (E), 467, 739; (IE), 709–711
Germination Rates and Flowering Dates (E), 12, 343
Hamilton Depression Rating Scale (E), 176
Ice Breaking Guess (BE), 702–705; (IE), 701–702
Impact Craters (BE), 693–694, 699–700, 708
Interbreeding (E), 624
International Astronomical Union (E), 181
IQ Tests/Scores (E), 134–135, 137, 139, 174, 339, 470, 724; (IE), 160, 168, 228, 644, 689
Language Skills (E), 394–395, 580
Maze Navigation (E), 546
Mouth Volume (JC), 701
Musical Composition Identification (IE), 504
National Oceanic and Atmospheric Administration (IE), 151–152
Neurological Research (E), 580
Ovulation and Determining Sexual Orientation (E), 11, 466–467
Paper Airplane Designs (E), 851, (JC), 755
Perfect Pitch (E), 736
Pi Calculations (E), 683
Planets (E), 103, 181
Predicting Abalone Age (E), 238
Psychology Experiments (IE), 322, 327
Richter Scale (IE), 46
Speed of Light (E), 543
Tank Draining (E), 252
Temperatures (E), 134, 136, 140, 145, 177, 238, 248–249, 343, 543, 649; (IE), 817
Tracking Fruit Flies (BE), 663; (E), 110, 682–683
Walking in Circles (E), 12
Water Testing (BE), 323–324, 327; (E), 12, 145, 176, 212, 579, 624–626, 583, 739, 850, 851; (IE), 321, 329–330
Weighing Bears (E), 12

Service Industries and Social Issues
American Red Cross (E), 364
Annenberg Foundation (BE), 458
Annie E. Casey Foundation (IE), 828n
Beliefs about Marriage (E), 36, 464, 579
Bill and Melinda Gates Foundation (BE), 458–459
Carnegie Corporation (BE), 458
Center for Collaborative Education (BE), 458
Center for School Change (BE), 458
Firefighters (IE), 160
Kids Count Data (IE), 828n
Nonprofit and Philanthropic Organizations (E), 139, 390, 398, 463–464, 470, 515, 570–571, 616–617; (IE), 828n; (JC), 592
Open Society Institute (BE), 458
Paralyzed Veterans of America (E), 491, 515
Pew Charitable Trusts (BE), 458
Police (E), 40, 137, 582, 682, 685
Thefts (E), 176
Violence (E), 682

Sports
Archery (E), 579
Atlas Bodybuilding Company (E), 392
Baseball (BE), 505–507, 656–657, 659, 668; (E), 79–80, 109–110, 147, 179, 204–205, 293, 314, 336, 490–491, 617, 625, 652–653, 728; (IE), 154, 350; (JC), 285, 371
Basketball (E), 106, 292–293, 347, 362, 383, 575–576, 735; (IE), 359, 372
Bowling (E), 365, 850
Cycling (E), 11, 149, 246; (JC), 7, 13
Dog Racing (IE), 338
Exercise (general) (E), 337, 340, 650
Fishing (E), 580
Football (BE), 20–21, 22–23, 25; (E), 10, 77, 139, 250–251, 337, 383, 516, 572, 617, 849; (JC), 53
Frisbee Throws (E), 775–776
Golf (E), 78, 347, 388, 544, 546, 628, 849
Golf Laboratories, Inc. (E), 628
Gymnastics (IE), 322
Hockey (E), 79, 718–719
Horse Racing (E), 389, 849–850, 891
Indianapolis 544 (E), 13
Javelin (E), 857
Kentucky Derby (BE), 50; (E), 12, 108, 177; (IE), 49, 53, 56, 68
Marathons (E), 649–650, 653–654, 856
Olympics (BE), 111–112, 115, 120, 635–636; (E), 39, 138, 144, 211–212, 249, 627–628; (IE), 110–112, 630–633
Running (E), 76, 766; (JC), 53
Skating (E), 581, 775
Skiing (BE), 111–112, 115, 120; (E), 104, 138, 545
Skydiving (E), 341
Soccer (E), 385, 388, 397
Speed Skating (BE), 635–636; (E), 144; 630–633, 639–640
Super Bowl (BE), 22–23, 25; (E), 10, 77, 337, 516; (IE), 494
Swimming (E), 244–245, 340, 391, 627–628
Track and Field (E), 110–112, 211–212, 249, 627–628
Treating Sports-Related Injuries (E), 39

United States Golf Association (E), 544
Video pinball (E), 856–857
Weightlifting (E), 392, 858

Surveys and Opinion Polls

American Association for Public Opinion Research (JC), 374
Claims Made (JC), 299
Company Surveys (IE), 307
Consumer Polls (E), 177
Election Polls (E), 12, 579–580; (IE), 305
Gallup Polls (BE), 22–23, 25, 480; (E), 10–12, 314, 363, 396, 488–489, 491–492, 582, 621, 696, 739
Health/Nutritional Data (IE), 113–114, 129
International Polls (E), 314, 384–385
Internet and E-Mail Polls (E), 307, 314; (IE), 295, 307
National Health Interview Survey (E), 386
Public Opinion Polls (BE), 476, 478–482; (E), 315, 363, 384–386, 466, 489, 491–492, 582–583, 681, 686, 849; (IE), 223, 294–296, 306; (JC), 374, 595
Student Surveys (BE), 301–302; (E), 41, 314, 571; (IE), 301; (JC), 449
Telephone Surveys (BE), 297; (E), 12, 142–143, 315–316, 364, 572, 741; (IE), 295–296, 307, 309, 482; (JC), 357, 374, 379

Technology

Cell Phones (E), 177, 238–239, 293, 316, 344, 361, 384, 386, 489, 513, 618; (IE), 30, 281, 309; (JC), 475
Computers (BE), 5, 7, 393–394, 500; (E), 109, 173, 201–202, 241, 344, 362, 396, 545, 571, 575, 726–727; (IE), 281–282, 320
Digital Music (E), 110–111, 361
Downloading Files/Music (E), 780–781
E-Books (E), 144, 387
E-Mail (BE), 374, 377; (E), 76, 573, 618
iPods and MP3 Players (E), 135–137, 465; (JC), 63
Printers (E), 516
Product Instruction Manuals (IE), 306
Software (E), 388–389, 574–575; (IE), 326
Stereograms (E), 111, 622

Transportation

Accidents (BE), 381, 556–557; (E), 35, 146, 648; (IE), 366, 377–378, 548
Air Travel (BE), 58–59, 61, 500; (E), 38, 101–102, 181, 240, 347, 362, 387, 389, 394, 396, 516, 544, 572; (IE), 60, 355; (JC), 89, 303
Bicycle Safety (E), 149, 581
Car Horsepower/Weight (E), 80, 178, 180, 202–203, 246–247, 251, 646–647, 810; (IE), 752, 827
Commuting to Work (BE), 631, 633–634, 636–637, 640; (E), 623, 682
Driving and Texting (IE), 2
Driving Speeds/Reaction Times (BE), 556; (E), 137–138, 147, 242, 363, 583; (IE), 557; (JC), 123
Driving Tests (BE), 498–500; (E), 292
Drunk Driving (E), 110, 387
Highways/Bridges (BE), 187–189; (E), 210, 245–246
Motorcycles (BE), 231–232, 550–552; (IE), 306, 548–549, 559
National Highway Traffic Safety Administration (BE), 556; (E), 146, 580, 810–811; (IE), 585
Traffic, Parking, Safety (BE), 381, 590–592; (E), 41, 176, 273, 315, 390, 466, 489, 543, 570–570, 580, 648, 686, 737, 853–854, 891–893; (IE), 554, 585
Traffic Lights/Stop Signs (BE), 354–356; (E), 389; (IE), 348–350
Trains (E), 365, 383
Travel and Tourism (E), 105, 384, 386
U.S. Bureau of Transportation Statistics (BE), 58–59, 89; (E), 101, 544, 850
Vehicle Searches (BE), 670, 672–673

Appendix D: Tables and Selected Formulas

TABLE OF RANDOM DIGITS

Row										
1	96299	07196	98642	20639	23185	56282	69929	14125	38872	94168
2	71622	35940	81807	59225	18192	08710	80777	84395	69563	86280
3	03272	41230	81739	74797	70406	18564	69273	72532	78340	36699
4	46376	58596	14365	63685	56555	42974	72944	96463	63533	24152
5	47352	42853	42903	97504	56655	70355	88606	61406	38757	70657
6	20064	04266	74017	79319	70170	96572	08523	56025	89077	57678
7	73184	95907	05179	51002	83374	52297	07769	99792	78365	93487
8	72753	36216	07230	35793	71907	65571	66784	25548	91861	15725
9	03939	30763	06138	80062	02537	23561	93136	61260	77935	93159
10	75998	37203	07959	38264	78120	77525	86481	54986	33042	70648
11	94435	97441	90998	25104	49761	14967	70724	67030	53887	81293
12	04362	40989	69167	38894	00172	02999	97377	33305	60782	29810
13	89059	43528	10547	40115	82234	86902	04121	83889	76208	31076
14	87736	04666	75145	49175	76754	07884	92564	80793	22573	67902
15	76488	88899	15860	07370	13431	84041	69202	18912	83173	11983
16	36460	53772	66634	25045	79007	78518	73580	14191	50353	32064
17	13205	69237	21820	20952	16635	58867	97650	82983	64865	93298
18	51242	12215	90739	36812	00436	31609	80333	96606	30430	31803
19	67819	00354	91439	91073	49258	15992	41277	75111	67496	68430
20	09875	08990	27656	15871	23637	00952	97818	64234	50199	05715
21	18192	95308	72975	01191	29958	09275	89141	19558	50524	32041
22	02763	33701	66188	50226	35813	72951	11638	01876	93664	37001
23	13349	46328	01856	29935	80563	03742	49470	67749	08578	21956
24	69238	92878	80067	80807	45096	22936	64325	19265	37755	69794
25	92207	63527	59398	29818	24789	94309	88380	57000	50171	17891
26	66679	99100	37072	30593	29665	84286	44458	60180	81451	58273
27	31087	42430	60322	34765	15757	53300	97392	98035	05228	68970
28	84432	04916	52949	78533	31666	62350	20584	56367	19701	60584
29	72042	12287	21081	48426	44321	58765	41760	43304	13399	02043
30	94534	73559	82135	70260	87936	85162	11937	18263	54138	69564
31	63971	97198	40974	45301	60177	35604	21580	68107	25184	42810
32	11227	58474	17272	37619	69517	62964	67962	34510	12607	52255
33	28541	02029	08068	96656	17795	21484	57722	76511	27849	61738
34	11282	43632	49531	78981	81980	08530	08629	32279	29478	50228
35	42907	15137	21918	13248	39129	49559	94540	24070	88151	36782
36	47119	76651	21732	32364	58545	50277	57558	30390	18771	72703
37	11232	99884	05087	76839	65142	19994	91397	29350	83852	04905
38	64725	06719	86262	53356	57999	50193	79936	97230	52073	94467
39	77007	26962	55466	12521	48125	12280	54985	26239	76044	54398
40	18375	19310	59796	89832	59417	18553	17238	05474	33259	50595

Table Z
Areas under the standard Normal curve

Second decimal place in z										z
0.09	0.08	0.07	0.06	0.05	0.04	0.03	0.02	0.01	0.00	
									0.0000†	−3.9
0.0001	0.0001	0.0001	0.0001	0.0001	0.0001	0.0001	0.0001	0.0001	0.0001	−3.8
0.0001	0.0001	0.0001	0.0001	0.0001	0.0001	0.0001	0.0001	0.0001	0.0001	−3.7
0.0001	0.0001	0.0001	0.0001	0.0001	0.0001	0.0001	0.0001	0.0002	0.0002	−3.6
0.0002	0.0002	0.0002	0.0002	0.0002	0.0002	0.0002	0.0002	0.0002	0.0002	−3.5
0.0002	0.0003	0.0003	0.0003	0.0003	0.0003	0.0003	0.0003	0.0003	0.0003	−3.4
0.0003	0.0004	0.0004	0.0004	0.0004	0.0004	0.0004	0.0005	0.0005	0.0005	−3.3
0.0005	0.0005	0.0005	0.0006	0.0006	0.0006	0.0006	0.0006	0.0007	0.0007	−3.2
0.0007	0.0007	0.0008	0.0008	0.0008	0.0008	0.0009	0.0009	0.0009	0.0010	−3.1
0.0010	0.0010	0.0011	0.0011	0.0011	0.0012	0.0012	0.0013	0.0013	0.0013	−3.0
0.0014	0.0014	0.0015	0.0015	0.0016	0.0016	0.0017	0.0018	0.0018	0.0019	−2.9
0.0019	0.0020	0.0021	0.0021	0.0022	0.0023	0.0023	0.0024	0.0025	0.0026	−2.8
0.0026	0.0027	0.0028	0.0029	0.0030	0.0031	0.0032	0.0033	0.0034	0.0035	−2.7
0.0036	0.0037	0.0038	0.0039	0.0040	0.0041	0.0043	0.0044	0.0045	0.0047	−2.6
0.0048	0.0049	0.0051	0.0052	0.0054	0.0055	0.0057	0.0059	0.0060	0.0062	−2.5
0.0064	0.0066	0.0068	0.0069	0.0071	0.0073	0.0075	0.0078	0.0080	0.0082	−2.4
0.0084	0.0087	0.0089	0.0091	0.0094	0.0096	0.0099	0.0102	0.0104	0.0107	−2.3
0.0110	0.0113	0.0116	0.0119	0.0122	0.0125	0.0129	0.0132	0.0136	0.0139	−2.2
0.0143	0.0146	0.0150	0.0154	0.0158	0.0162	0.0166	0.0170	0.0174	0.0179	−2.1
0.0183	0.0188	0.0192	0.0197	0.0202	0.0207	0.0212	0.0217	0.0222	0.0228	−2.0
0.0233	0.0239	0.0244	0.0250	0.0256	0.0262	0.0268	0.0274	0.0281	0.0287	−1.9
0.0294	0.0301	0.0307	0.0314	0.0322	0.0329	0.0336	0.0344	0.0351	0.0359	−1.8
0.0367	0.0375	0.0384	0.0392	0.0401	0.0409	0.0418	0.0427	0.0436	0.0446	−1.7
0.0455	0.0465	0.0475	0.0485	0.0495	0.0505	0.0516	0.0526	0.0537	0.0548	−1.6
0.0559	0.0571	0.0582	0.0594	0.0606	0.0618	0.0630	0.0643	0.0655	0.0668	−1.5
0.0681	0.0694	0.0708	0.0721	0.0735	0.0749	0.0764	0.0778	0.0793	0.0808	−1.4
0.0823	0.0838	0.0853	0.0869	0.0885	0.0901	0.0918	0.0934	0.0951	0.0968	−1.3
0.0985	0.1003	0.1020	0.1038	0.1056	0.1075	0.1093	0.1112	0.1131	0.1151	−1.2
0.1170	0.1190	0.1210	0.1230	0.1251	0.1271	0.1292	0.1314	0.1335	0.1357	−1.1
0.1379	0.1401	0.1423	0.1446	0.1469	0.1492	0.1515	0.1539	0.1562	0.1587	−1.0
0.1611	0.1635	0.1660	0.1685	0.1711	0.1736	0.1762	0.1788	0.1814	0.1841	−0.9
0.1867	0.1894	0.1922	0.1949	0.1977	0.2005	0.2033	0.2061	0.2090	0.2119	−0.8
0.2148	0.2177	0.2206	0.2236	0.2266	0.2296	0.2327	0.2358	0.2389	0.2420	−0.7
0.2451	0.2483	0.2514	0.2546	0.2578	0.2611	0.2643	0.2676	0.2709	0.2743	−0.6
0.2776	0.2810	0.2843	0.2877	0.2912	0.2946	0.2981	0.3015	0.3050	0.3085	−0.5
0.3121	0.3156	0.3192	0.3228	0.3264	0.3300	0.3336	0.3372	0.3409	0.3446	−0.4
0.3483	0.3520	0.3557	0.3594	0.3632	0.3669	0.3707	0.3745	0.3783	0.3821	−0.3
0.3859	0.3897	0.3936	0.3974	0.4013	0.4052	0.4090	0.4129	0.4168	0.4207	−0.2
0.4247	0.4286	0.4325	0.4364	0.4404	0.4443	0.4483	0.4522	0.4562	0.4602	−0.1
0.4641	0.4681	0.4721	0.4761	0.4801	0.4840	0.4880	0.4920	0.4960	0.5000	−0.0

†For $z \leq -3.90$, the areas are 0.0000 to four decimal places.

Table Z (cont.)
Areas under the standard Normal curve

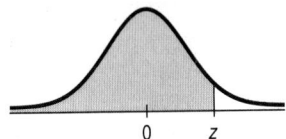

z	0.00	0.01	0.02	0.03	0.04	0.05	0.06	0.07	0.08	0.09
0.0	0.5000	0.5040	0.5080	0.5120	0.5160	0.5199	0.5239	0.5279	0.5319	0.5359
0.1	0.5398	0.5438	0.5478	0.5517	0.5557	0.5596	0.5636	0.5675	0.5714	0.5753
0.2	0.5793	0.5832	0.5871	0.5910	0.5948	0.5987	0.6026	0.6064	0.6103	0.6141
0.3	0.6179	0.6217	0.6255	0.6293	0.6331	0.6368	0.6406	0.6443	0.6480	0.6517
0.4	0.6554	0.6591	0.6628	0.6664	0.6700	0.6736	0.6772	0.6808	0.6844	0.6879
0.5	0.6915	0.6950	0.6985	0.7019	0.7054	0.7088	0.7123	0.7157	0.7190	0.7224
0.6	0.7257	0.7291	0.7324	0.7357	0.7389	0.7422	0.7454	0.7486	0.7517	0.7549
0.7	0.7580	0.7611	0.7642	0.7673	0.7704	0.7734	0.7764	0.7794	0.7823	0.7852
0.8	0.7881	0.7910	0.7939	0.7967	0.7995	0.8023	0.8051	0.8078	0.8106	0.8133
0.9	0.8159	0.8186	0.8212	0.8238	0.8264	0.8289	0.8315	0.8340	0.8365	0.8389
1.0	0.8413	0.8438	0.8461	0.8485	0.8508	0.8531	0.8554	0.8577	0.8599	0.8621
1.1	0.8643	0.8665	0.8686	0.8708	0.8729	0.8749	0.8770	0.8790	0.8810	0.8830
1.2	0.8849	0.8869	0.8888	0.8907	0.8925	0.8944	0.8962	0.8980	0.8997	0.9015
1.3	0.9032	0.9049	0.9066	0.9082	0.9099	0.9115	0.9131	0.9147	0.9162	0.9177
1.4	0.9192	0.9207	0.9222	0.9236	0.9251	0.9265	0.9279	0.9292	0.9306	0.9319
1.5	0.9332	0.9345	0.9357	0.9370	0.9382	0.9394	0.9406	0.9418	0.9429	0.9441
1.6	0.9452	0.9463	0.9474	0.9484	0.9495	0.9505	0.9515	0.9525	0.9535	0.9545
1.7	0.9554	0.9564	0.9573	0.9582	0.9591	0.9599	0.9608	0.9616	0.9625	0.9633
1.8	0.9641	0.9649	0.9656	0.9664	0.9671	0.9678	0.9686	0.9693	0.9699	0.9706
1.9	0.9713	0.9719	0.9726	0.9732	0.9738	0.9744	0.9750	0.9756	0.9761	0.9767
2.0	0.9772	0.9778	0.9783	0.9788	0.9793	0.9798	0.9803	0.9808	0.9812	0.9817
2.1	0.9821	0.9826	0.9830	0.9834	0.9838	0.9842	0.9846	0.9850	0.9854	0.9857
2.2	0.9861	0.9864	0.9868	0.9871	0.9875	0.9878	0.9881	0.9884	0.9887	0.9890
2.3	0.9893	0.9896	0.9898	0.9901	0.9904	0.9906	0.9909	0.9911	0.9913	0.9916
2.4	0.9918	0.9920	0.9922	0.9925	0.9927	0.9929	0.9931	0.9932	0.9934	0.9936
2.5	0.9938	0.9940	0.9941	0.9943	0.9945	0.9946	0.9948	0.9949	0.9951	0.9952
2.6	0.9953	0.9955	0.9956	0.9957	0.9959	0.9960	0.9961	0.9962	0.9963	0.9964
2.7	0.9965	0.9966	0.9967	0.9968	0.9969	0.9970	0.9971	0.9972	0.9973	0.9974
2.8	0.9974	0.9975	0.9976	0.9977	0.9977	0.9978	0.9979	0.9979	0.9980	0.9981
2.9	0.9981	0.9982	0.9982	0.9983	0.9984	0.9984	0.9985	0.9985	0.9986	0.9986
3.0	0.9987	0.9987	0.9987	0.9988	0.9988	0.9989	0.9989	0.9989	0.9990	0.9990
3.1	0.9990	0.9991	0.9991	0.9991	0.9992	0.9992	0.9992	0.9992	0.9993	0.9993
3.2	0.9993	0.9993	0.9994	0.9994	0.9994	0.9994	0.9994	0.9995	0.9995	0.9995
3.3	0.9995	0.9995	0.9995	0.9996	0.9996	0.9996	0.9996	0.9996	0.9996	0.9997
3.4	0.9997	0.9997	0.9997	0.9997	0.9997	0.9997	0.9997	0.9997	0.9997	0.9998
3.5	0.9998	0.9998	0.9998	0.9998	0.9998	0.9998	0.9998	0.9998	0.9998	0.9998
3.6	0.9998	0.9998	0.9999	0.9999	0.9999	0.9999	0.9999	0.9999	0.9999	0.9999
3.7	0.9999	0.9999	0.9999	0.9999	0.9999	0.9999	0.9999	0.9999	0.9999	0.9999
3.8	0.9999	0.9999	0.9999	0.9999	0.9999	0.9999	0.9999	0.9999	0.9999	0.9999
3.9	1.0000†									

†For $z \geq 3.90$, the areas are 1.0000 to four decimal places.

Table T
Values of t_α

Two-tail probability	0.20	0.10	0.05	0.02	0.01
One-tail probability	0.10	0.05	0.025	0.01	0.005
df					
1	3.078	6.314	12.706	31.821	63.657
2	1.886	2.920	4.303	6.965	9.925
3	1.638	2.353	3.182	4.541	5.841
4	1.533	2.132	2.776	3.747	4.604
5	1.476	2.015	2.571	3.365	4.032
6	1.440	1.943	2.447	3.143	3.707
7	1.415	1.895	2.365	2.998	3.499
8	1.397	1.860	2.306	2.896	3.355
9	1.383	1.833	2.262	2.821	3.250
10	1.372	1.812	2.228	2.764	3.169
11	1.363	1.796	2.201	2.718	3.106
12	1.356	1.782	2.179	2.681	3.055
13	1.350	1.771	2.160	2.650	3.012
14	1.345	1.761	2.145	2.624	2.977
15	1.341	1.753	2.131	2.602	2.947
16	1.337	1.746	2.120	2.583	2.921
17	1.333	1.740	2.110	2.567	2.898
18	1.330	1.734	2.101	2.552	2.878
19	1.328	1.729	2.093	2.539	2.861
20	1.325	1.725	2.086	2.528	2.845
21	1.323	1.721	2.080	2.518	2.831
22	1.321	1.717	2.074	2.508	2.819
23	1.319	1.714	2.069	2.500	2.807
24	1.318	1.711	2.064	2.492	2.797
25	1.316	1.708	2.060	2.485	2.787
26	1.315	1.706	2.056	2.479	2.779
27	1.314	1.703	2.052	2.473	2.771
28	1.313	1.701	2.048	2.467	2.763
29	1.311	1.699	2.045	2.462	2.756
30	1.310	1.697	2.042	2.457	2.750
32	1.309	1.694	2.037	2.449	2.738
35	1.306	1.690	2.030	2.438	2.725
40	1.303	1.684	2.021	2.423	2.704
45	1.301	1.679	2.014	2.412	2.690
50	1.299	1.676	2.009	2.403	2.678
60	1.296	1.671	2.000	2.390	2.660
75	1.293	1.665	1.992	2.377	2.643
100	1.290	1.660	1.984	2.364	2.626
120	1.289	1.658	1.980	2.358	2.617
140	1.288	1.656	1.977	2.353	2.611
180	1.286	1.653	1.973	2.347	2.603
250	1.285	1.651	1.969	2.341	2.596
400	1.284	1.649	1.966	2.336	2.588
1000	1.282	1.646	1.962	2.330	2.581
∞	1.282	1.645	1.960	2.326	2.576
Confidence levels	80%	90%	95%	98%	99%

Table X
Values of χ^2_α

Right-tail probability df	0.10	0.05	0.025	0.01	0.005
1	2.706	3.841	5.024	6.635	7.879
2	4.605	5.991	7.378	9.210	10.597
3	6.251	7.815	9.348	11.345	12.838
4	7.779	9.488	11.143	13.277	14.860
5	9.236	11.070	12.833	15.086	16.750
6	10.645	12.592	14.449	16.812	18.548
7	12.017	14.067	16.013	18.475	20.278
8	13.362	15.507	17.535	20.090	21.955
9	14.684	16.919	19.023	21.666	23.589
10	15.987	18.307	20.483	23.209	25.188
11	17.275	19.675	21.920	24.725	26.757
12	18.549	21.026	23.337	26.217	28.300
13	19.812	22.362	24.736	27.688	29.819
14	21.064	23.685	26.119	29.141	31.319
15	22.307	24.996	27.488	30.578	32.801
16	23.542	26.296	28.845	32.000	34.267
17	24.769	27.587	30.191	33.409	35.718
18	25.989	28.869	31.526	34.805	37.156
19	27.204	30.143	32.852	36.191	38.582
20	28.412	31.410	34.170	37.566	39.997
21	29.615	32.671	35.479	38.932	41.401
22	30.813	33.924	36.781	40.290	42.796
23	32.007	35.172	38.076	41.638	44.181
24	33.196	36.415	39.364	42.980	45.559
25	34.382	37.653	40.647	44.314	46.928
26	35.563	38.885	41.923	45.642	48.290
27	36.741	40.113	43.195	46.963	49.645
28	37.916	41.337	44.461	48.278	50.994
29	39.087	42.557	45.722	59.588	52.336
30	40.256	43.773	46.979	50.892	53.672
40	51.805	55.759	59.342	63.691	66.767
50	63.167	67.505	71.420	76.154	79.490
60	74.397	79.082	83.298	88.381	91.955
70	85.527	90.531	95.023	100.424	104.213
80	96.578	101.879	106.628	112.328	116.320
90	107.565	113.145	118.135	124.115	128.296
100	118.499	124.343	129.563	135.811	140.177

Table F

Numerator dF

Denominator df	1	2	3	4	5	6	7	8	9	10	11	12	13	14	15	16	17	18	19	20	21	22
1	4052.2	4999.3	5403.5	5624.3	5764.0	5859.0	5928.3	5981.0	6022.4	6055.9	6083.4	6106.7	6125.8	6143.0	6157.0	6170.0	6181.2	6191.4	6200.7	6208.7	6216.1	6223.1
2	98.50	99.00	99.16	99.25	99.30	99.33	99.36	99.38	99.39	99.40	99.41	99.42	99.42	99.43	99.43	99.44	99.44	99.44	99.45	99.45	99.45	99.46
3	34.12	30.82	29.46	28.71	28.24	27.91	27.67	27.49	27.34	27.23	27.13	27.05	26.98	26.92	26.87	26.83	26.79	26.75	26.72	26.69	26.66	26.64
4	21.20	18.00	16.69	15.98	15.52	15.21	14.98	14.80	14.66	14.55	14.45	14.37	14.31	14.25	14.20	14.15	14.11	14.08	14.05	14.02	13.99	13.97
5	16.26	13.27	12.06	11.39	10.97	10.67	10.46	10.29	10.16	10.05	9.96	9.89	9.82	9.77	9.72	9.68	9.64	9.61	9.58	9.55	9.53	9.51
6	13.75	10.92	9.78	9.15	8.75	8.47	8.26	8.10	7.98	7.87	7.79	7.72	7.66	7.60	7.56	7.52	7.48	7.45	7.42	7.40	7.37	7.35
7	12.25	9.55	8.45	7.85	7.46	7.19	6.99	6.84	6.72	6.62	6.54	6.47	6.41	6.36	6.31	6.28	6.24	6.21	6.18	6.16	6.13	6.11
8	11.26	8.65	7.59	7.01	6.63	6.37	6.18	6.03	5.91	5.81	5.73	5.67	5.61	5.56	5.52	5.48	5.44	5.41	5.38	5.36	5.34	5.32
9	10.56	8.02	6.99	6.42	6.06	5.80	5.61	5.47	5.35	5.26	5.18	5.11	5.05	5.01	4.96	4.92	4.89	4.86	4.83	4.81	4.79	4.77
10	10.04	7.56	6.55	5.99	5.64	5.39	5.20	5.06	4.94	4.85	4.77	4.71	4.65	4.60	4.56	4.52	4.49	4.46	4.43	4.41	4.38	4.36
11	9.65	7.21	6.22	5.67	5.32	5.07	4.89	4.74	4.63	4.54	4.46	4.40	4.34	4.29	4.25	4.21	4.18	4.15	4.12	4.10	4.08	4.06
12	9.33	6.93	5.95	5.41	5.06	4.82	4.64	4.50	4.39	4.30	4.22	4.16	4.10	4.05	4.01	3.97	3.94	3.91	3.88	3.86	3.84	3.82
13	9.07	6.70	5.74	5.21	4.86	4.62	4.44	4.30	4.19	4.10	4.02	3.96	3.91	3.86	3.82	3.78	3.75	3.72	3.69	3.66	3.64	3.62
14	8.86	6.51	5.56	5.04	4.69	4.46	4.28	4.14	4.03	3.94	3.86	3.80	3.75	3.70	3.66	3.62	3.59	3.56	3.53	3.51	3.48	3.46
15	8.68	6.36	5.42	4.89	4.56	4.32	4.14	4.00	3.89	3.80	3.73	3.67	3.61	3.56	3.52	3.49	3.45	3.42	3.40	3.37	3.35	3.33
16	8.53	6.23	5.29	4.77	4.44	4.20	4.03	3.89	3.78	3.69	3.62	3.55	3.50	3.45	3.41	3.37	3.34	3.31	3.28	3.26	3.24	3.22
17	8.40	6.11	5.19	4.67	4.34	4.10	3.93	3.79	3.68	3.59	3.52	3.46	3.40	3.35	3.31	3.27	3.24	3.21	3.19	3.16	3.14	3.12
18	8.29	6.01	5.09	4.58	4.25	4.01	3.84	3.71	3.60	3.51	3.43	3.37	3.32	3.27	3.23	3.19	3.16	3.13	3.10	3.08	3.05	3.03
19	8.18	5.93	5.01	4.50	4.17	3.94	3.77	3.63	3.52	3.43	3.36	3.30	3.24	3.19	3.15	3.12	3.08	3.05	3.03	3.00	2.98	2.96
20	8.10	5.85	4.94	4.43	4.10	3.87	3.70	3.56	3.46	3.37	3.29	3.23	3.18	3.13	3.09	3.05	3.02	2.99	2.96	2.94	2.92	2.90
21	8.02	5.78	4.87	4.37	4.04	3.81	3.64	3.51	3.40	3.31	3.24	3.17	3.12	3.07	3.03	2.99	2.96	2.93	2.90	2.88	2.86	2.84
22	7.95	5.72	4.82	4.31	3.99	3.76	3.59	3.45	3.35	3.26	3.18	3.12	3.07	3.02	2.98	2.94	2.91	2.88	2.85	2.83	2.81	2.78
23	7.88	5.66	4.76	4.26	3.94	3.71	3.54	3.41	3.30	3.21	3.14	3.07	3.02	2.97	2.93	2.89	2.86	2.83	2.80	2.78	2.76	2.74
24	7.82	5.61	4.72	4.22	3.90	3.67	3.50	3.36	3.26	3.17	3.09	3.03	2.98	2.93	2.89	2.85	2.82	2.79	2.76	2.74	2.72	2.70
25	7.77	5.57	4.68	4.18	3.85	3.63	3.46	3.32	3.22	3.13	3.06	2.99	2.94	2.89	2.85	2.81	2.78	2.75	2.72	2.70	2.68	2.66
26	7.72	5.53	4.64	4.14	3.82	3.59	3.42	3.29	3.18	3.09	3.02	2.96	2.90	2.86	2.81	2.78	2.75	2.72	2.69	2.66	2.64	2.62
27	7.68	5.49	4.60	4.11	3.78	3.56	3.39	3.26	3.15	3.06	2.99	2.93	2.87	2.82	2.78	2.75	2.71	2.68	2.66	2.63	2.61	2.59
28	7.64	5.45	4.57	4.07	3.75	3.53	3.36	3.23	3.12	3.03	2.96	2.90	2.84	2.79	2.75	2.72	2.68	2.65	2.63	2.60	2.58	2.56
29	7.60	5.42	4.54	4.04	3.73	3.50	3.33	3.20	3.09	3.00	2.93	2.87	2.81	2.77	2.73	2.69	2.66	2.63	2.60	2.57	2.55	2.53
30	7.56	5.39	4.51	4.02	3.70	3.47	3.30	3.17	3.07	2.98	2.91	2.84	2.79	2.74	2.70	2.66	2.63	2.60	2.57	2.55	2.53	2.51
32	7.50	5.34	4.46	3.97	3.65	3.43	3.26	3.13	3.02	2.93	2.86	2.80	2.74	2.70	2.65	2.62	2.58	2.55	2.53	2.50	2.48	2.46
35	7.42	5.27	4.40	3.91	3.59	3.37	3.20	3.07	2.96	2.88	2.80	2.74	2.69	2.64	2.60	2.56	2.53	2.50	2.47	2.44	2.42	2.40
40	7.31	5.18	4.31	3.83	3.51	3.29	3.12	2.99	2.89	2.80	2.73	2.66	2.61	2.56	2.52	2.48	2.45	2.42	2.39	2.37	2.35	2.33
45	7.23	5.11	4.25	3.77	3.45	3.23	3.07	2.94	2.83	2.74	2.67	2.61	2.55	2.51	2.46	2.43	2.39	2.36	2.34	2.31	2.29	2.27
50	7.17	5.06	4.20	3.72	3.41	3.19	3.02	2.89	2.78	2.70	2.63	2.56	2.51	2.46	2.42	2.38	2.35	2.32	2.29	2.27	2.24	2.22
60	7.08	4.98	4.13	3.65	3.34	3.12	2.95	2.82	2.72	2.63	2.56	2.50	2.44	2.39	2.35	2.31	2.28	2.25	2.22	2.20	2.17	2.15
75	6.99	4.90	4.05	3.58	3.27	3.05	2.89	2.76	2.65	2.57	2.49	2.43	2.38	2.33	2.29	2.25	2.22	2.18	2.16	2.13	2.11	2.09
100	6.90	4.82	3.98	3.51	3.21	2.99	2.82	2.69	2.59	2.50	2.43	2.37	2.31	2.27	2.22	2.19	2.15	2.12	2.09	2.07	2.04	2.02
120	6.85	4.79	3.95	3.48	3.17	2.96	2.79	2.66	2.56	2.47	2.40	2.34	2.28	2.23	2.19	2.15	2.12	2.09	2.06	2.03	2.01	1.99
140	6.82	4.76	3.92	3.46	3.15	2.93	2.77	2.64	2.54	2.45	2.38	2.31	2.26	2.21	2.17	2.13	2.10	2.07	2.04	2.01	1.99	1.97
180	6.78	4.73	3.89	3.43	3.12	2.90	2.74	2.61	2.51	2.42	2.35	2.28	2.23	2.18	2.14	2.10	2.07	2.04	2.01	1.98	1.96	1.94
250	6.74	4.69	3.86	3.40	3.09	2.87	2.71	2.58	2.48	2.39	2.32	2.26	2.20	2.15	2.11	2.07	2.04	2.01	1.98	1.95	1.93	1.91
400	6.70	4.66	3.83	3.37	3.06	2.85	2.68	2.56	2.45	2.37	2.29	2.23	2.17	2.13	2.08	2.05	2.01	1.98	1.95	1.92	1.90	1.88
1000	6.66	4.63	3.80	3.34	3.04	2.82	2.66	2.53	2.43	2.34	2.27	2.20	2.15	2.10	2.06	2.02	1.98	1.95	1.92	1.90	1.87	1.85

Table F (cont.) α = .01

Numerator dF

Denominator df	23	24	25	26	27	28	29	30	32	35	40	45	50	60	75	100	120	140	180	250	400	1000
1	6228.7	6234.3	6239.9	6244.5	6249.2	6252.9	6257.1	6260.4	6266.9	6275.3	6286.4	6295.7	6302.3	6313.0	6323.7	6333.9	6339.5	6343.2	6347.9	6353.5	6358.1	6362.8
2	99.46	99.46	99.46	99.46	99.46	99.46	99.46	99.47	99.47	99.47	99.48	99.48	99.48	99.48	99.48	99.49	99.49	99.49	99.49	99.50	99.50	99.50
3	26.62	26.60	26.58	26.56	26.55	26.53	26.52	26.50	26.48	26.45	26.41	26.38	26.35	26.32	26.28	26.24	26.22	26.21	26.19	26.17	26.15	26.14
4	13.95	13.93	13.91	13.89	13.88	13.86	13.85	13.84	13.81	13.79	13.75	13.71	13.69	13.65	13.61	13.58	13.56	13.54	13.53	13.51	13.49	13.47
5	9.49	9.47	9.45	9.43	9.42	9.40	9.39	9.38	9.36	9.33	9.29	9.26	9.24	9.20	9.17	9.13	9.11	9.10	9.08	9.06	9.05	9.03
6	7.33	7.31	7.30	7.28	7.27	7.25	7.24	7.23	7.21	7.18	7.14	7.11	7.09	7.06	7.02	6.99	6.97	6.96	6.94	6.92	6.91	6.89
7	6.09	6.07	6.06	6.04	6.03	6.02	6.00	5.99	5.97	5.94	5.91	5.88	5.86	5.82	5.79	5.75	5.74	5.72	5.71	5.69	5.68	5.66
8	5.30	5.28	5.26	5.25	5.23	5.22	5.21	5.20	5.18	5.15	5.12	5.09	5.07	5.03	5.00	4.96	4.95	4.93	4.92	4.90	4.89	4.87
9	4.75	4.73	4.71	4.70	4.68	4.67	4.66	4.65	4.63	4.60	4.57	4.54	4.52	4.48	4.45	4.41	4.40	4.39	4.37	4.35	4.34	4.32
10	4.34	4.33	4.31	4.30	4.28	4.27	4.26	4.25	4.23	4.20	4.17	4.14	4.12	4.08	4.05	4.01	4.00	3.98	3.97	3.95	3.94	3.92
11	4.04	4.02	4.01	3.99	3.98	3.96	3.95	3.94	3.92	3.89	3.86	3.83	3.81	3.78	3.74	3.71	3.69	3.68	3.66	3.64	3.63	3.61
12	3.80	3.78	3.76	3.75	3.74	3.72	3.71	3.70	3.68	3.65	3.62	3.59	3.57	3.54	3.50	3.47	3.45	3.44	3.42	3.40	3.39	3.37
13	3.60	3.59	3.57	3.56	3.54	3.53	3.52	3.51	3.49	3.46	3.43	3.40	3.38	3.34	3.31	3.27	3.25	3.24	3.23	3.21	3.19	3.18
14	3.44	3.43	3.41	3.40	3.38	3.37	3.36	3.35	3.33	3.30	3.27	3.24	3.22	3.18	3.15	3.11	3.09	3.08	3.06	3.05	3.03	3.02
15	3.31	3.29	3.28	3.26	3.25	3.24	3.23	3.21	3.19	3.17	3.13	3.10	3.08	3.05	3.01	2.98	2.96	2.95	2.93	2.91	2.90	2.88
16	3.20	3.18	3.16	3.15	3.14	3.12	3.11	3.10	3.08	3.05	3.02	2.99	2.97	2.93	2.90	2.86	2.84	2.83	2.81	2.80	2.78	2.76
17	3.10	3.08	3.07	3.05	3.04	3.03	3.01	3.00	2.98	2.96	2.92	2.89	2.87	2.83	2.80	2.76	2.75	2.73	2.72	2.70	2.68	2.66
18	3.02	3.00	2.98	2.97	2.95	2.94	2.93	2.92	2.90	2.87	2.84	2.81	2.78	2.75	2.71	2.68	2.66	2.65	2.63	2.61	2.59	2.58
19	2.94	2.92	2.91	2.89	2.88	2.87	2.86	2.84	2.82	2.80	2.76	2.73	2.71	2.67	2.64	2.60	2.58	2.57	2.55	2.54	2.52	2.50
20	2.88	2.86	2.84	2.83	2.81	2.80	2.79	2.78	2.76	2.73	2.69	2.67	2.64	2.61	2.57	2.54	2.52	2.50	2.49	2.47	2.45	2.43
21	2.82	2.80	2.79	2.77	2.76	2.74	2.73	2.72	2.70	2.67	2.64	2.61	2.58	2.55	2.51	2.48	2.46	2.44	2.43	2.41	2.39	2.37
22	2.77	2.75	2.73	2.72	2.70	2.69	2.68	2.67	2.65	2.62	2.58	2.55	2.53	2.50	2.46	2.42	2.40	2.39	2.37	2.35	2.34	2.32
23	2.72	2.70	2.69	2.67	2.66	2.64	2.63	2.62	2.60	2.57	2.54	2.51	2.48	2.45	2.41	2.37	2.35	2.34	2.32	2.30	2.29	2.27
24	2.68	2.66	2.64	2.63	2.61	2.60	2.59	2.58	2.56	2.53	2.49	2.46	2.44	2.40	2.37	2.33	2.31	2.30	2.28	2.26	2.24	2.22
25	2.64	2.62	2.60	2.59	2.58	2.56	2.55	2.54	2.52	2.49	2.45	2.42	2.40	2.36	2.33	2.29	2.27	2.26	2.24	2.22	2.20	2.18
26	2.60	2.58	2.57	2.55	2.54	2.53	2.51	2.50	2.48	2.45	2.42	2.39	2.36	2.33	2.29	2.25	2.23	2.22	2.20	2.18	2.16	2.14
27	2.57	2.55	2.54	2.52	2.51	2.49	2.48	2.47	2.45	2.42	2.38	2.35	2.33	2.29	2.26	2.22	2.20	2.18	2.17	2.15	2.13	2.11
28	2.54	2.52	2.51	2.49	2.48	2.46	2.45	2.44	2.42	2.39	2.35	2.32	2.30	2.26	2.23	2.19	2.17	2.15	2.13	2.11	2.10	2.08
29	2.51	2.49	2.48	2.46	2.45	2.44	2.42	2.41	2.39	2.36	2.33	2.30	2.27	2.23	2.20	2.16	2.14	2.12	2.10	2.08	2.07	2.05
30	2.49	2.47	2.45	2.44	2.42	2.41	2.40	2.39	2.36	2.34	2.30	2.27	2.25	2.21	2.17	2.13	2.11	2.10	2.08	2.06	2.04	2.02
32	2.44	2.42	2.41	2.39	2.38	2.36	2.35	2.34	2.32	2.29	2.25	2.22	2.20	2.16	2.12	2.08	2.06	2.05	2.03	2.01	1.99	1.97
35	2.38	2.36	2.35	2.33	2.32	2.30	2.29	2.28	2.26	2.23	2.19	2.16	2.14	2.10	2.06	2.02	2.00	1.98	1.96	1.94	1.92	1.90
40	2.31	2.29	2.27	2.26	2.24	2.23	2.22	2.20	2.18	2.15	2.11	2.08	2.06	2.02	1.98	1.94	1.92	1.90	1.88	1.86	1.84	1.82
45	2.25	2.23	2.21	2.20	2.18	2.17	2.16	2.14	2.12	2.09	2.05	2.02	2.00	1.96	1.92	1.88	1.85	1.84	1.82	1.79	1.77	1.75
50	2.20	2.18	2.17	2.15	2.14	2.12	2.11	2.10	2.08	2.05	2.01	1.97	1.95	1.91	1.87	1.82	1.80	1.79	1.76	1.74	1.72	1.70
60	2.13	2.12	2.10	2.08	2.07	2.05	2.04	2.03	2.01	1.98	1.94	1.90	1.88	1.84	1.79	1.75	1.73	1.71	1.69	1.66	1.64	1.62
75	2.07	2.05	2.03	2.02	2.00	1.99	1.97	1.96	1.94	1.91	1.87	1.83	1.81	1.76	1.72	1.67	1.65	1.63	1.61	1.58	1.56	1.53
100	2.00	1.98	1.97	1.95	1.93	1.92	1.91	1.89	1.87	1.84	1.80	1.76	1.74	1.69	1.65	1.60	1.57	1.55	1.53	1.50	1.47	1.45
120	1.97	1.95	1.93	1.92	1.90	1.89	1.87	1.86	1.84	1.81	1.76	1.73	1.70	1.66	1.61	1.56	1.53	1.51	1.49	1.46	1.43	1.40
140	1.95	1.93	1.91	1.89	1.88	1.86	1.85	1.84	1.81	1.78	1.74	1.70	1.67	1.63	1.58	1.53	1.50	1.48	1.46	1.43	1.40	1.37
180	1.92	1.90	1.88	1.86	1.85	1.83	1.82	1.81	1.78	1.75	1.71	1.67	1.64	1.60	1.55	1.49	1.47	1.45	1.42	1.39	1.35	1.32
250	1.89	1.87	1.85	1.83	1.82	1.80	1.79	1.77	1.75	1.72	1.67	1.64	1.61	1.56	1.51	1.46	1.43	1.41	1.38	1.34	1.31	1.27
400	1.86	1.84	1.82	1.80	1.79	1.77	1.76	1.75	1.72	1.69	1.64	1.61	1.58	1.53	1.48	1.42	1.39	1.37	1.33	1.30	1.26	1.22
1000	1.83	1.81	1.79	1.77	1.76	1.74	1.73	1.72	1.69	1.66	1.61	1.58	1.54	1.50	1.44	1.38	1.35	1.33	1.29	1.25	1.21	1.16

Table F (cont.)

$\alpha = .05$

Numerator df

Denominator df	1	2	3	4	5	6	7	8	9	10	11	12	13	14	15	16	17	18	19	20	21	22
1	161.4	199.5	215.7	224.6	230.2	234.0	236.8	238.9	240.5	241.9	243.0	243.9	244.7	245.4	245.9	246.5	246.9	247.3	247.7	248.0	248.3	248.6
2	18.51	19.00	19.16	19.25	19.30	19.33	19.35	19.37	19.38	19.40	19.40	19.41	19.42	19.42	19.43	19.43	19.44	19.44	19.44	19.45	19.45	19.45
3	10.13	9.55	9.28	9.12	9.01	8.94	8.89	8.85	8.81	8.79	8.76	8.74	8.73	8.71	8.70	8.69	8.68	8.67	8.67	8.66	8.65	8.65
4	7.71	6.94	6.59	6.39	6.26	6.16	6.09	6.04	6.00	5.96	5.94	5.91	5.89	5.87	5.86	5.84	5.83	5.82	5.81	5.80	5.79	5.79
5	6.61	5.79	5.41	5.19	5.05	4.95	4.88	4.82	4.77	4.74	4.70	4.68	4.66	4.64	4.62	4.60	4.59	4.58	4.57	4.56	4.55	4.54
6	5.99	5.14	4.76	4.53	4.39	4.28	4.21	4.15	4.10	4.06	4.03	4.00	3.98	3.96	3.94	3.92	3.91	3.90	3.88	3.87	3.86	3.86
7	5.59	4.74	4.35	4.12	3.97	3.87	3.79	3.73	3.68	3.64	3.60	3.57	3.55	3.53	3.51	3.49	3.48	3.47	3.46	3.44	3.43	3.43
8	5.32	4.46	4.07	3.84	3.69	3.58	3.50	3.44	3.39	3.35	3.31	3.28	3.26	3.24	3.22	3.20	3.19	3.17	3.16	3.15	3.14	3.13
9	5.12	4.26	3.86	3.63	3.48	3.37	3.29	3.23	3.18	3.14	3.10	3.07	3.05	3.03	3.01	2.99	2.97	2.96	2.95	2.94	2.93	2.92
10	4.96	4.10	3.71	3.48	3.33	3.22	3.14	3.07	3.02	2.98	2.94	2.91	2.89	2.86	2.85	2.83	2.81	2.80	2.79	2.77	2.76	2.75
11	4.84	3.98	3.59	3.36	3.20	3.09	3.01	2.95	2.90	2.85	2.82	2.79	2.76	2.74	2.72	2.70	2.69	2.67	2.66	2.65	2.64	2.63
12	4.75	3.89	3.49	3.26	3.11	3.00	2.91	2.85	2.80	2.75	2.72	2.69	2.66	2.64	2.62	2.60	2.58	2.57	2.56	2.54	2.53	2.52
13	4.67	3.81	3.41	3.18	3.03	2.92	2.83	2.77	2.71	2.67	2.63	2.60	2.58	2.55	2.53	2.51	2.50	2.48	2.47	2.46	2.45	2.44
14	4.60	3.74	3.34	3.11	2.96	2.85	2.76	2.70	2.65	2.60	2.57	2.53	2.51	2.48	2.46	2.44	2.43	2.41	2.40	2.39	2.38	2.37
15	4.54	3.68	3.29	3.06	2.90	2.79	2.71	2.64	2.59	2.54	2.51	2.48	2.45	2.42	2.40	2.38	2.37	2.35	2.34	2.33	2.32	2.31
16	4.49	3.63	3.24	3.01	2.85	2.74	2.66	2.59	2.54	2.49	2.46	2.42	2.40	2.37	2.35	2.33	2.32	2.30	2.29	2.28	2.26	2.25
17	4.45	3.59	3.20	2.96	2.81	2.70	2.61	2.55	2.49	2.45	2.41	2.38	2.35	2.33	2.31	2.29	2.27	2.26	2.24	2.23	2.22	2.21
18	4.41	3.55	3.16	2.93	2.77	2.66	2.58	2.51	2.46	2.41	2.37	2.34	2.31	2.29	2.27	2.25	2.23	2.22	2.20	2.19	2.18	2.17
19	4.38	3.52	3.13	2.90	2.74	2.63	2.54	2.48	2.42	2.38	2.34	2.31	2.28	2.26	2.23	2.21	2.20	2.18	2.17	2.16	2.14	2.13
20	4.35	3.49	3.10	2.87	2.71	2.60	2.51	2.45	2.39	2.35	2.31	2.28	2.25	2.22	2.20	2.18	2.17	2.15	2.14	2.12	2.11	2.10
21	4.32	3.47	3.07	2.84	2.68	2.57	2.49	2.42	2.37	2.32	2.28	2.25	2.22	2.20	2.18	2.16	2.14	2.12	2.11	2.10	2.08	2.07
22	4.30	3.44	3.05	2.82	2.66	2.55	2.46	2.40	2.34	2.30	2.26	2.23	2.20	2.17	2.15	2.13	2.11	2.10	2.08	2.07	2.06	2.05
23	4.28	3.42	3.03	2.80	2.64	2.53	2.44	2.37	2.32	2.27	2.24	2.20	2.18	2.15	2.13	2.11	2.09	2.08	2.06	2.05	2.04	2.02
24	4.26	3.40	3.01	2.78	2.62	2.51	2.42	2.36	2.30	2.25	2.22	2.18	2.15	2.13	2.11	2.09	2.07	2.05	2.04	2.03	2.01	2.00
25	4.24	3.39	2.99	2.76	2.60	2.49	2.40	2.34	2.28	2.24	2.20	2.16	2.14	2.11	2.09	2.07	2.05	2.04	2.02	2.01	2.00	1.98
26	4.23	3.37	2.98	2.74	2.59	2.47	2.39	2.32	2.27	2.22	2.18	2.15	2.12	2.09	2.07	2.05	2.03	2.02	2.00	1.99	1.98	1.97
27	4.21	3.35	2.96	2.73	2.57	2.46	2.37	2.31	2.25	2.20	2.17	2.13	2.10	2.08	2.06	2.04	2.02	2.00	1.99	1.97	1.96	1.95
28	4.20	3.34	2.95	2.71	2.56	2.45	2.36	2.29	2.24	2.19	2.15	2.12	2.09	2.06	2.04	2.02	2.00	1.99	1.97	1.96	1.95	1.93
29	4.18	3.33	2.93	2.70	2.55	2.43	2.35	2.28	2.22	2.18	2.14	2.10	2.08	2.05	2.03	2.01	1.99	1.97	1.96	1.94	1.93	1.92
30	4.17	3.32	2.92	2.69	2.53	2.42	2.33	2.27	2.21	2.16	2.13	2.09	2.06	2.04	2.01	1.99	1.98	1.96	1.95	1.93	1.92	1.91
32	4.15	3.29	2.90	2.67	2.51	2.40	2.31	2.24	2.19	2.14	2.10	2.07	2.04	2.01	1.99	1.97	1.95	1.94	1.92	1.91	1.90	1.88
35	4.12	3.27	2.87	2.64	2.49	2.37	2.29	2.22	2.16	2.11	2.07	2.04	2.01	1.99	1.96	1.94	1.92	1.91	1.89	1.88	1.87	1.85
40	4.08	3.23	2.84	2.61	2.45	2.34	2.25	2.18	2.12	2.08	2.04	2.00	1.97	1.95	1.92	1.90	1.89	1.87	1.85	1.84	1.83	1.81
45	4.06	3.20	2.81	2.58	2.42	2.31	2.22	2.15	2.10	2.05	2.01	1.97	1.94	1.92	1.89	1.87	1.86	1.84	1.82	1.81	1.80	1.78
50	4.03	3.18	2.79	2.56	2.40	2.29	2.20	2.13	2.07	2.03	1.99	1.95	1.92	1.89	1.87	1.85	1.83	1.81	1.80	1.78	1.77	1.76
60	4.00	3.15	2.76	2.53	2.37	2.25	2.17	2.10	2.04	1.99	1.95	1.92	1.89	1.86	1.84	1.82	1.80	1.78	1.76	1.75	1.73	1.72
75	3.97	3.12	2.73	2.49	2.34	2.22	2.13	2.06	2.01	1.96	1.92	1.88	1.85	1.83	1.80	1.78	1.76	1.74	1.73	1.71	1.70	1.69
100	3.94	3.09	2.70	2.46	2.31	2.19	2.10	2.03	1.97	1.93	1.89	1.85	1.82	1.79	1.77	1.75	1.73	1.71	1.69	1.68	1.66	1.65
120	3.92	3.07	2.68	2.45	2.29	2.18	2.09	2.02	1.96	1.91	1.87	1.83	1.80	1.78	1.75	1.73	1.71	1.69	1.67	1.66	1.64	1.63
140	3.91	3.06	2.67	2.44	2.28	2.16	2.08	2.01	1.95	1.90	1.86	1.82	1.79	1.76	1.74	1.72	1.70	1.68	1.66	1.65	1.63	1.62
180	3.89	3.05	2.65	2.42	2.26	2.15	2.06	1.99	1.93	1.88	1.84	1.81	1.77	1.75	1.72	1.70	1.68	1.66	1.64	1.63	1.61	1.60
250	3.88	3.03	2.64	2.41	2.25	2.13	2.05	1.98	1.92	1.87	1.83	1.79	1.76	1.73	1.71	1.68	1.66	1.65	1.63	1.61	1.60	1.58
400	3.86	3.02	2.63	2.39	2.24	2.12	2.03	1.96	1.90	1.85	1.81	1.78	1.74	1.72	1.69	1.67	1.65	1.63	1.61	1.60	1.58	1.57
1000	3.85	3.00	2.61	2.38	2.22	2.11	2.02	1.95	1.89	1.84	1.80	1.76	1.73	1.70	1.68	1.65	1.63	1.61	1.60	1.58	1.57	1.55

Table F (cont.)

$\alpha = .01$

Denominator dF

	23	24	25	26	27	28	29	30	32	35	40	45	50	60	75	100	120	140	180	250	400	1000
1	248.8	249.1	249.3	249.5	249.6	249.8	250.0	250.1	250.4	250.7	251.1	251.5	251.8	252.2	252.6	253.0	253.3	253.4	253.6	253.8	254.0	254.2
2	19.45	19.45	19.46	19.46	19.46	19.46	19.46	19.46	19.46	19.47	19.47	19.47	19.48	19.48	19.48	19.49	19.49	19.49	19.49	19.49	19.49	19.49
3	8.64	8.64	8.63	8.63	8.63	8.62	8.62	8.62	8.61	8.60	8.59	8.59	8.58	8.57	8.56	8.55	8.55	8.55	8.54	8.54	8.53	8.53
4	5.78	5.77	5.77	5.76	5.76	5.75	5.75	5.75	5.74	5.73	5.72	5.71	5.70	5.69	5.68	5.66	5.66	5.65	5.65	5.64	5.64	5.63
5	4.53	4.53	4.52	4.52	4.51	4.50	4.50	4.50	4.49	4.48	4.46	4.45	4.44	4.43	4.42	4.41	4.40	4.39	4.39	4.38	4.38	4.37
6	3.85	3.84	3.83	3.83	3.82	3.82	3.81	3.81	3.80	3.79	3.77	3.76	3.75	3.74	3.73	3.71	3.70	3.70	3.69	3.69	3.68	3.67
7	3.42	3.41	3.40	3.40	3.39	3.39	3.38	3.38	3.37	3.36	3.34	3.33	3.32	3.30	3.29	3.27	3.27	3.26	3.25	3.25	3.24	3.23
8	3.12	3.12	3.11	3.10	3.10	3.09	3.08	3.08	3.07	3.06	3.04	3.03	3.02	3.01	2.99	2.97	2.97	2.96	2.95	2.95	2.94	2.93
9	2.91	2.90	2.89	2.89	2.88	2.87	2.87	2.86	2.85	2.84	2.83	2.81	2.80	2.79	2.77	2.76	2.75	2.74	2.73	2.73	2.72	2.71
10	2.75	2.74	2.73	2.72	2.72	2.71	2.70	2.70	2.69	2.68	2.66	2.65	2.64	2.62	2.60	2.59	2.58	2.57	2.57	2.56	2.55	2.54
11	2.62	2.61	2.60	2.59	2.59	2.58	2.58	2.57	2.56	2.55	2.53	2.52	2.51	2.49	2.47	2.46	2.45	2.44	2.43	2.43	2.42	2.41
12	2.51	2.51	2.50	2.49	2.48	2.48	2.47	2.47	2.46	2.44	2.43	2.41	2.40	2.38	2.37	2.35	2.34	2.33	2.33	2.32	2.31	2.30
13	2.43	2.42	2.41	2.41	2.40	2.39	2.39	2.38	2.37	2.36	2.34	2.33	2.31	2.30	2.28	2.26	2.25	2.25	2.24	2.23	2.22	2.21
14	2.36	2.35	2.34	2.33	2.33	2.32	2.31	2.31	2.30	2.28	2.27	2.25	2.24	2.22	2.21	2.19	2.18	2.17	2.16	2.16	2.15	2.14
15	2.30	2.29	2.28	2.27	2.27	2.26	2.25	2.25	2.24	2.22	2.20	2.19	2.18	2.16	2.14	2.12	2.11	2.11	2.10	2.09	2.08	2.07
16	2.24	2.24	2.23	2.22	2.21	2.21	2.20	2.19	2.18	2.17	2.15	2.14	2.12	2.11	2.09	2.07	2.06	2.05	2.04	2.03	2.02	2.02
17	2.20	2.19	2.18	2.17	2.17	2.16	2.15	2.15	2.14	2.12	2.10	2.09	2.08	2.06	2.04	2.02	2.01	2.00	1.99	1.98	1.98	1.97
18	2.16	2.15	2.14	2.13	2.13	2.12	2.11	2.11	2.10	2.08	2.06	2.05	2.04	2.02	2.00	1.98	1.97	1.96	1.95	1.94	1.93	1.92
19	2.12	2.11	2.11	2.10	2.09	2.08	2.08	2.07	2.06	2.05	2.03	2.01	2.00	1.98	1.96	1.94	1.93	1.92	1.91	1.90	1.89	1.88
20	2.09	2.08	2.07	2.07	2.06	2.05	2.05	2.04	2.03	2.01	1.99	1.98	1.97	1.95	1.93	1.91	1.90	1.89	1.88	1.87	1.86	1.85
21	2.06	2.05	2.05	2.04	2.03	2.02	2.02	2.01	2.00	1.98	1.96	1.95	1.94	1.92	1.90	1.88	1.87	1.86	1.85	1.84	1.83	1.82
22	2.04	2.03	2.02	2.01	2.00	2.00	1.99	1.98	1.97	1.96	1.94	1.92	1.91	1.89	1.87	1.85	1.84	1.83	1.82	1.81	1.80	1.79
23	2.01	2.01	2.00	1.99	1.98	1.97	1.97	1.96	1.95	1.93	1.91	1.90	1.88	1.86	1.84	1.82	1.81	1.81	1.79	1.78	1.77	1.76
24	1.99	1.98	1.97	1.97	1.96	1.95	1.95	1.94	1.93	1.91	1.89	1.88	1.86	1.84	1.82	1.80	1.79	1.78	1.77	1.76	1.75	1.74
25	1.97	1.96	1.96	1.95	1.94	1.93	1.93	1.92	1.91	1.89	1.87	1.86	1.84	1.82	1.80	1.78	1.77	1.76	1.75	1.74	1.73	1.72
26	1.96	1.95	1.94	1.93	1.92	1.91	1.91	1.90	1.89	1.87	1.85	1.84	1.82	1.80	1.78	1.76	1.75	1.74	1.73	1.72	1.71	1.70
27	1.94	1.93	1.92	1.91	1.90	1.90	1.89	1.88	1.87	1.86	1.84	1.82	1.81	1.79	1.76	1.74	1.73	1.72	1.71	1.70	1.69	1.68
28	1.92	1.91	1.91	1.90	1.89	1.88	1.88	1.87	1.86	1.84	1.82	1.80	1.79	1.77	1.75	1.73	1.71	1.71	1.69	1.68	1.67	1.66
29	1.91	1.90	1.89	1.88	1.88	1.87	1.86	1.85	1.84	1.83	1.81	1.79	1.77	1.75	1.73	1.71	1.70	1.69	1.68	1.67	1.66	1.65
30	1.90	1.89	1.88	1.87	1.86	1.85	1.85	1.84	1.83	1.81	1.79	1.77	1.76	1.74	1.72	1.70	1.68	1.68	1.66	1.65	1.64	1.63
32	1.87	1.86	1.85	1.85	1.84	1.83	1.82	1.82	1.80	1.79	1.77	1.75	1.74	1.71	1.69	1.67	1.66	1.65	1.64	1.63	1.61	1.60
35	1.84	1.83	1.82	1.82	1.81	1.80	1.79	1.79	1.77	1.76	1.74	1.72	1.70	1.68	1.66	1.63	1.62	1.61	1.60	1.59	1.58	1.57
40	1.80	1.79	1.78	1.77	1.77	1.76	1.75	1.74	1.73	1.72	1.69	1.67	1.66	1.64	1.61	1.59	1.58	1.57	1.55	1.54	1.53	1.52
45	1.77	1.76	1.75	1.74	1.73	1.73	1.72	1.71	1.70	1.68	1.66	1.64	1.63	1.60	1.58	1.55	1.54	1.53	1.52	1.51	1.49	1.48
50	1.75	1.74	1.73	1.72	1.71	1.70	1.69	1.69	1.67	1.66	1.63	1.61	1.60	1.58	1.55	1.52	1.51	1.50	1.49	1.47	1.46	1.45
60	1.71	1.70	1.69	1.68	1.67	1.66	1.66	1.65	1.64	1.62	1.59	1.57	1.56	1.53	1.51	1.48	1.47	1.46	1.44	1.43	1.41	1.40
75	1.67	1.66	1.65	1.64	1.63	1.63	1.62	1.61	1.60	1.58	1.55	1.53	1.52	1.49	1.47	1.44	1.42	1.41	1.40	1.38	1.37	1.35
100	1.64	1.63	1.62	1.61	1.60	1.59	1.58	1.57	1.56	1.54	1.52	1.49	1.48	1.45	1.42	1.39	1.38	1.36	1.35	1.33	1.31	1.30
120	1.62	1.61	1.60	1.59	1.58	1.57	1.56	1.55	1.54	1.52	1.50	1.47	1.46	1.43	1.40	1.37	1.35	1.34	1.32	1.30	1.29	1.27
140	1.61	1.60	1.58	1.57	1.57	1.56	1.55	1.54	1.53	1.51	1.48	1.46	1.44	1.41	1.38	1.35	1.33	1.32	1.30	1.29	1.27	1.25
180	1.59	1.58	1.57	1.56	1.55	1.54	1.53	1.52	1.51	1.49	1.46	1.44	1.42	1.39	1.36	1.33	1.31	1.30	1.28	1.26	1.24	1.22
250	1.57	1.56	1.55	1.54	1.53	1.52	1.51	1.50	1.49	1.47	1.44	1.42	1.40	1.37	1.34	1.31	1.29	1.27	1.25	1.23	1.21	1.18
400	1.56	1.54	1.53	1.52	1.51	1.50	1.50	1.49	1.47	1.45	1.42	1.40	1.38	1.35	1.32	1.28	1.26	1.25	1.23	1.20	1.18	1.15
1000	1.54	1.53	1.52	1.51	1.50	1.49	1.48	1.47	1.46	1.43	1.41	1.38	1.36	1.33	1.30	1.26	1.24	1.22	1.20	1.17	1.14	1.11

Denominator df

Table F (cont.)

$\alpha = .01$

Denominator dF

Denominator df	1	2	3	4	5	6	7	8	9	10	11	12	13	14	15	16	17	18	19	20	21	22
1	39.9	49.5	53.6	55.8	57.2	58.2	58.9	59.4	59.9	60.2	60.5	60.7	60.9	61.1	61.2	61.3	61.5	61.6	61.7	61.7	61.8	61.9
2	8.53	9.00	9.16	9.24	9.29	9.33	9.35	9.37	9.38	9.39	9.40	9.41	9.41	9.42	9.42	9.43	9.43	9.44	9.44	9.44	9.44	9.45
3	5.54	5.46	5.39	5.34	5.31	5.28	5.27	5.25	5.24	5.23	5.22	5.22	5.21	5.20	5.20	5.20	5.19	5.19	5.19	5.18	5.18	5.18
4	4.54	4.32	4.19	4.11	4.05	4.01	3.98	3.95	3.94	3.92	3.91	3.90	3.89	3.88	3.87	3.86	3.86	3.85	3.85	3.84	3.84	3.84
5	4.06	3.78	3.62	3.52	3.45	3.40	3.37	3.34	3.32	3.30	3.28	3.27	3.26	3.25	3.24	3.23	3.22	3.22	3.21	3.21	3.20	3.20
6	3.78	3.46	3.29	3.18	3.11	3.05	3.01	2.98	2.96	2.94	2.92	2.90	2.89	2.88	2.87	2.86	2.85	2.85	2.84	2.84	2.83	2.83
7	3.59	3.26	3.07	2.96	2.88	2.83	2.78	2.75	2.72	2.70	2.68	2.67	2.65	2.64	2.63	2.62	2.61	2.61	2.60	2.59	2.59	2.58
8	3.46	3.11	2.92	2.81	2.73	2.67	2.62	2.59	2.56	2.54	2.52	2.50	2.49	2.48	2.46	2.45	2.45	2.44	2.43	2.42	2.42	2.41
9	3.36	3.01	2.81	2.69	2.61	2.55	2.51	2.47	2.44	2.42	2.40	2.38	2.36	2.35	2.34	2.33	2.32	2.31	2.30	2.30	2.29	2.29
10	3.29	2.92	2.73	2.61	2.52	2.46	2.41	2.38	2.35	2.32	2.30	2.28	2.27	2.26	2.24	2.23	2.22	2.22	2.21	2.20	2.19	2.19
11	3.23	2.86	2.66	2.54	2.45	2.39	2.34	2.30	2.27	2.25	2.23	2.21	2.19	2.18	2.17	2.16	2.15	2.14	2.13	2.12	2.12	2.11
12	3.18	2.81	2.61	2.48	2.39	2.33	2.28	2.24	2.21	2.19	2.17	2.15	2.13	2.12	2.10	2.09	2.08	2.08	2.07	2.06	2.05	2.05
13	3.14	2.76	2.56	2.43	2.35	2.28	2.23	2.20	2.16	2.14	2.12	2.10	2.08	2.07	2.05	2.04	2.03	2.02	2.01	2.01	2.00	1.99
14	3.10	2.73	2.52	2.39	2.31	2.24	2.19	2.15	2.12	2.10	2.07	2.05	2.04	2.02	2.01	2.00	1.99	1.98	1.97	1.96	1.96	1.95
15	3.07	2.70	2.49	2.36	2.27	2.21	2.16	2.12	2.09	2.06	2.04	2.02	2.00	1.99	1.97	1.96	1.95	1.94	1.93	1.92	1.92	1.91
16	3.05	2.67	2.46	2.33	2.24	2.18	2.13	2.09	2.06	2.03	2.01	1.99	1.97	1.95	1.94	1.93	1.92	1.91	1.90	1.89	1.88	1.88
17	3.03	2.64	2.44	2.31	2.22	2.15	2.10	2.06	2.03	2.00	1.98	1.96	1.94	1.93	1.91	1.90	1.89	1.88	1.87	1.86	1.86	1.85
18	3.01	2.62	2.42	2.29	2.20	2.13	2.08	2.04	2.00	1.98	1.95	1.93	1.92	1.90	1.89	1.87	1.86	1.85	1.84	1.84	1.83	1.82
19	2.99	2.61	2.40	2.27	2.18	2.11	2.06	2.02	1.98	1.96	1.93	1.91	1.89	1.88	1.86	1.85	1.84	1.83	1.82	1.81	1.81	1.80
20	2.97	2.59	2.38	2.25	2.16	2.09	2.04	2.00	1.96	1.94	1.91	1.89	1.87	1.86	1.84	1.83	1.82	1.81	1.80	1.79	1.79	1.78
21	2.96	2.57	2.36	2.23	2.14	2.08	2.02	1.98	1.95	1.92	1.90	1.87	1.86	1.84	1.83	1.81	1.80	1.79	1.78	1.78	1.77	1.76
22	2.95	2.56	2.35	2.22	2.13	2.06	2.01	1.97	1.93	1.90	1.88	1.86	1.84	1.83	1.81	1.80	1.79	1.78	1.77	1.76	1.75	1.74
23	2.94	2.55	2.34	2.21	2.11	2.05	1.99	1.95	1.92	1.89	1.87	1.84	1.83	1.81	1.80	1.78	1.77	1.76	1.75	1.74	1.74	1.73
24	2.93	2.54	2.33	2.19	2.10	2.04	1.98	1.94	1.91	1.88	1.85	1.83	1.81	1.80	1.78	1.77	1.76	1.75	1.74	1.73	1.72	1.71
25	2.92	2.53	2.32	2.18	2.09	2.02	1.97	1.93	1.89	1.87	1.84	1.82	1.80	1.79	1.77	1.76	1.75	1.74	1.73	1.72	1.71	1.70
26	2.91	2.52	2.31	2.17	2.08	2.01	1.96	1.92	1.88	1.86	1.83	1.81	1.79	1.77	1.76	1.75	1.73	1.72	1.71	1.71	1.70	1.69
27	2.90	2.51	2.30	2.17	2.07	2.00	1.95	1.91	1.87	1.85	1.82	1.80	1.78	1.76	1.75	1.74	1.72	1.71	1.70	1.70	1.69	1.68
28	2.89	2.50	2.29	2.16	2.06	2.00	1.94	1.90	1.87	1.84	1.81	1.79	1.77	1.75	1.74	1.73	1.71	1.70	1.69	1.69	1.68	1.67
29	2.89	2.50	2.28	2.15	2.06	1.99	1.93	1.89	1.86	1.83	1.80	1.78	1.76	1.75	1.73	1.72	1.71	1.69	1.68	1.68	1.67	1.66
30	2.88	2.49	2.28	2.14	2.05	1.98	1.93	1.88	1.85	1.82	1.79	1.77	1.75	1.74	1.72	1.71	1.70	1.69	1.68	1.67	1.66	1.65
32	2.87	2.48	2.26	2.13	2.04	1.97	1.91	1.87	1.83	1.81	1.78	1.76	1.74	1.72	1.71	1.69	1.68	1.67	1.66	1.65	1.64	1.64
35	2.85	2.46	2.25	2.11	2.02	1.95	1.90	1.85	1.82	1.79	1.76	1.74	1.72	1.70	1.69	1.67	1.66	1.65	1.64	1.63	1.62	1.62
40	2.84	2.44	2.23	2.09	2.00	1.93	1.87	1.83	1.79	1.76	1.74	1.71	1.70	1.68	1.66	1.65	1.64	1.62	1.61	1.61	1.60	1.59
45	2.82	2.42	2.21	2.07	1.98	1.91	1.85	1.81	1.77	1.74	1.72	1.70	1.68	1.66	1.64	1.63	1.62	1.60	1.59	1.58	1.58	1.57
50	2.81	2.41	2.20	2.06	1.97	1.90	1.84	1.80	1.76	1.73	1.70	1.68	1.66	1.64	1.63	1.61	1.60	1.59	1.58	1.57	1.56	1.55
60	2.79	2.39	2.18	2.04	1.95	1.87	1.82	1.77	1.74	1.71	1.68	1.66	1.64	1.62	1.60	1.59	1.58	1.56	1.55	1.54	1.53	1.53
75	2.77	2.37	2.16	2.02	1.93	1.85	1.80	1.75	1.72	1.69	1.66	1.63	1.61	1.60	1.58	1.57	1.55	1.54	1.53	1.52	1.51	1.50
100	2.76	2.36	2.14	2.00	1.91	1.83	1.78	1.73	1.69	1.66	1.64	1.61	1.59	1.57	1.56	1.54	1.53	1.52	1.50	1.49	1.48	1.48
120	2.75	2.35	2.13	1.99	1.90	1.82	1.77	1.72	1.68	1.65	1.63	1.60	1.58	1.56	1.55	1.53	1.52	1.50	1.49	1.48	1.47	1.46
140	2.74	2.34	2.12	1.99	1.89	1.82	1.76	1.71	1.68	1.64	1.62	1.59	1.57	1.55	1.54	1.52	1.51	1.50	1.48	1.47	1.46	1.45
180	2.73	2.33	2.11	1.98	1.88	1.81	1.75	1.70	1.67	1.63	1.61	1.58	1.56	1.54	1.53	1.51	1.50	1.48	1.47	1.46	1.45	1.44
250	2.73	2.32	2.11	1.97	1.87	1.80	1.74	1.69	1.66	1.62	1.60	1.57	1.55	1.53	1.51	1.50	1.49	1.47	1.46	1.45	1.44	1.43
400	2.72	2.32	2.10	1.96	1.86	1.79	1.73	1.69	1.65	1.61	1.59	1.56	1.54	1.52	1.50	1.49	1.47	1.46	1.45	1.44	1.43	1.42
1000	2.71	2.31	2.09	1.95	1.85	1.78	1.72	1.68	1.64	1.61	1.58	1.55	1.53	1.51	1.49	1.48	1.46	1.45	1.44	1.43	1.42	1.41

Table F(cont.)

Denominator dF

α = .01	23	24	25	26	27	28	29	30	32	35	40	45	50	60	75	100	120	140	180	250	400	1000
1	61.9	62.0	62.1	62.1	62.1	62.2	62.2	62.3	62.3	62.4	62.5	62.6	62.7	62.8	62.9	63.0	63.1	63.1	63.1	63.2	63.2	63.3
2	9.45	9.45	9.45	9.45	9.45	9.46	9.46	9.46	9.46	9.46	9.47	9.47	9.47	9.47	9.48	9.48	9.48	9.48	9.49	9.49	9.49	9.49
3	5.18	5.18	5.17	5.17	5.17	5.17	5.17	5.17	5.17	5.16	5.16	5.16	5.15	5.15	5.15	5.14	5.14	5.14	5.14	5.14	5.14	5.1
4	3.83	3.83	3.83	3.83	3.82	3.82	3.82	3.82	3.81	3.81	3.80	3.80	3.80	3.79	3.78	3.78	3.78	3.77	3.77	3.77	3.77	3.76
5	3.19	3.19	3.19	3.18	3.18	3.18	3.18	3.17	3.17	3.16	3.16	3.15	3.15	3.14	3.13	3.13	3.12	3.12	3.12	3.11	3.11	3.11
6	2.82	2.82	2.81	2.81	2.81	2.81	2.80	2.80	2.80	2.79	2.78	2.77	2.77	2.76	2.75	2.75	2.74	2.74	2.74	2.73	2.73	2.72
7	2.58	2.58	2.57	2.57	2.56	2.56	2.56	2.56	2.55	2.54	2.54	2.53	2.52	2.51	2.51	2.50	2.49	2.49	2.49	2.48	2.48	2.47
8	2.41	2.40	2.40	2.40	2.39	2.39	2.39	2.38	2.38	2.37	2.36	2.35	2.35	2.34	2.33	2.32	2.32	2.31	2.31	2.30	2.30	2.30
9	2.28	2.28	2.27	2.27	2.26	2.26	2.26	2.25	2.25	2.24	2.23	2.22	2.22	2.21	2.20	2.19	2.18	2.18	2.18	2.17	2.17	2.16
10	2.18	2.18	2.17	2.17	2.17	2.16	2.16	2.16	2.15	2.14	2.13	2.12	2.12	2.11	2.10	2.09	2.08	2.08	2.07	2.07	2.06	2.06
11	2.11	2.10	2.10	2.09	2.09	2.08	2.08	2.08	2.07	2.06	2.05	2.04	2.04	2.03	2.02	2.01	2.00	2.00	1.99	1.99	1.98	1.98
12	2.04	2.04	2.03	2.03	2.02	2.02	2.01	2.01	2.01	2.00	1.99	1.98	1.97	1.96	1.95	1.94	1.93	1.93	1.92	1.92	1.91	1.91
13	1.99	1.98	1.98	1.97	1.97	1.96	1.96	1.96	1.95	1.94	1.93	1.92	1.92	1.90	1.89	1.88	1.88	1.87	1.87	1.86	1.86	1.85
14	1.94	1.94	1.93	1.93	1.92	1.92	1.92	1.91	1.91	1.90	1.89	1.88	1.87	1.86	1.85	1.83	1.83	1.82	1.82	1.81	1.81	1.80
15	1.90	1.90	1.89	1.89	1.88	1.88	1.88	1.87	1.87	1.86	1.85	1.84	1.83	1.82	1.80	1.79	1.79	1.78	1.78	1.77	1.76	1.76
16	1.87	1.87	1.86	1.86	1.85	1.85	1.84	1.84	1.83	1.82	1.81	1.80	1.79	1.78	1.77	1.76	1.75	1.75	1.74	1.73	1.73	1.72
17	1.84	1.84	1.83	1.83	1.82	1.82	1.81	1.81	1.80	1.79	1.78	1.77	1.76	1.75	1.74	1.73	1.72	1.71	1.71	1.70	1.70	1.69
18	1.82	1.81	1.80	1.80	1.80	1.79	1.79	1.78	1.78	1.77	1.75	1.74	1.74	1.72	1.71	1.70	1.69	1.69	1.68	1.67	1.67	1.66
19	1.79	1.79	1.78	1.78	1.77	1.77	1.76	1.76	1.75	1.74	1.73	1.72	1.71	1.70	1.69	1.67	1.67	1.66	1.65	1.65	1.64	1.64
20	1.77	1.77	1.76	1.76	1.75	1.75	1.74	1.74	1.73	1.72	1.71	1.70	1.69	1.68	1.66	1.65	1.64	1.64	1.63	1.62	1.62	1.61
21	1.75	1.75	1.74	1.74	1.73	1.73	1.72	1.72	1.71	1.70	1.69	1.68	1.67	1.66	1.64	1.63	1.62	1.62	1.61	1.60	1.60	1.59
22	1.74	1.73	1.73	1.72	1.72	1.71	1.71	1.70	1.69	1.68	1.67	1.66	1.65	1.64	1.63	1.61	1.60	1.60	1.59	1.59	1.58	1.57
23	1.72	1.72	1.71	1.70	1.70	1.69	1.69	1.69	1.68	1.67	1.66	1.64	1.64	1.62	1.61	1.59	1.59	1.58	1.57	1.57	1.56	1.55
24	1.71	1.70	1.70	1.69	1.69	1.68	1.68	1.67	1.66	1.65	1.64	1.63	1.62	1.61	1.59	1.58	1.57	1.57	1.56	1.55	1.54	1.54
25	1.70	1.69	1.68	1.68	1.67	1.67	1.66	1.66	1.65	1.64	1.63	1.62	1.61	1.59	1.58	1.56	1.56	1.55	1.54	1.54	1.53	1.52
26	1.68	1.68	1.67	1.67	1.66	1.66	1.65	1.65	1.64	1.63	1.61	1.60	1.59	1.58	1.57	1.55	1.54	1.54	1.53	1.52	1.52	1.51
27	1.67	1.67	1.66	1.65	1.65	1.64	1.64	1.64	1.63	1.62	1.60	1.59	1.58	1.57	1.55	1.54	1.53	1.53	1.52	1.51	1.50	1.50
28	1.66	1.66	1.65	1.64	1.64	1.63	1.63	1.63	1.62	1.61	1.59	1.58	1.57	1.56	1.54	1.53	1.52	1.51	1.51	1.50	1.49	1.48
29	1.65	1.65	1.64	1.63	1.63	1.62	1.62	1.62	1.61	1.60	1.58	1.57	1.56	1.55	1.53	1.52	1.51	1.50	1.50	1.49	1.48	1.47
30	1.64	1.64	1.63	1.63	1.62	1.62	1.61	1.61	1.60	1.59	1.57	1.56	1.55	1.54	1.52	1.51	1.50	1.49	1.49	1.48	1.47	1.46
32	1.63	1.62	1.62	1.61	1.60	1.60	1.59	1.59	1.58	1.57	1.56	1.54	1.53	1.52	1.50	1.49	1.48	1.47	1.47	1.46	1.45	1.44
35	1.61	1.60	1.60	1.59	1.58	1.58	1.57	1.57	1.56	1.55	1.53	1.52	1.51	1.50	1.48	1.47	1.46	1.45	1.44	1.43	1.43	1.42
40	1.58	1.57	1.57	1.56	1.56	1.55	1.55	1.54	1.53	1.52	1.51	1.49	1.48	1.47	1.45	1.43	1.42	1.42	1.41	1.40	1.39	1.38
45	1.56	1.55	1.55	1.54	1.53	1.53	1.52	1.52	1.51	1.50	1.48	1.47	1.46	1.44	1.43	1.41	1.40	1.39	1.38	1.37	1.37	1.36
50	1.54	1.54	1.53	1.52	1.52	1.51	1.51	1.50	1.49	1.48	1.46	1.45	1.44	1.42	1.41	1.39	1.38	1.37	1.36	1.35	1.34	1.33
60	1.52	1.51	1.50	1.50	1.49	1.49	1.48	1.48	1.47	1.45	1.44	1.42	1.41	1.40	1.38	1.36	1.35	1.34	1.33	1.32	1.31	1.30
75	1.49	1.49	1.48	1.47	1.47	1.46	1.45	1.45	1.44	1.43	1.41	1.40	1.38	1.37	1.35	1.33	1.32	1.31	1.30	1.29	1.27	1.26
100	1.47	1.46	1.45	1.45	1.44	1.43	1.43	1.42	1.41	1.40	1.38	1.37	1.35	1.34	1.32	1.29	1.28	1.27	1.26	1.25	1.24	1.22
120	1.46	1.45	1.44	1.43	1.43	1.42	1.41	1.41	1.40	1.39	1.37	1.35	1.34	1.32	1.30	1.28	1.26	1.26	1.24	1.23	1.22	1.20
140	1.45	1.44	1.43	1.42	1.42	1.41	1.41	1.40	1.39	1.38	1.36	1.34	1.33	1.31	1.29	1.26	1.25	1.24	1.23	1.22	1.20	1.19
180	1.43	1.43	1.42	1.41	1.40	1.40	1.39	1.39	1.38	1.36	1.34	1.33	1.32	1.30	1.27	1.25	1.23	1.22	1.21	1.20	1.18	1.16
250	1.42	1.41	1.41	1.40	1.39	1.39	1.38	1.37	1.36	1.35	1.33	1.31	1.30	1.28	1.26	1.23	1.22	1.21	1.19	1.18	1.16	1.14
400	1.41	1.40	1.39	1.39	1.38	1.37	1.37	1.36	1.35	1.34	1.32	1.30	1.29	1.26	1.24	1.21	1.20	1.19	1.17	1.16	1.14	1.12
1000	1.40	1.39	1.38	1.38	1.37	1.36	1.36	1.35	1.34	1.32	1.30	1.29	1.27	1.25	1.23	1.20	1.18	1.17	1.15	1.13	1.11	1.08

Denominator df

Selected Formulas

Range = Max − Min
IQR = Q3 − Q1
Outlier Rule-of-Thumb: $y < Q1 - 1.5 \times IQR$ or $y > Q3 + 1.5 \times IQR$

$$\bar{y} = \frac{\sum y}{n}$$

$$s = \sqrt{\frac{\sum(y - \bar{y})^2}{n - 1}}$$

$$z = \frac{y - \mu}{\sigma} \text{ (model based)}$$

$$z = \frac{y - \bar{y}}{s} \text{ (data based)}$$

$$r = \frac{\sum z_x z_y}{n - 1}$$

$$\hat{y} = b_0 + b_1 x \quad \text{where } b_1 = r\frac{s_y}{s_x} \text{ and } b_0 = \bar{y} - b_1 \bar{x}$$

$$P(\mathbf{A}^C) = 1 - P(\mathbf{A})$$

$$P(\mathbf{A} \text{ or } \mathbf{B}) = P(\mathbf{A}) + P(\mathbf{B}) - P(\mathbf{A} \text{ and } \mathbf{B})$$

$$P(\mathbf{A} \text{ and } \mathbf{B}) = P(\mathbf{A}) \times P(\mathbf{B}|\mathbf{A})$$

$$P(\mathbf{B}|\mathbf{A}) = \frac{P(\mathbf{A} \text{ and } \mathbf{B})}{P(\mathbf{A})}$$

If $\mathbf{A}$ and $\mathbf{B}$ are independent, $P(\mathbf{B}|\mathbf{A}) = P(\mathbf{B})$

$E(X) = \mu = \sum x P(x)$ $\quad Var(X) = \sigma^2 = \sum (x - \mu)^2 P(x)$
$E(X \pm c) = E(X) \pm c$ $\quad Var(X \pm c) = Var(X)$
$E(aX) = aE(X)$ $\quad Var(aX) = a^2 Var(X)$
$E(X \pm Y) = E(X) \pm E(Y)$ $\quad Var(X \pm Y) = Var(X) + Var(Y)$
 $\quad$ if X and Y are independent

Geometric: $\quad P(x) = q^{z-1} p \quad \mu = \frac{1}{p} \quad \sigma = \sqrt{\frac{q}{p^2}}$

Binomial: $\quad P(x) = {}_n C_x p^x q^{n-x} \quad \mu = np \quad \sigma = \sqrt{npq}$

$\hat{p} = \frac{x}{n} \quad \mu(\hat{p}) = p \quad SD(\hat{p}) = \sqrt{\frac{pq}{n}}$

Sampling distribution of $\bar{y}$:
(CLT) As n grows, the sampling distribution approaches the Normal model with

$$\mu(\bar{y}) = \mu_y \quad SD(\bar{y}) = \frac{\sigma}{\sqrt{n}}$$

Inference:

Confidence interval for parameter = **statistic ± critical value × SE(statistic)**

$$\text{Test statistic} = \frac{\text{statistic} - \text{parameter}}{\text{SD}(\text{statistic})}$$

Parameter	Statistic	SD(statistic)	SE(statistic)
p	$\hat{p}$	$\sqrt{\dfrac{pq}{n}}$	$\sqrt{\dfrac{\hat{p}\hat{q}}{n}}$
$p_1 - p_2$	$\hat{p}_1 - \hat{p}_2$	$\sqrt{\dfrac{p_1 q_1}{n_1} + \dfrac{p_2 q_2}{n_2}}$	$\sqrt{\dfrac{\hat{p}_1 \hat{q}_1}{n_1} + \dfrac{\hat{p}_2 \hat{q}_2}{n_2}}$
μ	$\bar{y}$	$\dfrac{\sigma}{\sqrt{n}}$	$\dfrac{s}{\sqrt{n}}$
$\mu_1 - \mu_2$	$\bar{y}_1 - \bar{y}_2$	$\sqrt{\dfrac{\sigma_1^2}{n_1} + \dfrac{\sigma_2^2}{n_2}}$	$\sqrt{\dfrac{s_1^2}{n_1} + \dfrac{s_2^2}{n_2}}$
μ_d	$\bar{d}$	$\dfrac{\sigma_d}{\sqrt{n}}$	$\dfrac{s_d}{\sqrt{n}}$
σ_ε	$s_e = \sqrt{\dfrac{\sum(y - \hat{y})^2}{n-2}}$	(dividing $n-k-1$ in multiple regression)	
β_1	b_1	(in simple regression)	$\dfrac{s_e}{s_x \sqrt{n-1}}$
μ_ν	$\hat{y}_\nu$	(in simple regression)	$\sqrt{SE^2(b_1) \cdot (x_\nu - \bar{x})^2 + \dfrac{s_e^2}{n}}$
y_ν	$\hat{y}_\nu$	(in simple regression)	$\sqrt{SE^2(b_1) \cdot (x_\nu - \bar{x})^2 + \dfrac{s_e^2}{n} + s_e^2}$

Pooling: For testing difference between proportions: $\hat{p}_{pooled} = \dfrac{y_1 + y_2}{n_1 + n_2}$

For testing difference between means: $s_p = \sqrt{\dfrac{(n_1 - 1)s_1^2 + (n_2 - 1)s_2^2}{n_1 + n_2 - 2}}$

Substitute these pooled estimates in the respective SE formulas for both groups when assumptions and conditions are met.

Chi-square: $\chi^2 = \sum \dfrac{(Obs - Exp)^2}{Exp}$

ANNOVA: $SS_T = \sum\sum (\bar{y}_j - \bar{\bar{y}})^2;$ $MS_T = SS_T/(k-1)$
$SS_E = \sum\sum (y_{ij} - \bar{y}_j)^2;$ $MS_E = SS_E/(N-k)$
$F_{k-1, N-k} = MS_T / MS_E$